Electron Microscopic Cytochemistry and Immunocytochemistry in Biomedicine

Edited by

Kazuo Ogawa, M.D., Ph.D.
Department of Anatomy
Kyoto University
Kyoto, Japan

Tibor Barka, M.D.
Department of Cell Biology and Anatomy
Mount Sinai School of Medicine
New York

CRC Press
Boca Raton Ann Arbor London Tokyo

Cover — Peripheral blood cells (erythrocytes, platelets, monocyte, and three lymphocytes) fixed in whole blood; SEM micrograph. See also Figure 3 in Chapter 12. (Courtesy of Dr. Bruce Wetzel.)

Library of Congress Cataloging-in-Publication Data

Electron microscopic cytochemistry and immunocytochemistry in biomedicine/editors, Kazuo Ogawa, Tibor Barka.
p. cm.
Includes bibliographical references and index.
ISBN 0-8493-6012-9
1. Electron microscopic immunocytochemistry. 2. Cytochemistry.
3. Diagnosis, Electron microscopic. I. Ogawa, Kazuo, 1928- .
II. Barka, Tibor.
[DNLM: 1. Enzymes—analysis. 2. Histocytochemistry—methods.
3. Immunohistochemistry—methods. 4. Microscopy, Electron—methods.
QS 531 E38]
RB46.7.E43 1992
611′.0181—dc20
NDLM/CDLC
for Library of Congress 92-9241
CIP

Direct all inquiries to CRC Press, Inc., 2000 Corporate Blvd., N.W., Boca Raton, Florida, 33431.

International Standard Book Number 0-8493-6012-9

Library of Congress Card Number 92-9241

Printed in the United States of America 1 2 3 4 5 6 7 8 9 0

Printed on acid-free paper

PREFACE

As conceived, this book has two major objectives. First, it intends to provide an up-to-date guide for cytochemical and immunocytochemical methods that are considered reliable by present standards, with emphasis on electron microscopic applications of these methods. The book is multi-authored; each chapter is written by an expert who contributed to the development of the particular method and/or has extensive experience in its application. We felt that bringing in the critical viewpoint that only experts can provide outweighs the difficulty of assuring an entirely uniform presentation by a multitude of authors. Although the emphasis of the first part of the book is on techniques, the authors were free to discuss and illustrate applications of such methods, and they have done so in varying degrees.

A purely methodological text would have failed to reflect the major contributions of electron microscopic cytochemistry and immunocytochemistry to various fields of biomedicine. We consider cytochemistry and immunocytochemistry not an extension of descriptive morphology, but rather a part of an armamentarium employed in solving fundamental problems of biology. Keeping with this premise, the second objective of this book is to review selected topics in biomedicine — again by renowned experts — from a cytochemical point of view. It is our hope that, because of its dual scope, this treatise will be of value not only for practitioners of cytochemistry and immunocytochemistry but also for those of medicine.

Kazuo Ogawa
Tibor Barka

THE EDITORS

Kazuo Ogawa, M.D., Dr. Med. Sci., is Chairman and Professor of the Department of Anatomy, Faculty of Medicine at Kyoto University (Kyoto, Japan) and a Visiting Professor at Kanazawa Medical University (Kanazawa, Japan), Toho University School of Medicine (Tokyo, Japan), Shanghai Medical University (Shanghai, China), and China Medical University (Shenyang, China). He also served a Visiting Professorship at Yale University School of Medicine (New Haven, Connecticut) from 1969 to 1970.

Dr. Ogawa obtained his M.D. degree in 1954 from Kyoto University and his Dr. Med. Sci. degree in 1959 from Kobe University School of Medicine. In 1991 he received an Honoris Causa Doctoris degree from the Poznan Academy of Medicine (Poland). He has also served fellowships at the Rockefeller Foundation, the National Institutes of Health, and the Roosevelt International Cancer Foundation.

Dr. Ogawa is also a member of the Japan Society of Histochemistry and Cytochemistry, the Japanese Society of Electron Microscopy, the Japanese Association of Anatomists, the Clinical Electron Microscopy Society of Japan, the American Histochemical Society, the American Society of Cell Biology, the American Association of Anatomists, the New York Academy of Sciences, the American Association for the Advancement of Science, the International Brain Research Organization. He is an honorary member of the Ibero-American Society of Histochemistry and Cytochemistry, the Chinese Electron Microscopy Society, and the Polish Histochemical and Cytochemical Society.

Dr. Ogawa is a member of the Executive Committee of the International Federation of Societies for Electron Microscopy, and he was President of the International Federation of Societies for Histochemistry and Cytochemistry. He also was a member of the Japanese government's Science Council of Japan.

Dr. Ogawa is on the editorial board of *Acta Histochemica et Cytochemica, Microscopy Research and Technique, Cell Research,* and *Folia Histochemica et Cytochemica;* he also has served as an editor for various academic journals, such as *Denshi Kenbikyo (Electron Microscopy), Histochemistry, Journal of Histochemistry and Cytochemistry,* and *Biomedical Research.*

Among other awards, he has received the Japanese Society of Electron Microscopy Seto Prize for research in electron microscopic enzyme cytochemistry, the Pioneer's Award in histochemistry from the International Federation of Societies for Histochemistry and Cytochemistry, the Japan Society of Histochemistry and Cytochemistry Distinguished Scientist Award, and the Poznan Academy of Medicine Martinkowskie Medal.

Dr. Ogawa has presented over 70 invited lectures at international meetings and seminars and over 60 invited lectures at national meetings and seminars. He has published approximately 380 research papers and has edited or written more than 60 books. His current major research interests include the development of electron microscopic cytochemistry and its application to biomedicine with special reference to nematolysosomes and analytical color fluorescence electron microscopy.

Tibor Barka, M.D., is Professor of Cell Biology and Anatomy and Professor of Pathology at the Mount Sinai School of Medicine (New York). He obtained his M.D. degree in 1950 from the University of Debrecan Medical School (Debrecan, Hungary).

Dr. Barka is a member of the American Society of Cell Biology, the American Association of Pathologists, and the American Association of Anatomists; he is an honorary member and past president of the Histochemical Society.

Dr. Barka has a long-standing interest in the development and application of histochemical and cytochemical methods. He has authored several books and chapters on histochemistry as well as more than 135 research papers.

CONTRIBUTORS

Yoshihiro Akimoto, Ph.D.
Department of Anatomy
Kyorin University School of Medicine
Tokyo, Japan

Tsugio Amemiya, M.D.
Department of Ophthalmology
Nagasaki University School of Medicine
Nagasaki, Japan

Nobukazu Araki, Ph.D.
Department of Anatomy
Ehime University School of Medicine
Ehime, Japan

Tibor Barka, M.D.
Department of Cell Biology and Anatomy
Mount Sinai School of Medicine
New York, New York

Eveline Baumgart, M.D.
Department of Anatomy and Cell Biology II
University of Heidelberg
Heidelberg, Germany

Mary E. Bentfeld-Barker, Ph.D.
Department of Medicine
University of California – San Francisco
San Francisco, California

Virginia H. Black, Ph.D.
Department of Cell Biology
New York University School of Medicine
New York, New York

Ellis Denise White Bradley, B.Sc.
Department of Pathology
University of Mississippi Medical Center
Jackson, Mississippi

Stephen W. Carmichael, Ph.D., D.Sc.
Department of Anatomy
Mayo Clinic
Rochester, Minnesota

Gwen V. Childs, Ph.D.
Department of Anatomy and Neurosciences
University of Texas Medical Branch
Galveston, Texas

Bertalan Csillik, M.D., Ph.D., D.Sc.
Department of Anatomy
Albert Szent György Medical University
Szeged, Hungary

Shigeo Daikoku, M.D., Ph.D.
Department of Anatomy
University of Tokushima School of Medicine
Tokushima, Japan

Martin Dym, Ph.D.
Department of Anatomy and Cell Biology
Georgetown Medical School
Washington, D.C.

H. Dariush Fahimi, M.D.
Department of Anatomy and Cell Biology
University of Heidelberg
Heidelberg, Germany

Tullio Faraggiana, M.D.
Department of Pathology
Università ''La Sapienza''
Rome, Italy

Marilyn G. Farquhar, Ph.D.
Division of Cellular and Molecular Medicine and Center for Molecular Genetics
University of California – San Diego
La Jolla, California

Kazushi Fujimoto, Ph.D.
Department of Anatomy
Kyoto University
Kyoto, Japan

Toyoshi Fujimoto, M.D., D.M.S.
Department of Anatomy
Kyoto University
Kyoto, Japan

Atsuko Fujioka, M.D.
Kepartment of Anatomy
Kinki University
Osaka, Japan

Hisao Fujita, M.D., Ph.D.
Department of Anatomy
Osaka University Medical School
Osaka, Japan

Osamu Fukushima, M.D.
Department of Anatomy
Jikei University School of Medicine
Tokyo, Japan

Reinhart Gossrau, M.D.
Department of Anatomy
Free University of Berlin
Berlin, Germany

Barbara A. Hamkalo, Ph.D.
Department of Molecular Biology and Biochemistry
University of California – Irvine
Irvine, California

Hiroshi Hasegawa
Kanzakigawa Laboratory
Shionogi Research Laboratories
Osaka, Japan

Ken Hashimoto, M.D.
Department of Dermatology
Wayne State University
Detroit, Michigan

Kimiko Hayashi, Ph.D.
Department of Pathology
Robert Wood Johnson Medical School
University of Medicine and Dentistry of New Jersey
Piscataway, New Jersey

Masando Hayashi, M.D.
Department of Pathology
Robert Wood Johnson Medical School
University of Medicine and Dentistry of New Jersey
Piscataway, New Jersey

L. Hendricks
Division of Cellular and Molecular Medicine and Center for Molecular Genetics
University of California – San Diego
La Jolla, California

Guillermo A. Herrera, M.D.
Department of Pathology
University of Mississippi Medical Center
Jackson, Mississippi

Kei-Ichi Hirai, M.D., D.Sc.
Department of Anatomy
Kanazawa Medical University
Kanazawa, Japan

Hiroshi Hirano, M.D., Ph.D.
Department of Anatomy
Kyorin University School of Medicine
Tokyo, Japan

Setsuji Hisano, Ph.D.
Department of Anatomy
University of Tokushima School of Medicine
Tokushima, Japan

Su-Ming Hsu, M.D.
Department of Pathology
University of Texas
Houston, Texas

Yasuhiko Ibata, M.D.
Department of Anatomy
Kyoto Prefectural University of Medicine
Kyoto, Japan

Kenichirou Inomata
Department of Anatomy
Tokyo University School of Medicine
Tokyo, Japan

Hisanobu Kaiya, M.D.
Department of Neurology and Psychology
Central Hospital of Self-Defence Forces
Tokyo, Japan

Shinsuke Kanamura, M.D.
Department of Anatomy
Kansai Medical University
Osaka, Japan

Kioko Kawai, M.D.
Department of Pathology
Nagasaki University School of Medicine
Nagasaki, Japan

Hayato Kawakami, Ph.D.
Department of Anatomy
Kyorin University School of Medicine
Tokyo, Japan

Toshihiro Maeda, D.Sc.
Department of Anatomy
Shiga University of Medical Science
Shiga, Japan

Takashi Makita, Ph.D., V.D.
Department of Veterinary Anatomy
Yamaguchi University
Yamaguchi, Japan

Nenad Markovic, M.D., Ph.D.
Department of Hematology
Institute of Oncology
Novi Sad, Yugoslavia
and
Department of Pharmacology
Medical College of Pennsylvania
Philadelphia, Pennsylvania

Olivera Markovic-Saljinska, M.D., Ph.D.
Research Department
University Clinical Center
Belgrade, Yugoslavia
and
Department of Pathology and Laboratory Medicine
Albert Einstein Medical Center
Philadelphia, Pennsylvania

Hiroshi Mayahara, Ph.D., D.Sc.
Drug Safety Research Laboratory
Takeda Chemical Industries
Takatsuki, Japan

Vinci Mizuhira, M.D.
Kanzakigawa Laboratory
Shionogi Research Laboratories
Osaka, Japan
and
Department of Anatomy and Cell Biology
Tokyo Medical and Dental University
Tokyo, Japan

Masahiro Mizuno, M.D., Ph.D.
Department of Biology
University of California – San Diego
La Jolla, California

Akira Mizutani, M.D.
Aichi Prefectural Colony
Aichi, Japan

Paul K. Nakane, M.D., Ph.D.
Department of Anatomy
Nagasaki University School of Medicine
Nagasaki, Japan

Tohru Nakano, M.D.
Department of Anatomy
Kyoto University
Kyoto, Japan

Sandya Narayanswami, Ph.D.
Department of Molecular Biology and Biochemistry
University of California – Irvine
Irvine, California

Gang Ning, M.D.
Department of Anatomy
Kyoto University
Kyoto, Japan
and
Department of Cell Biology
China Medical University
Shenyang, China

Fumiaki Nishiyama, Ph.D.
Department of Pathology
Kyorin University School of Health Sciences
Tokyo, Japan

Toru Noda, M.D., Ph.D.
Division of Cellular and Molecular
Medicine and Center for
Molecular Genetics
University of California – La Jolla
La Jolla, California
and
Department of Anatomy
Kyoto University
Kyoto, Japan

Mitsuru Notoya
Kanzakigawa Laboratory
Shionogi Research Laboratories
Osaka, Japan

Takuro Ogata, M.D., Ph.D.
Department of Surgery
Kochi Medical School
Kochi, Japan

Kazuo Ogawa, M.D., Ph.D.
Department of Anatomy
Kyoto University
Kyoto, Japan

Kazushige Ogawa, D.V.M., Ph.D.
Department of Veterinary Anatomy
Iwate University
Morioka, Japan

Andrea Onetti-Muda, M.D.
Dipartimento Biopatologie Umana
Università ''La Sapienza''
Rome, Italy

Julia M. Polak, D.Sc., M.D.
Department of Histochemistry
Royal Postgraduate Medical School
London, England

John M. Robinson, Ph.D.
Department of Anatomy
The Ohio State University
Columbus, Ohio

Takuma Saito, M.D., Ph.D.
Department of Anatomy
Jichi Medical School
Tochigi, Japan

Masahiro Sakai, V.M.D., Ph.D.
Department of Anatomy
Kyoto University
Kyoto, Japan

Harumichi Seguchi, M.D., Ph.D.
Department of Anatomy
Kochi Medical School
Kochi, Japan

Mitsuru Setoyama
Department of Dermatology
Kagoshima University
Kagoshima, Japan

Carlos A. Suárez-Quian, Ph.D.
Department of Anatomy and Cell Biology
Georgetown Medical School
Washington, D.C.

Naonori Sugai, M.D.
Second Department of Anatomy
Fukushima Medical College
Fukushima, Japan

Yoichiro Takashima, M.D., D.M.S.
Department of Anatomy
Ehime University School of Medicine
Ehime, Japan

Kuniaki Takata, Ph.D.
Department of Anatomy
Kyorin University School of Medicine
Tokyo, Japan

Teruo Tanaka, D.D.S.
First Department of Oral Anatomy
Kyushu University
Fukuoka, Japan

Isao Tamura
Dosaka EM Co., Ltd.
Kyoto, Japan

Kazayuki Taniguchi, D.V.M., Ph.D.
Department of Veterinary Anatomy
Iwate University
Morioka, Japan
and
Department of Legal Medicine
St. Marianna University School of Medicine
Kanagawa, Japan

Elba A. Turbat-Herrera, M.D.
Department of Pathology
University of Alabama – Birmingham
Birmingham, Alabama

Satoki Ueno, M.D.
Department of Ophthalmology
Ohtsu Red Cross Hospital
Ohtsu, Japan

Susan Van Noorden, M.A.
Department of Histopathology
Royal Postgraduate Medical School
London, England

A. Velasco
Division of Cellular and Molecular Medicine and Center for Molecular Genetics
University of California – San Diego
La Jolla, California

Guo-Ying Wang, D.M.S.
Department of Anatomy
The First Medical College of PLA
Guangdong, People's Republic of China

Toshimitsu Watabiki, M.D.
Department of Legal Medicine
Kansai Medical University
Osaka, Japan

Mark C. Willingham, M.D.
Department of Pathology
Medical University of South Carolina
Charleston, South Carolina

Kouta Yamamoto, M.D.
Department of Gastroenterology
Shizuoka Municipal Hospital
Shizuoka, Japan

TABLE OF CONTENTS

PART I: METHODOLOGY

PART I

METHODOLOGY

ELECTRON MICROSCOPE LABORATORY
DEPARTMENT OF VETERINARY PATHOLOGY
UNIVERSITY OF MISSOURI, COLUMBIA

SAMPLE ____________________
MAGNIFICATION ____________________
PREPARATION ____________________
____
INVESTIGATOR ____________________
NEG. NO. __________ DATE __________

Chapter 1

ELECTRON MICROSCOPIC CYTOCHEMISTRY OF ENZYMES

1.1 An Historical Perspective

Kazuo Ogawa

Progress in science is in reality progress in methodology.[1] At a glance, scientific progress appears to be the accumulation of scientific information. However, if one considers the method by which results or data are obtained, it can easily be understood that this progress is in fact the development of the methods by which the results are obtained. If we lack the methodology to solve a problem, we can never obtain a solution to it. All methods have their own limitations and possibilities. Results obtained by a method can only be understood if the limitations and subsequent reproducibility of that method are taken into account. In other words, if the method or approach used is not appropriate, the question at hand can never be solved. Alternatively, if a new method is established, it can solve many hitherto unsolved problems and further open entirely new fields of study and deepen the understanding of Nature.

In the field of biomedicine, the most fundamental requirement is to understand the structure and function of biological materials at all levels, from individual organizations to macromolecules, or even molecules.[2] When pursuing the problem of structure and function, classical morphologists describe an exact form or morphology (*morph*) and presume its function, while biochemists perform functional research and assume functions in relation to morphology. Structure and function are, however, essentially inseparable. Histochemical, as well as cytochemical, approaches imply a primary goal of not only determining exact morphology, but also elucidating a direct correlation of structure to function *in situ*. This scientific approach takes as its most basic tenet the understanding of the correlation of structure and function *in situ,* even though the level of observation becomes ever finer from the tissue (*histos*) to the cell, further to the cell organelle, the membrane system, the macromolecule, and even down to the level of the molecule. Thus, a generation ago, when the primary object of morphological observation was at the tissue level, the above-mentioned approach was classified as light microscopic histochemistry. Later, after the introduction of electron microscopy and related techniques to the field of morphology, correlation of functions, especially biochemical functions, with subcellular fine structures such as organelles, membranes, macromolecules, and molecules, became the main theme in biomedicine. Subsequently, during the past several decades, electron microscopic cytochemistry and immunocytochemistry developed rapidly as one of the major fields in functional morphology.

Palade[3] retrospectively summarized, in 1971, how electron microscopic techniques were applied as a basic research tool in biomedicine. Ogawa (1971)[4] has reviewed historical changes in the understanding of the fine structures of mitochondria in the early days of electron microscopy, concluding that the understanding of the fine structures of mitochondria was established between the years of 1952 and 1954, followed in early 1956 by an elucidation of mitochondrial fine structure which is still valid today. Mitochondria were first observed under the electron microscope in 1945 by Keith R. Porter. However, correct understanding of the internal fine structure of mitochondria was possible only after the development and improvement of such techniques as fixation, embedding, and ultrathin sectioning.

0-8493-6012-9/93/$0.00 + $.50

A review of the literature of electron microscopic cytochemistry discloses a series of advances in methodology. Phosphotungstic acid was applied on striated muscle in 1945 by Hall et al.[5] and Gomori's silver technique was tried in 1951 on collagens of tendon by Dettmer et al.[6] The Feulgen reaction for the detection of deoxyribonucleic acid was applied on *Staphylococcus aureus* in 1954.[7] Glycogen was observed as early as 1940 by Huseman and Ruska,[8] but the amylase technique was first tried by Morgan and Mowry[9] only in 1951.

Electron microscopic enzyme cytochemistry was started in 1955 by Sheldon et al.,[10] who tried to demonstrate nonspecific acid phosphatase activity in mouse intestinal epithelium. However, their method was drastic in every way: they used as a fixative osmium tetroxide, which destroys enzymatic activity even during a short period of fixation, and also they incubated specimens in the reaction medium for up to 16 h at 37°C, which is much too long. Their approach was far from adequate[11] and is entirely different from that which is now in use. Following the trial of acid phosphatase activity, the electron microscope was used to demonstrate the activities of many other types of enzymes; alkaline phosphatase in 1956,[12] succinate dehydrogenase in 1957,[13] adenosine triphosphatase in 1958,[14] cholinesterase in 1959,[15,16] and so on. However, the preservation of fine structures and enzyme activity was extremely poor in these early trials and these experiments were not considered successful. The late 1950s was the Dark Age of electron microscopic enzyme cytochemistry. In the early 1960s, more reproducible techniques were developed in both electron microscopy per se and electron microscopic enzyme cytochemistry, as reviewed by Ogawa,[17] and electron microscopic enzyme cytochemistry was established as a major research area not only in electron microscopy, but also in molecular cytology, as seen in this book. It must be remembered here that alkaline phosphatase activity was first demonstrated under the light microscope in 1939 by H. Takamatsu (Kyoto University) and independently by G. Gomori (University of Chicago), thus opening the field of light microscopic histochemistry. Who could possibly have imagined then that the localization of the same enzyme activity would be performed under the electron microscope only a few decades later!

Pearse (1966)[18] referred to the 1962 Oxford Conference on Electron Cytochemistry at the 1966 International Symposium on Electron Microscopy and Cytochemistry held in Leiden in 1966. However, in the early days of electron microscopic cytochemistry, other symposia on electron microscopic cytochemistry were held:

1. Kyoto, April 5, 1963, titled "Histochemistry at the Level of Electron Microscope", held at the 4th annual meeting of The Japanese Histochemical Society (report published in *Archivum Histologicum Japonicum,* 23, 383, 1963)
2. Washington, D.C., April 8, 1963, titled "Applications of Cytochemistry to Electron Microscopy", held at the 14th annual meeting of the Histochemical Society (report published in the *Journal of Histochemistry and Cytochemistry,* 12, 57, 1964)

Since then numerous related symposia on electron microscopic cytochemistry have been held worldwide by various societies of histochemistry, cytochemistry, electron microscopy, anatomy and cell biology, and biochemistry, and the foundation of electron microscopic cytochemistry of today was established.

The primary aim of electron microscopic cytochemistry and immunocytochemistry is topographic visual chemistry and an elucidation of the functions of cellular fine structures *in situ,* allowing a precise understanding of the fine structural localization of biological activity, enzyme activity, antigenicity, functional groups, etc. of the cell. In other words, it is the objective of electron microscopic cytochemistry and immunocytochemistry to visualize the intracellular localization of chemical reactions, biological activities, antigenicity, etc. However, the cytological significance of their subcellular localization should not be overlooked. Electron microscopic cytochemistry and immunocytochemistry may appear to be developing, improving, or establishing new techniques for the *in situ* visualization of chemical reactions, biological activities, antigenicity, etc. However, the visualization methods or techniques available at present are still insufficient to solve many biological problems at hand. Therefore, the field may look static. However, the significance of the elucidation of subcellular localization of chemical reactions lies in eventually understanding the functional significance of cellular structures and in knowing the intracellular sites or compartments of chemical reactions or functions. Electron microscopic cytochemistry represents not only a method of topographic visual chemistry, but also a technique for pursuing the dynamic state of subcellular organelles such as mitochondria, lysosomes, peroxisomes, Golgi apparatus, endoplasmic reticulum, and other membrane systems, under various conditions by using cytochemical markers specific for each organelle. It can also examine the biosynthetic pathway of intracellular substances, e.g., the mechanism of membrane biosynthesis, vesicular transport, and intracellular trafficking. The ultimate aim of electron microscopic cytochemistry and immunocytochemistry is a dynamic understanding of the correlation of cellular structures to functions, and, thus, it represents a major field of functional and dynamic morphology.

REFERENCES

1. **Ogawa, K.,** Introduction, in *New Histochemistry,* Ogawa, K., Takeuchi, T., and Mori, T., Eds., Asakura Book Co., Tokyo, 1975, 2 (in Japanese).
2. **Ogawa, K.,** Introduction to electron microscopic cytochemistry. I, *Electron-Microscopy,* 10, 127, 1975 (in Japanese).
3. **Palade, G. E.,** Albert Claude and the beginnings of biological electron microscopy, *J. Cell Biol.,* 50, 5D, 1971.
4. **Ogawa, K.,** Structure of mitochondria, in *Mitochondria,* Hagiwara, B., Ed., Asakura Book Co., Tokyo, 1971, 51 (in Japanese).
5. **Hall, C. E., Jakus, M. A., and Schmitt, F. O.,** The structure of certain muscle fibrils as revealed by the use of electron stains, *J. Appl. Phys.,* 16, 459, 1945.
6. **von Dettmer, N., Neckel, I., and Ruska, H.,** Elektronenmikroskopische Befunde an versilberten kollagenen Fibrillen, *Z. Wiss. Mikrosk.,* 60, 290, 1951.
7. **Bradfield, J. R. G.,** Electron microscopic observations of bacterial nuclei, *Nature,* 173, 184, 1954.
8. **Huseman, E. and Ruska, H.,** Die Sichtbarmachung von Molekülen des p-Jodbenzoylglycogens, *Naturwissenschaften,* 28, 534, 1940.
9. **Morgan, C. and Mowry, R. W.,** Demonstration of glycogen in the human liver by the electron microscope, *Proc. Soc. Exp. Biol. Med.,* 76, 850, 1951.
10. **Sheldon, H., Zetterqvist, W., and Brandes, D.,** Histochemical reactions for electron microscopy: acid phosphatase, *Exp. Cell Res.,* 9, 592, 1955.
11. **Ogawa, K.,** Story of "pH" at the dawn of electron microscopic cytochemistry, *Acta Histochem. Cytochem.,* 22, 699, 1989.
12. **Brandes, D., Zetterqvist, W., and Sheldon, H.,** Histochemical techniques for electron microscopy: alkaline phosphatase, *Nature,* 177, 382, 1956.
13. **Barrnett, R. J. and Palade, G. E.,** Histochemical demonstration of the sites of activity of dehydrogenase system with the electron microscope, *J. Biophys. Biochem. Cytol.,* 3, 577, 1957.
14. **Essner, E., Novikoff, A. B., and Masek, B.,** Adenosine-triphosphatase and 5-nucleotidase activities in the plasma membrane of liver cells as revealed by electron microscopy, *J. Biophys. Biochem. Cytol.,* 4, 711, 1958.
15. **Barrnett, R. J. and Palade, G. E.,** Enzymatic activity in the M band, *J. Biophys. Biochem. Cytol.,* 6, 163, 1959.
16. **Lehrer, G. M. and Ornstein, L.,** A diazo coupling method for electron microscopic localization of cholinesterase, *J. Biophys. Biochem. Cytol.,* 6, 399, 1959.
17. **Ogawa, K.,** Electron microscopic cytochemistry, in *Enzyme Histochemistry,* Takeuchi, T., Shimizu, N., and Ogawa, K., Eds., Asakura Book Co., Tokyo, 1967, 429 (in Japanese).
18. **Pearse, A. G. E.,** The present and future of electron cytochemistry, *J. Histochem. Cytochem.,* 14, 768, 1966.

I. FIXATION

Atsuko Fujioka

A. INTRODUCTION

The aim of fixation is to preserve cell structure, with minimum change from the living state in various respects. When the goal is merely to study cellular ultrastructure, tissues are fixed to preserve morphology and to protect the cells from destruction by dehydration and embedding. In enzyme cytochemistry, however, the preservation of enzyme activity and definite localization of enzymes in reference to the ultrastructure of cells are required. As chemical fixation is usually accompanied by a decrease in enzyme activity, it seems difficult to achieve a satisfactory preservation of ultrastructure that is compatible with enzyme activity. High activity may be detected in fresh, unfixed specimens, but the localizations of enzymes in reference to the ultrastructure are harmed in such specimens. Also, unfixed specimens have the following disadvantages: (1) soluble enzymes move from the original site to another; (2) substrates or trapping agents do not readily penetrate through the membrane because of the membrane permeability barrier; and (3) handling is troublesome. It has been suggested that inactivation of enzymes up to the level of 75% is still compatible with precise localization of the surviving enzymes. The proper choice of fixative and various conditions of fixation (duration, pH, temperature, buffer, fixation mode) permit the compatibility of the preservation of enzyme activity and ultrastructure.

At present, aldehyde, osmium tetroxide and, to a lesser extent, potassium permanganate are used as fixatives for ultrastructural preservation in conventional electron microscopy. The use of the last two fixatives is limited in enzyme cytochemistry, because they oxidize enzymes, thereby destroying their active sites, although they are excellent fixatives which bring about good preservation of cellular fine structure and increase the electron density of the cellular components. It is generally accepted that aldehydes offer the best compromise for the preservation of both ultrastructure and enzyme activity. In the conventional cytochemical procedure, tissues are fixed in aldehyde prior to the detection of enzyme activity, and this is followed by osmium tetroxide.

B. ALDEHYDES

The use of aldehydes for electron microscopy was developed later than the use of osmium tetroxide or potassium permanganate. The introduction of aldehydes, especially glutaraldehyde, as a fixative for electron microscopy greatly contributed to the preservation of fragile components, such as microtubules and cellular membranes, as well as to enzyme cytochemistry.

Formalin was the first aldehyde fixative to be used for enzyme cytochemistry. Although it was useful for the retention of the activity of a wide range of enzymes, it did not always yield satisfactory preservation of ultrastructure. Sabatini et al.[1] applied various mono- and dialdehydes as fixatives to biological tissues and compared the ability of these reagents to preserve enzyme activity and ultrastructure. Glutaraldehyde and acrolein were excellent in ultrastructural preservation, but led to a significant decrease or loss of some enzyme activity. High retention of enzyme activity was found in tissues fixed by inferior fixatives, such as glyoxal and hydroxyadipaldehyde. Formaldehyde occupied an intermediate position in the preservation of both ultrastructure and enzyme activity. Each aldehyde differs in its ability to preserve enzyme activity and cellular morphology. In general, retention of enzyme activity is inversely proportional to the excellence of morphological preservation. There are also differences in the resistance of various enzymes to fixation. In general, hydrolytic enzymes are more resistant than oxidative enzymes. Thus, in enzyme cytochemistry, the suitable choice of aldehyde fixatives of different reactivities with enzymes enables the detection of a wide range of enzymes, from highly resistant enzymes to readily inactivated ones. Glutaraldehyde alone, or its mixture with formaldehyde, is usually used as the fixative prior to cytochemical reactions. Fixation with formaldehyde alone is employed for enzymes of lesser resistance.

1. Glutaraldehyde

Glutaraldehyde [$CHO(CH_2)_3CHO$] is a five-carbon dialdehyde with an aldehyde group at each end of a hydrocarbon straight chain. It has a molecular weight of 100.12.

Glutaraldehyde is usually supplied as a 25% aqueous solution with a pH of 3 to 6. It contains a number of impurities, such as glutaric acid, acrolein, glutaraldoxime, ethanol, and various polymers, which have an effect on the pH, osmolality, and aldehyde concentration of the fixative solution. It is recommended that the glutaraldehyde solution is purified by vacuum distillation, filtering through activated charcoal, ion exchange, or fractionation on Sephadex G-10. Ampules of purified glutaraldehyde solution in oxygen-free condition are commercially available.

Purified glutaraldehyde solution exhibits an absorption peak at 280 nm indicating the presence of monomeric glutaraldehyde. With time, another peak appears at 235 nm, especially under the condition of room temperature or alkaline pH, of during tissue fixation. This is attributed to dimer formation. Although the reaction mechanism of glutaraldehyde with proteins has not been determined, it has been suggested that the success of glutaraldehyde as a fixative depends on the simultaneous presence of a large range of glutaraldehyde oligomers, including the dimer, in the glutaraldehyde solution. Long polymeric links of glutaraldehyde may permit bridging between proteins across relatively large distances, thus stabilizing protein aggregates. The compounds obtained by cross linking with glutaraldehyde are irreversible and resistant to treatment with acids, urea, semicarbazide, or heat, in contrast to formaldehyde which is unable to form such polymeric links.

Glutaraldehyde primarily cross links ϵ-amino groups of lysine side chains in proteins to stabilize tissue fine structure. Also, α-amino groups of amino acid, SH groups of cysteine, and N-terminal amino groups of some peptides are involved in the reaction. Amino groups in phospholipids react with glutaraldehyde as well, but most lipids in tissues do not react with glutaraldehyde. Post-osmification is necessary to avoid extraction of lipids during dehydration. It is supposed that the reaction of proteins with glutaraldehyde is associated with little effect on conformation and biological activity of the proteins because the reactive amino groups are usually abundant on the protein surface.

2. Formaldehyde

Formaldehyde (HCOH) is a monoaldehyde that shows the simplest molecular formula of the aldehyde fixatives. It is a colorless, irritant gas readily soluble in water. It is supplied commercially as a 37 to 40% aqueous solution, known as formalin. Formalin contains a small amount of formic acid and a considerable amount of methanol to prevent polymerization of the formaldehyde. Formaldehyde that is freshly prepared by the dissociation of paraformaldehyde powder is recommended for enzyme cytochemistry.

The preparation of a 4% buffered formaldehyde solution using paraformaldehyde powder is as follows: (1) dissolve 2 g of paraformaldehyde powder in 25 ml of water by heating it to 60 to 70°C and stirring; (2) add one to two drops of 1 *N* NaOH while stirring until the solution clears (a slight milkiness may persist); (3) cool the solution and bring the volume to 50 ml with a buffer, pH 7.4.

Formaldehyde cross links between protein end-groups, such as amino, imino, amido, guanidino, thiol, phenolic, imidazolyl, and indolyl. The cross linking proceeds in two steps.

At first, formaldehyde forms additive compounds with an active hydrogen atom in functional groups described above. Then, the additive compounds, in turn, cause a condensation reaction with the other compounds with an active hydrogen atom resulting in cross linking by methylene bridges. In this function of forming intermolecular or intramolecular links, formalin acts as a polymerizing fixative. The cross linkages of formaldehyde with proteins are labile. The reaction is reversible, and formaldehyde, for the most part, is removable by washing with water.

Formaldehyde is not adequate for the preservation of lipids and carbohydrate, except for glycogen.

3. Comparison of Glutaraldehyde with Formaldehyde

Glutaraldehyde and formaldehyde show a similar reactivity with proteins, but there are differences in their ability to preserve cellular structure and enzyme activity. The former is superior to the latter in the preservation of ultrastructure. It has been suggested that glutaraldehyde completes fixation about 100 times more rapidly when fixing albumin, relative to formaldehyde. The cross linkages of proteins introduced by glutaraldehyde are stable and irreversible, in contrast to formaldehyde which forms labile cross linking of proteins. Glutaraldehyde resists treatment with boiling water or dilute acid, while the majority of the formaldehydes can be removed by simple washing with water. The number of cross linkages introduced by fixatives is proportional to the preservation of the ultrastructure, but it is inversely related to the preservation of enzyme activity. For most enzymes studied, formaldehyde preserves about twice as much activity as glutaraldehyde.

The penetration of glutaraldehyde into tissues is slower than that of formaldehyde. This may be due to the differences in their molecular size. The rate of penetration is calculated from the formula $d = Kt$, where d = depth of penetration, K = a constant specific for the fixative, and t = time. It has been suggested that the K value is 0.34 for glutaraldehyde and 2 for formaldehyde in the liver. Fixatives must penetrate into a tissue and complete fixation before autolysis of the tissue. The mixture of glutaraldehyde with formaldehyde is usually recommended as a fixative for the preservation of ultrastructure, for both the advantages of the formation of stable and rapid cross linkages with proteins by the former and the fast penetration into tissues by the latter.

C. FACTORS AFFECTING THE QUALITY OF FIXATION AND THE PRESERVATION OF ENZYME ACTIVITY

Factors affecting the quality of fixation and the preservation of enzyme activity should be noted upon the preparation of the fixative solution. They included pH, buffer, osmolality, and the concentrations of the solution.

1. pH

The characteristic structures and functions of cells greatly depend on proteins. Changes in intracellular pH caused by the pH of the fixative may bring about serious changes in the consistency of cytoplasm, selectivity of cell membranes, and the activity of enzymes because the physicochemical nature of proteins is affected by the pH of the solution in which they are placed. Also, the pH is responsible for the interaction of aldehydes with proteins. An increase in fixative pH results not only in an increase in binding capacity of aldehydes with proteins, but also in the rapid polymerization of glutaraldehyde.

The average pH values of most animal tissues is 7.4, and it is generally recommended that pH is kept in the range of 7.2 to 7.4. However, the average pH of plant cells is lower than that of animal cells, and some secretory vesicles show better preservation of ultrastructure under an acidic pH. The pHs vary from cell to cell, organelle to organelle, and from moment to moment. Strictly speaking, the optimal pH should be empirically determined for various cells or organelles.

2. Buffer

Although the cytoplasm of living cells consists of a kind of buffer, it exhibits only a poor capacity to maintain a constant pH in the presence of a weakly buffered or unbuffered fixative. The maintenance of a constant pH during fixation by the proper buffer vehicle is necessary for the preservation of fine structure and enzyme activity. The concentration of buffer needs to be at least 0.1 *M* for fixatives in order to provide an adequate buffering capacity during fixation.

The type of buffer used has an important effect on the stainability and morphological appearance of cells and organelles, as well as on enzyme activity due to specific ion effects and to the differential extraction by various buffers. Also, the type of buffer affects the protein cross linking rate of aldehydes.[2,3] Available buffers in the range of pH 7.2 to 7.4 include collidine, phosphate, cacodylate, HEPES (*N*-2-hydroxyethylpiperazine-*N′*-2-ethanesulfonic acid), MOPS [3-(*N*-morpholino)propane-sulfonic acid], PIPES [piperazine-*N,N′*-bis(2-ethane-sulfonic acid)], and veronal acetate buffers. The last is not appropriate for aldehyde fixatives because of its reactivity with the aldehydes. Collidine buffer, as well as phosphate buffers, have an excellent capacity for stabilization of the fixative pH, but the former is considered to extract cellular materials excessively. The use of phosphates as the buffer vehicle should be avoided in the cytochemical detection of phosphatases, because it precipitates polyvalent cations, which constitute a trapping agent in the incubation medium. Also, it has been proposed that glucose-6-phosphate dehydrogenase is completely inhibited by phosphate buffers. Cacodylate buffer is frequently employed for enzyme cytochemistry to avoid the presence of extracellular phosphate. Some inhibitory effects on enzyme activity, however, have been pointed out owing to arsenic in the cacodylate. Alkaline phosphatase activity in hepatocytes is an example.[4] The reaction products demonstrating the activity were confined to bile canalicular membranes when cacodylate buffer was employed as the vehicle of aldehyde fixative. However, when PIPES buffer was used as the vehicle for aldehyde fixatives, the activity was detected on entire plasma membranes of the hepatocytes. HEPES buffer also resulted in the same distribution. These buffers, which are zwitterionic buffers introduced by Good et al.,[5] are known to allow a higher retention of cellular materials, especially of proteins and phospholipids, and superior preservation of cellular structure. Many studies have shown that PIPES as well as HEPES buffers are superior to other buffers. However, even these buffers are not without producing some artifacts, depending on the kind of cells or organelles. One example is the appearance of multivesicular myeline figures in the rat cerebral cortex. There is no single buffer with superiority over the others. The type of buffer used should be determined by the aim of the study. The preparations of major buffers are shown below.

0.1 *M* Good buffer (HEPES, pH 6.8 to 8.2; PIPES, pH 6.1 to 7.5; MOPS, pH 6.5 to 7.9) — A 0.2 *M* solution of HEPES, PIPES, or MOPS is prepared in distilled water. The desired pH is obtained by adding 0.2 *M* NaOH to 50 ml of the solution. In the case of PIPES, the powder is not dissolved before pH 5.5 to 6.0. After the required pH has been reached, the solution is diluted to a total volume of 100 ml with distilled water.

0.1 *M* Cacodylate buffer, pH 5.0 to 7.4 — (1) 0.2 *M* sodium cacodylate (42.8 g of $Na(CH_3)_2AsO_2{\cdot}3H_2O$ in a total volume of 1000 ml of distilled water); (2) 0.2 *M* HCl (dilute 16.8 ml of conc. HCl to 1000 ml with distilled water). The desired pH can be obtained by adding solution (2) to 50 ml of solution (1) and diluting to a total volume of 100 ml.

0.1 *M* Phosphate buffer, pH 5.7 to 8.0 — (1) 0.2 *M* monobasic sodium phosphate (27.6 g of $NaH_2PO_4{\cdot}H_2O$ in a total volume of 1000 ml of distilled water); (2) 0.2 *M* dibasic

TABLE 1
Osmolalities (mOsm) of Buffers and Glutaraldehyde (GLA) Fixatives (pH 7.4)

Buffer	Buffer only	Buffer + 0.5% GLA	Buffer + 1.0% GLA	Buffer + 2% GLA
0.1 *M* Phosphate	220	259	306	401
0.1 *M* Cacodylate	188	235	264	395
0.1 *M* PIPES	211	265	310	408
0.1 *M* HEPES	112	144	207	315

TABLE 2
Total Osmolalities (mOsm) of 1% Glutaraldehyde Fixatives in 0.1 *M* Buffers with Various Concentrations of Sucrose (pH 7.4)

Buffer	Sucrose concentration 1%	2%	4%	6%
0.1 *M* Phosphate	344	375	422	520
0.1 *M* Cacodylate	305	336	409	470
0.1 *M* PIPES	328	341	395	455
0.1 *M* HEPES	217	255	305	370

sodium phosphate (53.6 g of $Na_2HPO_4 \cdot 7H_2O$ in a total volume of 1000 ml of distilled water). Adjust the pH by adding solution (2) to solution (1) and dilute with the same volume of distilled water as that of the buffer solution.

3. Osmolality

Osmolality of the fixative solution affects the appearance of cell structures. A hypertonic solution may cause shrinkage of cells and organelles. On the other hand, a hypotonic solution may bring about not only swelling of these structures but also leakage of enzymes from them. A somewhat hypertonic solution (400 to 500 mOsm or more) is usually used as the fixative solution. However, each cell type, as well as each organelle, has its own optimal osmolality. In addition, various cell types exhibit different sensitivities to changes in the osmolality of the fixative solution. The optimal osmolality for each cell type should be determined by trial and error. The osmolality can be adjusted by the addition of sucrose, or sometimes dextran, polyvinyl pyrrolidone (PVP), or glucose to a solution consisting of the fixative and the buffer. The osmolalities of available fixative solutions are shown in Tables 1 and 2.

4. Effects of Added Substances

The addition of nonelectrolytes, such as sucrose, dextran, or PVP, or electrolytes to the fixative solution is known to prevent the swelling of cells and to minimize the extraction of certain cellular elements, which may be attributed to their osmotic effects. $CaCl_2$, an electrolyte, binds to polypeptides or phospholipids, thereby stabilizing cellular membranes. It is effective for the stabilization of cytoskeletal elements, as well. Some additions have been attempted to achieve preservation of enzyme activity. The addition of substrates to the fixative solutions, as well as to washing buffers, is useful for the preservation of some enzymes, such as aspartate aminotransferase, ATPase, or glucose-6-phosphatase. The substrate may protect the active site of the enzyme by competing with the fixative.[6] Glutaraldehyde added to the incubation medium was employed for the detection of β-glucuronidase in erythrocyte ghosts.[7] The addition of an enzyme activator, such as Mg^{2+}, can be effective in preventing the depression of enzyme activity. Adenylate cyclase, guanylate cyclase, and K^+-dependent *p*-nitrophenyl phosphatase activities become cytochemically detectable by adding 5% dimethyl sulfoxide (DMSO) to the fixative solutions as well as to the incubation media, although their activities are too weak to be sufficiently detected after fixative without DMSO.[8]

D. METHODS OF FIXATION

Major modes of fixation include vascular perfusion, immersion, and dripping on the surface of the organ. Fixation by vascular perfusion shows superiority in most cases over the other methods in the preservation of ultrastructure as well as enzyme activities.

1. Vascular Perfusion

The fixative solution in vascular perfusion is supplied to tissues via the blood vessels. This method enables the rapid and uniform penetration of fixatives into the tissue and prevents the diffusion of enzymes and dislocation of intra- or intercellular substances by delay of fixation. Vascular perfusion with a low concentration of glutaraldehyde or its mixture with paraformaldehyde, for a short duration at low temperature, is recommended for enzyme cytochemistry, because it creates mild fixation conditions. The optimal duration and concentration are determined by experience, depending on the type of tissues and the ability of the enzymes to be detected. In the author's experience, vascular perfusion through the abdominal aorta with 1% glutaraldehyde for 3 minutes at room temperature has shown compatible preservation of fine structure and enzyme activity of adenylate cyclase or Ca^{2+}-ATPase in renal proximal tubular cells. Some enzymes may require fixation in lower concentrations of glutaraldehyde or in paraformaldehyde alone for the retention of detectable activity. Acid phosphatase in elongated lysosomes, nematolysosomes, was not cytochemically demonstrated until 2% paraformaldehyde and 0.25% glutaraldehyde were used in 30 m*M* PIPES buffer with 10% DMSO as the fixative.[9]

Tissues of a given organ are perfused through the most effective route. For example, the portal vein for the liver and the abdominal aorta for the kidney are used. When several organs in an animal must be perfused simultaneously, a cannula is introduced into the aorta via the left ventricle. This route is used for the brain also. An isotonic buffered saline or Ringer solution prior to a fixative solution is sometimes perfused with the aim of flushing out the blood. Half a minute to a few minutes are enough for washing out. Long washing may cause a deficiency of oxygen in tissues, thereby harming cellular structures. One percent of sodium nitrate as vasodilator and/or heparin in a dose of 0.5 to 1.0 ml/kg body weight as an anticoagulant can be added to the perfusate. The use of paraformaldehyde or 1% or less of glutaraldehyde as the fixative can eliminate the process of the washing out by isotonic buffer or Ringer solution.

The hydrostatic pressure of the perfusate should be identical to the normal pressure of the blood flowing in the tissue. The normal blood pressure in the rat is 100 to 120 mm Hg on the average. A hydrostatic pressure equivalent to that of blood pressure is obtained when the perfusate is supplied from a height that is about 13 times as high as the pressure in mm Hg of the blood. Thus, perfusate is supplied from a height of 100 to 150 cm for perfusion of the body, and 25 to 35 cm for the liver. The rate of flow is 5 to 10 ml min^{-1}. The simplest apparatus using the hydrostatic pressure was proposed by Palay et al.[10] Also, a pump is available for the perfusion of solutions.[11] The other useful apparatuses have been described by Rossi[12] and Furness et al.[13]

Procedure of Vascular Perfusion

1. Open the thoracic or abdominal cavity of an anesthetized animal and carefully expose the vessel supplying blood to the organ or the tissue to be perfused. The other main vessels should be ligatured. In some cases, opening of the thoracic cavity and artificial respiration via cannulae inserted in the trachea is performed.
2. Place a loose ligature around the supplying vessel.
3. Insert a needle or a cannula into the vessel. The needle or cannula is connected via polyethylene tubes, with inserted drip chambers for gross control of flow rates, to elevated flasks containing fixative and rinsing solution. The two polyethylene tubes are joined via a three-way valve just before the needle or cannula.[14] This apparatus is in principle the same as that by Palay et al.[10]
4. Tighten the ligature around the vessel and insert the needle or cannula immediately.
5. Widely open the superior or inferior vena cava or the right atrium of the heart with a cut from a pair of fine scissors, and start the perfusion. The perfusion starts with the rinsing solution and is switched over to the fixative when the rinsing solution comes out from the opened vein.

2. Immersion Fixation

Vascular perfusion is limited to use in human tissues. Such tissues are removed from the body, cut into small blocks in the fixative solution, and then immersed in the same solution. Immersion fixation is easy to perform, but it has some disadvantages, including the fact that uniform fixation cannot be obtained over all the small blocks because the fixative penetrates slowly from surface to center and the alteration of the blood pressure in removal of the tissues yields some artifacts, such as a closed lumen in the renal proximal tubule. Tissues for cytochemical use are cut into 1 mm^3 or less and usually fixed in 1 to 2% glutaraldehyde or a mixture with about 2% paraformaldehyde for 1 to 2 h at 0 to 4°C.

3. Dripping Method

In the dripping method, a tissue or an organ is exposed carefully and fixed by dripping the fixative solution on the surface, without removing the tissue from the body. As the blood supply and cell metabolism are maintained until the moment of fixation, tissues can be fixed under almost living conditions. Closed lumens in the renal proximal tubules do not appear in the fixation.[15] The dripping method is particularly effective in the fixation of relaxed muscles. In general, a 1 to 3% solution of chilled glutaraldehyde or its mixture with paraformaldehyde is dripped on the tissue surface drop by drop for the first few minutes. Then the tissue is covered with a thin pad of cotton soaked with the fixative solution. Sometimes a few fresh drops should be added. After about 20 min, the tissue, which becomes hard by fixation, is removed from the body and is immersed in the same fixative solution. Only the surface layers of the tissue show satisfactory preservation.

REFERENCES

1. **Sabatini, D. D., Bensch, K., and Barrnett, R. J.,** Cytochemistry and electron microscopy. The preservation of cellular ultrastructure and enzymatic activity by aldehyde fixation, *J. Cell Biol.*, 17, 19, 1963.
2. **Coetzee, J. and van der Merwe, C. F.,** Extraction of substances during glutaraldehyde fixation of plant cells, *J. Microsc.*, 135, 147, 1984.
3. **Coetzee, J. and van der Merwe, C. F.,** Penetration rate of glutaraldehyde in various buffers into plant tissue and gelatin gels, *J. Microsc.*, 137, 129, 1985.
4. **Yamamoto, K. and Ogawa, K.,** Effects of NaOH-PIPES buffer used in aldehyde fixative on alkaline phosphatase activity on rat hepatocytes, *Histochemistry,* 77, 339, 1983.
5. **Good, N. E., Winget, G. D., Winter, W., Connolly, T. N., Izawa, S., and Singh, R. M. M.,** Hydrogen ion buffers for biological research, *Biochemistry,* 5, 467, 1966.
6. **Papadimitriou, J. M. and van Duijn, P.,** Effects of fixation and substrate protection of the isoenzymes of aspartate aminotransferase studied in a quantitative cytochemical model system, *J. Cell Biol.*, 47, 71, 1970.
7. **Prosperi, E. and Raap, A. K.,** Substrate protection during the fixation of β-glucuronidase, *Histochem. J.*, 14, 689, 1982.
8. **Fujimoto, K., Toibana, M., and Ogawa, K.,** New cytochemical method for adenylate cyclase and guanylate cyclase with dimethyl sufloxide, *Acta Histochem. Cytochem.*, 14, 687, 1981.
9. **Araki, N., Matsubara, H., Takashima, Y., and Ogawa, K.,** Three-dimensional distribution of acid phosphatase-positive thread-like structures "nematolysosomes" in rat pancreatic exocrine cells, *Acta Histochem. Cytochem.*, 22, 585, 1989.
10. **Palay, S. L., McGee-Russell, S. M., Gordon, S., and Grillo, M. A.,** Fixation of neural tissue for electron microscopy by perfusion with solution of osmium tetroxide, *J. Cell Biol.*, 12, 385, 1962.
11. **Veerman, A. J. P.,** Perfusion fixation using a cushioning chamber coupled to a peristaltic pump, *Stain Technol.*, 49, 111, 1974.
12. **Rossi, G. L.,** Simple apparatus for perfusion fixation for electron microscopy, *Experientia,* 31, 998, 1975.
13. **Furness, J. B., Heath, J. W., and Costa, M.,** Aqueous aldehyde (Faglu) methods for the fluorescence histochemical localization of catecholamines and for ultrastructural studies of central nervous tissue, *Histochemistry,* 57, 285, 1978.
14. **Maunsbach, A. B.,** The influence of different fixatives and fixation methods on the ultrastructure of rat kidney proximal tubule cells. I. Comparison of different perfusion fixation methods and of glutaraldehyde, formaldehyde and osmium tetroxide fixatives, *J. Ultrastr. Res.*, 15, 242, 1966.
15. **Maunsbach, A. B., Madden, S. C., and Latta, H.,** Variations in fine structure of renal tubular epithelium under different conditions of fixation, *J. Ultrastr. Res.*, 6, 511, 1962.

II. MICROWAVE FIXATION METHOD FOR HISTO- AND CYTOCHEMISTRY

Vinci Mizuhira, Hiroshi Hasegawa, and Mitsuru Notoya

A. INTRODUCTION

Microwave irradiation (MWI), introduced by Mayers[1] as a fixation method for biological specimens, has been widely utilized by Mizuhira et al.[2] Microwave energy has been applied to biomedical research in three different ways: (1) basic studies concerning the effect of microwave energy on biological materials, (2) theoretical and practical studies on the microwave fixation method, and (3) studies on the biochemical stabilization of enzymes in the central nervous tissues by MWI, especially using extremely strong irradiation of more than 10 kW to fix brain tissue within milliseconds.[2] In this section, we describe the histo- and cytochemical applications of the MWI method. The theoretical aspects and basic experimental data have been described by Mizuhira.[2]

B. INSTRUMENTS AND EQUIPMENT

In the authors' laboratory, a National NEM-335 domestic microwave range, 2.45 GHz/500 W, fitted with a high-sensitivity thermometer (Resistance thermometer [Pt], Omron) and a millisecond twin timer controlled with a microcomputer (Figure 8), has been used. The same kind of product is available from Nisshin EM Co. & Ltd. (Model MWF-2*). Microwave ovens of different manufacture may be used successfully with care. Glass or plastic bottles of 30-ml capacity containing 10 ml of fixative were arranged in a water bath (large Petri dish). A Yokogawa digital high-sensitivity thermometer Model 2455-04 was used to measure the liquid (fixative) temperature just after MWI.

C. FIXATION METHODS USING MICROWAVE IRRADIATION

1. Fixatives for Conventional Histology

In general, buffered fixatives were used for prefixation in electron microscopy and for conventional fixation in light microscopy, as listed in Table 1, and for postfixation in electron microscopy, as shown in Table 2. The type of buffer used depended on the purpose that was intended, for example:

1. For kidney, 0.075 *M* sodium cacodylate buffer was used.
2. $CaCl_2$ and $MgCl_2$ can protect the membraneous structures from the MWI, but are omitted when trying to detect Ca^{2+} ion localization.
3. Tannic acid, 0.1 to 0.3%, is very effective for fixing the soluble peptides and proteins with aldehyde (Figure 3).[4,5]
4. Sodium cacodylate buffer was selected in order to add $CaCl_2$ to a fixative, instead of phosphate buffer, as phosphate buffer makes precipitates with calcium.
5. Formaldehyde was prepared from paraformaldehyde crystals just before use, and thus is referred to as (*para*)formaldehyde; this prefixative solution was also used for conventional light microscopy and gave excellent results.

* Nisshin EM Co. & Ltd., 23-9, Araki-Cho, Shinjuku-ku, Tokyo 160, Japan.

TABLE 1
Prefixation for Electron Microscopy

2.0% (*para*)formaldehyde
0.5% Glutaraldehyde
0.1 *M* Sodium cacodylate buffer
2–3 m*M* of $CaCl_2$, 1 m*M* of $MgCl_2$
0.1% Tannic acid, at pH 7.2–7.4

TABLE 2
Postfixative

1% Osmium tetroxide
0.1 *M* Sodium cacodylate buffer, at pH 7.2–7.4

Maintaining constant conditions is important in the MWI fixation method and care should be taken to use:

1. The same number of fixing bottles each time.
2. An equal amount of fixative in each bottle (10 ml bottle, for example).
3. A suitable arrangement of bottles in the Petri dish — peripheral position from the center of each dish.
4. A constant amount of water, at room temperature, in each dish (about 80% volume, almost the same as the water level of fixative in each bottle). If the number of fixing bottles is fewer than usual, use dummy bottles containing water instead of fixative. Constant conditions are essential for reliable results.
5. Fresh tissue is cut as soon as possible into small blocks of about 0.3 to 1 cm^3 and put into the fixing bottles.
6. The temperature (for example, 37°C) of the control box is set, then the irradiation is switched on. Usually, there is a temperature difference between the liquid of the central bottle (temperature-measuring bottle, Figure 8, 1) and the peripheral fixing bottles, about 4 to 5°C, and this must be corrected before starting the irradiation, that is, the temperature control setting should be 37°C + 5°C = 40 to 42°C (Figure 8B). Under this condition, the irradiation time is about 25 to 30 s. Of course, the irradiation time can be set first, for example, at 30 s, which would give a final temperature of about 37°C for the fixative (see Figure 1).
7. The temperature of the liquid (fixative in the bottle) after MWI should be kept below 35°C (maximum 40°C), as can be seen from Figure 1. Our basic experiments showed that if the temperature of the fixative rose above 40°C, membraneous ultrastructures would be broken and consequently become unsuitable for high-resolution electron microscopic observations. However, for light microscopy, temperatures of up to about 50°C are acceptable, because the histological resolution is lower than in electron microscopy.
8. MWI is usually done at room temperature (Figure 1A), but can be started at 5°C (ice cold) using a chilled water bath and chilled fixative, as shown in Figure 1B, in the case of enzymo- or immunocytochemical fixation. (*Note:* the chilled water does not contain the ice blocks or slush.)
9. After MWI, the samples are left in the glass bottle with the same fixative at room temperature for about 30 to 180 min for either light or electron microscopy. The chemical linking between the proteins and aldehyde proceeds during this time. For cytochemical fixation, this time is shortened as necessary.

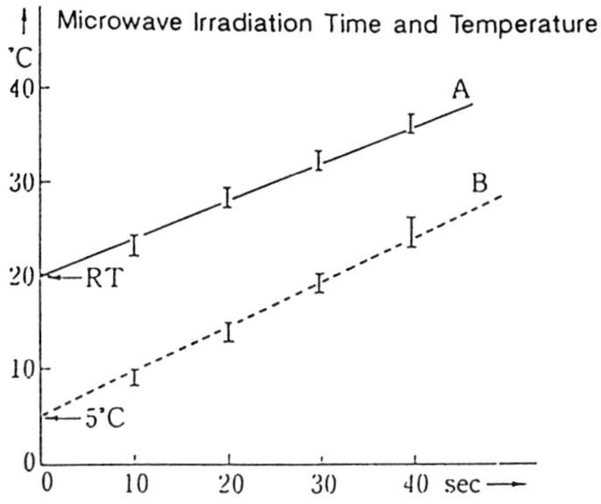

FIGURE 1. Correlation of the liquid temperature (fixative in a bottle) and microwave irradiation time (A) with water bath at room temperature; (B) with chilled water bath; RT, room temperature.

TABLE 3
Staining Mixture for Light Microscopy

0.13% Methylene blue aqueous solution	8 ml
0.13% Basic fuchsin aqueous solution	8 ml
0.2 *M* Phosphate buffer (pH 7.2)	14 ml
100% Ethanol	10 ml

Note: The newly prepared staining mixture should be left at room temperature for 4–5 d, and should be used after filtration.

10. For light microscopy, fixed tissue blocks are embedded in Technovit 7100 (Kulzer & Co. GmbH) resin or paraffin after the dehydration procedure. Resin-embedded blocks are cut with a glass knife with a JB-4 microtome into 0.5- to 1-μm-thick sections, then stained with a methylene blue/basic fuchsin/alcohol mixture (Table 3) for 5 min or stained with MWI. This procedure yields samples in which the structures, including mitochondria, can be clearly seen.
11. For electron microscopy, the blocks are postfixed with osmium tetroxide (Table 2) for 60 min or MWI for about 20 s, and left in the glass bottle for 30 min at room temperature, then embedded in Epon in the usual manner.

The following should be attended:

1. If the glass bottles are directly exposed to MWI without the water bath, the surface liquid temperature rises from 50 to 60°C, although the temperature at the bottom of the bottle remains below 40°C. Using the water bath cooling system keeps the temperature about the same at both the surface and the bottom of the bottle.
2. Some tissues, such as cartilage, bone, tooth, and lung, require special irradiation schedules because water or air (lung) caught in them can sometimes expand rapidly and explosively during the irradiation. Such tissues are subjected to 5-s irradiation periods given four or five times with an interval of several seconds using a computer control box.[2] A chilled water bath, without ice blocks or slush, is also required.

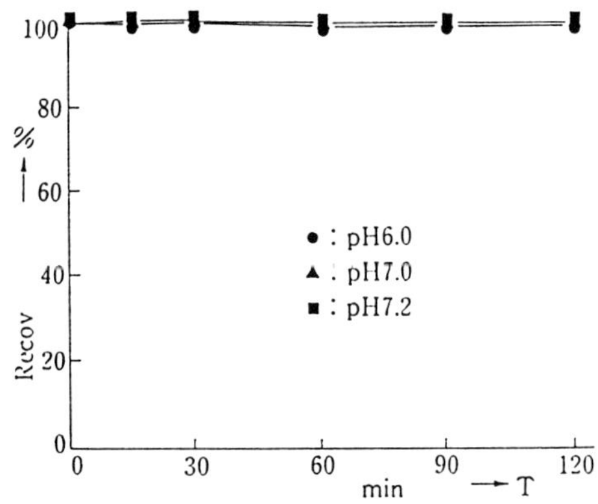

FIGURE 2. Alkaline phosphatase enzymatic activities in the rat proximal convoluted tubular epithelial cell, brush border.

2. Fixatives for Histo- and Cytochemistry

The fixative used varied; the choice depended on the nature of the target (enzymes, antigens, etc.). If necessary, MWI was applied under chilled conditions.

Enzymocytochemistry and immunocytochemistry[2,6] — We used the fixation shown in Table 1, but sometimes $CaCl_2$, $MgCl_2$, or glutaraldehyde was omitted and the concentration of formaldehyde was increased (3 to 4%). After MWI of the tissue blocks, they were kept immersed for 0 to 60 min, frozen in liquid nitrogen-frost or liquid propane, and then cut into thick slices (40 μm thick) using a Microslicer (Dosaka Co., Kyoto). To demonstrate alkaline phosphatase (ALPase) activity, incubation was carried out at pH 9.4 in a medium containing 20 m*M* β-glycerophosphate, 28 m*M* Tris-HCl buffer, 3.9 m*M* $MgSO_4$, 2.0 m*M* lead citrate, and 8.0% sucrose.[7] After incubation, sections were treated with 1% yellow ammonium sulfate for 2 min for light microscopy, and postfixed with osmium tetroxide for electron microscopy. The activity of the enzyme was so strong against the MWI that the reaction products could be detected even after embedding in resin (Figure 2).[2,7] For acid phosphatase (ACPase) activity, the tissue slices were incubated at pH 5.0 in a medium containing 3.0 m*M* β-glycerophosphate, 45.5 m*M* acetate buffer, 10 m*M* lead nitrate, and 7.3% sucrose.[7,9] ACPase enzymatic activity could be very clearly detected in the usual way in lysosomal granules in the liver, kidney, and other samples (Figure 3). However, this enzyme seemed very weak against the MWI. In the case of ACPase, $CaCl_2$ was added to the fixative, and intermittent irradiation was used on the sample in a chilled water bath.[2,7] For glucose-6-phosphatase (G-6-Pase) activity, incubation was performed at pH 6.7 in a medium containing 1.5 m*M* glucose-6-phosphatase, 80 m*M* Tris-maleate buffer, 3.6 m*M* lead nitrate, and 6.0% sucrose. G-6-Pase activity was detected clearly in liver or kidney, but this enzyme is also weak against MWI, and the same care is needed as that in the case of ACPase.[2,7] Ca^{2+}-ATPase activity was also clearly detected after using the selenium or lead salt method.[11,12]

Lectin-binding sites — Binding sites are clear in both light and electron microscopy and can resist MWI.[2]

Immunocytochemistry — Excellent reports are available on the applications of MWI for immunocytochemistry.[14,15]

Autoradiography (ARG) — To demonstrate ^3H-thymidine activity, a mouse was sacrified by cervical dislocation 150 min after intravenous injection. The testis, duodenum, and spleen were rapidly removed and cut into small pieces in the fixative (Table 1) and then subjected to MWI. Tissue blocks were embedded in Technovit for micro-ARG and in Epon for electron microscope-ARG. Microscope images for both light and electron microscopy were excellent, with very clear distribution of the developed silver grains localized only in the nuclei and not in the background.[2,5-7]

For radioactive-labeled peptide (^{125}I-peptide, S-6820), a 100% recovery rate was noted for 0.1 to 0.3% tannic acid with aldehyde (Figure 4).[2,4,6,7] ^{125}I-peptide (S-6820) 130 KBq/g, was injected intravenously into rats and their kidneys were removed after 10, 30, or 60 min. Thick sections (1.0 μm) mounted on glass slides were covered with Konica NR-M2 photographic emulsion. After exposure for 4 weeks, slides were developed in Konicadol-X developer, fixed in Super Fuji-fix, and stained with a mixture of methylene blue/basic fuchsin for micro-ARG.[2,6,7] Epon-embedded tissues were cut into 50- to 70-nm ultrathin sections with an ultramicrotome and mounted on a collodion-precoated copper grid.[5] The grids were coated with Konica NR-H2 emulsion, and exposed for 4 to 5 weeks, then developed in freshly prepared Konicadol-X developer at 20°C for 1 to 3 min. Under electron microscopy, developed silver grains were observed along the brush border and endocytotic vesicles in the proximal convoluted tubules 10 min after injection (Figure 5). The grains were located on lysosomes at 30 to 60 min.[2,6,7]

Electron probe X-ray microanalysis — This method works especially for Ca^{2+} ions in membraneous ultrastructures. Fresh tissues, e.g., rat brain, medulla oblongata, skeletal muscles (intercostal muscle), cardiac muscles, and aorta, were fixed by the MWI method using the fixative given in Table 1, except that $CaCl_2$ and $MgCl_2$ were omitted and 30 to 60 m*M* of K-oxalate (usually, 60 m*M*) was used as a prefixative. The samples were postfixed with 1% osmium tetroxide containing 1.5 to 2.0% K-antimonate for 180 min after the pH had been adjusted by adding a small amount of acetic acid or KOH at pH 7.4. During the postfixation procedure, calcium oxalate in the tissue was substituted with antimonic acid ($[Sb(OH)_6]^-$) for complete conversion to calcium antimonate, $Ca[Sb(OH)_6]_2$ (Figures 6B and 7B).[2,13] Analysis using EDAX-PV-9800 gave the chemical binding molar ratio (AT,%) and elemental molecular weight ratio (WT,%). The calculated theoretical values are shown in Table 4. EDX analysis (the electron probe X-ray microanalysis by the aid of energy dispersive X-ray [EDX] detector, usually simply called EDX analysis) yielded the values expected from the theoretical calculation, as shown in Figures 6 and 7 (compare Figures 6A, B and 7A, B, C). The calibration and resolution of the EDX detector must be checked before the experiments.

D. STAINING METHODS USING MICROWAVE IRRADIATION

This method is very fast and reliable for both light and electron microscopy.[16-19] We improved the technique using a computerized microwave oven (Figure 8). A resistance thermometer (Pt) was set in the center of the turntable in a glass bottle (Figure 8, 1) containing 10 ml of water to measure the liquid temperature. Two to four medium-sized Petri dishes with lids were arranged on the turntable.

1. For Light Microscopic Staining

1. Tissue sections, deparaffinized or resin embedded, are mounted on a glass slide.
2. A metallic mesh or slender glass rods are laid on the bottom of the dish.
3. Some water is put into the dish (to maintain the moisture in the dish chamber).
4. The glass slides are laid in sections on the mesh or glass rods.
5. The Petri dishes are covered with their lids.
6. The staining solution is poured on the sections (glass slide, Table 3).

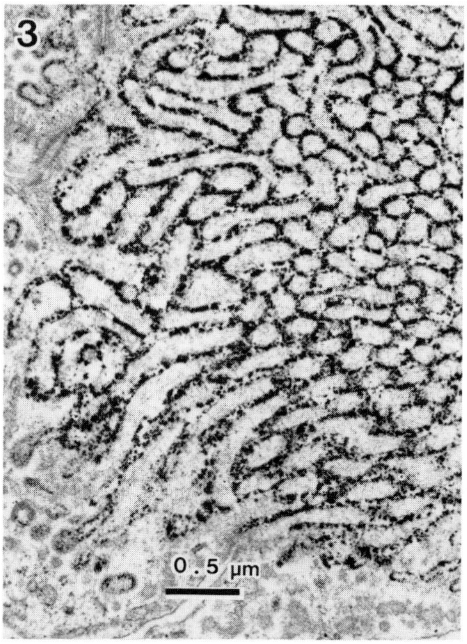

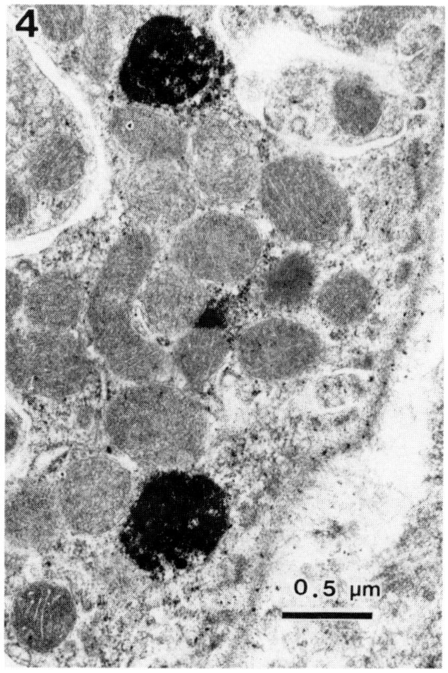

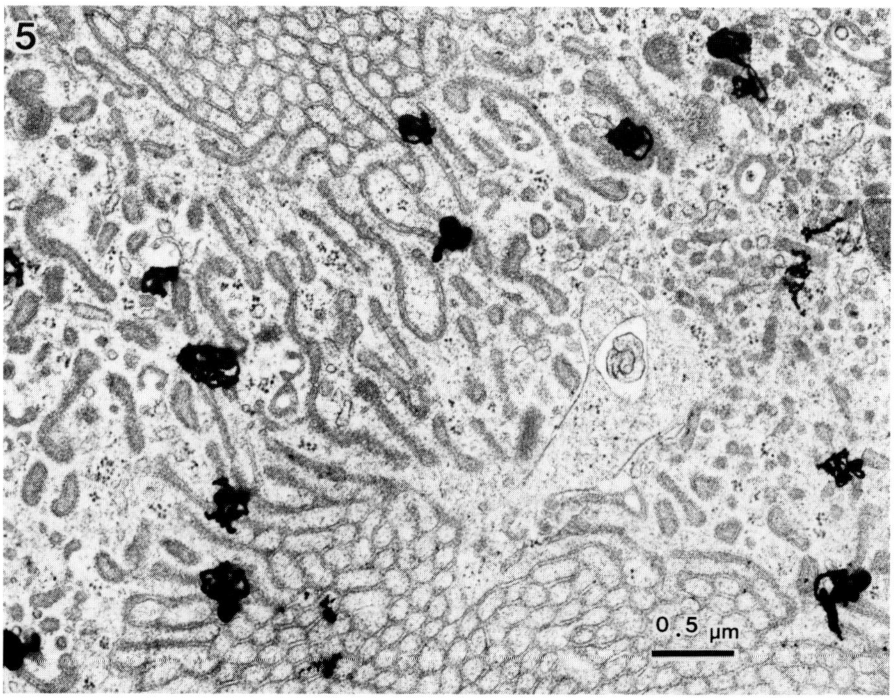

FIGURE 3. Acid phosphatase enzymatic activities in the rat tubular cell.

FIGURE 4. Fixation effect of tannic acid (TA) against ^{125}I-labeled peptide. A remarkable fixation effect is shown here. TA (0.3%) was used, and the labeled peptide (S-6820) showed a 100% recovery rate (Recov), with no correlations between the pH range and time (T, sec). In a case of 0.1% TA, the data were almost the same. This is very important because coexistence of the aldehyde group with TA is necessary, if without TA the reaction would be reversible.

FIGURE 5. ^{125}I-labeled peptide (S-6820) intravenously after 10 min, fixed with TA-containing fixative, electron microscope ARG. Developed silver grains appeared in the brush border area, then moved to the basal region, and were incorporated into the lysosomal granules at 30 to 60 min after injection.

TABLE 4
Theoretical Values of KCl and $Ca[Sb(OH)_6]_2$

In the case of KCl, K(1):Cl(1)
AT(%) = 50:50; WT(%) = (MW 35.43):(MW 39.10) = 47.55:52.45
In the case of $Ca[Sb(OH)_6]_2$, Ca(1):Sb(2)
AT(%) = 33.3:66.7; WT(%) = (MW 40.08):(121.7 × 2) = 14.1:85.9

TABLE 5
Sato's Lead Hydroxide Solution

Lead nitrate	1.0 g
Lead acetate, trihydrate	1.0 g
Lead citrate, trihydrate	1.0 g
Trisodium citrate, dihydrate	2.0 g
Deionized water	82.0 ml
1 *N* NaOH	18.0 ml

7. The temperature (for example, 50 or 70°C) and irradiation time (for example, 5 min) are set, with allowances for a temperature difference of about 20 to 25°C between the center bottle (temperature-measuring bottle) and the staining solution on the glass slide; that is, if the temperature is set at 70°C, the staining solution on the glass slide would reach 45 to 50°C (this relationship should be determined before the staining).
8. Finally, the microwave irradiation is switched on with the microcomputer set to run it off when the preset temperature is reached; these conditions keep the staining solution at 45 to 50°C during the irradiation time.

2. For Electron Microscopic Staining

1. Ultrathin sections are mounted on a grid.
2. A filter paper is laid on the bottom of the dish containing some water (to maintain moisture during staining with uranyl acetate); small staining dishes of plastic bottle caps are arranged in the Petri dish (on a wet filter paper).
3. Next, 2% uranyl acetate water solution is poured into the staining dish.
4. A section of the grid is sunk into the uranyl solution.
5. The Petri dish is covered.
6. The temperature and time are set as for the case of light microscopic staining.
7. The sections (grids) are rinsed with distilled or deionized water.
8. Soda-lime granules are placed at the bottom of the Petri dish and the staining dishes are placed on the soda-lime granules.
9. Sato's lead hydroxide solution (Table 5) is poured into each staining dish and the grids are laid in them.
10. MWI is switched on as described above; the samples are rinsed with deionized water, coated with carbon by the vacuum evaporation technique, and observed under an electron microscope.

Excellent results were obtained with good staining with electron opacity (Figures 9 to 11). Observations revealed fine intracellular structures, such as the cell skeleton (Figure 9), fine filaments, micro- or neurotubules, membraneous ultrafine structures, basement membranes, mitochondrial fine structures — elemental particles (ATPsynthetase) arranged on the mitochondrial crystae (Figure 11) — and correlations between the micro- or neurotubules and the outer surfaces of mitochondria — many filaments or tubules are attached to their outer surfaces (Figures 9 and 10), nuclear fine structures.[19]

Immunocytochemical and *in situ* hybridization staining also can be done in the same manner.[16,20,21]

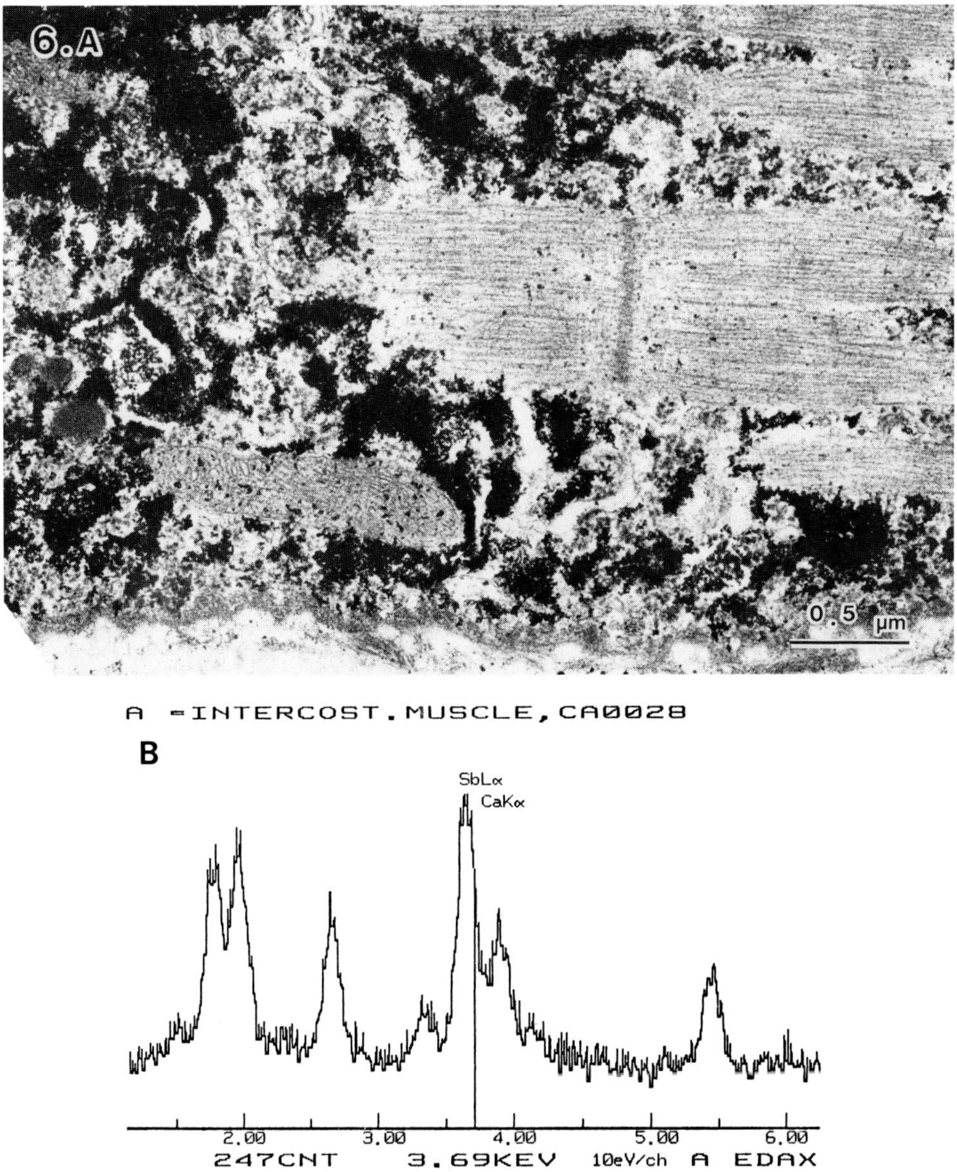

INTE-% :
LABEL = INTERCOST.MUSCLE,CA0028
08-FEB-91 07:46:05
80.670 LIVE SECONDS

ELEM	CPS	AT % ELEM
SB L	28.791	66.401
CA K	13.276	33.599
TOTAL		100.000

INTE-% :
LABEL = INTERCOST.MUSCLE,CA0028
08-FEB-91 07:46:47
80.670 LIVE SECONDS

ELEM	CPS	WT % ELEM
SB L	28.791	85.721
CA K	13.276	14.279
TOTAL		100.000

FIGURE 6. Demonstration of Ca^{2+} ion localization in the sarcoplasmic endsack of intercostal muscle. (A) Electron opaque precipitates of calcium antimonate appeared very clearly in the T-tubules (T, arrows), rat intercostal muscle. 60 m*M* of K-oxalate was used in a prefixative, followed by substitution of osmium with 2% K-antimonate. (B) Analyzed spectrum with EDAX-PV-9800: a Superquant program was used. AT and WT(%) values indicate coincidence with the calculated theoretical values, indicating the presence of calcium in these precipitates (see Table 4).

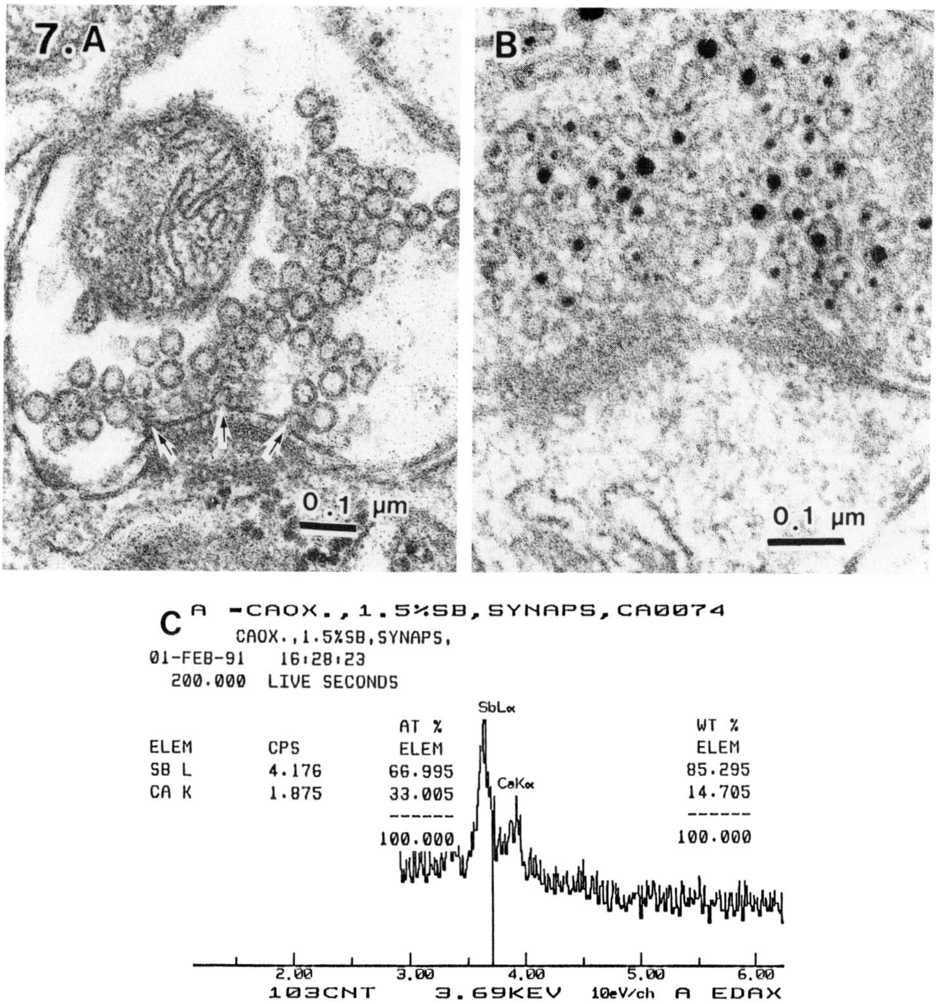

FIGURE 7. (A) Electron micrograph of an extremely fine instantaneous image of the secretion of synaptic vesicles to the postsynaptic membrane (arrows), rat medulla oblongata, MWI fixed. (B) Calcium ion localization in the synaptic membranes of a rat brain: 40 m*M* of K-oxalate was used for the prefixation. (C) EDX-analyzed spectrum obtained from the area of synaptic vesicles in Figure 7B. Calculated data were coincided with calculated theoretical values and also those obtained from other precipitates, such as from the intercostal muscle (see Figure 6B).

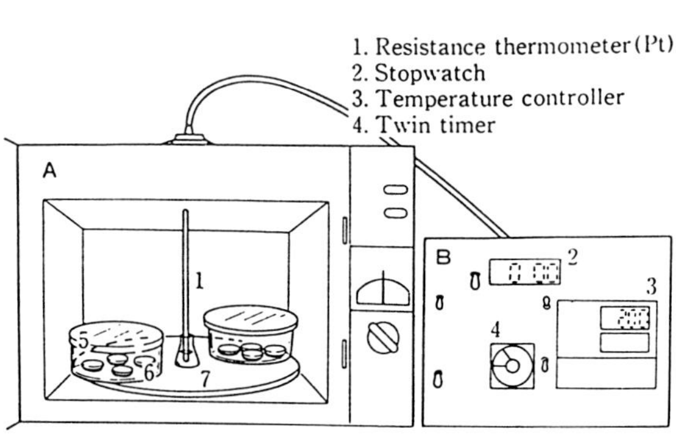

FIGURE 8. Schematic drawing of the staining for biological sections using MWI. (A) Microwave oven; (B) computerized control box. **5**, Glass or plastic moisture chamber with lid, filter paper was laid on the bottom of the chamber with some water; **6**, plastic caps for staining of grids, after uranyl acetate solution had been poured on the top of cap (in the case of lead staining, soda-lime granules were laid on the bottom of the chamber instead of a filter paper); **7**, turntable.

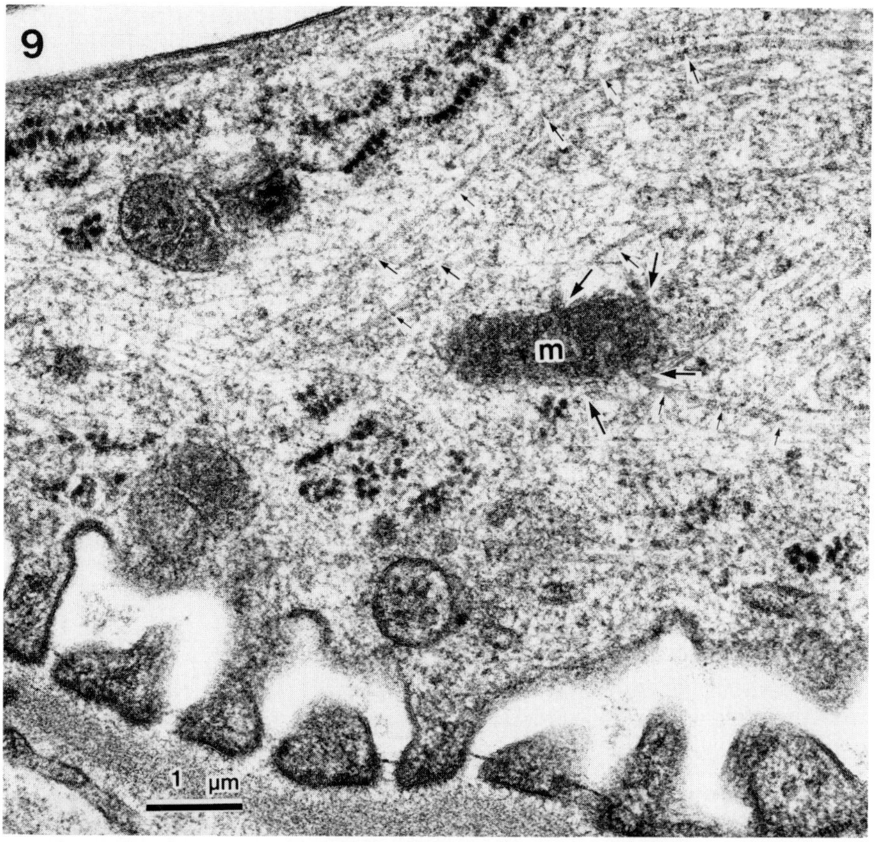

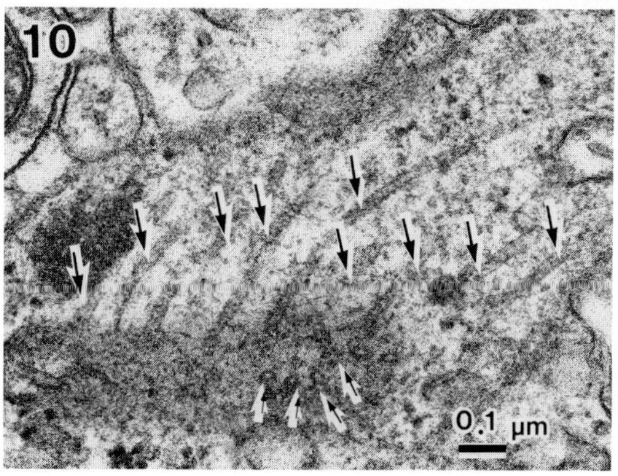

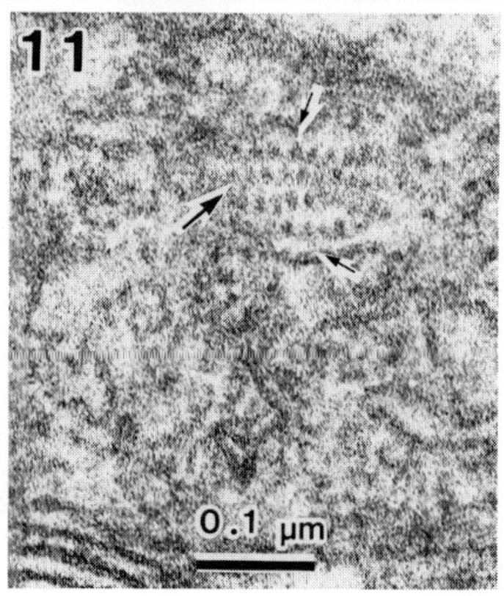

FIGURE 9. Rat kidney, fixed with MWI and stained by MWI. Part of a podocyte is very clearly stained showing every cell organellae, and beautiful intracellular cell skeletons, cell membranes, basement membranes, and mitochondria. In mitochondria (m) in Figures 9 and 10 (rat brain, axon), there seem to be distinct connections with cell skeletons, microtubules, and microfilaments (arrows), which show a very electron opaque nature.

FIGURE 10. Detail of rat brain, showing many neurotubules running parallel to each other in the axon; they finally become attached to the outer surface of the mitochondrial membrane (arrows). Small arrows indicate their cross-sectioned neurotubules attached to the mitochondrial surface. This section was fixed with MWI and was also stained by MWI with both uranyl acetate (5 min) and lead citrate (3 min, Table 5).

FIGURE 11. High magnifications reveal further complicated and precise fine structures of mitochondria. A triple-layered crystae membrane with fine particles can be seen (arrows). These granular fine structures should correspond to the elemental or ATPsynthetase particles, usually only observed under negative staining with broken mitochondrial fractionation. Such ultrastructural appearances can only be obtained by a combination MWI fixation using tannic acid containing fixative, and MWI staining of ultrathin sections.

REFERENCES

1. **Mayers, C. P.,** Histological fixation by microwave heating, *J. Clin. Pathol.*, 23, 273, 1970.
2. **Mizuhira, V., Notoya, M., and Hasegawa, H.,** New tissue fixation method for cytochemistry using microwave irradiation. I. General remarks, *Acta Histochem. Cytochem.*, 23, 501, 1990.
3. **Heuser, J. E., Reese, T. S., Dennis, M. J., Jan, Y., Jan, L., and Evans, L.,** Synaptic vesicle exocytosis captured by quick freezing and correlated with quantal transmitter release, *J. Cell Biol.*, 81, 275, 1979.
4. **Mizuhira, V. and Futaesaku, Y.,** New fixation method for biological membranes using tannic acids, *Acta Histochem. Cytochem.*, 5, 233, 1972.
5. **Mizuhira, V., Shiihashi, M., and Futaesaku, Y.,** High-speed electron microscope autoradiographic studies of diffusible compounds, *J. Histochem. Cytochem.*, 29, 143, 1981.
6. **Mizuhira, V., Notoya, M., and Hasegawa, H.,** Demonstrations of steroid and peptide hormone receptors by electron microscope autoradiography, in, *Using Autoradiography and Correlation Imaging in Vitro and in Vivo,* Stumph, W. E. and Solomon, H., Eds., Academic Press, Orlando, FL, in press, 1992.
7. **Notoya, M., Hasegawa, H., and Mizuhira, V.,** New tissue fixation method for cytochemistry by the aid of microwave irradiation. II. Details, *Acta Histochem. Cytochem.*, 23, 525, 1990.
8. **Mayahara, H., Hirano, H., Saito, T., and Ogawa, K.,** The new lead citrate methods for the ultracytochemical demonstration of activity of non-specific alkaline phosphatase, *Histochemistry,* 11, 88, 1967.
9. **Gomori, G.,** *Principles and Practice, Microscopic Histochemistry,* University of Chicago Press, Chicago, 1952, 193.
10. **Wachstein, M. and Meissel, E.,** On the histochemical demonstration of glucose-6-phosphatase, *J. Histochem. Cytochem.*, 4, 592, 1956.
11. **Kortije, K. H., Freihofer, D., and Rahman, H.,** Cytochemical localization of high-affinity Ca-ATPase activity in synaptic terminals, *J. Histochem. Cytochem.*, 38, 895, 1990.
12. **Ando, T., Fujimoto, K., Mayahara, H., Miyajima, H., and Ogawa, K.,** A new one-step method for the histochemistry and cytochemistry of Ca^{2+}-ATPase activity, *Acta Histochem. Cytochem.*, 14, 705, 1981.
13. **Mizuhira, V., Hasegawa, H., and Notoya, M.,** Demonstration of membrane associated calcium ions of X-ray microanalysis after microwave fixation. Proc. 22th Annu. Meeting Jpn. Soc. Clin. Electron Microsc., Ohtsu, 261, 1990, *J. Clin. Electron Microsc.*, 23, 1030, 1990.
14. **Login, G. R., Schmitt, S. J., and Dvorak, A. M.,** Rapid microwave fixation of human tissues for light microscopic immunoperoxidase identification of diagnostically useful antigens, *Lab. Invest.*, 57, 585, 1989.
15. **Login, G. R., Gall, S. J., Morgan, E., Arizono, N., Schwartz, L. B., and Dvorak, A. M.,** Rapid microwave postembedding immunogold technique, *Lab. Invest.*, 57, 592, 1981.
16. **Estrada, J. C., Brinn, N. T., and Bossen, E. H.,** A rapid method of staining ultrathin sections for surgical pathology TEM with the use of the microwave oven, *Am. J. Clin. Pathol.*, 83, 639, 1985.
17. **Hafiz, S., Soencer, P. C., Lee, M., Gooch, H., and Uerden, B. D.,** Use of microwave for acid and alcohol fast staining, *J. Clin. Pathol.*, 38, 1073, 1985.
18. **Hinds, J. L.,** A rapid and reliable silver impregnation method of Pneumocystis carinti and fungui, *J. Histochem.*, 11, 27, 1988.
19. **Hasegawa, H., Mizuhira, V., and Notoya, M.,** A new electron staining method for ultrathin sections using microwave irradiation (MWI), *J. Clin. Electron Microsc.*, 24, 5, 1991.
20. **Coates, P. J., Hall, P. A., Butcher, M. G., and D'denne, A. J.,** Rapid technique of DNA-DNA *in situ* hybridization of formalin field tissue sections using microwave irradiation, *J. Clin. Pathol.*, 40, 865, 1987.
21. **Volkers, H. H., Van den Brink, W. J., Rook, and Van den Berg, F. M.,** Microwave label detection technique for DNA *in situ* hybridization, *Eur. J. Morphol.*, 29 (1), 59, 1991.

III. THE USE OF NONFROZEN SECTIONS FOR THE CYTOCHEMICAL DEMONSTRATION OF ENZYME ACTIVITIES

Isao Tamura and Hiroshi Mayahara

A. INTRODUCTION

1. Necessity of Using Tissue Sections in Histo- and Cytochemistry

The use of tissue sections, not tissue blocks, in histo- or cytochemical reactions is one of the most important strategies in successful histo- or cytochemical examinations.[1,2] When cytochemical reactions are performed using tissue blocks, nonspecific deposits of reaction products and uneven distribution of the areas showing a positive reaction are often observed. For instance, it is widely accepted that nonspecific alkaline phosphatase (ALPase) activity in the kidney is localized only in the apical brush border of the proximal tubule cells. However, some early studies suggested that both the apical brush border and the basal plasma membrane infoldings were positive for ALPase activity, while other reports indicated that only the brush border was positive.[3-5] To explain this discrepancy, Reale and Luciano[5] surmised that the ALPases on the brush border and the basal infoldings differ in sensitivity to chemical fixatives. Mayahara and Ogawa[2] clarified this issue using a one-step lead citrate method[6] for ALPase. When tissue sections of less than 40 μm thick were used in the cytochemical reaction, they observed a positive ALPase reaction only on the brush border, irrespective of the fixative used or the length of the fixation. On the contrary, when they used tissue blocks, 1 × 1 × 1 mm, they often observed lead deposits both on the brush border and on the basal infoldings of the plasma membrane, irrespective of the fixative employed or the length of fixation. Surprisingly, lead deposits on the basal infolding membranes were found even when tissue blocks were incubated in a substrate-free medium. These results clearly showed that when tissue blocks are used in enzyme cytochemistry, nonspecific deposition of the reaction product occurs. Another problem in using tissue blocks in enzyme cytochemistry is the uneven pattern of reaction and/or the lack of reaction in the central part of the tissue block. The cause of such artifacts can be explained as follows.

1. If the tissue block is large, the reaction medium cannot penetrate into the central part.
2. Since the components of the incubation medium vary in their ability to penetrate the tissue, the proper ratio of the components that is essential for a proper reaction is not maintained in the central part of the tissue blocks. In the case of hydrolases, for instance, a relatively low concentration of the capturing molecules in the central part of the tissue blocks may result in diffusion of the reaction product, and thus diffusion artifacts.
3. In the case of phosphatases, for example, ions such as phosphate ions which may remain in the central part of the tissue block may react with the capture reagent, such as lead ions in the reaction medium, resulting in the formation of nonspecific (lead) deposits.

2. The Importance of Using Nonfrozen Sections

Tissue sections can be cut by frozen sectioning, nonfrozen sectioning, or sectioning after embedding. In the case of frozen sectioning, the ultrastructure of the cells tends to be altered by ice crystal formation. Such alterations are often seen as intracytoplasmic empty portions and waves of the membranous structures in electron micrographs.[7] It has been pointed out that nonfrozen sections provide a truer picture of morphology and sharper localization of reaction products than frozen sections.[8-10]

B. INSTRUMENTS FOR CUTTING NONFROZEN SECTIONS

Commercial instruments currently available for cutting nonfrozen sections include the Tissue Sectioner®,[11] Vibratome®,[12] and Microslicer®.[13] The Tissue Sectioner® is not widely used at present. Its main drawback is the lack of uniformity of section thickness, especially when the tissue is soft. This may be because the cutting movement of the blade is a simple chopping movement, and thus soft tissues are easily crushed.[12] The Tissue Sectioner® has another drawback in that cutting very thin (less than 20 μm thick) sections is almost impossible.[12]

The Vibratome®* was developed to overcome the drawbacks of the Tissue Sectioner®. The Vibratome® cuts tissue sections with a horizontally vibrating blade to avoid crushing the tissue during sectioning. The tissue block(s) is fixed on the stage with cyanoacrylate instant glue, and the stage is fixed with a small vice in a water bath. An appropriate buffer solution is poured into the water bath, and cutting is performed in the buffer, which lubricates the blade and cools the tissue block(s) during sectioning. The Vibratome® offered great advantages for nonfrozen sectioning. However, it has the following drawbacks:

1. The cutting speed is very slow; it often takes hours to obtain enough sections for cytochemistry.
2. Sections of less than 20 μm in thickness are almost impossible to cut when the tissue is not fixed.
3. When the components of the tissue are not uniform, sections easily fall apart.
4. Large sections are almost impossible to obtain.
5. The surface of the sections cut with a Vibratome® usually have chatter marks and thus cannot be used for light microscopy.

These drawbacks seem to originate from the use of electromagnetic vibration to shuttle the blade in the Vibratome® just as in a barber's clippers. The small width of vibration and the irregular up and down vibration of the arm which supports the blade holder seem to be responsible for the above-mentioned drawbacks.

C. DEVELOPMENT OF THE MICROSLICER® AND ITS APPLICATION

1. Development

The Microslicer®** was developed in 1981 by Ogawa et al. to overcome the drawbacks of the Vibratome®. In designing the Microslicer®, an attempt was made to develop a new blade vibration system different from that employed in the Vibratome®, because it was believed that the various drawbacks associated with the Vibratome® might be due to its blade vibration system.

The Vibratome® uses electromagnetic vibration as the blade driving system. The number of vibrations per second is determined *a priori* by that of the power source. When a Vibratome® is used at a relatively high ''amplitude'', a heavy noise and irregular up and down vibrations occur, and sections are often cut into pieces. In addition, when a Vibratome® is used at a low ''amplitude'', sections can be obtained only when the cutting speed is extremely slow. Such faults may originate from the use of the electromagnetic vibration system itself. Therefore, the electromagnetic vibration system was not chosen for the Microslicer®: a high-power micromotor (**M**) is used as the energy source for the blade driving

* Technical Products International, Inc., 13795 Rider Trail, Suite 104, St. Louis, Missouri 63045.

** Dosaka EM Co. Ltd., 1032-3, Ichihara-cho, Shizuichi, Sakyo-ku, Kyoto 601-11, Japan. This study was supported by Grant-in-Aid for Developmental Scientific Research from the Ministry of Education, Science and Culture, the Japanese Government, No. 487030.

FIGURE 3. A Microslicer®, type DTK-3000W.

and application of the Microslicer®, a representative apparatus for cutting nonfrozen sections, were reviewed. It is concluded that the Microslicer® seems to be the most reliable instrument for meeting the most severe requirements in cutting nonfrozen sections for enzyme cytochemistry, immunoelectron microscopy, and uses in various other fields of biomedical science.

REFERENCES

1. **Holt, S. J. and Hicks, R. M.,** The localization of acid phosphatase in rat liver cells as revealed by combined cytochemical staining and election microscopy, *J. Biophys. Biochem. Cytol.*, 11, 47, 1961.
2. **Mayahara, H. and Ogawa, K.,** The effect of thickness of specimen on the ultrastructural localization of alkaline phosphatase activity in the rat proximal convoluted tubule, *J. Histochem. Cytochem.*, 16, 721, 1968.
3. **Mölbert, R. G., Duspiva, F., and von Deimling, O. H.,** The demonstration of alkaline phosphatase in the electron microscope, *J. Biophys. Biochem. Cytol.*, 7, 387, 1960.
4. **Mizutani, A. and Barrnett, R. J.,** Fine structural demonstration of phosphatase activity at pH 9, *Nature,* 206, 1001, 1965.
5. **Reale, E. and Luciano, L.,** Effect of fixation on the alkaline phosphatase activity of mouse proximal convoluted tubule, *J. Histochem. Cytochem.*, 15, 413, 1967.
6. **Mayahara, H., Hirano, H., Saito, T., and Ogawa, K.,** The new lead citrate method for the ultracytochemical demonstration of activity of nonspecific alkaline phosphatase (orthophosphoric monoester phosphohydrolase), *Histochemistry,* 11, 88, 1967.

7. **Mayahara, H.,** On the Microslicer for non-frozen sectioning, in *Histo- and Cytochemistry 1982,* Ogawa, U., Ed., The Japan Society of Histochemistry and Cytochemistry, Gakusaikikaku, Tokyo, 1982, 43 (in Japanese).
8. **Novikoff, A. B., Quintana, N., Villaverde, H., and Forschirm, R.,** Nucleoside phosphatase and cholinesterase activities in dorsal root ganglia and peripheral nerve, *J. Cell Biol.,* 29, 525, 1966.
9. **Seligman, A. M., Plapinger, R. E., Wasserkrug, H. L., Deb. C., and Hanker, J. S.,** Ultrastructural demonstration of cytochrome oxidase activity by the Nadi reaction with osmiophilic reagents, *J. Cell Biol.,* 34, 787, 1967.
10. **Karnovsky, M. J.,** The ultrastructural basis of capillary permeability studied with peroxidase as a tracer, *J. Cell Biol.,* 35, 213, 1967.
11. **Smith, R. E. and Farquhar, M. G.,** TC-2 Tissue sectioner, *RCA Sci. Instrum. News,* 10, 13, 1965.
12. **Smith, R. E.,** Comparative evaluation of two instruments and procedures to cut non-frozen sections, *J. Histochem. Cytochem.,* 18, 590, 1970.
13. **Mayahara, H., Fujimoto, K., Noda, T., Tamura, I., and Ogawa, K.,** The "Microslicer", a new instrument for making non-frozen sections, *Acta Histochem. Cytochem.,* 14, 211, 1981.
14. **Mayahara, H., Ando, T., Ishikawa, Y., and Ogawa, K.,** Ultracytochemical study of the ouabain-sensitive, K-dependent p-nitrophenylphosphatase (Na-K-ATPase) activity in the developing rat kidney, *Acta Histochem. Cytochem.,* 15, 421, 1982.
15. **Inomata, K., Mayahara, H., Fujimoto, K., and Ogawa, K.,** Ultrastructural localization of ouabain-sensitive, potassium-dependent p-nitrophenylphosphatase (Na^+, K^+-ATPase) activity in the central nervous system of the rat, *Acta Histochem. Cytochem.,* 16, 277, 1983.
16. **Yamada, H., Mori, S. Ueda, S., Kawata, M., and Sano, Y.,** Improvement of technique on immunohistochemical demonstration of bioactive substances in the central nervous system, *Acta Histochem. Cytochem.,* 20, 629, 1987.
17. **Tsuto, T., Obata-Tsuto, H. L., Iwai, N., Takahashi, T., and Ibata, Y.,** Fine structure of neurons synthesizing vasoactive intestinal peptide in the human colon from patients with Hirschsprung's disease, *Histochemistry,* 93, 1, 1989.
18. **Ibata, Y., Okumura, H., Makino, S., Kawakami, F., Morimoto, N., and Chihara, K.,** Light and electron microscopic immunocytochemistry of GRF-like immunoreactive neurons and terminals in the rat hypothalamic arcuate nucleus and median eminence, *Brain Res.,* 370, 136, 1986.
19. **Ibata, Y., Takahashi, Y., Okamura, H., Kubo, T., and Kawakami, F.,** Fine structure of NPY-containing neurons in the lateral geniculate nucleus and their terminals in the suprachiasmatic nucleus of the rat, *Brain Res.,* 439, 230, 1988.
20. **Ichitani, Y., Morimoto, N., Okuyama, I., Hasegawa, M., Nagatsu, I., and Ibata, Y.,** Fine structure of tyrosine hydroxylase-like immunoreactive neurons and terminals of the hypothalamic arcuate nucleus and median eminence in young and aged rats, *J. Electron Microsc.,* 39, 396, 1990.
21. **Edwards, F. A., Konnerth, A., Sakmann, B., and Takahashi, T.,** Instruments and techniques. A thin slice preparation for patch clamp recordings from neurons of the mammalian central nervous system, *Eur. J. Physiol.,* 414, 600, 1989.
22. **Takahashi, T.,** Membrane currents in visually identified motoneurons of neonatal rat spinal cord, *J. Physiol.,* 423, 27, 1990.
23. **Ohshima, M., Itoh, H., Matsuoka, M., Murakami, T., and Ohashi, Y.,** Analysis of stress induced or salicylic acid-induced expression of the pathogenesis-related 1a protein gene in transgenic tobacco, *Plant Cell,* 2, 95, 1990.

IV. ULTRATHIN CRYOSECTION TECHNIQUES

Takuma Saito and Osamu Fukushima

A. INTRODUCTION

The aim of electron microscopic enzyme histochemistry is to localize the sites of enzyme activities on cellular ultrastructures in order to understand the total and precise metabolism of each cell. Routine enzyme histochemical techniques demonstrate enzyme activities in tissue sections of less than 40 μm in thickness, which allows the penetration of constituent chemicals in an incubation medium necessary to avoid false-negative and -positive reactions. The sections then are treated with osmic acid, dehydrated, and embedded in epoxy resins.

This routine enzyme histochemical technique limits us to the observation of a single enzyme activity from one Epon block containing a 40-μm section. But ultracryosections permit the demonstration of several different enzymes in a single cell or even within a single cell organelle. Another advantage of cryosection enzyme histochemistry is the avoidance of osmification, dehydration by organic solvents, plastics embedding, and heat polymerization, procedures which tend to modify and denature the biological materials. A further advantage of this techniques is the easy penetration of all constituent chemicals, such as substrate, activator, and capture agents, into the tissue during incubation.

The technique of ultrathin cryosectioning was first introduced by Fernandez-Moran,[1] and it was greatly improved by Bernhard and Leduc[2] and later by Christensen[3] and Tokuyasu.[4] These developments in the technique of ultramicrotomy, as well as the available supply of commercial ultracryomicrotomes, resulted in a wide histochemical application in a variety of fields in electron microscopy, such as immunohistochemistry,[5,6] X-ray microanalysis,[7,8] autoradiography,[9] affinity histochemistry,[10] and enzyme histochemistry.[11]

Although ultrathin cryosections were widely applied in histochemistry, their application for *enzyme* histochemistry has been rather limited and slow in development. It was first tried by Leduc and colleagues,[12] who made ultrathin cryosections from glutaraldehyde-fixed, gelatin-embedded kidney cortex. The ultrathin cryosections were retrieved with dimethylsulfoxide, rinsed in distilled water, air dried, and incubated in the media for the demonstration of adenosine triphosphatase. The sections were then rinsed in distilled water and air dried. Subsequently, several investigators employed this method for the localization of guanosine triphosphatase, 5′-nucleotidase, alkaline phosphatase, ATPase, DNase, and RNase activities using lead salt methods[13-18] and for the demonstration of peroxidase activity using the diaminobenzidine (DAB) method.[19] Bauer and Sigarlakie[20,21] successfully applied this method for localization of acid and alkaline phosphatase activities in yeast cells, which have a thick and rigid cell wall preventing the easy penetration of the incubation medium into the cell.

B. METHODS

1. Instruments and Equipment

For our observations, an LKB 8800 Ultrotome III with an LKB 14800 Cryokit was used for cryomicrotomy. However, a Reichert-Jung Ultracut S with cryosectioning system FC-4E and Reichert-Jung Cryo Super Novais were also available. For liquid nitrogen production, we utilized a mini P.S.A NL-1 (Iwatani Plantec Co., Tokyo), and for slushed nitrogen formation, a vacuum evaporated JEE 4C (JEOL Co., Tokyo).

Precipitates in the incubation mixture were removed by filtering with a disposable syringe filter unit Dismic-25, of cellulose nitrate, having 0.45-μm pore size (No. 25A5045 AS, Advantec Toyo, Tokyo). It was found convenient to use a wire loop (Nisshin EM, Tokyo) to transfer grids from medium to medium.

2. Chemicals and Reagents

A 25% glutaraldehyde solution of electron microscopy grade (Nakarai Chemical, Kyoto), paraformaldehyde (Merck, Darmstadt), and sodium cacodylate (Wako Chemical, Osaka, or Sigma Chemical Co., St Louis) were used for the preparation of fixatives.

For covered negative staining, a 0.5% Formvar solution (Nisshin EM, Tokyo), or 2% collodion (Oken Shoji, Tokyo), and sodium silicotungstate (TAAB Labs., England), or uranyl acetate (Merck, Darmstadt) were used.

Enzyme substrates were from Sigma Chemical Co., and other reagents were of analytical grade (Nakarai Chemical Co., Kyoto, Wako Chemicals Co., Osaka). Lead citrate was supplied by K & K Labs., New York.

C. GENERAL PROCEDURE OF ULTRACRYOMICROTOMY

For the demonstration of enzyme activity in ultrathin cryosections, the tissues have to be fixed. It is also possible to make ultrathin cryosections from unfixed tissue blocks, but it is suspected that such enzyme protein is lost from the ultrathin cryosections during the procedure of thawing, rinsing, and incubation. Regarding immunohistochemistry on ultracryosections, Slot et al.[22] stated that insufficient fixation leads to an uncontrolled loss of proteins, resulting in negative staining. To prevent this artificial loss and dislocation of enzymes, we usually use 2% glutaraldehyde in 0.1 *M* of a cacodylate buffer (pH 7.2) containing 8% sucrose or Karnovsky's fixative, a mixture of 5% glutaraldehyde with 4% formaldehyde, for 30 to 60 min at 0 to 4°C as a fixative. Perfusion fixation is possible with both fixatives. In the case of perfusion fixation, about 30 min of immersion fixation is employed after the 5- to 10-min perfusion fixation. The weaker fixation with low aldehyde concentration and short duration preserves a higher enzyme activity in the tissue. However, the activity is sometimes lower because of a leakage of the enzyme from the insufficiently fixed tissue. After fixation and rinsing in a cacodylate (0.1 *M*) buffer (pH 7.2) with 8% sucrose overnight, the trimmed tissue blocks are infiltrated with a series of sucrose-containing media, from 0.6, 1.2, 1.8, and 2.3 *M* sucrose in a 0.1 *M* cacodylate buffer (pH 7.2) for 10 to 20 min, at 4°C. To ensure a sufficient penetration of sucrose, 40-μm sections are used. High concentrations of sucrose provide a cryoprotection effect and also help the sectioning quality at low temperature. After infusion, the tissue blocks are frozen in slushed liquid nitrogen to prevent the insulating effect due to bubbling. To prepare slushed nitrogen, low vacuum (10 Torr) is applied to liquid nitrogen together with tooth picks in a polyethylene beaker. After bubbling, slushed nitrogen foams rise from the beaker. However, they are easily melted down when the slush nitrogen is brought to atmospheric pressure. After repeated slush formation and melting down of the liquid nitrogen, a solid nitrogen block appears at the bottom of the liquid nitrogen. This block remains intact for quite a long time under atmospheric pressure, providing −210°C liquid nitrogen without bubbling. The best preservation without ice crystal damage is seen at the peripheral part of the blocks (1 × 1 × 1 mm). The center of the tissue block usually shows a wide-spread intercellular space and a wide variety of holes in the cytoplasm.

The cutting temperature is −90 to −80°C for the specimen and −80 to −70°C for the glass knife. The interference color on the dry cut ultrathin cryosections is opal-like, of gold, purple, blue, and green. The ultrathin cryosections show severe compression but one can observe the interference color for judging the thickness of sections on the back of the glass knife. Babai[13] discussed the importance of section thickness for the precise localization of enzyme activities, indicating that it is very hard to demonstrate enzyme activities in ultrathin, 30- to 60-nm cryosections. We also recommend the use of blue- or green-colored sections for the demonstration of enzyme activity.

The ultrathin cryosections are retrieved on the surface of a drop of gelatin (0.75%):sucrose (2 *M*) mixture from the cryochamber, and thawed at room temperature to spread the sections on the surface of the mixture and to transfer them to copper grids coated with Formvar film. A binocular microscope is helpful in placing and orienting the sections on the grid for examination. The gelatin:sucrose solution on the wire loop should be picked up just before complete freezing when it is still transparent in the soft gel of the elastic bulb. Ultrathin cryosections may break up on contact, if they are picked up with a solid, frozen gelatin-sucrose bulb. On the other hand, the sections may adhere to the glass knife if picked up without gel formation. The dry ultrathin cryosections usually show some compression and have to be spread with the aid of the surface tension of the thawed gelatin-sucrose drop. If the surface tension is too great, the sections might be damaged or torn apart. As this damage of ultrathin cryosection during picking up tends to occur on the surface of old gelatin-sucrose drops, it is recommended to renew the solution at least every 2 to 3 weeks. In damaged sections, not only the otherwise excellent ultrastructure but also the enzyme activity might be disturbed.

The ultrathin cryosections are transferred to Formvar-coated grids. Before Formvar coating, the grids should be covered with a concentrated Formvar solution, to protect them from the undesirable reaction in the incubation media during incubation, and also from the ammonium sulfide used for color development in the succeeding steps.

Ultrathin cryosections on the grids are kept facing down, on the surface of 1% gelatin and 0.3% sucrose and then moved to a 0.1 *M* cacodylate buffer, pH 7.2, containing 8% sucrose. For incubation, the grids may be placed on the bottom of the incubation vial or on the surface of the reaction mixture. To avoid the precipitate formation on the surfaces of cryosections, the media which had a precipitate or milky appearance are filtered using a cellulose nitrate disposable filter unit (Advantec Toyo, Dismic-25).

D. DEMONSTRATION OF ALKALINE PHOSPHATASE AND 5′-NUCLEOTIDASE ACTIVITIES IN ULTRACRYOSECTION

The results obtained by using cryosections for the demonstration of phosphatase activities are illustrated by Figures 1 and 3. In the jejunum of a rat, alkaline phosphatase activity was demonstrated as dark deposits on the microvilli, on the interdigitating lateral plasma membranes, and in the Golgi apparatus in the supranuclear region. Some intracellular vacuoles also showed deposits of the reaction products in the subapical region and also around the Golgi apparatus. Small tubular alkaline phosphatase-positive structures were also seen in the cytoplasm (Figure 1).

Acid phosphatase activity was seen positive in the lysosomes, in the Golgi apparatus, as well as in the vacuoles and around the Golgi apparatus (Figure 3). Other organelles, such as mitochondria and endoplasmic reticulum, were not defined in the ultracryosections. The clear vacuoles near the Golgi apparatus appeared to be artifacts of ultracryotomy. These observations indicate that the ultrastructure of the cell organelles are difficult to discern without proper staining of the ultracryosections, and without such morphological information, the precise localization of enzyme activity cannot be established. Whenever alkaline phosphatase or acid phosphatase is observed positive in the Golgi lamellae, the metabolic meaning of this positive reaction is far from satisfactory, as it is difficult to orientate the positive cisternae without information as to the negative Golgi lamellae, vacuoles, and vesicles, which indicate the positive Golgi lamellae as *cis* or *trans,* and it is also difficult to evaluate them without knowing the relationship between the endoplasmic reticulum and the lysosomes. To solve those problems, a visualization of the ultrastructure surrounding the histochemical reaction products in the ultracryosections becomes important.

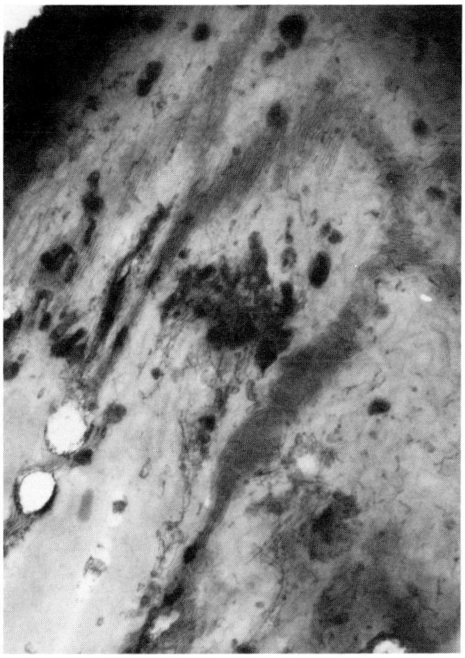

FIGURE 1. Alkaline phosphatase activity in the jejunum without electron staining. The reaction product is seen on microvilli, lateral plasma membrane, Golgi complex, and intracellular vacuoles as dark deposits. (Magnification × 6,800.)

E. STAINING TO VISUALIZE ULTRASTRUCTURE IN ULTRACRYOSECTIONS

1. Electron Staining Methods

There are two electron staining techniques available to visualize ultrathin cryosections: positive staining and negative staining. In the positive staining, ultrathin cryosections are stained with 2% neutral uranyl acetate in 0.15 *M* potassium oxalate, and then embedded in 0.05% uranyl acetate in 2% polyethylene glycol and 0.2% methyl cellulose.

Conventional negative staining usually gives an excellent negative image but is limited to patchy spots from puddles of stain solution in the wide unstained field of the ultracryosections. This problem may be overcome by "covered" negative staining to give uniform results on wide ultracryosections. Figures 2 and 3 show positive staining images of rat kidney cortex.

2. The Procedure of Covered Negative Staining

The procedure for making the covering film consists of the following: a thin collodion or Formvar film is made by dripping a 0.5 to 1% collodion solution in amyl acetate onto the surface of a glass slide floating on distilled water, or by dipping a glass slide into a 0.5% Formvar solution. In the case of collodion, an interfering silver or silver-gray film is picked up on a clean glass slide and dried vertically in a 500-ml beaker covered with aluminum foil. The film on the glass slide is scored into 3-mm squares a little bit larger than the copper grids. Then the small square of collodion film is gently separated from the slide, and transferred on to 4% neutral silicotungstate or 2% uranyl acetate using a small wire loop.

A small piece of collodion or Formvar sheet film is transferred to negative stain solutions by a wire loop. The technique is to apply a small piece of Formvar or collodion sheet film over a drop of the negative staining solutions covering the cryosection mounted on the grid;

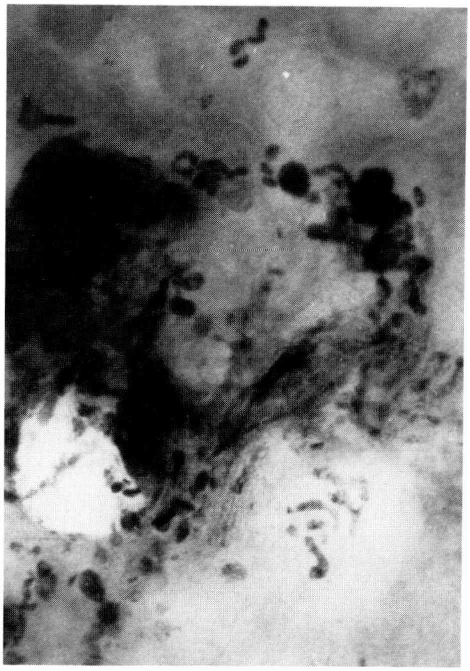

FIGURE 2. Acid phosphatase activity in the jejunum of a rat, without electron staining. Enzyme activity is seen in round lysosomes, and also in cisternae that appear to be part of the Golgi apparatus. However, it is difficult to identify, and may be *trans* cisternae of the Golgi apparatus or a part of GERL.

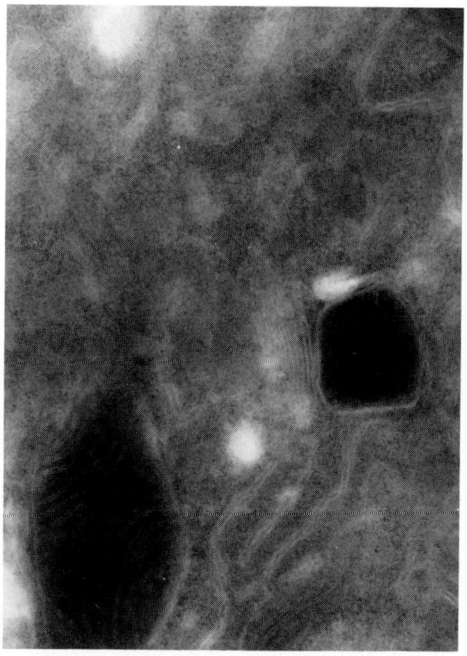

FIGURE 3. The negative-staining image of kidney proximal convoluted tubule. Mitochondria and endoplasmic reticulum are clearly identified. Golgi apparatus is made up of parallel cisternae. Stained with 4% silicotungstate. (Magnification × 27,000.)

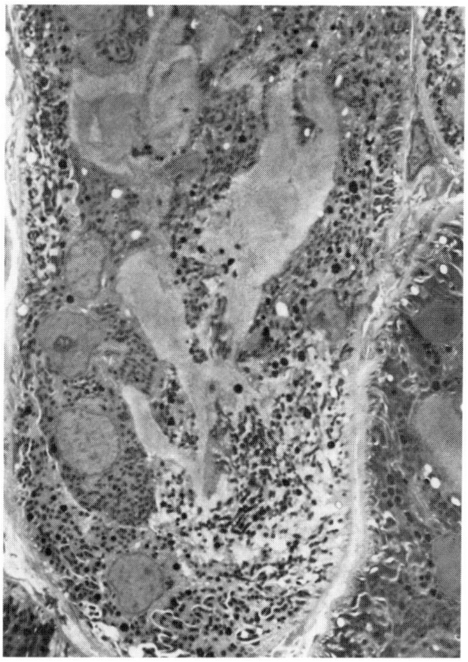

FIGURE 4. The negative-staining image of kidney cortex. The section was stained uniformly with silicotungstate using new covered negative staining. Nuclei are seen round, and numerous mitochondria are scattered in the cytoplasm. (Magnification × 1,400.)

after removing the extra negative staining solution by a small piece of filter paper, dry the preparation under a tungsten bulb. During this covering procedure, the Formvar sheet film spreads the negative staining solution over the ultracryosection.

With this new "covered" negative staining technique, the ultrastructure of ultracryosections is revealed,[14] showing an excellent, uniform negative image of the biological membranes and cytoskeletons.[23] The technique is easy to perform and specimens can be viewed at both high and low magnifications (Figures 2 and 3).

F. NEGATIVE STAINING IN ULTRACRYOSECTION ENZYME CYTOCHEMISTRY

Negative staining with uranyl acetate, phosphotungstic acid, silicotungstate, or ammonium molybdate applied to ultracryosections alter the demonstration of enzyme activities. The histochemical reaction products of enzyme activities were no longer seen after the application of negative staining. These results were also confirmed by light microscopic enzyme histochemistry. Leduc and colleagues,[12] Bauer and Sigarlakie,[21] and Babai[13] already reported that the reaction product, lead phosphate, in the lead methods for demonstrating enzyme activities, could not be detected in ultrathin cryosections after negative staining. Our light microscopic observations also showed that there was no reaction product of acid phosphatase activity by means of Gomori's method in the cryosections (1 μm in thickness) after immersion into 4% silicotungstate (Figure 4). With negative staining treatment the brown color did not develop, even after as short a treatment as 5 min. This suggests that the lead phosphate is dissolved in the negative staining solutions. However, by converting the lead phosphate to lead sulfide, a less soluble stable compound (Fukushima et al.[11]), the reaction product was retained in the ultracryosections after negative staining (Figure 4).

TABLE 1
Method for Negative Staining: Enzyme Histochemistry on Ultrathin Cryosections

1. Fixation: 2% glutaraldehyde, for 30 min, at 0–4°C
2. Rinse in cacodylate-buffered sucrose (8%)
3. Sucrose infusion: up to 2.3 *M* sucrose
4. Freezing with slushed nitrogen
5. Ultrathin cryosectioning
6. Section mounting on grids
7. Rinse in cacodylate-buffered sucrose
8. Incubation for demonstration of enzyme activity
9. Rinse in distilled water
10. Treatment with 1% yellow ammonium sulfide for 1 min
11. Rinse in distilled water
12. Negative staining with 4% neutral silicotungstate
13. Air drying
14. Observation with electron microscope

G. GENERAL PROCEDURES OF ENZYME HISTOCHEMISTRY ON ULTRACRYOSECTION WITH NEGATIVE STAINING

Before incubation, the sections on grids are washed in a cacodylate buffer containing 8% sucrose to remove the gelatin-sucrose mixture, and are then incubated in the medium for the demonstration of enzyme activity at 37°C. Cloudy incubation media, such as the 5′-nucleotidase medium of Wachstein and Meisel,[24] are filtrated through a 0.45-μm nitrocellulose Millipore filter before use to avoid the precipitates on the surface of the cryosections.

After incubation, the sections are well rinsed in distilled water for at least 15 min at room temperature, and then are treated with 1% yellow ammonium sulfide for 1 min. With this procedure, the lead phosphate is converted to brown lead sulfide, permitting light microscopical observations and the application of negative staining. After brief rinsing in distilled water, the sections are stained negatively with 4% neutral silicotungstate (Table 1).

The results obtained using cryosections are illustrated by a few examples. The localization of phosphatase activities were investigated in the epithelial cells of a rat jejunum. The reaction product of acid phosphatase[25] was restricted to the lysosomal matrix and the Golgi apparatus. As negative cisternae were seen on one side of the positive Golgi apparatus, the acid phosphatase-positive cisternae are the *trans* Golgi lamellae (Figure 5) or GERL (Figure 6). Thiamine pyrophosphatase[26] was also positive in several cisternae of the Golgi apparatus at the *trans* face. As negative cisternae are observed on the ultrathin cryosections, the penetration failure of constituent drug of incubation medium to negative Golgi membranes are unlikely, because most of the Golgi cisternae are cut in the ultrathin cryosections (Figure 7). Alkaline phosphatase[27] activity is also observed in the Golgi appartus. The site of the activity may be necessary to confirm comparison with cisternae of the acid phosphatase and thiamine pyrophosphatase positive ones using serial cryosections (Figure 8).

The histochemical method of DAB is applicable to cryosection cytochemistry. The demonstration of cytochrome oxidase[28] is used for the diagnosis of muscle disease, detecting the defect in mitochondria.[29] The diagnostic base of mitochondrial myopathy is on the establishment of histochemical technique to detect cytochrome oxidase activity in all normal mitochondria. However, the problem to detect mitochondrial enzyme activity is difficulty of penetration of incubation mixture path through double membrane of mitochondria. The failure to demonstrate enzyme activity can be overcome by applying cryosections. Figure 9 is a light microscopy of skeletal muscle. Positive reaction product of DAB is seen on mitochondria throughout the cross section. The cryosections are cut at right angles to the

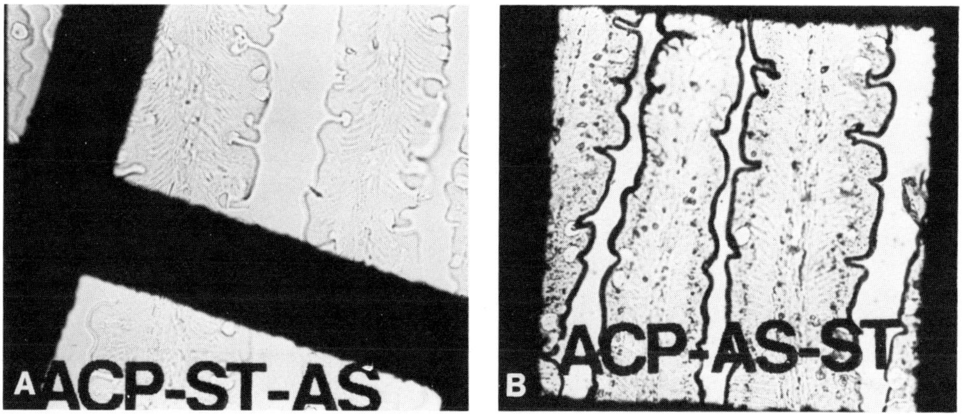

FIGURE 5. Semithin cryosections (1 μm in thickness) were incubated in an acid phosphatase (ACP) medium and then treated with negative staining solution of silicotungstic acid for 5 min. Reaction products disappeared after the sections were immersed in 4% silicotungstate (ST) just after incubation (A). Negative staining was applied after the conversion of lead phosphate of acid phosphatase reaction product into lead sulfide by treating the section with 1% yellow ammonium sulfide (AS) before negative staining in 4% silicotungstate (B). The reaction product remained positive on the microvilli and supranuclear region. (Magnification × 300.)

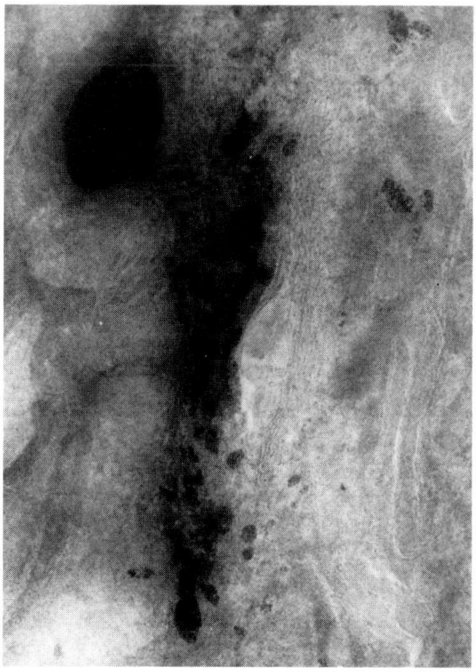

FIGURE 6. Acid phosphatase activity in kidney proximal convoluted tubule cell. After the ammonium sulfide treatment, the sections were stained with 4% silicotungstate. The activity is observed in the lysosomes and in the cisternae of the Golgi apparatus. (Magnification × 15,000.)

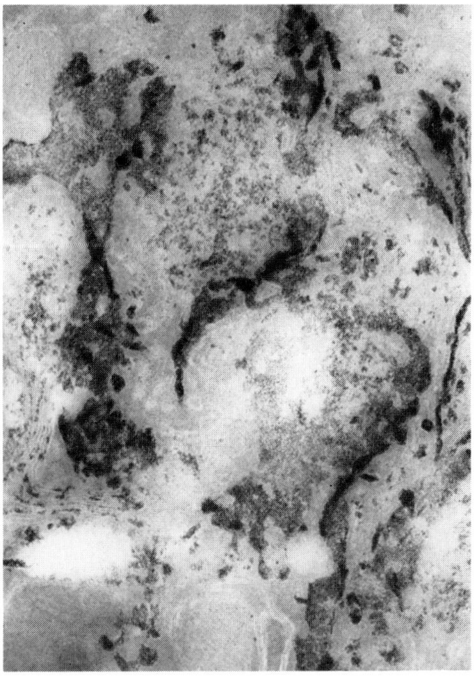

FIGURE 7. Acid phosphatase activity in jejunum epithelium, stained with 4% silicotungstate. The GERL cisternae are clearly seen along with the Golgi cisternae. Lysosomes are also seen positive with acid phosphatase. (Magnification × 32,000.)

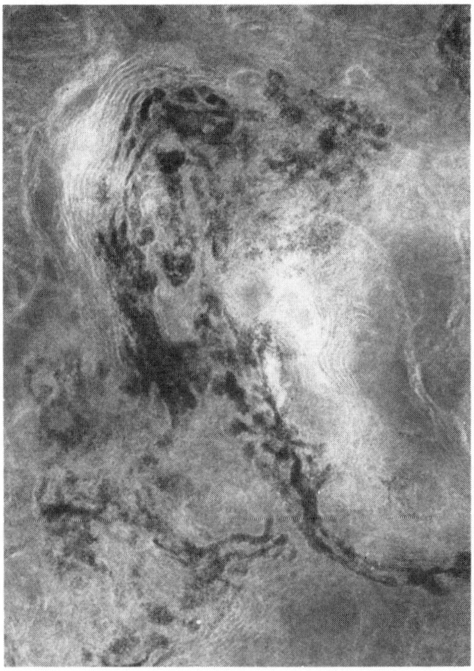

FIGURE 8. Thiamine pyrophosphatase activity in jejunum epithelium, stained with 4% silicotungstate. The activity is seen on the inner lamellae of the Golgi apparatus. The outer and intermediate lamellae are negative with thiamine pyrophosphatase. (Magnification × 44,000.)

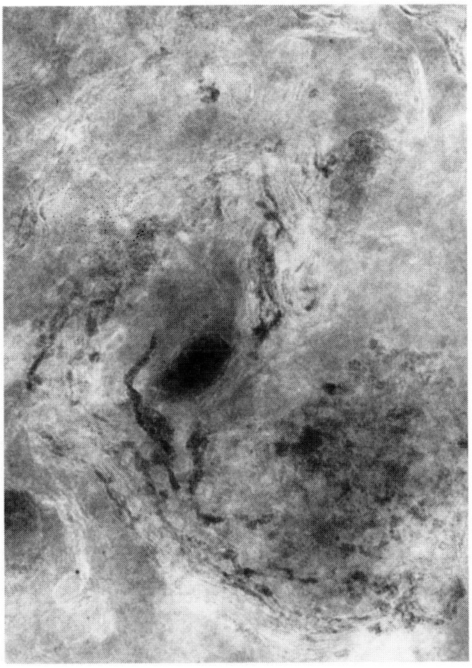

FIGURE 9. Alkaline phosphatase activity in jejunum epithelium, stained with 4% silicotungstate. The activity is seen on the Golgi apparatus. The outer and intermediate lamellae are negative with alkaline phosphatase. (Magnification × 33,000.)

surface after re-embedding in epoxy resin, as indicated in Figure 5. All mitochondria in the cryosection are shown positive for cytochrome oxidase (Figure 10).

H. CONCLUDING REMARKS

As ultracryosections are susceptible to mechanical distortions during sectioning, mounting on a grid, and histochemical incubation, care should be made to avoid the dislocation of enzyme proteins in the tissues. Bauer and Sigarlakie[20] sometimes observed a diffusion of acid phosphatase reaction product in the neighboring cytoplasm around the lysosomes. Some reaction products of adenosine triphosphatase, guanosine triphosphatase, and acid phosphatase activities were also found in the nucleus. However, gentle handling of ultracryosections and proper incubation procedures diminish these artifacts, and ultrathin cryosectioning has several advantages when applied for the localization of enzyme activity at the ultrastructural level.

Of the two electron staining methods applicable to ultrathin cryosections, negative[2] and positive stainings,[6] negative staining is simple and easy, and reliable on ultracryosections, especially with the new covered negative staining;[14] it gives uniform, good contrast of membrane structures and cytoskeleton elements.

One advantage is that the incubation of cryosections is much more easy than that of conventional 40-μm sections. Histochemical observations of cytochrome oxidase activity on ultracryosections cut perpendicular to the surface, showing the reaction products in all mitochondrial crystal spaces, proved the reliability of the method.

Serial ultracryosections permit the demonstration of different enzyme activities in the same cell and even in the same organelle.

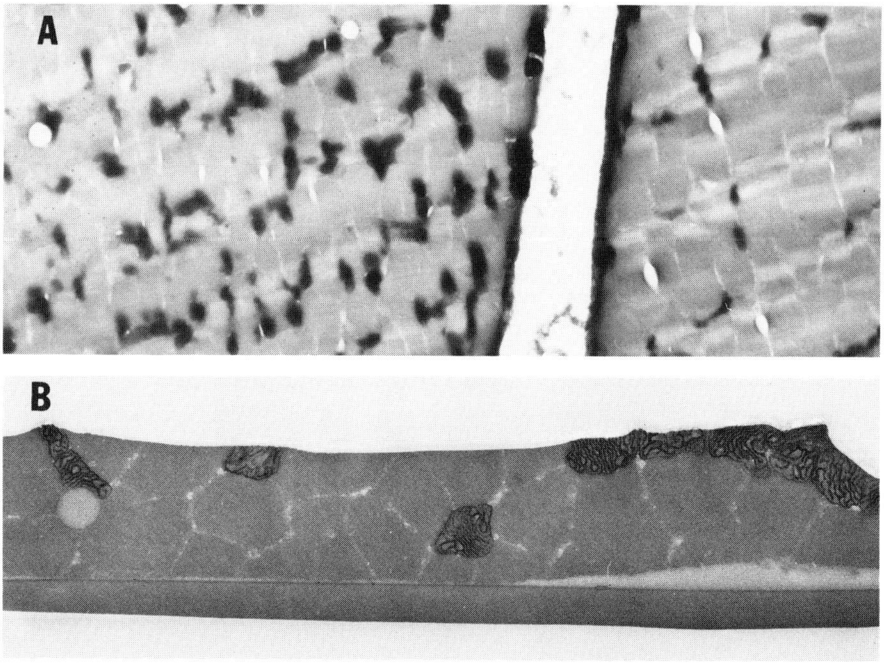

FIGURE 10. (A) Cytochrome *c* oxidase activity in rat anterior tibial muscle. The sections were embedded in 2% polyethylene glycol and 0.2% methyl cellulose before drying. The reaction product is seen in all the mitochondria in the section. (Magnification × 4,600.) (B) Cytochrome *c* oxidase activity in rat anterior tibial muscle. Semithin cryosections (1 μm in thickness) are re-embedded in Epon after demonstration of cytochrome *c* oxidase activity. The section surface is smooth. The activity is observed in all the mitochondria in the outer compartment in the section, as well as on the surface cut into half. Without electron staining. (Magnification × 16,000.)

ACKNOWLEDGMENTS

The authors wish to thank Professors K. Ogawa and T. Barka for their advice, discussion, and critical reading of the manuscript. The authors also acknowledge Professor H. Yamashita for his encouragement throughout this work. We also express our gratitude to Miss S. Komuro, Miss K. Onozaki, and Miss C. Ishijima for their excellent technical assistance.

REFERENCES

1. **Fernandez-Moran, H.,** Application of the ultrathin freezing-sectioning technique to the study of cell structures with the electron microscope, *Ark. Fys.*, 4, 471, 1952.
2. **Bernhard, W. and Leduc, E. H.,** Ultrathin frozen sections. I. Methods and ultrastructural preservation, *J. Cell Biol.*, 34, 757, 1967.
3. **Christensen, A. K.,** Frozen thin sections of fresh tissue for electron microscopy, with a description of pancreas and liver, *J. Cell Biol.*, 51, 7721, 1971.
4. **Tokuyasu, K. T.,** A technique for ultracryotomy of cell suspensions and tissues, *J. Cell Biol.*, 57, 551, 1973.
5. **Griffiths, G., McDowall, A., Back, R., and Dubochet, J.,** On the preparation of cryosections for immunocytochemistry, *J. Ultrastruct. Res.*, 89, 65, 1984.

6. **Tokuyasu, K. T.,** Application of cryoultramicrotomy to immunocytochemistry, *J. Microsc.,* 143, 139, 1986.
7. **Appleton, T. C.,** A cryostat approach to ultrathin "dry" frozen sections for electron microscopy: a morphological and X-ray analytical study, *J. Microsc.,* 100, 49, 1974.
8. **Wroblewski, R., Anniko, M., and Sakai, T.,** A unique striated muscle: further morphological and X-ray microanalytical investigation of the stapedius muscle of the guinea pig using thin and thick cryosections, *J. Submicrosc. Cytol.,* 16, 479, 1984.
9. **Futaesaku, Y. and Mizuhira, V.,** Negative-staining autoradiography: a new technique for ultracryotomy utilizing an interposed film, *J. Histochem. Cytochem.,* 34, 1085, 1986.
10. **Fujimoto, K., Araki, N., Ogawa, K.-S., Kondo, S., Kitaoka, T., and Ogawa, K.,** Ultracytochemistry of calmodulin binding sites in myocardial cells by staining of frozen thin sections with colloidal gold-labeled calmodulin, *J. Histochem. Cytochem.,* 37, 249, 1989.
11. **Fukushima, O., Yamashita, H., Sakai, T., and Saito, T.,** Application of ultrathin cryo-sections to enzyme histochemistry, *Acta Histochem. Cytochem.,* 22, 131, 1989.
12. **Leduc, E., Bernhard, W., Holt, S. J., and Tranzer, J. P.,** Ultrathin frozen sections. II. Demonstration of enzymic activity, *J. Cell Biol.,* 34, 773, 1967.
13. **Babai, F.,** Application of cryoultramicrotomy to enzyme cytochemistry, in *Electron Microscopy of Enzymes,* Hayat, M. A., Ed., Vol. 5, Van Nostrand Reinhold, New York, 1977, 162.
14. **Kuhlmann, W. D. and Viron, A.,** Cross-linked albumin supporting matrix in ultrathin cryo microtomy, *J. Ultrastruct. Res.,* 41, 385, 1972.
15. **Leung, T. K. and Babai, F.,** Detection ultracytochemique de la 5′-nucleotidase sur les coupes a congelation ultrafines, *J. Microsc. (Paris),* 21, 111, 1974.
16. **Ryder, T. and Bowen, I. D.,** The use of X-ray microanalysis to investigate problems encountered in enzyme cytochemistry, *J. Microsc.,* 101, 143, 1974.
17. **Vorbrodt, A. and Bernhard, W.,** Essais de localisation au microscopie electronique de l'activite phosphatasique nucleaire danc coupes a congelation ultrafines, *J. Microsc. (Paris),* 7, 195, 1968.
18. **Zotikov, L. and Bernhard, W.,** Localization au microscope electronique de l'activite de certaines nucleases dans des coupes a congelation ultrafines, *J. Ultrastruct. Res.,* 30, 642, 1970.
19. **Kuhlmann, W. D. and Viron, A.,** Diaminobenzidine cytochemistry in cryo-ultra cryotomy for the detection of peroxidases, *Histochemistry,* 54, 331, 1977.
20. **Bauer, H. and Sigarlakie, E.,** Cytochemistry on ultrathin frozen sections of yeast cell, *J. Mirosc.,* 99, 205, 1973.
21. **Bauer, H. and Sigalakie, E.,** Localization of alkaline phosphatase in Saccharomyces cerevisiae by means of ultrathin frozen sections, *J. Ultrastruct. Res.,* 50, 208, 1975.
22. **Slot, J. W., Posthuma, G., Chang, L.-Y., Crapo, J. D., and Geuze, H. J.,** Quantitative aspects of immunogold labeling in embedded and nonembedded sections, *Am. J. Anat.,* 185, 271, 1989.
23. **Sakai, T., Fukushima, O., and Saito, T.,** Negative staining for ultrathin cryosection, in *Proc. XIth Int. Cong. Electron Microscopy,* Imura, T., Maruse, S., and Suzuki, T., Eds., Japan Society of Electron Microscopy, Tokyo, 1986, 2221.
24. **Wachstein, M. and Meisel, M.,** Histochemistry of hepatic phosphatases at a physiologic pH, *Am. J. Clin. Pathol.,* 27, 13, 1957.
25. **Gomori, G.,** An improved histochemical technique for acid phosphatase, *Stain Technol.,* 25, 81, 1950.
26. **Novikoff, A. B. and Goldfischer, S.,** Nucleoside diphosphatase activity in the Golgi apparatus and its usefulness for cytological studies, *Proc. Natl. Acad. Sci. U.S.A.,* 47, 802, 1961.
27. **Mayahara, H., Hirano, H., Saito, T., and Ogawa, K.,** The new lead citrate method for the ultracytochemical demonstration of activity of non-specific alkaline phosphatase (orthophosphoric monoester phosphohydrolase), *Histochemie,* 11, 88, 1967.
28. **Seligman, A. M., Plapinger, R. E., Wasserkrug, H. L., Deb, C., and Hanker, J. S.,** Ultrastructural demonstration of cytochrome oxidase activity by the Nadi reaction with osmiophilic reagents, *J. Cell Biol.,* 34, 787, 1967.
29. **Oguro, N., Yanagisawa, M., Onozaki, K., and Saito, T.,** An improved method for electron microscopic demonstration of cytochrome C oxidase activity in cells in monolayer culture, *Acta Histochem. Cytochem.,* in press, 1991.

V. CRYOELECTRON MICROSCOPY

Masahiro Sakai

A. INTRODUCTION

Cryoelectron microscopy has been considered as a potential method for preserving materials in a state close to native conditions. By applying a rapid freezing technique,[1] one can investigate fine molecular details of biological specimens without artifacts caused by sample preparation procedures. In particular, cryoelectron microscopic observations are quite effective for the analysis of three-dimensional particle arrangements of isolated, frozen-hydrated samples such as ribosomal subunits,[2] microtubules,[3] gap junctions,[4] and viruses[5] by negative staining techniques with image reconstruction and/or X-ray diffraction. However, there have been very few reports concerning enzyme cytochemistry using cryoelectron microscopic techniques, because of the lack of useful methods.

In this chapter, we introduce an approach utilizing the cryoelectron microscope for enzyme cytochemistry by demonstration of Ca^{2+}-ATPase activity in isolated gap junctions. In principle, this specific method will be applicable for a wide variety of cell types and enzymes.

Gap junctions form specialized regions of intercellular contacts at which communication between cells occur. They are built up of pairs of plasma membranes connected with membrane proteins, called "connexons".[6] The connexons are arranged in a hexagonal lattice in the plane of the plasma membrane, and protrude into the extracellular space to create a 2- to 3-nm-wide gap between the membranes.[7,8] The connexon itself is an assembly of six identical transmembrane subunits,[9] pairs of which span both membrane bilayers and form a narrow continuous cell-to-cell aqueous pathway, a hydrophilic channel about 2 nm in diameter, through the membrane.[7] This generally well-accepted model of gap junction structure is based on physiological, electron microscopic, and X-ray diffraction data.[10-12] However, three-dimensional structural variations of gap junctions and differences in distribution of connexon particles related to cellular functions have been reported in various tissues.[13-15] These discrepancies may be derived from differences of experimental protocols or specimen conditions. Actually, it has been proved that electron beam irradiation damage,[16] detergents used for isolation procedures,[17] and staining methods[18] contribute to variations of gap junction image. This is why the cryoelectron microscope was applied to examine isolated gap junctions under near-to-native state by keeping the specimen temperature at −160°C during observation.

On the other hand, from the viewpoint of enzyme cytochemistry, the activities of three enzymes (Ca^{2+}-ATPase, adenylate cyclase, and guanylate cyclase) have been demonstrated in the gap junction of various tissues.[19] However, these enzymes were detected mostly in fixed specimens and observed in plastic resin thin sections by conventional electron microscope, so we examined the localization of Ca^{2+}-ATPase activity cytochemically on isolated, non-fixed heart gap junctions by cryoelectron microscope.

B. METHODS AND INSTRUMENTS

Cardiac gap junctions from four female or male Wistar rats, 7 to 8 weeks old and weighing 200 to 300 g, were isolated by the method of Manjunath et al.[20] The final gap junction pellet was suspended in 5 m*M* Tris, pH 10, and isolated gap junction specimens were deposited on Formvar-coated copper grids by adhesion. The grids were briefly washed with distilled water and floated on a drop of the incubation medium for the demonstration of Ca^{2+}-ATPase[21] for 10 min at 37°C. The reaction medium consisted of 250 m*M* glycine-25 m*M* KOH buffer, pH 9.0; 3 m*M* ATP-2Na (Sigma Chem. Co., St. Louis); 5 m*M* $CaCl_2$;

4 m*M* lead citrate; and 2.5 m*M* levamisol as a nonspecific alkaline phosphatase inhibitor. As cytochemical control, isolated gap junctions adhered to a Formvar-coated grid were incubated in a reaction medium lacking $CaCl_2$ or including 0.1 m*M* quercetin as an inhibitor of Ca^{2+}-transport ATPase. After washing on a drop of distilled water several times, the specimens on grids were rapidly frozen with liquid nitrogen without drying. The grid was then placed in a special grid capsule and loaded on the tip of the cryotransfer holder by using special specimen exchange instruments.

The procedure of specimen setting on the cryotransfer holder was carried out in the cryotransfer holder tip-housing box filled with liquid nitrogen. The samples loaded on the tip of the cryotransfer holder were inserted in a JEOL 2000 ES electron microscope equipped with a cryopump system for maintaining high vacuum instead of a diffusion-pump system when frozen-hydrated specimen are examined in an electron microscope. The localization of Ca^{2+}-ATPase activity on isolated heart gap junction was detected at −160°C at 200 kV accelerating voltage. Some samples were air dried after incubation of Ca^{2+}-ATPase activity, and examined by a conventional electron microscope. Furthermore, instead of the reaction procedure for isolated gap junctions applied onto Formvar-coated grids, the final gap junction suspension was mixed with a threefold or larger volume of the incubation medium, for the demonstration of Ca^{2+}-ATPase activity, for 20 min at room temperature. The mixture was pelleted (8000 × *g* for 30 min), washed with buffer solution, repelleted, and fixed with 2% glutaraldehyde in 0.1 *M* phosphate buffer, pH 7.4, for 30 min at 4°C. The pellet was processed for routine electron microscope observation. The ultrathin sections were stained with uranyl acetate and examined in a JEOL 1200 EX electron microscope at 100 kV accelerating voltage.

C. RESULTS

The activity of Ca^{2+}-ATPase in isolated heart gap junction was observed as electron-dense deposits of lead phosphate by cryoelectron microscope (Figure 1). The reaction products of Ca^{2+}-ATPase were detected on the entire surface of the isolated gap junction plaque with a quite uniform distribution pattern, clearly showing each connexon particle, so that the image obtained by enzyme reaction is similar to that of negatively stained gap junctions. The size of the reaction product which penetrated into the central region of the connexons was 6 to 7 nm in diameter; connexons were observed as 17- to 20-nm particles delineated roundly by the reaction products. By conventional electron microscopic observation, Ca^{2+}-ATPase activity could also be detected on the whole surface of isolated gap junctions, but reaction products were observed to be distributed with different sizes and patterns. That is to say, some reaction products seemed to be located along the arrangement of connexons with similar feature to those seen in cryoelectron microscopic specimens, others were aggregated to create small clumps without any relation to the connexon arrangement (Figure 2). The size of the connexons in the images obtained by the enzyme reaction products varied from connexon to connexon over a range of 9 to 13 nm in diameter, which was a little smaller than that of cryoelectron microscopic images. In control experiments, the exclusion of $CaCl_2$ from the incubation medium resulted in the complete disappearance of reaction products (Figure 3), and 0.1 m*M* quercetin revealed a complete inhibition of Ca^{2+}-ATPase activity of isolated gap junctions (Figure 4).

It is well known that the distribution pattern of connexon particles in gap junctions is quite different, depending on the cell type and on cellular functional state, even in the same cells.[13,14] In addition, the electron beam irradiation damage[16] and sample preparation, including staining[18] and detergent treatment,[17] which are required for the gap junction isolation procedure, are other important factors affecting the gap junction structure. We have demonstrated that the connexon arrangement of isolated, negatively stained gap junction reveals

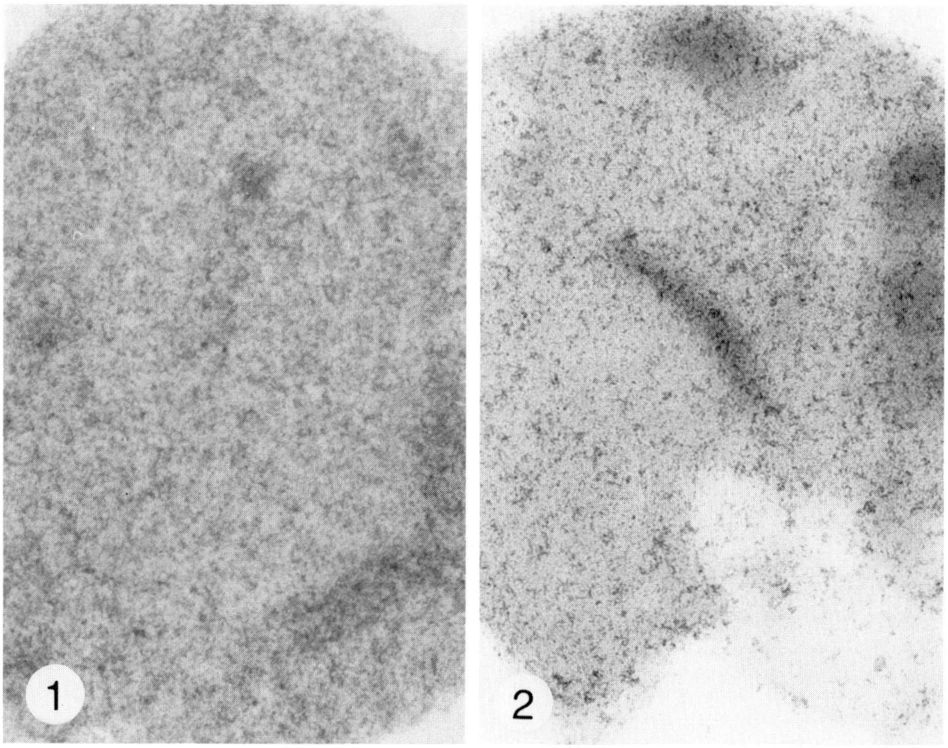

FIGURE 1. Ca^{2+}-ATPase activity on an isolated heart gap junction under a cryoelectron microscope. Electron-dense deposits of reaction products are observed over the whole area of the gap junction with a uniformly distributed pattern. (Magnification × 114,000.)

FIGURE 2. Ca^{2+}-ATPase activity on an isolated gap junction under a conventional electron microscope. Compared with Figure 1, the distribution pattern of reaction products is not uniform. (Magnification × 114,000.)

quite a different pattern between cryo- and conventional images, and that the average diameter of connexons in cryo-images was a little bigger than that of the conventional images.[22] It is suggested from this result that the change of localization of reaction products for Ca^{2+}-ATPase activity may be caused by alternations of the connexon arrangement.

In the plastic resin ultrathin sections, the reaction products of Ca^{2+}-ATPase activity on isolated gap junctions were much coarser and aggregated, forming bigger clumps than those observed on isolated gap junctions adhered to grids (Figures 5 and 6). They were distributed without any close relationship to connexon particles. Fujimoto et al.[19] also reported a different localization of reaction product of adenylate cyclase in isolated heart gap junctions between fixed and unfixed samples by conventional electron microscope.

From these results, it is obvious that the more treatment increases before and after incubation for the enzyme reaction, the greater the probability of false localization. In this respect, observation of isolated gap junctions under the cryoelectron microscope is ideal because no procedures are required after incubation. However, since isolated samples from living tissues may not represent their native structure, further improvements and developments of cryotechniques or other new methods are essential for the application of the cryoelectron microscope to the field of enzyme cytochemistry.

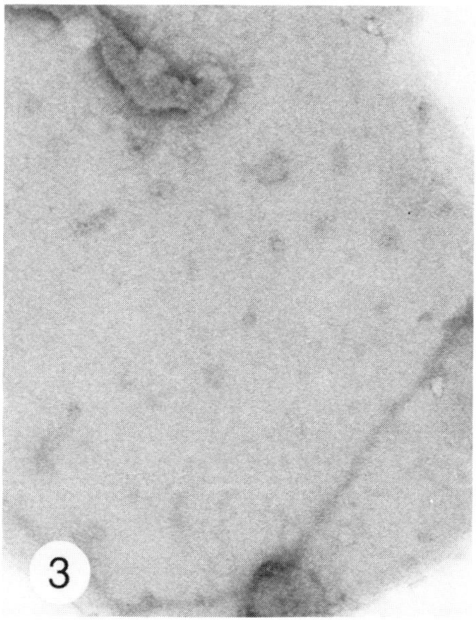

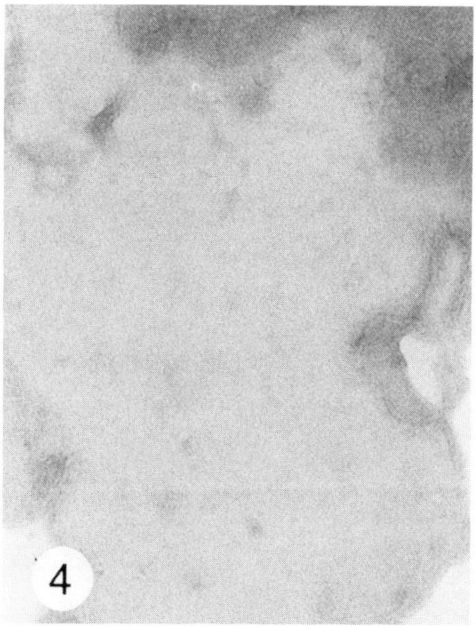

FIGURE 3. Cytochemical control incubated in the reaction medium without Ca^{2+} ions. No reaction products are observed on the gap junction. (Magnification × 120,000.)

FIGURE 4. Cytochemical control incubated in the reaction medium including 0.1 m*M* quercetin. The enzyme activity completely inhibited. (Magnification × 120,000.)

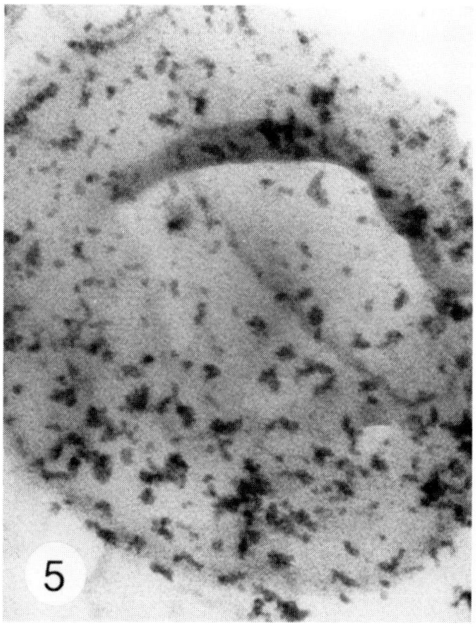

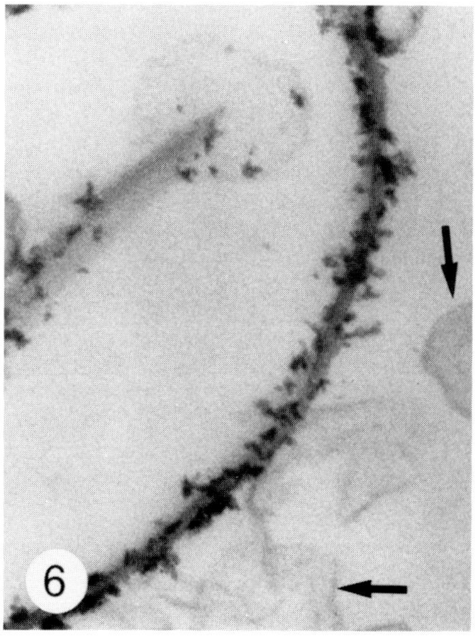

FIGURE 5. Ca^{2+}-ATPase activity of an isolated gap junction in an on-face view in plastic resin thin section under a conventional electron microscope. Reaction products were seen as aggregated clumps. (Magnification × 162,000.)

FIGURE 6. Ca^{2+}-ATPase activity of isolated gap junction in a transverse view. Reaction products were observed on both surfaces of gap junction but not on the other membrane fragments (arrows). (Magnification × 158,000.)

REFERENCES

1. **Heuser, J. E.,** Preparing biological samples for stereomicroscopy by the quick-freeze, deep-etch, rotary-replication technique, in *Methods in Cell Biology,* Vol. 22; *Three-Dimensional Ultrastructures in Biology,* Turner, J., Ed., Academic Press, New York, 1981.
2. **Milligan, R. A. and Unwin, P. N. T.,** Location of exit channel nascent protein 80S ribosome, *Nature,* 308, 32, 1986.
3. **Mandelkow, E.-M., Rapp, R., and Mandelkow, E.,** Microtubule structure studied by quick freezing: cryo-electron microscopy and freeze fracture, *J. Microsc.,* 141, 361, 1985.
4. **Makowski, L.,** Structural domains of gap junctions: implications for the control of intercellular communication, in *Gap Junctions,* Bennett, M. V. L. and Spray, D. C., Eds., Cold Spring Harbor Laboratory, New York, 1985, 13.
5. **Adrian, M., Dubochet, J., Lepault, J., and McDowell, A. W.,** Cryo-electron microscopy of viruses, *Nature,* 308, 32, 1984.
6. **Goodenough, D. A.,** In vitro formation of gap junction vesicles, *J. Cell Biol.,* 68, 220, 1976.
7. **Bennett, M. V. L.,** Function of electric junction in embryonic and adult tissue, *Fed. Proc.,* 32, 65, 1973.
8. **Revel, J. P. and Karnovsky, M. J.,** Hexagonal array of subunits in intercellular junctions of the mouse heart and liver, *J. Cell Biol.,* 33, c7, 1967.
9. **Caspar, D. L. D., Goodenough, D. A., Makowski, L., and Philips, W. C.,** Gap junction structures. I. Correlated electron microscopy and X-ray diffraction, *J. Cell Biol.,* 74, 605, 1977.
10. **Loewenstein, W. R.,** Permeability of membrane junctions, *Ann. N.Y. Acad. Sci.,* 137, 441, 1966.
11. **Mokawski, L., Caspar, D. L. D., Philips, W. C., and Goodenough, D. A.,** Gap junction structures. II. Analysis of the X-ray data, *J. Cell Biol.,* 74, 629, 1977.
12. **McNutt, N. S. and Weinstein, R. S.,** The ultrastructure of the nexus. A correlated thin section and freeze-cleave study, *J. Cell Biol.,* 47, 666, 1970.
13. **Page, E., Karrison, T., and Upshaw-Earley, J.,** Freeze-fractured cardiac gap junctions: structural analysis by three methods, *Am. J. Physiol.,* 244, H525, 1983.
14. **Peracchia, C. and Dulhunty, A. F.,** Structural changes with functional incoupling, *J. Cell Biol.,* 70, 419, 1976.
15. **Schellens, J. P. M., Blange, T., and Groot, K.,** Gap junction structure in rat liver parenchymal cells after in vivo ischemia, *Virchows Arch. B,* 53, 347, 1987.
16. **Baker, T. S., Caspar, D. L. D., Hollingshead, C. J., and Goodenough, D. A.,** Gap junction structures. IV. Asymmetric feature revealed by low-irradiation microscopy, *J. Cell Biol.,* 96, 204, 1983.
17. **Zampighi, G. and Unwin, P. N. T.,** Two forms of isolated gap junctions, *J. Mol. Biol.,* 135, 451, 1979.
18. **Henderson, D., Eibl, H., and Weber, K.,** Structure and biochemistry of mouse hepatic gap junction, *J. Mol. Biol.,* 132, 193, 1979.
19. **Fujimoto, K., Narimoto, K., Ogawa, K. S., Kondo, S., and Ogawa, K.,** Review: cytochemical aspect of gap junction: enzyme activity, hydrophilic channel and charge, *Acta Histochem. Cytochem.,* 18, 467, 1985.
20. **Manjunath, C. K., Going, G. E., and Page, E.,** Cytoplasmic surface and intramembrane components of rat heart gap junctions, *Am. J. Physiol.,* 246, H865, 1984.
21. **Ando, T., Fujimoto, K., Mayahara, H., Miyajima, H., and Ogawa, K.,** A new one-step method for the histochemistry and cytochemistry of Ca^{2+}-ATPase activity, *Acta Histochem. Cytochem.,* 14, 705, 1981.
22. **Sakai, M., Tanaka, T., and Ogawa, K.,** Ultrastructure and Ca^{2+}-ATPase activity of gap junctions under cryo-electron microscopic observation, *Acta Histochem. Cytochem.,* 21, 535, 1988.

Chapter 1

ELECTRON MICROSCOPIC CYTOCHEMISTRY OF ENZYMES

1.3 Basic Techniques for Major Groups of Enzymes

1.3.1 Oxidoreductases

TABLE OF CONTENTS

0-8493-6012-9/93/$0.00 + $.50

I. SUCCINATE DEHYDROGENASE

Kioko Kawai

A. INTRODUCTION

This flavoprotein enzyme is one of the important dehydrogenases and a main component of the tricarboxylic (TCA) cycle with which we are concerned.[1] This enzyme does not require a coenzyme and catalyzes the reaction between succinate and fumarate as follows:

$$\text{succinate} + \text{acceptor} \rightleftharpoons \text{fumarate} + \text{acceptor}$$

Succinate dehydrogenase (SDH) is localized only in mitochondria and is considered as a marker enzyme of the mitochondria. It can be demonstrated as succinate tetrazolium reductase[1-6] or succinate ferricyanide reductase.[7-10]

The two following methods have been reported for the ultrastructural demonstration of this enzyme.

B. COPPER-FERROCYANIDE METHOD

A method for the ultrastructural localization of SDH, employing ferricyanide as an electron acceptor, was introduced by Ogawa et al.[2] Ferricyanide can accept electrons from the respiratory chain. Ferrocyanide, formed as a result of reduction of ferricyanide by enzymatic activity, can be trapped by copper ions to produce insoluble, electron-dense copper ferrocyanide (Figure 1).

A modification of the original method of Ogawa et al.[2] was made by Kerpel-Fronius and Hajos,[3] who used sodium-potassium tartrate as a chelator.

The medium for succinate-ferricyanide reductase (Ogawa et al.[2]) is given below:

		Final concentration
Sodium succinate	100 mg	18 m*M*
0.1 *M* Phosphate buffer (pH 7.0)	13 ml	65 m*M*
0.1 *M* Sodium citrate	0.6 ml	3 m*M*
30 m*M* Copper sulfate	2.0 ml	30 m*M*
Double-distilled water	2.4 ml	
5 m*M* Potassium ferricyanide	2.0 ml	0.5 m*M*
Sucrose	2–3 g	0.29–0.44 m*M*

The ingredients are mixed, with constant stirring, in the order indicated. The final pH is adjusted to pH 7.0; total volume: 20 ml.

C. TETRAZOLE METHOD

Tetrazolium salts have been used as electron acceptors for the ultrastructural localization of oxidoreductase. The medium for succinate tetrazolium reductase is given below (Seligman[8]):

Sodium succinate	0.05 *M*
Distyryl ditetrazolium (DS-NBT)	0.5–1.0 mg/ml
Phosphate buffer (pH 7.0)	0.05 *M*
Sucrose	2.5%

Ingredients are added and rinsed in the above order.

Several different tetrazolium salts, such as tetra-NBT (TNBT), thiocarbanyl-NBT (TC-NBT), and DS-NBT have been employed to demonstrate SDH activity at the ultrastructural

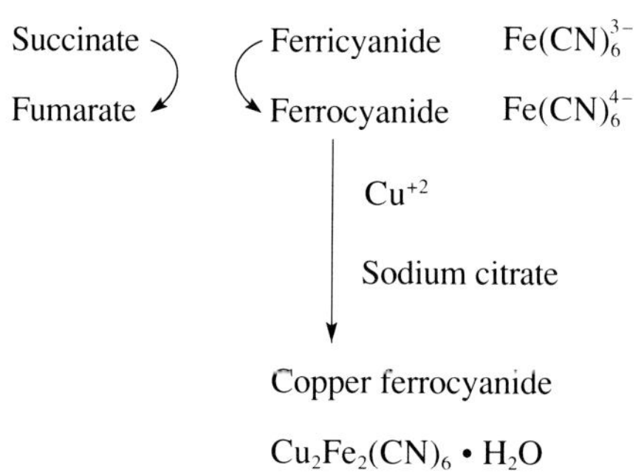

FIGURE 1. The mechanism of the reaction of the copper-ferrocyanide method is shown. Copper-ferrocyanide is the final reaction product.

TABLE 1
Procedures

1. Prefixation of tissues
2. After fixation, the tissues are washed for 30 to 60 min in three changes of buffer
3. Tissue sectioning
4. Incubation of specimens
5. After incubation, the specimens are washed in three changes of buffer
6. The specimens are stored in buffer at 4°C overnight
7. Postfixation in 1% OSO_4 for 60 min
8. Dehydration through graded alcohols
9. Embedding
10. Staining with uranyl acetate/or lead citrate

level.[1,6] TC-NBT penetrates into tissues to a depth of about 30 μm. Diformazan deposits from TC-NBT are observed to be uniformly distributed along both surfaces of the mitochondrial cristae with encroachment of the intracristal spaces.[8] DS-NBT is more reliable, and more readily osmicated than TC-NBT.[7,8]

D. GENERAL PROCEDURES

The basic procedures are summarized in Table 1. Some modification should be attempted depending on the tissues employed.

Prefixation — Fresh tissues are preferable for the ultracytochemical demonstration of the enzyme. However, it is not enough to preserve the cellular structure, and a brief fixation is desirable to obtain a clear discrete localization.[11,12] There have been reports concerning the fixation of tissues; 0.25% glutaraldehyde and 1% paraformaldehyde for 30 min at 4°C; 6.25% hydroxyadipaldehyde for 3 to 30 min at 4°C; 2% paraformaldehyde for 2 to 10 min at 4°C; 4% paraformaldehyde for 5 min at 4°C; and 5% buffered formalin for 5 min at 4°C. SDH activity was not inhibited significantly by these fixation procedures.

Tissue sectioning — Fresh or fixed tissues are cut into 40- to 60-μm-thick sections by an automicrochopper (Model KO-1, Horo shinichi Company, Osaka), Vibratome (Oxford Laboratories, U.S.), Microslicer (D.S.K., Dosaka, EMCO Ltd., Kyoto), or Tissue Sectioner (Sorvall, TC-2). Fresh or fixed tissues are also cut into small pieces (0.5 m^3) with a razor.

Incubation of specimens — Specimens are placed in the incubating medium as mentioned above for 10 to 60 min at room temperature, or 10 to 60 min at 37°C, with mechanical shaking. Dimethyl sulfoxide (DMSO) could be added to the medium to accelerate penetration into the tissues.[13] Phenazine methosulfate (PMS) has been thought to serve as a direct electron transfer and result in the reduction of potassium ferricyanide, which prevents the false

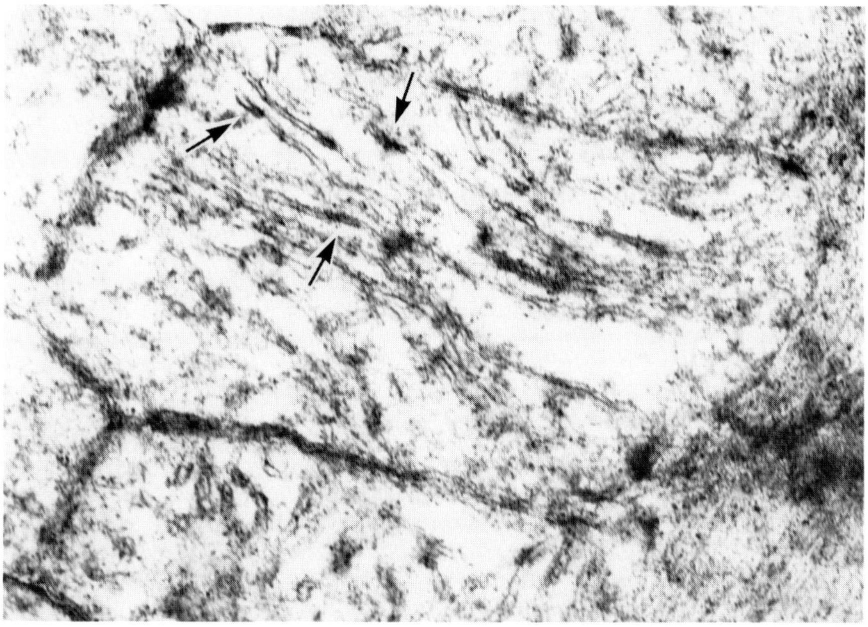

FIGURE 2. The reaction products of SDH are observed in intracristal spaces (arrows) of rat heart muscle.

localization of enzyme activity.[14,15] Adding PMS to the medium increases the electron density and reaction products.[14,15] The optimal concentration of PMS is 0.5 to 0.8 m*M*.[11,12] Incubation should be carried out in darkness when PMS is added to the solution.

Postfixation, dehydration, and embedding — After being rinsed in a buffer at room temperature, the sections are fixed in cold osmium tetroxide for 1 h at room temperature or at 4°C, dehydrated through graded alcohols and propylene oxide and embedded in Epon, GMA (glycol methacrylate), or Spurr medium.

Control media — To confirm the specificity of the reaction, the following[2,6] controls are recommended: (1) substrate-free medium — the sections are incubated in a medium from which the substrate is omitted; (2) medium containing various inhibitors — sodium malonate (36 m*M*), antimycin A (1 to 30 μg/ml), rotenone (33 to 330 μ*M*) are added to the incubating medium for SDH. The required amount of antimycin A is dissolved in 0.2 ml absolute alcohol, and rotenone is dissolved in 0.2 ml acetone. In these cases, the amount of double-distilled water should be decreased to 2.2 ml.

E. RESULTS

The reaction products are localized in the intracristal and outer spaces and in the unit membrane structure of the inner membrane system, including that of the cristae (Figure 2). No deposit is seen in the mitochondrial matrix.

Sodium malonate at a concentration of 18 m*M* inhibited the succinate ferricyanide reductase activity. When antimycin A is added to the incubating medium, the enzymatic reaction is inhibited. Positive reaction is not as evident when tetrazolium salts are used as electron acceptors. Formazan deposits as the reaction products are low in contrast and are ill defined. On the other hand, the reaction products of copper ferrocyanide are high in contrast and are well demarcated. The copper ferrocyanide method is superior to the tetrazolium salt method. Prolonged incubation time results in a reaction product filling the intracristal spaces, whereas a short incubation time results in a product which appears mainly at opposite points or both sides of the cristae.[4]

REFERENCES

1. **Pearse, A. G. E.,** Electron histochemistry, in *Histochemistry,* 3rd ed., Vol. 2, Churchill Livingstone, Edinburgh, 1972, 1286.
2. **Ogawa, K., Saito, T., and Mayahara, H.,** The site of ferricyanide reduction by reductases within mitochondria as studied by electron microscopy, *J. Histochem. Cytochem.,* 16, 49, 1968.
3. **Kerpel-Fronius, S. and Hajos, F.,** The use of ferricyanide for the light and electron microscopic demonstration of succinic dehydrogenase activity, *Histochemie,* 14, 343, 1968.
4. **Kalina, M., Weavers, B., and Pearse, A. G. E.,** Ultrastructural localization of succinate dehydrogenase in mouse liver mitochondria; a cytochemical study, *J. Histochem, Cytochem.,* 10, 124, 1971.
5. **Bogitsh, B. J.,** The cytochemical localization of succinate dehydrogenase in the spleen mitochondria of the gastropod Biomphalaria glabrata. A comparison of two methods, *J. Histochem. Cytochem.,* 23, 75, 1975.
6. **Makita, T.,** Ultracytochemistry of oxidoreductase, in *Shinkosososhikikagaku,* Ogawa, K. and Takeuchi, T., Eds., Asakura, Tokyo, 1980, 150 (in Japanese).
7. **Seligman, A. M., Ueno, H., Morizono, Y., Wasserkrug, H. L., Katzoff, L., and Hanker, J. S.,** Electron microscopic demonstration of dehydrogenase activity with a new osmiophilic ditetrazolium salt (TC-NBT), *J. Histochem. Cytochem.,* 15, 1, 1967.
8. **Seligman, A. M., Nir, I., and Plapinger, R. E.,** An osmiophilic disyryl ditetrazolium salt (DS-NBT) for ultrastructural dehydrogenase activity, *J. Histochem. Cytochem.,* 19, 273, 1971.
9. **Kalina, M., Plapinger, R. E., Hoshino, Y., and Seligman, A. M.,** Nonosmiophilic tetrazolium salts that yield osmiophilic, lipophobic formazans for ultrastructural localization of dehydrogenase activity, *J. Histochem. Cytochem.,* 20, 685, 1972.
10. **Spector, G. J.,** The ultrastructural cytochemistry of lactic dehydrogenase, succinic dehydrogenase, dihydronicotinamide adenine dinucleotide diaphorase and cytochrome oxidase activities in hair cell mitochondria of the guinea pig cochlea, *J. Histochem. Cytochem.,* 23, 216, 1975.
11. **Kawai, K.,** Ultracytochemical demonstration of dehydrogenases, in *Soshikisaibokagaku,* Japan Society of Histochemistry and Cytochemistry, Eds., Gakusaikikaku, Tokyo, 1987, 47 (in Japanese).
12. **Kawai, K., Tsuchiyama, H., and Shigematsu, K.,** The procedure for ultracytochemical demonstration of dehydrogenases in the rat adenal cortex using copper ferrocyanide, *Acta Histochem. Cytochem.,* 14, 408, 1981.
13. **Makita, T. and Sandborn, E. B.,** The effect of dimethyl sulfoxide (DMSO) in the incubation medium for the cytochemical localization of succinate dehydrogenase, *Histochemie,* 26, 305, 1971.
14. **Johnson, A. B.,** The use of phenazine methosulfate for improving the histochemical localization of ketose reductase, *J. Histochem. Cytochem.,* 15, 207, 1967.
15. **Hashimoto, R., Kaluza, J. S., and Berstone, M. S.,** The effect of menadione and phenazine methosulfate on the tetrazolium reduction system under histochemical conditions, *J. Histochem. Cytochem.,* 12, 794, 1964.

II. ALCOHOL DEHYDROGENASE

Toshimitsu Watabiki

A. INTRODUCTION

Alcohol dehydrogenase (ADH) occurs in a variety of organisms, including different types of bacteria, yeasts, fungi, plants, insects, fishes, amphibians, birds, and mammals.[1] Mammalian ADHs are dimeric zinc metalloenzymes with subunits of 40,000 Da.[2] In mammals, ADH is localized mainly in the liver, but also occurs in other organs, such as the gastrointestinal tract, lung, kidney, retina, brain, and testis.[3,4]

ADH exhibits a broad substrate specificity[3] and catalyzes the following reactions:

$$\underset{\text{Primary alcohol}}{R\text{–}CH_2OH} + NAD^+ \rightleftarrows \underset{\text{Aldehyde}}{R\text{–}CHO} + NADH + H^+$$

$$\underset{\text{Secondary alcohol}}{\overset{R}{\underset{R'}{>}}CHOH} + NAD^+ \rightleftarrows \underset{\text{Ketone}}{\overset{R}{\underset{R'}{>}}CO} + NADH + H^+$$

Hepatic ADH is the principal enzyme in ethanol elimination converting ethanol to acetaldehyde. Retina ADH in the visual cycle catalyzes the oxidation of retinol, vitamin A alcohol to retinal, vitamin A aldehyde.[5]

The localization of mouse liver ADH activity has been studied histochemically by the tetrazolium salt method,[6] and ultracytochemically by the copper-ferrocyanide method.[7,8]

B. LIGHT MICROSCOPIC HISTOCHEMISTRY: THE TETRAZOLIUM SALT METHOD

Principle of the tetrazolium salt method — If a tissue section containing cells with endogenous ADH activity is placed in the incubation medium containing NAD, substrate, tetrazolium salt, and phenazine methosulfate (PMS) as the intermediate acceptor, NADH which dissociated from the ADH-NADH complex is reoxidized nonenzymatically, directly by PMS, bypassing the NADH-diaphorase activity. The reduced PMS then reduces the tetrazolium salt to colored formazan. Therefore, the final reaction product, colored water-insoluble formazan, precipitates at or near the site of ADH activity in the sections.

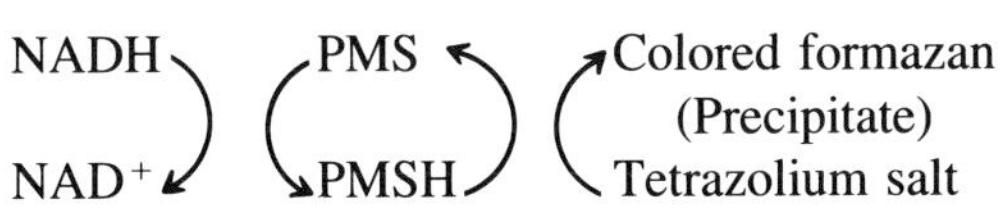

Chemicals — Nitro blue tetrazolium (NBT) $C_{40}H_{30}Cl_2N_{10}O_6$, mol wt 817.7 (Sigma Chemical Co); tetranitro blue tetrazolium (TNBT) $C_{40}H_{28}Cl_2N_{12}O_{10}$, mol wt 907.7 (Sigma); β-nicotinamide adenine dinucleotide (β-NAD) $C_{21}H_{27}N_7O_{14}P_2$, mol wt 663.4 (Sigma); phenazine methosulfate (PMS) $C_{12}H_8N_2{\cdot}(CH_3)_2SO_4$, mol wt 306.3 (Sigma); *n*-butyl alcohol: *n*-butanol $CH_3(CH_2)_2CH_2OH$, mol wt 74.1; pyrazole $C_3H_4N_2$, mol wt 68.08.

Procedures — (Watabiki et al.[6])

1. Cut fresh tissues into small pieces (5 × 5 × 2 mm), embed in M-1 Embedding Matrix (Lipshaw), and then freeze by immersion in dry ice/acetone.
2. Cut 12- to 14-μm-thick sections in a cryostat at −15 to −20°C. Place the sections on clean glass slides cooled in the cryochamber, attach them to the slides by finger thawing, and air dry in the cryostat.
3. Fix in 2 to 4% glutaraldehyde in 0.1 *M* phosphate buffer (pH 7.4) for 1 min at 0 to 4°C.
4. Wash in 0.1 *M* phosphate buffer (pH 7.4) for 1 h at 0 to 4°C.
5. Incubate the sections for 35 min in a medium containing TNBT or for 60 min in an incubating medium containing NBT in the dark at 37°C:

   a. Incubation medium for alcohol-tetrazolium reductase:

0.1 *M* Phosphate buffer, pH 7.4 or 0.1 *M* cacodylate buffer, pH 7.4	10 ml
20% Ethanol or 5% *n*-butanol	3 ml
NBT or TNBT (16.4 mg/6 ml)	6 ml
Distilled water	1 ml
NAD	60 mg
PMS	3 mg

   b. Control media
   Substrate-free medium
   Medium containing inhibitor such as pyrazole (20 m*M*)

6. Wash in distilled water
7. Mount in glycerine jelly

Results — Purplish-blue (NBT) or brownish-black (TNBT) formazan deposits indicate ADH activity. The ADH activity in mouse liver is localized mainly in the centrilobular zone (Figure 1). *n*-Butanol is a better substrate than ethanol; TNBT is a more sensitive indicator than NBT. When PMS is used, the incubation should be carried out in darkness, to avoid the photosensitive reduction of this compound. In some sections incubated in substrate-free medium, few formazan deposits are noted. This reaction, called "nothing dehydrogenase", is probably caused by nonspecific reduction of tetrazolium by sulfhydryl groups. The ADH activity is significantly inhibited by pyrazole.

C. ELECTRON MICROSCOPIC CYTOCHEMISTRY: THE COPPER-FERROCYANIDE METHOD

Principle of the copper-ferrocyanide method — NADH that has dissociated from the ADH-NADH complex is oxidized directly by ferricyanide, and ferricyanide itself is reduced to ferrocyanide. The ferrocyanide thus formed reacts with cupric ion complexed with citrate which prevents copper ferricyanide formation, so that it is trapped in the form of insoluble, electron-opaque copper ferrocyanide at or near the site of ADH activity.

$$\text{NADH} \rightarrow \text{NAD}^+ \quad ; \quad \text{Ferricyanide } Fe(CN)_6^{3-} \rightarrow \text{Ferrocyanide } Fe(CN)_6^{4-} + 2\,Cu^{2+} \rightarrow \text{Copper ferrocyanide } Cu_2Fe(CN)_6$$

Chemicals — Cacodylic acid sodium salt $(CH_3)_2AsO_2Na{\cdot}3H_2O$, mol wt 214.03; copper(II)sulfate-pentahydrate $CuSO_4{\cdot}5H_2O$, mol wt 249.68 (Merck); potassium ferricyanide:

potassium hexacyanoferrate(III) $K_3[Fe(CN)_6]$, mol wt 329.26; sodium citrate: tri-sodium citrate-dihydrate $Na_3C_6H_5O_7{\cdot}2H_2O$, mol wt 294.11.

Procedures (mouse liver) — (Watabiki et al.[7,8]):

1. Perfuse the liver through the portal vein at a pressure of 15 cm H_2O with Ringer's solution at room temperature, for 30 s, followed by 4% glutaraldehyde in 0.1 *M* cacodylate buffer (pH 7.4) containing 0.23 *M* sucrose for 10 min at 0 to 4°C.
2. Remove the liver, cut into small blocks, and immerse in the same fixative for 20 min at 0 to 4°C.
3. Wash overnight in 0.1 *M* cacodylate buffer (pH 7.4) containing 0.23 *M* sucrose at 0 to 4°C.
4. Cut 40-μm-thick sections with a freezing microtome or a Microslicer® (Dosaka EM, Kyoto).
5. Incubate the sections in the following incubation medium for 25 min at 37°C:

	(Final concentration)
0.1 *M* Cacodylate buffer, pH 7.4	5.0 ml (50 m*M*)
0.1 *M* Sodium citrate	0.4 ml (4 m*M*)
30 m*M* Copper sulfate	1.0 ml (3 m*M*)
Distilled water	1.1 ml
Sucrose	0.8 g (234 m*M*)
5 m*M* Potassium ferricyanide	1.0 ml (0.5 m*M*)
NAD	30 mg (4.5 m*M*)
20% Ethanol or 5% *n*-butanol	1.5 ml (506 m*M*; 83 m*M*)

6. Wash sections in 0.1 *M* cacodylate buffer (pH 7.4) containing 0.23 *M* sucrose for 10 min.
7. Postfix in 1% OsO_4 in 0.1 *M* cacodylate buffer for 1 h at 0 to 4°C.
8. Dehydrate via graded alcohols and propylene oxide.
9. Embed in Spurr's resin that has been poured into a silicone flat-embedding mold.
10. Cut ultrathin sections with an ultramicrotome.
11. Observe under an electron microscope without poststaining.
12. Cytochemical controls are similar to those of light microscopic histochemistry.

Results — When the 40-μm-thick sections in a flat block (Figure 2) are examined under the stereomicroscope, dark brown copper ferrocyanide deposits showing ADH activity are observed mainly in the centrilobular zone (Figure 3). Electron microscopically, the deposits are found mainly in the cytoplasmic matrix (Figure 4). In particular, they are seen abundantly in the cytoplasmic matrix immediately adjacent to profiles of the rough-surfaced endoplasmic reticulum (Figure 4). In addition, reaction products are also observed on the inner and outer membranes of mitochondria and cristae mitochondriales (Figure 5).

Comments — (1) Perfusion fixation of mouse liver: a 20-gauge needle is inserted into the portal vein and then both Aorta caudalis and V. cava caudalis are cut open at the level of the kidney. The total amount of fixative used for each mouse is approximately 10 ml perfused for 10 min. (2) ADH activity is better preserved if after a 7-min perfusion fixation the liver is perfused with 0.1 *M* cacodylate buffer (pH 7.4) containing 0.23 *M* sucrose for 10 min at 0 to 4°C, removed, and cut into small blocks, which are then washed in the same buffer for 60 min. Subsequent treatments follow stage 4 above. (3) The flat-embedding mold is convenient in trimming under a stereomicroscope (Figure 2) and obtaining light micrographs (Figure 3).

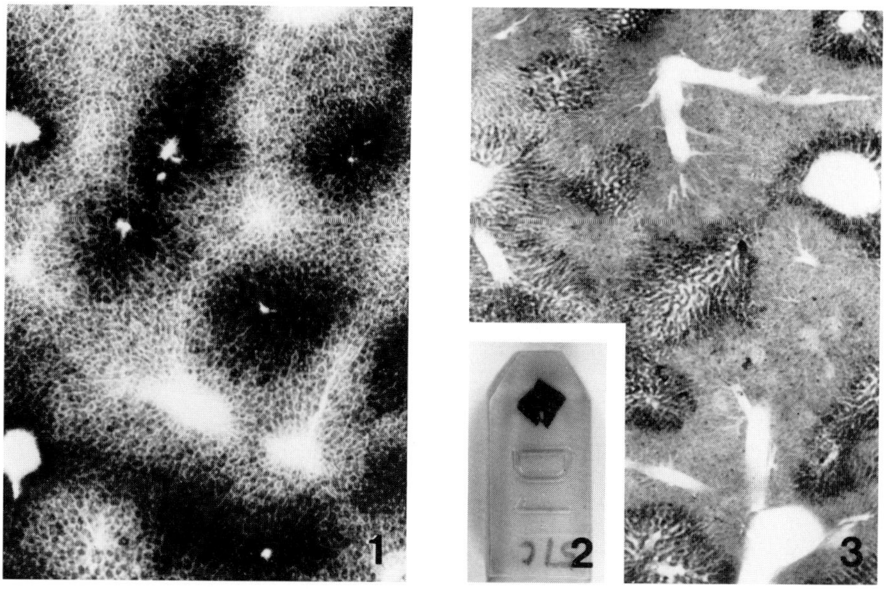

FIGURE 1. Mouse liver ADH activity demonstrated by the TNBT method; substrate ethanol. The TNBT formazan (brownish-black) deposits showing the activity are found mainly around the central vein of the liver lobule. Section, 12 μm thick; fixation, 1 min. (Magnification × 96.)

FIGURE 2. Mouse liver ADH activity demonstrated by the copper-ferrocyanide method; substrate ethanol. Prior to the preparation of ultrathin sections, a flat block is trimmed to expose the site of the activity. Fixation, 30 min.

FIGURE 3. A section, 40 μm thick, in the block is observed under a stereomicroscope, and the dark-brown copper ferrocyanide deposits show the ADH activity, in the same site as in Figure 1. (Magnification × 72.)

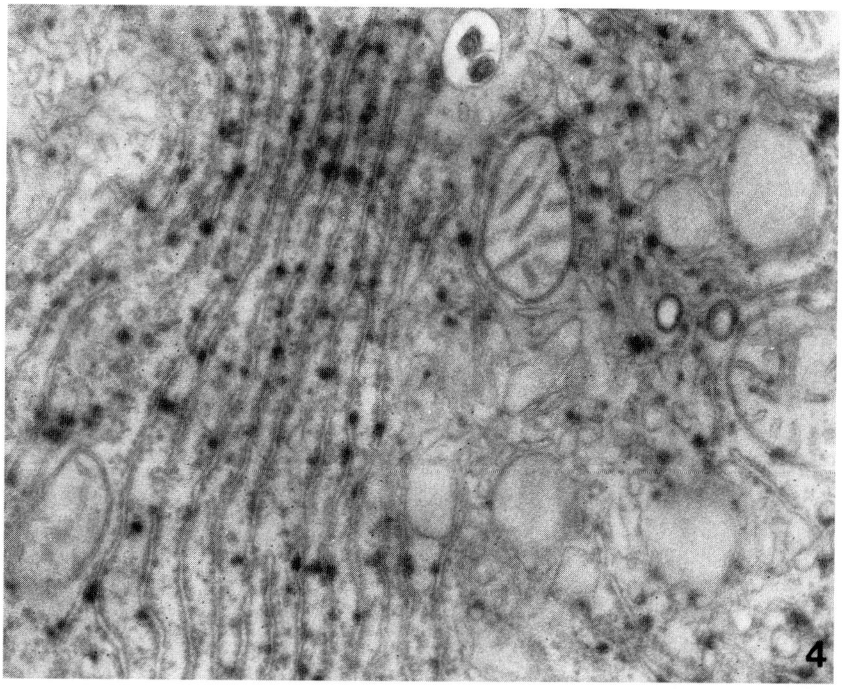

FIGURE 4. Mouse liver ADH activity demonstrated by the copper-ferrocyanide method; substrate ethanol. The copper ferrocyanide deposits showing the activity are seen in the cytoplasmic matrix immediately adjacent to both the rough- and smooth-surfaced endoplasmic reticulum. Unstained. (Magnification × 45,000.)

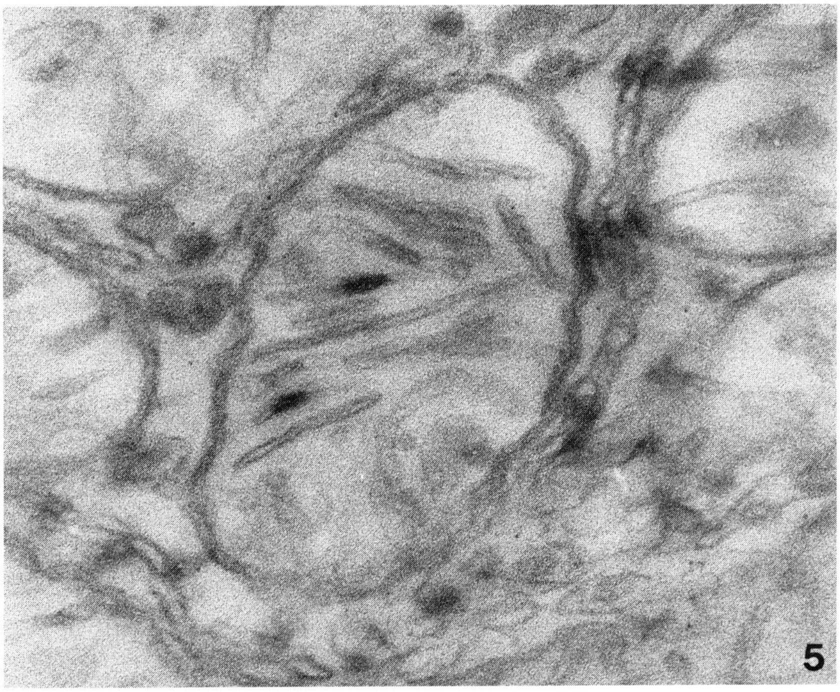

FIGURE 5. Mouse liver ADH activity on the inner and outer membranes of mitochondria and cristae mitochondriales. Unstained. (Magnification × 99,000.)

REFERENCES

1. **Brändén, C.-I., Jörnvall, H., Eklund, H., and Furugren, B.,** Alcohol dehydrogenases, in *The Enzymes,* Boyer, P. D., Ed., Academic Press, New York, 1975, 103.
2. **Bosron, W. F. and Li, T.-K.,** Catalytic properties of human liver alcohol dehydrogenase isoenzymes, *Enzyme,* 37, 19, 1987.
3. **Pietruszko, R.,** Mammalian liver alcohol dehydrogenases, *Adv. Exp. Med. Biol.,* 56, 1, 1975.
4. **Von Wartburg, J.-P. and Bühler, R.,** Biology of disease: alcoholism and aldehydism: new biomedical concepts, *Lab. Invest.,* 50, 5, 1984.
5. **Juliá, P., Farrés, J., and Parés, X.,** Purification and partial characterization of a rat retina alcohol dehydrogenase active with ethanol and retinol, *Biochem. J.,* 213, 547, 1983.
6. **Watabiki, T., Tokiyasu, T., Ishida, N., and Ogawa, K.,** Double staining procedure for histochemical localization of alcohol and aldehyde dehydrogenase activities in the mouse liver, *Acta Histochem. Cytochem.,* 22, 401, 1989.
7. **Watabiki, T.,** Ultracytochemical study of alcohol dehydrogenase activity in normal mouse hepatic cells, *Acta Histochem. Cytochem.,* 12, 368, 1979.
8. **Watabiki, T., Ishida, N., and Ogawa, K.,** Effect of ethanol administration on alcohol dehydrogenase activity in mouse hepatic cells: ultracytochemical study, *Acta Histochem. Cytochem.,* 14, 476, 1981.

III. β-HYDROXYBUTYRATE DEHYDROGENASE

Kioko Kawai

A. INTRODUCTION

This enzyme catalyzes the reaction of β-hydroxybutyric acid and acetoacetic acid. The enzyme is very firmly attached to the mitochondrial membranes and has a special relationship to the components of the respiratory chain.[1]

The reaction is as follows:

$$\text{D-3-hydroxybutyrate} + \text{NAD} \rightleftharpoons \text{acetoacetetic acid} + \text{NADH} + \text{H}^+$$

Most tissues contain very low levels of hydroxybutyrate dehydrogenase, but this enzyme is reversibly inactivated by aerobic conditions. It is also activated by thiol compounds.

The copper-ferrocyanide method has been reported for the ultrastructural demonstration of this enzyme.[2,3]

B. COPPER-FERROCYANIDE METHOD

Medium for β-hydroxybutyrate ferricyanide reductase[2] —

		Final concentration
Sodium D-L-β-hydroxybutyrate	100 mg	40 m*M*
NAD	132 mg	10 m*M*
0.1 *M* Sodium citrate	0.6 ml	3 m*M*
30 m*M* Copper sulfate	2.0 ml	30 m*M*
Double-distilled water	2.4 ml	
5 m*M* Potassium ferricyanide	2.0 ml	0.5 m*M*
Sucrose	2.0–3.0 g	
0.1 *M* Phosphate buffer, pH 7.0	13 ml	65 m*M*
Phenazine methosulfate (PMS)	0.9–9.2 mg	

The total volume of the medium is 20 ml. The final pH is adjusted to 7.0. The medium should be freshly prepared by adding the ingredients with constant stirring in the above-mentioned order. The solution is green and completely clear. It is stable for several hours. PMS is dissolved in 0.2 ml acetone. Incubation should be carried out in the dark when PMS is added to the solution.

Procedure —

1. Sections, 40 to 60 μm thick, are cut from fresh or fixed tissues using a Vibratome (Oxford Laboratories, U.S.) or a Microslicer (D.S.K., EMCO Ltd., Kyoto).
2. If fixation was used, the tissues are rinsed for 30 to 60 min in three changes of 0.1 *M* phosphate buffer, pH 7.0.
3. Sections are incubated in the freshly prepared medium at 37°C or at room temperature for 10 to 60 min with continuous shaking. The incubation is carried out in the dark if PMS was added to the incubating medium.
4. After rinsing in buffer at room temperature, the sections are fixed in 2% OsO_4 for 1 h at 4°C, dehydrated through graded alcohols and propylene oxide, and embedded in Epon, glycol methacrylate, or Spurr's resin.
5. The ultrathin sections are stained with uranyl acetate and/or lead citrate and examined under a transmission electron microscope.

To confirm the specificity of the enzyme reaction, sections should be (1) incubated in a medium from which the substrate was omitted, and (2) incubated in a medium containing various inhibitors, such as sodium malonate (36 m*M*) and antimycin A (1 to 30 μg/ml).

Results — The reaction product is localized in the intracristal and outer spaces in mitochondria of liver and heart.[1,2] These localizations are similar to those of succinate ferricyanide reductase.

REFERENCES

1. **Wise, J. B. and Lehninger, A. L.,** The stability of mitochondrial D-B-hydroxybutyric dehydrogenase and its relationship to the respiratory chain, *J. Biol. Chem.*, 237, 1363, 1962.
2. **Ogawa, K., Saito, T., and Mayahara, H.,** The site of ferricyanide reduction by reductases within mitochondria as studied by electron microscopy, *J. Histochem. Cytochem.*, 16, 49, 1968.
3. **Makita, T.,** Ultracytochemistry of oxidoreductase, in *Shinkosososhikigagaku,* Ogawa, K. and Takeuchi, T., Eds., Asakura, Tokyo, 1980, 151 (in Japanese).

IV. CYTOCHROME OXIDASE

Kei-Ichi Hirai and Guo-Ying Wang

A. INTRODUCTION

Cytochrome oxidase is the terminal enzyme of the respiratory (electron transfer) chain in aerobe mitochondria. The respiratory chain is present in the mitochondrial inner membrane, including the crista membrane. Cytochrome oxidase catalyzes the following reaction:

$$2 \text{ ferrocytochrome } c + 1/2 \ O_2 \rightleftharpoons 2 \text{ ferricytochrome } c + H_2O \qquad (1)$$

In 1968, Seligman et al.[1] developed a method using 3,3′-diaminobenzidine (DAB) to demonstrate cytochrome oxidase activity. In this method, 2 mol of DAB reduce 1 mol of ferricytochrome *c* to form ferrocytochrome *c*, which is then oxidized by cytochrome oxidase.[2] These reduction-oxidation cycles stimulate the oxidation of DAB and form insoluble DAB oxides as follows:

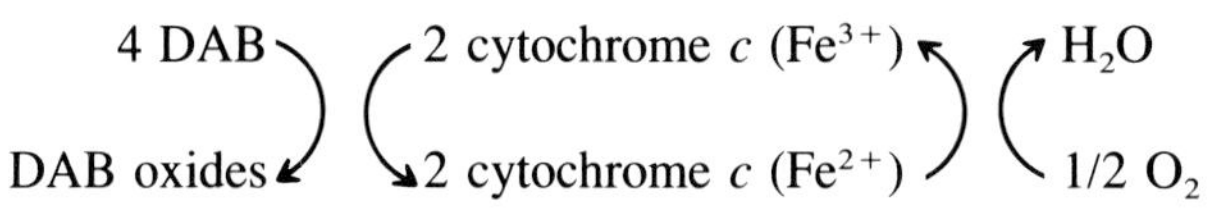

(2)

Because an electron spin resonance (ESR) signal (g = 2.0028) is detected on DAB oxides, but not on DAB free base, the reaction products may be free radical polymers formed by radical polymerization and condensation.[3]

Light microscopically, the reaction results in a brownish staining which represents the distribution of mitochondria. Since the reaction products may be converted with osmium tetroxide into electron-opaque deposits consisting of osmium black, the site(s) of the enzymatic reaction can be localized in cells or tissues electron microscopically.

The electron transfer chain exists in the mitochondrial inner membrane, while cytochrome *c* moieties are localized on the outer surface of the inner membrane and mobilized. The reaction products are accumulated in the intramitochondrial membrane spaces, including the cristal spaces.

B. METHODS

The DAB method is generally used to demonstrate cytochrome oxidase activity.[1] Because DAB cannot penetrate the plasma membrane of living cells, tissues are fixed with glutaraldehyde prior to histochemical incubation.[5] Glutaraldehyde should be spectrophotometrically pure, since impurities may inactivate the enzyme. Seligman's original incubation medium is buffered by Tris-hydrochloride, which can be replaced by phosphate, HEPES or PIPES buffer, or by phosphate-buffered saline (PBS). Catalase is added to the medium to eliminate peroxide contaminants which cause nonspecific peroxidatic reactions.[6]

C. EQUIPMENT AND CHEMICALS

1. A perfusion system for experimental animal tissues and a thermocontrolled shaker or rotator
2. Chemicals: 3,3′-diaminobenzidine-tetrahydrochloride (DAB·4HCl) salt (Wako Pure Chemical Co., Ltd., Tokyo; Sigma Chemical Co.); cytochrome *c* (Sigma, type IV); osmium tetroxide; glutaraldehyde (20 to 70%) of electron microscopic grade with a purity ratio of 0.2 (A235/A280) or less; catalase (Sigma, C-40); Ca^{2+}, Mg^{2+}-free PBS tablet (Flow Lab.).

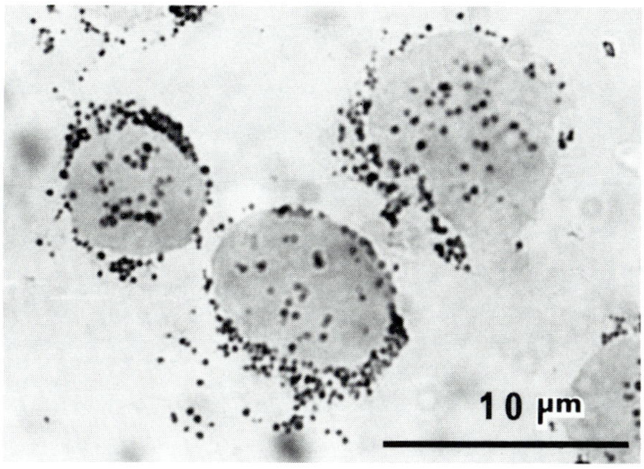

FIGURE 1. DAB reaction for cytochrome oxidase of cultured mouse myeloid leukemic cells. Semidry smear of cells was fixed in ice-cold 2% glutaraldehyde solution, for 10 min. Incubation was for 60 min at 37°C. (Magnification × 3000.)

3. DAB medium:

0.1 *M* Phosphate buffer or PBS (pH 7.4)	10 ml
DAB·4HCl Salt	10 mg
Cytochrome *c*	10 mg
Catalase	10 mg

4. Inhibitors: 2 m*M* KCN, 10 m*M* NaN_3

D. PROCEDURES

1. Use perfusion or immersion fixation with 2% glutaraldehyde in PBS, pH 7.4, at 4°C strictly for 10 min.
2. Wash tissues with PBS for 10 min at 4°C using reperfusion or immersion.
3. Dice and wash the tissues in three changes of PBS, for 20 min each at 4°C.
4. Cut 40-µm-thick slices.
5. Incubate tissue slices in the DAB medium for 60 min at 37°C with agitation using a rotator or shaker.
6. Wash the slices in three changes of PBS, for 20 min each, at 4°C.
7. Postfix the slices with 1% OsO_4 dissolved in PBS for 60 min at 4°C.
8. Wash the slices in three changes of 0.1 *M* sodium acetate, for 10 min each, at 4°C.
9. Block stain the slices with 0.5% uranyl acetate in distilled water for 20 min at 4°C.
10. Wash the slices with 50% ethanol for 10 min, and further dehydrate and embed them in a resin.

E. RESULTS

Mitochondria are stained black owing to their cytochrome oxidase activity. If smears of cells or semithin sections are used the mitochondria can be counted (Figure 1).[7] The mitochondria remain unstained if the cells were not fixed prior to incubation, if the glutaraldehyde was impure, or if the tissues were fixed for longer periods of time (Figure 2).[5] However, the histochemical staining can be performed using frozen sections of unfixed

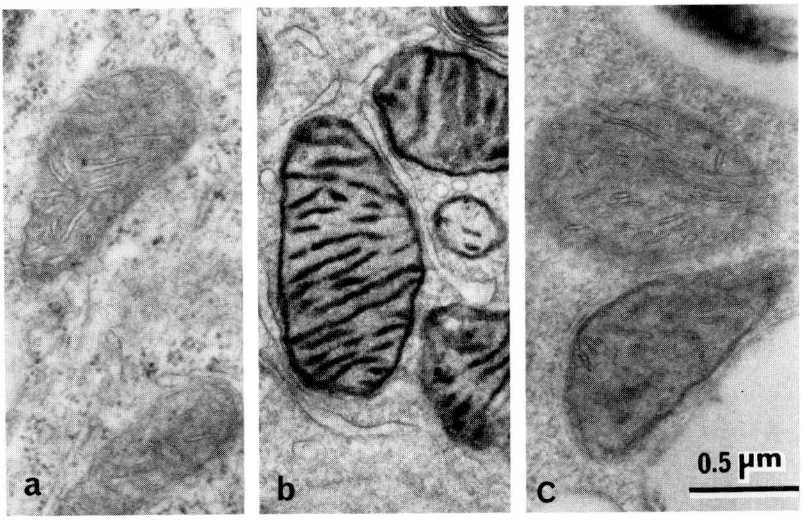

FIGURE 2. Effect of fixation on cytochrome oxidase activity of mouse lung alveolar type II cells. (a) Unfixed intact cell: (b) cells fixed in pure glutaraldehyde (A235/A280 = 0.22); (c) cells fixed in impure glutaraldehyde (A235/A280 = 0.46).

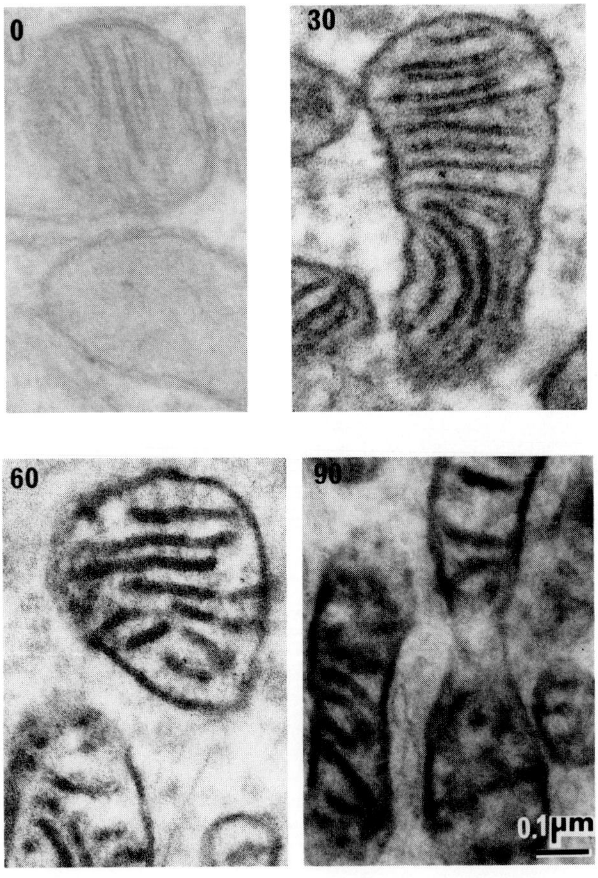

FIGURE 3. Time-dependent accumulation of oxidized DAB deposits in mitochondria of rat dorsal root ganglion cells[9] incubated for the demonstration of cytochrome oxidase activity. Slices of ganglion, 100 μm in thickness, were incubated in the DAB medium at 37°C for 0, 30, 60, and 90 min, respectively.

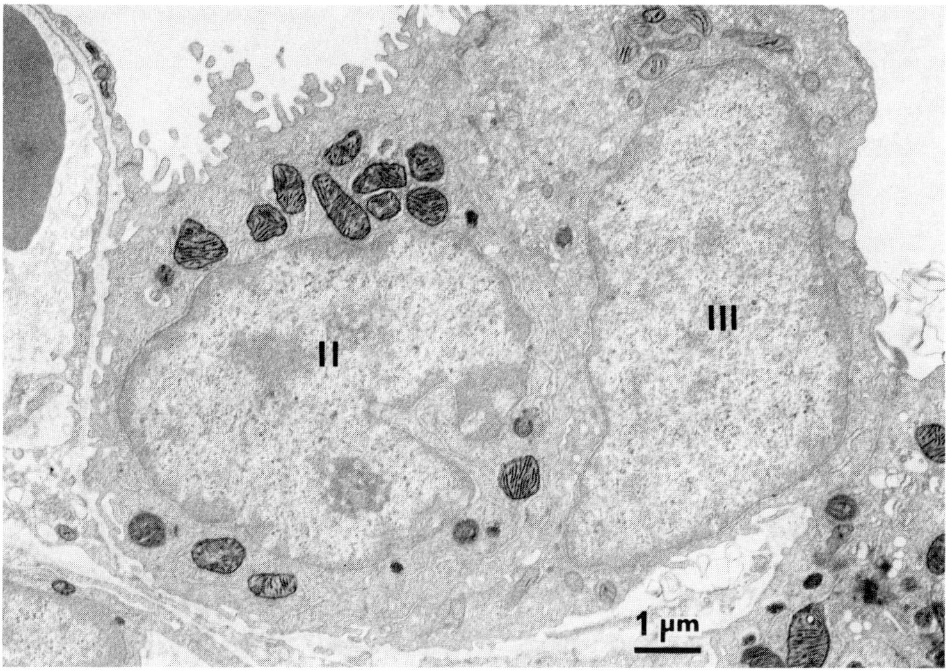

FIGURE 4. Cytochrome oxidase activity in alveolar type II and type III cells in rat lung.[5] The lung was instilled with cold 2% glutaraldehyde, pH 7.4, for 8 min, sliced in the fixative solution within 2 min, and washed.

tissues, since the plasma membrane becomes sufficiently damaged to allow the penetration of DAB.

Electron microscopically, the reaction product is localized along the mitochondrial inner membrane including the crista membrane, and is frequently seen accumulated in the intermembrane spaces including cristal spaces. The deposit accumulation is time dependent (Figure 3), and the rate varies among cell types (Figure 4).[5,8-10] These results suggest heterogenous activity among cells and even among the mitochondria of the same cell.[11]

Erythrocytes do not reveal DAB staining. Occasionally, however, these cells may be stained owing to a peroxidatic activity of hemoglobin which means an insufficiency in the reaction medium. It is supposed that this medium is contaminated somewhat by peroxides and causes peroxidase staining as a nonspecific reaction. Recently, cytochrome oxidase activity has been demonstrated in the lung,[5,8,12] leukocytes,[7] osteoclasts,[11] nerve system,[9,13-15] cardiac muscle,[10,16] and spermatozoa.[17]

REFERENCES

1. **Seligman, A. M., Karnovsky, M. J., Wasserkrug, H. L., and Hanker, J. S.,** Nondroplet ultrastructural demonstration of cytochrome oxidase activity with a polymerizing osmiophilic reagent, diaminobenzidine (DAB), *J. Cell Biol.*, 38, 1, 1968.
2. **Anderson, W. A., Bara, G., and Seligman, A. M.,** The ultrastructural localization of cytochrome oxidase via cytochrome c, *J. Histochem. Cytochem.*, 23, 13, 1975.
3. **Hirai, K.-I.,** Comparison between 3,3′-diaminobenzidine and auto-oxidized 3,3′-diaminobenzidine in the cytochemical demonstration of oxidative enzymes, *J. Histochem. Cytochem.*, 19, 434, 1971.

4. **Evans, W. H. and Graham, J. M.,** *Membrane Structure and Function,* IRL Press, Oxford, 1989, chap. 5.
5. **Hirai, K.-I., Ogawa, K., Wang, G.-Y., and Ueda, T.,** Varied cytochrome oxidase activities of the alveolar type I, type II and type III cells in rat lungs: quantitative cytochemistry, *J. Electron Microsc.,* 6, 449, 1989.
6. **Angermuller, S. and Fahimi, H. D.,** Selective cytochemical localization of peroxidase, cytochrome oxidase and catalase in rat liver with 3,3′-diaminobenzidine, *Histochemistry,* 71, 33, 1981.
7. **Hirai, K.-I., Nagata, K., Maeda, M., and Ichikawa, Y.,** Changes in ultrastructures and enzyme activities during differentiation of myeloid leukemia cells to normal macrophages, *Exp. Cell Res.,* 124, 269, 1979.
8. **Hirai, K.-I. and Ogawa, K.,** Cytochemical quantitation of cytochrome oxidase activity in rat pulmonary alveolar epithelial cells and possible defect in type I cells, *J. Electron Microsc.,* 35, 19, 1986.
9. **Wang, G.-Y., Hirai, K.-I., and Odashima, S.,** Quantitative cytochemical studies of cytochrome oxidase activity in rat dorsal root ganglion cells, *J. Electron Microsc.,* 39, 231, 1990.
10. **Perotti, M. E., Anderson, W. A., and Swift, H.,** Quantitative cytochemistry of the diaminobenzidine cytochrome oxidase reaction product in mitochondria of cardiac muscle and pancreas, *J. Histochem. Cytochem.,* 31, 351, 1983.
11. **Noda, K.,** Localization of cytochrome c oxidase activity in osteoclasts treated with calcitonin during experimental tooth movement, *Acta Histochem. Cytochem.,* 21, 301, 1988.
12. **Hirai, K.-I., Uyeda, T., and Ogawa, K.,** Electron-cytochemical studies on the differentiation of mouse lung alveolar epithelial cells with special reference to changes in mitochondria, *Acta Histochem. Cytochem.,* 17, 197, 1984.
13. **Darriet, D., Der, T., and Collins, R. C.,** Distribution of cytochrome oxidase in rat brain: studies with diaminobenzidine histochemistry *in vitro* and [^{14}C]cyanide tissue labeling *in vivo, J. Cereb. Blood Flow Metab.,* 6, 8, 1986.
14. **Wong-Riley, M. T. T. and Kageyama, G. H.,** Localization of cytochrome oxidase in the mammalian spinal cord and dorsal root ganglia, with quantitative analysis of ventral horn cells in monkey, *J. Comp. Neurol.,* 245, 41, 1986.
15. **Kugler, P., Vogel, S., Volk, H., and Schiebler, T. H.,** Cytochrome oxidase histochemistry in the rat hippocampus. A quantitative methodological study, *Histochemistry,* 89, 269, 1988.
16. **Millis, R. M., Stephens, T. A., Harris, G., Anonye, C., and Reynolds, M.,** Histochemical assessment of cytochrome oxidase activity for monitoring ischemic muscle injury, *Exp. Neurol.,* 88, 265, 1985.
17. **Hrudka, F.,** Cytochemical and ultracytochemical demonstration of cytochrome c oxidase in spermatozoa and dynamics of its changes accompanying ageing or induced by stress, *Int. J. Androl.,* 10, 809, 1987.

V. PEROXIDASE

Kei-Ichi Hirai

A. INTRODUCTION

Peroxidases catalyze the following reaction:

$$H_2O_2 + AH_2 \rightleftharpoons 2H_2O + A \tag{1}$$

where H_2O_2 is the first substrate (hydrogen acceptor), which may be replaced by CH_3OOH, C_2H_5OOH, $BaO_2Na_2O_2$, and so on; AH_2 is the second substrate (hydrogen donor) to be oxidized. Several peroxidases have been biochemically established which include, in animals, neutrophil myeloperoxidase,[1,2] eosinophil peroxidase,[3] lactoperoxidase,[4] thyroid peroxidase,[5] uterine peroxidase,[6] and salivary peroxidase;[7] and in plants, horseradish peroxidase[8] and turnip peroxidase.[9] Activity of a selenium-containing peroxidase, glutathione peroxidase, has not been demonstrated histochemically.

Several dyes have been used as hydrogen donors for the demonstration of peroxidase activity. These include 3,3′-diaminobenzidine (DAB),[10] 3,3′, 5,5′-tetramethyl-benzidine (TMB),[11,12] *p*-phenylenediamine-pyrocatechol (PPD-PC),[13] chloro-naphthol,[14] and 3-amino-9-ethylcarbazole.[15] For the electron microscopic demonstration of endogenous peroxidase activity in animal cells DAB is generally used.[10] The reaction is

$$H_2O_2 + \text{DAB free base} \rightleftharpoons 2H_2O + \text{DAB oxide} \tag{2}$$

where DAB is oxidized and forms DAB oxide, a radical polymer.[16] The DAB oxide deposits are insoluble and form electron-opaque osmium black after treatment with osmium tetroxide.

B. METHOD

A DAB-H_2O_2 method employing a neutral pH medium is recommended. The constituents of the reaction medium are the same as those used for the demonstration of catalase activity. The Tris-hydrochloride buffer used in the original medium of Graham and Karnovsky[10] is not suitable for the fixation and osmium treatment steps. This buffer may be replaced by PBS, phosphate or a Good (PIPES, HEPES) buffer, all of which are convenient for the whole range of procedures from fixation to osmium treatment. A short fixation time is recommended for some labile peroxidases, such as those of megakaryocytes or mouse lung alveolar type II cells. In order to eliminate the cytochrome oxidase reaction which frequently occurs when this neutral incubating medium is used, 1 m*M* KCN or 5 m*M* NaN_3 as inhibitor is added to the medium. The most important determinants of satisfactory results are the purity of the glutaraldehyde fixative and fixation time.

C. EQUIPMENT AND CHEMICALS

1. A perfusion system for experimental animal tissues and a thermocontrolled shaker or rotator
2. Chemicals: 3,3′-diaminobenzidine-tetrahydrochloride (DAB·4HCl) salt (Wako Pure Chemical Co., Ltd., Tokyo; Sigma Chemical Co.); 30% H_2O_2; osmium tetroxide; glutaraldehyde (20 to 70%), electron microscopic grade with purity ratio of 0.2 (A235/A280) or less; catalase (Sigma, C-40); Ca^{2+}, Mg^{2+}-free PBS tablet (Flow Lab.) or 0.1 *M* phosphate buffer, pH 7.0 to 7.6.

3. Preincubation medium:

DAB·4HCl Salt	10 mg
PBS	10 ml
100 m*M* KCN	0.1 ml (1 m*M*)*

4. DAB-H_2O_2 Reaction medium (pH 7.0 to 7.6):

DAB·4HCl Salt	10 mg (0.1%) (20 mg for megakaryocytes)
PBS	10 ml
100 m*M* KCN**	0.1 ml (1 m*M*)
1% H_2O_2	0.1 ml (0.01%)

5. Inhibitors: 1% H_2O_2;[17] 50 m*M* NaN_3; 50 m*M* 3-amino-1,2,4-triazole. Nonspecific denaturants are 90% methanol-3% H_2O_2;[18] 0.1% phenylhydrazine;[19] methanol-nitroferricyanide-acetic acid;[20] and 0.05 *M* periodic acid.[21]

D. PROCEDURES

1. Fixate with 1/2-strength Karnovsky fixative consisting of 2% paraformaldehyde and 2.5% glutaraldehyde[22] or 2% glutaraldehyde in PBS for 10 min at 4°C.

   - For experimental animals, perfusion fixation is recommended. First, perfuse cold PBS for 5 min to wash out the blood; second, place in a fixative solution for 10 min and reperfuse with PBS. Slice tissue at 40 μm thick.
   - Isolate white blood cells from heparinized venous blood by a Ficoll-Hyperque method and suspend them in a cold fixative solution for 7 min. Centrifuge the cells at 1000 × *g* for 3 min in conical tubes and wash in three changes of PBS.
   - Fix megakaryocytes for 1 h in a 1% tannic acid fixative containing 2% paraformaldehyde and 0.5% glutaraldehyde in 0.1 *M* phosphate buffer or PBS.[23] Wash cells with PBS by centrifugation three times.
   - Dice and immerse tissues in a cold fixative solution for 10 min with continuous agitation in a rotator or shaker.

2. Preincubate tissue slices in the preincubation medium for 10 min at room temperature to allow the penetration of DAB. Centrifuge cell suspensions for incubation and washing.
3. Incubate the slices in the neutral pH DAB-H_2O_2 medium for 60 min at 37°C. Incubate megakaryocytes for 60 min in a pH 7.6 medium containing 0.2% DAB.
4. Wash the slices in three changes of a cold PBS, for 20 min each.
5. Postfix the tissues with 1% OsO_4 dissolved in PBS for 60 min at 4°C.
6. Wash the osmium-treated tissue slices in three changes of 0.1 *M* sodium acetate, for 10 min each at 4°C.
7. Block stain the slices with 0.5% uranyl acetate in distilled water for 20 min at 4°C.

* Final concentrations are given in parentheses.
** Not used for the demonstration of the enzymes in megakaryocytes and mouse lung alveolar type II cells.

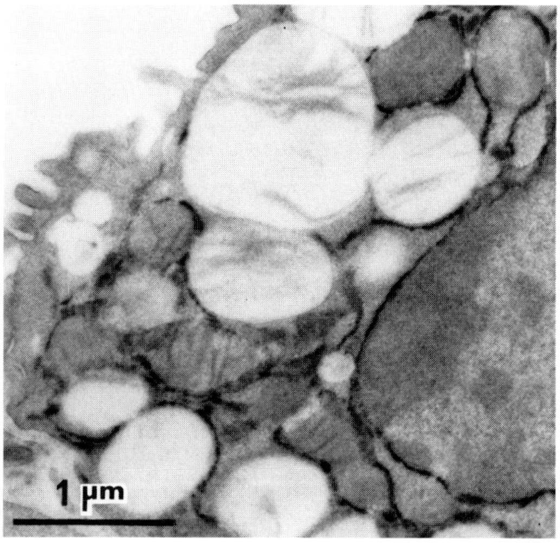

FIGURE 1. Peroxidase activity in the mouse lung alveolar type II cell. Lung was instilled with 2% pure glutaraldehyde and fixed for 10 min, and incubated for 60 min in a DAB-0.001% H_2O_2 medium (pH 7.6). (From Hirai, K.-I. et al., *Acta Histochem. Cytochem.*, 18, 255, 1985. With permission.)

8. Wash the slices with 50% ethanol for 10 min and further dehydrate and embed in a resin. At the 70% alcohol dehydration step, mix free cells suspended in 2 ml of the alcohol solution in a conical tube with 2 ml of 1% agarose warmed at 50°C and centrifuge immediately at 1200 × *g* for 5 min. Cool the tube and dice the solidified agarose containing the cells. Dehydrate and embed them in the usual manner.
9. Control: Incubate tissue slices in an H_2O_2-free DAB medium which contains 0.5 mg/ml catalase to eliminate endogenous peroxide in the medium.

E. RESULTS

Peroxidase activity in epithelial cells is commonly visualized as electron-opaque DAB deposits in the lumen of the endoplasmic reticulum and nuclear envelope of uterine endometrium nonciliated surface cells of rat and human,[24,27] and in mouse lung alveolar type II cells (Figure 1).[28,29] Pathogen-free rats are devoid of peroxidase activity in the respiratory tract epithelial cells, and start to show enzyme activity in tracheal ciliated cells, mucous cells, and glandular duct cells when infected with microorganisms (Figure 2).[30-31] Mucous cells, secretory granules, and Golgi apparatus show a positive reaction. Peroxidase activity has also been reported in laryngeal, thyroid,[32,33] lacrimal,[34,35] submandibular,[36,37] parotid,[38] mammary,[39] nasal,[40] and sublingual glands (Figure 3),[37] and epidermis,[41] conjunctiva,[42] colon,[43] and kidney collecting tubules.[44]

Peroxidase activity has been demonstrated in the azurophil granules of neutrophils, the specific granules and small granules (primary lysosomes) of eosinophils (Figure 4),[45,46] in mast cells,[47] monocytes,[48] and macrophages (Figure 5) and others.[49-50]

Horseradish peroxidase is widely used as a tracer in transport studies, and much effort has been spent on developing means to increase the sensitivity of detection of DAB deposits.[56-66]

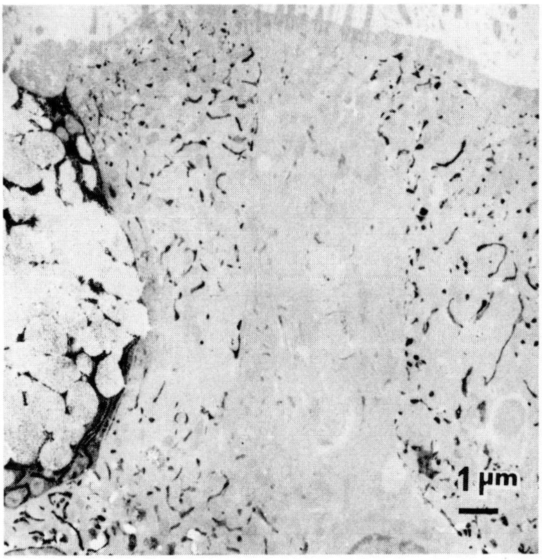

FIGURE 2. Peroxidase activity of the tracheal epithelium of a rat infected with corynebacteria. Pathogen-free rat tracheal epithelial cells are devoid of this activity. Experimental conditions were as described in legend to Figure 1. (From Sakai, J. et al., *Acta Histochem. Cytochem.*, 22, 421, 1989. With permission.)

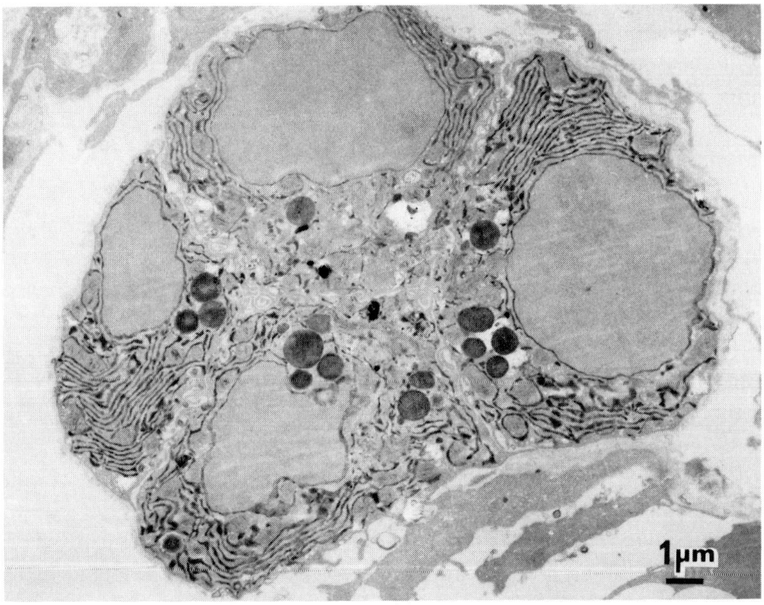

FIGURE 3. Peroxidase activity of the rat sublingual gland of a 23-d-old fetus. (From Asano, T., *Acta Anat. Nippon*, 63, 432, 1988. With permission.)

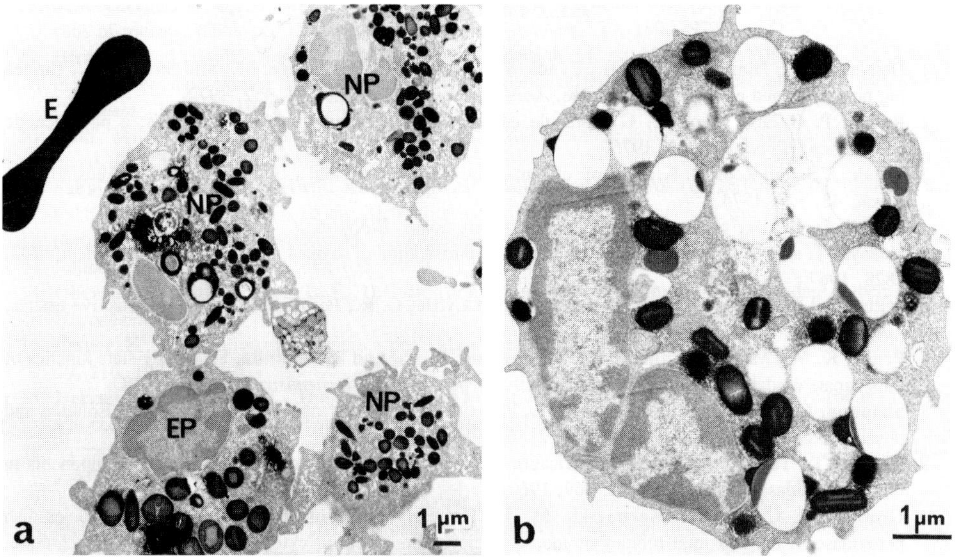

FIGURE 4. Peroxidase activity in human blood cells. (a) Neutrophils (NP), eosinophils (EP), and erythrocytes (E). (b) Eosinophil engaged in phagocytosis shows fusion of the specific granules to the phagosomes (arrowhead) and that of the small granules (arrow).

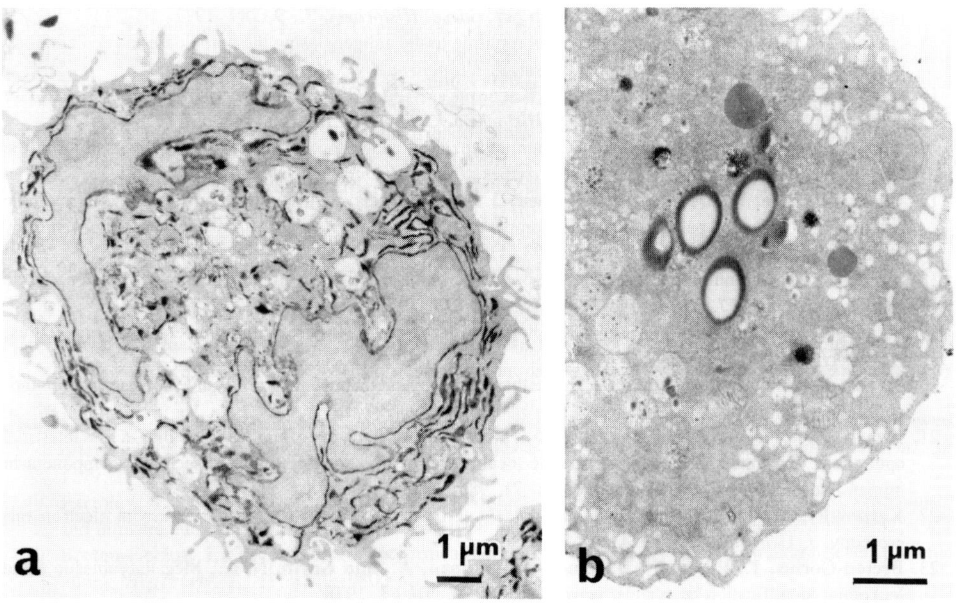

FIGURE 5. Peroxidase activity of rat peritoneal macrophage. (a) Resident macrophages were fixed for 10 min in 2% glutaraldehyde, and incubated for 1 h in a DAB-0.01% H_2O_2 medium (pH 7.6). (b) The enzyme activity was inhibited by 1% (>0.1%) H_2O_2, but catalase activity in peroxisomes remained unchanged.

54. **Shibata, A., Bennett, J. M., Castoldi, G. L., Catovsky, D., Flandrin, G., Jaffe, E. S., Katayama, I., Nanba, K., Schmalzl, F., Yam, L. T. et al.,** Recommended methods for cytological procedures in haematology. International Committee for Standardization in Haematology (ICSH), *Clin. Lab. Haematol.*, 7, 55, 1985.
55. **Berman, E. L., Carter, H. W., Ambrose, W. W., and Hanker, J. S.,** Studies of epsinophil medusa cells by electron imaging modes, *Scanning Electron Microsc.*, 1, 311, 1983.
56. **Bugelski, P. J., Gennaro, D. E., Poste, G., and Hoffstein, S. T.,** A new cytochemical method for ultrastructural detection of liposomes in tissues in vivo, *J. Histochem. Cytochem.*, 37, 843, 1989.
57. **Shu, S. Y., Ju, G.,and Fan, L. Z.,** The glucose oxidase-DAB-nickel method in peroxidase histochemistry of the nervous system, *Neurosci. Lett.*, 85, 169, 1988.
58. **Beach, T. G. and McGeer, E. G.,** Retrograde filling of pyramidal neurons in postmortem human cerebral cortex using horseradish peroxidase, *J. Neurosci. Meth.*, 23, 187, 1988.
59. **De Gennaro, L. D.,** The effects of lead nitrate on the central nervous system of the chick embryo. II. Electron microscopy and histochemistry: peroxisomes, *Growth*, 51, 213, 1987.
60. **De Bruijn, W. C., Van der Meulen, J., Brederoo, P., and Daems, W. T.,** Pt-staining of peroxidatic reaction products at the ultrastructural level, *Histochemistry*, 84, 492, 1986.
61. **Zaborszky, L. and Leranth, C.,** Simultaneous ultrastructural demonstration of retrogradely transported horseradish peroxidase and choline acetyltransferase immunoreactivity, *Histochemistry*, 82, 529, 1985.
62. **Liposits, Z., Setalo, G., and Flerko, B.,** Application of the silver-gold intensified 3,3′-diaminobenzidine chromogen to the light and electron microscopic detection of the luteinizing hormone-releasing hormone system of the rat brain, *Neuroscience*, 13, 513, 1984.
63. **Wainer, B. H. and Rye, D. B.,** Retrograde horseradish peroxidase tracing combined with localization of choline acetyltransferase immunoreactivity, *J. Histochem. Cytochem.*, 32, 439, 1984.
64. **Chan, K. Y. and Haschke, R. H.,** A re-evaluation of the retrograde axonal transport of horseradish peroxidase isoenzymes by semiquantitative cytochemical analysis, *Neuroscience*, 6, 1145, 1981.
65. **Carson, K. A., Lucas, W. J., Gregg, J. M., and Hanker, J. S.,** Facilitated ultracytochemical demonstration of retrograde axonal transport of horseradish peroxidase in peripheral nerve, *Histochemistry*, 67, 113, 1980.
66. **Gallyas, F., Gorcs, T., and Merchenthaler, I.,** High-grade intensification of the end-product of the diaminobenzidine reaction for peroxidase histochemistry, *J. Histochem. Cytochem.*, 30, 183, 1982.

VI. CATALASE

Kei-Ichi Hirai

A. INTRODUCTION

Catalase acts as a scavenger of hydrogen peroxide (H_2O_2) and forms oxygen gas and water as follows:

$$2H_2O_2 \rightleftharpoons 2H_2O + O_2 \qquad (1)$$

The catalytic activity of catalase has an optimum pH of 6.8. Catalase also has a peroxidatic activity towards alcohols, phenol, nitrite, aldehydes, formic acid, etc., in the presence of H_2O_2 as follows:

$$H_2O_2 + AH_2 \rightleftharpoons 2H_2O + A \qquad (2)$$

A purified catalase in association with four subunits shows catalytic activity but is generally devoid of peroxidatic activity. When catalase is treated with urea, formamide, or alkaline, the enzyme moiety dissociates into four to six subunits. During these dissociations, catalase loses its catalytic activity and exhibits peroxidatic activity (Table 1).[1-3] The latter activity is demonstrated histochemically.

In 1968 and 1969, several DAB-H_2O_2 methods were described; the high alkaline method (pH 9.0) of Novikoff and Goldfischer,[4,5] the pH 8.5 method of Fahimi,[6,7] and the methods of Hirai[8] employing a neutral pH DAB-H_2O_2 medium for formamide-treated specimens. Today, an alkaline DAB-H_2O_2 medium is generally used.

DAB (3,3′-diaminobenzidine-tetrahydrochloride) acts as hydrogen donor for the peroxidatic action of catalase. DAB is oxidized by catalase, and forms DAB oxide, a radical-polymer.[9]

$$H_2O_2 + \text{DAB free base} \rightleftharpoons 2H_2O + \text{DAB oxide} \qquad (3)$$

The deposits of oxidized DAB are insoluble and form osmium black after an osmium tetroxide treatment.

B. METHODS

A DAB-H_2O_2 method using an alkaline medium is recommended.[4-7] In order to specifically demonstrate catalase activity, the following should be considered (Table 2).

1. Tissues should be fixed thoroughly for at least 4 h; the mitochondria of short-term-fixed tissues are frequently stained by this method because of cytochrome oxidase activity toward DAB or nonspecific peroxidatic reaction of mitochondrial heme proteins.
2. Glutaraldehyde must be pure; this is the most critical factor, as impurities or methanol added as a stabilizer to commercial glutaraldehyde may produce inconclusive results.
3. The fixed tissues should be washed thoroughly to remove endogenous cytochrome *c*, which causes mitochondrial peroxidation of DAB.
4. The pH of the reaction medium should be high, ranging from 8.5 to 10.5, generally 9 to 9.7, to minimize peroxidase activity, the pH optimum of which is usually neutral.
5. The concentration of H_2O_2 should be relatively high: it is sometimes difficult to discriminate catalase from peroxidase in a DAB-H_2O_2 reaction; 1% H_2O_2 inactivates peroxidases but not catalase,[10] particularly in leukocytes which possess a high peroxidase activity, pretreatment with 1% H_2O_2 is essential.

TABLE 1
Dissociation of Hepatic Catalase Molecules and Changes in the Enzyme Activities

Treatments	Mol wt	Sedimentation coefficient	Catalytic activity	Peroxidatic activity
Native	250 kDa	11.2 S	+	−
Alkaline	1/3	3–3.8	−	+
6 *M* Formamide	1/3	3.8	−	+
2 *M* Urea	1/3	3.6	−	+
Acid	1/2	4.25	−	−

TABLE 2
Differences between Catalase and Peroxidase

	Catalase	Peroxidase
Long-term fixation	Resistant	Sensitive Resistant (EPO)
High concentration of glutaraldehyde	Resistant	Relatively sensitive
pH of reaction medium	Alkaline	Weakly acid-neutral
1% H_2O_2	Resistant	Sensitive
50 m*M* 3-Amino-1,2,4-triazole	Sensitive	Sensitive
Distribution	Peroxisome Cytosol	Nuclear envelope, endoplasmic reticulum, Golgi apparatus, secretory granule, azur granule, specific granule

C. EQUIPMENT AND CHEMICALS

1. A perfusion system for experimental animal tissues and a thermocontrolled shaker or rotator
2. Chemicals: 3,3′-diaminobenzidine-tetrahydrochloride (DAB·4HCl) salt (Wako Pure Chemical Co., Ltd., Tokyo; Sigma Chemical Co.); 30% H_2O_2; 2-amino-2-methyl-1,3-propanediol; osmium tetroxide; glutaraldehyde (20 to 70%), a special grade for electron microscopy with a purity ratio of 0.2 or less (A235/A280); catalase (Sigma, C-40); Ca^{2+}, Mg^{2+}-free PBS tablet (Flow Lab.).
3. Preincubation medium (neutral pH):

30% H_2O_2	1 ml (1%)
PBS	29 ml

4. DAB-H_2O_2 reaction medium (pH 9.7):

DAB·4HCl Salt	30 mg (0.1%)
Distilled water	12 ml
0.1 *M* 2-Amino-2-methyl-1,3-propandiol	15 ml
30% H_2O_2	1 ml (1%)
0.1 *N* NaOH	2.1 ml

5. Inhibitors: 50 m*M* 3-amino-1,2,4-triazole, 20 m*M* NaN_3

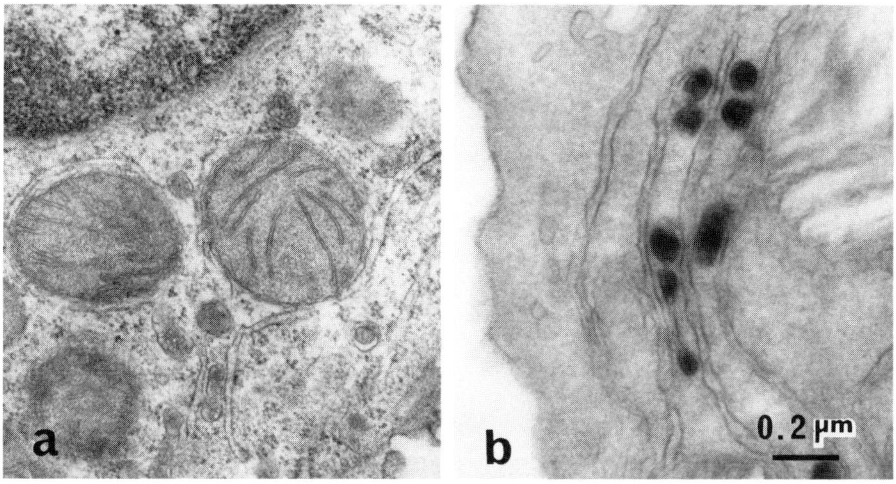

FIGURE 1. Peroxisome (a) and the peroxidatic activity of catalase of mouse lung alveolar type II cell (b). Tissues (b) were perfusion-fixed for 2 h, and incubated for 1 h in a DAB-0.01% H_2O_2 medium (pH 9.7). (From Hirai, K.-I. et al., *Exp. Mol. Pathol.*, 39, 129, 1983. With permission.)

D. PROCEDURES

1. Use perfusion or immersion fixation with 2% glutaraldehyde in PBS, pH 7.4 for 10 min at 4°C.
2. Dice and refix tissues for 4 to 24 h in the same fixative solution at 4°C with agitation using a rotator or shaker.
3. Wash tissues in three changes of PBS, for 20 min each, at 4°C.
4. Make 40-μm-thick slices.
5. Preincubate tissue slices in a preincubation medium for 10 min at room temperature to inactivate peroxidase or mitochondrial heme enzymes.
6. Incubate tissue slices in a DAB-H_2O_2 medium for 60 min at 37°C with agitation.
7. Wash the slices in three changes of a cold PBS, for 20 min each.
8. Postfix the slices with 1% OsO_4 dissolved in PBS for 60 min at 4°C.
9. Wash the osmium-treated slices in three changes of 0.1 *M* sodium acetate for 10 min each, at 4°C.
10. Block stain the slices with 0.5% uranyl acetate in distilled water for 20 min at 4°C.
11. Wash the slices with 50% ethanol for 10 min and embed them in a resin.
12. Control: incubate tissue slices in an H_2O_2-free DAB medium which contains 0.5 mg/ml catalase to eliminate endogenous peroxide in the medium.

E. RESULTS

Peroxisomes — The peroxidatic activity of catalase is demonstrated in the peroxisome matrix (Figure 1). The peroxisome core found in rodent hepatocytes is devoid of the activity (Figure 2). No reaction is found in the endoplasmic reticulum. However, DAB deposits have been demonstrated on the endoplasmic reticulum membrane and polysomes closely associated with peroxisomes.[11,12] Peroxidatic activity of catalase in the cytoplasmic background has been reported in the hepatocytes of sheep,[13] monkeys and guinea pigs (Figure 3),[14] and in the earthworm.[15] Peroxisomal catalase activity has been demonstrated in human hepatocytes,[16-20] lung alveolar type II cells (Figure 4),[21] and kidney.[20] Animal tissues containing peroxisomes include liver,[22,23] kidney,[25] lung,[26-28] small intestine,[29] cardiac muscle,[30] skeletal muscle,[31] pancreas,[32] adrenal glands,[33] nasal mucosa,[32] testis,[34] adipose tissue,[35]

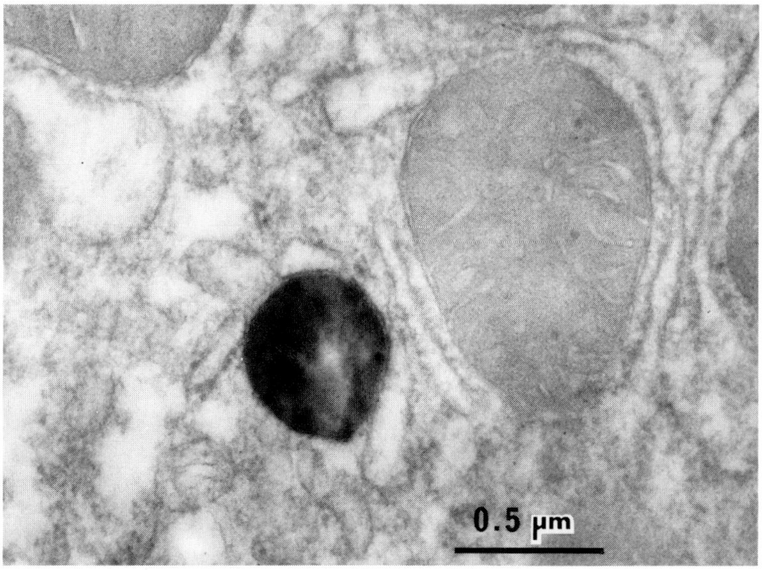

FIGURE 2. Localization of the peroxidatic activity of catalase in the rat hepatic peroxisome. Core is devoid of deposits. Tissues were immersion-fixed for 1 h with 2% glutaraldehyde, and incubated for 1 h in a DAB-0.01% H_2O_2 medium (pH 9.7).

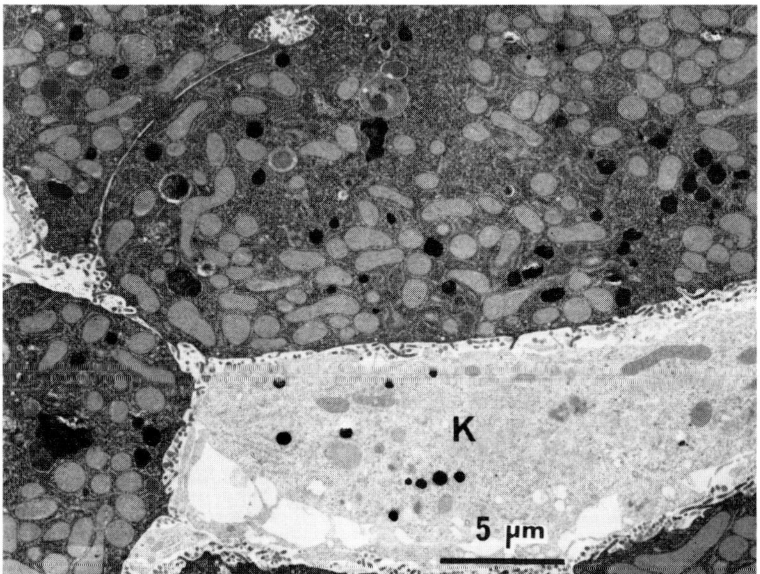

FIGURE 3. Distribution of the peroxidatic activity of catalase in guinea pig liver. Parenchymal cells exhibit two types of distribution of the activity in the peroxisomes and the cytoplasmic background. In Kupffer cells (K) the activity is localized in the peroxisomes. Tissues were fixed for 4 h, and incubated for 1 h in a DAB-0.01% H_2O_2 medium (pH 9.7).

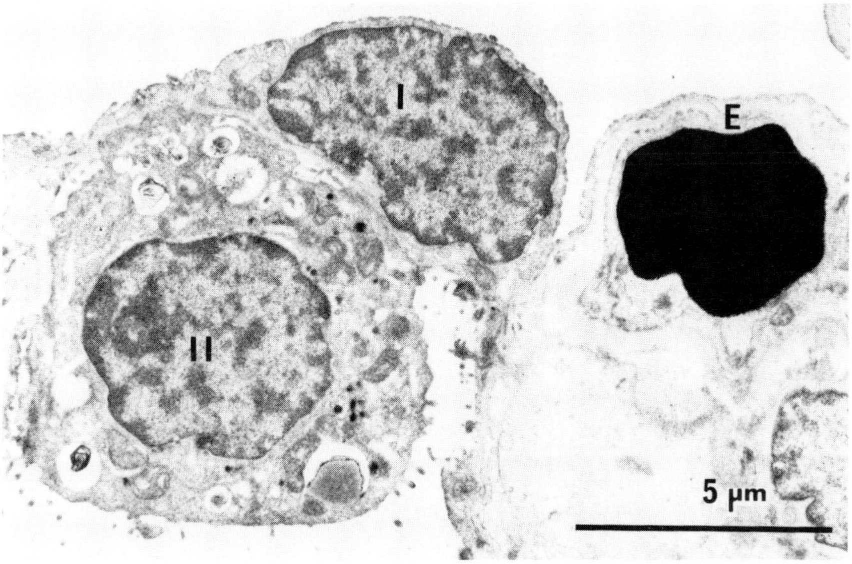

FIGURE 4. Localization of the peroxidatic activity of catalase in the human lung. Biopsies were immersion-fixed for 1 h with 2% glutaraldehyde-PIPES, and incubated for 1 h in a DAB-1% H_2O_2 medium (pH 9.7). (From Moriguchi, K. et al., *Exp. Mol. Pathol.*, 40, 262, 1984. With permission.)

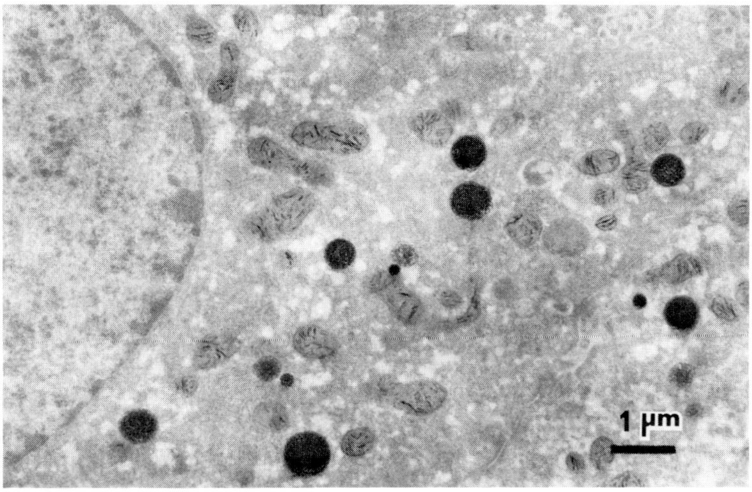

FIGURE 5. Artifact stain in the mitochondria. Rat liver was immersion-fixed for 1 h in a 2% paraformaldehyde-2.5% glutaraldehyde solution, and incubated for 1 h in a DAB-0.01% H_2O_2 medium (pH 9.7).

spinal cord,[36] salivary glands,[32,37] Hader's gland,[38] Ebner gland,[32] fibroblast,[39] and preputial gland.[33]

Erythrocytes — In tissue slices, erythrocytes in the vessels are always stained, although inhibitors do not affect the reaction. It is assumed that as a nonspecific peroxidatic reaction of hemoglobin, this erythrocytic stain is a good marker for evaluating the suitability of tissues or of the thorough penetration of the reaction reagents (Figure 4). An artefactual stain occurs in mitochondria when either fixation or washing is insufficient (Figure 5).

REFERENCES

1. **Inada, Y., Kurozumi, T., and Shibata, K.,** Peroxidase activity of hemoproteins. I. Generation of activity by acid or alkali denaturation of methemoglobin and catalase, *Arch. Biochem. Biophys.*, 93, 30, 1961.
2. **Kurozumi, T., Inada, Y., and Shibata, K.,** Peroxidase activity of hemoproteins. III. Activation of methemoglobin and catalase by formamide and guanidine, *Arch. Biochem. Biophys.*, 94, 464, 1961.
3. **Samejima, T. and Shibata, K.,** Denaturation of catalase by formamide and urea related to the subunit make-up of the molecule, *Arch. Biochem. Biophys.*, 93, 407, 1961.
4. **Novikoff, A. B. and Goldfischer, S.,** Visualization of microbodies for light and electron microscopy, *J. Histochem. Cytochem.*, 16, 507, 1968.
5. **Novikoff, A. B. and Goldfischer, S.,** Visualization of peroxisomes (microbodies) and mitochondria with diaminobenzidine, *J. Histochem. Cytochem.*, 17, 675, 1969.
6. **Fahimi, H. D.,** Cytochemical localization of peroxidatic activity in rat hepatic microbodies (peroxisomes), *J. Histochem. Cytochem.*, 16, 547, 1968.
7. **Fahimi, H. D.,** Cytochemical localization of peroxidatic activity of catalase in rat hepatic microbodies (peroxisomes), *J. Cell Biol.*, 43, 275, 1969.
8. **Hirai, K.-I.,** Light microscopic study of the peroxidatic activity of catalase in formaldehyde-fixed rat liver, *J. Histochem. Cytochem.*, 17, 585, 1969.
9. **Hirai, K.-I.,** Comparison, between 3,3′-diaminobenzidine and autooxidized 3,3′-diaminobenzidine in the cytochemical demonstration of oxidative enzymes, *J. Histochem. Cytochem.*, 19, 434, 1971.
10. **Wisse, E., Role, F., De Prest, B., Meulen, J. V. D., Emeis, J. J., and Deams, W. T.,** Peroxidatic reaction of Kupffer and parenchymal cells compared in rat liver, in *Electron Microscopy and Cytochemistry,* Wisse, E., Daems, W. Th., Molenaar, I., and Duijin, P. von, Eds., North-Holland Publishing, Amsterdam, 1973, 119.
11. **Legg, P. G. and Wood, R. L.,** New observations on microbodies. A cytochemical study on CPIB-treated rat liver, *J. Cell Biol.*, 45, 118, 1970.
12. **Wood, L. and Legg, P. G.,** Peroxidase activity in rat liver microbodies after amino-triazole inhibition, *J. Cell Biol.*, 45, 576, 1970.
13. **Roels, G.,** Cytochemical demonstration of extraperoxisomal catalase. I. Sheep liver, *J. Histochem. Cytochem.*, 24, 713, 1976.
14. **Roels, F., de Coster, W., and Goldfischer, S.,** Cytochemical demonstration of extraperoxisomal catalase. II. Liver of rhesus monkey and guinea pig, *J. Histochem. Cytochem.*, 25, 157, 1977.
15. **Fischer, E. and Horvath, I.,** Evidence of the presence of extraperoxisomal catalase in chloragogan cells of the earthworm, Lumbricus terrestris L., *Histochemistry,* 56, 165, 1978.
16. **Roels, F. and Cornelis, A.,** Heterogeneity of catalase staining in human hepatocellular peroxisomes, *J. Histochem. Cytochem.*, 37, 331, 1989.
17. **Lazarow, P. B., Small, G. M., Santos, M., Shio, H., Moser, A., Esterman, A., Black, V., and Dancis, J.,** Zellweger syndrome amniocytes: morphological appearance and a simple sedimentation method for prenatal diagnosis, *Pediatr. Res.*, 24, 63, 1988.
18. **King, B. F.,** Morphologic and cytochemical identification of microperoxisomes in human and non-human primate placental trophoblast, *Histochemistry,* 74, 115, 1982.
19. **Goldfischer, S., Moore, C. L., and Johnson, A. B.,** Peroxisomal and mitochondrial defects in the cerebro-hepato-renal syndrome, *Science*, 182, 62, 1973.
20. **Roels, F. and Goldfischer, S.,** An improved method for the cytochemical demonstration of peroxisomes in human liver and kidney, *J. Histochem. Cytochem.*, 25, 244, 1977.
21. **Moriguchi, K., Higashi, N., Kitagawa, S., Takase, K., Ohya, N., Uyeda, T., and Hirai, K.-I.,** Differentiation of human pulmonary alveolar epithelial cells revealed by peroxisome changes in pulmonary proteinosis, *Exp. Mol. Pathol.*, 40, 262, 1984.
22. **Essner, E.,** Observation on hepatic and renal peroxisomes (microbodies) in the developing chick, *J. Histochem. Cytochem.*, 18, 80, 1970.
23. **Saito, T., Iwata, K., and Ogawa, K.,** Changes of peroxisomes in the regenerating hepatic parenchymal cells in the rat, *Acta Histochem. Cytochem.*, 6, 212, 1975.
24. **Fahimi, H. D., Gray, B. A., and Herzog, V.,** Cytochemical localization of catalase and peroxidase in sinusoidal cells of rat liver, *Lab. Invest.*, 34, 192, 1976.
25. **Goeckermann, J. A. and Vigil, E. L.,** Peroxisome development in the metanephric kidney of mouse, *J. Histochem. Cytochem.*, 23, 957, 1975.
26. **Petrik, P.,** Fine structural identification of peroxisomes in mouse and rat bronchiolar and alveolar epithelium, *J. Histochem. Cytochem.*, 19, 339, 1971.
27. **Schneeberger, E. E.,** A comparative cytochemical study of microbodies (peroxisomes) in great alveolar cells of rodents, rabbit and monkey, *J. Histochem. Cytochem.*, 20, 180, 1972.

28. **Hirai, K.-I., Yamauchi, M., Witschi, H., and Cote, M. G.,** Disintegration of lung peroxisomes during differentiation of type II cells to type I cells in butylated hydroxytoluene-administered mice, *Exp. Mol. Pathol.,* 39, 129, 1983.
29. **Novikoff, P. M. and Novikoff, A. B.,** Peroxisomes in absorptive cells of mammalian small intestine, *J. Cell Biol.,* 53, 532, 1982.
30. **Herzog, V. and Fahimi, H. D.,** Microbodies (peroxisomes) containing catalase in myocardium: morphological and biochemical evidence, *Science,* 185, 271, 1974.
31. **Hand, A. R.,** Peroxisomes (microbodies) in striated muscle cells, *J. Histochem. Cytochem.,* 22, 207, 1974.
32. **Hand, A. R.,** Morphologic and cytochemical indentification of peroxisomes in the rat parotid and other exocrine glands, *J. Histochem. Cytochem.,* 21, 131, 1973.
33. **Goldenberg, J.-P., Huttinger, M., Ludwig, R., Kramar, R., and Bock, P.,** Catalase-positive particles (''microperoxisomes'') from rat preputial gland and bovine adrenal cortex, *Exp. Cell Res.,* 93, 438, 1975.
34. **Reddy, J. and Svoboda, D.,** Microbodies (peroxisomes) in the interstitial cells of rodent testes, *Lab. Invest.,* 26, 657, 1972.
35. **Pavelka, M., Goldenberg, H., Huttinger, M., and Kramar, R.,** Enzymic and morphological studies on catalase positive particles from brown fat of cold adapted rats, *Histochemistry,* 50, 47, 1976.
36. **Citkowritz, E. and Holtzman, E.,** Peroxisomes in dorsal root ganglia, *J. Histochem. Cytochem.,* 21, 34, 1973.
37. **Strum, J. M.,** Unusual peroxidase-positive granules in the developing rat submaxillary gland, *J. Cell Biol.,* 51, 575, 1971.
38. **Bock, P., Goldenberg, H., Huttinger, M., Kolar, M., and Kramar, R.,** Preparation and characterization of catalase-positive particles ('microperoxisomes') from Harder's gland of the rat, *Exp. Cell Res.,* 90, 15, 1975.
39. **Bock, P.,** Cytochemical demonstration of catalase positive particles (peroxisomes?) in fibroblasts, *Z. Zellforsch.,* 144, 539, 1973.

VII. MONOAMINE OXIDASE

Toshihiro Maeda

A. INTRODUCTION

The action of monoamine oxidase (MAO) (amine:oxygen oxidoreductase, deaminating [flavin-containing]; EC 1.4.3.4) on primary amines can be described as follows:

$$\text{E-FAD} + RCH_2NH_2 + H_2O \rightleftharpoons \text{E-FADH}_2 + RCHO + NH_3$$

$$\text{E-FADH}_2 + O_2 \rightleftharpoons \text{E-FAD} + H_2O_2$$

FIGURE 1. Principles of MAO histochemistry.

Aldehyde and hydrogen peroxide are the main products of MAO activity which are available for histochemical reaction. Reliable enzyme histochemical methods until now have always been based upon either aldehyde or hydrogen peroxide utilization. Historically, tetrazolium methods using the reduction ability of aldehyde, represented by Glenner's method, have long been most popular. However, the other method coupled with hydrogen peroxide has the advantage of using common or preferred substrates for the enzyme. This enables the MAO isozymes to be differentiated and so is an improvement. Thus, methods depending on aldehyde production, for which substrate availability is restricted, have become less meaningful in terms of chemical microscopic anatomy.

Two hydrogen peroxide methods based on different detection principles have been recently improved. One is a coupled peroxidation method[1,2] and the other is a cerium-based method.[3] The principles of histochemical staining are schematically summarized in Figure 1.

Both methods have advantages and disadvantages. The coupled peroxidation method consists of a one-step procedure in the presence of exogenous peroxidase (horseradish peroxidase: HRP). It can be visualized by the colored oxidation product of diaminobenzidine (DAB). However, the reaction product may diffuse, although the deposits have been shown to be confined to the inside of cellular structures, such as axon terminals or neuroglial processes in which the enzyme reaction is initiated.

On the other hand, the enzyme reaction product (H_2O_2) is captured directly in the cerium-based method, which may result in minimized diffusion artefacts. However, deposits of cerium perhydroxide, being the electron microscopically visible compound, are invisible in normal light. As a consequence, the staining intensity cannot be checked during the initial incubation, although this disadvantage was partially overcome by a second-step visualization procedure based on the conversion of cerium perhydroxide to cerium hydroxide and oxygen radicals at 55 to 60°C. The reactive oxygen oxidizes DAB into a visible polymer.[4]

B. METHODS

The coupled peroxidation and the cerium-based methods can both be employed at the level of the electron as well as the light microscope. Taking together the advantages and disadvantages of each method as described above, the coupled peroxidation and the cerium-based methods are recommended for neuroanatomical and cytochemical applications, respectively.

C. INSTRUMENTS AND EQUIPMENT

Perfusion fixation, which is mandatory for electron microscopic study of the central nervous system, requires a peristaltic pump. A refrigerated shaker is recommended, and other common tools for electron microscopy, including a Vibratome or a Microslicer are also required.

D. CHEMICALS AND REAGENTS

3,3′-Diaminobenzidine (DAB), Sigma grade II; horseradish peroxidase (HRP), Toyobo grade III (Toyobo, Tokyo). When HRP of another source or grade is used, the concentration may have to be changed. Nickel(II) (hexa)ammonium sulfate [$Ni(NH_3)_6SO_4$] may be obtained from Nacalai Tesque, Kyoto, Japan. It can also be prepared according to the instruction for copper(II) tetrammonium sulfate [$Cu(NH_3)_6SO_4$] with nickel(II) sulfate-hepta-hydrate and ammonium hydroxide (Hecht 1951: Preparative anorganische Chemie. Springer, Berlin). Tyramine HCl, sodium azide, 3-amino-1,2,4-triazole, clorgyline: *N*-methyl-*N*-propargyl-3-(2,4-dichlorophenoxy)propylamine, and other reagents are commercially available. A specific inhibitor of MAO-B, deprenyl, is not available commercially.

E. PROCEDURES

1. Coupled Peroxidation Method (for Neuroanatomical Research)[1]

1. Perfuse transcardially with 0.01 *M* phosphate-buffered saline (PBS) (pH 7.4), to wash out the blood (100 ml/2 min in the case of rat brain), followed by ice-cold fixative (250 ml/6 min in the case of rat brain after ligation of the descending aorta).
2. Remove brain and cut into sections of 30 to 60 μm on a Vibratome or a Microslicer.
3. Collect sections in ice-cold 0.1 *M* PBS (pH 7.4).
4. Incubate sections at 4°C with gentle agitation for 3 to 48 h. Preparation of the incubating medium: dissolve 5 mg DAB-4HCl in 10 ml of 0.05 *M* Tris-HCl buffer, pH 7.6, with constant stirring; add 10 mg HRP, 7.5 mg tyramine HCl, 60 mg nickel ammonium sulfate, and 6.5 mg sodium azide.
5. Rinse sections in ice-cold PBS.
6. Postfix sections with 1% OsO_4 in 0.1 *M* phosphate buffer, pH 7.4, at 4°C for 2 h.
7. Dehydrate and embed sections in epoxy resin (Luveak-812).

Notes: (1) A mixture of 2 to 4% formaldehyde and 2 to 3% glutaraldehyde in 0.1 *M* phosphate buffer, pH 7.4, can be used as a fixative. (2) To selectively observe type A or B of the enzyme, pretreat sections before step 4 with deprenyl (10^{-7} to 10^{-5} *M*) or clorgyline (10^{-6} to 10^{-4} *M*), respectively, at 4°C for 5 to 30 min, and rinse in PBS. Tyramine can be replaced by other substrates such as phenylethylamine or serotonin for type A or B enzyme, respectively. (3) Embedding sections in resin between the silicon-coated glass slide and cover glass allows trimming and ultrathin sectioning after light microscopic examination.

2. Cerium-Based Method (for Cytochemical Research)[3]

1. Fix tissues by immersion in a mixture of 0.1% glutaraldehyde and 1.5% formaldehyde in 1.0 *M* cacodylate buffer, pH 7.4, containing 0.15 *M* sucrose at 4°C for 60 min.
2. Rinse tissues in 0.1 *M* cacodylate buffer, pH 7.4, containing 0.15 *M* sucrose at 4°C for several hours.
3. Cut into sections of 4 to 10 μm (Cryostat) or 40 μm (Microslicer).
4. Immerse sections in a preincubation mixture containing 4.2 mg 3-amino 1,2,4-triazole and 2.5 g sucrose in 50 ml of 0.1 *M* Tris-maleate buffer, pH 7.5.

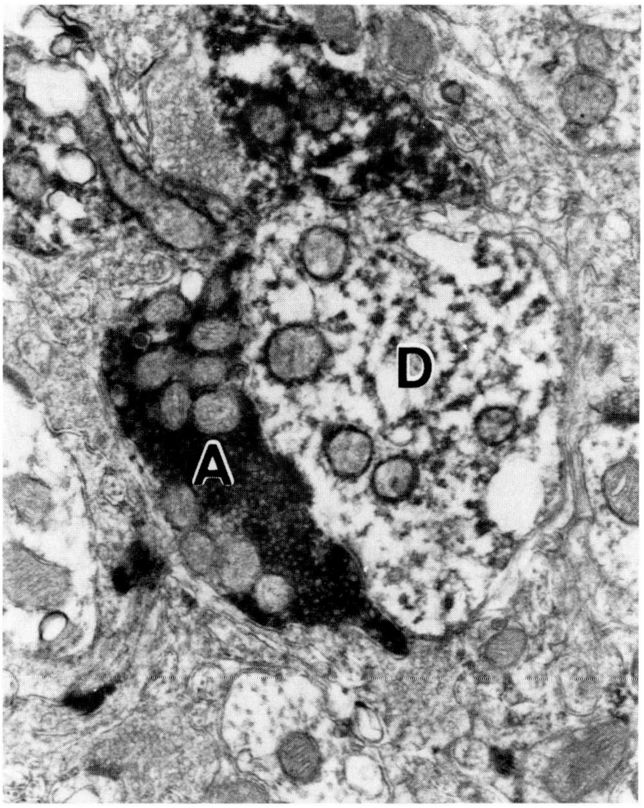

FIGURE 4. Electron micrograph of aminergic terminal (A) synapsed with the dendrite of amine-containing neuron (D). Note reaction products precipitated not only on the outer membrane of mitochondria but also in the cytoplasm. Locus coeruleus of rat, DAB-Ni. (Magnification × 26,000.)

REFERENCES

1. **Kishimoto, S., Kimura, H., and Maeda, T.** Histochemical demonstration for monoamine oxidaze (MAO) by a new coupled peroxidation method, *Cell Mol. Biol.*, 29, 61, 1983.
2. **Maeda, T., Imai, H., Arai, R., Tago, H., Nagai, T., Sakumoto, T., Kimura, H., Onteniente, B., and Kimura, H.,** An improved coupled peroxidatic oxidation method of MAO histochemistry for neuroanatomical research at light and electron microscopic levels, *Cell. Mol. Biol.*, 33, 1, 1987.
3. **Fujimoto, T., Inomata, K., and Ogawa, K.,** A certium method for the ultracytochemical localization of monoamine oxidase activity, *Histochem. J.*, 14, 87, 1982.
4. **Halbhuber, K.-J., Gossrau, R., Moller, U., Hulstaert, C. E., Zimmermann, N., and Feuerstein, H.,** The cerium perhydroxide-diaminobenzidine (Ce-H_2O_2-DAB) procedure. New methods for light microscopic phosphatase histochemistry and immunohistochemistry, *Histochemistry,* 90, 289, 1988.
5. **Arai, R., Kimura, H., and Maeda, T.,** Topographic atlas of monoamine oxidase-containing neurons in the rat brain studied by an improved histochmical method, *Neuroscience,* 19, 905, 1986.
6. **Kitahama, K., Sallanon, S., Lin, J. S., Maeda, T., and Jouvet, M.,** Type B monoamine-oxidase-containing cells and fibers in the cat hypothalamus demonstrated by an improved enzyme histochemical method, *J. Comp. Neurol.*, 285, 218, 1989.
7. **Maeda, T., Arai, R., and Kitahama, K.,** Extrinsic and intrinsic aminergic inputs to noradrenaline neurons in the locus coeruleus of rat: an electron microscopic study by monoamine oxidase enzyme histochemistry, *Biomed. Res.*, 10 (Suppl. 3), 259, 1989.

Chapter 1

ELECTRON MICROSCOPIC CYTOCHEMISTRY OF ENZYMES

1.3 Basic Techniques for Major Groups of Enzymes

1.3.2 Transferases

TABLE OF CONTENTS

0-8493-6012-9/93/$0.00 + $.50

I. ORNITHINE CARBAMOYLTRANSFERASE

Akira Mizutani

A. INTRODUCTION

Ornithine carbamoyltransferase (EC 2.1.3.3.) (OCT) catalyzes the carbamoylation of ornithine to citrulline by carbamoyl phosphate and, therefore, it participates in the urea cycle of the ammonia-urea metabolism.[1]

$$\text{L-ornithine} + \text{carbamoyl phosphate} \underset{\text{OCT}}{\overset{\uparrow}{\rightleftarrows}} \text{L-citruline} + \text{orthophosphate}$$

Brown and Cohen[2] have described the wide distribution of OCT in the livers of vertebrates, principally in ureotelic animals, and confirmed its significance in urea formation. Congenital deficiency of OCT results in severe hyperammoniemia. The fact that the transferase activity is connected with the liberation of inorganic phosphate led to the development of a Gomori-type histochemical method for the demonstration of OCT activity.[3-5] It is emphasized that (1) OCT is the only enzyme in the urea cycle for which a histochemical method is available, and (2) that the enzyme is localized in the mitochondrial matrix, specifically in that of the hepatocytes.

B. RECOMMENDED METHOD[3,4]

1. Chemicals and Reagents

Fixative — Formaldehyde solution (4%, pH 7.2) is recommended as a fixative. It is prepared freshly through depolymerization of paraformaldehyde, and buffered to pH 7.2 either by 0.05 *M* cacodylate-HCl or PIPES buffer solution.

Incubating Medium:

0.1 *M* Tris-maleate buffer (pH 7.4)	5 ml
Distilled water	4 ml
Dextran (mol wt 100,000 to 200,000)	0.1 g
1% Lead nitrate	1 ml
L-Ornithine-HCl	5 mg
Carbamoyl phosphate (Na or Li salt)	3 mg

Mix the reagents in above-mentioned order, and use immediately after adding the carbamoyl phosphate. Carbamoyl phosphate is unstable and should be stored dry below −20°C.

2. Methods

1. Perfusion fixation of the liver through the portal vein is recommended. In the case of the rat and guinea pig, perfusion is carried out under pressure equivalent to 35 to 40 cm of water, and 400 ml of fixative are allowed to drip for 30 min. In the case of small animals, such as newborns, transparenchymal perfusion may be used. If perfusion fixation is impossible, e.g., in the case of biopsy specimens, thin tissue slices less than 1 mm thick are immersed in the fixative for 6 to 8 h with continuous stirring. "Adequate but not excessive" fixation is one of the most critical factors of this method.
2. Wash dissected tissue blocks in cold buffer solution (Tris-maleate, Cacodylate or PIPES buffers, pH 7.4, containing 8% sucrose) for at least 2 h.

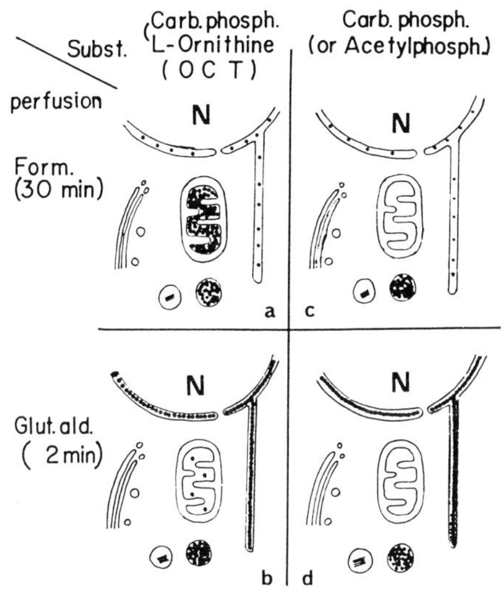

FIGURE 1. Schematic representation of OCT activity in a hepatocyte. With formaldehyde fixation, and with complete medium (a), the reaction product occurs mostly in the mitochondrial matrix (Figure 2). The reaction in the lysosomes, endoplasmic reticulum, and nuclear envelope is not specific since almost the same amount of reaction product is noted in control preparation (c) incubated in a medium without L-ornithine. With glutaraldehyde fixation and incubation in the complete medium (b), the reaction is slight in the mitochondria. The intensity of reaction in other organelles is moderate and similar to that seen in the control preparation (d).

3. Cut frozen or unfrozen sections 30 to 40 μm thick and float the sections onto the buffer solution.
4. Transfer free-floating sections into the incubating medium for 10 to 15 min at room temperature. Longer incubation should be avoided or, if necessary, the medium should be renewed, since carbamoyl phosphate is unstable and decomposes rapidly.
5. After incubation, wash the sections in buffer solution, immerse in a diluted solution of yellow ammonium sulfide, and rinse again. For light microscopy, mount the stained sections in glycerin jelly. This procedure can be omitted if electron microscopic observations are planned.
6. Postfix the sections in 1% osmium tetroxide in Veronal-acetate buffer, pH 7.4, for 1 h, and then follow the routine procedure for electron microscopy.

3. Results

Under the light microscope, deposits of the reaction product are seen as fine granules throughout the cytoplasm of the hepatocytes; the reaction tends to be more intense at the periportal than the centrolobular areas.

Under the electron microscope, the deposits of the final reaction product of OCT activity, lead phosphate or lead sulfide, are seen distributed in the matrix of the mitochondria of only the hepatocytes, but not of any other cells such as Kupffer cells, epithelium of bile ducts, or endothelium of blood vessels. The precipitates are also observed in lysosomes and, occasionally, in the endoplasmic reticulum and nuclear envelope. However, these reactions are proved to be nonspecific by control experiments (Figures 1 and 2).

4. Control Experiments

In order to support the specificity of the reaction, the following controls are indispensible: (1) omiting L-ornithine from the incubating medium or replacing it with D-ornithine;

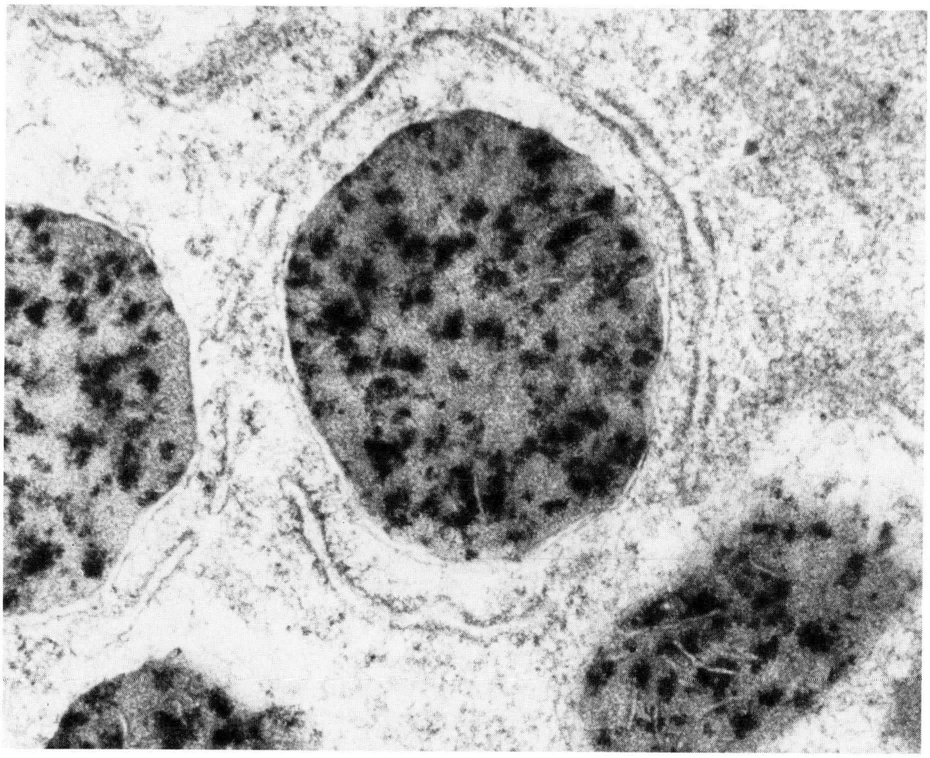

FIGURE 2. Hepatocyte of rat. Reaction products are noted in the matrix of mitochondria. (Magnification × 72,000.)

(2) pretreating the fixed sections with *p*-chloromercuribenzoate (5 m*M*, pH 7.4, containing 8% sucrose) for 15 min.

5. Comments

As already discussed, one of the most critical factors of the successful demonstration of OCT activity is the "adequate but not excessive" fixation of tissues. In order to preserve both OCT activity and tissue fine structure, tissue slices less than 1 mm thick should be fixed less than 8 h. The perfusion procedure is recommended. Glutaraldehyde, one of the most excellent fixatives, markedly reduces OCT activity, but, on the other hand, preserves the activities of other enzymes which hydrolyze carbamoyl phosphate.

Carbamoyl phosphate, one of the substrates of OCT, may also be hydrolyzed enzymatically and/or nonenzymatically, independently of OCT activity. Acid phosphatase in lysosomes and "carbamoyl phosphatase" in the endoplasmic reticulum and nuclear envelope seem to participate in the hydrolysis. The latter enzyme is similar to glucose-6-phosphatase in its histochemical behavior, and resembles acyl phosphatase biochemically, but has not been identified.[6,7]

REFERENCES

1. **Reichard, P.,** Ornithine carbamoyl transferase from rat liver, *Acta Chem. Scand.*, 11, 523, 1957.
2. **Brown, G. W., Jr. and Cohen, P. P.,** Comparative biochemistry of urea synthesis. III. Activities of ureacycle enzymes in various higher and lower vertebrates, *Biochem. J.*, 75, 82, 1960.
3. **Mizutani, A.,** Cytochemical demonstration of ornithine carbamoyl transferase activity in liver mitochondria of rat and mouse, *J. Histochem. Cytochem.*, 16, 172, 1968.
4. **Mizutani, A. and Fujita, H.,** Intramitochondrial localization of ornithine carbamoyltransferase activity in the guinea pig liver, *J. Electron Microsc.*, 18, 17, 1969.
5. **Merker, H.-J. and Spors, S.,** Electron microscopic demonstration of ornithine carbamoyltransferase in rat liver, *Histochemie*, 17, 83, 1969.
6. **Mizutani, A. and Fujita, H.,** Cytochemical studies on hydrolysis of carbamoyl phosphate in rat liver, *Acta Histochem. Cytochem.*, 1, 79, 1968.
7. **Mizutani, A.,** Lead ions in enzyme histochemistry, with special reference to OCT and related enzymes, *Acta Histochem. Cytochem.*, 12, 157, 1979.

II. CHOLINE ACETYLTRANSFERASE

Hisanobu Kaiya

A. INTRODUCTION

Choline acetyltransferase (ChAc, EC 2.3.1.6.) is an enzyme catalyzing the reaction between acetyl-CoA (coenzyme A) and choline:

$$\text{acetyl-CoA} + \text{choline} \rightleftharpoons \text{CoA} + O\text{-acetylcholine}$$

ChAc exists mainly in the nervous system and is involved in the synthesis of acetylcholine, an important neurotransmitter. Therefore, ChAc is specifically detected in the cholinergic nervous system such as the neuromuscular junctions, motor neuron-Renshaw cells, pre- and postganglionic sympathetic fibers, motor nuclei of cranial nerves, septohippocampal tract, habenulointerpeduncular tract, striatal interneurons, and globus pallidus-cortical projection. Apart from the nervous system, the enzyme is found in the placenta, corneal epithelium, and erythrocyte and platelet membranes. These distributions of ChAc outside the nervous system suggest that this enzyme plays a role in transmembranous transportation.

Methods for the cytochemical demonstration of ChAc activity at the level of electron microscope, utilizing basically the same approach as used for light microscopic histochemistry, have been reported. The ferricyanide reduction method, devised for the electron microscope cytochemical demonstration of the enzyme, will be described here.

B. PRINCIPLE OF THE REACTION

The histochemical method for the demonstration of ChAc activity is based on the reduction of ferricyanide to ferrocyanide by CoA-SH (Figure 1), generated in the transferase reaction catalyzed by ChAc. The ferrocyanide formed is trapped in the form of insoluble, electron-opaque cobalt ferrocyanide at or near the site of the enzyme (Figure 2). The method of Ago and Ogawa,[1] which provides reproducible results, is described here.

C. PROCEDURES

1. Fix in buffered 4% formaldehyde containing 0.06 mM acetyl-CoA and 5 mM choline chloride for 5 to 30 min at 0 to 4°C. The length of fixation depends on the tissue. Acetyl-CoA and choline chloride are added to the fixative to protect the enzyme.
2. Rinse in three changes of phosphate-buffered saline.
3. Incubate small blocks (40 μm) of fresh Vibratome or Microslicer sections of the tissue for 30 to 60 min at room temperature in a medium containing 60 mM phosphate buffer, pH 7.0; 0.6 mM acetyl-CoA; 10 mM choline chloride; 3 mM sodium citrate; 3 mM cobalt chloride; 0.5 mM potassium ferricyanide; 1 mM tetraisopropyrophosphoramide (a choline esterase inhibitor); and 440 mM sucrose. Adjust the final pH to 6.8 to 7.0.
4. Fix in 1% buffered osmium tetraoxide for 30 to 60 min at 0 to 4°C.
5. Process for routine electron microscopy.

As is usually the case, the fixation employed in the cytochemical demonstration of ChAc represents a compromise between maintenance of cellular ultrastructure and loss of enzyme activity. Since glutaraldehyde fixative causes a greater loss of ChAc activity than fixative with formaldehyde,[2] the latter is generally recommended. However, if formaldehyde fixative causes an unacceptable loss of enzyme activity, an attempt can be made to incubate unfixed,

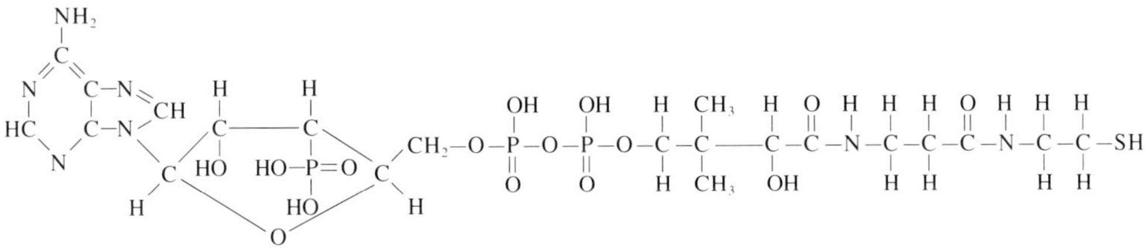

FIGURE 1. The chemical structure of coenzyme A.

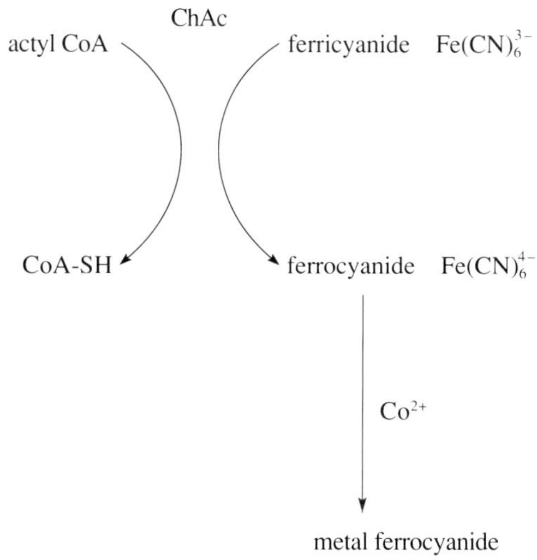

FIGURE 2. The principle of the ferricyanide reduction method.

fresh tissue and directly incubate in the reaction medium for 15 to 20 min, and then fix the tissue by osmium tetraoxide.

As a control, tissues are incubated in a medium without substrate, such as acetyl choline or acetyl-CoA, or in the presence of a SH-chain inhibitor, such as 3 m*M* *N*-ethyl-maleinimide. ChAc inhibitors such as *N*-hydroxyethyl-4-(1-naphthylvinyl) pyridinium bromide, 1 to 10 m*M*, or tetramethyl ammonium chloride, 40 m*M*, may also be used. After addition of the inhibitors, the pH of the medium should be readjusted.

D. RESULTS

Deposits of the reaction product indicating ChAc activity are observed in the plasma membrane, mitochondria, Golgi apparatus, and endoplasmic reticulum.[1,3,4] Especially in the rat brain, the ChAc activity is located to the synaptic membrane and synaptic vesicles (Figures 3, 4, and 5).[3,4]

E. DISCUSSION

The main problem with this method is its specificity. The method is based on the generation of CoA from acetyl-CoA in the presence of choline. However, reaction can occur with substrates other than choline, i.e., carnitine. Furthermore, endogenous oxidoreductases and CoA-hydrolyzing enzymes in the tissue can also facilitate the histochemical reaction. For this reason the above-mentioned controls, including the addition of a choline esterase inhibitor to the incubation medium, which prevents the hydrolysis of acetyl CoA, are indispensable.

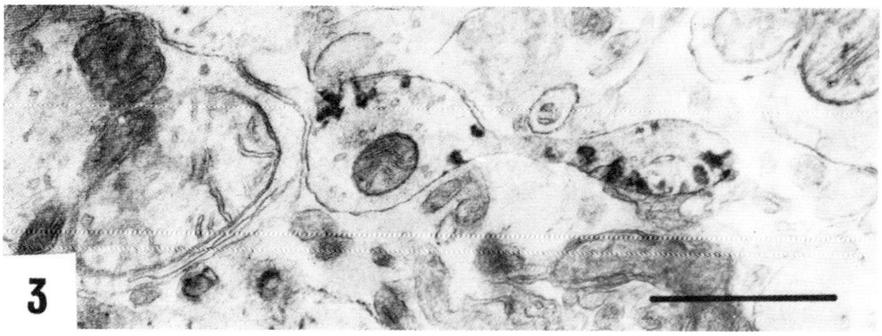

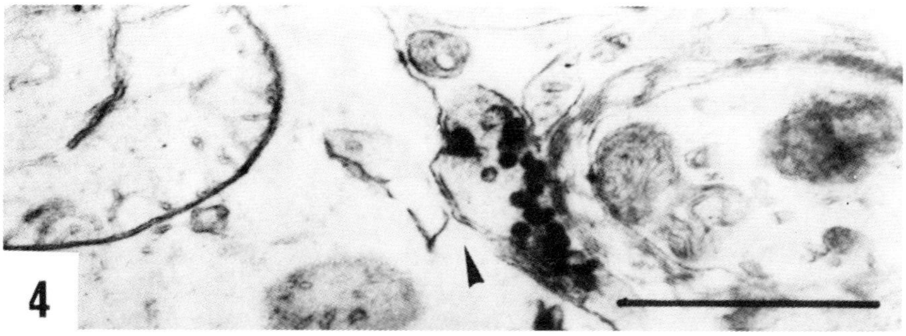

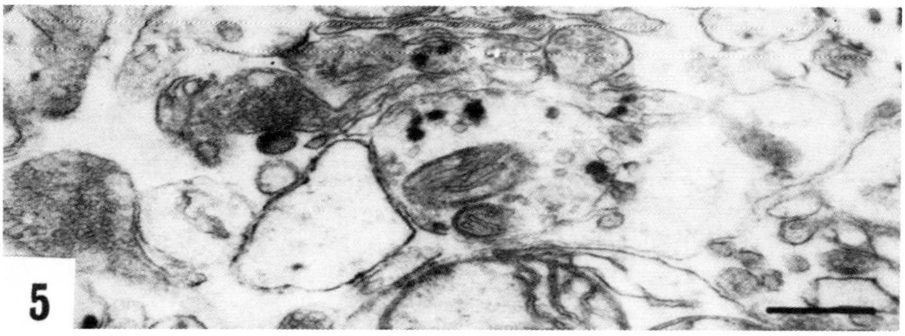

FIGURE 3. Varicose boutons with ChAc-positive synaptic vesicles. (Magnification × 26,000.)

FIGURE 4. An axo-dendritic synapse with round synaptic vesicles, most of which show ChAc activity. A large mitochondrion is seen in the postsynaptic dendrite. A cisternal structure is seen in the vicinity of the postsynaptic membrane (arrow). (Magnification × 33,000.)

FIGURE 5. An axo-spinous synapse. The bouton which contains ChAc-positive round synaptic vesicles and small mitochondria makes a junction with the dendritic spine, in which the plasma membrane also shows reaction products of ChAc. (Magnification × 28,000.)

REFERENCES

1. **Ago, Y. and Ogawa, K.,** Fine structural localization of choline acetylase activity in the nervous tissue of the rat, in *Electron Microscopy 1974,* Vol. 2, *Biological,* Sanders, J. V. and Goodchild, D. J., Eds., The Australia Academy of Science, Canberra, 1974, 148.
2. **Feigenson, M. E. and Barrnett, R. J.,** Combined pharmacological and fine-structural studies of choline o-acetyltransferase at the myoneural junction, *Brain Res.,* 119, 155, 1977.
3. **Kaiya, H., Iwata, T., Namba, M., Ago, Y., Mayahara, H., and Ogawa, K.,** Ultracytochemical observations of choline acetyltransferase neurons in the rat neostriatum, *Exp. Brain Res.,* 34, 395, 1979.
4. **Kaiya, H., Iwata, T., Namba, M., Ago, Y., Mayahara, H., and Ogawa, K.,** Ultracytochemistry of choline acetyltransferase in the neostriatum of the rat, *Acta Histochem. Cytochem.,* 12, 391, 1979.

III. GLYCOSYLTRANSFERASE

Tsugio Amemiya

A. INTRODUCTION

In glycogen, sugars are linked to other sugars by glycosidic bonds that are formed by specific glycosyltransferases. A glycosyltransferase reaction proceeds by donating the glycosyl unit from the nucleotide derivative to the nonreducing end of an acceptor sugar.[1] That is, glycosyltransferase catalyzes the following reaction: UDPG + primer → UPD + glycosyl-α-(1 → 4) primer. According to this reaction, polyglucose is synthesized in the cell if tissue containing glycosyltransferase is incubated in a substrate mixture containing UDPG and primer. This is the principle of the histochemical demonstration of glycosyltransferase.

B. LIGHT MICROSCOPIC DEMONSTRATION

Materials and reagents — Cryostat; uridine 5′- diphosphate glucose (UDPG); glucose-6-phosphate (G-6-P); reduced glutathion; α-amylase; β-amylase (all from Sigma Chemical Co., St. Louis, MO).

1. Methods[2,3]

1. Cut frozen sections of fresh unfixed tissue.
2. Incubate free-floating sections in the incubating medium at room temperature or at 37°C for 1 h.
3. Place in dilute Gram's iodine glycerol (I:KI:H_2O/1:2:3000) until red-brown color appears.
4. Mount sections on slides, cover with iodine glycerol (Gram's iodine solution:glycerol/1:10), and seal with paraffin.

The incubating medium is composed of 100 mg UDPG, 100 mg G-6-P, 10 mg primer, 2 mg reduced glutathion, 2 mg MgCl, 10 ml Tris-buffer or glycine-buffer (pH 8.4 to 9.0), and 15 ml distilled water.

The reaction can be visualized with periodic acid Schiff's reaction (PAS). Although the iodine reaction is specific and suitable, the PAS reaction is useful if the iodine reaction is unclear.

The control incubating medium is made without UDPG. The staining process is the same as that described above.

2. Results

The polysaccharide synthesized from UDPG in tissue sections during the incubation is stained red-brown with iodine and purple with PAS. The newly synthesized polysaccharide can be digested almost completely by α-amylase in 15 to 30 min, but it is not digested completely by β-amylase in 30 min.

Control sections incubated in a medium without UDPG show neither red-brown staining with iodine nor purple staining with PAS.

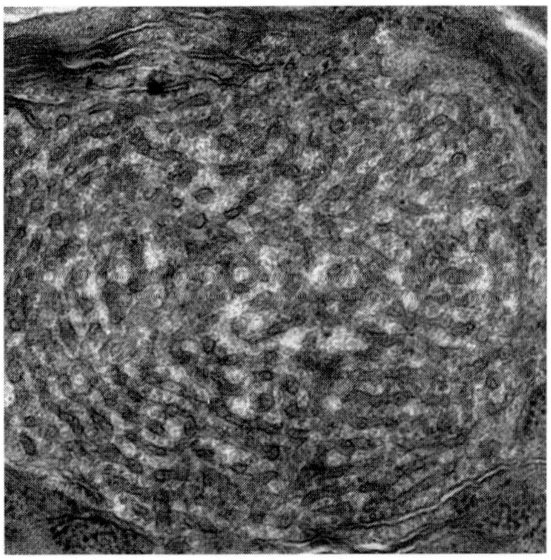

FIGURE 1. Electron micrograph of the paraboloid of a chick retina incubated in the medium for glycosyltransferase. Polyglucose particles are located in the cytoplasmic matrix. The cytoplasmic matrix is expanded by overproduced polyglucose particles, and the structure is flattened. (Magnification × 40,000.)

C. ELECTRON MICROSCOPIC DEMONSTRATION

1. Methods[2-4]

1. Incubate a very small fragment of unfixed tissue in the substrate medium for 30 min.
2. Fix in 6.5% glutaraldehyde for 30 min.
3. Immerse some of the specimens in α-amylase or β-amylase solution (digestion test).
4. Postfix specimens with and without amylase treatment in 1% osmium for 30 min.
5. Dehydrate in a series of ethanols as quickly as possible.
6. Embed in Epon 812.

To avoid the dissolution of the newly synthesized polyglucose in the water, steps 2 to 5 must be performed rapidly.

The incubating medium for the electron microscopic demonstration of glycosyltransferase is the same as that for light microscopy. The control medium contains no UDPG or G-6-P.

Before ultrathin sections are made, 1-μm slices are cut and stained with PAS, and the sites to be studied by electron microscopy are identified. The Epon-embedded sections also are stained red-brown with iodine and purple with PAS, but staining requires more than 6 h. Ultrathin sections are made and observed under an electron microscope.

2. Results

In sections incubated in the medium for glycosyltransferase, newly synthesized polyglucose particles are located in the cytoplasmic matrix (Figure 1). Most of the particles are less than 20 nm in diameter. These particles are rather well stained with lead citrate and fill in the cytoplasmic matrix.

In sections treated with α-amylase after incubation, no staining with iodine or PAS is found and no synthesized polyglucose particles are seen in the cytoplasmic matrix. On the other hand, sections treated with β-amylase after incubation show faint staining with iodine or PAS. Sections treated with β-amylase still contain synthesized polyglucose particles in the cytoplasmic matrix seen in electron micrographs.

Sections incubated in the medium without UDPG show no staining with iodine or PAS and a vacant cytoplasmic matrix, e.g., an expanded cytoplasmic matrix in electron micrographs.

D. LOCALIZATION

Glycosyltransferase is usually expected to coexist with phosphorylase in the same cell. Generally speaking, glycosyltransferase activity is weaker than phosphorylase activity, and occasionally it is not demonstrated by histochemical methods. For instance, glycosyltransferase and phosphorylase are located in the corneal epithelium, but glycosyltransferase is found mainly in the superficial half of the epithelium and phosphorylase mainly in the deep layer.

Localization of glycosyltransferase in various tissues is as follows.[3,5]

Skeletal muscle	Striated muscle fibers (rabbit) (intense)
Tongue	Striated muscle fibers (rabbit) (intense)
Esophagus	Striated muscle fibers (rabbit) (intense)
Stomach	No staining (rabbit)
Intestine	No staining (rabbit)
Liver	All liver cells (rabbit) (less intense)
Pancreas	No staining (rabbit)
Trachea	Epithelium, smooth muscle (rabbit) (faint), tracheal cartilage (rabbit)
Lung	No staining (rabbit)
Heart	Muscle fiber (rabbit) (less intense)
Kidney	Epithelium of proximal straight tubules (rabbit) (weak)
Urinary bladder	No staining (rabbit)
Brain	No staining (rabbit)
Bone	Cytoplasm of some cartilage cells (rabbit)
Skin	Sheath cells of hair follicles (rabbit) (faint)
Cornea	Epithelium (rabbit) (weak)
Retina	Paraboloid (chick accessory cone) (weak); pigment epithelium (moderate), photoreceptor inner segment (moderate); Müller cells (intense); outer and inner plexiform layers, ganglion cell layer, and nerve fiber layer (rat) (weak)
Blood vessel	Endothelium, smooth muscle (chick) (weak)

REFERENCES

1. **Leloir, L. F., Olavarría, J. M., Goldemberg, S. H., and Carminatti, H.,** Biosynthesis of glycogen from uridine diphosphate glucose, *Arch. Biochem. Biophys.*, 81, 508, 1959.
2. **Takeuchi, T. and Glenner, G. G.,** Histochemical demonstration of a pathway for polysaccharide synthesis from uridine-diphosphoglucose, *J. Histochem. Cytochem.*, 8, 227, 1960.
3. **Takeuchi, T. and Glenner, G. G.,** Histochemical demonstration of uridine diphosphate glucose-glycogen transferase in animal tissues, *J. Hitochem. Cytochem.*, 9, 304, 1961.
4. **Takeuchi, T., Sasaki, M., and Matsumoto, O.,** An electron histochemical observation of polysaccharide synthesized from uridine diphosphoglucose by the enzyme activity, *Acta Histochem. Cytochem.*, 3, 173, 1970.
5. **Amemiya, T. and Koyama, S.,** Electron histochemical study on the UDPG-glycogen glycosyltransferase and phosphorylase systems in the retina, *Acta Soc. Ophthalmol. Jpn.*, 74, 1032, 1970.

IV. GLUTAMIC-OXALACETIC TRANSAMINASE

Akira Mizutani

A. INTRODUCTION

Glutamic-oxalacetic transaminase (GOT), registered as aspartate aminotransferase (EC 2.6.1.1.), catalyzes the following reaction:

$$\text{L-aspartate} + \alpha\text{-ketoglutarate} \underset{\uparrow\ \text{GOT}}{\rightleftarrows} \text{L-glutamate} + \text{oxalacetate}$$

Two kinds of principles have been established for the histochemical demonstration of GOT activity, that is, the metal salt method and the azo dye method. The latter, however, is not adequate for electron microscopic observations, since the reaction products do not have satisfactory electron density.

Oxalacetate, one of the products of enzymatic transamination, can combine with lead ions and precipitate within alkaline ranges. These precipitates are detected under the electron microscope as the sites of transaminase activity.

B. RECOMMENDED METHOD[1,2]

1. Chemicals and Reagents

Fixatives may be either (1) 1% glutaraldehyde and 3.7% formaldehyde in 0.05 *M* imidazole-nitric acid buffer (pH 7.2), or (2) the same fixative with glutaraldehyde omitted.

Solutions for tissue rinse may be (1) 0.3 *M* (10.3%) sucrose buffered at pH 7.3 by 0.05 *M* imidazole-nitric acid buffer, or (2) the same solution with 20 m*M* L-aspartate added.

Preparation of the incubating medium is as follows:

L-aspartic acid	26.6 mg (final concentration 20 m*M*)
α-ketoglutaric acid	5.8 mg (final concentration 4 m*M*)
imidazole	34 mg (final concentration 50 m*M*)

Mix the above reagents in 3.5 ml distilled water and adjust to pH 7.5 by 1 *N* NaOH. Free aspartic acid, hardly soluble in water at a lower pH, becomes soluble at pH 7.5. Sodium aspartate may be used in place of free acid; its final concentration should be 20 m*M*. Add distilled water to 5 ml. Further, add 5 ml of 12 m*M* lead nitrate drop by drop with continuous stirring. Add 1.03 g of sucrose. The final pH must be 7.2 to 7.4. If the solution is turbid, filtrate and use immediately.

2. Method

1. Tissue fixation: Perfusion fixation with Fixative (1) through the thoracic descending aorta is recommended. Perfuse 200 ml of the fixative within 2 to 3 min for the adult rat.
2. Remove small slices of tissue and fix in cold Fixative (2) for another 30 min. If the perfusion fixation is not available, fix small fresh tissue slices in the cold Fixative (2) overnight.
3. Rinse the tissue in Rinse (1) for 1 h.
4. Incubate frozen or nonfrozen sections, 40 to 50 μm thick in the incubating medium at room temperature for 15 to 30 min.

5. Rinse the sections in Rinse (2) for 5 min.
6. Postfix in 1% OsO_4 buffered with Veronal to pH 7.2 to 7.4.
7. Briefly rinse in Rinse (1).
8. Dehydrate the sections in graded ethanols adjusted to pH 7.4 by imidazole buffer. Further dehydrate in propylene oxide and mount in the embedding medium. (For light microscopic observation, immerse the sections in diluted solution of yellow ammonium sulfide after step 5 and mount in glycerin jelly.)

3. Results

With the electron microscope, the reaction product (lead oxalacetate) is found to be localized in the cristae and surface membrane of the mitochondria, the limiting membrane of microbodies, and the nuclear envelopes of the hepatocytes of the rat.

4. Control Experiments

Omit α-ketoglutaric acid from the incubating medium or replace L-aspartic acid by its D-isomer.

5. Comments

The most important and critical factor of this method is to maintain a slightly alkaline pH of the solutions, such as the incubating medium, washing water, and graded ethanols for dehydration. The precipitates of lead oxalacetate are stable under this condition.

REFERENCES

1. **Lee, S. H. and Torack, R. M.,** Electron microscopic studies of glutamic oxalacetic transaminase in rat liver cell, *J. Cell Biol.*, 39, 716, 1968.
2. **Lee, S. H.,** Ultrastructural localization of glutamic oxalacetic transaminase activity in cardiac muscle fiber and cardiac mitochondrial fraction of the rat, *Histochemie,* 19, 99, 1969.

TABLE 3
Cu,Zn-SOD in Cell Organelles of Rat Hepatocyte

Nucleus	0.046 ± 0.017
Cytoplasm	0.282 ± 0.030
Mitochondria	0.033 ± 0.019
Rough ER[a]	0
Smooth ER	0.002 ± 0.002
Golgi complex	0
Lysosomes	0.021 ± 0.010
Peroxisomes	0.002 ± 0.000
Total	0.386 ± 0.028 mg/g

[a] ER, Endoplasmic reticulum

Data from Reference 12.

chromatography and Western blot analysis. Ultracryosections are the best choice of preparation if instruments for ultracryotomy are available. The next choice is fixation with 2% paraformaldehyde, with or without addition of 0.2 to 0.5% glutaraldehyde. Sections are treated with 3% H_2O_2 for 10 min to inactivate endogenous peroxidase and then with 10% normal goat serum for 30 min to diminish nonspecific binding. The sections are incubated with rabbit anti-Cu,Zn-SOD diluted (1:500) with PBS at pH 7.2 for several hours. The sections are then incubated at room temperature with 1% anti-rabbit goat IgG Fab fraction labeled with horseradish peroxidase for 60 min. After that the reacted sections are incubated in 0.02% 3,3′-diaminobenzidine containing 0.005% H_2O_2 for 5 min. Postfixation with 2% OsO_4 is for 60 min at 4°C. Reaction could be examined by backscatter electron image of reaction products with Wet Sample Scanning Electron Microscope (WET-SEM).[19]

Protein A-gold methods — The isolation of Cu,Zn-SOD from tissue homogenate or erythrocytes and the preparation of monoclonal or polyclonal antisera are similar to that of the ABC method. Ultracryosections are again the best choice of preparation. Alternatively, sliced, briefly fixed tissues are dehydrated and embedded in Lowicryl and polymerized at −20°C. The ultracryosection or embedded sections are immunolabeled with rabbit anti-rat (or swine, monkey, or human) Cu,Zn-SOD and 10 nm protein A-gold. Counterstained with or without uranyl acetate, the reacted sections are examined after slight carbon coating. A large dose (60,000 to 300,000 U/rat) of swine SOD is intravenously injected or administered via a gastric tube to the stomach of rats and then localized with protein A-gold method in target organs.[23] Human SOD is encapsulated with multilayered liposome composed of dipalmitoylphosphatidylcholine, cholesterol, and stearylamine. The ratio of human SOD to liposome in weight is 1:7 and the mean diameter of liposomes is 260 to 300 nm.[30,36] The liposomal human SOD is localized in various organs of rats with the protein A-gold method.

C. RESULTS

Cerium method — The dense reaction products are localized in the cytoplasm and the nuclear matrix of hepatocyte, gastric gland parietal cells, and cells of adrenal cortex, cardiac muscle, and many other organs. The size of reaction products on the surface of mitochondria is fairly large; more improvement is required to localize the SOD activity at the electron microscopic level.

ABC method — With this method, various tissues of baboon,[19] Japanese monkey, and rat[14-16] were examined for Cu,Zn-SOD localization (see Table 5). The reaction was located mainly in the cytoplasm and occasionally in the nucleus. The reaction in the nucleus was

TABLE 4
Effect of Fixatives for ABC Staining in Rat Hepatocytes

	Cytoplasm	Nucleus
Unfixed (fresh)	+ + +	± ~ +
Glutaraldehyde (2–4%)	+ +	+ +
Glutaraldehyde (2%) + paraformaldehyde (4%)	+ ~ + +	± ~ +
Paraformaldehyde (4%)	+ ~ + +	± ~ +

Key: + + +: intense reaction; + +: positive reaction; +: moderate reaction; ±: slight reaction.

variable and there were at least two types of hepatocytes: one with positive SOD in the nucleus and the other without SOD in the nucleus. The variation appeared partially due to fixation (Table 4). The reaction in the nuclear matrix also could be caused by diffusion of cytosolic SOD into the nucleus through nuclear pores.[12,19]

Protein A-gold method — Similar to the ABC method, polyclonal antiserum resulted in a heavier reaction than the monoclonal one. The slight fixation with 2% paraformaldehyde and 0.5% glutaraldehyde resulted in augmentation of immunolabeling in the nuclear matrix as compared to unfixed preparations. The postembedding procedure, after Lowicryl embedding, was as good as ultracryosections. Protein A-gold particles were distributed throughout the cytoplasm (Figure 1). The gold particles in the nucleus were rather clustered in the matrix. Although it has been accepted that tumor cells contain a significantly lower level of SOD,[24,25] immunocytochemical localization of SOD in a cell of Yoshida sarcoma was similar to that in intact cells (Figure 2). Since clinical application of both Cu,Zn-SOD and Mn-SOD has been reported by many investigators,[26-28] it will be essential to localize administered SODs immunocytochemically at the electron microscopic level. To distinguish administered SOD from endogenous SODs, human SOD or swine SOD was administered to rats. Endocytosis of bovine Cu,Zn-SOD to cultured hepatocyte has been described via specific receptors of target cells.[29] In the rat lung and kidney,[30] the protein A-gold particles were concentrated in dark spots within lysosomes (Figure 3). Lysosomal localization of Cu,Zn-SOD has been noticed by some other investigators.[31] It is also interesting to note that peroxisomes contain only little, if any, SOD activity.

D. DISCUSSION

Localization of SOD in regenerating cells, such as in regenerating liver after partial hepatoectomy and regenerating endothelial cells of aorta after ballooning, as well as in injured gastric mucous epithelium, also provides useful information for the clinical application of immunocytochemistry. An insignificant increase of SOD localization has been detected in all regenerating cells so far examined in this laboratory. The extracellular Cu,Zn-SOD was evident in the intima of regenerating aorta.[32] Elastin complex and fibrous components in the extracellular matrix adjacent to endothelial cells and smooth muscle cells had much protein A-gold labeling of SOD (Figure 4). As the clearance of administered SOD occurs within a few hours, several attempts to prolong its decomposition, such as chemical modification of SOD with dextran or polyethylene glycol[33] and encapsulation of SOD with many types of liposome,[34,35] have been reported. Immunolabeling of human Cu,Zn-SOD in lyposomes[36] could be visualized *in situ* (Figure 5).

TABLE 5
Distribution Immunolabeling (ABC Method) of Cu,Zn-SOD at the Electron Microscopic Level

	Baboon				Japanese monkey				Swine				Rat			
	C	N	M	EC	C	N	M	EC	C	N	M	EC	C	N	M	EC
Liver																
Hepatocyte	+ −	±	−	−	L+ +	±	±	−	+ +	±	−	−	L+ +	±	−	−
Kupffer cell	+ + +	+	±	−	L+ + +	+	−	−	+ +	±	−	−	L+ +		−	−
Kidney																
Glomerulus	+ −	+	−	−	L+	±	−	−	+	±	−	−	L+	±	−	−
Proximal tubules	+ + +	+	−	−	L+ + +	+	−	−	+	−	−	−	L+	+	−	−
Adrenal glands																
Z. glomerulosa	+ −	±	−	−	+ +	±	−	−	+ +	±	−	−	+ +	+	−	−
Z. fasciculata	+ + +	+	−	−	+ + +	+	−	−	+ + +	+	−	−	+ + +	+	−	−
Z. reticularis	+	±	−	−	+	+	−	−	+	−	−	−	+	±	−	−
Medulla	+	−	−	−	+	−	−	−	+	−	−	−	+	−	−	−
Gastric gland																
Chief cell	+	−	−	−	+	−	−	−	+	+	−	−	+	±	−	−
Parietal cell	+ + +	±	−	−	+ +	−	−	−	+ +	±	−	−	+ +	±	−	−
Mucous cell	+	±	−	−	+	±	−	−	+ +	±	−	−	+	±	−	−
Aorta (intima)																
Endothelial cell	+ +	+	−	+	+ +	+	−	L+	+ +	+	−	+	+ +	+	−	L+
Smooth muscle	+	+	−	+	+	+	−	L+	+	+	−	+	+	+	−	L+
Lung																
Epithelial cell	+	±	−	+	L+	±	−	±	+ +	+	−	+	L+ +	±	−	±

Note: There were no essential differences between species. C, cytoplasm; N, nucleus; M, mitochondria; EC, extracellular substances; L, liposomal human SOD and administered swine SOD; + +, intense reaction; +, moderate; ±, slight; −, negative.

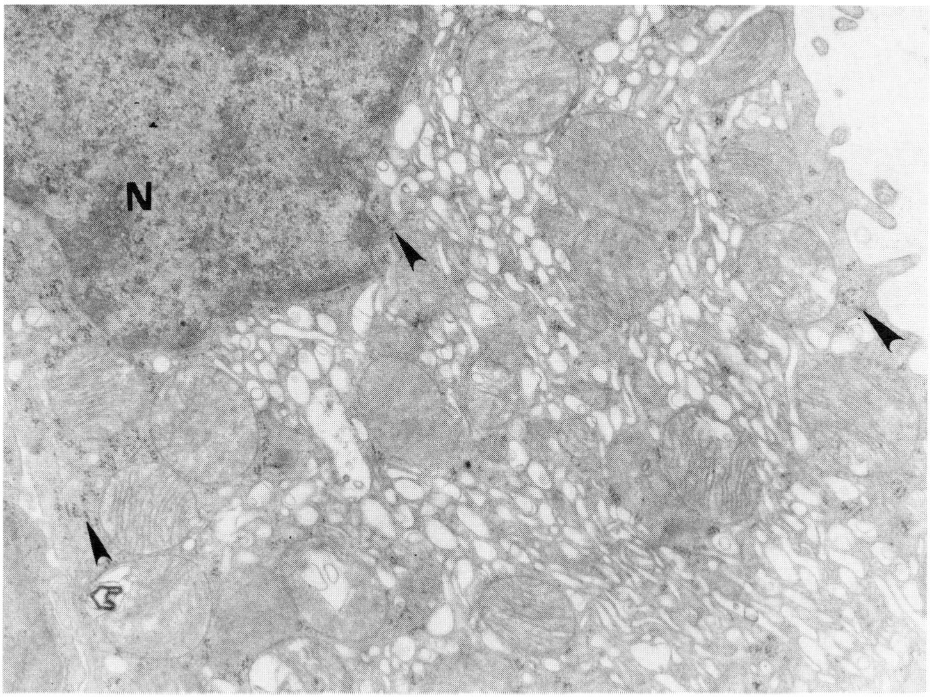

FIGURE 1. Immunolabeling of Cu,Zn-SOD on parietal cells in gastric glands of rat. Protein A-gold particles (arrow heads) are localized mainly in the cytoplasm. N, nucleus.

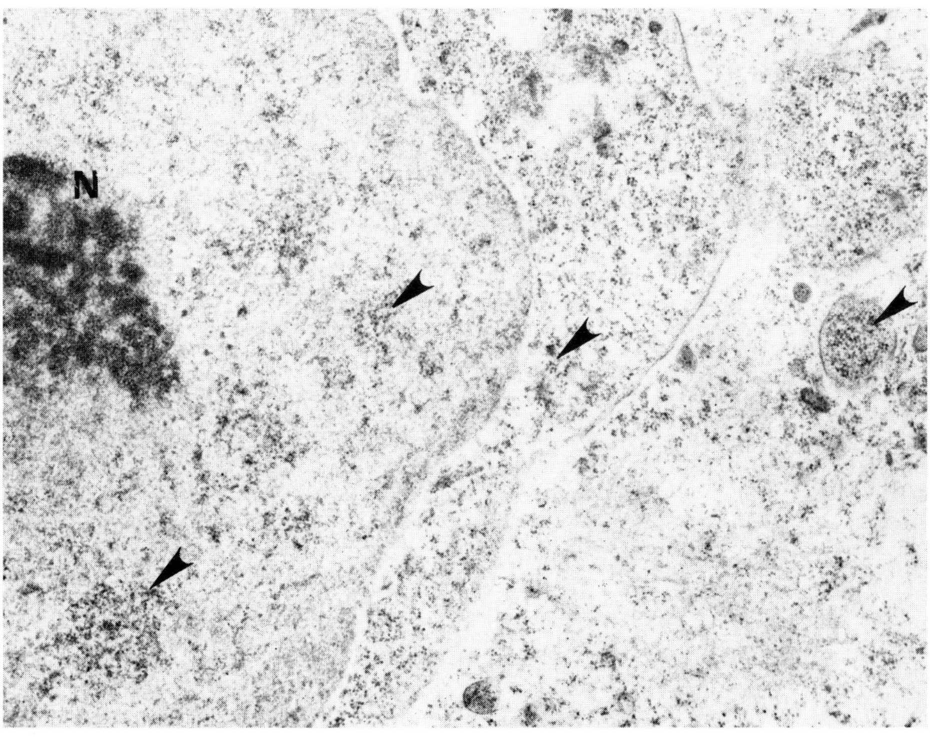

FIGURE 2. Immunolabeling of rat Cu,Zn-SOD on a tumor cell of Yoshida sarcoma. Protein A-gold particles of 10 nm were localized both in the cytoplasm and also in the nuclear matrix (arrow heads). N, nucleus.

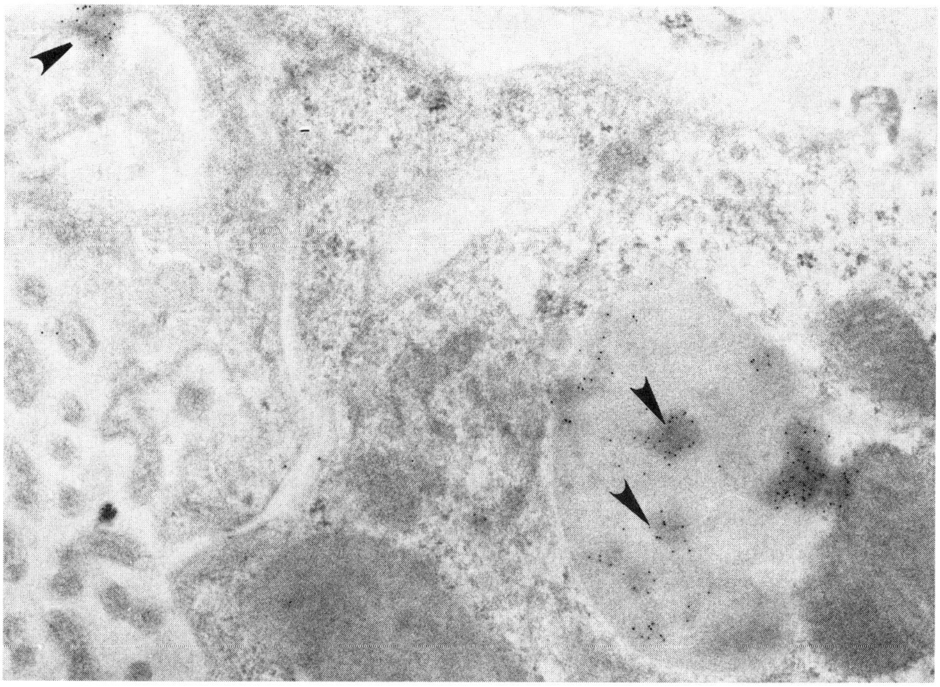

FIGURE 3. Immunolabeling of administered swine Cu,Zn-SOD in the epithelial cell of proximal convoluted tubule of rat kidney. Protein A-gold particles are concentrated on dense spots in the tip of microvilli and in lysosomes (arrow heads).

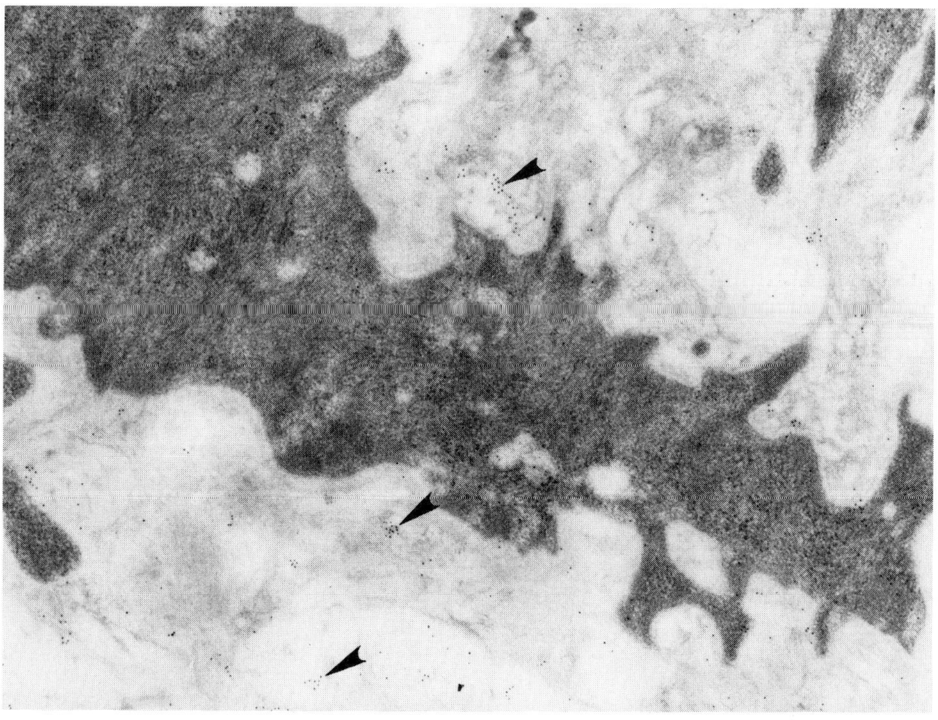

FIGURE 4. Extracellular Cu,Zn-SOD in the intima of regenerating rat aorta. Elastin complex, fibrous components, and basement membrane-like substance in the vicinity of smooth muscle and endothelium has many colloidal gold particles (arrow heads).

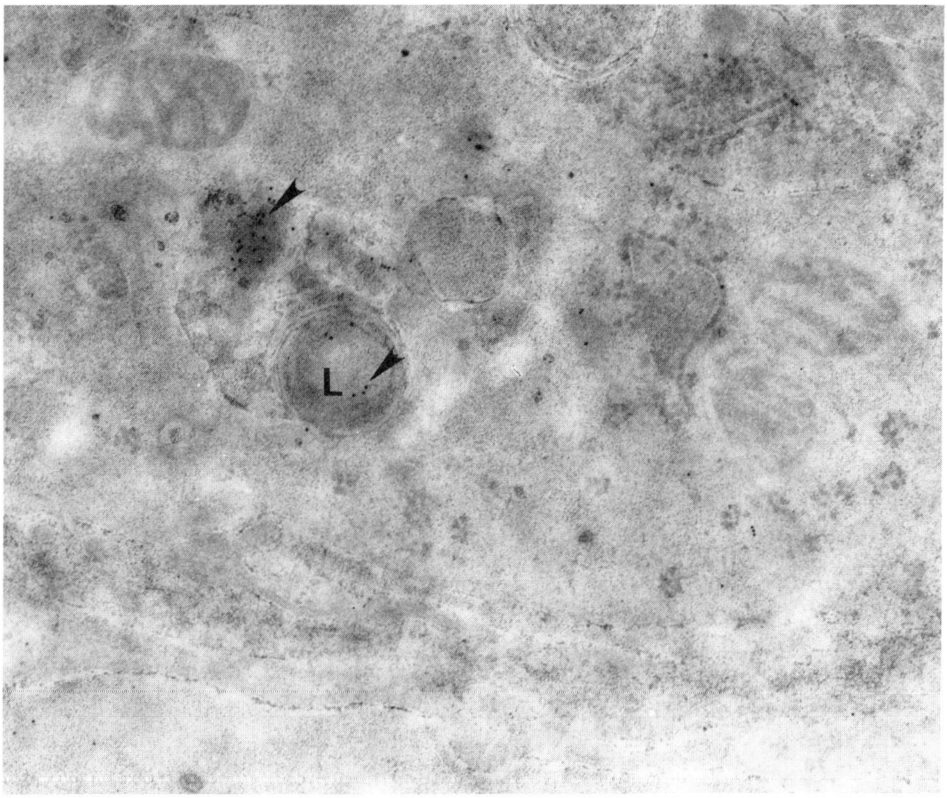

FIGURE 5. Liposomal human Cu,Zn-SOD in the epithelial cell of rat lung was immunolabeled with protein A-gold method (arrow heads). The liposome was made of dipalmitoylphosphatidylcholine, cholesterol, and stearylamine.

From a cytochemical point of view, the effect of SOD on Na-K-ATPase[37,38] is an interesting subject. The addition of swine Cu,Zn-SOD to the incubation medium of Na-K-ATPase, and to the solution of purified Na-K-ATPase itself, and other related projects, are being investigated in this laboratory.

REFERENCES

1. **Maccord, J. M. and Fridovich, I.,** Superoxide dismutase. An enzyme function or erythrocuprein (hemocuprein), *J. Biol. Chem.,* 244, 6049, 1969.
2. **Fridovich, I.,** Superoxide dismutase, *Annu. Rev. Biochem.,* 44, 147, 1975.
3. **Hassan, H. M., Dougherty, H., and Fridovich, I.,** Inhibitors of superoxide dismutase; a cautionary tale, *Arch. Biochem. Biophys.,* 199, 349, 1980.
4. **Freeman, B. A. and Crapo, D.,** Biology of disease: free radicals and tissue injury, *Lab. Invest.,* 47, 412, 1982.
5. **Wong, G. H. W. and Goeddel, D. V.,** Induction of manganous superoxide dismutase by tumor necrosis factor. Possible protective mechanism, *Science,* 242, 941, 1988.
6. **Balentine, J. D.,** *Pathology of Oxygen Toxicity,* Academic Press, London, 1982, 359.
7. **Marklund, S. L., Holme, E., and Hellener, L.,** Superoxide dismutase in extracellular fluid, *Clin. Chem. Acta,* 126, 41, 1982.

8. **Tainer, J. A., Getzoff, E. D., Richardson, J. S., and Richardson, D. C.,** Structure and mechanism of copper-zinc superoxide dismutase, *Nature,* 306, 284, 1983.
9. **Peters-Joris, C., Vandervoorde, A. M., and Bardhuin, P.,** Subcellular localization of superoxide dismutase in rat liver, *Biochem. J.,* 150, 31, 1975.
10. **Weisinger, R. A. and Fridovich, I.,** Mitochondrial superoxide dismutase. Site of synthesis and intramitochondrial localization, *J. Biol. Chem.,* 248, 4793, 1973.
11. **Weisinger, R. A. and Fridovich, I.,** Superoxide dismutase: organelle specificity, *J. Biol. Chem.,* 248, 3582, 1973.
12. **Chang, L.-Y., Slot, J. W., Geuze, H. J., and Crapo, J. D.,** Molecular immunocytochemistry of the Cu Zn superoxide dismutase in rat hepatocytes, *J. Cell Biol.,* 197, 2169, 1988.
13. **Asayama, K., Uchida, N., Dobashi, K., Hayashibe, H., Kato, K., and Kawamori, A.,** Immunohistochemical localization of copper zinc and manganese superoxide dismutases in murine tissues: comparison with the results in rat, *J. Active Oxygens Free Radicals,* 2, 122, 1991.
14. **Slot, J. W., Geuze, H. J., Freeman, B. A., and Crapo, J. D.,** Intracellular localization of the copper-zinc and manganese superoxide dismutase in rat liver parenchymal cells, *Lab. Invest.,* 55, 363, 1985.
15. **Thaets, L. G., Crouch, R., and Spicer, S. S.,** Immunolocalization of copper-zinc superoxide dismutase. II. Rat, *J. Histochem. Cytochem.,* 33, 803, 1985.
16. **Asayama, K. and Burr, I. M.,** Rat superoxide dismutase: purification, labeling, immunoassay and tissue concentration, *J. Biol. Chem.,* 260, 2212, 1985.
17. **Thaete, L. G., Crouch, R., and Spicer, S. S.,** The immunolocalization of copper-zinc superoxide dismutase in canine tissue, *J. Histochem. Cytochem.,* 31, 1299, 1983.
18. **Makita, T., Ishida, T., Kawata, M., and Lee, K. T.,** Cytochemical and immunohistochemical localization of superoxide dismutase (SOD) in the intima of swine and rat aorta, *Ann. N.Y. Acad. Sci.,* 598, 510, 1990.
19. **Makita, T., Ishida, T., and Ogino, K.,** Immunocytochemical localization of human Cu-Zn-SOD in the parietal cells of gastric gland of the baboon, in *Medical, Biochemical and Chemical Aspects of Free Radicals,* Hayashi, E., Niki, M., Kondo, T., and Yoshikawa, T., Eds., Elsevier, Amsterdam, 1989, 683.
20. **Crouch, R. K., Redmond, T. M., Gandy, S. E., Coles, W. A., Buse, W. G., and Simon, J. A.,** Localization of SOD in the cornea and pancreas by light and electron microscopy, in *Oxyradicals and their Scavengers: Biological and Medical Aspects,* Vol. 2, Greenwald, R. G. and Cohen, G., Eds., Elsevier/North-Holland, Amsterdam, 1983, 11.
21. **Rest, R. F. and Spitznagel, J. K.,** Subcellular distribution of superoxide dismutase in human neutrophils, *Biochem. J.,* 166, 145, 1977.
22. **Crapo, J. D., McCord, J. M., and Fridovich, I.,** Superoxide dismutases: preparation and assay, *Meth. Enzymol.,* 53, 382, 1978.
23. **Makita, T. and Lindblom, R.,** Electron microscopic localization of Cu-Zn-SOD of swine in the hepatocyte and parietal cells, Proc. 4th Int. Congr. Cell Biol., Int. Fed. Cell Biol., Montreal, August 14 to 19, 1988, 422.
24. **Van Balgooy, J. N. A. and Roberts, E.,** Superoxide dismutase in normal and malignant tissues in different species, *Comp. Biochem. Physiol. B,* 62, 263, 1979.
25. **Westman, G. and Marklund, S. L.,** Copper- and zinc-containing superoxide dismutase and manganese-containing superoxide dismutase in human tissues and human malignant tumors, *Cancer Res.,* 41, 2962, 1981.
26. **Marklund, S. L.,** Superoxide dismutase in human tissues, cells and extracellular fluid: clinical implications, in *Free Radicals, Aging, and Degenerative Diseases,* Johnson, J. E., Jr., Ed., Alan R. Liss, New York, 1986, 509.
27. **McBrien, D. C. H. and Slater, T. F., Eds.,** *Free Radicals, Lipid Peroxidation and Cancer,* Academic Press, New York, 1982, 447.
28. **Michelson, A. M., McCord, J. M., and Fridovich, I., Eds.,** *Superoxide and Superoxide Dismutases,* Academic Press, New York, 1977, 568.
29. **Dini, L. and Rotilio, G.,** Electron microscopic evidence for endocytosis of superoxide dismutase by hepatocytes using protein-gold adducts, *Biochem. Biophys. Res. Commun.,* 162, 940, 1989.
30. **Makita, T. and Ishida, T.,** Immunocytochemical localization of liposomal human superoxide dismutase (L-H-SOD) in the rat kidney, Second Japan-Korea Anatomy Joint Meeting, 1990, 101.
31. **Galler, B. L. and Weige, D. R.,** Rat liver Cu,Zn-superoxide dismutase: subcellular localization in lysosomes, *J. Biol. Chem.,* 257, 8945, 1982.
32. **Makita, T.,** Electron microscopic localization of Cu-Zn-SOD in the intima of rat aorta by colloidal gold method, *J. Clin. Electron Microsc.,* 22, 568, 1989.
33. **Miyata, K., Nakagawa, Y., Nakamura, M., Ito, T., Sugo, K., Fujita, T., and Tomoda, K.,** Altered properties of Serratia superoxide dismutase by chemical modification, *Agric. Biol. Chem.,* 52, 1575, 1988.

34. **Freeman, B. A., Young, S. L., and Crapo, J. D.,** Liposome-mediated augmentation of superoxide dismutase in endothelial cells prevent oxygen injury, *J. Biol. Chem.*, 258, 12534, 1983.
35. **Turrens, J. F., Crapo, J. D., and Freeman, B. A.,** Protection against oxygen toxicity by intravenous injection of liposome-entrapped catalase and superoxide dismutase, *J. Clin. Invest.*, 73, 87, 1984.
36. **Makita, T., Ishida, T., and Lindblom, R.,** Localization of administered liposomal human SOD and porcine SOD in the aorta of rats, Proc. 3rd Joint Meet. Japan Soc. Histochem. and Cytochem. and Histochem. Soc., Seattle, 1990.
37. **Makita, T., Ishida, T., Asahina, T., Kodaka, R., and Higo, Y.,** Effect of administration of swine Cu-Zn-SOD to electron microscopic localization of Na-K-ATPase in gastric gland, in *Medical, Biochemical and Chemical Aspects of Free Radicals,* Hayashi, E., Nikki, M., Kondo, T., and Yoshikawa, T., Eds., Elsevier, Amsterdam, 1989, 687.
38. **Makita, T., Ishida, T., Kawata, M., and Lindblom, R.,** Addition of Cu-Zn-SOD to the incubation medium for Na-K-ATPase, in *Medical, Biochemical and Chemical Aspects of Free Radicals,* Hayashi, E., Nikki, M., Kondo, T., and Yoshikawa, T., Eds., Elsevier, Amsterdam, 1989, 631.

Chapter 1

ELECTRON MICROSCOPIC CYTOCHEMISTRY OF ENZYMES

1.3 Basic Techniques for Major Groups of Enzymes

1.3.3 Hydrolases

TABLE OF CONTENTS

0-8493-6012-9/93/$0.00 + $.50

I. ESTERASES

Reinhart Gossrau

A. INTRODUCTION

The nonspecific esterases represent a subgroup of carboxylester hydrolases (members are carboxylesterase, arylesterase, lipase, acetylcholinesterase, cholinesterase, phospholipases) to which the carboxylesterases (EC 3.1.1.1; B-esterase, Ali-esterase, carboxylic ester hydrolase, monobutyrase, cocaine esterase, procaine esterase, methylbutyrase, organophosphate-sensitive esterase), arylesterases (EC 3.1.1.2; A-esterase, paraoxonase, arom-esterase, arylesterase, arylester hydrolase, organophosphate-resistent esterase, E-600 esterase, DFP-esterase), and acetylesterase (EC 3.1.1.6; C-esterase, acetic ester hydrolase) belong. These enzymes catalyze the hydrolysis of a wide range of synthetic aliphatic or aromatic esters of acetic acid or lower fatty acids as substrates, according to the following general reaction scheme:[1-3]

$$RCOOR' + H_2O \rightleftharpoons RCOOH + R'OH$$

Consequently, different types of substrates were also tried for the ultracytochemical detection of nonspecific esterases using different reaction principles such as azoindoxyl coupling, S-coupling (methods using the osmiophilic reagent principle), as well as metal salt methods or procedures using radioactive substrates. These methods are listed in Table 1.[4]

B. METHODS

Since not all of the procedures given in Table 1 are of equal value, in the following only the important and most reproducible methods are described for the ultracytochemical detection of nonspecific esterases in human, animal, and plant tissues.

In the azoindoxyl procedure of Holt and Hicks[5] unsubstituted indoxyl acetate serves as substrate. The enzymatically liberated indoxyl reacts with hexazotized pararosaniline yielding an osmiophilic azoindoxyl dye, i.e., upon treatment with OsO_4, azoindoxyl-OsO_4 complexes or chelates are formed. This reaction principle was modified by Kawashima[6] who replaced the hexazotized pararosaniline by tetrazotized BAXD ([*N*,*N*′-bis {*p*-aminophenyl}-1,3-xylendiamine]). Further ultracytochemical procedures based on simultaneous S-coupling are the diazothioether-OsO_4 methods developed by Seligman et al.[7,8] Using 2-naphthyl thiolacetate as substrate and Fast Blue BB for S-coupling of the thionaphthol liberated by the respective nonspecific esterase results in osmiophilic diazothioether molecules which form osmium black upon treatment with OsO_4; in an indentical procedure, 2-thiolacetoxybenzaldehyde is employed as substrate.[7,8] This substrate can also be used to visualize ultracytochemically nonspecific esterases in a metal salt procedure via the amplification effect of cupric ferrocyanide (Hatchett's Brown) with diaminobenzidin and again upon treatment of the generated diaminobenzidin brown with OsO_4, osmium black is formed (osmiophilic polymer generation method).[9,10] Another metal salt procedure was reported by Vatter et al.[11] In this reaction, thionitrophenol is freed from 2-nitrophenylthiol ester substrates and traps aurous ions, whereby electron-opaque deposits are generated. Another metal salt method was described by von Deimling and Madreiter[12,13] with 8-hydroxyquinoline or 8-mercaptoquinoline ester substrates. In the presence of bismuth a bismuth-quinoline complex, bismuth oxinate, is formed; this is then transformed into bismuth sulfide.

TABLE 1
Methods Employed for the Electron Microscopical Demonstration of Esterases

Substrate	Reagents used	Introduced by
1-Naphthyl acetate	Hexazonium pararosaniline	Lehrer and Ornstein[20]
Thiolacetic acid	Pb^{2+}	Barnett and Palade[21]
	^{3}H-DFP	Barnard et al.[22]
Indoxyl esters	Hexazotized pararosaniline, OsO_4	Holt and Hicks[5]
Indoxyl esters	Tetrazotized BAXD	Kawashima[6]
2-Naphthyl thiolacetate	Fast Blue BB, OsO_4	Seligman et al.[8]
2-Thiolacetoxy benzanilide	Fast Blue BB, OsO_4	Seligman et al.[8]
2-Thiolacetoxy benzanilide	Cu^{2+}, $Fe(CN)_6^{3-}$, diaminobenzidine, OsO_4	Hanker et al.[10]
2-Nitrophenyl esters	Au^{3+}	Vatter et al.[11]
Naphthol AS-D acetate	Hexazotized lead-phthalocyanine	Livingstone et al.[23]
8-Hydroxiquinoline acetate	Bi^{3+}	von Deimling and Madreiter[12]
8-Mercaptoquinoline acetate	Bi^{3+}	von Deimling and Madreiter[13]

After von Deimling, O. and Böcking, A., *Histochem. J.*, 8, 215, 1976.

TABLE 2
Inhibitors and Activators to Differentiate between Subgroups of Nonspecific Esterases and Cholinesterases

Enzyme	Other name	EC Number	Inhibitor	Activator
Carboxylester hydrolase	B-Esterase Aliesterase	3.1.1.1.	10 μ*M* E 600 0.1 m*M* Eserine	1 m*M* Cysteine
Arylester hydrolase	A-Esterase Aromesterase	3.1.1.2	10 m*M* E 600 0.1 m*M* *p*-Chlormercuribenzoic acid	1 m*M* Cysteine
Acetylester hydrolase	C-Esterase	3.1.1.6	10 m*M* 2-Phenylpropionic acid	
Acetylcholinester hydrolase	Specific cholinesterase	3.1.1.7	10 μ*M* DFP 10 μ*M* Eserine 0.1 m*M* E 600	
Cholinester hydrolase	Nonspecific cholinesterase	3.1.1.8	0.1 μ*M* DFP 10 μ*M* Eserine 0.1 μ*M* E 600 15–20 μ*M* NaF	

According to Wohlrab, F. and Gossrau, R., *Katalytische Enzymhistochemie. Grundlagen und Methoden für die Elektronenmikroskopie,* Fisher, Jena, 1992.

C. EQUIPMENT AND REAGENTS

The required equipment include a Vibratome, tissue sectioner, freezing-microtome, incubator, and hood. Chemicals and reagents are indoxyl acetate, 5-Br-4-Cl-3-indoxyl acetate, 2-naphthylthiol acetate, 2-thiolacetoxybenzanilide, 2-nitrophenylthiol acetate, 2-nitrophenylthiol butyrate, 8-hydroxyquinoline acetate, 8-mercaptoquinoline acetate, Fast Blue BB, pararosaniline, sodium aurous thiosulfate, tetrazotized BAXD, bismuthyl nitrate, 3,3′-diaminobenzidine, OsO_4, thiocarbohydrazide; inhibitors and activators (see Table 2).

D. PROCEDURES

- *Test tissue*: Rat kidney.
- *Tissue pretreatment:* Samples can be fixed by immersion in 2% glutardialdehyde in 0.1 *M* cacodylate buffer, pH 7.2, containing 5% sucrose, for 1 to 2 h at 4°C or in the mixture according to Karnovsky,[14] consisting of 3% glutardialdehyde and 2% depolymerized paraformaldehyde in 0.1 *M* cacodylate buffer, pH 7.2; alternatively, fixation by perfusion is also possible. Afterwards, 20- to 40-μm tissue chopper or Vibratome sections are prepared, or sections are cut using a freezing microtome.

Metal salt method with 8-hydroxyquinoline or 8-mercaptoquinoline esters according to von Deimling and Madreiter:[12,13]

- *Incubation medium*:

  · 24 ml 0.1 *M* Cacodylate buffer, pH 7.3
  · 23 ml Distilled water
  · 3 ml 0.05 *M* bismuth stock solution

  After thorough mixing cool to 7 to 14°C. Immediately before incubation, add 56 mg (0.3 m*M*) 8-hydroxyquinoline acetate or 61 mg 8-mercaptoquinoline acetate (the commercially available substrates are often not sufficiently pure). Incubate for 20 to 30 min at 9 to 11°C.
- *Post-treatment*: Rinse for 2 h in cold cacodylate buffer containing sucrose to remove soluble bismuth. Place the sections overnight at 4°C in diluted ammonium sulfide (two drops of a 20% solution in 10 ml buffer) to convert the bismuth oxitocinate into insoluble bismuth sulfide. Rinse several times in cold buffered sucrose solution for 3 h. Postfix in 2% OsO_4 in 0.1 *M* cacodylate buffer, pH 7.3, for 1 h at 4°C. Dehydrate in ethanol and embed in a resin, e.g., Epon.
- *Preparation of the bismuth stock solution:* Dissolve 1.525 g bismuthyl nitrate, ($[BiO]NO_3 \cdot H_2O$; mol wt, 305.03) and 6.0 g L-tartaric acid in 100 ml distilled water; add approximately 2 g NaOH for pH adjustment to 7.3. The clear solution is stable for several weeks at 4°C. The substrate is dissolved in 0.5 ml dimethylsulfoxide or acetone.

Detection of nonspecific esterases with the azoindoxyl-BAXD method according to Kawashima[6] follows the procedure below.

- *Incubation medium:*

  · 4 mg 5-Br-4-Cl-3-Indoxyl acetate (dissolved in 0.1 ml acetone)
  · 2.4 ml Distilled water
  · 0.85% NaCl
  · 0.2 ml 50 m*M* tetrazotized BAXD

  Incubate 1.5 to 3 h at 4 to 10°C.
- *Synthesis of BAXD* (*N,N'*-bis[*p*-aminophenyl]-1,3-xylendiamin) according to Plapinger et al.:[15] Dissolve 380 mg BAXD in 10 ml 20% HCl, add 10 ml freshly prepared 4% sodium nitrite solution. Shake and adjust pH with 1 *N* NaOH to 7.0.

- *Post-treatment:* Rinse 3 times for 20 min in 0.85% NaCl at 4°C; postfix in 2% OsO_4 for 90 to 120 min at 37°C. Rinse several times in 0.85% NaCl. Dehydrate in ethanol and embed in a resin, e.g., Epon.

The diazothioethyl-OsO_4 method, according to Seligman et al.,[7,8] is as follows:

- *Incubation medium* with 2-naphthylthiolacetate (NTA):
  - 10 ml 0.1 Phosphate buffer, pH 7.4 or 5
  - 0.1 ml 2.5% NTA in acetone
  - 5 mg Fast Blue BB
  - 0.75 g Sucrose

  For a medium with 2-thiolacetoxybenzanilide (TAB), use 0.1 ml 2.5% TAB in acetone instead of NTA. The substrate solution is added to the phosphate buffer-sucrose mixture with swirling; then Fast Blue BB is added. After vigorous shaking the medium is filtered and used immediately.
- *Incubation:* 30 to 60 min at 37°C with a change of the medium after 30 min.
- *Post-treatment:* Wash in cold 0.22 *M* sucrose for 30 min, osmificate for 20 to 30 min at 50°C, according to Seligman et al.[7] Wash with several changes of 0.22 *M* sucrose solution. Dehydrate in graded alcohols and embed in a resin, e.g., Araldite or Epon.

In case of low contrast and resolution, it might be useful to perform a further reaction step, i.e., the OTO (OsO_4-thiocarbohydrazide-OsO_4) method according to Seligman et al.[7]

Visualization of nonspecific esterases using the osmiophilic polymer generation method according to Hanker et al.[9,10] and Hanker[16] is as follows:

- *Incubation medium:* 2.5 mg 2-thiolacetoxybenzanilide (TAB) dissolved in 0.1 ml acetone with the aid of a vortex agitator; under magnetic stirring add:
  - 7.9 ml 0.8% Potassium acetate
  - 0.25 ml 0.1 *N* Acetic acid
  - 0.6 ml 0.1 *M* Sodium citrate
  - 1.25 ml 30 m*M* Cupric sulfate (dropwise)
  - 1.2 ml 5 m*M* Sodium ferricyanide (sodium hexacyanoferrate[III]) dropwise

  Adjust the pH to 5.5 to 5.6, cool to 0 to 4°C, and filter.
- *Post-treatment:* Rinse quickly in distilled water at 0 to 4°C; transfer sections into a freshly prepared solution of 3,3′-diaminobenzidine in 10 ml 0.05 *M* acetate buffer, pH 5.6, for 20 min. Rinse four times for 5 min each in distilled water at 4°C. Osmicate with 2% OsO_4 in 0.1 *M* cacodylate buffer, pH 7.2, for 15 to 20 min at 30 to 35°C. Rinse several times in distilled water, dehydrate in ethanol, and embed in resin, e.g., Epon.

The axoindoxyl-HPR (hexazotized pararosaniline) method, according to Holt and Hicks,[5] is as follows:

- *Incubation medium:*
  - 0.03 *M* Citrate buffer, pH 6.0
  - 1 m*M* Indoxyl acetate
  - 1 m*M* HPR

  Incubate for 2 to 3 h at 0°C.

- *Post-treatment:* Rinse in 0.1 *M* cacodylate buffer, pH 7.3; postfix in 1% OsO_4 in 0.1 *M* cacodylate buffer, pH 7.3, for 1 h. Dehydrate in ethanol and embed in a resin, e.g., Epon.
- *Preparation of hexazotized pararosaniline — Solution A:* dissolve 1.65 g pararosaniline·HCl in 50 ml 2 *N* HCl, heat, cool, and filtrate; *Solution B:* 3.3% sodium nitrate. To obtain hexazotized pararosaniline, equal volumes of the two solutions are mixed immediately before use, whereby the nitrate must be added to the pararosaniline. The resulting hexazotized pararosaniline should be light yellow. The nitrate solution should not be stored in the refrigerator for longer than one week.

The metal salt method with nitrophenylthiol esters, according to Vatter et al.,[11] is described below.

- *Stock solutions:*
  - 0.175 *M* Sodium maleate buffer, pH 7.4
  - 0.25 *M* NaCl
  - 0.025 *M* KCl
  - 0.025 *M* $CaCl_2$
  - 0.005 *M* $MgCl_2$ (adjusted to pH 7.4 with 1 *N* NaOH)
  - 0.025 *M* 2-Nitrophenylthiol acetate or 2-nitrophenylthiol butyrate (dissolved in dimethyl sulfoxide)
  - 0.02 *M* Sodium aurous thiosulfate in the ion-containing maleate buffer
- *Incubation medium:*
  - 1.58 ml Maleate buffer, pH 7.4
  - 0.12 ml Substrate solution
  - 0.30 ml Goldthiosulfate solution

  Incubate for 15 to 30 min at room temperature.
- *Post-treatment:* Rinse in maleate buffer; postfix in 1% OsO_4 for 20 min at 4°C; treat with 2% uranyl acetate for 45 min at 4°C.
- *Controls*
  - Basic controls: incubation in substrate-free media; incubation without trapping or coupling reagents; inactivation by heat at 60°C for 30 to 60 min; incubation of tissues with known nonspecific esterase activity; use of different methods for the ultracytochemical localization of these enzymes
  - Controls with inhibitors and activators: use of specific inhibitors for carboxylic esterases; reversible inhibitors, e.g., NaF can be added directly to the incubation medium, when they do not interact with other components. When irreversible inhibitors, e.g., DFP or E 600, are used, the sections are preincubated for 30 to 60 min. Further details about inhibitors and activators for the differentiation between various subgroups of nonspecific esterases, but also to discriminate between these enzymes and the lipase as well as cholinesterases, are given in Table 2 according to Wohlrab and Gossrau[19] (see also Reference 1 and Table 1, lipases)
  - Ultracytochemical controls: When metal salt methods are used, aldehyde-fixed tissue samples should be analyzed also without contrast procedures, e.g., without OsO_4, uranyl acetate, or lead citrate.

E. RESULTS

Reaction products can be found with all methods in many cell compartments, i.e., endoplasmic reticulum, lysosomes, Golgi apparatus, mitochondria, and nuclear envelope, and also in the cytoplasm as well as in lipid droplets, e.g., in proximal renal tubular cells, hepatocytes, or small intestinal enterocytes. From the results obtained with inhibitors, it is assumed that the activity in the endoplasmic reticulum, cytoplasm, and perinuclear space may be attributed to B-esterases, whereas the reaction in the Golgi apparatus, mitochondria, lysosomes, and lipid droplets is believed to be caused by A-esterases.[4,17-19]

It is important to bear in mind that the substrates have different rates of hydrolysis; e.g., indoxyl acetate is cleaved at a lower rate than the 5-Br-4-Cl-3-indoxyl derivative,[1,2] so that false-negative data may result when using the unsubstituted indoxyl substrate. Furthermore, the results may be affected by the different types of trapping reagents and different pH values employed for a given method.[1,2]

Since all substrates may be hydrolyzed also by acetylcholinesterase, cholinesterase, and lipase, the participation of these enzymes must be excluded by suitable controls, e.g., using appropriate inhibitors and activators (see Tables 1 and 2).

Using the azoindoxyl methods, the OsO_4-containing reaction product appears to be amorphous, as is that generated by the osmiophilic polymer generation method for the nonspecific esterases. With the diazothioether-OsO_4 procedure, the reaction product is also amorphous, but lipophilic like that obtained with the metal salt procedure with quinoline esters. In addition, the reaction products generated by the diazothioether-OsO_4 method may form more or less large droplets and tend to diffuse. Therefore, localization of the reaction product to certain cell structures might be difficult or sometimes even impossible. In the quinoline acetate technique the substrates have limited stability, e.g., spontaneous substrate hydrolysis occurs at pH 7.3 and 25°C, and bismuth can be bound to cell structures nonspecifically (metallophilia) and may induce nonenzymatic substrate hydrolysis. The nitrophenylthiol procedure results in a fine granular reaction product.[4,17,18]

The diazothiolether-OsO_4 and osmiophilic polymer generation methods have the drawback that the substrates, 2-naphthylthiolacetate and 2-thiolacetoxybenzanilide, diffuse poorly into the tissue.[7]

REFERENCES

1. **Pearse, A. G. E.,** *Histochemistry. Theoretical and Applied,* Churchill Livingstone, Edinburgh, 1972, chap. 17.
2. **Lojda, Z., Gossrau, R., and Schiebler, T. H.,** *Enzyme Histochemistry. A Laboratory Manual,* Springer, Berlin, 1979.
3. *Enzyme Nomenclature,* Academic Press, New York, 1984.
4. **von Deimling, O. and Böcking, A.,** Esterases in histochemistry and ultracytochemistry, *Histochem. J.,* 8, 215, 1976.
5. **Holt, S. J. and Hicks, R. M.,** The importance of osmiophilia in the production of stable azoindoxyl complexes of high contrast for combined cytochemistry and electron microscopy, *J. Cell Biol.,* 29, 361, 1966.
6. **Kawashima, T.,** A cytochemical method for indoxyl acetate esterase, *Acta Histochem. Cytochem.,* 1, 147, 1968.
7. **Seligman, A. M., Ueno, H., Wasserkrug, H. L., and Hanker, J. S.,** Esterase method for light and electron microscopy via the formation of osmiophilic diazothioethers, *Ann. Histochim.,* 11, 115, 1966.
8. **Seligman, M. L., Ueno, H., Hanker, J. S., Kramer, S. P., Wasserkrug, H. L., and Seligman, A. M.,** Cytochemical localization of pancreatic lipase with light and electron microscopy, *Exp. Mol. Pathol. Suppl.,* 3, 21, 1966.

9. **Hanker, J. S., Anderson, W., and Bloom, F.,** Osmiophilic polymer generation: catalysis by transition metal compounds in ultrastructural cytochemistry, *Science,* 175, 991, 1972.
10. **Hanker, J. S., Yates, P. E., Clapp, D. H., and Anderson, W. A.,** New methods for the demonstration of lysosomal hydrolases by the formation of osmium blacks, *Histochemie,* 30, 201, 1972.
11. **Vatter, A. E., Reiss, O. K., Newman, J. K., Lindquist, K., and Groeneboer, E.,** Enzymes of the lung. I. Detection of esterase with a new cytochemical method, *J. Cell Biol.,* 38, 80, 1968.
12. **von Deimling, O. and Madreiter, H.,** Esterase II, a new cytochemical method for the electron microscopical demonstration of non-specific esterase in animal tissue, *Histochemie,* 29, 83, 1972.
13. **von Deimling, O. and Madreiter, H.,** Esterase III, demonstration of a yet histochemically unknown thiolesterase by means of light and electron microscopy, *Histochemie,* 29, 340, 1972.
14. **Karnovsky, J. M.,** A formaldehyde-glutaraldehyde fixative of high osmolality for use in electron microscopy, *J. Cell Biol.,* 27, 137A, 1965.
15. **Plapinger, R. E., Linus, S. L., Kawashima, T., Deb, Ch., and Seligman, A. M.,** Preparation and structure-activity relationship of reagents for cytochrome oxidase activity: potential for light and electron microscopy, *Histochemie,* 14, 1, 1968.
16. **Hanker, J. S.,** Osmiophilic reagents in electronmicroscopic histochemistry, *Prog. Histochem. Cytochem.,* 127, 1979.
17. **Luppa, H. and Andrä, J.,** The histochemistry of carboxylester hydrolases: problems and possibilities, *Histochem. J.,* 15, 111, 1979.
18. **Shnitka, T. K.,** Esterase — nonspecific esterases, in *Electron Microscopy of Enzymes. Principles and Methods,* Vol. 3, Hayat, M. A., Ed., van Nostrand Reinhold, New York, 1974, 1.
19. **Wohlrab, F. and Gossrau, R.,** *Katalytische Enzymhistochemie. Grundlagen und Methoden für die Elektronenmikroskopie,* Fischer, Jena, 1992.
20. **Lehrer, G. M. and Ornstein, L.,** A diazo coupling method for the electron microscopic localization of cholinesterase, *J. Biophys. Biochem. Cytol.,* 6, 163, 1959.
21. **Barrnett, R. J. and Palade, G. E.,** Enzyme activity in the M-band, *J. Biophys. Biochem. Cytol.,* 6, 163, 1959.
22. **Barnard, E. A., Budd, G. C., and Ostrowski, K.,** Autoradiographic methods in enzyme cytochemistry. IV. The cellular and ultrastructural localization of organophosphate-sensitive esterases in mouse liver and kidney, *Exp. Cell Res.,* 60, 405, 1970.
23. **Livingstone, D. C., Coombs, M. M., Franks, L. M., Maggi, V., and Gahan, P. P.,** A lead phthalocyanin method for the demonstration of acid hydrolases in plant and animal tissues, *Histochemie,* 18, 48, 1969.

II. ACETYLCHOLINESTERASE AND CHOLINESTERASE

Bertalan Csillik

A. APPLICATIONS

Ever since the beginnings of neurohistochemistry, the ultimate goal of cholinesterase "staining" was to locate, at the molecular level, sites where the transmitter substance, acetylcholine, is inactivated.[1] However, electron microscopic enzyme cytochemistry, though undoubtedly approaching this objective, still fails to completely achieve this ultimate goal. Essentially, there are at least three major avenues to visualize the localization of acetylcholinesterase (AChE) at the ultrastructural level.

The first one of these includes numerous modifications of the original Koelle technique, which is based on the hydrolysis of acetylthiocholine by AChE. The resulting thiocholine can be captured with Cu^{2+}, Pb^{2+}, Fe^{3+}, etc., ions which, without any further treatment, yield electron-dense precipitates[3] — more or less accurately localized at the sites of the enzyme activity (Figure 1). The problem is, however, that the reaction product tends to form microcrystals, incorrectly identified with "groups of enzyme molecules" by several authors. On the other hand, osmic acid decolorization[4] often results in a false-negative reaction, erroneously identified with "lack of enzyme activity". Finally, especially if the end product (e.g., Cu-thiocholine) is "visualized" by H_2S treatment, substantivity of the resulting compound (CuS) will determine its fine structural localization. Therefore, the closer we get to the level of molecular organization, the more problems will be its reliability — quite similar to Heisenberg's law of infidelity. In the neuromuscular junction (Figure 2a) the acetylthiocholine reaction outlines both pre- and postsynaptic membranes.[5] In this context, it should be recalled that, even at the light microscopic level, if the thiocholine reaction is carried out at a physiological pH, it invariably results in nuclear staining (which is undoubtedly false); therefore, the reaction is traditionally performed at a slightly acidic pH (which is unquestionably aphysiological). The infidelity paradigm is thus even more exaggerated.

No less problematic is the azo-dye approach. Indoxyl acetate, naphthyl acetate, and their derivates, are hydrolyzed and coupled at physiological pH; the resulting stains are, however, lipophilic, therefore substantivity is again decisive for the fine structural localization. Osmium black, which is the end product of the procedure, is homogeneous and has no tendency to be translocated at the fine structural level. Therefore, the localization, though more diffuse and less accurate, at least does not induce temptations to interpret the staining at the molecular level. In the neuromuscular junction (Figures 2b and 2c), the azo-dye reaction is located in the synaptic cleft.[5]

Third, thiolacetic acid is also unspecifically hydrolyzed by AChE. The complex mechanism of the thiolacetic acid reaction has been described and analyzed in detail.[6,7] Accordingly, there is no reason to ascribe more fidelity to the molecular localization of the thiolacetic acid reaction than to that of the aforementioned ones. Thiolacetic acid outlines the postsynaptic membrane (Figure 2d) in the neuromuscular junction.[5] Essences of the fine structural localization of AChE with the above four techniques, with substrates and capturing moieties (in parentheses) are shown in Figure 3.

More unequivocal is the fine structural localization of AChE within the endoplasmic reticulum (ER) of cholinergic nerve cells. Here, the reaction end product of each of the available techniques is located within the cisternal lumina of the rough ER. Also the 35-nm globular units, dispersed more or less regularly within the cisterns, stand out clearly as nonreacting particles, irrespective of the paradigm of the reaction (Figure 4). Therefore, the

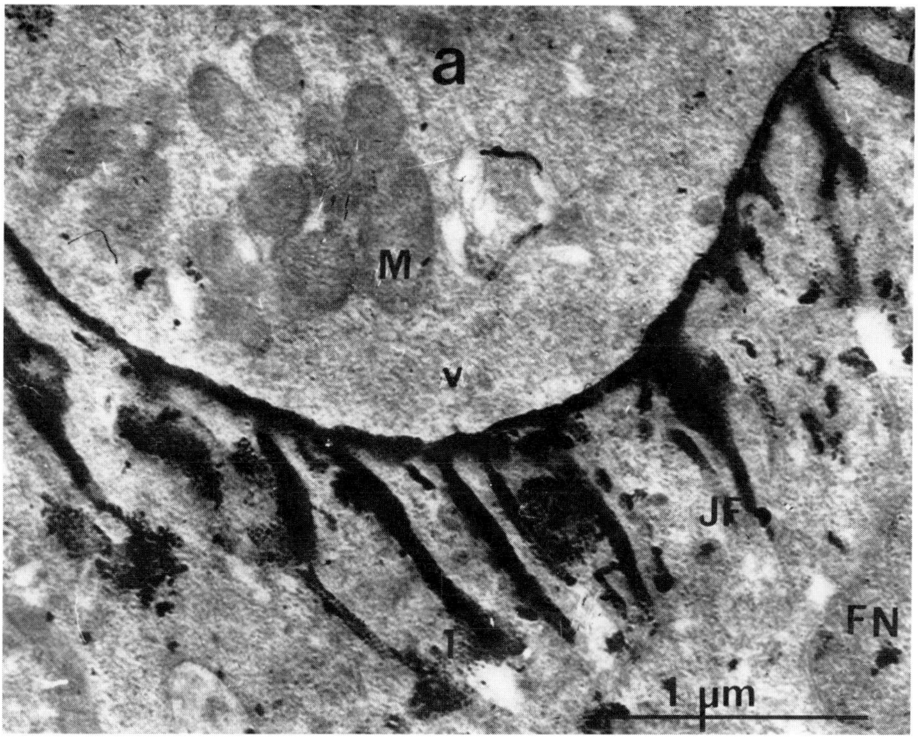

FIGURE 1. Acetylcholinesterase (AChE) reaction in the mammalian neuromuscular junction: portion of a motor end plate in the rat diaphragm. Note the regular profile of the axon terminal (a), containing mitochondria (M), and synaptic vesicles (v). AChE is displayed by the semicircular presynaptic membrane and by the postsynaptic membrane, thrown into numerous junctional folds (JF). Also the synaptic cleft contains reaction product. Pb-Cu-thiocholin technique. (From Csillik, B. et al., *Acta Histochem.*, 25, 58, 1966. With permission.)

reality of their structural entity, and, consequently, their plausible role in the establishment of the tertiary structure of enzyme protein molecules, as suggested two decades ago, seems to be feasible.[8]

Electron microscopic AChE reaction was found by numerous authors to be a classical tool to obtain final proof for the cholinergic nature of axons, both of the myelinated and the nonmyelinated variety. In this context, it is worth noting that the close structural intimacy of cholinergic and noncholinergic axons in the autonomic peripheral innervation apparatus has been first unequivocally proved by means of the electron microscopic AChE reaction (in the iris of the rat).[9]

In spite of all the limitations of the electron microscopic AChE reaction, a few examples selected from the recent literature will exemplify the widespread use of this methodological approach in various domains of neuroanatomy.

1. With regards to the motor endplate, two types of focal accumulations of AChE were shown to appear in noninnervated regenerating skeletal muscles of the rat.[10] Muscle fibers in the soleus muscle of the rat, injured by bupivacaine and free autografting, were allowed to regenerate within their old basal laminae. First, AChE gets concentrated at the sites of the former motor endplates. An additional kind of AChE accumulation appeared in regenerating muscles in regions remote from the sites where motor endplates were located in the muscles of origin. Both kinds of AChE

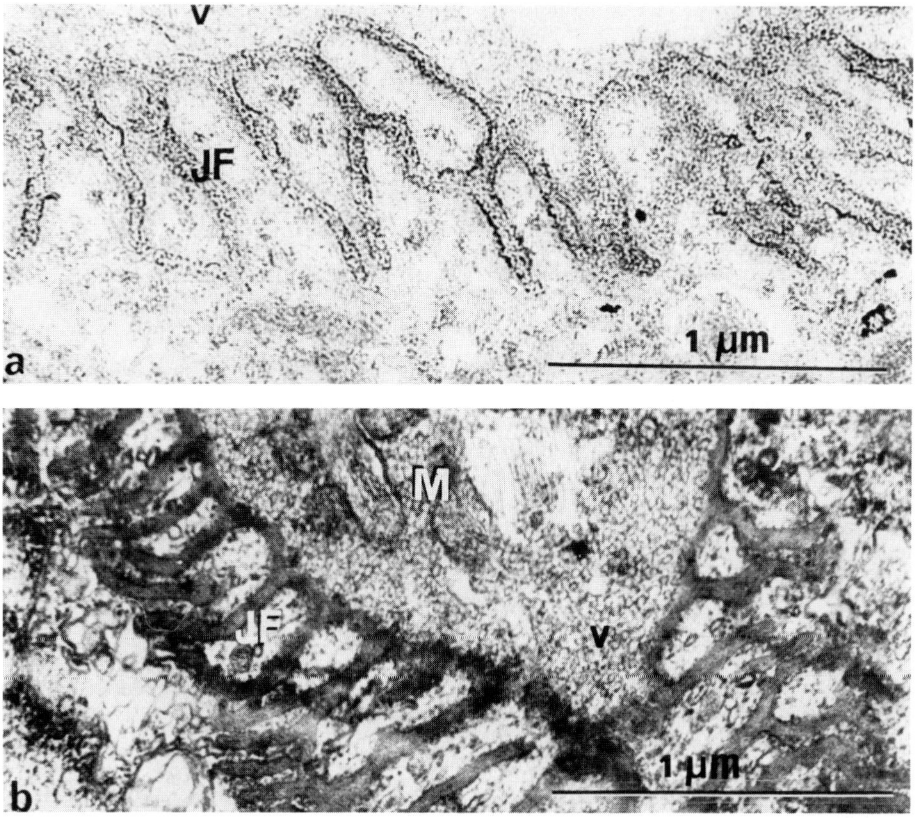

FIGURE 2. Comparison of the fine structural localization of four different enzyme histochemical methods, used to reveal AChE activity in the neuromuscular junction; rat diaphragm. Inhibition of nonspecific esterase with ethopropazine. v, Synaptic vesicles; M, mitochondrion in the axon terminal; JF, junctional folds. (a) Substrate: acetylthiocholine; coupling reaction: Cu^{2+} and Pb^{2+}. The enzyme reaction end product outlines both pre- and postsynaptic membranes. Rapid fixation in formaldehyde (10 min) and short incubation time (10 min) result in a poorly preserved ultrastructure but in a regular fine granular reaction of AChE on pre- and postsynaptic membranes and on the walls of the junctional folds, with virtually no reaction in the synaptic (or rather junctional) cleft. (b) Substrate: indoxyl acetate; endogeneous formation of indigo blue. Osmium black deposit is seen within the primary and secondary synaptic clefts (within the junctional folds). (c) Substrate: indoxyl butyrate; endogeneous formation of indigo blue. Osmium black deposit is seen within the primary and secondary synaptic clefts (within the junctional folds), the heaviest reaction is present in the "middle dense layer" of the synaptolemma. (d) Substrate: thiol acetic acid. Coarse granular reaction is confined to the postsynaptic membrane, including that of the junctional folds. A slight reaction is present in synaptic vesicles. (From Csillik, B. and Knyihár, E., *J. Cell Sci.*, 3, 529, 1968. With permission.)

accumulations ultrastructurally resembled the postsynaptic specialization of the motor endplate. In another study, lectin binding of neuromuscular junctions was studied as compared to the electron microscopic AChE reaction.[11] Lectin binding was found at neuromuscular junctions of muscles from newborn mice, when acetylcholine receptors and AChE were already concentrated at the endplate. As shown by ultrastructural histochemistry, accessible lectin-binding sites were confined to the crests of junctional folds in motor endplates of 4-day-old mice, whereas adult endplates exhibited lectin-binding sites in the entire synaptic cleft, including the junctional folds. Two neuromuscular hereditary diseases of the mouse, "wobbler" and "motor endplate disease"

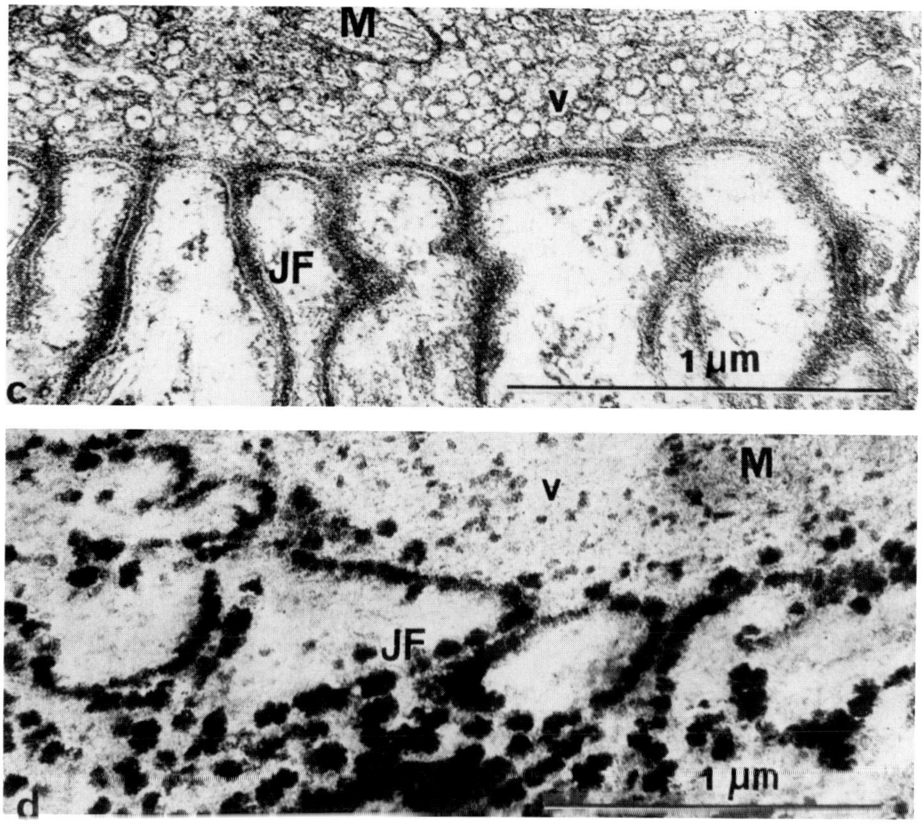

FIGURE 2 (continued).

were examined; at the ultrastructural level, diseased synapses showed reduced junctional folds but an unaffected lectin-binding capacity.

2. AChE activity in the perikarya was studied at the electron microscopic level in the magnocellular neurons of the hypothalamus.[12] AChE activity was particularly marked in the hypertrophied abnormal magnocellular neurons of homozygous Brattleboro rats which do not release significant amounts of vasopressin. It was assumed, therefore, that the increased AChE activity in osmotically stimulated animals cannot be a function of vasopressin. AChE activity was also detected in large multipolar neurons lying dorsolateral to the supraoptic nucleus, and in their fine axonal processes which project towards the supraoptic nucleus. AChE histochemistry was combined with choline acetyltransferase immunohistochemistry in order to characterize the ultrastructural distribution of AChE in identified cholinergic and noncholinergic neurons in the ferret brain.[13] In these cells intense staining for AChE was localized within the cisternae of the rER, in the nuclear envelope and Golgi apparatus, and along the plasma membranes of the soma and dendrites. In contrast, the distribution of AChE in noncholinergic neurons was restricted mainly to the cisternae of the ER and the nuclear envelope. In the cerebellum of the chicken, AChE was ultrastructurally distributed principally in the cisternae of rER and in a part of the nuclear envelope of the Purkinje, the Golgi, and some of the basket and granule cells, and in a portion of the saccules of the Golgi apparatus of the Purkinje cell.[14] In dendrites and the initial axon of Purkinje cells, the

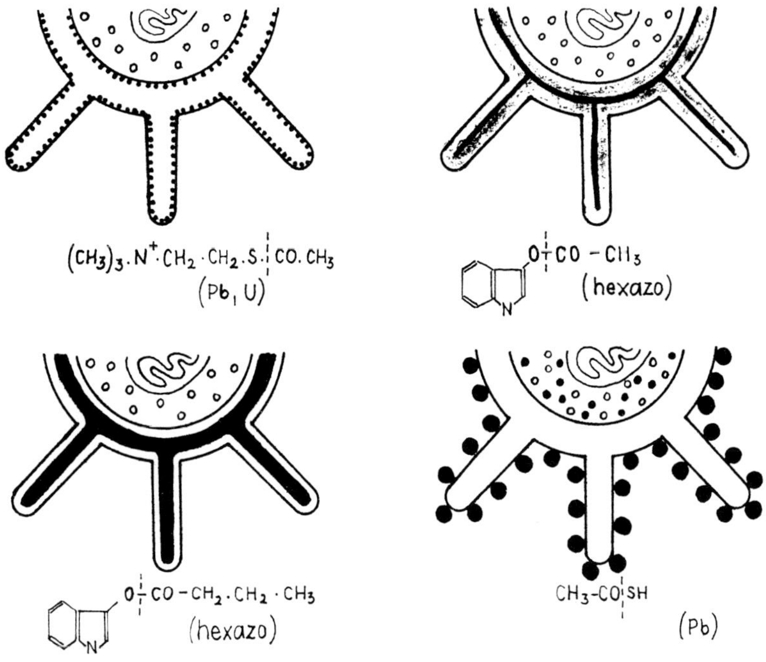

FIGURE 3. Schematic diagram of fine structural localization of AChE when using acetylthiocholine, indoxyl acetate, indoxyl butyrate, and thiol acetic acid substrates, respectively. Chemical formulas of substrates and the capturing agents are shown (the latter in parentheses). (From Csillik, B. and Knyihár, E., *J. Cell Sci.*, 3, 529, 1968. With permission.)

smooth ER also showed AChE activity. Although dendritic terminals of the Golgi cells contained AChE reaction products, the axon terminal did not. Some of the afferent terminal fibers forming the cerebellar glomerulus exhibited weakly positive AChE reaction. In order to study the distribution of enkephalin-containing nerve fibers and terminals in Meynert's nucleus basalis, cells of the latter were stained for choline acetyltransferase and AChE.[15] Electron microscopic observations of histochemically characterized neurons revealed that intensely AChE-stained structures in the region of Meynert's nucleus received sparse synaptic inputs from enkephalin-immunoreactive terminals. Synaptic contacts onto proximal AChE-positive dendrites were sparse, but the density increased on more distal regions of the dendrites.

3. Two important studies were done on alterations of AChE reaction of nerve cells at the electron microscopic level after nerve transection. AChE activity was investigated in synaptic areas of the cat spinal trigeminal nucleus (pars interpolaris and pars caudalis) ipsilateral and contralateral to complete retrogasserian rhizotomy.[16] Using the Karnovsky-Roots technique, enzymatic activity was initially reduced in synaptic clefts of injured afferent terminals and subsequently was enhanced throughout the extracellular space, including within synaptic clefts of possibly reinnervated sites. In another paper, neuronal surface changes were studied in the dorsal vagal motor nucleus of the guinea pig in response to axotomy; AChE was used as a cytochemical marker of neuronal membranes.[17] Vagal motoneurons did not show a loss of enzymatic activity after nerve transection; rather, a redistribution seemed to occur, with intensified staining of the plasmalemma. The newly formed processes were consistently found to be AChE-positive.

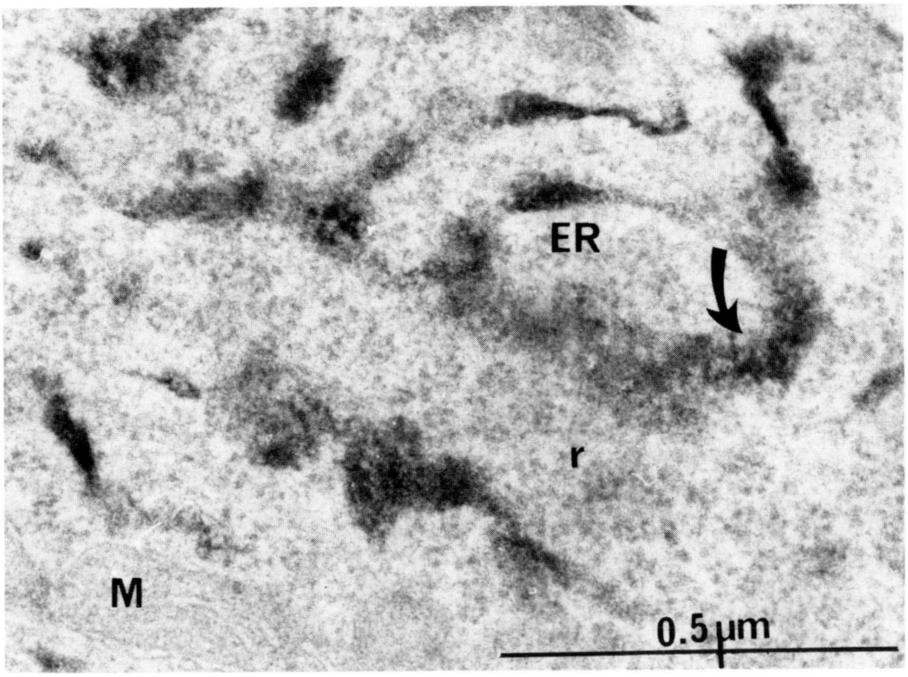

FIGURE 4. AChE reaction in the rough endoplasmic reticulum (ER) of a cholinergic autonomic nerve cell in the feline superior cervical ganglion. M, Mitochondrion; r, ribosomes, devoid of reaction end product. Within the cisterns, the end product of the enzyme reaction outlines nonreacting globular units, especially in tangential (grazing) sections of the cisterns (arrow). Measuring ~33 nm in diameter, such "endoplasmic units" are instrumental in carving out the tertiary molecular conformation of the enzyme protein molecules. (From Csillik, B. and Knyihár, E., *Acta Biochim. Biophys. Acad. Sci. Hung.*, 3, 165, 1968. With permission.)

4. Ultrastructural localization of AChE led to new discoveries with regards to cholinergic innervation of visceral organs. In the chicken pancreas, AChE-positive nerve bundles were found to run along the pancreaticoduodenal artery.[18] Their branches proceed into interlobular connective tissue, from a plexus around the interlobular secretory ducts and small arteries, and penetrate the exocrine parenchyma. Intrapancreatic ganglia showing a strong AChE activity were detected within the interlobular connective tissue or between acini. The exocrine pancreas was richly innervated with AChE-positive terminals. The endocrine pancreas was supplied with fewer nerves, though B and D cells were richly innervated by AChE-positive nerves. On the other hand, ultrastructure of cholinergic innervation in the cirrhotic liver in guinea pigs, studied by a modified Karnovsky and Roots AChE method, has revealed numerous AChE-positive nerve fibers traversing the collageneous bundles in the fibrotic areas, and cholinergic terminals forming close contacts with fibroblasts.[19] Finally, while studying nerve fibers supplying the Duvernoy's gland, a venom-secreting oral gland, of the Japanese colubrid snake *Rhabdophis tigrinus,* AChE-reactive fibers were found to abundantly supply the gland, especially the lobule consisting of secretory units.[20] Nerve fibers were frequently seen to terminate near capillaries subjacent to the secretory unit.
5. Electron microscopic AChE reaction is being used, however, not only in neurobiology. The appearance and distribution of AChE and butyrylcholinesterase (BuChE) in 12 human thyroid cancers and three normal thyroids were examined by electron micro-

scopic study.[21] In normal thyroid tissue samples examined, no activities of AChE and BuChE were detected. On the ultrastructural level, AChE reaction product was revealed in the perinuclear space, in the endoplasmic reticulum, and in the Golgi complex of some, but not in all, cells in less differentiated regions of the tumors. Recently, even helminthology seems to rely upon electron microscopic cholinesterase techniques. The teguments of *Orthocoelium scoliocoelium* and *Paramphistomum cervi* were examined using histochemical techniques and electron microscopy, which revealed that they comprise an outer surface syncytium underlain by a thick subsyncytial zone and musculature.[22]

B. METHODOLOGY

Out of the multitude of techniques developed during the last three decades for the electron microscopic localization of acetylcholinesterase and (pseudo)cholinesterase,[23-33] all of which can be used — though to different extents — to locate fine structural distribution of these enzymes, we found particularly useful the modification of our original technique[28,29] based on the classic Koelle reaction which yields convincing patterns both in the central and the peripheral nervous system. Prerequisite for the distinction between "true" acetylcholinesterase, "pseudo" cholinesterase, and the various nonspecific esterases, all of which would hydrolyze the substrate (acetylthiocholine or butyrylthiocholine) in the application of a (more or less) specific enzyme inhibitor *before* the actual incubation would take place; in addition, an appropriate concentration of this same inhibitor has to be maintained in the incubation solution as well. Accordingly, the technique recommended is as follows:

1. Equipment and Reagents

Required equipment includes a transcardial perfusion system (gravitational or motoric), a Vibratome, and manual surgical instruments (scalpels, pincets, scissors, tweezers).

Reagents required in the technique are

- Paraformaldehyde
- Glutaraldehyde
- Ethopropazine hydrochloride
- Acetyl thiocholine
- Butyryl thiocholine
- Sodium acetate, 0.1 *N* (aqueous solution)
- Acetic acid, 0.1 *N*
- Glycine, 3.8% (aqueous solution)
- $CuSO_4$, 0.1 *M* (aqueous solution)
- $Pb(NO_3)_2$, 0.5% (aqueous solution)

2. Method

1. Deep anesthesia is induced by an intraperitoneal injection of Nembutal (10 mg/kg) or by 10 ml/kg of a 4% aqueous solution of chloral hydrate. Transcardial perfusion fixation through the aorta is performed with a cold (4°C) buffered (pH 7.2) paraformaldehyde solution, containing 0.1% glutaraldehyde (note that glutaraldehyde preserves the ultrastructure but depresses enzyme activity). Alternatively, small samples of biopsy material (not exceeding 10 mm^3) should be subjected to immersion fixation.
2. Postfixation in glutaraldehyde-free buffered 4% paraformaldehyde (pH 7.2) at 4°C, overnight.
3. Cut Vibratome sections, 50 to 80 μm.

4. Preincubate in 0.007% ethopropazine hydrochloride (dissolved in distilled water) for 30 min.
5. Incubation at 25°C for 30 min in the following medium:

A. Acetylthiocholine	7 mg
Acetate buffer, 0.1 *N*, pH 6.2	2.5 ml
$CuSO_4$, 0.1 *M*	0.1 ml

Centrifuge at 3000 rpm for 10 min. Use the supernatant and add:

B. Glycine, 3.8%	0.1 ml
$Pb(NO_3)_2$, 0.5%	0.1 ml
Ethopropazine	0.175 mg

Vibratome sections should be completely covered by the incubation fluid. If necessary, use multiples of the above solution.
6. Rinse in 10% sucrose.
7. Postfix in 1% osmic acid buffered by Na-cacodylate (pH 6.8) at 4°C for 1 h.
   Important — Only cacodylate-buffered osmic acid prevents decolorization, which is common if any other combination of osmic acid is used. An alternative possibility to prevent decolorization is to perform conversion of the end product into CuS + PbS by means of a H_2S-saturated Ringer solution. However, the sulfides are prone to decolorization by osmic acid[4] and even Dalton's much-advertised fluid (dichromate-containing osmic acid), though undoubtedly the best of all the available possibilities, often fails to preserve the electron-dense end product, resulting in "patchy" localization.
8. Dehydrate in graded alcohols.
9. Process through propylene oxide.
10. Embed in Araldite or Durcupan ACM.
11. Thin sections, obtained on an ultramicrotome, should be stained with Reynold's lead citrate (and, if necessary, also with uranyl acetate).
12. Semithin sections, 0.5 to 1 μm, can be readily studied by phase contrast or by simple conventional light microscopy, without any further staining. If necessary, an alkaline toluidine blue solution might enhance the background staining.

3. Evaluation

Sites with increased electron density represent localization of acetylcholinesterase. In case of a too-heavy reaction (1) the pH of the incubation medium should be lowered (into the more acidic regions) or (2) the temperature of the incubation should be lowered, even down to 4°C. On the other hand, if there is no sign of a positive reaction, or only a faint electron density is observed, try to (1) use higher pH (as long as nuclear staining does not appear, pH can be elevated without any danger of artifacts), (2) use prolonged incubation time (up to 1 h but not longer), or (3) use an incubator at 37°C. A final possible cause of failure or a very faint reaction might be overfixation. Fixation time in aldehyde should be reduced to 1 h and the experiment repeated. Tissue preservation might be inferior, but intensity of the enzyme reaction will increase.

If ethopropazine pretreatment is omitted, sites of pseudocholinesterase activity will also show up in the electron micrograph. If a butyrylthiocholine substrate is used, these sites will give an enhanced reaction.

4. Controls

1. Most important is that, in the absence of the substrate (acetylthiocholine or butyrylthiocholine), no reaction, i.e., no electron density, should be present. Do not forget, however, that a background ''staining'' is not only permissible but, actually, desirable.
2. By increasing the concentration of ethopropazine (by a factor of 10), every kind of cholinesterase enzyme reaction should be eliminated. In case of a dubious effect, use diisopropyl fluorophosphonate (DFP) or eserine in high concentration. Lack of staining is proof of the enzymatic origin of the reaction observed in the experimental specimen.
3. *Beware of artifacts!* An electron microscopic enzyme reaction is acceptable only if it is visible also at the light microscopic level. Any other ''EM reaction'' is prone to be the result of a simple (or complex) precipitation. On the other hand, patchy enzyme localization can be traced back too often to osmic acid decolorization. This might be especially important if studying biopsy material from pathological samples. The ''rule of thumb'' is: electron histochemistry of cholinesterases is *not a direct methodology* — it is a technique which helps the investigator to draw conclusions. Therefore, the investigator should not *believe* anything he sees — or does not see! He has to *prove* the presence of the enzyme as well as its absence.

C. SUMMARY

In conclusion, it appears that the AChE reaction, at the electron microscopic level, finds widespread utilization in biology. Two important restrictions should, however, always be kept in mind: first, AChE positivity is *not* a sign for the presumed cholinergic character of a nerve cell and/or terminal — a serious setback, which was proclaimed nearly two decades ago;[2] and, second, that the presence or absence of an electron microscopic AChE reaction product at a special location might be either true or false. Therefore, the electron microscopic AChE reaction should never be taken at its face value. Remember Claude Bernard's century-old maxim: *La science ne consiste pas en faits, mais dans les consequences que l'on en tire.**

REFERENCES

1. **Koelle, G. B. and Friedenwald, J. S.,** A histochemical method for localizing cholinesterase activity, *Proc. Soc. Exp. Biol.*, 70, 617, 1949.
2. **Koelle, G. B.,** Cytological distribution and physiological functions of cholinesterases, in *Handbuch der Experimentellen Pharmakologie Ergänzungswerk,* Vol. 15, Koelle, G. B., Ed., Springer-Verlag, Berlin, 1963, 187.
3. **Csillik, B., Joó, F., Kása, P., and Sávay, Gy.,** Pb-Thiocholine techniques for the electron histochemical localization of acetylcholinesterase, *Acta Histochem.*, 25, 58, 1966.
4. **Reale, E. and Luciano, L.,** A probable source of errors in electron histochemistry, *J. Histochem. Cytochem.*, 12, 713, 1964.
5. **Csillik, B. and Knyihár, E.,** On the effect of motor nerve degeneration on the fine structural localization of esterases in the mammalian motor end-plate, *J. Cell Sci.*, 3, 529, 1968.
6. **Koelle, G. B. and Horn, R. S.,** Acetyl disulfide $(CH_3COS)_2$, a major active component in the thiolacetic acid histochemical method for acetylcholinesterase, *J. Histochem. Cytochem.*, 16, 743, 1968.
7. **Koelle, G. B., Davis, R., and Devlin, M.,** Acetyl disulfide $(CH_3COS)_2$, and bis-(thioacetoxy) aurate (I) complex, $Au(CH_3COS)_2^-$, histochemical substrates of unusual properties with acetylcholinesterase, *J. Histochem. Cytochem.*, 16, 754, 1968.

* Science consists not of facts, but of the conclusions one draws from them.

8. **Csillik, B. and Knyihár, E.,** Endoplasmic units: non-reacting globular intracisternal particles of the endoplasmic reticulum, revealed by an electron histochemical enzyme reaction as a negative stain, *Acta Biochim. Biophys. Acad. Sci. Hung.*, 3, 165, 1968.
9. **Csillik, B., Knyihár, E., and Halász, N.,** Ultrastructural localization of acetylcholinesterase in autonomic post-ganglionic axons, *J. Neuro-Visc. Relat.*, 31, 3, 1968.
10. **Sketelj, J., Crne, N., and Brzin, M.,** Two types of focal accumulations of acetylcholinesterase appear in noninnervated regenerating skeletal muscles of the rat, *J. Neurosci. Res.*, 20, 90, 1988.
11. **Kaupmann, K., Heimann, P., and Jockusch, H.,** Dolichos biflorus agglutinin receptors in mouse muscle. I. Developmental expression in relation to synaptic acetylcholinesterase and to neuromuscular disease, *Eur. J. Cell Biol.*, 46, 411, 1988.
12. **Morris, J. F.,** Differential distribution of acetylcholinesterase activity among vasopressin- and oxytocin-containing supraoptic magnocellular neurons, *Neuroscience*, 28, 109, 1989.
13. **Henderson, Z.,** Acetylcholinesterase on the dendrites of central cholinergic neurons: an electron microscopical study in the ferret, *Neuroscience*, 28, 95, 1989.
14. **Siripholvat, V., Watanabe, T., and Tomita, T.,** Histochemical and cytochemical localization of acetylcholinesterase in the cerebellar cortex of the chicken, *Jikken Dobutsu*, 37, 437, 1988.
15. **Martinez-Murillo, R., Blasco, I., Alvarez, F. J., Villalba, R., Solano, M. L., Montero-Caballero, M. I., and Rodrigo, J.,** Distribution of enkephalin-immunoreactive nerve fibres and terminals in the region of the nucleus basalis magnocellularis of the rat: a light and electron microscopic study, *J. Neurocytol.*, 17, 361, 1988.
16. **Wilson, F. R., Westrum, L. E., and Broderson, S. H.,** Ultrastructural changes in acetylcholinesterase activity in the deafferented spinal trigeminal nucleus, *Synapse*, 2, 341, 1988.
17. **Engel, A. K. and Kreutzberg, G. W.,** Neuronal surface changes in the dorsal vagal motor nucleus of the guinea pig in response to axotomy, *J. Comp. Neurol.*, 275, 181, 1988.
18. **Hiramatsu, K., Watanabe, T., and Fujioka, T.,** The distribution of acetylcholinesterase (AChE)-positive nerves in chicken pancreas demonstrated by light and electron microscopy, *Arch. Histol. Cytol.*, 51, 159, 1988.
19. **Akiyoshi, H.,** Ultrastructure of cholinergic innervation in the cirrhotic liver in guinea pigs. Neurohistochemical and ultrastructural study, *Virchows Arch. B*, 57, 81, 1989.
20. **Yoshie, S., Ogawa, T., and Fujita, T.,** Histochemical, immunohistochemical, and ultrastructural characteristics of nerves in the Duvernoy's gland of the Japanese colubrid snake, Rhabdophis tigrinus, *Arch. Histol. Jpn.*, 51, 459, 1988.
21. **Topilko, A. and Caillou, B.,** Acetylcholinesterase and butyrylcholinesterase activities in human thyroid cancer cells, *Cancer*, 61, 491, 1988.
22. **Sharma, P. N. and Hanna, R. E.,** Ultrastructure and cytochemistry of the tegument of Orthocoelium scoliocoelium and Paramphistonum cervi (Trematoda: Digenra), *J. Helminthol.*, 62, 331, 1988.
23. **Bell, C.,** Use of the direct-coloring thiocholine technique for demonstration of intracellular neuronal cholinesterases, *J. Histochem. Cytochem.*, 14, 567, 1966.
24. **Birks, R. I. and Brown, L. M.,** A method for locating the cholinesterase of a mammalian myoneural junction by electron microscopy, *J. Physiol.*, 152, 5P, 1960.
25. **Fränkö, O., Rechardt, L., and Hänninen, L.,** Electron microscopic demonstration of cholinesterases in nervous tissue, *Histochemie*, 8, 369, 1967.
26. **Joó, F., Sávay, G., and Csillik, B.,** A new modification of the Koelle-Friedenwald method for the histochemical demonstration of cholinesterase activity, *Acta Histochem.*, 22, 40, 1965.
27. **Karnovsky, M. J. and Roots, L.,** A "direct-coloring" thiocholine method for cholinesterases, *J. Histochem. Cytochem.*, 12, 219, 1964.
28. **Kása, P. and Csillik, B.,** Electron microscopic localization of cholinesterase by a copper-lead-thiocholine technique, *J. Neurochem.*, 13, 1345, 1966.
29. **Kása, P. and Csillik, B.,** AChE synthesis in cholinergic neurons: electron histochemistry of enzyme translocation, *Histochemie*, 12, 175, 1968.
30. **Koelle, G. B., Davis, R., and Gromadzki, C. G.,** Electron microscopic localization of cholinesterases by means of gold salts, *Ann. N.Y. Acad. Sci.*, 144, 613, 1967.
31. **Koelle, G. B. and Gromadzki, C. G.,** Comparison of the gold-thiolcholine and gold-thiolacetic acid methods for the histochemical localization of acetylcholinesterase and cholinesterases, *J. Histochem. Cytochem.*, 14, 443, 1966.
32. **Lewis, P. R. and Shute, C. C. D.,** An electron-microscopic study of cholinesterase distribution in the rat adrenal medulla, *J. Microsc.*, 89, 181, 1969.
33. **Pearse, A. G. E.,** *Histochemistry. Theoretical and Applied*, 3rd ed., Churchill Livingstone, Edinburgh, 1972.
34. **Csillik, B.,** Synaptochemistry of acetylcholine metabolism in a cholinergic neuron, *Int. Rev. Neurobiol.*, 18, 69, 1975.

III. LIPASES

Reinhart Gossrau

A. INTRODUCTION

Compared with many other enzymes, the ultracytochemical visualization of lipases (EC 3.1.1.3; triacylglycerol lipase, triglyceride lipase, tributyrase) remains difficult until now. For this reason, the available ultracytochemical lipase detection methods have their limitations, and therefore all will be described here.

Lipases catalyze the following reaction:

$$\text{triacylglycerol} + H_2O \rightleftharpoons \text{diacylglycerol} + \text{fatty acid}$$

The procedures tried to detect lipases at the electron microscopic level in human, animal, and plant tissues are either metal salt or simultaneous S-coupling reactions, originally developed for light microscopy or for both light and electron microscopy.

B. METHODS

The metal salt methods (Tween methods) use Tweens as substrates. These commercial detergents are water-soluble esters of long-chain saturated or unsaturated fatty acids with polysorbitol, polymannitol, or polyglycols; the remaining hydroxyl groups are etherified with ethylene oxide chains of varying lengths. In the indirect two-step procedure, calcium ions trap the fatty acids liberated by lipases as insoluble calcium soaps which are then converted by lead ions into electron-dense lead compounds. In the direct one-step procedure, lead ions are present in the incubation medium instead of calcium ions. Originally, Gomori[1,2] proposed the saturated Tween 20 (laurate) and 40 (palmitate) and recommended especially Tween 60 (estearic ester of a polymannitol) and the unsaturated Tween 81 (stearic ester of polyglycol) as lipase substrates. To demonstrate the ''true lipase'' Gomori used Tween 80 (oleic ester of polymannitol) or G-6486 (a ricinoleic aid ester); furthermore, Tween 85 (polyoxyethylene sorbitan trioleate) was tried by George and Iype[3] (see also Reference 4). At present, the oleate (Tween 80) is mostly used for lipase ultracytochemistry.

The simultaneous coupling method for lipase ultracytochemistry goes back to the observation of Kramer et al.[5,6] that pancreatic and serum lipase hydrolyze alkanoates of 1- and 2-naphthols (see also Reference 7: localization of acid lipase with Triton X-100 1-naphthyl palmitate micelles), the liberated naphthols being then used for the visualization of lipase.

In the presence of bile salts, pancreatic lipases hydrolyze preferentially 1- or 2-naphthol nonanoate (C_9). Based on this observation, Abe et al.[8] used several naphthol AS alkanoates (see also Reference 9: naphthol AS octanoate, C_8; decanoate, C_{10} and nonanoate, C_9; and the corresponding naphthol AS-BI or naphthol AS-LC fatty esters). In the presence of taurocholate naphthol AS nonanoate (pelargonic acid; dissolved in dimethylacetamide) with Fast Blue BB (or hexazotized pararosaniline[10]) for simultaneous coupling (formation of a blue or red azo-dye) was found to be the best substrate for the light microscopic demonstration of pancreatic lipase. However, for electron microscopic lipase cytochemistry the azo-dye did not possess sufficient electron opacity and was lipid soluble.[11]

Therefore, Seligman et al.[11] used the principle of osmiophilic reagents introduced by Hanker et al.[12] (see also Reference 13) and the nonanoate of 2-mercaptobenzanilide (2-thiolnonanoyl benzanilide) synthesized by Hanker et al.[14] as substrate. Upon enzymatic hydrolysis, the thiol, 2-mercaptobenzanilide, is liberated and S-couples with the diazonium salt Fast Blue BB to form an insoluble diazothioether dye. This compound can then be osmicated and an insoluble osmium black-containing pigment is formed (diazothioether-OsO_4 method).

C. EQUIPMENT AND REAGENTS

Equipment needed include a vibratome, tissue sectioner or freezing microtome, incubator (or, even better, shaking water bath), and a hood.

Required chemicals and reagents are glutardialdehyde, paraformaldehyde, sodium taurocholate, Tween 60 and 80, 2-thiononanoylbenzanilide, dimethylacetamide, Fast Blue BB, Tris(hydroxy)methyl aminomethane, maleic acid, calcium chloride, lead nitrate, and osmic acid.

D. PROCEDURES

- *Test tissue:* Rat pancreas
- *Tissue pretreatment:* Samples are fixed in 1.5 to 5% glutardialdehyde in 0.1 *M* cacodylate buffer, pH 7.2, according to Karnovsky,[15] or in a 1:4 diluted Karnovsky solution, or in 4% formaldehyde prepared from depolymerized paraformaldehyde in the same buffer for 1 to 4 h at 4°C, followed by rinsing in the same buffer for 5 to 10 h at 4°C. Afterwards 20- to 50-μm sections are cut with a tissue chopper, Vibratome, or freezing microtome, and collected in cacodylate buffer.

The two-step Tween method, according to Nagata[16] (see also Reference 17), is described below.

- *Preincubation medium (F):*

  - 2.5 ml 0.2 *M* Tris buffer, pH 7.2
  - 22.5 ml distilled water
  - 0.21 g sodium fluoride

  Preincubate for 0.5 to 1 h at 37°C. This is then followed by incubation in Medium A or B.

- *Incubation media:*

  - Medium A
    1.0 ml 5% Tween 80 solution
    2.5 ml 0.2 *M* Tris-maleate buffer, pH 7.2
    1.0 ml 10% Aqueous calcium chloride
    20.5 ml Distilled water
  - Medium B
    1.0 ml 5% Tween 80 solution
    2.5 ml 0.2 *M* Tris-maleate buffer, pH 7.2
    1.0 ml 10% Aqueous calcium chloride
    2.5 ml 2.5% Aqueous sodium taurocholate
    18.0 ml Distilled water
  - Medium C
    1.0 ml 5% Tween 80 solution
    2.5 ml 0.2 *M* Tris-maleate buffer, pH 7.2
    1.0 ml 0.08 to 0.02% Aqueous lead nitrate solution
    20.5 ml Distilled water

- Medium D
  2.5 ml 0.2 *M* Tris-maleate buffer, pH 7.2
  1.0 ml 10% Aqueous calcium chloride
  21.5 ml Distilled water
- Medium E
  1.0 ml 5% Tween 80 solution
  2.5 ml 0.2 *M* Tris-maleate buffer, pH 7.2
  1.0 ml 10% Aqueous calcium chloride
  0.099 g Quinine hydrochloride
  20.5 ml Distilled water

Incubate for 2 to 3 h at 37°C.

- *Post-treatment:* Wash sections incubated in medium A, B, C, D, and E in 2% ethylene diamine tetraacetate (EDTA) in 0.1 *M* cacodylate buffer, pH 7.2, for 3 min to remove calcium. Wash in cacodylate buffer for 5 min; immerse sections into 0.15% lead nitrate solution for 10 min to convert calcium soaps into lead soaps; wash in cacodylate buffer for 5 min; postfix with 1% OsO_4 in cacodylate buffer for 1 h at 0°C; wash in cacodylate buffer for 5 min; dehydrate rapidly, and embed in Epon; post-stain with uranyl acetate, followed by lead citrate.

The one-step Tween method, according to Pais[18] (also Gossrau, unpublished data, and see Reference 19), is as follows:

- *Incubation medium:*

  - 1.0 ml 5% Tween 80 or Tween 60 solution
  - 2.5 ml 0.2 *M* Tris-maleate buffer, pH 7.2
  - 1.0 ml 0.08 to 0.02% Aqueous lead nitrate solution
  - 20.5 ml Distilled water

  Incubate for 3 to 4 h at 37°C in a shaking water bath.

- *Post-treatment:* rinse in cold 0.1 *M* cacodylate buffer, ph 7.4, and postfix in ice-cold 2% OsO_4 in cacodylate buffer for 2 h. Dehydrate rapidly in acetone and embed in Epon. Stain some sections with uranyl acetate, followed by lead citrate.

The diazothioether osmium procedure, according to Seligman et al.,[11] is as follows:

- *Incubation medium:*

  - 10.0 ml 0.2 *M* Tris-HCl buffer, pH 7.4
  - 1.0 ml 2.5% Sodium taurocholate (25 mg/ml)
  - 0.1 ml 2-Thiononanoylbenzanilide in dimethylacetamide (40 mg/ml)
  - 10.0 mg Fast Blue BB

  The 0.1-ml 2-thiolnonanoylbenzanilide solution is added with swirling to the buffer taurocholate solution, yielding a turbid solution. After addition of Fast Blue BB, the solution is filtered. Incubate for 2 to 4 h at 37°C, with changes of the medium at 30-min intervals.

TABLE 1
Some Inhibitors and Activators of Lipase and Nonspecific Esterase

Lipase activator	Lipase inhibitor	Esterase inhibitor
10 m*M* Sodium taurocholate	3.5 m*M* Eserine	100 m*M* Eserine
	10 m*M* Quinine hydrochloride	100 m*M* Sodium arseniate
		10 m*M* Sodium fluoride
		10 m*M* Atoxyl
		1 μm Disopropylfluorophosphate
		10 μm Trichlorphone
		10 μm Diethyl 2-nitrophenyl phosphate

Data according to Pais, M. S. S., in *CRC Handbook of Plant Cytochemistry,* Vol. 1, Vaughn, K. D., Ed., CRC Press, Boca Raton, FL, 1987, 123.

- *Post-treatment:* rinse in 15% sucrose and expose to OsO_4 vapor for 30 min at 60°C according to Seligman et al.;[20] rinse in 15% sucrose for a few min; dehydrate in ethanol and embed in resins, e.g., Araldite or Epon.
- *Controls*

  - Basic controls: incubation with substrate-free media; incubation without trapping or coupling reagents; inactivation by heat at 60°C for 30 to 60 min; incubation of tissues with known lipase activity; use of both methods for the ultracytochemical detection of lipases.
  - Controls with inhibitors and activators: For the discrimination between lipase and the co-reacting nonspecific esterases, some compounds are given in Table 1.
  - Ultracytochemical controls: When the metal salt techniques with Tweens are used, aldehyde-fixed tissue samples should be analyzed also without contrast procedures, i.e., without uranyl acetate or lead nitrate.

E. RESULTS

Using the one-step Tween procedure according to Blanchette-Mackie and Scow,[19] granular and laminar precipitates are found in vacuoles and vesicles of capillary endothelial cells and in the subendothelial space of fat tissue. As described by Nagata[16] and Murata et al.,[17] the two-step procedure with taurocholate (medium B) shows varying amounts of lead soaps in the cisternae of the endoplasmic reticulum and dictyosomes of the Golgi apparatus, and in membranes of the zymogen (secretory) granules in the acinar cells of the exocrine pancreas. Incubation media without the lipase activator taurocholate (medium A) yields precipitates limited to the membranes of the zymogen granules and in the acinar lumen, as well as precipitates ubiquitous in the plasma membrane of the acinar cells. Media free of the substrate Tween 80 (medium D) and media containing lead instead of calcium (medium C), as well as incubation with the lipase inhibitor quinine hydrochloride (medium E), or preincubation in the presence of fluoride (medium F) do not lead to formation of deposits in the specimen.

With the diazothioether-osmic acid method, osmium black deposits can be detected in the endoplasmic reticulum, cisternae of the Golgi apparatus, around the secretory granules, and in the lumen of the pancreatic acini, but not within the zymogen granules.

Problems connected with the ultracytochemical visualization of lipase are also caused by the action of the enzyme at interfaces on substrates that are insoluble in water (lipid-water interface[21]) and by overlapping (co-reaction) with nonspecific esterases.[21-24] The use of natural lipase substrates (lipids, oils) is not possible because they are not sufficiently

IV. ALKALINE PHOSPHATASE

Kouta Yamamoto

A. INTRODUCTION

Although alkaline phosphatase (EC 3.1.3.1.) is present in almost every mammalian tissue, its physical functions and physiological substrates still remain unknown. Nonspecific alkaline phosphatase, an orthophosphoric monoester phosphohydrolase, hydrolyzes orthophosphoric monoesters into phosphoric acid and alcohol at alkaline pHs (optimal pH 7.6 to 9.9). This enzyme, localized in the plasma membrane, is a zinc-containing enzyme with a molecular weight of 100,000 to 200,000. It requires magnesium for catalytic activity and hydrolyzes various monoesters of phosphoric acid including ATP, AMP, and pyrophosphate. Therefore, the involvement of this enzyme in the cytochemical detection of the activities of other specific phosphatases, such as ATPase and AMPase, should be kept in mind.

There are two fundamentally different methods for the cytochemical detection of alkaline phosphatase activity: a metal capture method and an azo-dye method. The former is based on the precipitation of phosphate ions liberated by the enzyme action as a metal salt, and the latter on the coupling of liberated alcohol with diazonium salt to give an insoluble azo-dye at the sites of enzyme activity. However, the azo-dye methods do not yield reaction products of intrinsic electron density, and they are not particularly reliable above pH 8.5 because of the formation of massive precipitates in the incubating medium. Hanker et al.[1] applied a sulfate-containing substrate (2-naphtylthiolphosphoryl dichloride) to reduce osmium tetroxide forming so called ''osmium black'' in an azo-dye technique for electron microscopic observations (Figure 1). Although this procedure is theoretically interesting, the reaction product showed coarse deposits, which were not suitable for ultrastructural cytochemistry. The lead capture method described later is easily adapted for both light and electron microscopic examinations and the metal capture method is discussed below. The azo-dye techniques have been reviewed in detail by Burnstone.[2]

B. METHODS

Since the metal capture method was developed independently by Gomori[3] and Takamatsu,[4] various capture reagents, including calcium, cadmium, lead, and cerium ions have been proposed for the histocytochemical demonstration of alkaline phosphatase activity. Calcium phosphate has a low electron density and has to be converted to cobalt, silver, or lead phosphate. This conversion, however, may result in diffusion of reaction products. Cadmium phosphate is soluble in osmic acid and is not suitable for the electron microscopic demonstration of the enzyme. Hugon and Borgers[5] attempted to use lead nitrate at high alkaline pH without a chelating agent such as sodium citrate. Lead phosphate has sufficient electron density, and for light microscopic observations can be converted to brown lead sulfide (Figure 2). However, the lead nitrate medium is unstable at alkaline pH. Because of precipitate formation, the medium has to be filtered before use, and its exact composition becomes unpredictable. Although this method employs a medium having a pH of 8.2, hydrolysis of the substrate by enzymes other than alkaline phosphatases, such as ATPase and pyrophosphatase, active at this pH, cannot be avoided.

Recently, cerium ions have been intensively used for the histochemical demonstration of various phosphatases.[6,7] This capture reagent overcomes some of the problems inherent in the lead-based reactions, including nonspecific lead depositions and coarse granularity of the reaction products. However, cerium-based incubation media are turbid over pH 8.5, and cannot be used for the visualization of the reaction products under the light microscope.

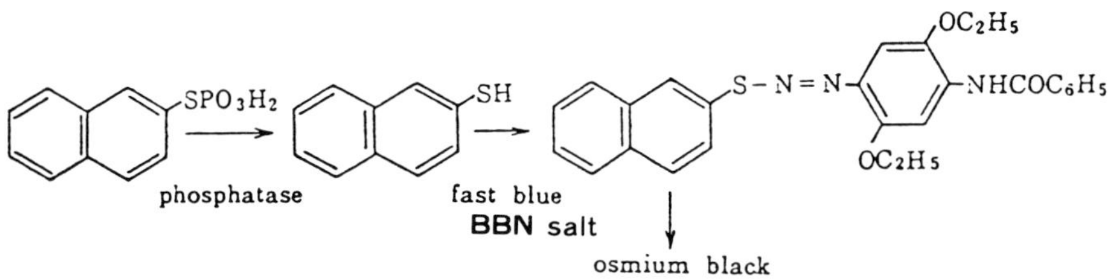

FIGURE 1. The azo-dye technique applied for electron microscopy.

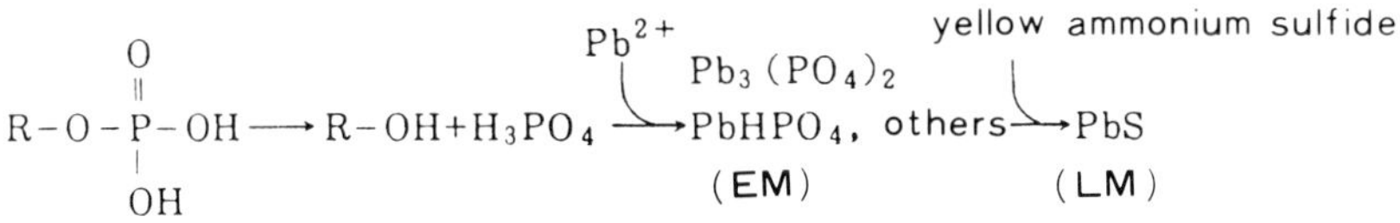

FIGURE 2. Histocytochemical detection of alkaline phosphatase activity (lead capture method).

Therefore, the cerium-based method has no advantage for the histocytochemical demonstration of alkaline phosphatase activity.

Mayahara et al.[8] introduced lead citrate as a capture reagent for the ultracytochemical detection of this enzyme activity at high alkaline pH. As discussed,[9] this method is the most reliable and provides the best results, for the following reasons: (1) it is a one-step method, (2) the incubating medium is stable and clear without the addition of a chelating agent at high alkaline pH, (3) the detection of the enzyme activity takes place near its pH optimum, and (4) the concentration of lead ions in the medium, which are known to inhibit the enzyme, is relatively low compared with the lead-based medium (12 m*M*) described by Tranzer.[10]

C. CHEMICALS AND REAGENTS

Sodium β-glycerophosphate was obtained from Sigma Chemical Co., St. Louis, MO, and lead citrate from K & K Lab., New York. All other reagents were of analytical grade.

Lead Citrate Method of Mayahara et al.[8]

Incubation medium	Final concentration			
Sucrose	8%	0.8 (g)	2.0	4.0
Sodium β-glycerophosphate	20 m*M*	60 (mg)	150	300
Distilled water		2.0 (ml)	5.0	10.0
0.2 *M* Tris-HCl buffer (pH 8.5)	28 m*M*	1.4 (ml)	3.5	7.0
0.015 *M* Magnesium sulfate	3.9 m*M*	2.6 (ml)	6.5	13.0
0.5% Lead citrate	2.0 m*M*	4.0 (ml)	10.0	20.0
Final pH 9.0–9.4	Total volume	10 (ml)	25	50

Weigh and add the reagents in the order shown in the table, except for the lead citrate, and dissolve completely using a magnetic stirrer. Add the lead citrate solution very slowly, with continuous stirring. Adjust the final pH to 9.0 to 9.4 with 1 *N* NaOH or 1 *N* HCl. The incubation medium is completely clear and can be used without filtration. However, if the pH falls below 8.5, a visible precipitate is formed. If a substrate is degenerated, it frequently causes precipitation in the medium. In the case of tissues containing high enzyme activity, such as small intestine, the concentration of the substrate should be reduced to 3 to 5 m*M*.

Stock solutions of Tris-HCl buffer and magnesium sulfate are recommended to obtain consistent reproducible results. Stock solutions are kept in a refrigerator.

Lead citrate is not easily soluble in water and it is dissolved by the addition of NaOH as follows: to 0.25 mg lead citrate in 48.25 ml of distilled water add 1.75 ml of 1 *N* NaOH. Place the solution into a small air-tight bottle, and shake vigorously. Complete dissolution requires about 20 to 30 min. Great care should be taken to exclude CO_2 from the medium, otherwise lead carbonate will be precipitated. The alkaline lead citrate solution should be prepared immediately before use. The solution cannot be stored.

Recently, Yu Xiu Shi et al.[11] reported that the use of 175 to 350 m*M* AMP (2-amino-2-methyl-1-propanol) buffer instead of Tris-HCl buffer provided finer localization of reaction products and higher preservation of alkaline phosphatase activity compared with the original lead citrate method. It seems to help to increase the chances of detecting weak alkaline phosphatase activity.

D. PROCEDURE

1. Fixation — Tissues cut into slices less than 1 mm thick are fixed for 1 h at 0 to 4°C by immersion in a mixture of 2 to 4% formaldehyde and 0.5 to 2% glutaraldehyde in 0.1 *M* sodium cacodylate buffer, pH 7.2, containing 8% sucrose. Perfusion fixation is preferable, which is performed for 5 to 10 min at room temperature (to avoid air embolism by small air bubbles liberated from cooled glutaraldehyde) with buffered glutaraldehyde. Following fixation, the tissues are rinsed from 1 h to overnight at 0 to 4°C in the same buffer containing 8% sucrose.
2. Cytochemical incubation — For light microscopy, 10- to 15-μm frozen sections are prepared with an electrofreezing microtome or in a cryostat. Prior to freezing, the tissues are soaked in 20% buffered glycerin for 30 min. For electron microscopy, 40- to 50-μm nonfrozen sections are cut with a Microslicer (Dosaka EM Co., Kyoto, Japan) or a vibratome. Sections are incubated for 10 to 30 min at room temperature or at 37°C. The length of incubation and the temperature depend on the intensity of reaction. A 5-min incubation at 0 to 4°C is recommended for small intestine, while 60-min incubation at 37°C is necessary for the detection of liver alkaline phosphatase.
3. For light microscopy, the sections are briefly rinsed in distilled water following the incubation and then treated with 1% yellow ammonium sulfide for 2 min. They are briefly rinsed in distilled water and mounted in glycerine jelly.
4. For electron microscopy, the sections are postfixed for 30 to 60 min with 1% buffered osmium tetroxide after a brief rinse in cold 0.1 *M* cacodylate buffer containing 8% sucrose. They are processed by routine procedures for electron microscopy. Ultrathin sections are counterstained briefly with uranyl acetate and examined under an electron microscope.

As cytochemical controls, sections are incubated either in the absence of substrate or in the complete medium containing 2.5 m*M* levamisole (Sigma Chemical Co., St. Louis). Sections are preincubated with 2.5 m*M* levamisole in 0.1 *M* cacodylate buffer for 30 min at 0 to 4°C prior to the cytochemical incubation to ensure complete inhibition. 12.5 m*M* L-phenylalanin[12] is used as control for alkaline phosphatase of small intestine, but does not completely inhibit the enzyme activity if β-glycerophosphate is used as substrate. Although the precise mechanism of inhibition still remains uncertain, levamisole is the most suitable inhibitor of nonspecific alkaline phosphatase, for the following reasons:[13] (1) it is a potent inhibitor, and complete inhibition may be achieved by very low concentrations (0.1 m*M*);

(2) specific other phosphatases are not inhibited; (3) its inhibitory action is independent of pH, substrate, or fixation; (4) the inhibitor can be used in both metal capture and azo-dye coupling methods, and (5) it is organ specific. Levamisole is an effective inhibitor of the alkaline phosphatase of various tissues, including liver, kidney, and bone, whereas the alkaline phosphatases of small intestine, one variant of HeLa cell, and the Regan isoenzyme are insensitive to this inhibitor. Because of its specificity as a potent inhibitor of alkaline phosphatases, levamisole is recommended also for cytochemical detection of other specific phosphatases in order to suppress coexisting alkaline phosphatase activity.

E. RESULTS

The reaction product of alkaline phosphatase activity is seen as brown precipitates of lead sulfide under the light microscope. Under the electron microscope, deposits of lead phosphate appear as fine granules which are generally localized to the plasma membrane. Although some reaction product may be present in lysosomes, deposits in other organelles or diffusion in the cytoplasm should be regarded as possible artifacts. In such cases, fixation conditions, incubation time, and substrate concentration should be carefully examined. Intense enzyme stainings are observed in the brush border of renal proximal convoluted tubules, bile canaliculi of hepatocytes, and striated border of small intestine in different species.

F. COMMENTS

1. Effect of NaOH-PIPES Buffer used in Aldehyde Fixative on Alkaline Phosphatase Activity

In general, a buffer is regarded as a substance which merely stabilizes the acid-base balance. In fact, when choosing the buffer for fixation, little attention is usually given to the influence of the specific components of buffers on the preservation of fine structure and the retention of enzyme activity. For the cytochemical detection of phosphatase activity, cacodylate buffer is frequently used as a vehicle for aldehyde fixatives, although it contains arsenic which reacts with the SH-groups of certain enzyme proteins and inhibits enzyme activity, including alkaline phosphatase. Yamamoto and Ogawa[14] reported that alkaline phosphatase activity in rat hepatocytes showed remarkable changes in localization when PIPES was used as a buffer of fixation for enzyme cytochemistry. The reaction product indicating alkaline phosphatase activity was confined to the bile canaliculi membrane when 0.1 *M* cacodylate buffer was used, while deposits were observed on the entire membranes of rat hepatocytes when PIPES was used. This difference may be explained by an inhibitory effect of cacodylate on hepatic alkaline phosphatase. These extraordinary cytochemical findings were confirmed by Yu Xiu Shi et al.[15] Recently, PIPES buffer has been utilized extensively for the cytochemical detection of various oxidases in peroxisomes with satisfactory preservation of both ultrastructure and enzyme activity. Therefore, its application for the cytochemical detection of various enzymes can be expected.

2. Localization of Intracellular Alkaline Phosphatase Activity

It is generally accepted that membrane proteins are synthesized on the rough endoplasmic reticulum, processed in the Golgi complex, and finally transported to the cell membrane. However, using conventional cytochemical technique, alkaline phosphatase activity was demonstrated exclusively in the plasma membrane, with no activity in other intracellular organelles, although a few exceptional localizations have been reported in human bladder tumor cell lines and in human ovarian cancer cells. Tokumitsu et al.[16] recently reported the detection of intracellular alkaline phosphatase activity in the endoplasmic reticulum, Golgi complex, Golgi-derived vesicles, and mitochondria, as well as in lysosomes and plasma membrane of cultured human cancer cells, by treating cells with 0.005% saponin before and during the incubation. This effect might be due to the exposition of latent enzyme activity by changing the structural arrangement of membranes. Although its application is limited at present only to cultured cells, the saponin treatment will be useful to understand the significance of intracellular alkaline phosphatases.

REFERENCES

1. **Hanker, J. S., Seaman, A. R., Weiss, L. P., Bergman, R. A., and Seligman, A. M.,** New cytochemical principle for light and electron microscopy, *Science,* 146, 1039, 1964.
2. **Burstone, M. M.,** Phosphatases, in *Enzyme Histochemistry, and Application in the Study of Neoplasmas,* Academic Press, New York, 1962, 160.
3. **Gomori, G.,** Microtechnical demonstration of phosphatase in tissue sections, *Proc. Soc. Exp. Biol.,* 42, 23, 1939.
4. **Takamatsu, H.,** Histochemical study of phosphatase, *J. Jpn. Pathol.,* 29, 492, 1939.
5. **Hugon, J. and Borgers, M.,** A direct lead method for the electron microscopic visualization of alkaline phosphatase activity, *J. Histochem. Cytochem.,* 14, 429, 1966.
6. **Robinson, J. M. and Karnovsky, M. J.,** Ultrastructural localization of several phosphatase with cerium, *J. Histochem. Cytochem.,* 31, 1197, 1983.
7. **Hulstaert, C. E., Kalicharan, D., and Hardonk, M. J.,** Cytochemical demonstration of phosphatases in the rat liver by a cerium-based method in combination with osmium tetroxide and potassium ferrocyanide postfixation, *Histochemistry,* 78, 71, 1983.
8. **Mayahara, H., Hirano, H., Saitoh, T., and Ogawa, K.,** The new lead citrate method for the ultrachemical demonstration of activity of non-specific alkaline phosphatase, *Histochemie,* 11, 88, 1967.
9. **Doty, S. B.,** Problem inherent in obtaining the alkaline phosphatase reaction, *J. Histochem. Cytochem.,* 28, 66, 1980.
10. **Tranzer, J. P.,** Utilisation de citrate de plomb pour la mise en évidence de la phosphatase alcaline au microscope électronique, *J. Microsc.,* 4, 409, 1965.
11. **Yu Xiu Shi, Araki, A., Fujimoto, T., and Ogawa, K.,** Use of AMP (2-amino-2-methyl-1-propanol) buffer for histochemical demonstration of alkaline phosphatase activity, *Acta Histochem. Cytochem.,* 21, 321, 1988.
12. **Rufo, M. B. and Fishman, W. H.,** L-Homoarginine, a specific inhibitor of liver-type alkaline phosphatase, applied to the recognition of liver-type enzyme activity in rat intestine, *J. Histochem. Cytochem.,* 20, 336, 1972.
13. **Borgers, M.,** The cytochemical application of new potent inhibitors of alkaline phosphatases, *J. Histochem. Cytochem.,* 21, 812, 1973.
14. **Yamamoto, K. and Ogawa, K.,** Effects of NaOH-PIPES buffer used in aldehyde fixative on alkaline phosphatase activity in rat hepatocytes, *Histochemistry,* 77, 339, 1983.
15. **Yu Xiu Shi, Araki, N., Fujimoto, T., and Ogawa, K.,** Use of glycine buffer for the lead citrate method of non-specific alkaline phosphatase cytochemistry, *Acta Histochem. Cytochem.,* 21, 81, 1988.
16. **Tokumitsu, S., Tokumitsu, K., and Fishman, W. H.,** Intracellular alkaline phosphatase activity in cultured human cancer cells, *Histochemistry,* 73, 1, 1981.

V. ACID PHOSPHATASE

Nobukazu Araki

A. INTRODUCTION

Acid phosphatases (ACPases, EC 3.1.3.2) are nonspecific phosphomonoesterases which catalyze the hydrolysis of almost any phosphomonoester to give phosphoric acid and the corresponding alcohol or phenol. They are distinguished from a similar group of enzymes, the alkaline phosphatases, primarily by their pH optima in the acid range. ACPases are widespread throughout nature, not only in higher animal tissues but also in bacteria, protozoa, and plants.

The ACPases can be differentiated into many types by their origin, molecular weight, pH optima, sensitivity to inhibitors or activators, and relative activity towards various substrates. However, there is no united view among biochemists and cytochemists on the precise classification of ACPases. Cytochemically, the ACPases are divided into two groups by their localization. One, the lysosomal ACPase, is mostly situated in lysosomes and trans Golgi cisternae. This enzyme is inhibited by tartrate and sodium fluoride. The other, nonlysosomal ACPase, is mainly localized in the plasma membrane and endoplasmic reticulum. The nonlysosomal ACPase activity is resistant to tartrate and sodium fluoride.

Since the discovery of lysosomes as ACPase-positive intracellular organelles by de Duve et al.,[1] ACPase has been widely used as a lysosomal marker enzyme by both biochemists and cytochemists. ACPase cytochemistry is one of the most popularized techniques. Most cytochemical studies on ACPase have dealt with lysosomal ACPases. The number of cytochemical studies on nonlysosomal ACPases is far fewer than that of lysosomal ACPase.

Principally, the cytochemical demonstration of ACPase activity is based on two methods. One is a metal salt method depending on the precipitation of released phosphoric acid as a metal-phosphate complex such as lead phosphate.[2,3] The lead phosphate precipitate is converted into a brown lead sulfide by ammonium sulfide treatment for light microscopy. Also, the metal-phosphate is visible under the electron microscope, owing to the high electron density of the metal. The other method is a coupling azo-dye method[4] which depends on the precipitation and coloration of the phenolic part (naphthol) of hydrolyzed substrates by coupling with a diazonium compound. The naphthol liberated from substrates by enzyme hydrolysis couples with diazonium salts to produce a colored insoluble azo-dye which can be visualized under a light microscope. Hexazonium pararosanilin is also useful as a coupler.[5] However, the coupling azo-dye method is not suitable for electron microscopic cytochemistry, because the azo-dye is soluble in reagents ordinarily used in processing tissues for electron microscopy, and it does not have enough electron density to be visualized under the electron microscope. Although a few applications of the coupling azo-dye method to electron microscopy have been attempted,[6-8] the coupling azo-dye method has not been popular in electron microscopic cytochemistry. Therefore, only the metal salt method is described below as a reproducible technique.

B. METHODS

The metal salt method for the histochemical demonstration of ACPase was originally developed by Gomori.[2] The composition of the improved Gomori's medium[3] is given in Table 1. Several modifications of Gomori's method have been described for electron microscopic cytochemistry. For example, Barka and Anderson[5] recommended the use of Tris-maleic acid buffer instead of acetate buffer. Various substrates such as cytidine-5′-monophosphate,[9] *p*-nitrophenyl phosphate,[10] and phosphorylcholine[11] have been applied to the

TABLE 1
The Composition of the Gomori's ACPase Incubation Medium

0.05 *M* Acetate buffer, pH 4.7–5.0	50 ml
Sucrose	4.0 g
Lead nitrate	50 mg
3% Sodium β-glycerophosphate	5 ml

Total volume 55 ml; final pH 5.0.

metal salt method. The substrate modifications may be suitable for detection of specific tissue ACPases or isozymes. Miyayama et al.[10] demonstrated that nonlysosomal ACPase localized on the plasma membrane and rough endoplasmic reticulum in mouse kidney tubule epithelial cells by using *p*-nitrophenyl phosphate. Addition of dimethyl sulfoxide (DMSO) to the incubation medium and the rinse solution was recommended for the demonstration of ACPase activity in culture cells in which lysosomal ACPase often exhibits a cytochemical latency.[12,13] DMSO may increase the permeability of membranes to cytochemical substrates. As trapping agents, aluminum nitrate[14] and cerium chloride[15] are substitutes for lead nitrate. Recently, the cerium-based method of Robinson and Karnovsky[15] is increasingly used for the electron microscopic cytochemistry of ACPases as well as other phosphatases. The incubation medium of Robinson and Karnovsky[15] consists of 0.1 *M* acetate buffer (pH 5.0), 1 m*M* sodium β-glycerophosphate, 2 m*M* cerium chloride, and 5% sucrose. Although various modifications may be applicable to the corresponding tissues or isozymes for special purposes of the studies, the Gomori's medium given in Table 1 seems to be sufficient in general if the specimen is processed under adequate conditions for ACPase cytochemistry.

Besides the composition of the incubation medium, there are other important factors to obtain precise enzyme localization without any artifacts. Hitherto, rather high concentrations (2 to 5%) of glutaraldehyde (GLA) have been customarily employed as a fixative for ACPase cytochemistry. Even if tissues fixed with a high concentration of GLA might be capable of giving adequate ACPase reaction in some lysosomes, GLA significantly inactivates some ACPase activity. Such an enzyme inactivation during tissue processing results in a false-negative reaction. Recently, we have demonstrated ACPase activity of elongated lysosomes (nematolysosomes) in rat pancreatic exocrine cells by employing 2% paraformaldehyde (PFA) and 0.25% GLA as a fixative.[16] In earlier studies,[17,18] it had been stated that these lysosomes were ACPase negative. Piperazine-*N,N′*-bis-2-ethanesulfonic acid (PIPES) buffer also seems to be preferable to cacodylate in the fixative and the rinse solution for the preservation of enzyme activity.[16] Such being the case, minimizing enzyme inactivation during tissue processing may allow the demonstration of ACPase activity where it had not been cytochemically detectable. The procedure described below is an adequate method for the fine localization of lysosomal ACPase activity. This method seems to detect the highest enzyme activity without any artifacts.

C. EQUIPMENT AND REAGENTS

Microslicer (Dosaka EM, Kyoto), Vibratome (Lancer, Ltd.), or cryostat is needed for the procedure.

Reagents include glutaraldehyde, paraformaldehyde, PIPES, sucrose, DMSO, acetic acid, sodium acetate, lead nitrate (Sigma, Nakalai, etc.), sodium β-glycerophosphate (Sigma, Merk, Wako, etc.), osmium tetroxide, ethanol, propylene oxide, embedding medium (Spurr's resin, Epon 812, et al.), and uranyl acetate (Merk).

D. PROCEDURES

1. *Fixation:* Fix tissues with 2% PFA and 0.25 to 1% GLA in 30 m*M* PIPES buffer, pH 7.4, containing 8% sucrose and 10% DMSO (optional) for 1 h at 4°C. If possible, fixation by perfusion with 2% PFA and 0.25% GLA for 5 min, followed by immersion with the same fixative for 1 h at 4°C, is advisable to preserve both ultrastructure and maximal enzyme activity.
2. *Rinse:* After the fixation, rinse the tissues thoroughly with 30 m*M* PIPES buffer, pH 7.4, containing 8% sucrose and 10% DMSO (optional) for 20 to 60 min at 4°C. Leaving the tissues in the buffer for longer than overnight is not recommended, because it may cause a decrease in the enzyme activity.
3. *Slice:* Prepare nonfrozen sections of 40 to 50 μm thickness with a Microslicer or Vibratome. Frozen sections cut in a cryostat or with a freezing microtome also can be used if it is difficult to cut nonfrozen sections of adequate thickness. The use of thin (<40 μm) frozen sections seems to give better results than the use of thick (>>50 μm) nonfrozen sections. Usually, the process of freeze-thawing tends to cause ultrastructural damage and enzyme release from lysosomes by cryoinjury. However, such damage might be significantly prevented by the addition of 10% DMSO, a cryoprotective reagent, to the rinse buffer.
4. *Preparation of the incubation medium:* Make up the incubation medium given in Table 1 just prior to incubation. It is very important to prepare a clear incubation medium without precipitation. The formation of precipitates alters the effective substrate and lead ion concentrations in the medium. This may result in a diffusion of reaction product from lysosomes and nonenzymatic reactions such as nuclear staining. Medium without precipitate formation can be prepared as follows: dissolve the sucrose and lead nitrate completely in 50 m*M* acetate buffer. While stirring, add substrate solution (3% sodium β-glycerophosphate) slowly to 50 m*M* acetate buffer containing sucrose and lead nitrate. The addition of substrate solution raises the pH of the medium. Since the rise of pH over 5.1 during or after the addition of substrate solution would produce heavy precipitation in the medium, the pH of the medium should be checked and carefully adjusted by HCl during the addition of the substrate. The use of acetate buffer at pH 4.7 facilitates the preparation of a medium without precipitate having a final pH of 5.0, containing substrate, lead, and buffer concentrations, as originally proposed by Gomori.[19] Alternatively, the substrate solution may be adjusted to pH 5.0 before mixing with the acetate buffer containing lead and sucrose.[20] De Jong et al.[20] recommended the use of HCl rather than acetic acid for adjusting the pH because chloride ions have an enhancing effect on the trapping efficiency in the incubation medium.
5. *Incubation:* Incubate tissue sections in a glass vial containing the incubation medium for 15 to 30 min at room temperature or 37°C with agitation.
6. *Cytochemical control:* Incubate some sections either in the absence of the substrate or in the full incubation medium containing an ACPase inhibitor. As lysosomal ACPase inhibitor, 10 m*M* sodium fluoride (NaF) or 10 m*M* L-(+)-tartrate is usually employed in the medium. These inhibitors have little effect on nonlysosomal ACPase. In contrast, nonlysosomal ACPase is inhibited by 1 μ*M* *p*-chloromercuribenzoate, although lysosomal ACPase is less sensitive to it.
7. *Rinse:* After the incubation, quickly rinse in cold 30 m*M* PIPES buffer, pH 7.4, containing 8% sucrose. This rinse solution should not contain DMSO because DMSO reacts with osmium.
8. *Postfixation:* Postfix with 1% osmium tetroxide in 30 m*M* PIPES buffer, pH 7.4, for 30 min at 4°C.

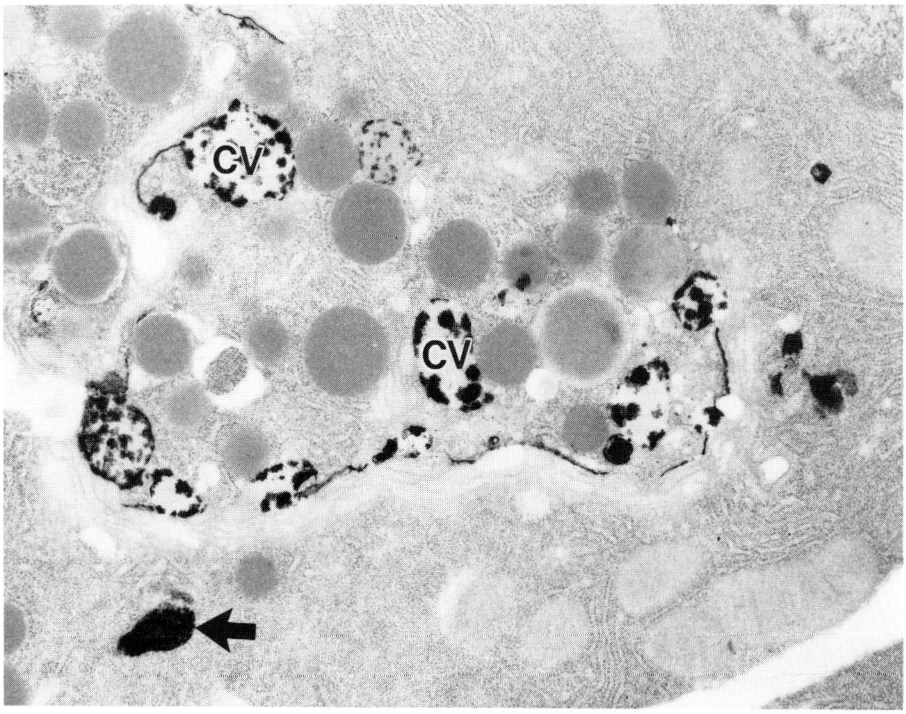

FIGURE 1. ACPase activity of the rat pancreatic exocrine cell. The tissue was fixed with a mixture of 2% PFA and 0.25% GLA in 30 m*M* PIPES buffer, pH 7.4, containing 8% sucrose and 10% DMSO, and incubated with Gomori's medium for 30 min at room temperature. Uranyl acetate staining. The reaction product is seen in lysosomes (arrow), *trans*-most cisterna of the Golgi complex and condensing vacuoles (CV). (Magnification × 19,000.)

9. *Dehydration and embedding:* After the postfixation, dehydrate the tissue through an ethanol series and propylene oxide, and embed in a resin as usual procedure.
10. *Sectioning, electron staining, and observation:* Cut ultrathin sections and mount them on specimen grids. Electron staining with uranyl acetate and/or lead citrate may be applied to the ultrathin sections. Combination of ACPase cytochemistry and 1- to 2-μm semithin section observation by 200-kV TEM is very useful for the three-dimensional structure and distribution of lysosomes in the cell.[16,21]

E. RESULTS

The reaction product of ACPase activity by the Gomori method localizes on primary lysosomes, secondary lysosomes, multivesicular bodies, some coated vesicles, and the *trans* side of Golgi apparatus including *trans*-Golgi network and GERL. In some secretory cells such as pancreatic exocrine cells, condensing vacuoles and secretory granules also show ACPase activity (Figure 1). Occasionally, ACPase activity is demonstrable in the lumen of endoplasmic reticulum. Recently, the presence of ACPase-positive elongated lysosomes, termed ''nematolysosomes''[16,21] or ''tubular lysosomes'',[22,23] was revealed in a variety of cell types. The distribution of nematolysosomes throughout the cell can be clearly revealed in semithin sections under 200-kV TEM (Figure 2). Much diffusion of the reaction product in the cytoplasm or nucleus should be regarded as a sign of the inadequacy either of the fixation to immobilize the enzyme or of phosphate trapping during the incubation. A little diffusion of reaction product around lysosomes can be minimized by shortening the incu-

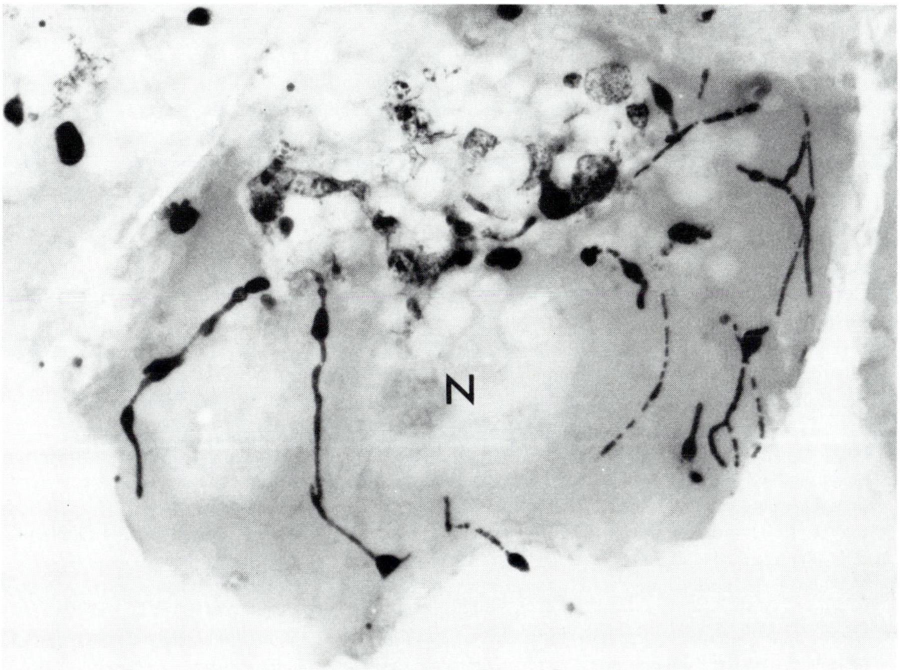

FIGURE 2. Demonstration of nematolysosomes showing ACPase activity in the rat pancreatic exocrine cell. The same sample as in Figure 1 was cut into 2-μm-thick sections, and viewed at 200 kV. ACPase-positive nematolysosomes are elongated from near the basolateral plasma membrane to the Golgi area. N, nucleus. (Magnification × 7800.)

bation time. When the reaction product localizes on some lysosomes, but not all throughout the cell, it is difficult to conclude an absence of ACPase from the lysosome with no reaction product. The heterogeneity in cytochemical reaction of ACPase among lysosomes may be ascribed to a difference in the content of the lysosomes or the lysosomal membrane. However, careful adequate procedures and a more critical eye are required for the judgment of the negative reaction.

It is likely that nonlysosomal ACPase is cytochemically demonstrated on the plasma membrane and other sites by using other substrates such as *p*-nitrophenyl phosphate. However, when using β-glycerophosphate, the reaction product seen on the plasma membrane is usually considered as a result of hydrolysis by other phosphatases. Reactions in extralysosomal sites must be assessed by control experiments with inhibitors.

REFERENCES

1. **de Duve, C., Pressman, B. C., Gianetto, R., Wattiaux, B., and Appelmans, F.,** Tissue fractionation studies. VI. Intracellular distribution pattern of enzymes in rat liver tissue, *Biochem. J.*, 60, 604, 1955.
2. **Gomori, G.,** Distribution of acid phosphatase in the tissues under normal and under pathologic conditions, *Arch. Pathol.*, 32, 189, 1941.
3. **Gomori, G.,** in *Microscopic Histochemistry, Principle and Practice*, University of Chicago Press, Chicago, 1952, 189.

4. **Burstone, M. S.,** Histochemical demonstration of acid phosphatase using naphthol AS phosphates, *J. Natl. Cancer Inst.*, 21, 523, 1958.
5. **Barka, T. and Anderson, P. J.,** Histochemical methods for acid phosphatase using hexazonium pararosanilin as coupler, *J. Histochem. Cytochem.*, 10, 741, 1962.
6. **Smith, R. E. and Fishman, W. H.,** *p*-(Acetoxymercuric) aniline diazotate, a reagent for visualizing the naphthol As-BI product of acid hydrolase action at the level of the light and electron microscopy, *J. Histochem. Cytochem.*, 17, 1, 1969.
7. **Tice, L. W. and Barrnett, R. J.,** Diazophthalocyanins as reagents for fine structural cytochemistry, *J. Cell Biol.*, 25, 23, 1965.
8. **Livingston, D. S., Fisher, S. W., Greenoak, G. E., and Maggi, V.,** A new procedure for the ultrastructural localization of acid phosphatases, *Histochemie,* 24, 159, 1970.
9. **Novikoff, A. B.,** Lysosome in the physiology and pathology of cells: contribution of staining methods, in *Ciba Foundation Symposium on Lysosomes,* de Reuck, A. V. S. and Cameron, M. P., Eds., Churchill Livingstone, London, 1963, 36.
10. **Miyayama, H., Solomom, R., Sasaki, M., Lin, C., and Fishman, W. H.,** Demonstration of lysosomal and extralysosomal sites for acid phosphatase in mouse kidney tubule cells with *p*-nitrophenyl-phosphate lead-salt technique, *J. Histochem. Cytochem.*, 23, 439, 1975.
11. **Serrano, J. A., Shannon, W. A., Jr., Sternberger, N. J., Wasserkrung, H. L., Serrano, A., and Seligman, A. M.,** The cytochemical demonstration of prostatic acid phosphatase using a new substrate, phosphorylcholine, *J. Histochem. Cytochem.*, 24, 1046, 1976.
12. **Brunk, U. T. and Ericsson, J. L. E.,** The demonstration of acid phosphatase in *in vitro* cultured tissues cells. Studies on the significance of fixation, toxicity and permeability, *Histochem. J.*, 4, 349, 1972.
13. **Mayahara, H. and Chang, J. P.,** Electron microscopic study of acid phosphatase activity in cultured human cystic fibrosis fibroblasts, *Acta Histochem. Cytochem.*, 11, 449, 1978.
14. **Berry, J. P., Hourdry, J., Sternberg, M., and Galle, P.,** Aluminium phosphate visualization of acid phosphatase activity: a biochemical and X-ray microanalysis study, *J. Histochem. Cytochem.*, 30, 86, 1982.
15. **Robinson, J. M. and Karnovsky, M. J.,** Ultrastructural localization of several phosphatases with cerium, *J. Histochem. Cytochem.*, 31, 1197, 1983.
16. **Araki, N., Matsubara, H., Takashima, Y., and Ogawa, K.,** Three-dimensional distribution of acid phosphatase-positive thread-like structures ''nematolysosomes'' in rat pancreatic exocrine cells, *Acta Histochem. Cytochem.*, 22, 585, 1989.
17. **Oliver, C.,** Cytochemical localization of acid phosphatase and trimetaphosphatase activities in exocrine acinar cells, *J. Histochem. Cytochem.*, 28, 78, 1980.
18. **Oliver, C.,** Characterization of basal lysosomes in exocrine acinar cells, *J. Histochem. Cytochem.*, 31, 1209, 1983.
19. **Waters, S. E. and Butcher, R. G.,** Studies on the Gomori acid phosphatase reaction: the preparation of the incubation medium, *Histochem. J.*, 12, 191, 1980.
20. **De Jong, A. S. H., Hank, T. J., Vanduijin, P., and Daems, W. Th.,** A new dynamic model system for the study of capture reactions for diffusable compounds in cytochemistry. II. Effect of the composition of the incubation medium on the trapping of phosphate ions in acid phosphatase cytochemistry, *Histochem. J.*, 11, 145, 1979.
21. **Araki, N. and Ogawa, K.,,** Regulation of intracellular lysosomal movements by the cytoskeletal system in rat alveolar macrophages, *Acta Histochem. Cytochem.*, 20, 659, 1987.
22. **Swanson, J., Burke, E., and Silverstein, S. C.,** Tubular lysosomes accompany stimulated pinocytosis in macrophages, *J. Cell Biol.*, 104, 1217, 1987.
23. **Krstić, R.,** Tubular lysosomes in rat and gerbil pinealocytes, *Histochemistry,* 88, 203, 1988.

VI. GLUCOSE-6-PHOSPHATASE

Shinsuke Kanamura

A. INTRODUCTION

Glucose-6-phosphatase is a multifunctional enzyme distributed in a variety of cell types of various organs. The enzyme has (1) phosphohydrolase activity (i.e., glucose-6-phosphate + $H_2O \rightarrow$ glucose + Pi), and (2) phosphotransferase activity (i.e., glucose-6-phosphate + glucose $^{14}C \rightleftharpoons$ glucose ^{14}C-6-phosphate + glucose). Although the enzyme shows a multiplicity of functions, it is suggested that, *in vivo,* this enzyme is specific for glucose-6-phosphate, and hydrolysis of glucose-6-phosphate is its sole function. The enzyme is bound tightly to the endoplasmic reticulum membrane, constituting a part of the membrane. Therefore, a complete purification of the enzyme has not been carried out.

Cytochemical demonstration of the activity of the enzyme is based on the capture of inorganic phosphate, released by the phosphohydrolase activity, by lead or cerium ions.

$$\text{glucose-6-phosphate} \rightarrow \text{glucose} + \text{phosphate}$$

$$\text{phosphate} + Pb^{2+} \text{ or } Ce^{3+} \rightarrow \text{lead phosphate or cerium phosphate} \downarrow$$

This enzyme cannot be demonstrated immunocytochemically. Chiquoine first demonstrated glucose-6-phosphatase activity by use of lead as the capture agent,[1] and then Wachstein and Meisel improved Chiquoine's method.[2] Tice and Barrnett demonstrated the activity ultrastructurally,[3] using the Wachstein and Meisel medium to which 0.4 *M* sucrose was added. Further, Robinson and Karnovsky,[4] Hulstaert et al.,[5] and Jonges et al.[6] used cerium ion as the capture agent. Hildebrand and Schleicher[7] and Jonges et al.[6] quantified the histochemical reaction for the enzyme by computerized densitometry using image analysis, and by microphotometry using the cerium-diaminobenzidine method, respectively. Teutch developed a microchemical method for the determination of glucose-6-phosphatase activity.[8]

B. METHODS

Because the enzyme is sensitive to glutaraldehyde,[9] prefixation for a short time is carried out. Generally, fresh frozen sections are used for light microscopic observation. The washing of tissue blocks after prefixation should also be brief.[16] Enzyme activity decreases to about 30% by immersion in a buffer at 4°C overnight or for 24 h.

The Wachstein-Meisel medium, and its modifications,[2,4,5,10-15] are used for demonstration of the activity. The incubation media are divided into several categories, containing lead or cerium as the capture agent, and media containing Tris-maleate or cacodylate as the buffer. The cerium-based media appear more suitable for the ultrastructural demonstration of the activity than the lead-based media, because of the uniformity and fine granularity of the reaction product. Tris-maleate is known to have some inhibitory effect on the enzyme and sometimes to cause poor preservation of morphology. However, when a medium containing low concentrations of Tris-maleate is used to avoid the inhibitory effect, lead tends to precipitate. Media containing cacodylate appear more reasonable. One may use any of these incubation media for cell types that contain high enzyme activity (see Results). However, results may vary in cell types containing low enzyme activity depending on the medium used.

For light microscopic observation, lead phosphate or cerium phosphate precipitated on the activity site is converted to black lead sulfide by dilute yellow ammonium sulfide[2] or visualized by the cerium-diaminobenzidine method.[6] Under the electron microscope, lead phosphate and cerium phosphate are visible as electron-dense reaction products.

C. EQUIPMENT AND REAGENTS

Instruments and equipment used included (1) equipment for perfusion fixation, (2) a Microslicer, Vibratome, or freezing microtome, and (3) a cryostat. Chemicals and reagents included (1) glutaraldehyde and osmium tetroxide, (2) glucose-6-phosphate (sodium or potassium salt), (3) cacodylate buffer or Tris-maleate buffer, and (4) lead nitrate or cerium chloride.

D. PROCEDURES

1. Electron Microscopy

1. Prefixation
   a. *Vascular perfusion:* For the liver, the perfusion is done via the portal vein. Physiologic saline (mouse, 3 to 4 ml/min; rat, 10 ml/min) is perfused for 1 min. Then, 2% glutaraldehyde containing 0.1 *M* sodium cacodylate, pH 7.2, is perfused at room temperature for 2 min at the same flow rate. When the perfusion fixation is done successfully, the liver becomes hard and pale. Keep the fixed liver at 0 to 4°C to prevent the inactivation of the enzyme.
   b. *Transparenchymal perfusion:* This method can be used only for the liver, and is useful for livers of very small animals, newborn animals, or embryos in which the vascular perfusion is difficult to do. The same equipment used for the vascular perfusion can be employed. A syringe with a needle is also useful. A hypodermic needle (e.g., 0.4 to 0.6 mm in diameter for newborn mouse and rat) is inserted into the parenchyma of the liver at the edge of the left or quadrate lobe. The jejunum with a part of the peritoneum is cut with scissors to open a branch of the superior mesenteric vein, and the perfusion is started immediately for 1 min at a perfusion pressure of 50 to 70 mmHg for a newborn mouse with physiologic saline, and then with 2% glutaraldehyde containing 0.1 *M* sodium cacodylate, pH 7.2, at room temperature for 1.5 to 3 min. The liver tissue adjacent to the perfusion needle cannot be used for the cytochemical analysis.
   c. *Immersion fixation:* This method is useful for cell types located on the surface of an organ, or for human materials. For the demonstration of the enzyme activity in intestinal epithelial cells, the intestinal mucosa is immersed in 3% glutaraldehyde containing 0.1 *M* sodium cacodylate, pH 7.2, at 4°C for 30 min.[11] Human materials sectioned at about 1 mm with a razor blade are immersed in 2% glutaraldehyde containing 0.1 *M* cacodylate, pH 7.2, at 4°C for 30 to 45 min, and then the tissue is cut into thinner sections during the fixation.
2. Wash the perfusion-fixed tissues for 5 to 15 min in 0.1 *M* cacodylate buffer, pH 7.2, containing 8% sucrose at 4°C. However, tissues fixed by immersion should be immersed into the buffer at 4°C for about 1 h.
3. Cut the tissues into 30-μm-thick sections by Microslicer, Vibratome (at 0 to 4°C), or freezing microtome.
4. Immediately incubate the sections at room temperature in the reaction media (Table 1) for 15 min in cell types having high activity (see Results), and for 1 h with a change of the medium in cell types containing moderate or low activity.

TABLE 1
Incubation Media

Lead-Based Media

Tris-Maleate as the Buffer

	Glucose-6-phosphate (m*M*)	Tris-maleate (m*M*)	$Pb(NO_3)_2$ (m*M*)	pH	Sucrose
Wachstein-Meisel[2]	1.68	80	3.63	6.7	
Lazarus and Barden[10]	6.4	20	2	6.7	
Hugon et al.[11]	3.6	60	3	6.5	7.5%
Teutsch[12]	10.0	40	3.6	6.5	300 m*M*

Cacodylate as the Buffer

	Glucose-6-phosphate (m*M*)	Cacodylate (m*M*)	$Pb(NO_3)_2$ (m*M*)	pH	Sucrose (m*M*)
Rosen[13]	4	50	3	6.5	350
Leskes et al.[14]	1	50	2	6.6	
Watanabe et al.[15]	3.7	80	3.6	6.7	230

Cerium-Based Media

	Glucose-6-phosphate (m*M*)	Buffer	$CeCl_3$ (m*M*)	pH	Sucrose
Robinson-Karnovsky[4]	1	0.1 *M* Cacodylate	2	6.6	5%
Hulstaert et al.[5]	2	80 m*M* Tris-maleate	1	6.5	
Asaka et al.[25]	3.7	80 m*M* Cacodylate	3.6	6.7	230 m*M*

Note: The medium of Watanabe et al.[15] or Asaka et al.[25] is made as follows: 18.5 ml of 10 m*M* glucose-6-phosphate is mixed with 20 ml of 0.2 *M* sodium cacodylate, pH 6.7, and 4.0 g of sucrose; then, 3.0 ml of 60 m*M* lead nitrate or cerium chloride is added little by little under continous stirring; finally, 8.5 ml of distilled water is added and the mixture is used immediately. To inhibit alkaline phosphatase activity, if necessary, (A) sections are preincubated in 0.1 *M* sodium cacodylate, pH 7.4, containing 50 m*M* EDTA and 0.23 *M* sucrose at room temperature for 30 min, or (B) sections are preincubated in 0.1 *M* sodium cacodylate, pH 7.4, containing 1 to 10 m*M* levamisole and 0.23 *M* sucrose at room temperature for 15 min, and then incubated in the reaction medium containing an equimolar amount of the inhibitor.

5. Wash the sections in 0.1 *M* cacodylate buffer, pH 7.2, containing 8% sucrose. The sections can be used for light microscopic observations after treatment as described in steps 5 and 6 under "Light Microscopy".
6. Postfix the sections in buffered 1 or 2% osmium tetroxide solution at 4°C for a few hours, dehydrate, and embed in resin.
7. Control experiments are carried out as described in Table 2.

2. Light Microscopy

1. Cut 12 (10 to 15)-μm-thick fresh frozen sections in a cryostat.
2. Incubate the sections in the reaction media (see Electron Microscopy, step 4).
3. Wash in physiologic saline.

TABLE 2
Control Experiments

	Reaction product in the endoplasmic reticulum and nuclear envelope or staining reaction in the section
Incubation in the reaction medium without the substrate	Negative
Incubation in the reaction medium without the substrate and containing equal moles of phosphates shown below	
Glucose-1-phosphate	Negative
Fructose-1,6-diphosphate	Negative
β-Glycerophosphate	Negative
Ribose-5-phosphate	Weakly positive
Fructose-6-phosphate	Weakly positive
Incubation in the reaction medium after preincubation in 0.1 *M* acetate buffer, pH 5.0, at 37°C for 15 min (to inhibit G6Pase activity)	Negative
Incubation in the reaction medium containing ions shown below after preincubation in 0.25 *M* sucrose containing equal moles of these ions (to inhibit G6Pase activity)	
1 m*M* Cu^{2+}	Negative
10 m*M* Zn^{2+}	Negative
10 m*M* CN^-	Negative
10 m*M* F^-	Weakly positive or negative

Note: Reaction product on the plasma membrane or in lysosomes, which may be seen in these control experiments, is probably related to nonspecific phosphatase or acid phosphatase activity.

4. Immerse the sections in 10% neutral formalin at room temperature for 5 min, and wash in distilled water.
5. Immerse the sections incubated in media containing lead into diluted (yellow) ammonium sulfide until no further change in color occurs. When media containing cerium are used, the sections are incubated at room temperature for 10 min in 50 m*M* Tris-maleate buffer, pH 8.0, including 10 m*M* 3,3′-diaminobenzidine (DAB), 10 m*M* sodium azide, and 4% H_2O_2 (add H_2O_2 solution to the buffer containing DAB and azide, and use the solution immediately).[6]
6. Wash in distilled water and mount in glycerol jelly.
7. Control experiments are carried out as given in Table 2.

E. RESULTS

No specific inhibitor has been found for the enzyme. Therefore, control experiments shown in Table 2 should be carried out. Although all of the control experiments should be done for safety, experiments using glucose-1-phosphate, fructose-1,6-diphosphate, ribose-5-phosphate, and fructose-6-phosphate may be omitted.

Generally, nonspecific deposits of lead are rare in sections incubated in the reaction media for 15 min. However, when sections are incubated for 1 h, nonspecific deposits of lead are sometimes observed. These are seldom seen if cerium is used. Deposits of reaction product caused probably by the activities of nonspecific phosphatase or acid phosphatase are also sometimes encountered on the plasma membrane or in lysosomes.

Glucose-6-phosphatase activity has been shown to be present in a variety of cell types of various organs. The enzyme activity is exclusively localized in the endoplasmic reticulum and nuclear envelope in all cell types. The activity is high in the hepatocyte,[17] proximal convoluted tubule cell,[18] and jejunal epithelial cell,[19] and moderate to low in cell types of other organs.

In the liver of the mouse and rat, the activity is higher in the periportal zone than in the perivenular zone, while the activity appears to be homogeneous throughout the lobule in the human and monkey livers. In the newborn mouse, the activity is evenly distributed throughout the lobule, and the adult type of sublobular distribution of the activity arises during postnatal development.[20] Teutch noticed a ''septal gradient'' (gradient in the lobular periphery) in the activity, in addition to the well-known ''periportal-perivenular gradient'', in female rat.[21] However, the activity appears to be homogeneous throughout the lobule in the male rat, although the two gradients become manifest by starvation. Teutch further confirmed microchemically the occurrence of the two gradients in male and female rats.[8] Thus, there are species and sex differences in the pattern of sublobular distribution of glucose-6-phosphatase.

Enzyme activity is moderate in the nerve cell of large type,[22] pancreatic B cell,[10] principal cell in the corpus and cauda of ductus epididymis,[23] epithelial cell of the seminal vesicle,[24] and nonpigmented epithelial cell in tips of ciliary processes in the pars plicata of ciliary body.[25] The activity is moderate to low in various cell types: the nerve cell of small type, neuroglial cell, acinar cell, myoepithelial cell, intercalated duct cell and striated duct cell of the salivary gland; follicular epithelial cell of the thyroid; tracheal epithelial cell; pancreatic A and D cells; secretory epithelial cell of the lactating mammary gland; epithelial cell of the adrenal cortex; endothelial cell; fibroblast; chondrocyte; osteoblast and osteocyte; smooth, striated, and cardiac muscle cells; cells in the reticuloendothelial system; heterophilic leukocyte; monocyte; lymphocyte and blood platelet; cells in the erythropoiesis and granulopoiesis; cells in the seminiferous epithelium; and uterine epithelial cell.

The physiological role of the enzyme is to release glucose into the blood in the liver. The enzyme is thought to be related to the reabsorption of glucose in the proximal convoluted tubule cell, and to the absorption of nutrients in the jejunal epithelial cell. However, the function of moderate to low activity in various cell types of other organs is not fully known. Relatively high activity in (1) the principal cell of the ductus epididymis, (2) epithelial cell of the seminal vesicle, (3) skeletal muscle cell, or (4) nonpigmented epithelial cell in the tip of ciliary processes in the pars plicata of ciliary body, is shown to be related to (1) release of glucose into the epididymal fluid,[23] (2) release of fructose into the seminal fluid,[24] (3) release of glucose into the blood in starved mice, or (4) release of glucose into the aqueous humor.[25]

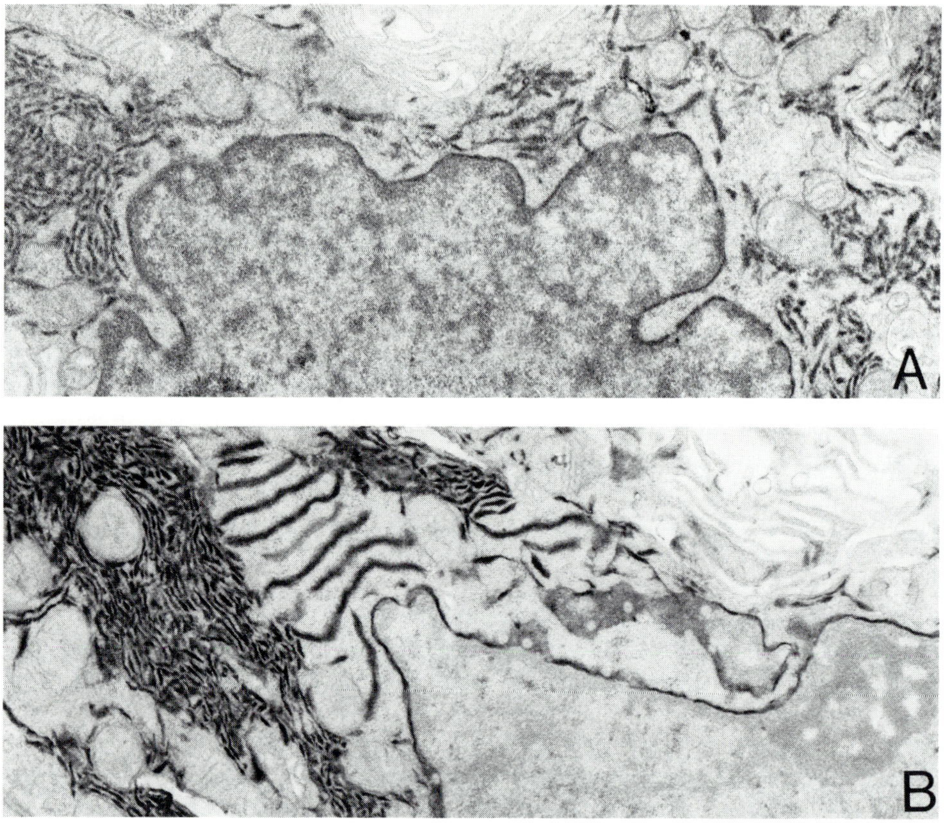

FIGURE 1. Demonstration of glucose-6-phosphatase activity in nonpigmented epithelial cells in tips of ciliary processes in the region containing large pockets within the pars plicata of rabbit ciliary epithelium. A section (30 μm) cut from tissue fixed by immersion in 2% buffered glutaraldehyde was incubated in the lead-based medium of Watanabe et al.[15] (A) or cerium-based medium of Asaka et al.[25] (B) at room temperature for 1 h with a change of the medium. The reaction product is abundant in the endoplasmic reticulum and nuclear envelope. Note that the reaction product is more uniform and more finely granular in B than in A. (Magnification × 20,000.)

REFERENCES

1. **Chiquoine, A. D.,** The distribution of glucose-6-phosphatase in the liver and kidney of the mouse, *J. Histochem. Cytochem.,* 1, 429, 1953.
2. **Wachstein, M. and Meisel, E.,** On the histochemical demonstration of glucose 6-phosphatase, *J. Histochem. Cytochem.,* 4, 592, 1956.
3. **Tice, L. and Barrnett, R.,** The fine structural localization of glucose-6-phosphatase in rat liver, *J. Histochem. Cytochem.,* 10, 754, 1962.
4. **Robinson, J. M. and Karnovsky, M. J.,** Ultrastructural localization of several phosphatases with cerium, *J. Histochem. Cytochem.,* 31, 1197, 1983.
5. **Hulstaert, C. E., Kalicharan, D., and Hardonk, M. J.,** Cytochemical demonstration of phosphatases in the rat liver by a cerium-based method in combination with osmium tetroxide and potassium ferrocyanide postfixation, *Histochemistry,* 78, 71, 1983.
6. **Jonges, G. N., Van Noorden, C. J. F., and Gossrau, R.,** Quantitative histochemical analysis of glucose-6-phosphatase activity in rat liver using an optimized cerium-diaminobenzidine method, *J. Histochem. Cytochem.,* 38, 1413, 1990.

7. **Hildebrand, R. and Schleicher, A.,** Image analysis of the histochemical demonstration of glucose-6-phosphatase activity in rat liver, *Histochemistry,* 86, 181, 1986.
8. **Teutsch, H. F.,** A new sample isolation procedure for microchemical analysis of functional liver cell heterogeneity, *J. Histochem. Cytochem.,* 34, 263, 1986.
9. **Kanamura, S.,** Sensitivity of glucose-6-phosphatase activity to glutaraldehyde, *Histochemistry,* 41, 257, 1975.
10. **Lazarus, S. S. and Barden, H.,** Specificity and ultrastructural localization of pancreatic B cell glucose-6-phosphatase, *Diabetes,* 14, 146, 1965.
11. **Hugon, J. S., Borgers, M., and Maestracci, D.,** Glucose 6-phosphatase and thiamine pyrophosphatase activities in the jejunal epithelium of the mouse, *J. Histochem. Cytochem.,* 18, 361, 1970.
12. **Teutsch, H. F.,** Improved method for the histochemical demonstration of glucose-6-phosphatase activity, *Histochemistry,* 57, 107, 1978.
13. **Rosen, S. I.,** The localization of glucose-6-phosphate hydrolysing enzyme in hepatocytes, red blood cells and leukocytes in the liver of the newborn rat, *J. Anat.,* 105, 579, 1969.
14. **Leskes, A. P., Siekevitz, P., and Palade, G.,** Differentiation of endoplasmic reticulum in hepatocytes. I. Glucose 6-phosphatase distribution in situ, *J. Cell Biol.,* 49, 264, 1971.
15. **Watanabe, J., Kanamura, S., Kanai, K., Oka, M., and Asada-Kubota, M.,** Cytochemical glucose 6-phosphatase activity in the cells of mouse pancreas and submandibular gland, *Histochem. J.,* 15, 999, 1983.
16. **Kanamura, S.,** Optimal postfixation washing time for ultrastructural demonstration of glucose 6-phosphatase activity, *J. Histochem. Cytochem.,* 21, 1086, 1973.
17. **Kanamura, S.,** Demonstration of glucose 6-phosphatase activity in hepatocytes following transparenchymal perfusion fixation with glutaraldehyde, *J. Histochem. Cytochem.,* 19, 386, 1971.
18. **Kanamura, S.** Ultrastructural localization of glucose 6-phosphatase activity in proximal convoluted tubule cells of rat kidney, *Histochemie,* 28, 288, 1971.
19. **Hugon, J. S., Maestracci, D., and Ménard, D.,** Glucose 6-phosphatase activity in the intestinal epithelium of the mouse, *J. Histochem. Cytochem.,* 19, 515, 1971.
20. **Kanamura, S.,** Postnatal changes in the localization of glucose 6-phosphatase activity within the liver lobule of the mouse, *Anat. Rec.,* 181, 635, 1975.
21. **Teutsch, H. F.,** Sex-specific regionality of liver metabolism during starvation; with special reference to the heterogeneity of the lobular periphery, *Histochemistry,* 81, 87, 1984.
22. **Stephans, H. R. and Sandborn, E. B.,** Cytochemical localization of glucose-6-phosphatase activity in the central nervous system of the rat, *Brain Res.,* 113, 127, 1976.
23. **Kanai, K., Kanamura, S., Watanabe, J., Asada-Kubota, M., and Oka, M.,** Effect of castration and testosterone replacement on high glucose 6-phosphatase activity in principal cells of the mouse epididymis, *Anat. Rec.,* 207, 289, 1983.
24. **Kanai, K., Kanamura, S., and Watanabe, J.,** High and testosterone-dependent glucose 6-phosphatase activity in epithelium of mouse seminal vesicle, *J. Histochem. Cytochem.,* 34, 1207, 1986.
25. **Asaka, Y., Watanabe, J., Kanai, K., and Kanamura, S.,** High glucose 6-phosphatase activity in non-pigmented epithelial cells of rabbit ciliary body, *J. Histochem. Cytochem.,* 39, 1113, 1991.

VII. 5′-NUCLEOTIDASE

John M. Robinson

A. INTRODUCTION

The enzyme 5′-nucleotidase (5′-NT, E.C. 3.1.3.5) is present in numerous cell types and has often been used, in cell fractionation studies, as a marker enzyme for the plasma membrane. Another important feature of the glycoprotein 5′-NT is that its active site is on the extracellular side of the plasmalemma and it is thus considered an ectoenzyme.[1] In addition, the products of the enzyme are localized to the exterior of the cell. While 5′-NT is an ectoenzyme, the ratio of enzyme on the cell surface compared to that in an intracellular pool may vary among different cell types. A steady-state ratio of cell surface and intracellular 5′-NT may occur due to membrane internalization and membrane recycling.[2] It is also noteworthy that 5′-NT belongs to that growing list of proteins that are anchored to the cell surface by a glycosyl-phosphatidylinositol linkage.[3]

The function of the ecto-5′-NT is to dephosphorylate 5′-mononucleotides, thus releasing the nucleoside and inorganic phosphate. This reaction is important as a source of the nucleoside adenosine. The enzyme plays a major role in regulating cellular uptake by converting the nontransportable 5′-nucleotide into the transportable nucleoside.

The enzyme cytochemical detection of 5′-NT relies upon reactions for demonstrating inorganic phosphate, which is released by hydrolysis of the 5′-mononucleotide. The free phosphate can be demonstrated by means of metal precipitation reactions. These metal-phosphate precipitates can then be imaged directly in the electron microscope, provided that they are electron dense. Earlier work utilized lead as the capture agent for this reaction by methods which were variations of that of Wachstein and Meisel.[4] Such lead-based reactions are capable of demonstrating 5′-NT as well as a variety of other phosphatases. However, these reactions suffer from certain limitations: (1) lead may be inhibitory to the enzyme, and (2) nonspecific deposition of lead phosphate often occurs and can lead to problems in identification of real sites of enzyme activity.

Cerium-based reactions for 5′-NT, as well as a number of other phosphatases, have been developed.[5-8] The introduction of cerium as a capture agent for phosphatases has led to numerous improvements, not the least of which is the virtual absence of nonspecific deposits of reaction product.

B. METHODS

The method of choice for the enzyme cytochemical localization of sites of 5′-NT uses cerium as the capture agent. Several incubation media have been described for the topographical detection of 5′-NT.[7,9] These different media are very similar, varying only slightly with regard to concentration of components and whether Mg^{2+} (an activator of 5′-NT) is derived from $MgCl_2$, $MgSO_4$, or $Mg(NO_3)_2$. Typically, adenosine 5′-monophosphate has been used as the substrate although guanosine 5′-monophosphate has also been used successfully.[10]

A cytochemical incubation medium which has the following components can be recommended:

100 m*M* Tris-maleate buffer
2 m*M* $CeCl_3$
2 m*M* $MgCl_2$
1 m*M* Adenosine 5′-monophosphate
1 m*M* Levamisole
145 m*M* Sucrose
Final pH 7.2 to 7.4

This medium should be clear; however, slight precipitation may occur if there is decomposition of the substrate during storage. Such precipitates can be removed by filtration. Passage of the medium once or twice through a small-pore (e.g., 0.2 μm) filter should suffice. Levamisole, a potent inhibitor of alkaline phosphatase,[11] should be included routinely in the reaction medium. Alkaline phosphatase can utilize a variety of phosphate-containing substrates and displays some activity at pH 7.2 to 7.4. Inhibition of alkaline phosphatase with levamisole thus provides for a high degree of specificity, since 5′-NT is not inhibited by this agent.

Proper controls for 5′-NT localization are essential. As previously indicated, nonspecific phosphatases such as ecto-alkaline phosphatase may contribute to the total reaction product observed. It is useful to compare the accumulation of reaction product with different substrates (e.g., β-glycerophosphate; *p*-nitrophenylphosphate) in the presence or absence of levamisole with that obtained with adenosine 5′-monophosphate as substrate in the presence or absence of levamisole (see Figure 1). Additional controls consist of (1) omission of substrate and (2) omission of Mg^{2+} and addition of the divalent cation chelator ethylenediaminetetraacetic acid; in the former case, one would expect no reaction in the absence of substrate (see Figure 2) and in the latter situation, reaction product would be expected to be absent or much reduced.

Adequate fixation is a prerequisite for the enzyme cytochemical demonstration of 5′-NT. That is, the fixation must be gentle enough to preserve enzyme activity yet sufficient for good morphological detail. A variety of fixation protocols has been used for the demonstration of 5′-NT in numerous cells and tissues. There does not appear to be a consensus for a "best fixative", rather the most useful fixation conditions should be determined for the experimental system under consideration. It should be noted, however, that fixation at reduced pH (1% glutaraldehyde, 0.1 *M* Tris-maleate buffer [pH 6.0]) has been shown to be important in preservation of 5′-NT activity in leukocytes.[12] This fixation procedure may be useful for other cells as well.

Inhibitors are also useful for establishing the specificity of the 5′-NT reaction and its location. That the majority of 5′-NT is an ectoenzyme on most cells can be demonstrated by the use of the diazonium salt of sulfanilic acid. This nonspecific cross-linking agent will interact with exposed proteins at the cell surface (and inhibit enzyme function) but will not cross the lipid bilayer.[13] Another nonpenetrating reagent which inhibits 5′-NT activity is the lectin concanavalin A. Furthermore this lectin inhibits the cerium-based cytochemical detection of 5′-NT in the enteric ganglia and in smooth muscle cells of the guinea pig.[14] Another useful reagent for the study of 5′-NT activity is the synthetic adenine nucleotide analog, α,β-methyleneadenosinediphosphate, which is a potent inhibitor of adenosine 5′-monophosphate catabolism.[15]

Penetration of cerium and the phosphorylated substrate into cells is a problem. While this is not a hindrance in localizing 5′-NT which is displayed on the cell surface, it is a concern in detecting an intracellular pool which may exist in some cells. Methods have been devised to circumvent this situation in cerium-based cytochemical reactions. One approach has been to use very mild detergent treatment on fixed cells for permeabilization;[16] alternatively, fixed material is frozen and thawed prior to the cytochemical incubation.[17]

The cerium-based reaction for 5′-NT should be carried out routinely at physiological temperature with constant agitation. The optimal incubation time may vary among different cell types and should be determined experimentally. It is usually advisable to remove the incubation medium and replace it with fresh medium at regular intervals (e.g., every 30 min). Following the reaction, the material should be washed with the cytochemical buffer. Some investigators routinely employ a second aldehyde fixation step at this point (e.g., 2 to 3% glutaraldehyde in 0.1 *M* cacodylate buffer [pH 7.2]).[8] This is followed by washes

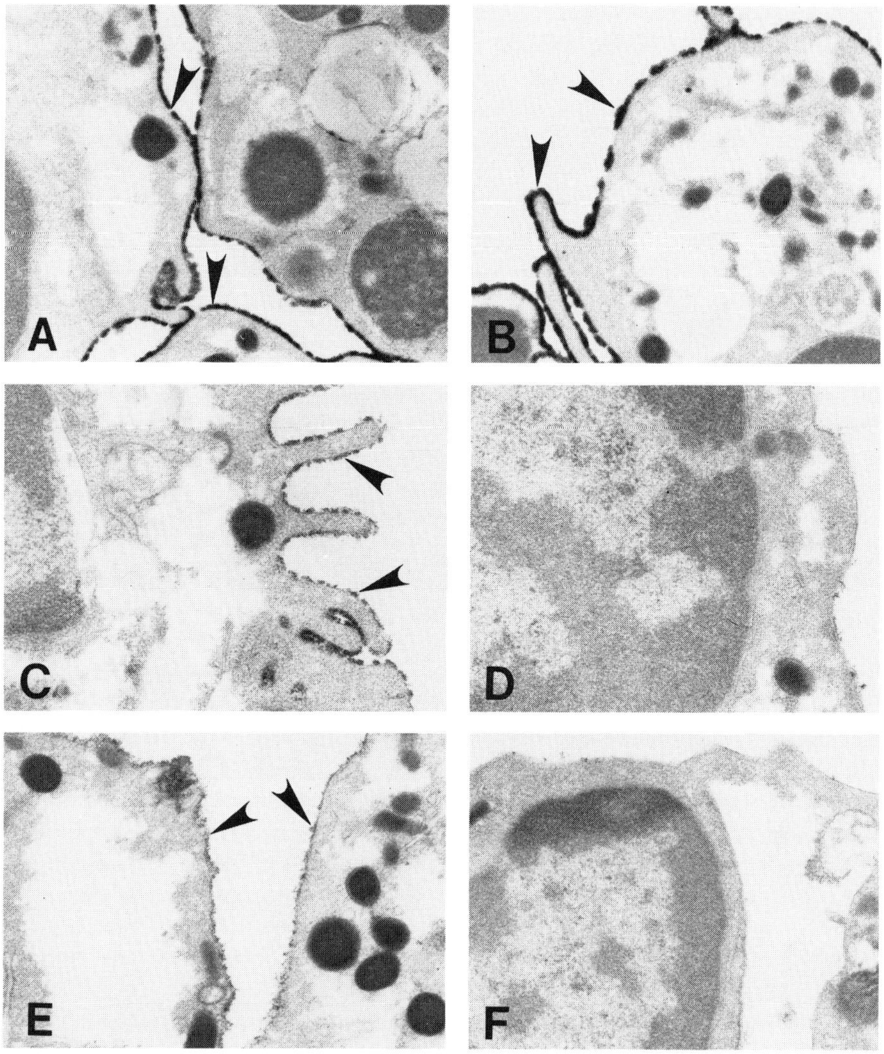

FIGURE 1. Electron micrographs of portions of guinea pig neutrophils incubated in the basic 5′-NT medium in the absence of levamisole (*Panels A,C,E*) or in the presence of 1 m*M* levamisole (*Panels B,D,F*). The substrates [illegible] phosphate (*Panels E,F*). Reaction product is present on the cell surface with each substrate tested in the absence of levamisole (*arrowheads, A,C,E*); however, in the presence of levamisole, reaction product only occurs with adenosine 5′-monophosphate as substrate (*arrowheads, B*). These results illustrate that levamisole inhibits hydrolysis of β-glycerophosphate and *p*-nitrophenylphosphate but does not inhibit the hydrolysis of adenosine 5′-monophosphate, showing that this cytochemical reaction is specific for 5′-NT.

and postfixation in osmium. Ferrocyanide-reduced osmium can be used following cerium reactions; this leads to enhanced membrane contrast in electron microscopy.[9] The material can then be processed for embedding by routine procedures and embedded in any of the commonly used plastic resins.

It is often very useful to visualize the cytochemical reaction at the light microscope level. It is the preferred way to screen samples rapidly to determine if the reaction has been successful; light microscopy can greatly facilitate location of appropriate areas of a block for further trimming. Cerium phosphate reaction products, while electron dense, are not readily visible by bright-field light microscopy. This drawback has been overcome by the

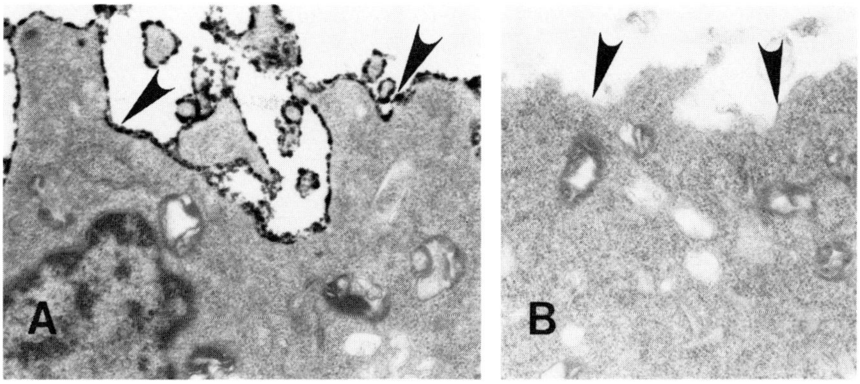

FIGURE 2. Electron micrographs of P815 cells which had been incubated in a cerium-based medium with adenosine 5′-monophosphate as substrate (*Panel A*) or in the absence of substrate (*Panel B*). Reaction product is evident at the surface of cells incubated with substrate (*arrowheads A*) but is lacking in the absence of substrate (*arrowheads B*).

development of methods to convert cerium phosphate into a colored reaction product.[18-20] It is beyond the scope of this chapter to compare these various procedures; however, it should be noted that visualization of 5′-NT at the light microscope level with cerium as the primary capture agent has been achieved.[21,22] In this case, the cerium is converted to lead phosphate by incubation with alkaline lead citrate; the lead phosphate is then converted to lead sulfide which is visible by bright field microscopy. This conversion in 1-μm-thick resin sections is facilitated by conducting the reaction in a microwave oven.[22]

C. EQUIPMENT AND REAGENTS

The cytochemical demonstration of 5′-NT can be achieved in both tissues and isolated cells. There are no special instruments or equipment required beyond those found in a well-equipped microscopy facility. A brief list of some of this equipment includes a tissue chopper or Vibratome for obtaining relatively thin tissue sections to aid in the penetration of the cytochemical medium. A centrifuge for working with isolated cells is essential in this situation. Embedding ovens, microtomes, optical and transmission electron microscopes (TEM) of the standard variety are sufficient for the localization of 5′-NT.

All of the chemicals and reagents employed in the cytochemical localization of 5′-NT are readily available from chemical and biochemical suppliers.

D. PROCEDURES

The procedures for localization of 5′-NT are given in more detail above in the section on Methods. Listed below is an abbreviated outline for carrying out this reaction.

Enzyme cytochemical detection of 5′-NT with cerium:

1. Initial aldehyde fixation of cells or tissue
   - Fixation conditions optimal for the sample
   - Wash in buffer to stop fixation
2. Preparation of fresh cytochemical incubation medium
   - Incubation in cytochemical medium
   - Wash in cytochemical buffer to stop reaction
   - Wash in second fixation buffer
3. Second aldehyde fixation of cells or tissue
   - Fixation for a convenient time (e.g., 1 h)
   - Wash in fixation buffer

4. Postfixation in osmium or ferrocyanide-reduced osmium
   - Fixation for about 1 h
   - Wash in fixation buffer
5. Dehydration of samples
6. Embedding of samples in plastic resin
   - Infiltration of resin
   - Polymerization of resin
7. Trimming of blocks and cutting sections
8. Microscopic examination of cerium phosphate
   - Light microscopic examination — visualization following conversion of cerium phosphate to colored product
   - Electron microscopic examination — visualization by direct examination by TEM
9. Photomicroscopy and data analysis

E. RESULTS

Localization of 5′-NT with cerium should lead to the formation of a fine amorphous reaction product (i.e., cerium phosphate) at the site of enzyme activity. Nonspecific deposition of electron-dense material should not be evident. There is, however, the possibility that metal precipitation reactions used for the localization of sites of enzyme activity have potential pitfalls. The problems of nonspecific deposits with lead as a capture agent are well known. This difficulty has been largely eliminated with the introduction of cerium as a capture agent. One notable problem is redistribution of reaction product. This is probably not a major concern with enzymes within membrane-bound organelles (e.g., lysosomes) where enzyme-generated reaction product should be confined by the surrounding membrane. Cell surface enzymes (e.g., 5′-NT) may, however, present a problem. In this case, barriers to diffusion of reaction product are less obvious. Blok et al.,[7] using a cerium-based reaction, report that latex beads (used in phagocytosis experiments) have deposits of reaction product following localization of 5′-NT in murine macrophages. Cramer et al.,[23] using a lead-based reaction for the localization of 5′-NT in rat macrophages, show reaction product on heat-killed bacteria which were bound to the cell surface during phagocytosis experiments. With guinea pig neutrophils which are actively attached to glass coverslips prior to fixation the cytochemical detection of 5′-NT, reaction product can be detected on the substratum near the cell surface (see Figure 3). On the other hand, Hulstaert et al.,[9] using a cerium-based method, show precise localization of 5′-NT to the bile canaliculus in rat liver without appreciable reaction product along the intervening plasma membrane regions. In any case, one should be aware of the possibility of the redistribution of reaction product when detecting 5′-NT by metal precipitation reactions.

Where possible, it is advisable to make side-by-side comparisons of the enzyme cytochemical distribution of 5′-NT activity with immunocytochemical detection of the protein itself. This is suggested since each technique provides different information (i.e., estimates of activity of the enzyme vs. distribution of the protein) and secondly the immunocytochemistry provides an additional control for potential artifacts which may be associated with enzyme cytochemical methods. Such a parallel study of the distribution of the ecto-5′-NT in rat kidney shows that the localization of enzyme activity is congruent with that of the antigen at the light microscope level.[24] Likewise, comparative enzyme cytochemical and immunocytochemical localization of 5′-NT has been carried out in rat salivary glands at the electron microscope (EM) level.[25] In this case, the distribution of 5′-NT demonstrated by cerium ions is essentially the same as that revealed by the antibody labeling technique. In other studies EM immunocytochemistry has shown by immuno-ferritin[26] and immuno-gold[27] methods that the predominant site for 5′-NT in liver is the bile canaliculus, the same site as previously demonstrated by cerium-based enzyme cytochemistry.[9] Immunocytochemical

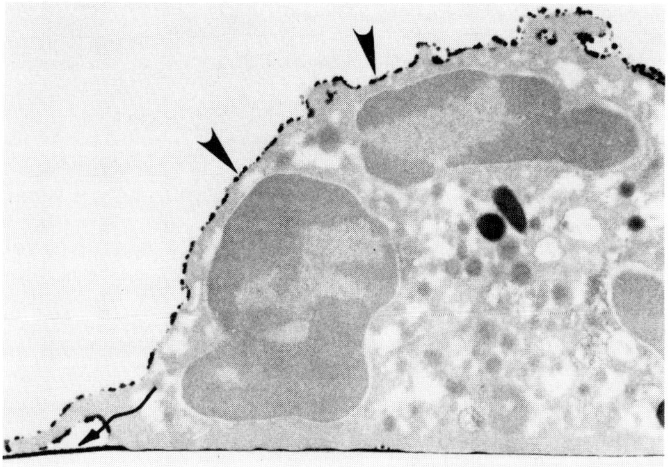

FIGURE 3. Electron micrograph of a guinea pig neutrophil which had attached to a glass coverslip under cell culture conditions. Following fixation, the sample was incubated for the localization of 5′-NT with cerium as the capture agent. The cell was subsequently sectioned perpendicular to the glass substratum. Note that reaction product is present on the cell surface (*arrowheads*). In addition there are deposits of electron-dense material near the cell-substratum interface, but slightly away from the cell membranes (*curved arrow*), suggesting that some redistribution of reaction product may occur under these conditions.

results should also be viewed with some caution, especially with regard to quantitation. For example, the immunolabeling of 5′-NT with colloidal gold probes in ultrathin cryosections of hepatoma cells has been shown to have a relatively low efficiency.[28] Nonetheless, cytochemical and immunocytochemical studies of the distribution of 5′-NT in a variety of cell types has provided valuable insight into the distribution and role of this enzyme.

REFERENCES

1. **DePierre, J. W. and Karnovsky, M. L.,** Ecto-enzymes of the guinea pig polymorphonuclear leukocytes. I. Evidence for an ecto-adenosine monophatase, -adenosine triphosphatase, and -p-nitrophenylphosphatase *J. Biol. Chem.*, 249, 7111, 1974.
2. **Stanley, K. K., Edwards, M. R., and Luzio, J. P.,** Subcellular distribution and movement of 5′-nucleotidase in rat cells, *Biochem. J.*, 186, 59, 1980.
3. **Low, M. G. and Saltiel, A. R.,** Structural and functional roles of glycosyl-phosphatidylinositol, *Science,* 239, 268, 1988.
4. **Wachstein, M. and Meisel, E.,** Histochemistry of hepatic phosphatases at physiological pH, *Am. J. Clin. Pathol.*, 27, 13, 1957.
5. **Veenhuis, M., Van Dijken, J. P., and Harder, W.,** A new method for the cytochemical demonstration of phosphatase activities in yeasts based on the use of cerous ions, *FEMS Microbiol. Lett.*, 9, 285, 1980.
6. **Robinson, J. M. and Karnovsky, M. J.,** Cytochemical localization of the ectoenzyme 5′-nucleotidase by a new cytochemical method, *J. Histochem. Cytochem.*, 30, 575, 1982.
7. **Blok, J., Onderwater, J. J. M., De Water, R., and Ginsel, L. A.,** A cytochemical method for the demonstration of 5′-nucleotidase in mouse peritoneal macrophages, with cerium ions used as trapping agent, *Histochemistry,* 75, 437, 1982.
8. **Robinson, J. M. and Karnovsky, M. J.,** Ultrastructural localization of 5′-nucleotidase in guinea pig neutrophils based upon the use of cerium as capturing agent, *J. Histochem. Cytochem.*, 31, 1190, 1983.
9. **Hulstaert, C. E., Kalicharan, D., and Hardonk, M. J.,** Cytochemical demonstration of phosphatases in the rat liver by a cerium-based method in combination with osmium tetroxide and potassium ferrocyanide postfixation, *Histochemistry,* 78, 71, 1983.

10. **Falcon, J., Besse, C., Guerlotté, J., and Collin, J.-P.,** 5′-Nucleotidase activity in the pineal organ of the pike. An electron-microscopic study, *Cell Tissue Res.*, 251, 495, 1988.
11. **Borgers, M.,** The cytochemical application of new potent inhibitors of alkaline phosphatase, *J. Histochem. Cytochem.*, 21, 812, 1973.
12. **Uusitalo, R. J. and Karnovsky, M. J.,** Surface localization of 5′-nucleotidase on the mouse lymphocyte, *J. Histochem. Cytochem.*, 25, 87, 1977.
13. **Karnovsly, M. L.,** Ectoenzymes: their modulation and similarity to certain enzymes of intracellular membranes, in *Cellular Biology of Ectoenzymes*, Kreutzberg, G. W., Ed., Springer-Verlag, Berlin, 1986, 3.
14. **Anderson Forman, C. and Gustafsson, L. E.,** Cytochemical localization of 5′-nucleotidase in the enteric ganglia and in smooth muscle cells of the guinea pig *J. Neurocytol.*, 14, 551, 1985.
15. **Pearson, J. D., Carleton, J. S., and Gordon, J. L.,** Metabolism of adenine nucleotides by ectoenzymes of vascular endothelial and smooth-muscle cells in culture, *Biochem. J.*, 190, 421, 1980.
16. **Robinson, J. M.,** Improved localization of intracellular sites of phosphatases using cerium and cell permeabilization, *J. Histochem.*, 33, 749, 1985.
17. **Kalicharan, D., Hulstaert, C. E., and Hardonk, M. J.,** Prevention of penetration hindrance in cerium-based glucose-6-phosphatase cytochemistry by freezing tissue in melting nitrogen, *Histochemistry*, 82, 287, 1985.
18. **Zimmermann, N. and Halbhuber, K.-J.,** Light microscopical localization of enzymes by cerium-based methods. I. Detection of acid phosphatase by a new cerium-lead-technique (Ce-Pb-method), *Acta Histochem.*, 76, 97, 1985.
19. **Okada, T., Kobayashi, T., Hakoi, K., and Seguchi, H.,** A simple method for light microscopical visualization of phosphohydrolase activities in materials processed for cerium-based ultracytochemistry, *Acta Histochem. Cytochem.*, 20, 105, 1987.
20. **Halbhuber, K.-J., Gossrau, R., Möller, U., Hulstaert, C. E., Zimmermann, N., and Feuerstein, H.,** The cerium perhydroxide-diaminobenzidine (Ce-H_2O_2-DAB) procedure. New methods for light microscopic phosphatase histochemistry and immunohistochemistry, *Histochemistry*, 90, 289, 1988.
21. **Werner, J. A. and Schünke, M.,** Cerium-induced light-microscopic demonstration of 5′-nucleotidase activity in the lymphatic capillaries of the proximal oesophagus of the rat, *Acta Histochem.*, 85, 15, 1989.
22. **Kalicharan, D. and Hulstaert, C. E.,** Light microscopical visualization of phosphatase activities demonstrated with the cerium method on resin sections of the kidney with the microwave oven, *Histochemistry*, 90, 485, 1989.
23. **Cramer, E. M., Werb, Z., and Bainton, D. F.,** Pitfalls in ecto-5′-nucleotidase enzyme cytochemistry as demonstrated by the immunogold-labeling technique on macrophages, *Histochem. J.*, 20, 108, 1988.
24. **Dawson, T. P., Gandi, R., Le Hir, M., and Kaissling, B.,** Ecto-5′-nucleotidase: localization in rat kidney by light microscopic histochemical and immunohistochemical methods, *J. Histochem. Cytochem.*, 37, 39, 1989.
25. **Yamashina, S., Katsumata, O., Wada, I., and Kato, K.,** Electron microscopic localization of 5′-nucleotidase in rat salivary glands. Comparative enzyme and immunohistochemical studies, *Histochemistry*, 84, 231, 1986.
26. **Matsuura, S., Eto, S., Kato, K., and Tashiro, Y.,** Ferritin immunoelectron microscopic localization of 5′-nucleotidase on rat liver cell surface, *J. Cell Biol.*, 99, 166, 1984.
27. **Geuze, H. J., Slot, J. W., Strous, G. J., Luzio, J. P., and Schwartz, A. L.,** A cycloheximide resistant pool of receptors for asialoglycoproteins and mannose-6-phosphate residues in the Golgi complex of hepatocytes, *EMBO J.*, 3, 2687, 1984.
28. **Howell, K. E., Reuter Carlson, U., Devaney, E., Luzio, J. P., and Fuller, S. D.,** One antigen, one gold? A quantitative analysis of immunogold labelling of plasma membrane 5′-nucleotidase in frozen thin sections, *Eur. J. Cell Biol.*, 44, 318, 1987.

VIII. PHOSPHODIESTERASE

Satoki Ueno

A. INTRODUCTION

It is well accepted that 3′,5′-cyclic adenosine monophosphate (cAMP) and guanosine monophosphate (cGMP) function as intracellular second messengers together with Ca ions to regulate a variety of important cellular functions after stimulation by exogenous factors such as hormones, neurotransmitters, or light. In some cells, one of the cyclic nucleotides is thought to be predominant, in other cells cAMP and GMP seem to work competitively to form some feedback mechanisms.

Levels of cAMP or cGMP in the cells are generally regulated and balanced by the activity of the synthesizing enzymes, adenylate cyclase or guanylate cyclase, and/or the degrading enzymes, cyclic nucleotide-phosphodiesterase (PDEase). These enzyme activities have been studied not only biochemically, but also histochemical demonstrations have been tried, to elucidate cellular regulatory mechanisms.

B. METHODS

1. Conventional Histochemical Approach

A histochemical attempt to demonstrate the activity of cyclic nucleotide-dependent PDEase was initially proposed by Shanta et al.[1] in 1966, based on the method for biochemical assay and Gomori's histochemical phosphate method. The fundamental principle of this procedure is summarized in Table 1. Cyclic monophosphate is hydrolyzed to 5′-mononucleotide by PDEase. In this process, no release of inorganic phosphate occurs. To capture an inorganic phosphate by a heavy-metal reagent for visualization of PDEase activity histo- and cyto-chemically, 5′-mononucleotide should be dephosphorylated by 5′-nucleotidase (5Nase) activity. For this purpose, the incubation medium contained snake venom as an exogenous 5Nase source.

The procedure of Shanta et al.[1] is based on light microscope utilizing unfixed cryostat sections. The sections are incubated in a medium which contains the following: cAMP, 1.44 m*M*; Tris-maleate buffer, 50 m*M* adjusted to pH 7.62; $MgCl_2$, 10 m*M* lead acetate, 2 m*M*; snake venom, 1 mg added to the above mixture. Following incubation, the sections are rinsed in distilled water, then placed in dilute yellow ammonium sulfide for 1 to 2 min for visualization.

Thereafter, several electron microscopical studies have been performed, and some of them proposed minor modifications to the procedure of Shanta et al. Of them, the procedure of Florendo et al.[2] is considered to be representative. The cytochemical procedure is as follows. The tissues are fixed in cold 2% glutaraldehyde in 0.05 *M* cacodylate-nitrate buffer containing 0.25 *M* dextrose. After fixation, tissues are cut into 1-mm cubes or strips, and washed in the same buffer for 3 to 5 h. The slices are first preincubated for 30 min at room temperature in the solution containing Tris-maleate buffer, 60 m*M*, adjusted to pH 7.4; $MgCl_2$, 2 m*M*; sucrose, 0.25 *M*; and snake venom, 5 mg/ml. The slices are immediately transferred to the incubation medium containing Tris-maleate buffer (60 m*M*, adjusted to pH 7.4), cAMP (either 0.75, 1.5, or 3.0 m*M*), lead nitrate (2 m*M*), and snake venom (3.0 mg/ml). The incubation is carried out at 37°C for 30 min. After incubation, the slices are rinsed three times in 0.05 *M* cacodylate buffer, and postfixed in 2% osmium tetroxide using the same buffer, dehydrated in ethanol, and embedded.

However, the method used possessed some fundamental disadvantages. First, since 5Nase is a relatively large molecule, uniform penetration into the tissues can be a problem.[3]

TABLE 2
Incubation Medium for 5Nase

Reagent		Conc (m*M*)
0.1 *M* Tris-maleate buffer, pH 7.4	8.0 ml	80
Distilled water	0.5 ml	
Sucrose	0.8 g	250
Levamisole	6.0 mg	2.5
60 m*M* $MnCl_2$	0.5 ml	3
5′-AMP or 5′-GMP (sodium salt)	10.4 or 12.2 mg	3
20 m*M* Lead nitrate	1.0 ml	2

Final volume: 10 ml, pH 7.4.

TABLE 3
Incubation Medium for PDEase

Reagent		Conc (m*M*)
0.1 *M* Tris-maleate buffer, pH 7.4	8.0 ml	80
Sucrose	0.8 g	250
Levamisole	6.0 mg	2.5
60 m*M* $MgSO_4$	0.5 ml	3
60 m*M* $MnCl_2$	0.5 ml	3
cAMP or cGMP	7.5 or 8.0 mg	3
20 m*M* Lead nitrate	1.0 ml	2

Final volume: 10 ml, pH 7.4.

ization of intrinsic 5Nase activity, the sections are incubated in the PDEase cytochemical medium at 37°C for 5 to 30 min.

After incubation, the sections are rinsed in distilled water, treated with 1% ammonium sulfide for 2 min for visualization of the reaction product, and mounted in glycerine jelly. For electron microscopy, the sections are rinsed briefly in 0.1 *M* cacodylate buffer, pH 7.4, containing 0.25 *M* sucrose and postfixed with 1% osmium acid solution for 30 min at 0 to 4°C. The sections are then dehydrated through a graded ethanol series and propylene oxide, and embedded in Spurr's epoxy resin.

The following controls should be carried out:

1. Incubate the sections in the standard medium containing 2 m*M* 3-isobutylmethylxanthine (IBMX) or 2 m*M* theophylline, which are potent inhibitors of cyclic nucleotide PDEase.[11,12]
2. Omit $MnCl_2$ and/or $MgSO_4$ from the incubation medium to observe the effect of activator ions.
3. Remove the substrate, cAMP or cGMP, from the standard medium.
4. Incubate the sections in the medium containing 1 m*M* $NiCl_2$, an inhibitor of 5Nase.[13] If the intrinsic 5Nase is inhibited, no phosphate is released and no reaction products are formed.

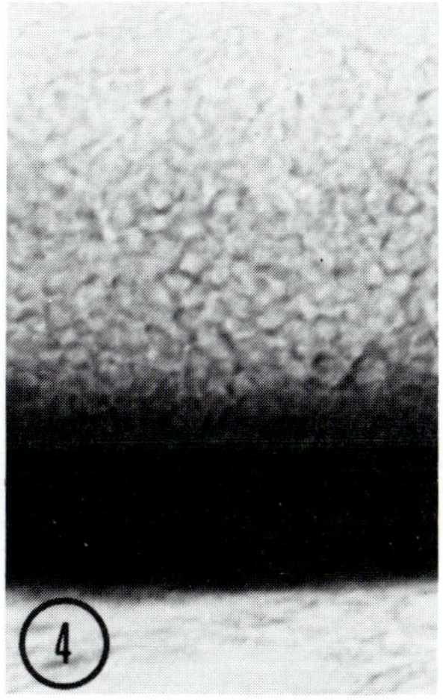

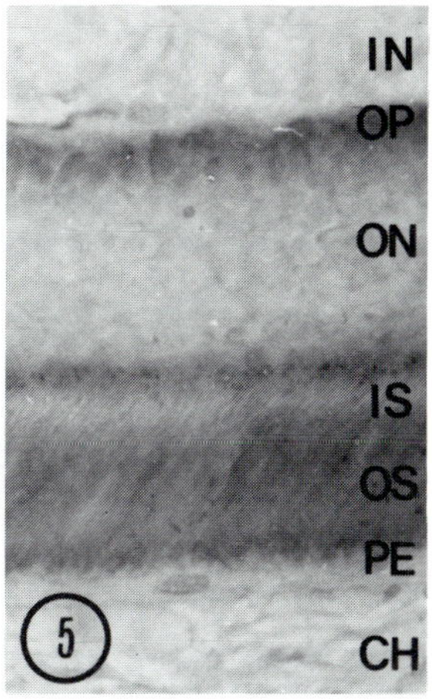

FIGURE 3. Light micrograph of intrinsic 5Nase activity, using 5′GMP as substrate, of rat retina. An intense staining is seen in all photoreceptor cells from the outer plexiform layer (OP) to the tip of the outer segments (OS) as well as in the pigment epithelium (PE). IN, Inner nuclear layer; ON, outer nuclear layer; IS, inner segments; CH, choroid. (Magnification × 500.)

FIGURE 4. Light micrograph of cGMP-PDEase activity. The intense activity of the enzyme is limited to the pigment epithelium (PE) and to the outer segments (OS). The inner segments (IS) are faintly stained. (Magnification × 500.)

FIGURE 5. When cAMP is used as substrate, the staining of the outer segments (OS) is much weaker than shown in Figure 4. In contrast, moderate reaction is seen in the outer synaptic plexiform layer (OP). Pigment epithelium (PE) also shows a moderate reaction. (Magnification × 500.)

E. RESULTS

In the rod photoreceptor and retinal pigment epithelial (RPE) cells, a high level of intrinsic 5Nase activity was observed. The intensity of the reaction was not altered by using 5′AMP or 5′GMP as substrate, and the reaction was uniformly distributed from the synaptic ending of photoreceptors in the outer plexiform layer to RPE cell membranes (Figure 3).

The method revealed differences in localization depending on the substrate, cAMP and cGMP. When cGMP was used as substrate, the activity was restricted to the inner and outer segments of photoreceptors and RPE. The activity was much more intense uniformly in the outer segments, whereas a weaker reaction was observed in the inner segments under conditions of illumination. In contrast to the localization of the 5Nase activity, no notable cGMP-dependent PDEase activity was detectable in the outer plexiform and nuclear layers (Figures 4 and 5). When cAMP was used as a substrate, the reaction in the outer segments was much weaker than when cGMP was used. It is noteworthy that moderate activity was noted in the outer plexiform layer (Figure 6). This reaction was not observed when cGMP was used or when snake venom was utilized. In the control study, both of these reactions were drastically inhibited by the addition of 2 m*M* IBMX to the standard incubation medium. When 10 m*M* NaF or 1 m*M* NiCl, inhibitors of 5′Nase, was added, the reaction was almost completely abolished.

The pattern of reaction under the electron microscope following standard incubation and in the control studies paralleled what was observed using a light microscope. Deposits of reaction product of cGMP-PDEase activity were localized on the disk membranes of the outer segments (Figure 7).

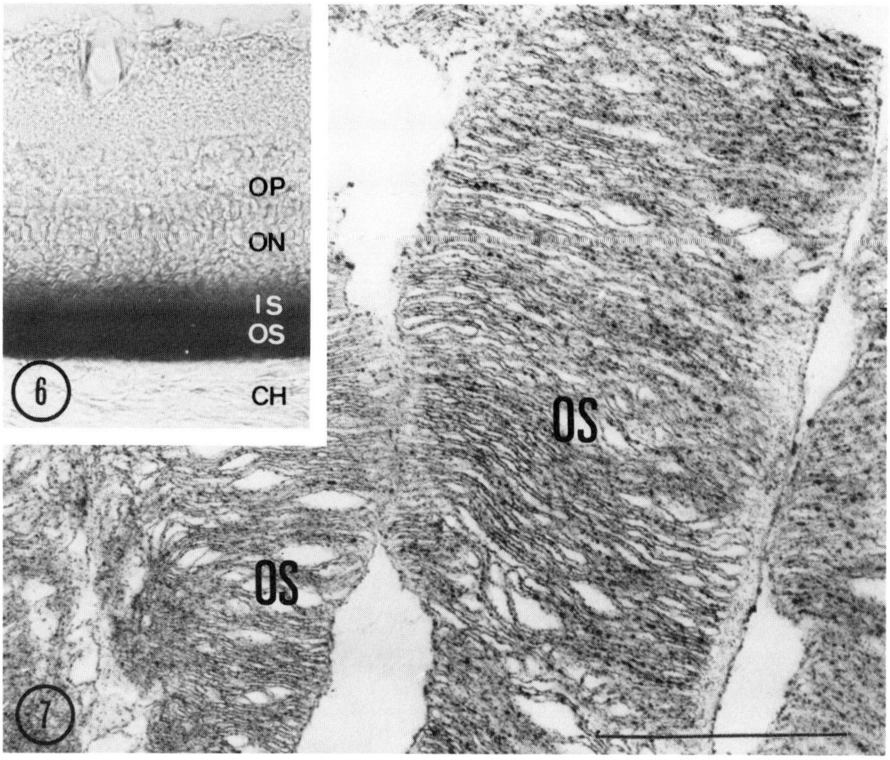

FIGURE 6. Light micrograph of cGMP-PDEase activity in the whole rat retina. The reaction is restricted to inner (IS) and outer segments of photoreceptors. (Magnification × 300.)
FIGURE 7. Electron micrograph of cGMP-PDEase activity in the outer segments. Intense deposits of the reaction product are located on the disk membranes. Conspicuous deposits of reaction product are observed on the plasma membranes of the outer segments (OS). (Magnification × 32,000.)

REFERENCES

1. **Shanta, T. R., Woods, W. D., Waitzman, M. B., and Bourne, G. H.,** Histochemical method for localization of cyclic 3′,5′-nucleotide phosphodiesterase, *Histochemie*, 7, 177, 1966.
2. **Florendo, N. T., Barnett, R. J., and Greengard, P.,** Cyclic 3′,5′-nucleotide phosphodiesterase: cytochemical localization in cerebral cortex, *Science*, 173, 745, 1971.
3. **Adinolfi, A. M. and Schmidt, S. Y.,** Cytochemical localization of cyclic nucleotide phosphodiesterase activity at developing synapses, *Brain Res.*, 76, 21, 1974.
4. **Roof, D. J. and Heuser, J. E.,** Surface of rod photoreceptor disk membranes: integral membrane components, *J. Cell Biol.*, 95, 487, 1982.
5. **Roof, D. J., Korenbrot, J. I., and Heuser, J. E.,** Surface of rod photoreceptor disk membranes: light-activated enzymes, *J. Cell Biol.*, 95, 501, 1982.
6. **Ueno, S., Bambauer, H. J., Umar, H., and Ueck, M.,** A new histo- and cytochemical method for demonstration of cyclic 3′,5′-nucleotide phosphodiesterase activity in retinal rod photoreceptor cells of the rat, *Histochemistry*, 81, 445, 1984.
7. **Ueno, S., Bambauer, H. J., Umar, H., and Ueck, M.,** Localization and function of cyclic guanosine monophosphate-phosphodiesterase activity in the retinal rods of the rat by means of a newly developed cytochemical method, *Cell Tis. Res.*, 238, 453, 1984.
8. **Ueno, S. and Ueck, M.,** Localization of cyclase and cyclic nucleotide phosphodiesterase activities in retinal pigment epithelial cells of the guinea pig, *Acta Histochem. Cytochem.*, 21, 45, 1988.

9. **Ueno, S. and Ueck, M.,** Cyclic nucleotide phosphodiesterase activity: histochemical and cytochemical methods, in *Methods in Enzymology: Initiation and Termination of Cyclic Nucleotide Action,* Corbin, J. D. and Johnson, R. A., Eds., Academic Press, San Diego, 1988, 477.
10. **Poeggel, G., Luppa, H., and Ludwig, A.,** Histochemistry of cyclic nucleotide phosphodiesterase in nervous tissue, immunohistochemical investigations, *Acta Histochem. Cytochem.,* 20, 381, 1987.
11. **Butcher, R. W. and Surtherland, E. W.,** Adenosine 3′,5′-phosphate in biological materials. I. Purification and properties of cyclic 3′,5′-nucleotide phosphodiesterase and use of the enzyme to characterize adenosine 3′,5′-phosphate in human urine, *J. Biol. Chem.,* 237, 1244, 1962.
12. **Peytreman, A., Nicholson, W. E., Liddle, G. W., and Sutherland, E. R.,** Effects of methylxanthines on adenosine 3′,5′-monophosphate and corticosterone in the rat adrenal, *Endocrinology,* 92, 525, 1973.
13. **Campbell, D. M.,** Determination of 5-nucleotidase in blood serum, *Biochem. J.,* 82, 34, 1962.

IX. THIAMINE PYROPHOSPHATASE

Kenichirou Inomata

A. INTRODUCTION

The enzyme thiamine pyrophosphatase (TPPase) catalyzes the reaction:

$$\text{thiamine pyrophosphatase} + H_2O \rightleftarrows \text{thiamine monophosphate} + \text{orthophosphate}$$

In histochemical studies, the enzyme is used as a marker of the Golgi apparatus. Differentiation between TPPase and nucleotide diphosphatase is difficult; thiamine pyrophosphate is also hydrolyzed in the endoplasmic reticulum, which has a high nucleotide diphosphatase activity, and conversely, nucleotide diphosphates are split in the Golgi apparatus. TPPase is relatively firmly bound to the Golgi apparatus. Only scanty information is available on the physiologic role of the enzyme. However, the detection of this enzyme also gives information on the functional state of the Golgi apparatus.

B. METHODS

The advantages of the lead citrate method are as follows:

1. The localization of the reaction product at the electron microscopic level is the best as compared with the other enzyme cytochemical methods.
2. The size of the reaction product seems to be the finest in the lead citrate method because of the lead-chelating effect of citrate in the reaction medium. This means that the highest resolution at the electron microscopic level is available by this method.

C. EQUIPMENT AND REAGENTS

For light microscopy, frozen sections, 10 to 14 μm thick, are cut with an electron freezing microtome. The sections for electron microscopy, 30 to 50 μm thick, are cut with a Microslicer or a Vibratome without freezing.

All the reagents must be added dropwise, in the order indicated, under stirring. The complete medium is clear. 50 m*M* KOH must be freshly prepared before use.

Reagents		Final conc (m*M*)
Distilled water	2.0 ml	
Sucrose	0.8 g	240.0
0.2 *M* Tris-HCl buffer, pH 8.5	1.4 ml	28.0
15 m*M* $MgCl_2$	2.6 ml	3.9
Thiamine pyrophosphate chloride	9.0 mg	2.0
0.5% Lead citrate	4.0 mg	2.0
Levamisole	6.0 mg	2.5
Final volume	10.0 ml	pH 9.2

D. PROCEDURES

1. Tissue Preparations

Tissues are fixed either by immersion or perfusion. In immersion fixation, tissue slices, less than 1 mm thick, are fixed in cold 2% paraformaldehyde, 0.25 to 1% glutaraldehyde in 0.03 *M* PIPES buffer, pH 7.2, for 30 to 50 min. In perfusion fixation, tissues are perfused with cold 1 to 2% paraformaldehyde, 0.25% glutaraldehyde in the same buffer for 15 min.

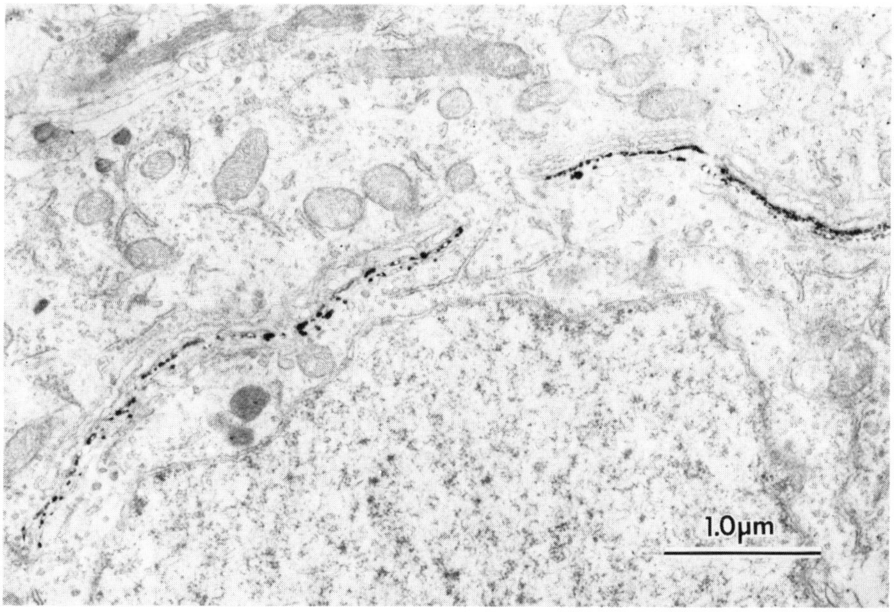

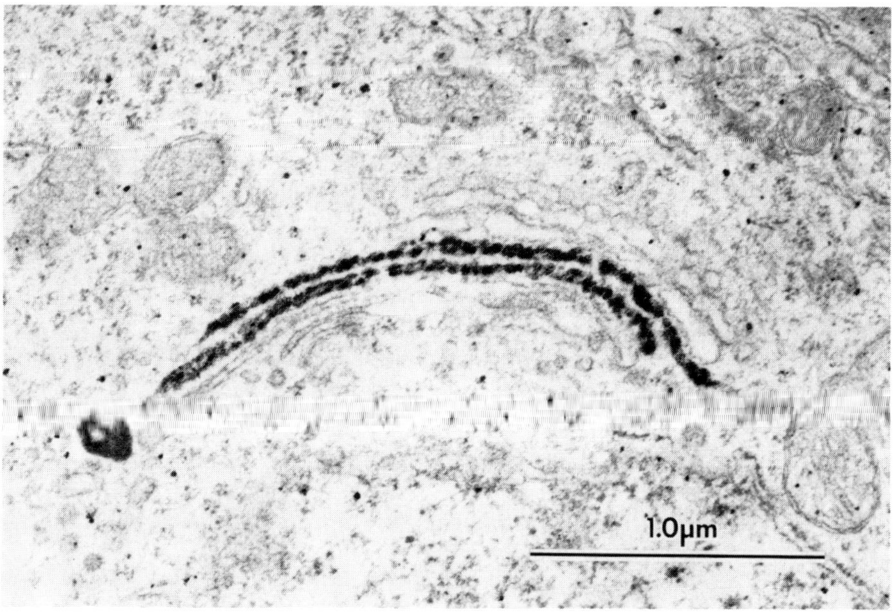

FIGURES 1 and 2. TPPase activity in the Golgi complex. Reaction products are localized in Golgi cisternae.

Excised tissues are subsequently fixed by immersion fixation for 15 min at 0 to 4°C in the same fixative. After fixation, the tissues are washed for several hours or overnight in 0.03 *M* PIPES buffer, pH 7.2, containing 8% sucrose. After washing, 10- to 14-μm-thick frozen sections are cut with an electron-freezing microtome for light microscopy. For electron microscopy, 30- to 50-μm-thick sections are cut with a Microslicer or a Vibratome without freezing.

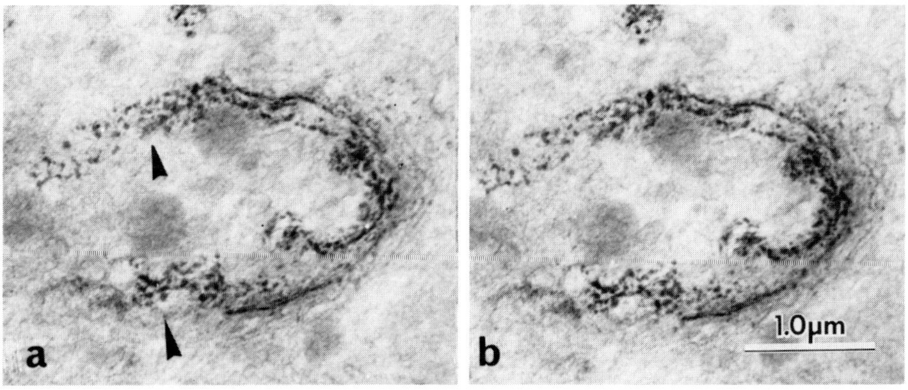

FIGURE 3. Stereopair pictures of TPPase activity in a 3.0-μm-thick section, using high-voltage electron microscope. Tilting angle ±4°. The TPPase-positive saccules are arranged in a regular hexagonal or polygonal network (arrowhead).

2. Cytochemical Procedures

Tissue sections are incubated in the medium for 30 min at 37°C. After incubation, frozen sections are washed in distilled water, treated with a 1% ammonium sulfide for 2 min for visualization of the reaction product mounted in glycerine jelly, and observed under a light microscope. For electron microscopy, nonfrozen sections are washed in 0.03 *M* PIPES buffer, pH 7.2, containing 8% sucrose, dehydrated with alcohol, embedded in the low-viscosity resin of Spurr. Ultrathin sections are cut with an ultratome, poststained with uranyl acetate, and observed with an electron microscope. The results are shown in Figures 1 to 3.

As a cytochemical control, the procedure should be carried out in a medium without substrate. The localization of the reaction product, lead phosphate, obtained by the lead citrate method is along the inner surface of Golgi apparatus in a variety of cells and tissues.

In the vascular epithelium, reaction products indicating the localization of TPPase activity are localized on the basolateral plasma membrane of vascular endothelial cells.

REFERENCES

1. **Ogawa, K., Sakai, M., and Inomata, K.,** Recent findings on ultracytochemistry of thiamin phosphatases, in *Thiamin,* Sable, H. Z. and Gubler, C. L., Eds., The New York Academy of Sciences, New York, 1982, 186.
2. **Inomata, K., Yoshioka, T., Nasu, F., and Mayahara, H.,** Ultracytochemical studies of capillary endothelial cells in the rat central nervous system, *Acta Anat.,* 118, 243, 1984.
3. **Tomiyasu, K., Nasu, F., Inomata, K., and Arii, Y.,** High-voltage electron microscopic observation of Golgi complex in the rat small ganglion cell — thiamine metabolism, *Acta Histochem. Cytochem.,* 23, 127, 1990.

X. Na^+,K^+-ADENOSINE TRIPHOSPHATASE

Hiroshi Mayahara and Kenichirou Inomata

A. INTRODUCTION

All cells expend metabolic energy to concentrate potassium in their cytoplasm and deplete intracellular sodium. As a result of pumping mechanisms, the potassium concentration within the cells is about one order of magnitude greater than that in the extracellular fluid, and the sodium concentration is about one order of magnitude less. These concentration gradients are of central importance for the cells' functions in the body. Sodium and potassium ions are pumped in opposite directions with the simultaneous hydrolysis of adenosine triphosphate (ATP) by a complex of proteins that spans the plasma membrane. This enzyme complex is named (Na^+ + K^+)-adenosine triphosphatase, sometimes shortened to Na^+,K^+-ATPase. In this section, cytochemical methods for the detection of Na^+,K^+-ATPase activity are reviewed.

B. METHODS FOR CYTOCHEMICAL DETECTION OF Na^+,K^+-ATPase ACTIVITY

Repeated attempts have been made to cytochemically localize the activity of this enzyme using ATP as a substrate.[1-3] These methods are all based on the same principle and use lead (nitrate) as a capture reagent which reacts with phosphate ions liberated by the enzymatic hydrolysis of ATP, and forms water-insoluble precipitates at, or close to, the site of enzymatic activity. However, none of these methods has been broadly applied because of poor reproducibility. After intensive efforts to find the reason for the failure to localize Na^+,K^+-ATPase activity, it became clear that the enzyme activity is easily lost by weak fixation or by the presence of low concentrations of heavy-metal ions, which are used as the capture reagent.[2]

Biochemically, it became clear that Na^+,K^+-ATPase hydrolyzes *para*-nitrophenylphosphate (NPP) in the presence of potassium and magnesium ions, and that this reaction is inhibited by a specific inhibitor of Na^+,K^+-ATPase, ouabain.[4-6] In 1972, Ernst[7] developed a cytochemical method to detect the K^+-dependent *p*-nitrophenylphosphatase (K-NPPase) activity of Na^+,K^+-ATPase using NPP as the substrate and $SrCl_2$ as a capture reagent. However, this method is a two-step method which is often accompanied by diffusion artifacts.[8]

C. THE LEAD CITRATE METHOD

In 1980, Mayahara et al.[8] presented a new, one-step lead citrate method for the light and electron microscopic localization of the K^+-dependent phosphatase activity of the Na^+,K^+-ATPase complex. The advantages of this method are the use of lead citrate for the direct capture of the liberated phosphate ions from NPP, the inclusion of 25% (v/v) dimethyl sulfoxide (DMSO) as an activator,[9] and a glycine buffer to stabilize the incubation medium. This one-step method allows for a more accurate localization of the reaction products. The use of DMSO not only amplifies the K^+-NPPase activity but also shifts the optimum pH to around pH 9.0, where the incubation is performed allowing a higher sensitivity.[9] The use of a glycine buffer results in extremely fine reaction products due to the chelation of lead; furthermore, chelation of heavy-metal ions[10] reportedly enhances Na^+,K^+-ATPase activity. In carefully controlled biochemical studies, it was shown that enough K^+-NPPase activity remains in properly fixed specimens in the presence of 4 mM lead citrate in the incubation medium, for the successful cytochemical localization of K^+-NPPase activity.[8]

1. Instruments and Equipment

For light microscopic cytochemistry, 10- to 14-μm-thick frozen sections are cut with a freezing microtome. Sections for electron microscopic cytochemistry, 30 to 40 μm thick, are cut with a Microslicer (Dosaka EM Co. Ltd., Kyoto) or a Vibratome (Technical Products International Inc., St. Louis) without freezing.

2. Chemicals and Reagents

Stock solutions	Volume of stock solutions	Final concentrations
1.0 *M* Glycine-KOH buffer, pH 9.0	2.5 ml	Glycine, 250 m*M*; K^+, 25 m*M*
1% Lead citrate in 50 m*M* KOH	4.0 ml	Lead citrate, 4.0 m*M*; K^+, 20 m*M*
Dimethyl sulfoxide	2.5 ml	25%, v/v
0.1 *M* *para*-Nitrophenyl phosphate (NPP), Mg salt	1.0 ml	NPP, 10 m*M*; Mg^{2+}, 10 m*M*
Levamisole	6.0 mg	2.5 m*M*

Final solution, 10.0 ml, pH 8.8.

All the reagents must be added in the order in which they appear from top to bottom, dropwise under constant stirring. The complete medium is a crystal-clear, light-yellow solution which is stable for hours. The 50 m*M* KOH must be freshly prepared before use. The magnesium salt of NPP (Sigma) is not easily dissolved and may require about 30 min of rigorous shaking. The sodium salt of NPP can be used, but this is not recommended because Na^+ inhibits activation of the enzyme by K^+. If the sodium salt is used, 10 m*M* $MgCl_2$ must be added to the incubation medium. The di-Tris salt of NPP cannot be used as it forms precipitates in the medium. Instead of levamisole, tetramisole (5.0 m*M*) or bromotetramisole (0.5 m*M*) can be used. These are added as specific inhibitors of nonspecific alkaline phosphatase (ALPase). It must be noted that these inhibitors very effectively inhibit the activity of ALPase isozymes in various tissues and organs, but they do not inhibit the intestinal ALPase activity.

3. Cytochemical Controls

In order to substantiate the specificity of the cytochemical procedures, the following control test should be carried out:

1. To demonstrate the K^+ dependency of the NPPase activity, K^+ is removed from the original incubation medium. This is achieved by substituting a glycine-NaOH buffer for the glycine-KOH, and dissolving the lead citrate in 50 m*M* NaOH.
2. To demonstrate the inhibition of the reaction by ouabain, ouabain (10 m*M*) is added to the incubation medium. A glass rod may be used to break up the ouabain crystal for quicker dissolution.
3. To compare the localization of nonspecific alkaline phosphatase (ALPase) activity with that of K^+-NPPase, ALPase activity is demonstrated by removing levamisole and K^+ from the incubation medium or by the lead citrate method for ALPase by Mayahara et al.[11] with sodium β-glycerophosphate as substrate.

D. PROCEDURES

Tissue preparation — Tissues are fixed either by immersion or perfusion. In immersion fixation, tissue slices less than 1 mm thick are fixed in a cold mixture of 2% paraformaldehyde and 0.5% glutaraldehyde in 0.05 *M* PIPES buffer, pH 7.2, for 30 to 60 min. In perfusion fixation, tissues are perfused with a cold mixture of 2% paraformaldehyde and 0.25% glutaraldehyde in the same buffer, pH 7.2, for 5 min. Excised tissues are subsequently fixed by immersion fixation for 15 min at 0 to 4°C in the same fixative. After fixation, the tissues are washed overnight in 0.05 *M* PIPES buffer, pH 7.2, containing 8% sucrose and 10% (v/v) DMSO. After this, frozen sections for light microscopy, 10 to 14 μm thick, are cut with a freezing microtome. The sections for electron microscopy, 30 to 40 μm thick, are cut with a Microslicer or a Vibratome without freezing.

Cytochemical procedures — Tissue sections are incubated in the reaction medium for K^+-NPPase. The appropriate incubation time depends on the intensity of the enzyme activity of the specimen; 5 to 15 min at room temperature for the kidney distal tubules and 10 to 25 min at 37°C for the retinal pigment epithelium are recommended. After incubation, frozen sections are washed in distilled water, treated with 1% ammonium sulfide for 2 min for visualization, washed in distilled water, mounted in glycerine jelly, and observed under a light microscope.

For electron microscopy, nonfrozen sections are washed in 0.05 *M* PIPES buffer, pH 7.2, containing 8% sucrose but no DMSO, dehydrated with alcohol and embedded in Spurr's low-viscosity resin. Ultrathin sections are cut with an ultratome, poststained with uranyl acetate, and observed under an electron microscope.

E. RESULTS

The reaction product, lead phosphate, obtained by the lead citrate method for the localization of K^+-NPPase activity, is found along the inner, cytoplasmic surface of the plasma membrane in a variety of cells and tissues. See References 12 to 16 for detailed information.

F. CONCLUSION

The methods for detecting Na^+,K^+-ATPase activity cytochemically are reviewed with special reference to the lead citrate method. The advantages of the lead citrate method are as follows:

1. The localization of the reaction products at the electron microscopic level seems to be the best among the enzyme cytochemical methods available, perhaps because diffusion artifacts which may occur during the conversion step in earlier methods is avoided using the direct capturing system.
2. The highest sensitivity among the available methods has been obtained with the lead citrate method.[11] This is achieved by using DMSO, which stimulates K^+-NPPase activity several-fold and at the same time enables incubation to occur at the optimum pH of the enzyme.
3. The use of glycine buffer, a metal chelator, protects against loss of the enzyme activity by chelating the heavy-metal ions used as the capturing agent, thereby increasing the sensitivity of this method.
4. The size of the reaction products seems to be the finest with the lead citrate method, perhaps because of the lead-chelating effect of glycine and citrate in the reaction medium. This means that this method offers the highest resolution available at the electron microscopic level.

The disadvantage of this method is that it does not allow a quantitative analysis since the reaction product is amplified at the sites of K^+-NPPase activity, as it is with all other cytochemical methods.

REFERENCES

1. **McClurkin, I. T.,** A method for the cytochemical demonstration of sodium-activated adenosine triphosphatase, *J. Histochem. Cytochem.*, 12, 654, 1964.
2. **Marchesi, V. T. and Palade, G. E.,** The localization of Mg-Na-K-activated adenosine triphosphatase on red cell ghost membranes, *J. Cell Biol.*, 35, 385, 1967.
3. **Uusitalo, R. and Palkama, A.,** Localization of sodium-potassium stimulated adenosine triphosphatase activity in the rabbit ciliary body using light and electron microscopy, *Ann. Med. Exp. Biol. Fenn.*, 48, 84, 1970.
4. **Judah, J. D., Ahmed, K., and McLean, A. E. M.,** Ion transport and phosphoproteins of human red cells, *Biochim. Biophys. Acta*, 65, 472, 1962.
5. **Fujita, M., Nakao, T., Tashima, Y., Mizuno, N., Nagano, K., and Nakao, M.,** Potassium-ion stimulated *p*-nitrophenylphosphatase activity occurring in a highly specific adenosine triphosphatase preparation from rabbit brain, *Biochim. Biophys. Acta,* 117, 42, 1966.
6. **Bader, H. and Sen, A.,** (K^+)-Dependent acyl phosphatase as part of the (Na^+ + K^+)-dependent ATPase of cell membranes, *Biochim. Biophys. Acta,* 118, 116, 1966.
7. **Ernst, S. A.,** Transport adenosine triphosphatase cytochemistry. II. Cytochemical localization of ouabain-sensitive, potassium-dependent phosphatase activity in the secretory epithelium of the avian salt gland, *J. Histochem. Cytochem.*, 20, 23, 1972.
8. **Mayahara, H., Fujimoto, K., Ando, T., and Ogawa, K.,** A new one-step method for cytochemical localization of ouabain-sensitive potassium-dependent *p*-nitrophenylphosphatase activity, *Histochemistry,* 67, 125, 1980.
9. **Guth, L. and Albers, R. W.,** Histochemical demonstration of (Na^+-K^+)-activated adenosine triphosphatase, *J. Histochem. Cytochem.*, 22, 320, 1974.
10. **Spechet, S. C. and Robinson, J. D.,** Stimulation of the (Na^+-K^+)-dependent adenosine triphosphatase by amino acids and phosphatidylserine: chelation of trace metal inhibitors, *Arch. Biochem. Biophys.*, 154, 314, 1973.
11. **Mayahara, H., Hirano, H., Saito, T., and Ogawa, K.,** The new lead citrate method for the demonstration of activity of non-specific alkaline phosphatase (orthophosphoric monoester phosphohydrolase), *Histochemie,* 11, 88, 1967.
12. **Mayahara, H. and Ogawa, K.,** Ultracytochemical localization of ouabain-sensitive, potassium-dependent *p*-nitrophenylphosphatase activity in the rat kidney, *Acta Histochem. Cytochem.*, 13, 90, 1980.
13. **Mayahara, H., Ando, T., Ishikawa, Y., and Ogawa, K.,** Ultracytochemical study of the ouabain-sensitive, K-dependent *p*-nitrophenylphosphatase (Na^+, K^+-ATPase) activity in the developing rat kidney, *Acta Histochem. Cytochem.*, 15, 421, 1982.
14. **Inomata, K., Mayahara, H., Fujimoto, K., and Ogawa, K.,** Ultrastructural localization of ouabain-sensitive, potassium-dependent *p*-nitrophenylphosphatase (Na^+-K^+-ATPase) activity in the central nervous system of the rat, *Acta Histochem. Cytochem.*, 16, 277, 1983.
15. **Ueno, S., Mayahara, H., Tsukahara, I., and Ogawa, K.,** Ultracytochemical localization of ouabain-sensitive, potassium-dependent *p*-nitrophenylphosphatase activity in guinea pig retina. II. Neurons and Müller cells, *Acta Histochem. Cytochem.*, 14, 186, 1981.
16. **Mayahara, H. and Ogawa, K.,** Histochemical localization of Na^+, K^+-ATPase, in *Methods in Enzymology,* Vol. 156, Fleischer, S. and Fleischer, B., Eds., Academic Press, New York, 1988, 417.
17. **Stahl, W. L. and Baskin, D. G.,** Histochemistry of ATPases, *J. Histochem. Cytochem.*, 38, 1099, 1990.

XI. H^+,K^+-ATPase AND LYSOSOMAL H^+-ATPase

Kazushi Fujimoto

A. INTRODUCTION

It is possible to classify H^+-adenosine triphosphatase (proton-translocating ATPase) according to its intracellular localization: (1) the transport-type H^+-ATPase resides in the plasma membrane and the lysosomal membrane, and (2) the ATP-synthetic or mitochondrial type resides in the mitochondria, bacteria, and chromaffin granules. The ATP-synthetic type, depending on the biological system or the physiological state, works in either the ATP-hydrolysis direction, driving ion transport or ATP synthesis. This is in contrast to the membrane-type H^+-ATPase, whose major function is to participate in ion transport. This chapter focuses on the recent progress in elucidating the enzyme cytochemistry of the ion transport-type H^+-ATPases, particularly gastric H^+,K^+-ATPase and lysosomal H^+-ATPase.

B. CYTOCHEMISTRY OF GASTRIC H^+,K^+-ATPase

The morphological changes that occur in parietal cells along with stimuli of various types of acid secretion are well demonstrated. The conspicuous ultrastructural feature is that active secretion is paralleled by an increase in both the number and the size of microvilli lining the intracellular canaliculi, and a decrease in the smooth cytoplasmic membrane components, called "tubulovesicles", while in a nonsecreting parietal cell the tubulovesicles are prominent. From the biochemical point of view, the gastric H^+,K^+-ATPase system has been identified as the pumping mechanism for the transport of H^+ ions across the secretory membranes of the parietal cells. It is considered of interest to localize the H^+,K^+-ATPase within the parietal cells and to study its distribution in relation to the ultrastructure that undergoes change during acid secretion.

H^+,K^+-ATPase activity can be described by several partial reactions, including an ATP phosphotransferase activity, a phosphoprotein intermediate, and a K^+-stimulated phosphoprotein phosphatase activity to hydrolyze *p*-nitrophenylphosphate (*p*-NPP). Like the enzyme histochemical methods used for detecting Na^+,K^+-ATPase, a few histochemical methods have been used in attempts to demonstrate H^+,K^+-ATPase activity using *p*-NPP as a substrate. Crucial factors required for validating the *p*-NPPase reaction corresponding to H^+,K^+-ATPase are

1. Stimulation by K^+
2. Ouabain insensitivity: unlike Na^+,K^+-ATPase, gastric H^+,K^+-ATPase is ouabain insensitive
3. Inhibition by nolinium bromide, which is known to interfere with the K^+-dependent phosphatase step of H^+,K^+-ATPase

1. Methods

Recently, Fujimoto et al.[1] introduced a lead citrate method for the demonstration of H^+,K^+-ATPase activity using *p*-NPP as a substrate and dimethyl sulfoxide (DMSO). Detailed biochemical studies have suggested that DMSO enhances ouabain-insensitive, K^+-dependent *p*-NPPase (gastric K^+-NPPase) and that it shifts the optimal pH for the K^+-NPPase reaction from neutral pH to pH 9.0. The effect of DMSO may also arise from an increase in the permeability of the incubation medium, since DMSO causes partial solubilization of lipids from cell membranes.

2. Reagents and Equipment

1. Buffer: dimethyl sulfoxide (DMSO), hydrochloric acid, sodium cacodylate, sucrose
2. Fixatives: glutaraldehyde (20 to 25% solution, electron microscopic grade), paraformaldehyde
3. Incubation medium: DMSO, glycine, lead citrate (ICN Pharmaceuticals, Inc.), levamisole (an inhibitor of nonspecific alkaline phosphatase; Sigma Chemical Co.), *p*-nitrophenylphosphate (Mg salt; Sigma Chemical Co.), nolinium bromide (a specific inhibitor of gastric H^+-ATPase; a gift from Norwich Eaton Pharmaceuticals, Inc.), ouabain (a specific inhibitor of Na^+,K^+-ATPase; Sigma Chemical Co.), potassium chloride, potassium hydroxide, sodium chloride, sodium hydroxide, vanadate
4. For light microscopy: ammonium sulfide, glycerin
5. For electron microscopy: epoxy resin, osmium tetroxide

All reagents should be of the highest purity available.

Preparation of solutions — The following solutions are prepared by standard methods: 0.2 *M* cacodylate buffer, pH 7.4; 0.1 *M* cacodylate buffer, pH 7.4, containing 0.25 *M* sucrose; and 0.01 *M* cacodylate buffer, pH 7.4, containing 10% DMSO and 0.25 *M* sucrose (DMSO buffer).

1. 1.0 *M* glycine-0.1 *M* KOH buffer, pH 9.0: dissolve 7.5 g glycine in 60 ml distilled water, adjust the pH to 9.0 with 10 ml of 1 *M* KOH, and make up to 100 ml with distilled water. Glycine buffer is stable for 2 to 3 days at room temperature.
2. 10.0 m*M* lead citrate: dissolve 0.2 g lead citrate in 19.0 ml distilled water, add 1.0 ml of 1 *N* KOH, and shake vigorously for 30 min to make a clear solution. Lead citrate solution is unstable.

A Vibratome (Oxford, California) or a Microslicer (Dosaka EM Co., Kyoto, Japan) is required for making nonfrozen sections.

3. Procedures

a. Preparation of Tissue Samples

1. Fixation: tissues are cut into slices less than 1 mm thick and are fixed for 30 to 60 min at 4°C in a mixture of 2% formaldehyde (freshly prepared by depolymerizing paraformaldehyde just before use) and 0.25% glutaraldehyde in 0.1 *M* cacodylate buffer, pH 7.4, containing 10% v/v DMSO and 0.25 *M* sucrose (washing buffer). Higher concentrations of glutaraldehyde should be avoided because these significantly reduce the enzymatic activity of H^+,K^+-ATPase. To preserve enzyme activity, 5 to 10% DMSO should be added to the fixatives.
2. Washing: fixed tissues are washed in several changes of DMSO buffer for a total of 1 to 2 h at 4°C.
3. Sectioning: the tissues are cut into sections less than 40 μm thick and are collected in DMSO buffer. To preserve the ultrastructure, nonfrozen sections should be used for electron microscopy.
4. Preincubation: prior to the cytochemical reaction, tissue sections are preincubated in DMSO buffer containing 10 m*M* ouabain for 30 min at 4°C to inhibit Na^+,K^+-ATPase activity.

b. Incubation

1. Preparation of incubation medium: prepare a complete incubation medium according to the formula listed below.

1.0 *M* Glycine-KOH buffer, pH 9.0	2.5 ml
3.1% *p*-Nitrophenylphosphate (Mg salt)	1.0 ml
1.1 *M* KCl	0.5 ml
10 m*M* Lead citrate in 50 m*M* KOH	4.0 ml
Dimethyl sulfoxide	2.0 ml
Levamisole	6.0 mg
Ouabain	73 mg

   The components of the incubation medium should be mixed in the order shown above, with gentle stirring. The resulting medium should be clear.
2. Incubation: a suitable time and temperature for the enzyme reaction will have to be determined by experiments. Incubation for 10 to 30 min at room temperature will probably suffice. In order to ensure adequate diffusion of reagents into the tissue sections, the medium should be gently agitated during incubation.
3. For light microscopy, sections are washed with distilled water, treated with 1% ammonium sulfide for 1 min, rinsed again in distilled water, and mounted on a glass slide with glycerin.
4. For electron microscopy, incubated sections are washed well in several changes of 0.1 *M* cacodylate buffer, pH 7.4, containing 0.25 *M* sucrose, and postfixed with 0.1 *M* cacodylate-buffer 1% osmium tetroxide for 10 min. Postfixed sections are rinsed in distilled water, dehydrated through a graded ethanol series, and embedded in epoxy resin. The ultrathin sections are stained with uranyl acetate and observed under an electron microscope.

c. Histo- and Cytochemical Controls

1. Inhibition: prior to the cytochemical reaction, sections are preincubated in DMSO buffer with 0.5 m*M* nolinium bromide or 0.25 m*M* vanadate for 30 to 60 min at 4°C; the sections are then incubated in media containing the respective inhibitors at the same concentrations. Since vanadate sensitivity is characteristic of ATPases that form a phosphorylated intermediate as part of their reaction sequence, it would appear that gastric H^+,K^+-ATPase, which forms a phosphorylated intermediate, is vanadate sensitive.
2. K^+ dependency: sections are incubated in a medium that lacks K^+ ions by replacing K^+ with Na^+.
3. Cytochemical demonstration of Na^+,K^+-ATPase: to examine the cytochemical differences between gastric K^+-NPPase and ouabain-sensitive K^+-NPPase corresponding to the second (dephosphorylation) step of Na^+,K^+-ATPase, ouabain-sensitive K^+-NPPase activity on parietal cells is demonstrated with Mayahara's medium containing 0.5 m*M* nolinium bromide.

4. Results

Light microscopic findings — Guinea pig stomachs were minced and fixed for 60 min at 4°C in a mixture of 2% formaldehyde and 0.25% glutaraldehyde in 0.1 *M* cacodylate

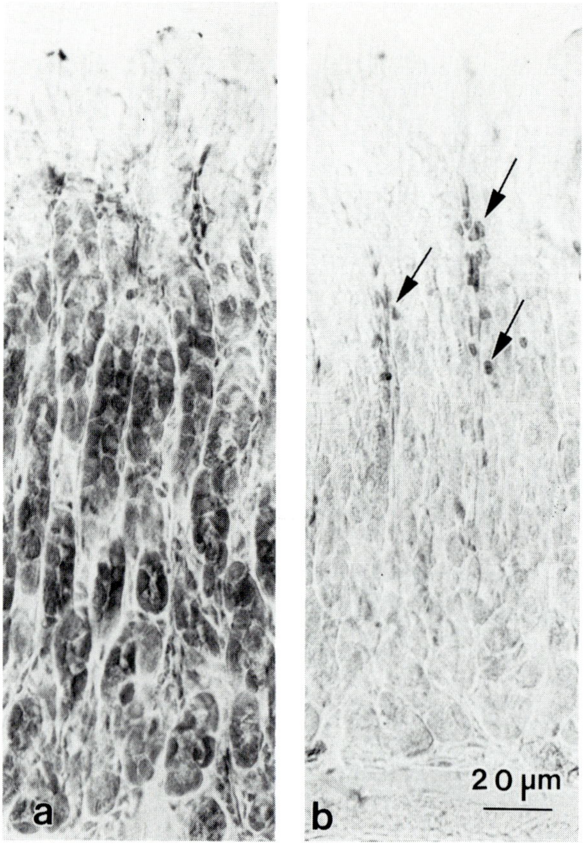

FIGURE 1. Light microscopic demonstration of gastric K^+-NPPase activity in guinea pig gastric mucosa. (a) Intense reaction is localized only in the parietal cells. (b) Incubation without K^+. The reaction in the parietal cells is almost abolished. Endothelial cells of the capillary (arrows) show nonspecific reaction. (Magnification × 500.)

buffer. Frozen sections were incubated in standard incubation medium for 45 min at room temperature. An intense K^+-NPPase reaction was observed in the parietal cells of the fundic glands, and there was a diffuse cytoplasmic staining (Figure 1a). Surface mucous, mucous neck, and chief cells were entirely free of reaction product.

Electron microscopic findings — Reaction products indicating gastric K^+-NPPase activity were found on the membranes lining the luminal cell surfaces, intracellular canaliculi (Figure 2, IC), and tubulovesicles (Figure 2, T). In the gastric glands of guinea pigs fed a normal diet, there was a variation in the ultrastructure of parietal cells associated with acid secretion. The cytoplasm of a typical secreting parietal cell was devoid of tubulovesicles, whereas the intracellular secretory canalicular lumen was filled with numerous microvilli. The cell membrane of the apical surfaces and secretory canaliculi exhibited a positive cytochemical reaction for gastric K^+-NPPase. However, the basolateral plasma membrane appeared free of reaction product. A typical, nonsecreting cell revealed generalized features which included extensive tubulovesicles whose membranes were intensely stained. Short, stubby microvilli, which projected into both the glandular and the intracellular canalicular lumen, displayed the same intense reaction as the region corresponding to the tubulovesicles.

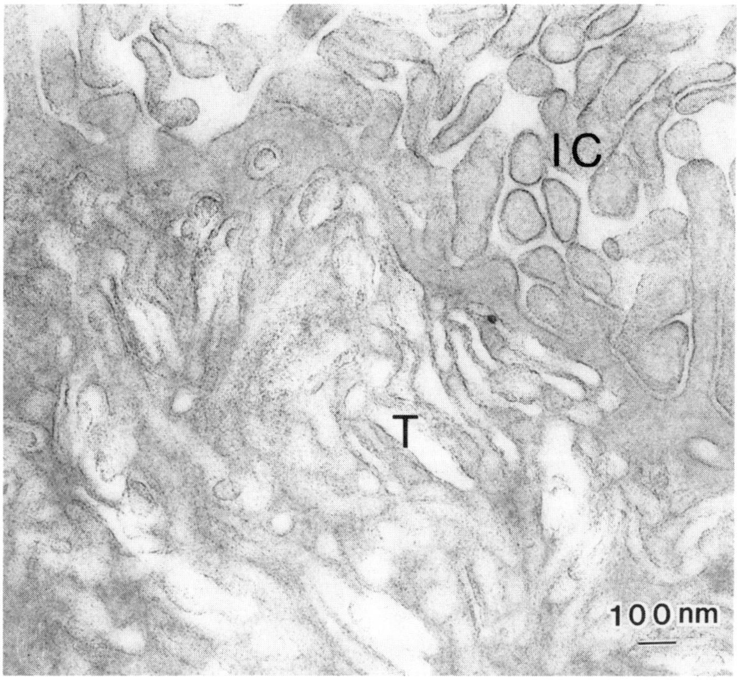

FIGURE 2. Gastric K^+-NPPase in a nonsecreting parietal cell. The tubulovesicles and microvilli are positive. (Magnification × 50,000.)

Cytochemical controls

1. Inhibitor sensitivity: nolinium bromide and vanadate completely inhibited the reaction of parietal cells.
2. Effect of omission of K^+: when K^+ was omitted from the incubation medium, no reaction was seen in the parietal cells (Figure 1b).
3. Cytochemical demonstration of ouabain-sensitive K^+-NPPase: ouabain-sensitive K^+-NPPase activity was demonstrated on the basolateral membranes of parietal cells.

These results showed that the basolateral membranes of parietal cells were not positive for gastric K^+-NPPase reaction; however, ouabain-sensitive K^+-NPPase activity was associated with the basolateral membrane. In other words, it appears that gastric H^+,K^+-ATPase and Na^+,K^+-ATPase are localized on the luminal membrane, including the tubulovesicles and the basolateral membrane, respectively, so there would appear to be a functional differentiation in the membranes of parietal cells.

C. CYTOCHEMISTRY OF LYSOSOMAL H^+-ATPase

The acidic interior of lysosomes provides a favorable environment for the digestive activity of the lysosomal enzymes. This is a reflection not only of the acidic pH optima of most lysosomal hydrolases, but also of the denaturing effects of low pH. In the last few years, an ATP-dependent proton pump, called lysosomal H^+-ATPase, has been postulated to play an important role in maintaining the pH gradient.

Histochemical detection of lysosomal H^+-ATPase in rat hepatocytes was reported by Luo et al.[2] Their method was based on the neutral pH *p*-NPPase method of Noda et al.,[3] which was introduced to detect Na^+,K^+-ATPase activity at neutral pH. Their improvement on the method of Noda et al. for the histochemical demonstration of lysosomal H^+-ATPase activity was to incorporate the use of 4,4-diisothiocyanatostilbene-2,2′-disulphonic acid (DIDS) and *N*-ethylmaleimide (NEM), which are regarded as inhibitors of lysosomal H^+-ATPase, but which have little or no effect on other lysosomal enzyme activity.

1. Procedure

1. Under anesthesia, animals are perfused with 2% formaldehyde and 0.25% glutaraldehyde in 0.1 *M* cacodylate buffer, pH 7.4, for 10 to 15 min. Tissues are then removed and cut into small pieces which are washed for 12 to 24 h in 0.1 *M* cacodylate buffer.
2. Fixed tissues are cut into 40-μm sections with a Microslicer or a Vibratome.
3. Sections are incubated for 5 to 30 min in the standard medium, which consists of 5 m*M* *p*-NPP (Mg salt), 50 m*M* KCl, 2.5 m*M* levamisole, 25% DMSO, and 10 m*M* lead citrate in 25 m*M* Tricine-KOH buffer, pH 7.4.
4. For light microscopy, incubated sections are rinsed in distilled water, treated with 1% ammonium sulfide for 1 min, rinsed in distilled water again, and mounted on glass slides with glycerin.
5. For electron microscopy, sections are rinsed in 0.1 *M* cacodylate buffer, fixed with 1% osmium tetroxide for 30 min at 4°C, rinsed in distilled water, dehydrated through a graded ethanol series, and embedded in epoxy resin.

2. Results

Following our attempt to validate the localization of lysosomal H^+-ATPase, our results, based on the method of Noda et al.[3] for neutral pH *p*-NPPase activity, were

1. When the standard incubation medium of Noda et al. was used, enzyme activity was found on lysosomal matrices as well as on lysosomal membranes.
2. When NaF at a concentration of 5 to 10 m*M* was added to the standard incubation medium, the enzyme activity of the lysosomal matrices was greatly reduced; however, that of the lysosomal membranes mostly remained. This finding implies that the *p*-NPPase reaction of the lysosomal matrices is due to acid phosphatase activity, and that acid phosphatase activity may not contribute to the reaction of the lysosomal membranes.
3. NEM (10 m*M*) or DIDS (10 m*M*), which are well-known inhibitors of lysosomal H^+-ATPase, inhibited the reaction of lysosomal membranes, but had little or no effect on the reaction of lysosomal matrices.

The information gathered provides evidence that lysosomal membrane-bound *p*-NPPase activity may represent H^+-ATPase activity.

REFERENCES

1. **Fujimoto, K., Ogawa, K. S., and Ogawa, K.,** Gastric K^+-stimulated *p*-nitrophenylphosphatase cytochemistry, *Histochemistry,* 84, 600, 1986.
2. **Luo, S.-Q., Sakai, M., Zhong, C.-S., and Ogawa, K.,** Histochemical and cytochemical studies of neutral pH *p*-NPPase on lysosomal membrane in rat liver, *Acta Histochem. Cytochem.,* 22, 187, 1989.
3. **Noda, T., Li, G.-J., and Ogawa, K.,** Histo- and cytochemical demonstration of *p*-NPPase in neutral pH with lead citrate method, *Acta Histochem. Cytochem.,* 21, 423, 1988.

XII. Ca^{2+}-ADENOSINE TRIPHOSPHATASE

Kazushi Fujimoto

A. INTRODUCTION

Calcium ion (Ca^{2+}) is recognized as the key ion in regulating cellular functions. For instance, muscle contraction is evoked by Ca^{2+} release from the sarcoplasmic reticulum or from membrane-associated calcium pools (probably mitochondria). In addition, in cellular responses to different stimuli, Ca^{2+} may function either by acting as an intracellular second messenger itself or by affecting other second messenger systems. It is well known that the Ca^{2+} concentration in cytoplasm is less than 10^{-5} to 10^{-7} *M*, in contrast to the much higher Ca^{2+} levels (about 10^{-3} *M*) found in extracellular fluids and in intracellular organelles such as mitochondria. This nonequilibrium of Ca^{2+} distribution is maintained mainly by the Ca^{2+}-transporting systems of the plasma membrane and intracellular membranes. Cell membranes contain both high- and low-affinity Ca^{2+}-transporting systems. The low-affinity systems correspond to various transporting modes (channels, exchangers, electrophoretic uniporters). The high-affinity systems correspond to the Ca^{2+}-activated adenosine triphosphatase (Ca^{2+}-ATPase) system.

The cytochemical method most often used for Ca^{2+}-ATPase studies was introduced by Ando et al.[1] This one-step method is applicable to both light and electron microscopic demonstration of the different types of Ca^{2+}-ATPase, and it reflects their biochemical and physiological properties. The method employs a medium containing ATP, $CaCl_2$, levamisole (an inhibitor of alkaline phosphatase), and lead as the capture agent. The procedure of Ando et al.,[1] which is now in general use for light and electron microscopy of Ca^{2+}-ATPase, is given below. However, so that the most appropriate level of Ca^{2+}-ATPase reaction can be selected for study, it is advisable to change the Ca^{2+} concentration of the incubation medium and the composition of the fixative solutions when necessary.

B. CHEMICALS AND REAGENTS

1. Buffer: hydrochloric acid, sodium cacodylate, sucrose
2. Fixatives: glutaraldehyde (20 to 25% solution, electron microscopic grade), paraformaldehyde
3. Incubation medium: adenosine triphosphate Na_2-salt (ATP; Sigma Chemical Co., St. Louis), calcium chloride, lead citrate (K and K Lab., New York), levamisole (an inhibitor of nonspecific alkaline phosphatase; Sigma Chemical Co.), potassium chloride, potassium hydroxide, sodium chloride, sodium hydroxide
4. For light microscopy: ammonium sulfide, glycerin
5. For electron microscopy: epoxy resin, osmium tetroxide
6. For cytochemical controls: adenosine diphosphate (ADP; Sigma Chemical Co.), adenosine monophosphate (AMP; Sigma Chemical Co.), β-glycerophosphate (β-GP; Sigma Chemical Co.), *p*-chloromercuric benzoate (PCMB), *p*-nitrophenyl phosphate (*p*-NPP; Sigma Chemical Co.), quercetin (an inhibitor of Ca^{2+}-ATPase; Sigma Chemical Co.)

All reagents should be of the highest purity available.

C. PREPARATION OF SOLUTIONS

The following solutions are prepared by standard methods: 0.2 *M* cacodylate buffer, pH 7.4; 0.1 *M* cacodylate buffer, pH 7.4, containing 0.25 *M* sucrose (washing buffer);

2% osmium tetroxide in distilled water; 100, 10, 1, or 0.1 m*M* calcium chloride in distilled water.

1. 1.0 *M* Glycine-NaOH buffer, pH 9.0: dissolve 7.5 g glycine in 60 ml distilled water, adjust the pH to 9.0 with 10 ml of 1 *N* NaOH and make up to 100 ml with distilled water. Glycine buffer is stable for 2 to 3 d at room temperature.
2. 10 m*M* Lead citrate solution: dissolve 0.2 g lead citrate in 19 ml with distilled water, add 1.0 ml of 1 *N* NaOH, and shake vigorously for 30 min to make a clear solution. The lead citrate solution should be freshly made immediately before use.

Additionally, according to our experience, NaOH solution must also be fresh, since a faint precipitate is occasionally seen when lead citrate is dissolved in NaOH solution prepared more than 1 week before use.

D. PROCEDURE

1. Preparation of Tissue Samples for Histochemical Incubation

1. Fixation: tissues are cut into slices less than 1 mm thick and are fixed for 60 min. It should be noted that the degree of inhibition of Ca^{2+}-ATPase activity in particular tissues may vary depending on the fixative used. Therefore, each tissue or cell preparation must be independently evaluated for fixative sensitivity. For instance, Ca^{2+}-ATPase reaction is found in the smooth muscle cells of rat intestine fixed with 2% formaldehyde alone (Figure 1a). However, no reaction is localized in the smooth muscle cells when tissues are fixed with a mixture containing 2% formaldehyde and 0.25% glutaraldehyde (Figure 1b). This is in contrast to the Ca^{2+}-ATPase reaction observed in the bile canaliculi of hepatocytes, which is insensitive to changes in the concentration of glutaraldehyde (Figures 1c and d). These observations emphasize the necessity for assessing enzyme localization by light microscopy of formaldehyde-fixed tissues.
2. Washing: fixed tissues are washed in several changes of washing buffer for a total of 1 to 2 h at 4°C.
3. Sectioning: fixed tissues are cut into sections less than 40 μm thick. To preserve the ultrastructure, nonfrozen sections should be used for electron microscopy.
4. Preparation of incubation medium: prepare the complete incubation medium according to the formula below:

1.0 *M* Glycine-NaOH buffer, pH 9.0	2.5 ml
Adenosine triphosphate (Na salt)	18 mg
Distilled water	2.5 ml
1–100 m*M* $CaCl_2$	1.0 ml
Levamisole	6.02 mg
10 m*M* Lead citrate	4.0 ml

The components of the incubation medium should be mixed in the order shown above, with gentle stirring. The resulting medium should be clear.

5. Incubation: a suitable time and temperature for enzyme reaction will have to be determined by experiments. Incubation for 10 to 30 min at room temperature will probably suffice.
6. For light microscopy, sections are washed in several changes of distilled water, treated with 1% ammonium sulfide for 1 min, rinsed again in distilled water, and mounted on glass slides with glycerin.

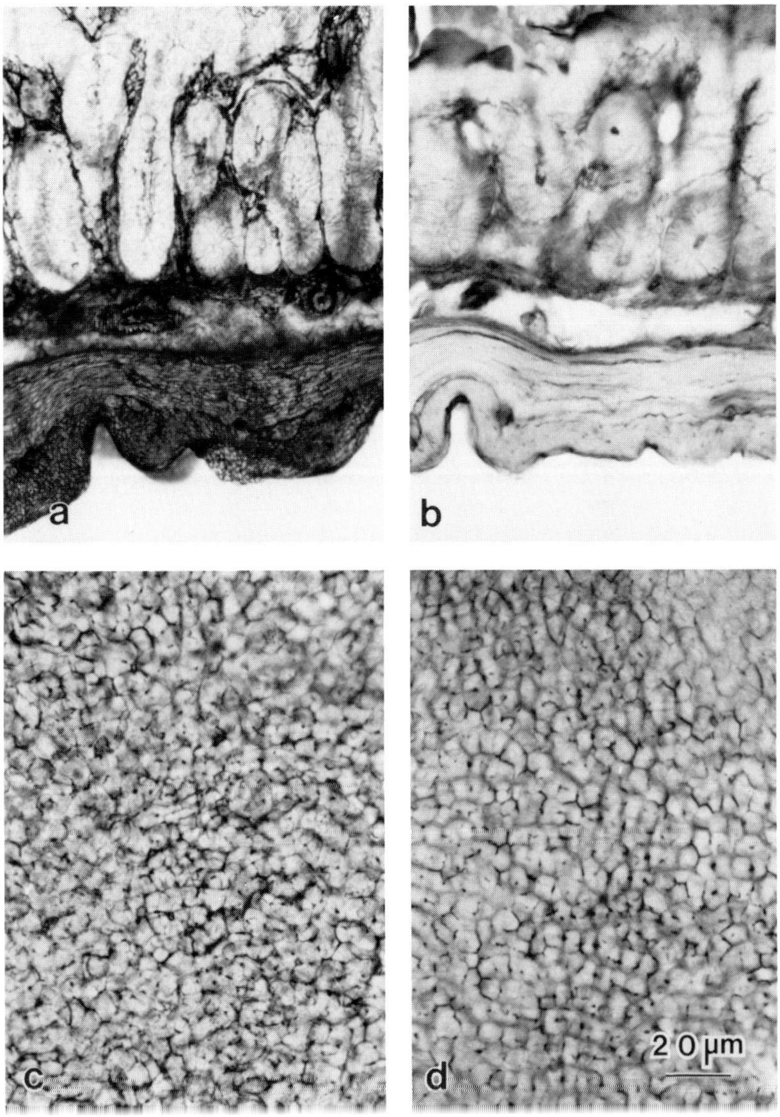

FIGURE 1. Light microscopic demonstration of Ca^{2+}-ATPase activity in the rat small intestine (a, b) and the rat liver (c, d). (a, c) Fixed with 2% formaldehyde for 60 min; (b, d) fixed with a mixture of 2% formaldehyde and 0.25% glutaraldehyde for 60 min. (Magnification × 450.)

7. For electron microscopy, incubated sections are rinsed in several changes of washing buffer and postfixed with 0.1 *M* cacodylate-buffered 1% osmium tetroxide for 10 min. Postfixed sections are rinsed in distilled water, dehydrated through a graded ethanol series, and embedded in epoxy resin. The resulting ultrathin sections are stained with uranyl acetate and observed under an electron microscope.

2. Histo- and Cytochemical Controls

1. Ca^{2+} dependency: sections are incubated in a Ca^{2+}-free medium or a medium in which $MgCl_2$ has been substituted for $CaCl_2$.

2. Substrate specificity: sections are incubated in a substrate-free medium and/or a medium in which various substrates (ADP, AMP, *p*-NPP, or β-GP) have been substituted for ATP.
3. Effect of inhibitors: sections are preincubated in a washing buffer containing 0.1 m*M* quercetin, an inhibitor of Ca^{2+}-ATPase activity, or 10 m*M* PCMB for 30 min at room temperature; the sections are then incubated in reaction media containing the respective inhibitors at the same concentrations.

E. RESULTS

Cytochemical controls — The elimination of $CaCl_2$ or ATP from the incubation medium results in complete loss of the cytochemical reaction in various tissues. Substitution of ADP, AMP, *p*-NPP, or β-GP results in no reaction. Preincubation in quercetin or PCMB followed by incubation in the incubation medium containing an inhibitor partly reduces the enzyme reaction in liver, kidney, heart, and trachea.[1,2]

Effect of changes of Ca ion concentration on the cytochemical reaction — The outcome of the Ca^{2+}-ATPase reaction is greatly influenced by the concentration of Ca^{2+} in the incubation medium. At 0.1 m*M* Ca^{2+}, in rat cardiac muscle cells fixed with 2% formaldehyde for 30 min, intense reaction is seen only on the myofilaments, probably on thick filaments (Figure 2a). At 1 m*M* Ca^{2+}, intense reactions are localized on the myofilaments and a weak reaction is seen in mitochondrial matrices (Figure 2b). An increase in Ca^{2+} concentration from 1 m*M* to 10 m*M* results in a sudden decrease of the reaction on the myofilaments, whereas, at 10 m*M* Ca^{2+}, the reaction in the mitochondrial matrices is predominant (Figure 2c).

Cytochemical localization of Ca^{2+}-ATPase activity — A cytochemical study using Ando's method has revealed many interesting localizations of Ca^{2+}-ATPase activity in various cells. It is possible to classify the Ca^{2+}-ATPase activity according to its intracellular localization: (1) plasma membrane, (2) mitochondria, (3) filamentous structures, (4) gap junction, and (5) other cell organelles.

- *Plasma membranes* — Applications of Ando's method have revealed depositions of the reaction product, indicating Ca^{2+}-ATPase activity on the plasma membranes of a variety of cells, including hepatocytes,[1] epithelial cells of the trachea,[3] aortic endothelial cells,[4] nonpigment ciliary epithelial cells,[5] toad urinary bladder cells,[6] peripheral nerve fiber cells,[7] and parietal cells.[8]
- *Mitochondria* — Using Ando's standard medium, mitochondrial Ca^{2+}-ATPase activity has been demonstrated by electron microscopy.[2,5,9] This activity is readily abolished even by a brief fixation with low concentrations of glutaraldehyde, but can be detected cytochemically after 2 to 4% formaldehyde fixation alone. Although an intense reaction was seen in the mitochondrial matrices, the outcome of the reaction was greatly influenced by changes in the concentration of Ca^{2+} in the incubation medium.
- *Gap junctions* — Ca^{2+}-ATPase activity has been demonstrated on the gap junctions of various cells, including cardiac muscle cells,[2] nonpigment ciliary epithelial cells,[5] and smooth muscle cells.[4] The reaction product is usually associated with the cytoplasmic side of the gap junctional membrane (Figure 3). The localization of enzyme activity in the gap junctions of various cell types has been reviewed in detail[10] (see Chapter 9.2).
- *Cytoskeletal system* — An intense Ca^{2+}-ATPase activity reaction has been demonstrated on the axoneme and basal bodies in the ciliated cells of rat trachea.[3] On closer inspection of the axoneme, the reaction products were found to be located between ciliary membranes and doublets, along both the outer and inner surfaces of doublets,

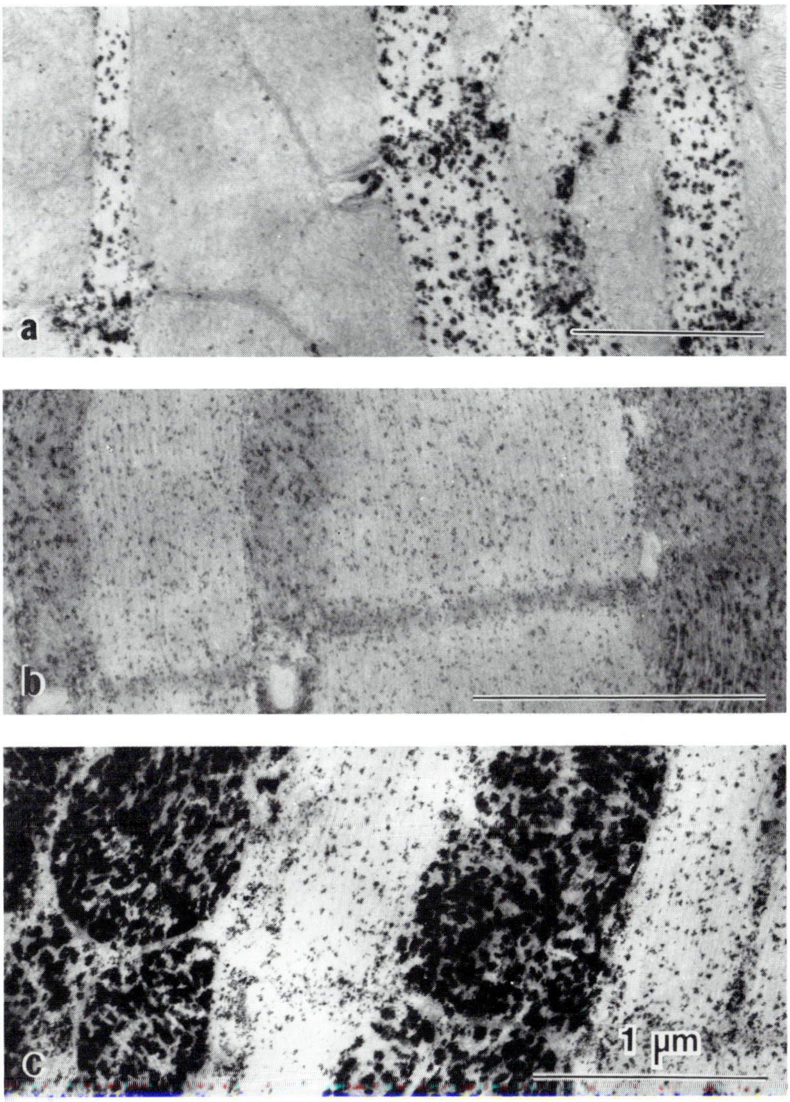

FIGURE 2. Effect of Ca^{2+} concentration on Ca^{2+}-ATPase activity in the rat cardiac muscle. The outcome of the reaction is influenced by changing the concentration of Ca^{2+} in the incubation medium. (a) 0.1 m*M*; (b) 1 m*M*; (c) 10 m*M* Ca^{2+}. (Magnifications: a, × 26,000; b, × 40,000; c, × 32,000.)

between doublets and the central sheath, on the periphery of the central pair, and between the central tubules. Ando et al.[3] considered that the Ca^{2+}-ATPase reaction on the axoneme described above indicated ATPase activity similar to that of dynein ATPase, which is related to the sliding of microtubules.

Ca^{2+}-ATPase reaction products have also been found in the following intracellular organelles of normal tissues; (1) heart — sarcoplasmic reticulum and transverse tubules of myocardial cells;[2] (2) liver — lysosomes of hepatocytes;[1] (3) intestine — Golgi apparatus of absorptive epithelial cells (Figure 4); (4) stomach — smooth endoplasmic reticulum of parietal cells;[8] (5) nerve — postsynaptic densities, synaptic vesicles, Golgi apparatus, and smooth endoplasmic reticulum;[11] and (6) retina — disks of photoreceptor cells.[9,12]

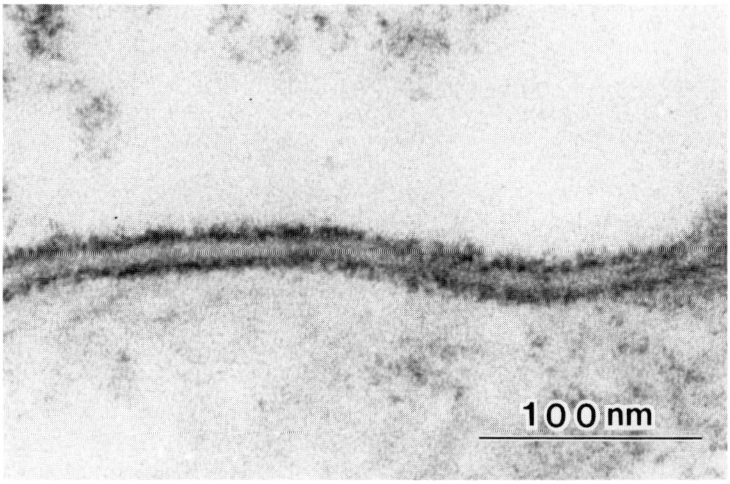

FIGURE 3. Ca^{2+}-ATPase activity on the gap junctions of rat myocardial cells. (Magnification × 300,000.)

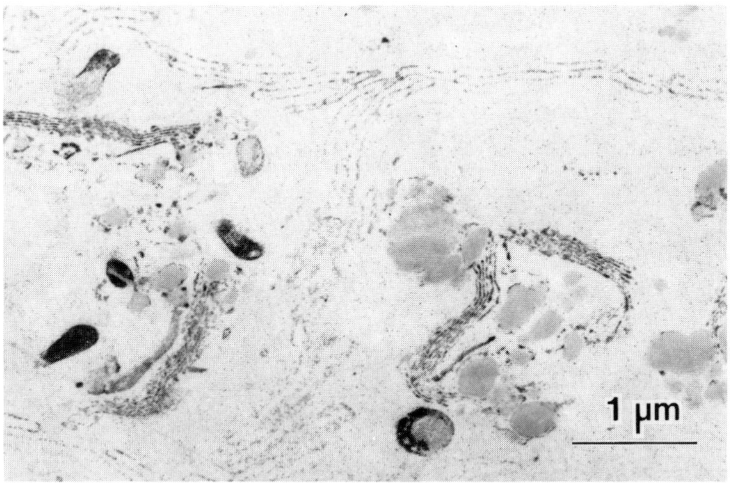

FIGURE 4. Ca^{2+}-ATPase activity in the Golgi apparatus of rat absorptive epithelial cells. (Magnification × 17,000.)

REFERENCES

1. **Ando, T., Fujimoto, K., Mayahara, H., Miyajima, H., and Ogawa, K.,** A new one-step method for the histochemistry and cytochemistry of Ca^{++}-ATPase activity, *Acta Histochem. Cytochem.*, 14, 705, 1981.
2. **Fujimoto, K. and Ogawa, K.,** Enzyme cytochemical study of rat cardiac muscle. II. Ca^{++}-ATPase and ouabain-sensitive, K^+-dependent *p*-nitrophenylphosphatase, *Acta Histochem. Cytochem.*, 15, 338, 1982.
3. **Ando, T., Mayahara, H., Fujimoto, K., Miyajima, H., and Ogawa, K.,** Ultracytochemical localization of Ca^{++}-ATPase activity on the epithelial cells of rat trachea, *Acta Histochem. Cytochem.*, 15, 812, 1981.
4. **Ogawa, K. S., Fujimoto, K., and Ogawa, K.,** Ultracytochemical studies of adenosine nucleotidases in aortic endothelial and smooth muscle cells — Ca^{++}-ATPase and Na^+, K^+-ATPase, *Acta Histochem. Cytochem.*, 19, 601, 1986.
5. **Yasubuchi, Y., Ueno, S., Tsukahara, I., and Ogawa, K.,** Ultracytochemical localization of Ca^{++}-ATPase activity in the epithelial cells of guinea pig ciliary body, *Acta Histochem. Cytochem.*, 18, 1, 1985.
6. **Davis, W. L., Jones, R. G., and Goodman, D. B. P.,** Electron microscopic cytochemical localization of Ca^{++}-ATPase in toad urinary bladder, *J. Histochem. Cytochem.*, 35, 39, 1987.
7. **Mata, M., Staple, J., and Fink, D. J.,** Cytochemical localization of Ca^{++}-ATPase in peripheral nerve, *Brain Res.*, 445, 47, 1988.
8. **Ogawa, K. S., Fujimoto, K., and Ogawa, K.,** Cytochemical localization of Ca^{++}-ATPase and Na^+, K^+-ATPase in acid-secreting parietal cell and non-secreting parietal cell, *Acta Histochem. Cytochem.*, 20, 197, 1987.
9. **Ueno, S., Bambauer, H. J., Umar, H., Ueck, M., and Ogawa, K.,** Ultracytochemical study of Ca^{++}-ATPase and K^+-NPPase activities in retinal photoreceptors of the guinea pig, *Cell Tissue Res.*, 237, 479, 1984.
10. **Fujimoto, K., Narimoto, K., Ogawa, K. S., Kondo, S., and Ogawa, K.,** Cytochemical aspects of gap junction: enzyme activity, hydrophilic channel and charge, *Acta Histochem. Cytochem.*, 18, 467, 1985.
11. **Mata, M. and Fink, D. J.,** Ca^{++}-ATPase in the central nervous system: an EM cytochemical study, *J. Histochem. Cytochem.*, 37, 971, 1989.
12. **Davis, W. L., Hagler, H. K., Farmer, G. R., Martin, J. H., and Bridges, G.,** Electron microscopic cytochemical localization of Ca^{++}-ATPase in the rod outer segments of the toad *Bufo marinus*, *Anat. Rec.*, 221, 761, 1988.

XIII. *p*-NITROPHENYL PHOSPHATASE

Harumichi Seguchi

A. INTRODUCTION

p-Nitrophenyl phosphatase (*p*-NPPase) is the general term used by histochemists to designate a group of enzymes which catalyze the hydrolysis of the substrate *p*-nitrophenyl phosphate (*p*-NPP). The chemical reaction proceeds as indicated below.

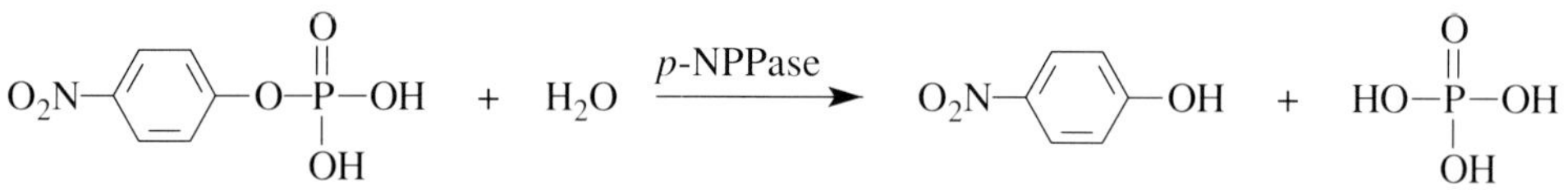

p-Nitrophenyl phosphate *p*-Nitrophenol Phosphoric acid

p-NPP has proved so far an excellent substrate for use in phosphatase research, since its hydrolysis yields an end-product, *p*-nitrophenol, which can be measured spectrophotometrically and allows the quantitative determination of enzymatic activity under cytochemical conditions, and also because it is hydrolyzed by a variety of phosphatases. This latter characteristic is also a handicap for the determination of enzyme specificity. However, it can be overcome by pH adjustments and by employing appropriate activators and inhibitors. The specificity of the reaction can also be assessed by careful examinations of the cytochemical localization of the reaction product.

The most relevant *p*-NPPases are classified as follows:

1. Nonspecific alkaline phosphatase (EC 3.1.3.1)
2. Nonspecific acid phosphatase (EC 3.1.3.2)
3. Specific alkaline *p*-NPPase (*p*-NPP phosphohydrolase, EC 3.1.3.41)
4. Acid *p*-NPPase (low-molecular-weight acid phosphatase)
5. Ouabain-sensitive, K^+-dependent *p*-NPPase or Na^+,K^+-ATPase (EC 3.6.1.3)
6. H^+-Transporting, K^+-stimulated *p*-NPPase or H^+,K^+-ATPase
7. Ouabain-insensitive, K^+-independent *p*-NPPase

See Chapter 3 for discussions on nonspecific alkaline phosphatase (EC 3.1.3.1); nonspecific acid phosphatase (EC 3.1.3.2); ouabain-sensitive, K^+-dependent *p*-NPPase Na^+,K^+-ATPase (EC 3.6.1.3); and H^+-transporting, K^+-stimulated *p*-NPPase H^+,K^+-ATPase.

B. SPECIFIC ALKALINE *p*-NPPase (*p*-NPP PHOSPHOHYDROLASE, EC 3.1.3.41)

This enzyme was first isolated and partially purified from cultures of wild-type *Saccharomyces cerevisiae* by Gorman and Hu,[1] and from commercial baker's yeast by Attias et al.[2]

The crude extract is heated and the supernant is then subjected to acetone fractionation followed by DEAE-cellulose column chromatography, $(NH_2)_2SO_4$ fractionation, and gel-exclusion chromatography on Sephadex G-100. The yield has been reported to be 23% for a 200-fold purification. The molecular weight was estimated to be 60,000 by gel chromatography.

This enzyme was found to catalyze the hydrolysis of *p*-NPP at a pH optimum of 8.0 to 8.2. The enzymatic activity is strongly stimulated by Mg^{2+}; Ca^{2+}, Zn^{2+}, and Be^{2+} are potent inhibitors, and inorganic phosphate is a competitive inhibitor. Ethylene diamine tetraacetate (EDTA) decreases the enzymatic activity by lowering the Mg^{2+} in the medium.[3] This enzyme is also stimulated by Co^{2+}, Zn^{2+}, Ni^{2+}, and Mn^{2+}, but to a lesser extent. The activation by Mg^{2+} is strongly antagonized by Ca^{2+}, Zn^{2+}, and Mn^{2+}; Na^+ and K^+ do not have any specific effect. The enzyme activity is inhibited by fluoride, arsenate, and *p*-chloromercuribenzoate (PCMB). Ouabain has no effect.[4]

This enzyme has not been found in any other living organisms.

C. ACID *p*-NPPase

This is a low-molecular-weight, nonlysosomal acid phosphatase (≦30,000). It differs from lysosomal acid phosphatase in that it does not hydrolyze the acid phosphatase substrate β-glycerophosphate. It is only specific for *p*-NPP, phenyl phosphate, and a few other substrates. The optimum pH is 5.5.

See also Article V (Araki) in this subsection.

1. Method

The usual preparative techniques can be used for light (LM) and electron microscopy (EM). Incubation is performed in the reaction medium of Miyayama et al.[5] for acid phosphatase. The composition of the incubation medium is set out below:

	LM	EM
p-NPP	1.5 m*M*	3.0 m*M*
Lead nitrate	2.0 m*M*	3.6 m*M*
Acetate buffer	0.05 *M*	0.1 *M*

The final pH is adjusted to pH 5.8; incubation is performed for 15 to 30 min at 37°C.

2. Results

Microvilli, lysosomes, basal infoldings of the plasma membrane, endoplasmic reticulum, and nuclear envelope of the epithelial cells of the renal tubule of mouse kidney give a positive reaction.[5] The physiological significance of this enzyme is still unknown.

D. OUABAIN-INSENSITIVE, K^+-INDEPENDENT *p*-NPPase

Cytochemical studies on the K^+-*p*-NPPase revealed the existence of a ouabain-insensitive, K^+-independent *p*-NPPase activity which exhibits tissue and cell specificity. In 1981, Ueno et al.[6] reported the presence of *p*-NPPase activity in the Müller's cells in the retina. This activity was insensitive to ouabain and did not require K^+ ions in the reaction medium. They observed the reaction product in the endoplasmic reticulum and in the nuclear envelope of these cells and, though the functional significance and the characteristics of this enzyme have not been clarified yet, these authors proposed its use as a marker enzyme for the Müller's cells.[6]

1. Histochemical Method

a. EM Procedure

1. Perfusion fixation with 2% paraformaldehyde and 0.25% or 0.5% glutaraldehyde in 0.1 *M* cacodylate buffer, pH 7.4, containing 6 to 8% sucrose, at 0 to 4°C, for 5 min is recommended. The dissected tissue is then immersed in the same fixative for a further 10 min.
2. Wash in the same buffer solution supplemented with 10% dimethyl sulfoxide (DMSO) for a minimum of 2 h (usually overnight).

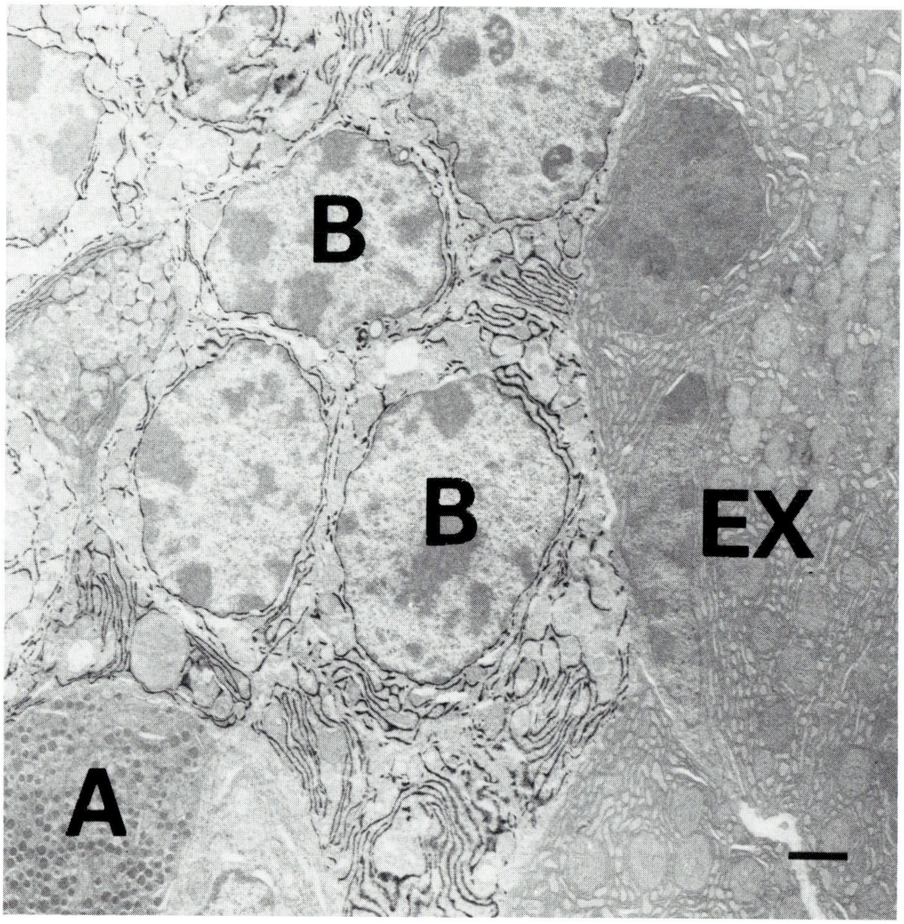

FIGURE 1. K^+-independent *p*-NPPase activity in guinea pig pancreatic islets. The reaction product of the ouabain-insensitive enzyme is observed in the cisternae of the endoplasmic reticulum and of the nuclear envelope in B cells (B). A, A cell; Ex, exocrine cell. (Bar = 1 μm.)

3. Cut 40-μm sections of preferably nonfrozen tissue with a Vibratome or a Microslicer.
4. Incubation in Ueno's modified reaction medium[6] of the Mayahara histochemical method[7] for non-K^+-*p*-NPPase, at room temperature or at 37°C for 5 to 30 min.

The composition of the modified incubation medium is

		Final conc (m*M*)
1 *M* Glycine-NaOH buffer, pH 9.0–9.4	2.5 ml	250
Levamisole	6.02 g	2.5
1% Lead citrate (dissolved in 50 m*M* NaOH)	4.0 ml	4.0
DMSO or distilled water	2.5 ml	
0.1 *M* *p*-NPP (Mg salt)	1.0 ml	10.0
Final pH 9.0–9.4		
Final volume	10.0 ml	

Notes: (1) In this incubation medium, K^+ ions have been removed from the reaction medium for K^+-*p*-NPPase and, therefore, KOH has to be substituted for NaOH in the buffer.

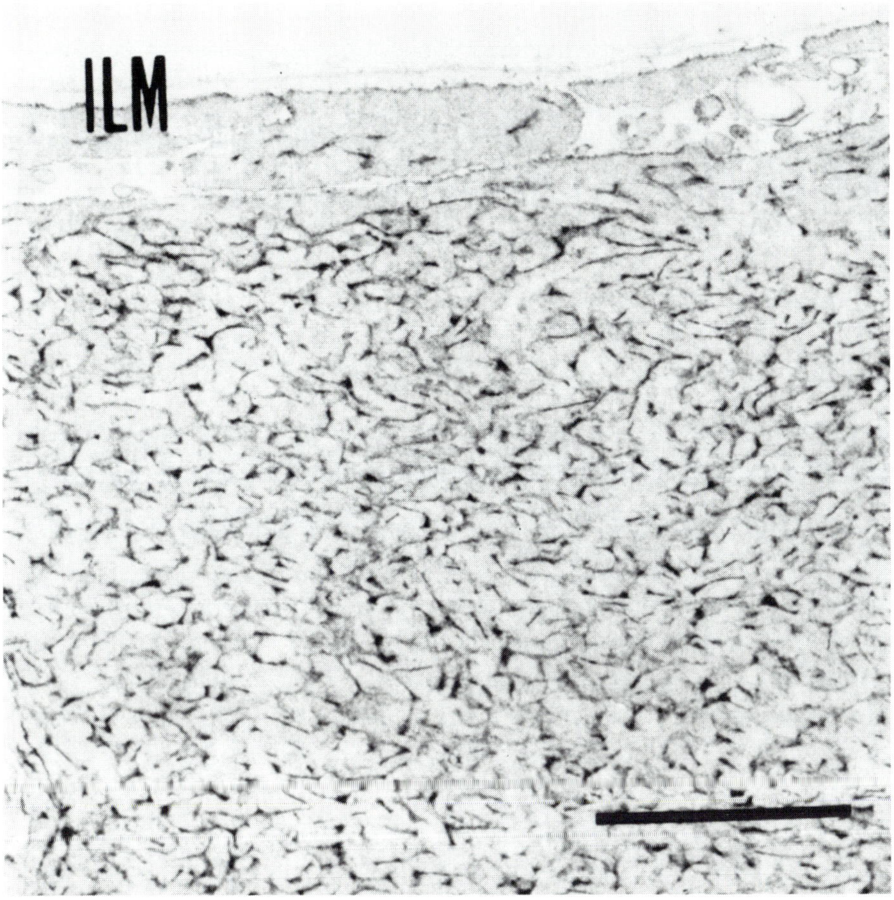

FIGURE 2. The inner limiting membrane (ILM) of the guinea pig retina. Ouabain-insensitive, K^+-independent *p*-NPPase activity is observed in the endoplasmic reticulum of Müller's cells. (Bar = 1 μm.) (Courtesy of Dr. S. Ueno.)

(2) Lead citrate is also dissolved in NaOH instead of KOH. (3) Unlike K^+-NPPase activity, this enzyme activity is not enhanced by DMSO. Therefore, the reaction is not affected by the use of 2.5 ml of distilled water instead of DMSO. (4) This incubation medium was originally developed to serve as control for the K^+-*p*-NPPase reaction, in order to see the effect of omission of K^+ ions from the medium.

Following incubation, wash in 0.1 *M* cacodylate buffer for 5 min (DMSO reacts with osmium and precipitates). Postfix with 1% OsO_4 in 0.1 *M* cacodylate buffer, for 30 to 60 min. Dehydrate in alcohol, and embed in epoxy resin.

b. LM Procedure

After incubation, as set out above, the sections are washed in distilled water for 1 min, followed by color development in 1% ammonium sulfide for 2 min, washing, and mounting in glycerin jelly.

2. Results

Ueno et al.[9] reported a high consistent reaction in the Müller's cells of the retina (Figures 1 and 2).

Notes: (1) If non-K^+-*p*-NPPase activity can be detected when K^+ ions are omitted from the incubation medium for K^+-*p*-NPPase, it should also be detectable when ouabain is added to the reaction medium. However, histochemically, the effect of ouabain is not complete and, therefore, the interpretation of the results is sometimes difficult. In such cases, good results can be obtained by previous incubation of the sections in buffer containing 10 m*M* ouabain.[8,9] (2) In rat tissues, however, the effect of ouabain is incomplete, and therefore, it is recommended to use guinea pig tissues as the test material.

REFERENCES

1. **Gorman, J. A. and Hu, A. S. L.,** The separation and partial characterization of L-histidinol phosphatase and an alkaline phosphatase of *Saccharomyces cerevisiae, J. Biol. Chem.,* 244, 1645, 1969.
2. **Attias, J., Bonnet, J. L., and Sauvagnargues, J. C.,** Separation et etude partielle de deux phosphatases alcalines de levure, *Biochim. Biophys. Acta,* 212, 315, 1970.
3. **Attias, J. and Bonnet, J. L.,** A specific alkaline *p*-nitrophenylphosphatase activity from baker's yeast, *Biochim. Biophys. Acta,* 269, 422, 1972.
4. **Attias, J. and Durand, H.,** Further characterization of a specific *p*-nitrophenylphosphatase from baker's yeast, *Biochim. Biophys. Acta,* 321, 561, 1973.
5. **Miyayama, H., Solomon, R., Sasaki, M., Lin, C. W., and Fishman, W. H.,** Demonstration of lysosomal and extralysosomal sites for acid phosphatase in mouse kidney tubule cells with *p*-nitrophenylphosphate lead-salt technique, *J. Histochem. Cytochem.,* 23, 439, 1975.
6. **Ueno, S., Mayahara, H., Tsukahara, I., and Ogawa, K.,** Ultracytochemical localization of ouabain-sensitive, potassium-dependent *p*-nitrophenylphosphatase activity in guinea pig retina. II. Neurons and Müller cells, *Acta Histochem. Cytochem.,* 14, 186, 1981.
7. **Mayahara, H., Fujimoto, K., Ando, T., and Ogawa, K.,** A new one-step method for the cytochemical localization of ouabain-sensitive, potassium-dependent *p*-nitrophenylphosphatase activity, *Histochemistry,* 67, 125, 1980.
8. **Ueno, S., Bambauer, H. J., Umar, H., Ueck, M., and Ogawa, K.,** Ultracytochemical study of Ca^{++}-ATPase and K^+-NPPase activities in retinal photoreceptors of the guinea pig, *Cell Tissue Res.,* 237, 479, 1984.
9. **Ueno, S., Mayahara, H., Ueck, M., Tsukahara, I., and Ogawa, K.,** Ultracytochemical localization of ouabain-sensitive, potassium-dependent *p*-nitrophenylphosphatase activity in the lacrimal gland of the rat, *Cell Tissue Res.,* 234, 497, 1983.

XIV. TRIMETAPHOSPHATASE

Nobukazu Araki

A. INTRODUCTION

The cytochemical demonstration of trimetaphosphatase (TMPase) activity, an acid inorganic polyphosphatase (EC 3.6.1.2) activity, is based on metal chelation with a ring polymer of phosphoric acid (trimetaphosphate) used as substrate.[1-4] The acid TMPase has a pH optimum of pH 4.0, and is not activated by divalent cations.[3] The intracellular distribution of TMPase is similar to that of acid phosphatase, the most well-known marker enzyme for lysosomes. Although the physiological role of the acid inorganic polyphosphatase remains unknown, it has been proposed that the TMPase is a consistent and reliable cytochemical marker for lysosomes.[5] In spite of the popularity of the cytochemical techniques for acid phosphatase, as a lysosomal marker enzyme, the cytochemical demonstration of acid phosphatase may give a false-negative reaction as the result of technical inadequacies such as overly prolonged fixation. The cytochemical demonstration of TMPase activity has several advantages over that of acid phosphatase. It is less sensitive to fixation and nonspecific staining is less frequent than in the case of acid phosphatase.[5,6] Hence, it has been frequently reported that some organelles are reactive for TMPase but not for acid phosphatase.[7-15] Therefore, considerable interest in using TMPase as a new cytochemical marker, which may define a novel type of lysosome, has arisen. Relying not only on the demonstration of acid phosphatase activity, but also on the localization of TMPase for the cytochemical identification of lysosomes, would be more sensible.

The staining technique for TMPase is based on Berg's chelate removal method.[4] The general scheme for the cytochemical reaction is as follows:

Substrate-metal chelate and capturing anion (soluble)

↓ enzyme

product (soluble) + salt of metal and anion (insoluble)

The addition of lead acetate to a solution of trimetaphosphate forms a lead chelate with a ring structure of trimetaphosphate. Since the lead-trimetaphosphate chelate is soluble in the form of an anion in the incubation medium, the substrate-metal chelate doubles in function as a capturing anion, forming an insoluble salt of metal. TMPase cleaves one of the chelating sites in a stable lead trimetaphosphate by enzymatic hydrolysis, and releases lead ions together with capturing anions (pyrophosphate and orthophosphate). The solubility product of lead ions in the medium is eventually exceeded, and the lead ions form salts with the capturing anions. As lead pyro- and orthophosphate are insoluble, they precipitate *in situ* as the reaction product.

B. METHODS

In earlier light microscopic studies by Berg,[1-4] TMPase was demonstrated at a variety of pH values, and the activity was not suggested to be lysosomal. Berg's method for acid TMPase was adapted to electron microscopic cytochemistry by Doty et al.,[5,15] and the enzyme was proven to be useful as a cytochemical marker for the electron microscopic localization of lysosomes.[5] The incubation medium for electron microscopic demonstration of TMPase is given in Procedure 4.

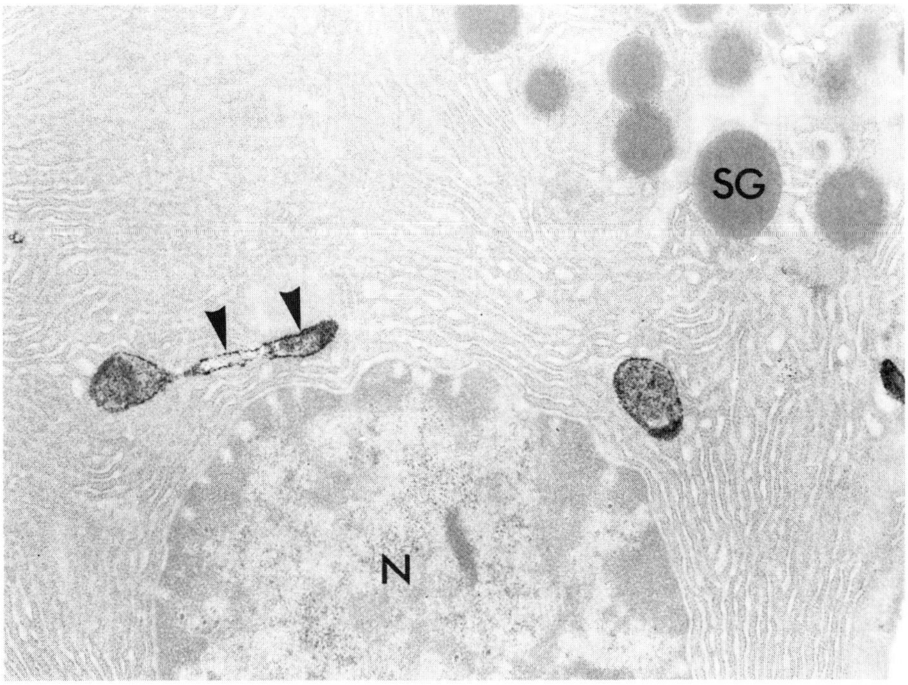

FIGURE 2. The localization of TMPase activity in rat pancreatic acinar cells. The tissue was incubated in a medium containing 2 m*M* cerium chloride instead of lead acetate. The reaction product is localized on lysosomes. An elongated lysosome showing TMPase activity is seen (arrows); N, nucleus; SG, secretory granule. (Magnification × 19,600.)

REFERENCES

1. **Berg, G. G.,** Histochemical demonstration of acid inorganic polyphosphatase, *J. Histochem. Cytochem.*, 3, 22, 1955.
2. **Berg, G. G.,** Histochemical demonstration of acid trimetaphosphatase and tetrametaphosphatase, *J. Histochem. Cytochem.*, 8, 92, 1960.
3. **Berg, G. G. and Gordon, L. H.,** Presence of trimetaphosphatase in the intestinal mucosa and properties of the enzyme, *J. Histochem. Cytochem.*, 8, 85, 1960.
4. **Berg, G. G.,** The staining of triphosphatases by the chelate removal method, *J. Histochem. Cytochem.*, 12, 341, 1964.
5. **Doty, S. B., Smith, C. E., Hand, A. R., and Oliver, C.,** Inorganic trimetaphosphatase as a histochemical marker for light and electron microscopy, *J. Histochem. Cytochem.*, 25, 1381, 1977.
6. **Petty, H. R., Hermann, W., and McConnell, H. M.,** Cytochemical study of macrophage lysosomal inorganic trimetaphosphatase and acid phosphatase, *J. Ultrastruct. Res.*, 90, 80, 1985.
7. **Oliver, C.,** Cytochemical localization of acid phosphatase and trimetaphosphatase activities in exocrine acinar cells, *J. Histochem. Cytochem.*, 28, 78, 1980.
8. **Oliver, C.,** Characterization of basal lysosomes in exocrine acinar cells, *J. Histochem. Cytochem.*, 31, 1209, 1983.
9. **Shannon, W. A., Jr. and Zellmer, D. M.,** A new species of lysosomes in rabbit polymorphonuclear leukocytes, *J. Histochem. Cytochem.*, 29, 1099, 1981.
10. **Maxwell, M. H.,** The distribution and localisation of acid trimetaphosphatase in developing heterophils and eosinophils in the bone marrow of the fowl and the duck, *Cell Tissue Res.*, 235, 171, 1984.

11. **Beaudoin, A. R., Grondin, G., Lord, A., and Pelletier, C.,** β-NADPHase- and TMPase-positive "snake-like" tubules in the exocrine pancreas: cytochemical and immunocytochemical studies, *J. Histochem. Cytochem.*, 33, 569, 1985.
12. **Jamur, M. C., Vugman, I., and Hand, A. R.,** Ultrastructural and cytochemical studies of acid phosphatase and trimetaphosphatase in rat peritoneal mast cells developing in vivo, *Cell Tissue Res.*, 244, 557, 1986.
13. **Kobayashi, T., Okada, T., and Seguchi, H.,** Ultracytochemical study of the trimetaphosphatase activity in the spiral ligament of guinea pig inner ear using cerium, *J. Histochem. Cytochem.*, 36, 959, 1988.
14. **Seguchi, H., Kobayashi, T., Zhang, S. X., and Okada, T.,** Biochemical and cytochemical study of trimetaphosphatase in guinea pig tissues, *Acta Histochem. Cytochem.*, 22, 734, 1989.
15. **Doty, S. B. and Schofield, B. H.,** Electron microscopic localization of hydrolytic enzymes in osteoclasts, *Histochem. J.*, 4, 245, 1972.
16. **Payer, A. F., Battle, C. L., and Peake, R. L.,** Use of osmium-ferrocyanide treatment for improved lysosomal acid trimetaphosphatase reaction and subcellular detail in thyroid follicular cells, *J. Histochem. Cytochem.*, 28, 183, 1980.
17. **Tiffon, Y., Buat, M. L., Landemore, G., and Izard, J.,** Cytochemical localization of acid phosphatase and trimetaphosphatase activities in Kurloff cells, *Biol. Cell*, 56, 85, 1986.
18. **Kobayashi, T., Seguchi, H., Nishiyama, S., and Okada, T.,** Localization of trimetaphosphatase activity in epithelium of the guinea pig cochlear duct, *Equilibrium Res. Suppl.*, 5, 65, 1989.
19. **Araki, N., Takashima, Y., and Seguchi, H.,** unpublished data, 1990.
20. **Araki, N., Matsubara, H., Takashima, Y., and Ogawa, K.,** Three-dimensional distribution of acid phosphatase-positive thread-like structures "nematolysosomes" in rat pancreatic exocrine cells, *Acta Histochem. Cytochem.*, 22, 585, 1989.

XV. THE SULFATASES

Mary E. Bentfeld-Barker

A. INTRODUCTION

The sulfatases are a diverse group of enzymes which catalyze the hydrolysis of various sulfate esters according to the general reaction:

$$R \cdot O \cdot SO_3^- + H_2O \rightarrow R \cdot OH + H^+ + SO_4^{2-}$$

They are divided into several subclasses, according to the type of sulfate ester hydrolyzed. (1) Arylsulfatases, found in all animal species studied, some higher plants, and some microorganisms, catalyze the hydrolysis of arylsulfates, i.e., sulfates containing a phenol or phenol derivative. Although they are active with several artificial substrates, their physiological substrates are unknown. (2) Steroid sulfatases hydrolyze the sulfate esters of steroid alcohols or phenols, and are found in mammalian liver as well as in digestive glands of mollusks. (3) Glycosulfatases, also found in mollusks as well as in bacteria, hydrolyze carbohydrate sulfates, including such complex molecules as cellulose polysulfates and disaccharides produced by the degradation of chondroitin sulfate. Several other minor sulfatases have been described but are limited mostly to bacteria and fungi. For a detailed description of these enzymes, the reader is referred to the exhaustive studies of Roy and colleagues.[1-3]

For electron microscopic cytochemistry only the arylsulfatases (ARS) are of interest, since no methods presently exist for demonstration of the other sulfatases. ARS are further classified into types I and II. Type I ARS have relatively low substrate specificity, show an alkaline pH optimum, are insensitive to sulfate or phosphate ions, but strongly inhibited by sulfite and cyanide. Type I ARS include ARS C, found in the endoplasmic reticulum of mammalian liver and kidney,[3] as well as several enzymes found in microorganisms. Type II ARS are widespread in animal tissues and in microorganisms. They are further divided into ARS A and B, both of which rapidly and specifically hydrolyze *p*-nitrocatecholsulfate, but other substrates only very slowly. These enzymes are both inhibited by phosphate and sulfate as well as sulfite, but not by cyanide or fluoride. They have an optimum pH of about 5.0, and are generally considered to be lysosomal enzymes.[3]

The cytochemical localization of ARS is based upon the formation of an insoluble precipitate at the site of enzyme action. Although in principle either the anion or the cation liberated in the reaction could be visualized, in practice precipitation of the sulfate anion by a heavy metal has been more successful. The most widely used methods utilize either lead or barium to visualize the enzyme reaction product.

B. METHODS FOR LOCALIZATION OF TYPE II ARYLSULFATASE

ARS A and B both have the same substrate specificity and approximately the same pH optimum and, therefore, differentiation of these two enzymes cytochemically is not readily accomplished. The genes for the two enzymes are known to be on different chromosomes, and the clinical syndromes produced by the absence of one enzyme or the other are strikingly different.[4] However, which of the two (or possibly both) enzymes is demonstrated by the methods to be described is uncertain. In some tissues ARS A has been reported to have a pH optimum of 4.2, while that for ARS B is about 6,[5] and therefore it has been assumed that the majority of reaction product is due to ARS B.[6]

1. Tissue Fixation

The fixative of choice appears to be glutaraldehyde, in spite of the fact that it may inhibit more than 70% of the enzyme activity. Maximum inhibition occurs within the first

5 min, but activity can be somewhat restored by thorough washing in buffer before incubation.[7] Phosphate buffers should be avoided since phosphate ions inhibit arylsulfatase. Cells in suspension can be fixed in 1 to 3% glutaraldehyde in cacodylate buffer for 10 to 30 min and then thoroughly washed for 1 to 2 h. Solid tissues can be fixed by perfusion, followed by immersion for up to 30 to 60 min. They should then be cut into 40- to 60-μm sections on a tissue chopper or Vibratome and washed thoroughly for several hours or overnight.

2. Incubation

The incubation medium consists of the enzyme substrate, here *p*-nitrocatechol sulfate, (2-hydroxy-5-nitrophenyl-sulfate), an appropriate buffer, and the heavy-metal capturing agent. The substrate concentration should be sufficient to saturate the enzyme and the pH of the buffer should be adjusted to the pH of optimum activity of the enzyme.

The two most commonly used incubation media for demonstrating ARS are those of Hopsu-Havu et al.[8] and Goldfischer.[9] Hopsu-Havu's method utilizes either barium or lead ions as capturing agents, while Goldfischer's uses only lead. The two methods give similar results although barium may produce a "cleaner" reaction since it does not precipitate nonspecifically in the reaction medium nor does it inhibit the enzyme. Lead, on the other hand, may form nonspecific complexes and inhibit enzyme activity to some extent. Nevertheless, lead may be preferable since lead sulfate can be easily converted to lead sulfide which is visible in the light microscope. Tissue reactivity can then be assessed before subsequent processing for electron microscopy.

Before the cells or tissue sections are incubated in the complete medium, they should be washed several times in the same buffer used in making up the medium. The osmolality of the wash buffer should be adjusted with sucrose.

The medium used by Goldfischer's method is as follows:

p-Nitrocatechol sulfate	15 to 30 mg
Veronal-acetate buffer (pH 5.4)	5.0 ml
Lead nitrate (24%)	0.16 ml
(Adjust pH to 5.5 with 0.1 *N* HCl.)	

Hopsu-Havu's medium is:

p-Nitrocatechol sulfate in 3 ml distilled H_2O	120 mg
Acetate buffer (0.1 *M*), pH 5.5	3.0 ml
Barium chloride (5%) or lead nitrate (8%)	3.0 ml
(Adjust pH to 5.0 to 5.5 with 0.2 *M* acetic acid.)	

The duration of incubation is 10 to 90 min at 30 to 37°C.

3. Controls

In order to ensure that heavy metal precipitates are not nonspecifically adsorbed to tissue components, some incubations must always be carried out without substrate. Nor is it sufficient to perform this control incubation once for a given tissue. Since the reaction can at times be somewhat capricious, it is important to include a sample without substrate every time the incubation is done. Some specimens should also be processed without postfixation with OsO_4, since this too can sometimes cause nonspecific precipitates. Finally, when analyzing the sections in the electron microscope, some sections should be left unstained, as excess staining can mask the reaction product, or even remove it from the section.[10]

For a more detailed discussion of these methods the reader should consult the excellent review by Hopsu-Havu and Helminen.[11]

4. Arylsulfatase A

A method to specifically demonstrate arylsulfatase A (ARS-A) electron microscopically has been published.[12] However, to date it has been utilized only on cultured human fibroblasts. For details, the reader should consult the original paper.

5. Arylsulfatase C

Although the enzyme is abundant in the rough endoplasmic reticulum of mammalian liver, and has been thoroughly studied biochemically, a method for the cytochemical demonstration of ARS-C has only recently been reported.[13] Since its optimum pH is basic (around 8.0) and neither ARS A nor B hydrolyzes the substrate used to any great extent, there should be little confusion about the interpretation of results. The present author has no experience with this procedure, and the reader is referred to the original paper.

6. Subsequent Processing

After incubation in any of the above media, tissues can be postfixed in OsO_4, dehydrated, and embedded according to standard procedures. Washing and conversion of lead sulfate into lead sulfide is described in detail below.

C. EQUIPMENT AND REAGENTS

All incubations and washing steps can be carried out in standard vials or tubes. If cells in suspension are used, they should be kept in 5- to 15-ml centrifuge tubes so they can be spun down before each change of medium or wash buffer. Other necessary equipment includes:

1. Tissue chopper (Sorvall) or Vibratome (Oxford) to cut solid tissues into 40- to 60-μm sections
2. Oven or shaking water bath
3. Rotator or shaking device — More uniform reaction will be achieved if the specimens can be agitated during the incubation period; if the rotator cannot be accommodated in the oven it is preferable to incubate with agitation at room temperature for a longer period.
4. Beckman microfuge to pack cells in suspension for embedding

Fixatives are listed below:

1. Glutaraldehyde — 1.5% in 0.1 *M* sodium cacodylate-HCl, pH 7.4, +1% sucrose (keeps for up to 4 weeks refrigerated)

   30 ml 10% Glutaraldehyde (from any reliable EM supplier)
   100 ml 0.2 *M* Sodium cacodylate stock
   50 ml Glass-distilled H_2O
   pH to 7.4 with HCl and bring total volume to 200 ml
   Add 2 g sucrose

2. Palade's osmium tetroxide — mix just before use:

   4 ml Acetate-veronal stock
   4 ml 0.1 *M* HCl
   2 ml Distilled H_2O
   pH to 7.6
   Add 10 ml 2% OsO_4
   Add 1 g (5%) sucrose

The incubation medium (Goldfischer) is made up as follows:

p-Nitrocatechol sulfate (Sigma Chemical Co.)	25 mg
Acetate-veronal buffer, pH 5.4	5.0 ml
24% $Pb(NO_3)_2$	0.16 ml

Dissolve *p*-nitrocatechol sulfate in the buffer and add the lead nitrate drop by drop with constant stirring. The lead will precipitate and the pH will rise. Adjust the pH to 5.5 with 0.1 *N* HCl. When it reaches about 5.7 the medium will begin to clear. It should be bright yellow and completely clear. Add 5% sucrose.

Hopsu-Havu's medium can be made up in the same way. The barium chloride should not precipitate.

Buffers are made up as follows:

1. Acetate-veronal stock (100 ml): 1.943 g sodium acetate (mol wt, 136; final concentration, 0.14 *M*); 2.886 g sodium veronal (barbital; since this is a barbiturate, it may be subject to special handling and storage restrictions) (mol wt, 206.18; final concentration, 0.14 *M*); dissolve and dilute to 100 ml.
2. Acetate-veronal buffer (for incubation medium): 20 ml acetate-veronal stock, 20 ml 0.1 *N* HCl; mix and dilute to 100 ml. Adjust pH to 5.4.
3. Acetate-veronal wash buffer: 20 ml acetate-veronal stock, 20 ml 0.1 *N* HCl; mix and dilute to 100 ml; adjust pH to 7.4; add 7% sucrose.
4. 0.2 Sodium cacodylate stock: 21.4 g sodium cacodylate; add distilled H_2O to 500 ml.
5. Sodium cacodylate wash buffer: 50 ml sodium cacodylate stock, 40 ml distilled H_2O; adjust pH to 7.4 with 0.1 *N* HCl; bring volume to 100 ml with H_2O and add 7% sucrose.

Ammonium sulfide is also required, and can be made up as follows: 50 ml acetate-veronal buffer, pH 5.4, 1.0 ml ammonium sulfide (concentrated); mix in well-ventilated fume hood just before use. For the in block stain (Kellenberger), use 20 ml acetate-veronal stock, 28 ml 0.1 *N* HCl, and 52 ml H_2O. Adjust pH to 6.0; add 0.5 g uranyl acetate. Store in refrigerator and protect from light. Add 4% sucrose to aliquot on day of use.

D. RECOMMENDED PROCEDURE

1. Cytochemistry

1. Fix tissues with 1.5% glutaraldehyde in 0.1 *M* sodium cocadylate buffer. Cultured cells, blood cells, or any isolated cells in suspension should be fixed for 30 min. For perfusion of solid tissues, 3% glutaraldehyde may be used. Perfusion should continue for 2 to 5 min. Tissues should be cut into small blocks and immersed in fresh fixative for another 30 to 60 min. They should then be cut into 40- to 60-μm sections.
2. Wash tissues thoroughly in sodium cacodylate wash buffer. Two hours of washing with at least five changes of buffer is minimum. Washing can be continued overnight.
3. Wash tissues three to five times (5 min each) in acetate-veronal buffer, pH 5.4, with 7% sucrose. Cacodylate ions can inhibit the reaction and they must be removed.
4. Incubate in complete medium for up to 3 h at room temperature or 30°C.
5. Wash four to five times in acetate-veronal buffer as in Step 3, until the wash is no longer yellow.
6. Convert reaction product (lead sulfate) to lead sulfide by immersion in 2% ammonium sulfide for 10 min at room temperature. Do this in a fume hood.

7. Wash for 2 h with at least four to five changes of veronal acetate buffer, pH 5.4, with 7% sucrose (as in Steps 3 and 5).
8. Wash three times (5 min each) in acetate-veronal wash buffer, pH 7.4, with 7% sucrose.

2. Subsequent Processing

1. If convenient, specimens may be returned to glutaraldehyde overnight, or they may be immediately postfixed.
2. Postfix in 1% OsO_4 in acetate-veronal buffer at 4°C for 1 to 2 h.
3. Stain with Kellenberger's in block uranyl acetate stain for 30 to 60 min at room temperature in the dark; it is best to rinse specimens once in the stain and then add a fresh batch to let them incubate.
4. Dehydrate and embed according to standard procedures for tissue being studied.
5. Stain sections briefly with lead citrate; leave some sections unstained.

E. INTERPRETATION OF RESULTS

1. Specificity of Reaction

The amount of enzyme reaction can be assessed in tissue sections before embedding by simply mounting a section in glycerol and observing it under the light microscope. Reaction product appears as dark-brown or black deposits. Embedded tissues should also be sectioned at 0.5 to 1.0 μm and examined at the light microscope level before cutting thin sections. The general distribution of the reaction product can be determined in this way (Figure 1). It is also possible to discover whether the reaction is generally uniform. Unfortunately, it is not uncommon for the distribution and/or amount of reaction product to vary from one block to another. Such inconsistencies are apparently due to minute differences in the degree of fixation experienced by different cells in the tissues. Such variations are most pronounced in immersion-fixed tissue.[11]

The presence of nonspecific deposits is best judged on unstained thin sections in the electron microscope. Since the method given localizes a lysosomal enzyme, the presence of dense deposits on nuclei or mitochondria is indicative of nonspecific precipitation (Figure 2). This is a common occurrence when lead is used as the capturing agent. Again the amount of such precipitation may also vary from one block to another. Washing the tissue after incubation and after converting with ammonium sulfide is critical for removing nonspecific precipitates. The "in block" staining with uranyl acetate also helps to clean up random precipitates.

2. Localization of Reaction Product

Lysosomes are the most morphologically heterogeneous organelles in the cell. Reaction product for ARS can be found in all components of the lysosomal system: Golgi[14] and/or GERL cisternae, coated vesicles, some endosomes,[15] multivesicular bodies,[16] primary lysosomes,[17] such as secretion granules of leukocytes, and pleomorphic secondary lysosomes, which include autophagic and heterophagic vacuoles[18,19] (Figures 3 to 5). It has also been demonstrated within the cisternae of the endoplasmic reticulum (Figure 4). This localization is assumed to represent newly synthesized enzyme, which has not yet been packaged into lysosomes. As discussed above, it is not thought to be ARS-C, which is limited to endoplasmic reticulum but does not readily hydrolyze *p*-nitrocatechol sulfate.

When reaction product is demonstrated in leukocyte or macrophage granules, it is often confined to the periphery of the granule or to only a few granules in the cell (Figure 6). Therefore, it is impossible to know whether the enzyme is also thus confined, or is distributed more uniformly. The absence of reaction product cannot be interpreted to necessarily indicate

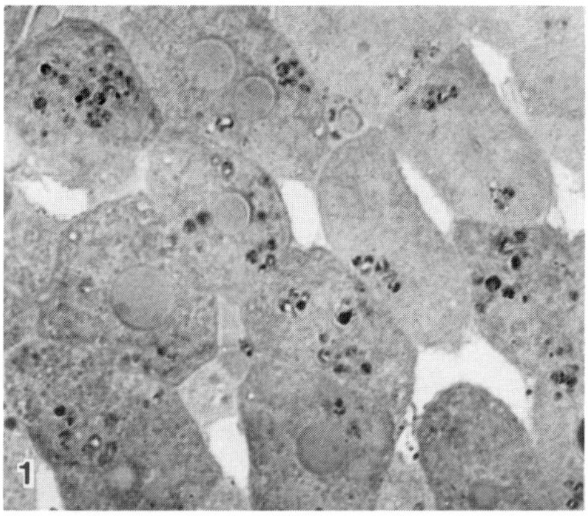

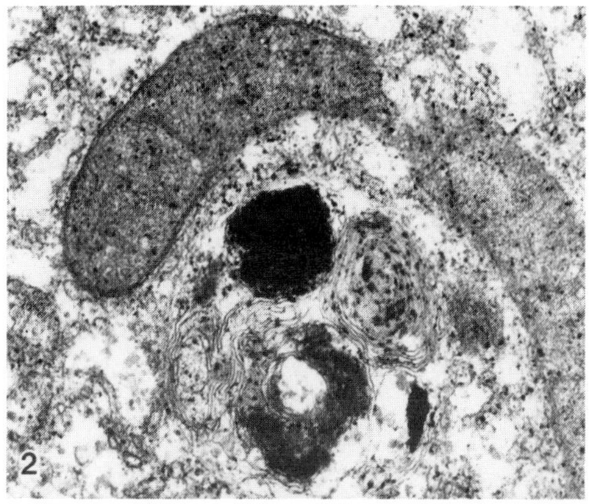

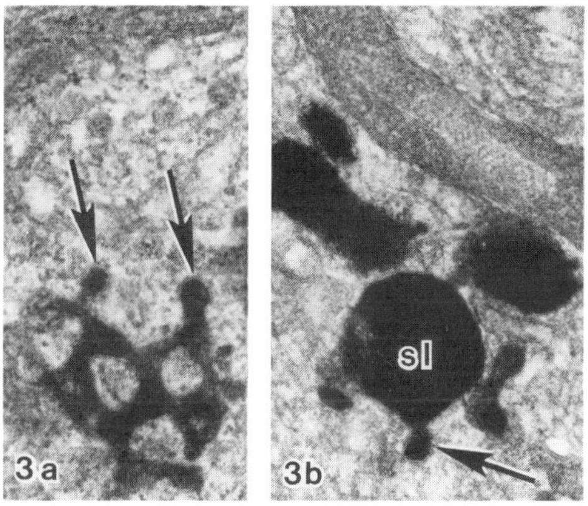

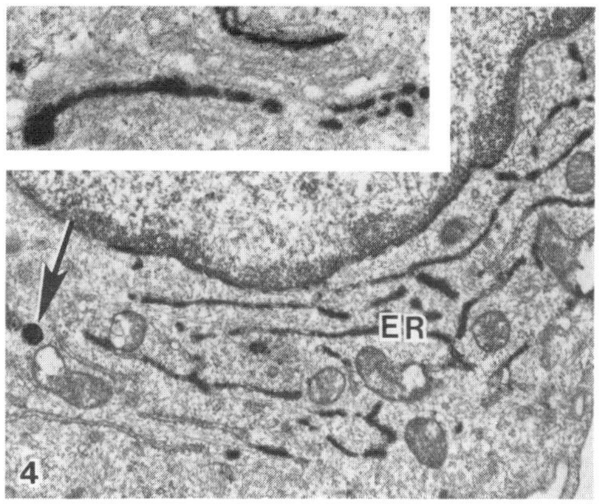

FIGURE 1. A 1-μm section of rat liver incubated in Goldfischer's medium and stained with toluidine blue. Reaction product is seen as small dark spots on the tissue. (Magnification × 2,200.)

FIGURE 2. Electron micrograph of rat liver incubated for demonstration of ARS. Micrograph shows a considerable amount of nonspecific background staining on mitochondria and general cytoplasmic areas. The "washed-out" appearance is due to the in-block uranyl acetate stain, which removes glycogen. (Magnification × 24,400.)

FIGURE 3. Electron micrographs of rabbit alveolar macrophages. (a) Reaction product of ARS (arrows) is seen in GERL and coated vesicles in continuity with it (magnification × 50,000); (b) reactive coated vesicles are continuous with a secondary lysosome (sl) (magnification × 47,900). (Reprinted from Nichols, B. A., *J. Exp. Med.*, 144, 920, 1976. With permission from Rockefeller University Press.)

FIGURE 4. Macrophage from rabbit peritoneal exudate. ARS reaction product is shown in the endoplasmic reticulum (ER) and a lysosome (arrow). The inset shows reaction in the Golgi cisternae of another cell. (Magnification × 19,100; inset × 41,600.) (Reprinted from Nichols, B. A. et al. *J. Cell Biol.*, 50, 498, 1971. With permission from Rockefeller University Press.)

absence of enzyme. Granule contents may inhibit the cytochemical reaction by preventing the substrate or the capture agent from penetrating the interior of the granule.

If equivocal results are obtained, or every block has an unacceptable level of nonspecific deposits, it is necessary to check the purity of all reagents used, including the water. Trace ions in some distilled water can cause unpredictable results. The use of a positive control tissue such as liver or kidney is worth the extra processing. Another important variable to examine is the time and manner of fixation. ARS cannot be demonstrated cytochemically in unfixed tissue, nor in tissue that is too well fixed.[11] The optimum fixation lies somewhere in the middle and may have to be determined empirically. Finally, it may be necessary to fractionate the tissue and examine it biochemically. Methods exist for identification of all ARS in this manner.[1-3]

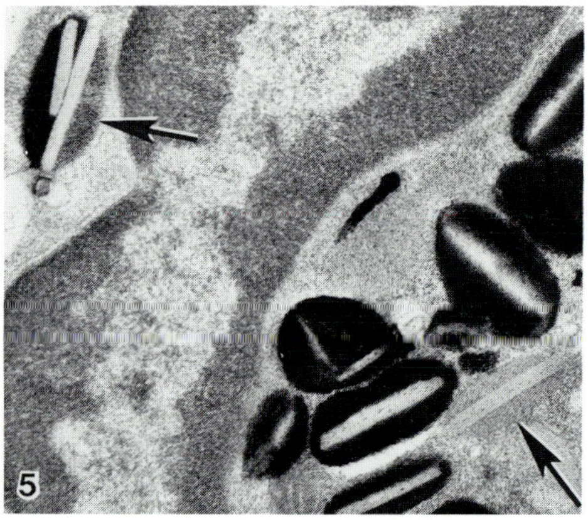

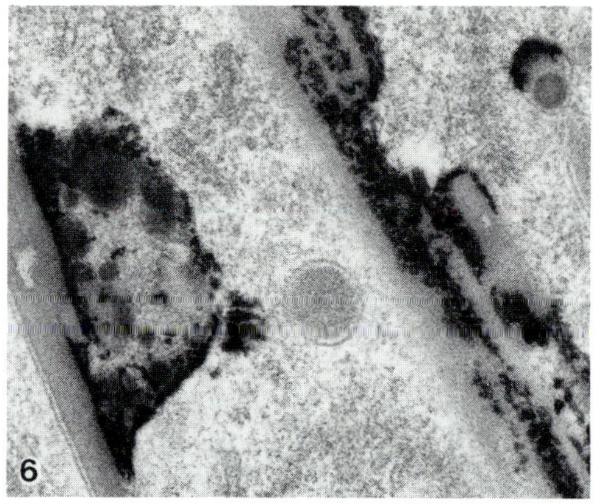

FIGURE 5. A rat eosinophilic myelocyte showing ARS reaction product in most of the granules. Arrows indicate granules containing either no reaction product, or reaction product in only part of the granule. (Magnification × 30,500.)

FIGURE 6. A mouse bone marrow macrophage. ARS reaction product is restricted mostly to the periphery of the large granules. Other granules in the same cell had no reaction. (Magnification × 38,300.)

F. CLINICAL SIGNIFICANCE OF THE ARYLSULFATES

Although the physiological substrates of these enzymes are presently unknown, the genes for the three ARS discussed in this chapter have been localized to three separate chromosomes: ARS-A to chromosome 22, ARS-B to chromosome 5, and ARS-C to the X chromosome.[4,12] Deficiency of any one enzyme produces a distinct metabolic disease. A defect in the lysosomal type II ARS leads to storage diseases: metachromatic leukodystrophy in the absence of ARS-A, and the Maroteaux-Lamy syndrome in the case of ARS-B deficiency. Furthermore, several other clinical syndromes have been described which are due to a lack of some of the other minor lysosomal sulfatases. Lack of ARS-C is associated with x-linked ichthyosis.[20] Fortunately, these genetic diseases are relatively rare, but all of them produce serious debilitating, and sometimes life-threatening, conditions.

REFERENCES

1. Roy, A. B., The synthesis and hydrolysis of sulfate esters, *Adv. Enzymol.*, 22, 205, 1960.
2. **Roy, A. B.,** The hydrolysis of sulfate esters, in *The Enzymes,* Vol. 5, 3rd ed., Boyer, P. D., Ed., Academic Press, New York, 1971, 1.
3. **Nicholls, R. G. and Roy, A. B.,** Arylsulfatase, in *The Enzymes,* Vol. 1, 3rd ed., Boyer, P. D., Ed., Academic Press, New York, 1971, 21.
4. **Holtzman, E.,** *Lysosomes,* Plenum Press, New York, 1989.
5. **Lynch, S. M., Austen, K. F., and Wasserman, S. I.,** Release of arylsulfatase A but not B from rat mast cells by noncytolytic secretory stimuli, *J. Immunol.,* 121, 1394, 1978.
6. **Bentfeld-Barker, M. E. and Bainton, D. F.,** Cytochemical localization of arylsulfatase B in rat basophils and mast cells, *J. Histochem. Cytochem.,* 28, 1055, 1980.
7. **Arborgh, B., Ericsson, J. L. E., and Helminen, H.,** Inhibition of renal acid phosphatase and arylsulfatase activity by glutaraldehyde fixation, *J. Histochem. Cytochem.,* 19, 449, 1971.
8. **Hopsu-Havu, V. K., Arstila, A. U., Helminen, H. J., and Kalimo, H. O.,** Improvements in the method for the electron microscopic localization of arylsulphatase activity, *Histochemie,* 8, 54, 1967.
9. **Goldfischer, S.,** The cytochemical demonstration of lysosomal arylsulfatase activity by light and electron microscopy, *J. Histochem. Cytochem.,* 13, 520, 1965.

10. **Kalimo, H. O., Helminen, H. J., Arstila, A. U., and Hopsu-Havu, V. K.,** The loss of enzyme reaction products from ultrathin sections during the staining for electron microscopy, *Histochemie,* 14, 123, 1968.
11. **Hopsu-Havu, V. K. and Helminen, H.,** Sulfatases, in *Electron Microscopy of Enzymes: Principles and Methods,* Vol. 2, Hayat, M. A., Ed., Van Nostrand Reinhold, New York, 1974, chap. 5.
12. **Chang, P. L. and Moudgil, G.,** A specific ultrastructural stain for arylsulfatase A activity in human cultured fibroblasts, *J. Histochem. Cytochem.,* 32, 617, 1984.
13. **Kawano, J. and Aikawa, E.,** Ultrastructural localization of arylsulfatase C activity in rat kidney, *J. Histochem. Cytochem.,* 35, 523, 1987.
14. **Rhodes, H., Mezitis, S. G. E., Gonatas, N. K., and Fleischer, B.,** Selective effect of nerve growth factor on some Golgi and lysosomal enzyme activities of rat pheochromocytoma (PC12) cells, *Arch. Biochem. Biophys.,* 272, 175, 1989.
15. **Kelly, B. M., Yu, C., and Chang, P. L.,** Presence of a lysosomal enzyme, arylsulfatase-A, in the prelysosome-endosome compartments of human cultured fibroblasts, *Eur. J. Cell Biol.,* 48, 71, 1989.
16. **Jost-Vu, M., Hamilton, R. L., Hornick, C. A., Belcher, J. D., and Havel, R. J.,** Multivesicular bodies isolated from rat hepatocytes: cytochemical evidence for transformation into secondary lysosomes by fusion with primary lysosomes, *Histochemistry,* 85, 457, 1986.
17. **Nichols, B. A., Bainton, D. F., and Farquhar, M. G.,** Differentiation of monocytes: origin, nature, and fate of their azurophil granules, *J. Cell Biol.,* 50, 498, 1971.
18. **Nichols, B. A.,** Normal rabbit alveolar macrophages. II. Their primary and secondary lysosomes as revealed by electron microscopy and cytochemistry, *J. Exp. Med.,* 144, 920, 1976.
19. **Pipe, R. K. and Moore, M. N.,** Arylsulphatase activity associated with phenanthrene induced digestive cell deletion in the marine mussel *Mytilus edulis, Histochem. J.,* 18, 557, 1986.
20. **Kolodny, E. H.,** Current concepts in genetics: lysosomal storage diseases, *N. Engl. J. Med.,* 294, 1217, 1976.

XVI. NUCLEOTIDYL CYCLASES

Kazushige Ogawa and Kazuyuki Taniguchi

A. NATURE OF CYCLASES

In order to precisely grasp the methods for cytochemical localization of the activity of cyclases (adenylate cyclase and guanylate cyclase), it is necessary to understand the nature of these enzymes. Therefore, a brief description, as well as the principle for cytochemical detection, are given here.

Cells in a multicellular organism communicate with one another through a variety of extracellular signal molecules, such as neurotransmitters and hormones, in order to modulate their development, growth, and functions. On the other hand, adenosine 3′,5′-monophosphate (cyclic AMP) is generally recognized as an intracellular regulatory molecule (second messenger) generated by the transduction of information from these extracellular signal molecules, and modulates significant metabolic processes through cyclic AMP-dependent protein kinase. Its generation and degradation are controlled by the enzymes-adenylate cyclase and cyclic AMP phosphodiesterase, respectively. Its intracellular level is thought to be mainly regulated by the adenylate cyclase system located in the cell membrane, which is a target for extracellular signaling molecules and is composed of three kinds of proteins, that is, receptor, guanosine triphosphate-binding regulatory protein (G-protein), and catalytic unit (adenylate cyclase). As for G-protein, it is composed of three subunits, α, β, and γ. Up to now, at least two kinds of G-proteins, stimulatory G-protein (Gs) and inhibitory G-protein (Gi), are recognized as the regulators for adenylate cyclase. Representative target receptors for Gs and Gi are β-adrenergic receptors and α_2-adrenergic receptors, respectively. Currently, the signal transduction to activate adenylate cyclase is thought to be as follows:[1,2] when an agonist (for example, noradrenaline, adenosine, and glucagon) binds to the receptor, the receptor-agonist complex is involved in a transient change in receptor-Gs interaction, allowing guanosine 5′-triphosphate (GTP) to replace guanosine 5′-diphosphate (GDP) on the α subunit (Gsα) in the presence of Mg^{2+}. As a consequence of the GTP binding, Gsα dissociates from β and γ subunits (G$\beta\gamma$). This allows Gsα to bind tightly to an adenylate cyclase molecule and results in the activation of adenylate cyclase. Together with intrinsic GTPase, Gsα hydrolyzes GTP to GDP to release inorganic phosphate within less than a minute. As a result, Gsα dissociates from adenylate cyclase to change it into an inactive form, and then connects with G$\beta\gamma$ to reform Gs, an inactive form. On the other hand, Gi is composed of a different α subunit (Giα) and the same G$\beta\gamma$ as Gs. Activation and inactivation of Gi are processed similarly to those of Gs, as described above. Giα binds to an adenylate cyclase molecule to cause direct inhibition of adenylate cyclase. G$\beta\gamma$ inhibits adenylate cyclase indirectly by binding to free Gsα to prevent it from activating adenylate cyclase.

Guanosine cyclic 3′,5′-monophosphate (cyclic GMP), an alternative cyclic nucleotide recognized in various animal cells, is also regarded as an intracellular messenger, and is thought to modulate cellular metabolic processes mostly through cyclic GMP-dependent protein kinase. However, its functional roles are not as well known as those of cyclic AMP, with a few exceptions, such as those in smooth muscle relaxation and Na^+ channel regulation of retinal rod cells. The generation of cyclic GMP is controlled by guanylate cyclase and its degradation is regulated by cyclic GMP phosphodiesterase. Guanylate cyclase is present in two different forms, a soluble form and a particulate form. The former is not coupled to plasma membrane receptors, although cyclic GMP levels often increase in accordance with the activation of the inositol-phospholipid pathway induced by the extracellular signal molecules such as acetylcholine. The latter is thought to be coupled to the receptors, for example, the receptor of atrial natriuretic factor.

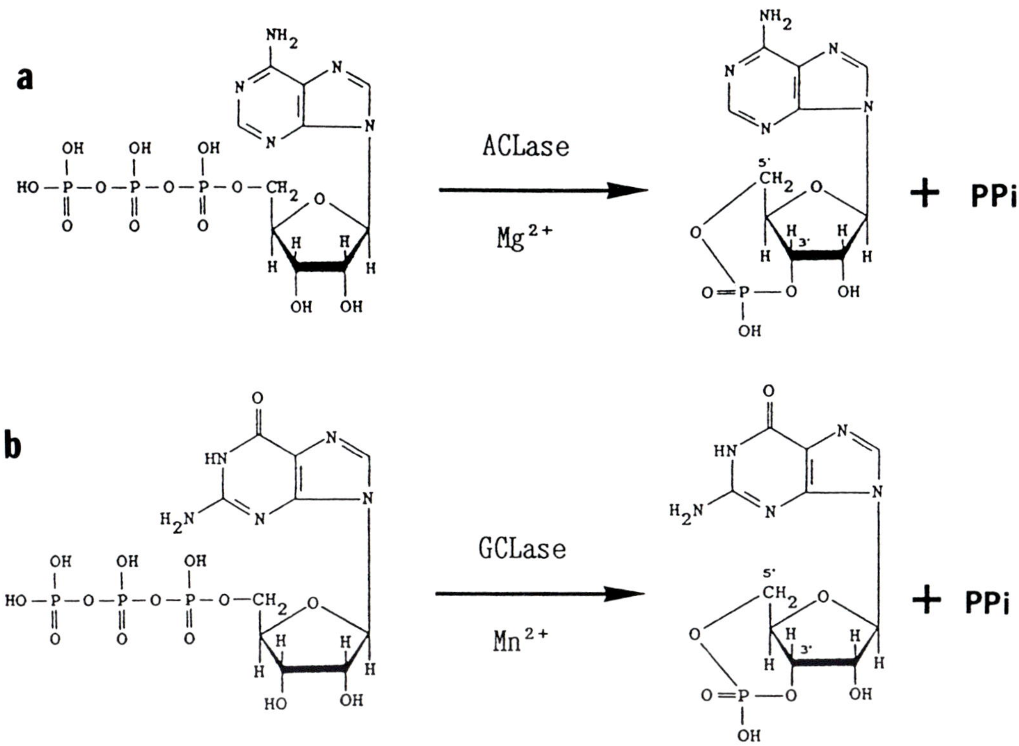

FIGURE 1. The catalytic reactions by ACLase (a) and GCLase (b); PPi, pyrophosphate.

B. GENERAL METHODOLOGICAL CONSIDERATIONS

Adenylate cyclase (EC 4.6.1.1, ATP-pyrophosphate lyase [cyclizing]) hydrolyzes adenosine 5′-triphosphate (ATP) at the first polyphosphate bond with Mg^{2+} as a cofactor and catalyzes the formation of cyclic AMP and pyrophosphate (PPi) (Figure 1). Reik and collaborators[3] first introduced a method for the cytochemical demonstration of adenylate cyclase (ACLase), fundamentally the same method as the one for ATPase introduced by Wachstein and Meisel,[4] and used ATP as substrate and lead ion as capture agent in the reaction medium. The method is based on the concept that heavy metal cations (lead ions) in the reaction medium capture PPi liberated by the enzymatic hydrolysis of substrate, which results in the formation of highly insoluble precipitates at or near the catalytic site of the enzyme. However, the use of ATP as the physiological substrate of ACLase may be subject to involvement of the activity of other ATP-hydrolyzing enzymes such as phosphohydrolases and pyrophosphohydrolases, especially ATPases. Adenylyl imidodiphosphate (AMP-P[NH]P) is an analog of ATP, and is synthesized by replacement of an oxygen with an -NH group between β- and γ-phosphorus atoms of ATP (Figure 2).[5,6] Howell and Whitfield,[7] and Wagner and co-workers[8] introduced AMP-P(NH)P as substrate in the cytochemical reaction medium in order to eliminate the activity of ATPases because it is an effective substrate for ACLase but it is not hydrolyzed by ATPases on account of the inherent stability of the P-N-P bond within it. Therefore, in almost all of the recent cytochemical studies, this newly synthesized substrate has been used as a specific one for ACLase. There is another serious criticism of the cytochemical demonstration of ACLase on the use of lead ion as capture agent: (1) lead ion is a potent inhibitor of ACLase and completely inhibits ACLase activity in millimolar concentration as usually used in cytochemical demonstrations;[9] (2) lead ion causes nonenzymatic hydrolysis of AMP-P(NH)P.[10] In order to overcome these problems,

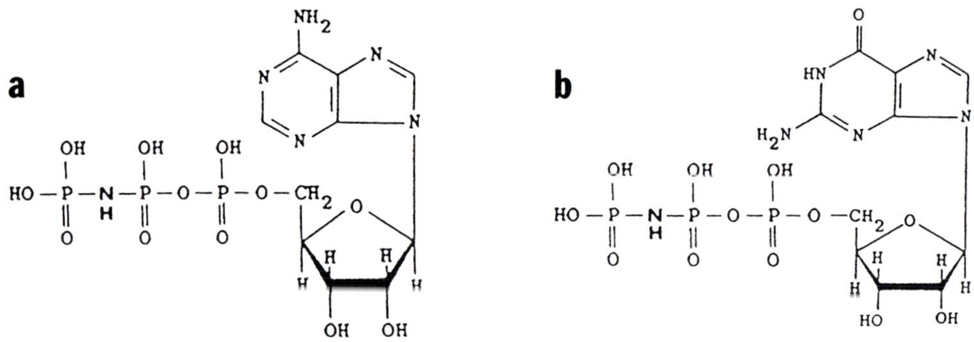

FIGURE 2. Structures of adenylyl imidodiphosphate (a) and guanylyl imidodiphosphate (b).

alternative capture agents, cobalt ion,[11] strontium ion,[12] and cerium ion,[13] have been introduced for the cytochemical demonstration of ACLase.

However, most metals inhibit enzyme activities and show similar drawbacks and advantages to lead in the cytochemical method. At present, the lead-based method is regarded as the most reliable one, for the following reasons: when maleate is used as a buffer reagent in the presence of lead nitrate in the reaction medium, it protects the enzyme and substrate from lead ion due to its weak chelation with lead ions. Wollenberger and Schulze[14] demonstrated that the free lead concentration is reduced to about one fiftieth (4×10^{-5} *M*) of the initial concentration in a medium of 80 m*M* maleate and 2 m*M* lead nitrate, the concentration usually used in cytochemical incubation media. Under this condition, ACLase activity is less inhibited in most tissues and retains enough residual activity to be detected cytochemically. Further, Cutler[15] demonstrated that lead ions do not cause nonenzymatic hydrolysis of AMP-P(NH)P under conditions similar to those described above, and suggested that impurities cause the formation of nonenzymatic lead salt precipitates in the tissue.

Guanylate cyclase (EC 4.6.1.2, GTP-pyrophosphate lyase [cyclizing]) catalyzes the formation of cyclic GMP and PPi from GTP in the physiological substrate containing Mn^{2+} as a cofactor (Figure 1). Its catalytic action is similar to that of ACLase. Therefore, the strategy for the cytochemical localization of guanylate cyclase (GCLase) is similar to that for ACLase: (1) the use of guanylyl imidodiphosphate (GMP-P[NH]P)[6] (Figure 2) instead of GTP as substrate prevents GTPases from hydrolyzing it and suppresses nonspecific reactions; (2) the use of Tris-maleate buffer together with lead nitrate as a capture agent minimizes the free lead concentration in the reaction medium.

In the following, we describe one of the most reliable cytochemical methods for cyclases, recently improved by Fujimoto et al.,[16] as well as our modification and some comparisons with other methods.

C. REAGENTS AND EQUIPMENT

All chemicals and reagents required for the cytochemical demonstration of cyclases need to be of analytical grade. Distilled and deionized water is used to prepare all solutions. Chemicals and reagents are as follows: paraformaldehyde; glutaraldehyde; sodium cacodylate; sucrose; dimethyl sulfoxide; osmium tetroxide; Tris (hydroxymethyl) aminomethane; maleic acid; 5′-adenylyl imidodiphosphate (Sigma Chem. Co. or Boehringer Mannheim); 5′-guanylyl imidodiphosphate (Sigma Chem. Co. or Boehringer Mannheim); theophylline; isobutyl methylxantine; magnesium sulfate; manganese chloride; levamisole (Sigma Chem. Co.); lead nitrate; sodium fluoride; sodium azide; ammonium sulfide solution (colorless); *d,l,*-isoproterenol-HCl; acetylcholine chloride; methylene blue.

An electrofreezing microtome or a cryostat and a Microslicer[17] (Dosaka EM Co., Kyoto) or a Vibratome are necessary to prepare frozen and nonfrozen tissue sections, respectively. For the cytochemical reactions, a water bath equipped with a shaker is convenient.

D. A GENERAL INCUBATION PROCEDURE FOR CYCLASES

1. Fixation

The concentration of paraformaldehyde (PFA) and glutaraldehyde (GLA) in the fixative and a way of fixation should be decided by trial and error. Our recommended method is as follows: tissues are fixed at 4°C for 30 to 60 min by immersion or perfusion with the standard fixative consisting of 2% PFA, 0.25% GLA, 8% sucrose, and 5% v/v dimethyl sulfoxide (DMSO) in 0.1 *M* cacodylate buffer, pH 7.4. In case of immersion fixation, tissues should be cut into slices less than 1 mm in thickness. Although both ACLase and GCLase show very low basal activities, as a rule, this fixative preserves sufficient ACLase and GCLase activities for ultrastructural localization. In some tissues, however, these enzymes are inactivated even by this weak fixation. In such cases, it is necessary to carry out the fixation for only a very short period or with a buffered solution containing 2 to 4% PFA only,[18,19] which results in a less satisfactory preservation of ultrastructure. Fixation with GLA inactivates the enzymes much more than with PFA. DMSO is thought to enhance penetration of solutions into tissues, and is recommended to be added to the fixative for the better preservation of ultrastructure.[20,21] In some cases, DMSO added to the incubation medium, as well as to the fixative, preserves the basal activities of cyclases.[16]

2. Washing

Fixed tissues are rinsed for 60 min at 4°C in 3 to 4 changes of the washing buffer solution, 0.1 *M* cacodylate buffer containing 8% sucrose and 5% (v/v) DMSO.

3. Sectioning

In order to avoid nonspecific deposition of reaction products, tissues should be cut into sections of a suitable thickness. For light microscopy, tissues are generally frozen and sectioned at 5 to 15 μm in thickness with an electrofreezing microtome or a cryostat. For electron microscopy, nonfrozen sections, about 40 μm, preferably not more than 50 μm in thickness, should be cut with a Microslicer or a Vibratome.

4. Incubation

The reaction media must be made up immediately before use as described in part F (see below). The optimal incubation time and temperature are decided by trial and error. We recommend to incubate tissue sections for 30 to 60 min at 37°C with changes of the reaction medium every 20 min. It is important to keep the reaction medium stirred gently during incubation.

5. Further Treatment

For light microscopy, the incubated tissue sections are washed in three changes of distilled water for a total of more than 20 min, then treated with 1% ammonium sulfide for 2 min, washed again in distilled water, and mounted with glycerine-jelly.

For electron microscopy, the sections are treated as described below:

1. The incubated tissue sections are washed in three to four changes of 0.1 *M* cacodylate buffer containing 8% sucrose for 60 min. This process needs to be carried out carefully because DMSO forms a precipitate with osmium tetroxide (OsO_4) if it remains in the tissue.

2. The sections are postfixed with 1% OsO_4 in 0.1 *M* cacodylate buffer at 4°C for less than 10 min. Postosmification longer than 10 min may cause diffusion artifacts because of the solubility of the reaction product in OsO_4.[16,22] In some cases, omission of postosmification may be helpful in identifying the reaction products and in excluding their possible reaction with OsO_4. It is also possible to use 2% GLA fixation instead of postosmification.
3. After washing in distilled water, the sections are dehydrated in graded ethanol series and embedded in epoxy resin.
4. Ultrathin sections are cut with an ultramicrotome, stained with uranyl acetate or left unstained, and viewed under an electron microscope.

E. INCUBATION MEDIA, CONTROL STUDIES, AND RESULTS

1. Incubation Media[16]

ACLase

	Final conc (m*M*)
Tris-maleate buffer, pH 7.4	80.0
AMP-P(NH)P[a]	0.5
Levamisole[b]	2.5
Theophylline[c] (or isobutyl methylxantine)	2.0
$MgSO_4$	4.0
NaF[d]	10.0–20.0
DMSO	5.0% v/v
Sucrose	8.0%
Lead nitrate[e]	2.0

Final pH 7.4

GLCase

Tris-maleate buffer, pH 7.4	80.0
GMP-P(NH)P[a]	0.5
Levamisole[b]	2.5
Theophylline[c] (or isobutyl methylxantine)	2.0
$MnCl_2$	4.0
NaN_3[f]	5.0
DMSO	5.0% v/v
Sucrose	8.0%
Lead nitrate[e]	2.0

Final pH 7.4

[a] The ingredients should be added and dissolved with stirring in the order listed. If the reaction medium is cloudy, this usually indicates that the substrate is old and/or impure. Commercially available AMP-P(NH)P preparations may require purification according to the method of Kvinnsland[23] or Cutler.[15] AMP-P(CH_2)P is another synthetic analog of ATP used successfully by Mayer et al.[24] for the localization of ACLase activity.

[b] Levamisole is a potent inhibitor of nonspecific alkaline phosphatase which causes nonspecific hydrolysis of the substrate.[25]

[c] Both theophylline and isobutyl methylxantine are potent inhibitors of phosphodiesterases which may cause the nonspecific hydrolysis of the substrate.

[d] NaF is an inhibitor of Mg^{2+}-ATPase and acid phosphatase, and also an activator of ACLase.

[e] There are alternative capture agents including cobalt nitrate,[11] strontium chloride,[12] and cerium chloride[13] for the heavy metal capturing method of cyclases. Each metal has inherent advantages and disadvantages compared with lead nitrate. The advantage of the cobalt-based method is that cobalt activates ACLase in some tissues;

the disadvantage is twofold: (1) NaF cannot be used in the reaction medium because of the formation of a precipitate, and (2) the electron density of the cobalt salt (the reaction product) is not high enough for electron microscopy. The strontium-based method has the advantage that it inhibits ACLase less than lead, and that the incubation can be carried out at alkaline pH optimal for ACLase using AMP-P(NH)P as a substrate. Its disadvantage is that the initial reaction product needs to be converted into a lead salt because it is soluble both at neutral pH and in OsO_4 solution. This process may cause diffusion of the reaction products. The advantage of the cerium-based method is that the reaction product (cerium salt) is generally observed as a very fine and uniform electron-dense precipitate; however, it has the disadvantage that the cerium salt cannot be observed at the light microscopic level and no method has been developed for light microscopy.

f NaN$_3$, an activator of both soluble and particulate GCLase,[26] is also an inhibitor of GTPases,[27] and it is sometimes added to the reaction medium.

2. Control Studies

As cytochemical controls, tissue sections are incubated in a substrate-free medium for the demonstration of both cyclases, or in a medium excluding Mg ions for ACLase and Mn ions for GCLase, respectively.

In addition, activators or inhibitors of cyclases may be used to evaluate the cytochemical reaction. The cytochemical reaction may also provide valuable information about the existence of receptors coupled to cyclase if neurotransmitters and hormones are used as activators or inhibitors. For example, as activation for ACLase activity, 0.1 to 1.0 m*M* *d,l*-isoproterenol-HCl, and for GCLase, 0.1 to 1.0 m*M* acetylcholine chloride and 0.1 to 0.5 m*M* methylene blue, are used. The latter is a free radical scavenger generally recognized as a potent inhibitor of soluble GCLase,[28] and also acts as an inhibitor of particulate GCLase in some instances.[29] The activators or inhibitors are dissolved in HEPES-buffered Krebs-Ringer solution (for general use) or in HEPES-buffered Dulbecco's modified Eagle medium (for tissues of the vascular system) and perfused for 5 min at room temperature intravascularly or intracardially, prior to the perfusion fixation.[18] Fixative, washing buffer, and incubation medium containing the same reagents may be used in the subsequent steps. In some cases, the first perfusion step is omitted or only the basal perfusate is used for comparison in evaluating the effectiveness of the agents. High concentrations of agents are usually employed in order to assure that an effective concentration of agents reaches the target cells (Figures 3 and 4). However, since the cytochemical method is essentially qualitative, small changes in the amount of reaction products caused by the activation or inhibition of the enzyme cannot be established. As an alternative of perfusion of activator, nonfrozen tissue sections cut with a Microslicer are incubated for 10 to 20 min in HEPES-buffered Krebs-Ringer solution containing the above-mentioned activator prior to the fixation step. It is very important to keep in mind that the activation of cyclase through the receptor requires intact plasma membrane. The receptor does not activate cyclase in fixed membranes.

3. Results

With the method of Fujimoto et al.,[16] both ACLase and GCLase are localized, in general, on the plasma membrane and/or on the cytoplasmic side of the plasma membrane (Figures 5 to 8). They are also localized on the gap junction (Figure 5), Golgi apparatus, endoplasmic reticulum, sarcoplasmic reticulum (Figure 6), and centriole (Figure 7). Using other reaction media, the reaction products are observed on the outside of the plasma membrane in many cases.

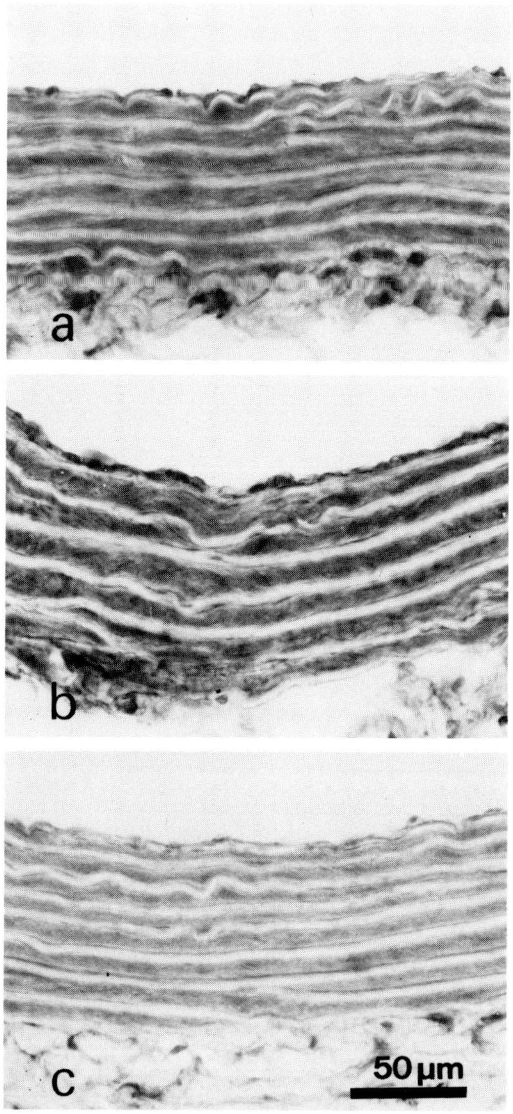

FIGURE 3. Light micrographs of ACLase activity in the rat thoracic aorta. (a) Basal ACLase activity, present in the endothelium and smooth muscle cell; (b) the effect of isoproterenol (1 m*M*) on ACLase activity: the activity in both endothelium and smooth muscle cell is greater than the basal activity; this indicates the existence of ACLase coupled to β-adrenergic receptors. (c) The effect of propranolol (1 m*M*) on ACLase activity. The activity in both endothelium and smooth muscle cell is reduced by propranolol; this suggests circulating catecholamines may contribute to the basal activity observed and the latter is blocked by propranolol.

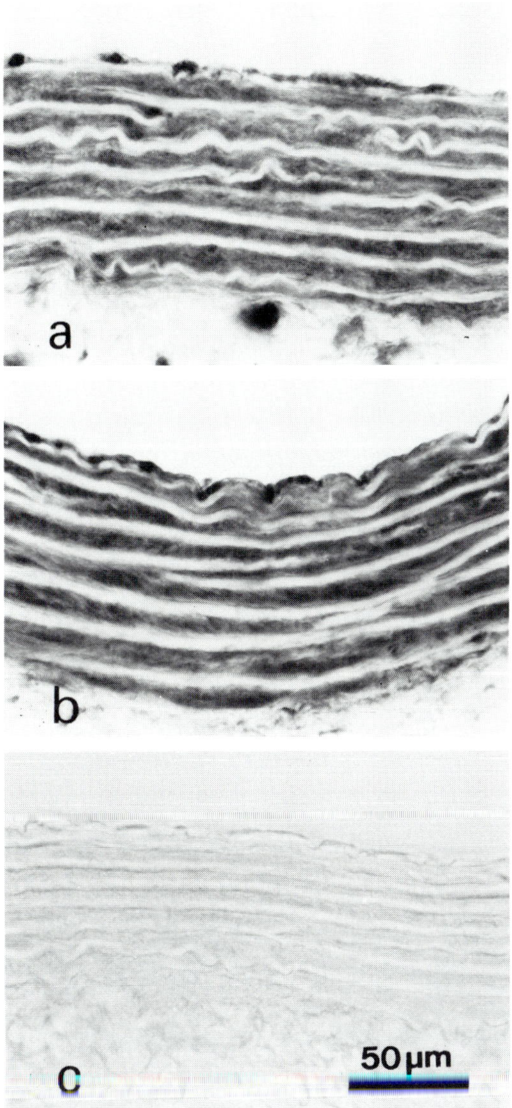

FIGURE 4. Light micrographs of GCLase activity in the rat thoracic aorta. (a) Basal GCLase activity is observed in the endothelium and smooth muscle cell; (b) GCLase activity affected by 1 m*M* acetylcholine — intense reaction is present in the endothelium and smooth muscle cell; (c) control preparation incubated without substrate in the GCLase medium — no deposits of the reaction product are seen in the endothelium and smooth muscle cell.

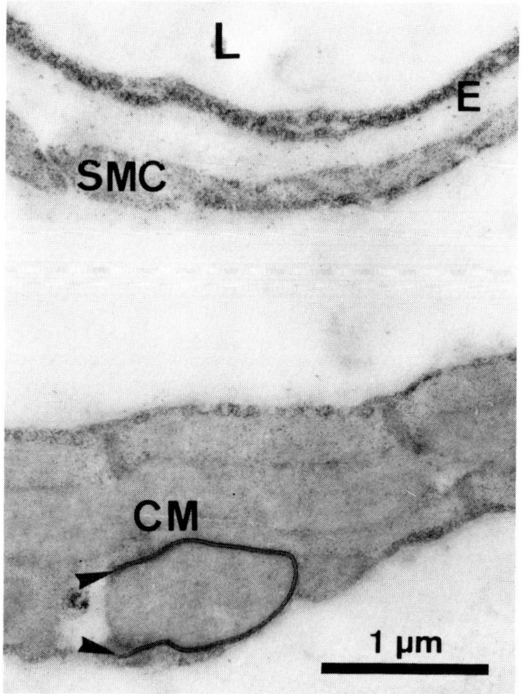

FIGURE 5. Electron micrograph of ACLase activity stimulated by isoproterenol in the rat heart. Reaction products are observed on the sarcolemma, sarcolemmal vesicles, and gap junctional membrane (arrowheads) in the cardiac muscle (CM). They are also observed on the vesicles of the venular endothelium (E) and on the plasma membrane and caveolae of the venular smooth muscle cell (SMC); L, lumen.

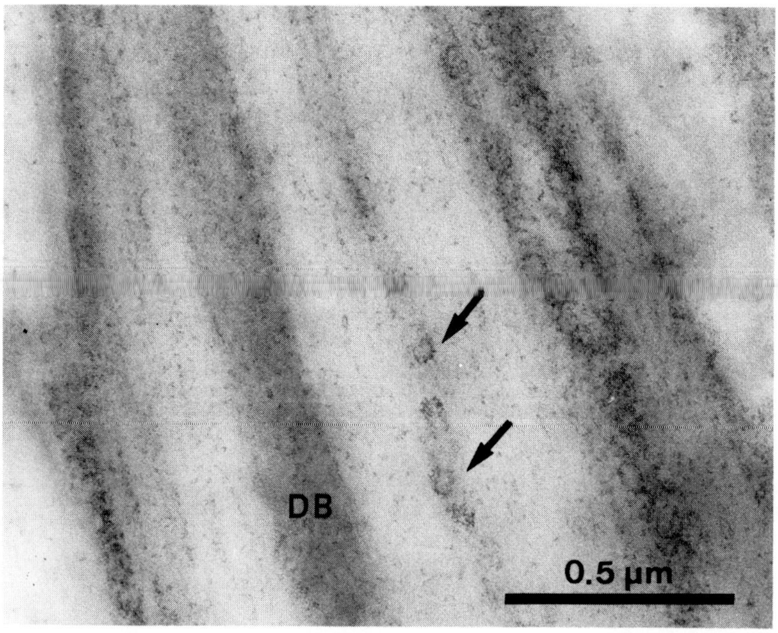

FIGURE 6. Electron micrograph of GCLase activity in the rat aortic smooth muscle cell. Reaction products are localized on the cytoplasmic side of the membrane covering caveolae, sarcoplasmic reticulum (arrows), and dense bodies (DB).

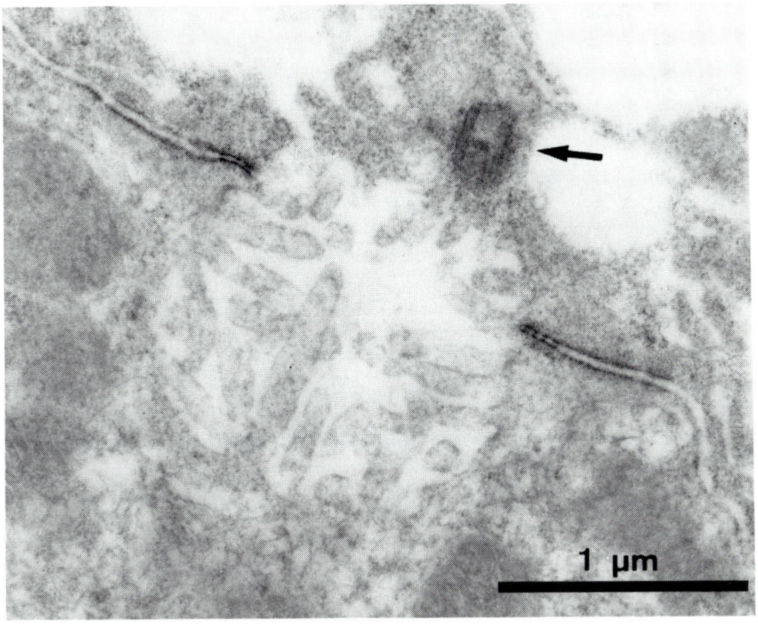

FIGURE 7. Electron micrograph of ACLase activity in the guinea pig gastric gland. Reaction products are observed on the lateral apical membrane composing the cell junction between parietal cell and chief cell and on the centriole (arrow) of the chief cell.

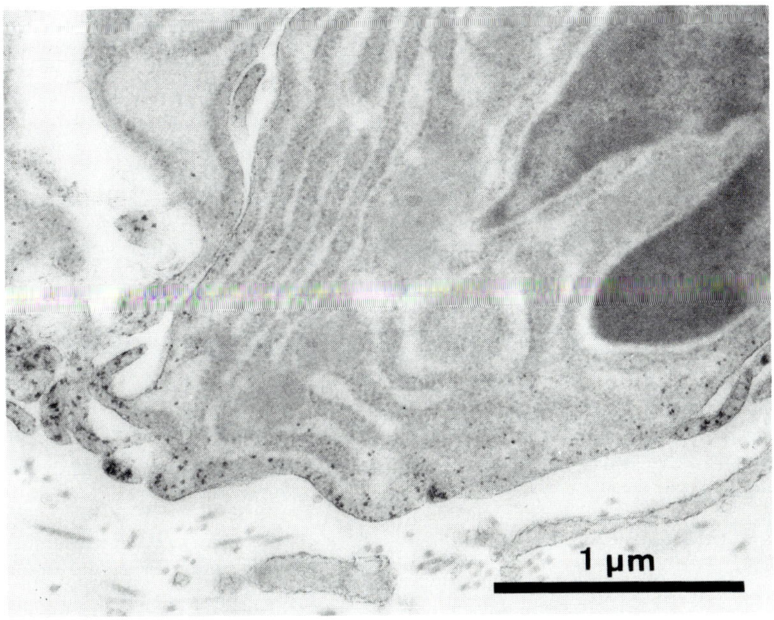

FIGURE 8. Electron micrograph of GCLase activity in the chief cell of guinea pig gastric gland. Reaction products are localized on the cytoplasmic side of the basal and lateral membrane.

REFERENCES

1. **Neer, E. J. and Clapham, D. E.,** Roles of G protein subunits in transmembrane signalling, *Nature,* 333, 129, 1988.
2. **Schramm, M. and Selinger, Z.,** Message transmission: receptor controlled adenylate cyclase system, *Nature,* 225, 1350, 1984.
3. **Reik, L., Petzold, G. L., Higgins, J. A., Greengard, P., and Barrnett, R. J.,** Hormone-sensitive adenyl cyclase: cytochemical localization in rat liver, *Science,* 168, 382, 1970.
4. **Wachstein, M. and Meisel, E.,** Histochemistry of hepatic phosphatases at physiological pH with special reference to the demonstration of bile canaliculi, *Am. J. Clin. Pathol.,* 27, 13, 1957.
5. **Yount, R. G., Babcock, D., Ballantyne, W., and Ojala, D.,** Adenylyl imidodiphosphate, an adenosine triphosphate analog containing a P-N-P linkage, *Biochemistry,* 10, 2484, 1971.
6. **Yount, R. G.,** Adenylylimidodiphosphate and guanylylimidodiphosphate, *Meth. Enzymol.,* 38, 420, 1974.
7. **Howell, S. L. and Whitfield, M.,** Cytochemical localization of adenyl cyclase activity in rat islets of Langerhans, *J. Histochem. Cytochem.,* 20, 873, 1972.
8. **Wagner, R. C., Kreiner, P., Barrnett, R. J., and Bitensky, M. W.,** Biochemical characterization and cytochemical localization of a catecholamine-sensitive adenylate cyclase in isolated capillary endothelium, *Proc. Natl. Acad. Sci. U.S.A.,* 69, 3175, 1972.
9. **Nathanson, J. A. and Bloom, F. E.,** Lead-induced inhibition of brain adenyl cyclase, *Nature,* 255, 419, 1975.
10. **Kempen, H. J. M., de Pont, J. J. H. H. M., Bonting, S. L., and Stadhouders, A. M.,** The cytochemical localization of adenylate cyclase: fact or artifact?, *J. Histochem. Cytochem.,* 26, 298, 1978.
11. **Rechardt, L. and Härkönen, M.,** Electron microscopical demonstration of adenylate cyclase activity in nervous tissue, *Histochemistry,* 51, 113, 1977.
12. **Zajic, G. and Schacht, J.,** Cytochemical demonstration of adenylate cyclase with strontium chloride in the rat pancreas, *J. Histochem. Cytochem.,* 31, 25, 1983.
13. **Rechardt, L. and Hervonen, H.,** Cytochemical demonstration of adenylate cyclase activity with cerium, *Histochemistry,* 82, 501, 1985.
14. **Wollenberger, A. and Schulze, W.,** Cytochemical studies on sarcolemma: Na^+-K^+-adenosine triphosphatase and adenylate cyclase, in *Recent Advances in Studies on Cardiac Structure and Metabolism,* Vol. 9, Roy, P. E. and Dhalla, N. S., Eds., University Park Press, Baltimore, 1976, 101.
15. **Cutler, L. S.,** Cytochemical methods for the localization of adenylate cyclase, *J. Histochem. Cytochem.,* 31, 85, 1983.
16. **Fujimoto, K., Toibana, M., and Ogawa, K.,** New cytochemical method for adenylate cyclase and guanylate cyclase with dimethyl sulfoxide, *Acta Histochem. Cytochem.,* 14, 687, 1981.
17. **Mayahara, H., Fujimoto, K., Noda, T., Tamura, I., and Ogawa, K.,** The "Microslicer", a new instrument for making non-frozen sections, *Acta Histochem. Cytochem.,* 14, 211, 1981.
18. **Ogawa, K. S., Fujimoto, K., and Ogawa, K.,** Cytochemical studies of nucleotidyl cyclases in aortic endothelium and smooth muscle cell, *Acta Histochem. Cytochem.,* 19, 521, 1986.
19. **Ogawa, K. S., Fujimoto, K., and Ogawa, K.,** Cytochemical localization of adenylate cyclase and guanylate cyclase in the parietal cell of guinea pig gastric gland, *Acta Histochem. Cytochem.,* 20, 185, 1987.
20. **Sandborn, E. B., Stephens, H., and Bendayan, M.,** The influence of dimethyl sulfoxide of cellular ultrastructure and cytochemistry, *Ann. N.Y. Acad. Sci.,* 243, 122, 1975.
21. **Slezak, J. and Geller, S. A.,** Cytochemical demonstration of adenylate cyclase in cardiac muscle. Effect of dimethyl sulfoxide, *J. Histochem. Cytochem.,* 27, 774, 1979.
22. **Poeggel, G., Luppa, H., Bernstein, H.-G., and Weiss, J.,** Histochemistry of adenylate cyclase, *Int. Rev. Cytol.,* 89, 35, 1984.
23. **Kvinnsland, S.,** Adenylate cyclase cytochemistry: a methodological evaluation, *Histochem. J.,* 11, 669, 1979.
24. **Mayer, D., Ehemann, V., Hacker, H.-J., Klimek, F., and Bannasch, P.,** Specificity of cytochemical demonstration of adenylate cyclase in liver using adenylate-(β,γ-methylene) diphosphate as substrate, *Histochemistry,* 82, 135, 1985.

25. **Borgers, M.,** The cytochemical application of new potent inhibitors of alkaline phosphatases, *J. Histochem. Cytochem.,* 21, 812, 1973.
26. **Kimura, H., Mittal, C. K., and Murad, F.,** Activation of guanylate cyclase from rat liver and other tissues by sodium azide, *J. Biol. Chem.,* 250, 8016, 1975.
27. **Murad, F., Arnold, W. P., Mittal, C. K., and Braughler, J. M.,** Properties and regulation of guanylate cyclase and some proposed functions of cyclic GMP, *Adv. Cyclic Nucleotide Res.,* 11, 175, 1979.
28. **Gruetter, C. A., Kadowitz, P. J., and Ignarro, L. J.,** Methylene blue inhibits coronary arterial relaxation and guanylate cyclase activation by nitroglycerin, sodium nitrite, and amyl nitrite, *Can. J. Physiol. Pharmacol.,* 59, 150, 1981.
29. **Ohlstein, E. H. and Berkowitz, A.,** Cyclic guanosine monophosphate mediates vascular relaxation induced by atrial natriuretic factor, *Hypertension,* 7, 306, 1985.

XVII. β-GLUCURONIDASE

Masando Hayashi and Kimiko Hayashi

A. SIMULTANEOUS COUPLING METHOD USING HEXAZONIUM PARAROSANILIN

1. Introduction

β-Glucuronidase catalyzes the hydrolysis of β-glycoside linkages of a variety of glucuronides. According to de Duve and associates,[1] the principal localization of β-glucuronidase is lysosomal. The simultaneous coupling method using naphthol AS-BI glucuronide and hexazonium pararosanilin has confirmed that the distribution of β-glucuronidase activity is associated with lysosomes in the majority of tissues.[2,3] It is to be noted that all lysosomal enzymes are not localized in the same pattern. A comparative study of the localization of three lysosomal enzymes (acid phosphatase, β-glucuronidase, and *N*-acetyl-β-glucosaminidase) shows pronounced differences among these enzymes in many tissues.[4]

Reactions involved in the simultaneous coupling method are as follows. At the sites of β-glucuronidase activity in tissues, naphthol AS-BI glucuronide liberates naphthol AS-BI, a hydrolysis product of extremely low solubility and high substantivity. Naphthol AS-BI is quickly converted to a more insoluble amorphous azo dye upon reaction with hexazonium pararosanilin included in the staining medium. Hexazonium pararosanilin is a capture reagent that can couple with naphthol rapidly at around pH 5. The azo dye is practically insoluble in ethanol and epoxy resins, and it is visible as a bright red stain in the light microscope and as a moderately opaque amorphous deposit in the electron microscope.[5] Presented here is our recent modification that employs glutaraldehyde for postfixation and eliminates the use of osmium tetroxide.

2. Equipment and Reagents

Required equipment include an unfrozen-tissue sectioner or freezing microtome, a water bath, a pH meter, and a refrigerator.

The reagents are naphthol AS-BI β-D-glucuronic acid (naphthol AS-BI β-D-glucuronide), Sigma Chemical Co., St. Louis, MO, product number N 1875; *p*-rosanilin, acridinfrei, distributed by Roboz Surgical Instrument Co., Washington, D.C., product number Chroma 11455; D-glucosaccharo-1,4-lactone: Chugai Pharmaceutical Co., Tokyo.

Stock solutions — (1) Substrate stock solution: 0.5 m*M* naphthol AS-BI glucuronide sodium salt in 0.2 *M* acetate buffer, pH 5.2. To 28 mg of naphthol AS-BI β-D-glucuronic acid in a 100-ml cylinder, add 1.2 ml of 0.05 *M* sodium bicarbonate and dissolve completely, either using a magnetic stirrer or by gentle shaking. Then add 0.2 *M* acetate buffer, pH 5.2, to 100 ml. The solution can be stored for months at 4°C. (2) Pararosanilin solution: dissolve 1 g *p*-rosanilin in 20 ml distilled water and 5 ml hydrochloric acid (HCl, approximately 37%). Filter through a No. 1 filter paper and store at room temperature. (3) 4% Sodium nitrite solution. Prepare before use.

Working staining medium — Mix 0.3 ml each of pararosanilin solution and 4% sodium nitrite solution in a 20-ml beaker. About 1 min later, pour 10 ml of the substrate stock solution into the dye mixture. Adjust pH to 5.2 with 1 *N* NaOH and add distilled water to a final volume of 20 ml. Filter through a No. 1 filter paper. The final concentration of the substrate is 0.25 m*M* and that of the diazo reagent is 1.8 m*M*.

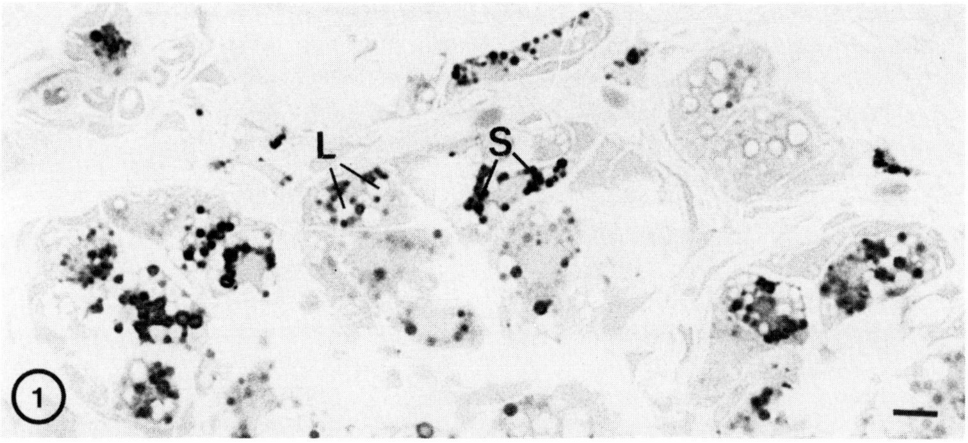

FIGURE 1. Light micrograph of acinar cells of a rat *preputial gland* stained for β-glucuronidase activity with the simultaneous coupling method, using naphthol AS-BI glucuronide and hexazonium pararosanilin. Note that a number of perinuclear secretory granules (S) are darkly stained for the enzyme activity, but lipid droplets (L) are not stained. (Bar = 10 μm.)

3. Procedures

1. Fix small tissue specimens (smaller than 3 × 3 × 3 mm) in a fixative containing 4% paraformaldehyde, 7.5% sucrose, and 0.1 *M* phosphate or cacodylate buffer, pH 7.4, for 24 h at 4°C. Wash the specimens in three changes of 0.1 *M* phosphate or cacodylate (if the specimens are fixed in the fixative prepared with cacodylate) buffer, pH 7.4, containing 7.5% sucrose for 3 h at 4°C.
2. Cut 40-μm-thick sections using an unfrozen-tissue sectioner or a freezing microtome, and keep them in cold 7.5% sucrose.
3. Treat the sections with 70, 100, and 70% ethanols, each for 1 min, at −15 to −20°C, rinse in two changes of cold 7.5% sucrose, and rinse in 0.1 *M* acetate buffer, pH 5.2. (Note that ethanol at room temperature completely inactivates β-glucuronidase activity in a few minutes.)
4. Incubate sections in the working staining medium prepared before use. Rat liver sections may be incubated for 1 h at 37°C and rat preputial gland sections for 30 min at room temperature (about 24°C). For controls, incubate control sections in the medium containing substrate and 0.5 m*M* D-glucosaccharo-1,4-lactone (mol wt, 192). This medium can be prepared by adding 10 μl of 1000× lactone solution (4.8 mg glucosaccharo lactone in 100 μl H_2O) to 10 ml of the working staining solution. Some control sections may be incubated in substrate-free medium.
5. Rinse in 0.1 *M* phosphate buffer, pH 7.4, and fix in 2% glutaraldehyde in 0.1 *M* phosphate buffer, pH 7.4, for 20 min at room temperature.
6. Rinse in 0.1 *M* phosphate buffer, pH 7.4, dehydrate in a graded series of ethanol, and embed in epoxy resin mixture.
7. Examine thin sections in the electron microscope without post-staining.

4. Results and Comments

Preputial glands are the richest source of β-glucuronidase in the rat. Acinar cells of the glands contain a number of red-stained perinuclear secretion granules and unstained lipid droplets (Figure 1). In the electron microscope, the red azo dye stain can be easily recognized

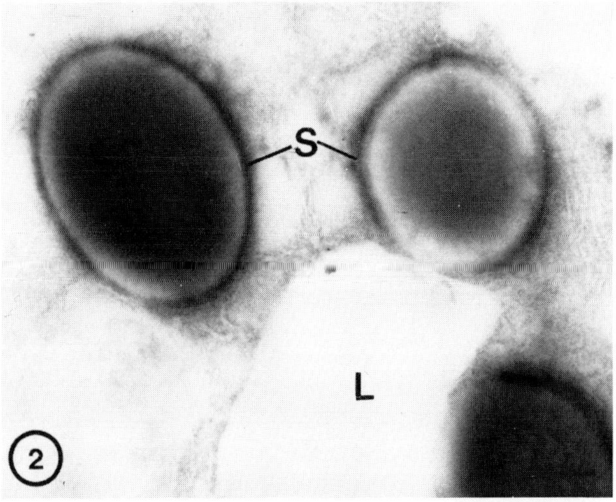

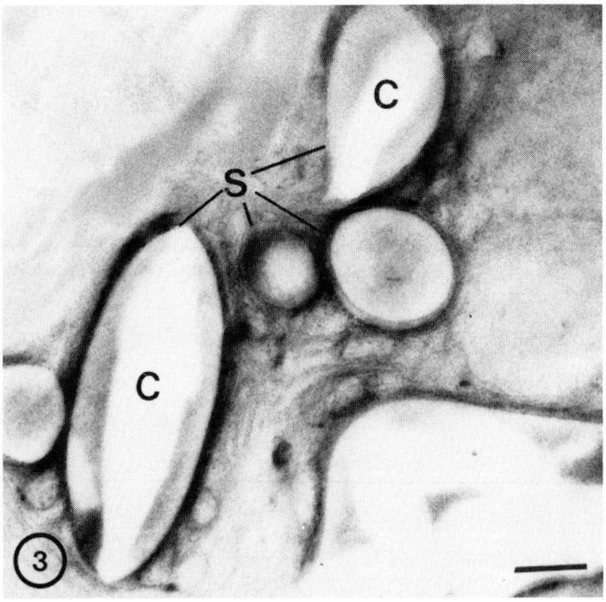

FIGURE 2. Electron micrograph of tissue shown in Figure 1. Note that opaque reaction product is present in the peripheral and central areas of the matrix of secretory granules (S). Lipid droplet (L) is not labeled.
FIGURE 3. Electron micrograph of tissue shown in Figure 1. Note secretory granules (S) containing unlabeled crystalloids (C).

as a moderately opaque, amorphous deposit on semithin (purple to blue) sections viewed under 80 kV. Opaque deposit is found frequently as a peripheral ring in the matrix of secretory granules, and the matrix of the granules is labeled with varying degree of opacity (Figure 2). Some secretory granules contain crystalloids, which are generally not labeled. These granules contain an opaque deposit in parts of the matrix (Figure 3) or in the entire matrix surrounding a crystalloid (Figure 4). Opaque deposit is found in the Golgi complex and condensing vacuoles (Figure 5). Lipid droplets are not labeled. No opaque deposit is found in control sections incubated in the media containing substrate and 0.5 m*M* glucosaccharolactone.

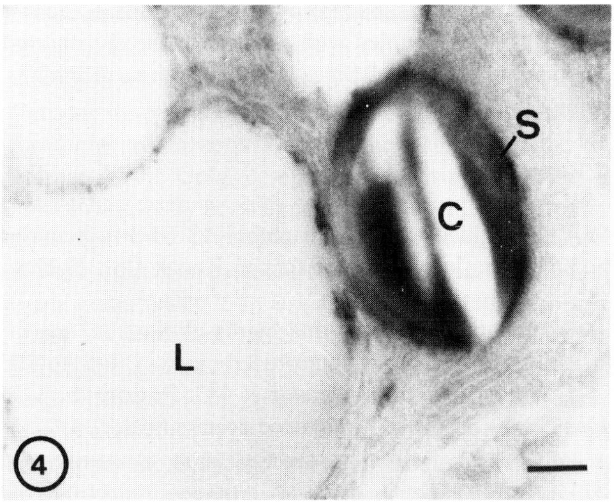

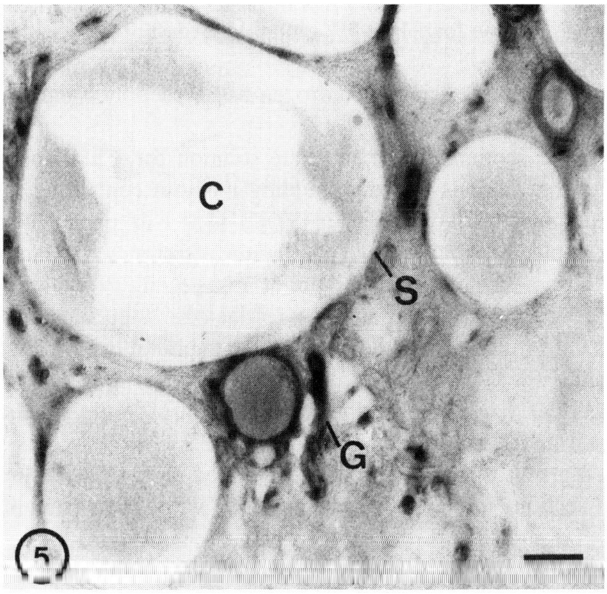

FIGURE 4. Electron micrograph — note a secretory granule in which almost the entire matrix surrounding a crystalloid (C) is labeled. Lipid droplets are not labeled.
FIGURE 5. Electron micrograph — note the presence of opaque deposit in the Golgi complex (G) and condensing vacuoles. (Bar = 0.5 μm.)

Subcellular localization of β-glucuronidase in the preputial gland as visualized with this technique is therefore almost identical to that of acid phosphatase, demonstrated by a lead phosphate method,[6] and to that of β-glucuronidase, demonstrated by an immunocytochemical method.[7]

B. POST-COUPLING *p*-NITROBENZENE DIAZONIUM TETRAFLUOROBORATE METHOD

1. Introduction

The procedures involve two steps: tissue sections are first incubated in the substrate, naphthol AS-BI glucuronide, solution without including a coupling agent. During incubation,

XVIII. PROTEASES

Reinhart Gossrau

A. INTRODUCTION

There is still much confusion created by the indiscriminate use of the terms peptidase, protease, and proteinase with different meanings. Based on the most recent recommendations in the latest edition of *Enzyme Nomenclature*[1] and in the monograph of McDonald and Barrett[2] (see also Reference 3), the whole group of enzymes is designated proteases, which consists of the exopeptidases and endopeptidases, also called proteinases. The exopeptidases attack peptide bonds at the ends of polypeptide molecules. Dependent on the site of cleavage and the size of the fragment liberated, the exopeptidases are further subdivided into aminopeptidases, dipeptidyl peptidases, tripeptidyl peptidases, carboxypeptidases, and peptidyl peptidases. The terms dipeptidase or tripeptidase refer to the size of the cleaved peptide molecule. Exopeptidases are also the omega peptidases, which remove terminal residues without a free α-amino or α-carboxyl group or terminal residues linked through an isopeptide bond.

In contrast, endopeptidases (or proteinases) attack peptide bonds inside substrate molecules, and are classified according to their catalytic mechanism as serine, cysteine, aspartic, and metalloproteinases (or endopeptidases), and unclassified proteases with unknown catalytic mechanisms (Table 1).

All peptidases (proteolytic enzymes) hydrolyze peptide bonds according to the following general reaction scheme:

$$R\text{–}CO\text{–}NH\text{–}R' + H_2O \rightarrow R\text{–}COOH + R'\text{–}NH_2$$

Of the classified known proteases, only a few can be visualized by ultracytochemical procedures, and most of the proteases that can be demonstrated in the electron microscope belong to the group of exopeptidases.

They are the aminopeptidases microsomal alanyl aminopeptidase (formerly aminopeptidase M or N, EC 3.4.11.2), glutamyl aminopeptidase (formerly aminopeptidase A, EC 3.4.11.7), the omega peptidase γ-glutamyl transpeptidase (formerly γ-glutamyl transferase, EC 2.3.2.2) and the dipeptidyl peptidases dipeptidyl peptidase I (EC 3.4.14.1), dipeptidyl peptidase II (EC 3.4.14.2), and dipeptidyl peptidase IV (EC 3.4.14.5). Cathepsin B (EC 3.4.22.1) and some not yet further characterized endopeptidases and ester proteinases, to which possibly also tissue kallikrein belongs, are the only proteinases that can be localized, with certain limitations, using ultracytochemical procedures.

In contrast to other enzymes which can be visualized by methods based on several reaction principles[4] especially one is used and is the most popular for the ultracytochemical detection of proteases. This principle was first reported by McDonald et al.[5] and then described in detail by Smith in collaboration with van Frank.[6] The method is based on the discovery that the leaving (detector group) 4-methoxy-2-naphthylamine couples rapidly with hexazotized pararosaniline, forming an azo dye which is able to bind osmic acid resulting in an amorphous, electron-dense reaction product.

Consequently, all protease substrates for electron microscopy represent either unsubstituted amino acid 4-methoxy-2-naphthylamides or unsubstituted or N-terminal substituted peptide 4-methoxy-2-naphthylamides, and are used in the simultaneous coupling method followed by osmification of the resulting azo dye according to the following reaction scheme:

TABLE 1
Classification of Proteases

I. Exopeptidases
  1. Aminopeptidases (aminoacylpeptidase hydrolases, EC 3.4.11)
  2. Dipeptidyl peptidase (dipeptidylpeptide hydrolases, EC 3.4.14)
  3. Tripeptidyl peptidase (tripeptidylpeptide hydrolases, EC number not yet assigned)
  4. Carboxypeptidases (serine-, EC 3.4.16; metallo-, EC 3.4.17; cysteine-, EC 3.4.18)
  5. Peptidyl dipeptidases (peptidyldipeptide hydrolases, EC 3.4.15)
  6. Dipeptidases (dipeptide hydrolases, EC 3.4.13)
  7. Omega peptidases (EC 3.4.19)

II. Endopeptidases or proteinases (EC 3.4.21 to 3.4.24, 3.4.99)
  1. Serine proteinases (EC 3.4.21)
  2. Cysteine proteinases (EC 3.4.22)
  3. Aspartic proteinases (EC 3.4.23)
  4. Metalloproteinases (EC 3.4.24)
  5. Proteinases of unknown catalytic mechanism (EC 3.4.99)

Based on References 1 and 2.

1. Amino acid-4-methoxy-2-naphthylamide or peptide-4-methoxy-2-naphthylamide + H_2O $\rightarrow$ amino acid or peptide + 4-methoxy-2-naphthylamine
2. 4-Methoxy-2-naphthylamine + hexazotized pararosaniline $\rightarrow$ azo dye
3. Osmiophilic azo dye + OsO_4 $\rightarrow$ OsO_4-containing azo dye

A further reaction principle was reported in conjunction with the ultracytochemical demonstration of γ-glutamyl transpeptidase by Rutenburg et al.[7] Originally described for the light microscopic visualization of the enzyme, this simultaneous azo-coupling procedure was modified for the electron microscopic localization of γ-glutamyl transpeptidase by bridging osmium to the copper chelate of the azo dye via thiocarbohydrazide (TCH; OTO reaction) according to the following scheme:

1. γ-Glutamyl-4-methoxy-2-naphthylamide $\rightarrow$ glutamic acid + 4-methoxy-2-naphthylamine
2. 4-Methoxy-2-naphthylamine + Fast Blue BB $\rightarrow$ azo dye
3. Azo dye + Cu^{2+} $\rightarrow$ copper-chelated azo dye
4. Copper-chelated azo dye + thiocarbohydrazide $\rightarrow$ TCH-containing copper-chelated azo dye
5. TCH-containing copper-chelated azo dye + OsO_4 $\rightarrow$ osmificated copper-chelated azo dye

A third procedure was described by Seligman and co-workers[8] for microsomal alanyl aminopeptidase and γ-glutamyl transpeptidase using again 4-methoxy-2-naphthylamine substrates but freshly diazotized 4-aminophthalhydrazide (4-APH) for simultaneous coupling of one molecule of 4-methoxy-2-naphthylamine (instead of Fast Blue BB) yielding an osmiophilic azo dye according to the following reaction scheme:

1. Amino acid-4-methoxy-2-naphthylamide $\rightarrow$ amino acid + 4-methoxy-2-naphthylamine
2. 4-Methoxy-2-naphthylamine + diazotized 4-aminophthalhydrazide $\rightarrow$ azo dye
3. Azo dye + OsO_4 $\rightarrow$ osmium black

B. METHODS

Historically, the first attempts to localize proteases ultracytochemically were those of Rutenburg et al.[7] and Smith,[9] who used either Fast Blue BB or Fast Blue B for the demonstration of γ-glutamyl transpeptidase and various dipeptidyl peptidases. In these reports, the azo dyes of 4-methoxy-2-naphthylamine and the Fast Blue compounds were compared with and without the subsequent use of the OTO method (a metal-to-metal bridging process), originally developed by Hanker et al.[10,11] Sufficiently increased electron density of the reaction product was reported and found to be insoluble when bridging to osmic acid was performed before resin embedding of the tissue. Because the azo dye appeared to be slightly diffusable or lipophilic, and tended to coalesce into droplets before chelation, Seligman and co-workers[8] introduced 4-aminophthalhydrazide diazotate as coupling agent in protease histochemistry. Although the resulting dye produces a sufficiently osmiophilic end product, which is lipophobic, its substantivity was questioned.

Therefore, in the following, primarily the simultaneous azo-coupling method with hexazonium pararosaniline or New Fuchsine and subsequent osmification of the azo dye is given. Because all proteases are visualized basically using the same tissue pretreatment, reaction principle, and post-treatment, the methods for these enzymes are presented together; only the incubation media are given separately.

However, because they were not yet consequently applied to other proteases, and critically compared with the azo dye-osmium procedure of Smith and van Frank[6] using hexazotized pararosaniline or hexazotized New Fuchsine, the techniques using the OTO principle after simultaneous coupling with Fast Blue BB and copper chelation of the resulting azo dye, as well as the osmium black procedure with diazotized 4-aminophthalhydrazide, are also presented.

C. EQUIPMENT AND REAGENTS

Equipment includes a Vibratome, tissue sectioner, freezing microtome, incubator, shaking water bath, and hood.

Chemicals and reagents include:

- α-Glutamyl-4-methoxy-2-naphthylamide (substrate for glutamyl aminopeptidase)
- Alanyl-4-methoxy-2-naphthylamide (substrate for microsomal alanyl aminopeptidase)
- γ-Glutamyl-4-methoxy-2-naphthylamide (substrate for γ-glutamyl transpeptidase)
- Glycyl-arginyl-4-methoxy-2-naphthylamide (substrate for dipeptidyl peptidase I)
- Lysyl-alanyl-4-methoxy-2-naphthylamide (substrate for dipeptidyl peptidase II)
- Glycyl-prolyl-4-methoxy-2-naphthylamide (substrate for dipeptidyl peptidase IV)
- CBZ-alanyl-arginyl-arginyl-4-methoxy-2-naphthylamide or CBZ-valyl-lysyl-lysyl-arginyl-4-methoxy-2-naphthylamide (substrate for cathepsin B)
- CBZ-glycyl-glycyl-arginyl-4-methoxy-2-naphthylamide, CBZ- or D-valyl-leucyl-arginyl-4-methoxy-2-naphthylamide or CBZ-alanyl-arginyl-arginyl-4-methoxy-2-naphthylamide (substrates for not yet classified endopeptidases, esterproteases, or kallikrein-type proteases)

All substrates may be obtained from Bachem, Bubendorf, Switzerland or Enzyme System Products, Livermore, CA. Also required are glycyl-glycine, pararosaniline, New Fuchsine, Fast Blue BB, thiocarbohydrazide (Polysciences, Warrington, PA), 4-aminophthalhydrazide (Aldrich, Steinheim, Germany), ethylene diamine tetraacetic acid (EDTA), cacodylate, sodium salt, osmic acid, diprotin A, amastatin, bestatin, leupeptin (Bachem, Bubendorf, Switzerland; Enzyme System Products, Livermore, CA), diisopropylfluorophosphate (DFP), Tris, 1,10-phenantroline, puromycin (Serva, Heidelberg, Germany) (see also Table 2), agar, and agarose type VII (Sigma Chemical Co.). Further details concerning protease inhibitors are given by Lojda et al.[21]

TABLE 2
Activators and Inhibitors of Proteases Detectable by Ultracytochemistry

Protease	Activator	Inhibitor
Microsomal alanyl aminopeptidase	—	1 m*M*, 1,10-Phenantroline 1 μ*M* Bestatin 1 μ*M* Amastatin
Glutamyl aminopeptidase	1 m*M* $CaCl_2$	1–5 m*M* EDTA 1 μ*M* Amastatin
Glutamyl transpeptidase	5 m*M* Glycyl-glycin	—
Dipeptidyl peptidase I	1 m*M* MEA	1 m*M* pCMB 1 m*M* Iodoacetate
Dipeptidyl peptidase II	—	1 m*M* Puromycin
Dipeptidyl peptidase IV		1 m*M* DFP 1 m*M* E 600 1 μ*M* Diprotin A
Cathepsin B	1 m*M* MEA 1 m*M* DTE	5 μ*M* Leupeptin

EDTA, ethylene diamine tetraacetic acid; MEA, mercaptoethylamine HCl; pCMB, *p*-chloromercuribenzoate; DTE, dithiothreitol; DFP, diisopropylfluorophosphate.

D. PROCEDURES

1. Test Tissues

Appropriate test tissue includes small intestine, kidney, liver, rat or mouse submandibular gland.

2. Tissue Pretreatment

Tissues can be fixed by immersion[12,13] in 1.5% purified glutaraldehyde in 0.05 *M* cacodylate or phosphate buffer, pH 7.2, containing 2% sucrose for 2.5 to 3 h at 4°C; or in a mixture according to Karnovsky[14] consisting of 3% glutaraldehyde and 2% formaldehyde (freshly prepared from paraformaldehyde) in 0.1 *M* cacodylate buffer, pH 7.2; or in a modified 4-times diluted mixture of Karnovsky.[14] Afterwards, the samples are rinsed in 0.1 *M* phosphate or cacodylate buffer, pH 7.2, containing 7% sucrose overnight at 4°C, embedded in 3% agar, cut at 30 to 40 μm using a tissue chopper or Vibratome, collected in centrifuge tubes in 0.1 *M* cacodylate buffer, pH 7, centrifuged at 4000 rpm for 5 min, and the buffer pipetted off. Furthermore, it might be useful to work with unfixed cryostat sections mounted on semipermeable membranes for the electron microscopic localization of proteases associated with the plasma membrane and Golgi apparatus, and fix the membrane without the use of gels briefly according to Karnovsky;[14] this is a modification of a procedure originally described for glucose-6-phosphatase ultracytochemistry by Bremer and colleagues.[15]

Fixation can also be carried out according to Kugler[16] by perfusion via the left ventricle: rinse for 2 min with a solution (osmolarity 330 to 340 mOsm) consisting of 137 *M* NaCl, 2.7 *M* KCl, 10 *M* Na_2HPO_4, 1.5 *M* KH_2PO_4, 0.5 *M* $MgCl_2$, 1 g/l glucose, 10 g/l sucrose, 5 g/l procain · HCl, and 0.5 ml/l thrombophob (5.000 U.S.P. units heparin, sodium salt per ml, pH 7.3); perfuse afterwards for 1 min with a mixture (330 to 340 mOsm) of 0.5% glutaraldehyde, 0.75% freshly prepared formaldehyde (made up from paraformaldehyde), and 10 *M* $CaCl_2$ in 0.05 *M* cacodylate buffer, pH 7.3; rinse (300 to 310 mOsm) with a solution of 50 g/l sucrose in 0.1 *M* cacodylate buffer, pH 7.4; prepare small tissue pieces, collect samples in this solution, and embed in 4% agarose type VII in 0.1 *M* cacodylate buffer, pH 7.4, at 37°C. After cooling of the embedded tissue samples to 4°C, 40-μm-thick sections are cut with a tissue chopper or Vibratome, and then collected and rinsed in 0.5 *M* cacodylate buffer, pH 7.4, containing 50 g/l sucrose.

Azo dye-osmium procedures for proteases with 4-methoxy-2-naphthylamine substrates according to Smith and van Frank,[6] Gossrau,[12] and Schröder and Gossrau[13] (see also References 16 and 17).

3. Incubation Media

1. Microsomal alanyl aminopeptidase

   - 5 mg Alanyl-4-methoxy-2-naphthylamide (dissolved in 0.25 ml *N,N′*-dimethylformamide)
   - 10 ml 0.1 *M* Cacodylate buffer, pH 7.0
   - 0.25 ml 2% Hexazotized pararosaniline or hexazotized New Fuchsine
   - Mix well, adjust the pH to 6.5 with NaOH, and filter

2. Glutamyl aminopeptidase

   - 7.5 mg α-Glutamyl-4-methoxy-2-naphthylamide (dissolved in 0.15 ml *N,N′*-dimethylformamide and 0.35 ml 0.4 *M* Tris stock solution)
   - 9.15 ml 0.1 *M* Cacodylate buffer, pH 7.0
   - 0.1 ml 1 *M* $CaCl_2$
   - 0.25 ml 2% Hexazotized pararosaniline or hexazotized New Fuchsine
   - Mix well, adjust the pH to 6.5 with NaOH, and filter

3. γ-Glutamyl transpeptidase

   - 5 mg γ-Glutamyl-4-methoxy-2-naphthylamide (completely dissolved in 0.5 ml *N,N′*-dimethylformamide)
   - 10 ml 0.1 *M* Cacodylate buffer, pH 7.0
   - 0.5 mg Glycyl-glycin
   - 0.25 ml 2% Hexazotized pararosaniline or hexazotized New Fuchsine
   - Mix well, adjust the pH to 6.5 with NaOH, and filter

4. Dipeptidyl peptidase I

   - 5 mg Glycyl-arginyl-4-methoxy-2-naphthylamide (dissolved in 0.25 *N,N′*-dimethylformamide)
   - 10 ml 0.1 *M* Cacodylate buffer, pH 5.5
   - 0.25 ml 2% Hexazotized pararosaniline or hexazotized New Fuchsine
   - Mix well, adjust the pH to 5.5 with NaOH, and filter

5. Dipeptidyl peptidase II

   - 5 mg Lysyl-alanyl-4-methoxy-2-naphthylamide (dissolved in 0.25 ml *N,N′*-dimethylformamide)
   - 10 ml 0.1 *M* Cacodylate buffer, pH 5.5
   - 0.25 ml 2% Hexazotized pararosaniline or hexazotized New Fuchsine
   - Mix well, adjust the pH to 5.5 with NaOH, and filter

6. Didpeptidyl peptidase IV

   - 5 mg Glycyl-prolyl-4-methoxy-2-naphthylamide (dissolved in 0.25 ml *N,N'*-dimethylformamide)
   - 10 ml 0.1 *M* Cacodylate or phosphate buffer, pH 7.0
   - 0.25 ml 2% Hexazotized pararosaniline or hexazotized New Fuchsine
   - Mix well, adjust the pH to 6.5 with NaOH, and filter

7. Cathepsin B

   - 5 mg CBZ-Arginyl-arginyl-4-methoxy-2-naphthylamide, or CBZ-alanyl-alanyl-arginyl-4-methoxy-2-naphthylamide, or CBZ-valyl-lysyl-lysyl-arginyl-4-methoxy-2-naphthylamide (dissolved in 0.5 ml *N,N'*-dimethylformamide)
   - 10 ml 0.05 *M* Phosphate buffer, pH 6.0
   - 0.25 ml 2% Hexazotized pararosaniline or hexazotized New Fuchsine
   - Mix well, adjust the pH to 6.0 with NaOH, and filter

8. Esterproteases, kallikrein-type proteases

   - 5 mg D- or CBZ-Valyl-leucyl-arginyl-4-methoxy-2-naphthylamide, CBZ-glycyl-glycyl-arginyl-4-methoxy-2-naphthylamide, or CBZ-alanyl-arginyl-arginyl-4-methoxy-2-naphthylamide (dissolve in 0.25 ml *N,N'*-dimethylformamide)
   - 10 ml 0.1 *M* Cacodylate buffer, pH 6.5
   - 0.25 ml 2% Hexazotized pararosaniline or hexazotized New Fuchsine
   - Mix well, adjust the pH to 6.5 with NaOH, and filter

Incubate the test tissue in incubation medium made up as described above, for 60 to 120 min at 37°C.

4. Post-Treatment

Rinse several times in 0.1 *M* cacodylate buffer, pH 7. Osmicate with 1.5 or 2% OsO_4 in 0.1 *M* cacodylate buffer, pH 7.2 to 8 (adjust with NaOH) for 30 to 120 min at room temperature. Rinse in 0.1 *M* cacodylate buffer, pH 7.6, and dehydrate in graded ethanol or acetone. Embed in a resin, e.g., Araldite or Epon, according to Spurr.[18]

Alternatively, post-fix sections in a mixture of 2.5% glutaraldehyde and 2.3% formaldehyde (freshly prepared from paraformaldehyde) in 0.05 *M* cacodylate buffer, pH 7.2, for 15 min at room temperature. Rinse in 0.1 *M* cacodylate buffer, pH 7.2, for 15 to 30 min. Osmicate in 2% aqueous OsO_4 for 14 to 16 h at room temperature. Dehydrate in graded acetone, and embed in Araldite.

5. Preparation of Hexazotized Pararosaniline or Hexazotized New Fuchsine (see also Reference 19)

Solution A: dissolve 0.4 g pararosaniline-HCl or New Fuchsine in 8 ml distilled water and 2 ml concentrated (36%) HCl, mix well until dissolved and filter. *Solution B:* 4.0% sodium nitrate. To obtain hexazotized pararosaniline or hexazotized New Fuchsine, equal volumes of both solutions are mixed immediately before use. The nitrate must be added to the pararosaniline or New Fuchsine. The resulting hexazotized pararosaniline should be light yellow, the hexazotized New Fuchsine yellow-brown. The nitrate solution should not be stored in the refrigerator for longer than 1 week.

E. OTHER TECHNIQUES

The OTO procedure for the ultracytochemical demonstration of proteases using γ-glutamyl transpeptidase as an example, according to Rutenberg et al.,[7] is given below. The electron microscopical visualization for the other proteases can be tried with the same reaction principle and the methods given above, but the hexazotized pararosaniline or New Fuchsine should be replaced by Fast Blue BB.

1. Incubation medium

- 2.0 ml γ-Glutamyl-4-methoxy-2-naphthylamide (1.25 mg/ml; dissolve 12.5 mg substrate in 0.25 ml dimethylsulfoxide and 0.25 ml 1 *N* NaOH and add distilled water to a volume of 10 ml)
- 5.0 ml 0.1 *M* Phosphate buffer, pH 7.4
- 13.0 ml Distilled water
- 10 mg Glycyl-glycine
- 10 mg Fast Blue BB
- 1.5 mg Sucrose

Incubate for 30 to 40 min at 25°C.

2. Post-treatment

- Rinse for 2 min in 7.5% sucrose
- Immerse for 2 min in 0.1 *M* cupric sulfate
- Rinse for 2 min in 7.5% sucrose
- Immerse for 15 min at 37°C in 1% thiocarbohydrazide containing 7.5% sucrose
- Wash three times in 7% sucrose at 37°C, for 5 min each
- Treat with 0.8 g OsO_4 dissolved in a few drops of distilled water
- Dehydrate in graded alcohols
- Embed in Araldite

The simultaneous azo dye method with 4-aminophthalhydrazide for the electron microscopic localization of microsomal alanyl aminopeptidase and γ-glutamyl transpeptidase, according to Seligman et al.,[8] is described below. As is the case with the OTO technique, this procedure can also be tried for other proteases given above, whereby diazotized 4-aminophthalhydrazide (4-APH) is used instead of hexazotized pararosaniline or New Fuchsine.

1. Incubation medium

- Microsomal alanyl aminopeptidase
- 0.5 ml Alanyl-4-methoxy-2-naphthylamide (8 mg/ml; dissolve 12.5 mg in 0.25 ml dimethylsulfoxide and 0.25 ml 1 *N* NaOH and add distilled water to a volume of 10 ml)
- 5 ml 0.1 *M* Phosphate buffer, pH 7.4
- 2.0 ml Distilled water
- 2.5 ml Freshly diazotized 4-APH (2.5 mg/ml)
- *p*-Glutamyl transpeptidase
- 1.0 ml γ-Glutamyl-4-methoxy-2-naphthylamide (1.25 mg/ml; for preparation, see alanyl-4-methoxy-2-naphthylamide)
- 5.0 ml 0.1 *M* Phosphate buffer, pH 7.4

- 5 mg Glycyl-glycine
- 1.5 ml Distilled water
- 300 mg Sucrose
- 2.5 ml Freshly diazotized 4-APH (2.5 mg/ml)

Incubate for 1 to 2 h with a change of the medium after 1 h.

2. Post-treatment

- Rinse for 30 min with several changes of 0.05 *M* phosphate buffer, pH 7.4, containing 7.5% sucrose
- Osmicate in 2% OsO_4 for 1 h at 45°C
- Dehydrate in graded alcohols
- Embed in Araldite

3. Diazotation of 4-aminophthalhydrazide (4-APH; if not available commercially, it may be synthesized according to Seligman et al.[8])

- Add 25 mg 4-APH to 0.5 ml cold 16.5% (v/v; 2 *N*) HCl followed by 0.5 ml cold 2% (20 mg/ml) $NaNO_2$ in a 15-ml graduated centrifuge tube.
- Mix the suspension and allow to stand in the cold for 5 min until complete solution.
- Add 1.0 ml 8.2% (1 *M*) sodium acetate to this reddish solution, adjust the pH to about 5.0 by dropwise addition of 4% (1 *N*) NaOH, and bring up to 10 ml with cold distilled water.
- The final concentration of the 4-APH diazotate in the incubation medium is 0.625 mg/ml.

F. CONTROLS

1. Basic controls

- Incubation in substrate-free media
- Incubation without hexazotized pararosaniline, or hexazotized New Fuchsine, or Fast Blue BB or 4-aminophthalhydrazide
- Incubation of sections in cupric sulfate and rinsing in sucrose
- Inactivation of the tissue by heat at 60°C for 30 to 60 min
- Incubation of tissues with known activity
- Localization of the respective protease

2. Controls with inhibitors and activators (see Table 2; for further activators and reading, see also References 20 and 21)

- *Microsomal analyl aminopeptidase:* preincubate in 0.02% (1 m*M*) 1,10-phenantroline (dissolved in a small amount of *N,N'*-dimethyl formamide) in 0.1 *M* phosphate buffer, pH 6.5 to 7.5, for 15 min at 37°C, and incubate further in the incubation medium containing the same concentration of phenantroline; incubate in the presence of 1 μ*M* bestatin or amastatin. Amastatin is an inhibitor of microsomal alanyl aminopeptidase and glutamyl aminopeptidase and bestatin is an inhibitor of microsomal alanyl aminopeptidase and arginyl aminopeptidase[20,21] (for further reading, see References 22 and 23).

- *Glutamyl aminopeptidase:* incubate in the absence of $CaCl_2$ and in the presence of 1 to 5 m*M* EDTA (instead of $CaCl_2$), or use 1 μ*M* amastatin (for details of the inhibition effect and references see microsomal alanyl aminopeptidase).
- *Dipeptidyl peptidase I:* preincubate with 1 m*M* iodoacetate or *p*-chloromercuribenzoate in 0.1 *M* cacodylate buffer, pH 5.0 or 5.5, for 15 min at 37°C.
- *Dipeptidyl peptidase II:* incubate in the presence of 1 m*M* puromycin.
- *Dipeptidyl peptidase IV:* incubate in the presence of 1 m*M* diisopropyl fluorophosphate (0.02%), which inhibits the enzyme completely. 1 m*M* E 600 (diethyl-*p*-nitrophenyl phosphate) or 1 m*M* Pb^{2+} ions inhibit incompletely. Addition of 1 μ*M* diprotin A to the incubation medium also results in a complete inhibition.
- *Cathepsin B:* incubate in the presence of 5 μ*M* leupeptin, or preincubate with 1 m*M* iodoacetate in 0.05 *M* phosphate buffer, pH 5.5 or 6.0, for 5 min.

G. RESULTS

Compared with the light microscopic visualization of proteases and many other enzymes, the electron microscopic demonstration methods for proteases have been less frequently applied up until now. This limited application of protease ultracytochemistry is explained by the comparatively strong inhibition of the proteases by aldehydes, the co-reaction of proteases different from those under investigation, as well as the reduced stability and moderate electron density of the osmicated final reaction product.

Especially at sites with low protease activity, such as plasma membranes of capillary endothelial cells, fixation either by immersion or perfusion can inhibit the respective enzyme nearly completely, e.g., glutamyl aminopeptidase or microsomal alanyl aminopeptidase. Then, the remaining activity is too low for a positive reaction. In this situation semithin sections, as an intermediate step (prepared from the same blocks from which the ultrathin sections are finally cut), can be helpful, showing whether there are at least minimum amounts of sufficiently osmicated azo dye. Its color should be black-brown or even better black; brown-yellow or only yellow final reaction products indicate insufficient osmication of the azo dyes from 4-methoxy-2-naphthylamine and hexazotized pararosaniline or hexazotized New Fuchsine.[6,13]

Glutaraldehyde alone at concentrations between 2 and 4% allows the most intense osmication. However, due to these relatively high concentrations, and probably to its effective cross-linking capacities with proteins and protein-containing molecules such as glycoproteins, to which all proteases belong,[21] it is deleterious to many protease activities. Therefore, concentrations below 2% or 1.5% glutaraldehyde can be useful. However, it is also important not to remain under a certain minimum concentration of glutaraldehyde in order to ensure sufficient osmication of the azo dye; below 0.5% glutaraldehyde, effective osmication is no longer guaranteed.

Formaldehyde, freshly prepared from paraformaldehyde, does not cross-link, and can be effectively used in concentrations ranging from 1 to 4%. Compared with glutaraldehyde, formaldehyde is less inhibitory to all kinds of proteases when used in these concentrations. A good compromise appears to be a diluted Karnovsky[14] mixture containing 1% glutaraldehyde and 1% formaldehyde (made up from paraformaldehyde) for protease ultracytochemistry. The periodate-lysine-formaldehyde fixative of McLean and Nakane appears to be useful also.[24]

The fixation time, as well as the type of buffer in the fixative and rinsing solution, depend upon the protease to be investigated. However, as a general rule, shorter fixation times with lower concentrations of fixative are preferable using the buffer and pH at which the respective protease is most active. For example, 1- to 10-min perfusion fixation followed by rinsing first in the buffer used to dilute the fixative containing 7% sucrose, and then with the buffer to be used for the demonstration of the respective protease, or 1- to 2-h immersion fixation are generally good for structural preservation and retention of sufficient protease activity; however, also, 1 to 2 h of perfusion with subsequent washing might be useful, depending on the protease and tissue studied.[13,16,17,21]

Another problem in protease ultracytochemistry is substrate specificity. One substrate can be cleaved by several proteases, e.g., lysyl-prolyl-4-methoxy-2-naphthylamide, alanyl-prolyl-4-methoxy-2-naphthylamide, and phenyl-prolyl-4-methoxy-2-naphthylamide are cleaved by dipeptidyl peptidase II and IV; glycyl-arginyl-4-methoxy-2-naphthylamide by dipeptidyl peptidase I and microsomal alanyl aminopeptidase in lysosomes and plasma membranes. CBZ-alanyl-arginyl-arginyl-4-methoxy-2-naphthylamide is hydrolyzed over a wide pH range by cathepsin B and by the not-yet-defined and -classified endopeptidases in the microvillous membrane of small intestinal enterocytes and proximal renal tubular cells. Moreover, one protease can hydrolyze more than one substrate, e.g., microsomal alanyl aminopeptidase cleaves alanyl-4-methoxy-2-naphthylamide, arginyl-4-methoxy-2-naphthylamide, leucyl-4-methoxy-2-naphthyl-amide, methionyl-4-methoxy-2-naphthylamide, and glycyl-4-methoxy-2-naphthylamide. Furthermore, a substrate may be cleaved successively by two different proteases, e.g., alanyl-phenyl-prolyl-4-methoxy-2-naphthylamide. In this case, microsomal alanyl aminopeptidase first cleaves off alanine, and the resulting phenyl-prolyl-4-methoxy-2-naphthylamide is attacked by dipeptidyl peptidase IV, whereby phenyl-proline and 4-methoxy-2-naphthylamine are formed.[21]

However, often the situation is less complicated because the cleaving proteases have different localizations, e.g., dipeptidyl peptidase IV is mainly associated with plasma membranes and dipeptidyl peptidase II with lysosomes. It is important to use the most sensitive and most discriminating substrate; however, the most sensitive substrate is not necessarily the most discriminating. For example, lysyl-prolyl-4-methoxy-2-naphthylamide and alanyl-prolyl-4-methoxy-2-naphthylamide are more sensitive substrates for dipeptidyl peptidase IV than glycyl-prolyl-4-methoxy-2-naphthylamide, but they are also cleaved by dipeptidyl peptidase II, even at pH 8.[21]

Therefore, specific controls using suitable inhibitors are important. However, even the application of inhibitors has limitations because most of them are only group specific, e.g., diisopropyl fluorophosphate which inhibits serine proteases such as dipeptidyl peptidase IV or EDTA; or phenantroline, which inhibits metallopeptidases, e.g., glutamyl aminopeptidase or microsomal alanyl aminopeptodase, but also the thiol peptidase dipeptidyl peptidase I. Except for diprotin A, which is said to be a specific inhibitor of dipeptidyl peptidylpeptidase IV, the microbial inhibitors are also only group specific. For example, amastatin inhibits microsomal alanyl aminopeptidase and glutamyl aminopeptidase, bestatin microsomal alanyl aminopeptidase and arginyl aminopeptidase; or leupeptin inhibits cathepsin B, plasmin, trypsin, and papain (see References 20 and 23).

Furthermore, it is possible that the inhibitors can react with compounds of the incubation media such as the coupling agents (diazonium salts), hexazotized New Fuchsine, hexazotized pararosaniline, Fast Blue BB, or diazotized 4-aminophthalhydrazide.

A special situation exists for dipeptidyl peptidase I and cathepsin B. As thiol (cysteine) proteases, they require thiol reagents, e.g., the presence of dithiotreitol in the incubation media. However, these reducing agents lead to a decomposition of the diazonium salts which can result in false-negative data which also occurs when the incubation is performed for these proteases in the absence of thiol reagents.[17,21]

4-Methoxy-2-naphthylamine is considered the best leaving group due to its ability to form osmiophilic complexes with hexazotized pararosaniline, originally shown by McDonald et al.[5] and Smith and van Frank,[6] and with hexazotized New Fuchsine[13] or the 4-amino-phthalhydrazide diacetate.[8] An alternative leaving group might be *p*-nitrothiophenol after its cleavage from amino acid or peptide ester substrates[25-27] in metal capture methods (reaction of the thiol of the liberated *p*-nitrothiophenolate with ionic gold to form an electron-dense gold *p*-nitrothiophenolate complex).[26-28] However, data on these substrates are limited, and thiol-containing substrates for cysteine (thiol) proteases, e.g., cathepsin B, do not yield positive results.[17,21]

Theoretically, hexazotized pararosaniline or hexazotized New Fuchsine binds three molecules of 4-methoxy-2-naphthylamine. The resulting azo dye is said to react with three molecules of osmium, provided that the osmification is carried out at alkaline pH for a sufficiently long time.[5,17] Practically, however, brown color can appear during the dehydration procedure, especially in the higher concentrations of acetone or ethanol. Then, the tissue looks comparatively pale and yellow-brown when viewed in semithin sections.[5,13] Whether the discoloring is caused by a loss of osmic acid alone, the osmicated azo dye, or both, has to be clarified. In every case, this can lead to false-negative results at sites with low staining intensity and low electron density. According to Sannes,[17] hexazotized pararosaniline (and hexazotized New Fuchsine) are lipid soluble and tend to make lipids more electron dense than is usually the case in routinely postosmicated tissues. Therefore, it can be difficult to differentiate between lipids and osmicated reaction products.[9]

Furthermore, the osmicated azo dye derived from hexazotized pararosaniline or New Fuchsine often appears to be amorphous and therefore it can be critical to differentiate between primary osmophilia of a certain structure, e.g., the lysosomal matrix, and the osmicated reaction product. This is true for telolysosomes in general,[17] or for the secondary lysosomes in the principal and clear cells of the rat epididymis or in the proximal renal tubular cells.[13] However, at other subcellular sites, but also in certain lysosomes, the osmicated azo dye as a final stain appears as nonhomogeneous electron-dense particles.

In comparison with hexazotized New Fuchsine, hexazotized pararosaniline, when used for the ultracytochemical visualization of plasma membrane or Golgi-associated proteases, such as glutamyl aminopeptidase or dipeptidyl peptidase IV, yields less and coarser granular stain which may occur in addition artificially in the cytoplasm or mitochondria.[13]

According to Hanker,[29] the OTO reaction (bridging osmium through thiocarbohydrazide [TCH] to copper-chelated azo dye), developed by Rutenburg and co-workers[7] for γ-glutamyl transpeptidase ultracytochemistry, has disadvantages, since the initial azo dye reaction product is lipophilic and tends to form coalescent masses which are large and not precise enough for high-resolution electron microscopy before chelation with copper may be affected. Furthermore, the copper chelate may not be sufficiently insoluble in organic solvents to withstand embedding in resins without loss and translocation.

Because of the lipophilic characteristics of the azo dye, the resulting droplet distribution of the deposits, and improper relation to the underlying ultrastructures, bridging of the chelated copper (copper-azo dye chelate) via TCH to osmium only magnifies these faults.[8] Therefore, Seligman et al.[8] developed a technique using diazotized 4-aminophthalhydrazide (4-APH). When coupled simultaneously to 4-methoxy-2-naphthylamine, a water-insoluble lipophobic, osmiophilic nondroplet azo dye results. However, when this procedure is applied to the electron microscopic demonstration of microsomal alanyl aminopeptidase and γ-glutamyl transpeptidase, e.g., in the proximal tubular cells of the kidney, the membranes of the endoplasmic reticulum and of cytoplasmic vesicles were more intensely stained than the microvillous plasma membranes where these enzymes primarily occur.[7,8,13] Furthermore, the 4-APH diazotate is comparatively unstable at neutral and especially at slightly alkaline pH, so that artificial precipitates may occur, and the incubation medium has to be renewed after 30 to 60 min.

Therefore, both the OTO method and the 4-APH procedure appear to deliver less reliable results than the azo dye-osmium procedure using hexazotized pararosaniline and especially hexazotized New Fuchsine.

The ultracytochemical azo dye-osmium methods with hexazotized pararosaniline and hexazotized New Fuchsine are useful to localize microsomal alanyl aminopeptidase and γ-glutamyl transpeptidase in the microvillous membranes of jejunal enterocytes and proximal renal tubular cells, or in certain capillary endothelial cells where also glutamyl aminopeptidase and dipeptidyl peptidase IV can be found. Additionally, the latter enzymes are present in the plasma membrane of renal glomerular endothelial cells, glutamyl aminopeptidase in the plasma membrane of smooth muscle cells, and together with dipeptidyl peptidase IV, in Golgi elements of enterocytes and endosomes of these cells and of proximal renal tubulocytes. Dipeptidyl peptidase IV exists in the plasma membrane of lymphocytes and alveolar macrophages. Other plasma membranes which show γ-glutamyl transpeptidase activity are those of the microvilli (stereocilia) of the principal cells in the head of the epididymis and decidual cells (for further localizations see References 17, 21, 30, and 31).

Dipeptidyl peptidase I and II are either present in the lysosomes of the same cells, e.g., macrophages, proximal renal tubular cells, hepatocytes, or principal cells of the epididymis, or in different cells such as dipeptidyl peptidase I in the acinar cells and dipeptidyl peptidase II in the islet cells of the pancreas. Cathepsin B can best be detected in the lysosomes of macrophages, epididymal principal cells, or proximal renal tubular cells (for further localizations, see References 17 and 21). The not-yet-classified kallikrein-type endopeptidase is present in the secretion granules of the tubular cells in the rat and mouse submandibular gland.[12,13,30]

REFERENCES

1. *Enzyme Nomenclature,* Academic Press, Orlando, 1984.
2. **McDonald, J. K. and Barrett, A. J.,** *Mammalian Proteases. A Glossary and Bibliography,* Academic Press, London, 1986.
3. **McDonald, J. K.,** An overview of protease specificity and catalytic mechanisms: aspects related to nomenclature and classification mechanisms, *Histochem. J.,* 17, 737, 1985.
4. **Wohlrab, F. and Gossrau, R.,** *Katalytische Enzymhistochemie. Grundlagen und Methoden für die Elektronenmikroskopie,* Fischer, Jena, 1992.
5. **McDonald, J. K., Callahan, P. X., Ellis, S., and Smith, R. E.,** Polypeptide degradation by dipeptidyl aminopeptidase I (cathepsin C) and related peptidases, in *Tissue Proteinases,* Barrett, A. J. and Dingle, J. T., Eds., North-Holland, Amsterdam, 1971, 69.
6. **Smith, R. E. and van Frank, R. M.,** The use of [illegible] of 4-methoxy-2-naphthylamine for the assay and subcellular localization of tissue proteinases, in *Lysosomes in Biology and Pathology,* Vol. 4, Dingle, J. T. and Deane, R. T., Eds., North-Holland, Amsterdam, 1975, 193.
7. **Rutenburg, A. M., Kim, H., Fischbein, J. W., Hanker, J. S., Wasserkrug, H. L., and Seligman, A. M.,** Histochemical and ultrastructural demonstration of γ-glutamyl transpeptidase, *J. Histochem. Cytochem.,* 17, 517, 1969.
8. **Seligman, A. M., Wasserkrug, H. L., Plapinger, R. E., Saito, T., and Hanker, J. S.,** Membraneous ultrastructural demonstration of aminopeptidase and γ-glutamyl transpeptidase activities with a diazonium salt that yields a lipophilic, osmiophilic azo dye, *J. Histochem. Cytochem.,* 18, 542, 1970.
9. **Smith, R. E.,** Subcellular localization of peptidases using aminoacyl and dipeptidyl derivatives of 4-methoxy-2-naphthylamine, *J. Cell Biol.,* 43, 135A, 1969.
10. **Hanker, J. S., Seaman, A. R., Weiss, L. P., Ueno, H., Bergman, R. A., and Seligman, A. M.,** Osmiophilic reagents: new cytochemical principle for light and electron microscopy, *Science,* 146, 1039, 1964.
11. **Hanker, J. S., Deb, C., Wasserkrug, H. L., and Seligman, A. M.,** Staining tissue for light and electron microscopy by bridging metals with multidentate ligands, *Science,* 152, 1631, 1966.
12. **Gossrau, R.,** Investigation of proteinases in the digestive tract using 4-methoxy-2-naphthylamine (MNA) substrates, *J. Histochem. Cytochem.,* 29, 464, 1981.
13. **Schröder, J. and Gossrau, R.,** Elektronenmikroskopische Lokalisation von Proteasen mit direkten Azo-Farbstoffmethoden, *Acta Histochem. Suppl.,* 25, 147, 1982.

14. **Karnovsky, M. J.,** A formaldehyde-glutaraldehyde fixative of high osmolality for use in electron microscopy, *J. Cell Biol.,* 27, 137A, 1965.
15. **Bremer, U., Hacker, H. J., and Bannasch, P.,** Electron microscopical demonstration of glucose-6-phosphatase in native cryostat sections fixed with glutaraldehyde through semipermeable membranes, *Histochemistry,* 65, 41, 1979.
16. **Kugler, P.,** On angiotensin-degrading aminopeptidases in the rat kidney, *Adv. Anat. Embryol. Cell Biol.,* 76, 1, 1982.
17. **Sannes, P. L.,** The histochemical and cytochemical localization of proteases, *Prog. Histochem. Cytochem.,* 18, 1, 1988.
18. **Spurr, A. R.,** A low-viscosity epoxy resin embedding medium for electron microscopy, *J. Ultrastruct. Res.,* 26, 31, 1969.
19. **Lojda, Z., Gossrau, R., and Schiebler, T. H.,** *Enzyme Histochemistry. A Laboratory Manual,* Springer, Berlin, 1979.
20. **Katanuma, N., Umezawa, H., and Holzer, H., Eds.,** *Protease Inhibitors. Medical and Biological Aspects,* Springer, Berlin, 1983.
21. **Lojda, Z., Gossrau, R., and Stoward, P. J.,** Proteases, in *Histochemistry. Theoretical and Applied,* Stoward, P. J. and Pearse, A. G. E., Eds., Churchill Livingstone, Edinburgh, 1991, 281.
22. **Umezawa, H.,** *Enzyme Inhibitors of Microbial Origin,* University Park Press and University Tokyo Press, Baltimore and Tokyo, 1972.
23. **Gossrau, R.,** Protease inhibitors (antiproteases): new trends in cytochemical protease research, *Histochem. J.,* 17, 522, 1985.
24. **McLean, I. W. and Nakane, P. K.,** Periodate-lysine-paraformaldehyde fixative. A new fixative for immunoelectronmicroscopy, *J. Histochem. Cytochem.,* 12, 1077, 1974.
25. **Castillo, M. J., Nakajima, K., Zimmerman, M., and Powers, J. C.,** Sensitive substrate for human leukocyte and porcine pancreatic elastase: a study of the merits of various chromophoric and fluorogenic leaving groups in assays of serine proteases, *Anal. Biochem.,* 99, 53, 1979.
26. **Clark, J. M., Aiken, B. M., Vaughan, D. W., and Kagan, H. M.,** Ultrastructural localization of elastase-like enzyme in human neutrophils, *J. Histochem. Cytochem.,* 28, 90, 1980.
27. **Clark, J. M., Aiken, B. M., Vaughan, D. W., and Kagan, H. M.,** Elastase-like enzyme in human neutrophils localized by ultrastructural cytochemistry, *J. Cell Biol.,* 84, 102, 1980.
28. **Vater, A. E., Reiss, O., Newman, J., Lindquist, K., and Groeneboer, E.,** Enzymes of the lung. I. Detection of esterase with a new cytochemical method, *J. Cell Biol.,* 38, 80, 1968.
29. **Hanker, J. S.,** Osmiophilic reagents in electronmicroscopic histochemistry, *Prog. Histochem. Cytochem.,* 12, 1, 1979.
30. **Gossrau, R.,** Localization and regulation of proteolytic enzymes in salivary glands, in *Molecular and Cellular Regulation of Enzyme Activity,* Part II, Barth, A., Haschen, R., Laibach, F. H., Possin, H., and Schowen, R. L., Eds., Martin-Luther-Universität, Halle-Wittenberg, 1984, 105.
31. **Gossrau, R.,** Cytochemistry of membrane proteases, *Histochem. J.,* 17, 737, 1985.

Chapter 1

ELECTRON MICROSCOPIC CYTOCHEMISTRY OF ENZYMES

1.3 Basic Techniques for Major Groups of Enzymes

1.3.4 Lyases – Carbonic Anhydrase

Naonori Sugai

TABLE OF CONTENTS

0-8493-6012-9/93/$0.00 + $.50

A. INTRODUCTION

Carbonic anhydrase (EC 4.2.1.1) was first discovered by Meldrum and Roughton (1933)[1] in human erythrocytes as an enzyme capable of catalyzing the hydration of carbon dioxide and dehydration of bicarbonate ions. Darvenport and Fisher[2] in 1939 found the enzyme in the gastrointestinal mucosa of animals (rat, cat, rabbit), which was the first discovery of the enzyme in tissues other than erythrocytes. Since then, the enzyme has been found in a large number of animal tissues, in some plants, and bacteria. Today, seven biochemically characterized isozymes of carbonic anhydrase (CA)[3] are known to occur in higher vertebrates, and the biochemical, physiological, pharmacological, and morphological studies on the enzyme are still carried out in increasing volume.

A histochemical method to detect CA activity had been initially devised by Kurata,[4] who used manganese in the incubation medium for capturing the reaction products. Kurata's method has undergone a series of improvements, particularly by Häussler[5] and Hansson.[6] Häussler substituted cobalt for manganese in the medium, which thus improved Kurata's method. Hansson's method, which diminished most of the criticism against the specificity of the CA histochemical reaction, depends upon enzymatic dehydration of the substrate H_2CO_3 and the formation of an alkaline-insoluble cobalt complex which is later converted into visible CoS. The histochemical reaction is accelerated by floating sections on the surface of the incubation medium, where a rapid CO_2 diffusion occurs. In addition, a relatively minor but important modification was made by Rosen,[7] who allowed the incubation medium to stand for a short period of time before use. This resulted in a moderate increase of alkalinity on the surface of the incubation medium, which caused a retardation of the noncatalyzed reaction. At present, Hansson's modified method is generally used; it is reliable and specific for CA.

B. HISTOCHEMICAL METHODS FOR THE DEMONSTRATION OF CARBONIC ANHYDRASE ACTIVITY AT THE ELECTRON MICROSCOPIC LEVEL

Two electron microscopic methods for the demonstration of CA activity have been reported to date, a preembedding staining method and a postembedding staining method.

Preembedding method[8] — After being reacted for CA activity, the tissue sections are embedded in Epon or Epon-Araldite and cut into thin sections for electron microscopical observations. In this method, the tissue sections are osmicated before embedding in resin, which thus allows an observation of excellent intracellular localization of CA activity at the ultrastructural level. This method will be discussed mainly in this chapter.

Postembedding method[9] — Semithin sections (0.1 to 0.2 µm) of the tissues embedded in a water-soluble resin (JB 4, Acrytron E, etc.) are reacted for CA activity and observed under an electron microscope. With this method, detailed ultrastructural observation are difficult because of the omission of osmication.

C. EQUIPMENT AND REAGENTS

In addition to standard histological laboratory equipment and supplies, a magnetic stirrer and a cryostat are needed. Reagents include sodium bicarbonate, cobalt sulfate, OCT compound (Tissue Tek II), acetazolamide (diamox), and potassium cyanate (KCNO).

D. ELECTRON MICROSCOPIC PROCEDURES FOR CA HISTOCHEMISTRY

1. Fixation

Fixed tissues or cells are usually used for CA histochemistry, because the reaction of frozen sections of unfixed tissues results in inconsistent and irregular histochemical stainings, owing to diffusion or loss of the enzyme. While fixation leads to a loss of CA activity, it preserves CA activity sufficiently to demonstrate its localization. Acetone, ethanol, or formaldehyde used for the demonstration of CA activity at the light microscopic level is unsuitable for CA electron microscopy, because of poor preservation of morphological structures and loss or diffusion of the enzyme. Although fixation with 1 to 2% glutaraldehyde decreases the enzyme activity to 3 to 5% of the original value, it is routinely used for CA electron microscopy, because, in spite of reduction of the activity, exquisite localization of enzyme activity is obtained with glutaraldehyde-fixed tissues or cells.[8]

The fixative we routinely use is 1 to 2% glutaraldehyde, with or without 3 to 4% formaldehyde (made freshly from paraformaldehyde at time of use), in 0.05 to 0.1 *M* cacodylate buffer, pH 7.4.

The best morphological preservation and reaction product localization are obtained using tissues fixed by perfusion. After perfusion, the dissected tissues are cut into slices or small blocks, and immersed in the same fixative as used in perfusion for 1 to 2 h at room temperature.

2. Preparation of Tissue and Tissue Sections

After fixation, the tissue slices or blocks are rinsed three times for 20 to 30 min at room temperature in 0.1 *M* cacodylate buffer, pH 7.4. Then, the tissues are cryoprotected with the method proposed by Barthel and Raymond,[10] modified as follows. The tissues are kept for 1 to 3 h at room temperature in each of a series of sucrose solutions of increasing concentration. The sucrose solutions are prepared by mixing 5 and 20% sucrose in 0.1 *M* cacodylate buffer, pH 7.4, in ratios of 2:1, 1:1, 1:2. Finally, the tissues are transferred into a 20% sucrose solution overnight at 4°C. The tissues are then infiltrated for 30 min to 3 h at room temperature in a mixture of 2 parts 20% sucrose in cacodylate buffer and 1 part OCT embedding medium, prior to embedding in the same solution used for infiltration. The embedding tissue blocks are frozen with dry ice, and sectioned in a cryostat at 6 to 18 μm thickness. Sections more than 20 μm thick should not be used.

3. Preparation of the Incubation Medium

Hanssons' medium contains 1.9 m*M* $CoSO_4$, 55.9 m*M* H_2SO_4, 3.7 m*M* KH_2PO_4, and 158.7 m*M* $NaHCO_3$. To make up this medium, 380 mg of $NaHCO_3$ is added to 20 ml of water in a 50-ml Erlenmeyer flask. When the sodium bicarbonate is dissolved, 8.5 ml of a freshly made reagent solution is added. The final incubating solution is stirred for 4 min on a magnetic stirrer, [illegible] for 1 min to increase the surface alkalinity of the incubation medium and to reduce noncatalyzed reaction, and then used for incubation within 1 min. The reagent solution is prepared from stock solution as follows:

0.1 *M* $CoSO_4$	5 ml
0.5 *M* H_2SO_4	30 ml
0.067 *M* KH_2PO_4	15 ml
H_2O	30 ml
Total volume	80 ml

4. Incubation and Examination under the Light Microscope

Frozen sections, while still frozen, are floated on the incubation medium. The reaction vessel is an open 90-mm Petri dish containing 8 to 10 ml of the medium. After incubation for 8 to 16 min, the sections are transferred with a glass rod or fresh wood applicator stick to the washing solution of 8% sucrose in 0.1 *M* cacodylate buffer, pH 7.4, for 20 to 30 s.

They are then transferred to a dish containing 2% ammonium sulfide for 1 to 3 min, and washed again in the sucrose solution. All the procedures are performed at 0 to 4°C. The sections are transferred from the sucrose washing solution and mounted, while wet, in water for light microscopic examination. Tissue sections are examined under the light microscope, and those that show the best histochemical reaction and morphological preservation are processed to the osmication step.

5. Osmication

To examine the detailed localization of CA activity at the ultrastructural level, osmication is indispensable. However, when the sections are treated with neutral or slightly alkaline-buffered OsO_4, as is done in the usual electron microscopic technique, the cobalt sulfide reaction product is solubilized and lost. To avoid this loss, some investigators elected to omit the osmication step,[7] while others used a very brief osmication to minimize loss of the cobalt precipitate. A different approach was employed by others,[11] who converted the cobalt to a lead precipitate, which is insoluble in osmic acid.

We have found[8] that buffering the osmium tetroxide to pH 6.0 or lower minimized the loss of the cobalt sulfide precipitate, while still providing ultrastructural preservation indistinguishable from that seen in tissues not incubated for the histochemical reaction. The tissue sections are routinely osmicated by 1% osmium tetroxide buffered with phosphate to pH 5.0 or 6.0 for 10 to 20 min at 4°C.

6. Block Staining and Embedding

After the tissue sections are postfixed in the osmium solution, they are treated with 1% uranyl acetate in 0.1 *M* maleate buffer, pH 5.2, for 10 to 20 min, dehydrated in ethanol, and embedded on glass slides in a drop of Epon or Epon-Araldite with a polymerized dummy specimen block resting on the specimen. The polymerized block with the specimen is removed by heating the slide and applying a lateral pressure.

7. Sectioning of Thin Tissue Sections for Electron Microscopic Observation

Light microscopical observations demonstrated that the cobalt sulfide reaction product was restricted to the surface of the sections nearest to the air interface. Electron microscopic examination of thin cross sections revealed that the reaction extended no deeper than 1 to 2 μm. Therefore, most of the observations of CA histochemical activity are made on the thin sections cut parallel or tangential to the plane of the tissue section.

The thin sections may be examined under the electron microscope, unstained or after uranyl acetate and lead citrate staining.

8. Inhibition Tests

The specificity of the histochemical reaction is supported by the following controls: (1) heating tissues to 97°C for 10 min.; (2) incubation in a medium without $NaHCO_3$; (3) addition of acetazolamide, 10^{-5} to 10^{-7} *M* to the medium; or (4) potassium cyanate (KCON) 5 to 6 $\times$ 10^{-3} *M*, added to the medium. No deposits of the reaction product should be seen in any of the control preparations.

E. RESULTS

Muther[12] has questioned the specificity of the histochemical techniques and suggested that the histochemical staining was caused, not by the enzymatic action of CA, but by the formation of an unknown reactive Co compound on the surface layer secondary to enzyme-independent alkalization of the incubation medium. He claimed that the inhibition of the histochemical staining by the sulfonamide inhibitors is caused by nonspecific interactions of the inhibitors with ingredients of the incubation medium. In addition, he maintained that potassium cyanate did not inhibit the histochemical staining.

However, several lines of evidence supporting the specificity of Hansson's method for CA activity have been reported.[13-17] In brief, Rosen[13] and Lönnerholm[14] demonstrated that the very low concentrations of the acetazolamide could abolish or partially inhibit the histochemical staining by inhibiting the enzyme. Sugai and Ito[8] have found that the histochemical reaction is inhibited by potassium cyanate. At present, Hansson's method is considered specific for the demonstration of CA activity. For a more detailed discussion of the specificity of CA histochemical methods, the readers are referred to Lönnerholm's review.[18]

Carbonic anhydrase is widely distributed throughout different tissues of vertebrates and it is found particularly in cells involved with the rapid exchange of carbon dioxide and bicarbonate. Its localization in the tissues or cells has been related to the transport of proton, bicarbonate, or other ions (Cl^-, Ca^{2+}), which have been implicated in such diverse processes as acid-base homeostasis (erythrocytes), respiration (lung and gill), calcification, bone resorption, reabsorption of bicarbonate (kidney), protection to epitheliums (gastric and intestinal mucosa), secretion of chloride ions (salt glands), and formation of aqueous humor (ciliary body), endolymph (semicircular canal and vestibular canal), cerebrospinal fluid (choroid plexus), sweat and pancreatic and gastric juices, and the synthesis of urea, glucose, and lipids.[3,19]

Seven biochemically distinct isozymes of CA have been identified and characterized in mammalian tissues and classified into I (low activity), II (high activity), III (liver and muscle), IV (membrane bound), V (mitochondrial), VI (secretory), and VII (salivary gland?) types, where types I, II, III, and VII are cytosolic.[3] All CA isozymes can be visualized at the ultrastructural level by Hansson's method, though none of the isozymes can be differentiated. Electron microscopically, the electron-dense Co precipitates indicating CA activity are seen in the cytoplasmic matrix, mitochondria, cell membranes, nucleus, and secretory granules.

The reaction of the cytoplasmic matrix has been demonstrated in a large number of cells, including the glial cells, gland cells of salivary and lacrimal glands, the epithelial cells of the stomach (surface mucous cells, mucous neck cells, parietal cells), Brunner's gland cells (see Figure 1), the epithelial cells of small intestine, cecum, and colon, small and great alveolar cells of the lung, Müller cells of the retina, the epithelial cells of turtle skin, the epithelial cells of proximal and distal tubules of the kidney, and capillary endothelial cells.

Recently, Wistrand and Knuuttila[20] have characterized biochemically the membrane-bound carbonic anhydrase in human kidney as type IV isozyme. A homologous isozyme was demonstrated to occur in the lungs of rat, rabbit, human, and cow as well.[21,22] Tashian[3] predicted earlier in his review that the membrane-bound CA isozyme would be found in tissues other than the kidney and lung. Recently, Carter et al.[23] described the ubiquitous distribution of type IV isozyme in a number of tissues, including the central nervous system, kidney, lung, liver, colon, pancreas, and fibroblasts. This wide distribution has been demonstrated also with histochemical methods. Membrane-bound cobalt precipitate due to CA activity has been shown with the electron microscopic technique in most of the tissues or cells, which indicate the reaction of cytoplasmic matrix. Deposits were seen either on the inner cytoplasmic or outer extracellular site of cell membrane (see Figures 2 and 3). The functional significance of this difference in localization has not been elucidated.

Mitochondrial CA isozyme, type V CA, has been demonstrated biochemically in the rat kidney and guinea pig skeletal muscle.[24,25] Mitochondrial localization also has been demonstrated cytochemically in a variety of cells, including glial cells, salivary or lacrimal gland cells, parietal cells, the epithelial cells of the small intestine and colon, etc. Cobalt precipitates have been observed on the mitochondrial cristae.

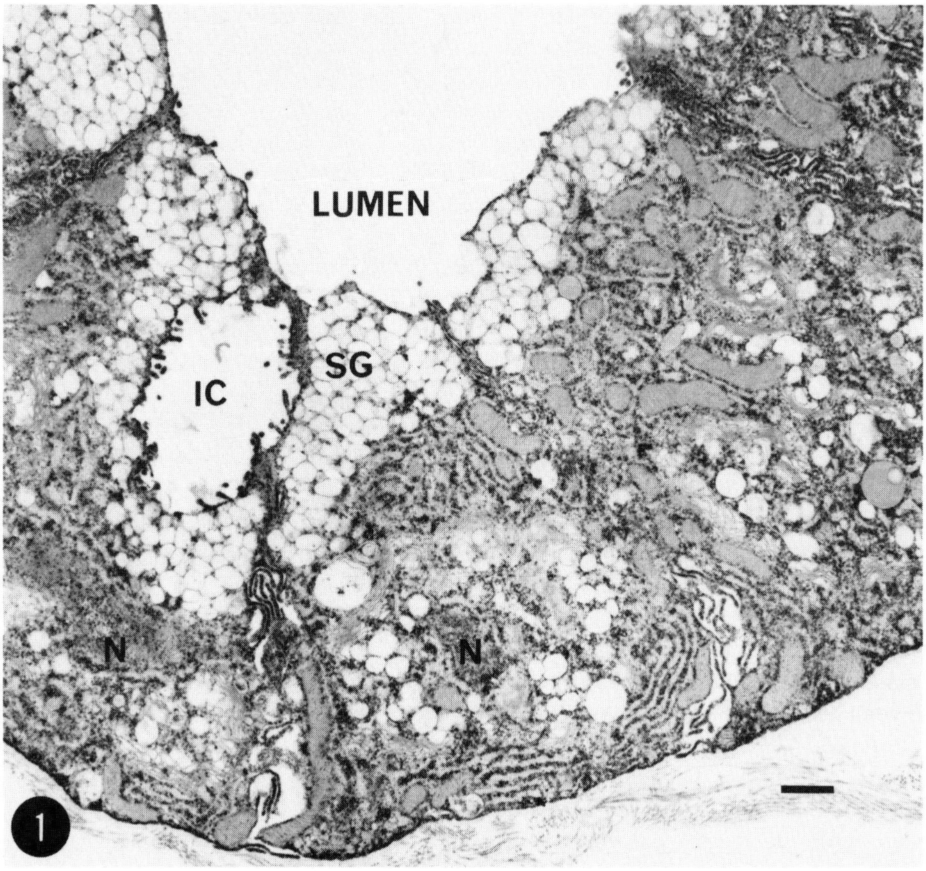

FIGURE 1. Electron micrograph of rat Brunner's gland fixed in the glutaraldehyde fixative, washed in a buffered sucrose solution, incubated in Hansson's medium, osmicated at pH 5.0, en bloc-stained with uranyl acetate and embedded with Epon-Araldite. Dense reaction product is present throughout the cytoplasmic matrix. The microvilli of intercellular canaliculi (IC) and the interdigitations of lateral cell membranes show reactivity. The cobalt precipitate is localized in the nucleus (N). However, the specificity of this nuclear staining is controversial. No reaction product is present in the secretory granules (SG). LUMEN, glandular lumen. (Bar = 1 μm.)

Scattered electron-dense CA reaction product has been demonstrated in the serous secretory granules of salivary[26] or lacrimal gland cells. In the rat lacrimal glands, dense cobalt precipitate was observed not only in the cytoplasmic matrix of the solitary cells, which are scattered among acinar cells with negative reactions, but also in the lumen of ducts (unpublished data). The latter indicates secretion of this enzyme into the tear fluid.

Histochemical reaction in the nucleus was varied considerably, though it was usually observed in cells with intense cytoplasmic reaction. In unfixed or weakly fixed tissues, no nuclear staining was found. The specificity of nuclear staining has been controversial.

In conclusion, the electron microscopic method can demonstrate the precise ultrastructural localization of CA activity involving all isoenzyme forms.

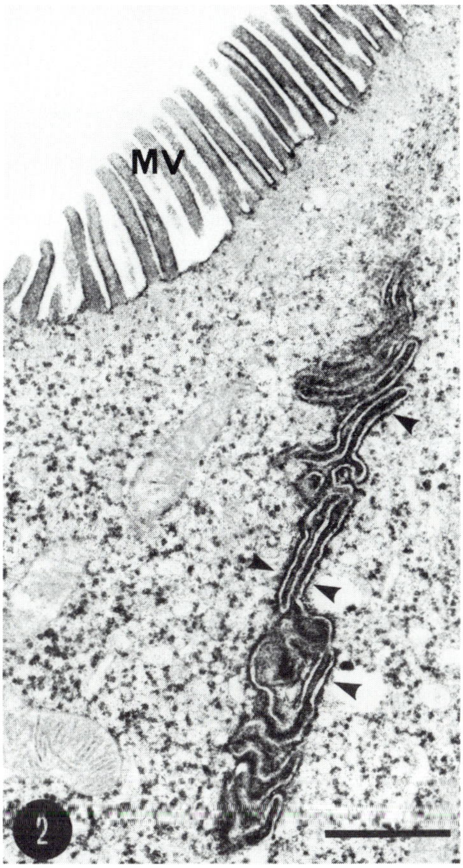

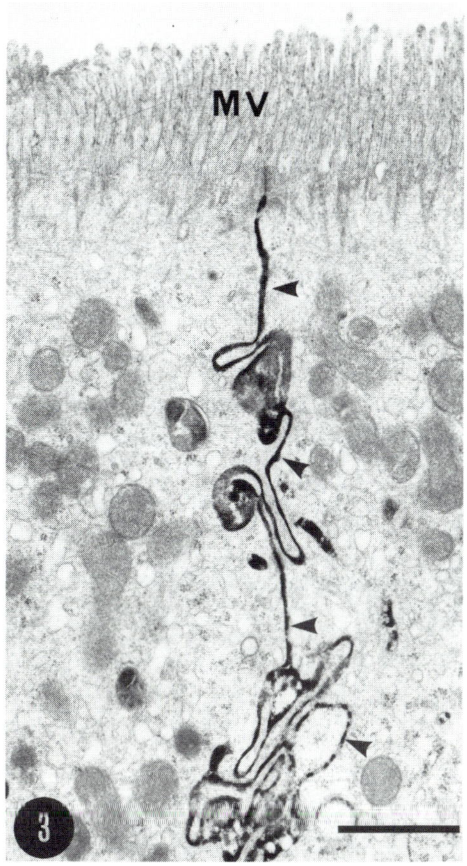

FIGURE 2. Electron micrograph of rat duodenal epithelial cells fixed in 2% glutaraldehyde-4% formaldehyde-0.05% picric acid fixative, incubated for 7 to 9 min in Hansson's medium, osmicated at pH 5.0 for 20 min. The dense reaction product is present on the inner surface of the microvillus (MV) and on lateral cell membranes (arrowheads). The reaction precipitate, which appears as small scattered dots, is localized in the cytoplasmic matrix. (Bar = 1 μm.)

FIGURE 3. Electron micrograph of epithelial cells of guinea pig duodenal epithelial cells fixed in 0.2% glutaraldehyde-4% formaldehyde-0.05% picric acid fixative, incubated in Hansson's medium for 5 to 6 min, osmicated at pH 5.0 for 15 min. Dense reaction product is present on the outer surface of lateral cell membranes (arrowheads). The reaction precipitate is excluded from microvilli (MV). (Bar = 1 μm.)

REFERENCES

1. **Meldrum, N. U. and Roughton, F. J. W.,** Carbonic anhydrase: its preparation and properties, *J. Physiol.,* 80, 113, 1933.
2. **Darvenport, H. W. and Fisher, R. B.,** Carbonic anhydrase in the gastro-intestinal mucosa, *J. Physiol.,* 94, 16P, 1983.
3. **Tashian, R. E.,** The carbonic anhydrases: widening perspectives on their evolution, expression and function, *BioEssays,* 10, 186, 1989.
4. **Kurata, Y.,** Histochemical demonstration of carbonic anhydrase activity, *Stain Technol.,* 28, 231, 1953.
5. **Häusler, G.,** Zur Technik und Spezifität des histochemischen Carbonanhydrasenachweises im Modelversuch und in Gewebsschnitten von Rattennieren, *Histochemie,* 1, 29, 1958.

6. **Hansson, H. P. J.,** Histochemical demonstration of carbonic anhydrase activity, *Histochemie,* 11, 112, 1967.
7. **Rosen, S.,** Carbonic anhydrase — alternative method, in *Electron Microscopy of Enzymes,* Vol. 3, Hayat, M. A. Ed., Van Nostrand Reinhold, New York, 1975, 109.
8. **Sugai, N. and Ito, S.,** Carbonic anhydrase, ultrastructural localization in the mouse gastric mucosa and improvements in the technique, *J. Histochem. Cytochem.,* 28, 511, 1980.
9. **Rinderstrale, Y.,** Intracellular localization of carbonic anhydrase in the frog nephron, *Acta Physiol. Scand.,* 98, 465, 1976.
10. **Barthel, L. K. and Raymond, P. A.,** Improved method for obtaining 3-μm cryosections for immunocytochemistry, *J. Histochem. Cytochem.,* 38, 1383, 1990.
11. **Yokota, S.,** Electron microscopic demonstration of carbonic anhydrase activity in mouse liver cells, *Histochemie,* 19, 225, 1969.
12. **Muther, T. F.,** A critical evaluation of the histochemical methods for carbonic anhydrase, *J. Histochem. Cytochem.,* 20, 319, 1972.
13. **Rosen, S.,** Observations on the specificity of newer histochemical methods for the demonstration of carbonic anhydrase activity, *J. Histochem. Cytochem.,* 20, 951, 1972.
14. **Lönnerholm, G.,** Carbonic anhydrase histochemistry. A critical study of Hansson's cobalt-phosphate method, *Acta Physiol. Scand. Suppl.,* 418, 1, 1974.
15. **Cassidy, M. M. and Lightfoot, F. G.,** Carbonic anhydrase, in *Electron Microscopy of Enzymes, Principles and Methods,* Vol. 3, Hayat, M. A., Ed., Van Nostrand Reinhold, New York, 1975, 77.
16. **Lönnerholm, G.,** Carbonic anhydrase in rat liver and rabbit skeletal muscle: further evidence for the specificity of the histochemical method of Hansson, *J. Histochem. Cytochem.,* 28, 427, 1980.
17. **Maren, T. H.,** Kinetics, equilibrium and inhibition in the Hansson histochemical procedure for carbonic anhydrase: a validation of the method, *Histochem. J.,* 12, 183, 1980.
18. **Lönnerholm, G.,** Histochemical localization of carbonic anhydrase in mammalian tissue, *Ann. N.Y. Acad. Sci.,* 429, 369, 1984.
19. **Maren, T. H.,** Carbonic anhydrase: chemistry, physiology and inhibition, *Physiol. Rev.,* 47, 595, 1967.
20. **Wistrand, P. J. and Knuuttila, K.-G.,** Renal membrane-bound carbonic anhydrase. Purification and properties, *Kidney Int.,* 35, 851, 1989.
21. **Whitney, P. L. and Briggle, T. V.,** Membrane-associated carbonic anhydrase purified from bovine lung, *J. Biol. Chem.,* 257, 12056, 1982.
22. **Zhu, X. L. and Sly, W. S.,** Carbonic anhydrase IV from human lung. Purification, characterization, and comparison with membrane carbonic anhydrase from human kidney, *J. Biol. Chem.,* 265, 8796, 1990.
23. **Carter, N. D., Fryer, A., Grant, A. G., Hume, R., Stange, R. G., and Wistrand, P. J.,** Membrane specific carbonic anhydrase (CA IV) expression in human tissues, *Biochim. Biophys. Acta,* 1026, 113, 1990.
24. **Dodgson, S. J. and Contino, L. C.,** Rat kidney mitochondrial carbonic anhydrase, *Arch. Biochem. Biophys.,* 260, 334, 1988.
25. **Storey, B. T., Lin, L. C., Tompkins, B., and Forster, R. E.,** II, Carbonic anhydrase in guinea pig skeletal muscle mitochondria, *Arch. Biochem. Biophys.,* 270, 144, 1989.
26. **Ikejima, T. and Ito, S.,** Carbonic anhydrase in mouse salivary glands and saliva: a histochemical, immunohistochemical, and enzyme activity study, *J. Histochem. Cytochem.,* 32, 625, 1984.

Chapter 2

ELECTRON MICROSCOPIC IMMUNOCYTOCHEMISTRY

2.1 An Historical Perspective

Tibor Barka

Historically, immunocytochemical methods were first developed and applied at the level of the light microscope, and important theoretical and practical considerations preceded the use of such methods and their modifications for the ultrastructural localization of cellular and extracellular antigens. The extension of light microscopic immunocytochemistry to the ultrastructural level posed no conceptual difficulties, but numerous technical problems had to be solved, including the simultaneous preservation of ultrastructure and immunoreactivity, penetration of macromolecular reagents into specimens destined for electron microscopic observations, and introduction of well-defined electron-dense markers. However, from the onset, the goal of electron microscopic immunocytochemistry, the specific demonstration of well-defined epitopes at the highest possible level of resolution, was clear.

Within the scope of this monograph, this brief historical overview aims at identifying crucial developments in electron microscopic* immunocytochemistry. In general, only publications** that first introduced a novel method or a significant modification are cited. For possible errors in order of priority or for omissions, which were assuredly unintentional, the author asks the reader's indulgence.

The first successful application of immunocytochemical techniques using fluorescent markers by Coons et al.[1,2] in 1941 to 1942 far preceded the wide-spread use of the electron microscope in cell biology. Fluorescent markers were not suitable for use in the transmission electron microscope, although they may be viewed under the scanning electron microscope.[3] The introduction of ferritin as an electron-dense label by Singer[4] and Singer and Schick[5] opened up the field of electron microscopic immunocytochemistry. Ferritin-labeled antibodies were applied most successfully to the ultrastructural localization of surface antigens, since difficulty of penetration of labeled antibody into fixed tissues made its use for the intracellular localization of antigens less effective. In order to overcome the problems of reagent penetration and nonspecific absorption, methods were developed for applying immunologic reagents to frozen sections[6] or directly to the surface of thin sections in conjunction with different embedding techniques such as cross-linked serum albumin,[7] methyl butyl methacrylate,[8] Epon, polyethylene glycol,[9] or Lowicryl.[10] Such "post-embedding" techniques were utilized not only with ferritin- or iron dextran-labeled antibodies[11] but also with cytochrome *c*-labeled antibody fragments[12] and with anbitodies labeled with heavy metals[13-15]

* Although the scanning electron microscope (SEM) has also been utilized for the localization of cell surface antigens with the use of large particulate markers (e.g., latex, bacteriophage T4, tobacco mosaic virus), ferritin, and enzyme reaction products, and attempts were made to demonstrate intracellular antigens with horseradish peroxidase-labeled antibody fragments (Nakane, P.K. and Hartman, A. L., Immunocytochemical localization of intracellular antigens with SEM, *Histochem. J.*, 12, 435, 1980), we will discuss only data obtained with the transmission electron microscope. For reviews of the use of SEM see: Molday, R. S. and Maher, P., A review of cell surface markers and labelling techniques, *Histochem. J.*, 12, 273, 1980; Hartman, A. L. and Nakane, P. K., Immunological methods in scanning electron microscopy, in *Techniques in Immunocytochemistry,* Vol. 2, Bullock, G. R. and Petrusz, P., Eds., Academic Press, New York, 1983, 103.

** Abstracts are excluded.

0-8493-6012-9/93/$0.00 + $.50

or enzymes. Of special interest was the application of ultracryotomy to electron microscopic immunocytochemistry[16,17] which, owing to improvements in techniques and in cryomicrotomes by manufacturers, is gaining wider acceptance.

A large volume of literature deals with the effect of fixatives on the immunocytochemical reactivity of various antigens. A compromise between preservation of immunoreactivity and ultrastructure needed special consideration of fixation in electron microscopic immunocytochemistry. Although some antigens were demonstrable in osmic acid-fixed tissues,[18,19] mixtures of paraformaldehyde with low concentrations of glutaraldehyde[20] proved to be the most widely acceptable fixative.

The introduction of enzyme-labeled antibodies by Avrameas and Uriel[21] and by Nakane and Pierce[22,23] in 1966 to 1967 represented the next milestone in the history of immunocytochemistry. Of the enzyme markers recommended, horseradish peroxidase offered several advantages, including its ready applicability to electron microscopic immunocytochemistry owing to the method of Graham and Karnovsky,[24] described coincidentally in 1966, for the ultrastructural localization of horseradish peroxidase activity.

In order to overcome the problems of conjugation and purification of enzyme-labeled antibodies and to increase the sensitivity of detection, ingenious unlabeled antibody enzyme methods were developed simultaneously by Sternberger and Cuculis[25] and by Mason et al.[26] in 1969. These methods, and their variants,[27-29] have been widely used in electron microscopic immunocytochemistry,[30] although, with the introduction of the biotin-avidin methods for light microscopic and the colloidal gold techniques for electron microscopic localization of antigens, their popularity is declining.

Overlapping with the widespread application of the peroxidase-antiperoxidase (PAP) method of Sternberger et al.,[27] colloidal gold was introduced as a particulate marker in electron microscopic immunocytochemistry by Faulk and Taylor[31] in 1971, and first applied in an indirect labeling technique by Romano et al.[32] in 1974. A great variety of ligands, such as immunoglobulins, lectins, protein A, avidin, toxins, and enzymes, has been used for complex formation with colloidal gold. Thus, colloidal gold markers became one of the most versatile tools, not only in immunology and immunocytochemistry, but in cytochemistry in general. The advantages of colloidal gold as an immunoelectron marker include the relative simplicity of preparing stabilized gold colloids of uniform, preselected particle sizes, high electron density, discreteness of particles facilitating quantification, and the relatively low degree of nonspecific interaction with sections of resin-embedded tissues, thus permitting its use in postembedding techniques.[33-35]

As an alternative to the unlabeled antibody enzyme methods, an important development was the introduction of biotin-avidin systems in immunocytochemistry. Biotin, a low-molecular weight vitamin, binds with very high affinity to avidin from egg white and to streptavidin from *Streptomyces avidini*. The association constant of avidin for biotin (about 10^{15} M^{-1}) is orders of magnitude larger than the association constants for antibody-antigen complexes. This high affinity, and the fact that biotinylation of antibodies does not significantly change their antigen binding capacity, provided the basis for the development of biotin-avidin indirect immunocytochemical techniques by Guesdon et al.[36] in 1979 and by Hsu et al.[37] in 1981. A ferritin-avidin-biotin complex was first used for the nonimmunological staining of biotinylated biological membranes in electron microscopy by Heitzmann and Richards[38] in 1974. Recent modifications of the biotin-avidin techniques, particularly those employing streptavidin, provide a convenient and sensitive means of detecting antigens at both the light and electron microscopic levels.[39]

The primary consideration in the application of immunocytochemical methods is the specificity and avidity of antibodies, their cross-reactivities, and nonspecific interactions with biological and embedding matrix components. The discovery of monoclonal antibodies by Köhler and Milstein[40] in 1975 had a great and lasting impact on the entire field of biology. Yet, monoclonal antibodies, perhaps because they are in general less sensitive than polyclonal antibodies, have been seldom used in electron microscopic immunocytochemistry.[41] Further efforts at increasing the sensitivity of techniques utilizing monoclonal antibodies in electron microscopic immunocytochemistry are certainly warranted.

In this brief historical survey, the author has not dealt with such important techniques as nucleic acid hybridization, including *in vitro* hybridization, using immunological or biotin-avidin-based systems, mainly because they are seldom applied at the electron microscopic level. Future, foreseeable developments of electron microscopic immunocytochemistry will lead to the discovery of more sensitive detection systems, the wider application of monoclonal antibodies and cryomicrotomy, and combinations of immunocytochemical techniques with methodologies of molecular biology. The future of electron microscopic immunocytochemistry is bright indeed.

REFERENCES

1. **Coons, A. H., Creech, H. J., and Jones, R. N.,** Immunological properties of an antibody containing a fluorescent group, *Proc. Soc. Exp. Biol. Med.*, 47, 200, 1941.
2. **Coons, A. H., Creech, H. J., Jones, R. N., and Berliner, E.,** The demonstration of pneumococcal antigen in tissues by the use of fluorescent antibody, *J. Immunol.*, 45, 150, 1942.
3. **Springer, E. L., Riggs, J. L., and Hackett, A. J.,** Viral identification by scanning electron microscopy of preparations stained with fluorescein labeled antibody, *J. Virol.*, 14, 1623, 1974.
4. **Singer, S. J.,** Preparation of an electron-dense antibody conjugate, *Nature*, 183, 1523, 1959.
5. **Singer, S. J. and Schick, A. F.,** The properties of specific stains for electron microscopy prepared by conjugation of antibody molecules with ferritin, *J. Biophys. Biochem. Cytol.*, 9, 519, 1961.
6. **Leduc, E. H., Scott, G. B., and Avrameas, S.,** Ultrastructural localization of intracellular immune globulins in plasma cells and lymphoblasts by enzyme-labeled antibodies, *J. Histochem. Cytochem.*, 17, 211, 1969.
7. **McLean, J. D. and Singer, S. J.,** A general method for specific staining of intracellular antigens with ferritin, *Proc. Natl. Acad. Sci. U.S.A.*, 65, 122, 1970.
8. **Kawarai, Y. and Nakane, P. K.,** Localization of tissue antigens on the ultrathin sections with peroxidase-labeled antibody method, *J. Histochem. Cytochem.*, 18, 161, 1970.
9. **Mazurkiewicz, J. E. and Nakane, P. K.,** Light and electron microscopic localization of antigens in tissues embedded in polyethylene glycol with peroxidase labeled antibody method, *J. Histochem. Cytochem.*, 20, 969, 1972.
10. **Carlemalm, E., Garavito, R. M., and Williger, W.,** Resin development for electron microscopy and an analysis of embedding at low temperature, *J. Microsc.*, 126, 123, 1982.
11. **Dutton, A. H., Tokuyasu, K. T., and Singer, S. J.,** Iron-dextran antibody conjugates: general method for simultaneous staining of two components in high-resolution immunoelectron microscopy, *Proc. Natl. Acad. Sci. U.S.A.*, 76, 3392, 1979.
12. **Kraehenbuhl, J. P., DeGrandi, P. B., and Campiche, M. A.,** Ultrastructural localization of intracellular antigen using enzyme-labeled antibody fragments, *J. Cell Biol.*, 50, 432, 1971.
13. **Pepe, F. A.,** The use of specific antibody in electron microscopy. I. Preparation of mercury-labeled antibody, *J. Biophys. Biochem. Cytol.*, 11, 515, 1961.
14. **Pepe, F. A. and Finck, H.,** The use of specific antibody in electron microscopy. II. The visualization of mercury-labeled antibody in the electron microscope, *J. Biophys. Biochem. Cytol.*, 11, 521, 1961.
15. **Sternberger, L. A., Donati, E. J., Cuculis, J. J., and Petrali, J. P.,** Indirect immunouranium technique for staining of embedded antigen in electron microscopy, *Exp. Mol. Pathol.*, 4, 112, 1965.
16. **Painter, R. G., Tokuyasu, T. K., and Singer, S. J.,** Immunoferritin localization of intracellular antigens: the use of ultracryotomy to obtain ultrathin sections suitable for direct immunoferritin staining, *Proc. Natl. Acad. Sci. U.S.A.*, 70, 1649, 1973.

17. **Tokuyasu, K. T. and Singer, S. J.,** Improved procedures for immunoferritin labeling of ultrathin frozen sections, *J. Cell Biol.*, 71, 894, 1976.
18. **Sternberger, L. A.,** Electron microscopic immunocytochemistry: a review, *J. Histochem. Cytochem.*, 15, 139, 1967.
19. **Beauvillain, J.-C., Tramu, G., and Dubois, M. P.,** Characterization by different techniques of adrenocorticotropin and gonadotropin producing cells in lerot pituitary *(Eliomys quercinus)*, *Cell Tissue Res.*, 158, 301, 1975.
20. **McLaughlin, B. J., Barber, R., Saito, K., Roberts, E., and Wu, J.-Y.,** Immunocytochemical localization of glutamate carboxylase in rat spinal cord, *J. Comp. Neurol.*, 164, 305, 1975.
21. **Avrameas, S. and Uriel, J.,** Methode dc marquage d'antigènes et d'anticorps avec des enzymes et son application en immunodiffusion, *C.R. Acad. Sci. Paris*, 262, 2543, 1966.
22. **Nakane, P. K. and Pierce, G. B.,** Enzyme-labeled antibodies: preparation and application for the localization of antigens, *J. Histochem. Cytochem.*, 14, 929, 1966.
23. **Nakane, P. K. and Pierce, G. B.,** Enzyme-labeled antibodies for the light and electron microscopic localization of tissue antigens, *J. Cell Biol.*, 33, 307, 1967.
24. **Graham, R. C. and Karnovsky, M. J.,** The early stages of absorption of injected horseradish peroxidase in the proximal tubules of mouse kidney: ultrastructural cytochemistry by a new technique, *J. Histochem. Cytochem.*, 14, 291, 1966.
25. **Sternberger, L. A. and Cuculis, J. J.,** Method for enzymatic identification of the immunocytochemical reaction without use of labeled antibodies, *J. Histochem. Cytochem.*, 17, 190, 1969.
26. **Mason, T. E., Phifer, R. F., Spicer, S. S., Swallow, R. A., and Dreskin, R. B.,** An immunoglobulin-enzyme bridge method for localizing tissue antigens, *J. Histochem. Cytochem.*, 17, 563, 1969.
27. **Sternberger, L. A., Hardy, P. H., Cuculis, J. J., and Meyer, H. G.,** The unlabeled antibody enzyme method of immunocytochemistry. Preparation and properties of soluble antigen-antibody complex (horseradish peroxidase-antihorseradish peroxidase) and its use in the identification of spirochetes, *J. Histochem. Cytochem.*, 18, 315, 1970.
28. **Vacca, L. L., Rosario, S. L., Zimmerman, E. A., Toashefsky, P., Ng, P.-Y., and Hsu, K. C.,** Application of immunoperoxidase techniques to localize horseradish peroxidase-tracer in the central nervous system, *J. Histochem. Cytochem.*, 23, 208, 1975.
29. **Mason, D. Y. and Sammons, R.,** Alkaline phosphatase and peroxidase for double immunoenzymatic labeling of cellular constituents, *J. Clin. Pathol.*, 31, 454, 1978.
30. **Moriarty, G. C. and Halmi, N. S.,** Electron microscopic study of the adrenocorticotropin-producing cell with the use of unlabeled antibody and the soluble peroxidase-antiperoxidase complex, *J. Histochem. Cytochem.*, 20, 590, 1972.
31. **Faulk, W. P. and Taylor, G. M.,** An immunocolloid for the electron microscope, *Immunochemistry,* 8, 1081, 1971.
32. **Romano, E. L., Stolinski, C., and Hughes-Jones, N. C.,** An antiglobulin reagent labelled with colloidal gold for use in electron microscopy, *Immunochemistry,* 11, 521, 1974.
33. **Horisberger, M. and Rosset, J.,** Colloidal gold, a useful marker for transmission and scanning electron microscopy, *J. Histochem. Cytochem.*, 25, 295, 1977.
34. **Roth, J., Bendayan, M., and Orci, L.,** Ultrastructural localization of intracellular antigens by the use of protein A-gold complex, *J. Histochem. Cytochem.*, 26, 1074, 1978.
35. **Batten, T. F. C. and Hopkins, C. R.,** Use of protein A-coated colloidal gold particles for immunoelectronmicroscopic localization of ACTH on ultrathin sections, *Histochemistry,* 60, 317, 1979.
36. **Guesdon, J.-L., Ternynck, T., and Avrameas, S.,** The use of avidin-biotin interaction in immunoenzymatic techniques, *J. Histochem. Cytochem.*, 27, 1131, 1979.
37. **Hsu, S.-M., Raine, L., and Fanger, H.,** Use of avidin-biotin-peroxidase techniques: a comparison between ABC and unlabeled antibody (PAP) procedures, *J. Histochem. Cytochem.*, 29, 577, 1981.
38. **Heitzmann, H. and Richards, F. M.,** Use of avidin-biotin complex for specific staining of biological membranes in electron microscopy, *Proc. Natl. Acad. Sci. U.S.A.*, 71, 3537, 1974.
39. **Childs, G. V. and Unabia, G.,** Application of a rapid avidin-biotin-peroxidase complex (ABC) technique to the localization of pituitary hormones at the electron microscopic level, *J. Histochem. Cytochem.*, 30, 1320, 1982.
40. **Köhler, G. and Milstein, C.,** Continuous culture of fused cells secreting antibody of pre-defined specificity, *Nature,* 256, 495, 1975.
41. **Cuello, A. C., Milstein, C., and Priestley, J. V.,** Use of monoclonal antibodies in immunocytochemistry with special reference to the central nervous system, *Brain Res. Bull.*, 5, 575, 1980.

Chapter 2

ELECTRON MICROSCOPIC IMMUNOCYTOCHEMISTRY

2.2 Preservation of Immunoreactivity and Ultrastructure

Masahiro Mizuno

TABLE OF CONTENTS

0-8493-6012-9/93/$0.00 + $.50

I. CRYOULTRAMICROTOMY

Masahiro Mizuno

A. INTRODUCTION

A variety of techniques have been developed to elucidate distribution of cellular substances at the ultrastructural level, and immunocytochemistry is now a major tool in medicine and cell biology for the study of cell structure. The ideal immunocytochemical method is one which meets the following requirements: (1) preservation of fine structure, (2) preservation of antigenicity, (3) easy accessibility of antigen by antibody, (4) immobilization of antigen at the original site, and (5) very little nonspecific reaction. In practice, it is difficult to satisfy all of the above requirements simultaneously, but cryoultramicrotomy is one of the best techniques currently available. Thus, immunogold labeling in combination with cryoultramicrotomy has been proven to be a powerful method for visualizing and localizing cell surface and intracellular proteins.

Pre- and postembedding methods, in which the antibody is applied before or after the tissue is embedded in a supporting medium, respectively, have several limitations with regard to the antibody-antigen reaction *in situ*. However, in immunocryoultramicrotomy, immunolabeling is performed directly on ultrathin frozen sections cut on a cryoultramicrotome, and, in principle, the antigen preservation is comparable to that of the light microscopic cryostat technique. This method has the following specific advantages: (1) good antigen preservation owing to the omission of the usual dehydration and resin embedding, (2) excellent preservation of membrane structures, (3) rapid evaluation of results, and (4) application of semithin (0.5 to 1.0 μm) sections to high-resolution immunofluorescence studies.

The drawbacks of the method are as follows: (1) difficulty in obtaining serial sections, (2) the cost of liquid nitrogen, and (3) difficulty in quantitative analysis. The last problem arises from differential penetration of immunoreagents into various cell structures resulting in different labeling efficiency. Regarding this problem, Slot et al.[1] reported that the embedding of tissues in polyacrylamide enabled a more uniform labeling of sections for quantitative analyses. On the other hand, Griffiths and Hoppeler[2] introduced a combined biochemical and morphometric approach for the quantitative estimation of immunolabeling of membrane antigens.

Until recently, mastering cryoultramicrotomy was a laborious task, but with recent improvements in both cryoultramicrotomes and procedures, this method has gained a measure of popularity; and although morphology in frozen thin sections was regarded as difficult to delineate, the introduction of polyvinyl alcohol (PVA) as an embedding material has brought frozen sections close to conventional images of plastic sections.[3]

Tokuyasu, the major pioneer of this method, has published several reviews.[4-7] In this section, practical aspects of the application of cryoultramicrotomy to immunocytochemistry according to Tokuyasu's method are described. The general steps in immunocytochemistry using cryoultramicrotomy are shown in Figure 1.

B. APPARATUS AND PRESTART REQUIREMENTS

The following apparatus and materials are required for cryoultramicrotomy.

1. Cryoultramicrotome — this comprises basically an ultramicrotome and a cryochamber which provides temperature stability and control; commercial cryoultramicrotomes are available from Reichert and RMC Inc.
2. Formvar- or parlodion-coated grids — carbon coating and ionization are preferable

Fixation

↓

2.3 *M* Sucrose or polyvinylpyrrolidone-sucrose mixture infusion

↓

Freezing in liquid nitrogen or Freon-22

↓

Cryosectioning

↓

On-grid immunostaining

↓

Embedding in polyvinyl alcohol

↓

Air drying

FIGURE 1. Steps in immunocytochemistry using cryoultramicrotomy.

3. Loops — two kinds of loops are used: platinum or stainless steel wire loops with a diameter of 2 to 3 mm, for picking up sections; Nichrome wire loops with a diameter a little larger than the grids, for grid transfer
4. Eyelash stick — a wooden, bamboo, or plastic rod with an eyelash attached for manipulation of sections
5. Slides — slides should be precleaned by storing in 10% HCl in 70% ethanol until needed; it is convenient to mark circles on them with a diamond pen for microscopic observation
6. 2.3 *M* sucrose in 0.1 *M* phosphate buffer, pH 7.4, containing 0.02% NaN_3
7. Polyvinylpyrrolidone (PVP)-sucrose mixture — a mixture of polyvinylpyrrolidone (10 to 30%) and sucrose is used as well as 2.3 *M* sucrose alone, as an infusion or suspension solution; to make a 100-ml solution containing x% of PVP, x g of PVP is dissolved in a mixture of 0.2x ml of 1.1 *M* $NaCO_3$ in a buffer and (100 − x) ml of 2.3 *M* sucrose in the same buffer
8. Polyvinyl alcohol solution (PVA) — 2.0 to 3.0% polyvinyl alcohol in distilled water containing 0.1 to 0.2% uranyl acetate is used as a thin embedding material
9. 2% neutral uranyl acetate — equal parts of 4% uranyl acetate and 0.3 *M* potassium oxalate, adjusted to pH 7.4 with 10% NH_4OH
10. 2% aqueous uranyl acetate

C. FIXATION

In cryoultramicrotomy, the only potential denaturing process is the initial fixation. As would be expected, different antigens vary in their susceptibility to fixation. Fortunately, experience has led to the selection of fixatives suitable for many systems. A simple fixation with 0.1 to 2.0% glutaraldehyde in combination with 2.0 to 4.0% formaldehyde for 0.5 to

1.0 h usually preserves both cell structure and antigenicity. However, some proteins lose their antigenicity even after fixation with a low concentration (0.1 to 0.2%) of glutaraldehyde. In this case, fixation with formaldehyde alone should be tried. The concentration of formaldehyde can be as high as 8%, and the fixation period as long as several hours or even overnight. Another method of dealing with this problem is the use of ethylacetimidate (EAI).[8] EAI, which blocks amino groups, reduces the degree of cross-linking. Hence, pretreatment with EAI and formaldehyde before glutaraldehyde fixation can improve the accessibility of some kinds of antigens as well as the retention of antigenicity. Further details are discussed in Article II of this section.

For single cells, postfixed cells are incubated in 5 to 10% gelatin at room temperature, sedimented in a centrifuge, and the solidified pellet at 4°C is then cut into small blocks.

Fixed tissues are stored either in a buffer solution, to which a small amount of formaldehyde (0.5%) should be added when glutaraldehyde is omitted in the fixative, or in liquid nitrogen after infiltration with sucrose.

For fluorescence microscopy, the tissues fixed with the glutaraldehyde mixture should be treated with 0.5 mg/ml sodium borohydride for 30 min to quench the background fluorescence before immunolabeling.

D. INFUSION

Prefreezing modifications of specimens are essential for optimum sectioning in cryoultramicrotomy. To prevent detectable damage during the freezing process, cryoprotective agents, such as sucrose, glycerol, dimethylsulfoxide, or polyethylene glycol, are employed. Sucrose infusion, as developed by Tokuyasu,[3] is the most widely used, and allows proper plasticity of frozen specimens.

Cut the fixed tissue into pieces up to 1 mm wide. The sugar-protected specimens can be cut relatively easily with surfaces up to 1 mm^2. Trimming of the specimen block is more easily done before sucrose infusion than after.

Transfer the pieces to 2.3 *M* sucrose in 0.1 *M* phosphate buffer and allow them to infiltrate for about 30 min. Infiltration of specimens up to 1 mm wide will usually take 30 to 60 min. Infusion is completed when the specimens are no longer floating on the top of the sucrose solution.

Recently, Tokuyasu[3] reported that a mixture of 10 to 30% polyvinylpyrrolidone (PVP) and 2.07 to 1.61 *M* sucrose can provide a still better postfreezing condition for the specimen as compared with sucrose alone. In many specimens infused with a PVP-sucrose mixture, extracellular spaces contain both sucrose and PVP, whereas cells contain chiefly sucrose with little PVP. PVP in extracellular spaces helps to make the sectioning smooth and uniform, especially in highly hydrated specimens like embryonic tissues. Since the viscosity of PVP-sucrose mixtures is high, infiltration usually takes 2 h

After infiltration, the specimen block is mounted onto a holder such as aluminum pins, and excess sucrose is removed.

E. FREEZING

Pick up the holder with forceps, plunge the head first into liquid nitrogen, and swirl vigorously. After that, transfer the mounted tissue to the precooled cryochamber.

Note that small specimens will be dehydrated quickly while being placed on the mounting block of the microtome. Supersaturation of sucrose, or formation of minute sucrose crystals resulting from dehydration, can seriously deteriorate the sectioning quality.

For specimens infiltrated in a mixture of PVP and sucrose, Freon-22 or Freon-12 is preferable to liquid nitrogen, because nitrogen gas around the specimen delays the freezing process, resulting in freezing damage.

F. SECTIONING

1. Knives

As with conventional ultramicrotomy, the quality of glass knives holds the key to good sectioning in this method. Normal 45° dry glass knives are also used in cryosectioning. A light casting of tungsten is said to enhance the cutting quality of glass knives and prolong their lives.[9]

It should be remembered that in cryoultramicrotomy, the best part of a glass knife is the central third; the knife angle is larger at the left third and flaws appear in the right third. The outer thirds may be used for surfacing the specimen of interest.

2. Semithin Sectioning

Semithin sectioning before ultrathin sectioning is not only useful in checking the target area of a tissue, but it is also an excellent technique for high-resolution immunofluorescence studies. It serves to detect intracellular distribution of the antigen in considerable detail at the light microscopic level. Moreover, cutting good semithin sections is a prerequisite for subsequent cryoultramicrotomy.

The temperature for sectioning is important and should be adjusted according to the section thickness. Semithin sections (0.5 to 1.0 μm) are usually cut at relatively high temperatures (−60 to −80°C) with a cutting speed of about 3 mm/s.

After the sections are cut, transfer them from the knife edge to an appropriate position using an eyelash stick. This stick is very useful in moving and collecting the sections.

Dip a small loop into 2.3 *M* sucrose and remove it quickly, letting a big droplet form at the tip. Then insert it quickly into the cryochamber, and attach sections on the sucrose droplet before it is completely solidified; bring the droplet out and thaw it at room temperature, then mount it on a slide. The sucrose droplet on the wire loop also acts as a magnifying lens for the section, allowing the condition of the section to be checked easily.

For confirmation of the area of interest before ultrathin sectioning, the semithin sections are stained with Toluidine Blue. When many semithin sections must be cut for immunofluorescence study, the slides should be kept in a humid chamber to avoid dehydration after (but not before) the sections are mounted on them.

3. Ultrathin Sectioning

If a semithin section examined under the light microscope reveals the area of interest, the cryochamber is cooled to a lower temperature fit for ultrathin sectioning. At lower temperatures, the material becomes more rigid and resistant to compression during ultrathin sectioning. The standard temperature for ultrathin sectioning is −90 to −110°C. Often section fragmentation occurs when the specimen is too rigid; to avoid this, the temperature should be raised or the section thickness should be decreased. When PVP-sucrose mixtures are employed, the temperature to cut ultrathin sections will be about 10°C colder than for 2.3 *M* sucrose-infused specimens.

Ultrathin sectioning is usually done at 50 to 100 nm, and under optimal conditions good sections are often obtained around 90 nm. The thickness of sections can be judged from their color. Gold sections are preferable, but often too fragile; blue-purple cellophane-like sections are good enough for most purposes. Tiny blocks are generally easier to cut, but subsequent manipulation of the section may be difficult.

Cutting speed should be kept slower than 2 mm/s, since too rapid sectioning causes severe compression damage to the section. Although slow sectioning may cause wrinkling of the sections, it can often be corrected by expanding the section on a sucrose droplet at room temperature.

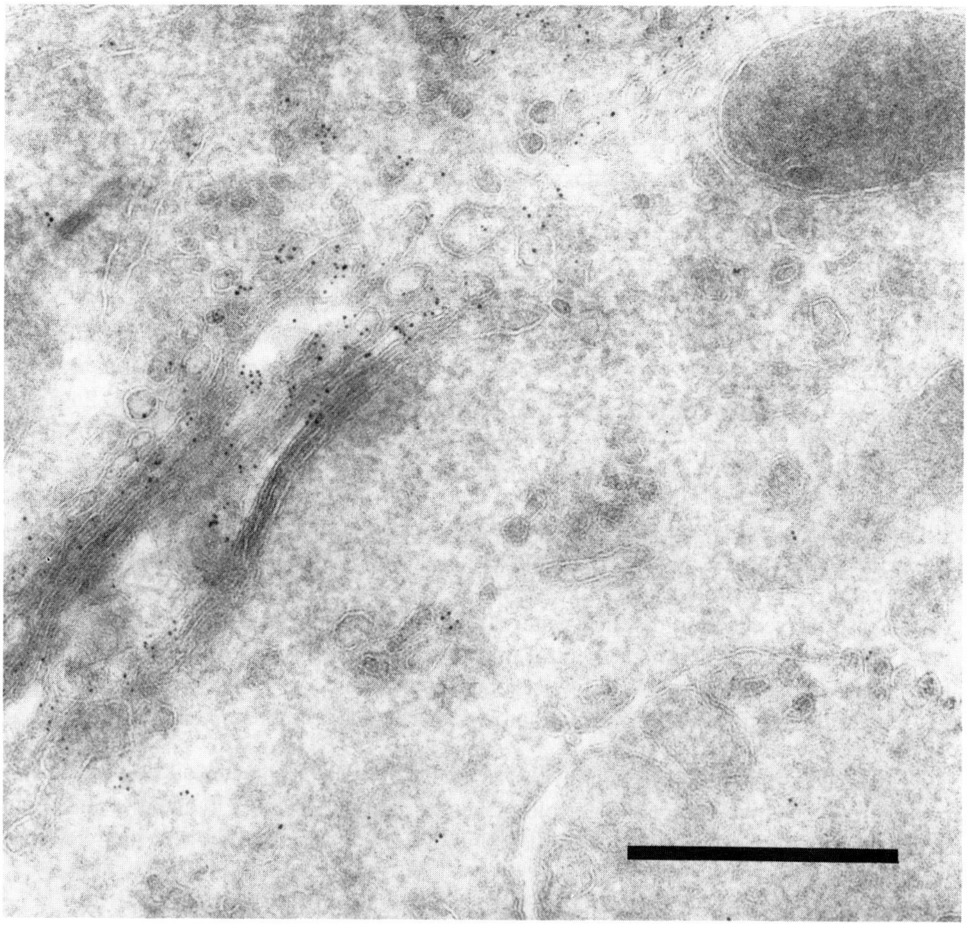

FIGURE 2. Ultrathin frozen section of Hep G2 cell, labeled for albumin with 5-nm gold particles. The fixed cells are resuspended in 10% gelatin, the suspension was centrifuged and cooled, and the pellet was processed for cryomicrotomy. Albumin is concentrated in the Golgi cisternae. (Bar = 0.5 μm.)

11. React the sections with 2% OsO_4 in a buffer or water for 10 min.
12. Rinse the sections in DW for 1 min, three times.
13. Stain the sections with 2% uranyl acetate for 10 min.
14. Rinse the sections briefly with DW.
15. Mount the sections in 3% PVA containing lead citrate (final concentration 0.002 to 0.01%) for 5 to 10 min.
16. Air-dry the sections.

Keller et al.[10] introduced the technique to use LR White or Epon 812 as an embedding medium for osmicated ultrathin frozen sections. The detail obtained is very similar to that of conventional plastic sections. However, because of high electron density of the cell as a whole, comparatively thick sections may not be suitable for observation. The steps before osmication are also the same as those described first.

11. React the immunostained and rinsed grids with 0.5 to 2.0% OsO_4 in 0.1 *M* sodium cacodylate buffer for 10 min.

12. Rinse the sections in 7% sucrose.
13. Stain the sections with 0.5% uranyl acetate in barbital/acetate buffer containing 5% sucrose for 20 min.
14. Dehydrate the sections through graded concentrations (40% to 100%) of ethanol.
15. Infiltrate the sections with LR White or Epon 812 in ethanol.
16. After infiltration, blot the grids between two sheets of Whatman harder filter paper to remove excess monomer.
17. Polymerize overnight at 60°C.

ACKNOWLEDGMENT

The author wishes to thank Professor K. T. Tokuyasu for critical reading of the manuscript.

REFERENCES

1. **Slot, J. W., Posthuma, G., Chang, L.-Y., Crapo, J. D., and Geuze, H. J.,** Quantitative aspects of immunogold labeling in embedded and in nonembedded sections, *Am. J. Anat.*, 185, 271, 1989.
2. **Griffiths, G. and Hoppeler, H.,** Quantitation in immunocytochemistry: correlation of immunogold labeling to absolute number of membrane antigens, *J. Histochem. Cytochem.*, 34, 1389, 1986.
3. **Tokuyasu, K. T.,** Use of poly(vinylpyrrolidone) and poly(vinyl alcohol) for cryoultramicrotomy, *Histochem. J.*, 21, 163, 1989.
4. **Tokuyasu, K. T.,** Immunochemistry on ultrathin frozen sections, *Histochem. J.*, 12, 381, 1980.
5. **Tokuyasu, K. T.,** Present state of immunocryoultramicrotomy, *J. Histochem. Cytochem.*, 31, 164, 1983.
6. **Tokuyasu, K. T.,** Immuno-cryoultramicrotomy, in *Immunolabelling for Electron Microscopy*, Polak, J. M. and Varndell, I. M., Eds., Elsevier, Amsterdam, 1984, 71.
7. **Tokuyasu, K. T.,** Application of cryoultramicrotomy to immunocytochemistry, *J. Microsc.*, 143(2), 139, 1986.
8. **Tokuyasu, K. T. and Singer, S. J.,** Improved procedures for immunoferritin labeling of ultrathin frozen sections, *J. Cell Biol.*, 71, 894, 1976.
9. **Griffiths, G., McDowall, A., Back, R., and Dubochet, J.,** On the preparation of cryosections for immunocytochemistry, *J. Ultrastruct. Res.*, 89, 65, 1984.
10. **Keller, G. A., Tokuyasu, K. T., Dutton, A. H., and Singer, S. J.,** An improved procedure for immunoelectron microscopy: ultrathin plastic embedding of immunolabeled ultrathin frozen sections, *Proc. Natl. Acad. Sci. U.S.A.*, 81, 5744, 1984.

II. FIXATION

Masahiro Mizuno

A. CHEMICAL FIXATION

The optimal fixative for immunocytochemistry must maintain original morphology, preserve antigenicity, immobilize antigens, and keep them accessible in the tissue. However, adequate structural preservation and retention of antigenicity are often conflicting goals, and the difficulty in fixation for immunoelectron microscopy is finding the point of compromise between the two. The choice of fixatives for a particular antigen is based on trial and error. A fixative may preserve some epitopes in a certain antigen, but may completely destroy reactivity of other epitopes.

In combination with glutaraldehyde, osmium tetroxide (OsO_4), which is classified as an oxidative fixative, provides optimal ultrastructural morphology in tissue sections for conventional electron microscopy. Actually, preservation of membrane structure greatly depends on the use of OsO_4 in hydrophobic resins. However, this fixative has powerful denaturing effects, and is not compatible with preservation of most biological activities in cytochemistry. Moreover, it is known that postfixation with OsO_4 destroys organization of F-actin.[1]

Currently, the most commonly used fixatives in immunoelectron microscopy are cross-linking fixatives. Formaldehyde, glutaraldehyde, acrolein, and imidates belong to this category. These fixatives confer rigidity upon the tissue by cross-linking proteins. Strong cross-linking is desirable for morphological preservation. However, extensive cross-linking impairs antigens by direct chemical effects and by structural distortion. For quick screening of antigenicity conversion under various fixation conditions, electrophobic protein blots or dot-immunobinding assay are useful.[2,3] The representative fixatives used in immunocytochemistry are discussed below.

1. Formaldehyde

Formaldehyde fixation is commonly used in light and electron microscopic immunocytochemistry, while glutaraldehyde fixation is used mainly for electron microscopy. Formaldehyde is the smallest aldehyde molecule and it provides fast diffusion and little alteration of proteins. It reacts with various groups of macromolecules in a cross-linking fashion. The most reactive sites of aldehydes are primarily free amino groups, which exist mainly in lysine residues. Hence, aldehydes will not severely affect antibody binding if the antigenic determinant does not comprise lysine residues.

This fixative is provided as a dry powder (paraformaldehyde) that should be dissolved immediately before use. The most common example of use is 3 to 4% formaldehyde either in 0.1 *M* sodium phosphate buffer, pH 7.3 to 7.4, or in 0.1 *M* sodium cacodylate buffer, pH 7.3 to 7.4. To make 100 ml of 4% formaldehyde in 0.1 *M* phosphate buffer, mix 40 ml of distilled water and 4 g of paraformaldehyde. Warm the solution to 60°C while stirring under a ventilation hood, then clear it by adding a few drops of 1 *N* NaOH. After the solution has cooled, add the following ingredients: distilled water (bring up to 50 ml) and 0.2 *M* phosphate buffer (bring up to 100 ml). Check the pH of the solution and adjust it to pH 7.3 to 7.4.

This fixative is often used in combination with other fixatives for immunoelectron microscopy. Formaldehyde fixation is reversible and in the case of formaldehyde alone, it is preferable to add 0.3 to 0.5% formaldehyde in the buffer for storing the fixed tissue.[4] When some epitopes are destroyed even by formaldehyde fixation, washing the fixed tissue thoroughly in buffer or treatment with borohydride[5] may be useful in regaining antigenicity.

It has been demonstrated that treatment of aldehyde-fixed materials with proteases such as trypsin, pepsin, papain, or pronase may cause an improvement in the intensity of immunolabeling and the number of labeled sites in light microscopic immunocytochemistry.[6-7] This is probably due to digestion of some cross-linked proteins, resulting in an increased permeability of antigens to antibodies. When such treatments are done, it is advisable to compare both protease-treated and nontreated sections, because in some cases the enzyme may digest the antigens. Such drastic treatments are not usually employed for immunoelectron microscopy because of the disruption of ultrastructure, but application of proteinase K to ultrastructural *in situ* hybridization using the preembedding method was reported with successful results.[8]

2. Glutaraldehyde

Glutaraldehyde, which is a dialdehyde, gives higher chemical stability and structural preservation than formaldehyde. Although glutaraldehyde surpasses all other fixatives in the ability to cross link and preserve ultrastructure, it often impairs antigens even in low concentrations because of steric hindrance of antigens to immunoreagents in tightly cross-linked tissues and because of the alteration of epitopes.

Glutaraldehyde fixation is also reversible, but at a much slower rate than formaldehyde. Thus, for practical use, it can be considered to be irreversible.

Penetration of glutaraldehyde into tissue is so slow that it is reasonable and popular to use this fixative in combination with other rapidly penetrating aldehydes (for example, formaldehyde[9] or acrolein[10]). A mixture of three aldehydes consisting of glutaraldehyde, formaldehyde, and acrolein is reported to reduce the tendency for shrinkage of the pineal gland when compared with glutaraldehyde alone.[11] A typical example for immunoelectron microscopy is 4% formaldehyde and 0.2% glutaraldehyde in 0.1 *M* phosphate buffer, pH 7.3 to 7.4. To make 100 ml of this fixative, mix 30 ml of distilled water and 4 g of paraformaldehyde. Warm the solution to 60°C while stirring, under a ventilation hood, then clear it by adding a few drops of 1 *N* NaOH. After the solution has cooled, add the following ingredients: 0.8 ml of 25% glutaraldehyde and distilled water (bring up to 50 ml), 0.2 *M* phosphate buffer (bring up to 100 ml). Check the pH of the solution and adjust it to pH 7.3 to 7.4.

The concentration of glutaraldehyde used in conventional microscopy is around 2.5%. However, in immunocytochemistry, many proteins, especially high-molecular-weight proteins,[3] cannot tolerate 0.5% glutaraldehyde. When the antigenicity is severely damaged even by a low concentration of glutaraldehyde, pretreatment with ethylacetimidate (EAI) may be effective.[4,12-13] EAI blocks a significant fraction of amino groups, and restricts the degree of the subsequent glutaraldehyde fixation.

According to Tokuyasu et al.,[13] a 5-min treatment with a mixture of 20 m*M* EAI and 8% formaldehyde is enough to protect the antigenicity of vinculin against a subsequent, strong fixation with a mixture of 4% formaldehyde and 4% glutaraldehyde.

For pretreatment with ethylacetimidate (EAI):

1. Fix the tissue with a mixture of 20 m*M* EAI and 3 to 8% formaldehyde in 0.1 *M* phosphate buffer, pH 7.4, for 5 min at room temperature. EAI should be dissolved immediately before use because it gradually hydrolyzes in solution. The pH is adjusted to 7.4 by adding 1 *N* NaOH. This procedure should be done in an ice box.
2. Subsequently, fix the tissue with a mixture of appropriate concentration of glutaraldehyde and 3 to 8% formaldehyde in 0.1 *M* phosphate buffer, pH 7.4, for 1 h at room temperature.

Some proteins easily diffuse from the original site under formaldehyde fixation alone. In such cases, glutaraldehyde fixation is required. For example, when fixed with 3% formaldehyde alone, the microvillar tip of the intestinal epithelial cells is labeled for α-actinin, but this labeling is reported to be an artifact caused by the noncovalent fixation.[14] The artifactual redistribution of α-actinin under mild fixation was also demonstrated in chicken heart muscle.[13]

It should be kept in mind that any free aldehyde groups remaining in the tissue may react with proteins (antibodies added in subsequent steps), resulting in high backgrounds and false positivity. For blocking the background, we add 10 to 50 m*M* glycine into the buffer. Pretreatment with 0.05% sodium borohydride ($NaBH_4$) in PBS is also effective in reducing nonspecific binding. Weber et al.[15] have demonstrated that the treatment of glutaraldehyde-fixed tissue culture cells with $NaBH_4$ improves detection of tubulin.

Generally speaking, one may first try a mixture of 0.1 to 1% glutaraldehyde and 2 to 4% formaldehyde and study the quality of ultrastructural preservation and the degree of immunolabeling. If necessary, up to 8% formaldehyde can be used. Whether the antigenicity of the protein of interest is sensitive or not can be determined by light microscopic immunohistochemistry. If the antigenicity of the protein is impaired by glutaraldehyde, a mixture with a lower concentration of glutaraldehyde and paraformaldehyde or paraformaldehyde alone should be employed. The periodate-lysine-paraformaldehyde solution described below is a good alternative. If the subject requires crucially fine structural preservation, EAI treatment may be necessary.

3. Periodate-Lysine-Paraformaldehyde Solution (PLP)

This fixative was originally introduced by McLean and Nakane[16] for fixation of glycoproteins, and was intended to act on carbohydrate side chains of glycoproteins. According to the original description of this fixative, periodate oxidizes the carbohydrates to aldehydes, and lysine cross-links the newly formed aldehyde groups intra- and intermolecularly. Formaldehyde is added to complement the fixation by periodate and lysine on the assumption that it does not undergo polymerization or form stable complexes with lysine.

The great advantage of PLP is excellent preservation of antigenicity of glycoproteins, and PLP gives a more intense labeling as compared to glutaraldehyde fixation. Thus, the use of PLP has enjoyed wide practice in immunocytochemistry, especially in the preembedding method. On the other hand, this fixative may be inferior to glutaraldehyde fixation in ultrastructural preservation.

The most commonly used combination is 0.01 *M* periodate, 0.075 *M* lysine, and 2% formaldehyde:

1. Stock solution A (0.1 *M* lysine 0.05 *M* phosphate buffer, pH 7.4)

Lysine hydrochloride	1.827 g
Distilled water	50 ml

Add 0.1 *M* $NaHPO_4$ to the above solution until pH is 7.4. Then double the volume by adding 0.1 *M* phosphate buffer, pH 7.4. This solution can be stored for 1 week at 4°C.

2. Stock solution B (8% formaldehyde)

Paraformaldehyde	8 g
Distilled water	100 ml

Warm the solution to 60°C while stirring under a ventilation hood, then clear it by adding a few drops of 1 *N* NaOH. This solution can be stored for 3 weeks at 4°C, but fresh solution is preferable.

3. Final solution: just before use, mix 3 parts of solution A with 1 part of solution B. Add solid sodium *m*-periodate ($NaIO_4$) to a final concentration of 0.01 *M* (21.4 mg/10 ml). Check the pH of the solution, and adjust it to pH 6.2. Fixation is done at 4°C.

When glutaraldehyde cannot be used, PLP solution is a good candidate to be tried, especially for fixing enzymes and peptides.

4. Zamboni's Fixative (Picric Acid-Formaldehyde)

Zamboni's buffered picric acid-formaldehyde fixative (a 15% saturated solution of picric acid and 2% formaldehyde in 0.1 *M* phosphate buffer, pH 7.2)[17] is mainly employed for fixation of polypeptides. Compared with formaldehyde alone, fine structure is better maintained in this fixative. Presumably the protein coagulative properties of picric acid may complement the cross-linking of formaldehyde. This fixative has also been used in combination with glutaraldehyde.[18] However, Zamboni's fixative is thought to be unsuitable for fixation of some proteins denatured by picric acid.

Zamboni's fixative is made up in the following manner:

1. Saturated aqueous solution of picric acid: dissolve more than 2.1 g of picric acid in 150 ml of distilled water and filter it.
2. 20% Formaldehyde: dissolve 20 g of paraformaldehyde in 100 ml of distilled water. Warm the solution to 60°C (in a ventilation hood) and add a few drops of 1 *N* NaOH until a clear solution is obtained. Cool and filter it.
3. Mix 150 ml of saturated aqueous solution of picric acid and 100 ml of 20% paraformaldehyde solution, and dilute it to 1000 ml with 320 mOsM phosphate buffer (3.31 g $NaH_2PO_4 \cdot H_2O$ and 33.7 g $Na_2HPO_4 \cdot 7H_2O$ in 1000 ml distilled water).

This solution is stable and can be stored for 1 year at room temperature. Fixation is done at 4°C.

5. Carbodiimide

Carbodiimide cross links carboxyl groups with amino acids, and its action is relatively moderate. Although it may allow detection of some antigenic sites which are lost by aldehyde fixation, this fixative is not used alone for electron microscopic study because of poor structural preservation. A combination of 0.35% glutaraldehyde and 1% 3-diethylaminopropyl carbodiimide is reported to be effective for preservation of morphology and preembedding detection of intracellular antigens in cultured cells.[19]

6. Acrolein

Acrolein is an aldehyde fixative with an ethylene bond which is highly reactive with amino groups such as histidine, tyrosine, and proline. Acrolein penetrates tissues rapidly, and facilitates fixation by other aldehydes. Hence, this fixative is effective in compensating for the slow penetration of glutaraldehyde into the tissue, especially when materials are immersion fixed. Its usefulness in fixation of several neuropeptides has been well documented.[20-21] Since acrolein is a very toxic agent, use of a ventilated hood and gloves is mandatory.

B. CRYOFIXATION

Conventional chemical fixation may cause not only impaired immunoreactivity of antigenic sites but artificial alteration of macromolecular structures.

Quick freezing followed by freeze substitution has been reported to provide satisfactory immunolabeling and morphology by avoiding artifacts induced by slow penetration of chemical fixatives. Since fixation and dehydration are performed at temperatures below the freezing and recrystallization points of cellular water, redistribution and loss of cellular components as well as shrinkage of samples should be minimal. This method is also reported to improve labeling accuracy through an increase in the specific intensity and a decrease in that of the background.

Freeze substitution following rapid freezing has opened new possibilities for postembedding immunocytochemistry, and its application to further studies is very promising. For further details, the reader is referred to the recent papers.[22-24]

REFERENCES

1. **Maupin-Szamier, P. and Pollard, T. D.,** Actin filament destruction by osmium tetroxide, *J. Cell Biol.*, 77, 837, 1978.
2. **Paiement, J. and Roy, L.,** Electrophoretic protein blots as aids in choosing fixatives for immunocytochemistry, *J. Histochem. Cytochem.*, 36, 441, 1988.
3. **Riederer, B. M.,** Antigen preservation tests for immunocytochemical detection of cytoskeletal proteins: influence of aldehyde fixatives, *J. Histochem. Cytochem.*, 37, 675, 1989.
4. **Tokuyasu, K. T. and Singer, S. J.,** Improved procedures for immunoferritin labeling of ultrathin frozen sections, *J. Cell Biol.*, 71, 894, 1976.
5. **Baschong, W., Baschong-Prescianotto, C., and Kellenberger, E.,** Reversible fixation for the study of morphology and macromolecular composition of fragile biological structures, *Eur. J. Cell Biol.*, 32, 1, 1983.
6. **Huang, S.-N.,** Immunohistochemical demonstration of hepatitis B core and surface antigens in paraffin sections, *Lab. Invest.*, 33, 88, 1975.
7. **Curran, R. C. and Gregory, J.,** Demonstration of immunoglobulin in cryostat and paraffin sections of human tonsil by immunofluorescence and immunoperoxidase techniques, *J. Clin. Pathol.*, 31, 974, 1978.
8. **Wolber, R. A., Beals, T. F., Lloyd, R. V., and Maassab, H. F.,** Ultrastructural localization of viral nucleic acid by in situ hybridization, *Lab. Invest.*, 59, 144, 1988.
9. **Karnovsky, M. J.,** A formaldehyde-glutaraldehyde fixative of high osmolarity for use in electron microscopy, *J. Cell Biol.*, 27, 137A, 1965.
10. **Sandborn, E., Koen, P. F., McNabb, J. D., and Moore, G.,** Cytoplasmic microtubules in mammalian cells, *J. Ultrastruct. Res.*, 11, 123, 1964.
11. **Schultz, R. L. and Witter, E. E.,** Improved procedures for pineal gland fixation for electron microscopy, *J. Pineal Res.*, 6, 267, 1971.
12. **Geiger, B., Dutton, A., Tokuyasu, K. T., and Singer, S. J.,** Immunoelectron microscopic studies of membrane-microfilament interactions: distribution of α-actinin, tropomyosin, and vinculin in intestinal epithelial brush border and chicken gizzard smooth muscle cells, *J. Cell Biol.*, 91, 614, 1981.
13. **Tokuyasu, K. T., Dutton, A. H., Geiger, B., and Singer, S. J.,** Ultrastructure of chicken cardiac muscle as studied by double immunolabeling in electron microscopy, *Proc. Natl. Acad. Sci. U.S.A.*, 78, 7619, 1981.
14. **Geiger, B., Tokuyasu, K. T., and Singer, S. J.,** Immunocytochemical localization of α-actinin in intestinal epithelial cells, *Proc. Natl. Acad. Sci. U.S.A.*, 76, 2833, 1979.
15. **Weber, K., Rathke, P. C., and Osborn, M.,** Cytoplasmic microtubular images in glutaraldehyde-fixed tissue cultured cells by electron microscopy and immunofluorescence microscopy, *Proc. Natl. Acad. Sci. U.S.A.*, 75, 1820, 1978.
16. **McLean, I. W. and Nakane, P. K.,** Periodate-lysine-paraformaldehyde fixative: a new fixative for immunoelectron microscopy, *J. Histochem. Cytochem.*, 22, 1077, 1974.

17. **Zamboni, L. and De Martino, C.,** Buffered picric acid-formaldehyde: a new rapid fixative for electron microscopy, *J. Cell Biol.*, 35, 148A, 1965.
18. **Somogyi, P. and Takagi, H.,** A note on the use of picric acid-paraformaldehyde-glutaraldehyde fixative for correlated light and electron microscopic immunocytochemistry, *Neuroscience*, 7, 1779, 1982.
19. **Willingham, M. C. and Yamada, S. S.,** Development of a new primary fixative for electron microscopic immunocytochemical localization of intracellular antigens in cultured cells, *J. Histochem. Cytochem.*, 27, 947, 1979.
20. **King, J. C., Lechan, R. M., Kugel, G., and Anthony, E. L. P.,** Acrolein: a fixative for immunocytochemical localization of peptides in the central nervous system, *J. Histochem. Cytochem.*, 31, 62, 1983.
21. **Smith, P. F. and Keefer, D. A.,** Acrolein/glutaraldehyde as a fixative for combined light and electron microscopic immunocytochemical detection of pituitary hormones in immersion-fixed tissue, *J. Histochem. Cytochem.*, 30, 1307, 1982.
22. **Ichikawa, M., Sasaki, K., and Ichikawa, A.,** Optimal preparatory procedures of cryofixation for immunocytochemistry, *J. Electron Microsc. Tech.*, 12, 88, 1989.
23. **Nicolas, M.-T., Bassot, J.-M., and Nicolas, G.,** Immunogold labeling of luciferase in the luminous bacterium *Vibrio harveyi* after fast-freeze fixation and different freeze-substitution and embedding procedures, *J. Histochem. Cytochem.*, 37, 663, 1989.
24. **Usuda, N., Ma, H., Hanai, T., Yokota, S., Hashimoto, T., and Nagata, T.,** Immunoelectron microscopy of tissues processed by rapid freezing and freeze-substitution fixation without chemical fixatives: application to catalase in rat liver hepatocytes, *J. Histochem. Cytochem.*, 38, 617, 1990.

III. EMBEDDING

Masahiro Mizuno

A. INTRODUCTION

Depending on whether staining is done before embedding or after, basic methods of electron microscopic immunocytochemistry fall into three types: preembedding technique, postembedding technique, and cryoultramicrotomy. Each has advantages and disadvantages with respect to preservation of fine structure and antigenicity, and immunoreaction (accessibility of antigen to antibody).

Cryoultramicrotomy was described in Article I. In this article, the preembedding and postembedding methods are outlined.

B. PREEMBEDDING METHOD

In this method, the specimens are generally prepared as isolated cells, monolayer cultures, tissue fragments, frozen sections cut on a cryostat, or unfrozen sections cut by a Vibratome or a Microslicer in a manner similar to that used for light microscopic immunocytochemistry. After immunostaining on permeabilized cells or thick sections, the specimens are processed for electron microscopic immunocytochemistry.

As with cryoultramicrotomy, the preembedding method is useful for antigens that cannot tolerate the dehydration and embedding procedures. This method has often been the choice to study the distribution of cell surface antigens and receptors, but is also widely applied to the detection of intracellular antigens. The preembedding method enables one to confirm the immunoreaction at the light microscopic level before getting to electron microscopic study. This ability is especially important for tissues with sparsely distributed antigens. Also, the use of osmium after immunostaining is suitable for the situation which needs adequate membrane contrast.[1]

An inherent problem of the preembedding method is the difficulty of penetration of markers and antibodies into cellular and organellar interiors: the impenetrability of plasma membranes and fixed cytosol matrix may limit the penetration, resulting in false-negative results. For improvement of the poor penetration, therefore, use of IgG fragments ($F_{(ab)}$, $F_{(ab')}$, $F_{(ab')2}$) and/or pretreatment with detergents is generally necessary. Triton X-100 and saponin are the best known and most commonly used detergents. Use of a high concentration of detergents causes membrane damage, and may affect the distribution of the antigen associated with the cell membrane. The loss of membrane structure is generally less in saponin treatment than in Triton X-100 treatment,[2,3] because the former selectively reacts with membrane-bound sterols, such as cholesterol, and keeps organelle membranes which have low sterol content intact. The freeze-thawing method[4] and the dehydration-rehydration method[5] are known as alternatives to the detergents for enhancing penetration.

The most popular marker for the preembedding method is horseradish peroxidase, and both the peroxidase-antiperoxidase (PAP) method and the avidin-biotin peroxidase complex (ABC) method can be applied. Piekut et al.[6] reported that in 80-μm-thick Vibratome sections, the PAP method allows only 8 to 9 μm of penetration under Triton-X 100 treatment at a concentration of 0.04 to 0.4%. When the horseradish peroxidase method is used, the drawback is that it does not always identify the accurate subcellular localization. This is mainly because the peroxidase reaction product formed by the diaminobenzidine (DAB) reaction diffuses away from the antigenic sites into the adjacent areas during subsequent processing of the tissue. The reaction product may be observed more than 100 nm away from the true antigenic sites.[7] Ferritin[3] or colloidal gold[8] labels can also be employed in the preembedding method. However, such particles do not easily penetrate intracellular organelles, and when large-sized markers are used, pretreatment with detergents, as described above, is usually necessary. Recently, the silver enhancement method of 1-nm gold particles has been applied to preembedding immunoelectron microscopy.[9,10] Chan et al.[10] has reported successful results using this method without detergents or freeze-thawing.

Briefly, the preembedding procedure involves the following steps. Details can be found in several references.[1,11,12]

1. Fixation of the tissue: Strong cross-linking fixatives (e.g., glutaraldehyde) fix both membranes and cytosol proteins firmly, and may cause a penetration problem in the preembedding method in addition to their effects on antigenicity. Fixation, therefore, should be appropriately light compared with the postembedding method. Preferably, periodate-lysine-paraformaldehyde (PLP) fixative or a mixture of formaldehyde and a low concentration of glutaraldehyde is used.
2. Sectioning of the specimen: Fixed unfrozen slices (30 to 70 μm) are obtained with a Vibratome or a Microslicer. In a cryostat, tissue specimens should be first cryoprotected by infiltrating with a graded concentration (10 to 20%) of sucrose solutions and sucrose-glycerol mixture. Then they are placed in a supporting medium (e.g., optimum cutting temperature [OCT] compound) and frozen with dry ice-acetone or methylbutane in liquid nitrogen. Frozen sections (6 to 10 μm) are cut in a cryostat, collected on coated slides and air dried for 30 min.
3. Pretreatment: When peroxidase is used, it may be necessary sometimes to treat the sections with methanol and 0.03% H_2O_2 in phosphate-buffered saline (PBS)[13] or with 0.02 *M* periodic acid in PBS,[11] to inhibit intrinsic peroxidase activity. This is particularly important for those cells that have intrinsic peroxidase activity, such as neutrophils and macrophages, as well as heme-containing cells. If penetration is a problem, a small amount of detergent (e.g., 0.02 to 0.1% Triton X-100 or 0.03 to 0.1% saponin for 30 min) may be employed.
4. Immunostaining

   - Rinse the sections in PBS three times, 5 min each time.
   - React the sections with blocking proteins (2 to 10% normal serum from the host species of the secondary antibody or 3% BSA) for 30 to 60 min.
   - Rinse the sections in PBS three times, 5 min each time.
   - Incubate the section with the primary antibody at 4°C in a moist chamber for 6 to 10 h.
   - Rinse the sections in PBS three times, 5 min each time.
   - Incubate the section with peroxidase-conjugated secondary antibody in a moist chamber for 4 to 6 h.

   Antibody concentration and incubation time and temperature should be varied for optimal immunocytochemical detection.
5. Postfixation by 2 to 3% glutaraldehyde for 10 min — The postfixation supplements the initial light fixation, stabilizing the immune complex and tissue components.
6. Administration of diaminobenzidine (DAB) to visualize peroxidase activity: After the PBS rinse, the slides are incubated with 5 mg of 3,3′-diaminobenzidine tetrahydrochloride (DAB) dissolved in 10 ml of 50 m*M* Tris-HCl, pH 7.6, for 10 to 30 min in the dark. Subsequently, 1% H_2O_2 in distilled water is added to the above DAB solution (to a final concentration of 0.01 to 0.05%) and the slides are further incubated for 3 to 5 min. Brown precipitates indicate the antigenic sites at the light microscopic level. *Note:* DAB is a potential carcinogen and should be handled with caution.

7. Osmication, dehydration, and embedding in epoxy resin for observation with electron microscopy — The sections are fixed for 1 h in 1 to 2% OsO_4 solution in 0.1 *M* cacodylate buffer, pH 7.4. The DAB deposit itself does not possess the electron density necessary for many microscopic examinations. The reason for osmication is that DAB is highly osmiophilic, and the bound osmium imparts electron density. Subsequently, the sections are dehydrated through a graded series of ethanol solutions and embedded in epoxy resins.
8. Ultrathin sectioning and observation.

C. POSTEMBEDDING METHOD

In the postembedding method, only the antigens exposed at the surface of resin-embedded sections are immunolabeled. Colloidal gold particles conjugated with immunoglobulin or protein A are the best markers in this method with respect to precise localization. Moreover, based on the uniformity of antibody penetration into all cell compartments exposed on the surface, quantitative analysis is possible. Since there is no penetration of immunolabeling markers in this approach, the size of markers is not expected to affect greatly the labeling efficiency. However, Yokota has shown that labeling density for catalase on Lowicryl K4M sections using protein A-gold decreases as the size of gold particles increases.[14] The reason for this is not clear, but Yokota speculated that the binding force of the antibody to the antigen or of the protein A to the antibody may not be enough to support the large-sized gold particles.

The resins used are divided into two major groups; epoxy resins (e.g., Epon, Araldite, and Spurr's resin) and acrylic resins (e.g., Lowicryl K4M and LR White).

1. Epoxy Embedding of Osmicated Tissues

The hydrophobic epoxy resins such as Epon, Araldite, and Spurr's resin have traditionally been used in immunocytochemistry.

Conventional epoxy embedding of osmicated tissues provides the required ultrastructural preservation, and lack of osmication may cause swelling, shrinking, and extraction of tissue components. However, these treatments irreversibly affect the antigenicity of most molecules. The high chemical reactivity and hydrophobic nature of epoxy resin, and its high denaturing temperature of polymerization, impair antigenicities of most cytoskeletal proteins. Furthermore, use of osmium tetroxide is usually not compatible with preservation of most biological activities in immunocytochemistry, except for some polypeptides. For this reason, various methods using etching agents, such as sodium metaperiodate, hydrogen peroxide, alcoholic sodium hydroxide, periodic acid, and so forth,[15] have been developed to increase antigen accessibility in the resins and to restore osmicated antigen. This etching process is crucial to determine the degree of both restoration of antigenicity and damage of cell structures. The concentration of etching agents and the time for reaction with them depend on the composition of resins, section thickness, and the degree of desired contrast. With improvement of the etching technique, there have been many reports of successful immunocytochemical detection of hormones,[15] enzymes,[15] neuropeptides,[15] and immunoglobulins.[17,18] Sodium metaperiodate treatment restores tissue antigenicity by removing osmium and enhances the hydrophilic nature of the resin by oxidizing the hydrophobic alkane side chains to alcohols, aldehydes, and acids. This agent is superior to hydrogen peroxide in many cases.[15]

Apart from the problem of fixation, satisfactory results in the epoxy-embedding procedure are closely related to the antigen type.[19] Some antigens (e.g., cytoskeletal proteins, membrane proteins, collagen proteins) are very labile to the embedding process, and epoxy resins have been found inadequate for their detection. On the other hand, the resins are compatible with immunolabeling for other antigens. Indeed, the advent of hydrophilic resins may have made this type of postembedding method obsolete for some applications, but in combination with freeze-substitution, the desirability of epoxy resins may be reevaluated.

2. Lowicryl K4M

The development of hydrophilic resins such as Lowicryl K4M and LR White has made possible the adequate preservation of cellular fine-structural details and the reactivity of cellular components with antibodies. As regards retention of antigenicity, there are many reports that demonstrate the superiority of hydrophilic resins to embedding in Epon or Araldite resins. The most popular of the hydrophilic resins is Lowicryl K4M, which is an acrylate-methacrylate mixture of low viscosity for low-temperature embedding. This method, introduced by Carlemalm et al.,[20] has been applied and improved for immunocytochemistry by Roth[21,22] and, in combination with colloidal gold techniques, by other researchers.

The original principles of the Lowicryl K4M embedding method are (1) dehydration at low temperature minimizes the detrimental effects of solvent exchange on protein conformation, (2) polymerization with long-wave ultraviolet light at low temperature avoids denaturation of antigenicity at high temperature, and (3) the hydrophilic nature of the resin reduces nonspecific binding of immunoreagents.

Polymerization is usually achieved by ultraviolet irradiation. Since osmium tetroxide in the specimen interferes with unsaturated bonds in the resin and precludes polymerization, it should be omitted. Omission of OsO_4 also contributes to antigenicity retention. Unlike the case with Epon and Araldite, the etching process prior to immunostaining is not required.

Compared to cryoultramicrotomy, labeling of Lowicryl K4M sections is generally less efficient, and antigenicities of some proteins may not be retained, especially membrane proteins, which are preserved in the former. However, it is definitely an advantage of the Lowicryl embedding method that once the blocks are made, the following process can be performed with little difficulty.

A minor drawback of the original Lowicryl K4M embedding method is the long time required for polymerization at $-35°C$. Altman et al.[23] have reported a modified Lowicryl K4M embedding procedure with adequate results. This modification has made it possible to shorten the time needed for polymerization with short-distance (10 cm) ultraviolet radiation. The method used by Altman et al. is described below.

a. Fixation

The principle of choice of fixatives is the same as with other embedding methods, however, osmium tetroxide cannot be employed, for the reason described above.

A mixture of 0.1 to 1.0% glutaraldehyde and 3 to 4% formaldehyde in 0.1 *M* sodium phosphate buffer, pH 7.4, is commonly used.

b. Dehydration and Infiltration

The subsequent dehydration and infiltration steps are performed at room temperature.

Dehydration (dimethylformamide)

50% Dimethylformamide	10 min
75% Dimethylformamide	10 min
90% Dimethylformamide	10 min

Infiltration

Lowicryl K4M:Dimethylformamide, 1:2	10 min
Lowicryl K4M:Dimethylformamide, 1:1	15 min
Lowicryl K4M	20 min
Lowicryl K4M	25 min

Lowicryl K4M resin is prepared just before use by gently mixing the cross-linker, the monomer, and the initiator in the following proportion: cross-linker A, 2 g; monomer B, 13 g; and initiator, 75 mg.

Careful attention should be paid when handling Lowicryl K4M resin, for it is extremely noxious, generating cyanogen gas when mixed and causing severe dermatitis. Therefore, all work with the resin must be done in a ventilation hood with gloves.

c. Embedding and Polymerization

The samples are embedded in the fresh Lowicryl K4M in BEEM capsules and polymerized with long-wave (360 nm) ultraviolet radiation at a distance of 10 cm for 45 min at 4°C. Under some conditions, complete polymerization may require more time. In order to facilitate polymerization, a cardboard box lined with aluminum foil should be used as a reflecting chamber.

d. Sectioning

Gold to silver sections are cut as with epoxy resins. Because of the hydrophilic nature of the blocks, the water level in the boat should be kept below normal, just sufficient to keep the knife edge wet. Once the surface of the block touches water, subsequent cutting becomes markedly difficult. If this happens, one should renew the surface of the block by cutting several thick sections with a dry knife, after which thin sections can again be cut.

e. Immunoreaction

The sections are immersed for 10 min in 3% BSA-PBS or 2 to 10% goat serum-PBS. They are incubated with the primary antibody for 30 to 60 min, washed in PBS (1 min $\times$ 8), and incubated with colloidal gold-conjugated IgG or protein A for 60 min. The sections are then thoroughly washed in PBS (1 min $\times$ 4), and rinsed in distilled water (1 min $\times$ 6).

f. Staining

Finally, the sections are stained with 2% aqueous uranyl acetate for 5 min and Millonig's lead acetate for 45 s.

The poor membrane contrast on Lowicryl K4M sections has often been pointed out. This is basically due to the necessity to avoid osmication of specimens before dehydration and embedding. Recently, uranyl acetate has been used before or during the dehydration step to improve membrane contrast and tissue preservation in the postembedding method.[24-26] Originally, uranyl acetate, which has properties of a fixative (it was first used en bloc to stabilize viral and bacterial DNA[27]) as well as a staining agent, has long been recognized to reduce extraction of membrane phospholipids[28] and increase membrane contrast.[29] Thus, the use of uranyl acetate as a postfixative in the Lowicryl K4M method is reported to improve morphology without impairing antigenicity.[24,25]

Another way to get better contrast on Lowicryl K4M sections has recently been developed by Roth et al.[30] The method, uranyl acetate (UA)/methyl cellulose (MC) adsorption staining, is as follows.

1. After immunolabeling, stain the grids with a mixture of 1.8% UA/0.2% MC, pH 4.5 to 6.0, for 5 min at room temperature.
2. Remove the excess UA/MC stain with a filter paper, to such an extent that the dried film of a staining mixture shows a silver-gold to gold-blue color.

This staining method is reported to be useful for increasing membrane contrast without obscuring colloidal gold particles.

Another acrylic embedding agent, LR White, is also hydrophilic, but the polymerization is carried out by heat (50°C). Unlike Lowicryl K4M, the resin is compatible with the use of osmic acid postfixation. For further details of LR White, see Newman and Hobot.[31]

REFERENCES

1. **Priestley, J. V.,** Pre-embedding ultrastructural immunocytochemistry: immunoenzyme techniques, in *Immunolabeling for Electron Microscopy,* Polak, J. M. and Varndell, I. M., Eds., Elsevier Science Publishers, Amsterdam, 1984, 37.
2. **Ohtsuki, I., Manzi, R. M., Palade, G. E., and Jamieson, J. D.,** Entry of macromolecular tracers into cells fixed with low concentrations of aldehydes, *Biol. Cell,* 31, 119, 1978.
3. **Willingham, M. C.,** An alternative fixation-processing method for preembedding ultrastructural immunocytochemistry of cytoplasmic antigens, *J. Histochem. Cytochem.,* 31, 791, 1983.
4. **Ribak, C. E., Vaughn, J. E., and Barber, R. P.,** Immunocytochemical localization of GABAergic neurones at electron microscopic level, *Histochem. J.,* 13, 555, 1981.
5. **Eldred, W. D., Zucker, C., Karten, H. J., and Yazulla, S.,** Comparison of fixation and penetration enhancement techniques for use in ultrastructural immunocytochemistry, *J. Histochem. Cytochem.,* 31, 285, 1983.
6. **Piekut, D. T. and Casey, S. M.,** Penetration of immunoreagents in Vibratome-sectioned brain: a light and electron microscope study, *J. Histochem. Cytochem.,* 31, 669, 1983.
7. **Courtoy, P. J., Picton, D. H., and Farquhar, M. G.,** Resolution and limitations of the immunoperoxidase procedure in the localization of extracellular matrix antigens, *J. Histochem. Cytochem.,* 31, 945, 1983.
8. **Yang, H.-Y., Lieska, N., Goldman, A. E., and Goldman, R. D.,** A 300,000-mol-wt intermediate filament-associated protein in baby hamster kidney (BHK-21) cells, *J. Cell Biol.,* 100, 620, 1985.
9. **Lah, J. J., Hayes, D. M., and Burry, R. W.,** A neutral pH silver development method for the visualization of 1-nanometer gold particles in pre-embedding electron microscopic immunocytochemistry, *J. Histochem. Cytochem.,* 38, 503, 1990.
10. **Chan, J., Aoki, C., and Pickel, V. M.,** Optimization of differential immunogold-silver and peroxidase labeling with maintenance of ultrastructure in brain sections before plastic embedding, *J. Neurosci. Meth.,* 33, 113, 1990.
11. **Amenta, P. S. and Martinez-Hernandez, A.,** Specific methods for electron immunohistochemistry, *Meth. Enzymol.,* 145, 133, 1987.
12. **Brown, W. J. and Farquhar, M. G.,** Immunoperoxidase methods for the localization of antigens in cultured cells and tissue sections by electron microscopy, *Meth. Cell Biol.,* 31, 553, 1989.
13. **Streefkerk, J. G. and van der Ploeg, M.,** Model studies on quantitative aspects of histochemical reactions on peroxidase with special reference to the effect of methanol on peroxidase activity, *Ann. Histochim.,* 21, 67, 1976.
14. **Yokota, S.,** Effect of particle size on labeling density for catalase in protein A-gold immunocytochemistry, *J. Histochem. Cytochem.,* 36, 107, 1988.
15. **Bendayan, M. and Zollinger, M.,** Ultrastructural localization of antigenic sites on osmium-fixed tissues applying the protein A-gold technique, *J. Histochem. Cytochem.,* 31, 101, 1983.
16. **Castel, M., Morris, J., Ben-Barak, Y., Timberg, R., Sivan, N., and Gainer, H.,** Ultrastructural localization of immunoreactive neurophysins using monoclonal antibodies and protein A-gold, *J. Histochem. Cytochem.,* 33, 1015, 1985.
17. **Hearn, S. A., Silver, M. M., and Sholdice, J. A.,** Immunoelectron microscopic labeling of immunoglobulin in plasma cells after osmium fixation and epoxy embedding, *J. Histochem. Cytochem.,* 33, 1212, 1985.

A. INTRODUCTION

Immunocytochemistry at the electron microscopic level first became possible by the introduction of ferritin-conjugated antibody by Singer (1959).[1] His choice of ferritin as a marker molecule was reasonable. Ferritin is a protein of about 650 kDa which is composed of an internal iron-containing core and an outer apoprotein shell; the core is visible under the electron microscope as a distinct particle of about 5 nm in diameter, while the outer shell of about 10 nm in diameter can be used for chemical conjugation with antibodies.

In recent years, the ferritin technique has been largely taken over by methods utilizing colloidal gold. Although colloidal gold has several advantages over ferritin, e.g., visibility, variable particle sizes, etc., some people prefer ferritin conjugates because they tend to give more signals than colloidal gold conjugates. Moreover, because the conjugation ratio of ferritin and immunoglobulin can be adjusted to 1:1, while that of colloidal gold and immunoglobulin is unknown, the ferritin method is superior in quantitation of antigenic sites.

In this article, methods to immunolabel ultrathin sections using the ferritin-conjugated antibody will be described in detail and, as an interesting variation, the ferritin bridge technique for cultured cells will be explained. One major problem with ferritin techniques is that commercial ferritin-conjugated antibodies are often not of adequate quality; to the author's knowledge, antibodies used for conjugation are immunoglobulin fraction of antisera and not affinity purified. Therefore, to overcome this problem, it is advisable to prepare the conjugates in the laboratory. Methods for conjugation will not be treated here, but several protocols using different chemicals for conjugation are available.[2-5] As described below, for those who do not want to prepare ferritin-conjugated antibody themselves, a combination of biotinylated immunoglobulin and streptavidin-ferritin, both of which are available commercially, can be used as an alternative. On the other hand, the ferritin bridge technique can be performed with reagents easily obtainable commercially.

B. IMMUNOLABELING OF ULTRATHIN CRYOSECTIONS USING FERRITIN AS A MARKER

Because ferritin-conjugated immunoglobulin is fairly large and does not penetrate easily into tissues, the material for immunolabeling must be either isolated or prepared to expose antigenic sites on the surface. Isolated cells and organelles are appropriate for surface labeling by preembedding procedures; for labeling of intracellular structure *in situ,* ultrathin sections are commonly used for postembedding procedures. In this article, a method to label ultrathin cryosections with the indirect immunoferritin technique is described.[6,7]

1. Equipment

1. Small loops: a loop slightly larger than electron microscope (EM) grids in diameter attached to a needle holder for easy handling; nichrome wire sold for electricity of 100 to 200 W is appropriate.
2. Anticapillary forceps: forceps which are made to take up a minimal amount of solution are available commercially.
3. Glass plate and Parafilm: a sheet of Parafilm is stuck on a glass plate with some moisture in between; it provides a clean surface for the immunolabeling procedure.

2. Chemicals and Reagents

1. Biotinylated anti-immunoglobulin antibody, available from Cappel, Zymed, Sigma, EY, and other companies; antibody must be affinity-purified for specific immunoreaction.

2. Streptavidin-ferritin, available from Bethesda Research Laboratories.
3. 2% Neutral uranyl acetate solution.
4. 2% Acidic uranyl acetate solution in distilled water.
5. Methyl cellulose solution: heat distilled water to about 60°C, add an appropriate amount of methyl cellulose powder (1500 cp) and mix with magnetic stirrer for a few minutes. Transfer the turbid solution directly into ice and leave there for 1 to 2 h. The solution should become transparent. Add distilled water to adjust the final concentration. This solution can be kept at 4°C for at least a few months. It must be centrifuged at 3000 rpm or at a higher speed for 5 min before use.
6. Polyethylene glycol solution: polyethylene glycol (Carbowax) of 1540 in molecular weight dissolved in distilled water.

3. Procedures

1. Fix tissues with an appropriate fixative; 3 to 8% formaldehyde (freshly depolymerized from paraformaldehyde) in 0.1 *M* sodium phosphate buffer, pH 7.4, is usually used as a basic fixative. Glutaraldehyde, even as little as 0.01%, improves morphology, but it may reduce or abolish the reactivity of antigens. The duration and temperature of fixation must be determined for a specific antibody; 3% formaldehyde for 60 min at room temperature (20 to 25°C) is a reasonable point to start; if a positive immunoreaction is seen, try a mixture of 3% formaldehyde and 0.1% glutaraldehyde. All procedures below are carried out at room temperature unless specified differently.
2. Prepare ultrathin cryosections as described elsewhere in this volume, and mount them on formvar-coated EM grids.
3. Rinse sections by floating the EM grids on phosphate-buffered saline (PBS; Ca^{2+}, Mg^{2+} free) containing 10 m*M* glycine (PBS-glycine) with the section side down. Solutions are placed on a Parafilm sheet. The grids can be transferred easily by a wire loop. They must be handled gently so that their upper side (opposite to the section side) do not get wet with splash; once wet with buffer or other solutions, the portion becomes contaminated and unusable for observation. PBS-glycine is used as a vehicle of proteins and antibodies up to step 8.
4. Pretreat with 3% bovine serum albumin, 2% gelatin, 5% normal serum (of animals used to raise the secondary antibody), or a combination of the proteins, for 10 min.
5. Rinse briefly by transferring the grids through 2 to 4 droplets of PBS-glycine.
6. Incubate with 10 to 30 μg/ml primary antibody solution for 30 to 60 min. Antibody solution as little as 10 μl suffices for a grid. The grids must be placed in a humid chamber during the incubation. To minimize carry-over from the foregoing rinse buffer to the antibody solution, transfer grids with anticapillary forceps at this step and at step 8.
7. Rinse for 10 min by transferring the grids through at least 7 droplets of PBS-glycine.
8. Incubate with appropriately diluted ferritin-conjugated secondary antibody for 30 to 60 min at room temperature or at 37°C. The optimal dilution is variable among different preparations and must be determined by trial and error.
9. Rinse as in step 7, but use PBS without glycine because glycine will react with glutaraldehyde in the next step.
10. Fix with 2% glutaraldehyde in 0.1 *M* sodium phosphate buffer, pH 7.4, for 10 min.
11. Rinse with 6 drops of distilled water for 3 min.
12. Treat with 2% neutral uranyl acetate in the dark for adsorptive staining for 10 min.
13. Rinse briefly with 4 drops of distilled water, each for a few seconds.

14. Place on a mixture of 0.2% methyl cellulose, 2% polyethylene glycol, and 0.01 to 0.02% acidic uranyl acetate in the dark for 10 min. The mixture of methyl cellulose and polyethylene glycol can be kept at 4°C for several weeks. Uranyl acetate must be added just before use. Sections should not be left in the mixture for long; severe precipitation can occur.
15. Pick up grids with wire loops and wick off excess amount of solution by touching filter paper to the loop. The thickness of the embedding is critical for preservation of ultrastructure; the interference color of gold to purple is appropriate.
16. Dry and observe under an electron microscope.

To use the biotin-streptavidin system for visualization, step 8 is to be done as described below:[8]

8a. Incubate with 10 to 30 μg/ml biotinylated secondary antibody for 30 to 60 min at room temperature or at 37°C.
8b. Rinse as in step 7.
8c. Incubate with an appropriate dilution of streptavidin-ferritin for 30 to 60 min at room temperature or at 37°C.

Double labeling is possible by using primary antibodies raised in two different animal species, e.g., rabbit and guinea pig, rabbit and mouse, etc. As a marker for secondary antibodies, colloidal gold of 10 nm or larger[9] and Imposil (iron-dextran),[10] which are clearly distinguishable from ferritin, may be used. An example of the double-labeling procedure is shown below. Steps 6 through 8 are done as follows.

6. Incubate with a mixture of rabbit anti-antigen I antibody and guinea pig anti-antigen II antibody, each at a concentration of 10 to 30 μg/ml, for 30 to 60 min.
7. Rinse for 10 min by transferring the grids through at least 7 droplets of PBS-glycine.
8. Incubate with a mixture of ferritin-conjugated goat anti-rabbit IgG antibody and colloidal gold-conjugated goat anti-guinea pig IgG antibody for 30 to 60 min at room temperature or at 37°C. Cross-reaction between the secondary antibodies must be minimal; it has to be checked beforehand that goat anti-rabbit IgG and anti-guinea pig IgG antibodies do not recognize guinea pig IgG and rabbit IgG, respectively.

4. Results

Ferritin molecules are seen as discrete particles on ultrathin frozen sections embedded in a mixture of methyl cellulose and polyethylene glycol. The density of the particles can be maximized by taking electron micrographs slightly off the right focus. Since the electron density of ferritin is lower than that of colloidal gold, it is not compatible with methods giving higher contrast to cellular structures. For example, ferritin becomes less visible if too much uranyl acetate is added in the last embedding step. Embedding in polyvinyl alcohol or LR White gives much higher contrast than the methyl cellulose method and thus is not appropriate for the ferritin method. Another precaution with the ferritin method is that some cells and tissues (e.g., liver, spleen, etc.) contain innate ferritin, which can be confused with the marker.

Labeling can be enhanced by the use of biotinylated secondary antibody and streptavidin-ferritin (Figure 1). Because of the three layers of reagents binding to each antigenic site, ferritin tends to form small clusters. By the double immunolabeling procedure, relative localizations of two antigens can be determined clearly (Figure 2).

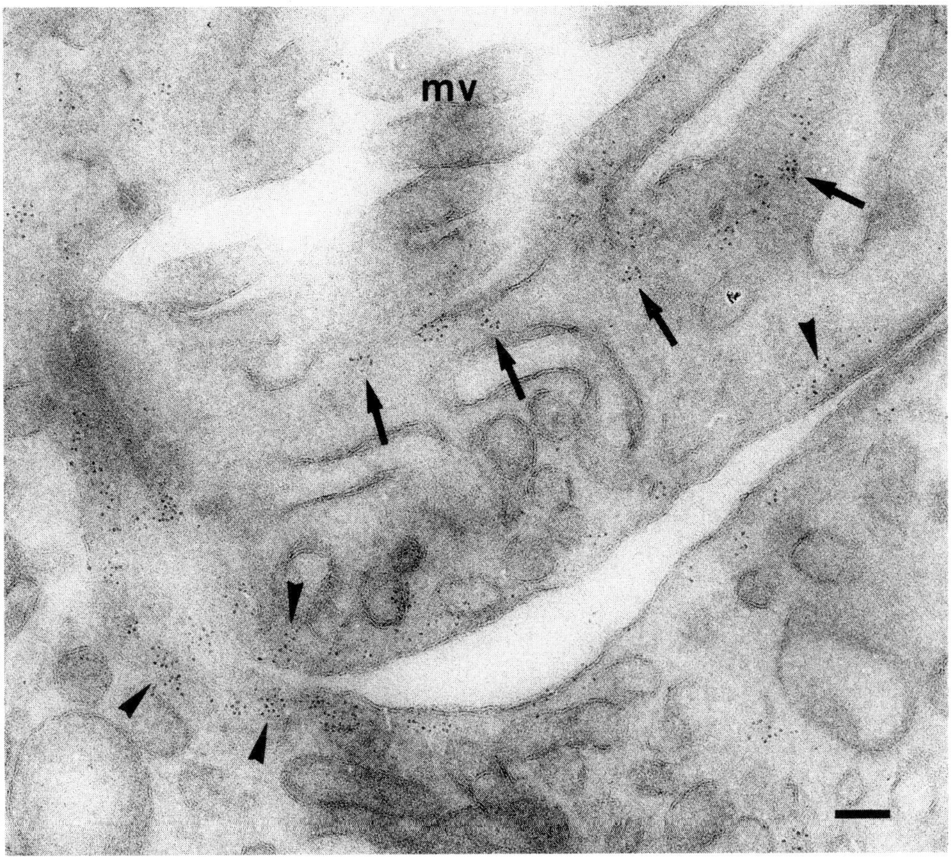

FIGURE 1. Localization of fodrin in an ultrathin frozen section of renal proximal tubule epithelium of the rat by the combination of biotinyl secondary antibody and streptavidin-ferritin. Fodrin is localized along the plasma membrane (arrowheads) as well as in the terminal web. Note that ferritin particles are mostly observed as small clusters when this procedure is used; mv, microvilli. (Bar, 100 nm; magnification × 74,000.)

C. IMMUNOLABELING OF ISOLATED CELLS WITH THE FERRITIN BRIDGE TECHNIQUE

This procedure is based on two technical improvements. First, the cells must be fixed so as to allow penetration of antibodies while still preserving adequate morphology. This is possible by utilizing a combination fixative, the reactivity of which is optimized by precise timing.[11] Saponin is generally used to make the cell membrane permeable. Second, immunolabeling is performed by four steps, three unlabeled antibodies and free ferritin, for better penetration and maximum enhancement.[12] A scheme of the immunolabeling procedure is shown in Figure 3.

1. Equipment and Reagents

Small glass coverslips, 9 × 9 mm, are convenient for culturing cells for immunolabeling. A drop of 20 μl of antibodies is sufficient for labeling cells on a coverslip. Coverslips can be placed in 24-well culture plates for culture, fixation, and rinses.

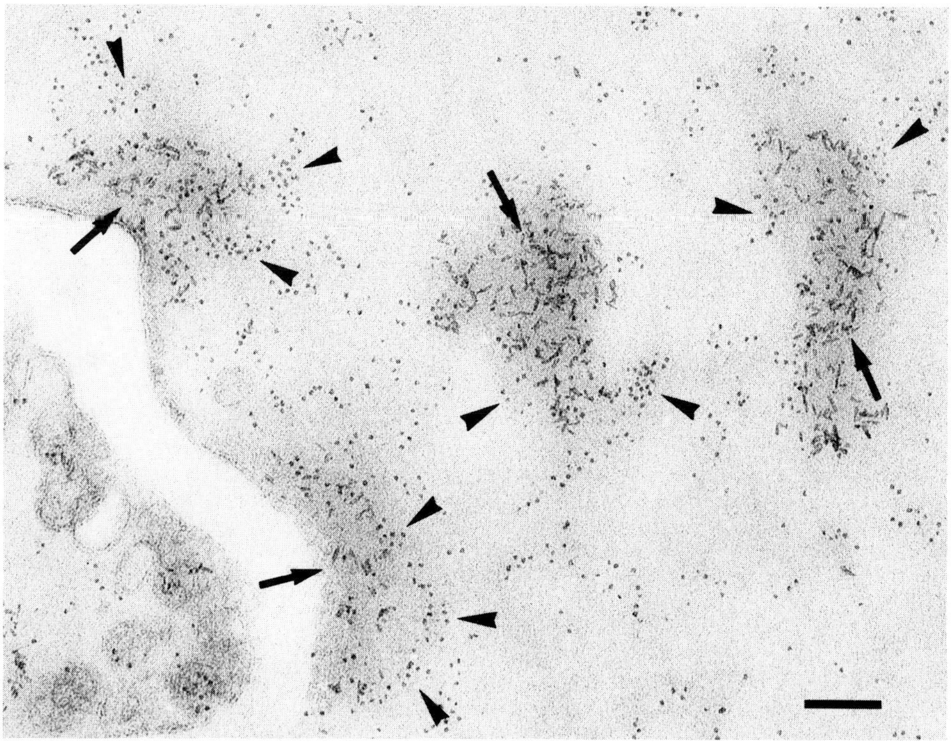

FIGURE 2. Double immunolabeling of α-actinin and filamin in an ultrathin frozen section of the chicken aortic smooth muscle. α-Actinin and filamin are marked with Imposil (elongated shape; arrows) and ferritin (round shape; arrowheads), respectively. While α-actinin is concentrated in the center of dense regions, filamin is mostly localized in its periphery. (Bar, 100 nm; magnification × 105,000.)

Chemicals and reagents include:

1. Ethyldimethylaminopropyl carbodiimide (EDC) — water-soluble and crosslinks proteins through adjacent carboxyl and amino groups
2. Unlabeled secondary antibody
3. Anti-ferritin antibody — it must be raised in the same species as the primary antibody; rabbit antibody is available from Sigma
4. Ferritin (ferritin obtained from horse spleen is available from various companies)

2. Procedures

1. Prepare the fixative solution by gently mixing solutions A and B *precisely* for 4 min at 23°C. While mixing, adjust the pH to 7.0 with 1 *N* NaOH. The time must be exact since the reaction proceeds in the fixative solution. Premixing for 4 min was found an optimal time for morphology and antibody penetration. *Solution A*: mix equal volumes of 100 m*M* sodium phosphate buffer, pH 7.0, and Ca^{2+}, Mg^{2+}-free PBS, add solid Tris base to be 1.4%, and adjust the pH to 7.0 with concentrated HCl. *Solution B*: 0.3% glutaraldehyde and 1.0% EDC in distilled water. The concentration of glutaraldehyde can be varied depending on cell types and antigens studied.
2. Rinse cells with PBS, and fix them with the fixative solution for 7 min at 23°C.
3. Rinse several times with PBS-glycine.

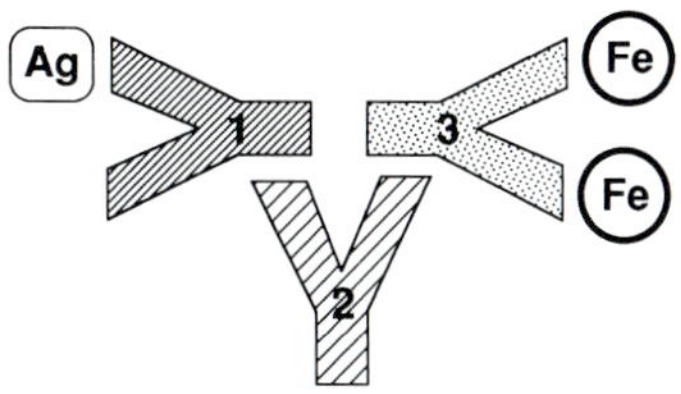

FIGURE 3. A scheme showing the principle of the ferritin bridge method. Ag, Antigen to be detected; Fe, ferritin; 1, primary rabbit antibody; 2, secondary goat anti-rabbit IgG antibody; 3, rabbit anti-ferritin antibody.

4. Permeabilize and pretreat with 0.02% saponin and 5 mg/ml normal goat globulin (NGG) in PBS for 10 to 30 min at room temperature (subsequent procedures are carried out at room temperature unless specified differently). This same solution (SGP) is used for dilution of antibodies and rinses.
5. Rinse several times with SGP.
6. Incubate with 50 μg/ml primary rabbit antibody or preimmune rabbit globulin (for control) for 60 min. After absorbing excess solution, cells on a coverslip can be inverted on a small droplet of the antibody placed on a flat surface of Parafilm.
7. Rinse well with SGP.
8. Incubate with 1 mg/ml goat anti-rabbit IgG antibody for 60 min.
9. The same as step 6.
10. Incubate with 100 μg/ml rabbit anti-ferritin antibody for 60 min.
11. The same as step 6.
12. Incubate with 100 μg/ml horse spleen ferritin for 60 min.
13. Rinse several times with PBS.
14. Postfix with 3% glutaraldehyde in PBS for 15 min.
15. Rinse well with PBS.
16. Osmicate with 1.5% OsO_4 in PBS for 30 min.
17. Dehydrate with an ethanol series and embed in epoxy resin. Coverslips are inverted on a mold filled with the resin and placed in an oven for polymerization.
18. Coverslips can be detached from the epon block by plunging them into liquid nitrogen and then into hot water.
19. Prepare ultrathin sections and counterstain them with uranyl acetate or with lead citrate followed by bismuth subnitrate.

3. Results

The morphology of cells fixed with the above protocol is acceptable for most purposes. The method has been most successfully used to localize cytoskeletal proteins (Figure 4). Membrane structures are perturbed because of the saponin treatment, but are still good enough to allow distinct identification of various organelles. For preservation of membranes, it is recommended to transfer samples directly from 100% ethanol to 100% Epon instead of using a gradual transition.[11]

ACKNOWLEDGMENT

The author is grateful to Dr. Mark C. Willingham for kindly providing electron micrographs for the ferritin bridge method.

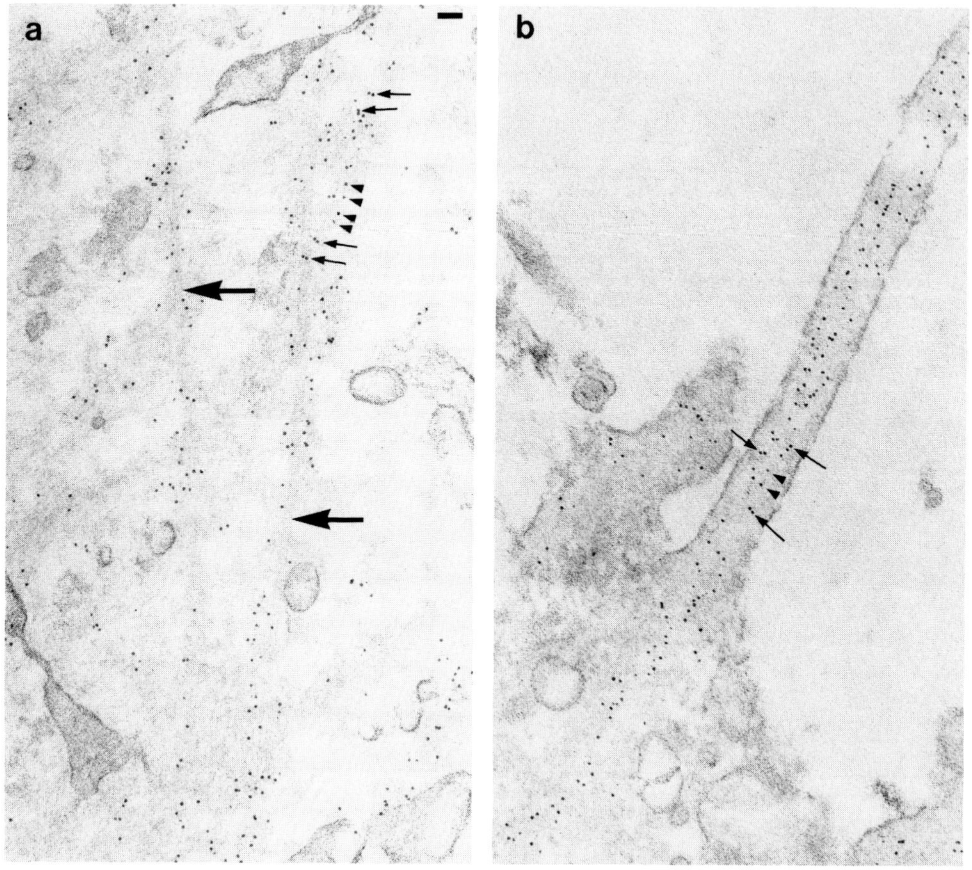

FIGURE 4. Ultrastructural localization of tubulin and actin in Swiss 3T3 cells by the ferritin bridge method. (a) Tubulin. Two microtubules (large arrows) are surrounded by ferritin particles, many of which appear as doublets (small arrows) characteristic of the attachment of two ferritin molecules to divalent anti-ferritin antibody. (b) Actin. The core microfilaments of a surface microvillus is shown labeled positively. (Bar, 60 nm; magnifications: a, × 54,000, b, × 60,600. Photographs courtesy of Dr. M. C. Willingham.)

REFERENCES

1. **Singer, S. J.,** Preparation of an electron-dense antibody conjugate, *Nature (London),* 183, 1523, 1959.
2. **Singer, S. J. and Schick, A. F.,** The properties of specific stains for electron microscopy prepared by conjugation of antibody molecules with ferritin, *J. Biophys. Biochem. Cytol.,* 9, 519, 1961.
3. **Tanaka, H.,** The ferritin-labeled antibody method: its advantages and disadvantages. A methodological review, *Acta Haematol. Jpn.,* 31, 125, 1968.
4. **Otto, H., Takamiya, H., and Vogt, A.,** A two stage method for cross-linking antibody globulin to ferritin by glutaraldehyde. Comparison between the one-stage and the two-stage method, *J. Immunol. Meth.,* 3, 137, 1973.
5. **Kishida, Y., Olsen, B. R., Berg, R. A., and Prockop, D. J.,** Two improved methods for preparing ferritin-protein conjugates for electron microscopy, *J. Cell Biol.,* 64, 331, 1975.
6. **Tokuyasu, K. T. and Singer, S. J.,** Improved procedures for immunoferritin labeling of ultrathin frozen sections, *J. Cell Biol.,* 71, 894, 1976.

7. **Tokuyasu, K. T.,** Immunochemistry of ultrathin frozen sections, *Histochem. J.*, 12, 381, 1980.
8. **Fujimoto, T. and Ogawa, K.,** Immunoelectron microscopy of fodrin in the rat uriniferous and collecting tubular epithelium, *J. Histochem. Cytochem.*, 37, 1345, 1989.
9. **Fujimoto, T. and Ogawa, K.,** Retrieving vesicles in secretion-induced rat chromaffin cells contain fodrin, *J. Histochem. Cytochem.*, 37, 1589, 1989.
10. **Dutton, A., Tokuyasu, K. T., and Singer, S. J.,** Iron-dextran antibody conjugates. General method for simultaneous staining of two components in high resolution immunoelectron microscopy, *Proc. Natl. Acad. Sci. U.S.A.*, 76, 3392, 1979.
11. **Willingham, M. C. and Yamada, S. S.,** Development of a new primary fixative for electron microscopic immunocytochemical localization of intracellular antigens in cultured cells, *J. Histochem. Cytochem.*, 27, 947, 1979.
12. **Willingham, M. C.,** Electron microscopic immunocytochemical localization of intracellular antigens in cultured cells: the EGS and ferritin bridge procedures, *Histochem. J.*, 12, 419, 1980.

Chapter 2

ELECTRON MICROSCOPIC IMMUNOCYTOCHEMISTRY

2.3 Major Visualization Techniques

2.3.2 Immunoelectron Microscopy by Enzyme-Labeled Antibodies

Paul K. Nakane

TABLE OF CONTENTS

0-8493-6012-9/93/$0.00 + $.50

A. INTRODUCTION

Horseradish peroxidase (HRP)-labeled antibodies were introduced in 1966,[1] using 3,3′-diaminobenzidine (DAB), a benzidine derivative, as substrate.[2] The many advantages of the immunoperoxidase technique in light microscopy (increased sensitivity, permanent staining, bright-field visualization), as well as its adaptability to electron microscopy, have rapidly established it as the method of choice in much immunohistochemical investigation.

For the localization of antigens within cells and tissues at the ultrastructural level using the immunoperoxidase method, two approaches have been employed. One is to perform immunochemical reactions first in fixed whole cells or in thick sections of tissues, embed them in plastic polymers, obtain ultrathin sections from the plastic blocks, and examine the sections with an electron microscope (preembedding method).[3] The other is to embed fixed cells and tissues in plastic polymers first, obtain ultrathin sections from the plastic block, perform the immunochemical reaction on the surface of the ultrathin sections, and examine the section with an electron microscope (postembedding method).[4] Each approach has its own demerits and merits. The main demerits are that with the preembedding method, one must assure that antibodies gain access to all immunoreactive antigens within cells and tissue sections; and with the postembedding method, one must assure that the plastic embedding does not interfere with the immunoreaction between antibodies and immunoreactive antigens on the surface of ultrathin sections. The merits are that with the preembedding method, one can gain a broad perspective on the distribution of antigens at the light microscopic level prior to observation at the ultrastructural level and antigens which will not withstand the harshness of the plastic polymerization reaction can be localized at the ultrastructural level; and with the postembedding method, one needs not be concerned with the size of labeled antibodies.

In this chapter, preembedding steps using peroxidase-labeled antibodies are outlined in the order of use. At each step, alternative modifications and some needed comments are inserted.

B. PROCEDURE

Step 1: Specimens

Antigens in most cells and tissues may be localized at the ultrastructural level by the enzyme-labeled antibody method. The procedure for specimen preparation varies, depending upon the type of materials examined. For preparation of biological materials the following steps are recommended:

a. Cells in suspension, e.g., blood cells and cultured cells in suspension. Deposit cells onto albumin-coated microscope slides* using a cytocentrifuge at 900 rpm for 5 min and air dry for 10 to 30 min depending upon the relative humidity of the room. (Fixation may be done during or after the centrifugation.) Go to Step 2a.

b. Cells attached to smooth surface, e.g., monolayer of cells in culture. It is recommended to culture cells in plastic wells adhered to glass slide such as LAB-TEK chamber. Go to Step 2b.

c. Excised tissues, e.g., surgical and biopsy specimens. Go to Step 2c.

d. Tissues *in situ*, e.g., organs of experimental animals. Go to Step 2d.

* Preparation of egg albumin-coated glass slides:[5] (1) clean glass microscope slides in 1% HCl-70% ethanol by immersion and agitation for several minutes; (2) dip slides in 100% acetone and allow to air dry; (3) combine one egg white and 1 ml concentrated (28%) NH_4OH with 500 ml distilled water, stir 10 minutes, and filter through a paper towel or six layers of gauze; (4) place cleaned slides in racks and dip into the above freshly made egg albumin solution for 1 minute. Drain off excess solution on a paper towel and dry in racks at 60°C for 2 hours to overnight. Up to 500 slides can be coated from the above egg mixture and they can be kept indefinitely.

Step 2: Fixation

Ideally, a fixative should preserve both antigenicity and morphology. The best fixative for a particular system is chosen empirically, considering the properties of the antigen and tissue as well as the resolution desired. The mildest fixatives, such as acetone and ethanol, are excellent for preserving antigenicity, but tissue morphology is little more than adequately preserved and is not suitable for electron microscopy. Formaldehyde solutions and picric acid containing Zamboni's solution[6] are superior in preserving morphology and can be used for electron microscopic preparations for certain antigens. However, conventional electron microscopic fixatives such as glutaraldehyde and OsO_4 are usually not suitable because they either interact strongly with proteins or oxidize proteins, respectively, and frequently denature protein antigens. A fixative designed for immunoelectron microscopy, periodate-lysine paraformaldehyde[7] (PLP) on the other hand, primarily stabilizes carbohydrate moieties with adequate preservation of most antigenicity and ultrastructure.

Preparation of PLP fixative is as follows. Stock solution A, 0.1 *M* lysine in 0.05 *M* sodium phosphate, pH 7.4:

1. Dissolve 1.827 g L-lysine HCl (Sigma) in 50 ml H_2O (0.2 *M* lysine HCl)
2. Adjust pH to 7.4 with 0.1 *M* $NaHPO_4$
3. Make up to 100 ml with 0.1 *M* sodium phosphate buffer, pH 7.4; osmolality should be approximately 300 mOs
4. This solution can be stored at 4°C for a maximum of 10 days

Stock solution B, 8% paraformaldehyde:

1. Mix 8 g paraformaldehyde (Matheson, Coleman, Bell) in 100 ml H_2O
2. Heat to 60°C with stirring
3. Slowly add 1 to 3 drops of 1 *N* NaOH until the solution clears
4. Filter through paper filter Whatman #2
5. Store at 4°C

Just before use, combine three parts A with one part B and add sodium *m*-periodate to 0.01 *M* (i.e., 21.4 mg $NaIO_4$/10 ml). Final composition of the fixative therefore is 0.01 *M* $NaIO_4$, 0.075 *M* lysine, 0.0375 *M* sodium phosphate buffer, and 2% paraformaldehyde. The final pH will be approximately 6.2 and osmolality around 700 mOs.

a. Fixation of cells on glass slides. Immerse the slide in PLP at 4°C for 10 min. Stop the fixation by washing in 0.005 *M* sodium phosphate buffer, pH 7.2, containing 10% sucrose (PB-sucrose). The fixation may be stopped also by PB-sucrose containing 4×10^{-5} *M* digitonin (Sigma Chemicals) to further improve membrane fixation. Digitonin may be dissolved by heating slightly or overnight at 4°C on a shaker. Go to Step 4.

b. Fixation of monolayer of cultured cells. Discard culture medium and immerse the slide in PLP at 4°C for 10 min. Stop the fixation by washing in PB-sucrose solution. The fixation may be stopped also by PB-sucrose containing 4×10^{-5} *M* digitonin (Sigma Chemicals) to improve membrane fixation (see Step 2a). Place the slide in a freezer (−20°C) overnight. Bring the slide to room temperature and wash with PB-sucrose. This freeze and thaw step is necessary for good penetration of labeled antibodies. Go to Step 4.

c. Fixation of tissue blocks by immersion in fixative. Slice tissues into thicknesses of less than 2 mm. Immerse the tissue slices in PLP at 4°C for 4 h under gentle shaking. Stop the fixation by washing in PB-sucrose. The fixation may be stopped also by PB-10% sucrose containing 4×10^{-5} *M* digitonin to improve membrane fixation (see Step 2a). Go to Step 3.

d. Fixation by perfusion for small animals (e.g., rats):

1. Lightly anesthetize experimental animals with ether.
2. Expose the thoracic cavity and abdominal viscera by bilateral incisions.
3. Insert an 18-gauge angiocath into the descending thoracic aorta and cut open the right atrium to allow outflow of blood.
4. Inject through the angiocath with 100 units heparin first then perfuse with 100 ml of balanced salt solution (5.6 m*M* dextrose; 0.4 m*M* potassium phosphate, monobasic; 1.3 m*M* sodium phosphate, dibasic, 7-hydrate; 0.03 m*M* phenol red; 1.4 m*M* calcium chloride, 2-hydrate; 5.4 m*M* potassium chloride; 0.14 *M* sodium chloride; 1.1 m*M* magnesium chloride; and 0.8 m*M* magnesium sulfate, 7-hydrate).
5. Once completely blanched and cleared of blood, perfuse with 150 ml of PLP or other suitable fixatives.
6. Excise desired tissue or organ, cut it into 2- to 3-mm cubes, and immerse in PLP at 4°C with gentle shaking for an additional 3 to 4 h. Wash the tissues overnight by two changes of PB-sucrose. Go to Step 3.

Step 3: Frozen Sections of Fixed Tissues

1. Wash in PB-sucrose.
2. Wash in 0.005 *M* phosphate buffer, pH 7.2, containing 15% sucrose for 4 h to overnight.
3. Wash in 0.005 *M* phosphate buffer, pH 7.2, containing 20% sucrose for 4 h.
4. Wash in 0.005 *M* phosphate buffer, pH 7.2, containing 20% sucrose and 5% glycerol for 1 h or wash in changes of 1% gum arabic in 30% sucrose solution.[8]
5. Embed in O.C.T. embedding medium (Ames) and quick-freeze in a mixture of dry ice and ethanol. We shape a piece of heavy aluminum foil into a boat over the end of a thick pencil, about 1 cm in diameter, place the tissue inside, fill with O.C.T., and freeze, being very careful not to let the ethanol seep into the boat.
6. Section the frozen tissue to about 6 μm in thickness in a cryostat. Place the frozen sections onto albumin-coated slides (see Step 1a). Sectioning will be somewhat difficult, but the high sucrose-glycerol content is required for good ultrastructural preservation.
7. Air dry for at least 30 min and wash in PB-sucrose.
8. Wash in three changes of PB-sucrose for 10 min each. Go to Step 4.

Step 4: Moist Chamber

Make a moist chamber. We usually use a clear cake dish or Petri dish as the moist chamber. Place a sheet of filter paper inside the dish and spray PB-sucrose to moisten the filter paper, then place the dish upside down onto a sheet of moist filter paper on a table. This way one can avoid accidental shaking of the moist chamber. Go to Step 5.

Step 5: First Antibody Application

Dry off excess PB-sucrose from around the section with gauze. With a capillary pipet apply about 20 μl of antibody in PB-sucrose. Mix the antibody solution thoroughly with residual PB-sucrose over tissue, being careful not to scratch the section. Place in moist chamber for 3 to 4 h. *Keep chamber moistened for the duration.* To assure the complete penetration of antibody through the thickness of tissue sections, the antibody should be either Fab or Fab′ fragments. If horseradish peroxidase-conjugated antibody (HRP-AB) is used, the antibody also should be either Fab or Fab′. To reduce the time required and damage to the tissues, apply HRP-AB at this step when the conjugate is at hand. Recently, varieties of antibodies conjugated with HRP are available from commercial sources, but when it is necessary to conjugate HRP, please refer to the method introduced by Wilson and Nakane.[9] If the HRP-AB is used, go to Step 9, otherwise go to Step 7.

Step 7: PB-Sucrose Wash

Rinse antibody solution off the slides with PB-sucrose. Wash in three changes of PB-sucrose for 10 min each. Go to Step 8.

Step 8: Second Antibody Application

Dry off excess PB-sucrose from around section with gauze. With a capillary pipet apply about 20 μl of HRP-AB in PB-sucrose which is against the IgG of animal from which the antibody used in Step 6 was obtained. The antibody in HRP-AB should be either Fab or Fab′. Mix HRP-AB solution thoroughly with residual PB-sucrose over tissue as in Step 6, again being careful not to scratch the section. Place the slide in a moist chamber for 3 to 4 h. *Keep chamber moistened for the duration.* Go to Step 9.

Step 9: PB-Sucrose Wash

Rinse HRP-AB solution off the slides with PB-sucrose. Wash in three changes of PB-sucrose for 10 min each. Go to Step 10.

Step 10: Postfixation

Incubate the slide in 2% glutaraldehyde in 0.1 *M* sodium phosphate buffer, pH 7.2, for 30 min, or with a capillary pipet apply about 40 μl of 2% glutaraldehyde in 0.1 *M* sodium phosphate as the HRP-AB was applied in Step 8, and place the slide in the moist chamber for 30 min. This step is to secure the antibody firmly at the site of antigens.

Step 11: PB-Sucrose Wash

Rinse antibody solution off the slides with PB-sucrose. Wash in three changes of PB-sucrose for 10 min each. Go to Step 12.

Step 12: DAB Penetration

Incubate the slide in DAB solution (see below) for 30 min. Note that this solution does not contain H_2O_2. This solution is used to assure even penetration of DAB throughout the thickness of the section, otherwise HRP near the surface of the tissue section will be stained darker than that deep in the section. Go to Step 13.

Preparation of DAB solution (adapted from Graham and Karnovsky[2]):

1. Dissolve 25 mg 3,3′ diaminobenzidine-4 HCl in 100 ml 0.05 *M* Tris-HCl buffer, pH 7.6, containing 10% sucrose. Filter if not completely colorless.
2. Adjust the pH to 7.6 with 0.05 *M* Tris-base or with HCl if necessary.

Step 13: DAB Reaction

Without washing off the DAB solution from Step 12, incubate the section in DAB solution containing H_2O_2* for 2 min. Color will develop, but the intensity is somewhat

* Preparation of DAB solution with H_2O_2: to 50 ml of the DAB solution made up in Step 12, add 0.05 ml 5% H_2O_2.

lighter than that experienced by the light microscopic immunoperoxidase histochemistry. If the color is developed longer, one will experience diffusion of reaction product from the site of antigen.

Step 14: PB-Sucrose Wash

Rinse the slides with PB-sucrose. Wash in three changes of PB-sucrose for 10 min each. At this step, the slide may be examined with a light microscope to see the general distribution of antigens. Do not let the slide dry while examining it. Go to Step 15.

Step 15: Osmication

Apply a drop of 2% OsO_4 in 1 *M* sodium phosphate buffer, pH 7.2, as in Step 13 and incubate in a moist chamber for 1 h. At this step, the DAB reaction product reduces and chelates OsO_4, leaving an electron-dense, insoluble precipitate at the site of enzyme action. Go to Step 16.

Step 16: PB-Sucrose Wash

Rinse the slides with PB-sucrose. Wash in three changes of PB-sucrose for 10 min each. Go to Step 17.

Step 17: Dehydration

Dehydrate the slides in 70, 80, 90, and 95% ethanol for 5 min each and three changes of 100% ethanol for 5 min each. Note that propylene oxide is not used for clearing, so that the 100% ethanol should be absolutely free of water. Go to Step 18.

Step 18: Embedding

Fill gelatin capsules (No. 0, Eli Lilly, Indianapolis, IN) with fresh Epon 812 with catalyst. Flip and place the Epon-filled capsule over each section on glass slide. There is no need for concern if small air bubbles are present, they will migrate upward slowly; however, too much air will push Epon out of the capsules when they are warmed. While keeping the capsules upside down on the slide, we normally place them on a sheet of glass plate. Go to Step 19.

Step 19: Polymerization

Place the glass plate with slide in a 37°C oven. Make sure the plate is horizontal, otherwise the Epon-filled capsules will migrate slowly on the glass slide toward the lower side. Polymerize the Epon by incubating at 37°C overnight and then at 60°C for 24 to 48 h. Go to Step 20.

Step 20: Capsule Removal

The polymerized Epon-filled capsule may be removed from the slide by heating. We usually heat the block briefly over a bunsen burner flame and push or pull the capsule sideways. With this maneuver, the tissue section is now embedded in Epon. If the tissue section is left on the glass slide, it is an indication that the absolute ethanol used in Step 17 contained water. Go to Step 21.

Step 21: Trimming and Sectioning

The immunostained section in the Epon block may be examined under an ordinal light or dissection microscope. Although the tissues are darkened by osmication in Step 15, immunostained areas usually can be recognized. Trim down the face of the block to desired areas and section by an ultramicrotome. The ultrathin sections should be slightly thicker (gold color) than normal ultrathin sections for easier recognition of oxidized DAB. Go to Step 22.

Step 22: Electron Microscope Observation

Examine the ultrathin sections without counterstaining. Counterstaining with lead and/ or uranium will make the oxidized DAB difficult to recognize.

C. METHODS TO ELIMINATE ENDOGENOUS PEROXIDASE ACTIVITY

Sometimes it is necessary to inactivate endogenous peroxidase activity in blood cells if antigens are thought to be situated in the cells, in order to avoid confusion with the specific immunostaining. For cells and tissues fixed by PLP fixative: (1) to eliminate peroxidase activity in white blood cells, add 0.005 *M* sodium azide (NaN_3) to the DAB solution with H_2O_2[10] at step 13; (2) to eliminate peroxidase activity in white and red blood cells,[11] immerse slide in a solution of 0.01 *M* $NaIO_4$ for 10 min then follow by washing with 0.1 mg/ml $NaBH_4$ solution for 10 min between Steps 4 and 5.

D. OTHER USAGES

More recently, a method to examine an immunoperoxidase-stained light microscopic section on a glass slide under the electron microscope has been introduced.[12] A backscattered electron image (BEI) of osmium which has chelated with the reaction product of HRP was formed by a scanning electron microscope. Specifically, first embed tissues fixed with 4% formaldehyde in 0.1 *M* phosphate buffer, pH 7.4, in JB-4. Tissue sections at 2 μm in thickness are obtained from the JB-4 block. Place the sections on carbon-coated glass slides, partially remove the JB-4 from the sections, and the sections are reacted with antibody followed with HRP-AB. The slides are incubated with DAB and H_2O_2. The DAB reaction product is chelated then with OsO_4 → thiocarbohydrazide → OsO_4. After washing and drying, the sections are observed with BEI mode. The reaction product of HRP can be recognized as black under a light microscope and as amorphous masses with a scanning electron microscope.

Another recent development is the use of HRP-labeled antibody for *in situ* hybridization. This also is carried out by first localizing specific mRNA or DNA by enzyme immunohisto *in situ* hybridization[13] (EIHISH) on cells and sections mounted on glass slides, then examined at the electron microscopic level, as in the preembedding method.[14]

REFERENCES

1. **Nakane, P. K. and Pierce, G. B.,** Enzyme-labeled antibodies: preparation and application for the localization of antigen, *J. Histochem. Cytochem.*, 4, 929, 1966.
2. **Graham, R. C., Jr. and Karnovsky, M. J.,** The early stages of absorption of injected horseradish peroxidase in the proximal tubules of mouse kidney: ultrastructural cytochemistry by a new technique, *J. Histochem. Cytochem.*, 14, 291, 1966.
3. **Martinez-Hernandez, A., Nakane, P. K., and Pierce, G. B.,** Intracellular localization of basement membrane antigen in parietal yolk sac cells, *Am. J. Pathol.*, 76, 549, 1974.
4. **Nakane, P. K.,** Application of peroxidase-labeled antibodies to the intracellular localization of hormones, in *Karolinska Symposia on Research Methods in Reproductive Endocrinology,* 3rd Symposium. In Vitro *Methods in Reproductive Cell Biology,* Acta Endocrinologia Separatum III, 1971, 190.

5. **Pearse, A. G. E.,** *Histochemistry, Theoretical and Applied,* Vol. 2, Little, Brown, Boston, 1972.
6. **Stefanini, M., DeMartino, C., and Zamboni, L.,** Fixation of ejaculated spermatozoa for electron microscopy, *Nature,* 216, 173, 1967.
7. **McLean, I. W. and Nakane, P. K.,** Periodate-lysine-paraformaldehyde fixative: a new fixative for immunoelectron microscopy, *J. Histochem. Cytochem.,* 22, 1077, 1974.
8. **Holt, S. J. and Withers, R. F. J.,** Studies in enzyme cytochemistry. V. An appraisal of indigogenic reactions for esterase localization, *Proc. R. Soc. London Ser. B,* 148, 520, 1958.
9. **Wilson, M. B. and Nakane, P. K.,** Recent development in the periodate method of conjugating horseradish peroxidase (HRPO) to antibodies, in *Immunofluorescence and Related Staining Techniques,* Knapp, W., Holubar, K., and Wick, G., Eds., Elsevier/North-Holland Biochemical Press, Amsterdam, 1978, 215.
10. **Yamada, G. and Nakane, P. K.,** Hepatitis B core and surface antigens in liver tissue: light and electron microscopic localization by the peroxidase-labeled antibody method, *Lab. Invest.,* 36, 649, 1977.
11. **Isobe, Y., Chen, S., Nakane, P. K., and Brown, W. R.,** Studies on translocation of immunoglobulins across intestinal epithelium. I. Improvements in the peroxidase-labeled antibody method for application to study of human intestinal mucosa, *Acta Histochem. Cytochem.,* 10, 161, 1977.
12. **Izumi, S. and Nakane, P. K.,** Immunohistochemical localization of antigens in tissues embedded in JB-4 by backscattered electron imaging utilizing colloidal gold-labeled and peroxidase-labeled antibodies, *Acta Histochem. Cytochem.,* 21, 237, 1988.
13. **Nakane, P. K., Moriuchi, T., Koji, T., Tanno, M., and Abe, K.,** Non-radioactive cDNA probes for in situ hybridization of mRNA, *Acta Paediatr. Jpn.,* 29, 500, 1987.
14. **Nakane, P. K., Matsumura, H., and Koji, T.,** In situ hybridization using T-T dimerized non-radioactive probes, in *Biological Aspects of Brain Tumors,* Tabuchi, K., Ed., Springer-Verlag, Tokyo, 1992, 52.

Chapter 2

ELECTRON MICROSCOPIC IMMUNOCYTOCHEMISTRY

2.3 Major Visualization Techniques

2.3.3 The Use of Avidin-Biotin Interaction in Immunocytochemistry and Immunoelectron Microscopy

Su-Ming Hsu

TABLE OF CONTENTS

0-8493-6012-9/93/$0.00 + $.50

A. INTRODUCTION

Although several immunohistochemical methods have been used successfully for detection of antigens by electron microscopy, these techniques are somewhat limited if antigenic epitopes are destroyed during fixation and/or tissue processing. An additional problem in dealing with immunoelectron microscopy is the accessibility of the antigens to antibodies because of limited penetration of antibodies. Careful selection of staining methods (preembedding vs. postembedding), fixatives, types of antibodies (Fab fragment vs. whole immunoglobulin molecule), source of antibodies (polyclonal vs. monoclonal), and methods of section etching must be taken into consideration for successful staining.

Optimal staining requires (1) that the tissue sections provide sufficient cytologic detail, allowing the determination of the type of cells stained and the distribution and relationship of these cells to nonstained cells and (2) that the antigenic determinants are maximally preserved, so that sensitivity of detection is guaranteed. These two aspects are generally not compatible. A tissue-processing technique that is suitable for morphologic study is generally detrimental to antigen preservation, and vice versa. None of the available tissue preparation methods meet both criteria. A compromise is often needed, and the selection of staining method depends on the aims of the study. The general consideration for immunoelectron microscopy which is discussed in several previous chapters is also applicable to methods employing the avidin-biotin interaction.

Any antibody tagged with biotin can be visualized with an avidin-conjugated or an avidin-biotin-complexed substance.[1-12] The substance can be an enzyme, a fluorescent dye, electron-dense particles (ferritin or colloid gold), or a radioactive isotope. Therefore, the avidin-biotin detection technique can be applied to the following: (1) antibody-antigen reactions (immunocytochemistry, immunohistochemistry, immunofluorescence, flow cytometry, enzyme-linked immunoassays, immunoelectron microscopy, or Western blotting), (2) lectin histochemistry, (3) receptor-ligand reactions, and (4) *in situ* hybridizations with biotin-labeled nucleic acid probes.

In this chapter, we describe the methods based on the avidin-biotin interaction. These methods can be used in cell and tissue preparations at the light microscopic as well as the ultrastructural levels. These methods are very sensitive because of the extraordinary affinity between avidin and biotin.

B. PRINCIPLES OF AVIDIN-BIOTIN TECHNIQUES

The following two avidin-biotin techniques are generally used: the labeled avidin-biotin (LAB) technique and the avidin-biotin complex (ABC) technique. In the LAB method, the biotin-labeled substance is visualized directly by means of an avidin conjugate, such as avidin-colloid gold, avidin-peroxidase, or avidin-ferritin conjugates. Although the use of an avidin-peroxidase conjugate yields a reasonable sensitivity, a major drawback is that the avidin conjugates usually cause a high background reaction as well as nonspecific staining of some tissues, such as the kidney.[4]

In the ABC method, the formation of an ABC, by which the biotin-labeled antibody is visualized, is based on the following biochemical characteristics of avidin and of biotin conjugates:[3]

1. Avidin has four binding sites for biotin conjugates; thus, avidin can form a link with two or more biotin conjugates (e.g., biotin-labeled peroxidase).
2. Several biotin molecules can be coupled to one peroxidase molecule; thus, each biotin-peroxidase conjugate can react with several avidin molecules.
3. With an optimal ratio of avidin to the biotin-peroxidase conjugate, the formation of a complex containing multiple peroxidase molecules is possible.

4. The sensitivity of the detecting system is increased when the number of peroxidase molecules present is increased.

The optimal ratio of avidin to biotin-peroxidase conjugate is determined empirically. The sensitivity is also affected greatly by the degree of biotinylation of the peroxidase. The ABC can be prepared in the laboratory of the user according to a previously published method.[3] In general, the amounts of avidin and biotin-peroxidase conjugate are about 10 and 2.5 μg per milliliter, respectively. It should be noted that the availability of free biotin-binding sites in the complex is created by incubation of excess amounts of avidin with respect to the biotin-peroxidase molecules. The same idea of enhancing sensitivity has also been applied recently to the formation of an avidin-biotin-alkaline phosphatase complex and an avidin-biotin-glucose oxidase complex. However, it has not been used for preparation of avidin-biotin-ferritin or avidin-biotin-colloid gold complexes.

1. Preparation of Biotin-Conjugated Proteins or Nucleic Acid Probes

Biotinylation of immunoglobulins (Igs), lectins, cytokines, or other proteins can be performed by the following procedure:

1. *N*-Hydroxysuccinimide biotin (Sigma) is dissolved in dimethyl sulfoxide (DMSO) at a concentration of 1 mg/ml (solution A).
2. The concentration of protein to be biotin labeled is adjusted to 1 mg/ml in 0.1 *M* $NaHCO_3$, pH 9.0 (solution B).
3. The above two solutions are mixed at a volume ratio of A:B about 1:8 at room temperature for 4 h.
4. The solution is dialyzed against phosphate buffer overnight.
5. The solution can be concentrated further by vacuum dialysis.

The following is a step-by-step description of the method for 3′ end-labeling DNA with biotin-14-dATP.

1. Mix the following components in a 1.5-ml microcentrifuge tube: 12.5 μl of 0.4 m*M* biotin-14-dATP, 2 pmol oligonucleotide, 10 μl of 5X TdT (terminal deoxynucleotidyl transferase) (Bethesda Research Lab., Gaithersburg, MD) reaction buffer. Add autoclaved distilled water up to 49 μl. The 5X TdT reaction buffer contains 500 m*M* potassium cacodylate (pH 7.2), 10 m*M* $CoCl_2$, and 1 m*M* dithiothreitol (DTT).
2. Add 1.0 μl (15 units) TdT. Mix by gentle pipetting.
3. Stop the reaction by placing the tube on ice and adding 10 μl of 0.1 *M* Na_2EDTA (pH 8.0).

2. Immunoelectron Microscopic Method with Cell Suspension

The following is a step-by-step description of the method for detecting surface antigen of cells with our ABC technique for immunoelectron microscopy (preembedding method).[13]

1. Wash cells with culture medium (e.g., RPMI 1640), spin down, and resuspend cells in culture medium. Fix the cells with low concentrations of cold buffered glutaraldehyde or paraformaldehyde for 1 to 5 min, if the target antigens can sustain the fixation. No fixatives should be applied if fixation may destroy the antigenic epitope.
2. Wash cells thoroughly with RPMI medium to remove fixative.
3. Incubate cells with 1% nonimmune or preimmune serum in RPMI medium. The serum used should preferably be derived from the same animal species as that of the secondary antibody.

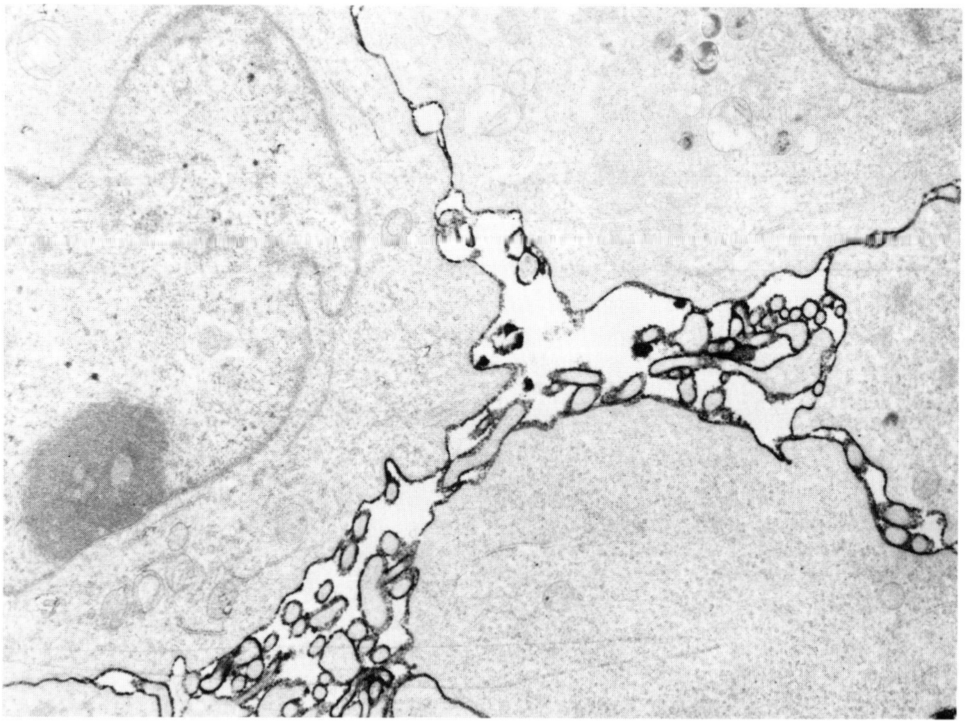

FIGURE 1. Example of immunoelectron microscopic staining in cultured cells. The cells (Reed-Sternberg cells, neoplastic cells of Hodgkin's lymphoma) were stained with anti-Leu M1, and developed with the ABC and DAB substrate.

4. Spin down cells and incubate with primary antibody at a proper dilution (usually 1 to 5 μg/ml) for 30 min to 2 h, depending on the affinity of the antibody.
5. Wash cells in RPMI medium three times.
6. Incubate with biotin-labeled secondary antibody at a dilution of 1 to 200 for 30 min.
7. Wash cells in RPMI medium three times.
8. Incubate cells with the ABC reagent for 1 h. The ABC reagent can be obtained from Vector Laboratories (Burlingame, CA) and prepared as instructed by the manufacturer.
9. Wash cells in RPMI medium three times.
10. Develop the cell-antibody reaction in a substrate such a diaminobenzidine (DAB)-hydrogen peroxide. The substrate is prepared as follows: dissolve 100 mg DAB (Sigma, St. Louis, MO) in 200 ml Tris-buffered saline (TBS); stir and filter; add 1 ml of 8% $NiCl_2$ (aqueous solution) and 25 μl of 3% hydrogen peroxide.[14]
11. Terminate the reaction by washing the cells in RPMI medium.
12. Fix cells with routine fixative (e.g., 1% glutaraldehyde) and spin down the cell pellet.
13. Submit the cell pellet to routine electron microscopic examination.

The above method is suitable for detection of cell surface antigens (Figure 1). For localization of intracytoplasmic antigen, staining methods may need some modifications. For example, light fixation of cells, which preserves the structure but does not destroy antigens, and treatment of cells with mild detergents (e.g., 0.01% Triton X) for increased permeability of antibody, may be applied. The selection of small-sized reagents (e.g., Fab fragments of antibody, or avidin-peroxidase conjugate) may also be helpful.

Use of the peroxidase reaction yields a homogeneous, electron-dense deposition of polymerized DAB at the antigen-antibody reaction sites. The reaction can be amplified by the addition of a heavy metal (e.g., nickel chloride) to the substrate solution without causing background staining.[14] Quantitation of the peroxidase reaction product may be difficult to achieve using the peroxidase method. For quantitation, one may choose to use avidin-ferritin or avidin-colloid gold conjugates, with which the reaction products are discrete particles.

3. Immunoelectron Microscopy of Tissue Sections

Different types of tissue sections, such as cryostat- or Vibratome-cut sections, paraffin sections, or 1-μm-thick sections cut from epon blocks may be used for immunostaining. Sections from fixed tissues preserve the ultrastructural details, but the target antigens may be destroyed. In contrast, the antigens are preserved in cryostat- and Vibratome-cut sections, but the morphologic detail is generally poor. In all types of sections, accessibility of antigens at deep locations may pose a problem for successful detection.

The use of small-sized reagents may be desirable. The biotin-conjugated Fab fragment of antibody is perhaps the smallest reagent, and the binding of this reagent to antigen can subsequently be detected with avidin-peroxidase conjugates. If accessibility of antigens is not the problem, the use of biotin-labeled antibodies followed by the ABC should give the highest sensitivity of detection as compared with that of other staining methods. The staining method to be used for tissue sections is generally the same as that for cell suspensions.

C. ADVANTAGES OF AVIDIN-BIOTIN TECHNIQUES

The use of the avidin-biotin immunocytochemical technique generally offers much greater sensitivity, specificity, and flexibility than is obtainable with the use of other immunoperoxidase techniques.[15]

1. Use of Biotin-Labeled Primary Antibody

For practical purposes, the method of choice for immunohistologic staining is a three-step method in which a biotin-labeled secondary antibody is applied, followed by the ABC or by avidin conjugates. The three-step method does not require direct conjugation of primary antibodies, which sometimes is not feasible because of the small quantities of antibody available. The sensitivity of the three-step ABC method is enhanced by amplification through the primary antibody-secondary antibody reactions. However, the secondary antibody may introduce some degree of nonspecificity or an unwanted reaction. For example, some commercially available anti-mouse Igs have a strong affinity to human epithelial cells. In addition, the three-step ABC method is not suitable for staining of mouse tissue with mouse monoclonal antibodies, or of human tissues with human antibodies, because the secondary antibody (e.g., biotin-labeled rabbit anti-mouse or human Ig) binds not only to the added primary antibody, but also to the interstitial or cytoplasmic Ig in tissue sections.

One can overcome this problem by using the biotin-labeled MAb or biotin-labeled human autoimmune antibody and then ABC, eliminating the need for a secondary antibody. This two-step ABC method is most useful in the detection of Ig in tissue sections, or in a system in which the primary antibody and the tissue sections have the same source. The two-step ABC method is less sensitive than the three-step method but yields much lower background staining.

2. Flexibility

Any protein, carbohydrate, or nucleic acid tagged with biotin can be visualized with an avidin-conjugated or an avidin-biotin-complexed substance. This offers flexibility in visualizing specific target substances in cells. Examples include (1) detection of specific carbohydrate moieties (lectin receptor) by use of a biotin-labeled lectin,[16] (2) detection of cytokine or hormone receptors by treatment of cells with biotin-labeled cytokines or hormones, and (3) localization of specific mRNA in cells by hybridizing of cells with biotin-labeled DNA probes.

The staining methods for detection of lectin, hormone, or cytokine receptors are very similar to that for detection of antigen by antibodies. Because carbohydrates are generally (but not always) resistant to destruction by fixation, ABC-lectin histochemical staining can be applied to formalin- or glutaraldehyde-fixed tissue sections. However, some lectin receptors may nevertheless be destroyed or masked by fixation.[10] Thus, staining results achieved with fixed sections can be quite different from those with unfixed sections or cells.

In contrast to lectin receptors, hormone and cytokine receptors are very labile and are easily destroyed by fixation. Unfixed tissue sections or cells are generally required for successful staining. The cells or tissues can be fixed gently, once the biotin-labeled hormone or cytokine has bound to receptors. The binding between cytokine/hormone and its receptors can then be visualized by the ABC method.

The staining technique involving biotin-labeled probes and *in situ* hybridization is generally more complicated and thus requires special attention. The staining method for the detection of interleukin-5 (IL-5) mRNA in cultured cells is described below.

1. Cytospin smears are prepared on gelatin-coated slides.
2. The smears are fixed in 4% paraformaldehyde in phosphate-buffered saline for 15 min at room temperature, then rinsed three times in 70% alcohol prepared with diethylpyrocarbonate-treated water (0.1% of DEPC in distilled water).
3. The smears are rinsed in 0.5X standard saline citrate (SSC) for 10 min at room temperature.
4. To decrease nonspecific interactions, the smears are treated with 0.25% v/v acetic anhydride prepared in 0.1 *M* triethanolamine (pH 7.2) for 5 min, and then rinsed for 5 min in 2X SSC.
5. The smears are covered with prehybridization solution consisting of 20% deionized formamide, 5X SSC, 10% dextran sulfate, 100 μg/ml of salmon sperm DNA, 100 μg tRNA, 1X Denhardt's, and 10 m*M* dithiothreitol. Smears are prehybridized for 2 h at 42°C.
6. The prehybridization solution is shaken off, and each smear is flooded with 100 μl of hybridization solution containing 1 ng/ml of the biotin-labeled probe. The hybridization solution is the same as the prehybridization solution, except for addition of the probe (biotin-labeled IL-5 oligonucleotide).
7. The smears are coverslipped and incubated overnight at 42°C in a sealed, humidified container.
8. After hybridization, the smears are rinsed once in 2X SSC for 10 min at room temperature, and once in 1X SSC for 60 min at 42°C.
9. The smears are incubated with ABC for 1 h at room temperature.
10. The smears are then developed in DNA-$NiCl_2$-H_2O_2 solution.
11. The slides are washed in TBS, counterstained with methyl green, and dehydrated through graded alcohols and xylene.

3. Double Staining

A combination of the biotin-avidin technique with other immunoenzymatic or immunofluorescence methods can also be used for double-staining purposes. For immunoelectron microscopy, one can detect one antigen by using the biotin-avidin and peroxidase system and at the same time visualize another by using the colloid gold-conjugate antibody method.

This method requires the use of various secondary antibodies which will not cross-react with each primary antibody. The following is a step-by-step method for localization of cytoplasmic Ig and surface CD38 antigen in myeloma (plasma) cells.

1. Wash myeloma cells with culture medium (e.g., RPMI 1640), spin down, and resuspend cells in culture medium containing 1% normal horse serum for 10 min.
2. Spin down cells and incubate with anti-CD38 mouse monoclonal antibody at a dilution of 1 μg/ml for 1 h.
3. Wash cells in RPMI medium three times.
4. Incubate with biotin-labeled horse anti-mouse antibody at a dilution of 1 to 200 for 30 min.
5. Wash cells in RPMI medium three times.
6. Fix cells with 0.4% paraformaldehyde for 5 min, and wash thoroughly with RPMI medium.
7. Treat cells with 0.01% Triton X for 5 min, and wash again.
8. Incubate cells with ferritin-labeled rabbit anti-human Ig and the ABC reagent simultaneously for 1 h.
9. Wash cells in RPMI medium three times.
10. Develop in substrate containing DAB-$NiCl_2$-H_2O_2.
11. Terminate the reaction by washing the cells in RPMI medium.
12. Fix cells with routine fixative (e.g., 1% glutaraldehyde) and spin down the cell pellet.
13. Submit the cell pellet to routine electron microscopic examination.

4. Stability of Avidin and Biotin Reagents

The conjugation of biotin to a protein (e.g., immunoglobulin) is a relatively mild chemical reaction; thus, biotinylation usually does not affect the biological or physiologic affinity of the protein. Therefore, the preparation of biotin-labeled reagents is suitable not only for antibody-antigen reactions, but also for cytokine/hormone-receptor binding assays. Once biotin is conjugated, the biotin-labeled proteins or oligonucleotides are very stable. They maintain the same reactivities for up to 5 years when stored at 4°C.

5. Small Size of Biotin-Labeled Antibodies

Biotin is a small molecule, and thus the biotin-labeled substance maintains a size comparable to that of the original molecule. Because the molecular size of the antibody is very important for its good penetration into tissues (e.g., in immunoelectron microscopy), it may be best to use the biotin-labeled Fab fragments of an antibody.

6. Minimal Background Staining

Because of the high sensitivity of the ABC method, one can use highly diluted primary and secondary antibodies. The use of highly diluted antibodies causes minimal background staining.[15] On some occasions, nonspecific staining may be derived from the unwanted binding of secondary antibody with tissue elements and the binding cannot be prevented by pretreatment of cells or tissue sections with nonimmune serum. In this case, one can pretreat the cells or tissue sections with a nonconjugated secondary antibody, which will saturate the nonspecific binding sites. The subsequent reaction with the ABC reagent recognizes only the biotin-labeled secondary antibody, which binds only to the primary antibody, and does not react with nonspecific binding sites.

D. PITFALLS IN THE USE OF AVIDIN-BIOTIN STAINING TECHNIQUES

General nonspecific background staining — Background staining seen in immunohistochemical stains with the ABC reagent can be blocked effectively by incubation of cells or tissue sections with 1% nonimmune or preimmune serum in TBS. The serum used should preferably be derived from the same animal species as that of the secondary antibody. In other staining systems (e.g., Western blotting or enzyme-linked immunosorbent assay [ELISA]), the background staining of antibody and ABC can be minimized effectively by preincubation of the cellulose paper or plastic wells with 1% solution of dry, mild powder and/or 0.01% Triton X.

Endogenous biotin — Biotin is a coenzyme of decarboxylase. It is present in many tissues, such as the liver, pancreas, kidney, and intestine. In fixed tissues, the biotin-containing proteins are likely to be destroyed by the fixation or embedding procedures; thus, the use of ABC does not produce unwanted binding to tissues or cells. In unfixed sections of liver, kidney, or pancreas, and in some cultured cells, the sections or cells have to be pretreated with avidin (25 μg/ml) for 15 min before biotin is added at a saturating concentration for 15 min, with an interval for washing in buffer. This pretreatment with avidin can successfully block endogenous biotin molecules.[17] Because avidin is a tetramer which has four binding sites, the sections require additional biotin treatment for saturation for all free biotin-binding sites associated with avidin, so that the added avidin will not react with biotin-labeled antibody or ABC. Biotin, when bound to avidin, will not react with a second avidin molecule.

Endogenous peroxidase — Several types of cells, including red blood cells, neutrophils, and eosinophils, contain peroxidase. One has to distinguish the staining due to added peroxidase (i.e., in the ABC) from that due to endogenous peroxidase. If the presence of an endogenous peroxidase reaction does not affect the accuracy of interpretation, blocking of endogenous peroxidase is not necessary. The endogenous peroxidase activity can be blocked by pretreatment of the sections or cells with 3% hydrogen peroxide in absolute methanol for 30 min. This treatment does not affect the preservation of antigens, if such antigens already escape destruction by the fixation and embedding processes.

Nonspecific binding with mast cells — ABC can react with mast cells through an unknown mechanism, and the reaction cannot be prevented by avidin-biotin pretreatment. This unwanted reaction presents a problem when antigen is localized in mast cells. One can prepare the ABC in a high-pH (8.5) solution or in high-ionic-strength buffer such as in 0.5 *M* NaCl. High-pH or high-ionic-strength solutions will not affect the ABC-biotin reaction, but can prevent or minimize unwanted binding to mast cells.

E. SUMMARY

The use of the ABC technique, in which an avidin-biotin interaction is used for immunohistochemical studies, is a simple and straightforward procedure. All required reagents are commercially available. The main advantages of this technique are its sensitivity, specificity, and flexibility. It is crucial for the success of this staining procedure that the antibodies, the types of tissues, and the fixatives that are used be selected properly. The method can easily be adapted for electron microscopy, flow cytometry, *in situ* hybridization, Western blotting, and ELISA.

REFERENCES

1. **Hsu, S.-M., Raine, L., and Fanger, H.,** A comparative study of the peroxidase-anti-peroxidase method and an avidin-biotin complex method for studying polypeptide hormones with radio-immunoassay antibodies, *Am. J. Clin. Pathol.,* 75, 734, 1981.
2. **Hsu, S.-M., Raine, L., and Fanger, H.,** The use of anti-avidin antibody and avidin-biotin-peroxidase complex in immunoperoxidase techniques, *Am. J. Clin. Pathol.,* 75, 816, 1981.
3. **Hsu, S.-M., Raine, L., and Fanger, H.,** Use of avidin-biotin-peroxidase complex (ABC) in immunoperoxidase techniques: a comparison between ABC and unlabeled antibody (PAP) procedures, *J. Histochem. Cytochem.,* 29, 577, 1981.
4. **Hsu, S.-M. and Raine, L.,** Use of avidin-biotin-peroxidase complex (ABC) in pathology, in *Advances in Immunohistochemistry,* DeLellis, R. A., Ed., Masson, Paris, 1984, 31.
5. **Hsu, S.-M.,** Immunoperoxidase techniques using the avidin-biotin system, in *Enzyme Mediated Immunoassays,* Ngo, T. T. and Lenhoff, H. M., Eds., Plenum Press, New York, 1985, 467.
6. **Hsu, S.-M.,** Immunohistochemistry in lymphomas. Review and outlook, *Hematol. Pathol.,* 2, 91, 1988.
7. **Hsu, S.-M.,** Avidin-peroxidase conjugates and complexes for histochemical localization *Meth. Enzymol.,* 184, 357, 1990.
8. **Hsu, S.-M. and Raine, L.,** Protein A, avidin and biotin in immunohistochemistry, *J. Histochem. Cytochem.,* 29, 1349, 1981.
9. **Hsu, S.-M. and Raine, L.,** Versatility of biotin-labeled lectins and avidin-biotin-peroxidase complex for localization of carbohydrates in tissue sections, *J. Histochem. Cytochem.,* 30, 157, 1982.
10. **Hsu, S.-M.,** The use of biotin-labeled lectin in formalin-fixed tissues, *J. Histochem. Cytochem.,* 31, 258, 1983.
11. **Morstyn, G., Hsu, S.-M., Kinsella, T., Gratzner, H., Russo, A., and Mitchell, J. B.,** Bromodeoxyuridine in tumours and chromosomes detected with a monoclonal antibody, *J. Clin. Invest.,* 72, 1844, 1983.
12. **Hsu, P.-L., Hsu, S.-M., and Appella, E.,** Immunological screening of a cDNA library using avidin-biotin-peroxidase method, *Gene Anal. Techn.,* 2, 30, 1985.
13. **Hsu, S.-M., Hsu, P.-L., Mulshine, J., Ho, Y.-S., Ge, Z.-H., and Huang, L.,** Ultrastructural and biochemical studies of Leu M1 antigens in granulocytes and H-RS cells in Hodgkin's disease, *J. Natl. Cancer Inst.,* 77, 363, 1986
14. **Hsu, S.-M. and Soban, E.,** Color modification of diaminobenzidine (DAB) precipitation by metallic ions and its application for double immuno-histochemistry, *J. Histochem. Cytochem.,* 30, 1079, 1982.
15. **Hsu, S.-M., Cossman, J., and Jaffe, E. S.,** A comparison of ABC, unlabeled antibody and conjugated immunohistochemical methods with monoclonal and polyclonal antibodies — an examination of germinal center of tonsils, *Am. J. Clin. Pathol.,* 80, 429, 1983.
16. **Hsu, S.-M. and Ree, H. J.,** Histochemical studies on lectin binding in reactive lymphoid tissues, *J. Histochem. Cytochem.,* 31, 538, 1983.
17. **Hsu, S.-M. and Raine, L.,** Unwanted binding associated with avidin-biotin-peroxidase complex in frozen tissue, *Am. J. Clin. Pathol.,* 77, 775, 1982.

A. INTRODUCTION

The principal aim of immunohistochemistry, originally developed by Coons,[1] was to determine the accurate localization of biologically active substances in order to understand cell structure and function. Much effort has been made to develop immunolabeling techniques for the localization of substances at the ultrastructural level. Gold labeling of antibody-antigen complexes is one of the most promising procedures, because gold has a high electron density, which is easily recognizable under the electron microscope. Colloidal gold techniques can be used not only in transmission electron microscopy for ultrathin sections, but also in scanning or freeze-etch electron microscopy to determine three-dimensional distribution of surface antigens. In this chapter, we describe immunohistological procedures using colloidal gold, and review the literature on their applications in transmission electron microscopy. Reviews on immunogold labeling procedures have appeared and can be consulted for further information.[2-8]

B. PREPARATION OF COLLOIDAL GOLD COMPLEXES

1. Colloidal Gold Particles

The optimal size of colloidal gold particles is in the range of 3 to 30 nm in diameter. Colloidal gold of different uniform particle sizes can be prepared for labeling of various antigens. We recommend the original method of Frens[9] to prepare gold particles larger than 15 nm, and the method of Mühlpfordt[10] for smaller ones. For further information on other methods, recent excellent reviews are available.[3,5]

Frens' method — The protocol for making 15-nm particles is as follows: (1) add 0.5 ml 2% tetrachloroauric acid ($HAuCl_4$) to 100 ml of distilled water, then heat it to boiling; (2) add quickly 2.0 to 2.3 ml 1% sodium citrate, and boil for 5 min while stirring; (3) cool the solution in an ice-water bath. The 1% tri-sodium citrate solution should be freshly prepared.

Mühlpfordt's method — Prepare 1% tri-sodium citrate (freshly prepared), 1% tannic acid (freshly prepared), and 2% $HAuCl_4$ solutions. Five-nanometer particles are made as follows: (1) add 0.5 ml 2% $HAuCl_4$ solution to 100 ml of distilled water and boil it; (2) add a mixture of 2.0 ml of sodium citrate solution and 0.2 ml of tannic acid solution, prepared just before use, to the boiling $HAuCl_4$ under vigorous stirring; (3) cool the solution for 15 min in an ice-water bath. As the ratio of tannic acid to sodium citrate increases, the gold particles become smaller.

The gold particles thus prepared can adsorb macromolecules, such as immunoglobulins or protein A, onto their surfaces. The exact mechanism of this adsorption is yet unknown. This property is the basis of immunogold labeling. The gold particles, however, are unstable and aggregate in the presence of salts (e.g., NaCl). With flocculation, the color of the suspension changes from red to blue or grey. In the extreme case, precipitation may occur. To prevent flocculation or precipitation, we add the minimal amount of protein A or immunoglobulin required to stabilize the gold colloid suspension. The adsorption of proteins by colloidal gold particles is pH dependent and the optimal pH is different for each protein. The recommended pH values for protein A and immunoglobulins are 6.9 and 9.0, respectively.

2. Preparation of Protein-Gold Complexes

a. Protein A-Gold Complex

Protein A-gold complex is widely employed in immunohistochemical labeling because of its ease of use. Protein A, a component of the cell wall of *Staphylococcus aureus,* adsorbed on gold colloid particles as a stabilizer, binds to the Fc portion of immunoglobulin G subclasses of various mammals. Using this property, Romano and Romano[11] and Roth et al.[12] introduced the protein A into immunogold labeling. The chemistry and use of protein A and protein A-gold complex have been reviewed.[5,13]

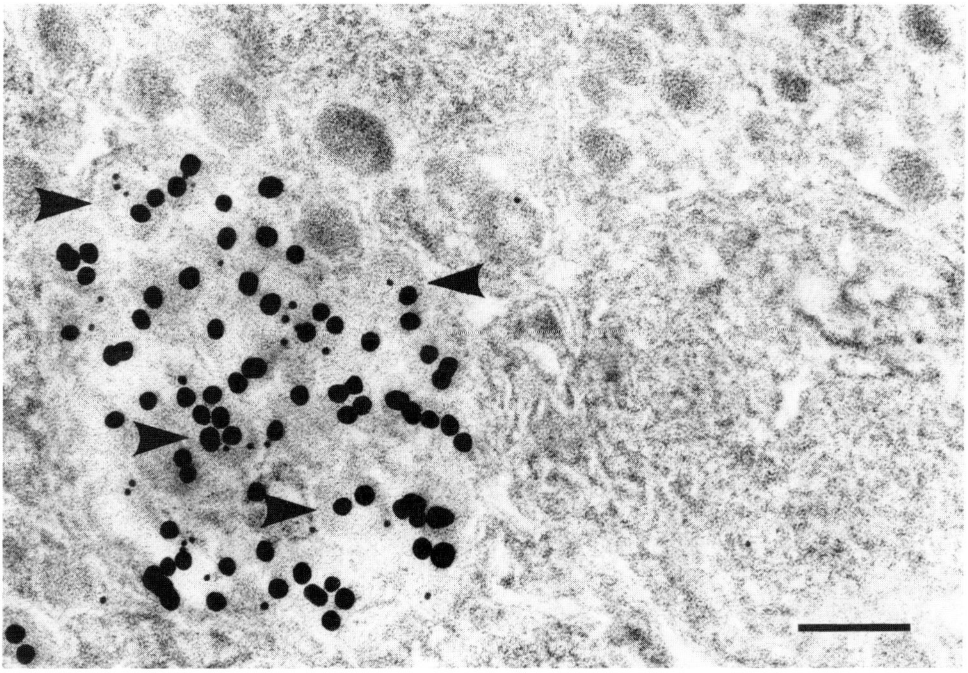

FIGURE 1. Double immunogold labeling with the gold-antibody complex method of a section of rat median eminence prepared by the freeze-drying technique. Vasopressin and corticoliberin are labeled with 15- and 5-nm gold particles, respectively. In a nerve terminal, several secretory granules (arrowheads) show colocalization of the two immunoreactivities. (Bar = 0.1 μm.)

Preparation of protein A-gold complex is described below. Stock solutions: protein A (Pharmacia), aliquots of 1 mg/0.2 ml in distilled water; 0.2 *M* K_2CO_3; 1% polyethylene glycol (PEG, mol wt 20,000); 0.02% PEG in 0.02 *M* phosphate-buffered saline (PBS). Protein A-colloidal gold is prepared as follows: (1) adjust the pH of the gold suspension (15-nm particles) to 6.9 by K_2CO_3 just before use; (2) add 0.2 ml protein A (1 mg) to 10 ml of the suspension; (3) 2 min later, add 0.2 ml 1% PEG; (4) centrifuge the mixture (25,000 × *g* for 15 min at 4°C) and discard the supernatant; (5) resuspend the sediment in 0.1 ml PEG-PBS and store at 4°C until use.

An example of the preparation of protein A-gold-antibody complex, using anti-vasopressin is as follows: (1) add 10 μl of 22.5-fold condensed affinity-purified anti-vasopressin antibody solution to 0.1 ml of the final protein A-gold suspension described above and mix well at room temperature; (2) incubate the mixture at 32°C for 2 h; (3) store at 4°C until use; and (4) dilute 10 to 20 times by PEG-PBS for staining. In double staining, we use two complexes, each labeled with gold particles of different sizes, and stain the sections consecutively with the two complexes. Details of the double staining were described previously.[6,14,15]

b. Antibody-Gold Complex

This complex directly labels tissue antigens. Romano et al.[16] and De Mey et al.[17] assessed this method as a sensitive and stable one in pre- and postembedding electron microscopic immunohistochemistry. We have employed it for the demonstration of different neuropeptides in neuronal perikarya and axon terminals, especially to determine intragranular colocalization of vasopressin, corticoliberin, oxytocin, and enkephalin (Figure 1).[6,18,19]

For optimal results, either affinity-purified antibodies or monoclonal antibodies are required.

The preparation of antibody-gold complex is as follows. Stock solutions: colloidal gold; 0.2 *M* K_2CO_3; antibody solution; 2 m*M* borate buffer, pH 9.0; 10% bovine serum albumin (BSA), pH 9.0; and a mixture of 20 m*M* Tris, 1% BSA, 150 m*M* NaCl, and 20 m*M* NaN_3 as a diluent.

The complex is made as follows: (1) adjust the pH of colloidal gold to 9.0 with K_2CO_3 just before making the complex, (2) add quickly 0.1 ml of antibody solution diluted with borate buffer to 1 ml of colloidal gold and mix well at room temperature, (3) let stand for 2 min, then add 0.1 ml of 10% BSA. Place the mixture for at least 1 min at room temperature and cool it on ice, (4) centrifuge for 30 min at 20,000 × *g*, remove the supernatant carefully, resuspend sediments with 0.1 ml of diluent, and store at 4°C until use.

C. TISSUE PREPARATION

Of crucial importance in electron microscopic immunohistochemistry is preservation of ultrastructure and antigenicity of bioactive substances. In addition to conventional tissue preparation procedures, we prefer to use freeze-substitution and freeze-drying techniques (Figure 1).[6,15] In the postembedding procedure, the fixed tissues are embedded in Epon or Spurr, and ultrathin-sectioned. In the preembedding procedure, the tissues are immunolabeled and embedded thereafter.

D. PRACTICE OF IMMUNOLABELING

To localize tissue antigens with the colloidal gold method, we incubate tissues or cells according to the following schedule.

1. Postembedding Procedure

Protein A-gold method — (1) Block nonspecific binding sites by incubating ultrathin sections, supported with uncoated nickel grids, in 1% egg albumin in PBS solution for 15 to 20 min at room temperature, (2) without washing, transfer the sections to the primary antiserum and incubate for 2 h at 32°C or overnight at 4°C in moist atmosphere, (3) wash sections with distilled water three times, 10 min each, (4) incubate in protein A-gold (1:10 to 20) for 2 h at 32°C, (5) wash with distilled water and counterstain with 1% uranyl acetate for 10 min and lead citrate for 1 min. The optimal dilution of the primary antiserum is determined empirically.

Protein A-gold-antibody complex method — (1) Block nonspecific binding sites as in Step 1 described above, (2) without washing, incubate sections in protein A-gold-antibody complex (1:10 to 20) for 2 h at 32°C or overnight at 4°C, (3) wash with distilled water and counterstain. In double labeling, the sections, washed as in Step 3, are incubated with the second antibody complexed with protein A-gold of a particle size different from that used for the first antigen. We suggest the use of a complex of smaller gold particles for the first staining because large gold particles are lost more easily from sections by repeated washing.

Antibody-gold complex method — (1) Immerse sections in 10% normal serum for 15 min at room temperature to block nonspecific binding sites, (2) without washing, incubate the sections in unlabeled primary antiserum for 2 h at 32°C or overnight at 4°C, (3) after washing with distilled water, incubate the sections in gold-labeled secondary antibody (1:10 to 20) for 2 h at 32°C, (4) wash with distilled water and counterstain. In the case of multiple labeling, we use different antibodies bound with different-sized gold particles. In this case, at the second step, the sections are incubated consecutively in the different complexes after washing with distilled water.

2. Combination of Pre- and Postembedding Techniques

A combination of conventional preembedding immunoperoxidase procedure for one antigen and the postembedding immunogold labeling for another is a useful technique.[20]

The procedures are as follows: (1) incubate Vibratome sections (30 μm thick) with the primary antiserum for 2 to 3 h, followed by incubation with biotinylated secondary antibody and avidin-biotin-peroxidase complex (Vectastain ABC Kit, Vector Laboratories, CA) for 1 h at 18°C, (2) visualize the immunoreaction with 0.007% diaminobenzidine and 0.03% H_2O_2 at 32°C, (3) osmicate the sections for 30 min at 4°C, dehydrate with an acetone series, embed in Araldite resin, and cut into ultrathin sections, (4) treat sections with 2% sodium metaperiodate for 15 to 30 min at 4°C and then wash with distilled water, (5) immunostain sections with either one of immunogold methods described above.

E. APPLICATIONS

Colloidal gold marking is very efficient, not only in electron microscopy, but also in light microscopy in which the aggregated gold particles show a distinct natural color. Owing to the high atomic number of gold, particles of colloidal gold exhibit extremely high density when viewed in the transmission electron microscope. Accordingly, even particles of 3 nm and less in diameter can be easily recognized and clearly discriminated from cellular constituents. The gold particles are capable of strong emission of secondary electrons and are, therefore, useful in scanning electron microscopy as well. Since gold emits characteristic signals, X-ray microanalysis can be performed on thin sections, and therefore quantitative X-ray microanalysis becomes possible.

The gold labeling may be applicable in indicating the dynamics of intra- and extracellular antigens. The use of a variety of proteins or peptides bound to gold particles further expands the field of investigation: lipoprotein[21,22] or cholera toxin B[23] is used for detection of receptors; enzymes[24,25] indicate the places of intracellular metabolism; anti-biotin antibody[26] or streptavidin[27] serves as a marker for *in situ* hybridization immunohistochemistry. Colloidal gold-wheat germ agglutinin complex is a valuable tool in detecting neuronal cell body retrograde transport.[28] Recently, 1-nm gold particles have become available for preembedding electron microscopic immunohistochemistry.[29] This procedure permits precise ultrastructural localization of a variety of intracellular antigens and, hence, is advantageous in detecting multiple antigens. The development of new microscopic equipment widens the field of application of gold labeling techniques. Reflection contrast microscope is a recent example.[30] Quantitative analyses of antigens are a problem which future studies will address.

REFERENCES

1. **Coons, A. H.,** Fluorescent antibody methods, in *General Cytochemical Methods,* Danielli, J. F., Ed., Academic Press, New York, 1958, 399.
2. **Faulk, W. P. and Taylor, G. M.,** An immunocolloid method for the electron microscope, *Immunochemistry,* 8, 1081, 1971.
3. **Horisberger, M.,** Colloidal gold: a cytochemical marker for light and fluorescent microscopy and for transmission and scanning electron microscopy, in *Scanning Electron Microscopy,* Vol. 2, Johari, O., Ed., SEM Inc., AMF O'Hare, Chicago, 1981, 9.
4. **De Mey, J. R.,** The preparation of immunoglobulin gold conjugates (IGS reagents) and their use as markers for light and electron microscopic immunocytochemistry, in *Immunohistochemistry,* Cuello, A. C., Ed., IBRO, 1983, 347.
5. **Roth, J.,** The colloidal gold marker system for light and electron microscopic cytochemistry, in *Techniques in Immunocytochemistry,* Vol. 2, Bullock, G. R. and Petrusz, P., Eds., Academic Press, London, 1983, 217.

6. **Hisano, S., Adachi, T., Maegawa, M., and Daikoku, S.,** Some improvement in tissue preparation and colloidal-gold immunolabeling for electron microscopy, *Am. J. Anat.*, 175, 245, 1986.
7. **Timms, B. G.,** Postembedding immunogold labeling for electron microscopy using "LR White" resin, *Am. J. Anat.*, 175, 267, 1986.
8. **de Harven, E. and Soligo, D.,** Scanning electron microscopy of cell surface antigens labeled with colloidal gold, *Am. J. Anat.*, 175, 277, 1986.
9. **Frens, G.,** Controlled nucleation for the regulation of the particle size in monodisperse gold suspensions, *Nature Phys. Sci.*, 241, 20, 1973.
10. **Mühlpfordt, H.,** The preparation of colloidal gold particles using tannic acid as an additional reducing agent, *Experientia*, 38, 1127, 1982.
11. **Romano, E. L. and Romano, M.,** Staphylococcal protein A bound to colloidal gold: a useful reagent to label antigen-antibody sites in electron microscopy, *Immunochemistry,* 14, 711, 1977.
12. **Roth, J., Bendayan, M., and Orci, L.,** Ultrastructural localization of intracellular antigens by the use of protein A-gold complex, *J. Histochem. Cytochem.*, 26, 1074, 1978.
13. **Langone, J. J.,** Protein A of *Staphylococcus aureus* and related immunoglobulin receptors produced by streptococci and pneumonococci, *Adv. Immunol.*, 32, 157, 1982.
14. **Adachi, T., Hisano, S., and Daikoku, S.,** Intragranular colocalization of immunoreactive methionine-enkephalin and oxytocin within the nerve terminals in the posterior pituitary, *J. Histochem. Cytochem.*, 33, 891, 1985.
15. **Hisano, S., Adachi, T., and Daikoku, S.,** Freeze-drying technique in electron microscopic immunohistochemistry, *J. Histochem. Cytochem.*, 33, 485, 1985.
16. **Romano, E. L., Stolinski, C., and Hughes-Jones, N. C.,** An antiglobulin reagent labelled with colloidal gold for use in electron microscopy, *Immunochemistry,* 11, 521, 1974.
17. **De Mey, J., Moeremans, M., Geuens, G., Nuydens, R., and De Brabander, M.,** High resolution light and electron microscopic localization of tubulin with IGS (immuno gold staining) method, *Cell Biol. Int. Rep.*, 5, 889, 1981.
18. **Hisano, S., Tsuruo, Y., Katoh, S., Daikoku, S., Yanaihara, N., and Shibasaki, T.,** Intragranular colocalization of arginine vasopressin and methionine-enkephalin-octapeptide in CRF-axons in the rat median eminence, *Cell Tissue Res.*, 249, 497, 1987.
19. **Kagotani, Y., Hisano, S., Tsuruo, Y., Daikoku, S., Okimura, Y., and Chihara, K.,** Intragranular co-storage of neuropeptide Y and arginine vasopressin in the paraventricular magnocellular neurons of the rat hypothalamus, *Cell Tissue Res.*, 262, 47, 1990.
20. **Kagotani, Y., Tsuruo, Y., Hisano, S., Daikoku, S., and Chihara, K.,** Axons containing neuropeptide Y innervate arginine vasopressin-containing neurons in the rat paraventricular nucleus. Dual electron microscopic immunolabeling, *Histochemistry,* 91, 273, 1989.
21. **Robenek, H. and Severs, N. J.,** Double labelling of lipoprotein receptors in fibroblast cell surface replicas, *J. Ultrastruct. Res.*, 87, 149, 1984.
22. **Severs, N. J. and Robenek, H.,** Detection of microdomains in biomembranes. An appraisal of recent developments in freeze-fracture cytochemistry, *Biochim. Biophys. Acta,* 737, 373, 1983.
23. **Llewellyn-Smith, I. J., Minson, J. B., Wright, A. P., and Hodgson, A. J.,** Cholera toxin B-gold, a retrograde tracer that can be used in light and electron microscopic immunocytochemical studies, *J. Comp. Neurol.*, 294, 179, 1990.
24. **Bendayan, M.,** Ultrastructural localization of nucleic acids by use of enzyme-gold complexes, *J. Histochem. Cytochem.*, 29, 531, 1981.
25. **Furuhashi, M., Nakahara, A., Fukutomi, H., Kominami, E., Grube, D., and Uchiyama, Y.,** Immunocytochemical localization of cathepsins B, H, and L in the rat gastro-duodenal mucosa, *Histochemistry,* 95, 231, 1991.
26. **Binder, M., Tourmente, S., Roth, J., Renaud, M., and Gehring, W. J.,** In situ hybridization at the electron microscope level: localization of transcripts on ultrathin sections of Lowicryl K4M-embedded tissue using biotinylated probes and protein A-gold complexes, *J. Cell Biol.*, 102, 1646, 1986.
27. **Wolber, R. A., Beals, T. F., Lloyd, R. V., and Maassab, H. F.,** Ultrastructural localization of viral nucleic acid by *in situ* hybridization, *Lab. Invest.*, 59, 144, 1988.
28. **Seeley, P. J. and Field, P. M.,** Use of colloidal gold complexes of wheat germ agglutinin as a label for neural cells, *Brain Res.*, 449, 177, 1988.
29. **Lah, J. J., Hayes, D. M., and Burry, R. W.,** A neutral pH silver development method for the visualization of 1-nanometer gold particles in pre-embedding electron microscopic immunocytochemistry, *J. Histochem. Cytochem.*, 38, 503, 1990.
30. **Cornelese-ten Velde, I. and Prins, F. A.,** New sensitive light microscopical detection of colloidal gold on ultrathin sections by reflection contrast microscopy. Combination of reflection contrast and electron microscopy in post-embedding immunogold histochemistry, *Histochemistry,* 94, 61, 1990.

Chapter 2

ELECTRON MICROSCOPIC IMMUNOCYTOCHEMISTRY

2.3 Major Visualization Techniques

2.3.5 Electron Microscopic Immunocytochemistry of Neuropeptides and Neurotransmitters

Yasuhiko Ibata

TABLE OF CONTENTS

0-8493-6012-9/93/$0.00 + $.50

A. INTRODUCTION

Electron microscopic immunocytochemistry is a powerful tool for the morphological investigation of the nervous system. It enables the examination of the fine structure of neuronal perikarya, axons, dendrites, as well as synapses. These components contain neuroactive substances, including neuropeptides and neurotransmitters, which are localized by using specific antibodies generated against these neuroactive substances.

In the morphological survey of the neuronal elements in the central nervous system, as well as in the peripheral nervous system, it is very important and attractive to elucidate the interaction of each of the particular neuronal elements. These elements often contain different neuroactive substances to comprise a neuronal circuit. For such purpose, double-labeling immunoelectron microscopic methods also allow the detection of different neuroactive substances in single sections by employing the use of two different antibodies.

In this text, I will describe some of the procedures used to detect neuropeptides and neurotransmitters by electron microscopic immunocytochemistry as well as a method for double-labeling immunoelectron microscopy for the detection of a different neuropeptide and neurotransmitter in the same ultrathin section, using the rat as the material.

B. APPLICATION TO NEUROPEPTIDES AND NEUROTRANSMITTERS

Application of electron microscopic immunocytochemistry to detect neuropeptides, neurotransmitters, and their synthesizing enzymes (e.g., tyrosine hydroxylase, TH) is carried out by either the peroxidase-antiperoxidase (PAP) method or the avidin-biotin peroxidase complex (ABC) method.[1-6]

Animals are routinely injected with colchicine (100 to 300 μg/rat) under Nembutal anesthesia into the lateral cerebral ventricle 48 h before sacrifice, to promote the storage of neuropeptides or enzymes in the neuronal perikarya. This treatment enables the easy detection of the immunoreaction in the neuronal perikarya.

1. Fixation

Animals are perfused via the left cardiac ventricle with a mixture of 4% paraformaldehyde, 0.2% picric acid, and 0.35% glutaraldehyde buffered with 0.1 *M* phosphate buffer (PB) at a pH of 7.4. The brain, spinal cord, or peripheral nervous tissues are removed and further fixed in the same solution, but without the glutaraldehyde, for an additional 24 to 48 h at 4°C. The tissues are stored in 0.1 *M* phosphate-buffered saline (PBS). Sections, 40 μm in thickness, are cut in a PBS bath using a Microslicer. They are incubated in 0.1 *M* PBS containing 1% bovine serum for 1 h prior to the immunohistochemical procedure.

Sections are incubated in specific antisera generated against neuropeptides (dilution: 3000 to 5000 times) or a synthetic enzyme (dilution: 10,000 times in the case of TH) for 72 to 96 h at 4°C. In the case of the PAP method, sections are incubated in an IgG solution (dilution: 200 times), specific for the species of the animal that was immunized with the primary antigen, for 2 h at room temperature, and thereafter they are incubated in PAP solution (dilution: 200 times) for 90 min at room temperature. In the case of the ABC method, they are incubated in a similar biotinylated IgG solution (dilution: 200 times) for 1 to 2 h at room temperature, then they are incubated in avidin-biotin peroxidase complex (ABC) solution (dilution: 200 times) for 1 to 2 h at room temperature.

2. Visualization of Immunoreactive Products

After immunoreaction following either the PAP or ABC method, the sections are exposed at room temperature to a 0.02% 3,3′-diaminobenzidine-4 HCl (DAB) solution containing H_2O_2 (0.045%) for 5 min. After confirmation by observation of the development of the final reaction product in the immunoreactive neuronal elements by light microscopy, the sections are fixed in a chilled 1% OsO_4 solution for 1 h, dehydrated in a graded series of acetone, and embedded in an Epon mixture.

3. Examination in an Electron Microscope

Semithin sections are examined under a light microscope, following which ultrathin sections (stained with uranyl acetate) are examined under an electron microscope. Examples of the fine structural localization of neuropeptides, neurotransmitters, and neurotransmitter-synthesizing enzymes by this procedure are shown in Figures 1 and 2.

C. APPLICATION FOR THE DETECTION OF DOPAMINE-CONTAINING NEURONAL ELEMENTS

The application of immunoelectron microscopy using a high concentration of glutaraldehyde as the fixative for the detection of dopamine-containing neuronal elements[7-8] is a requisite and is useful for detection of the fine structure of the dopamine neuron system. The method uses dopamine-glutaraldehyde-coupled immunogen to generate specific antisera against dopamine.

Animals are perfused via the left cardiac ventricle with a fixative containing 5% glutaraldehyde buffered at pH 7.4 with 0.1 *M* phosphate buffer. The brains are removed and tissue blocks are fixed in the same solution for an additional 2 h at 4°C and stored in PB overnight. Sections, 40 μm in thickness, are then cut in PB by a Microslicer, and immersed in 0.1 *M* PB containing 1% $Na_2S_2O_5$ and 1% sodium borohydrate, pH 7.4, for 30 min at room temperature. The sections are preincubated in 0.1 *M* PB containing 1% calf serum and 1% $Na_2S_2O_5$ for 1 h at room temperature.

Immunoreaction, visualization, and examination at the ultrastructural level are similar to those of the procedure outlined above. In this method, $Na_2S_2O_5$ is used to protect the oxidation of dopamine and the sodium borohydrate is used for the inactivation of glutaraldehyde. Prolongation of the fixation period results in less immunoreaction. Figure 3 shows dopamine-like immunoreactive elements visualized by this procedure.

D. DOUBLE IMMUNOLABELING FOR ELECTRON MICROSCOPIC IMMUNOCYTOCHEMISTRY

Several investigators have tried to design double-labeling immunoelectron microscopic procedures. I wish to introduce here a procedure in which one neuroactive substance, such as a neuropeptide, is detected as a silver-intensified end product. In this procedure, the silver-intensified end product is substituted with gold particles to render it more stable. The other substance, such as another neuropeptide, or a neurotransmitter, and/or one of its synthesizing enzymes, is visualized as an end product of the DAB reaction.[9,10]

1. Fixation

Fixation and sectioning is similar to the procedure described above.

2. Immunoreaction and Visualization of End Product

1. Sections are incubated in antiserum against substance A (dilution: 3000 to 5000 times) that also contains a low concentration (0.06%) of Triton 100 X for 72 h at 4°C. Triton 100 X enables the penetration of the chemical solutions (e.g., immunoglobulins) more easily into the sections.
2. Sections are immersed in the biotinylated IgG solution (dilution: 200 times) for 2 h at room temperature.
3. Sections are incubated in ABC solution for 2 h at room temperature.

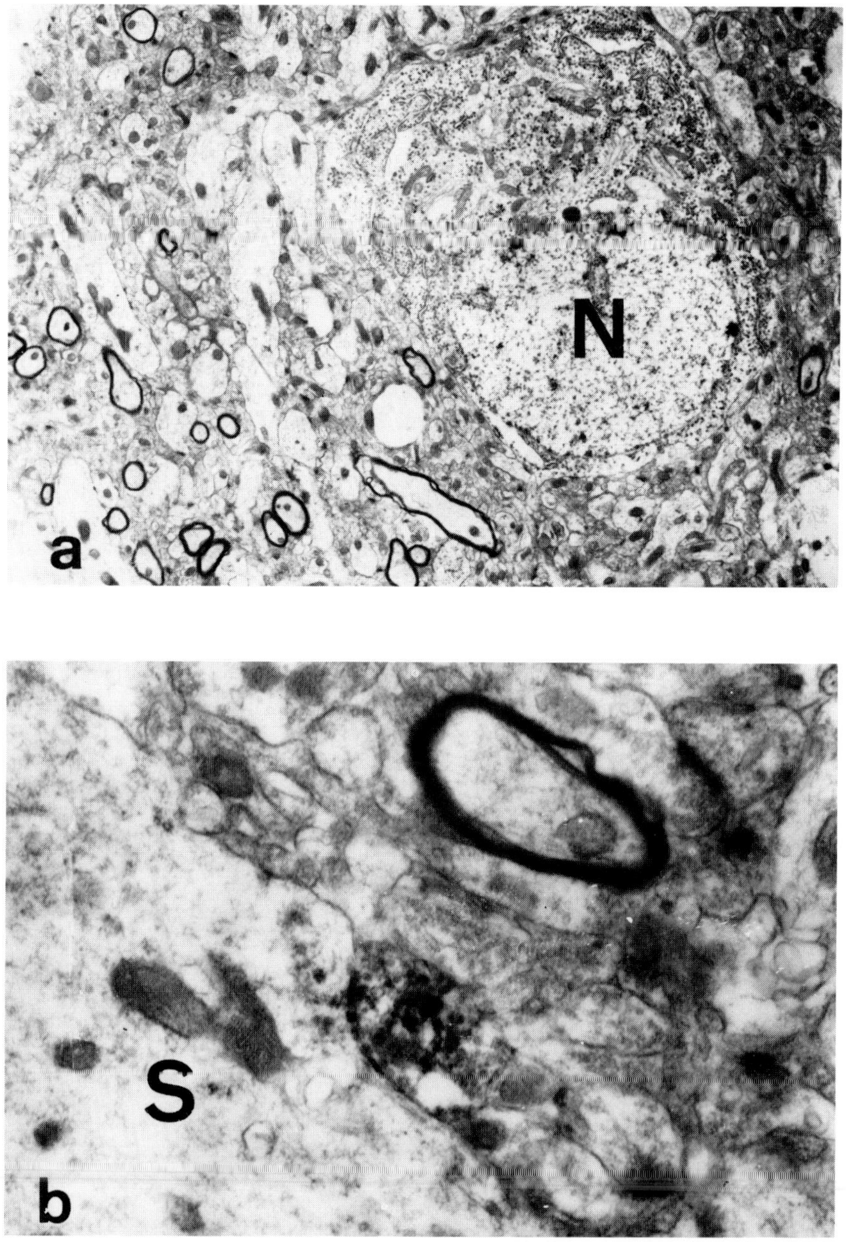

FIGURE 1. Gamma-aminobutyric acid (GABA)-like immunoreactive neuronal perikaryon (a) and a presynaptic region (b) of symmetrical axosomatic synapse in the hippocampal formation. Immunoreactive substance is localized throughout the cytoplasm. Presynaptic region contains synaptic vesicles and immunoreactive dense granules. N, Nucleus; S, postsynaptic pyramidal cell soma. Tyrosine hydroxylase (TH)-like immunoreactive neuronal perikaryon demonstrating strong immunoreactivity in the hypothalamic arcuate nucleus (c). This neuron contains well-developed cell organelles such as rough endoplasmic reticulum (ER) and mitochondria. (Magnifications: a, × 4,300; b, × 25,000; c, × 11,000.)

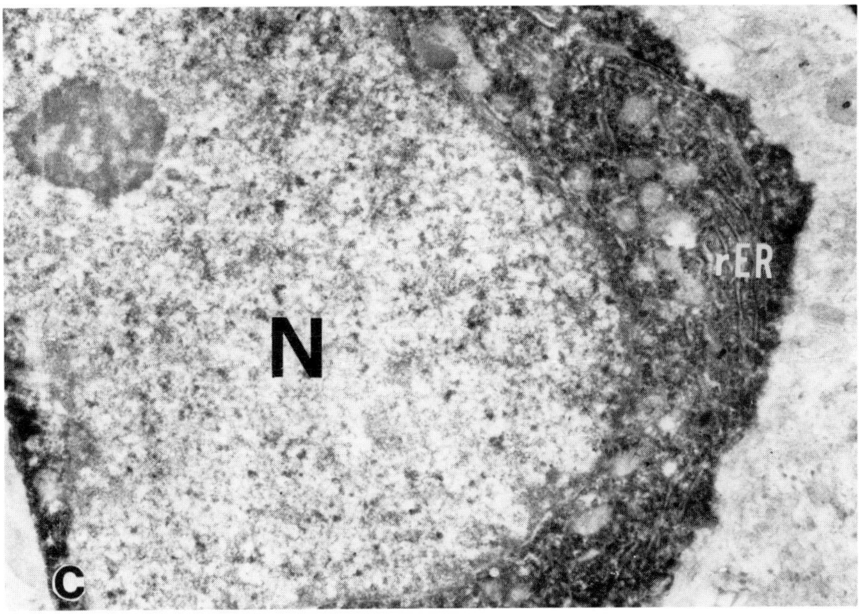

FIGURE 1 (continued).

4. Sections are then exposed to the DAB solution (8 min). After this step, the sections are reacted with antiserum against substance B as in step 5, since substance A is further intensified with silver and the silver-intensified end product is finally substituted with gold in steps 7 to 11.
5. The sections are incubated in antiserum against substance B (dilution: 500 times) containing 0.06% Triton X 100 for 24 h at room temperature.
6. Following exposure to antiserum B, the sections are immersed in biotinylated IgG solution (dilution: 200 times) for 6 h at room temperature.
7. Sections are immersed in 10% thioglycolic acid to suppress the argyrophil reaction of the tissue for 3 h at room temperature and rinsed in 2% sodium acetate for 30 min.
8. To silver intensify the DAB reaction product, sections are placed in the physical developer for 5 to 10 min at room temperature. The physical developer consists of a mixture of 1 part of stock solution A and 1 part of stock solution B, which are mixed just before use. Stock solution A: 5% sodium carbonate solution. Stock solution B: 0.2% ammonium nitrate, 0.2% silver nitrate, 1% tangstosilioic acid, and 1.9% formaldehyde.
9. Sections are rinsed in 1% acetic acid for 15 min followed by another rinse in 2% sodium acetate for 5 min.
10. To gold-tone the substituted silver, the sections are immersed in 0.05% $HAuCl_4 \cdot 4H_2O$ for 10 min at 4°C.
11. Sections are then rinsed in 2% sodium acetate for 5 min and fixed with a 3% sodium thiosulfate solution for 10 min. A final 10-min rinse in 2% sodium acetate is performed.
12. Incubation in the ABC solution for 12 h at room temperature is used to detect the immunoreaction of substance B.
13. Thereafter, the sections are exposed to the DAB solution (10 min) and are then examined in a light microscope to confirm the immunoreaction.
14. Finally, the sections are fixed in a 1% OsO_4 solution for 1 h, dehydrated, and embedded in the Epon mixture.

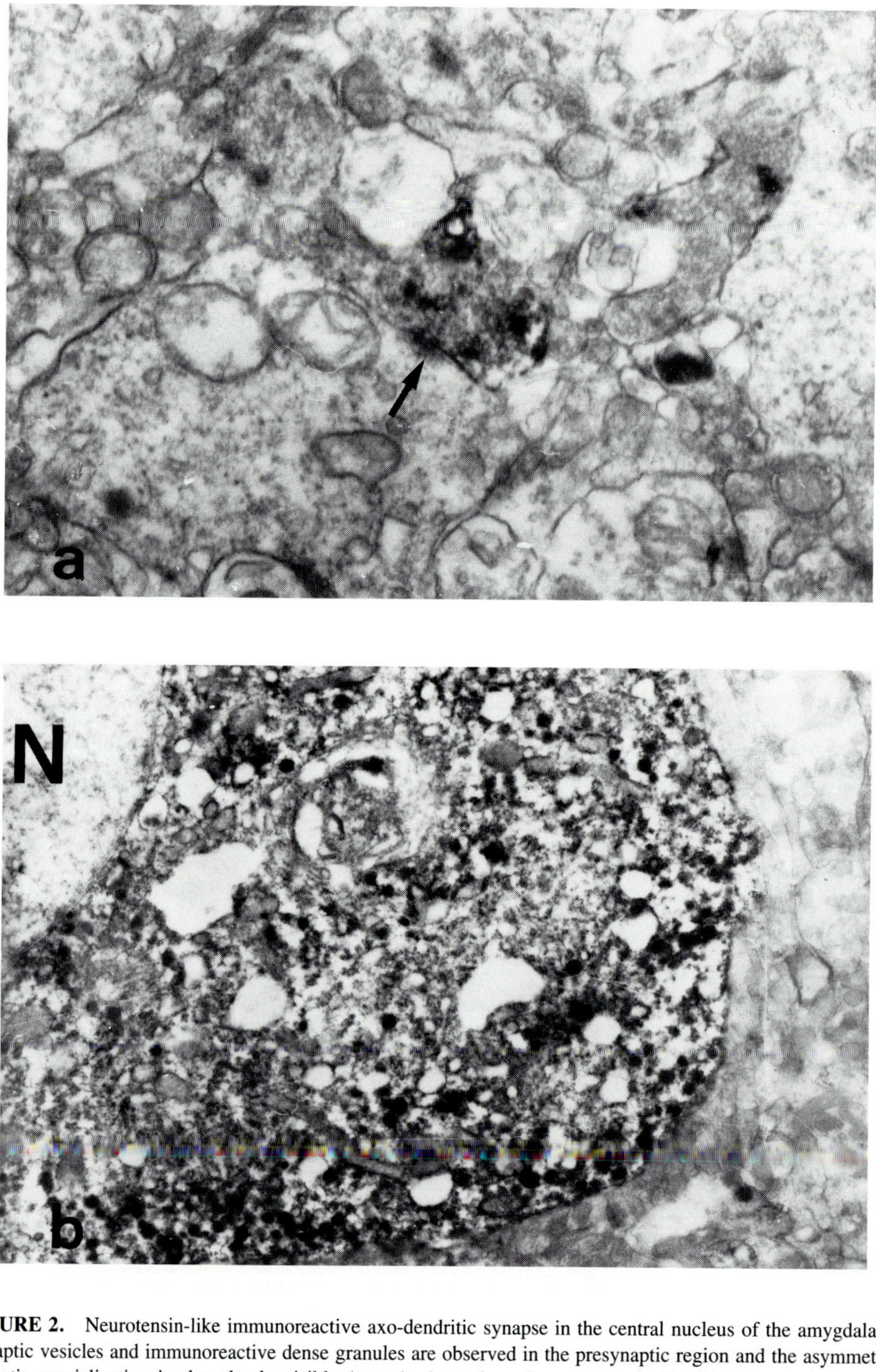

FIGURE 2. Neurotensin-like immunoreactive axo-dendritic synapse in the central nucleus of the amygdala (a). Synaptic vesicles and immunoreactive dense granules are observed in the presynaptic region and the asymmetrical synaptic specialization is also clearly visible (arrow). A portion of a vasopressin-like immunoreactive neuronal perikaryon located within the magnocellular portion of the hypothalamic paraventricular nucleus is shown in (b). This neuron contains immunoreactive dense granules as well as ER and mitochondria; N, nucleus. Vasopressin-like immunoreactive nerve endings demonstrating strong immunoreactivity in the external layer of the median eminence (c). Ps: Pericapillary space of the portal vessel. (Magnifications: a, × 28,000; b, × 13,000; c, × 12,000.)

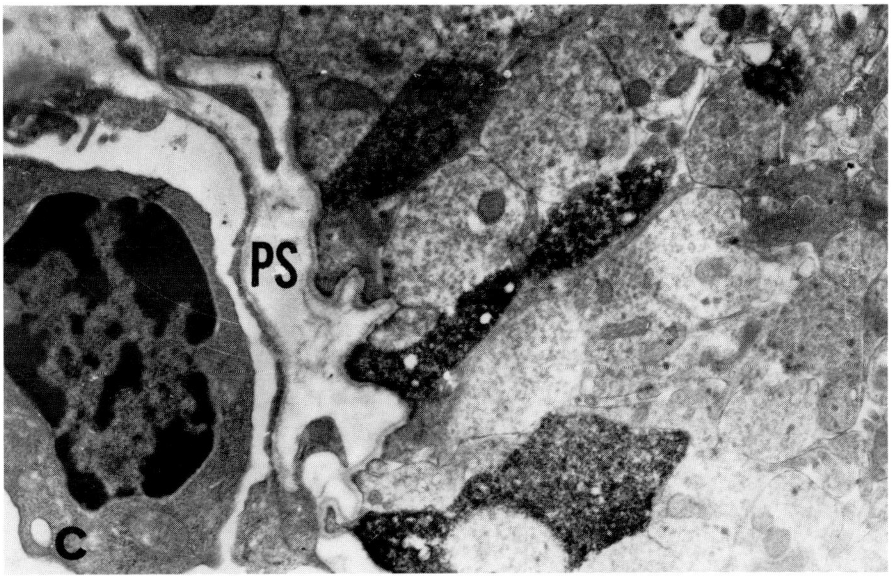

FIGURE 2 (continued).

3. Examination in an Electron Microscope

Ultrathin sections are stained with uranyl acetate and examined under an electron microscope. Deposits of the gold particles are found in the neuronal elements containing substance A, while substance B is visualized as osmium black of the end product after the DAB reaction. Several controls are prerequisite in the critical steps of the procedure.

Other methods of double-labeling electron microscopic immunocytochemistry are available to scrutinize the fine structural interaction of neuronal elements which contain different types of neuroactive substances.[10,11] Figure 4 shows neuronal elements revealed by this procedure.

Postembedding immunoelectron microscopy using different sizes of protein A gold, which specifically bind to the IgG, has been frequently applied for the detection of hormones in anterior pituitary cells.[12-14] These methods have not been applied to detect neuropeptide- and/or transmitter-containing elements in the nervous system, since the preexamination of the [illegible] under light microscopy using gold particles is not as easy as the use of a preembedding method.

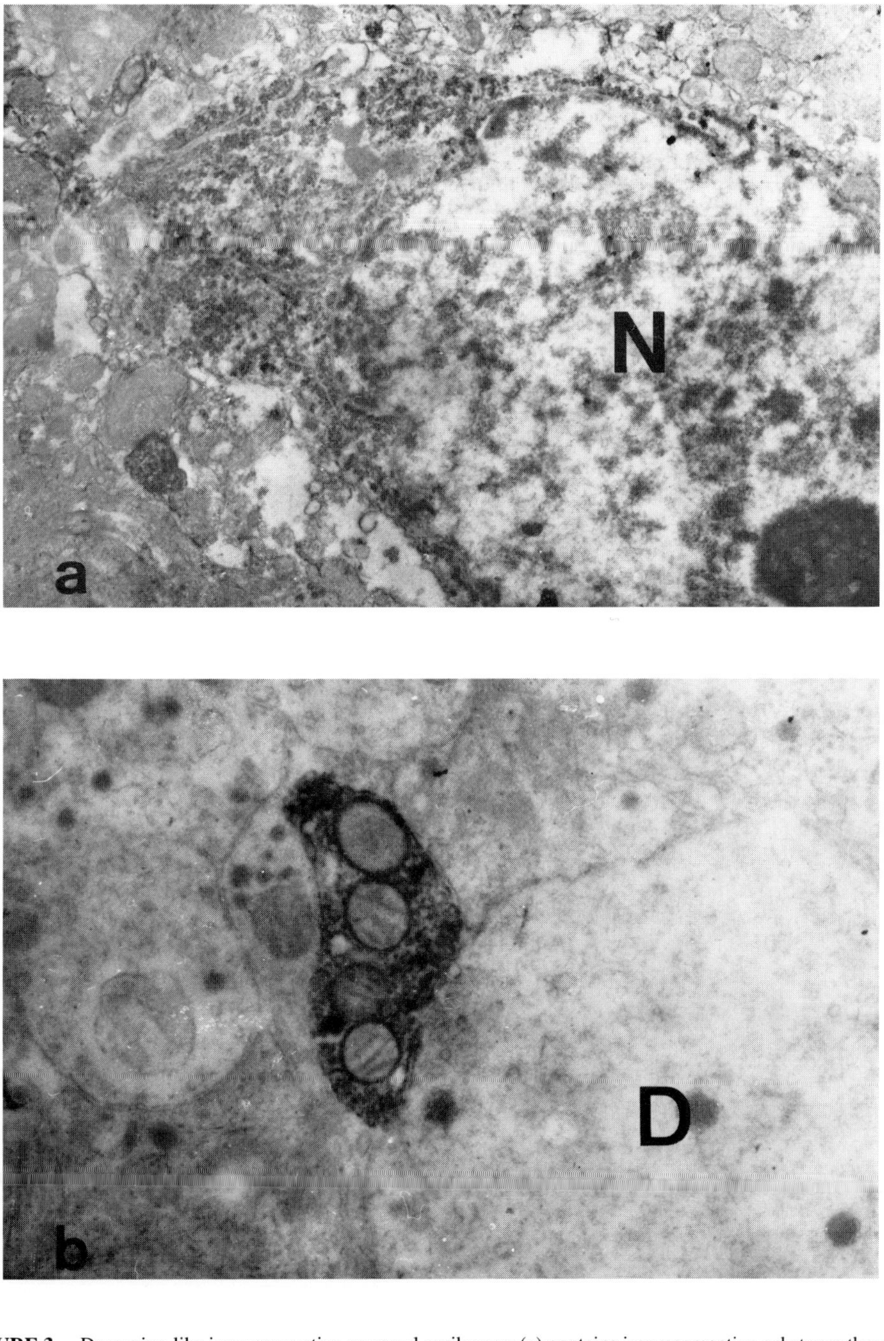

FIGURE 3. Dopamine-like immunoreactive neuronal perikaryon (a) contains immunoreactive substance throughout its cytoplasm. Dopamine-like immunoreactive axo-dendritic synapse: this presynaptic region contains a large number of synaptic vesicles and mitochondria (b); N, nucleus; D, dendrite. (Magnifications: a, × 10,000; b, × 27,000.)

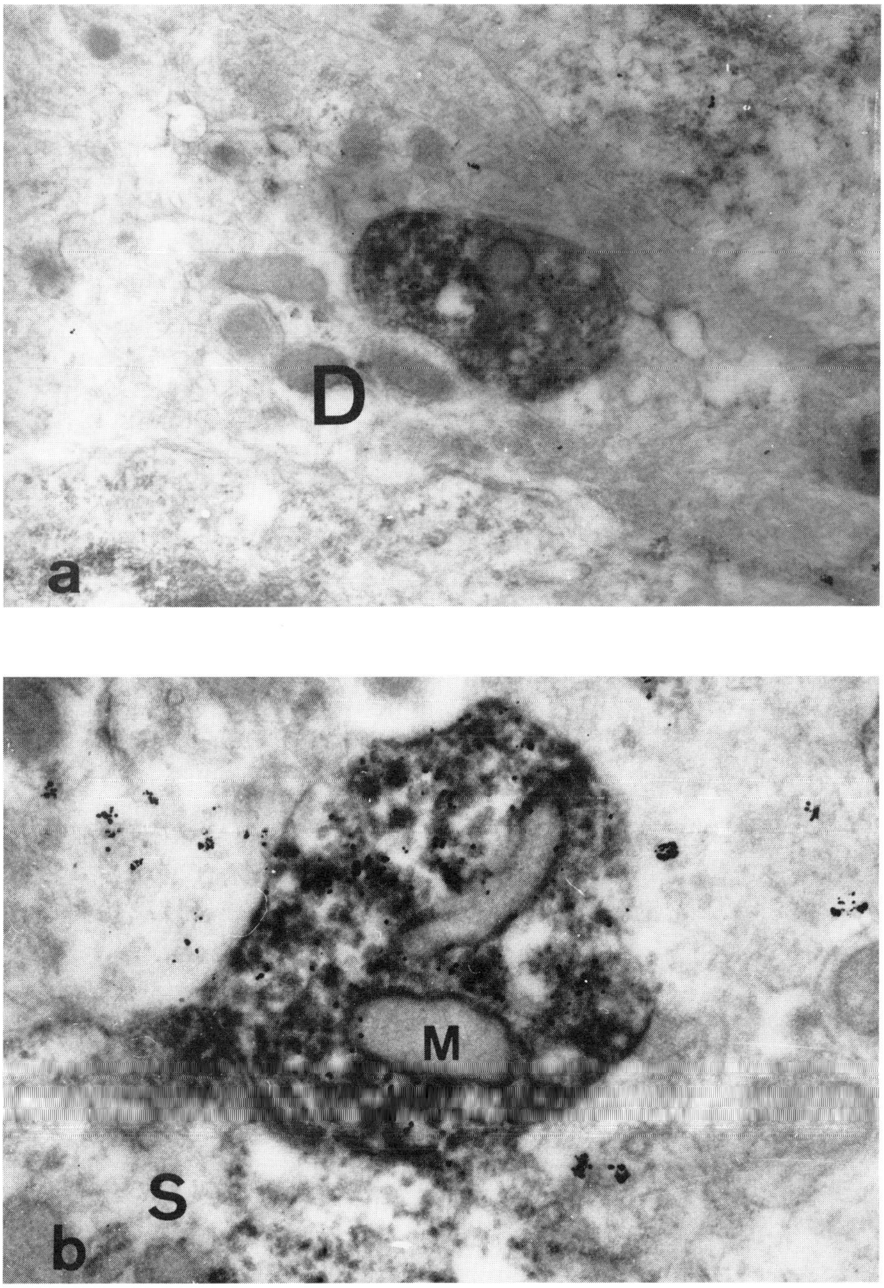

FIGURE 4. Vasoactive intestinal peptide (VIP)-like immunoreactive preterminal axon makes an axo-dendritic synaptic contact. In the presynaptic element, gold-toned, silver-intensified DAB particles are seen in addition to large numbers of synaptic vesicles (a). Immunoreactive dense granules and mitochondria are present. VIP-like immunoreactive presynaptic terminal in which gold particles as well as synaptic vesicles, immunoreactive dense granules, and mitochondria are observed forming an axo-somatic synapse on a portion of a vasopressin-like immunoreactive neuronal somata (b). The postsynaptic cell contains dense immunoreactive substances demonstrated by the DAB reaction alone. D, Dendrite; M, mitochondria; S, cell soma. (Magnifications: a, × 25,000; b, × 60,000.)

REFERENCES

1. **Pickel, V. M., Joh, T. H., Reis, D. J., Leeman, S. E., and Miller, R. J.,** Electron microscopic localization of substance P and enkephalin in axon terminals related to dendrites of catecholaminergic neurons, *Brain Res.,* 160, 387, 1979.
2. **Ibata, Y., Okamura, H., Makino, S., Kawakami, F., Morimoto, N., and Chihara, K.,** Light and electron microscopic immunocytochemistry of GRF-like immunoreactive neurons and terminals in the rat hypothalamic arcuate nucleus and median eminence, *Brain Res.,* 370, 136, 1986.
3. **Ibata, Y., Takahashi, Y., Okamura, H., Kubo, T., and Kawakami, F.,** Fine structure of NPY-containing neurons in the lateral geniculate nucleus and their terminals in the suprachiasmatic nucleus of the rat, *Brain Res.,* 439, 230, 1988.
4. **Chan-Palay, V., Zaborszky, L., Kohler, C., Goldstein, M., and Paly, S. L.,** Distribution of tyrosine-hydroxylase-immunoreactive neurons in the hypothalamus of rat, *J. Comp. Neurol.,* 227, 467, 1984.
5. **Potte, M., Beaudet, A., and Brauer, J. R.,** Light and electron microscopic study of tyrosine hydroxylase-immunoreactive neurons within the developing rat arcuate nucleus, *Brain Res.,* 439, 127, 1988.
6. **Ichitani, Y., Morimoto, N., Okuyama, I., Hasegawa, M., Nagatsu, I., and Ibata, Y.,** Fine structure of tyrosine hydroxylase-like immunoreactive neurons and terminals of the hypothalamic arcuate nucleus and median eminence in young and aged rats, *J. Electron Microsc.,* 39, 396, 1990.
7. **Sakai, M., Kani, K., Yoshida, M., and Nagatsu, I.,** Dopaminergic cells in the superior cervical ganglion of the rat: light and electron microscopic study using an antibody against dopamine, *Neurosci. Lett.,* 96, 157, 1989.
8. **Ibata, Y., Ichitani, Y., Okamura, H., Miyake, M., Okuyama, I., Okamoto, S., Matsumoto, Y., and Nagatsu, I.,** Light and electron microscopic immunocytochemistry of tuberoinfundibular dopamine system of the rat using antiserum against dopamine, *Biomed. Res.,* 10 (Suppl.), 47, 1989.
9. **Maegawa, M., Hisano, S., Tsuruo, Y., Katoh, S., Nakanishi, J., Chikamori-Aoyama, M., and Daikoku, S.,** Differential immunolabeling for electron microscopy of diverse peptidergic neurons, *J. Histochem. Cytochem.,* 35, 251, 1987.
10. **Liposits, Z. S., Sherman, D., Phelix, C., and Paull, W. K.,** A combined light and electron microscopic immunocytochemical method for the simultaneous localization of multiple tissue antigens. Tyrosine hydroxylase immunoreactive innervation of corticotropin releasing factor synthesizing neurons in the paraventricular nucleus of the rat, *Histochemistry,* 85, 95, 1986.
11. **van den Pol, A. N.,** Silver-intensified gold and peroxidase as dual ultrastructural immunolabels for pre- and postsynaptic neurotransmitters, *Science,* 228, 332, 1985.
12. **Childs, G. V., Ellison, D. G., and Garner, L. L.,** An immunocytochemist's view of gonadotropin storage in the adult male rat: cytochemical and morphological heterogeneity in serially sectioned gonadotropes, *Am. J. Anat.,* 158, 397, 1980.
13. **Bendyan, M.,** Double immunocytochemical labeling applying the protein A-gold technique, *J. Histochem. Cytochem.,* 30, 81, 1982.
14. **Fumagalli, G. and Zanini, A.,** In cow anterior pituitary, growth hormone and prolactin can be packed in separate granules of the same cell, *J. Cell Biol.,* 100, 2019, 1985.

Chapter 3

LECTIN HISTOCHEMISTRY AND CYTOCHEMISTRY

Hayato Kawakami and Hiroshi Hirano

TABLE OF CONTENTS

0-8493-6012-9/93/$0.00 + $.50

A. INTRODUCTION

Lectins are carbohydrate-binding proteins or glycoproteins other than enzymes or antibodies that are extracted from plants, animals, or microorganisms. In histochemistry and cytochemistry, labeled lectins have been widely used as a probe for the detection of carbohydrates. With lectins, cells and tissues may be clearly identified as a particular type and/or a group under the microscope, and changes in glycoconjugates during development or under pathological conditions, including malignant transformation, have been successfully visualized *in situ*.[1-6] Several comprehensive reviews concerning lectin histochemistry and cytochemistry are available, both from technical and from pathological points of view.[7-12]

B. SPECIFICITIES AND CHARACTERISTICS OF LECTINS

Lectins commonly used in histochemistry and cytochemistry are usually divided into several groups, depending on their preferential binding specificities (Table 1).[7,8] Some of the lectins display a strong anomeric specificity, whereas others react equally with both α and β forms of sugars. Many lectins react predominantly with terminal nonreducing sugars, while a few lectins, e.g., concanavalin A (ConA), bind to internal or external constituents of sugar chains. Certain groups of lectins have an affinity for either *O*-glycoside-type or *N*-glycoside-type sugar chains, whereas others have an affinity for both types (Table 2).[13]

In general, monosaccharide specificity of lectins does not always reflect the exact nature of the carbohydrate structure of each lectin receptor. Furthermore, in most histochemical preparations, lectins with the same nominal monosaccharide specificity may have a different affinity for the same cell or tissue. Therefore, these facts should be taken into consideration for evaluation of the detailed carbohydrate structure of lectin-binding sites.

C. A GUIDE TO SELECTION OF METHODS FOR LECTIN HISTOCHEMISTRY

Most of the methods for conventional immunohistochemistry are applicable for lectin histochemistry also. Contrary to the fact that ordinary protein antigens tend to lose their activity easily during histochemical preparation, the carbohydrate moiety seen by the lectin receptor usually maintains its reactivity even after chemical fixation or heat treatment. For routine light microscopic studies, 10% formalin-, Bouin's fluid-, or ethanol-fixed materials embedded in paraffin are recommended. Recently, various types of fixatives and procedures for lectin histochemistry have been thoroughly compared and described.[14]

For transmission electron microscopy, a variety of preparative procedures for lectin histochemistry have been used. In preembedding techniques, free cells or thick frozen sections of tissues are at first incubated with the lectin and then embedded, while in post-embedding techniques, ultrathin sections from resin-embedded specimens or ultrathin cryosections are prepared in advance and then exposed directly to the lectin conjugate. Scanning electron microscopy with secondary-electron images and/or backscattered electron images is useful, particularly for three-dimensional observation of lectin-binding sites (see Chapter 6).

Lectin-binding sites are visualized with different kinds of markers (fluorescent dyes, horseradish peroxidase [HRP], ferritin, colloidal gold, etc.), among which the most appropriate marker is selected depending on whether the binding sites are to be examined at the light or electron microscopic level, and whether they are located intracellularly or at the cell surface. In each case, both direct and indirect methods (Figure 1) are applicable. The direct method is rather simple from a technical point of view. With this method, the whole procedure can usually be carried out within a relatively short time. Detection of the binding sites by this method, however, is sometimes less sensitive as compared with that by the indirect

TABLE 1
Lectins Commonly Used in Histochemistry and Cytochemistry

Sources	Abbrev.	Carbohydrate specificities: αFuc	αMan	αGlc	αGlcNAc	βGlcNAc	αGal	βGal	αGalNAc	βGalNAc	αNeu	Complex carbohydrates
Aleuria aurantia	AAL	+										Fuc$_{1\alpha6}$GlcNAc
Anguilla anguilla	AAA (ESH)	+										anti-H
Lotus tetragonolobus	Lotus	+										anti-H
Ulex europeus I	UEA-I	+										anti-H, Fuc$_{1\alpha2}$Gal$_{1\beta4}$GalNAc
Ulex europeus II	UEA-II	+				+						anti-H
Galantus nivalis	GNA		+									Man$_{1\alpha3}$Man
Canavalia ensiformis	ConA		+ +	+	(+)							branched mannose
Lens culinaris	LCA		+ +	+	+							branched mannose with α-L-Fuc
Pisum sativum	PSA		+ +	+	+							branched mannose with α-L-Fuc
Vicia faba	VFA		+ +	+	+							branched mannose with α-L-Fuc
Griffonia simplicifolia II	GS-II				+ +	+ +						glycogen
Triticum vulgaris	WGA					+ +					+	β(1,4)GlcNAc oligomer
Datura stramonium	DSA					+						(Gal$_{1\beta4}$GlcNAc)$_2$Man
Lycopersicon esculentum	LEA					+						β(1,4)GlcNAc oligomer
Solanum tuberosum	STA					+						β(1,4)GlcNAc oligomer
Oryza sativa	OSA					+						β(1,4)GlcNAc oligomer
Phytolacca americana	PWM					+						β(1,4)GlcNAc oligomer
Cytisus sessilifolius	CSA					+						anti-H, GlcNAc$_{1\beta4}$GlcNAc
Laburnum alpinum	LAA					+						anti-H, GlcNAc$_{1\beta4}$GlcNAc
Artocarpus integrifolia	Jacalin						+ +					Gal$_{1\beta3}$GalNAc
Griffonia simplicifolia I	GS-I						+ +		+ +			A subunit: αGalNAc; B subunit: αGal
Vicia villosa	VVA						+		+ +			GalNAc$_{1\alpha3}$Gal
Maclura pomifera	MPA						+		+ +			Gal$_{1\beta3}$GalNAc
Glycine max	SBA						+		+ +	+ +		GalNAc$_{1\alpha3}$Gal
Viscum album	VAA							+ +				βGal > GalNAc

TABLE 3
A Summarized Guide to the Selection of Lectin-Labeling Method

		TEM		
Markers	**Light microscopy**	**Preembedding method**	**Postembedding method**	**SEM[a]**
Fluorescent dyes	◎	X	X	X
HRP, ALP	◎	○	△	△
Ferritin	X	○	○	○
Colloidal gold	○	○	◎	◎

[a] Both secondary- and backscattered-electron images are included.

method, and the optimum dilution of lectin conjugates used should be determined for each batch. The indirect method employing biotinylated lectins is a useful tool with high sensitivity and with low background. This method works well for both light and electron microscopy. The background binding of avidin to endogenous tissue biotin, if any, may be overcome by the use of unlabeled avidin.[15]

A summarized guide to the selection of labeling methods is shown in Table 3.

D. REAGENTS

Conjugated and unconjugated lectins are commercially available through E.Y. Lab. (U.S.), Vector Lab. (U.S.), Sigma (U.S.), Boehringer-Mannheim (Germany), Honen Corp. (Japan), etc. Conjugated lectins for the direct method are fairly easily prepared.[12,16,17] In the indirect method, in addition to biotinylated lectins, avidin conjugates are also commercially available. Hapten sugars for each lectin should be prepared at the required concentration (Table 1).

E. LIGHT MICROSCOPY

The fluorescence-labeling method (Figure 2) — The avidin-biotin indirect method is described below. Specimens (paraffin section, cryostat section, semithin frozen section, cell monolayers, etc.) are

1. Dipped in 1% bovine serum albumin and phosphate-buffered saline (BSA-PBS) for 10 min
2. Treated with biotinylated lectin (5 to 25 μg/ml 0.1% BSA-PBS) for 30 to 60 min
3. Washed with PBS three times
4. Treated with fluorescein isothiocyanate (FITC)- or Rhodamine-labeled streptavidin for 30 min
5. Washed with PBS three times
6. Treated with 4′6-diamidino-2-phenylindole · 2 HCl (DAPI) (0.5 to 1 μg/ml PBS) for 10 min for nuclear staining
7. Observed under a fluorescence microscope

The HRP(ABC)-labeling method (Figure 3) — Specimens (paraffin sections, frozen sections, etc.) are treated as described in steps 1 to 3 above. Then the sections are

4. Treated with ABC (avidin-biotinyl peroxidase complex, Vector Lab.) for 30 min
5. Washed with PBS three times

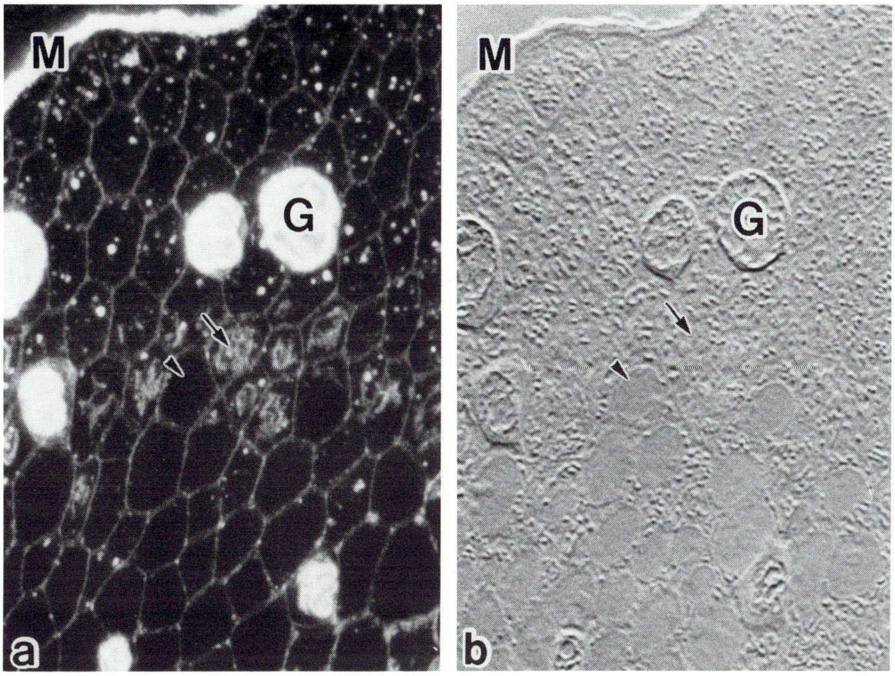

FIGURE 2. Light micrographs of a portion of the mouse jejunum epithelium.[18] (a) Semithin frozen section (1 μm) stained with biotinylated RCA and streptavidin-Texas Red (indirect method). Arrows show nucleus, and the arrowheads point out the supranuclear staining for RCA. (b) The corresponding Nomarski differential interference image. G, Goblet cell; M, microvilli. (Magnification × 1300.)

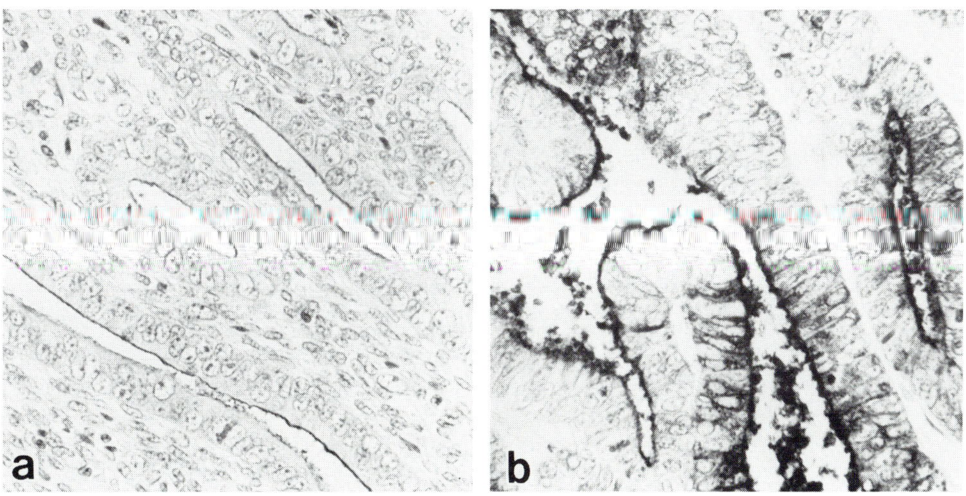

FIGURE 3. Light micrographs of the human endometrium. Stained with biotinylated PNA in combination with ABC complex (indirect method).[16] (a) Normal endometrium. PNA stains only the apical plasma membrane of the glandular cells. (b) Endometrial adenocarcinoma. The cytoplasm and the plasma membrane of some cancer cells are positively stained. (Magnification × 280.)

TABLE 4
An Example for the Evaluation of Lectin Staining

INCIDENCE \ INTENSITY	negative	weak	moderate	strong
50~100%	−	+	++	+++
10~50%	−	+	++	++
~10%	−	+/−	+	+
negative	−	−	−	−

Note: Staining intensity is graded as negative, weak, moderate, or strong. Incidence is also divided into four grades: negative, less than 10%, 10 to 50%, and more than 50%. According to these criteria, cases scoring + or more are regarded as positive.

Data from Reference 6.

6. Treated with 0.02% DAB (3,3′-diaminobenzidine)-0.005% H_2O_2 for 10 min
7. Poststained with hematoxylin or methylgreen
8. Observed under a light microscope

Endogenous peroxidase activity may be abolished, if necessary, by pretreatment with methanol and H_2O_2.[19] Meanwhile, the presence of the endogenous peroxidase activity in erythrocytes visualized under the microscope is useful as a positive control for the DAB reaction.

To estimate semiquantitatively the intensity of the lectin staining, we evaluate the staining pattern of each specimen on the basis of both staining intensity and incidence of positive cells. Intensity is graded on an arbitrary scale into four categories, and incidence is classified into four grades according to the percentage of positive cells (Table 4). Combining both intensity and incidence, the reactivity of each specimen is scored from − to + + + (Table 4). Specimens with a score greater than or equal to + are designated as positive cases, and the respective positive rate for each lectin staining is defined as the percentage of positive cases with respect to the total number of specimens.[6]

F. ELECTRON MICROSCOPY

The preembedding method for transmission electron microscopy (TEM) — The direct method is preferred when preembedding staining is done. HRP-labeled lectins are usually used in combination with thick frozen sections (20 to 50 μm) (Figures 4 and 5). Particulate markers, such as colloidal gold and ferritin, penetrate poorly into tissues and cells unless the specimens are pretreated in advance with detergents such as saponin or Triton X 100[20] and/or with freeze-thawing. These markers are particularly useful for the detection of carbohydrates distributed at the cell surface, and are also convenient to examine the rearrangement and redistribution of lectin receptors *in situ* (e.g., endocytosis) (Figures 6 and 7). In the preembedding method, homogenized cell subfractions can also be used for the study of the intracellular location of lectin receptors (Figure 8).

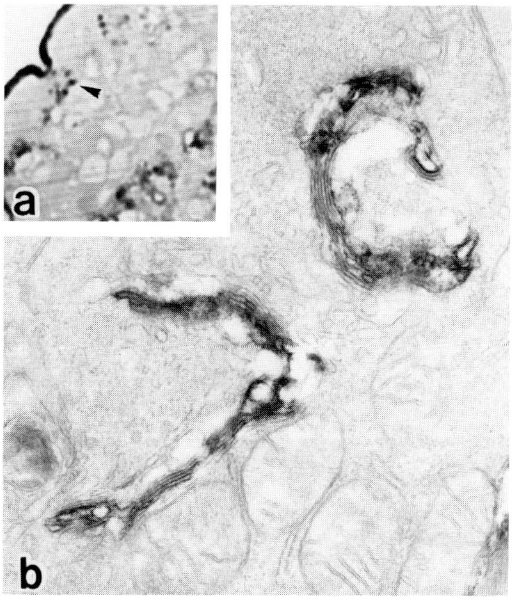

FIGURE 4. Light (a) and electron (b) micrographs of rat intestinal epithelium. Stained with HRP-labeled PNA (direct method). Preembedding staining. Thick frozen sections were incubated for 72 h, subjected to the DBA reaction, osmicated, embedded in Epon, then sectioned. Supranuclear staining in light microscopy (a) (arrowhead) corresponds to the positive staining in the Golgi apparatus in electron microscopy (b), especially the medial area of which is intensely stained. (Magnifications: a, × 850, b, × 23,000.)

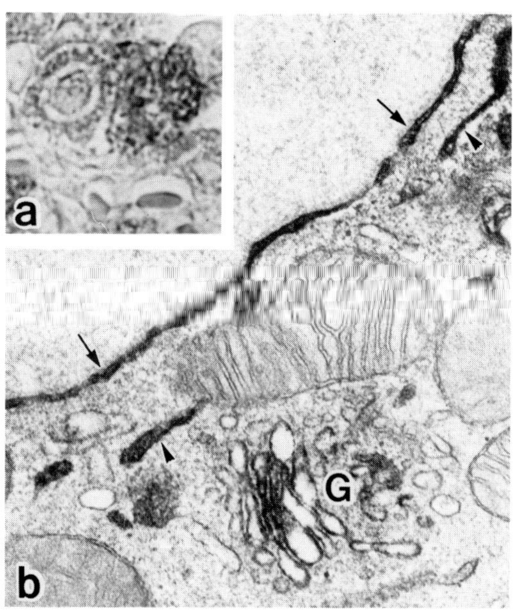

FIGURE 5. Light and electron micrographs of human gastric mucosa stained with ConA by the two-step method.[21] Preembedding staining. Thick frozen sections that had been incubated with unlabeled ConA (100 μg/ml) for 72 h were labeled with HRP (50 μg/ml). (a) Diffuse cytoplasmic staining in light microscopy corresponds to the positive staining in the endoplasmic reticulum (arrowheads), and (b) nuclear membranes (arrows) at the electron microscopic level. G, Golgi apparatus. (Magnifications: a, × 500, b, × 28,000.)

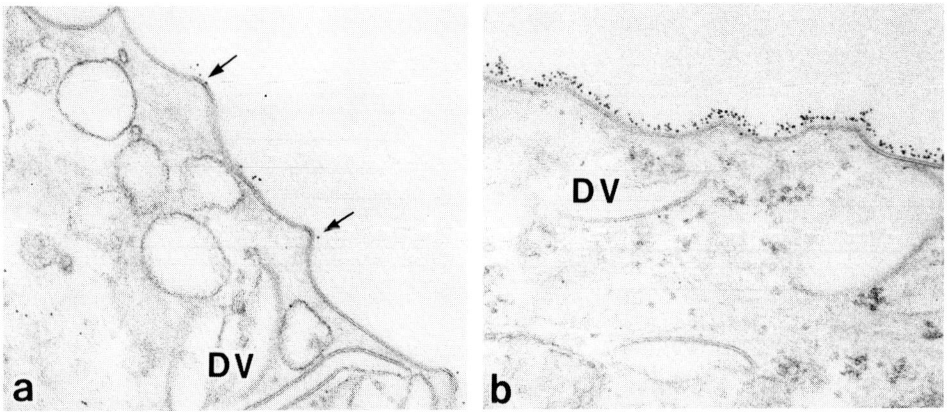

FIGURE 6. Electron micrographs of the apical region of superficial cells of the bladder mucosa from a normal rat.[22] Preembedding staining. Stained with ferritin-labeled lectins. (a) ConA. (b) RCA. ConA binding sites are restricted to the ridge regions (arrows) of the luminal membrane (a). On the other hand, RCA binds diffusely to the entire luminal surface. DV, discoid vesicle. (Magnification × 46,000.)

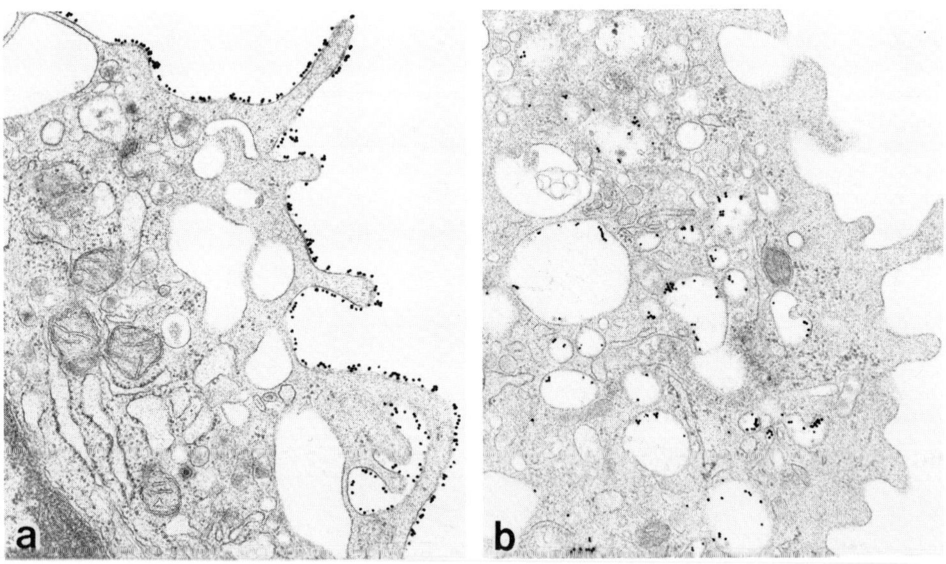

FIGURE 7. Electron micrographs of mouse peritoneal macrophage.[23] Preembedding staining. Stained with colloidal gold-labeled ConA. Two-step method was applied by use of unlabeled ConA in combination with colloidal gold-labeled HRP. (a) Incubation with ConA was carried out after glutaraldehyde prefixation. ConA binds evenly to the cell surface. (b) Incubated at 4°C, then at 37°C for 30 min. Bound ConA is incorporated into vesicular or vacuolar structures. (Magnifications: a, × 18,400; b, × 21,200.)

The procedure using the HRP method (Figures 4 and 5) is as follows. Specimens fixed with 2.5% glutaraldehyde, 4% formaldehyde, or an aldehyde mixture (2.5% glutaraldehyde + 2% formaldehyde) are

1. Cut into thick frozen sections (20 to 50 μm thick)
2. Treated with HRP-labeled lectin (10 to 50 μg lectin/ml PBS) at 4°C for 1 to 72 h

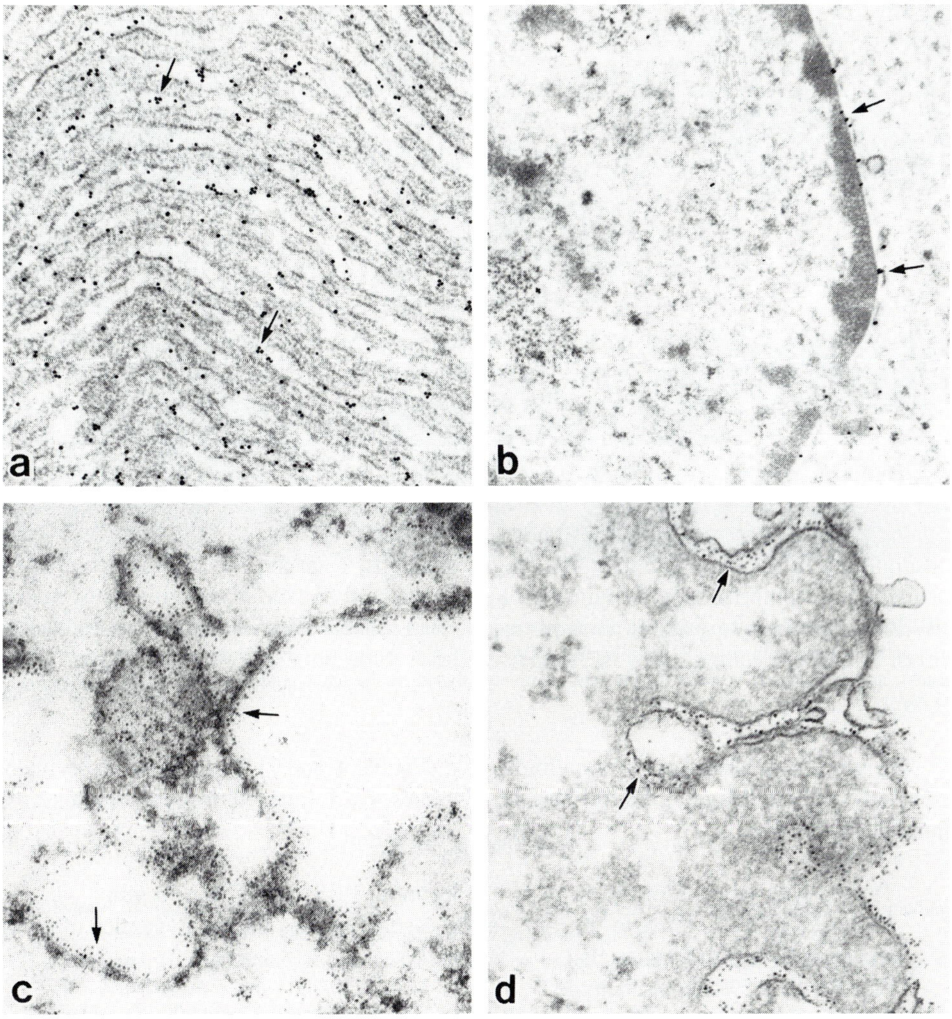

FIGURE 8. Electron micrographs to show intracellular stainings with ConA. Comparison between postembedding (a, b) and preembedding (c, d) methods. (a, b) Lowicryl K4M embedded [illegible] gland of rat [illegible] stained with biotinylated ConA and streptavidin-colloidal gold. (c, d) Subfractions of unfixed myeloma cells (c) and prefixed rat cerebrum (d) stained with ferritin-labeled ConA. Cisternal surfaces of the endoplasmic reticulum (a, c) and nuclear membranes (b, d) are positively stained (arrows). (Magnifications: a, b, × 16,600; c, × 55,200; d, × 47,800.)

3. Washed thoroughly with PBS
4. Treated with 0.05% DAB at room temperature for 10 min, followed by the addition of 1/20 volume of 0.1% H_2O_2 (final concentration = 0.005% H_2O_2)
5. Reacted at room temperature for 5 to 10 min
6. Washed with water
7. Postfixed with 1% OsO_4 for 1 h
8. Dehydrated, embedded in Epon, and processed for TEM without poststaining or with light lead staining

14. **Allison, R. T.,** The effects of various fixatives on subsequent lectin binding to tissue sections, *Histochem. J.,* 19, 65, 1987.
15. **Wood, G. S. and Warnke, R.,** Suppression of endogenous avidin-binding activity in tissues and its relevance to biotin-avidin detection systems, *J. Histochem. Cytochem.,* 29, 1196, 1981.
16. **Irimura, T. and Hirano, H.,** Application of the ferritin-labeling technique to the study of the molecular architecture of cell membranes, *Acta Histochem. Cytochem.,* 10, 230, 1977.
17. **Yokoyama, M., Nishiyama, F., Kawai, N., and Hirano, H.,** The staining of Golgi membranes with *Ricinus communis* agglutinin-horseradish peroxidase conjugate in mice tissue cells, *Exp. Cell Res.,* 125, 47, 1980.
18. **Takata, K. and Hirano, H.,** Use of fluorescein-phalloidin and DAPI as a counterstain for immunofluorescence microscopic studies with semithin frozen sections, *Acta Histochem. Cytochem.,* 23, 679, 1990.
19. **Streefkerk, J. G.,** Inhibition of erythrocyte pseudoperoxidase activity by treatment with hydrogen peroxide following methanol, *J. Histochem. Cytochem.,* 20, 829, 1972.
20. **Hedman, K., Pastan, I., and Willingham, M. C.,** The organelles of the trans domain of the cell. Ultrastructural localization of sialoglycoconjugates using *Limax flavus* agglutinin, *J. Histochem. Cytochem.,* 34, 1069, 1986.
21. **Ito, M., Takata, K., Saito, S., Aoyagi, T., and Hirano, H.,** Lectin-binding pattern in normal human gastric mucosa. A light and electron microscopic study, *Histochemistry,* 83, 189, 1985.
22. **Hirano, H., Nishiyama, F., and Yokoyama, M.,** Regional differences in cytochemical properties of the plasma membrane of rat bladder transitional epithelial cells, in *Recent Development of Electron Microscopy 1985,* Hashimoto, H., Kuo, K. H., Lee, K., and Ogawa, K., Eds., Japanese Society of Electron Microscopy, Tokyo, 1986, 275.
23. **Takata, K. and Hirano, H.,** Whole-cell-mount cytochemistry by the colloidal gold labeling method. Combined transmission and scanning electron microscopic study of ConA binding sites in mouse macrophages, *Histochemistry,* 81, 435, 1984.
24. **Carlemalm, E., Garavitom, R. M., and Villiger, W.,** Resin development for electron microscopy and an analysis of embedding at low temperature, *J. Microsc.,* 126, 123, 1982.
25. **Roth, J., Bendayan, M., Carlemalm, E., Villiger, W., and Garavito, M.,** Enhancement of structural preservation and immunocytochemical staining in low temperature embedding pancreatic tissue, *J. Histochem. Cytochem.,* 29, 663, 1981.
26. **Altman, L. G., Schneider, B. G., and Papermaster, D. S.,** Rapid embedded of tissues in Lowicryl K4M for immunoelectron microscopy, *J. Histochem. Cytochem.,* 32, 1217, 1984.
27. **Newman, G. R. and Hobot, J. A.,** Modern acrylics for post-embedding immunostaining techniques, *J. Histochem. Cytochem.,* 35, 971, 1987.

Chapter 4

IN SITU HYBRIDIZATION AT THE ELECTRON MICROSCOPE LEVEL

Barbara A. Hamkalo and Sandya Narayanswami

TABLE OF CONTENTS

0-8493-6012-9/93/$0.00 + $.50

A. INTRODUCTION

It is possible to localize specific nucleic acid sequences in specimens prepared for electron microscopy by combining *in situ* hybridization with colloidal gold detection. This technique, electron microscope (EM) *in situ* hybridization, will be referred to as EMISH.[1] DNA or RNA probes are labeled nonisotopically either by *in vitro* incorporation of nucleotides that are covalently linked to a ligand or by direct coupling to a ligand that can be detected subsequently via interaction with high-affinity ligands, such as antibodies coupled to colloidal gold particles. The availability of several different nonisotopic labels and colloidal gold particles in a large range of sizes enhances the flexibility and versatility of the technique. Localization of sequences in the EM is particularly applicable when very small structures are the target and when sequences are sufficiently closely linked so that they are indistinguishable in the light microscope.

B. METHODS

1. Probe Labeling

DNA probes can be labeled in a variety of *in vitro* reactions which incorporate modified nucleotides such as biotin-dNTPs or digoxigenin dUTP, both of which are commercially available. To date, most EM localizations have employed double-stranded DNA probes labeled by nick translation in the presence of bio-dUTP. Incorporation of this nucleotide is nearly equal to that of dTTP and the reactions are highly reproducible. Biotin-coupled nucleotides are available with variable carbon linkers that separate the biotin moiety from the base. C-11, C-16, and C-21 linkers give approximately the same signal, which is considerably greater than the C-4 nucleotide. Digoxigenin is incorporated at a somewhat lower level and does not give as large a signal as biotin in comparable experiments. This is probably due in part to differences in the efficacy of detection inherent in different antibodies. However, it is possible to amplify a small signal, as described in a later section. Oligonucleotide probes can be labeled with the same ligands by the activity of terminal transferase[2] or by the polymerase chain reaction.[3] An alternative to enzyme-based labeling involves the direct covalent attachment of the molecule *N*-acetoxy-2-acetylaminofluorene (AAF) to guanine residues of either DNA or RNA. Although modification is reproducible, AAF-labeled probes give a considerably smaller signal, which can be partially compensated for by increasing the probe concentration as well as by signal amplification. Despite the differences in efficiency of detection, the availability of several possible nonisotopic labeling schemes permits multiple sequence detection.

2. Specimen Preparation

To date, EMISH has been carried out primarily with whole mount preparations of mitotic chromosomes or interphase nuclei. Therefore, methods will be detailed for this type of preparation, although it is feasible to adapt the localization techniques for sectioned material. The simplest and most reproducible whole mount specimen preparation is by a modification of the Miller spreading technique[4] described below. After specimens are dried onto parlodion-carbon-coated gold electron microscope (EM) grids, the material is fixed to the film with glutaraldehyde. Fixed specimens can be stored at room temperature for several months without deterioration of morphology or hybridization signal.

3. *In Situ* Hybridization and Detection

Sequence localization involves denaturation of chromosomal DNA by one of several alternative treatments. Alkali denaturation described in *Procedures* below provides good morphological preservation concomitant with reproducible signals and very low background. Denaturation at 70°C in 2 × SSC/70% formamide gives about a twofold lower signal although morphological preservation is somewhat better. Double-stranded DNA probes in hybridization buffer are denatured by boiling followed by rapid quenching on ice. Prior to hybridization, specimens can be treated with RNase A to remove nascent RNA from transcribed sequences and/or Proteinase K to increase accessibility of target to probe. The necessity to include these steps must be determined empirically. Hybridization is carried out to two grids per small glass Reactivial by incubation in a water bath adjusted to the appropriate temperature. The temperature is determined by the composition of the hybridization buffer, the type of label, and, to a lesser extent, the base composition of the target. After hybridization and removal of unhybridized probe by rinsing the grids, hybrid sites are detected by incubation in either streptavidin-gold (for direct detection of bio-probes) or by sequential incubation in specific primary antibody and secondary antibody-gold. Two or more probes can be localized simultaneously by hybridization to a mixture of alternately labeled probes and detection with colloidal gold particles of different sizes (available from one to several hundred nanometers), each tagged to identify a single probe. Specimens are viewed in a standard transmission electron microscope without staining or heavy-metal shadowing since chromosomes and nuclei possess sufficient inherent electron density. Data are collected by photographing randomly selected fields. Although a qualitative analysis usually suffices for mapping, it is possible to quantitate the signal by simply counting the number of gold particles in a defined area.

C. EQUIPMENT AND REAGENTS

The techniques described below do not require unusual equipment or instruments. A few small, reusable accessories are required which are not necessarily standard laboratory possessions. These are listed with sources when possible: microcentrifugation chambers for deposition of specimens on EM grids — these are made in our shop and are diagrammed in Figure 1; gold EM grids — any EM supply catalog; single grid holders — E.F. Fullam #14890; Reactivials — Kontes #749001-000.

Standard chemicals and reagents are not listed.

Nick-translation kit — BRL
Bio-dNTP-Enzo — Sigma, BRL, Clontech
Digoxigenin-dUTP — Boehringer Mannheim
Photobiotin (Photoprobe) — Vector Labs
Terminal transferase — Ratliff Biochemicals (Los Alamos, NM)
N-Acetoxy-2-acetylaminofluorene — National Cancer Institute, Carcinogen Standard Reference Repository (Bethesda, MD)
Colcemid — Gibco
Nocodazole — Sigma
Glutaraldehyde — 8%, EM grade — any EM supply catalog
Bovine serum albumin (BSA) (nuclease-free, for hybridization buffer) — BRL
Colloidal gold sols (free or coupled) — Janssen Pharmaceutica, E-Y Labs, Electron Beam Sciences, Sigma, Polysciences
Streptavidin — BRL
BSA (for colloidal gold buffer) — Fraction V powder — Sigma
Fish gelatin — Sigma
Anti-biotin-Enzo — Vector
Anti-digoxigenin — Boehringer Mannheim
Anti-AAF — Dr. R. Baan; Netherlands Organisation for Applied Scientific Science Research; TNO Medical Biological Laboratory; P.O. Box 45; Lange Kleiveg 151; 2280 HV; Rijswijk; The Netherlands

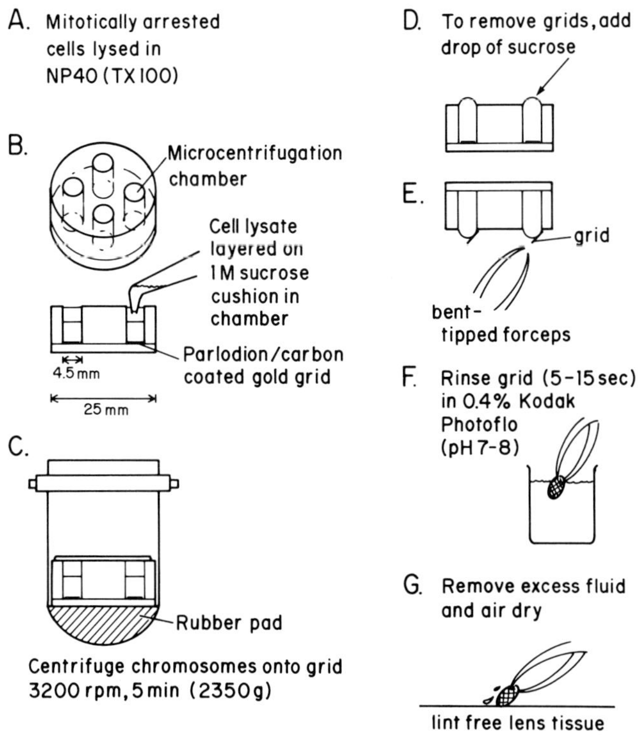

FIGURE 1. Schematic of the whole mount specimen preparation technique.

Biotin-conjugated antibodies — Jackson Immunoresearch
Silver intensification kits — Janssen Pharmaceutica; BioCell
Micro cleaning solution — E.F. Fullam

Stock solutions are made up as described below.

- *1% NP40* (per 100 ml)

  1 ml Nonidet P40
  99 ml Double-distilled H_2O
  Adjust pH to 8 to 9 with 0.1 *N* NaOH
  Millipore filter through 0.22 μm

- *20 × SSC*

  3 *M* NaCl
  0.3 *M* Sodium citrate
  pH 7.0

- *Alkali denaturation solution:* approximately 24 drops of 10 *N* NaOH in 100 ml 2 × SSC (final pH 12.0)

- *Hybridization buffer* (HB)

  50% Formamide
  10% Dextran sulfate
  1 m*M* trisodium EDTA (ethylene diamine tetraacetic acid)
  10 m*M* Tris, (pH 7.6)
  0.2% Ficoll
  1 mg/ml BSA
  40 μg/ml *E. coli* DNA
  0.6 *M* NaCl

Stock solutions of 0.5 *M* EDTA, 1 *M* Tris, 2% Ficoll, and 5 *M* NaCl are Millipore-filtered before use.

Coupling of protein to colloidal gold: this protocol is derived from de Waele et al.[5] It gives details for coupling streptavidin to 20-nm gold, but it can be adapted to the adsorption to colloidal gold of any other protein.

1. All glassware and test tubes used in this protocol are soaked and washed in a hot 1% solution of Micro cleaning solution, rinsed ten times with tap water, ten times with distilled water, and either autoclaved or baked at 200°C for 12 h. Polycarbonate tubes should be used for the centrifugation and storage of colloidal gold to prevent adsorption of the gold to the tubes and resultant flocculation.
2. Prior to protein adsorption, the minimum amount of protein required to protect the colloid from salt-induced flocculation is determined. The pH of 10 to 15 ml of a colloidal gold sol is adjusted with 0.2 *M* K_2CO_3 to 0.4 pH units above the isoelectric point (IEP) of the protein (IEP of streptavidin is 7.0, colloidal gold is adjusted to 7.4). Dilute buffer is used for pH adjustment because high salt concentrations cause unprotected colloids to flocculate. Colloid pH can be read directly with a gel-filled electrode. If one is unavailable, 10 drops (250 μl) of 1% PEG-6000 are added to 0.5-ml aliquots of gold before reading, in order to prevent flocculation on the electrode.
3. The appropriate protein concentration for coupling is determined by adding increasing amounts of protein (e.g., 0.1, 1, 2, 5 μg, etc.) to a series of ten tubes.
4. Water is added to each tube to bring the volume to 100 μl followed by the addition of 1 ml of colloid. After 10 min of incubation at 25°C in order to allow the protein to adsorb to the gold, 200 μl of 10% NaCl is added to each tube, followed by incubation at 25°C for at least 1 h.
5. The A^{520} of each sample is read and plotted against thc amount of protein. The plot will show a linear increase up to a certain protein concentration where the A^{520} plateaus. The amount of protein at the plateau plus 10% is the quantity of protein that should be added to 1 ml of gold in order to stabilize it against salt-induced flocculation.
6. A solution of 8% BSA is prepared prior to making the bulk preparation; BSA is used to quench the conjugation. The BSA is dissolved in 2 m*M* Na borate, pH 9 (Millipore filtered), dialyzed overnight at 4°C against the same buffer, and centrifuged in a refrigerated centrifuge at 100,000 × *g* for 1 h at 4°C in a swinging bucket rotor to clarify before use.
7. For the bulk preparation, the pH of 20 to 30 ml of gold solution is adjusted to 7.4 as above; an appropriate amount of streptavidin is added, followed by incubation for 2 min at 25°C with stirring.

8. The 8% BSA solution is then added to a final concentration of 1% and the gold is washed twice in 1% BSA buffer (1% BSA, 0.9% NaCl, 0.02 *M* sodium azide, 20 m*M* Tris, pH 8.2) in order to remove free streptavidin. Several 5-ml aliquots of the gold are placed in polycarbonate centrifuge tubes and the gold is pelleted by centrifugation at 23,400 × *g* in a refrigerated centrifuge in a fixed-angle rotor for 50 min at 4°C. Most of the supernatant is removed and 5 ml of fresh 1% BSA buffer is added. The pellet is resuspended and the wash is repeated. In each case, the pellet should be soft and loose in appearance. A hard compact pellet should be discarded, since it signifies that the gold preparation is massively aggregated and will fail to resuspend properly.
9. The second resuspended pellet is centrifuged as above and the soft pellet resuspended in one tenth the original volume of 1% BSA buffer. The gold preparation is stored at 4°C and is stable for several months.

D. PROCEDURES

1. Probe Labeling

Double-stranded DNA probes are nick-translated in the presence of modified nucleotide using a commercial nick-translation kit according to the supplier's instructions. Each reaction contains 2 μg of DNA in 100 μl, and, in addition to the other dNTPs, 5 μ*M* biodUTP or digoxigenin dUTP; 20 μCi ^{3}HdATP is included so that incorporation of biotin can be monitored.

1. A 20-μCi sample of ^{3}HdATP (17 Ci/mmol) is lyophilized in an Eppendorf tube for each reaction.
2. To each tube on ice, dNTPs and 10 μl of the PolI-DNAseI stock solution are added; the reactions are mixed well and incubated at 15°C for 90 min. The reaction is stopped by addition of 10 μl of STOP buffer (300 m*M* EDTA, pH 8.0) and 2 μl of 5 *M* NaCl.
3. In order to determine the specific activity, a 5-μl aliquot is removed from each reaction and spotted on a GF/C filter (Whatman). The filter is dried, placed in cold 5% trichloracetic acid (TCA) for at least 1 h, and washed with five changes of cold 5% TCA followed by two rinses in 95% ethanol and air drying. The filter is counted in a liquid scintillation counter and the specific activity expressed as cpm/μg DNA.
4. While the filters are drying, unincorporated nucleotides are removed from the reaction by Sephadex G50 chromatography using the spin column procedure.[6] A 100-μl reaction can be loaded on a 1-ml G50 column. The volume of the column eluate is measured and the amount of incorporated radioactivity determined by scintillation counting of a 5-μl aliquot of the eluate.
5. The amount of DNA recovered in the column eluate (μg) is determined by dividing the total radioactivity in the eluate by the specific activity of the probe preparation as determined in step 3. The probe is diluted to 4 μg/ml in either 100 m*M* NaCl, 10 m*M* Tris, pH 8.0, 1 m*M* EDTA, or distilled water and stored frozen.

The following alternative labeling schemes are not described in detail because, in our hands, they give either less reproducible or lower signals than nick-translated probes. Oligonucleotide probes can be labeled by the terminal transferase-directed tailing reaction as detailed in Moyzis et al.[2] Unlike DNA polymerase I, terminal transferase incorporates bio-dUTP less well than dTTP but incorporates biodCTP severalfold better than bio-dUTP. Thus, tailing reactions should use bio-dCTP for optimal labeling. Recently, Weier et al.[3] described an alternate labeling protocol which involves the extension of synthetic oligonucleotide primers via the polymerase chain reaction (PCR). This protocol permits a high level of incorporation of modified nucleotide in the course of generating a large amount of probe. Finally, photoactivatable forms of bio-dUTP are commercially available which can be directly coupled to a nucleic acid probe under a sun lamp.

Nonenzymatic labeling of probes can also be effected with the molecule AAF, as follows:

1. DNA is digested with DNAseI to a mean length of about 3 kb. It is denatured by boiling at 100°C for 5 min, and fast-cooled on ice.
2. DNA is diluted to 0.2 mg/ml in 1.6 m*M* sodium citrate, pH 7, 20% (v/v) ethanol. AAF is added from a stock solution in dimethyl sulfoxide (DMSO) (6 mg/ml) to a final concentration of 60 μg/ml (for approximately 5% modification). The final concentration of AAF determines the amount of substitution of the DNA; however, the final concentration of DMSO should not exceed 0.8% (v/v).
3. Reactions are run for 1 h at 37°C in the dark, during which time AAF covalently attaches to G residues in the DNA.
4. Unreacted AAF is removed by extracting the reactions six times with water-saturated diethylether.
5. The ratio of A^{305}/A^{260} is used to calculate the percent modification of the DNA, based on a standard curve.[7]
6. Probes are stored frozen in water at −20°C.

2. Specimen Preparation

Specimen preparation methods for EM *in situ* hybridization (EMISH) to whole-mount metaphase chromosomes involve the generation of large numbers of chromosomes from either established cell lines or diploid tissue, and the deposition of these chromosomes onto EM grids.[8] It is essential to begin with preparations containing large numbers of metaphase cells, since the small size of an EM grid limits the amount of material that can be examined in a single specimen and the large size of contaminating interphase nuclei reduces one's ability to analyze large numbers of individual chromosomes. Established adherent cell lines are an excellent source of metaphase chromosomes because they can be synchronized at metaphase by the use of agents such as Colcemid or Nocodazole and mitotic cells can be preferentially collected by shake off. If cell lines are unavailable, one can use diploid cells derived from blood or spleen cultures. Although these cells grow in suspension and do not divide very rapidly, it is possible to obtain sufficient numbers of dividing cells from both phytohemagglutinin (PHA) stimulated human peripheral blood blocked with methotrexate and from interleukin 2 stimulated mouse splenic lymphocytes after mitotic arrest with Colcemid.

A detailed protocol for a typical mouse cell line (L929) follows. The density of a culture which gives optimal metaphase arrest and the amount of time required for detergent lysis of arrested cells varies among cell lines and must be determined empirically.

1. Logarithmic cells are blocked in metaphase by treatment with Colcemid (50 to 80 ng/ml) or nocodazole (100 ng/ml) for 6 to 18 h.
2. Metaphase cells are selectively detached and collected by gently shaking the culture flask followed by pelleting in a 15-ml disposable conical centrifuge tube in a table top centrifuge (740 × *g*) and resuspension in 0.1 to 0.2 times the original volume of the same culture medium to a final concentration of about 1×10^5 cells/ml.
3. Cells are lysed by the addition of Nonidet P-40 (NP40) at 25°C. The stock solution typically used is 1% in double-distilled (dd) H_2O with the pH adjusted to 8 to 9 with 0.1 *N* NaOH, followed by filtering through a 0.22-μm Millipore filter. One drop of

the cell suspension is placed in one drop of 1% NP40 on a clean microscope slide and, after careful mixing, cell lysis is monitored by phase contrast microscopy. One should assess the mitotic index and determine the amount of time for at least 70% cell lysis and release of chromosomes that are visible as individual structures. When lysis is carried out with a 1:1 dilution of NP40, nearly 100% of L929 cells lyse within 1 min of detergent addition.

4. After optimal lysis conditions are determined, the chambers used for centrifuging the chromosomes onto EM grids are prepared (Figure 1). Chambers are made out of Plexiglass and each can be used to prepare four specimens at once. The chambers fit the 50-ml swinging buckets of any centrifuge. Chambers are filled with 1 *M* sucrose (pH 8.5) until a convex meniscus is formed; gold EM grids covered with a parlodion-carbon support film are rendered hydrophilic by placing in 95% ethanol for 1 min; they are then dipped briefly in 1 *M* sucrose and dropped into one of the four wells of the microcentrifugation chamber carbon film side up; they will fall to the bottom and rest on the Plexiglass surface. Form a cushion in the chambers by removing about half of the sucrose. Gold grids are used because copper grids release ions into solution during EMISH, resulting in a marked reduction in preparation quality.
5. When the chambers are ready, cells are lysed as determined in step 3 in a siliconized test tube. If necessary, mix the lysate carefully to separate released chromosomes. A 30- to 50-μm aliquot of the chromosome suspension is layered with a Pasteur pipette on top of the sucrose cushion in each well and preparations are centrifuged for 5 min at 2500 × *g* at 25°C in 50-ml swinging buckets in a table-top centrifuge.
6. After centrifugation, 1 *M* sucrose is added to each chamber to form a convex meniscus; upon inverting the chamber, grids float away from the bottom of the chamber and are held in the hanging droplet. Grids are picked from the droplets individually with forceps and rinsed in Photoflo solution (0.4% Kodak Photoflo 200, pH 8.5, Millipore filtered) to reduce surface tension and remove residual sucrose. Preparations are then air dried.
7. Prior to carrying out EMISH, each grid is checked by phase contrast microscopy (40×) in order to evaluate chromosome morphology and density. Optimal preparations should consist of well-spread groups of chromosomes, with few interphase nuclei. Approximately one metaphase per grid square is an optimal chromosome density.

3. Prehybridization Treatments

a. Fixation

1. The fixative is prepared fresh each day by diluting 1.25 ml of a newly opened ampoule of 8% glutaraldehyde into 100 ml of freshly prepared 2 × SSC in a beaker on a magnetic stirrer.
2. Grids are placed in single grid holders and immersed in the fixative, avoiding the formation of air bubbles on the grid surfaces. Air bubbles can be removed by carefully aspirating them with a Pasteur pipette.
3. Grids are fixed for 20 min at 25°C with gentle stirring. These conditions appear to be optimal, since increasing the concentration of glutaraldehyde or the time of fixation does not improve morphology. It is possible to fix for as little as 5 min, although chromosome morphology is less well preserved.
4. After fixation, individual grids are rinsed in the Photoflo solution and air dried. Fixed specimens can be stored at 25°C for at least 3 months without noticeable loss of signal or deterioration in chromosome morphology.

b. RNase and Protease Digestion

RNase digestion is typically included when transcribed sequences are localized, since nascent RNA can interfere with hybridization to template sequences.

1. After fixation, grids are incubated in 400 μg/ml of pancreatic RNAse in 2 × SSC for 30 min at 25°C. The grids are placed in 50-μl droplets of enzyme on the surface of a clean Petri dish.
2. RNAse is removed by washing the grids, in single grid holders, once in 2 × SSC for 10 min at 25°C, followed by a Photoflo rinse and air drying.

Proteinase K is employed to increase target sequence accessibility. Grids are placed in 50-μl droplets of 1 μg/ml autodigested Proteinase K in 2 × SSC for 30 min at 37°C in a Petri dish. Grids are washed, rinsed, and dried as above.

c. Denaturation

Denaturation is routinely performed using alkaline 2 × SSC. Denaturation should be carried out immediately prior to hybridization, since overnight storage of denatured preparations results in a reduced signal, presumably as a consequence of the reassociation of denatured DNA. Denaturation times used vary with the base composition of the target since (G + C)-rich DNAs (>60% G + C) require more extensive denaturation treatments (up to 30 min) to achieve a reproducible signal.

1. Denaturation is carried out at 25°C with the grids in single grid holders. The denaturation solution is prepared just prior to use by adding 10 *N* NaOH to 100 ml of 2 × SSC to a final pH of 12.0 (this is approximately 24 drops using a Pasteur pipette, or 1 ml).
2. Grids in single-grid holders are immersed in this solution for 2 min, avoiding the formation of air bubbles as noted above.
3. Grids are then rinsed in Photoflo to remove alkali and air dried. The air-drying step in the three preceding sections is not obligatory but represents a convenient stopping point during the procedure.

d.* In Situ *Hybridization

Hybridization of chromosomes or nuclei to dsDNA probes labeled with biotin, dinitrophenyl groups DNP, digoxigenin, or AAF is carried out in a standard hybridization buffer containing high salt, formamide, and dextran sulfate. Final probe concentrations typically are about 800 ng/ml for biotin, DNP, and digoxigenin, and 5- to 10-fold higher for AAF. Hybridizations are carried out at 30°C, except in the case of AAF-substituted probes, where hybridization is performed at 27°C (Landegent et al., 1984), due to the significantly reduced stability of hybrids containing AAF residues. In general, overnight hybridization is most convenient, although a good signal is obtained after 6 to 7 h with repetitive sequence targets. The conditions detailed are used with double-stranded probes, regardless of modification.

1. For 100 μl of hybridization solution, mix 79 μl of HB, 1 μl of 7 mg/ml yeast tRNA (phenol extracted, reprecipitated, and dissolved in dd H_2O) and 20 μl of probe DNA (4 μg/ml) in a siliconized Eppendorf tube.
2. This mixture is placed in a boiling water bath for 5 min to denature the DNA, followed by rapid cooling on ice for 2 min. Tubes are spun briefly in a microfuge (up to maximum speed and then down) in order to bring droplets of condensation to the bottom of the tube, and tubes are kept on ice prior to use.

3. A 50-μl aliquot of the hybridization solution is pipetted into a siliconized Reactivial.
4. Two grids, back to back (i.e., specimen side facing out), are placed in each vial using forceps. The vials are capped and incubated overnight in a water bath at the temperature selected for hybridization.
5. After hybridization, grids in individual grid holders are rinsed in a beaker with stirring three times for 20 min each in 100 ml of 2 × SSC at 25°C to remove unhybridized probe. Grids are kept moist between this and subsequent steps to retain antibody reactivity.

Incorporation of minor modifications in the protocol described above permits localization of multiple probes. Probes are added to the hybridization mixture simultaneously and denatured together. As a result, probes used for multiple localizations should be at higher stock concentrations (about 20 μg/ml) than the 4 μg/ml recommended for a nick-translated probe used alone. They can be stored without dilution after purification over Sephadex G50. Several probes can then be added to hybridization buffer without drastically changing the concentrations of buffer components. When AAF is used to label one probe, hybridization is performed at 27°C, which is optimal for AAF hybrid formation.

e. Signal Detection

Hybridized preparations are first reacted with specific antibodies to the ligands used to modify DNA probes followed by incubation with an appropriate secondary antibody immobilized on colloidal gold. In addition, streptavidin or avidin can be used to locate biotin-containing hybrids. Colloidal gold and colloidal gold complexed with antibodies or streptavidin/avidin in sizes from 1 to 40 nm are commercially available. Colloidal gold can be prepared in the laboratory with minimal equipment. The protocol involves reduction of gold chloride by appropriate reducing agents selected according to the size of colloid desired.[9]

Primary antibodies are diluted to an appropriate concentration in phosphate-buffered saline (PBS) supplemented with BSA and NaCl. The presence of BSA and high salt reduces nonspecific binding of antibodies to the grid film to virtually zero. Such nonspecific adsorption represented the major source of background observed during the development of this technique. Each antibody batch is titrated using a test system (in our case mouse satellite DNA and mouse metaphase chromosomes) in order to determine optimal antibody dilution and reaction time. The conditions given are suitable for multiple-step antibody sandwich reactions and multiple labeling.

1. Grids are placed in 50-μl droplets of affinity-purified primary antibody (e.g., for antibiotin antibodies, 0.8 μg/ml in PBS, 2 mg/ml BSA, 0.5 *M* NaCl) in Petri dishes with the specimen side up and incubated in a moist atmosphere for 4 h at 37°C. A plastic slide box lined with dampened paper towels forms a convenient moist chamber.
2. Unbound antibody is removed by rinsing grids at 25°C three times with 100 ml of PBS/NaCl for 10 min each in single grid holders in a beaker with stirring. Grids should not be allowed to dry between this step and the secondary antibody/gold reactions, described below.

Secondary antibody/gold reactions are carried out as follows.

1. 500 μl of the 15- to 20-nm gold stock is pelleted at 12,000 × *g* for 15 min at 4°C in a fixed-angle rotor in a refrigerated centrifuge. 5-nm gold stocks are pelleted at 23,400 × *g* for 45 min to 1 h in order to separate unabsorbed protein which will be in the supernatant. In both cases, the pellet should be soft and loose in appearance.

2. Most of the supernatant, containing released protein, is discarded, except for a small quantity directly over the gold. The remaining pellet is resuspended in the same volume of Millipore-filtered (8-μm followed by 0.22-μm filters) 1% BSA buffer. The BSA can be substituted with 1% fish gelatin with comparable results.
3. Gold is diluted in 1% BSA to a concentration determined empirically for each batch, followed by low-speed centrifugation of the diluted sample in a table-top centrifuge with a swinging bucket rotor (400 to 700 × *g* for 10 min at 25°C) in order to remove aggregates. Several 50-μl aliquots of the supernatant are carefully removed with a P200 Pipetman and placed in clean Petri dishes. Grids are individually placed in the 50-μl droplets and incubated in a moist chamber at 25°C. Although overnight incubation is convenient, the time can be shortened based upon that required to give a visible signal. Highly repeated sequences (e.g., mouse satellite) exhibit a good signal after as little as 2 h of incubation in gold.
4. Grids are rinsed free of unbound gold at 25°C in single grid holders three times, for 20 min each in 100 ml 1% BSA buffer in a beaker with stirring. They are then rinsed in Photoflo and air dried. Specimens can be viewed in the EM without further contrasting, since chromosomes and gold are sufficiently electron dense. However, greater contrast can be obtained by staining or heavy metal shadowing.

In general, 15- to 20-nm gold is used because the particles are readily identified even at low magnifications (2600× to 3300×). However, smaller or larger particles also can be used. The low electron density of 5-nm particles makes them most suitable for use when a large signal is expected, since large clusters of these particles are readily located even at low magnifications. One-nanometer particles are most useful when accessibility is a problem. Although they are invisible at the magnifications used to examine chromosomes, they can be identified by silver intensification (see below).

Signal detection schemes for multiple-probe labeling are basically the same as for single-probe labeling, except that precautions are necessary to avoid cross-reaction among immunological reagents. Avidin or streptavidin can be used to label bio-probes in conjunction with antibodies that recognize other modifications. Primary antibodies generated in enough different species are available so that differently substituted probes can be combined, at least pairwise. In this case, grids are incubated with a mixture of two or more primary antibodies, followed by incubation with a mixture of appropriately labeled secondary antibodies on different-sized gold particles.

The choice of gold particle sizes for multiple-probe labeling depends upon parameters such as signal size and the distance between sequences of interest. At the magnifications used to examine metaphase chromosomes, 5-nm gold is difficult to detect unless the signal is large. However, 10-, 15-, 20-, and 30-nm particles are all visible and distinguishable from each other in the EM. Although we have only carried out double labeling to date, the availability of four modified probes and different sizes of gold should permit simultaneous detection of four or more loci.

It is possible to enhance a small signal by one of two amplification approaches. (1) Antibody sandwich — antibody sandwich amplification is described for biotin-labeled probes but can be generalized by applying layers of primary and secondary antibody when other ligands are used. After hybridization as described, grids are incubated in antibiotin followed by incubation with biotin-coupled secondary antibody under the same conditions. These steps can be repeated one or more times if necessary. Amplified signals ultimately are detected with streptavidin gold. Although moderately repeated sequences (about 100 copies) are detectable without amplification, two cycles of amplification are sufficient to produce a signal which is visible in the light microscope.[10] One drawback of several rounds of signal amplification in EMISH is a loss of spatial resolution. (2) The second approach to signal amplification, silver intensification, involves increasing the size, hence visibility, of individual particles.[11,12] Layers of metallic silver are deposited directly on colloidal gold particles. Amplification by silver intensification is particularly useful when reduced accessibility dictates the use of very small particles and is imperative when 1-nm gold is used.[13] The procedure is rapid and results in a severalfold increase in particle size within about 10 min. Several different kits are available and any one can be used according to the supplier's instructions.

E. RESULTS

Prior to the application of EMISH to map an unknown, as a positive control it is imperative to localize a previously mapped sequence in order to evaluate specificity. A moderately (about 100 copies) or highly reiterated sequence which exists as tandem arrays represents an excellent test system in which to define conditions. Although the procedures described above are generally applicable, it may be necessary to alter, for example, denaturation time, probe concentration, or to include RNase or protease steps. The test probe should give a reproducible signal with respect to the appropriate chromosomal location on most, if not all, relevant chromosomes. There are several routine negative controls that should be run when a new probe and/or system is investigated. These include a series of experiments in which a single component is excluded and a series in which a nonspecific modified probe and an unlabeled specific probe are hybridized.

Under most circumstances, background labeling of the support film is very low. However, some probes appear to stick to the film more than others and give rise to significant gold deposition on the film. The cause of this nonspecific sticking is not apparent. Nevertheless, such background can be tolerated since there is an inherent repulsion between colloidal gold and chromatin so that, even if the support film shows significant labeling, chromosomes typically are surrounded by a halo free of background. One can be confident of a positive signal even with very few gold particles, as long as both sister chromatids are labeled. In our experience, this is nearly always the case.

One situation that can give rise to a lower than expected signal based on results with a comparable target is differential accessibility of sequences in condensed chromosomes. This often can be minimized by including a low concentration (about 0.2 m*M* EDTA) in the sucrose cushion through which chromosomes are centrifuged onto grids in order to partially decondense the chromosome.[12] When multiple simultaneous localizations are carried out, a common artifact is nonspecific labeling due to cross-hybridization between different probes cloned in vectors that share some common sequences. This artifact can be eliminated by isolating the cloned inserts for all but one probe prior to labeling and hybridization. In such a situation, the least abundant target is detected by the intact recombinant molecule so that networking among DNA molecules is permitted in order to increase the signal.

Figure 2 illustrates typical localization results with moderately or highly repetitive DNA probes. Figure 2a shows hybridization of a centromere-restricted repetitive DNA to mouse chromosomes. This probe labels the entire centromere region but the patchy distribution of gold particle clusters suggests that it actually occurs in several noncontiguous blocks. Figure 2b is an example of a low signal which has been amplified by two successive antibody treatments. Although small, the signal is reproducible on both sister chromatids. An interphase nucleus hybridized with a probe for 5S ribosomal DNA is shown in Figure 2c. Clusters of gold particles are distributed throughout the structure. Finally, Figure 2d represents a double-label hybridization in which two alternately labeled human chromosome 1 centromeric probes were localized to a small chromosome fragment derived from this chromosome. One sequence was labeled with biotin and detected with 20-nm gold and the other was coupled

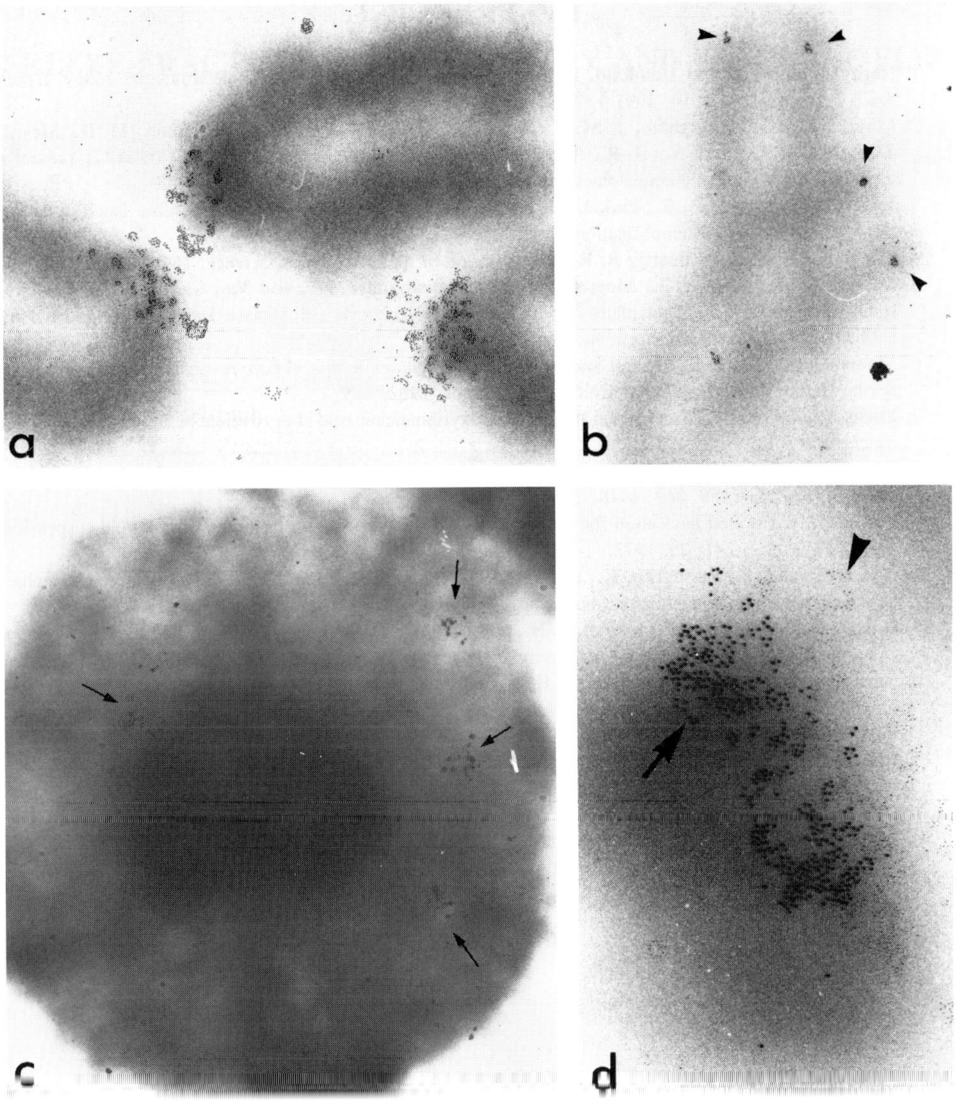

FIGURE 2. Examples of DNA sequence localization at the EM level. (a) Mouse chromosomes after hybridization with a biotin-probe and immunogold (15 nm) detection (× 114,100). (b) The ends of mouse chromosomes after hybridization with a biotin-labeled telomere oligomer and detection via antibiotin and streptavidin gold (× 114,100). (c) A frog cell nucleus hybridized to a biotin-5S RNA probe (arrows show hybrid sites) (× 99,100). (d) A chromosome fragment derived from human chromosome 1 labeled with biotin probe and 20-nm gold (arrow) and AAF probe and 5-nm gold (arrowhead) (× 271,900).

to AAF and labeled with 5-nm gold. The relative positions of these sequences would be indistinguishable in the light microscope.

The current methodology is not capable of reproducibly detecting single-copy sequences. However, considerable effort is directed towards modifying steps in the protocol to increase its sensitivity. Nevertheless, the current methodology is capable of rapidly and precisely locating reiterated sequences in whole mount preparations. It is particularly useful when small chromosomal structures or nuclei are the targets. In addition, the general methodology should be applicable to the localization of sequences in sectioned material.

A. INTRODUCTION

Freeze-fracture is a very useful method for investigating the macromolecular architecture of the cell membrane. Cells are frozen at the temperature of liquid nitrogen in the presence of a cryoprotectant (e.g., glycerol), then the frozen specimen is cracked with a knife. The fracture plane passes through the hydrophobic middle of lipid bilayers. As a result of this split there are two fracture surfaces: the fracture face on the exterior leaflet is the exoplasmic fracture face (EF face), whereas that on the interior leaflet is the protoplasmic fracture face (PF face). Both exposed faces are evaporated with platinum (Pt) and carbon (C) to create thin metal replicas. In the resulting replicas, smooth areas correspond to the fatty acid tails of phospholipids and the intramembranous particles (IMPs) represent proteins. Freeze-fracture electron microscopy reveals the precise macromolecular architecture of the membranes, so that the number and distribution of a wide variety of transmembrane macromolecules, revealed as IMPs, can be determined. Freeze-fracture when used alone, however, offers little information on the biochemical composition and function of cell membrane components.

The combination of freeze-fracture with cytochemical labeling of individual cell membrane molecules, freeze-fracture cytochemistry, would provide direct evidence of the chemical nature and topology of the cell membrane components. To date, four major approaches have been developed for freeze-fracture cytochemistry: label-etch,[1] label-fracture,[2] fracture-label,[3,4] and replica-label.[5-7]

In principle, the conventional freeze-fracture methods provide no information concerning the hydrophilic (actual) cell surfaces. Recently, Pinto da Silva and co-workers[8-10] have developed a new variant of freeze-fracture, called "fracture-flip", to provide extended views of the surfaces of cell membranes at close to macromolecular resolution. Fracture-flip proceeds as follows:

1. Freezing: fixed or unfixed cells are frozen.
2. Freeze-fracture: freeze-fracture splits the membranes into protoplasmic and exoplasmic halves.
3. Carbon-fixation: the exposed EF faces are evaporated with C. This produces a C cast with stabilized membrane halves.
4. Washing, flipping (inversion), and Pt-shadowing of the carbon cast: the C casts are thawed.

Contrary to conventional freeze-fracture, the C casts are not digested with acids or bases. Instead, they are repeatedly rinsed in distilled water to wash away unfractured cells. The exoplasmic halves of membranes remain attached to the C casts, which are transferred to electron microscope grids. The C casts on grids are flipped (inverted). Pt-shadowing creates the informative three dimensional images of membrane surfaces.[11-13] The most important advantage is that the combination of fracture-flip with cytochemical methods, fracture-flip cytochemistry, provides high-resolution views of the distribution of antigens and receptors on cell surfaces.[8,12,13] Table 1 lists the techniques for freeze-fracture and fracture-flip cytochemistry and their simplified protocols.

B. FREEZE-FRACTURE CYTOCHEMISTRY

1. Label-Etch

Pinto da Silva and Branton[1] developed freeze-etching as the first cytochemical approach to characterize freeze-fractured images. This procedure involves the cytochemical labeling of antigens or receptors on cell surfaces. The labeled cells, frozen in the absence of cryoprotectant, are freeze-fractured and the preparations allowed to etch at $-100°C$ for 1 min

TABLE 1
Protocols for Freeze-Fracture Cytochemistry and Fracture-Flip Cytochemistry

Conventional F-F	Label-etch	Label-fracture	Fracture-label	Replica-label: Replica-stain	Replica-label: Postshadow-label	Replica-label: Postreplication-label	Fracture-flip
Fixation	Fixation	Fixation	Fixation	Fixation	Fixation	Fixation	Fixation
	LABEL*	LABEL*					LABEL*
Cryoprotect		Cryoprotect	Cryoprotect	Cryoprotect	Cryoprotect	Cryoprotect	Cryoprotect
Freeze-fracture	F-F	F-F	F-F	F-F	F-F	F-F	F-F
	Etching						
Pt-Shadowing	Pt	Pt		Pt	Pt	Pt	
C-Evaporation	C	C		C		C	C
Thawing	Thawing	Thawing	Thawing	Thawing	Thawing	Thawing	Thawing
Digestion with bleach	Bleach						
Washing	Washing	Washing	Washing	Washing	Washing	Washing	Washing
			LABEL	LABEL	LABEL	LABEL	LABEL
			Washing	Washing	Washing	Washing	Washing
Pick up on grids	Pick up	Pick up	Embedding	Pick up	Embedding	Embedding	Pick up
EM Observation	EM	EM	Thin section	EM	Thin section	Thin section	Flip
			EM / CPD		EM	EM	Pt
			Pt/C				EM
			Bleach				
			EM				

LABEL*: Cytochemical labeling can be performed before or after fixation — F-F, freeze-fracture; Pt, platinum; C, carbon; CPD, critical-point dry.

or more. In freeze-etching, the distribution of the cytochemical marker over the membrane surface exposed by sublimation can be compared to that of the IMPs exposed by freeze-fracture. It was used first to show that IMPs corresponded to transmembrane proteins.[1,14,15] However, additional characterization of freeze-fracture replicas using freeze-etch has been lacking, mainly due to the technical difficulty in the cryofixation processing. In freeze-etch, it is usually necessary to freeze specimens without cryoprotectants. The zone of good freezing is therefore limited to a thickness of less than 10 μm, and only relatively narrow areas of the cell surface are available for observation.

2. Label-Fracture

In the label-fracture procedure, the cells in suspension are labeled with colloidal gold probes before being freeze-fractured and replicated.[2] In freeze-fracture cytochemistry, colloidal gold is generally chosen as the electron-dense marker. In contrast to other electron-dense macromolecules such as ferritin, colloidal gold is easily visualized on Pt/C replicas and is also easily prepared in many sizes.

Fixed, labeled cells are impregnated in 30% glycerol, frozen, freeze-fractured at −130°C, and immediately replicated by Pt/C evaporation. Contrary to conventional freeze-fracture replication, Pt/C-replicated specimens are not digested with acids or bases (e.g., chromic acid, bleach). Instead, they are repeatedly washed in distilled water to free cellular components and unbound gold particles. When observed under an electron microscope, gold particles bound to the outer half of the plasma membrane, which remains associated with the replica after the washing procedure, can be readily recognized through the replica. The technique allows the visualization at high resolution of cytochemical markers of cell surface sites over large cell surface areas.[2,16,17] This distribution is seen superimposed on the image of an EF face (Figure 1).

3. Fracture-Label

While label-fracture is restricted to isolated cells or monolayers, fracture-label permits the cytochemical characterization of the PF and EF faces of cell membranes (and also cross-fractured cytoplasm and intracellular organelles) in both isolated cells and tissue samples.[3,4] In the fracture-label procedure, the specimen is tissue bulk or cells embedded in bovine serum albumin gel. Cells and tissues are fixed with conventional aldehyde fixatives, impregnated in glycerol, frozen, freeze-fractured by grinding under liquid nitrogen, thawed, deglycerinated, and then exposed to a specific cytochemical reagent labeled with an electron-dense marker such as ferritin or colloidal gold. There are two variants of fracture-label, thin section[3] and critical-point drying.[4] In thin-section fracture-label, the labeled tissue samples are processed for resin embedding, and thin sections are cut perpendicular to the labeled faces (Figure 2). In critical-point drying fracture-label, the labeled samples are critical-point dried then replicated with Pt/C. The Pt/C-replicated specimens are digested with bleach. Thus, the critical-point drying fracture-label permits the *en face* observation of the label along the fracture plane. In the fracture-label procedure, the split membrane halves are exposed to aqueous environments. They therefore undergo reorganization into interrupted bilayers. The fracture-label replica is thus not comparable with conventional freeze-fracture replicas.[18]

Recently, the fracture-label technique was combined with backscattered electron imaging to visualize the *in vivo* distribution of lectin-colloidal gold-binding sites on freeze-fractured membranes in tissues and cells.[19] In this technique, labeled specimens were critical-point dried and covered with carbon alone. Lectin-gold labeling revealed by a backscattered electron imaging system showed a labeling pattern similar to that seen by the critical-point

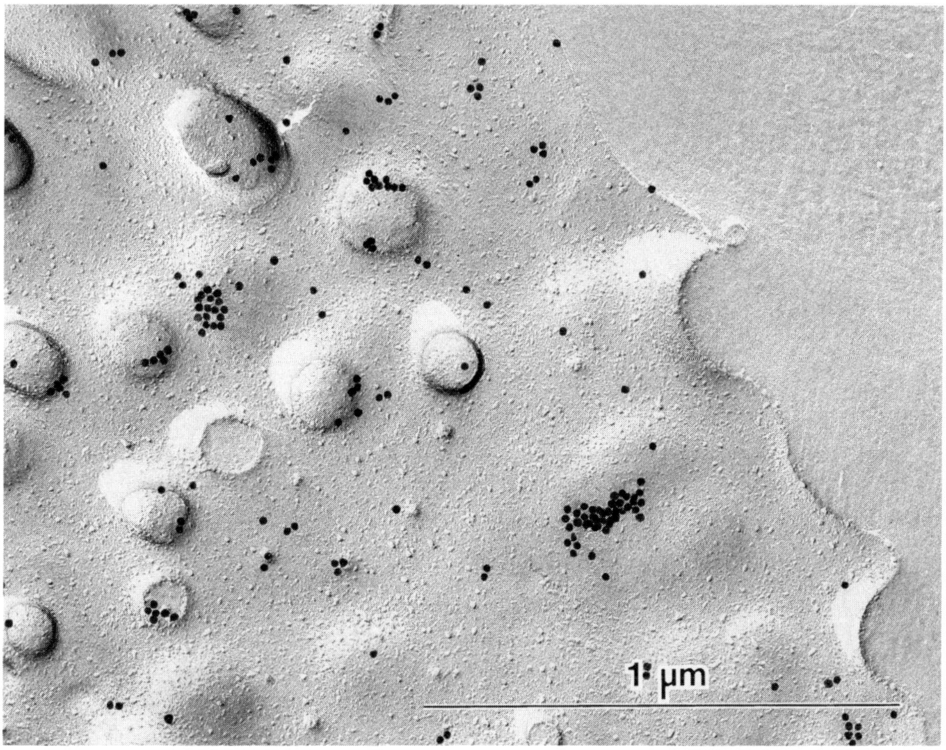

FIGURE 1. Label-fracture of WGA binding sites on a rat macrophage.

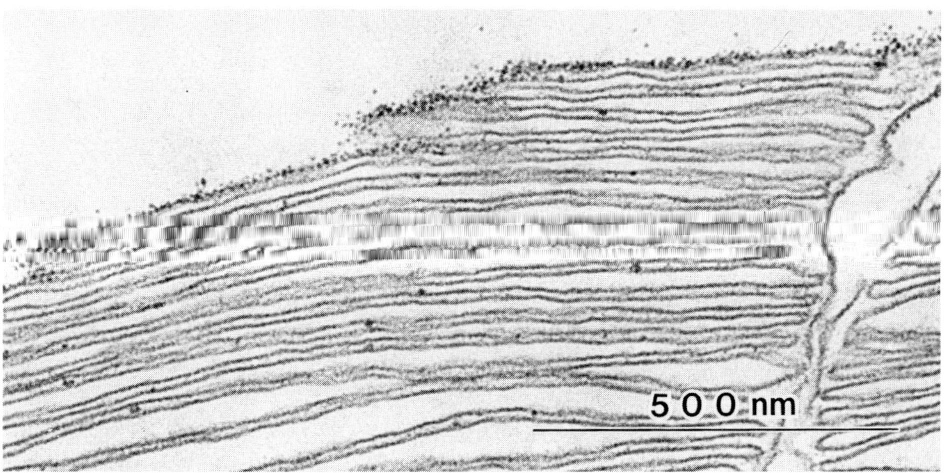

FIGURE 2. Fracture-label of cationized ferritin-binding sites on disk membranes of a rat photoreceptor cell.

drying fracture-label method. The backscattered electron imaging provides additional information concerning the distribution of labeling with respect to the three-dimensional organization of fractured faces.[19]

4. Replica-Label

As mentioned above, the rationale of label-fracture is that the cytochemical marker is retained in place because the exoplasmic (outer) halves of the membranes are stabilized by, and remain attached to, the Pt/C replica. It should therefore be possible to label the outer surfaces of membranes after the specimens are freeze-fractured, replicated, and washed. To date, two main approaches have been developed for replica-labeling.

Replica-staining label-fracture — Andersson-Forsman and Pinto da Silva[5] developed a new version of label-fracture, called ''replica-staining label-fracture'' for the cytochemical characterization of membrane surfaces. They showed that cytochemical labeling of cell surfaces can be performed using conventionally prepared Pt/C replicas. After freeze-fracture, Pt/C replication, and thawing, the replicas are washed with an appropriate physiological buffer or distilled water. The Pt/C replicas, along with their attached outer halves of membranes, are floated on droplets of cytochemical reagents. After cytochemical staining, the labeled replica fragments are rinsed several times in distilled water and are mounted on grids.

Postshadow labeling (sectioned replica label) — In the postshadow labeling procedure, tissue is conventionally freeze-fractured and then lightly shadowed with Pt, but C deposition is omitted.[6] Thus, the tissue is still accessible to cytochemical probes. After Pt shadowing, the specimens are thawed and exposed to cytochemical reagents. Subsequently, the labeled tissue and its attached replica are processed for resin embedding. Conventional transmission electron microscopic examination of sectioned, labeled replicas reveals that cytochemical markers can be seen associated with certain IMPs. Recently, Dinchuk et al.[7] introduced a new version of sectioned replica label, called ''postreplication labeling''. In this method, cells are conventionally freeze-fractured and shadowed with Pt and C. The Pt/C replicas are thawed, washed with a physiological buffer, exposed to cytochemical reagents, and are processed for resin embedding. In both methods for sectioned replica label, only relatively small areas of the replica are available for observation.

C. FRACTURE-FLIP AND FRACTURE-FLIP CYTOCHEMISTRY

To understand how and why fracture-flip works, it is necessary to explain the carbon-stabilizing and flipping steps in more detail. The practical procedure for fracture-flip is described here in instructive style, the interpretation is briefly discussed and, finally, some examples of the application of fracture-flip and fracture-flip cytochemistry are presented.

1. Standard Protocol

Conventional fracture-flip (as opposed to a modified fracture-flip, ''fracture-flip/Triton X-100'' described later) can reveal only outside surfaces of free cells, including isolated cells, cells in monolayer, and isolated cell organelles.

1. Fixation: specimens are chemically fixed with conventional fixatives (e.g., 2% formaldehyde and 2% glutaraldehyde in 0.1 *M* phosphate buffer, pH 7.4).
2. Washing and cryoprotection: fixed specimens are washed with the same buffer several times, and impregnated stepwise in 30% glycerol in 0.1 *M* phosphate buffer, pH 7.4.
3. Freezing (Figure 3, step 1): cryoprotected specimens are mounted on single- or double-replica specimen carriers, and frozen by plunging them into liquid nitrogen slush.

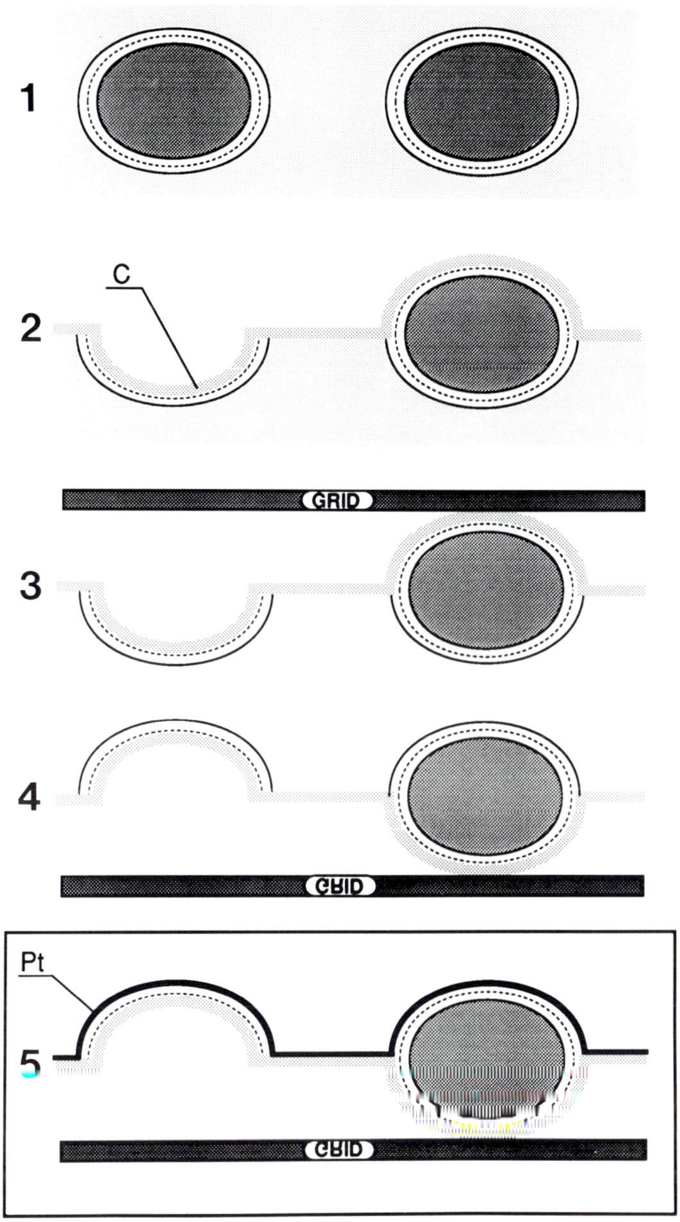

FIGURE 3. Procedure for fracture-flip and cytochemistry. The diagram illustrates the main steps.

4. Freeze-fracture: freeze-fracture is performed at −120°C to −130°C.
5. Carbon fixation (Figure 3, step 2): the fracture faces are stabilized by the unidirectional deposition of C from a standard electron-beam gun mounted at 90° to the specimen plane. A suitable thickness of C will have to be empirically determined, but 10 to 20 nm will probably suffice for C stabilization of membrane halves.
6. Thawing and washing: to release the C cast from the specimen carrier, the carrier is immersed gently in distilled water. After floating off, pieces of the C cast are transferred to distilled water using a platinum wire loop, and floated sequentially through four washes in distilled water.

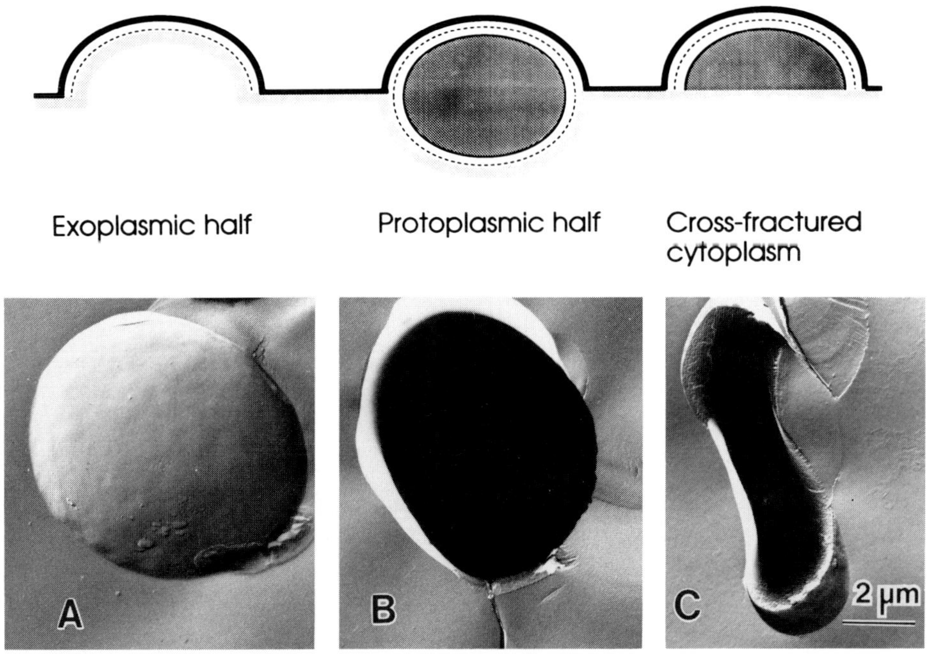

FIGURE 4. How to interpret images reproduced by fracture-flip. (A) to (C), Fracture-flipped human erythrocytes.

7. Pick up on grids (Figure 3, step 3) and flipping (Figure 3, step 4): after washing, the C casts are transferred to electron microscope grids. For most purposes, Formvar-coated 100- and/or 150-mesh copper grids are most suitable. These grids are bent at the edge so that the grid can be held parallel with the floating carbon cast. The Formvar-coated surface of the grids is pushed against the carbon cast and lifted from the water. The C casts on Formvar-coated grids are flipped. Therefore, the casts are mounted with the carbon side down. Excess water is removed by blotting with filter paper. Double-sided adhesive tape is attached to the surface of a flat-topped specimen stage, and 3-mm-wide strips of white paper are placed across the surface of the tape. The grids are placed over the paper with grid edges touching the adhesive tape.
8. Pt/C shadowing (Figure 3, step 5): Pt/C is evaporated unidirectionally at an angle of 45° for 10 s (2 nm thick). For rotary shadowing, the angle is 20 to 22°.
9. Observations: these casts (actually carbon-stabilized membrane halves) are observed under a transmission electron microscope. Electron micrographs are taken stereoscopically by tilting the specimens ±6°.
10. Cytochemistry: cytochemical labeling can be performed before or after freeze-fracture (Table 1).

2. Interpretation

In the fracture-flip technique, the carbon cast with carbon-stabilized membrane halves is not treated with acids or bases. Therefore, a considerable amount of fractured or cross-fractured cells remain attached to the carbon cast, either because they are attached to the carbon-stabilized protoplasmic half of their fractured membrane (Figure 4, protoplasmic half), or because they were cross-fractured and attached to the cytoplasm (Figure 4, cross-fractured cytoplasm). These cells are collapsed and shrivelled after subsequent processing. Figures 4B and C show shrivelled erythrocytes attached to the carbon cast.

3. Application of Fracture-Flip

a. Erythrocytes[8]

Freshly harvested erythrocytes were fixed in 1% glutaraldehyde in phosphate-buffered saline (PBS), pH 7.4, for 1 h at 4°C and washed twice in PBS. Fixed cells were impregnated in 30% glycerol, mounted on double-replica specimen carriers, frozen, and freeze-fractured at −120°C. The subsequent process was carried out as described above. Low-magnification views of an erythrocyte fracture-flip carbon are shown in Figure 4. The gently curved convex surface of erythrocytes is easily recognized (Figure 4A). Close observation at high magnification reveals the cell surfaces of erythrocytes to be covered with numerous particles ranging from 3 to 5 nm in diameter.[8]

b. Macrophages[11]

Macromolecular resolution views of the cell surfaces of macrophages during spreading have been studied using fracture-flip. Rat pulmonary macrophages were plated into culture dishes containing double-replica specimen carriers and allowed to spread on the carriers for 30 min at 37°C. This process is known to involve the assembly and endocytosis of numerous clathrin-coated vesicles. After incubation, the macrophages were fixed with 2% glutaraldehyde for 60 min. Fracture-flip was performed as described above. To immunocytochemically demonstrate fibronectin-binding sites, some specimens were immediately frozen without fixation after incubation. Immunolabeling was performed after freeze-fracture and carbon stabilization.

Cell surfaces of freshly harvested macrophages were covered by numerous particles ranging from 10 to 25 nm in diameter. These surface particles appeared to be randomly distributed over the cell surface. Incubation of adherent macrophages for 15 min at 37°C resulted in the formation of irregularly shaped patches (area 0.5 to 4 μm^2). After 30 min at 37°C, particle aggregation became more pronounced, with the particles restricted to circular depressions (Figure 5A). Intervening areas between the patches or the depressions were covered with small particles (10 nm in diameter). Immunoelectron microscopy of fibronectin-binding sites revealed that immunogold particles were exclusively confined to the intervening areas and the periphery of the particle patches (Figure 5B).[20]

c. Nuclear Membrane and Nuclear Pore Complexes[21]

Fracture-flip cannot reveal membrane surfaces in instances where the cell body remains attached to the carbon cast. In order to overcome this limitation, the use of fracture-flip combined with Triton X-100 extraction has been proposed.[21,22] In principle, fracture-flip/Triton X-100 makes it possible to visualize the exoplasmic luminal surfaces of cell organelles as well as the cytoplasmic surfaces of plasma membranes. The cytoplasmic surface of the outer nuclear membrane and exoplasmic surface of the inner nuclear membrane of liver cell nuclei have been studied using this technique. After freeze-fracture, carbon stabilization, and thawing, unfixed nuclei were treated with 2% Triton X-100 to dissolve unfractured portions of the outer nuclear membranes. To determine the precise localization of *N*-acetylglucosamine residues, after treatment with Triton X-100 some carbon casts were rinsed with 5 m*M* Tris-HCl buffer, pH 7.4, and incubated for 30 min at room temperature with wheat germ agglutinin (WGA)-colloidal gold (E-Y Laboratories, Inc.). After incubation with WGA-colloidal gold, the carbon casts were washed gently four times with Tris-HCl buffer, fixed in 1% glutaraldehyde for 10 min, then washed with distilled water. Flipping and Pt-shadowing of the carbon-stabilized membrane halves (cytoplasmic half of the outer nuclear membrane and exoplasmic half of the inner nuclear membrane) revealed the cytoplasmic surface of the outer and the exoplasmic surface of the inner nuclear membranes facing the perinuclear space.

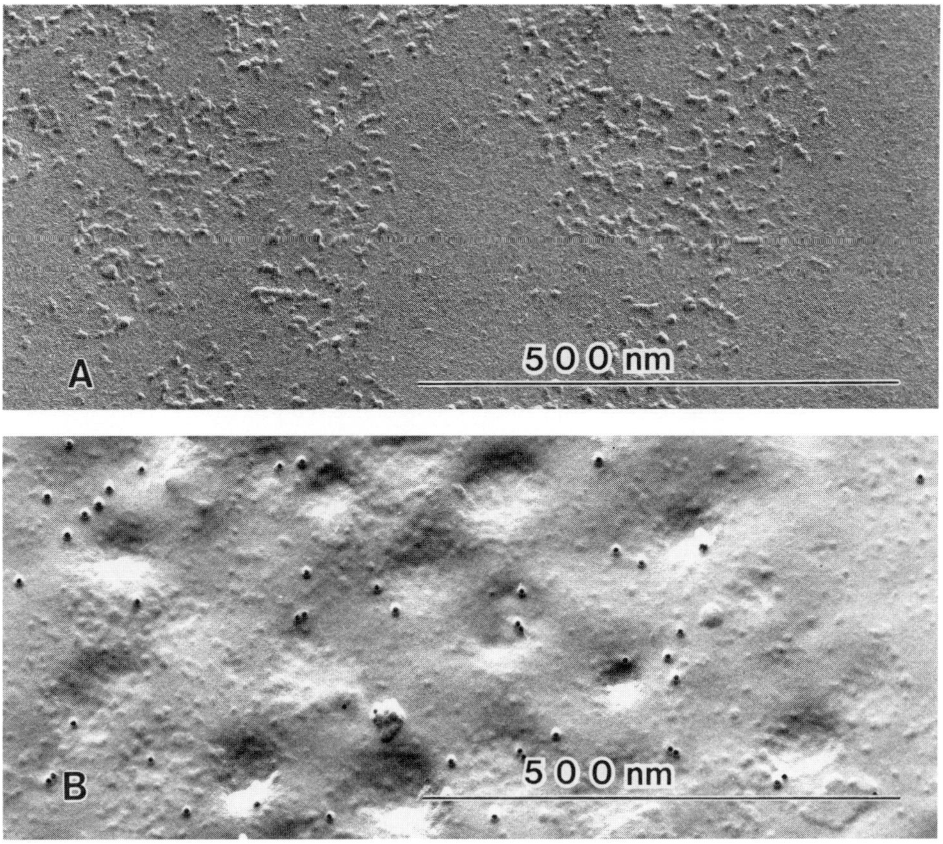

FIGURE 5. Adherent surface of macrophage warmed to 37°C for 30 min. (A) Membrane particles are clustered into patches. (B) Immunogold particles indicating localization of fibronectin-binding sites are associated with the periphery of the patches.

The cytoplasmic surface of the outer nuclear membrane was covered by numerous globular particles (30 to 35 nm length, 25 to 30 nm width), often aligned into curved chains (Figure 6A, arrows). Because of their size and characteristic arrays, these oblong particles were interpreted as ribosomes. Nuclear complexes were distinguishable from polysomes by their configuration delimiting a central pore (Figure 6A, corresponding arrowheads). The approximate dimensions of nuclear pore complexes revealed by fracture-flip are (1) outside diameter, 120 to 130 nm, (2) central pore, 40 to 45 nm, (3) thickness of the annular ring, 40 nm. On the exoplasmic surface of the inner nuclear membrane, the nuclear pore complexes appeared as prominent annular structures with an outer diameter of about 130 nm and an inner pore of 40 to 45 nm (Figure 6B). WGA-colloidal gold particles were found to be selectively associated with the inner pore of the nuclear pore complex (Figure 6C, arrowheads).

D. CONCLUSIONS

This chapter has reviewed freeze-fracture techniques combined with cytochemical labeling. Technically, these approaches are simple. However, interpretation is much more complex. In particular, fracture-label involves delicate points of interpretation. The reader is referred to the original articles or the reviews[18,23-25] for detailed interpretation of freeze-fracture cytochemistry.

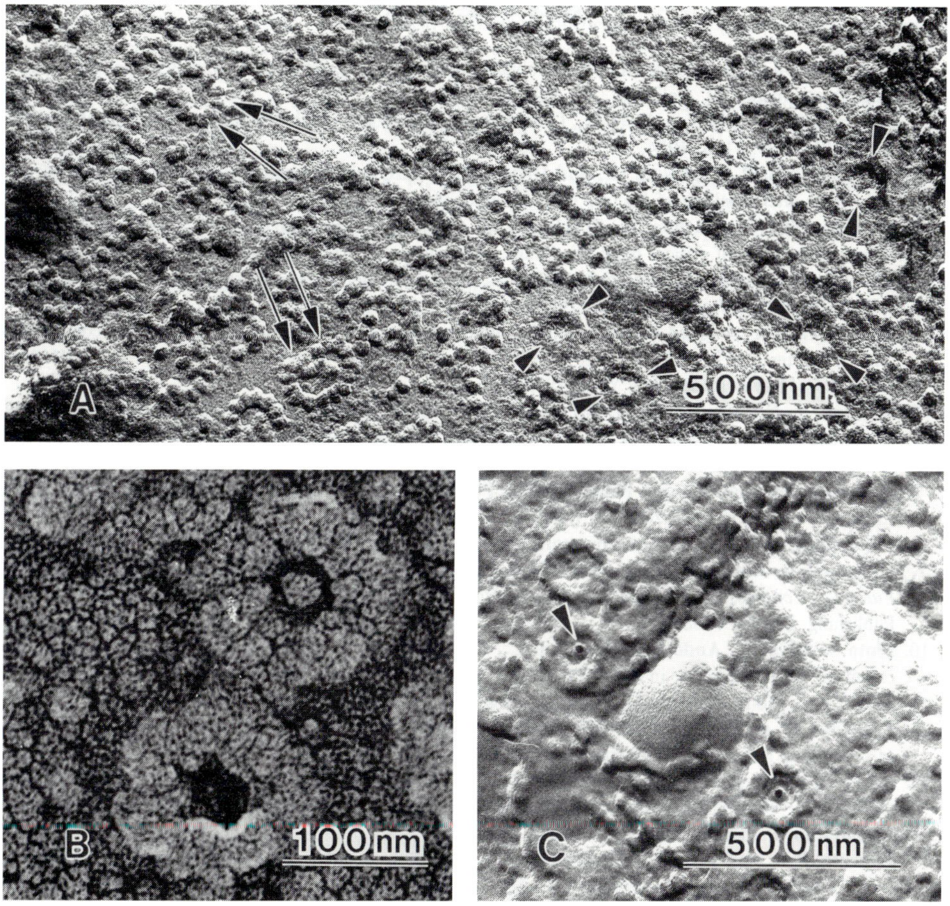

FIGURE 6. Fracture-flip/Triton X-100 and cytochemistry of rat liver nuclei. (B) Reversed prints of rotary-shadowed cast; (C) binding of WGA-gold to nuclear pore complexes.

Fracture-flip quickly produces macromolecular images of membrane surfaces, and requires only routine equipment. Most biological activities, including antigenicity, remain intact after freeze-fracture and C fixation. In this paper we show some examples of fracture-flip combined with cytochemical labeling after carbon stabilization. Fracture-flip and fracture-flip cytochemistry make it possible to study the precise macromolecular architecture of biomembranes by a combination of conventional freeze-fracture techniques and freeze-fracture cytochemistry.

The development and diversification of freeze-fracture techniques combined with cytochemistry continue steadily.[26-28] It is clear that freeze-fracture and fracture-flip cytochemistry will bridge the gap between biochemistry and more morphologically oriented disciplines in cell membrane research.

ACKNOWLEDGMENTS

This work was supported in part by Grants-in-Aid for Scientific Research from the Ministry of Education, Science and Culture of Japan (No. 02404020).

Chapter 6

ELECTRON MICROSCOPIC CYTOCHEMISTRY UNDER THE SCANNING ELECTRON MICROSCOPE

Kuniaki Takata, Fumiaki Nishiyama, Yoshihiro Akimoto, and Hiroshi Hirano

TABLE OF CONTENTS

0-8493-6012-9/93/$0.00 + $.50

A. INTRODUCTION

Scanning electron microscopic observation has several advantages over the conventional transmission electron microscopic observation of ultrathin sections, the following being the main ones: (1) specimens are relatively easy to prepare, because samples are ready to be examined after the critical-point drying and coating; (2) there is no laborious work of ultrathin sectioning required for transmission electron microscopy; (3) a wide area of a given specimen can easily be examined at a glance and can be recorded in a single micrograph; (4) when employed properly, a couple of scanning electron micrographs sometimes provide the information equivalent to that otherwise only obtained from many transmission electron micrographs; and (5) in addition, three-dimensional localization of the label can be demonstrated more easily.

Scanning electron microscopy has been used in conjunction with both enzyme cytochemistry and immunocytochemistry.[1-3] For the latter, colloidal gold labeling is very useful, just as it is in transmission electron microscopy. The recent introduction of the high-resolution scanning electron microscope with a field-emission gun and a sensitive backscattered-electron detector has made the combination of immunocytochemistry and scanning electron microscopy a versatile method.[4,5]

B. METHODS

Immunocytochemistry — A wide variety of substances have been used as immunocytochemical markers in scanning electron microscopy. As in transmission electron microscopy, colloidal gold, ferritin, and horseradish peroxidase (HRP) are commonly used (Figure 1).

Enzyme cytochemistry — For enzyme cytochemical demonstration in scanning electron microscopy, the methods using heavy metals (such as lead), described in detail in the preceding sections of this book, can be employed. The heavy metal deposits caused by an enzyme reaction are good objects for backscattered-electron imaging (Figure 1).[2]

Autoradiography — Silver grains in autoradiography exhibit strong contrast in backscattered-electron imaging. When applied in semithin plastic sections, good resolutions were obtained.[6]

C. EQUIPMENT AND REAGENTS

A scanning electron microscope (high-resolution type, preferably with a field-emission gun; equipped with secondary- as well as backscattered-electron detector), a critical-point dryer, and a carbon-coating machine.

For immunocytochemistry or lectin cytochemistry, the following conjugates (labeled probes) should be prepared or purchased: colloidal gold conjugates (conjugated to antibodies, protein A, streptavidin, lectins, etc.), ferritin-conjugates, and HRP-conjugates.

D. PROCEDURES

1. Labeling and Observation of Surface Antigens of Cultured Cells by the Colloidal Gold Method

Cells are cultivated on coverslips (diameter: 10 mm). If the transmission electron microscopic observation of the same specimen used as a whole-cell mount is also intended,[7,8] cells should be cultured on gold or platinum grids coated with Formvar and carbon. (These grids can be prepared as follows: the thin Formvar membrane is floated on water. Grids are placed on it. Coverslips are then placed over the grids. The whole membrane with grids and coverslips is retrieved by placing a sheet of vinyl over it, air dried, and then coated with carbon.) The use of coated grids as a substratum for cell culture is convenient when

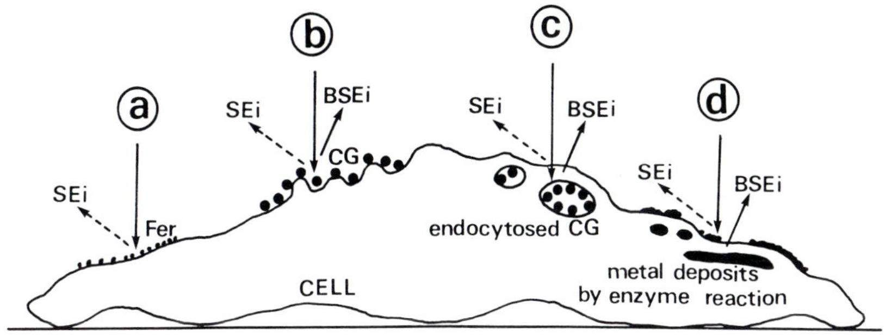

FIGURE 1. Schematic diagram showing typical labeling methods for cytochemistry in scanning electron microscopy. (**a**) Labeling with ferritin (Fer) conjugates, and observation by secondary-electron imaging (SEi). (**b**) Labeling with colloidal gold (CG) conjugates, and observation by backscattered-electron imaging (BSEi) or SEi. (**c**) Observation of endocytosed CG by BSEi. (**d**) Detection of metal deposits formed by enzyme reaction by BSEi. Note that when BSEi is used to detect the label (**b, c, d**), SEi provides the topographical image.

specimens are to be examined in the high-resolution scanning electron microscope, because only small specimens are accepted in these microscopes. Comparison between whole-mount scanning electron microscopic images and whole-mount transmission electron microscopic images is also useful.[7,8]

In the case of free cells, such as blood cells, it is convenient to attach such cells to polylysine-coated coverslips, and proceed in the same way as for cultured cells. To coat coverslips with polylysine, clean coverslips are immersed in 1 mg/ml aqueous polylysine solution for 30 min. If desired, they can be stored in polylysine solution at −20°C until used. After being rinsed with water, air dried, and sterilized by ultraviolet irradiation, the coated coverslips are ready to use. Cells are plated on polylysine-coated coverslips, and incubated for an appropriate time until an adequate number of cells has attached. If the use of polylysine-coated coverslips is not suitable for some reason, cells can be processed in suspension in test tubes, and dried in a special drying cell for suspension specimens. Such samples are then mounted on a strip of carbon tape with conductive adhesive for electron microscopic observation.

Cells on coverslips (or on grids on coverslips) are washed with phosphate-buffered saline (PBS) or medium. To avoid the possible alteration of surface ultrastructure by the adsorption of unwanted materials during fixation, the presence of serum or any other protein(s) during fixation should be avoided. The condition of fixation should be determined in consideration of the following: type of cell, type of antigen to be detected, type of antibody used, type of cytochemical label. We usually use a mixture of 3 to 4% formaldehyde and 0.1 to 0.5% glutaraldehyde-0.1 *M* sodium phosphate buffer, pH 7.4. PBS can be used instead of 0.1 *M* sodium phosphate buffer. Osmolarity of the fixatives should be carefully controlled, since it greatly alters the cell surface morphology. Periodate-lysine-paraformaldehyde (PLP) fixative is also useful. For lectin cytochemistry, we used 1.25% glutaraldehyde-PBS. Cells are fixed at room temperature for 10 to 30 min, washed with PBS, and then incubated with inert proteins such as bovine serum albumin (1%), gelatin (1%), or normal serum (5%) from the same species used to generate the secondary antibody to quench nonspecific binding.

For immunostaining, a drop of the primary antibody solution (about 100 μl) is placed on a sheet of Parafilm. A coverslip bearing the specimen is then floated on the antibody solution with the cell attached side down after the other side of the coverslip is wiped dry. Incubation is carried out at room temperature for 1 h, or at 4°C overnight. For the overnight incubation, coverslips are housed in a small chamber and sealed to avoid drying. Only the cell surface is accessible to antibodies, especially colloidal gold-labeled antibodies. Therefore, antigenic sites exposed on the cell surface can be demonstrated with this method. After the incubation with the primary antibody, the coverslips are placed in Petri dishes, and are washed with PBS three times for a total of 15 min. Coverslips are then floated on drops of the colloidal gold (10 to 20 nm in diameter)-labeled secondary antibody (or protein A) solution on Parafilm sheet in the same way as for the primary antibody. Specimens are then washed again with PBS as above. Specimens should be washed thoroughly because any adherent substances (such as serum or aggregated proteins) will be fixed in the subsequent refixation step. Note that it is only the surface of the specimen that is to be examined under the microscope, so the above possible source of contamination should be carefully avoided.

Specimens are refixed in 2.0 to 2.5% glutaraldehyde-PBS for 10 min at room temperature to fix the bound antibodies, and washed with PBS three times for a total of 15 min. When osmication is omitted, this refixation procedure with glutaraldehyde is also beneficial in that it strengthens the specimen for the subsequent electron microscopic observation. Specimens are then osmicated in 1% osmium tetroxide-0.1 *M* cacodylate buffer, pH 7.4, for 5 to 15 min at 4°C. They are subsequently rinsed with deionized water, dehydrated with a graded series of ethanols, and critical-point dried with liquid carbon dioxide used as a transition medium.

When only conventional secondary-electron imaging is used, dried specimens are coated with carbon, gold, or platinum as usual. Extreme care should be taken not to coat the specimen too much with metals, because colloidal gold particles are hard to distinguish on a heavily coated cell surface. For the noncoating observation, specimens can be treated after osmium treatment with 1% aqueous tannic acid solution for 10 min at room temperature as a conductive stain. Specimens are then thoroughly washed with deionized water, reosmicated, and processed as described above. Observation of colloidal gold particles with secondary-electron imaging at low accelerating voltages (2 to 5 kV) provides another way of visualization of both the surface topography and the colloidal gold label simultaneously.[9]

When backscattered-electron imaging is used to demonstrate the colloidal gold label, dried specimens are lightly coated with carbon. The degree of contrast of the backscattered-electron image is mainly dependent on the atomic number. Colloidal gold therefore exhibits a very high contrast against the cell, which is made up of mainly light elements such as hydrogen, carbon, and oxygen. Colloidal gold particles (10 to 20 nm in diameter) on the cell surface are easily detected at magnifications of 20,000 to 40,000 as bright dots on the backscattered-electron images, irrespective of the ultrastructural topography of the specimen surface. This specific detection of the cytochemical marker can only be achieved by the combination of colloidal gold labeling and backscattered-electron imaging.

Although backscattered-electron images give the unequivocal identification of colloidal gold label, they do not provide information on the surface topography. In order to identify the location of the label, secondary-electron images of the same specimen are used to demonstrate the surface ultrastructure. The electronic mixing of these two images at an appropriate ratio results in an excellent visualization of both the label and the ultrastructure. Also, the latest model of the scanning electron microscope can assign different colors to these two images, superimpose them electronically, and display them as a single image, which is very useful. To obtain secondary-electron images of better quality, deposition of metal in the specimen is recommended; however, this results in a decrease in the contrast of colloidal gold particles in the backscattered-electron images. Therefore, the conditions of osmication and conductive tannic acid treatment must be carefully chosen with respect to the size of the colloidal gold particles used.

When the use of smaller colloidal gold particles, such as 4 nm in diameter, is required to identify the microdomains on the cell surface, the osmium refixation step should be omitted. Omission of osmium treatment minimizes the adsorption of metal (in this case osmium), which drastically enhances the relative contrast of small colloidal gold particles against the cell surface. The contrast of the colloidal gold particles can also be enhanced by digital image processing[10] or by the silver enhancement method.[10,11]

If the purpose of the immunolabeling is simply to identify the cell types with the appropriate surface markers, larger colloidal gold particles (larger than 15 nm) can be used for easy observation. In this case, specimens can be treated with osmium/tannic acid/osmium. Although more osmium is deposited in such specimens, larger colloidal gold particles exhibit such high contrast that they can be distinguished in the backscattered-electron image.

The images with reversed polarity (negative images) are useful when samples are examined as a whole mount with a transmission electron microscope as well. The reversed images are remarkably similar to those of transmission electron images of the surface structure and subsurface cytosol.

To demonstrate the location of antigens on the surfaces of isolated organelles, basically the same procedure used for cultured cells can be employed. Samples are attached or adsorbed on coverslips or coated grids and processed as described above.

2. Observation of Antigens on the Specimen Surface by the Ferritin Labeling Method

Basically similar procedures are used as described for the colloidal gold labeling method. Only secondary-electron images can be used for the observation. Therefore, specimens with a relatively simple surface ultrastructure, such as flat cell surfaces[12] or bundles of filaments, have been used. Compared with the colloidal gold labeling method, a high density of label is usually obtained. In addition to the immunostaining, cationized ferritin can be used as well.

E. RESULTS

Labeling of mouse macrophages with concanavalin A (ConA)/HRP-colloidal gold[7,8,13] — Mouse peritoneal macrophages were cultured on Formvar/carbon-coated grids, and sequentially incubated with ConA and HRP-colloidal gold (17 nm in diameter). ConA binding sites are visualized as colloidal gold particles based on the affinity of ConA for mannose residues of HRP. Figure 2 shows the secondary-electron microscopic image of a specimen treated with osmium/tannic acid/osmium and lightly coated with platinum. Colloidal gold particles representing ConA-binding sites are seen as bright particles on the cell surface. The surface topography of the cell is well resolved as a result of the conductive staining with tannic acid and metal coating. Observation of the secondary electron images alone have intrinsic shortcomings, however. Over the nonflat surfaces, colloidal gold particles are hard to distinguish because of the ''edge effect''. In case ''charge-up'' of the specimen occurs, metal coating drastically improves the ease of observation and quality of the surface topographic image. However, since metal coating decreases the relative contrast of the colloidal gold image, such coating should be minimal.

Visualization of surface immunoglobulins (Igs) on the mouse lymphocyte cell surface by use of backscattered-electron images[4] — Mouse lymphocytes were labeled for their surface Igs with goat anti-mouse Igs-colloidal gold. After carbon coating, secondary-electron images, backscattered-electron images, and a mixture of the two were photographed (Figure 3). Basically, the secondary-electron images show the surface topography. In the backscattered-electron image, each colloidal gold particle on the specimen surface is unequivocally demonstrated as a bright (normal polarity) or a dark (reversed polarity) spot. Observation of the secondary-electron images and the backscattered-electron images side by side, or of the mixture of secondary- and backscattered-electron images, is recommended.

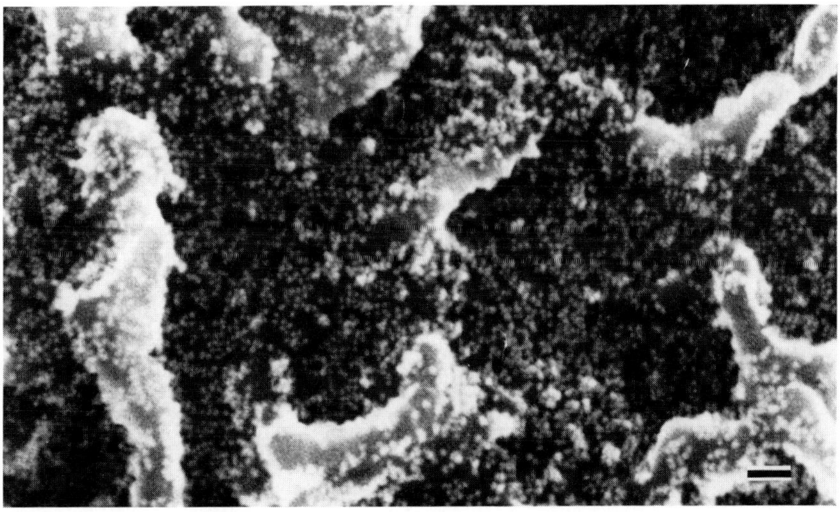

FIGURE 2. Portion of mouse peritoneal macrophage labeled with ConA and HRP-colloidal gold (17 nm in diameter). The secondary-electron image of a specimen lightly coated with platinum is shown. (Bar = 0.1 μm.)

Observation of endocytosis from backscattered-electron images[4] — The backscattered electron possesses much higher energy than does the secondary electron. It, therefore, can be used to detect endocytosed colloidal gold particles inside cytoplasmic vesicles. Such endocytosed colloidal gold particles exhibit vague and weak signals, and can thus easily be distinguished from the surface-bound colloidal gold particles, which give clear and bright images (Figure 4). As the sharpness of the endocytosed colloidal gold particles is a function of the accelerating voltage, it can be increased if a higher accelerating voltage is used. In combination with stereoscopic observation, backscattered-electron imaging of labeled colloidal gold particles can provide a method for the observation of the early endocytic events.

Labeling of luminal surfaces of organs[12] — Rat urinary bladders were removed, opened flat, and incubated with ferritin-labeled ConA for mannose residues on the luminal cell surface of the urinary bladder epithelium. Specimens were observed under scanning electron microscope after platinum coating. Ferritin particles are seen as small round particles on the surface by the secondary-electron imaging. Ferritin label, representing ConA-binding sites, is restricted to ridges, a microdomain of the luminal plasma membrane (Figure 5).

Observation of HRP-labeled thick sections — Treatment of HRP-labeled specimens with diaminobenzidine-H_2O_2, and then with osmium, results in the deposition of osmium in the positive sites, thus making the positive sites suitable for detection by backscattered-electron imaging. Figure 6 shows a thick section of an Epon-embedded specimen labeled with HRP-peanut agglutinin (PNA). The lectin-binding sites are clearly demonstrated in this backscattered-electron image.

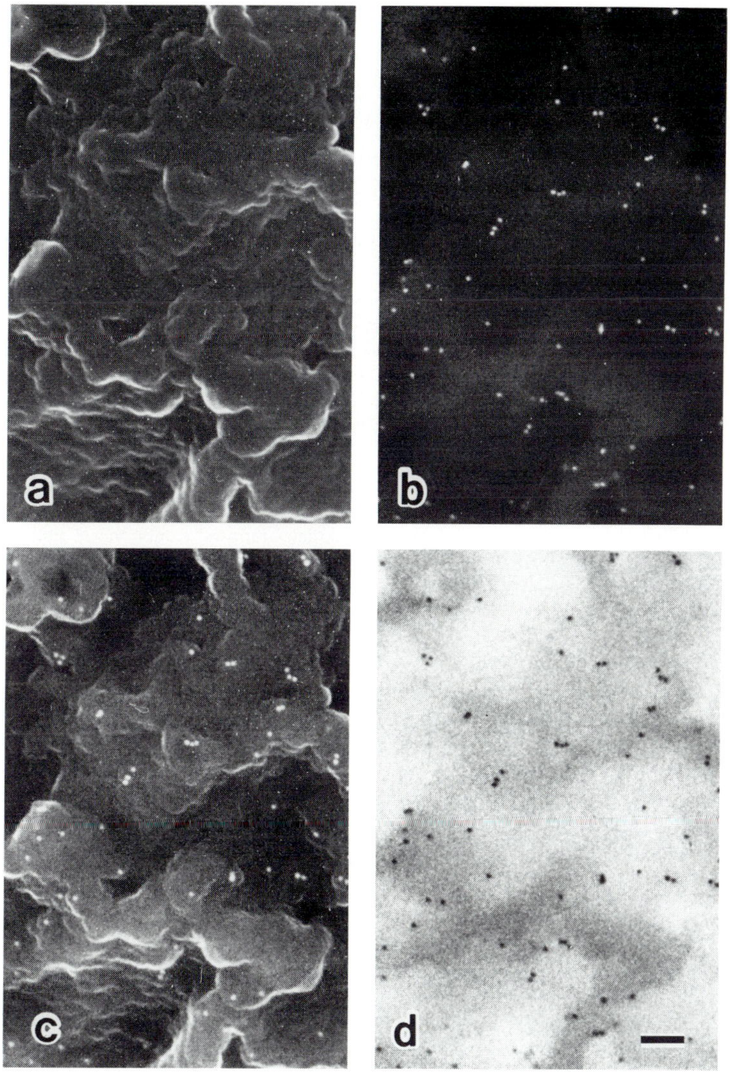

FIGURE 3. Portion of mouse lymphocyte labeled with anti mouse Igg colloidal gold (10 nm in diameter). (a) Secondary-electron image. (b) Backscattered-electron image. (c) Image obtained by mixing of the secondary- and backscattered-electron images. (d) Backscattered-electron image (reversed polarity). (Bar = 0.1 μm.)

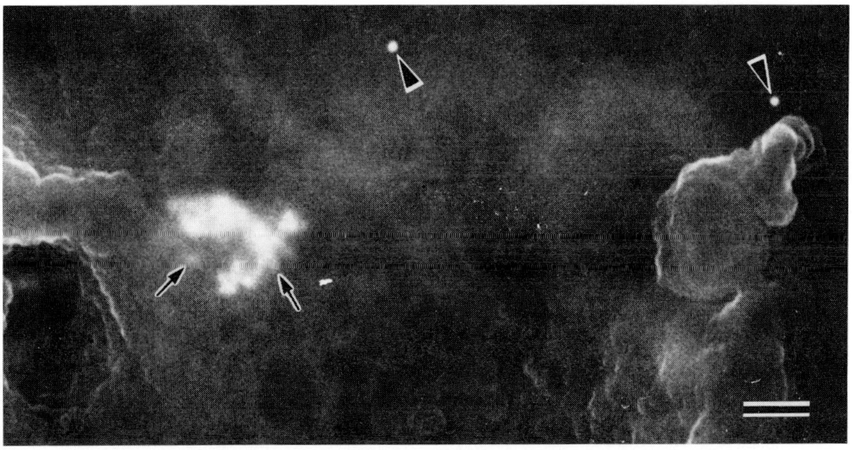

FIGURE 4. Observation of endocytosis of colloidal gold (15 nm in diameter)-labeled *Ricinus communis* agglutinin in a mouse macrophage. The endocytosed colloidal gold particles (arrows) exhibit vague signals compared with the surface-bound colloidal gold particles, which give clear and bright images (arrowheads). This image was obtained by mixing of backscattered- and corresponding secondary-electron images. Accelerating voltage: 15 kV. (Bar = 0.1 μm.)

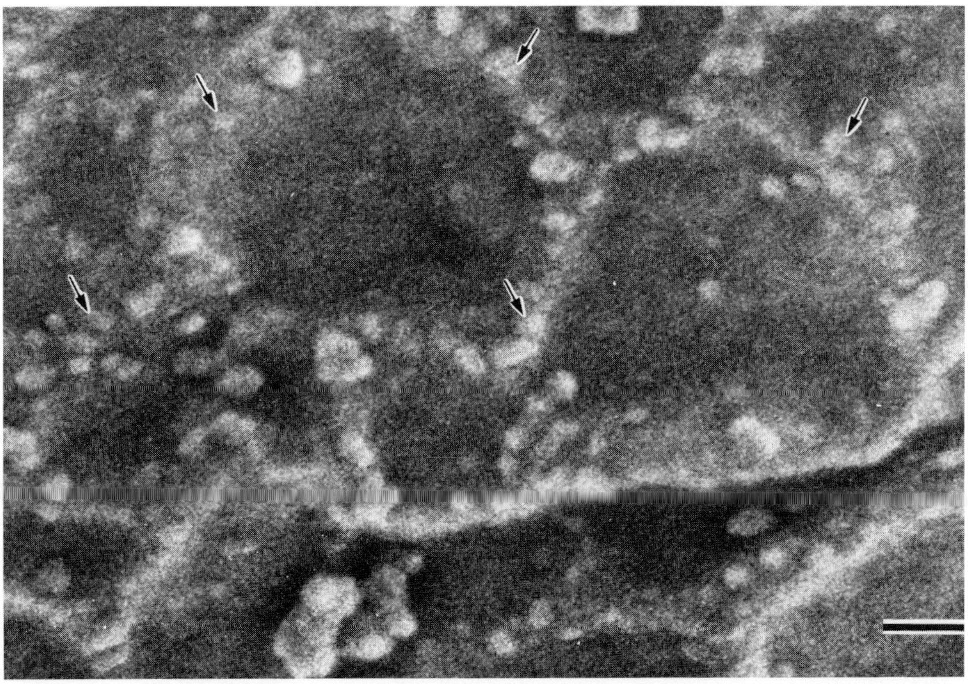

FIGURE 5. Luminal surface of rat urinary bladder. Ferritin label representing ConA-binding sites is seen on the ridges of the luminal plasma membrane (arrows). (Bar = 0.1 μm.)

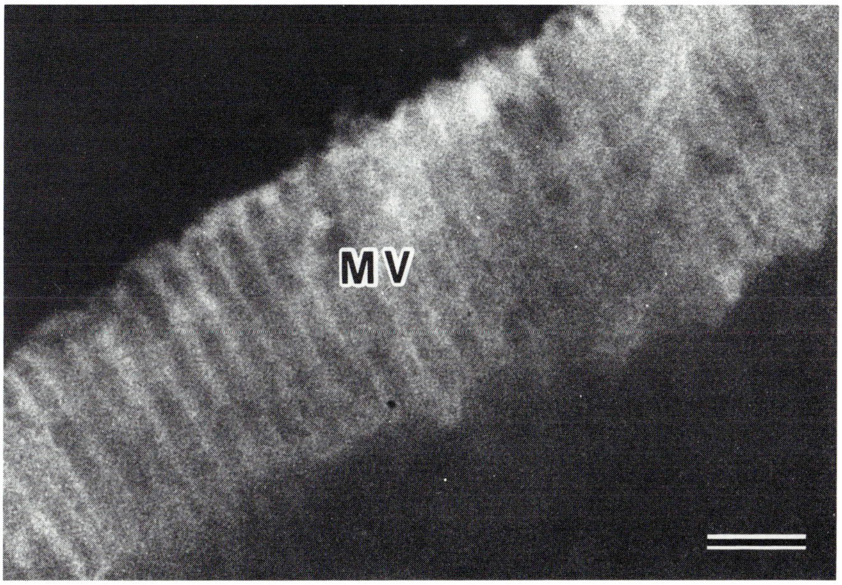

FIGURE 6. Rat intestinal epithelium stained with HRP-peanut agglutinin (PNA). A backscattered-electron image of a thick section of an Epon-embedded specimen. Plasma membrane of the microvilli (MV) is positive. (Bar = 0.5 μm.)

REFERENCES

1. **Molday, R. S.,** Labeling of cell surface antigens for SEM, in *Techniques in Immunocytochemistry,* Vol. 2, Bullock, G. R. and Petrusz, P., Eds., Academic Press, London, 1983, 117.
2. **Ushiki, T. and Fujita, T.,** Backscattered electron imaging. Its application to biological specimens stained with heavy metals, *Arch. Histol. Jpn.,* 49, 139, 1986.
3. **Horisberger, M.,** Colloidal gold for scanning electron microscopy, in *Colloidal Gold,* Vol. 1, Hayat, M. A., Ed., Academic Press, San Diego, 1989, 217.
4. **Takata, K., Akimoto, Y., Ogura, K., Yamagishi, S., and Hirano, H.,** Colloidal gold label observed with a high resolution backscattered electron imaging in mouse lymphocytes, *J. Electron Microsc.,* 37, 346, 1988.
5. **De Harven, E. and Soligo, D.,** Backscattered electron imaging of the colloidal gold marker on cell surfaces, in *Colloidal Gold,* Vol. 1, Hayat, M. A., Ed., Academic Press, San Diego, 1989, 229.
6. **Nanci, A., Zalzal, S., and Smith, C. E.,** Routine use of backscattered electron imaging to visualize cytochemical and autoradiographic reactions in semi-thin plastic sections, *J. Histochem. Cytochem.,* 38, 403, 1990.
7. **Takata, K. and Hirano, H.,** Whole-cell-mount cytochemistry by the colloidal gold labeling method. Combined transmission and scanning electron microscopic study of ConA binding sites in mouse macrophages, *Histochemistry,* 81, 435, 1984.
8. **Takata, K. and Hirano, H.,** Colloidal gold in high-voltage electron microscopy — Ruthenium red method and whole-cell mount, in *Colloidal Gold,* Vol. 2, Hayat, M. A., Ed., Academic Press, San Diego, 1989, 345.
9. **Pawley, J. and Albrecht, R.,** Imaging colloidal gold labels in LVSEM, *Scanning,* 10, 184, 1988.

A. INTRODUCTION

The electron microscopic images obtained from ultrathin sections of tissues are widely believed to express "true" cell structure. However, it is apparent that most cellular components have a three-dimensional expansion, and their three-dimensional structures are very difficult to conceive from two-dimensional images obtained from ultrathin sections. High-voltage electron microscopy (HVEM) provides an opportunity to observe thick sections which contain the three-dimensional structure of cellular components. There are fundamental differences in the structural imagery between ultrathin section and thick section. The images obtained from ultrathin sections are cut-face images of the three-dimensional structure, whereas that of thick sections can be understood as projected images of three-dimensional structures. Therefore, there are several things which we cannot expect in observation of thick section, such as a sharp image of the membranes or fine internal structures of organelles, which are usually observed clearly in ultrathin section. However, by increasing the thickness of the section, we can expect more chances to observe the continuity of the structure which appeared to be fragmented in the ultrathin section. This advantage helps in visualizing the expansion of cell organelles such as endoplasmic reticulum, mitochondria, Golgi apparatus, and specialized regions of the plasma membrane. Even smaller and thinner structures, like cytoskeletons and collagen fibers, would be visualized not as fragmented but preserved in their long structure.

However, it was difficult to observe significant structures in regular plastic thick sections prepared by conventional EM techniques, which were developed for ultrathin section, under HVEM. Numerous attempts were made to visualize the structures of the cell and to localize molecules on thick sections under HVEM. This chapter will focus on the cytochemical techniques which visualize cellular structures and localization of molecules under HVEM.

B. METHODS

1. The Golgi Method for Visualizing Cell Shape

Techniques for HVEM were often adapted from those of light microscopy. The Golgi impregnation method has been used for detecting extension of neurons in the central nervous system at the light microscopic level. One of the representative methods is the "rapid Golgi method". In this staining method one uses the triple impregnation procedure of soaking the slabs of aldehyde-perfused nerve tissue in 2.33% potassium dichromate, 0.19% osmium tetroxide, and 0.75% silver nitrate aqueous solutions.[1] There are many modifications of this kind of staining. Selective staining by electron-dense deposits in certain cells in nervous tissue shows the shape and extension of the whole cell in three-dimensional images. The study of the spine of the nerve cell was found to be especially difficult with light microscopy and conventional electron microscopy (EM) by observation of ultrathin section, because this structure is below the resolution of the light microscope and too large to be observed with ultrathin sections for EM. Hama et al.[2] performed a morphometric study using a computer-aided image analyzer extracting three-dimensional image of spines from thick sections of Golgi-impregnated nerve tissue.

2. Methods for Visualizing Plasma Membrane

a. Ruthenium Red Method

Ruthenium red was introduced as a stain for the visualization of mucopolysaccharides and glycoproteins under the electron microscope by Luft.[3] Electron-dense deposits could be observed on the plasma membrane of most cell types. The staining method is simple. By adding ruthenium red to both the prefixative and the postfixative solutions at concentrations of 0.05 to 0.06%, the plasma membrane could be stained for HVEM observation. The deposits in thick sections are electron dense enough to reveal the two-dimensional expansion of the plasma membrane easily with HVEM.[4,5] This staining method could be combined with immunogold or ligand labeling of the plasma membrane for studying ligand-binding sites in relation to the three-dimensional structure of the plasma membrane.[6] However, in order to use this counter staining, the labeling should be performed before the aldehyde fixation.

b. Ferri- or Ferrocyanide Osmium Method

Introduction of potassium ferricyanide, as well as potassium ferrocyanide, in osmium tetroxide postfixative was found to increase the electron density of glycogen particles and membrane contrast in tissues prepared for conventional EM.[7,8] However, occasionally, these methods were also found to form heavy electron-dense deposits on the plasma membrane and endoplasmic reticulum (ER) that then allows HVEM applications to the observation of thick sections. Although the staining mechanisms of these methods are not well understood, several techniques were suggested for obtaining electron-dense deposits.[9-11] For the ferrocyanide osmium method, 0.08% potassium ferrocyanide is included in 1% osmium tetroxide, 0.1 *M* cacodylate or phosphate buffer, pH 7.4, for the observation of the plasma membrane under the HVEM. Increased concentration of potassium ferrocyanide to 1.5% in a 1 to 2% osmium tetroxide unbuffered solution is identical to Karnovsky's postfixative that is widely used for increasing membrane contrast of most cellular components,[8] although this method decreases electron-dense deposits on the plasma membrane. For the ferricyanide-osmium technique, 0.8 to 8.0% potassium ferricyanide is dissolved in 1 to 2% osmium tetroxide aqueous or cacodylate-buffered solution. According to Jinguji and Ishikawa,[10] the staining is affected by selection of buffer and pH. Optimal staining was obtained with sodium cacodylate-HCl buffer, pH 7.0, without calcium ions. They also found that addition of calcium (0.01 to 0.2 *M*) to the postfixative shifts the staining of the intercellular space to the intracellular membrane compartment, such as ER, nuclear envelope, and mitochondria. This observation may reflect the similar staining pattern observed by Forbes et al. using the diaminobenzidine-ferrocyanide method.[12] Moreover, Forbes and Neil[13] claimed that three factors are of crucial importance in the postfixation with ferrocyanide-osmium solution. First, postfixation in this solution should be carried out without agitation. Second, postfixation at pH 10.0 tends to stain plasma membrane (transverse-axial tubular system, insertion of plasma membrane into cytoplasm), whereas it stains more intracellular membrane compartment (sarcoplasmic reticulum, a specialized ER of muscle cells) at pH 7.6. Third, the tissue postfixed with the ferrocyanide osmium has poor infiltration characteristics with Epon. This tendency can be overcome with gradual introduction of the resin (Poly/Bed 812) into the tissue, as follows:

2:1 Propylene oxide: Poly/Bed, overnight
1:1 Propylene oxide: Poly/Bed, 8 h the next day
1:2 Propylene oxide: Poly/Bed, overnight
Pure Poly/Bed, 6 h under vacuum

Electron-dense deposits are observed in the ER of plant cells,[9] on basolateral membranes of intestinal cells,[10,14] and intracellular canaliculi (apical membrane) of parietal cells of gastric epithelium[14] (Figure 1). A HVEM study could be made by these methods, but because of the overlapping of complex structures, observation of semithin sections with lower-voltage EM (for example, 200 kV) is sometimes more effective for structural analysis.

As mentioned above, the diaminobenzidine-ferrocyanide method also visualizes the plasma membrane. This method will be described later.

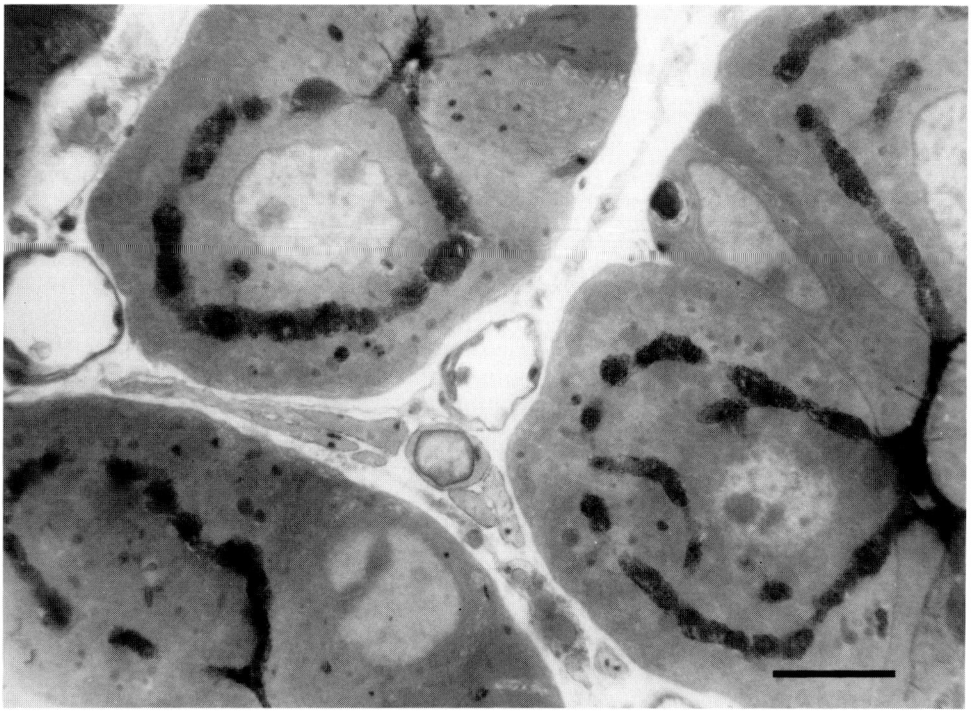

FIGURE 1. Gastric parietal cells of mouse were stained with 0.8% ferricyanide and 1% osmic acid. A 1-μm-thick section was observed using 200-kV EM. The intracellular canaliculi of a parietal cell, which is the intracellular extension of the apical membrane, was stained with the electron-dense material. This structure, which consists of microvilli, appear to surround the nuclei in the form of a huge network. (Bar = 5 μm.)

3. Methods for Visualizing Intracellular Organelles

a. Osmium-Impregnation Method

After Friend and Murray[15] introduced a method for visualizing the *cis*-side of the Golgi apparatus, this method was applied to HVEM. Extensive HVEM studies were made by Rambourg[16] and Rambourg and Clermont,[17] focusing on the intracellular distribution and three-dimensional visualization of Golgi apparatus (for a review see Rambourg and Clermont[17]). They reported different intracellular distributions of the Golgi apparatus in various cell types. Another osmium-impregnation method was described for the visualization of the Golgi apparatus by introducing zinc iodide into osmium solution (zinc iodide osmium; ZIO method). The availability of an aldehyde prefixation before osmium impregnation in this method improved structures of organelles and resulted in a more reliable morphological analysis under HVEM.[14,18,19]

In the original osmium impregnation method, fresh tissue blocks without aldehyde prefixation are incubated in unbuffered 2% aqueous osmium tetroxide solution for 48 h at 40°C. During the fixation, the fixative should be replaced after it turns black. Mostly, the *cis*-most cisterna of the Golgi apparatus is stained by electron-dense deposits. The structures revealed by this method are ribbon-like extending and often meshed by numerous small pores under HVEM. For the ZIO stain, tissue blocks prefixed with 1 to 2% glutaraldehyde, are soaked in ZIO staining solution, in the dark, overnight at 4°C after rinsing with Tris-HCl buffer. The composition of Tris-HCl buffer is 50 m*M* Tris, 0.057 *M* NaCl, 5 m*M* $CaCl_2$, and 15 m*M* $MgCl_2$, and the pH is adjusted to 7.4 with HCl. The ZIO solution is

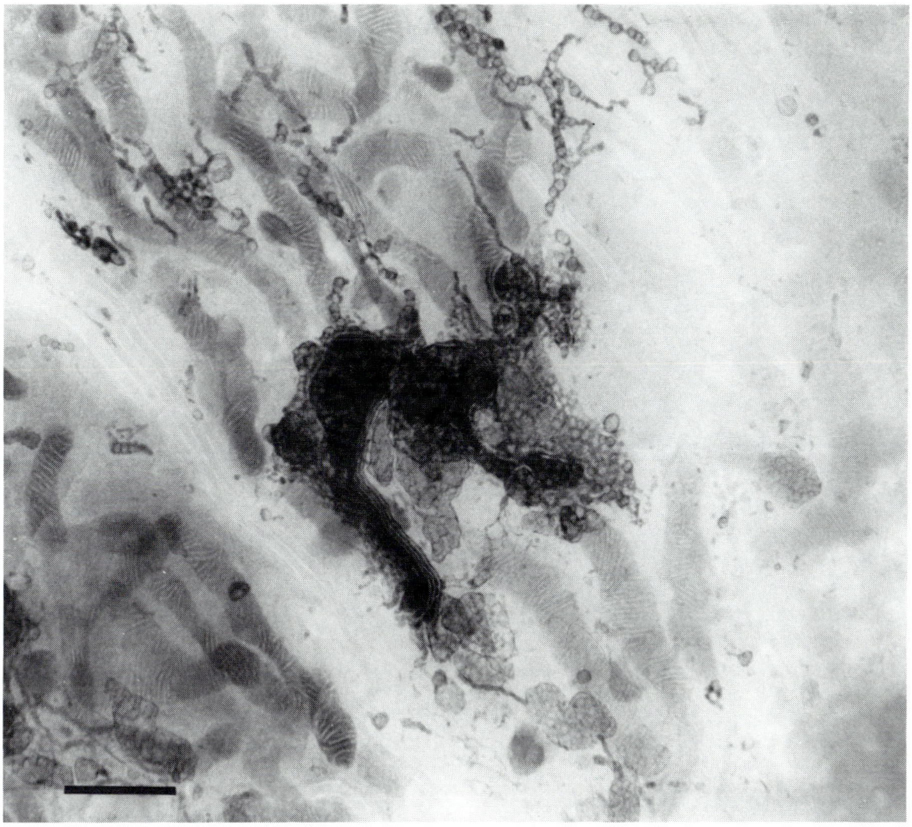

FIGURE 2. Duodenal epithelial cell was stained with the zinc-iodide osmium method; a 1-μm-thick section was observed using a 1000-kV EM. The three-dimensional Golgi apparatus is visualized as an electron-dense structure. Tubular structures with small lipid vesicles extended from the apical cytoplasm appear to connect to a mesh-like *cis*-Golgi cisterna. The same lipid vesicles can be identified in mesh-like cisternae. From the *trans*-side of the Golgi apparatus another tubular structure appears to extend to the basal cytoplasm containing large secretory granules. (Bar = 1 μm.)

prepared as follows: 3 g of zinc powder is sonicated in 20 ml of distilled water for 10 min and 1 g of iodine crystal is slowly added to the zinc solution and stirred. After cooling, this mixture is filtered and 4 ml of this solution is mixed with 2 ml of the above mentioned Tris-HCl buffer and 2 ml of 2% osmium tetroxide aqueous solution. After staining, the tissue blocks are treated with 0.5% aqueous uranyl acetate solution for 30 to 60 min at 40°C and processed for electron microscopy. Stain is found preferentially in Golgi apparatus of some cell types with high electron density, but other cell organelles like ER, mitochondria, or endomembrane structures, like tubulo-vesicular structure of gastric parietal cell (Figure 2), could be visualized under HVEM. Using this staining, the Golgi apparatus in the pancreatic exocrine cell was demonstrated to be an interconnected reticular structure composed of stacks of ribbon-like cisternae by observations of serial thick sections under HVEM.[18] Two distinctive tubular structures, which connect both *cis*- and *trans*-sides of Golgi apparatus, were also visualized in duodenal epithelial cells[14] (Figure 3).

b. Diaminobenzidine-Ferrocyanide Method

This method was introduced for staining the sarcoplasmic reticulum of muscle cell, but after a slight modification, it was used to demonstrate the T-system in muscle cells, and

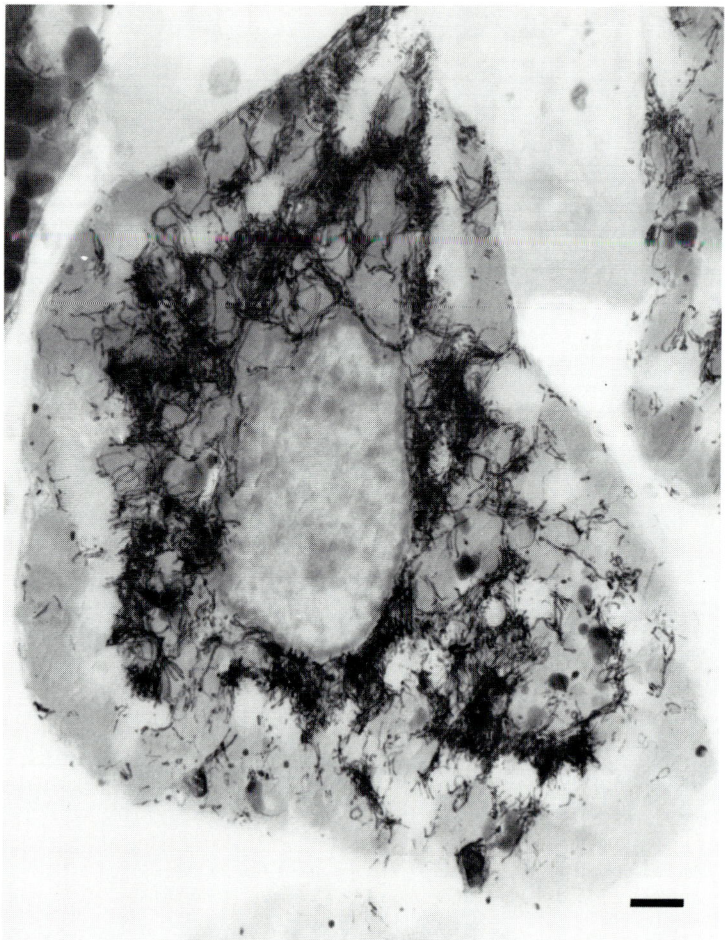

FIGURE 3. Gastric parietal cell was stained with the zinc-iodide-osmium method; a 1-μm-thick section was observed with 1000-kV EM. Together with the Golgi apparatus, the tubulo-vesicular system is revealed as numerous fibrous structures. These structures appear mostly to be located between the intracellular canaliculi and the nucleus. (Bar = 1 μm.)

the smooth ER and mitochondria in neurons.[20] This method is as follows: muscle tissue is fixed with 2.0% formaldehyde, 2.5% glutaraldehyde in 0.1 *M* cacodylate buffer, pH 7.3, for 2 h and rinsed in 10% sucrose in the same buffer overnight. After a brief wash with 0.05 *M* Tris-HCl buffer, pH 7.6, tissue blocks are incubated in 0.05% diaminobenzidine (DAB) and 0.01% H_2O_2 in Tris-HCl buffer, pH 7.3, for 2 h at room temperature and fixed again in the same aldehyde fixative overnight at 4°C. The tissue blocks are postfixed in 2% osmium tetroxide in cacodylate buffer, pH 7.3, containing 0.8% potassium ferrocyanide for 2 h at 4°C.

According to Forbes et al.,[12] sarcoplasmic reticulum staining obtained by the DAB-ferrocyanide method could also be achieved simply by postfixation with ferrocyanide-osmium solution when the tissue is prefixed with an aldehyde fixative containing calcium without phosphate. Conversely, prefixation with aldehyde fixative containing phosphate without calcium stains the plasma membrane like the T-system.

c. Ur-Pb-Cu Method

After glutaraldehyde fixation, tissue blocks are immersed in 5% uranyl acetate aqueous solution, pH 3.5, for 1 h. The tissue blocks are rinsed in distilled water and then incubated in lead and copper citrate solution for 1 h at 37°C. Lead and copper citrate solution is prepared as follows: mix 2 ml of 1 *M* lead nitrate aqueous solution with 0.5 ml of 10% copper sulfate. White precipitates are formed. To this solution, 19 ml of 4.6% sodium citrate and then 4 ml of 4% sodium hydroxide are added with stirring. After rinsing with distilled water, tissue is postfixed with 1% osmium tetroxide overnight, and processed for electron microscopy. The ER and Golgi apparatus are heavily stained by electron-dense precipitates.[21] This method can be used for HVEM.

d. Enzyme Cytochemistry

Reproducible and selective visualization of organelles can be achieved by several enzyme cytochemical techniques. Demonstration of phosphatase activity using a lead capture method is one of the applicable methods for HVEM observation. Sometimes these methods do not produce enough electron density for observation with HVEM, and the distribution of lead precipitates does not reveal all the structures where the enzyme is located. Therefore, relatively weak fixation which preserves the enzyme activities, longer incubation times, increase of concentration of the substrates in the incubation medium, or adding dimethyl sulfoxide (DMSO) to the incubation medium for enhancing penetration of the chemicals were recommended.[22,23] So far, glucose-6-phosphatase for ER,[24] nicotinamide adenine dinucleotide phosphatase[17,25] for medial Golgi cisternae, thiamine pyrophosphatase,[17,26] and acid phosphatase (cytidine monophosphatase)[17,23] for *trans*-cisternae of Golgi apparatus, and acid phosphatase[22,27] for lysosomes are reported for HVEM and 200-kV EM applications. Because of peculiar structures, thick sections of the Golgi apparatus and lysosome revealed very striking images that cannot be obtained from ultrathin sections.

4. Methods for Localizing Ligand-Binding Sites under HVEM

a. Preembedding Labeling Method or Labeling of Whole Cell Mounts

Basically, preembedding immunocytochemical techniques have an advantage, in that labeling is performed prior to the embedding process, which sometimes results in a loss of immunoreactivity. However, at the same time, it has a problem in allowing antibody access to intracellular antigenic sites. At present, immunogold labeling or lectin-gold labeling of whole cell mounts is mostly used as the preembedding labeling method for HVEM.

Cultured cells, grown on Formvar-coated grids, are labeled before or after fixation. For localizing intracellular antigenic sites, cells should be treated with 0.5% Triton X-100 in buffer before labeling. After postfixation with 1% osmium tetroxide and/or 1% uranyl acetate in buffer or water, and dehydration, the cells are critical-point dried and examined with HVEM. Fibronectin, tubulin, and microtubule-associated proteins have been localized with this method.[28,29] According to Ochs and Stearns,[28] ferritin conjugated with a secondary antibody could be substituted for protein-A gold particles. Ferritin particles are too small, however, to be recognized as individual particles in HVEM, although a massive binding of ferritin may be useful for visualizing three-dimensional structure.[28,30]

The membrane surface of free cells can be labeled and observed under HVEM. A suspension of free cells is labeled with cationized ferritin, before or after fixation, and postfixed, dehydrated, and embedded in Epon. Semithin sections (0.5 μm) are cut and observed in HVEM.[30]

b. Postembedding Labeling Method

The postembedding ligand labeling method cannot be applied to thick sections in the same manner as ultrathin sections, because labeling only the surface of a thick section does

Stereoview is the most exciting advantage of thick section observation. With the aide of a goniometer attached to HVEM, stereopaired electron micrographs can be taken by tilting the specimen. The more the tilting angle increases, the greater the stereo effect is, although there is a limit to the tilting angle that permits the recognition of a three-dimensional image from the stereopaired micrographs. Stereo effects can be experienced by observing stereopaired electron micrographs separately with both naked eyes. As this effect results from the difference of the distance between the same two points on two stereopaired images, it will be amplified by increasing the thickness of the section, the tilting angles, and the magnification of both EM and the print level with the same tilting angle. Therefore, the choice of these three factors that influence the stereo effect can be determined by the size of the object and the thickness of the section observed. A stereoscope is very helpful for the critical observation of stereopaired electron micrographs.

Several factors regarding the observation of serial sections must be considered. Before beginning the observation, each serial section should be placed in the specimen holder of HVEM in the same direction, because the axes of tilting angle for stereopaired electron micrographs must be the same direction in all serial sections. If a rotation specimen holder is available, correction of the direction would be very easy and accurate. After checking all serial sections with light micrographs, observation with HVEM should start from the middle section in the series and one should search for the structure being of the most interest and most easily identified by surrounding structures. Then, the observation of the same structure should be continued in the adjacent sections having direct connections to the first section observed. Conversely, if one starts the observation from the first section, the connections of the structure may be interrupted after observing only a few serial sections, because the structure of interest, which appeared in some sections, may not be presented in the rest of the serial sections.

Three-dimensional images of a structure from stereopaired prints are sometimes difficult to be shared with other people because these images are created in the sense of vision and do not exist objectively. Whole three-dimensional images of certain structures reconstructed from serial sections is also very difficult to demonstrate in printed form. Superimposing the images obtained from two to three serial sections may sometimes help in understanding the reconstructed three-dimensional structure.

Compared to HVEM (1000 kV or higher voltages), intermediate-voltage EM (400 to 800 kV) or lower-voltage EM (200 to 400 kV) are more popular. Although even 1- to 5-μm-thick sections are difficult to observe under 200-kV EM, one can still obtain important information from semithin sections (0.5 to 1.0 μm) with less three-dimensional effects. As most of the cytochemical techniques introduced here would be applicable to semithin sections, some useful information can be obtained from observation of semithin sections even with an intermediate-voltage EM.

ACKNOWLEDGMENT

High-voltage electron micrographs presented in this chapter were taken with H-1250 M at the National Institute for Physiological Sciences in Okazaki, Japan. The author is very grateful to Dr. Tatsuo Arii for technical help and suggestions.

REFERENCES

1. **Chan-Palay, V. and Palay, S. L.,** High voltage electron microscopy of rapid Golgi preparations. Neurons and their processes in the cerebellar cortex of monkey and rat, *Z. Anat. Entwickl.-Gesch.*, 137, 125, 1972.
2. **Hama, K., Arii, T., and Kosaka, T.,** Three-dimensional morphometrical study of dendritic spines of the granule cell in the rat dentate gyrus with HVEM stereo images, *J. Electron Microsc. Tech.*, 12, 80, 1989.
3. **Luft, J. H.,** Ruthenium red and violet II. Fine structural localization in animal tissue, *Anat. Rec.*, 171, 369, 1971.
4. **Kawakami, H. and Hirano, H.,** Three-dimensional reconstruction of the open canalicular system of human platelets by means of high-voltage electron microscopy, *J. Electron Microsc.*, 34, 38, 1985.
5. **Yamada, E. and Ishikawa, H.,** Dense tissue and special stains, in *Methods in Cell Biology*, Vol. 22, Turner, J. N., Ed., Academic Press, New York, 1981, 123.
6. **Takata, K., Arii, T., Yamagishi, S., and Hirano, H.,** Use of colloidal gold and ruthenium red in stereo high-voltage electron microscopic study of con A-binding sites in mouse macrophages, *Histochemistry*, 81, 441, 1984.
7. **De Bruijn, W. C.,** Glycogen, its chemistry and morphologic appearances in the electron microscope. I. A modified OsO_4 fixative which selectively contrasts glycogen, *J. Ultrastruct. Res.*, 42, 29, 1973.
8. **Karnovsky, M. J.,** Use of ferrocyanide-reduced osmium in electron microscopy, Proc. 14th Annual Meeting Am. Soc. Cell Biol., 1971, 146.
9. **Hepler, P. K.,** The structure of the endoplasmic reticulum revealed by osmium tetroxide-potassium ferricyanide staining, *Eur. J. Cell Biol.*, 26, 102, 1981.
10. **Jinguji, Y. and Ishikawa, H.,** An osmium-ferricyanide staining method for the intercellular space in the small intestinal epithelium, *J. Electron Microsc.*, 39, 59, 1990.
11. **Neiss, W. E.,** Electron staining of the cell surface coat by osmium-low ferrocyanide, *Histochemistry*, 80, 231, 1984.
12. **Forbes, M. S., Plantholt, B. A., and Sperelakis, N.,** Cytochemical staining procedures selective for sarcotubular systems of muscle: modifications and applications, *J. Ultrastruct. Res.*, 60, 306, 1977.
13. **Forbes, M. S. and Neil, E. E. V.,** Membrane systems of guinea pig myocardium: ultrastructure and morphometric studies, *Anat. Rec.*, 222, 362, 1988.
14. **Noda, T. and Ogawa, K.,** The structures of cellular components visualized under high voltage electron microscope, in *Recent Development of Electron Microscopy 1987*, Hashimoto, H., Kuo, K. H., Lee, K., and Ogawa, K., Eds., Nakanishi Printing, Kyoto, 1988, 183.
15. **Friend, D. S. and Murray, M. J.,** Osmium impregnation of the Golgi apparatus, *Am. J. Anat.*, 117, 135, 1965.
16. **Rambourg, A.,** Three-dimensional structure of the osmium-impregnated Golgi apparatus as seen in the high voltage electron microscope, *Am. J. Anat.*, 140, 27, 1974.
17. **Rambourg, A. and Clermont, Y.,** Three-dimensional electron microscopy: structure of Golgi apparatus, *Eur. J. Cell Biol.*, 51, 189, 1990.
18. **Noda, T. and Ogawa, K.,** Golgi apparatus is one continuous organelle in pancreatic exocrine cell of mouse, *Acta Histochem. Cytochem.*, 17, 435, 1984.
19. **Noda, T. and Ogawa, K.,** Golgi apparatus and its structural differentiation in duodenal epithelial cells, *Acta Histochem. Cytochem.*, 21, 343, 1988.
20. **Tsukita, S. and Ishikawa, H.,** Three-dimensional distribution of smooth endoplasmic reticulum in myelinated axons, *J. Electron Microsc.*, 25, 141, 1976.
21. **Thiéry, G. and Rambourg, A.,** A new staining technique for studying thick sections in the electron microscope, *J. Microsc. Biol. Cell*, 26, 103, 1976.
22. **Araki, N., Matsubara, H., Takashima, Y., and Ogawa, K.,** Three-dimensional distribution of acid phosphatase-positive thread-like structures ''nematolysosomes'' in rat pancreatic exocrine cells, *Acta Histochem. Cytochem.*, 22, 585, 1989.
23. **Mayahara, H., Ishikawa, T., Ogawa, K., and Chang, J. P.,** The three-dimensional structure of the Golgi complex in cultured fibroblasts. An ultracytochemical study with thin and thick sections, *Acta Histochem. Cytochem.*, 11, 239, 1978.

24. **Favard, P. and Carasso, N.,** The preparation and observation of thick biological sections in the high voltage electron microscope, *J. Microsc. (Oxford),* 97, 59, 1973.
25. **Rambourg, A., Clermont, Y., Hermo, L., and Segretain, D.,** Tridimensional architecture of the Golgi apparatus and its components in mucous cells of Brunner's glands of the mouse, *Am. J. Anat.,* 179, 95, 1987.
26. **Novikoff, P. M., Novikoff, A. B., Quintana, N., and Hauw, J.-J.,** Golgi apparatus, GERL and lysosomes of neurons in rat dorsal root ganglia, studied by thick section and thin section cytochemistry, *J. Cell Biol.,* 50, 859, 1971.
27. **Sakai, M. and Ogawa, K.,** Energy-dependent lysosomal wrapping mechanism, *Histochemistry,* 76, 479, 1982.
28. **Ochs, R. L. and Stearns, M. E.,** Colloidal gold immunolabeling of whole-mount erythrophore cytoskeletons: localization of tubulin and HMW-MAPs, *Biol. Cell,* 42, 19, 1981.
29. **Peters, D. M. P. and Mosher, D. F.,** Localization of cell surface sites involved in fibronectin fibrillogenesis, *J. Cell Biol.,* 104, 121, 1987.
30. **Kawakami, H. and Hirano, H.,** Rearrangement of the open-canalicular system of the human blood platelet after incorporation of surface-bound ligands. A high-voltage electron-microscopic study, *Cell Tissue Res.,* 245, 465, 1986.
31. **Gorbsky, G. and Borisy, G. G.,** Microtubule distribution in cultured cells and intact tissues: improved immunolabeling resolution through the use of reversible embedment cytochemistry, *Proc. Natl. Acad. Sci. U.S.A.,* 82, 6889, 1985.
32. **Hama, K. and Hirosawa, K.,** High voltage electron microscopy: a study of autoradiography, *J. Electron Microsc.,* 26, 187, 1977.
33. **Hirosawa, K. and Hama, K.,** High voltage electron microscope autoradiography. II. A study on the resolution of semi-thin sections, *J. Electron Microsc.,* 31, 405, 1982.
34. **Pylypas, S. P., Schwrath, R. B., and Johnson, R. B.,** Preparation of bone for high-voltage electron microscopic radioautography, *J. Electron Microsc. Tech.,* 14, 324, 1990.
35. **Katagiri, N., Hama, K., Katagiri, Y., Shimatani, Y., Hashimoto, Y., and Aikawa, E.,** High-voltage electron microscopic study on the axon of the dermal photoreceptor cell in the dorsal mantle of *Onchidium cerruculatum, J. Electron Microsc.,* 36, 38, 1990.

Chapter 8

ANALYTICAL COLOR FLUORESCENCE ELECTRON MICROSCOPY

Tohru Nakano and Gang Ning

TABLE OF CONTENTS

0-8493-6012-9/93/$0.00 + $.50

A. INTRODUCTION

Several signals or information can be obtained with the scanning electron microscope, such as secondary electrons, backscattered electrons, transmitted electrons, Auger's electrons, and X-rays. Cathodoluminescence (CL) is emitted as one of these signals under electron beam irradiation.[1] Although CL is essentially a similar phenomenon to fluorescence elicited by ultraviolet light, it has some intrinsic advantages over the latter:[1,2] (1) CL can be amplified electronically and measured quantitatively, (2) the depth of focus and the resolution in the CL mode are higher than in the conventional fluorescence microscope, and (3) some substances exhibit luminescence only under electron irradiation. However, applications of the CL technique to the biomedical field have been limited, because CL from organic chemicals is a very weak signal and easily fades under high beam current (beam effect).[3]

On the other hand, although several techniques of micro-area analysis based on electron microscopy have been developed, it is no exaggeration to say that no practical techniques are available for information on the chemical bonding states of constituent elements of biomolecules. CL reflects electronic configuration of the outer electron shell and is derived from loosely bound π-electrons in the case of organic compounds.[4] This suggests that the CL analysis technique may potentially provide information about chemical bonds, that cannot be obtained using conventional analytical electron microscopic techniques.

In recent years, we have developed an analytical color fluorescence electron microscope (ACFEM) based on a high-resolution scanning electron microscope (SEM).[5,6] The ACFEM enabled us to detect CL from biological specimens as a color image simultaneously with the corresponding secondary electron (SE) image and also to analyze the CL spectrum. Using the ACFEM, in combination with a cryo-SEM technique,[7] the distribution of several organic chemicals such as vitamin A ester[6] and cholesterol ester[8] and their dynamic changes could be detected *in situ* in various tissues, which suggests that the CL analysis by the ACFEM is a simple and effective new method and promises to open a productive new area in the field of histo- and cytochemistry.

B. METHODS

The application of CL in the biomedical field can be classified into two categories.[9] One is aiming at the identification of substances by CL analysis as described here. For this purpose, a CL-detector system combined with an SEM are required, and the observation should be performed at very low temperature for accurate and stable CL emission.[10] Although the relationship between molecular structure and CL yield has not been established yet,[3,11] CL spectrometry can be used to identify intracellular substances to a certain extent by comparing the CL spectrum of the specimen with those of purified compounds.[6,8,9] However, because the fluorescence spectra could shift due to several factors, such as energy transfer between chromophores and solvent polarity,[12] it is difficult to determine the origin of the CL in some cases by simply comparing the peak CL wavelength of the tissue with those of purified compounds. In such a case, observations of the changes of CL images in different physiological and pathological conditions and reference to biochemical data under the same conditions could help to identify the origin of the CL.[6,8] Once the correspondence between the organic substance and its CL spectra is elucidated, monochromatic CL images (obtained using a monochromator) at the peak wavelength of the CL spectrum will reveal the distribution of the CL-emitting substance in the specimen.

Another method of CL application to biomedicine is performed under an ultraviolet-elicited fluorescence microscope, for example, the immunocathodoluminescence method.[13-16] However, most fluorescent dyes used under a light microscope are easily damaged by electron irradiation, and their luminescence fades very quickly.[17] For this reason, although several experiments using fluorescent dyes[3,18] and immunocathodoluminescence[13-16] have

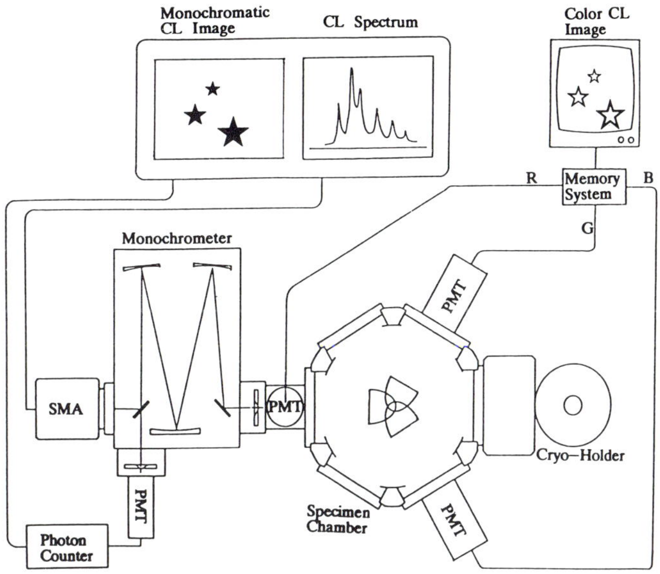

FIGURE 1. Schematic diagram of the ACFEM. CL, divided into three components (R, G, B), is reconstructed into a single color image. For CL spectrometry and monochromatic CL imaging, the monochromator is placed at a focus of an ellipsoidal mirror. PMT, photomultiplier tube; SMA, spectrometric multichannel analyzer.

been carried out, the resolution obtained does not seem to exceed that of the light microscope. When CL dyes that emit a stable CL and are resistant to electron beam damage are developed, the application of CL, including immunocathodoluminescence technique, will progress rapidly.

C. INSTRUMENTS

The ACFEM (Figure 1) was developed on the basis of a high-resolution SEM (ISI-DS 130; Akashi Beam Technology Corp., Tokyo).[5,6] This SEM has two specimen stages: a top stage for ultrahigh-resolution observation (3 nm with a tungsten filament and 2 nm with a LaB_6 filament) and a bottom stage for the observation of large specimens. The top stage is used to observe CL in color and to analyze CL spectra.

In order to maintain high performance at low acceleration voltages, which should be used to avoid the beam effect in the CL mode, a high-excitation objective lens was introduced. Three ellipsoidal mirrors were arranged at 120° to each other and combined so as to hold a single focus accurately. This set of three combined ellipsoidal mirrors has a small hole in the center for electrons to pass through to the specimens, and is placed below the upper objective lens pole-piece as a collection system for CL photons. Considering the angle distribution of CL emission and solid angle of the mirror, about 90% of CL photons should be detected. At the other focal points of the three ellipsoidal mirrors, red, green, and blue filters were placed. A photomultiplier tube (PMT) (R-1104; Hamamatsu Photonics Co., Hamamatsu, Japan) was placed next to each filter, and the CL, divided into three components, was amplified by the PMT and integrated to form a single color image through a memory system. For CL spectrometry, a monochromator (HR-320; Jobin-Yvon, U.S.) was placed at the focal point of an ellipsoidal mirror followed by a spectrometric multichannel analyzer (SMA; Prinston Instruments Inc., U.S.) and SMA control system. Monochromatic CL images at any wavelength could be obtained by attaching a PMT next to the monochromator.

D. PROCEDURES

1. Specimen Preparation

Fresh unfixed tissues should be used, because chemical fixatives such as glutaraldehyde decrease the intensity of CL emission. However, when the specimen is fragile and difficult to mount on a cryo-tip without chemical fixation, weak fixative (e.g., 2% formaldehyde) can be used for a short time. After cutting the tissues into small pieces (about 1 mm in thickness), they are mounted on small aluminum cryo-tips and rapidly frozen in liquid propane precooled with liquid nitrogen. The surface of the frozen specimen is cut by a cryoultramicrotome (Ultracut E, FC-4; C. Reichert Optishe AG, Vienna, Austria) at −100°C to expose intracellular structures and to make the surface flat. Because the CL emitted from the surface of the specimen is collected by a set of three ellipsoidal mirrors placed above the specimen, the surface of the specimen should be flat and smooth so that the CL would be separated and detected by the three mirrors (for red, green, and blue color components) equally. Cutting of the surface of the specimen by a cryoultramicrotome should not be deep, because the zone of good freezing is limited to no more than a few microns.

2. Observation under the ACFEM

After preparing the specimens, they are attached to the cryo-holder precooled by liquid nitrogen, and transferred into the ACFEM. During the process of transferring the specimens into the microscope in the air, the surfaces of the specimens become covered with ice crystals. However, they will be sublimated in the specimen chamber of the ACFEM within 20 min. The cryo-holder is kept at −120°C by liquid nitrogen, and observations by the ACFEM are performed under accelerating voltages of 5 to 9 kV. The procedure of observation under the ACFEM is as follows:

Color CL image and SE image — By comparing the color CL image with the SE image, correspondence between the intracellular structures and the CL-emitting substances is revealed. When observing the SE images, thin gold coating can be performed in the specimen chamber of the ACFEM.

CL spectrometry and analysis — The CL spectrum is recorded using the SMA system. If biochemical data on the chemical composition of the tissue are available, the CL spectra of each purified chemical found in the tissue should be investigated. A comparison between the CL spectra and intensity of the specimen and those of the chemicals can furnish much information as to the identity of the CL-emitting substances.

Monochromatic CL image using a monochromator — When there are multiple peaks in the CL spectrum, monochromatic CL images at each peak should be compared in the same field of the specimen. If the CL of each peak is derived from different substances, each monochromatic image will show the distribution of each corresponding substance.

E. RESULTS

1. Rat Adrenal Cortex

Strong blue-green CL was observed to be emitted from the adrenal cortex of the rat (data not shown here).[8] By comparing the color CL images with the SE image at high magnification, the CL was found to be emitted from the lipid droplets (LD) in the adrenocortical cells (Figure 2). The CL spectrum of rat adrenal cortex had two peaks at 320 nm and 430 nm (Figure 3). Comparing the two monochromatic CL images at each peak wavelength, it was found that every LD emitted CL of both peak wavelength components in the zonae glomerulosa and fasciculata (Figure 4); on the other hand, a few LD in the zona reticularis emitted CL with a peak of 430 nm only (Figure 5).

In order to elucidate the changes of CL emission from adrenal cortex in different experimental conditions, the effects of hypocholesterolemic drugs, adrenocorticotropic

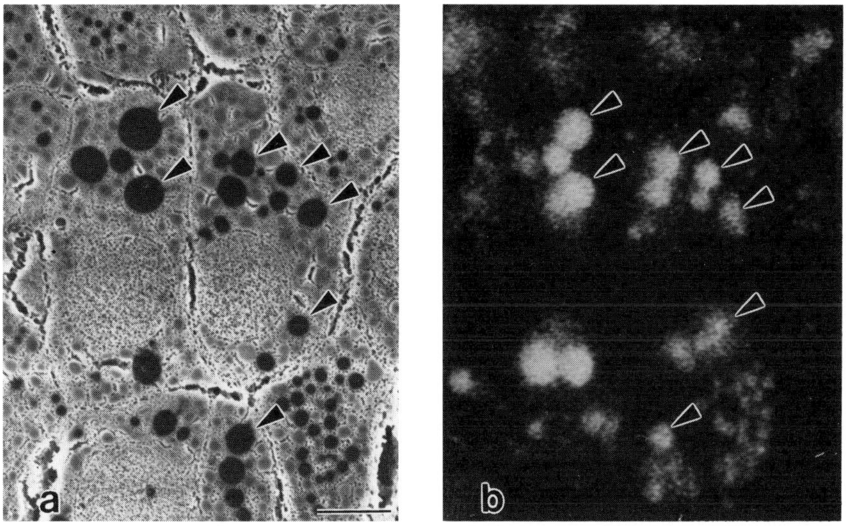

FIGURE 2. SE (a) and corresponding CL (b) images of the zona fasciculata cells of adrenal cortex of an untreated rat. CL is emitted from LD (arrowheads). (Bar = 5 μm.)

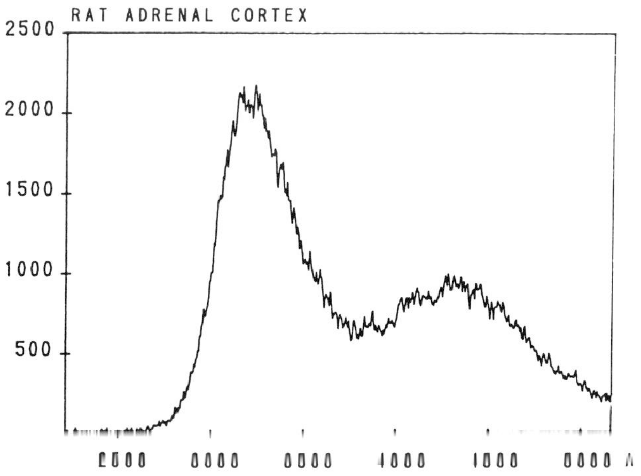

FIGURE 3. CL spectrum of rat adrenal cortex.

hormone (ACTH), and hypophysectomy on the CL of rat adrenal cortex were examined. 17-α-Ethinyl estradiol (EE) and 4-aminopyrazolo-pyrimidine (APP) were used as hypocholesterolemic drugs. Both of these drugs are known to cause a significant decrease of esterified cholesterol in the adrenal.[19-23] EE administration also results in a decrease of corticosterone concentration in plasma and adrenals.[19,20] However, chronic APP administration does not significantly affect the basal plasma corticosterone concentration, and it is reported to stimulate endogenous cholesterol synthesis in the adrenocortical cells affected by APP.[22,23] When the two monochromatic images of each peak wavelength were compared in the adrenal cortex of EE- and APP-treated rats, LD in all three zones of the cortex emitted CL of 430 nm alone, and CL of 320 nm disappeared (Figures 6 and 7). In these two groups, the total number of LD in every zone decreased; the APP-treated group contained more LD than the EE-treated group (Figures 6b and 7b).

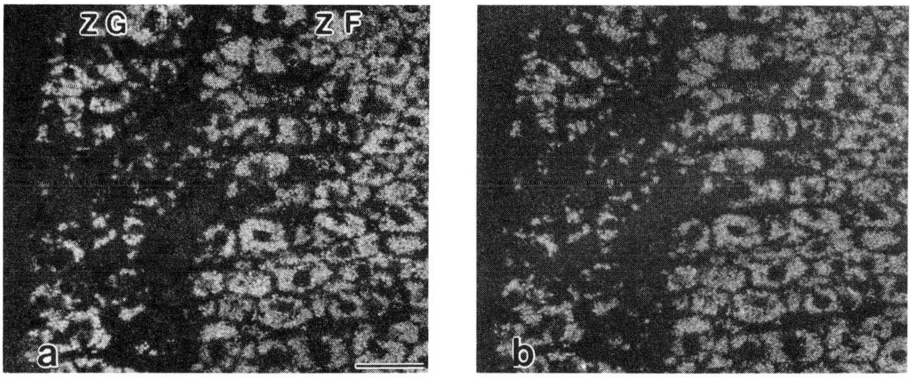

FIGURE 4. Monochromatic CL images of the zonae glomerulosa and fasciculata of the adrenal cortex of an untreated rat at each peak wavelength: 320 nm (a) and 430 nm (b). Every LD emits CL of both peak wavelengths. ZG, zona glomerulosa; ZF, zona fasciculata. (Bar = 30 μm.)

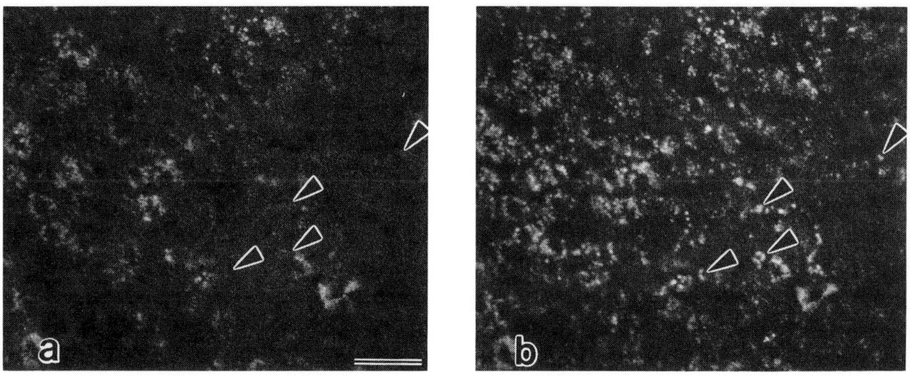

FIGURE 5. Monochromatic CL images of the zona reticularis cells of the adrenal cortex of an untreated rat at each peak wavelength: 320 nm (a) and 430 nm (b). Some LD (arrowheads) emit CL of 430 nm, but not 320 nm. (Bar = 30 μm.)

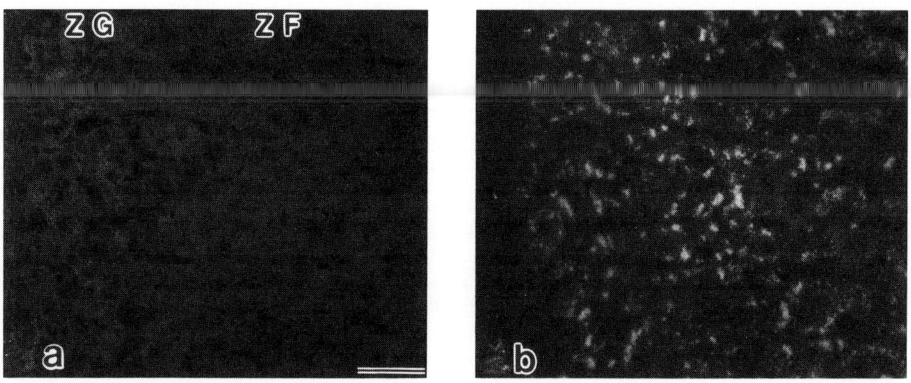

FIGURE 6. Monochromatic CL images of the adrenal cortex of an EE-treated rat at each peak wavelength: 320 nm (a) and 430 nm (b). The LD do not emit CL of 320 nm. (Bar = 30 μm.)

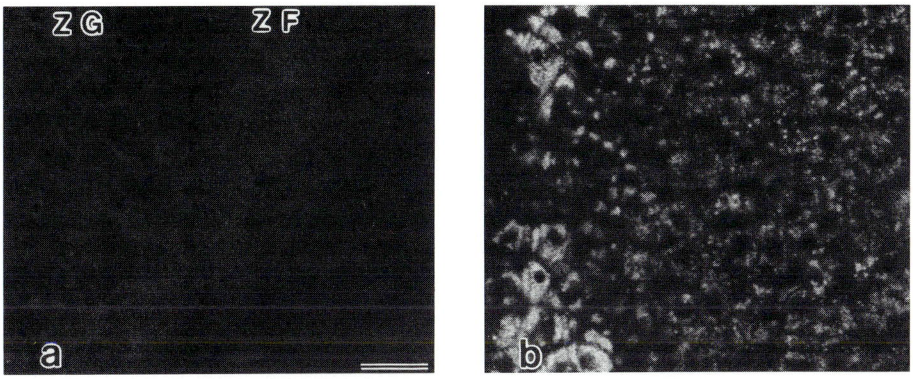

FIGURE 7. Monochromatic CL images of the adrenal cortex of an APP-treated rat at each peak wavelength: 320 nm (a) and 430 nm (b). No CL from LD is seen at 320 nm as in the EE-treated rat, while the number of the LD which emit CL of 430 nm is larger than that of the EE-treated adrenal cortex. (Bar = 30 μm.)

In the hypophysectomized rats, severe atrophy of both adrenals was observed. The volume of whole adrenal cortex decreased significantly, and the zonae fasciculata and reticularis appeared thin and dense, while the zonae glomerulosa and intermedia were found to be wider compared with the control (Figure 8, b and c). An SE image of the zona fasciculata cells in high magnification revealed that there were two types of LD: huge LD and densely packed smaller LD (Figure 8a). The monochromatic images at the two peak wavelengths revealed that CL from huge LD had a peak at 320 nm but not at 430 nm, and a diffuse CL derived from the smaller LD in the zonae fasciculata and reticularis was more intense at 320 nm (Figure 8, b and c).

Rats were killed at different times after the administration of ACTH. CL of 320 nm of the zonae fasciculata and reticularis cells apparently decreased at 5 min after the injection of the hormone (Figure 9a); it almost disappeared at 2 h after the injection (Figure 10a). The number of LD also decreased, but CL of 430 nm remained at all times (Figures 9b and 10b). CL of the zona glomerulosa did not disappear in either peak wavelength, while CL intensity at 320 nm seemed weaker than that of untreated animals (Figures 9a and 10a). At 4 h after the injection, CL of 320 nm in the zonae fasciculata and reticularis cells increased again, while the number of LD in these zones was still smaller than in untreated animals (Figure 11).

[illegible] synthesis could be separated by CL spectroscopy *in vitro*. However, it was difficult to differentiate these substances *in vivo,* because the wavelength peaks of these chemicals are close to each other and sometimes overlap. Therefore, rat adrenals under some experimental conditions, which are known to change chemical composition in adrenal cortex, were observed by the ACFEM. The above results clearly showed the changes of the CL of 320 nm. Comparison of the data shown here with previously reported biochemical data[19-26] suggested a close relationship between the CL of 320 nm and cholesterol ester.[8]

The conventional histochemical technique to detect cholesterol and its ester, developed by Schultz,[27] has several problems in both specificity and preservation of tissue structures. Especially, the poor preservation of the ultrastructure because of a concentrated acid treatment makes it difficult to apply this method to electron microscopic cytochemistry. The results presented here show that CL spectroscopy combined with the cryo-SEM method preserves intracellular structures and detects cholesterol ester specifically.

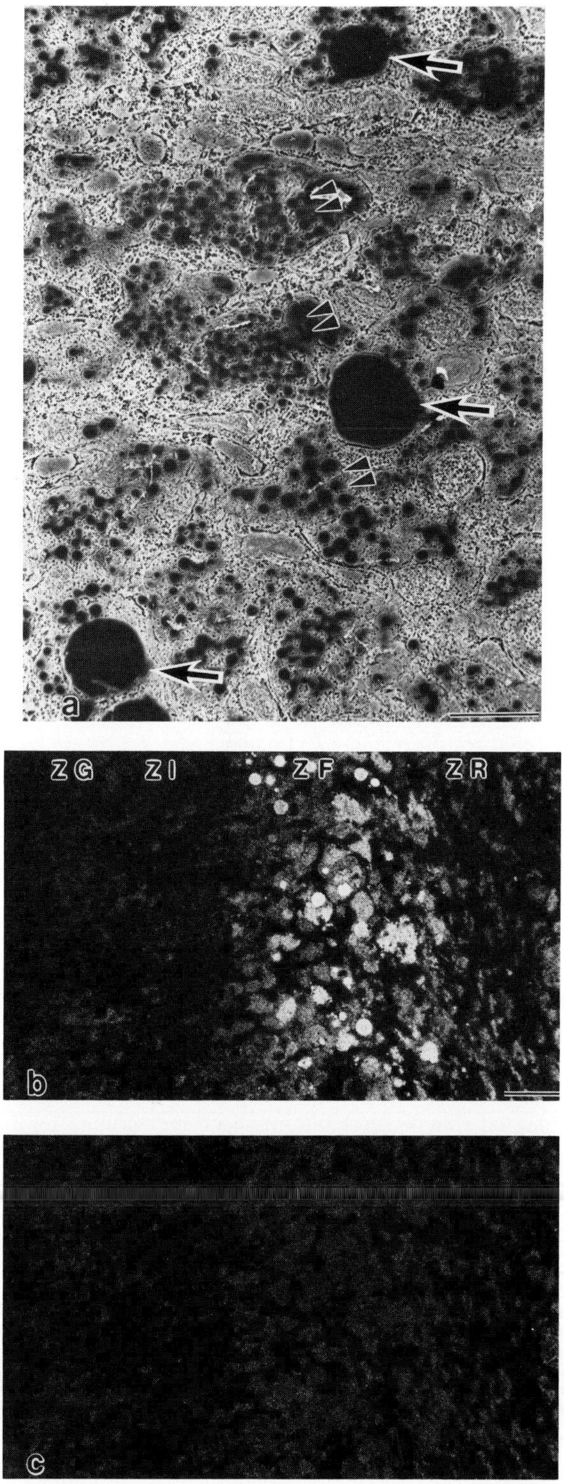

FIGURE 8. SE image (a) and monochromatic CL images of the adrenal cortex of a hypophysectomized rat at each peak wavelength: 320 nm (b) and 430 nm (c). There are two types of LD in the zona fasciculata cells: huge LD (arrows) and densely packed smaller LD (arrowheads). CL of 320 nm in the zonae fasciculata and reticularis is much more intense than that of 430 nm. ZI, zona intermedia; ZR, zona reticularis. (Bars = 10 μm (a) and 30 μm (b and c).)

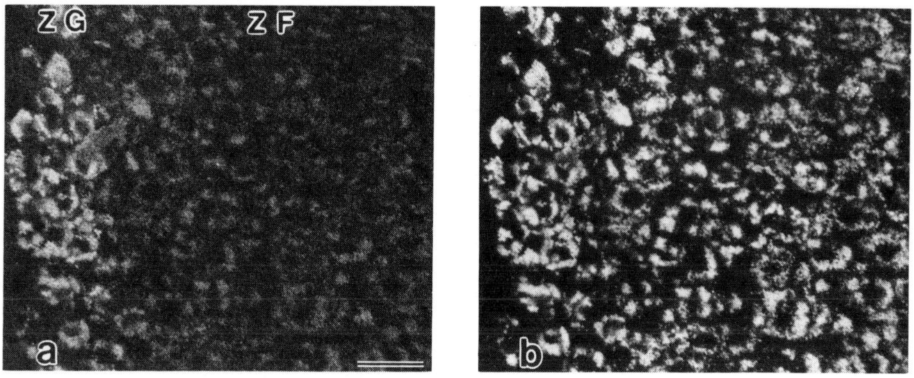

FIGURE 9. Monochromatic CL images of the adrenal cortex of an ACTH-treated rat 5 min after injection at each peak wavelength: 320 nm (a) and 430 nm (b). CL of 320 nm in the zonae fasciculata is apparently decreased. (Bar = 30 μm.)

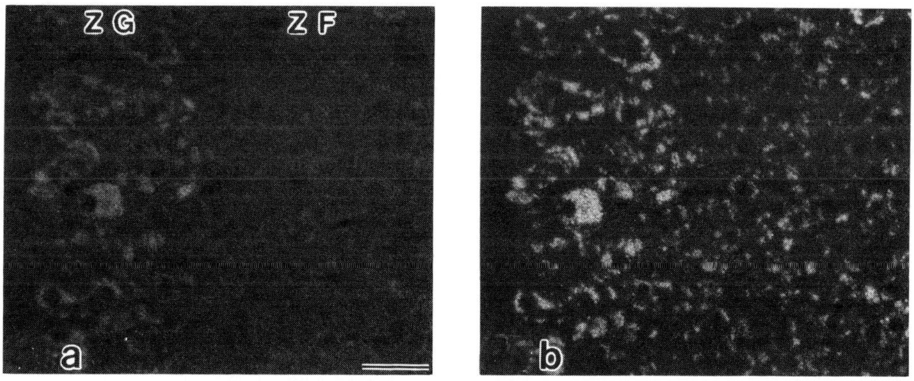

FIGURE 10. Monochromatic CL images of the adrenal cortex of an ACTH-treated rat 2 h after injection of the hormone at each peak wavelength: 320 nm (a) and 430 nm (b). CL of 320 nm in the zona fasciculata has almost disappeared. (Bar = 30 μm.)

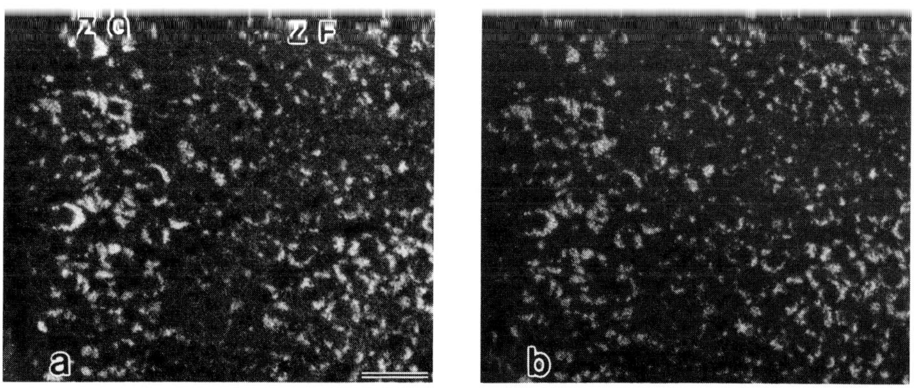

FIGURE 11. Monochromatic CL images of the adrenal cortex of an ACTH-treated rat 4 h after injection of the hormone at each peak wavelength: 320 nm (a) and 430 nm (b). CL of 320 nm in the zona fasciculata appears to be increased again. (Bar = 30 μm.)

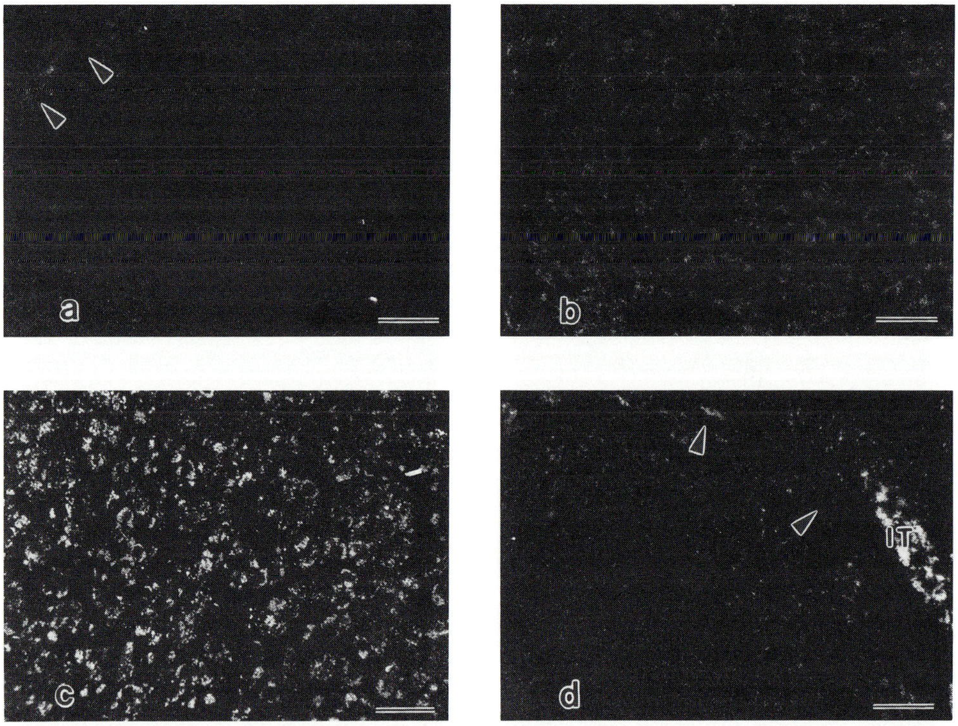

FIGURE 12. The CL images of corpora lutea in Wistar rats obtained at a wavelength of 320 nm: newly formed corpus luteum, estrus (a); diestrus 1 (b); later diestrus 2 (c); and 7-d pregnancy (d). Arrowheads indicate the edge of a corpus luteum. IT, interstitial tissue. (Bar = 40 μm.)

2. Rat Ovary

In the ovaries of rats, the CL can usually be found from LD in the cells of corpora lutea, interstitial tissues, theca interna of follicles, and occasionally from the membrana granulosa (Figures 12 and 13).

It is well known that laboratory rats usually have a 4- to 6-day estrous cycle through puberty (except for conception), and that the complicated structural and physiological changes of the ovary (ovarian cycle), which are essentially associated with the ovarian steroidogenesis, are synchronously repeated following the estrous cycle. The findings gained from ACFEM observations demonstrated that the CL from the LD in the corpus luteum also has regular changes in intensity throughout the estrous cycle, and CL at a wavelength of 320 nm always reveals a close correlation with progesterone levels in the plasma. The results of CL detected in the rat ovary following an estrous cycle by ACFEM, moreover, is coincident with an earlier study of cholesterol ester by conventional microscopy.[28] These facts suggested that the CL of 320 nm is derived from the cholesterol esters — a storage form of the physiological precursor of steroid hormones — in the LD of cells of the corpus luteum. In the review by Freeman,[29] rat corpora lutea were described as being functionally defective, except during pregnancy, because they only secrete a small amount of progesterone during a short post-ovulatory period of 24 to 48 h, as the rats are in the late estrus and diestrus 1, when a small amount of LD containing little cholesterol esters is retained in the cells of corpora lutea. As soon as the rats are in the stage of diestrus 2, hormonal secretion of corpora lutea begins to slow down and eventually ceases, the amount of LD increases, causing rapid storage of abundant cholesterol esters, and the cells producing progesterone no longer degenerate.

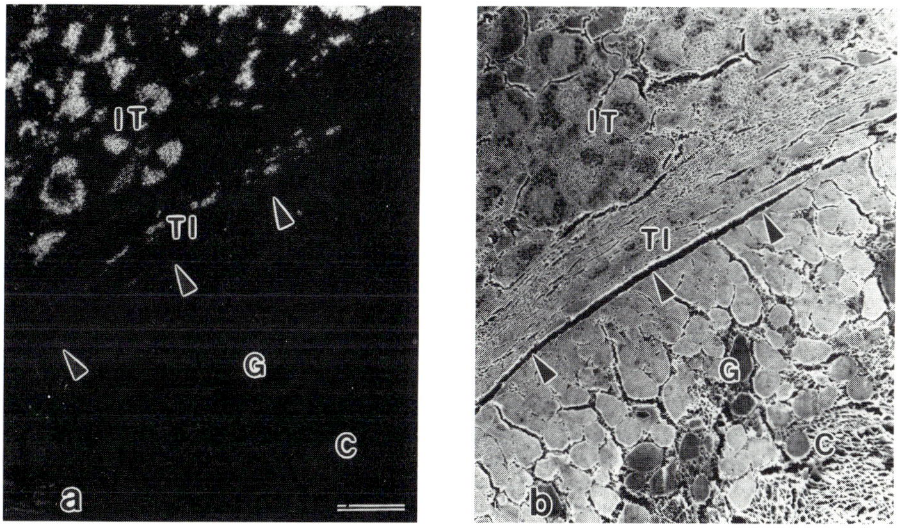

FIGURE 13. The CL image (a) and its secondary electron image (b) of a portion of a preovulatory follicle and surrounding interstitial tissue. C, Follicular cavity; G, membrana granulosa; TI, theca interna; IT, interstitial tissue. Arrowheads indicate the basement membrane. Note the consistent distribution of the CL and lipid droplets (dark grains) in cells in the two pictures. (Bar = 20 μm.)

However, in the case of pregnancy, the rats possess well-functioning corpora lutea which synthesize a considerably large amount of steroid hormones to maintain the early conception, during which the LD are smaller and contain little cholesterol esters. This dynamic metabolism of cholesterol esters in the corpus luteum is reflected by CL intensity at 320 nm (Figure 12). At estrus, very weak CL is barely detected only from the edge of newly formed corpora lutea, namely luteinizing theca interna. At diestrus 1, the CL becomes stronger than it is at estrus, and it is uniformly scattered around the corpora lutea in either the central or the edge portions. At late diestrus 2 and proestrus, the intensity of CL is enhanced significantly and it persists throughout the next estrous cycle, though an uneven distribution gradually arises. When ovaries of 7-day pregnant rats are examined, the weak CL of 320 nm appears to have a similar pattern of distribution to the CL observed from corpora lutea at estrus.

The CL can also be observed in various follicles, where it shows different intensities and regular distribution patterns which are closely related to the states of steroid hormonal secreting function of the follicles at a given developing stage, and cytochemically provides the dynamic information of steroidogenesis of different cells and organelles. The CL from the LD in follicles are located as masses in the theca interna and scattered as granules around the membrana granulosa (Figure 13).

REFERENCES

1. **Pfefferkorn, G., Bröcker, W., and Hastenrath, M.,** The cathodoluminescence method in the scanning electron microscope, *Scan. Electron Microsc.*, 1, 251, 1980.
2. **Bröcker, W. and Pfefferkorn, G.,** Applications of the cathodoluminescence method in biology and medicine, *Scan. Electron Microsc.*, 2, 125, 1979.

PART II

MAJOR CONTRIBUTIONS OF ELECTRON MICROSCOPIC CYTOCHEMISTRY AND IMMUNOCYTOCHEMISTRY TO BIOMEDICINE

Chapter 9

CELL ORGANELLES

9.1 Applications of Electron Microscopic Cytochemistry and Immunocytochemistry to the Study of the Plasma Membrane

Mark C. Willingham

TABLE OF CONTENTS

0-8493-6012-9/93/$0.00 + $.50

A. INTRODUCTION

In this article, I will describe examples of the types of experimental results that cytochemical methods can provide in the study of the plasma membrane and structures related to the plasma membrane. It is not my intention to include all possible membrane components, or to cover all applicable methods in such studies, but rather to concentrate on those with which I have had direct experience and those that have been especially useful in the last decade in clarifying some of the physiological aspects of plasma membrane function and morphology.

B. THE PLASMA MEMBRANE

1. The Physiological Nature of the Plasma Membrane

The plasma membrane of mammalian cells represents a highly specialized and vital part of cell structure, probably the earliest unique morphological unit of individual cells in evolution. This thin, fragile membrane mediates the interactions of the cell with its environment, both physically and chemically. Most individual mammalian cells are very small, perhaps 7 to 40 μm in diameter. As a result, the plasma membrane that each cell maintains encompasses roughly the surface of a sphere of this diameter, with some structures that might increase its actual surface area, such as microvilli. However, for some specialized cells, such as the syncytium of cells in skeletal muscle or placenta, or the extremely long neurons in the spinal cord, their single continuous plasma membrane may be quite large. It is this unit of membrane that the cell must maintain in good working order to keep the outside of the cell separated from the inside.

Even though mammalian cells are usually sequestered in warm, watery surroundings, the inside of the cell is very different from the outside in chemical composition. The inside of cells, that is, the cytoplasmic fluid compartment (i.e., cytosol), contains high levels of potassium, very low sodium, very low calcium, a slightly acidic pH, and large amounts of ATP and special soluble proteins. The outside has very little potassium, lots of sodium and calcium, and a slightly alkaline pH. The plasma membrane is the only barrier that separates these very different environments, and it must keep this balance of ions and other solutes steady for the cell to function. A large percentage of the energy of the cell, in fact, is used to maintain this balance.

One of the hallmark properties of the plasma membrane is that it is impermeable to most hydrophilic molecules, based mainly on its largely lipid composition. Small hydrophobic materials, however, frequently can dissolve in the lipid bilayer, and those that are small and slightly hydrophobic can readily cross the membrane, albeit with low efficiency. Charged, hydrophilic molecules and large molecules such as proteins have almost no permeability to the plasma membrane and their ability to enter and exit cells is based mainly on selective, often energy-dependent, carrier systems.

2. Plasma Membrane Morphology

From a morphologist's point of view, the plasma membrane initially might seem to be rather simple, but there are a number of morphological specializations and features that one can recognize by electron microscopy. Cells contact other cells and other objects, such as substrata (e.g., basement membrane), and these interactions often require specialized morphological features. Focal adhesions characterize many cells in their attachment to substrata in culture, and these have specialized proteins that associate with them, both inside and outside of the membrane. Gap junctions, tight junctions, and desmosomes are all specialized regions that some cells have that mediate interaction between adjacent cells. Migrating cells frequently extend thin lamellapodia ahead of their migration or extend ruffles above the cell, some of which slowly fall back onto the adjacent cell surface and trap bubbles of media as an intracellular pinocytic vesicle, a process originally described by the term "pinocytosis".[1] Other areas of the cell surface may have microvilli, thin villous extensions of the cell with an internal core of actin microfilaments. Some cells have stable populations of microvilli on their surface, while others have microvilli that vary in number and length. For example, at the time of mitosis, many cells become rounder and frequently show large numbers of microvilli on their surface. Many tumor cells that have lost their attachment to their normal substrata frequently have many microvilli.[2] Another surface specialization is the zeiotic bleb, structures that appear as rapidly expanding and contracting bubbles on the surface of cells.[3] In cell culture, with phase contrast microscopy, these structures appear very phase dense, and they contain extruded cytoplasm, occasionally containing intracellular organelles such as lysosomes or mitochondria. Some tumor cells in culture show intense blebbing activity. At the time of cytokinesis in most cells, the increased intracellular pressure produces blebs on the surface of the dividing cells. Interruption of the microfilaments that support the inside of the plasma membrane by agents such as the cytochalasins produce surface blebs that fail to contract.[3] This perhaps suggests that the mechanical integrity of the membrane is very dependent on the microfilamentous meshwork that lies just beneath its interior surface. This blebbing activity is physiologic and should not be confused with fixation artifacts created by osmotic effects in aldehyde-fixed cells.[4] This latter artifact is characterized by a lipid-rich blister of the surface membrane that contains no internal cytoplasm, and in phase contrast appears clear. Such structures are frequently observed in scanning micrographs of cells fixed for too long a time in low-osmolarity buffers.

Another group of surface specializations are the indentations of the plasma membrane. One group of uniform-sized, flask-shaped indentations are called caveolae.[5] The structure of these and their function have long been an issue of debate, and will be discussed further below. Another indentation is a very different structure, termed a coated pit.[6] Each cell has 500 to 1000 coated pits on its surface at any one moment, and these are very specialized and unique structures. Coated pits mediate the vast majority of all types of endocytosis in cells.[7] They are invaginations of the plasma membrane that have a unique cage-like coating on their intracellular surface, a coating composed mainly of a protein called clathrin.[8]

3. Two Functions Mediated at the Surface: Endocytosis and Exocytosis

Endocytosis is the vesicular-mediated entry of macromolecules and other large elements into the interior compartments of cells.[7] Many objects enter the cell in this way, but all have the property that they are either too large or too hydrophilic to cross the plasma membrane directly, and the coated pit produces vesicles that deliver materials to precise destinations in the cell interior, usually never allowing the materials out of the vesicular compartments inside into the cytosol. In this sense, the internal compartments derived from endocytosis and vesicular sorting events are still, functionally, outside of the cell, that is, outside of the cytoplasm. This also brings up another important aspect of cell membrane relationships. All of the membranes within the cell, including the plasma membrane, are at one time or another in physical continuity with each other, except for mitochondria and, in plant cells, chloroplasts. These latter structures are thought to have entered a symbiotic relationship with cells long ago in evolution, originally entering cells as parasites. Their membranes, therefore, do not interact directly with the other membranes of the cell. Indeed, one could imagine the major membrane systems of the cell as nothing more than internalizations and specializations of the plasma membrane itself, the functionally most important membrane structure that defines the existence of individual cells.

Another important morphologic event that occurs at the plasma membrane is the exit of vesicular materials from the internal compartments of the cell, a process called exocytosis.

Exocytosis is specialized into a constitutive pathway that constantly brings material from the Golgi system to the cell surface, and a triggered or concentrative storage release process mediated by specialized secretory granules.[9] The former exocytic process is not well understood because of its speed and the lack of clear markers for the specialized structures involved, but occurs in all cells. The latter process has been very well studied morphologically because of the large size and long life of the specialized secretory granules involved, but is not found in all cells. The constitutive pathway constantly replaces surface components removed by endocytosis or surface degradation, as well as releasing some vesicular materials into the surrounding environment. The secretory granule pathway mainly releases concentrated material in the lumen of these granules into the cell surroundings, but also may contribute some components of the plasma membrane.

4. The Chemical Makeup of the Plasma Membrane

The plasma membrane, similar to other cell membranes, is chemically characterized as a bilayer of laterally mobile phospholipids in which are buried a number of other materials. Some materials are embedded in the membrane, or even lie across the membrane, and are termed integral elements. Some molecules may be attached to other integral components of the membrane but are not integral themselves, and are termed peripheral elements. Such peripheral associations occur on both the inside as well as the outside of the plasma membrane. Since the plasma membrane is derived from synthesis through the Golgi system, an internal membrane system that has numerous glycosylation enzymes on its noncytoplasmic (lumenal) face, many surface elements, both lipid and protein, have carbohydrates added as part of their covalent structure on the extracellular face of the plasma membrane. This dense layer of carbohydrates is often characterized by terminal sialic acids, which impart a net negative charge to the plasma membrane at physiological pH, a property that can be exploited cytochemically. In addition, some of these carbohydrate structures form unique antigenic epitopes to which antibodies can be raised, examples of which are the human blood group antigens A and B. Another hallmark of the plasma membrane, as well as the trans-Golgi membrane system from which it is derived and with which it interacts, is a very high content of cholesterol. This cholesterol content can, in fact, be a cytochemically useful marker for these membrane systems. This cholesterol content also makes these membranes susceptible to permeabilization with cholesterol-like detergents such as saponins, a property which has been exploited for cytochemistry of fixed membranes (see below).[10]

5. Molecules of the Membrane with Physiological Importance

Buried in and attached to the plasma membrane are a large collection of receptors, that concentrate their ligands from the cell's environment and bind them to the cell surface.[7] Such receptors include those for some of the cell's nutrient requirements, such as the receptors for low-density lipoproteins, transferrin, or vitamins such as folate. Other receptors bind polypeptide hormones or growth factors that regulate cell behavior, such as the receptors for insulin or epidermal growth factor. Some pathologic processes such as viruses or protein toxins also bind to unique components of the membrane, components that represent, as a result, receptors for these enemies of the cell. Other receptors interact with substrate components such as fibronectin or laminin.

In addition to receptors, the plasma membrane also contains transport systems. Many of these transporters are energy-requiring transmembrane proteins that form selective carrier channels that allow the entry or exit of hydrophilic materials that would otherwise not be able to cross the membrane. Some of these are related to maintaining the ionic balance across the membrane, such as Na^+, K^+-ATPase, some are related to transport of nutrients into the cell, such as the glucose transporter, and some are responsible for removing toxic materials from the cell interior, such as the multidrug transporter (P-glycoprotein).[11] Cells invest a large amount of their energy requirements on these transport functions, partly because this provides the selectivity that allows the internal environment to be so radically different from the external.

6. Internal Structures Closely Associated with the Membrane

The inner surface of the plasma membrane is intimately associated with a meshwork of actin microfilaments and many other proteins. This meshwork forms a support for the delicate membrane lipid bilayer that prevents deformations of the membrane. As described above, in cases in which the microfilaments are disrupted, such as in treatment with cytochalasins, the surface forms zeiotic blebs, indicative of the high internal pressure within the cell. In cells that are in the process of dying in tissue culture, surface blebbing is an event that occurs just prior to death. The disorder of the microfilament mat with high internal pressure at cytokinesis produces extensive surface blebbing. When cells are microinjected mechanically with a micropipette, one can see surface blebbing as a result of high internal cytoplasmic pressure, followed by a rapid return to normal morphology. These and other observations suggest that the microfilament meshwork stabilizes and supports the membrane, which otherwise would be subject to drastic physical distortions.

Other specialized areas of the membrane mediate intercellular contracts, for example, as physical connections, as with desmosomes, as a transfer point for low-molecular-weight solutes, as with gap junctions, or as a lateral barrier to restrict one area of the cell surface from another, as with tight junctions. Other sites, such as focal adhesions, have proteins which mediate transmembrane anchoring of the cell surface to some external structure.

At one type of site on the cell surface the actin meshwork is interrupted by another intimately associated protein substructure, the clathrin coat. Except for some intercellular junctions, coated pits are probably the only spot on the membrane in which the actin meshwork is interrupted physiologically.[12] This spot allows the surface membrane to invaginate, and following an increase in the internal pressure of a functionally closed pit, the actin-free surface blebs inward within the cell, leading to a separation event that forms a separated internal vesicle, the process of endocytosis. Underneath caveolae, on the other hand, the base of these invaginations are surrounded by the dense microfilament meshwork, and even if these structures functionally close and increase their internal pressure, the surrounding meshwork may prevent their continued expansion. Perhaps this is the reason that caveolae do not appear to mediate an actual vesicle separation event. One would expect [illegible] membrane-to-membrane contact, as is seen for secretory granules in stimulated secretory exocytosis. So little is known about the morphologic mechanism of constitutive exocytosis, however, that the fusions sites and mechanisms for vesicle contact with the membrane are completely unknown.

C. CYTOCHEMICALLY EXPLOITABLE FEATURES OF THE PLASMA MEMBRANE

1. Surface Charge

Since the surface of mammalian cells has a net negative charge, materials that are positively charged and visible in the electron microscope are potential candidates for labeling the entire external surface of living or fixed cells. One such approach is the use of large polymeric electron-dense molecules, such as Ruthenium red, colloidal tannic acid, or colloidal lanthanum salts.[13] Examples of these will be shown below. In addition, a recently

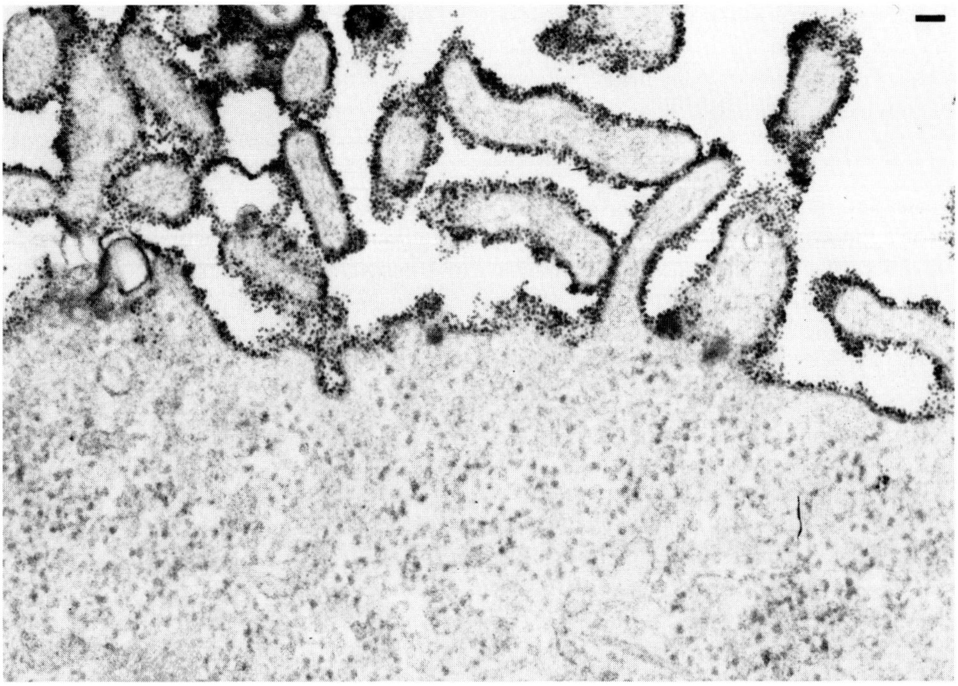

FIGURE 1. Cytochemical labeling of surface anionic sites using the polylysine-ferritin method. MDCK cells were fixed using glutaraldehyde, scraped from the cultured dish, then sequentially incubated in poly-L-lysine, followed by washing and incubation with horse spleen ferritin. The apical surface microvilli show extensive binding of ferritin on the external surface of the plasma membrane. (Magnification × 60,000; bar = 67 nm.)

described method[14] that involves the sequential incubation of cells with polylysine, followed by incubation with electron-dense ferritin, allows the labeling of the entire accessible plasma membrane (Figure 1). The principle of this method is simply to bridge the negatively charged cell surface to negatively charged ferritin using a multivalent positively charged intermediate polymer. This method is simple and predictable, and provides a clear, dense labeling of the exposed surface of any mammalian cell. Its major disadvantage is the necessity for two sequential steps. We have utilized it to show that both the basal and apical surfaces of polarized cells in culture were accessible to large molecules only when they were removed from the culture surface, exposing their basal surface.[15] While ferritin is directly visible only using electron microscopy, one could also use this method for surface labeling by following the ferritin incubation with a commercially available anti-ferritin antibody labeled with a cytochemical tag appropriate for light microscopy, as well. An alternative single-step procedure is the use of cationic ferritin, ferritin derivatized to have a net positive charge, to bind directly to the negative charge of the cell surface.[16] This method works well, but perhaps does not produce quite the same intense degree of labeling seen with the polylysine-ferritin method.

2. Surface Carbohydrates

Since the external surface of cells has an extensive layer of carbohydrates attached to both lipid and protein elements in the plasma membrane, one can target such carbohydrate groups directly for cytochemical labeling. Plant lectins are available that can interact extensively with the surface of mammalian cells, and these are also available with cytochemical

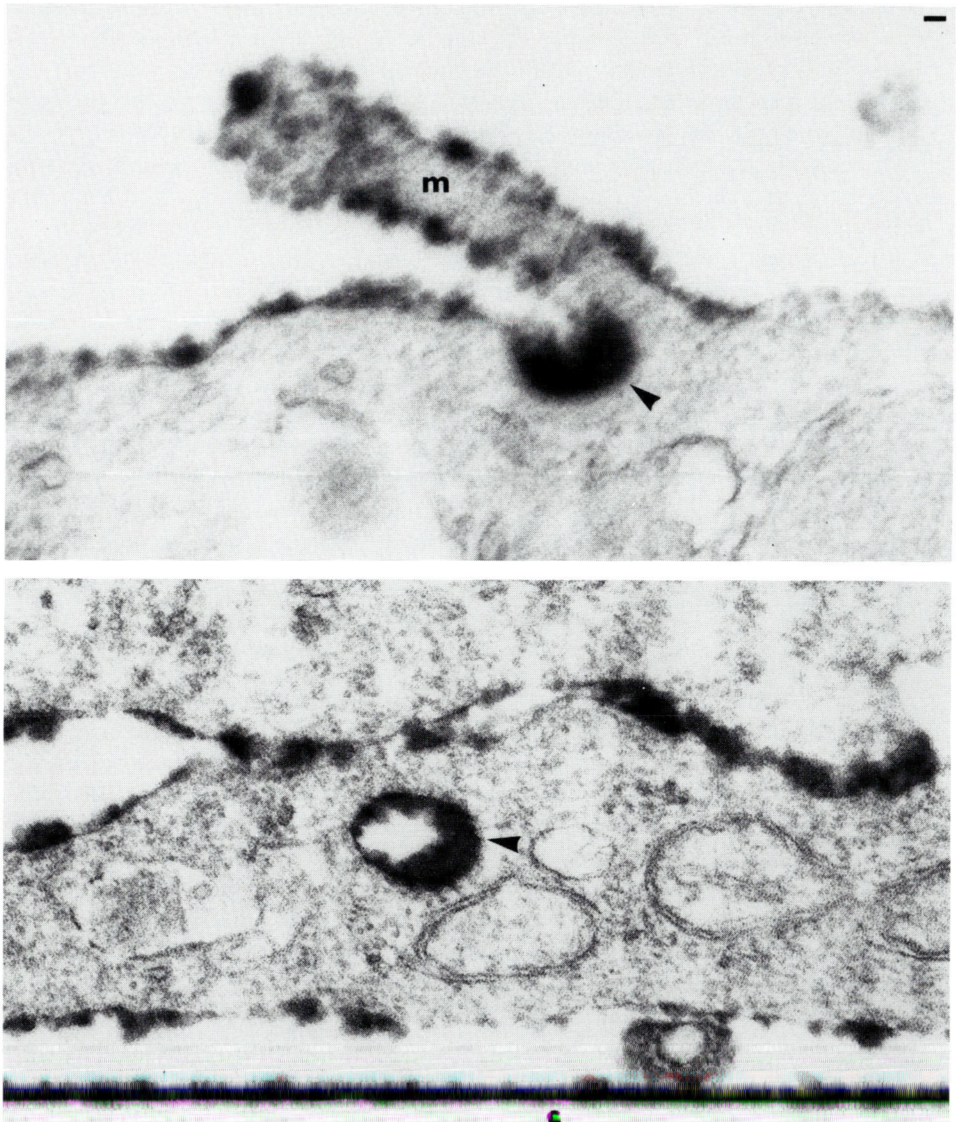

FIGURE 2. Labeling of the external face of the plasma membrane using the concanavalin A-horseradish peroxidase method. Swiss 3T3 cultured fibroblasts were incubated at 4°C with concanavalin A, then washed and incubated with horseradish peroxidase (HRP), and then fixed with glutaraldehyde. HRP was detected using diaminobenzidine to create the dense reaction product on the plasma membrane, and the cells were embedded and sections were prepared perpendicular to the substratum (S). The arrowheads point out the base of a coated pit next to a microvillus (m) in the upper panel; another pit can be seen in the lower panel in which the pit has been sectioned tangentially so that the connection to the cell surface lies out of the plane of section. In the lower panel, another cell process overlies the lower one, and reaction product can be seen between these two cell surfaces. (Magnification × 96,200; bar = 31 nm.)

labels attached, such as horseradish peroxidase (HRP) or ferritin. One such lectin is concanavalin A (con A) that binds to mannosyl residues on glycoproteins and glycolipids of the cell surface. An example of labeling using con A is shown in Figure 2, in which the surface of a cultured cell has been labeled using a sequential incubation of con A followed by HRP.[17] This method is a bridge approach in which the con A binds to the carbohydrates on the cell, and because it is multivalent in its carbohydrate-binding sites, the excess unreacted sites on the lectin then bind to the mannose-rich surface of the plant enzyme HRP.[18] The peroxidase activity is then demonstrated using diaminobenzidine (DAB) as a substrate, and it is this DAB reaction product that is visualized in Figure 2. This figure also shows the labeling of coated pits that are still accessible to the cell surface in these cells incubated at 4°C, a temperature at which endocytosis is blocked. The labeling of a tangentially sectioned base of a coated pit in the lower panel of Figure 2 demonstrates the ability of such a method to demonstrate continuity of a surface indentation with the outside of the cell when it is sectioned tangentially, and its connection with the outside is not included in the section.

In addition to lectins, one can also use the antigenic character of carbohydrate structures present on both glycolipids and glycoproteins of the cell surface to generate antibodies reactive with specific carbohydrate groups. An example of this is the antibody OVB1,[19] which reacts with a sialic acid-containing epitope on the surface of some tumor cells and on myelocytic cells in humans. Another carbohydrate epitope present on some tumor cells is the Lewis^y antigen, part of the structure of which is recognized by the recently isolated anti-tumor antibodies B1 and B3.[20] These carbohydrate epitopes represent an extremely high number of sites on each cell, perhaps 10 million sites per cell or more. The precision of antibody recognition for subtle changes in carbohydrate structure has suggested that many antibodies useful for therapy or diagnosis of human cancer may be found in these anti-carbohydrate epitope antibodies.

3. Surface Indentations and External Connections

As alluded to above, the cell surface has indentations that function in different ways. Clathrin-coated pits are the major sites of endocytosis on the cell surface. Caveolae, on the other hand, have not been clearly demonstrated to mediate endocytosis, in spite of numerous reports of receptor localization to these sites that appear to be inaccessible to the cell exterior at a later time. Recently, a new model for the role of caveolae[21] has suggested that they may represent invaginations that close and open to the cell exterior, but never actually separate from the cell surface by a membrane fusion, vesicular separation event. This model is supported by evidence from the study of receptors for folate, in which the release of folate into a form that can cross the intact membrane is favored by an acidic pH, a condition that would be likely to exist in functionally closed caveolae. These caveolae would then reopen to the cell exterior, where the folate receptors could then concentrate more folate from the surrounding medium. The times of the closed vs. open states of caveolae could mostly favor the open state, so that cytochemical studies examining the continuity of caveolae with the outside of the cell would usually find these structures to be open to the exterior.

The actual connection of an invagination of the surface with the cell exterior is a remarkably difficult thing to study. Many cytochemical markers used previously are large colloidal substances which may not permeate a narrow, but still intact, membranous connection to the cell surface. The difficulty in interpretation of tangentially sectioned membranes and connections to the surface out of the plane of section is much more complicated than one would imagine. This area still is one of great controversy.[22] However, there are some issues that cytochemistry and its limitations can or cannot address. Figure 2 demonstrates the continuity of a coated pit with the cell surface in a cell essentially frozen in time at 4°C using the impermeant cytochemical marker, con A-HRP. Clearly, the connections of such coated pits with the surface are frequently not present in each thin section.[23]

Figure 3 represents other impermeant cytochemical markers. Caveolae are frequently seen in sections tangential to the plasma membrane, appearing as small apparently isolated vesicles, leading to their interpretation as so-called "micropinosomes". However, when impermeant markers such as Ruthenium red, lanthanum, or con A-HRP are used in prefixed

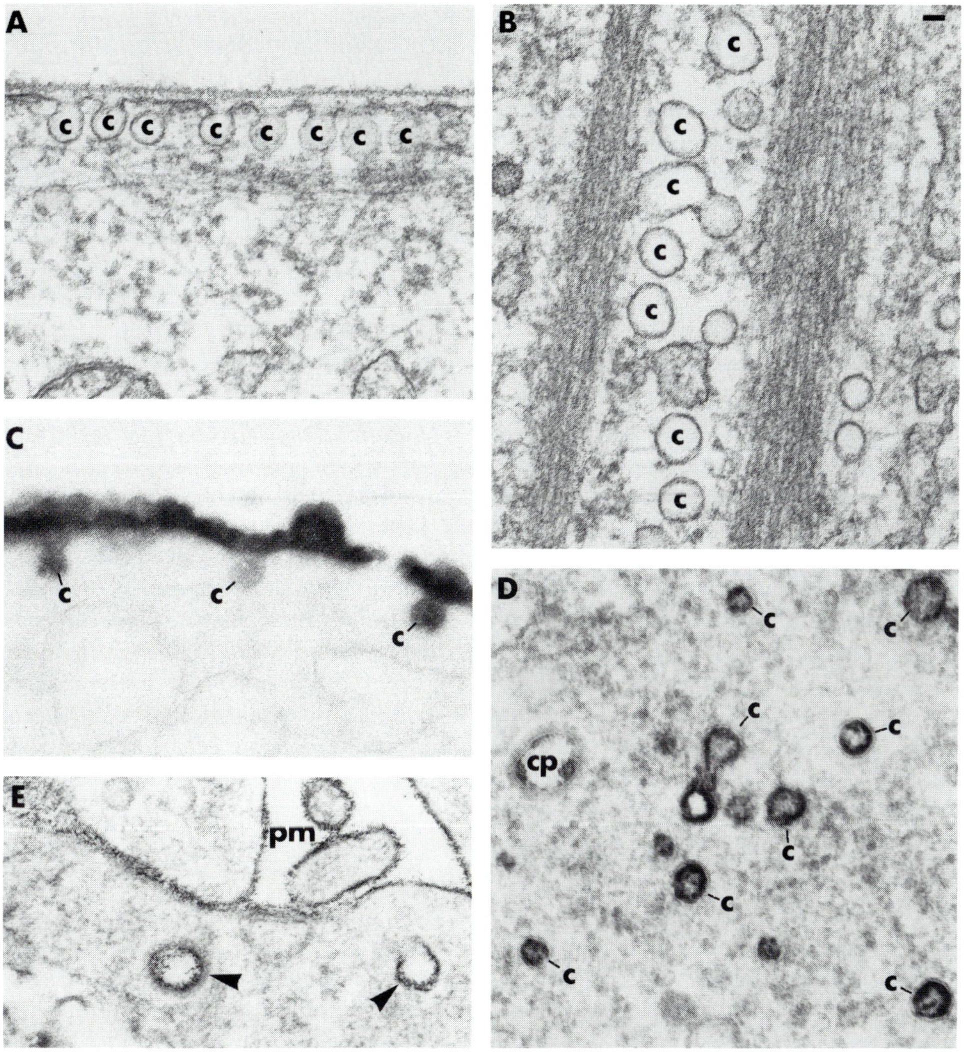

FIGURE 3. Impermeant markers of the plasma membrane and surface invaginations. Swiss 3T3 cultured cells were fixed for electron microscopy either using routine methods (A, B), or after surface labeling with concanavalin A-HRP (C), Ruthenium red (D), or lanthanum (E). A and B show images of caveolae (c) which are sectioned either perpendicular (A) or tangential (B) to the plasma membrane. C demonstrates that these structures are still in communication with the external surface using con A-HRP. In a section tangential to the membrane surface (D), Ruthenium red applied after fixation demonstrates that both a coated pit (cp) and caveolae (c) are still accessible to the cell exterior. In E, coated pits (arrowheads) tangentially sectioned show connection to the exterior of the plasma membrane (pm) using lanthanum labeling. (Magnification × 73,000; bar = 40 nm.)

cells, one can clearly see continuity of most of these structures with the cell surface (Figure 3).[13] A number of studies have suggested that these caveolae are always connected to the surface,[24] a belief this author shares. More complex, however, are those instances in which the neck regions of both caveolae and coated pits are so narrow that the impermeant markers cannot penetrate into the lumen of the invagination.

This problem clearly exists for coated pits when the cells are fixed at physiological temperature, then probed with impermeant markers.[13] Examples of such "cryptic" pits are easily found, and careful analysis of such structures using serial sections and special preservation techniques to preserve very high contrast for the membranes have convinced me that all structures with the appearance of coated "vesicles" of the size of the pits at the surface are actually still connected to the cell surface. The actual separation of these structures from the cell surface during endocytosis appears to occur only after the pits have functionally closed from the cell exterior, have swollen to a much larger size, and only then do they actually physically separate their membrane from the plasma membrane to become isolated intracellular vesicles.[25] Their appearance at that point is that of the "receptosome"[26] or "endosome",[27] a larger, clear vesicle with a fuzzy plaque at one edge that, based on immunocytochemical evidence using anti-clathrin, is believed to be the residue of the clathrin coat from the cell surface coated pit. Such a fast and complex morphologic event has proven to be remarkably difficult to study, but cytochemical markers have played a crucial part in this analysis.

4. Surface Protein Antigens

In addition to carbohydrate epitopes that exist on surface glycoproteins, there are also epitopes that are determined by the peptide itself. Some of these epitopes are based on the primary amino acid sequence in an unfolded region of the protein, and others are conformational epitopes present in the native state, but not in the denatured state of the protein. Figure 4 demonstrates two externally exposed epitopes of protein antigens detected using immunocytochemistry. One of these is the multidrug resistance transporter, P-glycoprotein (Figure 4A), detected on the apical surface of a polarized cell in tissue culture.[15] The other is a viral glycoprotein antigen, G-protein of the vesicular stomatitis virus, that is present both on the surface of viral particles and also on infected cells that produce these particles by budding at the plasma membrane.[28] These viruses infect cells by endocytosis of the viral particles mediated by coated pits at the cell surface, as shown in Figure 4B.

5. Surface Receptors

Figure 5 demonstrates different labeling strategies for receptors at the plasma membrane using cytochemical markers attached to the ligands for these receptors. Horseradish peroxidase,[29] ferritin,[17] and colloidal gold[30] have been successfully used to demonstrate morphologically ligand binding to cell surface receptors, and Figure 5 shows examples of ligands that concentrate in coated pits on the cell surface. There are other ligand-receptor systems, of course, that do not concentratively cluster in coated pits.

Figure 6 shows that not only ligands but antibodies to the receptors themselves can be used as markers for the distribution of surface receptors. The receptor for epidermal growth factor (EGF) is a transmembrane protein to which antibodies have been raised, some to the external domain near, but not within, the ligand-binding site, and others to the internal domain. In Figure 6A, a ligand label is demonstrated using ferritin coupled to EGF,[31] in Figure 6B is shown an immunocytochemical experiment using an antibody to the external domain of the receptor,[32] and in Figure 6C is shown an immunocytochemical experiment in fixed and permeabilized cells using an antibody to the internal domain of the receptor.[33] In all cases, the receptor is found clustered in coated pits with the ferritin label present on the side of the membrane on which the reactive site was located. Such experiments clearly show the high level of resolution achievable using electron microscopy and cytochemical labeling, considering that each single ferritin molecule approximately demonstrates reaction with a single receptor molecule.

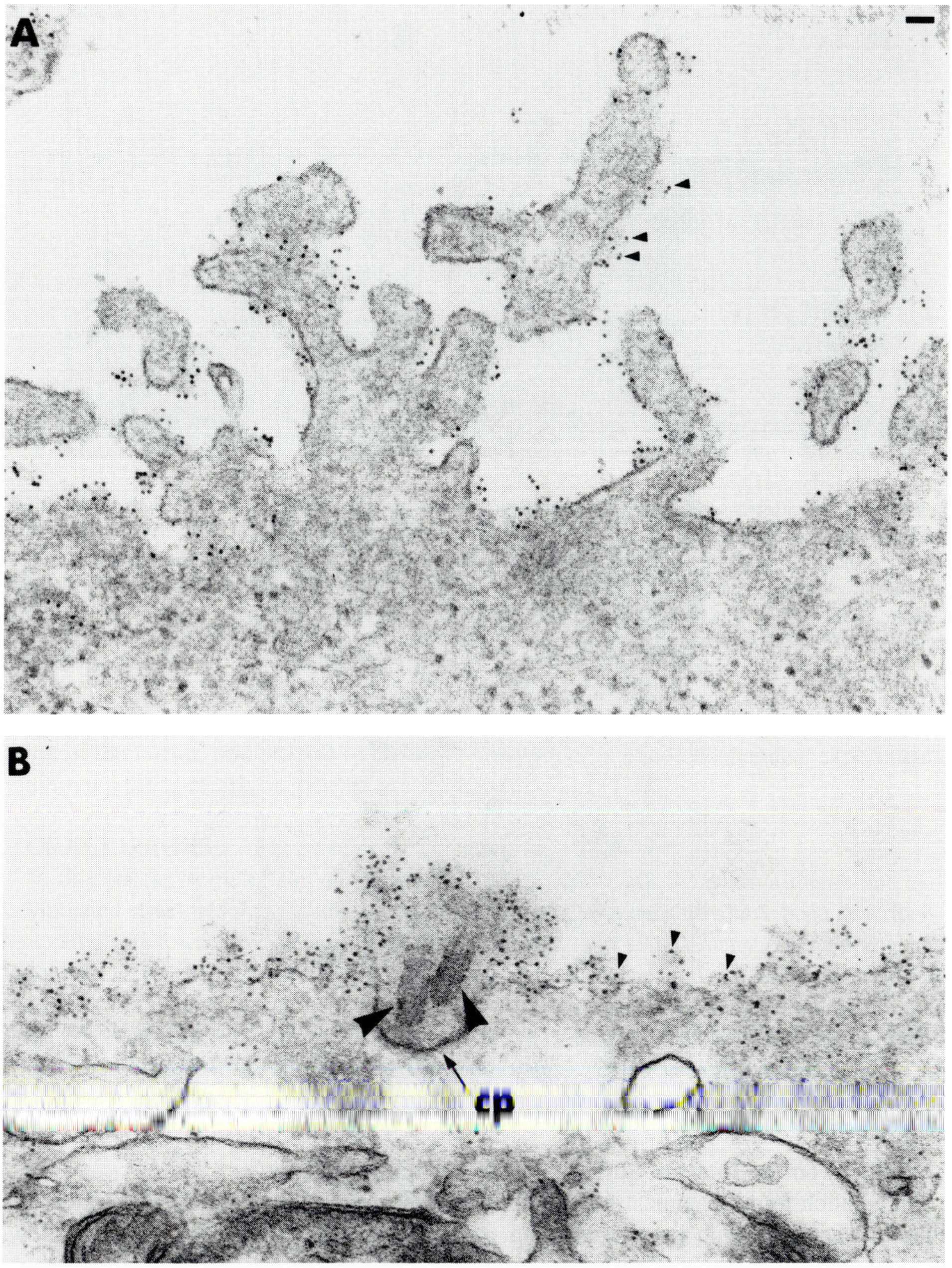

FIGURE 4. Immunocytochemical labeling of protein antigens on the external surface of the plasma membrane. MDCK-MDR cells (A) and VSV-infected 3T3 cells (B) were labeled using immunocytochemistry with the EGS and ferritin bridge procedures with antibodies to the multidrug transporter (A) or for VSV G-protein (B). The ferritin cores (small arrowheads) shown in A demonstrate the diffuse distribution of the multidrug transporter over the microvilli of the apical cell surface. In B, a coated pit (cp) contains two VSV virions (arrowheads) prior to their endocytosis into these cells that were previously infected and, therefore, have large amounts of G protein labeling on the rest of the cell surface (small arrowheads). (Magnification × 70,500; bar = 60 nm.)

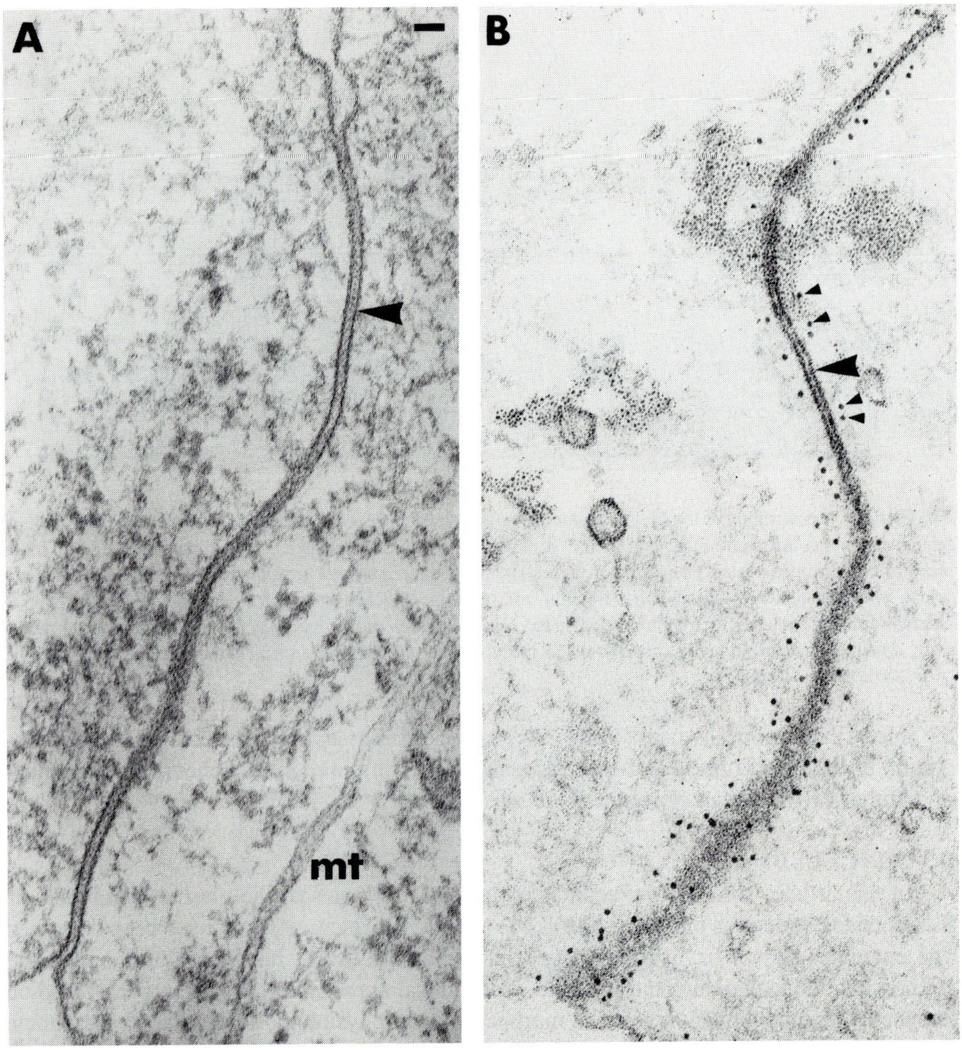

FIGURE 7. Localization of the RSV *src* gene product on the inner surface of the plasma membrane at a gap junction using immunocytochemistry. RSV-transformed NRK cells were fixed, permeabilized using saponin and incubated with anti-pp60*src* antibodies, which were then detected using the ferritin bridge method. In A, an image in a cell fixed using routine methods, the structure of a gap junction (large arrowhead) found between two cells is shown, whereas in B, ferritin cores (small arrowheads) demonstrate the location of the *src* gene product on the inner surface of the membrane in these regions. (Magnification × 100,000; bar = 40 nm.)

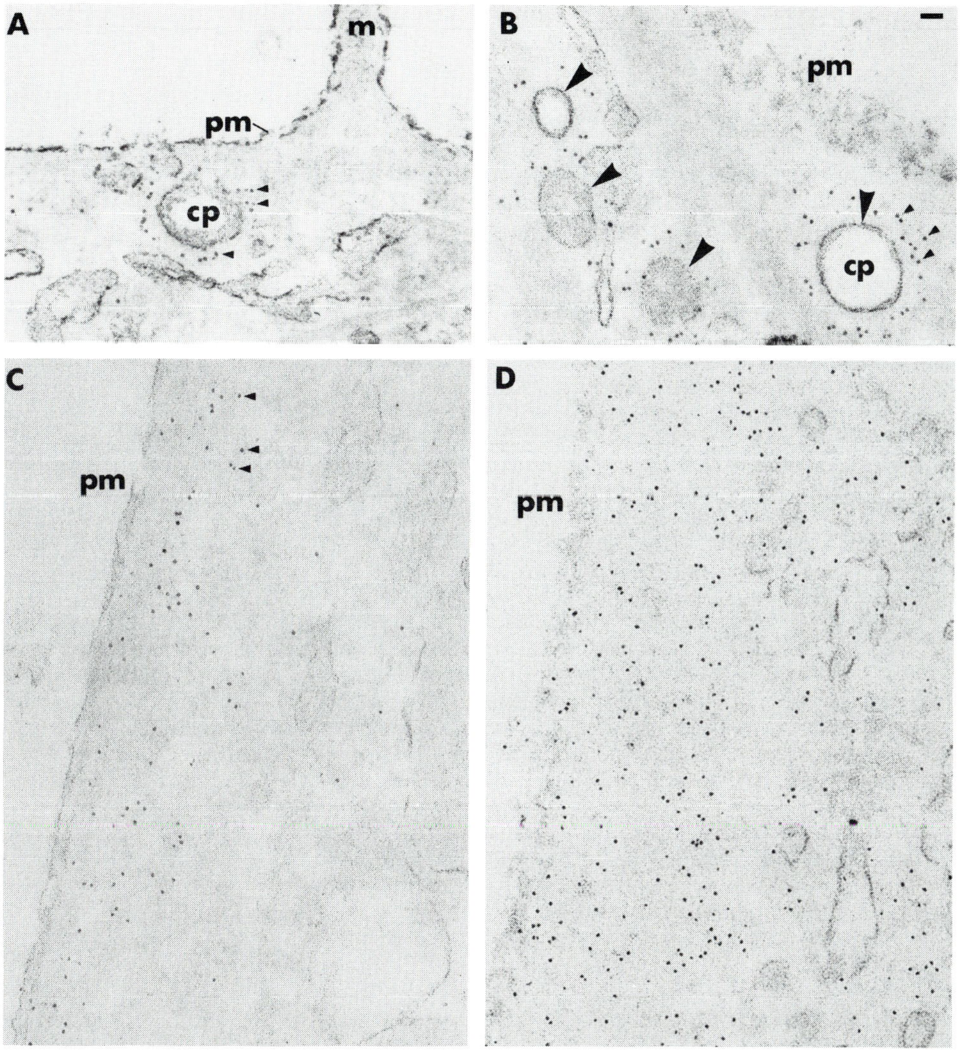

FIGURE 8. Electron microscopic immunocytochemical localization of antigens associated with the inner surface of the plasma membrane. A and B show localization of clathrin (bristle coats, small arrowheads) around coated pits (cp) (large arrowheads) of the plasma membrane (pm), sectioned either (A) perpendicular to or (B) tangential to the plane of the membrane. C demonstrates localization of actin in the microfilament meshwork just beneath the plasma membrane (pm), whereas D demonstrates that the actin-associated protein filamin is present not only in the meshwork, but also deeper in the cytoplasm. (Magnification × 73,000; bar = 40 nm.)

REFERENCES

1. **Lewis, W. H.,** Pinocytosis, *Bull. Johns Hopkins Hosp.,* 49, 17, 1931.
2. **Willingham, M. C. and Pastan, I.,** Cyclic AMP modulates microvillus formation and agglutinability in transformed and normal mouse fibroblasts, *Proc. Natl. Acad. Sci. U.S.A.,* 72, 1263, 1975.
3. **Godman, G. C., Miranda, A. F., Deitch, A. D., and Tanenbaum, S. W.,** Action of cytochalasin D on cells of established lines. III. Zeiosis and movements at the cell surface, *J. Cell Biol.,* 64, 644, 1975.
4. **Shelton, E. and Mowczko, W. E.,** Membrane blisters: a fixation artifact. A study in fixation for scanning electron microscopy, *Scanning,* 1, 166, 1978.
5. **Bruns, R. R. and Palade, G. E.,** Studies on blood capillaries. II. Transport of ferritin molecules across the wall of muscle capillaries, *J. Cell Biol.,* 37, 277, 1968.
6. **Roth, T. F. and Porter, K. R.,** Yolk protein uptake in the oocyte of the mosquito Aedes aegypti L, *J. Cell Biol.,* 20, 313, 1964.
7. **Pastan, I. and Willingham, M. C.,** The pathway of endocytosis, in *Endocytosis,* Pastan, I. and Willingham, M. C., Eds., Plenum Press, New York, 1985, chap. 1.
8. **Pearse, B. M. F.,** Clathrin: a unique protein associated with intracellular transfer of membrane by coated vesicles, *Proc. Natl. Aad. Sci. U.S.A.,* 73, 1255, 1976.
9. **Gumbiner, B. and Kelly, R. B.,** Two distinct intracellular pathways transport secretory and membrane glycoproteins to the surface of pituitary tumor, *Cell,* 28, 51, 1982.
10. **Willingham, M. C.,** Electron microscopic immunocytochemical localization of intracellular antigens in cultured cells: the EGS and ferritin bridge procedures, *Histochem. J.,* 12, 419, 1980.
11. **Gottesman, M. M. and Pastan, I.,** The multidrug transporter, a double-edged sword, *J. Biol. Chem.,* 263, 12163, 1988.
12. **Willingham, M. C., Yamada, S. S., Davies, P. J. A., Rutherford, A. V., Gallo, M. G., and Pastan, I.,** Intracellular localization of actin in cultured fibroblasts by electron microscopic immunocytochemistry, *J. Histochem. Cytochem.,* 29, 17, 1981.
13. **Willingham, M. C., Rutherford, A. V., Gallo, M. G., Wehland, J., Dickson, R. B., Schlegel, R., and Pastan, I.,** Receptor-mediated endocytosis in cultured fibroblasts: cryptic coated pits and the formation of receptosomes, *J. Histochem. Cytochem.,* 29, 1003, 1981.
14. **Skutelsky, E. and Bayer, E. A.,** A simple two-step labeling procedure for ultrastructural localization of cell surface anionic sites, *J. Histochem. Cytochem.,* 35, 1063, 1987.
15. **Pastan, I., Gottesman, M. M., Ueda, K., Lovelace, E., Rutherford, A. V., and Willingham, M. C.,** A retrovirus carrying an MDR1 cDNA confers multidrug resistance and polarized expression of P-glycoprotein in MDCK cells, *Proc. Natl. Acad. Sci. U.S.A.,* 85, 4486, 1988.
16. **Danon, D., Goldstein, L., Marikovsky, Y., and Skutelsky, E.,** Use of cationized ferritin as a label of negative charges on cell surface, *J. Ultrastruct. Res.,* 38, 500, 1972.
17. **Willingham, M. C., Maxfield, F. R., and Pastan, I.,** Alpha$_2$-macroglobulin binding to the plasma membrane of cultured fibroblasts, *J. Cell Biol.,* 82, 614, 1979.
18. **Bernhard, W. and Avrameas, S.,** Ultrastructural visualization of cellular carbohydrate components by means of concanavalin A, *Exp. Cell Res.,* 64, 232, 1971.
19. **Kurrasch, R. H., Rutherford, A. V., Rick, M. E., Gallo, M. G., Lovelace, E. T., Pastan, I., and Willingham, M. C.,** Characterization of a monoclonal antibody, OVB1, which binds to a unique determinant in human ovarian carcinomas and myeloid cells, *J. Histochem. Cytochem.,* 37, 57, 1989.
20. **Pastan, I., Lovelace, E. T., Gallo, M. G., Rutherford, A. V., Magnani, J. L., and Willingham, M. C.,** Characterization of monoclonal antibodies B1 and B3 that react with mucinous adenocarcinomas, *Cancer Res.,* 51, 3781, 1991.
21. **Rothberg, K. B., Ying, T., Kolhouse, J. F., Kamen, B. A., and Anderson, R. G. W.,** The glycophospholipid-linked folate receptor internalizes folate without entering the clathrin-coated pit endocytic pathway, *J. Cell Biol.,* 110, 637, 1990.
22. **Willingham, M. C. and Pastan, I.,** Do coated vesicles exist?, *Trends Biochem. Sci.,* 9, 93, 1984.
23. **Willingham, M. C. and Pastan, I.,** Formation of receptosomes from plasma membrane coated pits during endocytosis: analysis of serial sections with improved membrane labeling and preservation techniques, *Proc. Natl. Acad. Sci. U.S.A.,* 80, 5617, 1983.
24. **Bundgaard, M., Hagman, P., and Crone, C.,** The three-dimensional organization of plasmalemmal vesicular profiles in the endothelium of rat heart capillaries, *Microvasc. Res.,* 25, 358, 1983.
25. **Holtzman, E.,** *Lysosomes,* Plenum Press, New York, 1989, 52.
26. **Willingham, M. C. and Pastan, I.,** The receptosome: an intermediate organelle of receptor-mediated endocytosis in cultured fibroblasts, *Cell,* 21, 67, 1980.
27. **Willingham, M. C. and Pastan, I.,** Receptosomes, endosomes, CURL: different terms for the same organelle system, *Trends Biochem. Sci.,* 10, 190, 1985.

28. **Wehland, J., Willingham, M. C., Gallo, M. G., and Pastan, I.,** The morphologic pathway of exocytosis of the G protein of vesicular stomatitis virus in cultured fibroblasts, *Cell,* 28, 831, 1982.
29. **Willingham, M. C. and Pastan, I. H.,** Transit of epidermal growth factor through coated pits of the Golgi system, *J. Cell Biol.,* 94, 207, 1982.
30. **Dickson, R. B., Willingham, M. C., and Pastan, I.,** Alpha$_2$-macroglobulin absorbed to colloidal gold: a new probe in the study of receptor-mediated endocytosis, *J. Cell Biol.,* 89, 29, 1981.
31. **Willingham, M. C., Haigler, H. T., FitzGerald, D. J. P., Gallo, M. G., Rutherford, A. V., and Pastan, I.,** The morphologic pathway of binding and internalization of epidermal growth factor in cultured cells: studies on A431, KB and 3T3 cells using multiple methods of labeling, *Exp. Cell Res.,* 146, 163, 1983.
32. **Beguinot, L., Lyall, R. M., Willingham, M. C., and Pastan, I.,** Down regulation of the epidermal growth factor receptor in KB cells is due to receptor internalization and subsequent degradation in lysosomes, *Proc. Natl. Acad. Sci. U.S.A.,* 81, 2384, 1984.
33. **Beguinot, L., Werth, D., Ito, S., Richert, N., Willingham, M. C., and Pastan, I.,** Functional studies on the EGF receptor with an antibody that recognizes the intracellular portion of the receptor, *J. Biol. Chem.,* 261, 1801, 1986.
34. **Willingham, M. C., Jay, G., and Pastan, I.,** Localization of the ASV *src* gene product to the plasma membrane of transformed cells by electron microscopic immunocytochemistry, *Cell,* 18, 125, 1979.
35. **Willingham, M. C., Keen, J. H., and Pastan, I.,** Ultrastructural immunocytochemical localization of clathrin in cultured fibroblasts, *Exp. Cell Res.,* 132, 329, 1981.
36. **Wallach, D., Davies, P., Bechtel, P., Willingham, M. C., and Pastan, I.,** Cyclic AMP-dependent phosphorylation of the actin-binding protein, filamin, in *Advances in Cyclic Nucleotide Research,* Vol. 9, George, W. J. and Ignarro, L. J., Eds., Raven Press, New York, 1978, 371.

Chapter 9

CELL ORGANELLES

9.2 Intercellular Junctions

Kazushi Fujimoto

TABLE OF CONTENTS

0-8493-6012-9/93/$0.00 + $.50

A. INTRODUCTION

Cells in direct contact with neighboring cells are often linked to each other at specialized intercellular junctions. These junctions can be classified into three major functional groups: (1) tight junctions (occluding junctions), which occlude the intercellular space and thus restrict the movement of molecules through the spaces between the cells, or through the paracellular pathway; (2) adhering junctions (anchoring junctions), which mechanically attach cells to neighbors or to the extracellular matrix; and (3) gap junctions (communicating junctions), which permit ions and other small molecules to pass between cells.

These intercellular junctions were originally identified and named according to their morphology and function. However, recently, their biochemical composition and more precise physiological functions have begun to be elucidated.

B. TIGHT JUNCTIONS (ZONULA OCCLUDENS)

The zonula occludens, or tight junctions, represent the transepithelial permeability barrier of the paracellular pathway in most vertebrate epithelia. In thin-section electron microscopy, the tight junctions appear as a series of focal sites of apparent fusion between the outer membrane leaflets of neighboring cells. Freeze-fracture replicas of the tight junction reveal a band of variably spaced and anastomosing fibrils that circumscribe the cell at the apical-basolateral border. These fibrils correspond to the punctate membrane fusion, which presumably contribute to the intercellular adhesion and barrier capacity at these junctions.

Although the macromolecular structure of the tight junction is not well understood, two basic models exist: the subunit model[1] and the linear-membrane-micellar model.[2] Both models derive from the morphological analysis of freeze-fracture images of the junction. The two models can be summarized as follows. The subunit model envisages junctional strands as linear, anastomosing arrays of intercellular protein rivets, and the continuous strand seen in chemically fixed specimens as the product of lateral cross-linking of these proteins by aldehyde. The linear-membrane-micellar model views tight junctions as stabilized instances of linear membrane fusion intermediates and visualizes the strands as cylindrical inverted membrane micelles composed of a lipid continuum and varying complements of integral proteins.

Recently, two tight junction proteins of unknown function have been identified; ZO-1, the first protein to be localized to the tight junction, has been characterized as a high-molecular-weight (225 kDa), monomeric phosphoprotein peripherally associated with the junctional membrane.[3] More recently, another tight junction-associated protein, called cingulin, has been identified.[4] Cingulin, like ZO-1, is a peripheral membrane protein, but is present at about five times the quantity of ZO-1.[5] Immunoelectron microscopy demonstrates that, although both ZO-1 and cingulin are found only at tight junctions, immunogold particles associated with ZO-1 appear clustered at the points of junctional membrane contact, while those for cingulin are more diffusely distributed.[6] The mean distance of the immunogold particles from the tight junctional membrane in chicken intestinal epithelial cells is 67 and 23 nm for cingulin and ZO-1 immunogold particles, respectively.[6] These observations suggest that cingulin is close to, but not colocalized with, the tight junctional membrane itself.

The structure models of the tight junction, mentioned above, are structurally and physiologically plausible. Because so little is known about the actual composition of the integral components of the junction (obviously both ZO-1 and cingulin are peripheral proteins at the junctional region), both models remain vague in their predictions. Verification of the precise molecular structure of the tight junction is clearly dependent on further analysis of these peripheral proteins and other junctional components.

TABLE 1
Components of Adhering Junctions

Molecule	Ref.
Adherens junction	
Integral components	
E-Cadherin (uvomorulin)	12
A-CAM	13
Peripheral components	
Actin	7
Myosin	7
Tropomyosin	8
α-Actinin	7, 8
Vinculin	7, 8
400-kDa Protein (tenuin)	55
Desmosome	
Integral components	
Desmogleins	56
125-kDa Glycoprotein	17
Peripheral components	
Desmoplakins	56
Desmocalmin	18

C. ADHERING JUNCTIONS

Intercellular adhering junctions can be grouped into two categories: desmosomes and adherens junctions. In thin-section electron microscopy, adherens junctions and desmosomes are characterized by a prominent undercoat, through which actin and intermediate-sized filaments, respectively, are associated with the plasma membrane. Although it has been well characterized morphologically, these intercellular junctions have only recently begun to be described at the molecular level. Table 1 lists the known components of this type of junction.

1. Adherens Junctions

Adherens junctions are thought to play crucial roles in the morphogenesis of organs as well as in the cell adhesion mechanism. The adherens junctions are subclassified as cell-to-cell (the zonula adherens of polarized epithelia and the fascia adherens of cardiac muscle) and cell-to-substrate (adhesion plaque). Immunogold labeling of ultrathin sections of the chicken and human intestinal epithelium clearly showed that actin, myosin, and α-actinin are affiliated with the zonula adherens, and none of these proteins are associated with the desmosome.[7] In the adherens junctions, some unique proteins, such as vinculin and talain, were also detected. Vinculin was the first protein reported to be localized in the cell-to-cell and cell-to-substrate adherens junctions.[8] Talin was detected only in the cell-to-substrate junctions.[9] The other adherens junction-specific protein is pp60*src*, which is the product of the oncogene of RSV and is a tyrosine-specific protein kinase.[10] This protein is known to be localized in both types of adherens junctions.

It is well known that the intercellular junctions are part of a continuum of cell-cell interactions which involve cell adhesion molecules (CAMs). The existence of specific CAMs

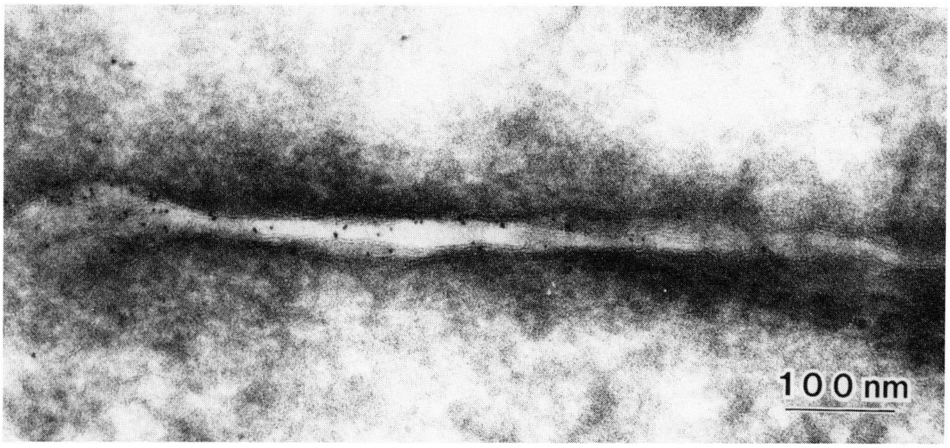

FIGURE 1. Frozen thin sections of mouse intestinal epithelium incubated with anti-E-cadherin (kindly supplied by Dr. M. Takeichi, Department of Biophysics, Faculty of Science, Kyoto University, Kyoto, Japan) were visualized using 5-nm IgG-gold particles. The specific localization of gold particles is in the intercellular space of the adherens junction.

with distinct tissue distributions has been postulated to explain the selective adhesion of cells that occurs *in vivo* or *in vitro*.[11] Many classes of CAMs have been identified in various cells and tissues. Immunoelectron microscopy has shown that E-cadherin (uvomorulin)[12] and *N*-cadherin (A-CAM)[13] are tightly associated with the adherens junction (Figure 1). Cadherins are integral membrane proteins responsible for Ca^{2+}-dependent cell-to-cell adhesion mechanisms. They are divided into E-cadherin, *P*-cadherin, *N*-cadherin, and L-CAM (related E-cadherin) subclasses.[11] Recent studies have clearly indicated that interfering with the normal interaction of E-cadherin at the cell surface inhibits the formation of the zonula adherens, the tight junctions, and the desmosomes.[14] Expression of L-CAM by transferring cDNA encoding the L-CAM in mouse sarcoma cells, which have little endogenous L-CAM, causes a morphological change accompanied by increases in both the adherens junction and gap junction areas.[15] It seems likely that the initial cell-cell contact involves interaction of specific CAMs, including those that eventually associate in adherens junctions. Subsequently, both tight junctions and gap junctions can develop.

2. Desmosomes

In desmosomes, the membranes of two cells are separated by a 20- to 30-nm space which often has a central dense line. The cytoplasmic aspect of the desmosome consists of a pair of dense attachment plaques immediately beneath each of the two adjacent lateral cell membranes. In desmosomes isolated from bovine muzzle epidermis, the major constituents have been identified on SDS-PAGE (Table 1). Some of these proteins (desmoplakin I, II, III, desmocalmin, etc.) were shown to localize in the dense plaque using immunocytochemical techniques.[16,17] In addition, bovine tongue epithelial cell desmosomes contain a 125-kDa glycoprotein that is immunologically identical with E-cadherin.[18] Desmocalmin, which can bind both keratins and calmodulin in calcium-dependent reactions,[17] is the most promising candidate for a role in the binding of intermediate filaments.

When mouse epidermal keratinocytes are grown in low calcium (<0.1 mM) medium, they do not form desmosomes. Following the addition of calcium to the low-calcium medium, these keratinocytes begin to form desmosomes within 5 min.[19] This experimental system is called "calcium-switch". Recently, using calcium-switch in MDCK cells, Pasdar and

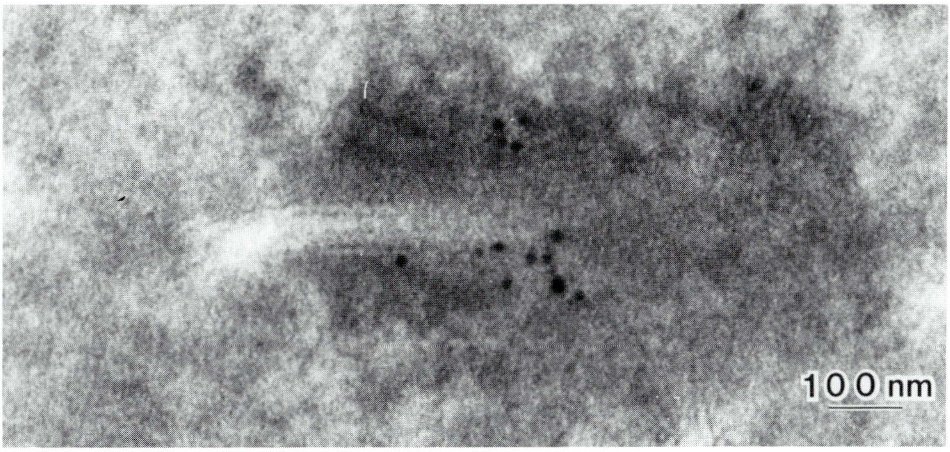

FIGURE 2. Frozen thin sections of rabbit cardiac muscle incubated with anti-type 2 protein kinase C (MBL Co, Nagoya, Japan) were visualized with 10-nm IgG-gold particles. Most of the gold label is located on the cytoplasmic plaque close to the plasma membrane.

Nelson[20,21] showed that in cells cultured in low-calcium medium (the absence of cell-to-cell contacts), desmoplakin I and II are found in Triton X-100/high salt-soluble and insoluble pools. The soluble type presented a diffuse perinuclear staining by immunofluorescence microscopy, and the insoluble type exhibited a punctate staining pattern in the cytoplasm. When the cells were transferred from low to normal calcium medium, soluble desmoplakin I and II changed rapidly to the insoluble type, so the proportion of the insoluble pool increased significantly. Immunofluorescence microscopy indicated that the soluble type was diffusely distributed in the cytoplasm, while the insoluble type was observed as punctate cytoplasmic staining that colocalized with intermediate filaments. This cytoplasmic punctate staining is thought to represent the assembly of desmosome components which move to the cell-to-cell contact areas. This information provides evidence that desmosome assembly may require physiological levels of calcium and cell-to-cell contact mediated by the calcium-dependent CAMs.

Despite the important role of calcium in the formation of desmosomes, little is known about the intracellular signal transduction mechanism leading to desmosome formation after switching to normal extracellular calcium levels. Recent work indicates that in some cells switching to normal calcium levels causes an increase in protein kinase C (PKC) activity, and 12-*O*-tetradecanoylphorbol-13-acetate (TPA), an activator of PKC, induces rapid cell-to-cell contact and redistribution of desmoplakins from the cytoplasm to the plasma membrane, with desmosome formation.[22] It is well known that PKC plays a crucial role in signal transduction for a variety of biologically active substances that activate many cellular functions and proliferation.[23] It thus seems possible that PKC is directly involved in desmosome formation and function. In support of this possibility, we demonstrated the immunoelectron microscopic localization of PKC with monoclonal antibodies against type 2 and type 3 PKCs in desmosomes. The immunogold label was located on the cytoplasmic dense plaque close to the desmosome membrane of normal rabbit myocardial cells (Figure 2), but was not seen in hepatocyte desmosomes. Our results suggest that PKC may play an important role in the function of desmosomes of some cells under physiological conditions. However, the substrate for PKC and how PKC is involved in the mechanisms of desmosome assembly remain to be elucidated.

D. GAP JUNCTIONS

Gap junctions are specialized regions of the plasma membranes comprised of closely packed aggregates of channels that span the membranes of adjacent cells to form a low-resistance intercellular pathway for the diffusion of small molecules (<1 to 2000 Da).

Gap junctions can be recognized by their characteristic morphology. In thin sections, the cross-sectional profile of the gap junction is displayed as two parallel plasma membranes, separated from each other by a narrow intercellular "gap" of 2 to 3 nm. The narrow gap is easily confirmed by infiltration with colloidal lanthanum[24] or horseradish peroxidase.[25] In freeze-fracture replication and negative staining, the gap junctions can be recognized as patchy arrays of closely packed 8-nm membrane particles at an average center-to-center distance of 10 nm. However, gap junctions do not all appear structurally and functionally identical. For instance, myocardial gap junctions are thicker in profile than those in liver, implying that in the heart either additional proteins are associated with the cytoplasmic surfaces or that the heart gap junction protein is larger. In addition, many studies have suggested physiological differences between gap junctions in different cell types.[26]

A high-resolution structural study using X-ray diffraction and low-dose Fourier microscopy suggests that the gap junction channel is composed of a hemichannel, called connexon, embedded in the plasma membrane of one cell joined with a connexon in the opposing cell membrane, and each connexon is an oligomer of six integral membrane protein subunits.[27]

When proteins of purified gap junctions of murine liver are separated by SDS-PAGE, two major proteins of 26 kDa (also described as 27 or 28 kDa) and 21 kDa, respectively, are found. While a number of different gap junction proteins have now been identified, hepatic gap junctions are unique in being the first demonstrated case where there are two distinct proteins. Beyer et al.[28] have introduced a system of nomenclature for gap junction proteins based on using the term "connexin" coupled with the molecular mass predicted by cDNA analysis. The cDNA for the rat liver 26-kDa component has been cloned and, based on its deduced formula weight of 32 kDa, has been designated connexin 32. The 21-kDa liver gap junction protein, which has a calculated molecular mass of 26 kDa, is referred to as connexin 26; the 43-kDa heart gap junction protein becomes connexin 43.

On the other hand, cytochemical studies of gap junctions is also progressing. The focus of this chapter is to describe the recent progress in cytochemistry of the gap junctions, and some implications of the structure and function of the gap junction are discussed.

1. Enzyme Cytochemistry

At present, cytochemical evidence for the localization of adenylate cyclase, guanylate cyclase, and Ca^{2+}-ATPase activities at the gap junctions of various cells have been reported.[29-38] The reaction products were observed on the inner leaflets (cytoplasmic sides) of the junctional membrane (see Figure 3 in Chapter 1, Section 1.3.3, article on Ca^{2+}-ATPase). When the isolated gap junctions were fixed prior to cytochemical incubation, the pattern of cytochemical staining was the same as *in situ*. However, in the unfixed gap junction the reaction products spanned the two junctional membranes.[33]

2. Cytochemical Distribution of Hydrophilic Channels

Cytochemical visualization of the hydrophilic channel in the gap junction has been attempted in our laboratory using alkali-metal ions.[39] The experimental approach attempted takes advantage of the fact that alkali-metal ions K^+, Rb^+, or Cs^+, can be exchanged freely between cells through gap junctions. However, since the cytochemical demonstration of water-soluble ions involves problems such as difficulty of visualization and diffusibility of ions, we devised the following protocol to resolve these difficulties: isolated gap junctions

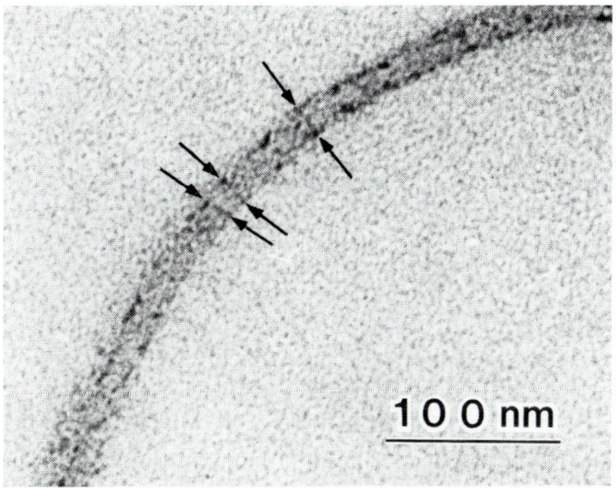

FIGURE 3. Cytochemical visualization of hydrophilic channels. Cesium precipitation is seen on the inner leaflets of the gap junctional membranes. In some areas precipitates are seen as narrow cross-bridges spanning two junctional membranes (corresponding arrows).

were incubated in 10 to 20 m*M* cesium chloride or rubidium chloride, buffered to pH 7.4 with 10 m*M* Tris for 30 min, then sedimented by centrifugation. The incubated gap junction pellets were frozen in liquid nitrogen. Freeze-substitution was performed by storing frozen samples in acetone at −80°C for 3 days. Samples were slowly warmed to room temperature and embedded in Epoxy resin. Ultrathin dry-cut sections were examined without staining under an electron microscope. Figure 3 is an electron micrograph showing the cross-sectional view of an isolated gap junction incubated with cesium. Electron-dense cesium deposits were seen mainly on the inner leaflets of junctional membranes. In some areas narrow cross-bridges of 1 to 2 nm in width were seen transversing two opposing gap junctional membranes (Figure 3, corresponding arrows).

3. Charges and Glycosyl Residues

It is well known that Ca^{2+} is the most effective uncoupler in causing the closure of hydrophilic channels. Peracchia et al. proposed the following hypothesis of an electrostatic mechanism for opening and closing channels.[40] Gap junction proteins have negatively charged sites, so an open channel may result from repulsion between negative charges of neighboring connexins, while if the repulsive charges are neutralized by binding divalent cations such as calcium ion to the connexin, a closing of the channels could be caused by the disappearance of the electrostatic repulsion.

Ultimately, a cytochemical study for the demonstration of anionic sites on the gap junctions would be required to assess this hypothesis. Recently, we have developed a simple method for visualizing the intracellular distribution of the anionic sites at the ultrastructural level by labeling frozen thin sections with cationized ferritin.[41] In the intercalated disc of cardiac muscles, cationized ferritin particles almost continuously labeled the intercellular surface of intermediate junction and desmosome, while in the gap junction a high affinity for cationized ferritin was seen on the cytoplasmic side of the gap junction. It is generally accepted that anionic sites on most biological membranes represent various macromolecules: proteins, glycoprotein, but most often, sialoglycoproteins, so we have attempted to stain frozen thin sections with lectin conjugated with ferritin. The results revealed that the intracellular surface of the intermediate junction and desmosome specifically bound the lectins, whereas no lectin binding was seen at the gap junction.[42] These results imply that the charged proteins and lipids of gap junctions contribute to the cytoplasmic surface charge.

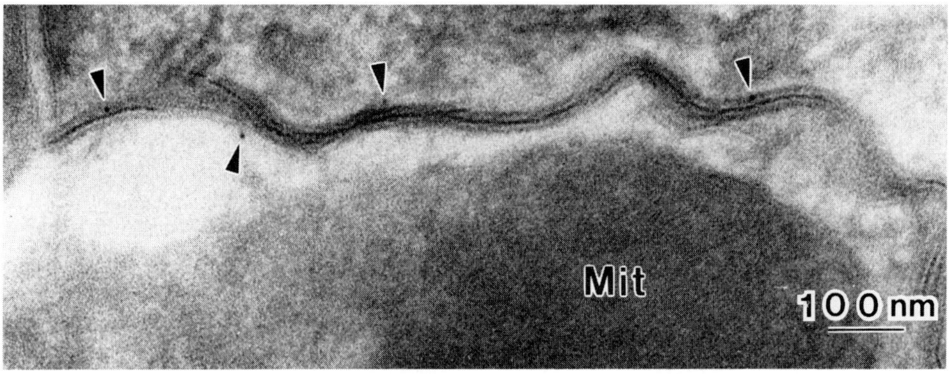

FIGURE 4. Frozen thin sections of rabbit liver incubated with anti-type 2 protein kinase C (MBL Co, Nagoya, Japan) and visualized with 10-nm IgG-gold particles. Electron microscopy revealed TPA-induced translocation of cytosolic PKC to the gap junctional membrane (arrowheads); Mit, mitochondrion.

4. Calmodulin and Protein Kinase C

Calmodulin-binding sites — Calmodulin, originally characterized as a calcium-dependent activator of cyclic nucleotide phosphodiesterase, is a Ca^{2+}-dependent modulator protein which mediates the effects of Ca^{2+} in a variety of cell reactions and processes.[43] We recently introduced biologically active colloidal gold-labeled calmodulin as a marker for identifying potential calmodulin-binding sites by electron microscopy and have stained frozen thin sections of rat cardiac muscle with this conjugate.[44] In the presence of Ca^{2+}, gold particles were found localized on the sarcoplasmic reticulum, mitochondria, and gap junctions. The Ca^{2+}-dependent binding of calmodulin to gap junction is of particular interest. Wojtczak has reported that calmodulin inhibitors can prevent uncoupling in cardiac muscle induced by Ca^{2+}.[45] Peracchia and Girsch[46] have reported the inhibitory effects of calmodulin inhibitors on uncoupling using liposome-incorporated lens gap junction channels. In addition, Welsh et al.[47] and Hertzberg and Gilula[48] have demonstrated the binding of calmodulin to lens and liver gap junction protein in gel overlay experiments. Cytochemical studies demonstrating the Ca^{2+}-dependent binding of calmodulin to gap junctions directly suggest that calmodulin is associated with the gap junctions in the presence of Ca^{2+} and directly or indirectly regulates gap junctional permeability.

Protein kinase C — Agonists and/or antagonists of cAMP-dependent protein kinase, protein kinase C (PKC), calmodulin/Ca^{2+}-dependent protein kinase, and several tyrosine kinases modulate gap junctional communication in various cell systems.[49] Recent studies have led to the conclusion that connexins themselves are substrates for protein kinases. Takeda et al.[50] have determined that connexin 32 is phosphorylated *in vitro* by PKC. They also showed that the addition of TPA, an activator of PKC, to intact hepatocytes rapidly increased connexin 32 phosphorylation at serine residues.[51] Recently, we demonstrated that the addition of TPA to intact hepatocytes leads to marked decreases in gap junction areas (within 20 min). Furthermore, immunolocalization of PKC by electron microscopy with two monoclonal antibodies against type 2 and type 3 PKCs showed that both epitopes are associated with the gap junctional membranes in TPA-treated hepatocytes (Figure 4, arrowheads). Although the physiological significance of PKC-catalyzed phosphorylation of connexin 32 is unknown, our results suggest that TPA-induced phosphorylation of connexin 32 causes breakdown of the gap junction. In fact, we detected internalization of the gap junctional membrane and annular gap junctions, which is one sequential step in the breakdown of the gap junction, in TPA-treated hepatocytes.

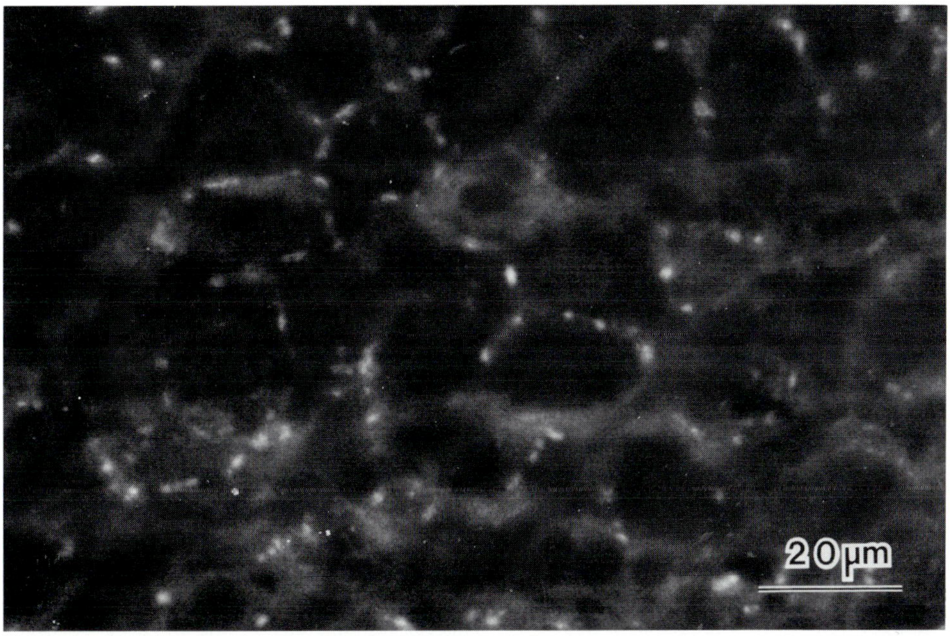

FIGURE 5. Immunofluorescent localization of connexin 32 in frozen sections of rat liver. Notice the punctate labeling for connexin 32 along the cell membranes. (Anti-connexin 32 was the kind gift of Dr. A. Takeda, Department of Laboratory Medicine, Ehime University School of Medicine, Ehime, Japan.)

5. Immunocytochemistry of Gap Junctional Proteins

The production of polyclonal and monoclonal antibodies against gap junction proteins and connexins provided an immunohisto- and cytochemical approach to the characterization of gap junctions. Figure 5 shows indirect immunohistochemistry of connexin 32 by labeling frozen sections of unfixed rat liver with a monoclonal antibody. The primary antibody was visualized with a FITC-labeled, goat anti-mouse secondary antibody. The staining consisted of numerous punctate or thread-like regions of intense fluorescence outlining the lateral aspects of the cells. This staining pattern is consistent with the size and distribution of gap junctions in the liver as revealed by numerous electron microscopic studies. Immunoelectron microscopy of connexin 32 using frozen thin sections of 0.5% formaldehyde fixed liver revealed that the monoclonal antibody was found directly on the gap junction in rat hepatocytes (Figure 6).

Immunohisto- and cytochemistry of connexins reveal a complex pattern of overlapping expression. Immunoelectron microscopy showed that connexin 26 colocalizes with connexin 32 in the same gap junctional plaques in liver.[52] However, connexin 26, unlike connexin 32, is not evenly distributed throughout the liver lobule. Immunolabeling of connexin 26 reveals a gradient with maximum immunoreactivity in the periportal zone and a faint reaction in the perivenous zone. Connexin 43 and 32 are both expressed in proximal renal cells, but the staining pattern does not overlap.[53] This suggests that both connexins may be expressed in the same cell but not in the same gap junctional plaque. The physiological significance of these different expression patterns is not known.

Quantitative approaches using immunofluorescence microscopy are useful for analyzing the alteration of gap junctions.[54] Immunofluorescence microscopy reveals small plaques at cell peripheries. Plaques can be counted and an estimate of their number per section area obtained. Obviously, a much larger number of samples can be efficiently observed than that

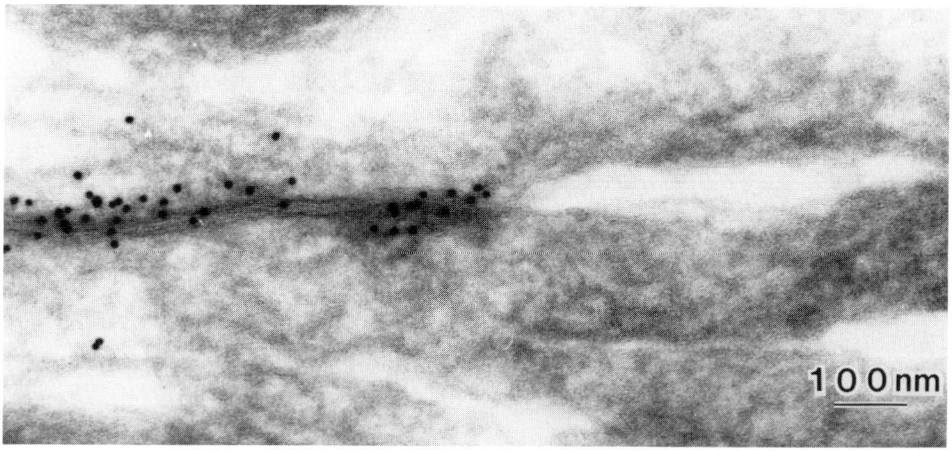

FIGURE 6. Immunoelectron microscopic localization of connexin 32 in rat hepatocytes. Immunogold particles decorate the cytoplasmic surfaces of the gap junction.

of quantitative analysis using thin-section or freeze-fracture electron microscopy. The limitation of immunofluorescence microscopic analysis is that any changes observed in the number of plaques may not necessarily reflect changes in the amount of gap junction protein, since plaques could aggregate or disperse, altering both size and number, with no concomitant change in the actual area of gap junction per cell or in the absolute amount of gap junction protein.

ACKNOWLEDGMENT

This work was supported in part by Grants-in-Aid for Scientific Research from the Ministry of Education, Science and Culture of Japan (No. 03454116).

REFERENCES

1. **Staehelin, L. A.,** Further observations on the fine structure of freeze-cleaved tight junctions, *J. Cell Sci.,* 13, 763, 1973.
2. **Pinto da Silva, P. and Kachar, B.,** On tight-junction structure, *Cell,* 28, 441, 1982.
3. **Stevenson, B. R., Siliciano, J. D., Mooseker, M. S., and Goodenough, D. A.,** Identification of ZO-1. a high molecular weight polypeptide associated with the tight junction (zonula occludens) in a variety of epithelia, *J. Cell Biol.,* 103, 755, 1986.
4. **Citi, S., Sabanay, H., Jakes, R., Geiger, B., and Kendrick-Jones, J.,** Cingulin, a new peripheral component of tight junctions, *Nature (London),* 333, 272, 19888.
5. **Citi, S., Sabanay, H., Kendrick-Jones, J., and Geiger, B.,** Cingulin: characterization and localization, *J. Cell Sci.,* 93, 107, 1989.
6. **Stevenson, B. R., Heintzelman, M. B., Anderson, J. M., Citi, S., and Mooseker, M. S.,** ZO-1 and cingulin: tight junction proteins with distinct identities and localizations, *Am. J. Physiol.,* 257, C621, 1989.
7. **Drenckhahn, D. and Dermietzel, R.,** Organization of the actin filament cytoskeleton in the intestinal brush border: a quantitative and qualitative immunoelectron microscope study, *J. Cell Biol.,* 107, 1037, 1988.
8. **Geiger, B., Dutton, A. H., Tokuyasu, K. T., and Singer, S. J.,** Immunoelectron microscopic studies of membrane-microfilament interactions: the distribution of α-actinin, tropomyosin, and vinculin in intestinal epithelial brush border and in chicken gizzard smooth muscle cells, *J. Cell Biol.,* 91, 614, 1981.

9. **Geiger, B., Volk, T., and Volberg, T.,** Molecular heterogeneity of adherens junctions, *J. Cell Biol.,* 101, 1523, 1985.
10. **Shriver, K. and Rohrschneider, L.,** Organization of pp60*src* and selected cytoskeletal proteins within adhesion plaques and junctions of Rous sarcome virus-transformed rat cells, *J. Cell Biol.,* 89, 525, 1981.
11. **Takeichi, M.,** The cadherins: cell-cell adhesion molecules controlling animal morphogenesis, *Development,* 102, 639, 1988.
12. **Boller, K., Vestweber, D., and Kemler, R.,** Cell-adhesion molecule uvomorulin is localized in the intermediate junctions of adult intestinal epithelial cells, *J. Cell Biol.,* 100, 327, 1985.
13. **Volk, T. and Geiger, B.,** A-CAM: a 135-kD receptor of intercellular adherens junctions. I. Immunoelectron microscopic localization and biochemical studies, *J. Cell Biol.,* 103, 1441, 1986.
14. **Gumbiner, B., Stevenson, B., and Grimaldi, A.,** The role of the cell adhesion molecule uvomorulin in the formation and maintenance of the epithelial junctional complex, *J. Cell Biol.,* 107, 1575, 1988.
15. **Mege, R. M., Matsuzaki, F., Gallin, W. J., Goldberg, J. I., Cunningham, B. A., and Edelman, G. M.,** Construction of epithelloid sheets by transfection of mouse sarcoma cells with cDNA for chicken cell adhesion molecules, *Proc. Natl. Acad. Sci. U.S.A.,* 85, 7274, 1988.
16. **Franke, W. W., Cowin, P., Schmelz, M., and Kapprell, H.-P.,** The desmosomal plaque and the cytoskeleton, in *Junctional Complexes of Epithelial Cells,* Ciba Found. Symp. 125, John Wiley & Sons, Sussex, 1987, 26.
17. **Tsukita, S. and Tsukita, S.,** Desmocalmin: a calmodulin-binding high molecular weight protein isolated from desmosomes, *J. Cell Biol.,* 101, 2070, 1985.
18. **Jones, J. C. R.,** Characterization of a 125 K glycoprotein associated with bovine epithelial desmosomes, *J. Cell Sci.,* 89, 207, 1988.
19. **Hennings, H. and Holbrook, K. A.,** Calcium regulation of cell-cell contact and differentiation of epidermal cells in culture, *Exp. Cell Res.,* 143, 127, 1983.
20. **Pasdar, M. and Nelson, W. J.,** Kinetics of desmosome assembly in Madin-Darby canine kidney epithelial cells: temporal and spatial regulation of desmoplakin organization and stabilization upon cell-cell contact. I. Biochemical analysis, *J. Cell Biol.,* 106, 677, 1988.
21. **Pasdar, M. and Nelson, W. J.,** Kinetics of desmosome assembly in Madin-Darby canine kidney epithelial cells: temporal and spatial regulation of desmoplakin organization and stabilization upon cell-cell contact. II. Morphological analysis, *J. Cell Biol.,* 106, 687, 1988.
22. **Sheu, H.-M., Kitajima, Y., and Yaoita, H.,** Involvement of protein kinase C in translocation of desmoplakins from cytosol to plasma membrane during desmosome formation in human squamous cell carcinoma cells grown in low to normal calcium concentration, *Exp. Cell Res.,* 185, 176, 1989.
23. **Nishizuka, Y.,** Studies and perspectives of protein kinase C, *Science,* 233, 305, 1986.
24. **Revel, J.-P. and Karnovsky, M. J.,** Hexagonal array of subunits in intercellular junctions of mouse heart and liver, *J. Cell Biol.,* 33, c7, 1967.
25. **Goodenough, D. A. and Revel, J.-P.,** The permeability of isolated and in situ mouse hepatic gap junction studies with enzymatic tracers, *J. Cell Biol.,* 50, 81, 1971.
26. **Zampighi, G.,** Gap junction structure, in *Cell-to-Cell Communication,* De Mello, W. C., Ed., Plenum Press, New York, 1987, 1.
27. **Makowski, L.,** Structural domains in gap junctions: implications for the control of intercellular communication, in *Gap Junctions,* Bennett, M. V. L. and Spray, D. C., Eds., Cold Spring Harbor Laboratory, Cold Spring Harbor, NY, [illegible]
28. **Beyer, E. C., Goodenough, D. A., and Paul, D. L.,** The connexins: a family of related gap junction proteins, in *Gap Junctions,* Herzberg, E. L. and Johnson, R., Eds., Alan R. Liss, New York, 1988, 167.
29. **Ando, T., Fujimoto, K., Mayahara, H., Miyajima, H., and Ogawa, K.,** A new one-step method for the histochemistry and cytochemistry of Ca^{2+}-ATPase activity, *Acta Histochem. Cytochem.,* 14, 705, 1981.
30. **Fujimoto, K. and Ogawa, K.,** Enzyme cytochemical study of rat cardiac muscle. I. Adenylate cyclase and guanylate cyclase, *Acta Histochem. Cytochem.,* 15, 325, 1982.
31. **Fujimoto, K. and Ogawa, K.,** Enzyme cytochemical study of rat cardiac muscle. II. Ca^{++}-ATPase and ouabain-sensitive, K^{+}-dependent p-nitrophenylphosphatase, *Acta Histochem. Cytochem.,* 15, 338, 1982.
32. **Fujimoto, K., Toibana, M., and Ogawa, K.,** New cytochemical method for adenylate cyclase and guanylate cyclase with dimethyl sulfoxide, *Acta Histochem. Cytochem.,* 14, 687, 1981.
33. **Fujimoto, K., Narimoto, K., Ogawa, K. S., Kondo, S., and Ogawa, K.,** Cytochemical aspects of gap junction: enzyme activity, hydrophilic channel and charge, *Acta Histochem. Cytochem.,* 18, 467, 1985.
34. **Kondo, S., Imamura, S., Fujimoto, K., and Ogawa, K.,** Calcium-activated adenosine triphosphatase and gap junctions in rat epidermis, *Acta Histochem. Cytochem.,* 21, 521, 1988.
35. **Narimoto, K., Okamura, H., Mori, T., Fujimoto, K., and Ogawa, K.,** Enzyme cytochemical and morphometrical analysis of gap junction in rabbit ovarian granulosa cells during follicle maturation and ovulatory process, *Acta Histochem. Cytochem.,* 18, 565, 1985.

36. **Ogawa, K. S., Fujimoto, K., and Ogawa, K.,** Cytochemical studies on nucleotidyl cyclases in aortic endothelium and smooth muscle cell, *Acta Histochem. Cytochem.*, 19, 521, 1986.
37. **Sakai, M., Tanaka, T., and Ogawa, K.,** Ultrastructure and Ca^{++}-ATPase activity of gap junctions under cryo-electron microscopic observation, *Acta Histochem. Cytochem.*, 21, 535, 1988.
38. **Ueno, S., Umar, H., Bambaner, H. J., and Ueck, K.,** Ultracytochemical localization of Ca^{++}-ATPase activity in the paraphysis of the frog, *Rana esculenta, Cell Tissue Res.*, 235, 3, 1984.
39. **Fujimoto, K., Ogawa, K. S., and Ogawa, K.,** Direct visualization of the hydrophilic channel in the gap junction under an electron microscope using alkali-metal ions, *Acta Histochem. Cytochem.*, 17, 453, 1984.
40. **Peracchia, C., Bernardini, G., and Peracchia, L. L.,** Uncoupling mechanism: a hypothesis, *J. Cell Biol.*, 83, 86a, 1979.
41. **Fujimoto, K., Ogawa, K. S., Kondo, S., and Ogawa, K.,** Distribution of anionic sites on intracellular organelles: a study by labeling frozen thin-section with cationized ferritin, *Acta Histochem. Cytochem.*, 18, 455, 1985.
42. **Fujimoto, K., Ogawa, K. S., Kondo, S., and Ogawa, K.,** Ultracytochemistry of gap junctions, in *Proc. XIth Int. Congress on Electron Microscopy,* Vol. 3, Imura, T., Maruse, S., and Suzuki, T., Eds., The Japanese Society of Electron Microscopy, Tokyo, 1986, 1869.
43. **Cheung, W. Y.,** Calmodulin plays a pivotal role in cellular regulation, *Science,* 207, 19, 1980.
44. **Fujimoto, K., Araki, N., Ogawa, K. S., Kondo, S., Kitaoka, T., and Ogawa, K.,** Ultracytochemistry of calmodulin binding sites in myocardial cells by staining of frozen thin sections with colloidal gold-labeled calmodulin, *J. Histochem. Cytochem.*, 37, 249, 1989.
45. **Wojtczak, J. A.,** Electrical uncoupling induced by general anesthetics: a calcium-independent process?, in *Gap Junctions,* Bennett, M. V. L. and Spray, D. C., Eds., Cold Spring Harbor Laboratory, Cold Spring Harbor, NY, 1985, 167.
46. **Peracchia, C. and Girsch, S. J.,** An in vitro approach to cell coupling: permeability and gating of gap junction channels incorporated into liposomes, in *Gap Junctions,* Bennett, M. V. L. and Spray, D. C., Eds., Cold Spring Harbor Laboratory, Cold Spring Harbor, NY, 1985, 191.
47. **Welsh, M. J., Aster, J. C., Ireland, M., Alcala, J., and Maisel, H.,** Calmodulin bindings to chick lens gap junction protein in a calcium-dependent manner, *Science,* 216, 642, 1982.
48. **Hertzberg, E. L. and Gilula, N. B.,** Liver gap junctions and lens fiber junctions: comparative analysis and calmodulin interaction, *Cold Spring Harbor Symp. Quant. Biol.*, 46, 639, 1982.
49. **Loewenstein, W. R.,** The cell-to-cell channel of gap junctions, *Cell,* 48, 725, 1987.
50. **Takeda, A., Hashimoto, E., Yamamura, H., and Shimazu, T.,** Phosphorylation of liver gap junction protein by protein kinase C, *FEBS Lett.*, 210, 169, 1987.
51. **Takeda, A., Saheki, S., Shimazu, T., and Takeuchi, N.,** Phosphorylation of the 27 kDa gap junction protein by protein kinase C in vitro and rat hepatocytes, *J. Biochem.*, 106, 723, 1989.
52. **Traub, O., Look, J., Dermietzel, R., Brümmer, F., Hülser, D., and Willecke, K.,** Comparative characterization of the 21-kD and 26-kD gap junction proteins in murine liver and cultured hepatocytes, *J. Cell Biol.*, 108, 1039, 1989.
53. **Beyer, E. C., Kistler, J., Paul, D. L., and Goodenough, D. A.,** Antisera directed against connexin 43 peptides react with a 43-kD protein localized to gap junctions in myocardium and other tissues, *J. Cell Biol.*, 108, 595, 1989.
54. **Sugiyama, Y. and Ohta, H.,** Changes in density and distribution of gap junctions after partial hepatectomy: immunohistochemical and morphometric studies, *Arch. Histol. Cytol.*, 53, 71, 1990.
55. **Tsukita, S., Itoh, M., and Tsukita, S.,** A new 400-kD protein from isolated adherens junctions: its localization at the undercoat of adherens junctions and at microfilament bundles such as stress fibers and circumferential bundles, *J. Cell Biol.*, 109, 2905, 1989.
56. **Steinberg, M. S., Shida, H., Gindice, G. J., Shida, M., Patel, N. H., and Blaschuk, O. W.,** On the molecular organization, diversity and functions of desmosomal proteins, in *Junctional Complexes of Epithelial Cells,* Ciba Foundation Symp., Bock, G. and Clark, S., Eds., John Wiley & Sons, Sussex, 1987, 3.

Chapter 9

CELL ORGANELLES

9.3 Cytoskeleton

Toyoshi Fujimoto

TABLE OF CONTENTS

0-8493-6012-9/93/$0.00 + $.50

A. INTRODUCTION

It is only two decades ago that the term ''cytoskeleton'' appeared in modern biology literature. In the short period of time since then, the cytoskeleton has become one of the main foci of cell biological research and has exerted a wide-spread influence on other fields of life science. Today, we know that the cytoskeleton basically consists of a few filamentous structures which are made of different sets of proteins and interact with each other to perform a variety of functions. We also realize that the cytoskeleton has a three-dimensional architecture from which each component is interlaced in the cytoplasm. It can be said for sure that a major portion of the present knowledge of the cytoskeleton, especially its overall organization, was brought about through cytochemical techniques.

Because of the enormous scope of this research area, it is apparently impossible, in this short article, to refer to all important work done with cytochemical methods. As a consequence, only a few selected topics will be discussed.

B. ACTIN, MYOSIN, AND RELATED PROTEINS

1. Actin

Actin and myosin, which had been known to constitute the myofilaments of the skeletal muscle fiber, later became identified in virtually all nonmuscle cells. Actin is a 42-kDa protein; its molecular structure is very similar even in cells of distant progeny. Actin polymerizes to form a filament of 5 to 7 nm in diameter. With the advent of modern fixation protocols for electron microscopy, it was discovered that numerous thin filaments exist in the cytoplasm. It was also noticed that filaments of different sizes are present in a cell. A cytochemical method was developed to distinguish actin filaments:[1] it was found that heavy meromyosin, or S1 fragment, specifically binds to actin filaments and forms arrowhead-like structures visible under an electron microscope (Figure 1). By this technique, the coexistence of actin filaments and intermediate filaments in a single cell was first established. At the same time, it was discovered that each actin filament owns innate polarity: one end is a pointed end and the other barbed. This latter discovery was very important in deciphering the functions of actin filaments *in situ,* because it indicates the direction of contractile forces exerted with myosin. For example, a bundle of actin filaments of a mixed polarity, as in the contractile ring of a dividing cell, can contract and shorten its length, whereas that of uniform polarity, such as the core filaments of the microvillus, maintains a constant length and may play mostly a structural role.

Since the amino acid sequence and hence the molecular structure of actin are very similar among different species, it had been difficult to raise antibody against the protein. A high-titer antiserum was finally obtained by injecting animals with the protein denatured with sodium dodecyl sulfate (SDS).[2] Immunofluorescence micrographs obtained by use of this antiserum were vivid and illustrative. It is not exaggerating to say that the very basic image of the cytoskeleton was established by these pictures. Later, a mushroom toxin, phalloidin or phallacidin, was discovered to have a high affinity for F-actin, and the toxins conjugated to fluorochromes, e.g., NBD-phallacidin, became popular as specific labeling reagents for actin filaments for fluorescence microscopy.

Recently, a novel approach combining *in situ* hybridization and immunoelectron microscopy was introduced and revealed that messenger RNA molecules coding for cytoskeletal proteins are associated with actin filaments but not with intermediate filaments or microtubules.[3] The result indicates that actin filaments play a role in transporting the message to specific areas of the cell.

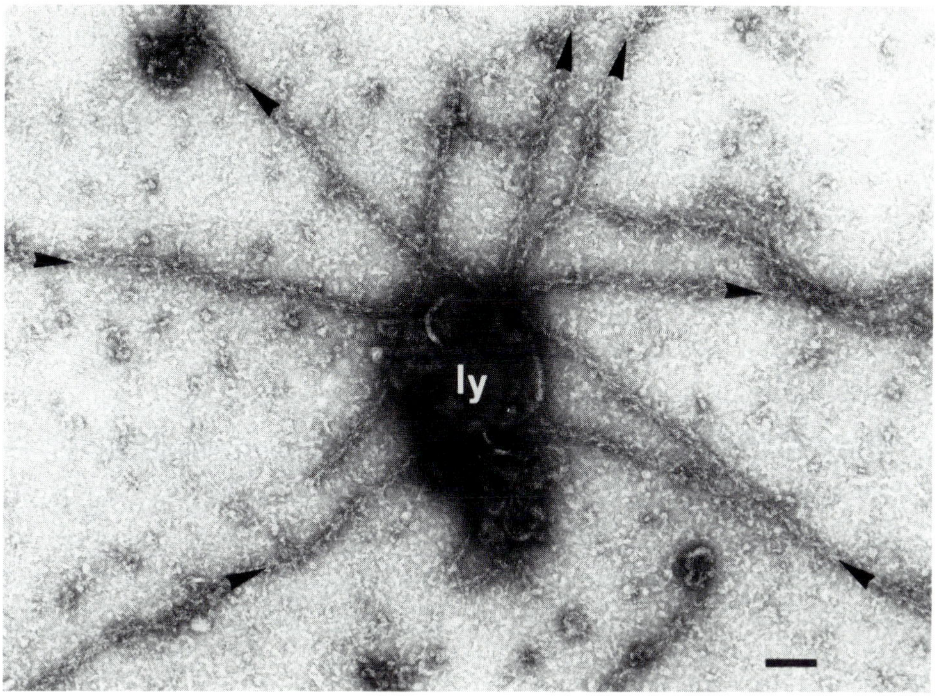

FIGURE 1. Negative staining of isolated actin filaments decorated with heavy meromyosin. Interaction of lysosomes (ly) with the filaments is shown. Polarity of filaments is indicated with arrowheads. (Bar = 100 nm; magnification × 70,000. Photograph courtesy of Dr. N. Araki.)

2. Myosin

In the striated muscle, myosin occurs as thick myofilaments interdigitated with actin filaments. Although it has been known for years that myosin exists in nonmuscle cells, there are still many questions about what kind of structures it forms and how the assembly process is regulated. Obviously, myosin in nonmuscle cells is far more plastic than that in striated muscle cells; probably it takes a filamentous form only under certain conditions. *In vitro* experiments showed that nonmuscle as well as smooth muscle myosin is only polymerized to form filaments when phosphorylated in the light chain.

In contrast to actin, myosin does not form a three-dimensional network of its own. Instead, myosin, probably forming a bipolar rod, is distributed along actin filaments where contractile force is to be generated. In cultured cells, myosin has been localized intermittently along stress fibers; α-actinin is also present in a similar manner, but the two proteins are segregated from each other, a distribution which is similar to the sarcomere of the skeletal muscle fiber. By using a battery of antibodies raised against different portions of myosin molecule, it was concluded that myosin in stress fibers is organized into bipolar filaments.[4] In the intestinal epithelium, myosin is thought to be localized in the terminal web and to interact with the rootlet of microvillar core filaments.[5]

In addition to conventional myosin as described above, there is now known to exist another kind of myosin which lacks the tail domain and is called myosin I or minimyosin; for distinction, the former type of myosin is often called myosin II. In *Acanthamoeba,* it was demonstrated by immunoelectron microscopy that myosin I is in the membrane while myosin II forms a short filament in the cytoplasm.[6] In mammalian tissues, it is only in the intestinal microvillar membrane that localization of myosin I is known for sure.

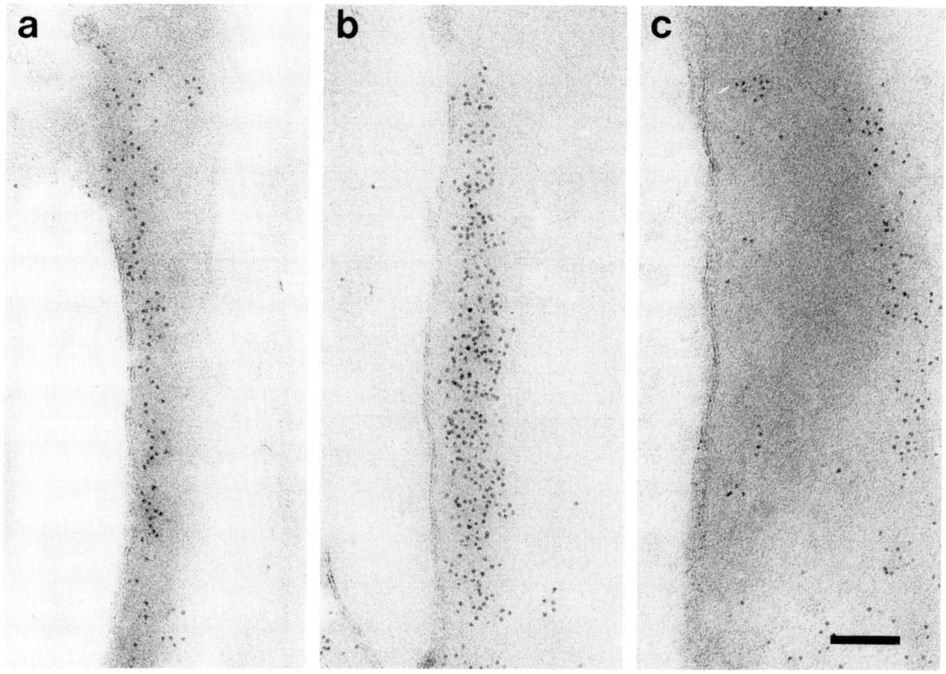

FIGURE 2. Localization of vinculin (a), α-actinin (b), and filamin (c) in the dense plaque of the aortic smooth muscle. The three antigens show different distributions in relation to the cell membrane and the dense region: vinculin is the closest to the cell membrane, α-actinin is found throughout the dense region, and filamin is farthest from the cell membrane and localized in the periphery of the dense region. (Bar = 100 nm; magnification × 92,000.)

3. Actin-Binding Proteins

Tens of proteins which bind to actin have been discovered and the number is still increasing. Immunocytochemistry has been particularly useful in determining precise localization of those proteins and thus in understanding their functions. One example which illustrates the importance of immunoelectron microscopy concerns the zonula adherens of intestinal epithelial cells. By immunolabeling ultrathin frozen sections, it was clearly demonstrated that vinculin was localized closer to the plasma membrane than α-actinin.[7] More recently, another protein, talin, was found even closer to the membrane. From these results, topography at the macromolecule level, that is, talin, vinculin, α-actinin, in this order from the junctional membrane, was established. A similar distribution of macromolecular localization in the dense body of the smooth muscle cell is shown in Figure 2.

Depending on the cellular environment, the localization of some actin-binding proteins are known to change. Gelsolin in platelets was found to become associated with the plasma membrane following thrombin stimulation, and then to return to the cytoplasm as in the resting condition. Immunoelectron microscopy and semiquantitative treatment of the results revealed the distributional change conclusively.[8]

It is important to localize various proteins in relation to the three-dimensional structure of the cytoskeleton. This goal has been achieved by replica techniques. Cells or tissues first immunolabeled are critical-point dried, or rapidly frozen, and deep etched, and replicas are prepared. Several proteins, including actin-binding protein,[9] gelsolin,[8] and fodrin,[10] have been observed to bind to actin filaments in different manners by this technique.

C. MICROTUBULES

1. Tubulin

The microtubule is a hollow tube of about 25 nm in outer diameter and 15 nm in inner diameter. Its wall is made of tubulin, a globular protein of about 55 kDa in molecular weight. The distribution of microtubules in cells was visualized by immunolabeling with anti-tubulin antibody. By high-resolution immunofluorescence microscopy, each microtubule can be seen as an independent filament in cultured cells; in most cases, microtubules are seen to be emanated from a perinuclear region and extend toward the cell periphery. The focus where microtubules seem to originate is called the microtubule-organizing center (MTOC), the component of which have been labeled by autoantibodies found occasionally in normal animals.

Under the electron microscope, microtubules can be discerned easily in ordinary resin sections if cells are fixed properly. They have an inherent polarity, as do actin filaments, and can be detected by two cytochemical techniques: the first one, called the hook-decoration method, attaches free tubulin molecules to the microtubules;[11] the added molecules form curved protofilament sheets on the original microtubular walls, and the polarity can be determined by observing the direction of curvature (clockwise vs. counterclockwise) in cross sections. The second method utilizes dynein which binds to microtubules at oblique angles and thus makes it possible to establish the direction.[12] Knowing the polarity of a microtubule is important, especially if it works as a track for intracellular transport using motor molecules, as described later.

Microtubules had been believed to be depolymerized rapidly in the cold. However, in recent years, a population of microtubules which are stable even in the cold were found to exist and were thus named ''cold-stable microtubules''. Although the molecular basis of this peculiar characteristic has not been fully understood, posttranslational modification of tubulin molecules may be responsible for the stability of the subpopulation; for example, detyrosination, acetylation, and phosphorylation of tubulin may be involved. Double-label immunoelectron microscopy with monospecific antibodies to tyrosinated and detyrosinated tubulin showed the following:[13] the two kinds of tubulin coexist in all microtubules; in most microtubules, the tyrosinated form overwhelms the detyrosinated one, whereas in the rest of the microtubules the proportion of the two forms is reversed. The meaning of this chemical modification *in vivo* is still controversial, and needs to be clarified to understand microtubular functions.

2. Microtubule-Associated Proteins

In addition to posttranslational modification, microtubule-associated proteins are likely to impart microtubules of different characteristics matched to local environment. They presumably stabilize microtubules and provide linkages with other cellular components. These proteins have been studied most extensively in neurons. The major microtubule-associated proteins in neurons are MAP1, MAP2, and tau. It was demonstrated by immunocytochemistry that they show differential distributions in a nerve cell: MAP2 is mostly in dendrites, tau is in axons, and MAP1 occurs equally in dendrites and axons. By immunoelectron microscopy, it was found that the microtubule-associated proteins intercalate adjacent microtubules[14] and/or between microtubules and intermediate filaments[15] (Figure 3).

Recently, the so-called ''motor'' molecules which drive various organelles along microtubules have been identified. Kinesin and cytoplasmic dynein are the two which have been best characterized so far. Kinesin is considered as an anterograde motor which transports various particles from the minus to the plus end of microtubules, while cytoplasmic dynein is a retrograde motor which works in the opposite direction. Immunofluorescence studies[16,17] with anti-kinesin antibodies showed either diffuse cytoplasmic, tubular, or punctate labeling

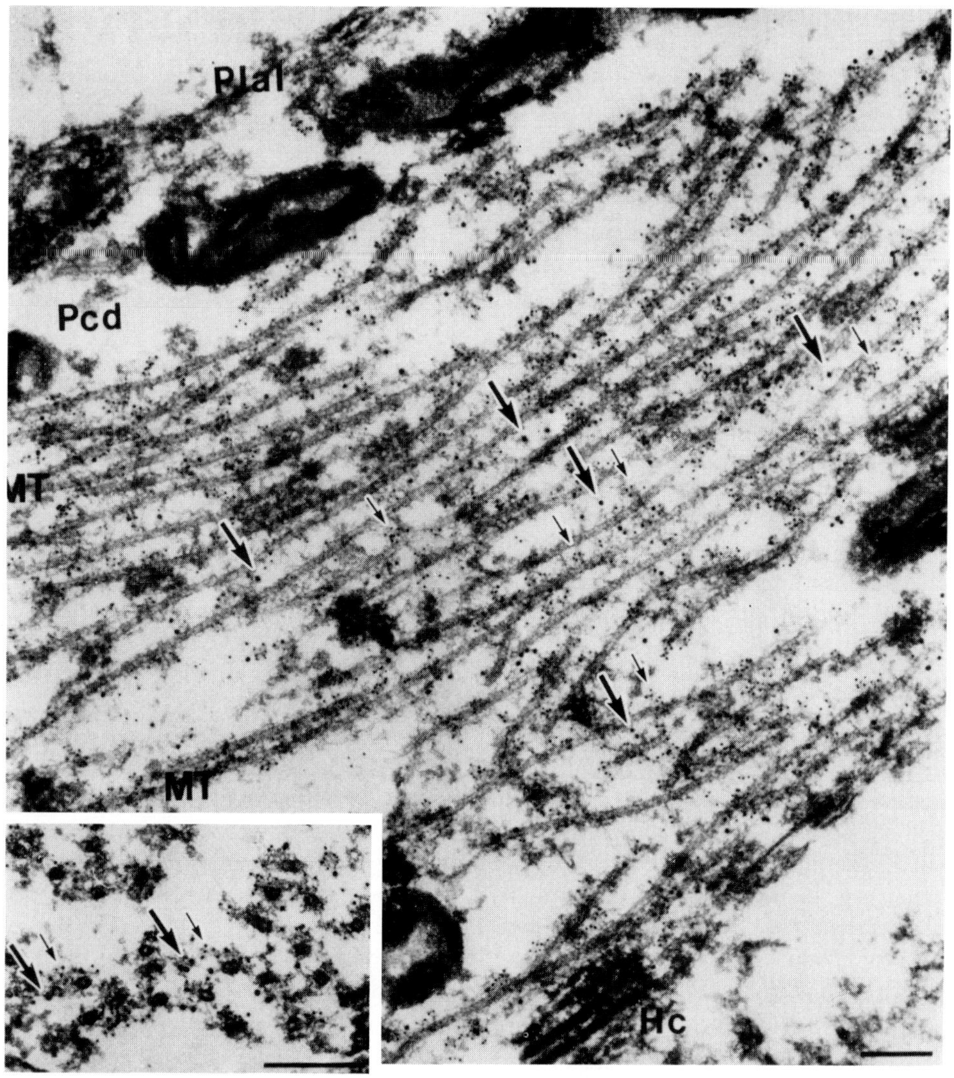

FIGURE 3. Double immunolabeling of MAP1A and MAP2 in a dendrite of a rat cerebellar Purkinje cell. MAP1A (marked with 10-nm gold; large arrows) and MAP2 (marked with 5-nm gold; small arrows) are shown to coexist on the fuzzy material between microtubules. MT, microtubules; Pcd, Purkinje cell dendrite; Hc, hypolemmal cisternae; Plal, plasmalemma. Bar, 200 nm; magnification × 46,800. Inset: cross section of a dendrite. (Bar = 200 nm; magnification × 70,200. Reproduced from Shiomura, Y. and Hirokawa, N., *J. Cell Biol.*, 104, 1575, 1987. With the permission of the Rockefeller University Press.)

patterns; but they commonly demonstrated that kinesin is extracted by permeabilization before fixation. Despite the variation as to its immunocytochemical localization, the latter results indicate that kinesin is associated with membrane organelles.

D. INTERMEDIATE FILAMENTS

Intermediate filaments are also called 10-nm filaments because of their diameter. In contrast to actin filaments and microtubules, intermediate filaments consist of different proteins in cells of different origin. Vimentin, desmin, prekeratin, glial acidic fibrillary protein, and neurofilament proteins form intermediate filaments in mesenchymal cells, muscle cells, epithelial cells, astrocytes, and nerve cells, respectively. In addition, nuclear lamins, which make a scaffold for the nuclear membrane, are found homologous to intermediate filament proteins. It has been thought that intermediate filaments are invariably stable and seldom go through depolymerization. However, in recent years, it was found that phosphorylation-dephosphorylation of the component proteins modulates the filaments' stability, and that the intermediate filament is depolymerized in a physiological environment.

Immunofluorescence microscopy revealed that in some cells, intermediate filaments are distributed in close association with microtubules, suggesting a close relationship between the two systems. As a practical medical application, the cell-type-specific expression of intermediate filament proteins is utilized to identify histological origin of tumors by immunocytochemistry.

In some cells, two different intermediate filament proteins are coexpressed. For example, in developing chicken embryonic heart muscles, intermediate filaments are first composed solely of vimentin, but in adults, they are made of desmin alone; in transition, it was found by immunolabeling that a single intermediate filament contains both proteins. Immunoelectron microscopy revealed that the two proteins are localized in a single intermediate filament and that copolymerization occurs *in situ*[18] (Figure 4). Immunocytochemistry in conjunction with the quick-freeze deep-etch technique showed that neurofilaments in mature axons are connected laterally with crossbridges, and that they are in fact made of one of the three neurofilament proteins.[19]

A number of proteins have been identified to bind to intermediate filaments, but their characterization has been slow, mainly due to their scarcity. It was shown by immunoelectron microscopy that plectin is localized at junction sites and branching points of intermediate filaments,[20] while epinemin is distributed randomly.[21] The functions of the associated proteins are not understood yet, but it is probable that they modify the nature of intermediate filaments to adapt to various cellular requirements.

E. MEMBRANE SKELETON

In the cytoplasmic surface of the erythrocyte plasma membrane, peripheral membrane proteins constitute an extensive protein network, and interact with the cytoplasmic portion of integral membrane proteins. Spectrin is the major constituent of the network and exists as a dimer of filamentous proteins, 240 and 220 kDa in molecular weight and 100 nm in length. In recent years, several methods to observe the erythrocyte membrane *en face* were introduced, and direct visualization of spectrin and other peripheral membrane proteins became possible.

It was suspected that spectrin analogs, and hence similar networks of proteins, might occur prevalently, but antibodies to spectrin did not react with other cells. It was several years later before it was known that proteins similar to spectrin exist in virtually all cells but that they do not cross-react immunologically with spectrin. They have been given several different names by discoverers; in this article, they are called "fodrin". Fodrin shares many characteristics with spectrin; it is a dimer of 240 and 235 kDa, 100 nm in length, and binds to other peripheral membrane proteins such as ankyrin, band 4.1, and adducin. Interestingly, fodrin binds to many proteins with higher affinity than spectrin, to name a few, actin, microtubules, intermediate filaments, and calmodulin.

By immunocytochemical techniques, fodrin has been found to occur beneath the plasma membrane of many cells. Based on this localization, and on the fact that analog of other erythrocyte proteins also exist in those cells, it is postulated that a protein network is formed in many cells as in the erythrocyte. However, in contrast to spectrin in the erythrocyte, fodrin in nonerythroid cells is not necessarily bound to the cytoplasmic surface of the plasma

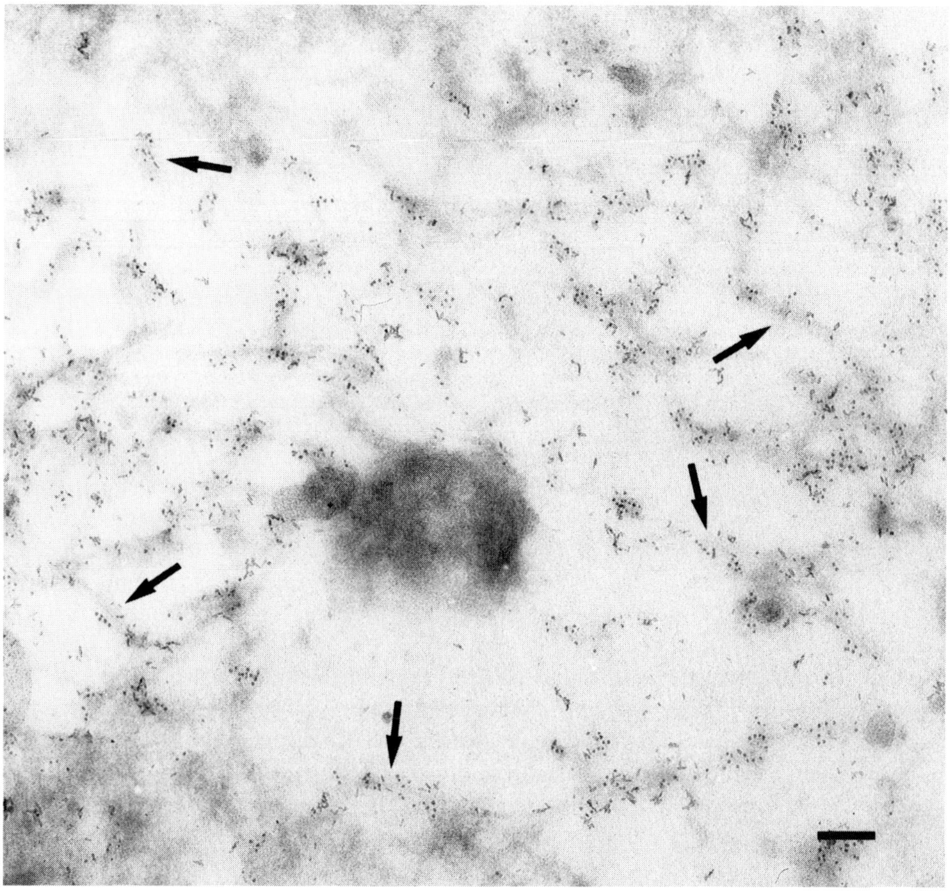

FIGURE 4. Double immunolabeling of vimentin and desmin in an ultrathin frozen section of the chicken aortic smooth muscle. Intermediate filaments are labeled for both vimentin (marked with Imposil; elongated shape) and desmin (marked with ferritin; round shape). Colocalization of the two proteins are most obvious in the regions indicated by arrows. (Bar = 100 nm; magnification × 79,000.)

membrane. Immunoelectron microscopy showed that fodrin in the neutrophil occurs as a cytoskeletal protein.[22] It was also revealed that in T lymphocytes[23] and keratinocyte[24] fodrin has a dual distribution in the cytoplasm and beneath the plasma membrane, depending on various conditions (Figure 5). In addition, translational redistribution of fodrin along the plasma membrane was shown in the chromaffin cell.[25] It is thought to indicate that an actin filament meshwork crosslinked with fodrin hinders secretory granules from approaching the plasma membrane in a resting state, but dissolves when secretion is stimulated. Incorporation of fodrin into the cell interior with vesicles was also demonstrated by immunoelectron microscopy.[26] These findings indicate that fodrin is concentrated beneath the plasma membrane in many cells, but that its organization may be variable and much more flexible than that of its erythrocyte counterpart.

Besides spectrin, other erythrocyte peripheral membrane proteins, or their analogs, have been known to occur in various cells and tissues. Ankyrin is a protein of 215 kDa, and it interconnects between spectrin and band 3 in the erythrocyte. That this linkage also occurs in nonerythroid cells was shown in the renal intercalated cells by immunoelectron microscopy.[27] Brain isoform of ankyrin, sharing some antigenicity with the erythrocyte isoform,

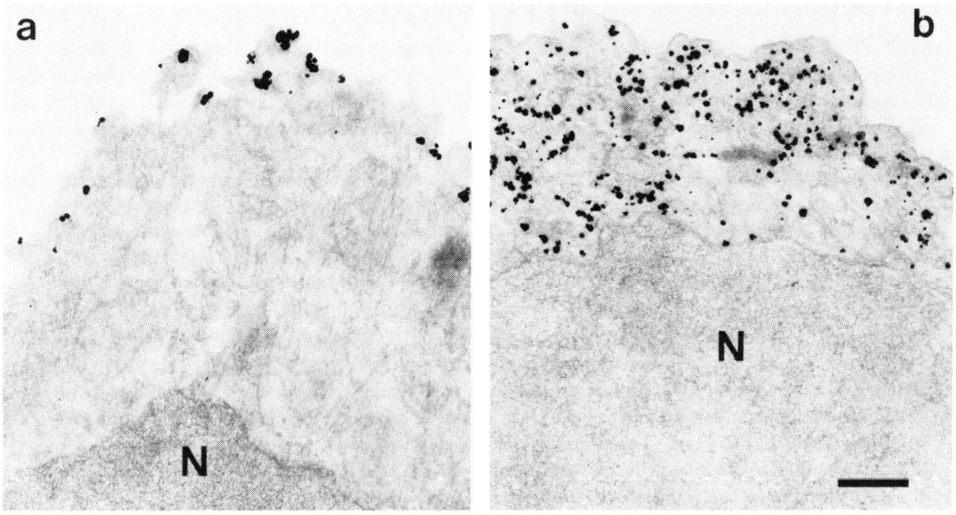

FIGURE 5. Immunolabeling of fodrin in a mouse keratinocyte cell line (PAM 212) cultured in standard Ca^{2+} medium (a) and in low-Ca^{2+} medium (b). Labeling was made with 1-nm colloidal gold and later developed with silver. While fodrin is localized beneath the cell membrane in standard Ca^{2+}, it is found in the cytoplasm in cells cultured in low-Ca^{2+} medium; N: nucleus. (Bar = 500 nm; magnification × 19,000. Photograph courtesy of Dr. K. Yoneda.)

was identified in many tissues. Immunocytochemistry showed differential distribution of the two isoforms in the nervous system; the brain isoform has a broad distribution, while the erythroid isoform is limited to a few sites, including the nodes of Ranvier.[28]

Band 4.1, a protein of about 80 kDa, forms a ternary complex with the tail of spectrin dimer and actin oligomer in the erythrocyte. It also binds to glycophorin, an integral membrane protein in the erythrocyte, only when the latter is associated with phosphatidylinositol 4,5-bisphosphate (PIP_2). This property makes band 4.1 an attractive target of research in relation to signal transduction in the plasma membrane. Band 4.1 in nonerythroid cells is known to occur in several forms generated by differential RNA splicing. Immunofluorescence microscopy of cultured endothelial cells revealed that distribution of band 4.1 is modulated by the extracellular matrix.[79] Notably, the protein occurs in the perinuclear region, but it is not known what the band 4.1-positive structure corresponds to at the ultrastructural level.

F. CONCLUSION

Even from this short and incomplete survey, it is obvious that cytochemistry has made an enormous contribution to cytoskeletal research. In the forthcoming years, by incorporating new techniques, especially those of molecular biology, cytochemistry will certainly remain as a major tool in this attractive field.

ACKNOWLEDGMENTS

The author is grateful to Drs. N. Araki, N. Hirokawa, and K. Yoneda for kindly providing their electron micrographs.

Chapter 9

CELL ORGANELLES

9.4 Contributions of Cytochemistry and Immunocytochemistry to our Understanding of the Organization and Function of the Golgi Apparatus

Marilyn Gist Farquhar, Linda Hendricks, Toru Noda, and Angel Velasco

TABLE OF CONTENTS

0-8493-6012-9/93/$0.00 + $.50

A. INTRODUCTION

There are few areas in the cell biology of organelles where the value of observations based on electron microscopic cytochemistry and immunocytochemistry are more evident than in the case of the Golgi apparatus or complex. In the case of the Golgi there are numerous instances where electron microscopic cytochemical and immunocytochemical findings dictated the directions of research in the field, and the insights gained provided the basis for the working hypotheses that were eventually tested and corroborated by alternative techniques such as cell fractionation, *in vitro* transport assays, etc. The reason for this should be clear: the Golgi is not a single uniform organelle comparable to the mitochondria, for example. Rather it is a collection of smooth membrane compartments of different composition and function but of similar size and density (Figure 1), which cannot be reliably separated by the usual cell fractionation procedures. Only by localization of various Golgi components *in situ* or its products in transit can one reliably resolve one subcompartment from the other. It will be the theme of this chapter that electron microscopic cytochemistry and immunocytochemistry have played a crucial role at virtually every juncture in the development of our current concepts of Golgi organization and function and of the traffic of membrane and products through this organelle.

B. CURRENT CONCEPTS OF GOLGI ORGANIZATION AND FUNCTION

The Golgi apparatus consists of a stack of cisternae and associated vesicles which constitute an obligatory station along the so-called exocytic or endoplasmic reticulum (ER) to plasmalemmal pathway. That is, secretory proteins destined for export from the cell, lysosomal enzymes destined for lysosomes, and integral membrane proteins destined for the plasmalemma and other organelles must pass through this organelle in order to be delivered to their final destination. Before proceeding with the main topic of this chapter, it is essential to briefly summarize our current concept of the organization of the Golgi apparatus and how membrane and product traffic is directed to and through it. The key features of our current model (reviewed by Farquhar[1,2] and diagrammed in Figure 2) are (1) the Golgi stack has a distinct polarity: it has a *cis* or entry face and a *trans* or exit face; (2) products enter at the *cis* face and move vectorially across the stack in the *cis* to *trans* direction, traversing the cisternae one by one, and are progressively posttranslationally modified in transit; (3) each cisterna or set of cisternae constitutes a separate subcompartment with a distinctive composition and functions imparted by resident membrane proteins (e.g., glycosyltransferases, receptors, transport systems, etc.); (4) transport from the ER to the Golgi and between Golgi subcompartments is carried out by vesicular carriers coming to and departing from the dilated rims of the cisternae; (5) all products that follow this route are mixed together in transit until they arrive on the *trans* side of the Golgi where sorting takes place and secretory proteins, lysosomal enzymes, and membrane proteins are packaged into separate vesicular containers.

The posttranslational modifications of proteins that take place largely or exclusively in the Golgi apparatus include trimming and terminal glycosylation (addition of galactose and sialic acid residues) of N-linked glycoproteins, sulfation, proprotein processing, and phosphorylation of lysosomal enzymes.

In addition to newly synthesized proteins, it has been established that the Golgi receives recycling membrane traffic from various locations such as the plasmalemma, lysosomes, and endosomes, and sorts and directs this recycling traffic to its correct destination (see Figure 2). Thus, another main function of the Golgi is in the sorting of protein and vesicular traffic. Further discussion of many of these concepts are included in the reviews cited above, and in several other recent reviews[3-9] dealing with specific aspects of Golgi organization and function.

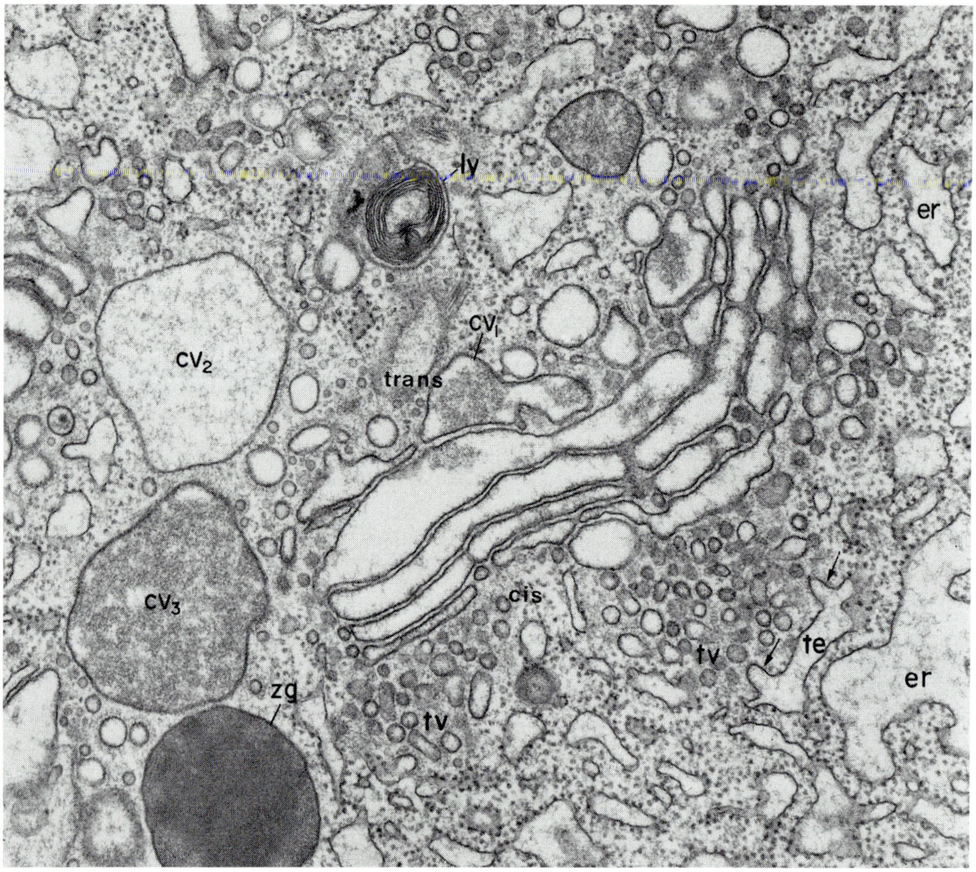

FIGURE 1. Golgi complex from an exocrine pancreatic cell (guinea pig) demonstrating its multiple subcompartments. Characteristic features are the presence of a stack of four to five slightly dilated Golgi cisternae associated with condensing vacuoles (cv_{1-3}) on the *trans* side, and a profusion of small peripheral Golgi vesicles, or transport vesicles (tv), along the *cis* side. These vesicles are assumed to bud (arrows) from the transitional elements (te) of the rough ER (er) and to transport secretory products to the *cis*-most Golgi cisternae. The condensing vacuoles gradually fill with secretory proteins (mostly pancreatic zymogens), undergo progressive concentration (cv_{1-3}), and eventually become mature zymogen granules (zg); ly, lysosome. (Magnification × 29,600. From Farquhar, M. G. and Palade, G. E., *J. Cell Biol.*, 91, 77, 1981. With permission.)

We would now like to trace the history of how these concepts evolved, and the role that insights obtained from cytochemical and immunocytochemical studies played in this development.

C. CONTRIBUTIONS OF ELECTRON MICROSCOPE CYTOCHEMISTRY

From the time it was described by Golgi in 1898[10] until the 1960s, the Golgi apparatus remained a mysterious organelle of unknown function whose very existence as a bona fide organelle was questioned.[11] During the 1960s a number of cytochemical and other techniques were introduced that were very useful for demonstrating the Golgi complex or subcompartments of this organelle at the electron microscope level. These included osmium impregnation, enzyme cytochemical demonstration of various phosphatases, and autoradiography. The information gained with these techniques provided many of the first insights into the detailed organization and functions of this organelle.

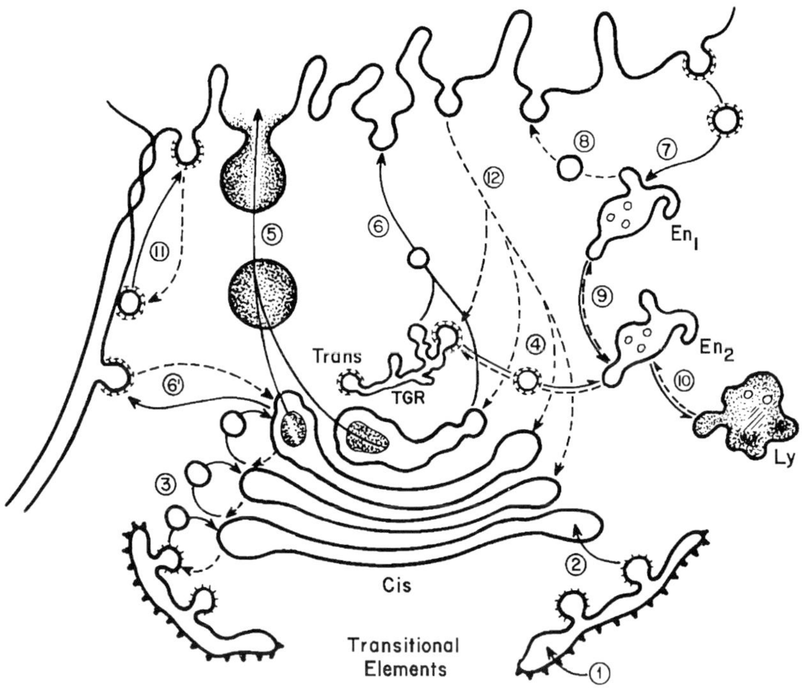

FIGURE 2. Diagram depicting the major routes of protein and membrane traffic along the exocytic (1–6) and endocytic (7–12) pathways. *Exocytic pathways:* secretory proteins, membrane glycoproteins, and lysosomal enzymes are synthesized on polyribosomes and are translocated to the ER lumen (**1**) where they spend a variable amount of time undergoing cotranslational and posttranslational processing (e.g., disulfide bonding, glycosylation, trimming, folding, and oligomerization). They exit the ER via small vesicles that serve to shuttle them to the *cis* side of the Golgi complex (**2**) and subsequently traverse the Golgi cisternae sequentially via vesicular carriers (**3**). Lysosomal enzymes bind to mannose-6-phosphate receptors (MPR) in the Golgi complex, are believed to be sorted from secretory and membrane proteins in the *trans*-Golgi cisterna or *trans*-Golgi reticulum network, and ferried via coated vesicle carriers to late endosomes (**4**). Membrane and secretory proteins are also sorted on the *trans* side of the Golgi complex in either the *trans*-most Golgi cisterna or *trans*-Golgi reticulum (TGR) and delivered by exocytosis along either the regulated pathway via secretion granules (**5**) or along the constitutive pathway via nonclathrin-coated vesicles (**6**) which fuse either with the apical (**6**) or basolateral (**6′**) domains of the plasmalemma. *Endocytic pathways:* several endocytic pathways have been outlined — to endosomes (**7**), to lysosomes (**9** and **10**), to the opposite cell surface via transcytosis (**11**), or to the Golgi (**12**). The latter is believed to represent Golgi or plasma membrane proteins that cycle between the plasma membrane and the Golgi. En_1, Early endosome; En_2, late endosome or "prelysosome"; Ly, lysosome. (Modified from Farquhar, M. G., *Annu. Rev. Cell Biol.*, 1, 447, 1985.)

1. Selective Impregnation of *Cis*-Cisternae by OsO_4

The very discovery of the Golgi apparatus as a distinct organelle by Golgi was dependent on its general impregnation with heavy metals (silver and, later, OsO_4) which made possible its identification by light microscopy. In 1965, Friend and Murray[12] introduced a modified osmium impregnation method suitable for demonstration of the Golgi by electron microscopy and discovered that the osmium was selectively deposited within *cis*-cisternae (Figure 3D). The basis of its selective deposition in *cis*-cisternae remains unknown to this day. One can speculate that there are components in the *cis*-cisternae, not present in other cisternae, that reduce the osmium. Irrespective of the fact that its biochemical basis remains unknown, the method itself has proved to be useful for distinguishing the *cis* from the *trans* side of the Golgi and, until quite recently, was the only *cis*-Golgi marker available.

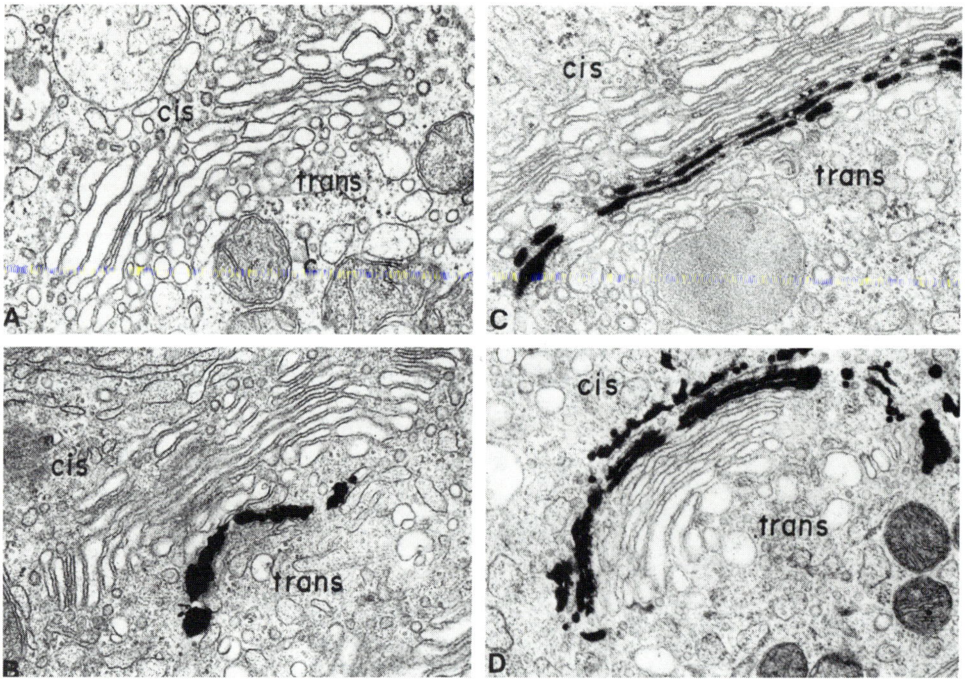

FIGURE 3. Differential cytochemical staining of Golgi cisternae of the rat epidydymis. (A) Unstained. The Golgi complex of this epithelium consists of eight to ten parallel stacked cisternae with numerous associated vesicles. (B) Acid phosphatase (β-glycerophosphatase) is detected in a single *trans*-cisterna which is set off from the rest of the stack, and which has the properties ascribed to GERL or the TGN. (C) Thiamine pyrophosphatase (TPPase) is detected in two or three of the *trans*-most cisternae in the stack. (D) Preparation impregnated with OsO_4. Osmium deposits are found in two of the *cis*-most cisternae in the stack. (Magnifications: A, × 34,400; B and C, × 25,800; D, × 20,600. From Friend, D. S., *J. Cell Biol.*, 41, 269, 1969. With permission.)

2. Phosphatase Cytochemistry Provides the First Indication of Compositional Heterogeneity Among Golgi Cisternae

TPPase and acid phosphatase are markers for *trans*-Golgi cisternae — The first evidence that the Golgi cisternae differ from one another came from cytochemical staining procedures that demonstrated qualitative differences in staining for several phosphatases. In 1961, Novikoff and Goldfischer[13] demonstrated that thiamine pyrophosphatase (TPPase) represents a cytochemical marker for the Golgi complex that can be used to characterize its form and distribution in different cell types. Subsequently, Novikoff and Novikoff[14,15] showed that within the Golgi, TPPase is restricted to one to two *trans*-cisternae (see Figure 3C). To this day, TPPase remains the most readily accessible marker for *trans*-Golgi cisternae and the most reliable general marker for demonstration of the polarity of the Golgi stack, because it is applicable to virtually all cell types. This enzyme is believed to function in the removal of phosphate from nucleotide diphosphates formed as products of the transfer of activated sugar nucleotide donors (e.g., UDP-gal) to maturing oligosaccharides.[16,17] It was established later that TPPase activity coincides with galactosyltransferase activity (see below). Novikoff and Novikoff[14,15] further showed that acid phosphatase (CMPase),* like TPPase, is also

* Both β-glycerophosphate and cytidine monophosphate have been used as substrates for acid phosphatase. In general the localizations with the two different substrates have been the same, but some differences have also been reported.[164]

restricted to one to two cisternae on the *trans* side of the stack, but CMPase and TPPase are present in different cisternae in many cell types (Figure 3B), with the CMPase-positive cisterna often being the first cisterna on the *trans* side of the Golgi stack and the TPPase-positive cisterna the second.

The GERL concept — Based on the knowledge that acid phosphatase is a lysosomal enzyme and that this marker was found in the *trans*-most Golgi cisterna, Novikoff and Novikoff[14,15] formulated the GERL (Golgi-ER-lysosome) hypothesis for the origin of lysosomal enzymes. GERL was visualized as a separate cisterna distinct from the Golgi cisternae in that it was straighter, thicker walled, and set off from the rest of the stacked cisternae, which was TPPase-negative and acid phosphatase-positive. Novikoff believed GERL constituted the site where lysosomal enzymes are concentrated and packaged. According to his original formulation, the GERL cisterna was believed to be devoted exclusively to the packaging and processing of lysosomal enzymes which received its newly synthesized lysosomal enzymes directly from the ER. Later on, it was suggested that some secretory components may have direct access to GERL, bypassing the Golgi stack.[18] We now know that the latter is not the case, as there is ample evidence that secretory, lysosomal, and membrane glycoproteins are mixed together in the same ER and Golgi cisternae, up to and including the *trans*-most cisterna (see below). Moreover, there is every reason to believe that the GERL cisterna is an integral part of the Golgi complex, since packaging of secretory proteins, which is a classical Golgi function, takes place in this cisterna (Figure 4) as well as in the *trans*-most cisterna of the stack (see Farquhar, 1991[2]) in many cell types. Also, the CMPase activity found there does not necessarily represent only lysosomal enzyme activity. It may function, at least in part, to degrade CMP generated from transfer of sialic acid to oligosaccharides via CMP-sialic donor. However, even though some of the original ideas about GERL's function were incorrect, the concept did prove to be very useful. It had the advantage that it provoked considerable controversy, called attention to some of the distinctive morphological and cytochemical features of the *trans*-most cisternae, and recognized the close relationship between this region of the Golgi and lysosomal enzyme trafficking.

NADPase is a middle-Golgi marker — Later, Smith[19] demonstrated that in at least some (but not all) cell types the middle cisternae stain selectively for nicotinamide adenine dinucleotide phosphatase (NADPase). This technique is useful for some purposes, but it has not been widely applied because the main application of cytochemical techniques has been to find a marker to indicate the polarity of this organelle, and middle Golgi markers are not useful for this purpose.

Phosphatase cytochemistry provides the first indication of a compositional difference within the same Golgi cisterna — The fact that secretion granules bud from the dilated rims of Golgi cisternae[11] suggested that the dilated rims must have a different protein composition than that of their centers. However, direct evidence for this concept is limited. The first and only direct information indicating that this is the case was provided by phosphatase cytochemistry: when 5′-nucleotidase[20] and adenylate cyclase[21] were demonstrated in unfixed Golgi fractions, the lead phosphate reaction product was concentrated along the dilated rims of individual Golgi cisternae (Figure 5). These findings also provided the first evidence that some plasma membrane proteins are indigenous to Golgi fractions, and thus the presence of these plasma membrane enzymes in Golgi fractions is not due exclusively to contamination with plasma membrane.

3. Heterogeneity of Golgi Lipid Composition

Heterogeneity has also been demonstrated in the lipid composition of Golgi membranes. Isolated Golgi membranes were shown to have a lipid composition which was quantitatively

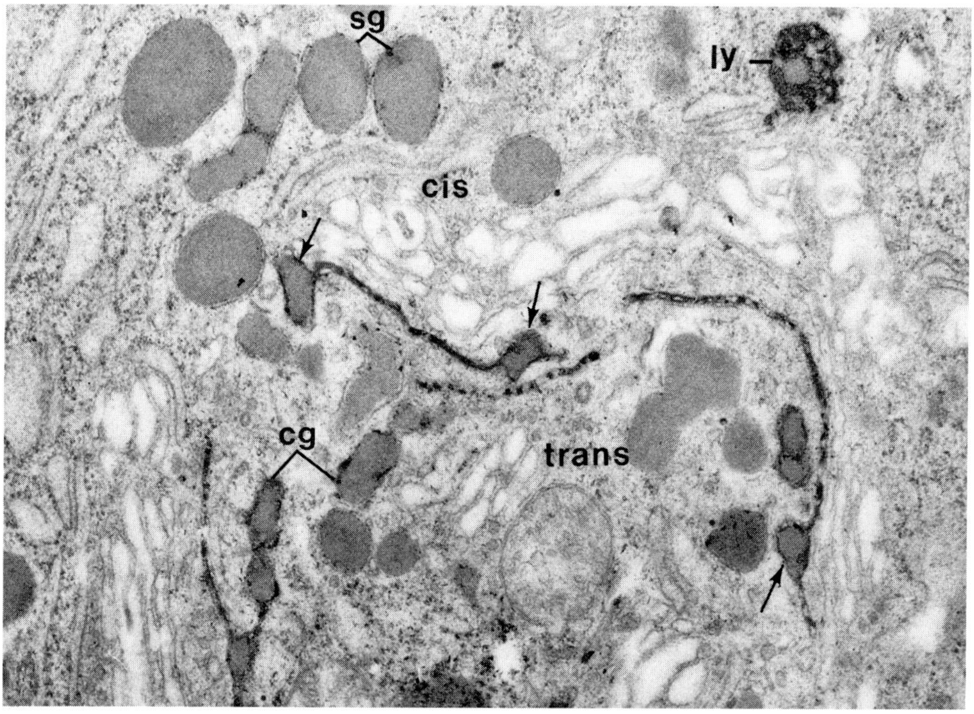

FIGURE 4. Golgi region of a prolactin-secreting cell from a lactating rat demonstrating condensing secretory granules (arrows) and acid phosphatase in the *trans*-most Golgi cisterna. The *trans*-most cisterna is less dilated than the rest and has the features sometimes ascribed to GERL and the TGN. Note that in some places (to the right), the reactive cisterna seems to be included in the regular stack, and in other places (to the left) it appears to be set off from the stack. Acid phosphatase (β-glycerophosphatase) reaction product is also seen around some of the immature or aggregating granules (ag) found on the *trans*-Golgi face, at the periphery of a few of the mature granules (sg), and in lysosomes (ly). (Magnification × 30,000. From Smith, R. E. and Farquhar, M. G., *J. Cell Biol.*, 31, 319, 1966. With permission.)

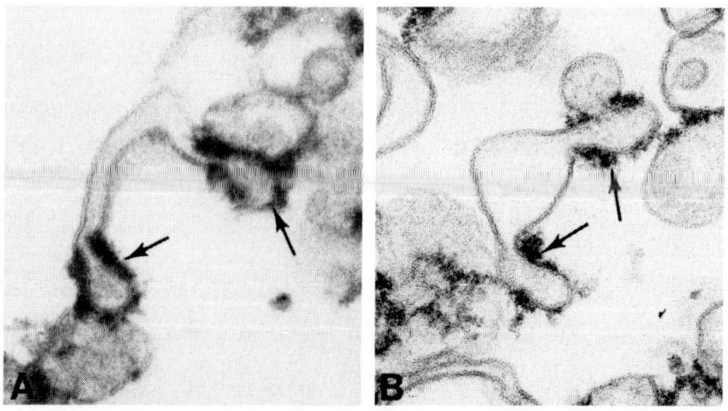

FIGURE 5. Compositional heterogeneity within a single Golgi cisterna. Golgi cisternae from heavy Golgi subfractions (GF_3) prepared by the method of Ehrenreich et al.[166] were reacted for 5′-nucleotidase (A) or adenylate cyclase (B) prior to fixation. Reaction product (lead phosphate) is concentrated on the dilated rims of the cisternae (arrows) and is absent from their central regions. (Magnification × 85,000. A, From Farquhar, M. G. et al., *J. Cell Biol.*, 60, 8, 1974; B, from Cheng, H. and Farquhar, M. G., *J. Cell Biol.*, 70, 660, 1976. With permission.)

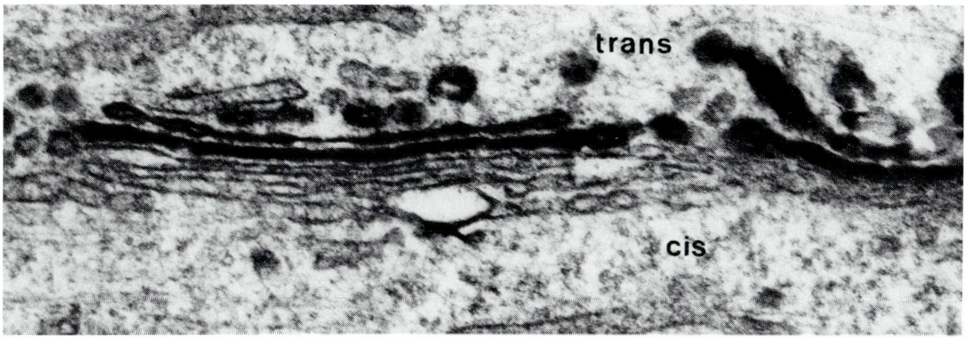

FIGURE 6. Electron microscopic localization of C6-NBD-Cer in the Golgi apparatus of CHO-K1 cells. Cells were fixed with 0.5% glutaraldehyde for 10 min, incubated for 30 min at 2°C with 5 nmol C6-NBD-Cer followed by incubation with 3.4 mg defatted bovine serum albumin/ml (to reduce background and give prominent labeling of the Golgi apparatus), photobleached in the presence of DAB, treated with OsO_4 and processed for electron microscopy. Only one or two cisternae, assumed to be *trans*-cisternae, and numerous small vesicles near the Golgi apparatus, are labeled. (Magnification × 90,000. From Pagano, R. E. et al., *J. Cell Biol.*, 109, 2067, 1989. With permission.)

different from either the ER or the plasma membrane.[22] The observation that membranes of different Golgi subfractions differentially bound digitonin (which forms insoluble complexes with cholesterol), suggested that cholesterol was heterogeneously distributed in the Golgi.[23,24] Subsequently, direct visualization by electron microscopy of filipin-cholesterol complexes in exocrine and endocrine pancreatic cells[25] indicated that a gradient (*cis* to *trans*) of cholesterol existed across the Golgi stacks. Filipin-cholesterol complexes (visualized either as protuberances or pits in freeze-fracture or as characteristic scalloped areas of the membrane in thin sections) were abundant in *trans*- (mature) cisternae, but were lacking in the *cis*- (forming) Golgi cisternae and ER cisternae.

More recently, fluorescent lipid derivatives that behave as analogs of their natural counterparts have been introduced to study lipid synthesis, sorting, and transport in vitally stained cells. Using a fluorescent analog of ceramide (*N*-[e-7-nitrobenz-2-oxa-13-diazol-4-yl-aminocaproyl]-D-*erythro*-sphingosine, C_6-NBD-Cer) as a probe, it has been demonstrated that the Golgi apparatus is the major site of ceramide incorporation and therefore of sphingomyelin synthesis in cultured fibroblasts[26,27] and Madin-Darby canine kidney (MDCK) cells.[28] Recently, Pagano and co-workers[29] found that this probe localizes selectively to the *trans*-cisternae of the Golgi complex of human skin fibroblasts (Figure 6). This was demonstrated using a cytochemical method which involves the photoconversion of the fluorescent analog into an electron-dense diaminobenzidine (DAB) product. These cytochemical findings together with complementary biochemical findings[30] established that the Golgi is an important site of sphingomyelin synthesis. In addition, the availability of a Golgi-specific lipid probe represents an exciting development, as it has provided a convenient *trans*-Golgi marker which is less likely to be restricted by cell, tissue, or species specificity than proteins and therefore should be broadly applicable to localization studies at both the light and ultrastructural level.

4. Autoradiography Provided the First Evidence that Sulfation and Terminal Glycosylation of N-Linked Glycoproteins Takes Place in the Golgi Complex

Sulfation — Shortly after the introduction of techniques for electron microscope autoradiography it was shown[31,32] that immediately after administration of radioactive sulfate *in vivo,* autoradiographic grains are concentrated over the Golgi cisternae and vacuoles in goblet cells and cartilage cells — cells that are known to make sulfated glycoproteins and proteoglycans. This constituted the first evidence that sulfation occurs exclusively in the Golgi. Later, Young[33] surveyed a variety of cells and found that many additional cell types also incorporate sulfate exclusively into the Golgi complex. Biochemical evidence subsequently corroborated that sulfotransferases are concentrated in Golgi fractions and established that sulfation is a very late posttranslational event (reviewed by Farquhar and Palade[11]) that occurs shortly before discharge. So far, no sulfotransferase has yet been localized *in situ.*

Glycosylation — It is now well established that one of the major functions of the Golgi apparatus is in the trimming and terminal glycosylation of N-linked glycoproteins.[34] Initial glycosylation occurs cotranslationally in the ER with the transfer of high mannose oligosaccharides to the nascent peptide. The latter consist of a 9-mannose unit which is subsequently trimmed to a 5-mannose unit prior to terminal glycosylation in the Golgi. Golgi enzymes are responsible for most of the trimming of these oligosaccharides and for the attachment of terminal sugars (*N*-acetylglucosamine, galactose, fucose, and sialic acid) to them. The very first evidence that the Golgi was involved in glycosylation came from electron microscope autoradiography. Neutra and Leblond[35,36] showed that when [^{3}H]glucose and [^{3}H]galactose were given *in vivo* the vast majority of the autoradiographic grains were localized over the Golgi of intestinal goblet cells, suggesting the primary role of this organelle in glycosylation and indicating its specific role in galactose incorporation into glycoproteins. A few years later Fleischer et al.[37] isolated Golgi fractions and showed they were enriched in galactosyltransferase. The latter enzyme has — to this time — remained the preferred biochemical marker for Golgi fractions. The same year, in studies on thyroid cells Whur and co-workers showed by autoradiography[38] that the addition of mannose to thyroglobulin occurs in the ER, and the addition of the terminal sugars (galactose and, later on, fucose and sialic acid[5]) occurs in the Golgi. This represented the first indication that there was division of labor between the ER and Golgi in glycosylation of N-linked glycoproteins. Biochemical evidence validating the autoradiographic observations was provided by the same laboratory.[39]

5. Autoradiography also Provided the First Indication of Vectorial Transport of Secretory Proteins and the First Quantitative Data on the Concentration of Secretory Proteins

Vectorial transport — An early contribution of EM autoradiography was the clear demonstration that newly synthesized secretory proteins move vectorially through the cell from rough ER to condensing vacuoles to mature secretion granules.[40,41] By carrying out pulse-chase experiments on exocrine pancreatic cells *in vivo,* it was demonstrated that autoradiographic grains appeared first (10 min chase) over the ER, at 30 min over condensing vacuoles, and after 60 min over mature secretion granules. This sequence was corroborated by simultaneous autoradiographic and cell fractionation experiments on pancreas incubated *in vitro,* which indicated that the label moved from rough microsomes to smooth microsomes to condensing vacuoles (granules).[42] Initially the autoradiographic data suggested that the Golgi might be bypassed, but later experiments on stimulated exocrine pancreatic cells demonstrated that the grains were seen over bona fide Golgi cisternae prior to appearing in condensing vacuoles.[43] The same sequence has also been demonstrated by autoradiography in other cell types such as the mammotroph of the anterior pituitary[44] (see Figure 7) except that in the latter cells, as in most other regulated secretory cell types, packaging (and concentration) of secretory proteins occurs in the dilated rims of the *trans*-most cisternae, rather than in condensing vacuoles.

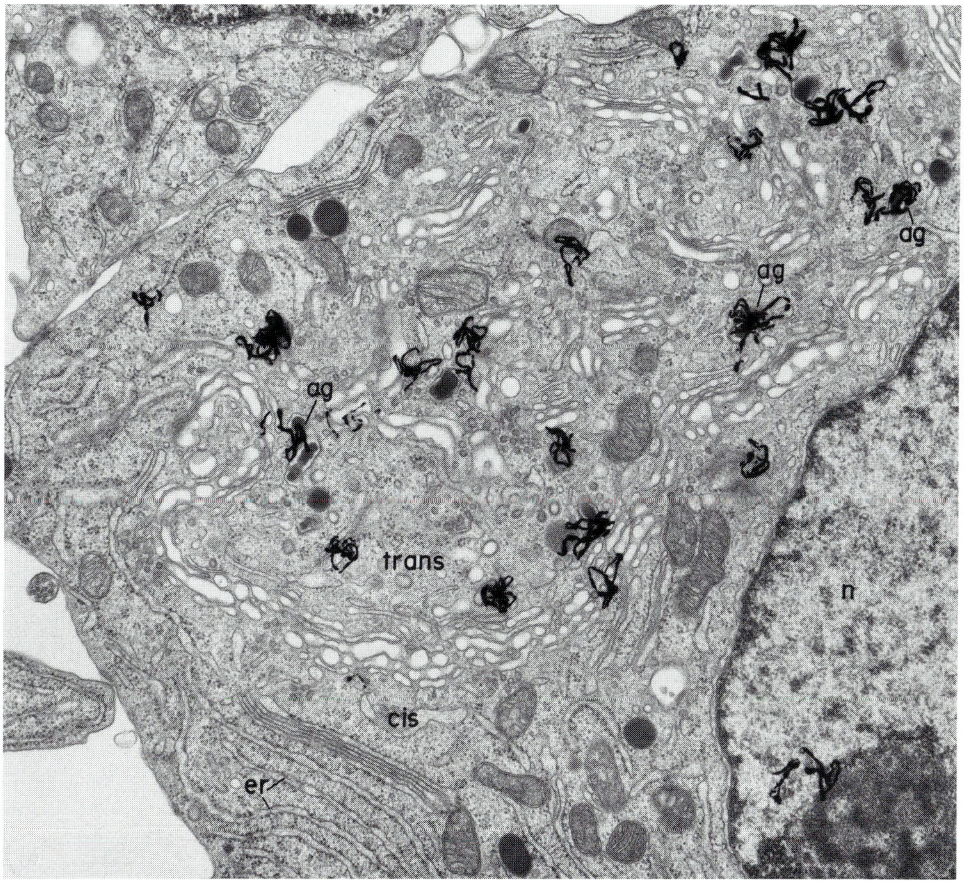

FIGURE 7. Concentration of secretory proteins within newly formed secretory granules. Autoradiogram of a prolactin cell (rat anterior pituitary) from a dissociated cell preparation pulsed *in vitro* for 5 min with [^{3}H]leucine and fixed after a 30-min chase. Grains are concentrated over immature or aggregating granules (ag) located on the *trans* side of Golgi stacks. When corrected for radiation spread,[45] the grain density (grains/unit area) of the immature granules at peak labeling is 50 to 200 times that of the rough ER, indicating that the secretory product (>85% prolactin) undergoes a ~200-fold concentration. (Magnification × 18,500. From Farquhar, M. G. et al., *Endocrinology*, 102, 296, 1978. With permission.)

Concentration Morphologic evidence in exocrine pancreatic cells, anterior pituitary cells, pancreatic β-cells, to mention a few, clearly suggested that considerable concentration of secretory proteins occurs in condensing vacuoles and in the dilated rims of the Golgi cisternae, because the content of these structures is denser than that of the ER and Golgi cisternae (Figures 1 and 4). A measure of the amount of concentration that can occur was provided by quantitative autoradiography of pituitary mammotrophs: up to 200 times more autoradiographic grains were found associated with the dilated rims of the Golgi cisternae than with the rough ER.[45] An estimate of the extent of concentration in the exocrine pancreas was also provided by immunogold labeling of secretory proteins.[46] However, to date the autoradiographic results constitute the most definitive quantitative data available on this topic, as immunogold labeling is semiquantitative at best.[47] Most limiting for this purpose is the fact that the labeling efficiency has been shown to vary from one intracellular compartment to another.[48]

6. Confirmation of the Compartments along the Exocytic Pathway was Provided by Peroxidase Cytochemistry

The full extent and connections of the exocytic pathway were subsequently visualized by peroxidase cytochemistry using the DAB method. In cells such as blood eosinophils[49] and parotid[50] or lacrimal[51] acinar cells, that synthesize peroxidase as one of their main secretory products, it was demonstrated that all ER and Golgi cisternae plus transitional elements and associated vesicles on the *cis* side of the Golgi, as well as condensing vacuoles and granules, lie along the secretory route. This corroborative evidence obtained via a different approach was very important, because although it may be hard to believe at present, the idea that there is compartmentalization of the secretory pathway was new and rather heretic at the time (early 1970s) that it was introduced. There were those who strongly believed that secretory proteins are transported through the cytosol prior to their secretion from the cell. Although the combined autoradiographic and cell fractionation approach provided the first indication of the vectorial routing of secretory proteins and of the role of clusters of smooth vesicles and transitional elements of the ER in ER-to-Golgi transport, neither technique had sufficient resolution to accurately define whether most, all, or only some of the pre-Golgi vesicles and Golgi cisternae were involved.

7. Electron-Dense Tracers Provided the First Evidence for Recycling of Membranes between the Plasmalemma and the Golgi Complex

In the mid-1970s, it was commonly believed that all macromolecules taken up by endocytosis were delivered to endosomes and lysosomes and that recycling occurred exclusively between the plasmalemma and these compartments.[52] This conclusion was reached based on the finding that when cultured cells such as fibroblasts and macrophages were incubated with electron-dense tracers such as horseradish peroxidase (HRP), the tracers were taken up primarily into these organelles. However, later on it was shown that when secretory cells such as neurons,[53] parotid cells,[54] anterior pituitary cells[55] (Figure 8), and plasma cells[56] were incubated with a variety of tracers such as dextran, HRP, HRP-tagged lectins, and cationized ferritins, these tracers were taken up into Golgi cisternae. In most cases uptake was primarily into the *trans*-cisternae, but in at least some cases the tracer was detected more widely across the Golgi stack and was even found to be packaged into granules[55] (Figure 8). From these observations it was concluded that (1) endocytosed macromolecules can reach the Golgi under some circumstances, and (2) since the tracers are packaged with the granules that are eventually discharged by exocytosis, membrane must recycle from the plasma membrane to the Golgi and back. Later on, further evidence for existence of recycling was obtained when it was shown that transferrin receptors tagged with antibodies at the plasma membrane reach the Golgi[57] and are resialylated.[58] Similar findings have also been obtained in the case of mannose 6-phosphate receptors.[59]

D. CONTRIBUTIONS OF IMMUNOELECTRON MICROSCOPY

The rapid advances in the development of techniques for immunoelectron microscopy that have been made in the last 10 years have made possible the application of these methods to Golgi research. The impact of information obtained by immunolocalization of proteins at the electron microscope level has been spectacular. It has opened new vistas and made possible the integration of biochemical and morphologic information that was not possible before. Some of the concepts on which information gathered by these approaches has had an impact are given below.

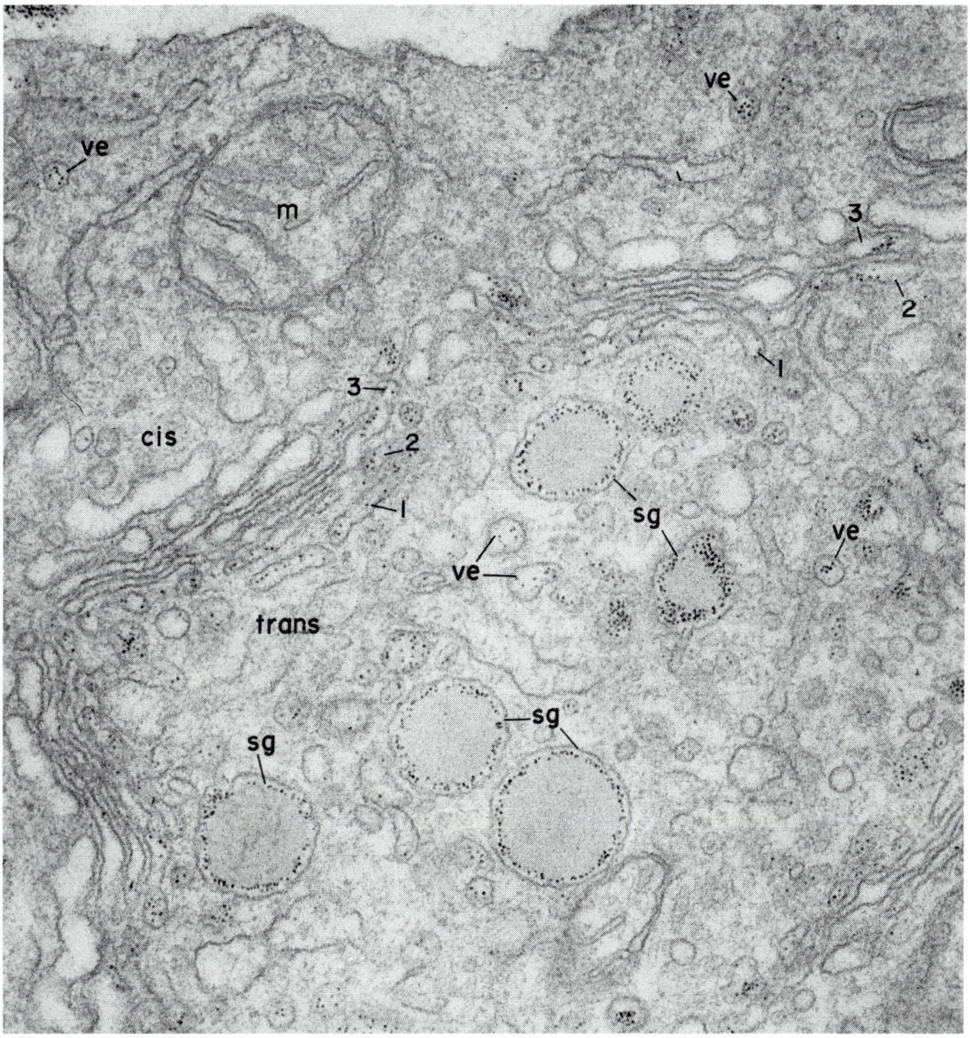

FIGURE 8. Recycling of membranes to the Golgi complex. Cultured somatotroph or growth hormone-secreting cell (rat) incubated 60 min in cationized ferritin (CF), 0.05 mg/ml. CF is taken up in small vesicles (ve), multiple Golgi cisternae, and secretion granules (sg). Note that the CF is most abundant in the three cisternae (**1–3**) on the *trans* side of the Golgi stack, indicating that the incoming vesicles carrying the CF fuse preferentially with the *trans*-cisternae. The tracer is packaged along with growth hormone into the forming granules and is located between the granule membrane and its dense contents. (Magnification × 67,200. From Farquhar, M. G., *J. Cell Biol.*, 77, R35, 1978. With permission.)

1. Direct Demonstration of Sequential Transport of Membrane Glycoproteins across the Golgi Stack

As mentioned earlier, it is now accepted that membrane, lysosomal, and secretory proteins all travel through the Golgi complex on their way to the plasma membrane along the so-called exocytic route. The assumption that membrane proteins follow the same route was initially based on the fact that several viral membrane glycoproteins acquire terminal sugars, a posttranslational modification known to occur exclusively in the Golgi. Viral glycoproteins have provided a very valuable model for studies on membrane biogenesis in general and for demonstration of sequential transport through intracellular compartments in particular. The first direct evidence that membrane proteins follow the secretory pathway came from immunocytochemical studies demonstrating that upon synchronization of expression, vesicular stomatitis virus (VSV) G-glycoprotein[60] and the Semliki Forest virus spike glycoprotein[61] appear first in the ER and then in the Golgi apparatus prior to delivery to the plasma membrane. The latter work was also supported by parallel biochemical studies demonstrating a temporal correlation between oligosaccharide maturation and transport of Semliki Forest virus spike proteins from the ER to the Golgi.[62] Subsequently, Bergmann and Singer,[63] using a temperature-sensitive mutant of VSV, carried out another immunocytochemical study in which it was determined that G-protein leaves the ER in 50- to 70-nm electron-dense vesicles which reach the Golgi complex about 3 min after the temperature shift. At this time the majority (70%) of the cisternae showed asymmetric labeling of only the *cis* or entry face, whereas by 4 min a broader distribution over the Golgi stack was observed. These experiments demonstrated (1) the *cis-trans* migration of membrane proteins through the Golgi and (2) the rapidity with which at least some membrane proteins traverse the stack. Confirmation of the passage of G-protein through the Golgi in transit to the plasma membrane was also obtained by autoradiography.[64] It was further demonstrated by Strous et al.[65] by double immunogold labeling of human hepatoma cells that secretory proteins (albumin and transferrin) following the constitutive pathway and membrane glycoproteins (VSV G-protein) destined for the plasma membrane travel through the same Golgi compartments and are delivered to the cell membrane in the same carrier vesicles.

2. Localization of Glycoprotein-Processing Enzymes in Golgi Subcompartments

a. Localization of Terminal Transferases to the **Trans-***Golgi*

In 1982, Roth and Berger[66] described the ultrastructural immunolocalization of UDP-gal:β-D-*N*-acetylglucosamine β1-4 galactosyltransferase in 2-3 *trans*-Golgi cisternae of HeLa cells by immunogold labeling of Lowicryl-embedded specimens (Figure 9A). The cisternae were identified as *trans*-cisternae by the simultaneous demonstration of TPPase activity in the same section. Galactosyltransferase had been shown some time before to be a reliable biochemical marker for Golgi fractions,[37] but it was not until the immunocytochemical study that this enzyme was established beyond any reasonable doubt to be restricted to the Golgi and to be concentrated in a specific *(trans)* subcompartment of the organelle in many cell types.[8] This finding represented the first direct demonstration of a functionally defined Golgi subcompartment and greatly strengthened the idea that the Golgi complex is organized into a series of functionally distinct subcompartments. Moreover, its spatial colocalization with TPPase activity suggested a possible concerted action of the two enzymes with TPPase (a nucleoside diphosphatase) serving to hydrolyze UDP generated in the transfer by galactosyltransferase of galactose from a UDP-galactose donor to oligosaccharides.[16,17] The localization of galactosyltransferase proved the ability of the immunocytochemical methods to resolve different Golgi subcompartments *in situ* and provided a key piece of evidence that made possible integration of morphologic and biochemical data[62,67] on the compartmentalization of oligosaccharide processing in the Golgi.

Not long after, Roth et al.[68] localized Galβ1,4-GlcNAc α2,6-sialyltransferase in both the *trans*-most cisterna of the stack and the transtubular network of rat hepatocytes (Figure 9B). The site of immunolabeling coincided with the cytochemical label for sialic acid demonstrated with *Limax flavus* lectin. It is disappointing that both the galactosyl- and sialyltransferase antibodies proved to be highly species specific (see Table 1). Thus, their application as markers for the *trans*-Golgi elements has necessarily been limited.

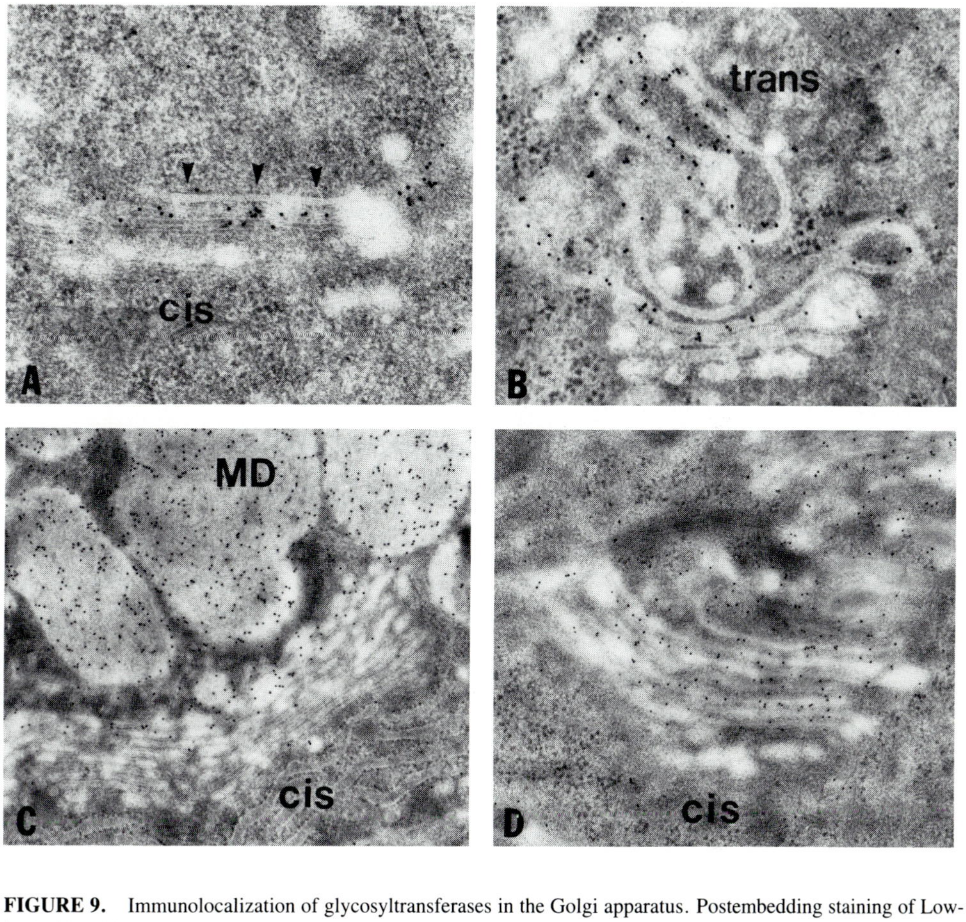

FIGURE 9. Immunolocalization of glycosyltransferases in the Golgi apparatus. Postembedding staining of Lowicryl K4M thin sections with the protein A gold technique. (A) Galactosyltransferase immunoreactivity is restricted to *trans*-cisternae of the Golgi apparatus in HeLa cells. The *trans*-most Golgi cisterna, or so-called "rigid lamella" (arrowheads), is unlabeled. (B) Immunolabeling for sialyltransferase is present over the *trans*-cisternae and the *trans*-Golgi network. (C) In the Golgi apparatus of intestinal goblet cells, immunolabeling for sialyltransferase obtained with polypeptide epitope-specific sialyltransferase antibody is restricted to *trans*-cisternae. It is also present in mucus droplets (MD). (D) Sialyltransferase immunoreactivity as detected with polypeptide epitope-specific antibody is present throughout the Golgi stack of absorptive enterocytes. (Magnifications: A, × 56,000; B and D, × 40,000; C, × 19,000. Figure provided by Dr. Jürgen Roth.)

b. Other Glycoprotein-Processing Enzymes

In addition to the *trans*-Golgi enzymes, β1,4-galactosyltransferase and α2,6-sialyltransferase, two other Golgi enzymes, α-mannosidase II[69] and GlcNAc transferase I,[70] have also been purified and localized (see Table 1). The latter was localized to the middle Golgi cisternae in hepatocytes and kidney cells. α-Mannosidase II was originally localized all across the stack in hepatocytes,[71] but a more restricted localization was reported by others.[1,72] These studies, together with data on the distribution of enzymatic activities in Golgi subfractions,[62] have provided the basis for our current model describing the compartmentalized organization of N-glycosylation reactions across the Golgi stack.[34]

By contrast, very little is known about the localization of steps in the biosynthesis of O-linked glycoproteins. This is due to the fact that data obtained by cell fractionation is conflicting[1] and the lack of availability of monospecific antibodies against the glycosyltransferases involved. The only enzyme participating in O-glycosylation that has been

TABLE 1
Antibodies to Golgi Glycoprocessing Enzymes

Antibody	Nature	Species/cell-type specificity	Antigen source	Antigen nature	Localization	Ref.
Anti-β-1-4 galactosyl-transferase	Polyclonal, rabbit	Human: HeLa, CaCo-2, fibroblasts	Galactosyltransferase purified from human milk	54-kDa O-linked trans-membrane glycoprotein	*trans*-Golgi	66, 164
Anti-GlcNAc I trans-ferase I	Monoclonal, mouse	Rabbit: LLC-RK1 and hepatocytes	GlcNAc I transferase purified from rabbit liver acetone powder	44-, 45-kDa hydrophobic doublet sometimes associated with 32-, 55-kDa bands	medial Golgi	70
Anti-α-2-6 sialyltrans-ferase	Polyclonal, rabbit	Rat: hepatocytes, goblet cells. Bovine: fibroblasts	α-2-6 Sialyltransferase purified from rat liver microsomes	47-kDa transmembrane glycoprotein	*trans*-Golgi, TGN	68
Anti-α-mannosidase II	Polyclonal, rabbit	Rodent (rat, mouse), human	α-Mannosidase II purified from rat liver Golgi membranes	124-kDa transmembrane glycoprotein	middle and *trans*-Golgi	69
Anti-α-mannosidase IA	Polyclonal, rabbit	Rat	α-Mannosidase IA purified from rat liver salt washed Golgi membranes	58-kDa transmembrane glycoprotein, not terminally glycosylated	?	165
Anti-α-1-3 GalNAc transferase	Polyclonal, rabbit	Human intestine	α-1-3GalNAc transferase purified from human group A, plasma	Unknown	*trans*-Golgi/TGN; mucus droplets	166

localized by immunocytochemistry is a terminal $\alpha 1,3$-*N*-acetylgalactosaminyltransferase which functions in the synthesis of the blood group A carbohydrate sequence.[73] Interestingly, this enzyme has been localized to the *trans*-Golgi region of Goblet cells of the intestine but was found to have a broader distribution across the Golgi stack in the intestinal epithelial cell. Indirect evidence on O-glycosylation was also obtained using lectin probes specific for O-linked sugars.[74] Obviously, the list of Golgi enzymes that have been localized to date is quite incomplete. This is due to the fact that relatively few Golgi enzymes have been purified, and hence relatively few antibodies are available. Clearly this represents a major challenge for the future.

c. *Cell Type Variations in the Distribution of Golgi Markers*

Immunocytochemical and cytochemical studies have also exposed some complications in our current dogmas concerning Golgi subcompartments, in that a number of findings to date indicate that there are variations in the distribution of Golgi markers among different cell types or in the same cell type under different conditions (see References 8 and 75 for further discussion of this problem). To begin with, it is clear that in some cell types, terminal transferases can be present at the plasma membrane. For example, sialyltransferase has been localized to *trans*-cisternae and mucus droplets (Figure 9C) and plasma membranes in intestinal goblet cells using polypeptide epitope-purified antibody.[75] Second, variations have been documented among different cell types in the distribution of $\alpha 2,6$-sialyltransferase and blood group A, *N*-acetylgalactosaminyltransferase within Golgi subcompartments.[73] In addition, the distribution of Golgi cytochemical markers can apparently vary according to the state of differentiation of the cells, as has been shown for intestinal epithelial cells.[73,76] In the intestine it is restricted to *trans*-cisternae of goblet cells (Figure 9C) whereas it is found throughout the Golgi stack in absorptive cells (Figure 9D).

Two important lessons have been learned from the immunocytochemical demonstration of glycosyltransferases that were not apparent from cell fractionation studies. One already referred to is the fact that the subcompartment organization of the Golgi is not as rigid as implied in most diagrams. In particular, the distribution of certain Golgi resident proteins across the Golgi stack can be different from one cell type to another. The other is the importance of working with well-characterized immunological reagents[75] to avoid misinterpretation of the immunocytochemical results as a consequence of the antibodies cross-reacting with other than protein epitopes (e.g., carbohydrate epitopes).[77]

3. Identification and Localization of Other Golgi Resident Proteins

a. *Immunoelectron Microscopy has been Used to Localize Golgi Antigens in Golgi Subcompartments*

A number of monoclonal and polyclonal anti-Golgi antibodies have been generated using a variety of immunogens that recognize antigens localized to specific Golgi subcompartments. Unfortunately, many of these reagents are of limited use for immunoelectron microscopy, primarily due to restrictions in their species and/or cell lineage specificity and their ability to recognize aldehyde-fixed antigens in a tissue section or cultured cell. The first of those that proved to be a very useful reagent resulted from immunization of rabbits with a Golgi membrane fraction prepared from rat liver.[78] The antisera, when affinity purified by absorption of unwanted reactivities on membrane fractions of other cellular components, recognized a 135-kDa membrane glycoprotein which was shown to be broadly distributed across the Golgi stacks. Subsequently, a monoclonal antibody to this protein was prepared using a similar immunogen and localized to all but the *cis*-most cisternae of the Golgi stacks.[79] Both of these reagents have been widely used to distinguish Golgi from other cellular compartments and to demonstrate a Golgi location for putative Golgi proteins in immuno-

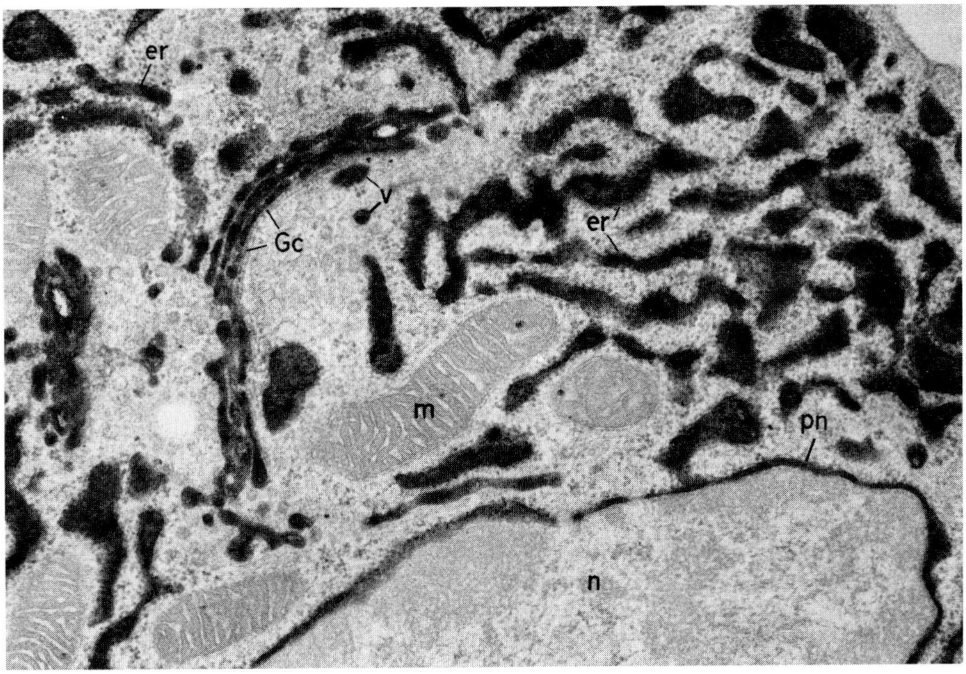

FIGURE 13. Immunocytochemical localization of IgG-containing compartments along the constitutive secretory pathway in a plasma cell. Spleen cells were harvested from a rat immunized against horseradish peroxidase (HRP) and lightly fixed; cryostat sections were incubated with HRP and subsequently reacted with diaminobenzine (DAB). Anti-HRP immunoglobulins are seen throughout the rough ER (er), including the perinuclear cisterna (pn), and in the stacked Golgi cisternae (Gc) and associated secretory vesicles (v). (Magnification × 16,700. From Ottosen, P. D. et al., *J. Exp. Med.*, 152, 1, 1980.

that passage along the constitutive secretory pathway occurs by default, whereas passage along the regulated pathway requires sorting.[113] Chung et al.[114] have isolated a series of ~25-kDa proteins which have the properties expected of sorting carriers for regulated secretory proteins. The 25-kDa proteins were shown to be concentrated in the Golgi region of AtT-20 and GH_3 pituitary cells by immunofluorescence, but they have not yet been localized at the EM level.

b. Sorting of Plasma Membrane Proteins for Polarized Delivery to the Apical and Basolateral Domains of the Plasmalemma

Polarized epithelial cells are known to possess apical and basolateral domains of distinctive protein and lipid composition.[115,116] Transport of integral plasma membrane proteins from the *trans*-Golgi to the cell surface occurs via vesicular carriers which have been identified by immunocytochemical studies as a class of 100 to 200-nm, noncoated vesicles[63,112,117,118] that have been isolated recently.[119] The vesicles appear to bud from regions of the TGN that are not coated with clathrin,[118,120] and they are considered to be the bulk-flow carriers for delivery of components to the plasma membrane. In polarized cells, vesicles with different compositions and targeting specificities must exist in order to maintain the differences in protein and lipid composition between apical and basolateral plasma membrane domains.[115,116,121] Once again, the use of viral proteins combined with immunoelectron microscopy proved to be fruitful to study sorting of membrane proteins for apical and basolateral delivery. Rindler et al.[122,123] set out to determine the intracellular route taken by

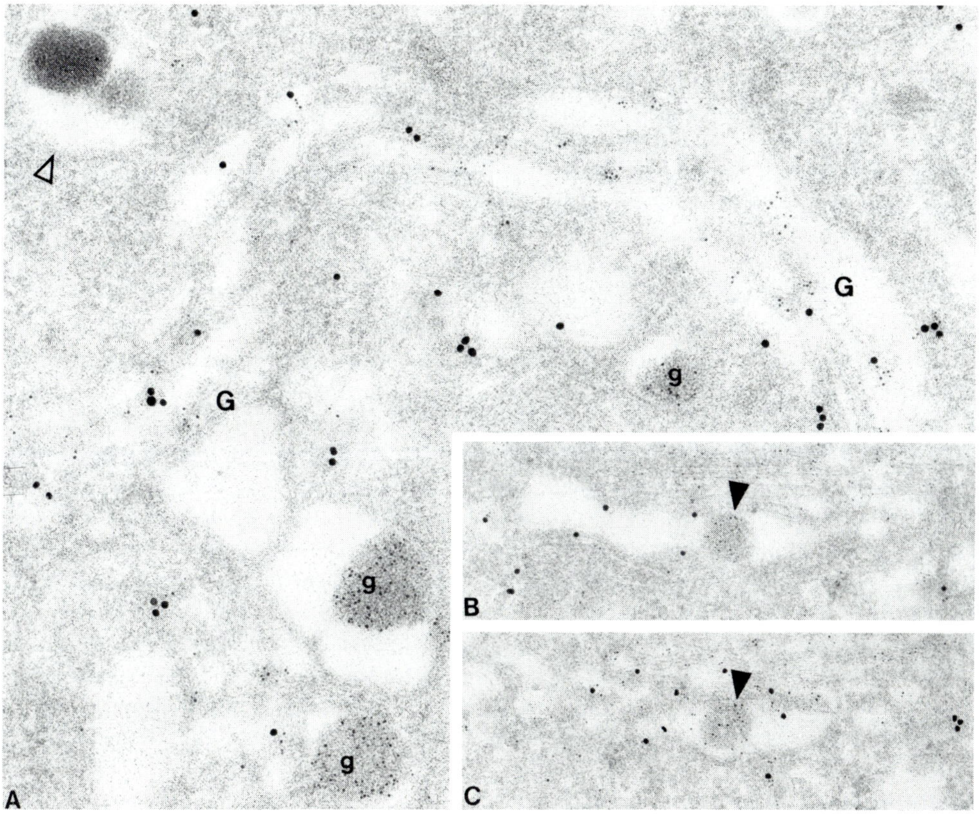

FIGURE 14. Colocalization of a membrane protein (influenza hemagglutinin) and a secretory protein (insulin) in Golgi cisternae. Isolated islets from neonatal rat pancreas were infected with influenza virus and doubly labeled with anti-proinsulin and anti-hemagglutinin (HA) antibodies followed by protein A gold. Tissue was fixed in glutaraldehyde and embedded in Lowicryl K4M at low temperature. Thin sections were labeled simultaneously with anti-proinsulin and anti-HA antibodies. (A) Proinsulin (5-nm gold) and HA (15-nm gold) immunoreactive sites codistribute throughout the cisternae of the Golgi stack. (G). Proinsulin is concentrated over three condensing secretory granule cores (g); (B) and (C) show two consecutive serial sections of a *trans*-most condensing Golgi cisterna. HA labeling (15-nm gold) is present on the cisternal membrane, while the dense secretory granule core (black arrowheads) is stained by proinsulin antibodies. (Magnification × 62,900; insets × 45,100. From Orci, L. et al., *Cell*, 51, 1039, 1987. With permission.)

newly synthesized VSV G-protein delivered to the basolateral plasmalemma, and influenza hemagglutinin, delivered to the apical plasmalemma in MDCK cells coinfected with both viruses. The two proteins were colocalized throughout the same individual Golgi stack, indicating that sorting occurs after traversing this organelle and before delivery to their respective plasma membrane surface. Biochemical evidence has also been obtained showing that VSV G-protein destined for the basolateral cell surface and influenza neuraminidase destined for the apical cell surface exist together in the same *trans*-Golgi compartment.[124] These studies led to the conclusion that the TGN must be the sorting site where membrane proteins destined for different domains of the plasma membrane are segregated into separate vesicle carriers.[3] It is clear, however, from other studies (reviewed in Reference 115) that the routes of delivery (direct or indirect) differ from one cell type to another.

In addition to membrane proteins, polarized cells are also able to segregate intracellularly newly synthesized secretory products and release them in a polarized fashion.[125] This process has not yet been followed by immunocytochemistry. The signals and receptors responsible for polarized sorting of both membrane and secretory proteins are under active investigation but have not yet been characterized, with the possible exception of apical proteins in MDCK cells. In the case of the latter, there is evidence that anchoring to glycosyl-phosphatidylinositol determines targeting,[115] in at least some cases.

c. Sorting of Lysosomal Enzymes from Other Proteins

It is well established that sorting of most lysosomal enzymes from other proteins comes about through the interaction between the mannose 6-phosphate recognition marker on lysosomal enzymes and mannose 6-phosphate receptors (MPR).[126,127] Two such receptors have been purified and sequenced, the 215-kDa MPR and the 46-kDa MPR. Once again, immunocytochemistry has provided important information crucial to our understanding of the trafficking of lysosomal enzymes and the recycling itinerary of their receptors. Lysosomal enzymes have been found to be most concentrated in lysosomes, but in several studies[106,128,129] they were also localized in the Golgi complex, indicating that they follow the same intracellular route as secretory proteins. The same localization applies to acid phosphatase,[130] which is atypical in that it does not contain the recognition marker, and its transport to lysosomes does not involve either MPR.

Regarding the distribution of the receptors, immunocytochemical studies from several laboratories have established that the 215-kDa MPR, the best studied of the two receptors, is found in three locations — in the Golgi (Figures 15 and 16), in coated vesicles, and in endosomal structures[131-135] — which were therefore assumed to represent the sorting site, the carriers, and the delivery site, respectively, for lysosomal enzymes.[131] More recently, the delivery site has been identified as a late endosome or prelysosome.[134-136] In the majority of cell types studied up to now, the receptor is concentrated within the TGR or TGN of the Golgi. However, for unknown reasons, variations in receptor distribution within the Golgi have been encountered in some cell lines in culture as determined by immunocytochemistry in combination with cell fractionation.[137]

It has been proposed that lysosomal enzymes bind to their receptors that cluster into coated pits exclusively in the TGR or TGN.[4,134,135] Several immunocytochemical findings lend credence to this concept. In one study, clathrin-coated buds showed a higher concentration of MPR than the remainder of the TGR/TGN membrane, which lacks a clathrin coat.[134] In another study, it was shown that membrane proteins in transit to the cell surface appear to be excluded from the coated buds.[118]

Relatively little work has been done to date on the 46-kDa receptor. However, immunofluoresence[138,139] and immunoelectron microscopic,[139] as well as cell fractionation[140] studies suggest that its distribution overlaps considerably with the 215-kDa MPR in that it is also found in the Golgi and in late endosomes. However, there are subtle differences in the distribution of the two receptors within the Golgi in the same cell type, suggesting that the two receptors play somewhat different roles in the sorting and trafficking of lysosomal enzymes.[139]

7. Characterization of Pre-Golgi Compartments

In the last few years there has been increasing interest in the nature and function of a possible intermediate or pre-Golgi compartment(s) situated between the ER and the *cis* side of the Golgi complex. The first evidence suggesting the existence of such a compartment was obtained from the immunocytochemical observations of Saraste and Kuismanen[141] in cultured BHK cells infected with Semliki Forest virus. When transport to the cell surface was arrested by incubating the cells at 15°C, the viral glycoprotein detected by immunoperoxidase accumulated in dilated, 300- to 400-nm, smooth membrane-bound vesicles and

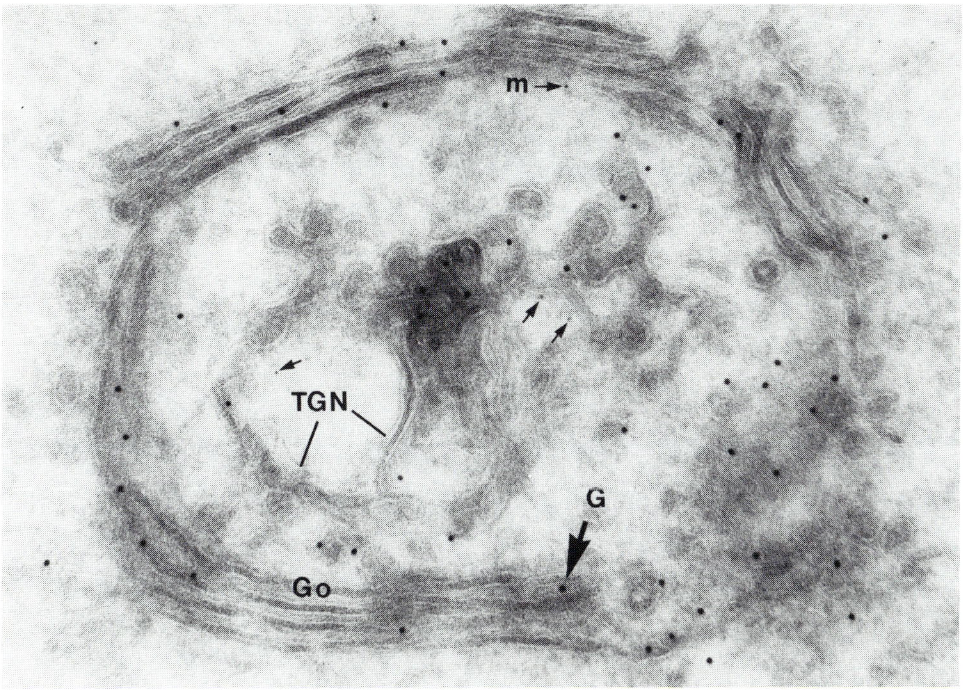

FIGURE 15. Double immunogold labeling for VSV G-protein and 215-kDa mannose 6-phosphate receptors (MPR). NRK cells were infected with a VSV temperature-sensitive mutant (Ts-045) at 37°C to prevent transport from the Golgi to the cell membrane and shifted to 18°C to allow transport through the Golgi. Cells were fixed in 8% paraformaldehyde in 200 μ*M* Pipes buffer (pH 7.0). Cryosections were prepared and labeled with anti-MPR and anti-G protein and detected with protein A gold of two different sizes. Under these circumstances (viral infection, 18°C), the TGN is significantly enlarged. Both the Golgi stack (Go) and the TGN contain high concentrations of G protein (G) (9-nm gold) and lower concentrations of MPR (5-nm gold), indicated by arrows. (Magnification × 60,200. From Griffiths, G. et al., *Cell,* 52, 329, 1988. With permission.)

protruding tubules. Tooze et al.[92] also described the budding of coronavirus particles into a smooth membrane compartment located between the ER and the Golgi stack. Proteins believed to represent immunocytochemical markers for pre-Golgi tubulo-vesicular structures located in proximity to the *cis*-Golgi (Figure 10) have been identified by Saraste et al.[86] and Schweizer et al.[illegible] In the latter case, the protein marker p53 has been used to identify the compartment in which VSV G-protein is arrested during a 15°C temperature block, providing evidence that such an intermediate compartment is a normal, preGolgi station in the secretory pathway.[87] Interest in such a compartment was enhanced by Pelham's proposal[142,143] that luminal residents of the ER continuously leave the ER and are recycled back from a so-called intermediate or ''salvage'' compartment via a putative KDEL receptor. KDEL, a carboxy-terminal sequence characteristic of some luminal residents of the ER, is believed to represent an ER retention signal. Additionally, Vaux and co-workers,[144] by generating anti-idiotype antibodies to KDEL, have isolated a 72-kDa protein that has many of the properties expected of a putative KDEL receptor. The immunofluorescence staining pattern of this protein is that anticipated for the salvage compartment, i.e., it does not totally correspond to either the ER or the Golgi complex and displays modifications during 15°C and brefeldin A incubations.[144] The putative receptor has not yet been localized at the EM level. At present, the possible structural and/or functional relationship between the

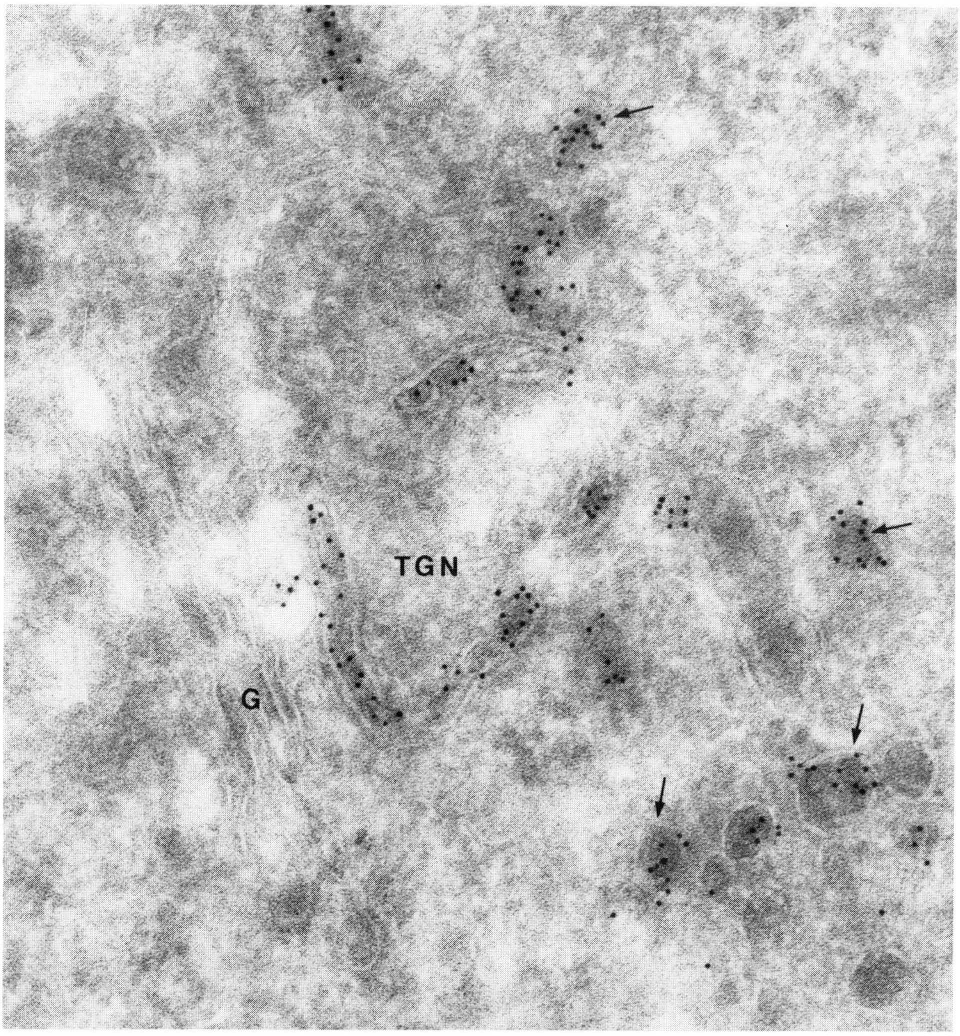

FIGURE 16. Profile of the *trans*-Golgi reticulum or network (TGR) located adjacent to a Golgi stack (G) demonstrating the localization of the 215-kDa mannose-6-phosphate receptor (MPR) by immunogold labeling of a human hepatoma (HepG2) cell. MPR are found in the TGR and in vesicles (arrowheads) which are thought to derive from the TGR. (Magnification × 98,700. From Geuze, H. J. and Morré, D. J., *J. Electron Microsc. Tech.*, 17, 24, 1991. With permission.)

intermediate or salvage compartment and the smooth membrane-bound ER subcompartments (transitional elements [Figure 1], tubular structures, smooth cisternae) present in certain cell types[145,146] (Figure 17) is not clear and remains to be determined. Immunoelectron microscopic studies are required to resolve this issue.

Effects of brefeldin A — Indirect evidence that a Golgi-ER recycling pathway normally operates in eukaryotic cells has been obtained in studies with brefeldin A. Initial morphological[147] and biochemical[148] reports had indicated that this agent causes a reversible inhibition of the intracellular transport of secretory proteins by promoting selective disruption of the Golgi complex. Lippincott-Schwartz and co-workers[150] have investigated the effects of brefeldin A on the morphology of cultured cells by immunocytochemistry using Golgi-

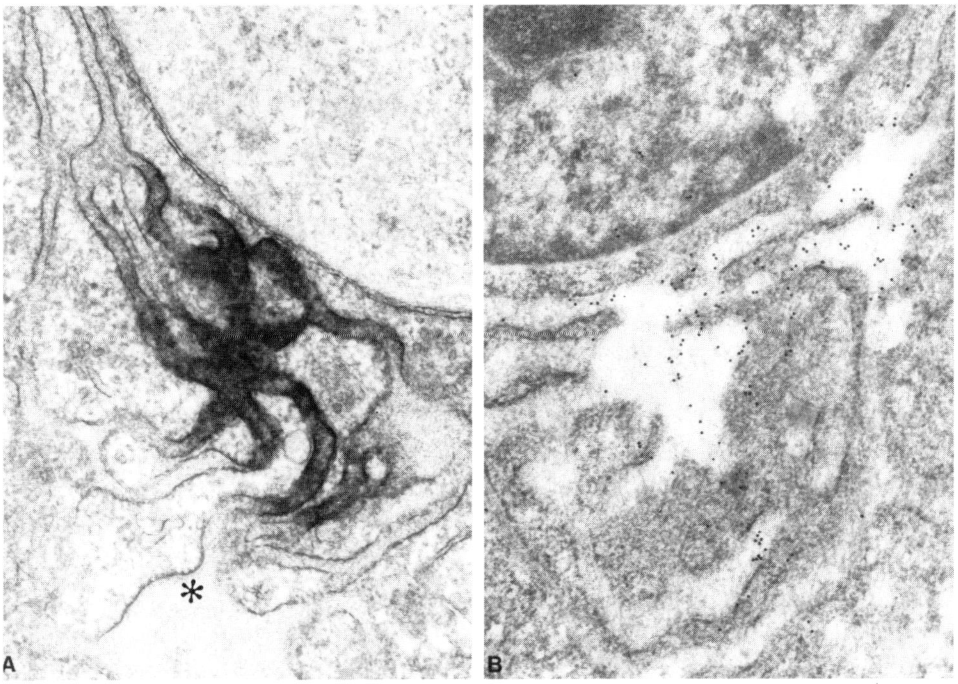

FIGURE 17. Accumulation of proteoglycan precursors in a smooth ER subcompartment of chondrocytes. In cultured chondrocytes the ER is clearly compartmentalized, as indicated by the detection of proteoglycan precursors by either immunoperoxidase (A) or with a protein A-gold postembedding technique on Lowicryl sections (B). The ER subcompartment where proteoglycan precursors accumulate is a specialized, smooth membrane-bound region consisting of anastomosing tubules with a content of lower density than that of the remaining rough ER. However, it is continuous with typical rough ER cisternae (*). (Magnifications: A, × 28,400; B, × 34,800. From Vertel, B. M. et al., *J. Cell Biol.*, 109, 1827, 1989. With permission.)

associated markers (p53,[85] α-mannosidase II,[69] and β-COP [coat protein][149]). They obtained results indicating that in the presence of brefeldin A, the Golgi complex is disorganized into a population of vesicles and tubules which fuses with the ER.[72] This retrograde movement of Golgi proteins is believed to involve all Golgi resident proteins (except those of the TGN) which rapidly redistribute to the ER and remain enzymatically active. The concept that they have developed, for which there is mostly indirect evidence to date, is that brefeldin A selectively disrupts the anterograde arm of vesicle cycling that normally exists between the ER and the *cis*-Golgi but does not affect the retrograde part of the cycle.[72] The time course of brefeldin A treatment and reversal,[72,84] as well as the energy, temperature, cytoskeletal, and nucleotide requirements for the retrograde movement,[150,151] have been studied by immunofluorescence. Of the markers used, only α-mannosidase II has been localized at the EM level in brefeldin-treated cells and has served to document the disruption of the Golgi cisternae.[72] Further immunoelectron microscopic studies are needed to verify the basis for these concepts and to precisely identify the compartments involved.

8. Coated Vesicles and Vesicle Coats

There is now considerable evidence that transport of membrane and secretory proteins along the exocytic pathway is mediated by vesicular carriers.[9,11] Morphologically, two populations of coated vesicular carriers can be distinguished. One type of vesicle associated with the *trans* side of the Golgi possesses characteristic cage-like cytoplasmic coats resembling clathrin cages. Verification of the clathrin nature of the coats was obtained by immunogold labeling[152] with an affinity-purified polyclonal antibody that recognized clathrin polymerized into coats. Clathrin-immunoreactive sites were not uniformly distributed throughout the Golgi apparatus: they were concentrated in the *trans*-cisternae, the *trans*-Golgi network, maturing secretory granules, and typical coated vesicles in polypeptide hormone-secreting cells. Other vesicles, such as those associated with the *cis*- and medial-Golgi cisternae, transitional elements, and ER, were coated but their coats were not organized into clathrin cages,[153] and they did not stain for clathrin.[154,155] This implied that there is a distinct population of nonclathrin-coated vesicles involved in ER-to-Golgi transport and intracisternal Golgi transport. These vesicles have recently been isolated and found to have a different protein composition from either Golgi or ER membranes.[156] Very recently, one of the main components of the nonclathrin coats has been identified as a protein known as β-COP (coat protein), a 110-kDa peripheral Golgi membrane protein.[149,157] Immunogold electron microscopy was used to confirm that β-COP is present on nonclathrin-coated vesicles and cisternae of the Golgi complex,[149] where it was concentrated along the dilated rims and budding profiles (Figure 18) of Golgi cisternae. Clathrin-coated vesicles function in receptor-mediated endocytosis[158] and also serve as carriers for lysosomal enzymes from the *trans*-Golgi/*trans*-Golgi network to endosomes.[4,118,131,132] It has been proposed that nonclathrin-coated vesicles serve as "bulk carriers" for membrane and secretory proteins.[7] A future contribution of electron microscopic immunocytochemistry to our understanding of transport through the Golgi complex may come from double immunogold labeling studies using probes of bulk vs. selective vesicle cargo to test the "bulk flow" hypothesis at the ultrastructural level.

9. GTP-Binding Proteins and Golgi Traffic

As already discussed, transport of proteins and lipids between the ER and Golgi, between adjacent Golgi cisternae, and between the Golgi and plasma membrane is believed to be mediated by vesicular carriers. Despite the generality of this mechanism, the principles governing the flow of vesicular traffic remain poorly understood at the molecular level. Identification in *Saccharomyces cerevisiae* of two small ras-related GTP-binding proteins, Sec4[159] and Ypt1,[160] which control distinct stages of the secretory pathway, as well as studies on the effects of the GTPγS, a nonhydrolyzable GTP analog, on different transport events in mammalian cell systems[9] have led to the hypothesis that small GTP-binding proteins may mediate vesicular traffic along the secretory pathway. The working model predicts that GTP hydrolysis triggers conformational switches promoting unidirectional movement of transport vesicles from a donor compartment to an acceptor compartment.[160,161]

Immunocytochemistry has provided important new information on the distribution of these proteins. The Ypt1 gene product and its mammalian counterpart, implicated in the regulation of vesicular traffic between the ER and Golgi, were localized by immunofluorescence to Golgi-like structures in yeast and to the Golgi region of mouse cells.[160] By immunoperoxidase localization at the EM level another small GTP-binding protein, rab6, which is highly homologous to both Ypt1 and Sec4,[162] was found to be associated with the cytoplasmic surface of the middle-Golgi cisternae in mammalian cells.[88] Three additional members of the ras family, rabs 2, 5, and 7, were localized by immunofluorescence and immunogold techniques in MDCK cells and found to be associated with different compartments — rab2 with an intermediate compartment between the ER and Golgi, rab5 with the cytoplasmic surface of the plasma membrane and early endosomes, and rab7 with late endosomes. ADP-ribosylation factor, another GTP-binding protein not closely related to either Ypt1 or Sec4, was recently localized by immunoferritin electron microscopy to the cytosolic face of predominantly *cis*-Golgi membranes.[163] Localization by immunoelectron

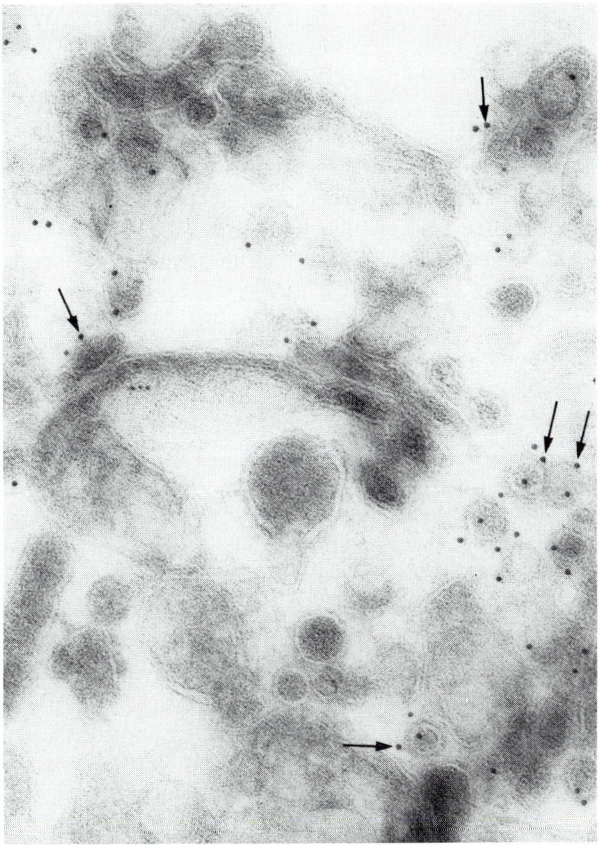

FIGURE 18. Immunogold labeling of β-COP in GTPγS-treated Golgi fractions. Golgi fractions were prepared from rat liver and incubated with an ATP-generating system in the presence of GTPγS and processed for ultrathin cryosectioning as described in Reference 149. Ultrathin cryosections were incubated with anti-β-COP followed by 9-nm protein A gold. β-COP was preferentially found on coated vesicles close to the Golgi cisternae (arrows) and on profiles of vesicles presumably budding from the Golgi cisternae. (Magnification × 85,000. Figure provided by Dr. Gareth Griffiths.)

microscopy of different GTP binding in association with different intracellular compartments has led to the conclusion that specific GTP binding proteins may be required at each step in vesicular transport, and each GTP-binding protein has a distinctive distribution and distinct function.

E. CONCLUDING REMARKS: PROSPECTS FOR THE FUTURE

In this review we have attempted to provide an overview of the key role electron cytochemical and immunocytochemical findings have played in reaching our current understanding of the organization and function of the Golgi apparatus. We have attempted to point out how at every juncture morphological findings obtained with these techniques have provided important insights that provided the basis for our working models. The question that arises is, what are the prospects for the continued contribution of cytochemistry and, particularly, immunoelectron microscopy in this field? The answer is that due to the very nature of this organelle, immunocytochemistry is such a powerful tool that it has become a required component of any comprehensive study of Golgi structure and function. In effect, the situation at present requires that biochemical (cell fractionation) data and immunochemical (immunoprecipitation, immunoblotting) data be supplemented with morphologic data obtained by immunocytochemical localizations wherever possible. This is because it provides information that not only complements and extends results obtained by biochemical and immunochemical methods, but also it provides information on the precise distribution of protein antigens in Golgi subcompartments that cannot be obtained by any other existing method, including current methods of cell fractionation involving separations based on physical parameters. Also, in spite of the present trend in the field, immunofluorescence cannot replace immunoelectron microscopy for this purpose. A so-called ''Golgi'' signal by immunofluorescence can be deceiving, as many organelles such as lysosomes and endosomes are concentrated in the Golgi region. Moreover, to reliably localize an antigen in a specific Golgi subcompartment is virtually impossible at the light microscope level. This means that immunoelectron microscopic studies will be essential for future progress in the field. It has been pointed out that very few Golgi resident proteins have been characterized to date. Only six glycosylation enzymes and a dozen or so proteins of unknown function have yet been purified and localized. Contrast this with the list of proteins known to be present in the Golgi, which includes not only glycosyltransferases and enzymes involved in trimming of oligosaccharides, but also ion and nucleotide-sugar transporters, proteins involved in sorting, and those involved in vesicle transport, among others. In addition, many studies designed to determine the nature of signals involved in membrane trafficking and sorting involve construction (by recombinant DNA technology) of chimeric proteins which are expressed and localized in cultured cells. To accurately assess the localization of these chimeras will require immunoelectron microscopy.

For immunoelectron microscopy to have its greatest impact there are two points that should be kept in mind. The first is that there are a variety of methods available at present for carrying out immunocytochemical localizations which include immunofluorescence, immunoperoxidase, and immunogold labeling — the latter either on ultrathin cryosections (preembedding) or Lowicryl-embedded ultrathin sections (postembedding). Each of these approaches has advantages and disadvantages and has already made its contribution to our understanding of the Golgi as detailed herein Since these methods are complementary, more than one approach should be tried in the hope of obtaining convergent results in order to reach the most reliable conclusions. The other point to be made is that although we have stressed the value of immunoelectron microscopy in this field, we believe that biochemical, molecular biological, immunochemical, and immunocytochemical approaches should be combined wherever feasible, because the synthesis of all three approaches is much more powerful than their additive value.

ACKNOWLEDGMENT

This paper was supported by National Institutes of Health grants DK17780 and CA46128 (to MGF).

REFERENCES

1. **Farquhar, M. G.,** Progress in unraveling pathways of Golgi traffic, *Annu. Rev. Cell Biol.,* 1, 447, 1985.
2. **Farquhar, M. G.,** Protein traffic through the Golgi, in *Intracellular Trafficking of Proteins,* Steer, C. and Hanover, J., Eds., Cambridge University Press, p. 430, 1991.

3. **Griffiths, G. and Simons, K.,** The *trans* Golgi network: sorting at the exit site of the Golgi complex, *Science,* 234, 438, 1986.
4. **Geuze, H. J. and Morré, D. J.,** Trans-Golgi reticulum, *J. Electron Microsc. Techn.,* 17, 24, 1991.
5. **Bennett, G. and Wild, G.,** Traffic through the Golgi apparatus as studied by radiography, *J. Electron Microsc. Tech.,* 17, 132, 1991.
6. **Mollenhauer, H. H. and Morré, D. J.,** Perspectives on Golgi apparatus form and funtion, *J. Electron Microsc. Tech.,* 17, 2, 1991.
7. **Pfeffer, S. R. and Rothman, J. E.,** Biosynthetic protein transport and sorting by the endoplasmic reticulum and Golgi, *Annu. Rev. Biochem.,* 56, 829, 1987.
8. **Roth, J.,** Localization of glycosylation sites in the Golgi apparatus using immunolabeling and cytochemistry, *J. Electron Microsc. Tech.,* 17, 121, 1991.
9. **Rothman, J. E. and Orci, L.,** Movement of proteins through the Golgi stack: a molecular dissection of vesicular transport, *FASEB J.,* 4, 1460, 1990.
10. **Golgi, C.,** Sur la structure des cellules nerveuses, *Arch. Ital. Biol.,* 30, 60, 1898.
11. **Farquhar, M. G. and Palade, G. E.,** The Golgi apparatus (complex) — (1954–1981) — from artifact to center stage, *J. Cell Biol.,* 91, 77, 1981.
12. **Friend, D. S. and Murray, M. J.,** Osmium impregnation of the Golgi apparatus, *Am. J. Anat.,* 117, 135, 1965.
13. **Novikoff, A. and Goldfischer, S.,** Nucleosidediphosphatase activity in the Golgi apparatus and its usefulness for cytological studies, *Proc. Natl. Acad. Sci. U.S.A.,* 47, 802, 1961.
14. **Novikoff, A. B.,** The endoplasmic reticulum: a cytochemist's view, *Proc. Natl. Acad. Sci. U.S.A.,* 73, 2781, 1976.
15. **Novikoff, A. B. and Novikoff, P. M.,** Cytochemical contributions of differentiating GERL from the Golgi apparatus, *Histochem. J.,* 9, 525, 1977.
16. **Brandan, E. and Fleischer, B.,** Orientation and role of nucleoside diphosphatase and 5′-nucleotidase in Golgi vesicles from rat liver, *Biochemistry,* 21, 4640, 1982.
17. **Fleischer, B.,** Mechanism of glycosylation in the Golgi apparatus, *J. Histochem. Cytochem.,* 31, 1033, 1983.
18. **Novikoff, A. B., Mori, M., Quintana, N., and Yam, A.,** Studies of the secretory process in the mammalian exocrine pancreas. I. The condensing vacuoles, *J. Cell Biol.,* 75, 148, 1977.
19. **Smith, C. E.,** Ultrrastructural localization of nicotinamide adenine dinucleotide phosphatase (NADPase) activity in the intermediate saccules of the Golgi apparatus in rat incisor ameloblasts, *J. Histochem. Cytochem.,* 28, 16, 1980.
20. **Farquhar, M. G., Bergeron, J. J. M., and Palade, G. E.,** Cytochemistry of Golgi fractions prepared from rat liver, *J. Cell Biol.,* 60, 8, 1974.
21. **Cheng, H. and Farquhar, M. G.,** I. Presence of adenylate cyclase activity in Golgi and other fractions from rat liver. II. Cytochemical localization within Golgi and ER membranes, *J. Cell Biol.,* 70, 660, 1976.
22. **Meldolesi, J., Jamieson, J. D., and Palade, G. E.,** Composition of cellular membranes in the pancreas of the guinea pig, *J. Cell Biol.,* 49, 130, 1971.
23. **Amar-Costesec, A., Wibo, M., Thinés-Sempoux, D., Beaufay, H., and Berthet, J.,** Analytical study of microsomes and isolated subcellular membranes from rat liver, *J. Cell Biol.,* 62, 717, 1974.
24. **Borgese, N. and Meldolesi, J.,** Localization and biosynthesis of NADH-cytochrome b_5 reductase, an integral membrane protein, in rat liver cells, *J. Cell Biol.,* 85, 501, 1980.
25. **Orci, L., Montesano, R., Meda, P., Malaisse-Lagae, F., Brown, D., Perrelet, A., and Vassalli, P.,** Heterogeneous distribution of filipin-cholesterol complexes across the cisternae of the Golgi apparatus, *Proc. Natl. Acad. Sci. U.S.A.,* 78, 293, 1981.
26. **Lipsky, N. G. and Pagano, R. E.,** Intracellular translocation of fluorescent sphingolipids in cultured fibroblasts: endogenously synthesized sphingomyelin and glucocerebroside analogues pass through the Golgi apparatus en route to the plasma membrane, *J. Cell Biol.,* 100, 27, 1985.
27. **Pagano, R. E. and Sleight, R. G.,** Defining lipid transport pathways in animal cells, *Science,* 229, 1051, 1985.
28. **van Meer, G., Stelzer, H. K., Wijnaendts-van-Resandt, R. W., and Simons, K.,** Sorting of sphingolipids in epithelial (Madin-Darby canine kidney) cells, *J. Cell Biol.,* 105, 1623, 1987.
29. **Pagano, R. E., Sepanski, M. A., and Martin, O. C.,** Molecular trapping of a fluorescent ceramide analogue at the Golgi apparatus of fixed cells: interaction with endogenous lipids provides a *trans*-Golgi marker for both light and electron microscopy, *J. Cell Biol.,* 109, 2067, 1989.
30. **Futerman, A. H., Stieger, B., Hubbard, A. L., and Pagano, R. E.,** Sphingomyelin synthesis in rat liver occurs predominantly at the *cis* and *medial* cisternae of the Golgi apparatus, *J. Biol. Chem.,* 265, 8650, 1990.
31. **Godman, G. C. and Lane, N.,** On the site of sulfation in the chondrocyte, *J. Cell Biol.,* 21, 353, 1964.

32. **Lane, N., Caro, L., Otero-Vilardebo, L. R., and Godman, G. C.,** On the site of sulfation in colonic goblet cells, *J. Cell Biol.,* 21, 339, 1964.
33. **Young, R. W.,** The role of the Golgi apparatus in sulfate metabolism, *J. Cell Biol.,* 57, 175, 1973.
34. **Kornfeld, R. and Kornfeld, S.,** Assembly of asparagine-linked oligosaccharides, *Annu. Rev. Biochem.,* 54, 631, 1985.
35. **Neutra, M. and Leblond, C. P.,** Synthesis of the carbohydrates in the Golgi of mucus in the Golgi complex as shown by electron microscope radioautography of goblet cells from rats injected with glucose-^{3}H, *J. Cell Biol.,* 30, 119, 1966.
36. **Neutra, M. and Leblond, C. P.,** Radioautographic comparison of the uptake of galactose-^{3}H and glucose-^{3}H in the Golgi region of various cells secreting glycoproteins or mucopolysaccharides, *J. Cell Biol.,* 30, 137, 1966.
37. **Fleischer, B., Fleischer, S., and Ozawa, H.,** Isolation and characterization of Golgi membranes from bovine liver, *J. Cell Biol.,* 43, 59, 1969.
38. **Whur, P., Herscovics, A., and Leblond, C. P.,** Radioautographic visualization of the incorporation of galactose-^{3}H and mannose-^{3}H by rat thyroids in vitro in relation to the stages of thyroglobulin synthesis, *J. Cell Biol.,* 43, 289, 1969.
39. **Herscovics, A.,** Biosynthesis of thryoglobulin. Incorporation of ^{14}C-galactose, ^{14}C-mannose and ^{3}H-leucine into soluble proteins by rat thyroids in vitro, *Biochem. J.,* 112, 709, 1969.
40. **Caro, L. G. and Palade, G. E.,** Protein synthesis, storage and discharge in the pancreatic exocrine cell, *J. Cell Biol.,* 20, 473, 1964.
41. **Jamieson, J. D. and Palade, G. E.,** Intracellular transport of secretory proteins in the pancreatic exocrine cell. II. Transport to condensing vacuoles and zymogen granules *J. Cell Biol.,* 34, 597, 1967.
42. **Palade, G.,** Intracellular aspects of the process of protein secretion, *Science,* 89, 347, 1975.
43. **Jamieson, J. D. and Palade, G. E.,** Synthesis, intracellular transport and discharge of secretory proteins in stimulated pancreatic exocrine cells, *J. Cell Biol.,* 50, 135, 1971.
44. **Farquhar, M. G., Reid, J. A., and Daniell, L.,** Intracellular transport and packaging of prolactin. A quantitative electron microscope autoradiography study of mammotrophs dissociated from rat pituitaries, *Endocrinology,* 102, 296, 1978.
45. **Salpeter, M. and Farquhar, M. G.,** High resolution autoradiographic analysis of the secretory pathway in mammotrophs of the rat anterior pituitary, *J. Cell Biol.,* 91, 240, 1981.
46. **Bendayan, M., Roth, J., Perrelet, A., and Orci, L.,** Quantitative immunocytochemical localization of pancreatic secretory proteins in subcellular compartments of the rat acinar cell, *J. Histochem. Cytochem.,* 28, 149, 1980.
47. **Howell, K. E., Reuter-Carlson, U., Devaney, E., Luzio, J. P., and Fuller, S. D.,** One antigen, one gold? A quantitative analysis of immunogold labeling of plasma membrane 5′-nucleotidase in frozen thin sections, *Eur. J. Cell Biol.,* 44, 318, 1987.
48. **Griffiths, G. and Hoppeler, H.,** Quantitation in immunocytochemistry: correlation of immunogold labeling to absolute number of membrane antigens, *J. Histochem. Cytochem.,* 38, 312, 1985.
49. **Bainton, D. F. and Farquhar, M. G.,** Segregation and packaging of granule enzymes in eosinophilic leukocytes, *J. Cell Biol.,* 45, 54, 1970.
50. **Herzog, V. and Miller, F.,** Die Lokalisation Endogener Peroxydase in der Glandular Parotis der Ratte, *Z. Zellforsch. Mikrosk. Anat.,* 107, 403, 1970.
51. **Hand, A. R. and Oliver, C.,** The Golgi apparatus: protein transport and packaging in secretory cells, in *Methods in Cell Biology,* Hand, A. R. and Oliver, C., Eds., Academic Press, New York, 1981, 137.
52. **Steinman, R. M., Brodie, S. E., and Cohn, Z. A.,** Membrane flow during pinocytosis. A stereologic analysis, *J. Cell Biol.,* 68, 665, 1976.
53. **Gonatas, N. K.,** The role of the neuronal Golgi apparatus in centripetal membrane vesicular traffic, *J. Neuropathol. Exp. Neurol.,* 41, 6, 1982.
54. **Herzog, V. and Farquhar, M. G.,** Luminal membrane retrieved after exocytosis reaches most Golgi cisternae in secretory cells, *Proc. Natl. Acad. Sci. U.S.A.,* 74, 5073, 1977.
55. **Farquhar, M. G.,** Recovery of surface membrane in anterior pituitary cells. Variations in traffic detected with anionic and cationic ferritin, *J. Cell Biol.,* 77, R35, 1978.
56. **Ottosen, P. D., Courtoy, P. J., and Farquhar, M. G.,** Pathways followed by membrane recovered from the surface of plasma cells and myeloma cells, *J. Exp. Med.,* 152, 1, 1980.
57. **Woods, J. W., Doriaux, M., and Farquhar, M. G.,** Transferrin receptors recycle to cis and middle as well as trans Golgi cisternae in Ig-secreting mouse myeloma cells, *J. Cell Biol.,* 103, 277, 1986.
58. **Snider, M. D. and Rogers, O. C.,** Intracellular movement of cell surface receptors after endocytosis: resialyation of asialotransferrin receptor in human erythroleukemia cells, *J. Cell Biol.,* 100, 826, 1985.
59. **Duncan, J. R. and Kornfeld, S.,** Intracellular movement of two mannose 6-phosphate receptors: return to the Golgi apparatus, *J. Cell Biol.,* 106, 617, 1988.

60. **Bergmann, J. E., Tokuyasu, K. T., and Singer, S. J.,** Passage of an integral membrane protein, the vesicular stomatitis virus glycoprotein, through the Golgi apparatus *en route* to the plasma membrane, *Proc. Natl. Acad. Sci. U.S.A.,* 78, 1746, 1981.
61. **Green, J., Griffiths, G., Louvard, D., Quinn, P., and Warren, G.,** Passage of viral membrane proteins through the Golgi complex, *J. Mol. Biol.,* 152, 663, 1981.
62. **Dunphy, W. G., Fries, E., Urbani, L. J., and Rothman, J. E.,** Early and late functions associated with the Golgi apparatus reside in distinct compartments, *Proc. Natl. Acad. Sci. U.S.A.,* 78, 7453, 1981.
63. **Bergmann, J. E. and Singer, S. J.,** Immunoelectron microscopic studies of the intracellular transport of the membrane glycoprotein (G) of vesicular stomatitis virus infected Chinese hamster ovary cells, *J. Cell Biol.,* 97, 1777, 1983.
64. **Bergeron, J. J. M., Kotwal, G. J., Levine, G., Bilan, P., Rachubinski, R., Hamilton, M., Shore, G. C., and Ghosh, H. P.,** Intracellular transport of the transmembrane glycoprotein G of vesicular stomatitis virus through the Golgi apparatus as visualized by electron microscope radioautography, *J. Cell Biol.,* 94, 36, 1982.
65. **Strous, G. J. A. M., Willemsen, R., Van Kerkhof, P., Slot, J. W., Geuze, H. J., and Lodish, H. F.,** Vesicular stomatitis virus glycoprotein, albumin, and transferrin are transported to the cell surface via the same Golgi vesicles, *J. Cell Biol.,* 97, 1815, 1983.
66. **Roth, J. and Berger, E. G.,** Immunocytochemical localization of galactosyltransferase in HeLa cells: codistribution with thiamine pyrophosphatase in trans-Golgi cisternae, *J. Cell Biol.,* 92, 223, 1982.
67. **Dunphy, W. G. and Rothman, J. E.,** Compartmental organization of the Golgi stack, *Cell,* 42, 13, 1985.
68. **Roth, J., Taatjes, D. J., Lucocq, J. M., Weinstein, J., and Paulson, J. C.,** Demonstration of an extensive trans-tubular network continuous with the Golgi apparatus stack that may function in glycosylation, *Cell,* 43, 287, 1985.
69. **Moremen, K. W. and Touster, O.,** Biosynthesis and modification of Golgi mannosidase II in HeLa and 3T3 cells, *J. Biol. Chem.,* 260, 6654, 1985.
70. **Dunphy, W. G., Brands, R., and Rothman, J. E.,** Attachment of terminal N-acetylglucosamine to asparagine-linked oligosaccharides occurs in central cisternae of the Golgi stack, *Cell,* 40, 463, 1985.
71. **Novikoff, P. M., Tulsiani, D. R. P., Touster, O., Yam, A., and Novikoff, A. B.,** Immunocytochemical localization of α-D-mannosidase II in the Golgi apparatus of the rat liver, *Proc. Natl. Acad. Sci. U.S.A.,* 80, 4364, 1983.
72. **Lippincott-Schwartz, J., Yuan, L. C., Bonifacino, J. S., and Klausner, R. D.,** Rapid redistribution of Golgi proteins into the ER in cells treated with brefeldin A: evidence for membrane cycling from Golgi to ER, *Cell,* 56, 801, 1989.
73. **Roth, J., Taatjes, D. J., Weinstein, J., Paulson, J. C., Greenwell, P., and Watkins, W. M.,** Differential subcompartmentalization of terminal glycosylation in the Golgi apparatus of intestinal absorptive and goblet cells, *J. Biol. Chem.,* 261, 14307, 1986.
74. **Roth, J.,** Cytochemical localization of terminal N-acetyl-D-galactosamine residues in cellular compartments of intestinal goblet cells: implications for the topology of O-glycosylation, *J. Cell Biol.,* 98, 399, 1984.
75. **Taatjes, D. J., Roth, J., Weinstein, J., and Paulson, J. C.,** Post-Golgi apparatus localization and regional expression of rat intestinal sialyltransferase detected by immunoelectron microscopy with polypeptide epitope-purified antibody, *J. Biol. Chem.,* 263, 6302, 1988.
76. **Pavelka, M. and Ellinger, A.,** Cytochemical characteristics of the Golgi apparatus, *J. Electron Microsc. Tech.,* 17, 11, 1991.
77. **Roth, J., Lentze, M. J., and Berger, E. G.,** Immunocytochemical demonstration of ectogalactosyltransferase in absorptive intestinal cells, *J. Cell Biol.,* 100, 118, 1985.
78. **Louvard, D., Reggio, H., and Warren, G.,** Antibodies to the Golgi complex and the rough endoplasmic reticulum, *J. Cell Biol.,* 92, 92, 1982.
79. **Burke, B., Griffiths, G., Reggio, H., Louvard, D., and Warren, G.,** A monoclonal antibody against a 135-K Golgi membrane protein, *EMBO J.,* 1, 1621, 1982.
80. **Baron, M. D. and Garoff, H.,** Mannosidase II and the 135-kDa Golgi-specific antigen recognized by monoclonal antibody 53FC3 are the same dimeric protein, *J. Biol. Chem.,* 265, 19928, 1990.
81. **Yuan, L., Barriocanal, J. G., Bonifacino, J. S., and Sandoval, I. V.,** Two integral membrane proteins located in the *cis*-, middle and *trans*-part of the Golgi system acquire sialylated N-linked carbohydrates and display different turnovers and sensitivity to cAMP-dependent phosphorylation, *J. Cell Biol.,* 105, 215, 1987.
82. **Chicheportiche, Y., Vassalli, P., and Tartakoff, A. M.,** Characterization of cytoplasmically oriented Golgi proteins with a monoclonal antibody, *J. Cell Biol.,* 99, 2200, 1984.
83. **Chicheportiche, Y. and Tartakoff, A. M.,** Monoclonal antibodies as markers of the endocytic and secretory pathways, *Eur. J. Cell Biol.,* 44, 135, 1987.

84. **Donaldson, J. G., Lippincott-Schwartz, J., Bloom, G. S., Kreis, T., and Klausner, R. D.,** Dissociation of a 110 kD peripheral membrane protein from the Golgi apparatus is an early event in brefeldin A action, *J. Cell Biol.,* 111, 2295, 1990.
85. **Schweizer, A., Fransen, J. A. M., Bächi, T., Ginsel, L., and Hauri, H.-P.,** Identification, by a monoclonal antibody, of a 53-kD protein associated with a tubulo-vesicular compartment at the *cis*-side of the Golgi apparatus, *J. Cell Biol.,* 107, 1643, 1988.
86. **Saraste, J., Palade, G. E., and Farquhar, M. G.,** Antibodies to rat pancreas Golgi subfractions: identification of a 58 kD cis-Golgi protein, *J. Cell Biol.,* 105, 2021, 1987.
87. **Schweizer, A., Fransen, J. A. M., Matter, K., Kreis, T. E., Ginsel, L., and Hauri, H.-P.,** Identification of an intermediate compartment involved in protein transport from endoplasmic reticulum to Golgi apparatus, *Eur. J. Cell Biol.,* 53, 185, 1990.
88. **Goud, B., Zahraoui, A., Tavitian, A., and Saraste, J.,** Small GTP-binding protein associated with Golgi cisternae, *Nature,* 345, 553, 1990.
89. **Gonatas, J. O., Mezitis, S. G. E., Stieber, A., Fleischer, B., and Gonatas, N. K.,** MG-160: A novel sialoglycoprotein of the medial cisternae of the Golgi apparatus, *J. Biol. Chem.,* 264, 646, 1989.
90. **Croul, S., Mezitis, G. E., Stieber, A., Chen, Y., Gonatas, J. O., Goud, B., and Gonatas, N. K.,** Immunocytochemical visualization of the Golgi apparatus in several species, including human, and tissues with an antiserum against MG-160, a sialoglycoprotein of rat Golgi apparatus, *J. Histochem. Cytochem.,* 38, 957, 1990.
91. **Machamer, C. E. and Rose, J. K.,** A specific transmembrane domain of a coronavirus E1 glycoprotein is required for its retention in the Golgi region, *J. Cell Biol.,* 105, 1205, 1987.
92. **Tooze, J., Tooze, S., and Warren, G.,** Replication of coronavirus MHV-A59 in sac-cells: determination of the first site of budding of progeny virions, *Eur. J. Cell Biol.,* 33, 281, 1984.
93. **Machamer, C., Rose, J., Mentone, S., and Farquhar, M. G.,** The E1 glycoprotein of an avian coronavirus is targeted to the cis Golgi complex, *Proc. Natl. Acad. Sci. U.S.A.,* 87, 6929, 1990.
94. **Orci, L., Ravazzola, M., Storch, M.-J., Anderson, R. G. W., Vassalli, J.-D., and Perrelet, A.,** Proteolytic maturation of insulin is a post-Golgi event which occurs in acidifying clathrin-coated secretory vesicles, *Cell,* 49, 865, 1987.
95. **Schnabel, E., Mains, R. E., and Farquhar, M. G.,** Proteolytic processing of pro-ACTH/endorphin begins in the Golgi complex of pituitary corticotropes and AtT-20 cells, *Mol. Endocrinol.,* 8, 1223, 1989.
96. **Fisher, J. M., Sossin, W., Newcomb, R., and Scheller, R.,** Multiple neuropeptides derived from a common precursor are differentially packaged and transported, *Cell,* 54, 813, 1988.
97. **Lepage-Lezin, A., Joseph-Bravo, P., Devilliers, G., Benedetti, L., Launay, J.-M., Gomez, S., and Cohen, P.,** Prosomatostatin is processed in the Golgi apparatus of rat neural cells, *J. Biol. Chem.,* 266, 1679, 1991.
98. **Brennan, S. O. and Peach, R. J.,** Calcium-dependent KEX2-like protease found in hepatic secretory vesicles converts proalbumin to albumin, *FEBS Lett.,* 229, 167, 1988.
99. **Herz, J., Kowal, R. C., Goldstein, J. L., and Brown, M. S.,** Proteolytic processing of the 600 kd low density lipoprotein receptor-related protein (LRP) occurs in a *trans*-Golgi compartment, *EMBO J.,* 9, 1769, 1990.
100. **Anderson, R. G. W. and Pathak, R. K.,** Vesicles and cisternae in the *trans* Golgi apparatus of human fibroblasts are acidic compartments, *Cell,* 40, 635, 1985.
101. **Schwartz, S. L., Strous, G. J. A. M., Slot, J. W., and Geuze, H.,** Immunoelectron microscopic localization of acidic intracellular compartments in hepatoma cells, *EMBO J.,* 4, 899, 1985.
102. **Anderson, R. G. W. and Orci, L.,** A view of acidic intracellular compartments, *J. Cell Biol.,* 106, 539, 1988.
103. **Araki, N., Lee, M., Takashima, Y., and Ogawa, K.,** Cytochemical demonstration of NPPase activity for detecting proton-translocating ATPase of Golgi complex in rat pancreatic acinar cells, *Histochemistry,* 93, 453, 1990.
104. **Yurko, M. A. and Gluck, S.,** Production and characterization of a monoclonal antibody to vacuolar H^+-ATPase of renal epithelia, *J. Biol. Chem.,* 99, 1511, 1987.
105. **Rodman, S. J., Stahl, P. D., and Gluck, S.,** Distribution and structure of the vacuolar H^+ATPase in endosomes and lysosomes from LLC-PK1 cells, *Exp. Cell. Res.,* 192, 445, 1991.
106. **Slot, J. W. and Geuze, H. J.,** Immunoelectron microscopic exploration of the Golgi complex, *J. Histochem. Cytochem.,* 31, 1049, 1983.
107. **Tartakoff, A. M. and Vassalli, P.,** Comparative studies of intracellular transport of secretory proteins, *J. Cell Biol.,* 79, 694, 1978.
108. **Tartakoff, A. M., and Vassalli, P.,** Plasma cell immunoglobulin secretion, *J. Exp. Med.,* 146, 1332, 1977.

109. **Geuze, H. J. and Slot, J. W.,** The subcellular localization of immunoglobulin in mouse plasma cells, as studied with immunoferritin cytochemistry on ultrathin frozen sections, *Am. J. Anat.*, 158, 161, 1980.
110. **Kelly, R. B.,** Pathways of protein secretion in eukaryotes, *Science,* 230, 25, 1985.
111. **Tooze, J., Tooze, S. A., and Fuller, S. D.,** Sorting of progeny coronavirus from condensed secretory proteins at the exit from the trans-Golgi network of AtT-20 cells, *J. Cell Biol.*, 105, 1215, 1987.
112. **Orci, L., Ravazzola, M., Amherdt, M., Perrelet, A., Powell, S. K., Quinn, D. L., and Moore, H.-P.,** The trans-most cisternae of the Golgi complex: a compartment for sorting of secretory and plasma membrane proteins, *Cell,* 51, 1039, 1987.
113. **Burgess, T. L. and Kelly, R. B.,** Constitutive and regulated secretion of proteins, *Annu. Rev. Cell Biol.*, 3, 243, 1987.
114. **Chung, K.-N., Walter, P., Aponte, G. W., and Moore, H.-P. H.,** Molecular sorting in the secretory pathway, *Science,* 243, 192, 1988.
115. **Rodriguez-Boulan, E. and Nelson, W. J.,** Morphogenesis of the polarized epithelial cell phenotype, *Science,* 245, 718, 1989.
116. **Simons, K. and Wandinger-Ness, A.,** Polarized sorting in epithelia, *Cell,* 62, 207, 1990.
117. **Rodriguez-Boulan, E., Pasiek, K. T., Salas, P. J. I., and Sabatini, D. D.,** Intracellular transport of influenza virus hemagglutinin to the apical surface of Madin-Darby canine kidney cells, *J. Cell Biol.*, 98, 308, 1984.
118. **Griffiths, G., Pfeiffer, S., Simons, K., and Matlin, K.,** Exit of newly synthesized membrane proteins from the trans cisternae of the Golgi complex to the plasma membrane, *J. Cell Biol.*, 101, 949, 1985.
119. **De Curtis, I. and Simons, K.,** Isolation of exocytic carrier vesicles from BHK cells, *Cell,* 58, 719, 1989.
120. **Wehland, J., Willingham, M. C., Gallo, M. G., and Pastan, I.,** The morphological pathway of exocytosis of the vesicular stomatitis virus G protein in cultured fibroblasts, *Cell,* 28, 831, 1982.
121. **Wandinger-Ness, A., Bennett, M. K., Antony, C., and Simons, K.,** Distinct transport vesicles mediate the delivery of plasma protein to the apical and basolateral domains of MDCK cells, *J. Cell Biol.*, 111, 987, 1990.
122. **Rindler, M. J., Ivanov, I. E., Plesken, H., and Sabatini, D. D.,** Polarized delivery of viral glycoproteins to the apical and basolateral plasma membranes of Madin-Darby canine kidney cells infected with temperature-sensitive viruses, *J. Cell Biol.*, 100, 136, 1985.
123. **Rindler, M. J., Ivanov, I. E., Plesken, H., Rodriguez-Boulan, E. J., and Sabatini, D. D.,** Viral glycoproteins destined for apical or basolateral plasma membrane domains traverse the same Golgi apparatus during their intracellular transport in doubly infected Madin-Darby canine kidney cells, *J. Cell Biol.*, 98, 1304, 1984.
124. **Fuller, S. D., Bravo, R., and Simons, K.,** An enzymatic assay reveals that proteins destined for the apical or basolateral domains of an epithelial cell line share the same late Golgi compartment, *EMBO J.*, 4, 297, 1985.
125. **Caplan, M. J., Stow, J. L., Newman, A. P., Madri, J., Anderson, H. C., Farquhar, M. G., Palade, G. E., and Jamieson, J. D.,** Dependence on pH of polarized sorting of secreted proteins, *Nature,* 329, 632, 1987.
126. **Kornfeld, S.,** Trafficking of lysosomal enzymes, *FASEB J.*, 1, 462, 1987.
127. **von Figura, K. and Hasilik, A.,** Lysosomal enzymes and their receptors, *Annu. Rev. Biochem.*, 55, 167, 1986.
128. **Geuze, H. J., Slot, J. W., Strous, G. J., Hasilik, A., and von Figura, K.,** Possible pathways for lysosomal enzyme delivery, *J. Cell Biol.*, 101, 2253, 1985.
129. **van Dongen, J. M., Barneveld, R. A., Geuze, H. J., and Galjaard, H.,** Immunocytochemistry of lysosomal hydrolases and their precursor form in normal and mutant human cells, *Histochem. J.*, 16, 941, 1984.
130. **Yokota, S., Himeno, M., and Kato, K.,** Immunocytochemical localization of acid phosphatase in rat liver, *Cell Struct. Funct.*, 14, 163, 1989.
131. **Brown, W. J., Goodhouse, J., and Farquhar, M. G.,** Mannose-6-phosphate receptors for lysosomal enzymes cycle between the Golgi complex and endosomes, *J. Cell Biol.*, 103, 1235, 1986.
132. **Brown, W. J., Constantinescu, E., and Farquhar, M. G.,** Redistribution of mannose-6-phosphate receptors induced by tunicamycin and chloroquine treatments, *J. Cell Biol.*, 99, 320, 1984.
133. **Geuze, H., Slot, J. W., Strous, G. J. A. M., and von Figura, K.,** Ultrastructural localization of the mannose-6-phosphate receptor in rat liver, *J. Cell Biol.*, 98, 2047, 1984.
134. **Geuze, H. J., Stoorvogel, W., Strous, G. J., Slot, J. W., Bleekemolen, J. E., and Mellman, I.,** Sorting of mannose-6-phosphate receptors and lysosomal membrane proteins in endocytic vesicles, *J. Cell Biol.*, 107, 2491, 1988.
135. **Griffiths, G., Hoflack, B., Simons, K., Mellman, I., and Kornfeld, S.,** The mannose-6-phosphate receptor and the biogenesis of lysosomes, *Cell,* 52, 329, 1988.

Chapter 9

CELL ORGANELLES

9.5 Lysosomes

Masahiro Sakai and Nobukazu Araki

TABLE OF CONTENTS

0-8493-6012-9/93/$0.00 + $.50

A. INTRODUCTION

It is well known that lysosomes contain a large number and variety of hydrolytic enzymes and are the main sites of intracellular digestion of both exogenous materials (heterophagy) and endogenous substances (autophagy). Recently, it became evident that lysosomes are polymorphic and their shape and size change considerably during the digestive process in several cell types.[1] Although the relationship between lysosomes and both heterophagy and autophagy has been studied extensively by morphological means, little attention has been paid to the details of the regulatory mechanism of the intracellular movement of lysosomes with regard to their function.

Heterophagy is defined as the cellular activity by which exogenous substances are taken into cells by endocytosis or pinocytosis and degraded later by lysosomal enzymes. It is currently agreed that cytoskeletal elements such as microfilaments and microtubules are necessary to complete the process of heterophagy and to maintain both cellular shape and distribution of organelles within the cell. It has been recently reported that dynamic changes of lysosomal structure occur during pinocytosis, resulting in the formation of ''tubular lysosomes''[2] or ''nematolysosomes''.[3] In the first part of this chapter, we describe how cytoskeletal elements behave when lysosomes reveal unusual changes in their shape and size by demonstrating the effects of drugs that destabilize actin filaments or the inhibition of microtubule formation. In addition, we describe the direct interaction between microfilaments and isolated lysosomes from alveolar macrophages investigated *in vitro*.

The term autophagy designates the intracellular process of sequestration of the cell's own components and subsequent degradation by the lysosomes. The cell organelles associated with the autophagic process, termed autophagic vacuoles, constitute membrane-bound organelles containing cytoplasmic components structurally intact or in various stages of disintegration. Although many morphological and biochemical studies have been carried out on cellular autophagy in different cell types under different conditions,[4] little is known concerning the mechanism that controls the formation of autophagic vacuoles or the role of autophagy in cell life.

In the second half of this chapter, an attempt to clarify the control mechanism of autophagy will be made by studying whether there are any detectable changes in cytoplasmic components, which are thought to be catabolized by autophagy, or in membranes, by which autophagic vacuoles form in mouse or rat livers, from the viewpoint of the distribution of electric charges and lectin-binding patterns.

B. MATERIALS AND METHODS

1. Heterophagy and Nematolysosomes

Alveolar macrophages were prepared from lungs of Wistar rats (male or female, 200 to 300 g body weight), and cultured in Medium 199 containing 10% newborn calf serum. For the induction of nematolysosome formation, cultured macrophages were exposed to rat peroxidase-anti-peroxidase immune complex (PAP; 0.25 mg/ml culture Medium 199) as a pinocytic marker at 4°C for 30 min, and then reincubated for 10 to 120 min in PAP-free medium at 37°C to allow pinocytosis. To investigate the role of actin filaments and microfilaments in the regulation of the nematolysosome formation, either actin filament destabilizers (cytochalasin B, cytochalasin D, 20 μ*M*) or antimicrotubular drugs (colchicine, nocodazole, 20 μ*M*) were added to the medium employed in studying PAP pinocytosis. Macrophages were fixed with 1% glutaraldehyde in cacodylate buffer, pH 7.4, containing 5% sucrose at room temperature. The transport of PAP in the cell was traced by the diaminobenzidine (DAB) reaction,[5] and nematolysosomes were detected by the demonstration of acid phosphatase (AcPase) activity cytochemically.[6] The specimens were dehydrated with graded ethanol series and embedded in Spurr's resin. Ultrathin sections and 1-μm semithin sections were prepared and observed under electron microscopes (JEOL 1200EX and 2000ES) at accelerating voltages to 100 or 200 kV.

In a separate experiment, in order to examine the direct relationship between lysosomes and microfilaments, lysosome-enriched fractions from alveolar macrophages were prepared and mixed with G-actin (0.1 to 0.5 mg protein) in an actin polymerization buffer for 30 min at 37°C.[7] The solution containing lysosomes and polymerized F-actin was dropped onto Formvar-coated copper grids. In some specimens F-actin filaments were decorated with heavy meromyosin (HMM) to examine their polarity. The specimens were negatively stained with 1% uranyl acetate in aqueous solution and observed under a JEOL 1200EX electron microscope.

2. Autophagy and Lysosomes

The following two methods were applied to examine the nature of membranes and matrices of autophagic vacuoles and lysosomes in hepatocyes.

Frozen ultrathin section labeling — Livers of Wistar rats (male or female, 200 to 300 g) or ddY mice (male or female, 25 to 35 g) were fixed by perfusion through the left ventricle with 2% formaldehyde and 0.25% glutaraldehyde in 0.1 *M* phosphate buffer, pH 7.4, for 10 min at room temperature. In some cases, autophagy was induced in hepatocytes by an injection of vinblastine (VBL; 50 mg/kg body weight) 30 min or 2 h prior to removal of the livers. Small fixed tissue blocks were immersed in 2.3 *M* sucrose solution for more than 2 h, and rapidly frozen with liquid nitrogen after placement of blocks in a specimen holder. Frozen thin sections of 90 to 100 nm in thickness were prepared with a Reichert-Jung Cryomicrotome at −90°C to −110°C, and recovered on Formvar-coated copper grids by the method of Tokuyasu.[8] Sections on the grids were rinsed with phosphate-buffered saline (PBS) several times to remove sucrose and then incubated for 30 min at room temperature in one of the following compounds: cationized ferritin (CF; 50 to 100 μg/ml PBS, pH 7.4), anionized ferritin (AF; 50 to 100 μg/ml PBS, pH 7.4), ferritin-labeled Concanavalin A (Con A-F; 50 to 200 μg/ml PBS, pH 7.4), wheat germ agglutinin (WGA-F; 50 to 200 μg/ml PBS, pH 7.4). After washing with PBS several times again, the sections were refixed with 2% glutaraldehyde in 0.1 *M* phosphate buffer, stained with 2% uranyl acetate, and embedded in a mixture of 0.2% methylcellulose and 2% polyethylene glycol.

Isolated lysosomes or autophagic vacuole labeling — Livers of Wistar rats (male or female, 200 to 250 g) or of rats in which autophagy was induced by VBL administration were minced in 0.3 *M* sucrose and homogenized (1:2 w/v) with an ultradisperser for 1 min. The mitochondria and lysosomes (ML) fraction was obtained by the centrifugation of 24,000 × *g* for 20 min from the postnuclear fraction of liver homogenate. The ML fraction suspended in 50% Metrizamide was overlaid with a discontinuous gradient composed of 26/24/20/15% of Metrizamide from bottom to top. After centrifugation at 80,000 × *g* for 2 h, the fractions bonding at 24/20% and 20/15% Metrizamide interfaces were collected and washed twice with PBS. The procedure for the isolation of lysosomes and autophagic vacuoles was based on the method of Marzella et al.[9] The final suspensions of isolated lysosomes and autophagic vacuoles were mixed with CF or AF at the concentration of 50 to 100 μg/ml suspension or with Con A-F or WGA-F at the concentration of 100 to 200 μg/ml suspension and kept for 30 min at room temperature. After incubation, they were pelleted at 15,000 × *g* for 10 min and washed with PBS several times. The pellets were then fixed with 2% glutaraldehyde in 0.1 *M* phosphate buffer and postfixed with 1% osmium tetraoxide for 30 min at 4°C. These samples were dehydrated with ethanol series and embedded in Spurr's resin. Ultrathin sections stained with 2% uranyl acetate were examined under a 1200 EX electron microscope at 100 kV accelerating voltage.

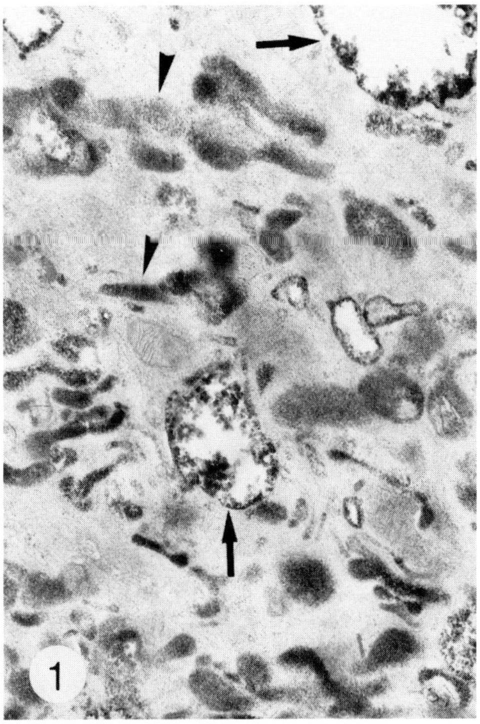

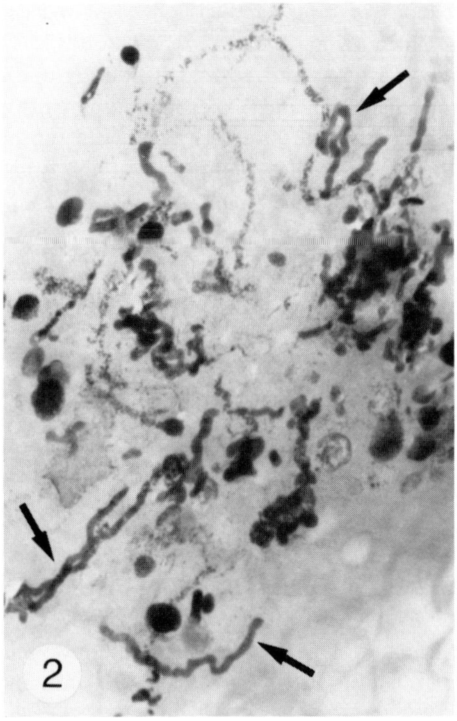

FIGURE 1. A macrophage incubated in PAP-free culture medium for 20 min at 37°C after preincubation in PAP-containing medium at 4°C; PAP detected by DAB reaction is observed in pinosomes (arrows) and a large number of elongated structures filled with PAP are conspicuous in the cytoplasm (arrowheads). (Magnification × 21,300.)

FIGURE 2. Cytochemical detection of AcPase activity in the cell in the same condition as Figure 1. AcPase-positive nematolysosomes reveal long thread-like appearances in the 1-μm-thick sections (arrows). (Magnification × 8,700.)

C. RESULTS

1. Heterophagy and Nematolysosomes

After 5 to 20 min incubation of macrophages in the PAP-free medium at 37°C following exposure to PAP at 4°C, PAP was detected in pinosomes of various size. Subsequently, during 20 to 60 min incubation at 37°C, an increasingly large number of thread-like elongated structures, which were entirely filled with PAP were observed in the cytoplasm (Figure 1). Since these structures revealed AcPase activity cytochemically, they were recognized as one type of lysosome. Although fragments of these structures were occasionally seen in the routine ultrathin sections, 1- to 2-μm-thick sections observed under a 200-kV electron microscope allowed us to see that they expand to 8 to 10 μm in maximum length (Figure 2). We have termed these thread-like lysosomes "nematolysosomes", emphasizing that they are not tubule structures,[2] because similar elongated lysosomes showing intense AcPase activity have been reported as mostly "tubular lysosomes" in some types of cell such as endothelial cells,[10] neurons,[11] peritoneal macrophages,[3] cells in the anterior pituitary,[12] smooth muscle cells,[13] striated muscle cells,[14] and pinealocytes.[15] The presence of AcPase-negative and trimetaphosphatase-positive basal elongated lysosomes has been also reported as tubular lysosomes in exocrine acinar cells of pancreas, parotid gland, and lacrimal glands.[16] By this unique characteristic of cytochemical staining of basal elongated lysosomes, they are believed to be a distinct population of lysosomes, but a recent study of Araki et al.[17] demonstrated that AcPase activity could be detected in basal elongated lysosomes in rat pancreatic exocrine acinar cells if the specimens were fixed with PIPES-buffered fixative containing 0.25% glutaraldehyde and 2% formaldehyde, suggesting that lysosomes in exocrine cells may move and transform to create nematolysosomes as they also do in alveolar macrophages. It is concluded that although the cell types possessing nematolysosomes or tubular lysosomes are limited so far, these lysosomes may represent a universal structure related to some activity of the cells.

Actin filament destabilizers (cytochalasin B or cytochalasin D) or antimicrotubular drugs (colchicine or nocodazole), added to the cultures had no effect on the binding of PAP to the plasma membranes of macrophages or PAP uptake into pinosomes. However, both types of drugs completely suppressed the appearance of nematolysosomes during PAP pinocytosis (Figure 3). These results suggest that both intact actin filaments and microfilaments are essential for the formation of nematolysosomes in alveolar macrophages. In macrophages treated with antimicrotubular drugs, another form of lysosomes clearly distinct from nematolysosomes was observed. These lysosomes, which had an extended or flattened structure that thinly wraps local cytoplasmic areas, and small spherical lysosomes are considered "wrapping lysosomes" (Figure 3, arrows). Such lysosomes, thought to be involved in the autophagic process, have been reported using different terms, such as the lysosomal wrapping mechanism (LWM),[18] microautophagy,[19] and lysosomophagy,[20] under various experimental conditions. The regulatory mechanisms of the formation of nematolysosomes and wrapping lysosomes by the cytoskeletal system seem to be different since LWM in histiocytes was not dependent on the presence of microtubules but required only intact actin filaments. This difference may be explained by speculating that nematolysosome formation showing a dynamic and directional lysosomal movement needs the association of microtubules and actin filaments, but that the transformation of wrapping lysosomes, regarded as a local and unidirectional lysosomal activity, is controlled by actin filaments.

Electron microscopic observations of negatively stained preparations of isolated lysosomes mixed with actin *in vitro* substantiate this hypothesis. It was found that actin filaments were preferentially assembled on lysosomes, and the actin filaments seemed to contact the lysosomal membranes at the sides of filaments (Figure 4). The polarity of actin filaments in relation to lysosomal membranes was not uniform, probably because of changes in lysosomal shapes. Furthermore, it has been reported that microtubules and filamentous elements, 5 to 10 nm in diameter, could be observed in the vicinity of nematolysosomes, likely in contact with lysosomal membranes, of alveolar macrophages fixed with 1% glutaraldehyde containing 1% saponin and 0.4% tannic acid.[2]

In conclusion, the intracellular lysosomal movements are definitely regulated by the cytoskeletal system, but the interaction of lysosomes and cytoskeleton may be different depending on the mode of function-related lysosomal movement or transformation.

2. Autophagy and Lysosomes

Labeling methods using lectin-binding ferritin, or cationized or anionized ferritin (CF or AF) and ultrathin sections prepared by cryomicrotomy were utilized in order to detect changes in membranes forming autophagic vacuoles and of cytoplasmic components which are destined for the autophagic process in normal or vinblastine-treated mouse or rat livers.

Such studies revealed that Con A-F particles were distributed mainly on the luminal surface of bile canaliculi and in matrices of lysosomes (Figure 5). In general, the distribution pattern of WGA-F on hepatocytes was similar to that of Con A-F. Binding of Con A-F and WGA-F to lysosomal matrices differed from lysosome to lysosome in the same cell; some lysosomes revealed only a moderate accumulation of Con A-F and WGA-F particles in their matrices, but others showed a dense distribution of the label. The considerable differences

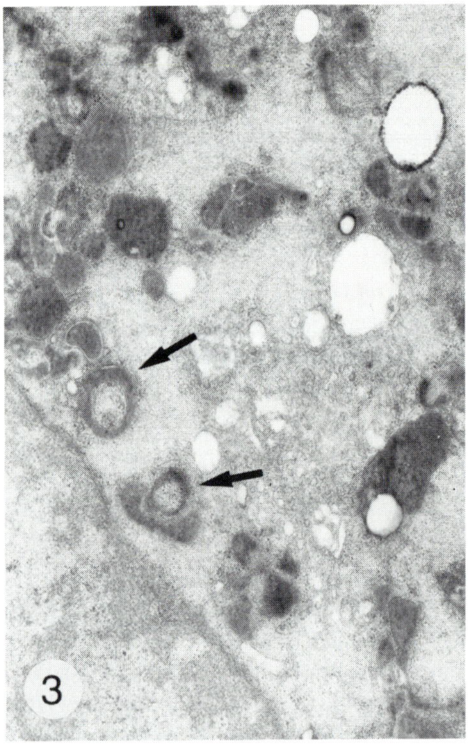

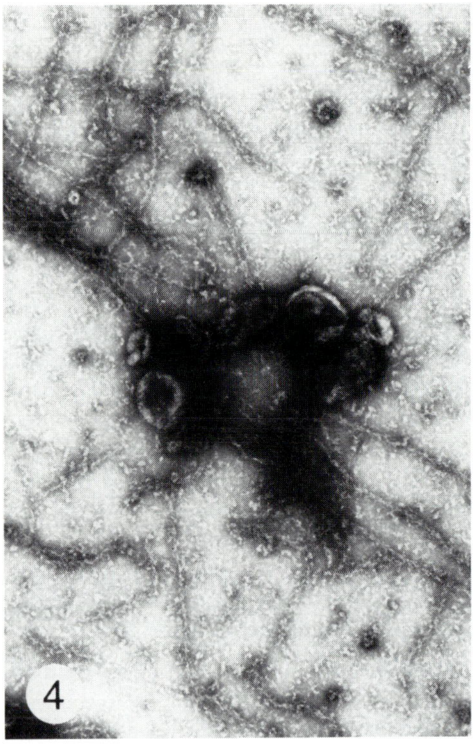

FIGURE 3. Nocodazole-treated macrophages incubated for 20 min at 37°C after preincubation with PAP in the presence of 20 μM nocodazole; DAB-positive endosomes are observed but nematolysosomes are not present in the cytoplasm. Wrapping lysosomes are frequently seen (arrows). (Magnification × 18,000.)

FIGURE 4. Negative-stained electron micrograph showing the interaction of actin filaments with isolated lysosomes from alveolar macrophages. Actin filaments decorated with HMM are attached to lysosomes at the side of filaments. (Magnification × 37,000.)

in the labeling of Con A-F and WGA-F on lysosomal matrices may indicate a heterogeneity of lysosomal contents even though lysosomal matrices reveal morphologically similar features. Whereas Con A-F and WGA-F particles did not display a preferential binding to the limiting membranes of autophagic vacuoles, within autophagic vacuoles they were frequently observed. Interestingly, the binding of Con A-F and WGA-F varied depending on the stage of the autophagic vacuoles; in matrices of newly formed vacuoles few Con A-F and WGA-F were seen but many ferritin particles were observed in vacuoles at advanced degradation stages. It is generally accepted that the digestion of sequestered macromolecules within autophagic vacuoles is performed by lysosomal enzymes. In addition, since lysosomes showed a preferential accumulation of Con A-F and WGA-F particles, most of these binding sites within autophagic vacuoles seem to derive from lysosomes and not from the engulfed cytoplasmic materials.

On frozen thin sections incubated with CF, ferritin particles were observed throughout the cytoplasm of hepatocytes. However, observation of a large number of preparations revealed a higher density of ferritin particles on the outer surface of plasma membranes, in lysosomal matrices, and on the autophagic vacuoles, compared to those in mitochondria, endoplasmic reticulum (ER) and Golgi apparatus. Autophagic vacuoles revealed dense accumulation of ferritin particles, not only on the inner surface of the limiting membrane, but also over sequestered cellular components (Figure 6). Quantification of CF particles bound to various cell membranes confirmed that the plasma membrane and the limiting membrane

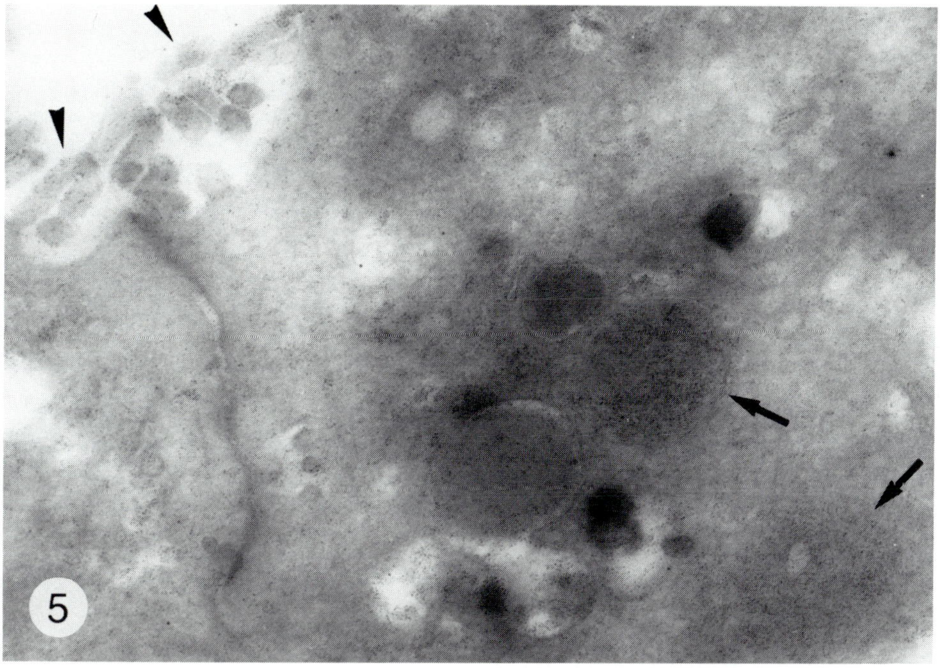

FIGURE 5. Peribilliary regions of hepatocytes incubated with Con A-F; ferritin particles are observed mainly on the outer surface of the canaliculi membranes (arrowheads) and in the lysosomal matrices (arrows). (Magnification × 48,000.)

of autophagic vacuoles have a concentration of anionic binding sites higher than those of ER and Golgi membranes.[21] On the other hand, AF particles were also observed throughout the cytoplasm, but there was no preferential localization on the plasma membranes and in lysosomes. Furthermore, autophagic vacuoles showed a relatively low AF binding to both membranes and matrices. These results suggested that the inner surface of the limiting membrane of autophagic vacuoles has a preponderance of negative over positive charges, and if the membrane of autophagic vacuoles was to originate from preexisting membranes of ER, Golgi apparatus, lysosomes etc., the membranes from which autophagic vacuoles derive must somehow acquire a higher negative charge.

When livers of Wistar rats were used for the isolation of lysosomes, lysosome-enriched fraction was recovered from the 24/20% Metrizamide interface. Most ingredients in this fraction were lysosomes of various shapes and sizes showing good preservation of ultrastructure, although a few mitochondria and fragments of ER were concluded. After incubation with ferritin-conjugated reagents, the outer surfaces of lysosomal membranes showed high bindings of both CF and WGA-F particles, but not of AF and Con A-F particles. Although CF particles were uniformly distributed on all lysosomal membranes (Figure 7), WGA-F showed a uniform distribution pattern, varying from lysosome to lysosome. These results indicate that the outer surfaces of lysosomal membranes contain negative charges that may be partially caused by *N*-acetylneuraminic acid (sialic acid) judging from its characteristic of binding of WGA. It is of interest to clarify the existence of one of the sugar residues on the cytoplasmic site of organelles such as lysosomes, because glycoproteins composed of cellular membranes usually display their sugar residues toward the outside of the cytoplasm. Nevertheless, these findings provide the fascinating hypothesis that sugar residues on lysosomal membranes facing the cytoplasm may play an important role as a trigger for multiple functions of lysosomes in the life of the cell.

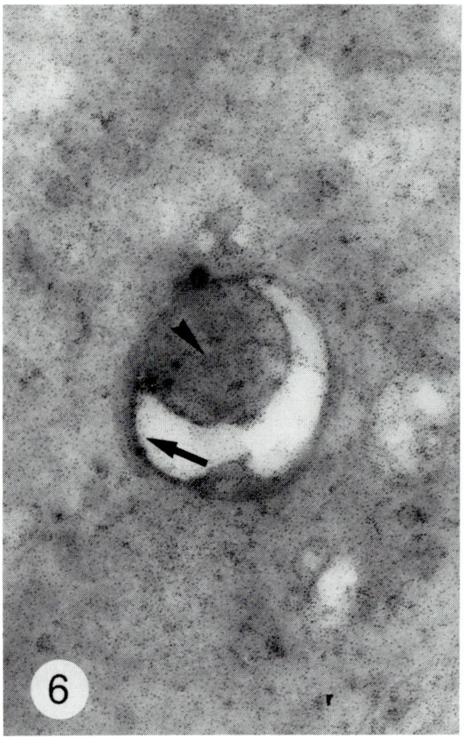

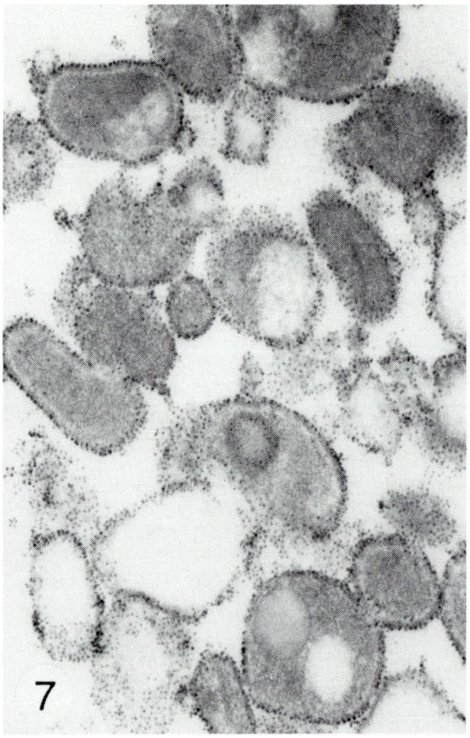

FIGURE 6. A frozen thin section of mouse hepatocyte incubated with cationized ferritin (CF). The dense accumulation of CF particles is observed not only on the inner surface of limiting membrane (arrows) but also over sequestered materials (arrowhead). (Magnification × 47,000.)

FIGURE 7. Isolated lysosomes from rat liver incubated with CF; CF particles are observed on the outer surface of all lysosomal membranes. (Magnification × 65,000.)

REFERENCES

1. **Sakai, M., Araki, N., and Ogawa, K.,** Lysosomal movement during heterophagy and autophagy: with special reference to nematolysosome and wrapping lysosome, *J. Electron Microsc. Tech.,* 12, 101, 1989.
2. **Araki, N. and Ogawa, K.,** Regulation of intracellular lysosomal movements by the cytoskeletal system in rat alveolar macrophages, *Acta Histochem. Cytochem.,* 20, 659, 1987.
3. **Swanson, J., Burke, E., and Silverstein, S. C.,** Tubular lysosomes accompany stimulated pinocytosis in macrophages, *J. Cell Biol.,* 104, 1217, 1987.
4. **Grinde, B.,** Autophagy and lysosomal proteolysis in the liver, *Experientia,* 41, 1089, 1985.
5. **Graham, R. C., Jr. and Karnovsky, M. J.,** The early stages of absorption of injected horseradish peroxidase in the proximal tubules of mouse kidney; ultrastructural cytochemistry by a new technique, *J. Histochem. Cytochem.,* 14, 291, 1966.
6. **Gomori, G.,** *Microscopic Histochemistry, Principle and Practice,* University of Chicago Press, Chicago, 1952, 189.
7. **Mehrabian, M., Bame, K. J., and Rome, L. H.,** Interaction of rat liver lysosomal membranes with actin, *J. Cell Biol.,* 99, 680, 1984.
8. **Tokuyasu, K. T.,** A technique for ultracryotomy of all suspensions and tissues, *J. Cell Biol.,* 57, 551, 1973.
9. **Marzella, L., Ahlberg, J., and Glaumann, H.,** Isolation of autophagic vacuoles from rat liver, *J. Cell Biol.,* 93, 144, 1982.

10. **DeBruyn, P. P. H., Michelson, S., and Becher, R. P.,** Endocytosis, transfer tubules, and lysosomal activity in myeloid sinusoidal endothelium, *J. Ultrastruct. Res.*, 53, 133, 1975.
11. **Broadwell, R. D., Oliver, C., and Brightman, M. W.,** Neuronal transport of acid hydrolases and peroxidase within the lysosomal system of organelles: involvement of agranular reticulum-like cisterns, *J. Comp. Neurol.*, 190, 519, 1980.
12. **Broadwell, R. D. and Oliver, C.,** An enzyme cytochemical study of the endocytic pathways in anterior pituitary cells of the mouse in vivo, *J. Histochem. Cytochem.*, 31, 325, 1983.
13. **Robinson, J. M., Okada, T., Castellot, J. J., Jr., and Karnovsky, M. J.,** Unusual lysosomes in aortic smooth muscles cells: presence in living and rapidly frozen cells, *J. Cell Biol.*, 102, 1616, 1986.
14. **Okada, T., Robinson, J. M., and Karnovsky, M. J.,** Cytochemical localization of acid phosphatase in striated muscle, *Histochemistry*, 85, 177, 1986.
15. **Kristic, R.,** Tubular lysosomes in rat gerbil pinealcytes, *Histochemistry*, 88, 203, 1985.
16. **Oliver, C.,** Characterization of basal lysosomes in eoxcrine acinar cells, *J. Histochem. Cytochem.*, 31, 1209, 1983.
17. **Araki, N., Matubara, H., Takashima, Y., and Ogawa, K.,** Three-dimensional distribution of acid phosphatase-positive thread-like structures "nematolysosomes" in rat pancreatic exocrine cells, *Acta Histochem. Cytochem.*, 22, 585, 1989.
18. **Mayahara, H. and Ogawa, K.,** Lysosomal wrapping mechanism observed in the autophagolysosome formation, in *Histochemistry and Cytochemistry*, Takeuchi, T., Ogawa, K., and Fujita, S., Eds., Japanese Society of Histochemistry and Cytochemistry, Kyoto, 1972, 29.
19. **Ahlberg, J., Marzella, L., and Glaumann, H.,** Uptake and degradation of proteins by isolated rat liver lysosomes: suggestion of a microautophagic pathway of proteolysis, *Lab. Invest.*, 47, 523, 1982.
20. **Thyberg, J., Blomgren, K., Hellgren, D., and Heiden, U.,** Lysosomophagy in cultured macrophages treated with the antimicrotubular drug nocodazole, *Eur. J. Cell Biol.*, 27, 279, 1982.
21. **Sakai, M., Luo, S. Q., Tanaka, T., and Ogawa, K.,** Distribution of electric charges and Concanavalin A binding sites on autophagic vacuoles and lysosomes in mouse hepatocytes, *Exp. Cell Res.*, 185, 143, 1989.

A. INTRODUCTION

The peroxisome is a ubiquitous cell organelle surrounded by a single membrane and a finely granular matrix containing in some cell types a crystalline electron-dense core (Figure 2).[1,2] The enzymes of peroxisomes and the specific functions of the organelle vary in different cells. In hepatocytes, they are involved in several important metabolic processes, such as the biosynthesis of ether glycerolipids, the catabolism of very long chain fatty acids and production of cholesterol and bile acids (Figure 1).[3] The discovery of a group of inherited diseases characterized by the morphologic absence or numerical decrease, as well as biochemical dysfunction, of peroxisomes established their vital importance in human metabolism.[4] In this chapter, after a brief review of the current methods for electron microscopic cytochemical and immunocytochemical investigation of peroxisomes, we present some of the major areas of active research on peroxisomes which have particularly benefited from the application of those ultrastructural techniques.

B. ELECTRON MICROSCOPIC CYTOCHEMICAL METHODS FOR PEROXISOMAL ENZYMES

From the large number of peroxisomal enzymes only a few can be detected by ultrastructural cytochemistry (Table 1). This table does not include the purely light microscopic methods and some of the older ultrastructural techniques for which new and better alternatives have been developed.

Catalase — The most common method for the detection of peroxisomes is based on the visualization of the peroxidatic activity of catalase with 3,3′-diaminobenzidine (DAB) (Figure 3).[5-8] The fixation of the tissue with an aldehyde is the prerequisite for detection of catalase with the DAB technique.[21,22] The fixation inhibits the catalatic activity of catalase while enhancing its peroxidatic activity.[21] Model studies with crystalline beef liver catalase have revealed that the peroxidatic activity is further augmented by higher temperatures in the incubation medium (37 or 45°C), alkaline pH (up to pH 10.5), and high hydrogen peroxide concentrations (0.15%).[23] Based on these observations, strategies were outlined for the optimal detection of catalase, used either as an ultrastructural tracer or as a marker of peroxisomes.[24] Moreover, the modifications of the DAB method permitting the differentiation between catalase, peroxidase, and cytochrome oxidase were described.[25]

Oxidases — There have been several methods for the light and electron microscopic localization of peroxisomal oxidases, but currently the cerium technique is the one most commonly used (for a review, see Reference 15). Cerium was introduced originally for the detection of hydrogen peroxide in granulocytes[26] and a modification of this technique was adopted for the localization of peroxisomal oxidases in yeast.[27] However, because of the sensitivity of oxidases in peroxisomes of mammals to fixation with high concentrations of aldehydes, this method was not often applied. It was only after the introduction of appropriate fixation conditions (e.g., 5-min perfusion with 0.25% glutaraldehyde in PIPES buffer) that convincing and reproducible results could be obtained.[9-11] Moreover, Angermüller and Fahimi[9] noticed that the type of buffer used in the incubation medium affected the pattern of localization in peroxisomes. Particularly, the Tris-maleate buffer used originally by Briggs et al.[26] seems to interfere with the localization of enzymes in the cores of peroxisomes.[9-11] Angermüller[15] has recently found that maleic acid in this buffer prevents the staining of the cores and has suggested that it may be responsible for chelating the copper atoms in the urate oxidase molecule. Indeed, if Tris is used without the maleate, the crystalline cores of peroxisomes are stained intensely, similar to the pattern when PIPES buffer is used (Figure 4). Similarly, when in the incubation medium for α-hydroxyacid oxidase (with argininic acid as substrate), the PIPES buffer is used, the cores stand out prominently (Figure 6),

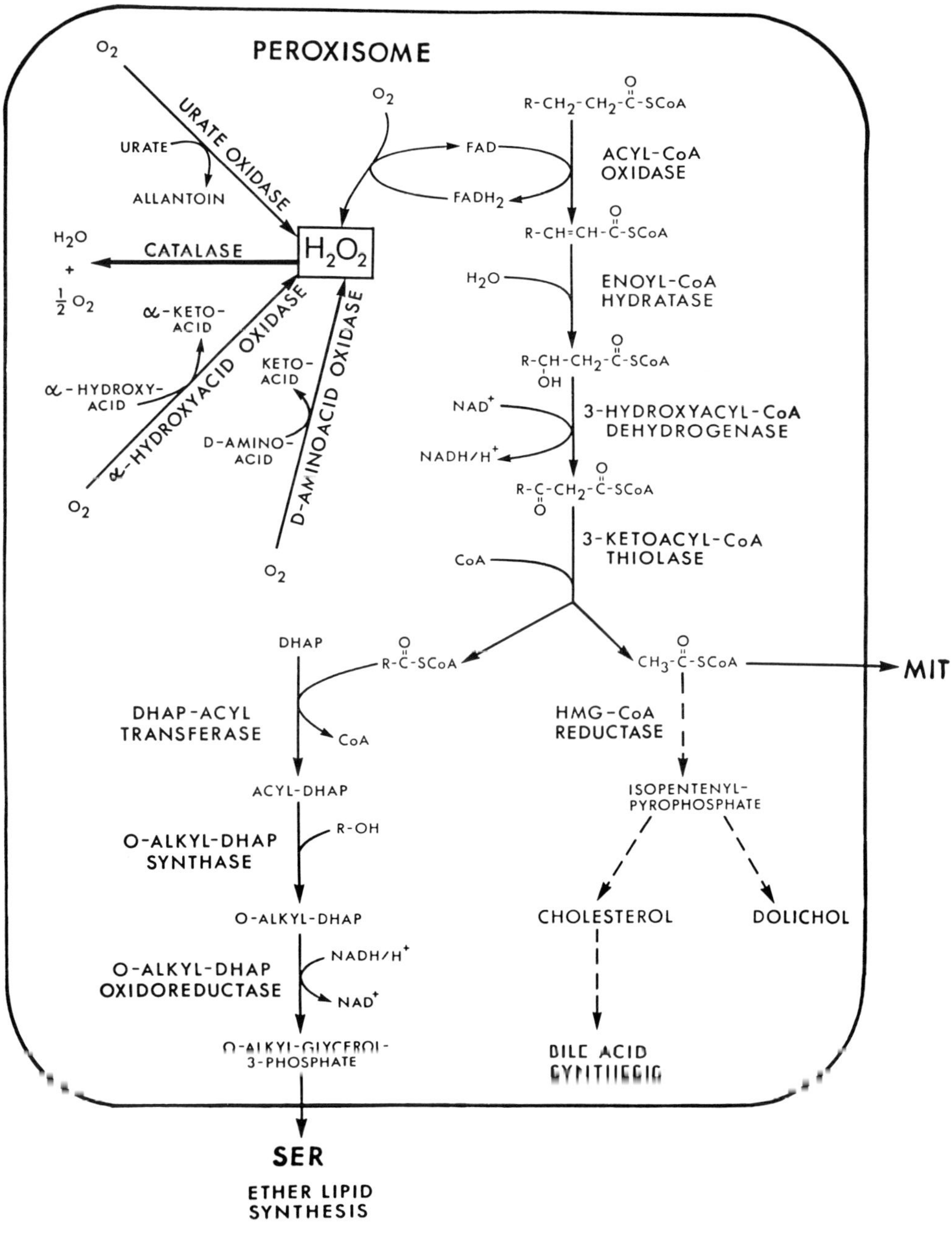

FIGURE 1. The metabolic pathways in rat liver peroxisomes.

whereas with Tris-maleate a uniform diffuse staining of the matrix is obtained (Figure 7). Although peroxisomes and most of their enzymes are uniformly distributed across the liver lobule, the D-amino acid oxidase activity is more pronounced in the periportal than in pericentral (perivenous) hepatocytes.[13] Moreover, the intensity of staining for this enzyme in peroxisomes varies not only in neighboring cells but even within one and the same cell (Figure 5).[13,14] The enzyme malate synthase, which is a marker of the glyoxylate cycle, has

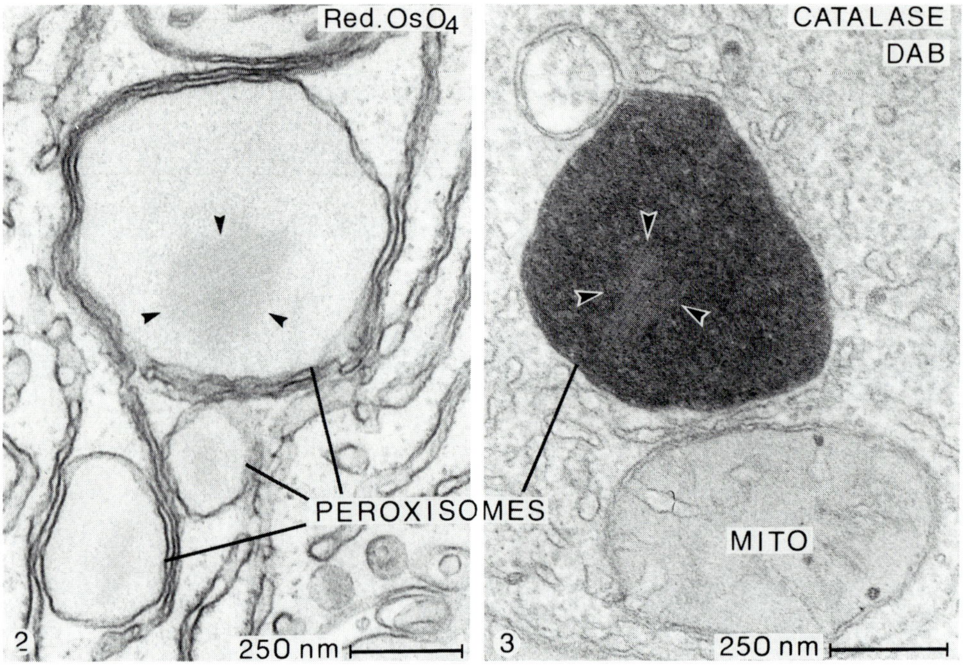

FIGURE 2. Rat liver peroxisomes after processing by the K-ferrocyanide reduced osmium procedure. Note the cores in some peroxisomes (arrowheads).

FIGURE 3. A rat liver peroxisome stained with the DAB method for catalase;[6] MITO: mitochondrion.

TABLE 1
Current Methods for the Electron Microscopic Cytochemical Localization of Peroxisomal Enzymes

Enzyme	Principal agent	Localization within the peroxisomes	Ref.
Catalase	DAB	Matrix	5–8
Urate oxidase	Cerium	Core	9
L-α-Hydroxyacid oxidase	Cerium	Matrix (marginal plates)	10–12
D-Amino acid oxidase	Cerium	Matrix (focal aggregation)	13–15
Xanthine oxidase	Cerium	Core	16,17
Polyamine oxidase	Cerium	Matrix	18
D-Aspartate oxidase	Cerium	Matrix	19
Malate synthase	K-Ferricyanide	Matrix	20

been localized in glyoxysomes of plant cells using a ferricyanide reduction method.[20] By applying this technique, malate synthase has also been found in a few animal cell types, such as the epithelial cells lining the toad urinary bladder.[28]

C. IMMUNOELECTRON MICROSCOPIC METHODS FOR PEROXISOMAL PROTEINS

Numerous peroxisomal proteins have been investigated by immunoelectron microscopy (Table 2). The first studies were carried out by Yokota and Nagata,[29] using ultrathin cryosections and ferritin-labeled antibodies. More convincing results were reported later by the

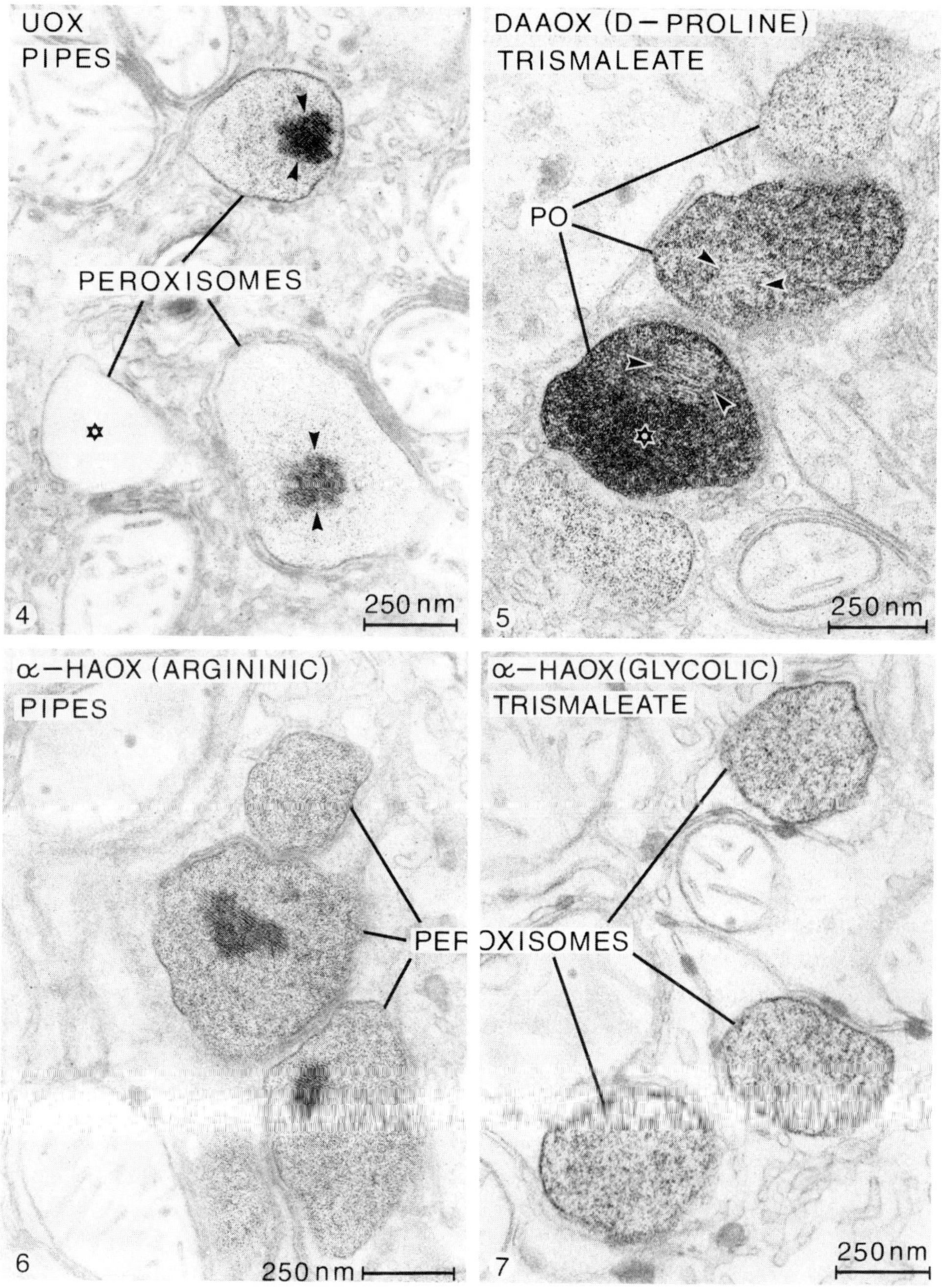

FIGURE 4. Rat liver peroxisomes stained for urate oxidase (UOX) using the cerium technique. Note the distinctly positive crystalline cores (arrowheads). One peroxisome without a core completely lacks cerium reaction product (*).

FIGURE 5. Rat liver peroxisomes stained for D-amino acid oxidase (DAAOX) with D-proline as substrate and Tris-maleate buffer, using the cerium technique. Note the heterogeneity in the intensity of staining and the focal aggregation of the reaction product in the matrix (*) which is distinct from the urate oxidase-positive crystalline core (arrowheads).

FIGURE 6. Peroxisomes (rat liver) stained for α-hydroxy acid oxidase with argininic acid as substrate and PIPES buffer. Note the diffuse reaction product in the matrix of all peroxisomes with focal condensation in the cores.

FIGURE 7. A similar preparation as in Figure 5, but with glycolic acid as substrate and Tris-maleate buffer. Only the diffuse staining in the matrix is observed but the focal condensation in the cores is absent.

TABLE 2
Peroxisomal Proteins Localized by Immunoelectron Microscopy (Mainly in Liver and Kidney)

Protein	Method used	Localization within the peroxisome	Ref.
Catalase	Preembedding and postembedding PAG[a]	Matrix	29–32
Urate oxidase	Postembedding PAG and preembedding PAG applied to isolated cores	Cores	33,34
Acyl-CoA oxidase	Preembedding and postembedding PAG	Matrix	32,35,36
Hydratase/dehydrogenase (bifunctional enzyme)	Postembedding PAG	Matrix	32,36,37
Thiolase	Postembedding PAG	Matrix	32,36,38
D-Amino acid oxidase	Postembedding PAG	Aggregated in matrix	38,39
α-Hydroxy acid oxidase	Postembedding PAG	Matrix (marginal plates)	12,36,38
Acyl-CoA synthase	Postembedding PAG	Po-Membrane	38,40
HMG-CoA reductase	Cryosection PAG	Matrix	41
Alanine-glyoxylate aminotransferase	Postembedding PAG	Matrix	42
Serine pyruvate aminotransferase	Postembedding PAG	Matrix	38,43
Po Membrane Proteins (22, 26, 70 kDa)	Postembedding PAG	Membrane	44,45
Sterol-carrier protein	Postembedding PAG	Matrix	46,47

[a] PAG, Protein A-gold technique.

application of the preembedding immunoelectron microscopy using peroxidase-labeled Fab fragments against catalase and acyl-CoA oxidase.[30,35] After the introduction of the protein A-gold technique, this method has been used extensively for the investigation of peroxisomal proteins.[31-47] Figures 8 to 10 show examples of the application of this method for characterization of distinct domains within peroxisomes, with catalase in the matrix (Figure 8), urate oxidase in the core region (Figure 9), and the 70-kDa peroxisomal membrane protein localized to the limiting membrane (Figure 10). This method has been applied to both ultrathin frozen sections,[41] as well as to the material embedded in various resins which preserve the antigenicity of peroxisomal proteins.[33,45] Immunoelectron microscopy has proven very useful for the characterization of peroxisomal fractions isolated in the course of biochemical studies (Figures 11 to 13). Figures 11 and 12 show the application of the preembedding immunolabeling with antibodies to 70-kDa membrane proteins and urate oxidase, respectively, to isolated peroxisomes or their cores. On the other hand, it is possible to apply the postembedding protein A-gold technique to fractions embedded in resins, as shown in Figure 13, with the localization of catalase. Because of the high specificity and sensitivity of the immunocytochemical procedures, this approach has become the method of choice for ultrastructural investigation of peroxisomes and analysis of their proteins.

D. CONTRIBUTIONS OF CYTOCHEMISTRY AND IMMUNOCYTOCHEMISTRY TO THE UNDERSTANDING OF THE BIOLOGY AND PATHOBIOLOGY OF PEROXISOMES

1. The Discovery of the Ubiquity of Peroxisomes and "Microperoxisomes"

In the original studies of de Duve and Baudhuin,[48] peroxisomes were considered to be present only in a few cell types such as hepatocytes and renal tubular epithelial cells.

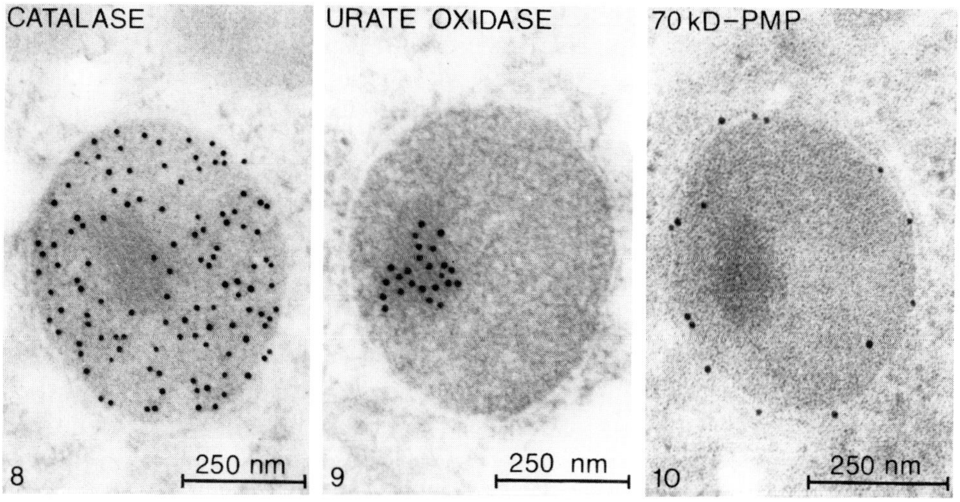

FIGURE 8. Immunocytochemical localization of catalase by the postembedding protein A-gold (PAG) technique. Catalase is localized in the matrix, sparing the core region.

FIGURE 9. Localization of peroxisomal urate oxidase with PAG. Urate oxidase is localized only in the electron-dense core.

FIGURE 10. The 70-kDa peroxisomal membrane protein (PMP) is confined to the limiting membrane, with localization by protein A-gold.

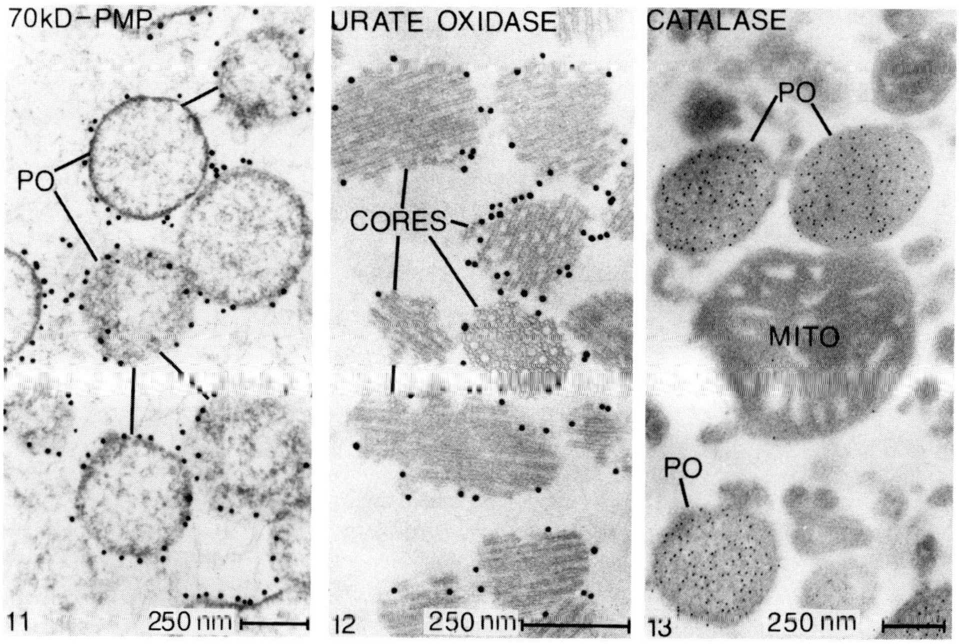

FIGURE 11. Localization of 70-kDa peroxisomal membrane protein in isolated peroxisomes by the preembedding PAG technique; PO: peroxisomes.

FIGURE 12. Localization of urate oxidase in isolated cores by the preembedding technique.

FIGURE 13. Localization of catalase in an isolated peroxisome fraction by the postembedding technique after embedding in LR-white; PO: peroxisomes, MITO: mitochondrion.

Application of the DAB technique for cytochemical localization of catalase led to the discovery of the ubiquity of this organelle.[2,49] Indeed, peroxisomes of different forms and sizes were found in almost all plant and animal cells, with a few rare exceptions, such as in mature red blood cells. Novikoff and Novikoff[50] suggested the term "microperoxisomes" for the designation of most of the newly found peroxisomes because of their smaller size (0.1 to 0.25 μm), their presumed frequent continuities with endoplasmic reticulum (ER), and their lack of crystalline inclusions of urate oxidase. We believe that there is little ground for further use of this term and that it should be abandoned because: (1) peroxisomes differ not only in their content of urate oxidase but also in that of several other oxidases (Figure 5),[15] (2) the presumed continuities with ER[51] have proven to be interconnections between peroxisomes.[14,45,52] Thus, the only justification for further use of the term "microperoxisome" would be for designation of small peroxisomes in contrast to the larger-sized ones. However, since no separate terms are used for small mitochondria or lysosomes, there is no reason to use a special term for designation of small peroxisomes.

In addition to catalase which serves for the identification of peroxisomes, the enzymes of peroxisomal lipid β-oxidation, particularly acyl-CoA oxidase, should be useful markers for the immunocytochemical detection of peroxisomes in different cell types.[53]

2. Xenobiotic-Induced Proliferation of Peroxisomes

Marked proliferation of peroxisomes is observed after the treatment of most rodents with clofibrate and several other hypolipidemic drugs (see Figure 14). Since long-term treatment with such drugs is associated with the development of malignant tumors in some rodent species (for a review see Reference 54), careful toxicologic screening, including morphometric analysis of peroxisome proliferation, have become quite important in recent years. In this context, the selective staining of peroxisomes with the cytochemical DAB method for catalase has been found very useful for the application of the automatic image analysis systems for the determination of morphometric parameters of peroxisomes.[55,56] In general, it can be stated that this approach has proven not only accurate, but also efficient and cost saving.

The xenobiotic-induced peroxisome proliferation is associated with the increase of the activity of some peroxisomal enzymes, particularly those of fatty acid β-oxidation, whereas the majority of peroxisomal proteins are unaffected or even reduced.[36,37] Figures 14 and 15 show the cytochemical localization of catalase and D-amino acid oxidase in livers of rats treated with BM 15766, a new hypocholesterolemic agent.[57] Whereas catalase is uniformly present in proliferated peroxisomes, the staining intensity for D-amino acid oxidase varies in different peroxisomes and is generally reduced in comparison to controls (Figure 5). The xenobiotic-induced alterations of peroxisomal proteins have also been investigated quantitatively by the protein A-gold technique.[36,37,53] The induction of peroxisomal bifunctional protein (hydratase/dehydrogenase) by bezafibrate,[58] is shown in Figures 16 and 17 with the marked increase of the labeling density of peroxisomes in the treated animal. The density of gold particles over the matrix of peroxisomes (gold particle/μm^2) is determined by measuring the surface area of the peroxisome in sections and counting the number of gold particles over it. The application of an automatic image analysis system for this time-consuming and tedious task was introduced by Beier and Fahimi[59] and it seems that quantitative immunoelectron microscopy, in conjunction with automatic image analysis, provides an efficient and reliable approach for the detection of modulations of peroxisomal proteins.[36]

It must be emphasized that most studies on peroxisome proliferation and formation of liver tumors after treatment with hypolipidemic drugs have been carried out in rodents and their relevance for humans has not been established.[60] Indeed, in a few morphometric studies

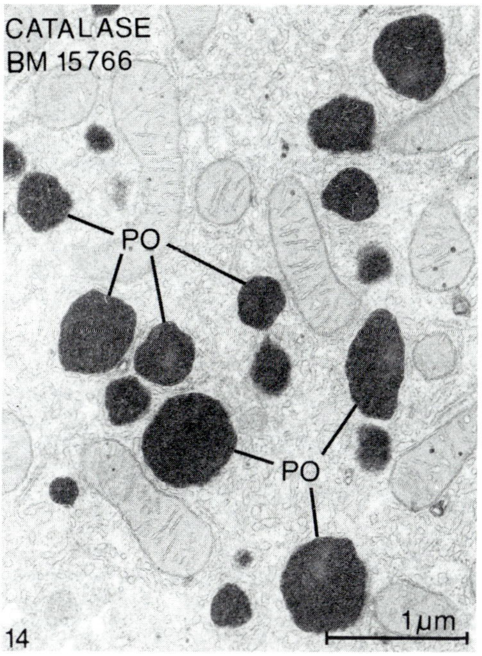

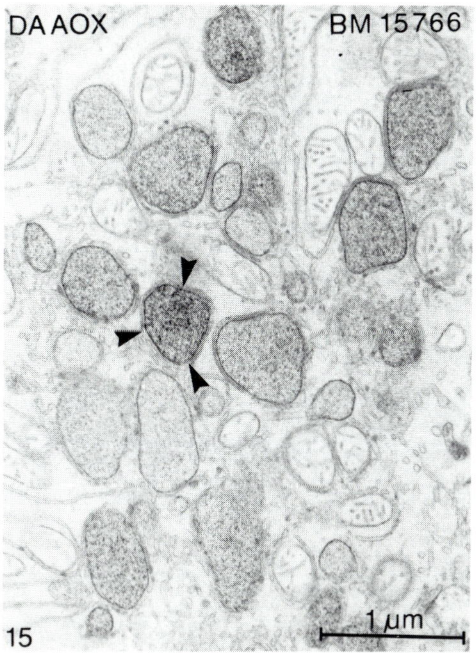

FIGURE 14. Peroxisome proliferation in the liver of a rat treated with a hypolipidemic drug (BM 15766)[57] and stained for catalase with DAB. Note the uniform staining in all peroxisomes.

FIGURE 15. Peroxisome proliferation in rat liver with staining for D-amino acid oxidase (DAAOX) with D-proline as substrate. Note the overall reduction in the intensity of staining in comparison to the control (Figure 5) and the heterogeneity of the reaction with darker- (arrowheads) and lighter-stained peroxisomes.

published hitherto, the proliferation of peroxisomes in livers of patients treated with fibrates has not been unequivocally demonstrated.[61] With the availability of the highly sensitive quantitative immunogold technique[36] and appropriate methods for processing of human tissues for immunocytochemistry of peroxisomal enzymes,[32] it should be rewarding to investigate biopsies from humans exposed to such drugs for the assessment of the risk factors.

3. Peroxisomal Diseases

In recent years a growing number of inherited metabolic diseases have been discovered which are characterized by the impairment of one or several peroxisomal functions.[illegible] Essentially two categories are distinguished: (1) in the first group, peroxisomes are either absent or markedly reduced in number and clinically multiple peroxisomal functions are deficient; (2) in the second group, peroxisomes are present and appear unaffected by regular ultrastructural criteria, but they lack one or several specific enzymes. The absence of peroxisomes in Zellweger syndrome, which is the main representative of the first category of these disorders, was discovered in human liver biopsies, which were processed for the electron microscopic cytochemical localization of catalase.[64] It is of interest that whereas biochemically catalase activity was detectable in homogenates, by cytochemistry peroxisomes could not be found.[64] Subsequent studies have revealed that indeed catalase and some other peroxisomal matrix proteins are synthesized normally and are present in the cytoplasm but their import into the matrix of peroxisomes is apparently defective. By immunocytochemistry, the 70-kDa peroxisomal membrane protein has been localized to small and large vesicles which have been called ''peroxisomal ghosts'' in the cytoplasm of cell lines from

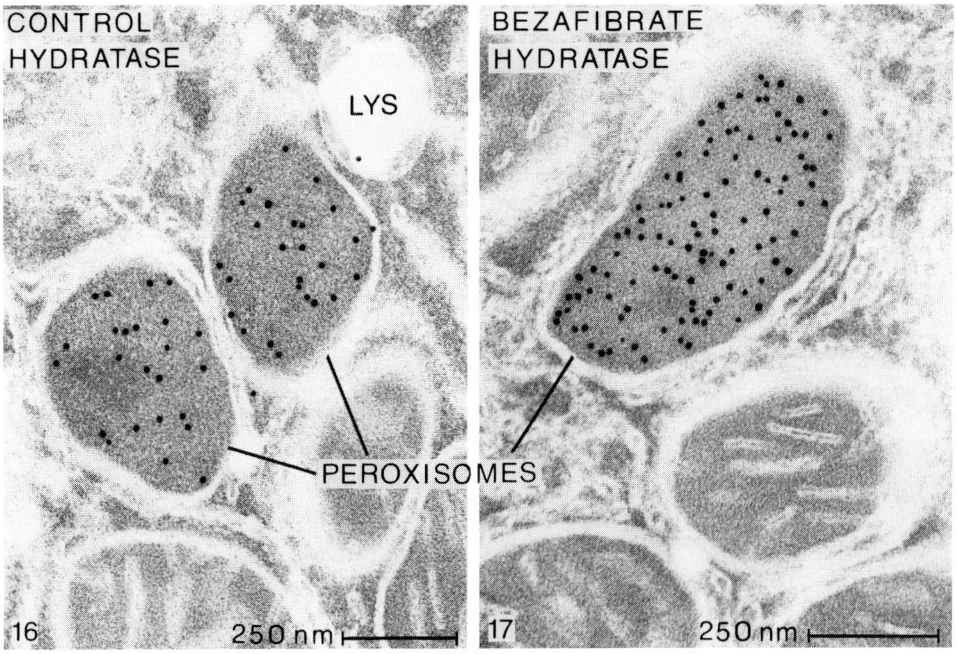

FIGURE 16. Rat liver peroxisomes immunostained for hydratase/dehydrogenase in a section from a control animal (see Figure 17); LYS: lysosome.

FIGURE 17. Rat liver section from a rat treated with bezafibrate, showing peroxisomes after immunocytochemical incubation for hydratase/dehydrogenase. Note the marked increase in labeling density.

Zellweger syndrome patients.[65,66] Further studies are now in progress to characterize the exact molecular mechanisms which are involved in the import of peroxisomal proteins and their possible malfunction in Zellweger syndrome and related disease (see also the next section on biogenesis of peroxisomes).

The application of the DAB-cytochemical technique to liver biopsies from the second category of peroxisomal deficiency diseases of diverse types (neonatal and adult types of adrenoleukodystrophy and peroxisomal acyl-CoA oxidase deficiency) have revealed unusually large peroxisomes with diminished catalase staining and marked heterogeneity in the intensity of the cytochemical reaction.[67,68] In general, it seems that the decreased metabolic capacity of peroxisomes is associated with an increase of their size and heterogeneity in the catalase staining.

Immunoelectron microscopy has been applied only in a few instances to the investigation of peroxisomal disorders, particularly for the study of the hyperoxaluria type 1, which is due to an inherited deficiency of the hepatic alanine/glyoxylate aminotransferase.[42] This enzyme is localized in normal human liver entirely in peroxisomes, but in carnivores it is mitochondrial and in rodents is present both in peroxisomes and in mitochondria. In some patients with a severe clinical course, residual enzyme activity is present, ranging up to 30% of controls, but immunoelectron microscopy has revealed that the immunoreactive enzyme protein is localized in mitochondria.[69] It has been suggested that this disease is due to a unique trafficking defect, in which peroxisomal alanine/glyoxylate aminotransferase is diverted from peroxisomes to mitochondria.[69]

Since most antibodies to rat peroxisomal proteins cross-react with the corresponding proteins in human liver and kidney[32,70] and the methods for fixation and processing of human biopsies are relatively simple, it is expected that in future, immunoelectron microscopy will become a routine procedure in the investigation of primary peroxisomal diseases, as well as in studies of secondary alterations of peroxisomes in human pathology.[71]

4. Biogenesis of Peroxisomes

It is now well established that all peroxisomal proteins are synthesized on free polyribosomes and are integrated posttranslationally into the various compartments of the organelle.[72] A minimal targeting signal for the import of proteins into peroxisomes consisting of a C-terminal tripeptide: serine-lysine-leucine (SKL) has been identified,[73] and it seems that the import signal is conserved between yeast, plants, and mammalian cells.[74] In spite of the great progress in biochemical elucidation of the mechanisms involved in peroxisomal biogenesis, the morphological basis of *de novo* formation of peroxisomes remained controversial for some time.[2,72] The early observations of morphological association of peroxisomes with the ER led to the suggestion that peroxisomes may arise by budding from the ER.[48] The introduction of DAB-cytochemistry, however, revealed that the catalase reaction product was confined to the matrix of peroxisomes and was absent from the ER.[6] Moreover, differential cytochemical staining with DAB for catalase and peroxidase and serial ultrathin sectioning showed that peroxisomes and ER formed two distinct and separate compartments in sinus-lining cells of rat liver.[75] By using computer-assisted three-dimensional reconstruction of ultrathin sections it was shown that in regenerating rat liver several peroxisomes were interconnected at times, forming a so-called "peroxisomal reticulum".[14] Similar structures were also identified in mouse liver[76] and in sebaceous glands.[52] The tail-like extensions of peroxisomal membrane, which were once considered to be the connections to the ER, were found recently to be interconnecting bridges between adjacent peroxisomes, containing peroxisomal membrane proteins.[45] These observations provide morphologic evidence in support of the concept of *de novo* formation of peroxisomes from preexisting ones.[72] On the other hand, in some cell types, such as bovine renal cortical tubular epithelial cells, peroxisomes exhibit lateral membrane contacts with ER which even persist after the isolation of peroxisomes.[77] This observation confirms the close functional cooperation between the two compartments, e.g., in the biosynthesis of ether-glycerolipids. Such lateral membrane contacts could also serve as bridges for the transport of phospholipids from their site of synthesis at the outer leaflet of ER to the peroxisomal membrane.[78]

ACKNOWLEDGMENTS

The authors thank Professors A. Völkl and T. Hashimoto for providing the antibodies used in this study. They also acknowledge the assistance of Ms. A. Achten, H. Hamberger, and Dr. K. Beier in word processing and assembling of the references. The technical assistance of Ms. P. Christmann and G. Krämer is gratefully acknowledged.

The experimental work presented here has been supported by grants from the Deutsche Forschungsgemeinschaft (Vo 317/3-2) and grants of the state of Baden-Württemberg (LFSP; Transportvorgänge im Endomembransystem).

54. **Reddy, J. K. and Lalwani, N. D.,** Carcinogenesis by hepatic peroxisome proliferators: evaluation of the risk of hypolipidemic drugs and industrial plasticizers to humans, *Crit. Rev. Toxicol.,* 12, 1, 1983.
55. **Beier, K. and Fahimi, H. D.,** Application of automatic image analysis for morphometric studies of peroxisomes stained cytochemically for catalase. I. Electron-microscopic application, *Cell Tissue Res.,* 246, 635, 1986.
56. **Beier, K. and Fahimi, H. D.,** Application of automatic image analysis for morphometric studies of peroxisomes stained cytochemically for catalase. II. Light-microscopic application, *Cell Tissue Res.,* 247, 179, 1987.
57. **Baumgart, E., Stegmeier, K., Schmidt, F. H., and Fahimi, H. D.,** Proliferation of peroxisomes in pericentral hepatocytes of rat liver after administration of a new hypocholesterolemic agent (BM 15766), *Lab. Invest.,* 56, 554, 1987.
58. **Fahimi, H. D., Reinicke, A., Sujatta, M., Yokota, S., Özel, M., Hartig, F., and Stegmeier, K.,** The short- and long-term effects of bezafibrate in the rat, *Ann. N.Y. Acad. Sci.,* 386, 111, 1982.
59. **Beier, K. and Fahimi, H. D.,** Automatic determination of labeling density in protein A-gold immunocytochemical preparations using an image analyzer, *Histochemistry,* 82, 99, 1985.
60. **Bentley, P., Waechter, F., Bieri, F., and Stäubli, W.,** Risks and benefits of agents which induce hepatic peroxisome proliferation, in *Peroxisomes in Biology and Medicine,* Fahimi, H. D. and Sies, H., Eds., Springer, Berlin, 1987, 309.
61. **Gariot, P., Barrat, E., Drouin, P., Genton, P., Pointel, J. P., Foliguet, B., Kolopp, M., and Debry, G.,** Morphometric study of human cell modifications induced by fenofibrate, *Metabolism,* 36, 203, 1987.
62. **Lazarow, P. B. and Moser, H. W.,** Disorders of peroxisome biogenesis, in *The Metabolic Basis of Inherited Disease,* Scriver, C. R., Beaudet, A. L., Sly, W. S., and Valle, D., Eds., McGraw-Hill, New York, 1988, 1479.
63. **Wanders, R. J. A., Heymans, H. S. A., Schutgens, R. B. H., Barth, P. G., Van den Bosch, H., and Tager, J. M.,** Peroxisomal disorders in neurology, *J. Neurol. Sci.,* 88, 1, 1988.
64. **Goldfischer, S. L., Moore, C. L., Johnson, A. B., Spiro, A. P., Valsamis, M. P., Wisniewski, H. K., Ritch, R. H., and Norton, W. T.,** Peroxisomal and mitochondrial defects in the cerebro-hepato-renal syndrome, *Science,* 182, 62, 1973.
65. **Santos, M. J., Imanaka, T., Shio, H., and Lazarow, P. B.,** Peroxisomal integral membrane proteins in control and Zellweger fibroblasts, *J. Biol. Chem.,* 263, 10502, 1988.
66. **Wiemer, E. A. C., Brul, S., Just, W. W., Van Driel, R., Brouwer-Kelder, E., Van den Berg, M., Weijers, P. J., Schutgens, R. B. H., Van den Bosch, H., Schram, A., Wanders, R. J. A., and Tager, J. M.,** Presence of peroxisomal membrane proteins in liver and fibroblasts from patients with the Zellweger syndrome and related disorders: evidence for the existence of peroxisomal ghosts, *Eur. J. Cell Biol.,* 50, 407, 1989.
67. **Roels, F., Pauwels, M., Poll-Thé, B. T., Scotto, J., Ogier, H., Aubourg, P., and Saudubray, J. M.,** Hepatic peroxisomes in adrenoleukodystrophy and related syndromes: cytochemical and morphometric data, *Virchows Arch. A,* 413, 275, 1988.
68. **Poll-The, B. T., Roels, F., Ogier, H., Scotto, J., Vamecq, J., Schutgens, R. B. H., Wanders, R. J. A., and Van Roermund C. W. T.,** A new peroxisomal disorder with enlarged peroxisomes and a specific deficiency of acyl-CoA oxidase (pseudo-neonatal adrenoleukodystrophy), *Am. J. Hum. Genet.,* 42, 422, 1988.
69. **Danpure, C. J., Cooper, P. J., Wise, P. J., and Jennings, P. R.,** An enzyme trafficking defect in two patients with primary hyperoxaluria type I: peroxisomal alanine/glyoxylate aminotransferase rerouted to mitochondria, *J. Cell Biol.,* 108, 1345, 1989.
70. **Litwin, J. A., Völkl, A., Müller-Höcker, J., and Fahimi, H. D.,** Immunocytochemical demonstration of peroxisomal enzymes in human kidney biopsies, *Virchows Arch. B,* 54, 207, 1988.
71. **Goldfischer, S. L. and Reddy, J. K.,** Peroxisomes (microbodies) in cell pathology, *Int. Rev. Exp. Pathol.,* 26, 45, 1984.
72. **Lazarow, P. B. and Fujiki, Y.,** Biogenesis of peroxisomes, *Annu. Rev. Cell Biol.,* 1, 489, 1985.
73. **Gould, S. J., Keller, G. A., Hosken, N., Wilkinson, J., and Subramani, S.,** A conserved tripeptide sorts proteins to peroxisomes, *J. Cell Biol.,* 108, 1657, 1989.
74. **Gould, S. J., Keller, G. A., Schneider, M., Howell, S. H., Garrard, L. J., Goodman, J. M., Distel, B., Tabak, H., and Subramani, S.,** Peroxisomal protein import is conserved between yeast, plants, insects and mammals, *EMBO J.,* 9, 85, 1990.
75. **Fahimi, H. D., Gray, B. A., and Herzog, V.,** Cytochemical localization of catalase and peroxidase in sinusoidal cells of rat liver, *Lab. Invest.,* 34, 192, 1976.
76. **Gorgas, K.,** Serial section analysis of mouse hepatic peroxisomes, *Anat. Embryol.,* 172, 21, 1985.
77. **Zaar, K., Völkl, A., and Fahimi, H. D.,** Association of isolated bovine kidney cortex peroxisomes with endoplasmic reticulum, *Biochim. Biophys. Acta,* 897, 135, 1987.
78. **Pagano, R. E.,** Lipid traffic in eukaryotic cells: mechanisms for intracellular transport and organelle-specific enrichment of lipids, *Curr. Opinion Cell Biol.,* 2, 652, 1990.

Chapter 9

CELL ORGANELLES

9.7 Mitochondria

Takashi Makita

TABLE OF CONTENTS

0-8493-6012-9/93/$0.00 + $.50

A. INTRODUCTION

There are numerous reviews on mitochondrial structure.[1-4] The mitochondria have four compartments: (1) the outer membrane, (2) the outer compartment or intermembrane space, (3) the inner membrane including cristae, and (4) the inner compartment or matrix. The overall structure is illustrated in Figure 1 and the subsidiary structure of each compartment will be discussed later. Table 1 is the list of enzymes so far localized in these four compartments.

Not all those enzymes were localized cytochemically at the electron microscopic level but many of them particularly related to the mitochondrial respiratory chain in the inner membrane (Figure 2) have been visualized at the ultrastructural level. In this short article, (1) succinate dehydrogenase (SDH) and its related enzymes, (2) ATPases, (3) glutamic oxaloacetic transaminase (GOT), (4) carnitine acetyltransferase (CAT), (5) Mn-superoxide dismutase (SOD), and some others will be briefly described as general subjects of electron microscopic enzyme cytochemistry. Other current topics on mitochondria will be discussed as perspectives in Section C.

B. EXAMPLES OF ENZYME CYTOCHEMISTRY OF MITOCHONDRIA

1. Succinate Dehydrogenase (SDH)

Succinate dehydrogenase (SDH, EC 1.3.99.1) is one of the typical marker enzymes of mitochondria. The cytochemistry of this enzyme is also applicable to other related enzymes such as cytochrome *c* reductase, β-hydroxybutyrate reductase, $NADH_2$ reductase, MAO, xanthin oxidase, malate reductase, lactate reductase, 3β-hydroxysteroid reductase, α-hydroxyacid oxidase, and some others.

Since the activities of these enzymes are easily abolished by aldehyde fixatives, slices of unfixed tissues are incubated in the medium of Ogawa et al.[5] or that of Kerpel-Fronius and Hajós.[8] The compositions of those media are described earlier in this volume.

Dimethyl sulfoxide (DMSO) can be added to the incubation medium (5% v/v) to increase the density of reaction and obtain a uniform reaction of mitochondria.[6,7] Otherwise, reaction in each mitochondrion is not uniform and it has been considered as a reflection of irregular level of SDH activity in each mitochondrion. The comparison of both incubation media was fully discussed elsewhere.[9] Since reaction product, cupric ferrocyanide, is soluble in alkaline solution, post-staining with alkaline lead citrate should be avoided.[10] Postfixation with glutaraldehyde and OsO_4 is essential.

As shown in Figures 3 to 7, reaction products are on the outer surface of the inner membrane. The composition of the reaction product can be identified with X-ray microanalysis. However, deposit between the outer and inner membranes is patchy and irregular among adjacent mitochondria, between adjacent crista, and even in a given crista. The distribution of the deposit may not reflect the precise site of SDH reaction in mitochondria.

Besides the copper ferricyanide methods mentioned above, SDH can be localized by the 2-(2′-benzothiozolyl)-5-styryl-3-(4-pathalhydrazidyl) tetrazolium chloride (BSPT) method.[11]

Incubation medium:	Tris-HCl buffer (0.1 *M*, pH 7.2)	5.0 ml
	BSPT dissolved in 0.25 ml of DMSO	2.5 mg
	Sodium succinate (0.1 *M*)	0.5 ml

The final pH is 7.0 to 7.2. Osmolarity is 350 mOsM. The medium has to be mixed immediately before use. Incubation time is 30 to 90 min at 37°C. Reaction products are uniform and linear in most cristae and some parts of the outer compartment. Occasionally the reaction product does not cover the entire crista but stops at the joint of the crista and

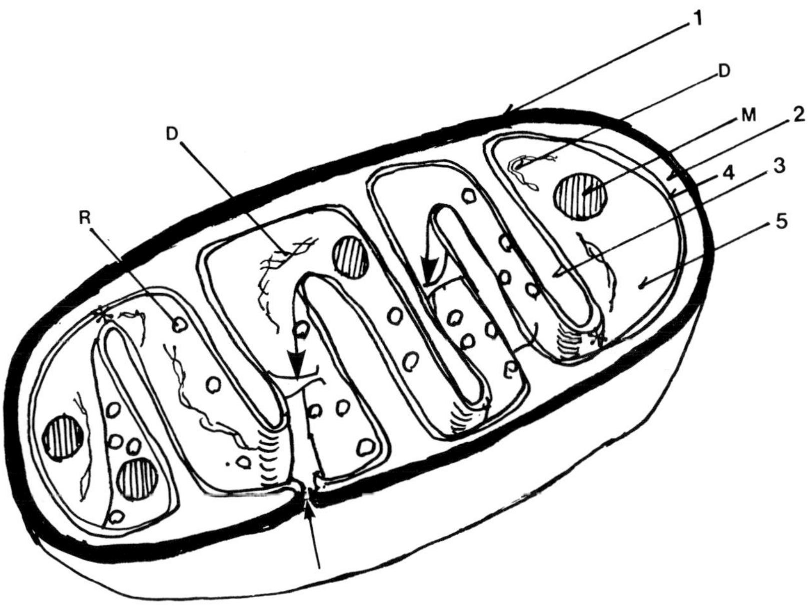

FIGURE 1. General subdivision and subsidiary structure of mitochondrion. 1, Outer membrane; 2, outer compartment or intermembrane space; 3, cristae; 4, inner membrane; 5, inner compartment or matrix; M, matrix granule; r, ribosomes (smaller than cytosolic ones in size); D, mitochondrial DNA; Arrowheads, bridge structure between cristae; Arrows, channel structure. The size of each substructure is not proportional to real size.

the inner membrane. This joint portion has been noticed as an SDH-free region, but whether it reflects any functional significance or not is still unknown.

Control experiments: incubation in the following media is informative: (1) substrate-free medium, (2) inhibitor-containing medium — 36 m*M* oxaloacetate, 36 m*M* malonate, or 36 m*M* malonate plus 20 μg/ml antimycin A. Treatments such as boiling sections for 5 to 10 min or fixation with 2% glutaraldehyde for 15 to 60 min will also abolish the SDH activity.

Application of SDH cytochemistry to other related enzymes with minor modification is available.

1. Cytochrome *c* reductase — instead of sodium succinate, add 7.4 mg (30 μM) of cytochrome *c* to the medium used in the copper ferrocyanide method.[5] The reaction is less intensive than that of SDH. The outer compartment has an even weaker reaction than the intracristal space (Figure 6).
2. $NADH_2$ reductase — instead of sodium succinate in the medium, add 10 mg (0.7 m*M*) $NADH_2$ to it. The reaction products are localized both in the outer compartment and intracristal space (Figure 7).
3. β-Hydroxybutyrate reductase — replace sodium succinate with 100 mg (40 m*M*) D-L-β-hydroxybutyrate and add 132 mg of NAD. Phenazine methosulfate (0.9 to 9.2 mg) and ubiquinon (UQ_{10}) dissolved in 0.2 ml of acetone (0.075 to 9.2 m*M*) can be added to the medium.

2. Mg-ATPase and Ca-ATPase

Compared with ATPase on plasma membrane, mitochondrial ATPase was difficult to localize at the electron microscopic level. Mg-ATPase has been localized in the matrix of

TABLE 1
Distribution of Enzymes in Mitochondrial Compartments

Outer membrane	Intermembrane space	Inner membrane plus crista	Matrix
NADH-cytochrome b_5 reductase	Adenylate kinase	Cytochrome *c* reductase	Mn-superoxide dismutase
Monoamine oxidase	Nucleoside diphosphokinase		Carnitine acetyltransferase
Phospholipase A	Nucleoside monophosphokinase		Ornitine carbamyl transferase
Nucleoside diphosphokinase		Cytochrome oxidase	Isocitrate dehydrogenase (NAD)
Phosphatidate phosphatase		Oligomycin-sensitive ATPase	Fumarase
Glycerophosphate acyltransferase		β-Hydroxybutyrate dehydrogenase	Glutamate dehydrogenase
Acyl-CoA synthetase		Carnitine acetyltransferase	Aspartate aminotransferase
Hexokinase 2		Choline dehydrogenase	β-Hydroxyacyl-CoA dehydrogenase
		α-Glycerophosphate dehydrogenase	RNA polymerase
			DNA polymerase
		Steroid 11 β-hydroxylase	
		Carnitine palmitoyltransferase	
		NAD(P) transhydrogenase	
		NADH dehydrogenase	
		Succinate dehydrogenase	

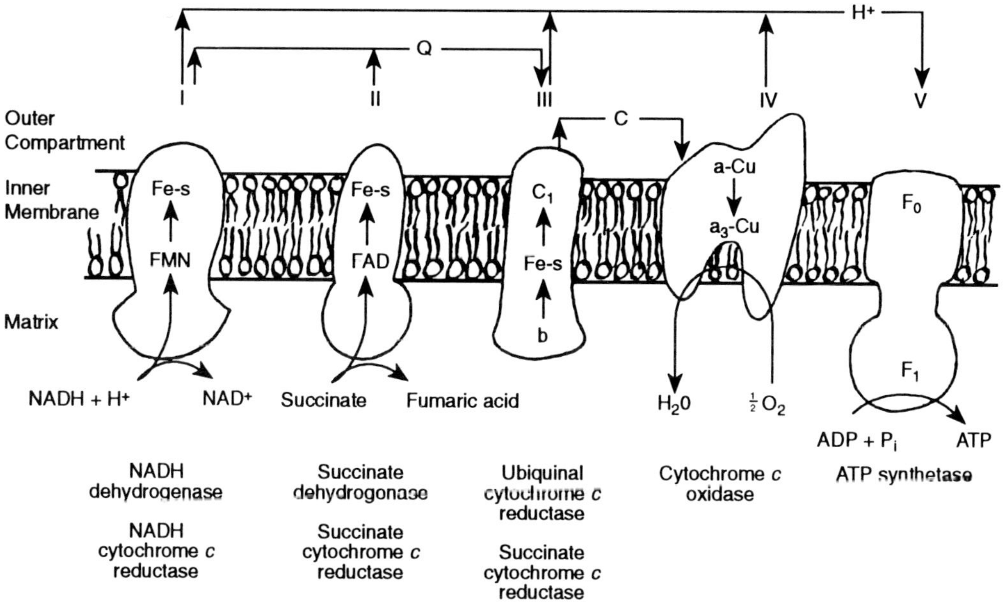

FIGURE 2. The inner mitochondrial membrane and enzymes related to its respiratory chain.

mitochondria.[12,13] In the course of investigation of creatine phosphokinase (CPK) cytochemistry, the following medium was found to result in stable and repeatable deposition of reaction product along the matrix side of inner membrane.[14]

Incubation medium:	Tris-HCl buffer (0.2 *M*, pH 9.0)	13 ml
	$MgSO_4 \cdot 7H_2O$	49 mg
	(or $CaCl_2$)	(23 mg)
	ATP-2Na	30-90 mg
	Lead citrate (0.5% in 50 m*M* NaOH)	5 ml
	DMSO	1 ml
	2-4-Dinitrophenol (DNP) (2 m*M*)	2 ml
Inhibitor:	Olygomycin or PCMB	10 m*M*

Since mitochondrial ATPase is partially abolished by glutaraldehyde, 2% paraformaldehyde with or without 0.5% glutaraldehyde is recommendable as a fixative. Incubation time ranges from 20 to 40 min. Incubation temperature is 37°C or room temperature. When sections are incubated at 37°C, reaction products fill the inner compartment and practically all mitochondria are positive to the reaction (Figure 8). Deposition of the reaction product indicates the partition or compartmentation of the intercristal space (Figure 9). It may imply the existence of branches of crista or a bridge structure spanning between cristae (Figure 1). When the reaction is suppressed with short incubation time and/or lowered incubation temperature, reaction products are lined on the matrix side of the inner membrane (Figure 10). As for the structure of mitochondrial ATPase and biosynthesis of it, some excellent reviews are available.[15-17]

Ca-ATPase location is quite similar to that of Mg-ATPase both in mitochondria and peroxisomes,[14] but Ca-ATPase activity appears weaker than Mg-ATPase under the condition we employed.

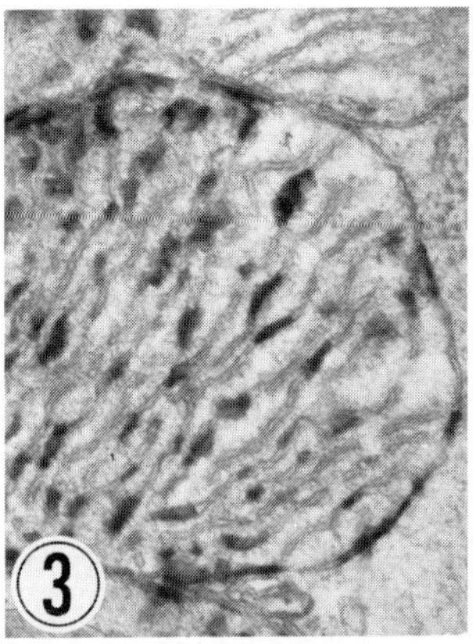

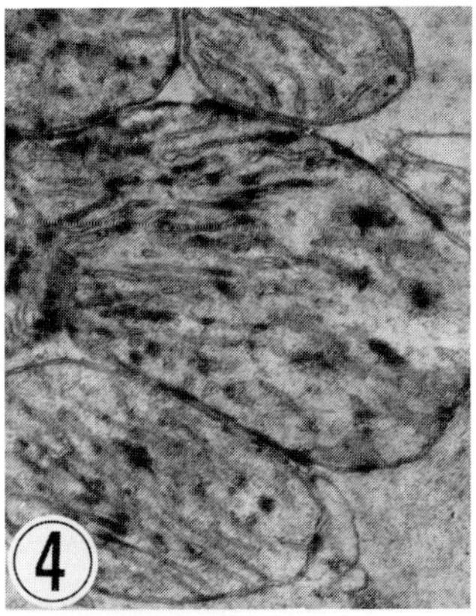

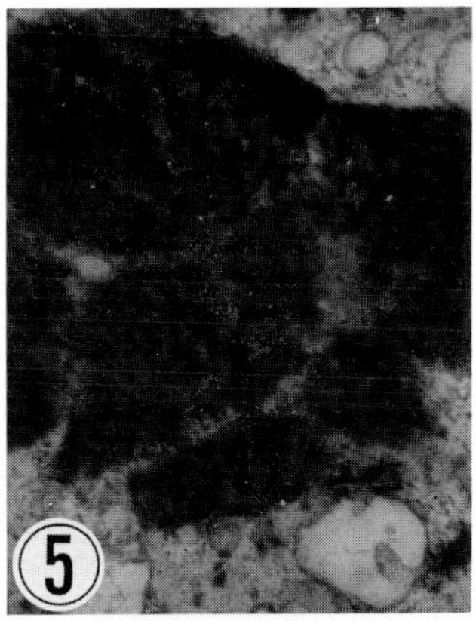

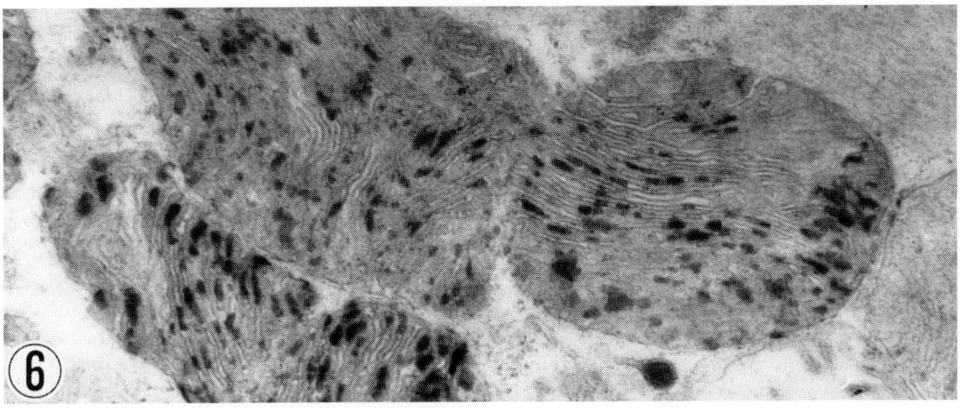

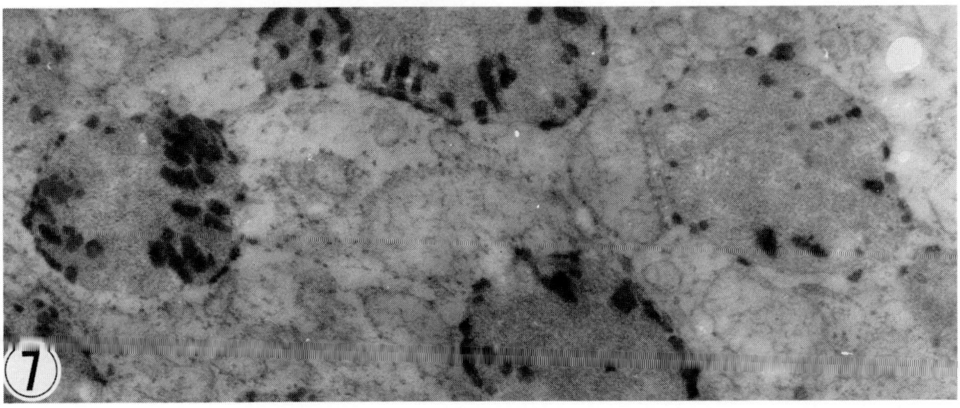

FIGURE 3. Succinate dehydrogenase (SDH) activity in mitochondria detected with copper ferrocyanide methods; unfixed rat cardiac muscle.

FIGURE 4. The intense and uniform reaction of SDH with addition of 5% DMSO to the incubation medium and incubation for 30 min.

FIGURE 5. The intense reaction of SDH with 5% DMSO-containing medium; incubation for 60 min.

FIGURE 6. Cytochrome *c* reductase activity; unfixed rat cardiac muscle. Incubation for 60 min.

FIGURE 7. $NADH_2$-reductase in unfixed hen oviduct — rather clumped deposits of reaction products are characteristic of this enzyme.

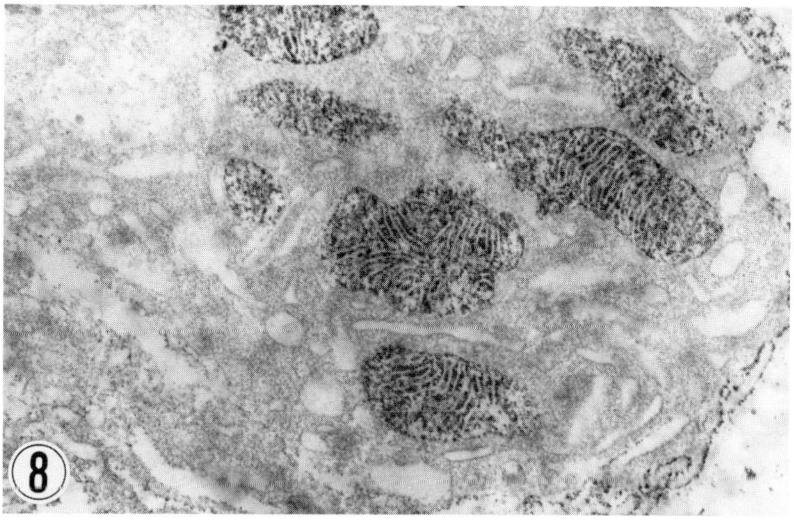

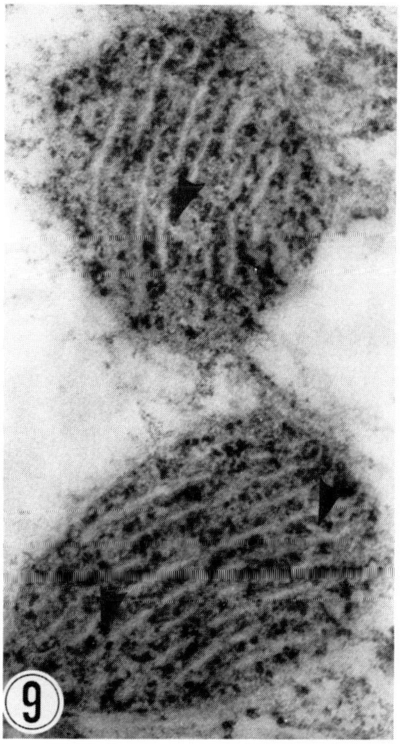

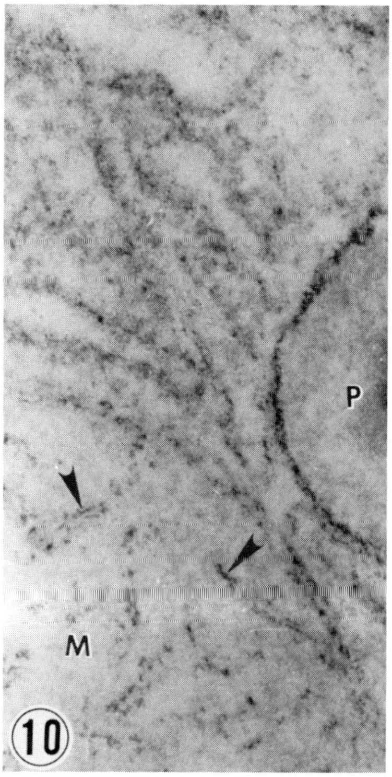

FIGURE 8. All mitochondria (rat liver) are reactive to Mg-ATPase reaction. The incubation medium (see text) gives stable and repeatable results without difficulties.

FIGURE 9. At a higher magnification (compare with Figure 8), reaction products fill the inner compartment and the mass of reaction product (ppb) superimpose the partition-like bridge structure spanning between cristae (arrowheads).

FIGURE 10. When the incubation time is diminished and incubation is at room temperature, reaction products are limited to the matrix side of the inner membrane (arrowheads); M, mitochondrion; P, peroxisome. Notice that the limiting membrane of peroxisome has Mg-ATPase activity.

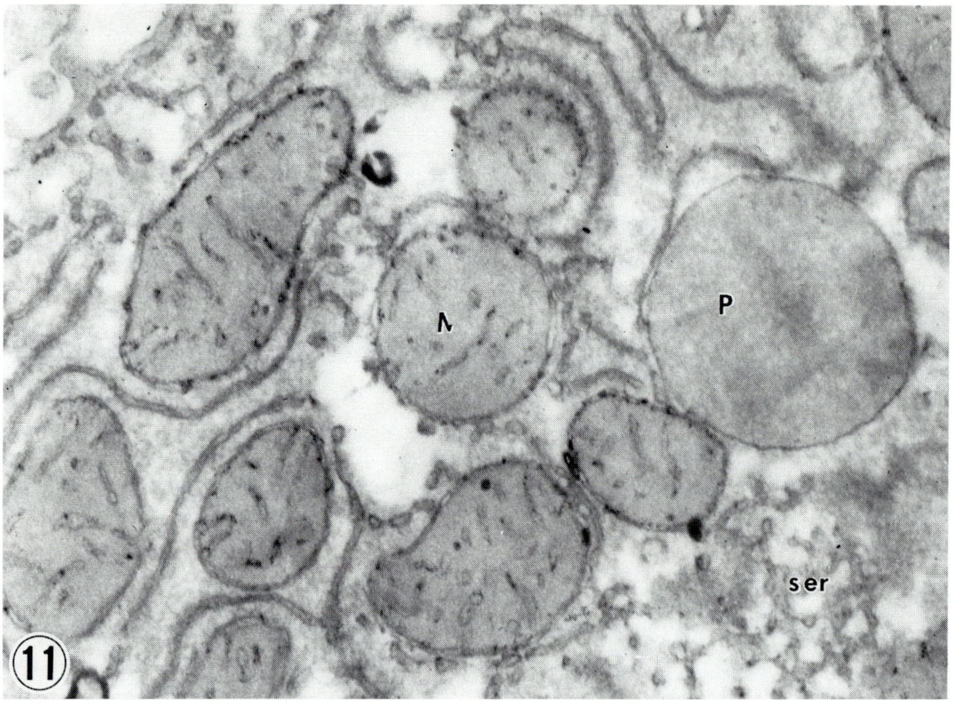

FIGURE 11. Glutamic oxaloacetic transaminase (GOT) in the mitochondria (M) of a hepatocyte of hamster. Though it is not illustrated in this picture, some part of the rough ER and limiting membrane of some peroxisomes also appear to be positive to GOT reaction; P, peroxisome; ser, smooth-surfaced endoplasmic reticulum.

3. Glutamic Oxaloacetic Transaminase (GOT)

GOT (EC 2.6.1.1.) has the highest activity of the transaminases in mammalian tissues. The cardiac muscle has the highest activity and other major organs such as skeletal muscle, kidney, brain, liver, spleen, testis, and prostate also contain a considerable level of the enzyme. This enzyme is involved in the following reaction:

$$\alpha\text{-ketoglutarate} + \text{L-aspartate} \rightleftarrows \text{L-glutamate} + \text{oxaloacetate}$$

Ultrastructural localization of GOT has been reported in the liver and other organs of rat,[18] mice,[22] and hamsters.[20] A method for the demonstration of GOT is described earlier in this volume.

As shown in Figure 11, GOT is localized in the cristal space and outer compartment. Occasionally GOT is seen on rough endoplasmic reticulum (ER) and peroxisomal membrane. Plasma membrane, smooth ER, and Golgi complex are devoid of GOT reaction.

Substrate protection on GOT[19,21] is one of the interesting subjects of GOT cytochemistry. Application of GOT cytochemistry for clinical investigations remains to be developed.

4. Carnitine Acetyltransferase (CAT)

CAT (EC 2.3.1.7.) is one of the members of the carnitine acyltransferases. This enzyme plays an important role in lipid metabolism and catalyzes the following reaction:

$$\text{acetyl-CoA} + \text{carnitine} \rightleftarrows \text{acetylcarnitine} + \text{CoA}$$

CAT is known to bind to the mitochondrial membrane. Slices of tissues, either unfixed or fixed with a mixture of 4% paraformaldehyde and 1% glutaraldehyde in cacodylate buffer (pH 7.4, 0.05 *M*), are incubated in either one of the following media.[23]

1. Uranyl-ferrocyanide method:

Potassium ferricyanide	5 mg
Uranyl acetate	2.5 mg
Acetyl CoA	2.0 mg
Carnitine	4.0 mg
Maleate buffer (0.05 *M*, pH 7.0)	2.5 mg
Dextrose	4.5% (w/v)
Inhibitor (mercuric chloride)	4×10^{-4} *M* (preincubate for 5 min)

2. Cadmium method:

Cacodylate buffer* (0.05 M, pH 7.0) with 4.5% (w/v) dextrose	2.5 ml
Cadmium nitrate	10.0 mg
Acetyl CoA	2.0 mg
Carnitine	4.0 mg

The reaction product (cadmium-CoA) is soluble. A high concentration of competitive inhibitor, acetyl carnitine (49 mg/12.5 ml of medium) is required as a specific inhibitor.

Localization of CAT (Figure 12) is essentially on the outer surface of the inner membrane of mitochondria,[24] though it has been reported to be localized biochemically both in the intermembrane space and in the matrix.[25] Some mitochondrial cristae have a central line[26] and those cristae have little, if any, reaction product of this enzyme.[27] The dystrophic muscle has many lipid droplets in the vicinity of mitochondria. Those mitochondria that engulf lipid droplets have altered arrangement of cristae and there are no reaction products (Figure 13) in such mitochondria.

5. Other Mitochondrial Enzymes

1. Cytochrome *c* oxidase (see also Chapter 1):

Incubation medium:[28]	Diaminobenzidine (DAB)	5 mg
	Cytochrome *c* (type II)	10 mg
	Sodium phosphate buffer (0.05 *M*, pH 7.4)	9 ml
	Catalase	20 μg/ml
	Sucrose	750 mg
Inhibitors:	Potassium cyanide	6.5 mg
	(or sodium azide)	(3-6 m*M*)

Reaction product in unfixed or 4% paraformaldehyde-fixed tissue is localized in the intermembrane space and intracristal space.

2. Superoxide dismutase (SOD) — SOD (EC 1.15.11.) is one of the free-radical scavengers and its clinical application and pathological study are progressively increased. There are three types of SOD and one of them, Mn-SOD, is exclusively localized in the matrix of mitochondria.[29] The appropriate method to localize Mn-SOD is an

* This buffer is not cacodylate/HCl, but cacodylate/nitric acid.

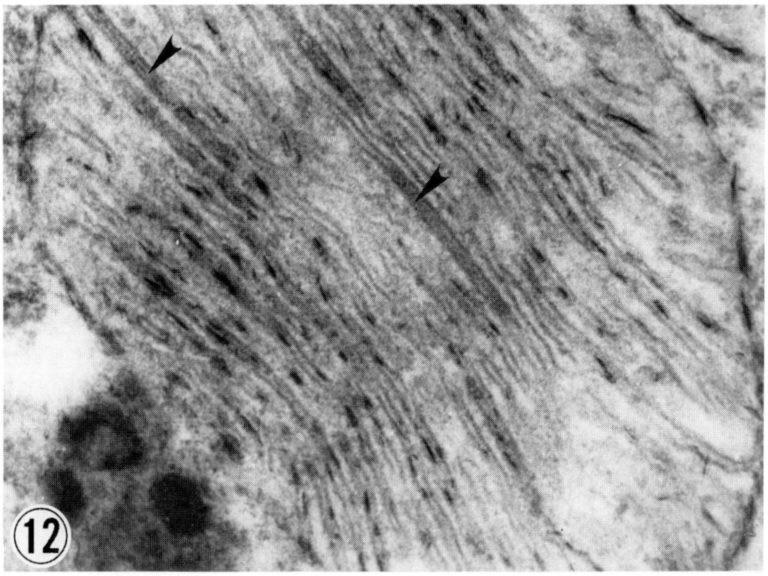

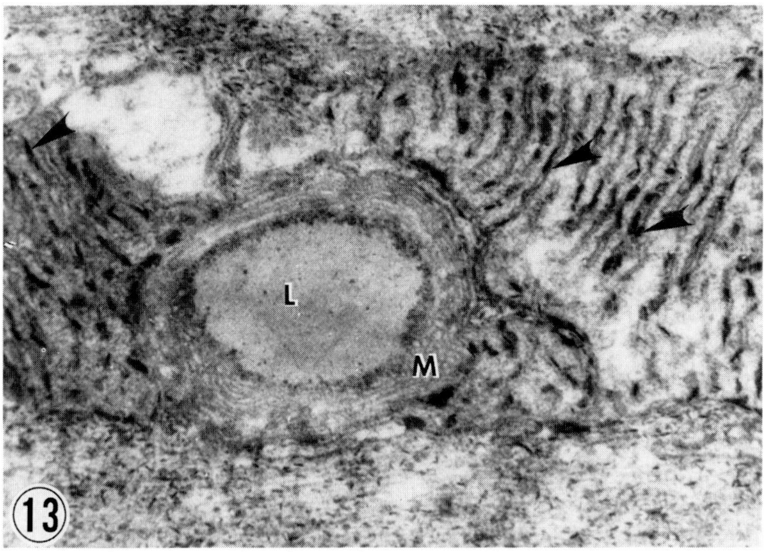

FIGURE 12. Carnitine acetyltransferase (CAT) activity in mitochondria of skeletal muscle of mice. Reaction products are on the outer surface of the inner membrane. Some cristae have a central line (arrowheads) and there are few reaction products, if any, on those cristae.

FIGURE 13. In dystrophic muscle of mice, many lipid droplets (L) are closely associated with mitochondria and cristae of those mitochondria are irregularly arranged (M). Such cristae have no reaction product of CAT despite the fact that reactions (arrowheads) are evident in neighboring mitochondria.

immunocytochemical procedure utilizing either avidin-biotin-peroxidase complex or protein A-gold. The sites of immunolabeled markers are uniformly distributed in the matrix of mitochondria without any clump formation or attaching to the inner surface of the inner membrane. (See also Chapter 1, Section 3, Article V.)

3. Monoamine oxidase (MAO), peroxidase, and ornitine carbamoyltransferase are all localized in the outer compartment, including the intracristal space. Alkaline phos-

phatase is often localized in the outer compartment but not in the cristae. This observation in this laboratory requires some other experiments to confirm it.

C. PERSPECTIVE

Although it would not be appropriate to try to review numerous subjects on mitochondria in this short article, the following topics should be interesting to those who are planning cytochemistry of mitochondria.

Transport of cytosolic substance into mitochondria — As an essential part of the argument on morphogenesis of mitochondria,[3,4] the precise localization of synthesis of proteins and other components or enzymes and their transportation to substructures of mitochondria have been key subjects.[30,31,37] One of the interesting concepts on this subject is the channel where the inner and outer mitochondrial membranes are fused, like in a nuclear pore, to allow the cytoplasmic substance to penetrate into the matrix of the mitochondria. Such channels are not easy to identify even at the electron microscopic level, but it will be important to localize certain enzymes, for example ATPase, at the channel. Those channels appear to increase in number in the mitochondria of the cardiac muscle of rats when an antitumor drug, adreamycin for example, is administered.

Pathology of mitochondria — The pathology of cell organelles is eventually the pathology of the membrane.[32] The deformation of cristae, increase of matrix granules, increase of channel structure, characterization of giant mitochondria, and specific inclusion bodies in the matrix are some of the subjects related to the pathology of mitochondria. The possibility that interdigitation of outer membranes of two adjacent mitochondria develops into dense bodies in degenerating mitochondria is an example of interesting suggestions.[33] Cytochemistry can detect the topographic change of mitochondrial enzyme activity. As shown in the case of carnitine acetyltransferase (Figure 13), deformed cristae and denatured membranes often lack marker enzyme.

Cytochemistry of isolated mitochondria — Despite the fact that progress of biochemistry of mitochondria largely depends on analysis of mitochondrial fractions, including naked mitochondria, cytochemistry of mitochondrial fraction has not been as fully investigated as that of mitochondria *in situ*. Observations of isolated and unfixed mitochondria in wet-chamber or wet-cell and cryosection of isolated mitochondrial fraction certainly represent a new field of mitochondrial cytochemistry. The difference between the low-energy orthodox and high-energy condensed conformation[34] in isolated mitochondria may account for functional stages in *in situ* mitochondria and cytochemical studies of these two conditions remain to be developed.

Technical problems associated with penetration of substrates into the mitochondria — Since mitochondria *in situ* are well guarded with two membranes, any substance to react with enzymes in mitochondria has to cross these two membranes. The antibody for immunocytochemical reaction and many other macromolecular substrates for cytochemical detection of enzymes have difficulty in reaching the site of reaction. Unfixed tissue slices or cryosections often result in inferior reaction intensity as compared with briefly fixed materials, most probably because of the self-guarding ability of living cell membranes to prevent exogenous substances from penetration. Treatment of membrane with detergents, such as saponin, to perforate it, freeze and thaw procedures, addition of DMSO to the incubation medium,[6,7] prolonged incubation time, elevated incubation temperature, and much more have been attempted to obtain regular and uniform reactions in mitochondria. All of those means, however, inevitably denature mitochondrial membranes. A uniform reaction is needed to evaluate the functional stage or/and pathological condition of mitochondria and also to minimize the denaturation of the mitochondria. This paradox is one of the main subjects of mitochondrial cytochemistry.

Some unveiled substructures of mitochondria — This article intentionally oversimplifies the structure of mitochondria, ignoring the variety of functional and structural differences of these organelles. As an excellent tool to detect pin-point localization of certain enzymes, cytochemistry must be utilized to visualize the difference between, for example, subsarcolemmal mitochondria and centrally located ones in cardiac muscle, newly formed young mitochondria and matured ones, apically located mitochondria and ones which are lined with basal infolding, and much more. The specific feature of mitochondria in sperm tail, differences among different organs of the same body, aging-related changes of mitochondria, and other topics are also being investigated cytochemically. In addition to the main mitochondrial substructures, cytochemists also should be concerned with some subsidiary structures, for example, the relationship between the cytoskeleton and the outer surface of the outer membrane of mitochondria,[35,36] bridge-like structures spanning between cristae (Figure 1), and the fine network of membranes in matrix which is similar to those in the matrix of peroxisomes,[14] mitochondrial ribosomes, and mitochondrial DNA are targets of recent cytochemistry of mitochondria. Since analyses of DNA sequences are a key subject of current biotechnology, autoradiography and molecular analysis of mitochondrial DNA will be a promising field of cytochemistry. Although it is beyond the scope of this article, mitochondrial cytochemistry includes X-ray microanalysis, pattern analysis, three-dimensional reconstruction, and cytoflowmetry in addition to the enzyme cytochemistry and immunocytochemistry summarized in this chapter.

REFERENCES

1. **Munn, E. A.,** *The Structure of Mitochondria,* Academic Press, London, 1974, 465.
2. **Tzagoloff, A.,** *Mitochondria,* Plenum Press, New York, 1982, 342.
3. **Attardi, G. and Schatz, G.,** Biogenesis of mitochondria, *Annu. Rev. Cell Biol.,* 4, 289, 1988.
4. **Schatz, G.,** The biogenesis of mitochondria, in *Mitochondria and Microsomes,* Lee, C. P., Schatz, G., and Dallner, G., Eds., Addison-Wesley, London, 1981, 45.
5. **Ogawa, K., Saito, T., and Mayahara, H.,** The site of ferricyanide reduction by reductases within mitochondria as studied by electron microscopy, *J. Histochem. Cytochem.,* 16, 49, 1968.
6. **Makita, T. and Sandborn, E. B.,** Fine structural localization of reductases and transferases in muscle mitochondria and the effect of dimethylsulfoxide (DMSO) in the incubation medium, *J. Histochem. Cytochem.,* 18, 686, 1970.
7. **Makita, T. and Sandborn, E. B.,** The effect of dimethylsulfoxide (DMSO) in the incubation medium for the cytochemical localization of succinate dehydrogenase, *Histochemie,* 26, 305, 1971.
8. **Kerpel-Fronius, S. and Hajós, F.,** The use of ferricyanide for the light and electron microscopic demonstration of succinic dehydrogenase activity, *Histochemie,* 14, 343, 1968.
9. **Ogawa, K. and Ago, Y.,** Oxidoreductase and related enzymes, *Cell,* 5, 18, 1973.
10. **Hanker, J. S.,** Oxidoreductases, in *Electron Microscopy of Enzymes. Principles and Methods,* Vol. 4, Hayat, M. A., Ed., Van Nostrand Reinhold, New York, 1975, 1.
11. **Kalina, M., Plapinger, R. E., Hoshino, Y., and Seligman, A. M.,** Nonosmophilic tetrazolium salts that yield osmophobic formazans for ultrastructural localization of dehydrogenase activity, *J. Histochem. Cytochem.,* 20, 685, 1972.
12. **Ogawa, K. and Mayahara, H.,** Intramitochondrial localization of adenosine triphosphatase activity, *J. Histochem. Cytochem.,* 17, 487, 1969.
13. **Ernst, S. A. and Philpott, C. W.,** Preservation of Na-K-activated and Mg-activated adenosine triphosphatase activities of avian salt gland and teleost gill with formaldehyde as fixative, *J. Histochem. Cytochem.,* 18, 251, 1970.
14. **Makita, T., Hakoi, K., and Araki, N.,** Cytochemical localization of Mg^{++}-ATPase and Ca^{++}-ATPase on the limiting membrane of rat liver peroxisomes, *Acta Histochem. Cytochem.,* 23, 601, 1990.
15. **Senior, E.,** The structure of mitochondrial ATPase, *Biochim. Biophys. Acta,* 301, 249, 1973.
16. **Tzagoloff, A., Rubin, M. S., and Sierra, M. F.,** Biogenesis of mitochondrial enzymes, *Biochim. Biophys. Acta,* 301, 71, 1973.

17. **Seabald, W.,** Biogenesis of mitochondrial ATPase, *Biochim. Biophys. Acta,* 463, 1, 1977.
18. **Lee, S. H.,** Ultrastructural localization of glutamic oxaloacetic transaminase activity in cardiac muscle fiber and cardiac mitochondrial fraction of the rat, *Histochemie,* 19, 99, 1969.
19. **Makita, T. and Sandborn, E. B.,** Substrate protection in the ultrastructural localization of aspartate aminotransferase activity in hamster tissues, Proc. 4th Int. Congr. Histochem., Helsinki Finland Histochem. Soc., 1972, 293.
20. **Makita, T., Matuura, H., Morimoto, M., Kiwaki, S., and Sandborn, E. B.,** Ultrastructural localization of glutamic oxaloacetic transaminase (GOT) activity in hamster tissues, *Arch. Histol. Jap.,* 35, 301, 1973.
21. **Papadimitrious, J. M. and Van Duijn, P.,** Effect of fixation and substrate protection on the histoenzymes of aspartate aminotransferase studied in a quantitative cytochemical model system, *J. Cell Biol.,* 47, 84, 1970.
22. **Papadimitrious, J. M. and Van Duijn, P.,** The ultrastructural localization of the isozymes of aspartate aminotransferase in murine tissues, *J. Cell Biol.,* 47, 84, 1970.
23. **Higgins, J. A. and Barrnett, R. J.,** Cytochemical localization of transferases activities: carnitine acetyltransferase, *J. Cell Sci.,* 6, 29, 1970.
24. **Makita, T. and Sandborn, E. B.,** A comparison of ultrastructural localization of carnitine acetyltransferase activity in mouse liver mitochondria with that in cardiac muscle, *Experentia,* 27, 184, 1971.
25. **Brdiczka, D., Gerbitz, K., and Pette, D.,** Localization and function of external and internal carnitine acetyltransferases in mitochondria of rat liver and pig kidney, *Eur. J. Biochem.,* 11, 234, 1969.
26. **Hall, J. D. and Crane, F. L.,** Intracristal rods. A new structure in beef heart mitochondria, *J. Cell Biol.,* 48, 420, 1971.
27. **Makita, T., Kiwaki, S., and Sandborn, E. B.,** Scanning electron microscopy and cytochemical localization of carnitine acetyltransferase activity in normal and dystrophic muscle of mice, *Histochem. J.,* 5, 335, 1973.
28. **Seligman, A. M., Karnovsky, M. J., Wasserkrug, H. L., and Hanker, J. S.,** Nondroplet ultrastructural demonstration of cytochrome oxidase activity with a polymerizing osmiophilic reagent, diaminobenzidine (DAB), *J. Cell Biol.,* 38, 1, 1968.
29. **Weisinger, R. A. and Fridovich, I.,** Mitochondrial superoxide dismutase. Site of synthesis and intramitochondrial localization, *J. Biol. Chem.,* 248, 4793, 1973.
30. **Grivell, L. A.,** Protein import into mitochondria, *Int. Rev. Cytol.,* 111, 107, 1988.
31. **Pon, L., Moll, T., Vestwever, D., Marshallsay, B., and Schatz, G.,** Protein import into mitochondria: ATP-dependent protein translocation activity in a submitochondrial fraction enriched in membrane contact sites and specific proteins, *J. Cell Biol.,* 109, 2603, 1989.
32. **Carafoli, E.,** Mitochondrial pathology: an overview, *Ann. N.Y. Acad. Sci.,* 488, 11, 1986.
33. **Makita, T. and Sasaki, K.,** A possible mode of formation of mitochondrial dense bodies in cardiac muscle of a dystrophic hamster, *Cytobios,* 25, 183, 1979.
34. **Hackenbrock, C. R.,** Ultrastructural basis for metabolically linked mechanical activity of mitochondria. II. Electron-transport linked ultrastructural transformation in mitochondria, *J. Cell Biol.,* 37, 345, 1968.
35. **Olson, G. E. and Winfrey, V. P.,** Mitochondria-cytoskeleton interactions in the sperm midpiece, *J. Struct. Biol.,* 103, 13, 1990.
36. **Bereiter-Hahn, J.,** Behavior of mitochondria in the living cell, *Int. Rev. Cytol.,* 122, 1, 1990.
37. **Pfanner, N. and Neupert, W.,** The mitochondrial protein import apparatus, *Annu. Rev. Biochem.,* 59, 1, 1990.

A. INTRODUCTION

1. An Historical Perspective

The earliest histochemical studies of the anterior pituitary gland recognized individual cell types by their affinity for different types of acid dyes or the periodic acid Schiff reaction.[1-3] The cells were then identified when they changed staining characteristics and/or size following stimulation. This was done by removing the target organ and observing changes in a specific cell type. For example, removal of the gonads caused enlargement of the blue, ovoid gonadotropes and removal of the thyroid produced enlargement of the darker blue, angular thyrotropes. The investigators used this evidence to conclude that the cell type that changed must be the source of the stimulatory hormone. This linked histochemistry to the identity of specific pituitary cells and introduced the concept that one pituitary cell type may produce only one hormone.

Because serum levels of the specific pituitary hormone were high following removal of the target organ, these early histochemical studies also provided the first correlation between cytological changes and enhanced secretion. The signs of enhanced secretory activity included enlargement, changes in distribution of the stainable material (presumably the hormone stores) and, in the case of gonadotropes and thyrotropes, formation of large vacuoles. Thus, these early studies also demonstrated that target organs may communicate with pituitary cells, either directly or indirectly. They led the way for subsequent studies of effects of steroids, thyroid hormone, or other peptides on pituitary cell function.

2. Contributions by the First Ultrastructural Studies

The early histochemical studies had set the stage for the electron microscopic analyses of the changes in the different cell types as the target organ feedback was removed or manipulated. Pioneering ultrastructural studies by Farquhar and Rinehart[4,5] demonstrated changes in the cytology of thyrotropes and gonadotropes coincident with removal of the target organ. These investigators showed that the pituitary cells were typical of protein hormone-secreting cells in that they contained rough endoplasmic reticulum, a prominent Golgi apparatus, and secretory granules. Indeed, there was strong evidence that the individual cell types could be differentiated by the size and shape of their secretory granules. However, there was some controversy because cells with morphological characteristics of more than one cell type were evident. Clearly, cytochemical confirmation of the hormone content of the granules was needed.

Furthermore, these early studies showed distinct changes in the distribution and size of the various protein-synthesizing organelles as the individual pituitary cells were stimulated. For example, following removal of the thyroid, cells with the smallest secretory granules enlarged and became filled with rough endoplasmic reticulum. The secretory granules were less numerous and dense material accumulated in the rough endoplasmic reticulum.[5] This provided evidence that the rough endoplasmic reticulum may have taken over the packaging function in these cells.

This evidence was later corroborated by cytochemical studies by Pelletier and Puviani,[6] who used autoradiography to show that attachment of a terminal sugar (fucose) to the glycoprotein hormones occurred primarily in the Golgi complex in thyrotropes or gonadotropes from normal animals. However, in thyrotropes or gonadotropes stimulated by removal of their target organs, the predominant labeling site for fucose was in the rough endoplasmic reticulum.[6] Thus, the addition of fucose to the glycoproteins during final packaging apparently had shifted from the Golgi complex to the rough endoplasmic reticulum.

In another early histochemical study, Farquhar[7] reported that pituitary cells have mechanisms for controlling internal stores of their hormones. Cytochemical labeling for acid phosphatase showed high activity in specific zones of the Golgi complex in actively secreting prolactin cells from suckling rats. However, when the young were removed and the stimulus was thereby stopped, the granules were merged with acid phosphatase-positive lysosomes where they were presumably degraded. This phenomenon was termed "crinophagy".

B. CONTRIBUTIONS BY IMMUNOCYTOCHEMISTRY

1. Advent of Immunolabeling Studies

By the early 1970s, the field of immunocytochemistry had begun to blossom and studies of the anterior pituitary cell types were proliferating. The earliest electron microscopic studies[8-14] corroborated many of the early ultrastructural studies in that they identified label for the hormones in specific types of secretion granules. The early studies linking size and shape of the secretion granule with the hormone content were accurate for the major subtypes of the cells.[11] However, cells with identical morphologies could store different hormones. Most of the studies followed the lead of the earliest pioneers and used the experimental models in which the target organ was removed. They confirmed that the immunocytochemically identified cell types enlarged and changed size and shape after manipulation of target organ feedback.

In this way, the immunocytochemists were able to correlate changes in the immunolabeled granule population with blood levels of secreted hormone and provide a more accurate view of the cytophysiological responses to stimuli.[10-14] For example, after adrenalectomy, corticotropes lost secretion granules immunolabeled for adrenocorticotropic hormone (ACTH) as serum levels of ACTH rose during the first 24 h.[10,11] Then, during the first 4 to 7 days, the immunolabeled corticotropes enlarged and both the Golgi complex and rough endoplasmic reticulum were expanded. This was accompanied by repopulation of the ACTH storage granules coinciding with increases in pituitary levels of ACTH. In later studies some of the same phenomena were seen following treatment with corticotropin-releasing hormone.[15]

In another example, removal of the thyroid produced enlarged thyrotropes.[13,14] Early studies had suggested that packaging had shifted from the rough endoplasmic reticulum to the Golgi complex. The immunolabeling evidence confirmed the depletion of thyroid-stimulating hormone (TSH) storage granules and the presence of stored TSH in both dense and diffuse regions in the rough endoplasmic reticulum.[14] Thus, the early immunolabeling studies confirmed both the identity of the cell that produced the hormone and evidence to support some of the hypotheses about changes in the sites of production and packaging after stimulation.

2. Discovery of Multihormonal Pituitary Cells

By the mid to late 1970s, immunocytochemists began to realize that their techniques were more powerful than they anticipated. Whereas the results were frequently confirmatory, they also began to challenge major dogma that were in vogue at the time. The first stated that each pituitary cell type contained only one hormone. The second assumed that all members of a given cell type were identical and would therefore respond in synchrony to a given stimulus.

Multihormonal cells that stored both follicle-stimulating hormone (FSH) and luteinizing hormone (LH) were identified immunocytochemically by Phifer et al.,[16] and Nakane,[17] thereby challenging the one cell, one hormone hypothesis. This was quickly followed by immunolabeling evidence for corticotropes that contained melanocyte-stimulating hormone (MSH) and adrenocorticotropin (ACTH).[18]

This became easier to interpret as our understanding of the chemistry of the hormones improved. The peptides produced by corticotropes were formed from a common precursor molecule called pro-opiomelanocortin which is split to produce a number of different types of opiocortin molecules. The multihormonal characteristics of corticotropes were further explained by the corelease of the peptides following stress or trauma.

With respect to the gonadotropins, LH and FSH have two subunits (alpha and beta). The alpha subunit is identical in both LH and FSH molecules. Therefore, only one gene is required for its transcription. Different genes transcribe the message for the beta subunit which is unique to LH and FSH. The pituitary cytochemists took advantage of that fact to discriminate between sites of LH and FSH storage.[11,16,19,20]

However, the existence of multihormonal gonadotropes were more difficult to explain physiologically because only one known releasing hormone had been identified for LH and FSH (gonadotropin-releasing hormone, or GnRH).[19,20] Thus, investigators wondered how nonparallel release was controlled. Studies in our laboratory agreed that 50 to 75% of gonadotropes stored both LH and FSH, depending on the physiological state of the rat.[19,20] The remaining cells appeared to store only one of the hormones which provided a potential mechanism behind monohormonal (or nonparallel) release. However, at the same time, it challenged the dogma that all gonadotropes are identical and operate in synchrony.

Further immunolabeling tests showed that the percentage of multihormonal gonadotropes increased to over 90% if the population was stimulated following removal of the gonads. Thus, these data suggested that the population is multipotential. This multipotential characteristic pointed to an efficient way of controlling the gonadotrope.

More recent studies have employed dual immunogold labels for these beta subunits to determine how the multihormonal gonadotropes process and package the hormones.[20] These dual labels have shown that LH and FSH can be in either the same or different granules. Monohormonal granules containing only LH predominate just before the time of the proestrous LH surge and monohormonal FSH granules are most frequent before the estrous FSH rise. Thus, these cytochemical data suggested that the packaging process is another site for the control of nonparallel secretion of LH and FSH.

The gonadotropes were also able to change storage patterns in response to GnRH stimulation (more became multihormonal; however, more of the granules were monohormonal).[19] This pointed to the fluidity of the cells and the presence of subtypes of cells and granules. Perhaps gonadotrope subtypes are activated to control different phases of the cycle.

3. Multihormonal Cells with Unusual Combinations of Hormones

Shortly after the initial immunolabeling studies of gonadotropes were published, another group of multihormonal cells was discovered.[12] These cells were more unusual and difficult to explain in that they stored gonadotropins with ACTH.[21] At the time they were discovered, there was no physiological rationale for their existence. We postulated that they were a residual stem or undifferentiated cell and may reflect a cell lineage present in the pituitary cell population. Tests of their frequency agreed that they were more abundant during development than in the adult, which supported this hypothesis.[22] However, today, our understanding of reproductive physiology has grown to include the recognition that ACTH stimulates the rise in progesterone which is needed to facilitate the LH surge and the rise in FSH.[23] Furthermore, studies have shown that the adrenal is the source of the progesterone, under the control of ACTH. This brings forth an important reason for the coexistence of gonadotropins with ACTH. One may postulate that this multipotential cell could be controlled independently after appropriate estrogen priming, to stimulate the production and secretion of progesterone by the adrenal. If this multihormonal cell could also respond to stress, then it could be diverted to a corticotrope under acute or chronic stress conditions. This may be one reason why stress interferes with ovulation and other normal events in reproductive system.

By the early 1980s, the cytochemical studies had provided clear evidence that multihormonal cells exist in the pituitary. In normal pituitaries, the following combinations were found: GH and PRL (growth hormone and prolactin, respectively),[24] ACTH and PRL,[25] ACTH and TSH,[26] and TSH and GH[27] or PRL.[27] As these subgroups were described, it became obvious that most anterior pituitary cell types exhibited heterogeneous storage patterns, so that one could no longer use granule size or shape as the sole criteria for identifying pituitary cell types. These facts were further supported by data from cytochemical studies of pituitary tumors[28] which showed different combinations of pituitary hormones and granules in the same tumor cells. The most frequent combination of hormones found were GH and PRL. Less frequent combinations included GH and TSH, or GH, PRL, and TSH. The least frequent combinations were GH and ACTH or ACTH and LH.

There are no cytochemically based studies that have provided concrete answers to the question of physiological significance of these cells. They may be residual stem cells, or subsets of multifunctional cells with the capacity to support more than one system. They may have unique combinations of receptors that also allow them to be multifunctional.

4. Immunolabeling for Regulatory Peptides in the Pituitary

As the studies of the known pituitary hormones developed, immunolabeling for other types of peptides was found in different cell types. The earliest examples of this were interpreted to be labeling at target sites of action for the neuropeptide. Sternberger and Petrali[29] found labeling for the neuropeptide gonadotropin-releasing hormone (GnRH) in granules of pituitary cells that were later confirmed as gonadotropes by Bauer and Childs.[30] Sternberger suggested that the labeling signified spare receptor sites stored in the secretory granule.

Similar labeling for thyrotropin-releasing hormone (TRH) was found in granules in thyrotropes and in cells enlarged by thyroidectomy.[32] Parallel studies tested the assumption that GnRH and TRH were derived from the neuroendocrine cells in the hypothalamus by growing the cells in culture.[30,31] Surprisingly, immunolabeling evidence showed that the GnRH or TRH peptides persisted in the absence of hypothalamic input. Furthermore, May et al.[31] demonstrated that these neuropeptides could be found in subsets of cells that also contained ACTH and gonadotropins at the end of the 21-d culture period. Thus, the cytochemical evidence raised questions that suggested the pituitary might be able to produce these secretagogues.

If the pituitary produced regulatory peptides such as these, this might indicate much broader functions for pituitary cells in the control of peripheral target cells. It may also reflect the presence of paracrine or autocrine regulatory systems in the pituitary itself. It is interesting to note that Nakane's early observation that immunolabeled gonadotropes and prolactin cells were frequently in juxtaposition[8,17] actually predicted such an interaction.

This particular interaction has recently been investigated most thoroughly by Houben and Denef.[33] They have used cell separation techniques to show that enriched aggregates of GnRH-stimulated gonadotropes do control prolactin cells. The exact mediating factor may be angiotensin II; however, additional candidate peptides are now being studied. In a recent review, Houben and Denef[33] described a number of studies of regulatory peptides identified by immunolabeling in the anterior pituitary. These include neuropeptides that were originally found in either the brain or gastrointestinal tract, the renin-angiotensin system, gonadal peptides that regulate gonadotropes, and growth factors and cytokines. In some cases, these peptides were localized in the same secretory granules with the pituitary hormone.

Whereas the immunolabeling evidence is strong, it must be interpreted with caution. First, one cannot use it to discriminate between sites of action on a target cell from sites of storage or production. Therefore, conclusions about the production of each of the peptides in the pituitary cells must first be based on assays for the mRNA for the peptide (or its precursor).

Furthermore, once the mRNA is assayed, there is often no specific consensus as to the exact cellular site of synthesis. Immunolabeling shows that many of these peptides appear to exist in a subset of several of the known cell types. This is also reviewed by Houben and

Denef.[33] For example, substance P appears in lactotropes, gonadotropes, and thyrotropes. Vasoactive intestinal peptide (VIP) appears in lactotropes and thyrotropes. Neuropeptide Y is found in gonadotropes, somatotropes, corticotropes, and some lactotropes. Thus, dual labeling protocols that localize the mRNA for the peptide (with *in situ* hybridization) and a pituitary hormone antigen (with immunolabeling) are needed to prove which, if any, of these cells represent the site of synthesis of the particular peptide. Different types of probes for mRNA, antigens, and receptors are being used to discriminate sites of synthesis from sites of action. After these important mapping studies are completed, one can then design more mechanistic studies to test potential paracrine or autocrine systems.

C. CONTRIBUTIONS BY AFFINITY CYTOCHEMISTRY

1. Cytochemical Labeling for Receptor Sites

One way to probe for receptor sites is to use a potent, labeled analog of the neuropeptide in question. This has been done in a number of laboratories and their studies are reviewed in previous papers from this laboratory.[34-36] At the ultrastructural level, the sites of the peptides can be identified by localizing labels such as tritium, biotin, ferritin, or colloidal gold. It is very important to prove that the attachment of the label does not interfere with potency of the molecule. In general, the least interference has been seen with radioactive labels (which are localized by autoradiography), or biotin (which is localized by labeled avidin compounds).

The earliest studies of receptor binding focused on the phenomenon of the rate and route of processing of the regulatory peptide-receptor complex.[34-36] This event has been named "receptor-mediated endocytosis". It begins by demonstration of rapid binding by the labeled hormone to the surface. Counts of the amount of label over the surface of the target cell show that the labeling index is highest during the first 3 min. Then, in some of the cells the label appears to cluster and it may be associated with pits in the membrane called "coated pits". At this point, one may also see label associated with small vesicles inside the cell.

The labeling over these vesicles increases with time as the labeling over the surface of the cell decreases. Then the labeling of larger vesicles, sometimes called receptosomes, increases. It has been postulated that these structures serve as a site for receptor recycling and transport back to the membrane. With further time, the labeling in the Golgi complex increases and label appears to be associated with granules and lysosomes.

Most of the cytochemical studies agreed that pituitary target cells for gonadotropin-releasing hormone (GnRH) or corticotropin-releasing hormone (CRH) internalized the regulatory peptides within 10 min of binding and then transported them to the Golgi complex.[34-36] There they became associated with lysosomes, or secretory granules. The workers have hypothesized that this final site represents a site for degradation of the regulatory peptide, and subsequent recycling of the receptor to the surface. Finally, some labeling of regions associated with rough endoplasmic reticulum or the nucleus suggests that the neuropeptide may regulate transcription or translation. However, the cytochemical evidence cannot determine if these peptides are intact and active at these sites. Furthermore, the ultrastructural cytochemical studies have not been able to link different rates or routes of internalization with a specific function inside the cell or specific mechanisms of action. Thus, these types of studies, like those of the multihormonal cells, are at this point, fairly descriptive.

There is value to the cytochemical studies of regulatory peptides when they are combined with immunolabeling for the hormone. This will help determine the identity of the target cells for the regulatory peptide.[36] Furthermore, questions about the significance of multihormonal target cells can be addressed. For example, the neuropeptide, arginine vasopressin (AVP), regulates the release of both ACTH and TSH from the anterior pituitary.[26] Dual cytochemical labeling studies have been used to prove that AVP binds to subsets of corticotropes and thyrotropes; AVP also increases the proportion of multihormonal cells that store both ACTH and TSH. Thus, the cytochemical studies have revealed a possible mechanism whereby AVP could control the release of both hormones during cold stress.

2. Cytochemical Labeling for mRNA

In situ hybridization for pituitary hormone mRNA is still in the early stages of application, especially at the electron microscopic level. At this point the application of this technology appears to be most useful when it is combined with immunolabeling to identify the pituitary cell that expresses the mRNA.[37] It is also useful when it is quantified, to show changes in the number of labeled cells or the amount of label per cell.

However, both of the above phenomena can be demonstrated most efficiently at the light microscopic levels. The electron microscopic studies are laborious and they have not provided striking new information. Recently, we showed label for βFSH mRNA over dilated rough endoplasmic reticulum that expanded after gonadectomy.[37] These data confirmed the 1954 morphological studies by Farquhar and Rinehart,[4] who showed a similar expansion in rough endoplasmic reticulum after removal of the gonads. As expected for a cell that exports the product of this mRNA, the labeling pattern for the cRNA-mRNA hybrids is almost exclusively on rough endoplasmic reticulum.

D. SUMMARY AND CONCLUSIONS

To summarize, the immunocytochemical studies of the pituitary have challenged the dogma that individual pituitary cells identified by tinctorial labeling produce and store only one hormone. They also have demonstrated heterogeneity in the pituitary cell population in terms of the types of hormones that are stored. Some cells appear to be monohormonal, others may store different combinations of hormones. This predicts functionally different subtypes that could be activated during states when more than one system is needed. Examples include gonadotropes that could release ACTH to help mediate the proestrous LH surge and cells that respond to cold stress by releasing both ACTH and TSH.

Some of the newest directions in research in the field are focused on the many new peptides found in pituitary cells. They may support distant target organs and some appear to be regulated by secretions from target organs.[33] If so, then this suggests much broader functions for the pituitary gland than previously believed.

In addition, many of these peptides are candidates for mediators of autocrine or paracrine interactions in the pituitary.[33] They may be local controlling agents that enable the pituitary cells to respond to outside influences. Future cytochemical studies that correlate cytophysiology with immunolabeling, *in situ* hybridization, and receptor labeling will help elucidate the functions of these various peptides.

REFERENCES

1. **Halmi, N. S.,** Some unsolved problems of anterior pituitary histophysiology, in *Cytologie de l'Adenohypophyse,* Benoit, J. and DaLage, C., Eds., Colloques Internationaux du Centre National de la Recherche Scientifique, Paris, Editions du CNRS, 1963, No. 123.
2. **Halmi, N. S.,** Adenohypophysis, in *Histology,* 2nd ed., Greep, R. O., Ed., McGraw-Hill, New York, 1966, chap 29.
3. **Purves, H. D.,** Morphology of the hypophysis as related to its function, in *Sex and Internal Secretions,* 3rd ed., Young, W. C., Ed., Williams & Wilkins, Baltimore, 1961.

4. **Farquhar, M. G. and Rinehart, J. F.,** Electron microscopic studies of the anterior pituitary gland of castrate rats, *Endocrinology,* 54, 516, 1954.
5. **Farquhar, M. G. and Rinehart, J. F.,** Cytologic alterations in the anterior pituitary gland following thyroidectomy: an electron microscope study, *Endocrinology,* 55, 857, 1954.
6. **Pelletier, G. and Puviani, R.,** Detection of glycoproteins and autoradiographic localization of $[^3H]$ fucose in the thyroidectomy cells of rat anterior pituitary gland, *J. Cell Biol.,* 45, 600, 1973.
7. **Farquhar, M. G.,** Processing of secretory products by cells of the anterior pituitary gland, *Mem. Soc. Endocrinol.,* 19, 79, 1971.
8. **Nakane, P. K.,** Application of peroxidase-labeled antibodies to the intracellular localization of hormones, *Acta Endocrinol. (Copenhagen) Suppl.,* 153, 190, 1971.
9. **Kawarai, Y. and Nakane, P. K.,** Localization of tissue antigen on the ultrathin sections with peroxidase-labeled antibody method, *J. Histochem. Cytochem.,* 18, 161, 1970.
10. **Moriarty, G. C. and Halmi, N. S.,** Electron microscopic study of the adrenocorticotropic cell with the use of unlabeled antibody and the peroxidase-antiperoxidase complex, *J. Histochem. Cytochem.,* 20, 590, 1972.
11. **Moriarty, G. C.,** Adenohypophysis: ultrastructural cytochemistry, a review, *J. Histochem. Cytochem.,* 21, 855, 1973.
12. **Moriarty, G. C.,** Electron microscopic immunocytochemical study of rat pituitary gonadotropes: a sex difference in morphology and cytochemistry of LH cells, *Endocrinology,* 97, 1215, 1975.
13. **Moriarty, G. C. and Tobin, R. B.,** Ultrastructural immunocytochemical characterization of the thyrotroph in rat and human pituitaries, *J. Histochem. Cytochem.,* 24, 1131, 1976.
14. **Moriarty, G. C. and Tobin, R. B.,** An immunocytochemical study of TSHβ storage in rat thyroidectomy cells with and without D or L thyroxine treatment, *J. Histochem. Cytochem.,* 24, 1140, 1976.
15. **Westlund, K. N., Aguilera, G., and Childs, G. V.,** Quantification of morphological changes in pituitary corticotropes produced by *in vivo* CRF stimulation and adrenalectomy, *Endocrinology,* 116, 439, 1985.
16. **Phifer, R. F., Midgley, A. R., and Spicer, S. S.,** Immunohistologic and histologic evidence that luteinizing hormone and follicle stimulating hormones are present in the same cell types in the human pars distalis, *J. Clin. Endocrinol.,* 36, 124, 1973.
17. **Nakane, P. K.,** Simultaneous localization of multiple tissue antigens using the peroxidase-labeled antibody method, *J. Histochem. Cytochem.,* 16, 557, 1968.
18. **Phifer, R. F., Orth, D. N., and Spicer, S. S.,** Immunohistologic evidence that β-melanocyte stimulating hormone and adrenocorticotropin are produced in the same human adenohypophyseal cell, Abstr., IV International Endocrine Congress, Washington, D. C., June 18, 1972.
19. **Childs, G. V.,** Functional ultrastructure of gonadotropes: a review, in *Current Topics in Neuroendocrinology,* Vol. 7, Pfaff, D., Ed., Springer-Verlag, Berlin, 1986, 49.
20. **Childs, G. V., Unabia, G., Tibolt, R., and Lloyd, J. M.,** Cytological factors that support non-parallel secretion of LH and FSH during the estrous cycle, *Endocrinology,* 121, 1801, 1987.
21. **Moriarty, G. C. and Garner, L. L.,** Immunocytochemical studies of cells in the rat adenohypophysis containing both ACTH and FSH, *Nature,* 265, 356, 1977.
22. **Childs, G. V., Ellison, D. G., and Ramaley, J. A.,** Adrenocorticotropin storage in corticotropes and a subpopulation of gonadotropes during the stress non-responsive period in the neonatal male rat, *Endocrinology,* 110, 1676, 1982.
23. **Brann, D. W. and Mahesh, V. B.,** Regulation of gonadotropin secretion by steroid hormones, *Front. Neuroendocrinol.,* 12, 165, 1991.
24. **Papka, R. E., Yu, S. M., and Nikitovitch-Winer, M. B.,** Use of immunoperoxidase and immunogold methods in studying prolactin secretion and application of immunogold labeling for pituitary hormones and neuropeptides, *Am. J. Anat.,* 175, 289, 1986.
25. **Losinski, N. E., Horvath, E., and Kovacs, K.,** Double-labeling immunogold electron-microscopic study of hormonal colocalization in nontumorous and adenomatous rat pituitaries, *Am. J. Anat.,* 185, 236, 1989.
26. **Childs, G. V., Westlund, K., and Unabia, G.,** Characterization of anterior pituitary target cells for arginine vasopressin: including cells that store adrenocorticotropin, thyrotropin-β and both hormones, *Endocrinology,* 125, 554, 1989.
27. **Newman, G. R., Jasani, B., and Williams, E. D.,** Multiple hormone storage by cells of the human pituitary, *J. Histochem. Cytochem.,* 37, 1183, 1989.
28. **Kovacs, K., Horvath, E., Asa, S. L., Stefaneanu, L., and Sano, T.,** Pituitary cells producing more than one hormone human pituitary adenomas, *Trends Endocrinol. Metab.,* 1, 104, 1989.
29. **Sternberger, L. A. and Petrali, J. P.,** Quantitative immunocytochemistry of pituitary receptors for luteinizing hormone releasing hormone, *Cell Tissue Res.,* 162, 141, 1975.
30. **Bauer, T. W. and Childs, G. V.,** Studies of immunoreactive gonadotropin releasing hormone (GnRH) in the rat anterior pituitary, *J. Histochem. Cytochem.,* 29, 1171, 1981.

31. **May, V., Wilber, J. F., U'Prichard, D. C., and Childs, G. V.,** Persistence of immunoreactive TRH and GnRH in long-term primary anterior pituitary culture, *Peptides,* 8, 543, 1987.
32. **Childs, G. V., Ellison, D. G., Yang, H.-Y., Kubek, M., Tobin, R. B., and Wilber, J. F.,** The effect of thyroidectomy, propylthiouracil and thyroxine on pituitary content and immunocytochemical staining of thyrotropin and thyrotropin releasing hormone, *J. Histochem. Cytochem.,* 29, 357, 1981.
33. **Houben, H. and Dendef, C.,** Regulatory peptides produced in the anterior pituitary, *Trends Endocrinol. Metab.,* 1, 398, 1990.
34. **Childs, G. V., Hazum, E., Amsterdam, A., Limor, R., and Naor, Z.,** Cytochemical evidence for different routes of GnRH processing by large gonadotropes and granulosa cells, *Endocrinology,* 119, 1329, 1986.
35. **Childs, G. V., Morell, J. L., and Aguilera, G.,** Cytochemical studies of CRF receptors in anterior lobe corticotropes: binding, glucocorticoid regulation and endocytosis of [Biotinyl-Ser1] CRF, *Endocrinology,* 119, 2129, 1986.
36. **Childs, G. V., Westlund, K. N., Tibolt, R. E., and Lloyd, J. M.,** Hypothalamic regulatory peptides and their receptors: their role in regulation at the adenohypophysial level, *J. Electron Microsc. Tech.,* 1991.
37. **Childs, G. V., Unabia, G., Weirman, M. E., Gharib, S. D., and Chin, W. W.,** Castration induces time-dependent changes in the FSHβ-mRNA-containing gonadotrope cell population, *Endocrinology,* 126, 2205, 1990.

Chapter 10

ENDOCRINE ORGANS

10.2 The Adrenal Gland

TABLE OF CONTENTS

0-8493-6012-9/93/$0.00 + $.50

I. THE ADRENAL CORTEX — A CYTOCHEMICAL PERSPECTIVE

Virginia H. Black

A. INTRODUCTION

Many different steroids have been extracted from adrenal glands of humans and experimental animals, but only a few of these are normally released from the gland and are hormonally active. All of these steroids are the product of the adrenal cortex. There are three major classes of adrenal corticosteroids: (1) glucocorticoids, which have pronounced effects on carbohydrate metabolism; (2) mineralocorticoids, which affect electrolyte balance and blood pressure; and (3) sex steroids, primarily androgens, whose effects are similar to those secreted by the testis. In most species, the adrenal cortex consists of three concentric zones, distinguishable by their morphology. The outer zona glomerulosa features clusters of cells or arched cords. The middle zona fasciculata is characterized by radiating straight cords of cells. The inner zona reticularis has interweaving cellular cords. This morphological zonation is paralleled by important functional zonation. Each cortical zone produces a different spectra of products. For example, aldosterone, a mineralocorticoid, is secreted by cells of the zona glomerulosa, whereas cortisol, a glucocorticoid, and dehydroepiandrosterone (DHEA), which can act as an androgen, are formed by cells of the zonae fasciculata and reticularis. The factors which determine the types of products secreted by the cells are not fully understood. The expression of only certain enzymes, the specificity of the enzymes for different intermediates in the pathway, and the organization of the enzymes within organelle membranes may all play a role in both the types of steroids produced and their relative proportions.

Steroid hormones are formed from cholesterol by removal of a six-carbon fragment of the side chain followed by dehydrogenation at carbon-3 and a series of hydroxylations at specific sites. A simplified scheme showing the biosynthesis of the principal adrenal steroids is presented in Figure 1. Variations on this theme undoubtedly occur and a thought-provoking discussion of the possibilities is given in two review articles by Lieberman and co-workers.[1,2] As will become clear in this review, precursors and intermediates traverse many organelles in their transit along the steroidogenic pathway (Figure 2). What governs their movement in ordered fashion is not entirely clear. Binding proteins are being discovered and may facilitate movement between organelles. Microdomains of particular sets of enzymes within the organelles may provide efficient pathways and influence the type of steroid produced.

Like other steroid-secreting cells (testicular interstitial cells, ovarian corpus luteal cells), adrenocortical cells have abundant mitochondria with tubular or vesicular cristae, a large amount of smooth endoplasmic reticulum (SER), numerous lipid inclusions, many peroxisomes, and a prominent Golgi complex. Despite their similarities, the cells of each of the zones do have characteristic features which vary among species. The fine structure of the cortical cells and the responses of their cellular organelles to stimulation and suppression have been extensively reviewed, most recently by Nussdorfer.[3]

Much of the information we have about the organization of the functional components of these subcellular structures comes from biochemistry. Hall[4] nicely summarized the information available from this perspective concerning subcellular localization and Lieberman and Prasad[2] have challenged some of the conventional interpretations of biochemical data. However, biochemistry by its very nature presents a homogenized view. Both Hall and Lieberman stressed that the organization of enzymes in the steroidogenic pathway within the cell might be more complex than currently appreciated. Hall suggested that microdomains of enzymes might exist within membranous organelles, which would facilitate movement

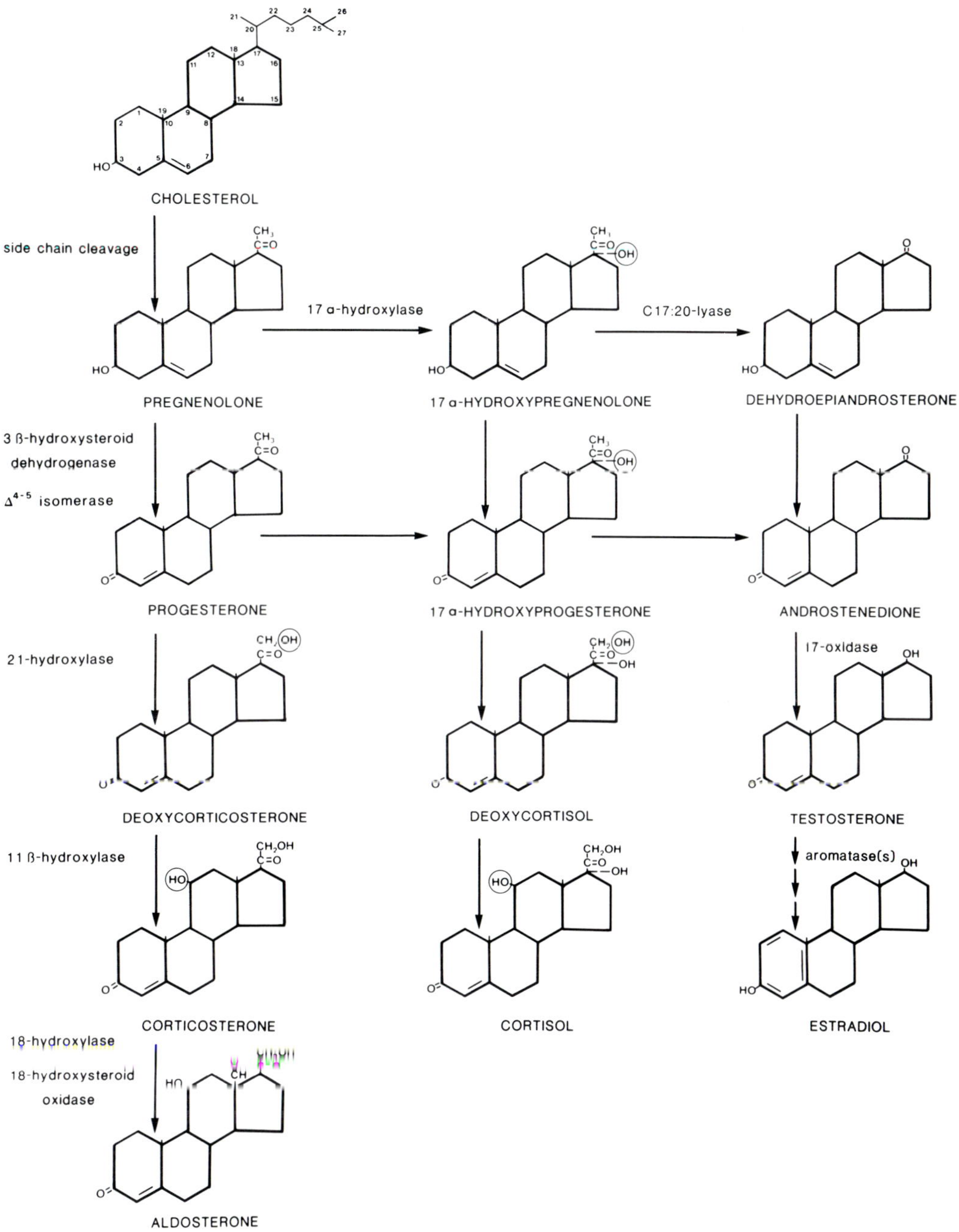

FIGURE 1. Diagram showing the pathways for biosynthesis of the principal steroid hormones from the precursor, cholesterol. For each hydroxylation step, the newly added hydroxyl group is circled.

of intermediates along the biosynthetic pathway.[4] Lieberman and co-workers suggested that there are specific associations of enzymes for each of the steroids produced (hormonads).[1] Such associations might or might not be confined to one organelle.

Cytochemical analysis allows localization *in situ* and avoids the problems of tissue dispersion, organelle breakage, and contamination of one fraction with another. It is not,

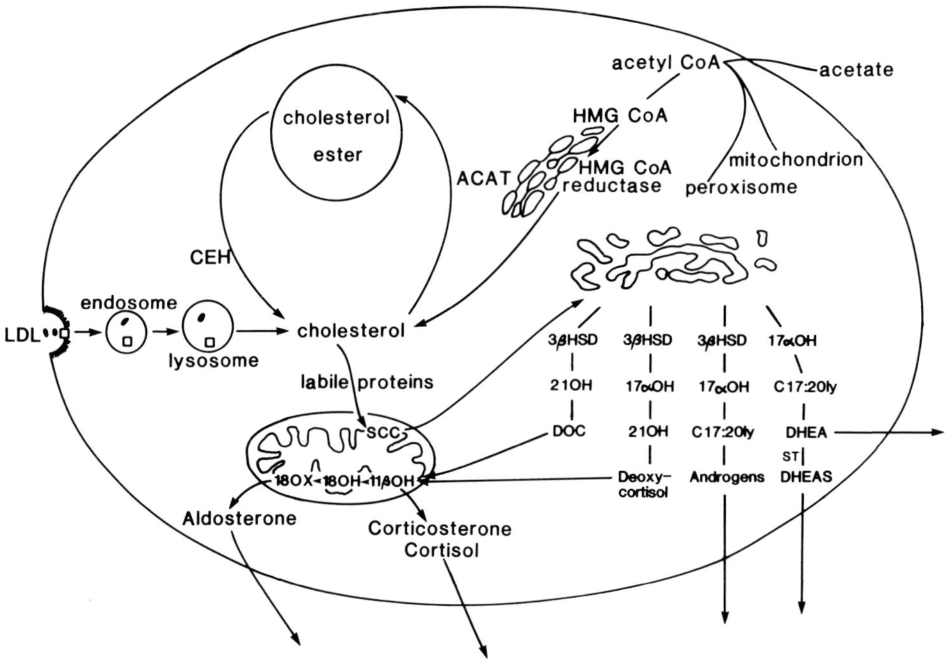

FIGURE 2. Diagram showing the sources of cholesterol, the steps involved in maintaining cholesterol homeostasis, and the flow of sterol and steroid intermediates along the biosynthetic pathway toward the major secreted steroid hormones. The localization of some steps by cytochemistry is given in Table 1. Other organellar localizations are derived solely from biochemical data; LDL, low-density lipoprotein; ACAT, acetyl CoA cholesterol:acyltransferase; SCC, cholesterol side-chain cleavage enzyme; 3βHSD, 3β-hydroxysteroid dehydrogenase; 17OH, 17α-hydroxylase; C17:20ly, cleavage between C17 and C20, C17:20 lyase; ST, sulfotransferase; 21OH, 21 hydroxylase; DOC, deoxycorticosterone; 11βOH, 11β-hydroxylase; 18OH, 18 hydroxylase; 18OX, 18 hydroxysteroid oxidase.

however, without its own limitations. Fixation and resolution present dual problems which are more stringent for analysis at the electron microscopic (EM) level than at the light microscopic (LM) level. It is difficult to preserve membrane structure at the low concentration of fixative required for retention of enzyme activity or protein antigenicity. The use of various combinations of paraformaldehyde and glutaraldehyde, both in low percentages, has gone a long way toward resolving this dual problem, particularly when the tissue is perfused with the fixative rather than simply being immersed in it. Another limitation of resolution has been finding techniques which will result in discrete electron-dense products. Diffusion and/or heavy deposition of reaction product can obscure membrane localization and protein distribution. The advent of immunochemical techniques, particularly the use of protein A-gold have helped to solve this. In addition, when protein A-gold-linked immunocytochemistry is employed at the electron microscope (EM) level, detailed topological analysis and quantitation are possible. Double labeling analysis offers a way to visualize the relationships between membrane-associated proteins and other enzymes in the biosynthetic pathway and would allow the hypotheses of specific enzyme domains to be tested.

The adrenal cortex has been extensively studied by cytochemical techniques, which provided considerable insight into its structure and function. Many of the LM analyses were elegantly summarized by Deane.[5] Early studies done at the EM level were discussed in the comprehensive review by Nussdorfer.[3] Although some of these studies are incorporated here for the sake of continuity and completeness, this review will focus on more recent contributions of cytochemistry to our understanding of the role of cellular organelles in steroidogenesis and of factors which contribute to functional zonation. In general, the review follows the path of steroidogenesis through the cell, presenting zonal differences where appropriate. Hopefully, it will also point to future avenues of research. Cytochemical localizations are summarized in Table 1.

The synthesis of corticosteroid hormones is influenced by a number of different factors. Classically, angiotensin II (AII) is the primary stimulator of mineralocorticoid synthesis and adrenocorticotropic hormone (ACTH) the primary stimulator of glucocorticoid production. However, an increasing number of other regulatory factors have been shown recently to modulate the function of these cells. Stimulation by most trophic hormones has both a short-term and a long-term component. The short-term response involves the transport of cholesterol to mitochondria for cleavage of its side chain. In the long term, increases in steroidogenic enzymes occur, with consequent increases in the volumes of organelles with which they are associated.[3] Many of the genes for these enzymes have been cloned and sequenced. Their regulation has been and is being extensively analyzed.[6,7] Thus far, the increase in these proteins following stimulation seems to be effected primarily via enhanced mRNA production. Most of this information has been derived from Northern blot analysis of mRNA prepared from cultured cells or from whole adrenal tissue. Analysis by *in situ* hybridization would allow questions regarding differential zonal responses and uniformity of response among zonal cell populations to be addressed, but little investigation has been done at the cellular level. While information currently available from this cytochemical approach is incorporated into the appropriate sections of this review, this area offers fertile ground for future investigation.

B. ORGANELLAR LOCALIZATION: STEROID SYNTHESIS AND TRANSPORT

1. Lipid Droplets/Cholesterol

Cholesterol is the basic building block of all steroid hormones. As indicated in Figure 2, it can be obtained by uptake of cholesterol from plasma lipoproteins or synthesized endogenously via HMG CoA reductase, a rate-limiting enzyme in cholesterol biosynthesis. It can be stored within the cell in esterified form or used immediately for steroid synthesis. The rate of its esterification and hydrolysis, its movement through the cell, and its entry into mitochondria for conversion to pregnenolone are limiting steps in steroid synthesis. Due to its importance to the function of these cells, its distribution has been extensively studied.

Digitonin reacts with all unesterified 3β hydroxysterols to form crystals which can be seen in the EM.[8] Free cholesterol is the most abundant of these. The localization of digitonin-cholesterol reaction product in lipid droplets of adrenocortical cells and in rectangular crystals occasionally seen in normal fasciculata cells is well established.[8,9] Free cholesterol was also localized in lipid droplets by examining cells after uptake of ^{3}H-cholesterol.[10] When combined with results from differential centrifugation, Moses and co-workers concluded that most adrenal cholesterol is in esterified form within lipid droplets.[10] They reasoned that some mechanism must be required for mobilizing it for use in steroid synthesis, the initial step of which occurs in the mitochondria.[10] Szabo and co-workers, using combined radiolabeling and digitonin techniques, found that treatment with cycloheximide caused increases in lipid droplet size and labeled cholesterol accumulation.[9] Such accumulation was enhanced by cotreatment with ACTH. In these cells digitonin-cholesterol product was found not only in the lipid droplets but also around mitochondria. Cycloheximide inhibits the conversion of cholesterol to pregnenolone. The authors suggested that inhibition occurred not at conversion of cholesterol ester to free cholesterol, for free cholesterol was, if anything, more abundant

TABLE 1 (continued)
Cytochemical Localizations in the Adrenal Cortex

Compound	Compartment	Species	Ref.
	Cytoskeleton		
Actin	Y-1 adrenal tumor cells	Mouse	128
	Cytosol, cortical and microvillar	Rat	129
Intermediate filaments			
Cytokeratins	Throughout cortex, decrease in neoplastic tissue	Human	131–133
Vimentin	Zona glomerulosa and zona fasciculata, increase in neoplastic tissue	Human	131–133
	Y-1 Adrenal tumor cells	Mouse	134
Amyloid	Cytosol	Human, elderly	135
	Calcium		
	Mitochondrial intracristal space and smooth endoplasmic reticulum	Rat	117, 118
	Mitochondrial matrix granules	Rat	119
	Intracellular canaliculi	Rat	149, 150
	Peptide Hormones		
Renin	Secretory granules	Rat	140–142
	Zona glomerulosa		
	Immunocytochemistry	Rat	139, 140
	In situ hybridization	Rat	139
Oxytocin and vasopressin	Definitive cortex	Human, fetal	143
(AVP)	All zones, zona glomerulosa most intense	Human, adult, cow, hamster, guinea pig	144
IGF	Definitive and fetal cortex	Human, fetal	145
bFGF	Throughout cortex	Rat, postnatal cortex	146
	Zona glomerulosa	Rat, adult	147
	Receptors		
AII	Zona glomerulosa	Rat	152
AVP	Zona glomerulosa	Rat, human, guinea pig	154
ACTH	All zones, plasma membrane and intracellular	Rat	155
PRL	Zona fasciculata	Mouse	156
GH	Zonae glomerulosa and fasciculata	Rat	157, 158
EGF	All zones, scattered cells, increases in neoplastic tissue	Human	159
LDL	All zones, zona fasciculata most intense	Rat	160
	Cultured cortical cells	Rat, human, normal	25
HDL	Cultured zonae glomerulosa and fasciculata cells	Rat	24

TABLE 1 (continued)
Cytochemical Localizations in the Adrenal Cortex

Compound	Compartment	Species	Ref.
	Microvasculature		
Peroxidase	Mononuclear phagocytes, inner cortical sinuses	Rat	164
Serum albumin-gold uptake	Endothelial cells, medulla, not cortex	Rat	165
Vitamin A	Fibroblasts, pericapillary space	Mouse	166
Actin, myosin, and adrenergic innervation, glyoxylic acid technique	Arteriolar vessels, endothelial and perivascular cells of cortical capillaries	Rat, guinea pig	168
	Nerves		
Vasopressin	Nerves, zona glomerulosa	Human, fetal	143
VIP	Nerves, zona glomerulosa	Rat	170
CRF	Nerves, zonae glomerulosa and reticularis	Sheep	171

in the treated cells. Instead, they concluded that the blockade was at the entry of cholesterol into mitochondria. Magalhães and Magalhães[11] found digitonin-cholesterol complexes in cytoplasmic lipid droplets and at the mitochondrial periphery as well as associated with Golgi vesicles.[11] In animals treated with aminoglutethimide, an inhibitor of cholesterol side-chain cleavage, they found a distinct increase in the number of cholesterol-digitonide complexes in the Golgi, localized to the inner face. This unusual observation suggests specific association of 3β-hydroxysterols with this organelle.

However, both the autoradiographic technique and the digitonin technique have limitations. The latter does not precipitate esterified sterols and the crystals formed disrupt cellular architecture, while the former suffers from possible diffusion and nonspecific adherence of the labeled free sterol to other cellular organelles. To circumvent these problems, Jones and Miyai used enzyme histochemistry to localize cholesterol.[12] Esterified cholesterol was hydrolyzed with cholesterol esterase and the free cholesterol oxidized to cholesterol oxidase. This reaction produces hydrogen peroxide which can oxidize diaminobenzidine (DAB) in the presence of horseradish peroxidase. This method allowed localization of esterified, free, or total cholesterol depending upon the enzymes used. Dense, amorphous reaction product occurred in cytoplasmic vacuoles, which could have been derived from endoplasmic reticulum or Golgi, in lipid droplets, and around mitochondria. This distribution was similar to that obtained with digitonin, but had the advantage of precise localization due to absence of membrane disruption.

2. Lysosomes

Membrane-bound bodies of varying sizes, with finely granular and sometimes heterogeneous content, were described in early fine structural studies of adrenocortical cells. Cytochemical identification of the smaller dense bodies as lysosomes, i.e., organelles containing the marker enzyme acid phosphatase, was made by Penney and Barrnett.[13] Later, Szabo and co-workers noted that most such structures gave a positive reaction for acid phosphatase and were thus classifiable as lysosomal in nature.[14] Positive reaction product

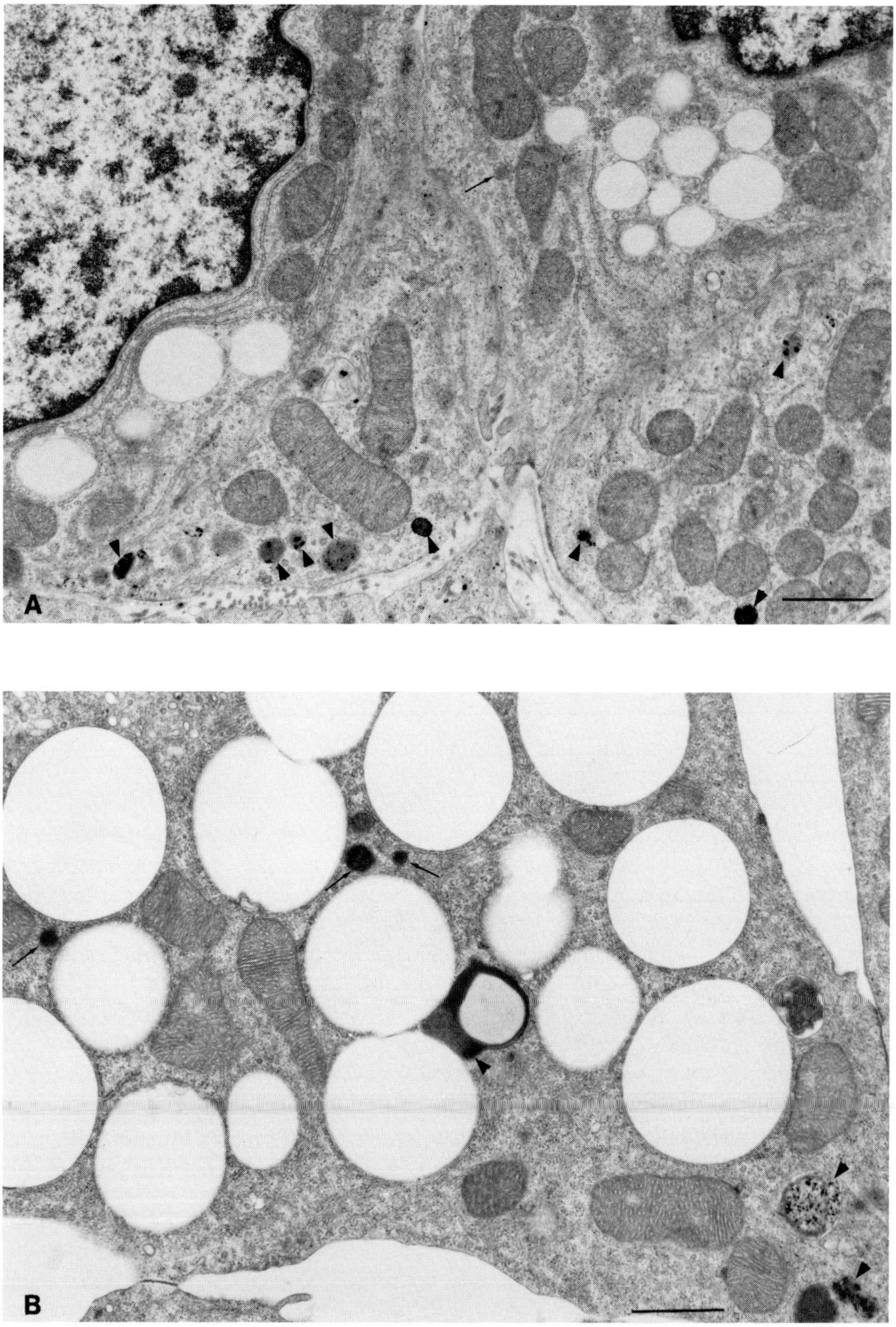

FIGURE 3. Electron micrographs of (A) zona glomerulosa and (B) zona fasciculata cells showing variable acid phosphatase reactivity among lysosomes (arrowheads) and demonstrating peroxisomes (arrows) in tissue reacted with DAB (B) and in tissue not reacted with DAB (A). (Bar = 1.0 μm.)

for acid phosphatase was also noted in the innermost Golgi cisternae and associated tubules and vesicles[11,15] (Figure 3), later identified as a specialized region of the Golgi complex involved in the transport of materials from the Golgi to lysosomes and other cellular components (GERL, Novikoff et al.[16]; TGN, *trans*-Golgi network, Griffiths and Simons[17]), as recently reviewed by Holtzman.[18] Changes in the number, size, and heterogeneity of cytochemically identified lysosomes have been reported following hypophysectomy and ACTH treatment, as well as during development, suggesting an important functional role for this organelle.[14,19-21]

A role for lysosomes in cholesterol metabolism at several levels has been suggested. Dietert and Scallen showed specific increases in large acid phosphatase-positive bodies with heterogeneous content following inhibition of cholesterol synthesis by three different compounds.[22] Reaction product seen adjacent to lipid droplets suggested involvement in cholesterol mobilization.[14] Lysosomes have also been strongly implicated in the release of cholesterol from plasma lipoproteins taken into the cell by receptor-mediated endocytosis,[23] and there is some cytochemical support for this pathway in adrenocortical cells.[24,25] In time-course studies, gold-labeled lipoprotein particles were initially seen bound to the plasma membrane of human and rat adrenocortical cells. At later time points the label was incorporated into lysosomes. Enhanced steroidogenesis occurred in the presence of the lipoprotein, suggesting that its cholesterol component had been utilized. However, because protein A-gold labels the protein and not the sterol portion of the particles, these studies support the uptake of the receptor/protein complex into lysosomes but cannot prove that sterol follows the same pathway. Detailed cytochemical analysis of this pathway using labels for both components, such as that done with ovarian corpora luteal cells,[25] is still lacking for the adrenocortical cell.

In addition to a role in cholesterol metabolism, lysosomes may be involved in the metabolism of sulfated steroids. Cholesterol sulfate can serve as a precursor for steroid synthesis and sulfated steroids may serve as intermediates in the steroidogenic pathway.[1] Although at least one steroid, dehydroepiandrosterone (DHEA), is secreted in its sulfated form, others are apparently desulfated prior to secretion. Rappay and co-workers demonstrated lysosomal arylsulfatase cytochemically in adrenocortical lysosomes and Golgi cisternae, suggesting a link between lysosomes and the sulfation and desulfation of steroids.[27]

Lysosomes in adrenocortical cells are not homogeneous in their enzyme composition and show zonal differences in enzyme content. Bacsy demonstrated that acid phosphatase and arylsulfatase were not distributed equally in the lysosomal population.[28] In a ''double-labeling'' study employing lead phosphate as the electron-dense product for acid phosphatase and barium sulfate as the detectable reaction product for arylsulfatase, he showed by X-ray microanalysis that the Ba/Pb ratio was not uniform among lysosomes. Both the zona fasciculata and zona glomerulosa contained populations with relatively high arylsulfatase activity compared to other tissues, but levels of this enzyme varied from one lysosome to another in proportion to acid phosphatase. In addition, the zona glomerulosa contained a unique population of lysosomes high in acid phosphatase, but with low, if any, arylsulfatase, a ratio closer to that seen in lysosomes of other tissues. The high level of sulfatase in adrenocortical lysosomes underscores the possibility of involvement in metabolism of sulfated steroid intermediates. Their zonal differences may reflect differences in steroid pathways and products.

Zonal differences with regard to lysosomes have also been noted in the fetal adrenal. The cells of the fetal zone were shown to contain large and intensely stained lysosomes, while those in the permanent cortex were only lightly stained.[29] It was suggested that lysosomes were involved in degradation of SER, abundant in fetal cortical cells, but less prominent in the relatively quiescent permanent cortical cells. An association of larger numbers of lysosomes with abundant SER was also observed in the developing and adult guinea pig adrenal cortex.[19,20] Their increased numbers following ACTH treatment could reflect increased membrane turnover in the stimulated cells.[21]

Biochemical analysis has shown lysosomes of adrenocortical cells to contain at least six acid hydrolases in addition to the above, with an unusually high content of cathepsin D.[30] The distribution of these enzymes among lysosomes and between zones has not been investigated, but lysosomes in adrenocortical cells are affected in several disease states, underscoring their role in several degradative processes in which these additional enzymes might be involved. In Cushing's syndrome they are closely related to well-developed SER and mitochondria in compact cells, but to lipid droplets in clear cells.[31] These relationships are similar to those seen in normal cells, but are amplified in the hyperplastic tissue. In Pompe's disease (type II A glycogenosis) characteristic intralysosomal accumulations of glycogen occur in the adrenal cortex and medulla, as well as in other tissues.[32] In this disease, smaller lysosomes can be seen within larger acid phosphatase-positive bodies, suggesting autophagocytosis. In Niemann-Pick disease caused by sphingomyelinase deficiency, lysosomes increase markedly and are filled with lipoidal material.[33]

3. Peroxisomes

Small granules bound by a single membrane and containing a less electron-dense, more coarsely granular content than lysosomes were described in early accounts of steroid-secreting cell fine structure. They were variously identified as lysosomes, secretory granules, or microbodies (peroxisomes).[15] The advent of diaminobenzidine for localization of catalase, the marker enzyme for peroxisomes, allowed their positive identification as peroxisomes.[15,34,35] However, unambiguous distinction between small primary lysosomes, lacking heterogeneous content, and peroxisomes requires sequential incubation of tissues for both marker enzymes (Figure 3).[21]

A significant functional role for peroxisomes in normal adrenocortical cells was indicated by their increase during differentiation and following ACTH treatment.[15,19] It was shown that these organelles contain a KCN-insensitive fatty acid β-oxidation system, capable of generating acetyl CoA, the building block for cholesterol synthesis, whose activity was also enhanced by ACTH.[36]

Roles for adrenocortical cell peroxisomes in cholesterol synthesis, utilization, and transport have been supported by several cytochemical studies. HMG CoA reductase is a key enzyme in cholesterol synthesis. Peroxisomes in adrenocortical cells were shown to increase following inhibition of cholesterol synthesis by mevinolin, a competitive inhibitor of HMG CoA reductase,[37] as well as after treatment with 4-aminopyrazolo-pyrimidine (4-APP)[38] or Nafenopin,[39] compounds known to decrease plasma lipoproteins and result in compensatory increase in HMG CoA reductase activity. They also increase in number in adrenocortical cells of thyroparathyroidectomized rats.[40] Hypothyroidism results in increased catalase activity in the adrenal and liver, correlating with the increase in peroxisomes and an increase in plasma cholesterol. The increases seen in peroxisomes following treatment with these agents could have several explanations, one of which could be an increase in enzymes associated with cholesterol synthesis. In hepatocytes, HMG CoA reductase has been localized in peroxisomes, as well as in its more accepted site in the endoplasmic reticulum, and the level of the peroxisomal enzyme is increased by several hypolipidemic drugs.[41,42] It would be of interest to see if HMG CoA reductase is a peroxisomal enzyme in adrenocortical cells and if its regulation is the same as, or different from, that in the endoplasmic reticulum.

Cleavage of the side chain of cholesterol to form pregnenolone is a rate-limiting step in steroid synthesis. Treatment with aminoglutethimide, an inhibitor of cholesterol side-chain cleavage, also resulted in an increase in peroxisomes in adrenocortical cells.[11] It is of interest to note that in these animals, no increase in lysosomes was detected. This contrasts with the results of Dietert and Scallen,[22] discussed above, showing that inhibition of cholesterol synthesis led to specific increase in large lysosomes, and may point to a difference in function between these two organelles with regard to cholesterol metabolism.

Carrier proteins have been implicated in the transport of cholesterol through the cell and into mitochondria, where side-chain cleavage takes place.[43] Recently, preliminary immunochemical data has been presented demonstrating localization of sterol carrier protein 2 (SCP2) in peroxisomes of rat adrenocortical cells.[44] It will be of interest to see if this is universal among species and zonal cell types and if SCP2 levels and/or distribution change with function in adrenocortical cells, as recently described for testicular interstitial cells.[45]

An association between peroxisomes and lipid metabolism in adrenocortical cells is clinically illustrated by the peroxisomal deficiency diseases, cerebro-hepato-renal (Zellweger's) syndrome (CHRS) and adrenoleukodystrophy (ALD). In these diseases morphologically demonstrable peroxisomes are decreased in number or absent and long-chain fatty acid metabolism is impaired. Adrenal glands from these patients have numerous intracellular lamellar profiles, attributed to accumulations of very-long-chain fatty acids esterified to cholesterol.[46-48] In ALD patients the profiles were not stained with the Schultz reaction for cholesterol, but did react with digitonin. The lack of staining could be due to the solid nature of the lamellar structures or indicate the presence of free 3β-hydroxysterols other than cholesterol (see cholesterol/digitonin above).[46,49,50] The most likely candidates would be cholesterol precursors, lanosterol, zymosterol or 7-dehydrocholesterol. The origin of the lamellae and the significance of their associated free sterol is uncertain. They may result from membrane lipid abnormalities with resultant membrane instability or from esterification of sterol not utilized in steroid synthesis to unmetabolized long-chain fatty acids, which are not incorporated into membranes.[46,51] Often there is no clinical evidence of adrenal hypofunction, suggesting compensatory activity by the cortical cells which may accentuate the accumulation of abnormal lipids.

The changes with function and disease states described above are most prominent in the inner cortical cells of the zona fasciculata and zona reticularis. However, peroxisomes are also present in zona glomerulosa cells and lamellar inclusions are present in zona glomerulosa cells of ALD patients.[19,49] This suggests a commonality of peroxisomal function among the different zonal cell types.

4. Endoplasmic Reticulum and Mitochondria

As reviewed by Hall,[4] the mitochondrial and microsomal fractions have been shown by biochemical analysis to contain many of the enzymes required for steroid synthesis (Figure 2). The localization of several has been confirmed cytochemically, but much remains to be done, particularly with regard to the topological distribution of enzymes in both organelles.[1,2,4]

a. 3β-Hydroxysteroid Dehydrogenase (3βHSD)

3βHSD acts in the conversion of pregnenolone to progesterone by dehydrogenation of a 3β-hydroxy-5-ene-steroid and isomerization of the 3-oxo-5-ene-steroid. The enzyme catalyzes both steps and its activity is required for production of all secreted steroid hormones, except for DHEA. Because of this, it serves as a marker of steroid-producing cells and has been the most widely studied enzyme. Biochemical assay of adrenal subcellular fractions reported 3βHSD activity in the microsomal fraction[52-54] or in both microsomal and mitochondrial fractions.[55-58] However, activity in the mitochondrial fraction was suggested to be due to microsomal contamination.[54,59] Because its relative activity and intracellular distribution with respect to subsequent enzymes in the steroidogenic pathway may influence the pattern of steroid products (see Figure 2), knowledge of both the organellar localization and topological distribution of 3βHSD are important.

In attempts to distinguish whether or not the 3βHSD found in the mitochondrial fraction was a contaminant, several groups measured the ratio of 3βHSD to 21-hydroxylase, using the latter as a marker of microsomes. Even though this ratio was consistently greater than in the microsomal fraction, the result was interpreted in two ways. One group said that it was due to specific retention in the mitochondrial fraction of microsomal fragments high in 3βHSD content.[59] The other groups recognized that the data were consistent with such an interpretation but, lacking evidence that 3βHSD was nonuniformly distributed in the ER, concluded that the mitochondrial localization reflected the state *in vivo*.[55,58]

Cytochemistry has been applied to this question, but the answer has not been definitive. Several techniques were developed at the histochemical level for detection of 3βHSD in steroid-secreting cells; one employing nitroblue tetrazolium was extended to the EM level.[60,61] When the reaction was run without an electron-transfer intermediate, product was localized largely in mitochondria.[60] This was undoubtedly due to mitochondrial NADPH diaphorase, for inclusion of an electron-transfer intermediate to bypass this enzyme resulted in decreased mitochondrial staining and the deposition of product adjacent to the ER.[61] However, the lipophilic nature of the reaction product called into question a preferential localization near lipid droplets. When analysis was carried out using copper ferrocyanide in the presence of electron-transfer intermediates, intense reaction product occurred adjacent to ER membranes with no preferential localization near lipid droplets. Reaction product was also seen adjacent to the inner mitochondrial membranes,[62-64] but its relative intensity was variable. When the overall deposition was intense both organelles were well labeled. When it was less intense, labeling in the ER predominated.[65-67] The dependence of localization on the strength of the reaction leaves open the possibility that residual NADPH diaphorase activity was responsible for the mitochondrial product. Furthermore, the fact that reaction product seen in association with the ER using ferrocyanide was not uniformly distributed[67] leaves open the chance that differential association of 3βHSD-enriched ER fragments with mitochondria could occur. The somewhat diffuse nature of the reaction product also prevented definitive membrane localization. Even the best results showed cytoplasmic staining "associated with the SER" and staining of the mitochondrial matrix "associated with the inner membrane".

When EM immunocytochemistry was employed no staining occurred in the mitochondria and dense reaction product occurring in association with the ER was fairly uniformly distributed.[68] However, because the immunoperoxidase method was used and because the ER elements are closely packed in the cell, definitive statements about the membrane association could not be made. A more detailed analysis would be possible using the immunogold technique.

Although the results with EM immunocytochemistry are suggestive of ER localization, they are not conclusive, for they only pertain to the particular antibody used. Chapman and Sauer found several differences in the activity of 3βHSD in mitochondrial and microsomal fractions.[58] They suggested that they contained related but distinct enzymes. Handler and Bransome had earlier suggested two forms of the enzyme but did not place the forms in different organelles.[69] Distinct enzymes in the two organelles, with slightly different activities and sensitivity to fixation could explain both the biochemical and cytochemical variations reported in the literature. In cattle and humans, only one gene for the protein has been reported.[70,71] However, in rats two cDNAs have been characterized.[72] Both cDNAs encode 372aa proteins which are of slightly different molecular weights. The proteins are 94% homologous and both have activity for 3βHSD as well as Δ^5-Δ^4 isomerase activity for

Δ^5-pregnene and Δ^5-androstene precursors. However, one is more active than the other and the two have different tissue distributions. It would be of interest to examine their intracellular localization as well.

Differences in the distribution of 3βHSD activity among the different zones of the cortex may account, in part, for zonal differences in steroid products. In most species cytochemically demonstrable 3βHSD is most intense in the zona fasciculata.[57,63,68,73] However, in the Japanese harvest mouse the zona glomerulosa was intensely stained.[74] This species exhibits some characteristics of zona glomerulosa hyperfunction[75] and the enhanced activity may relate to this functional state. In the human fetal adrenal 3βHSD biochemical activity is lower in the fetal than in the definitive cortex and in the human and guinea pig adult adrenal it is lower in the zona reticularis than in the zona fasciculata.[73,76] The lower levels are thought to favor the formation of DHEA or its sulfate by 17α-hydroxylase, present in both zona fasciculata and zona reticularis, but not in zona glomerulosa of the normal adrenal cortex (see below). It is not clear, however, whether these differences in activity are due to different levels of the protein or to the arrangement of the proteins within the intercellular membranes. In at least one species, the guinea pig, protein levels by immunoblotting are equal in the outer (ZG + ZF) and inner (predominantly ZR) zones, yet biochemical and cytochemical activity in the inner zone is 25 to 50% lower than in the outer zone.[73,77] Other factors, such as the surrounding lipid environment and the relationship of 3βHSD to other enzymes, may influence activity. Topological immunochemical data might be informative.

b. Steroid Hydroxylases

Many of the steps in steroidogenesis involve site-specific hydroxylation of the sterol ring structure (Figures 1 and 2). These reactions involve cytochrome P450s as their terminal oxidase and the specificity of the reaction is inherent, at least in part, in these proteins. To date, five cytochrome P450s with steroid hydroxylase activity have been isolated, three in the mitochondrial fraction and two in the microsomal fraction. The structure, synthesis, and regulation of these proteins has been extensively studied, but they are probably not the only P450s in the adrenal cortex. At least three of the known P450s are associated with several enzymatic steps ($P450_{scc}$, $P450_{11\beta}$, $P450_{17\alpha}$). Some of these may prove to be functions of other proteins, as recently shown for aldosterone biosynthesis (see below). Additional P450s may also be specific for intermediates in alternate pathways of steroid synthesis and metabolism,[1,2] some with clinical relevance.[78] Furthermore, cytochrome P450s as yet unidentified must be necessary for several steps in cholesterol biosynthesis. Candidates for alternate functions include cytochrome P450s immunochemically related to hepatic P450s, recently identified in the guinea pig adrenal.[79-81]

i. Mitochondria

Cholesterol Side-Chain Cleavage Enzyme and 11β- and 18-Hydroxylase

Cholesterol side-chain cleavage is an initial step in the synthesis of steroids from cholesterol, while 11β- and 18-hydroxylase are involved in the final steps of mineralo- and glucocorticoid synthesis. The association of these enzymes with the mitochondrial fraction was established by biochemical analysis of subfractionated adrenal homogenates.[82] Subsequently, it was reported that most of their activity could be recovered from an inner membrane fraction and that both enzymes had specific associations with the matrical side of the inner membrane.[83-85] Two cytochrome P450s associated with these enzymatic steps were purified from bovine mitochondria, one with activity for side-chain cleavage ($P450_{scc}$) and the other with activity for both 11β- and 18-hydroxylation ($P450_{11\beta}$).[86,87] An additional cytochrome P450 specifically capable of aldosterone synthesis ($P450_{aldo}$) has been purified recently from rat zona glomerulosa mitochondria.[88]

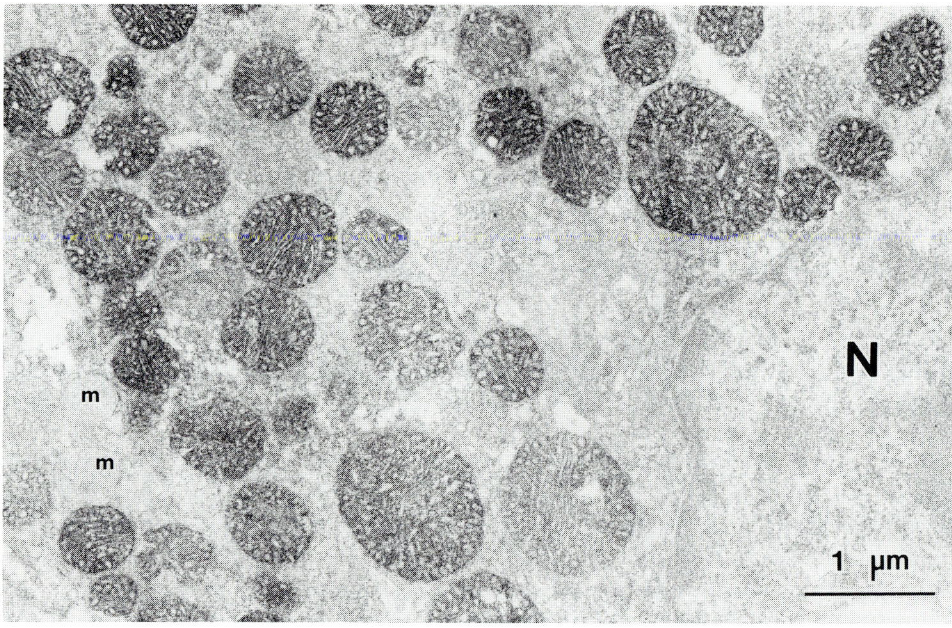

FIGURE 4. Immunoperoxidase localization of $P450_{11\beta}$ in the mitochondria of bovine zona fasciculata cells. Reaction product is located in the inner mitochondrial compartment and is most intense along the inner membrane. The intracristal space is devoid of staining. (Bar = 1.0 μm. From Mitani, F. et al., *J. Histochem. Cytochem.*, 30, 1066, 1982. With permission.)

The localization of 11β-hydroxylase was interpreted to be mitochondrial following specific localization of DOC in this organelle by EM radioautographic analysis.[89] Immunocytochemical analysis of $P450_{scc}$ and $P450_{11\beta}$ by Mitani and co-workers, employing highly specific polyclonal antibodies, firmly established the localization of these P450s to the inner mitochondrial compartment (Figure 4).[99] Two additional immunocytochemical studies using monoclonal antibodies have confirmed their finding: i.e., intense staining along the inner membranes of the mitochondria with no staining of other cell membranes such as endoplasmic reticulum, nuclear or plasma membranes.[91,92] These electron microscopic studies were done on bovine adrenal cortex. In each case it was noted that the staining intensity varied among the mitochondria: 50% were positive, 50% negative.

The relative intensity of the mitochondrial hydroxylases also varied among the different zones of normal adrenals and in pathological tissues. In the normal human adrenal histochemical staining for 11β-hydroxylase was most intense in the zonae fasciculata and reticularis.[93] Immunoreactivity for $P450_{scc}$ and $P450_{11\beta}$ showed a similar distribution, being weak in the zona glomerulosa and intense in the zonae fasciculata and reticularis in bovine and pig adrenals.[90-95] In adrenocortical adenomas of the clear cell type, both enzyme activity and immunostaining were weaker than in normal cells, while in the compact cell type they were stronger, probably indicating greater steroidogenic activity in the latter.[93,94] In hyperplastic adrenals from patients with Cushing's disease, in which cortisol production is increased, immunostaining for both $P450_{11\beta}$ and $P450_{scc}$ were intense in the inner cortical zones and in cortical micronodules.[94,95] Immunostaining was also intense in the outer zones of adrenals from patients with idiopathic hyperaldosteronism, correlating with increased steroidogenesis in this region. In this case, immunostaining in the attached normal adrenal tissue was reduced.

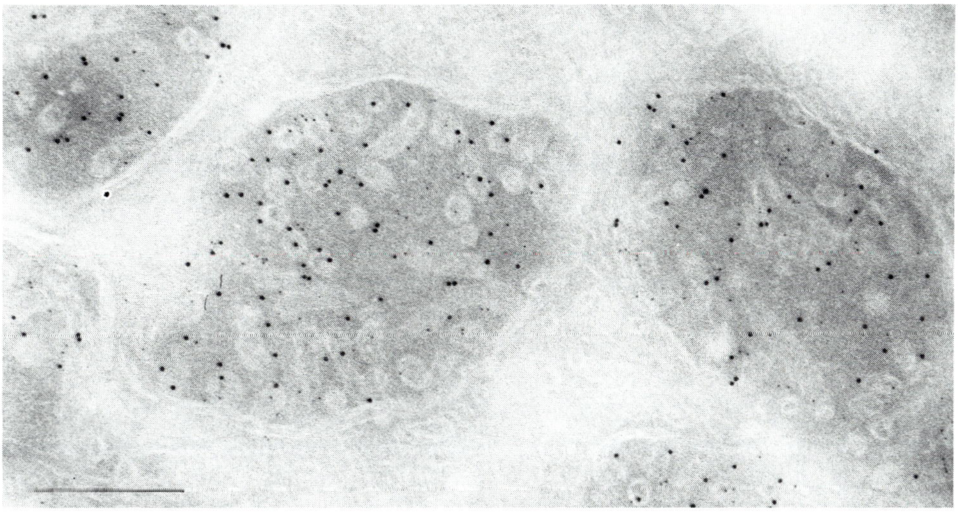

FIGURE 5. Ultrathin cryosections of bovine zona fasciculata cells immunogold labeled for demonstration of $P450_{scc}$ (large particles) and $P450_{11\beta}$ (small particles). (Bar = 0.25 mm. From Geuze, H. J. et al., *Histochemistry*, 86, 551, 1987. With permission.)

The 50:50 distribution of labeling for both $P450_{11\beta}$ and $P450_{scc}$ among the mitochondria was somewhat surprising, as was the almost negligible immunostaining of the zona glomerulosa which, although producing lower total steroid levels, requires both P450s to synthesize its characteristic product, aldosterone.[96] In addition, while the work cited above indicated that the staining was most intense on the matrical side of the inner membranes; in cases where the staining was particularly intense, the immunoperoxidase methodology used resulted in reaction product appearing in the matrical space.[90-92] Peroxidase immunochemistry also did not permit double labeling, necessary for simultaneous demonstration of their distribution among the mitochondria and for precise analysis of their topological relationship within the inner mitochondrial membrane.

Quantitative immunogold cytochemistry carried out in a collaborative study by Geuze et al. confirmed and extended these earlier observations.[97] Using the very sensitive protein A/gold/silver technique they were able to demonstrate $P450_{scc}$ and $P450_{11\beta}$ in all three zones of the adrenal. The zona glomerulosa was less intensely stained, probably due to its lower number of mitochondria. Using double labeling at the EM level, they found that all of the mitochondria were labeled and that each contained both enzymes (Figure 5). Most of the label occurred on the matrical side of the inner mitochondrial membrane, confirming earlier biochemical reports on the topological distribution of $P450_{scc}$[85,86] and demonstrating that both enzymes are present on the inner aspect of the inner mitochondrial membrane. Analysis of the relationship of these P450s to NADPH reductase and adrenodoxin, a nonheme iron protein intermediate in the reaction, would provide additional confirmation of the topology derived from biochemistry which suggests that they occur in coupled sets.

Recent studies have shown that there are two variants of $P450_{11\beta}$. In the bovine adrenal there is one gene ($P450_{11\beta}$) whose product performs 11β-, 18-, and 19-hydroxylations of deoxycorticosterone and is also capable of converting 18-hydroxycorticosterone to aldosterone.[87] However, in the rat and human adrenal there are two similar but distinct genes.[98-101] The products of both are capable of 11β-hydroxylation and 18-hydroxylation,[88,101] but only one can catalyze formation of aldosterone from 18-hydroxycorticosterone ($P450_{aldo}$), the other

($P450_{11\beta}$) cannot. Recent studies indicate that mRNA for $P450_{aldo}$ is highly expressed in an adrenal tumor of a patient suffering from primary aldosteronism.[101] However, it was undetectable in total adrenal mRNA from normal human adrenal.[98,100] In rat, $P450_{aldo}$ has been isolatable only from mitochondria of the zona glomerulosa.[88] Its expression in the normal human adrenal may have been masked by examination of total adrenal RNA rather than RNA specifically from the zona glomerulosa. *In situ* hybridization would allow definitive zonal localization of mRNA, while immunocytochemistry would be capable of elucidating the zonal and subcellular distribution of these two variant proteins *in situ*.

Adrenodoxin and Adrenodoxin Reductase

In the mitochondria, but not in the ER, a nonheme iron protein, adrenodoxin, mediates the transfer of electrons from adrenodoxin reductase to mitochondrial cytochrome P450s. This protein, like the P450s scc and 11β above, has been localized to the inner mitochondrial membrane by both the peroxidase and protein A-gold techniques.[101,102] Using the latter technique, adrenodoxin was homogeneously distributed among the mitochondria of the three cortical zones, although, as for the cytochromes P450, the labeling intensity was lower for the zona glomerulosa than for the zonae fasciculata and reticularis. In the zona glomerulosa, adrenodoxin immunolabeling was not only on the inner mitochondrial membrane but also in the mitochondrial matrix and over electron-dense intramitochondrial bodies.[103]

Adrenodoxin reductase has also been localized to the inner mitochondrial membrane,[102] but its precise distribution remains to be investigated with the protein A-gold technique. Its zonal distribution in the bovine and human adrenal, both normal and pathological, is similar to that described for the mitochondrial cytochrome P450s and adrenodoxin.[102] Additional topological information might be provided by double-labeling analysis of these mitochondrial proteins, as suggested above.

ii. Endoplasmic Reticulum

21-Hydroxylase and 17α-Hydroxylase/C17:20 Lyase

Two steroid hydroxylases have been localized to the microsomal fraction by biochemical analysis: 21 hydroxylase ($P450_{21}$), which is required for synthesis of mineralo- and glucocorticoids, and 17α-hydroxylase ($P450_{17\alpha}$), which is necessary for production of cortisol and androgens.[1,2,4] $P450_{17\alpha}$ has activity both for 17α-hydroxylation, involved in synthesis of cortisol and androgens, and for cleavage of the side chain between C17 and C20 (C17:20 lyase).[105] It is well recognized that the relative activities of these two P450s determine the type of products synthesized. The distribution of these proteins among the zones of the adrenal cortex has been analyzed by immunohistochemistry in both normal and pathological tissues and is generally consistent with their pattern of steroid production.

Immunodetectable $P450_{21}$ is present in all three cortical zones, but its relative intensity among the zones is species dependent. In the bovine adrenal cortex, immunostaining for $P450_{21}$ is most intense in the zona glomerulosa and inner zona reticularis, somewhat less intense in the zona fasciculata, and faint in the region between the glomerulosa and fasciculata.[106] In the guinea pig adrenal, immunostaining for $P450_{21}$ was most intense in the zona reticularis.[107] Immunodetectable $P450_{17\alpha}$, on the other hand, was present in the zonae fasciculata and reticularis, but not in the zona glomerulosa in pig, bovine, human, and guinea pig adrenal.[107,108] This distribution of $P450_{17\alpha}$ is consistent, for the most part, with the ability of the tissues to produce cortisol and androgens. However, in the guinea pig, zona glomerulosa cells can produce cortisol *in vitro* and possess $P450_{17\alpha}$, when assayed by ELISA, suggesting that the immunochemistry employed was not sufficiently sensitive in this species.[96,109]

The intensity of immunolabeling is altered in several disease states.[108] Immunostaining for $P450_{17\alpha}$ was particularly intense in the zonae fasciculata and reticularis of patients with

Cushing's disease, but negative in the hyperplastic zona glomerulosa and outer zona fasciculata of adrenals from individuals with idiopathic hyperaldosteronism. However, in some aldosteronomas, presumably derived from zona glomerulosa cells, the tumor cells contained immunoreactive $P450_{17\alpha}$. As in the case of normal guinea pig zona glomerulosa cells this was consistent with their production of cortisol and responsiveness to ACTH.[108]

The distribution of steroid hydroxylase gene expression has also been analyzed by *in situ* hybridization.[110] In adrenals of fetal sheep, mRNA for $P450_{21}$, $P450_{17\alpha}$ and $P450_{scc}$ was distributed throughout the cortex in young fetuses. However, with the appearance of two zones, labeling for $P450_{17\alpha}$ was limited to the zona fasciculata while mRNA for $P450_{21}$ and $P450_{scc}$ was found in both zona glomerulosa and zona fasciculata. Changes in the intensity of labeling for $P450_{17\alpha}$ during gestation (lower at midterm than in early or late term) were consistent with changes in the ability of the adrenal to synthesize cortisol.

In addition to the types of cytochrome P450 present in the cell, the precise pathway of synthesis followed may also be influenced by the arrangement of these proteins within the ER with respect to each other and to 3βHSD (Figure 2).[4] $P450_{17\alpha}$ has been demonstrated along the membranes of the ER in guinea pig adrenal, using the immunoperoxidase technique and a similar location was stated for $P450_{21}$ (Figure 6).[107] With the immunoperoxidase technique employed, distribution along the ER membrane appeared to be fairly uniform, at least for $P450_{17\alpha}$. However, the relationship of the microsomal cytochrome P450s to each other and their relationship with NADPH cytochrome P450 reductase and with other enzymes in the steroidogenic pathway, such as 3βHSD, remains to be examined.

Spironolactone (SL) Bodies

These lamellated structures were first described in zona glomerulosa cells of individuals treated with spironolactone, a compound which inhibits aldosterone action and destroys cytochrome P450s.[111,112] Their origin from the endoplasmic reticulum is suggested by the fact that they stain intensely for 3βHSD and NADH tetrazolium reductase[113] and that their lamellae connect to membranes of the ER.[33,113] SL bodies in aldosteronomas have been specifically stained by immunoperoxidase using anti-aldosterone antibody.[114] The rest of the cytoplasm was unreactive. This may have been due to extraction during tissue preparation or indicate preferential retention of aldosterone in the SER membranes making up the SL body. It has been suggested that SL may itself become incorporated into the membranes, interfering with normal catabolism. However, the formation of such bodies does not require SL, nor are they in all instances confined to zona glomerulosa cells. Similar structures have been observed in zona glomerulosa cells of individuals with type B sphingomyelinase deficiency and in zona reticularis cells of a female patient with a virilizing adenoma, both instances where no SL treatment was given.[114,115] Furthermore, they have been induced in rat zonae fasciculata and reticularis cells following treatment with cycloheximide and ACTH.[116] Their formation probably results from abnormal turnover of ER membrane, but their precise functional significance remains to be elucidated.

Calcium

Calcium is an important "third messenger" in stimulation of adrenocortical cells by either AII or ACTH. Significant fluxes of calcium occur in concert with hormonal stimulation. The predominant localization of calcium in the mitochondrial intracristal space and SER lumen was established by Aguas and Nickerson.[117,118] To facilitate calcium retention, they fixed tissue in potassium pyroantimonate-osmium tetroxide. Fixation with osmium rather than with glutaraldehyde retained calcium in tissues and potassium pyroantimonate precipitated the calcium *in situ* during the fixation process. Other intracellular sites included the nuclear heterochromatin and nucleolus, as well as Golgi vesicles and small vesicles near the cell surface. The latter deposits suggested vesicular uptake or release of calcium.

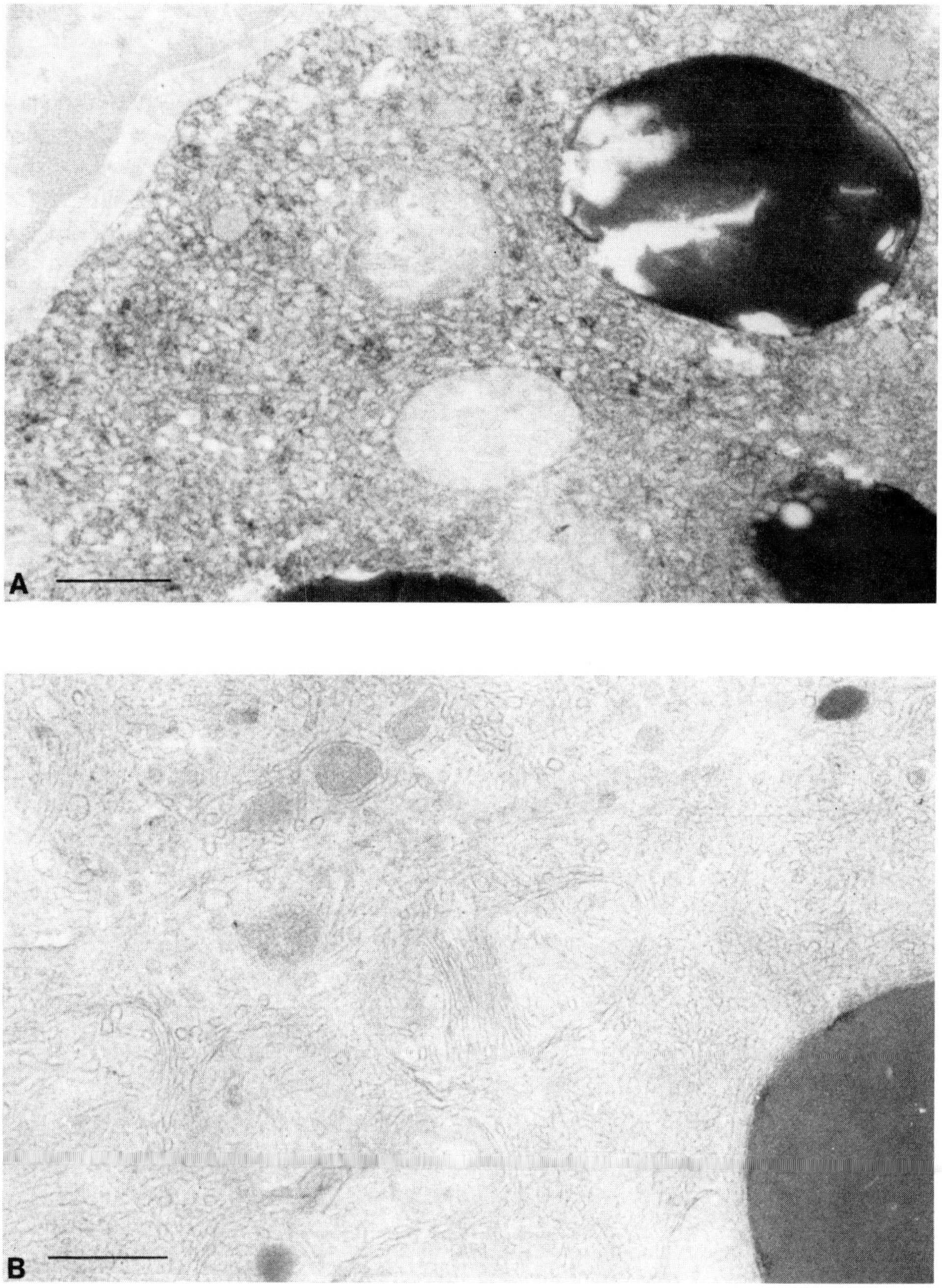

FIGURE 6. (A) Immunoperoxidase localization of $P450_{17\alpha}$ in guinea pig zona fasciculata cells. Staining is most intense along the ER surface. No reaction product occurs within the ER lumen. (B) Control. (Bar = 1.0 μm. From Shinzawa, K. et al., *J. Endocrinol.*, 119, 191, 1988. With permission.)

Saturation of the fixative with EGTA [Ethyleneglycol-bis-(β-aminoethyl ether)-*N,N,N′,N′*-tetraacetic acid], a calcium chelator, resulted in no nuclear staining and a significant decrease in deposition in all other compartments, indicating that most of the precipitate seen was due to formation of calcium-antimonate. Deposits were also seen in the extracellular space and within the capillary lumen. The volume of these precipitates was altered in regenerating rat adrenals and in normal rats following hormone treatment.[118] Hypophysectomy reduced deposition in both the mitochondrial intracristal space and endoplasmic reticulum cisternae. ACTH increased deposition in both compartments.

Matrix granules within adrenocortical mitochondria can be removed by EGTA treatment of thin sections, suggesting that they are sites of concentrated cations such as calcium.[119] Their decrease in number following ACTH treatment is consistent with their representing a site of sequestered calcium which responds to hormonally induced calcium fluxes.

5. Cytosol

a. Dehydroepiandrosterone (DHEA) Sulfotransferase

DHEA and DHEA sulfate are prominant products of the human adrenal gland. The enzyme responsible for sulfation of this steroid is DHEA sulfotransferase. Ultramicrochemical analysis demonstrated that DHEA sulfotransferase was localized to the zona reticularis of the human and guinea pig adrenal.[120,121] This zonal distribution was confirmed for the human adrenal by immunoperoxidase histochemistry.[122] However, staining within the zona reticularis cells was not uniform; it was most intense in the region of the cell containing abundant lipid droplets. EM immunocytochemistry confirmed that product was localized around lipid droplets in the human zona reticularis cells.[122] Although sulfotransferase is considered to be a cytoplasmic enzyme based on biochemical analysis, the diffuse nature of the immunoperoxidase product precluded finer localization to membranes vs. cytosol. Other sulfated precursors or intermediates may well have their own specific sulfotransferase. The identification and localization of these has not yet been pursued cytochemically.

Sulfotransferase can transfer sulfate to hydroxyl groups at either the 3β or 17β positions. DHEA is the preferred substrate being formed from 17α-hydroxypregnenolone.[120] The intracellular localization suggests association between DHEAS formation and enzymes of the early steps in steroidogenesis: cholesterol hydroxylase, side-chain cleavage, 17α-hydroxylase and 3βHSD.[122]

b. Glucose-6-Phosphate Dehydrogenase (G6PD)

G6PD, a ubiquitous enzyme in cells, provides NADPH for microsomal hydroxylases in steroid-secreting cells. ACTH stimulates its activity in concert with stimulation of the hydroxylases.[123] Appropriate localization in tissue requires an electron transfer intermediate which will bypass cellular NADPH-reductase, "diaphorase", localized in the mitochondria. To stabilize the soluble G6PD, polyvinyl alcohol is often included in the medium. In a procedure using PMS (phenazine methosulfate) as the intermediate and potassium ferricyanide as an electron acceptor, Berchtold showed fine deposits of copper ferrocyanide reaction product in the cytoplasm of rat adrenocortical cells.[124] Product indicative of NADPH diaphorase was found in the mitochondrial matrix when G6PD medium lacking PMS was used or if NADPH was used as the substrate. NADPH diaphorase reaction product was not uniform among the mitochondria, occurring in only 50% of them in either case. The cytoplasmic localization of G6PD has been confirmed many times, particularly in conjunction

further investigation of the trafficking of molecules for lysosomes vs. secretory granules upon leaving the Golgi through the TGN.

While renin appears to reside in a population of granules in the zona glomerulosa which may be synonymous with those having low acid phosphatase activity, the possibility of peptide hormone localization in similar granules found in cells of the zona fasciculata and zona reticularis remains to be examined. Several peptide hormones, recently demonstrated in the adrenal cortex by immunohistochemistry, are candidates worthy of investigation. Some are located in nerves (see below), but others appear to be in the parenchymal cells. The neuropeptides oxytocin and vasopressin have been identified in the definitive cortex of the human fetal adrenal.[143] Their presence in anencephalic fetuses suggested intraadrenal production and they are stored in secretory granules in other tissues. In the adult adrenal cortex of several species (rat, cow, hamster, guinea pig) immunostaining for these neuropeptides is most intense in the zona glomerulosa, but all zones are labeled.[144] The growth factors, insulin-like growth factor (IGF) and basic fibroblast growth factor (bFGF), have also been identified in adrenocortical cells. Their mode of storage and secretion requires investigation. IGF is present in both zones of the human fetal adrenal[145] and bFGF is present throughout the early postnatal cortex of the rat.[146] With further development, bFGF becomes confined to the zona glomerulosa,[146] the site in which it is also localized in the bovine adrenal.[147] Aside from the obvious mitogenic effect of the growth factors, the functions of these additional intraadrenal peptide hormones is uncertain. AVP is capable of stimulating cortical cells *in vitro*[148] and possible involvement in modulation of adrenal function, especially in stress-related changes, has been proposed.

8. Plasma Membrane

a. Cell Junctions/Contacts

Tracer molecules detectable by EM cytochemical methods have been used to delineate adrenocortical cell junctions.[149] While tight junctions, which preclude passage of such tracers, were found between endothelial cells, they did not occur between the parenchymal cells of the cortex. However, adrenocortical cells do possess extensive gap junctions,[149] which become larger as one progresses from zona glomerulosa to zona reticularis.[21] Cell contacts with a septate-like appearance have also been described.[150] These occur between interdigitating processes along the intercellular and pericapillary boundaries of cortical cells in rat adrenal. Deposits of calcium-antimonate are retained in the intercellular canaliculi formed by these microvillar contacts, suggesting a role in calcium sequestration.[150] Reaven and coworkers recently demonstrated that radiolabeled high-density lipoproteins (HDL) also accumulate in these channels.[151] They suggested that these cell surface compartments serve as storage sites for lipoproteins and/or facilitate their uptake.

b. Receptors

Both radioautography and immunocytochemistry have been used to analyze the distribution of receptors for hormones, growth factors, and lipoproteins in the adrenal cortex at the LM level. Receptors for angiotensin II (AII), the primary trophic hormone of zona glomerulosa cells, have been localized in the zona glomerulosa by autoradiography.[152] Atrial natriuretic peptide (ANP), a peptide hormone acting antagonistically to AII, i.e., inhibiting aldosterone secretion, also has receptors located in the zona glomerulosa.[153,154] ACTH can stimulate steroidogenesis by all cortical cells and its receptors have been demonstrated on the plasma membrane of all zonal cell types by immunoelectron microscopy. Labeling for ACTH occurred not only along the plasma membrane, but also intracellularly, suggesting that the hormone may be taken up by the cells following binding to plasma membrane receptors.[155] The significance of this tantalizing observation needs to be analyzed in more detail.[155]

Receptors for several other hormones of pituitary origin which affect adrenocortical function have also been demonstrated cytochemically. Anti-prolactin (PRL) binding was localized in the mouse zona fasciculata.[156] ^{125}I-labeled growth hormone binding occurred in the rat adrenal and the relative intensity of labeling among the cortical zones was shown to vary with hormonal treatment.[157,158] Expression of epidermal growth factor (EGF) receptors was immunochemically demonstrated in scattered cells in all zones in nonneoplastic adrenals, but increased in neoplastic tissue and was greater in adrenocortical carcinomas than in adenomas.[159] In adrenocortical adenomas, immunostaining was confined to compact cells; clear cells were unstained.

Lipoproteins are required by adrenocortical cells of many species for maximal steroidogenesis. Both low-density lipoprotein (LDL) and high-density lipoprotein (HDL) can serve as a source of cholesterol and the preference for one over the other seems to be species dependent. LDL receptors were demonstrated immunohistochemically in the rat adrenal.[160] Immunoreactivity was most intense in the zona fasciculata. Binding to plasma membrane of gold-labeled LDL was shown in monolayer cultures of human and rat adrenocortical cells,[24] but HDL binding was only observed on adrenocortical cells from the rat.[24,25] Label also occurred in coated vesicles and lysosomes. Labeling of both plasma membrane and intracellular sites was competed by unlabeled lipoprotein, providing evidence that at least the protein component of the particles followed this pathway. Radiolabeled HDL has also been shown to accumulate in microvillar channels present *in situ* in this species, but uptake by receptor-mediated endocytosis was not certain.[151]

c. *Alkaline Phosphatase*

Although the functional role of alkaline phosphatase remains ill-defined, its distribution has been extensively studied and it has been used as an index of cell function.[3] At the EM level, alkaline phosphatase has been demonstrated along the plasma membrane of adrenocortical cells of lower vertebrates and in normal human fetal and adult adrenals.[29,31,161] In fetal adrenals, staining was more intense in the cells of the fetal cortex than in those of the permanent cortex.[29] In Cushing's syndrome and primary aldosteronism, staining was more intense in compact cells than in more lipid-filled clear cells.[31] In these apparently more active cells, deposits occurred along surfaces of abundant microvilli. In the clear cells the plasma membrane, while labeled, lacked abundant microvilli. In the case of primary aldosteronism, alkaline phosphatase was also found in the outer mitochondrial membrane.[31]

C. MICROVASCULATURE

1. Phagocytosis

Early studies of the reticuloendothelial system included the cells lining sinusoids of the adrenal cortex and medulla.[162] This system was later redefined in terms of mononuclear phagocytes.[163] Several criteria, including EM cytochemistry for peroxidase, have been used to characterize a small population of cells in the sinuses of the adrenal cortex and medulla as being of monocytic origin.[164] These phagocytes are more abundant in the inner cortical region than in the outer cortex or medulla and can be increased by zymosan treatment.

Other analyses show, however, that adrenal endothelial cells, specifically those of the medulla, can take up gold-labeled albumin.[165] Gold particles are seen in coated pits and vesicles, endosomes, and inside lysosomes, suggesting receptor-mediated uptake, possibly via the "scavenger receptor". These cells are similar in this respect to cells lining the sinusoids of liver and bone marrow and could account for the ability of the adrenals to catabolize albumin. In the cortex, the cortical capillaries showed no sign of endocytosis. The gold-labeled albumin appeared to cross through the fenestrae in the endothelial cells to reach the perivascular space. Unfortunately, this paper did not comment on the macrophage population, discussed above, which exists in the inner cortical and medullary sinuses. Macrophages also have "scavenger receptors" and could contribute significantly to adrenal protein catabolism.

2. Vitamin A Storage

Storage of radiolabeled vitamin A has been demonstrated by autoradiography to occur in lipid droplets of fibroblast-like cells located in the pericapillary spaces of the mouse adrenal.[166] These cells are most prominent in the cortex. Their pericapillary location is similar to the vitamin A-storing cells in the liver.[167] The functional significance of vitamin A storage in the adrenal is unclear.

3. Contractile Proteins

The distribution of contractile proteins in the guinea pig and rat adrenal cortex was investigated by immunofluorescence using antibodies made against chicken actin and myosin.[168] Both proteins were found in arteriolar vessels, as expected; but, myosin-specific proteins were also detected in the endothelial and perivascular cells of cortical capillaries. Cortical capillary myosin proteins were much more prominant in the guinea pig than in the rat and were more abundant in the zonae glomerulosa and reticularis than in the zona fasciculata. Adrenergic nerves, detectable by the glyoxylic acid technique, showed a striking correspondence in distribution.[168] Taken together, these observations suggest that local regulation of blood flow may occur throughout the cortex, rather than solely at the level of the subcapsular plexus and central vein. In addition to altering the rate of flow, contraction of the capillary wall could also influence the transport of hormones into (or out of) the intercellular spaces by changing the dimensions of microvillar channels between cells or altering transendothelial channels.

D. NERVES

The adrenal cortex is well innervated with fibers involved in regulation of blood vessels, as noted above, and with fibers passing through to the adrenal medulla. However, the innervation of the parenchymal cells of the cortex and its significance to cortical function has been debated. There is evidence that nerves are involved in compensatory hypertrophy that occurs when one adrenal is removed and may be involved in stress-related responses.[169] Recently, several neuropeptides capable of influencing cortical cell function, have been localized within fibers in the cortex. Fibers containing vasopressin (AVP) and vasointestinal peptide (VIP) are most prominent in the zona glomerulosa,[143,170] while fibers immunoreactive for corticotrophin-releasing factor (CRF) penetrate the cortex in close contact with parenchymal cells and form a plexus in the zona reticularis adjacent to the medulla.[171] AVP and VIP are known stimulators of adrenocortical function *in vitro*,[148,172] but their precise functions, as well as the functions of CRF, remain to be clarified.

E. CONCLUSIONS

As can be seen from the discussion above, cytochemistry is a powerful technique, particularly when combined with biochemistry and experimental manipulation. It has contributed significantly to our understanding of adrenocortical structure and function, both normal and pathological. In the near future, the increasing availability of specific antibodies to proteins of interest and cDNA probes for their mRNAs will allow more detailed examination of zonal and intracellular distribution of proteins which are involved in steroidogenesis and of factors involved in their hormonal regulation. EM immunocytochemistry may prove uniquely capable of mapping enzyme relationships which could influence the specific path followed in synthesis of particular hormones and *in situ* hybridization will allow analysis of regulation at the level of individual cells.

ACKNOWLEDGMENTS

The author wishes to thank Drs. Steven Carmichael, Seymour Lieberman, Manuel Magalhães, Maria Magalhães, Peter Nickerson, and Gastone Nussdorfer for their critical reading of the manuscript and insightful discussion.

REFERENCES

1. **Lieberman, S., Greenfield, N. J., and Wolfson, A.,** A heuristic proposal for understanding steroidogenic processes, *Endocr. Rev.*, 5, 128, 1984.
2. **Lieberman, S. and Prasad, V. V. K.,** Heterodox notions on pathways of steroidogenesis, *Endocr. Rev.*, 11, 469, 1990.
3. **Nussdorfer, G. G.,** Cytophysiology of the adrenal cortex, *Int. Rev. Cytol.*, 98, 1, 1986.
4. **Hall, P. F.,** Cellular organization for steroidogenesis, *Int. Rev. Cytol.*, 86, 53, 1984.
5. **Deane, H. W.,** The anatomy, chemistry, and physiology of adrenocortical tissue, in *Handbuch der Experimentellen Pharmakologie Erganzungswerk,* Vol. 14, Eichler, O. and Farah, A., Eds., Deane, H. W., Subed., Springer-Verlag, Berlin, 1962, chap. 1.
6. **Waterman, M. R. and Simpson, E. R.,** Regulation of the biosynthesis of cytochromes P-450 involved in steroid hormone synthesis, *Mol. Cell. Endocrinol.*, 39, 81, 1985.
7. **Miller, W. L.,** Molecular biology of steroid hormone synthesis, *Endocr. Rev.*, 9, 295, 1988.
8. **Okros, I.,** Digitonin reaction in electron microscopy, *Histochemie,* 13, 91, 1968.
9. **Szabo, D., Dzsinich, Cs., and Okros, I.,** Ultrastructural localization of adrenal cholesterol by autoradiograph and digitonin reaction after cycloheximide-induced inhibition of corticosterone synthesis, *Histochemie,* 27, 43, 1971.
10. **Moses, H., Davis, W. W., Rosenthal, A. S., and Garren, L. D.,** Adrenal cholesterol: localization by electron-microscope autoradiography, *Science,* 163, 1203, 1969.
11. **Magalhães, M. C. and Magalhães, M. M.,** Ultrastructural alterations produced in rat adrenal by aminoglutethimide. A stereological and cytochemical study, *Endocrinology,* 90, 444, 1972.
12. **Jones, H. M. and Miyai, K.,** Ultrastructural localization of cholesterol by enzyme histochemistry, *Histochem. J.*, 13, 1017, 1981.
13. **Penney, D. P. and Barrnett, R. J.,** The fine structural localization and selective inhibition of nucleosidephosphatases in the rat adrenal cortex, *Anat. Rec.*, 152, 265, 1965.
14. **Szabo, D., Stark, E., and Varga, B.** The localization of acid phosphatase activity changes in the adrenal zona fasciculata of intact and hypophysectomized rats following ACTH administration, *Histochemie,* 10, 321, 1967.
15. **Black, V. H. and Bogart, B. I.,** Peroxisomes in inner adrenocortical cells of fetal and adult guinea pigs, *J. Cell Biol.*, 57, 345, 1973.
16. **Novikoff, P. M., Novikoff, A. B., Quintana, N., and Hauw, J. J.,** Golgi apparatus, GERL and lysosomes of neurons in rat dorsal root ganglion studied by thin section and thick section cytochemistry, *J. Cell Biol.*, 50, 859, 1971.
17. **Griffiths, G. and Simons, K.,** The trans-Golgi network: sorting at the exit site of the Golgi apparatus, *Science,* 239, 438, 1986.
18. **Holtzman, E.,** *Lysosomes,* Plenum Press, New York, 1989.
19. **Black, V. H. and Russo, J. J.,** Stereological analysis of the guinea pig adrenal: effects of dexamethasone and ACTH treatment with emphasis on the inner cortex, *Am. J. Anat.*, 159, 85, 1980.
20. **Black, V. H.,** The development of smooth-surfaced endoplasmic reticulum in adrenal cortical cells of fetal guinea pigs, *Am. J. Anat.*, 135, 381, 1972.
21. **Black, V. H., Robbins, E., McNamara, N., and Huima, T.,** A correlated thin-section and freeze-fracture analysis of guinea pig adrenocortical cells, *Am. J. Anat.*, 159, 453, 1979.

22. **Dietert, S. E. and Scallen, T. J.,** An ultrastructural and biochemical study of the effects of three inhibitors of cholesterol biosynthesis upon murine adrenal gland and testis. Histochemical evidence for a lysosome response, *J. Cell Biol.,* 40, 44, 1969.
23. **Brown, M. S., Kovanen, P. T., and Goldstein, J. L.,** Receptor-mediated uptake of lipoprotein-cholesterol and its utilization for steroid synthesis in the adrenal cortex, *Rec. Prog. Hormone Res.,* 35, 215, 1979.
24. **Toth, I. E., Szabo, D., Bacsy, E., Szalay, K., Sz., Hesz, A., and Szollar, L. G.,** Morphological evidence of lysosomal uptake of high-density lipoproteins by rat adrenocortical cells in vitro, *Mol. Cell. Endocrinol.,* 44, 185, 1986.
25. **Toth, I. E., Szabo, D., Szalay, K. Sz., Gyevai, A., Szollar, L. G., and Glaz, E.,** Colloidal gold-labeled lipoprotein binding and internalization in adrenocortical cells in vitro, *Clin. Biochem.,* 21, 101, 1988.
26. **Paavola, L. G., Strauss, J. F., III, Boyd, C. O., and Nestler, J. E.,** Uptake of gold and [^{3}H]cholesteryl linoleate-labeled human low density lipoprotein by cultured rat granulosa cells: cellular mechanisms involved in lipoprotein metabolism and their importance to steroidogenesis, *J. Cell Biol.,* 100, 1235, 1985.
27. **Rappay, Gy., Kondies, L., and Bacsy, E.,** Intracellular distribution of arylsulphatase activity in adrenal cortical cells of the rat, *Histochemie,* 34, 271, 1973.
28. **Bacsy, E.,** Enzymic heterogeneity of adrenocortical lysosomes: an X-ray microanalytical study, *Histochem. J.,* 14, 99, 1982.
29. **Murakoshi, M., Osamura, Y., Watanabe, K., and Kuroshima, Y.,** Enzyme histo-cytochemical studies in human fetal adrenal glands, *Tokai J. Exp. Clin. Med.,* 8, 89, 1983.
30. **Taguchi, S., Watabe, S., Kouyama, H., Tatsunami, S., and Yago, N.,** Characterization of porcine adrenocortical lysosomes, *J. Biochem.,* 89, 1411, 1981.
31. **Murakoshi, M., Osamura, Y., Watanabe, K., Nomoto, Y., and Sakai, H.,** Enzyme histo-cytochemical studies in human adrenocortical adenomas (II) Cushing's syndrome, *Tokai J. Exp. Clin. Med.,* 8, 293, 1983.
32. **Hui, K. S., Williams, J. C., Borit, A., and Rosenberg, H. S.,** The endocrine glands in Pompe's disease, *Arch. Pathol. Lab. Med.,* 109, 921, 1985.
33. **Elleder, M. and Smid, F.,** Adrenal changes in Niemann-Pick disease: differences between sphingomyelinase deficiency and type C, *Acta Histochem.,* 76, 163, 1985.
34. **Magalhaes, M. M. and Magalhaes, M. C.,** Microbodies (peroxisomes) in rat adrenal cortex, *J. Ultrastruct. Res.,* 37, 563, 1971.
35. **Beard, M. E.,** Identification of peroxisomes in the rat adrenal cortex, *J. Histochem. Cytochem.,* 20, 173, 1972.
36. **Russo, J. J. and Black, V. H.,** Hormone-dependent changes in peroxisomal enzyme activity in guinea pig adrenal, *J. Biol. Chem.,* 257, 3883, 1982.
37. **Rebuffat, P., Mazzocchi, G., and Nussdorfer, G. G.,** Effect of long-term inhibition of hydroxy-methylglutaryl coenzyme A reductase by mevinolin on the zona fasciculata of rat adrenal cortex, *Virchows Arch. B,* 54, 67, 1987.
38. **Mazzocchi, G., Robba, C., Meneghelli, V., and Nussdorfer, G. G.,** Longterm effects of the hypocholesterolaemic drug 4-aminopyrazolo-pyrimidine on the zona fasciculata of the rat adrenal cortex, *J. Anat.,* 149, 1, 1986.
39. **Mazzocchi, G., Robba, C., Rebuffat, P., Belloni, A. S., and Nussdorfer, G. G.,** Effects of the hypolipidemic drug Nafenopin on the zona fasciculata of the rat adrenal cortex: a correlated biochemical and stereological study, *Anat. Rec.,* 204, 245, 1982.
40. **Conran, R. M. and Nickerson, P. A.,** Thyroparathyroidectomy increases the activity of catalase in the adrenal gland of the rat, *J. Microsc. Cytol.,* 11, 305, 1982.
41. **Keller, G. A., Pazirandeh, M., and Krisans, S.,** 3-Hydroxy-3-methylglutaryl coenzyme A reductase localization in rat liver peroxisomes and microsomes of control and cholestyramine-treated animals: quantitative biochemical and immunoelectron microscopical analyses, *J. Cell Biol.,* 103, 875, 1986.
42. **Keller, G. A., Scallen, T. J., Clarke, D., Maher, P. A., Krisans, S. K., and Singer, S. J.,** Subcellular localization of sterol carrier protein-2 in rat hepatocytes: its primary localization to peroxisomes, *J. Cell Biol.,* 108, 1353, 1989.
43. **Vahouny, G. V., Dennis, P., Chanderbhan, R., Fiskum, G., Noland, B. J., and Scallen, T. J.,** Steroid carrier proteins (SCPs)-mediated transfer of cholesterol to mitochondrial inner membranes, *Biochem. Biophys. Res. Commun.,* 122, 509, 1984.
44. **van Amerongen, A., van Noort, M., van Beckhoven, J. R. C. M., Rommerts, F. F. G., Orly, J., and Wirtz, K. W. A.,** The subcellular distribution of the nonspecific lipid transfer protein (sterol carrier protein 2) in rat liver and adrenal gland, *Biochem. Biophys. Acta,* 1001, 243, 1989.
45. **Mendis-Handagama, S. M. L. C., Zirkin, B. R., Scallen, T. J., and Ewing, L. L.,** Studies on peroxisomes of the adult rat Leydig cell, *J. Androl.,* 1, 270, 1990.

46. **Powers, J. M., Schaumburg, H. H., Johnson, A. B., and Raine, C. S.,** A correlative study of the adrenal cortex in adreno-leukodystrophy — evidence for a fatal intoxication with very long chain saturated fatty acids, *Invest. Cell. Pathol.*, 3, 353, 1980.
47. **Powers, J. M., Moser, H. W., Moser, A. B., and Schaumburg, H. H.,** Fetal adrenoleukodystrophy: the significance of pathologic lesions in adrenal gland and testis, *Hum. Pathol.*, 13, 1013, 1982.
48. **Goldfischer, S., Powers, J. M., Johnson, A. B., Axe, S., Brown, F. R., and Moser, H. W.,** Striated adrenocortical cells in cerebro-hepato-renal (Zellweger) syndrome, *Virchows Arch. A*, 401, 355, 1983.
49. **Powers, J. M. and Schaumburg, H. H.,** Adreno-leukodystrophy (sex-linked Schilder's disease), *Am. J. Pathol.*, 76, 481, 1974.
50. **Powers, J. M. and Schaumburg, H. H.,** Adreno-leukodystrophy. Similar ultrastructural changes in adrenal cortical and Schwann cells, *Arch. Neurol.*, 30, 406, 1974.
51. **Singh, I., Moser, H. W., and Moser, A. B.,** Adrenoleukodystrophy: impaired oxidation of long-chain fatty acids in cultured skin fibroblasts and adrenal cortex, *Biochem. Biophys. Res. Commun.*, 102, 1223, 1981.
52. **Beyer, K. F. and Samuels, L. T.,** Distribution of steroid-3β-ol dehydrogenase in cellular structures of the adrenal gland, *J. Biol. Chem.*, 219, 69, 1956.
53. **Inano, H., Machino, A., and Tamaoki, B.,** Localization of the Δ^5-3β-hydroxysteroid dehydrogenase and 21-hydroxylase activities in the smooth-surfaced microsomes of adrenals, *Steroids*, 13, 357, 1969.
54. **Cowan, R. A., Giles, C. A., and Grant, J. K.,** The intramitochondrial distribution of the 3β-hydroxysteroid dehydrogenase-oxosteroid isomerase. A probable redistribution artifact, *J. Steroid Biochem.*, 5, 607, 1974.
55. **McCune, R. W., Roberts, S., and Young, P.,** Competitive inhibition of adrenal Δ^5-3β-hydroxysteroid dehydrogenase and Δ^5-3-ketosteroid isomerase activities by adenosine 3′,5′-monophosphate, *J. Biol. Chem.*, 245, 3859, 1970.
56. **Basch, R. S. and Finegold, M. J.,** 3β-Hydroxysteroid dehydrogenase activity in the mitochondria of rat adrenal homogenates, *Biochem. J.*, 125, 983, 1971.
57. **Snoeijing, R. and Hardonk, M. J.,** Discrepancy between the histochemically and biochemically determined 3β-hydroxysteroid dehydrogenase activity in the calf adrenal cortex, *Histochemie*, 33, 323, 1973.
58. **Chapman, J. C. and Sauer, L. A.,** Intracellular localization and properties of 3β-hydroxysteroid dehydrogenase/isomerase in the adrenal cortex, *J. Biol. Chem.*, 254, 6624, 1979.
59. **Moustafa, A. M. and Kortiz, S. B.,** Concerning the subcellular distribution of 3β-hydroxysteroid dehydrogenase/isomerase in the rat adrenal, *Proc. Soc. Exp. Biol. Med.*, 149, 823, 1975.
60. **Bradbury, S. and Steward, V. W.,** A note on the electron staining of diformazan deposits in tissue sections, *J. R. Microsc. Soc.*, 83, 467, 1964.
61. **Laffargue, P., Chamlian, A., and Adechy-Benkoel, L.,** Localisation probable en microscopie electronique de la 3β-hydroxysteroide deshydrogenase, de la glucose-6-phosphate deshydrogenase et de la NADH diaphorase dans le corps jaune ovarien de la femme, *J. Microsc.*, 13, 235, 1972.
62. **Kagawa, H. and Nagahama, Y.,** Ultrastructural localization of Δ^5-3β-hydroxysteroid dehydrogenase in the interrenal cells of the goldfish (Carassius auratus), *Cell Tissue Res.*, 212, 225, 1980.
63. **Kawai, K.,** Ultracytochemical localization of dehydrogenase in the rat adrenal cortex, *Acta Histochem. Cytochem.*, 14, 426, 1981.
64. **Kawai, K., Tsuchiyama, H., and Shigematsu, K.,** The procedure for ultracytochemical demonstration of dehydrogenases in the rat adrenal cortex using copper ferrocyanide, *Acta Histochem. Cytochem.*, 14, 408, 1981.
65. **Berchtold, J. P.,** Ultracytochemical demonstration and probable localization of 3β hydroxysteroid dehydrogenase activity with a ferricyanide technique, *Histochemistry*, 50, 171, 1977.
66. **Berchtold, J. P.,** Electron microscopic localization of 3β-hydroxysteroid dehydrogenase and NADH-ferricyanide reductase activities in amphibian interrenal cells, *Cell Tissue Res.*, 188, 142, 1978.
67. **Anderson, J. E., Thliveris, J. A., and Penner, S. B.,** Effect of fixation on the ultrastructural localization of Δ^5,3β-hydroxysteroid dehydrogenase in adrenocortical and Leydig cells of the rat, *Acta Histochem.*, 73, 79, 1983.
68. **Ishimura, K., Yoshinaga-Hirabayashi, T., Fujita, H., Ishii-Ohba, H., Inanao, H., and Tamaoiki, B.,** Light and electron microscopic immunocytochemistry on the localization of 3β-hydroxysteroid dehydrogenase/isomerase in the bovine adrenal cortical cells, *Histochemistry*, 89, 35, 1988.
69. **Handler, R. P. and Bransome, E. D.,** Guinea pig adrenal 3β-hydroxysteroid dehydrogenase: is there more than one enzyme?, *J. Clin. Endocrinol.*, 29, 1117, 1969.
70. **Zhao, H.-F., Simard, J., Labrie, C., Breton, N., Rheaume, E., Luu-The, V., and Labrie, F.,** Molecular cloning, cDNA structure and predicted amino acid sequence of bovine 3β-hydroxy-5-ene-steroid dehydrogenase/Δ^5-Δ^4 isomerase, *FEBS Lett.*, 259, 153, 1989.
71. **Luu-The, V., Lachance, Y., Labrie, C., Leblanc, G., Thomas, J. L., Strickler, R. C., and Labrie, F.,** Full length cDNA structure and deduced amino acid sequence of human 3β-hydroxy-5-ene steroid dehydrogenase, *Mol. Endocrinol.*, 3, 1310, 1989.

72. **Zhao, H.-F., Labrie, C., Simard, J., deLaunoit, Y., Trudel, C., Martel, C., Rheaume, E., Dupont, E., Luu-The, V., Pelletier, G., and Labrie, F.,** Characterization of rat 3β-hydroxysteroid dehydrogenase/ Δ^5-Δ^4 isomerase cDNAs and differential tissue-specific expression of the corresponding mRNAs in steroidogenic and peripheral tissues, *J. Biol. Chem.*, 266, 583, 1991.
73. **Davison, B., Large, D. M., Anderson, D. C., and Robertson, W. R.,** Basal steroid production by the zona reticularis of the guinea-pig adrenal cortex, *J. Steroid Biochem.*, 18, 285, 1983.
74. **Sugihara, H., Yun, K., Yonemitsu, N., Miyabara, S., Tsutsui, C., Nakashima, H., and Watanabe, F.,** Adrenal cortex of micromys minutus japonicus (Japanese harvest mouse). Its histochemical and fine structural study, *Acta Pathol. Jpn.*, 35, 323, 1985.
75. **Watanabe, F.,** Breeding experiments of wild micromys minutus japonicus and its cancer study, *Nagasaki Igakkai Zasshi*, 49, 200, 1974.
76. **O'Hare, M. J., Nice, E. C., and Neville, A. M.,** Regulation of androgen secretion and sulphoconjugation in the adult human adrenal cortex: studies with primary monolayer cell cultures, in *Adrenal Androgens*, Genazzani, A. R., Thijssen, J. H. H., and Siiteri, P. K., Eds., Raven Press, New York, 1980, 7.
77. **Black, V. H. and Stalvey, J.,** 3β-Hydroxysteroid dehydrogenase in guinea pig adrenal: an immunochemical analysis, unpublished results, 1991.
78. **Speiser, P. W., Agdere, L., Ueshiba, H., White, P. C., and New, M. I.,** Aldosterone synthesis in salt-wasting congenital adrenal hyperplasia with complete absence of adrenal 21-hydroxylase, *N. Engl. J. Med.*, 324, 145, 1991.
79. **Black, V. H., Barilla, J. R., Russo, J. J., and Martin, K. O.,** A cytochrome P450 immunochemically related to $P450_{c,d}$ (P450I) localized to the smooth microsomes and inner zone of the guinea pig adrenal, *Endocrinology*, 124, 2480, 1989.
80. **Black, V. H., Barilla, J. R., and Martin, K. O.,** Effects of age, adrenocorticotropin, and dexamethasone on a male-specific cytochrome P450 localized in the inner zone of the guinea pig adrenal, *Endocrinology*, 124, 2492, 1989.
81. **Black, V. H.,** Immunodetectable cytochromes P450I, II, and III in guinea pig adrenal — hormone responsiveness and relationship to capacity for xenobiotic metabolism, *Endocrinology*, 127, 1153, 1990.
82. **Yago, N. and Ichii, S.,** Submitochondrial distribution of components of the steroid 11β-hydroxylase and cholesterol sidechain-cleaving enzyme systems in hog adrenal cortex, *J. Biochem.*, 65, 215, 1969.
83. **Dodge, A. H., Christensen, A. K., and Clayton, R. B.,** Localization of a steroid 11β-hydroxylase in the inner membrane subfraction of rat adrenal mitochondria, *Endocrinology*, 87, 254, 1970.
84. **Curchill, P. F., de Alvare, L. R., and Kimura, T.,** Topological studies of the steroid hydroxylase complexes in bovine adrenocortical mitochondria, *J. Biol. Chem.*, 253, 4924, 1978.
85. **Churchill, P. F. and Kimura, T.,** Topological studies of cytochromes $P\text{-}450_{scc}$ and $P\text{-}450_{11\beta}$ in bovine adrenocortical inner mitochondrial membranes. Effects of controlled tryptic digestion, *J. Biol. Chem.*, 254, 10433, 1979.
86. **Suhara, K., Gomi, T., Sato, H., Itagaki, E., Takemori, S., and Datagiri, M.,** Purification and immunochemical characterization of the two adrenal cortex mitochondrial cytochrome P-450-proteins, *Arch. Biochem. Biophys.*, 190, 290, 1978.
87. **Watanuki, M., Tilley, B. E., and Hall, P. F.,** Cytochrome P-450 for 11β- and 18-hydroxylase activities of bovine adrenocortical mitochondria: one enzyme or two, *Biochemistry*, 17, 127, 1978.
88. **Ogishima, T., Mitani, F., and Ishimura, Y.,** Isolation of aldosterone synthase cytochrome P-450 from zona glomerulosa mitochondria of rat adrenal cortex, *J. Biol. Chem.*, 264, 10935, 1989.
89. **Magalhães, M. C., Magalhães, M. M., and Coimbra, A.,** An electron microscope autoradiographic study of DOC conversion to corticosterone in the rat adrenal cortex, *Cell Tissue Res.*, 151, 47, 1974.
90. **Mitani, F., Shimizu, T., Ueno, R., Ishimura, Y., Izumi, S., Komatsu, N., and Watanabe, K.,** Cytochrome $P\text{-}450_{11\beta}$ and $P\text{-}450_{scc}$ in adrenal cortex: zonal distribution and intramitochondrial localization by the horseradish peroxidase-labeled antibody method, *J. Histochem. Cytochem.*, 30, 1066, 1982.
91. **Sugano, S., Ohnishi, T., Hatae, N., Ishimur, K., Jujita, H., Yamano, T., and Okamoto, M.,** Monoclonal antibodies against bovine adrenal cytochrome $P\text{-}450_{11\beta}$ and cytochrome $P\text{-}450_{scc}$. Their isolation, characterization and application to immunohistochemical analysis of adrenal cortex, *J. Steroid Biochem.*, 23, 1013, 1985.
92. **Ishimura, K., Yoshinaga, T., Fujita, H., Sugano, S., Okamoto, M., and Yamano, T.,** Light and electron microscopic immunohistochemistry on the localization of cytochrome P-450 of the side chain cleavage system and of cytochrome P-450 of 11β-hydroxylase in the bovine adrenal cortical cells, *Arch. Histol. Jap.*, 48, 541, 1985.
93. **Murakoshi, M., Osamura, Y., Watanabe, K., Izumi, S., Kawakami, T., Nomoto, Y., and Sakai, H.,** Enzyme histochemical studies in human adrenocortical adenomas. I. Primary aldosteronism, *Tokai J. Exp. Clin. Med.*, 8, 155, 1983.

94. **Sasano, H., Okamoto, M., and Sasano, M.,** Immunohistochemical study of cytochrome P-45011β-hydroxylase in human adrenal cortex with mineralo- and glucocorticoid excess, *Virchows Arch. A,* 413, 313, 1988.
95. **Sasano, H. and Sasano, N.,** Immunohistochemical demonstration of cholesterol side-chain cleavage cytochrome P-450 in bovine and human adrenals, *Pathol. Res. Pract.,* 184, 337, 1989.
96. **Black, V. H.,** Lipoprotein requirements for secretion of ultraviolet-absorbing corticosteroids by guinea pig adrenocortical cells in vitro: inner versus outer cortices; zona glomerulosa versus zona fasciculata, *Endocrinology,* 120, 640, 1987.
97. **Geuze, H. J., Slot, J. W., Yanagivashi, K., McCracken, J. A., Schwartz, A. L., and Hall, P. F.,** Immunogold cytochemistry of cytochromes P450 in porcine adrenal cortex, *Histochemistry,* 86, 551, 1987.
98. **Nonaka, Y., Matsukawa, N., Morohashi, K.-I., Omura, T., Ogihara, T., Teraoka, H., and Okamoto, M.,** Molecular cloning and sequence analysis of cDNA encoding rat adrenal cytochrome P450 11β, *FEBS Lett.,* 255, 21, 1989.
99. **Imai, M., Shimada, H., Okada, Y., Matsushima-Hibiya, Y., Ogishima, T., and Ishimura, Y.,** Molecular cloning of a cDNA encoding aldosterone synthase cytochrome P450 in rat adrenal cortex, *FEBS Lett.,* 263, 299, 1990.
100. **Mornet, E., Dupont, J., Vitek, A., and White, P. C.,** Characterization of two genes encoding human steroid 11β-hydroxylase ($P450_{11\beta}$), *J. Biol. Chem.,* 264, 20961, 1989.
101. **Kawamoto, T., Mitsuuchi, Y., Ohnishi, T., Ichikawa, Y., Yokoyama, Y., Sumimoto, H., Toda, K., Miyahara, K., Kuribayashi, I., Nakao, K., Hosoda, K., Yamamoto, Y., Imura, H., and Shizuta, Y.,** Cloning and expression of a cDNA for human cytochrome $P450_{aldo}$ as related to primary aldosteronism, *Biochem. Biophys. Res. Commun.,* 173, 309, 1990.
102. **Mitani, F., Ishimura, Y., Izumi, S., and Watanabe, K.,** Immunohistochemical localization of adrenodoxin and adrenodoxin reductase in bovine adrenal cortex, *Acta Endocrinol.,* 90, 317, 1979.
103. **Hatano, O., Sagara, Y., Omura, T., and Takakusu, A.,** Immunocytochemical localization of adrenodoxin in bovine adrenal cortex by protein A gold technique, *Histochemistry,* 91, 89, 1989.
104. **Sasano, H., Sasano, N., Okamoto, M., and Nonaka, Y.,** Immunohistochemical demonstration of adrenodoxin reductase in bovine and human adrenals, *Pathol. Res. Pract.,* 184, 473, 1989.
105. **Zuber, M. X., Simpson, E. R., and Waterman, W. R.,** Expression of bovine 17α-hydroxylase cytochrome P-450 cDNA in nonsteroidogenic (COS 1) cells, *Science,* 234, 1258, 1986.
106. **Sasano, H., White, P. C., New, M. I., and Sasano, N.,** Immunohistochemistry of cytochrome P450 21-hydroxylase: microscopic examination of the enzyme in the bovine adrenal cortex and kidney, *Endocrinology,* 122, 291, 1988.
107. **Shinzawa, K., Ishibashi, S., Murakoshi, M., Watanabe, K., Konimani, S., Kawahara, A., and Takemori, S.,** Relationship between zonal distribution of microsomal cytochrome P450s ($P450_{17\alpha,lyase}$ and $P450_{c21}$) and steroidogenic activities in guinea pig adrenal cortex, *J. Endocrinol.,* 119, 191, 1988.
108. **Sasano, H., Mason, J. I., and Sasano, N.,** Immunohistochemical study of cytochrome $P450_{17\alpha}$ in human adrenocortical disorders, *Hum. Pathol.,* 20, 113, 1989.
109. **Black, V. H. and Brody, R. I.,** Differential ACTH response of immunodetectable HMGCoA reductase and cytochromes $P450_{17\alpha}$ and $P450_{21}$ in guinea pig adrenal outer zone cell types, zona glomerulosa and zona fasciculata, *Endocr. Res.,* 17, 193, 1991.
110. **Tangalakis, K., Coghlan, J. P., Connell, J., Crawford, R., Darling, P., Hammond, V. E., Haralam-** [illegible] steroid hydroxylases in ovine fetal adrenal glands, *Acta Endocrinol.,* 120, 225, 1987.
111. **Davis, D. A. and Medline, N. M.,** Spironolactone (Aldactone) bodies, *Am. J. Clin. Pathol.,* 54, 22, 1970.
112. **Kovacs, K., Horvath, E., and Singer, W.,** Fine structure and morphogenesis of spironolactone bodies in the zona glomerulosa of the human adrenal cortex, *J. Clin. Pathol.,* 26, 949, 1973.
113. **Aiba, M., Suzuki, H., Kageyama, K., Murai, M., Tazaki, H., Abe, O., and Saruta, T.,** Spironolactone bodies in aldosteronomas and in the attached adrenals, *Am. J. Pathol.,* 103, 404, 1981.
114. **Hsu, S. M., Raine, L., and Martin, H. F.,** Spironolactone bodies. An immunoperoxidase study with biochemical correlation, *Am. J. Clin. Pathol.,* 75, 92, 1981.
115. **Aiba, M., Kameya, T., Suzuki, H., Nakumura, H., Mazuno, Y., and Kanno, T.,** Enzyme histochemical and electron microscopic study of a virilizing adrenocortical adenoma, *Acta Pathol. Jpn.,* 28, 615, 1978.
116. **Dzsinich, Cs., Szabo, D., and Okros, I.,** Cycloheximide-induced ultrastructural changes in the adrenal cortex of the rat, *Experientia,* 25, 835, 1969.
117. **Aguas, A. P. and Nickerson, P. A.,** Subcellular distribution of calcium in zona fasciculata cells of the rat adrenal cortex, *Tissue Cell,* 3, 491, 1981.
118. **Nickerson, P. A.,** Cytochemical localization of calcium in mitochondria of regenerating rat adrenal cortex. A study of adrenal regeneration hypertension, *J. Submicrosc. Cytol.,* 19, 63, 1987.

119. **Suyama, A. T., Long, J., and Ramachandran, J.,** Ultrastructural changes induced by ACTH in normal adrenocortical cells in culture, *J. Cell Biol.*, 72, 757, 1977.
120. **Cameron, E. H. D., Jones, T., Jones, D., Anderson, A. B. M., and Griffiths, K.,** Further studies on the relationship between C_{19} and C_{21} steroid synthesis in the human adrenal gland, *J. Endocrinol.*, 45, 215, 1969.
121. **Jones, T. and Griffiths, K.,** Ultramicrochemical studies on the site of formation of dehydroepiandrosterone-sulphate in the adrenal cortex of the guinea pig, *J. Endocrinol.*, 42, 559, 1968.
122. **Kennerson, A. R., McDonald, D. A., and Adams, J. B.,** Dehydroepiandrosterone sulfotransferase localization in human adrenal glands: a light and electron microscopic study, *J. Clin. Endocrinol. Metab.*, 56, 786, 1983.
123. **McKerns, K. W.,** Mechanism of action of adrenocorticotrophic hormone through activation of glucose-6-phosphate dehydrogenase, *Biochim. Biophys. Acta,* 90, 357, 1964.
124. **Berchtold, J. P.,** Ultrastructural demonstration of glucose-6-phosphate dehydrogenase activity in steroid-secreting cells, *Histochemistry,* 63, 173, 1979.
125. **Strott, C. A.,** A pregnenolone-binding protein in the soluble fraction of guinea pig adrenal cortex, *J. Biol. Chem.*, 252, 464, 1977.
126. **Strott, C. A. and Lyons, C. D.,** Pregnenolone sulfate binding in the guinea pig adrenal cortex: comparisons with pregnenolone binding, *Biochemistry,* 17, 4557, 1978.
127. **Whitnall, M. H., Lee, Y. C., Driscoll, W. J., and Strott, C. A.,** Immunocytochemical localization of the 34KD pregnenolone-binding protein to fasciculata and reticularis cells and a novel 32KD protein specific for reticularis cells in guinea pig adrenal cortex, *J. Histochem. Cytochem.*, 38, 1607, 1990.
128. **Hall, P.,** On the mechanism of action of ACTH: the role of actin, *Endocr. Res.*, 10, 431, 1984.
129. **Loesser, K. E. and Malamed, S.,** A morphometric analysis of adrenocortical actin localized by immunoelectron microscopy: the effect of adrenocorticotropin, *Endocrinology,* 121, 1400, 1987.
130. **Cheitlin, R. and Ramachandran, J.,** Regulation of actin in rat adrenocortical cells by corticotropin, *J. Biol. Chem.*, 256, 3156, 1981.
131. **Nakano, M.,** Adrenal cortical carcinoma, *Acta Pathol. Jpn.*, 38, 163, 1988.
132. **Henzen-Logmans, S. C., Stel, H. V., van Muigen, G. N. P., Mullink, H., and Meijer, C. J. L. M.,** Expression of intermediate filament proteins in adrenal cortex and related tumours, *Histopathology,* 12, 359, 1988.
133. **Miettinen, M., Lehto, V.-P., and Virtanen, I.,** Immunofluorescence microscopic evaluation of the intermediate filament expression of the adrenal cortex and medulla and their tumors, *Am. J. Pathol.*, 118, 360, 1985.
134. **Almahbobi, G. and Hall, P. F.,** The role of intermediate filaments in adrenal steroidogenesis, *J. Cell Sci.*, 97, 679, 1990.
135. **Eriksson, L. and Westermark, P.,** Age-related accumulation of amyloid inclusions in adrenal cortical cells, *Am. J. Pathol.*, 136, 461, 1990.
136. **Gemmell, R. T., Laychock, S. G., and Rubin, R. P.,** Ultrastructural and biochemical evidence for a steroid-containing secretory organelle in the perfused cat adrenal gland, *J. Cell Biol.*, 72, 209, 1977.
137. **Rebuffat, P., Belloni, A. S., Mazzocchi, G., Vassamelli, P., and Nussdorfer, G. G.,** A stereological study of the trophic effects of the renin-angiotensin system on the rat adrenal zona glomerulosa, *J. Anat.*, 129, 561, 1979.
138. **Hashida, Y. and Yunis, E. J.,** Ultrastructure of the adrenal zona glomerulosa in children with renovascular hypertension, *Hum. Pathol.*, 3, 301, 1972.
139. **Deschepper, C. F., Mellon, S. H., Cumin, F., Baxter, J. D., and Ganong, W. F.,** Analysis by immunocytochemistry and in situ hybridization of renin and its mRNA in kidney, testis, adrenal, and pituitary of the rat, Proc. Natl. Acad. Sci. U.S.A., 83, 7552, 1986.
140. **Mizuno, K., Hoffman, L. H., McKenzie, J. C., and Inagami, T.,** Presence of renin secretory granules in rat adrenal gland and stimulation of renin secretion by angiotensin II but not by adrenocorticotrophin, *J. Clin. Invest.*, 82, 1007, 1988.
141. **Inagami, T., Mizuno, K., Naruse, M., Nakamaru, M., Naruse, K., Hoffman, L. H., and McKenzie, J. C.,** Active and inactive renin in the adrenal, *Am. J. Hypertens.*, 2, 311, 1989.
142. **Naruse, K., Murakoshi, M., Osamura, R. Y., Naruse, M., Hiroshi, T., Watanabe, K., Demura, H., Inagami, T., and Shizume, K.,** Immunohistological evidence for renin in human endocrine tissues, *J. Clin. Endocrinol. Metab.*, 61, 172, 1985.
143. **Ravid, R., Oosterbaan, H. P., Hansen, B. L., and Swaab, D. F.,** Localization of oxytocin, vasopressin and parts of precursors in the human neonatal adrenal, *Histochemistry,* 84, 401, 1986.
144. **Hawthorne, J., Nussey, S. S., Henderson, J. R., and Jenkins, J. S.,** Immunohistochemical localization of oxytocin and vasopressin in the adrenal glands of rat, cow, hamster and guinea pig, *Cell Tissue Res.*, 250, 1, 1987.

145. **Hill, D. J., Clemmons, D. R., Wilson, S., Han, V. K. M., Strain, A. J., and Milner, R. D. G.,** Immunological distribution of one form of insulin-like growth factor (IGF)-binding protein and IGF peptides in human fetal tissues, *J. Mol. Endocrinol.*, 2, 31, 1989.
146. **Grothe, C. and Unsicker, K.,** Immunocytochemical mapping of basic fibroblast growth factor in the developing and adult rat adrenal gland, *Histochemistry,* 94, 141, 1990.
147. **Grothe, C. and Unsicker, K.,** Immunocytochemical localization of basic fibroblast growth factor in bovine adrenal gland, ovary, and pituitary, *J. Histochem. Cytochem.*, 37, 1877, 1989.
148. **Gallo-Payet, N., Guillon, G., Balestra, M. N., and Jard, S.,** Vasopressin induces breakdown of membrane phosphoinositides in adrenal glomerulosa and fasciculata cells, *Endocrinology,* 119, 1042, 1986.
149. **Friend, D. S. and Gilula, N. B.,** Variations in tight and gap junctions in mammalian tissues, *J. Cell Biol.*, 53, 758, 1972.
150. **Friend, D. S. and Gilula, N. B.,** A distinctive cell contact in the rat adrenal cortex, *J. Cell Biol.*, 53, 148, 1972.
151. **Reaven, E., Spicher, M., and Aszhar, S.,** Microvillar channels: a unique plasma membrane compartment for concentrating lipoproteins on the surface of rat adrenal cortical cells, *J. Lipid Res.*, 30, 1551, 1989.
152. **Coutard, M. and Osborne-Pellegrin, M. J.,** Autoradiographic localization of ^{3}H-angiotensin II in rat adrenal glands, *Experientia,* 36, 251, 1980.
153. **Elliott, M. E. and Goodfriend, T. L.,** Inhibition of aldosterone synthesis by atrial natriuretic factor, *Fed. Proc.*, 45, 2376, 1986.
154. **Mantyh, C. R., Kruger, L., Brecha, N. C., and Mantyh, P. W.,** Localization of specific binding sites for atrial natriuretic factor in peripheral tissues of the guinea pig, rat, and human, *Hypertension,* 8, 712, 1986.
155. **Garcia-Caballero, T. and Morel, G.,** Immunocytochemical localization of ACTH-like immunoreactivity in rat adrenal glomerulosa and fasciculata cells, *Histochemistry,* 91, 205, 1989.
156. **McDonough, L. B. and Ewig, J. E.,** Immunocytochemical localization of prolactin binding sites in mouse adrenal gland, *Comp. Biochem. Physiol.*, 72A, 259, 1982.
157. **deKretser, D. M., Catt, K. J., Burger, H. G., and Smith, G. C.,** Radioautographic studies on the localization of ^{125}I-labelled human luteinizing and growth hormone in immature male rats, *J. Endocrinol.*, 43, 105, 1969.
158. **Mayberry, H. E. and Waddell, W. J.,** Distribution of ^{125}I-labeled human growth hormone in immature, hypophysectomized rats pretreated with ACTH or unlabeled human growth hormone, *Anat. Rec.*, 179, 107, 1974.
159. **Kamio, T., Shigematsu, K., Sou, H., Kawai, K., and Tsuchiyama, H.,** Immunohistochemical expression of epidermal growth factor receptors in human adrenocortical carcinoma, *Hum. Pathol.*, 21, 277, 1990.
160. **Fong, L. G., Bonney, E., Kosek, J. C., and Cooper, A. D.,** Immunohistochemical localization of low density lipoprotein receptors in adrenal gland, liver, and intestine, *J. Clin. Invest.*, 84, 847, 1989.
161. **Berchtold, J. P.,** Localisation ultrastructurale de l'activite phosphatasique alcaline dans les cellules interrenales d'un amphibien urodele, *Histochemie,* 26, 258, 1971.
162. **Ashchoff, L.,** Das reticulo-endotheliale system, *Ergeb. Inn. Med. Kinderheilkd.*, 26, 1, 1925.
163. **VanFurth, E., Cohn, Z. A., Hirsch, J. C., Spektor, W. C., and Langevoort, H. L.,** The mononuclear phagocyte system: a new classification of macrophages, monocytes and their precursor cells, *Bull. WHO,* 46, 845, 1972.
164. **Sunloff, S. V. and Papadimitriou, J. M.,** The mononuclear phagocytes of the rat adrenal, *Am. J. Pathol.*, 104, 258, 1981.
165. **Bumbasirevic, V., Pappas, G. D., and Becker, R. C.,** Endocytosis of serum albumin-gold conjugates by microvascular endothelial cells in rat adrenal gland: regional differences between cortex and medulla, *J. Submicrosc. Cytol. Pathol.*, 22, 135, 1990.
166. **Hirosawa, K. and Yamada, E.,** Localization of vitamin A-storing cells in the mouse adrenal gland: an electron microscopic autoradiographic study, *Am. J. Anat.*, 153, 233, 1978.
167. **Wake, K.,** "Sternzellen" in the liver: perisinusoidal cells with special reference to storage of vitamin A, *Am. J. Anat.*, 132, 429, 1971.
168. **Unsicker, K. and Groschel-Stewart, U.,** Distribution of contractile proteins and adrenergic nerves in the adrenal gland of guinea-pig, rat and ox as revealed by immunofluorescence and the glyoxylic acid technique, *Experientia,* 34, 102, 1978.
169. **Dallman, M. F.,** Control of adrenocortical growth in vivo, *Endocr. Res.*, 10, 213, 1985.
170. **Cunningham, L. A. and Holzwarth, M. A.,** Audioradiographic distribution of ^{125}I-VIP binding in the rat adrenal cortex, *Peptides,* 10, 1105, 1989.
171. **Rundle, S. E., Canny, B. J., Robinson, P. M., and Funder, J. W.,** Innervation of the sheep adrenal cortex: an immunohistochemical study with rat corticotrophin-releasing factor antiserum, *Neuroendocrinology,* 48, 8, 1988.
172. **Cunningham, L. A. and Holzwarth, M. A.,** Vasoactive intestinal peptide (VIP) stimulates adrenal aldosterone and corticosterone secretion, *Endocrinology,* 122, 2090, 1988.

II. CYTOCHEMISTRY OF THE ADRENAL MEDULLA

Stephen W. Carmichael

A. HISTORICAL INTRODUCTION

The adrenal medulla is a modified sympathetic ganglion. It has proven useful as a model for neurobiologic studies.[1] Whereas these cells have been studied extensively, the chromaffin vesicle has been more thoroughly investigated. Probably because the adrenal chromaffin vesicle was the first secretory vesicle to be isolated, it is the best characterized in biochemical studies. As reviewed by Winkler et al.[2] and others, the composition of the chromaffin vesicle can be almost completely defined in molecular terms. Still, cytochemical studies have made, and continue to make, valuable contributions to our understanding of the adrenal medulla. This review attempts to give a complete accounting of cytochemistry as applied to the adrenal medulla. Except for a few historical references, this review is restricted to ultrastuctural studies. Histochemical studies at the light microscopic level far outnumber cytochemical studies and have been covered elsewhere.[3] A summary of this review is given in tabular form (Table 1).

Studies using microscopy to examine the composition of the adrenal medulla have a long and interesting history. To summarize this history, the first histochemical study was published in 1856 by Gabriel Constant Colin, who observed that the adrenal medulla became blue when treated with iron sulfate.[4] In a more widely cited study, Alfred Vulpian showed that ferric chloride colored the adrenal medulla green.[5] This reaction was also noted by Rudolf Virchow.[6] The most famous histochemical reaction associated with the adrenal medulla, the chromaffin reaction, is named for a relatively specific reaction to chromate salts. It was apparently first observed by Bertholdus Werner as a brownish deposit after fixation in chromic acid or dichromate salts.[7] Jacob Henle was the first to give cytologic details of the color reaction.[8] Following this, there were several others who observed that there is something peculiar about the way in which the adrenal medulla reacted with metallic solutions, particularly with chromium. Yet, it was not until the end of the century that the term "chromaffin" was introduced to describe this reaction. The "chromaffin cell" was so named by Alfred Kohn.[9] Whereas other terms have been introduced, the term "chromaffin" has stuck and is used universally to describe the parenchymal cells of the adrenal medulla and related cells that exhibit the chromaffin reaction. Later, using energy-dispersive X-ray analysis at the electron microscopic level, Wood[10,11] definitively showed that chromium is indeed complexed with catecholamines in chromaffin vesicles.

In 1948 it was shown that the adrenal medulla contained not only epinephrine (adrenaline), but also norepinephrine (noradrenaline).[12] This discovery was confirmed by others and presented the obvious question as to whether these two catecholamines resided in the same cell, or in two different cell types. Histochemists rose to the challenge! Bänder[13] was apparently the first histochemist to attempt to correlate staining properties of chromaffin cells with their amine content. He noted cells with an affinity for acid fuchsin and other cells with an affinity for picric acid, with the former probably corresponding to norepinephrine-containing cells and the latter to epinephrine-containing cells. Eränkö,[14] observing autofluorescence after formaldehyde fixation, concluded that there are two different types of cells in the adrenal medulla, rather than one cell type in different secretory phases. The iodate reaction, introduced by Hillarp and Hökfelt,[15] led to the same conclusion. This was confirmed by Wood[16] using the classic chromaffin reaction.

Lever[17] was the first to publish an electron microscopic study of the adrenal medulla. He noted osmiophilic secretory vesicles that have become known as chromaffin vesicles. He also noted two types of cells, light and dark, based on cytoplasmic density. This

TABLE 1
Cytochemical Localizations in the Adrenal Medulla

Compound	Compartment	Species	Ref.
	Catecholamine-Synthesizing Enzymes		
Tyrosine hydroxylase	Cytosol, outside of chromaffin vesicles	Rat, frog	38
	Cytosol, some chromaffin vesicles	Rat	41
	Cytosol and chromaffin vesicles	Rat, PC12 cells	44
	Nerve fibers	Rat	45
Dopamine β-hydroxylase	Chromaffin vesicles	Cow	49
	Chromaffin vesicles	Rat, frog	38
	Chromaffin vesicles	Cow	50
	Chromaffin vesicles	Cow	51
Phenylethanolamine *N*-methyltransferase	Cytosol, some chromaffin vesicles	Cow	52
	Cytosol, some chromaffin vesicles	Rat, frog	38
	Periphery of chromaffin vesicles	Cow	53
	Phosphatases		
Triphosphatase and diphosphatase	Membranes of chromaffin cells, nerves, spaces between cells, endothelium	Rat, cat	54
	Not in spaces between cells	Rat	55
Monophosphatase	Membranes of chromaffin cells, chromaffin vesicles	Rat, cat	54
	Around blood vessels, not in adrenal medullary cells	Rat	55
Mg-ATPase	Cell membrane, cytosol	Rat, cat	54
	Plasma membrane	Hamster	18
	Cell membrane, cytosol	Hamster	57
	Cell membrane, cytosol, mitochondria, nerves	Mouse	58
	Cell membrane, cytosol, and chromaffin vesicles	Hamster	59
Ca-ATPase	Golgi, plasma membrane, nerves	Hamster	56
	Questioned presence in Golgi, not in chromaffin vesicles	Rat	55
	Golgi and chromaffin vesicles	Mouse	58
Acid phosphatase	Golgi and chromaffin vesicles	Rat	61
	Lysosomes, not Golgi	Rat	62
	Golgi and lysosomes	Rat	55
	Golgi, lysosomes, and chromaffin vesicles	Frog	63
	Golgi, lysosomes, and chromaffin vesicles	Amphibians	64
Thiamine pyrophosphatase	Golgi, axons, Schwann cells	Rat	55
Inosine diphosphatase	Golgi, axons, Schwann cells	Rat	55
	Cholinesterases		
Acetylcholinesterase	Plasma and nuclear membranes, axons, ganglion cells	Rat	65

TABLE 1 (continued)
Cytochemical Localizations in the Adrenal Medulla

Compound	Compartment	Species	Ref.
	Cholinesterases		
	Plasma and nuclear membranes, RER, SER, axons, axonal SER	Cow	70
	As above, plus some chromaffin vesicle membranes	Cow, pig, hamster	71
	Plasma membrane	Cow	119
	Plasma and nuclear membranes, RER, ganglion cells	Chick	69
	Plasma and nuclear membranes, chromaffin vesicle membrane, ER	Amphibians	67
	Plasma and nuclear membranes, ER, nerves	Amphibian	68
Cholinesterase	Nerves, chromaffin cells	Reptile, amphibians	66
	Other Enzymes		
Adenylate cyclase	Plasma membrane, chromaffin vesicles, endothelium	Hamster, cow	72
Catalase	Peroxisomes	Rat	73
	Neuropeptides		
Enkephalin	Chromaffin vesicles	Rat	76
	Chromaffin vesicles, RER, heterogeneous	Man	77
	Epinephrine-containing cells, $<10\%$ norepinephrine-containing cells	Cat	89
	Chromaffin vesicles	Cat, pig, horse, man, pheochromocytoma	78
[Met]enkephalin-Arg^6-Gly^7-Leu^8	Presynaptic nerve terminals	Rat	79
	Presynaptic nerve terminals, 20% of chromaffin cells, chromaffin vesicles	Rat	129
	Presynaptic nerve terminals, large-cored vesicles	Guinea pig	81
[Met^5]- and [Leu^5]enkephalin	Epinephrine-containing vesicles, ganglion cells, nerves	Rat	82
[Met]enkephalin	Epinephrine-containing vesicles	Frog	80
Neuropeptide Y	Norepinephrine-containing vesicles	Horse, cat, rat, guinea pig, mouse	83
	50% of epinephrine-containing cells, cytosol, nerves	Rat	84
	Epinephrine-containing vesicles, nerves	Rat, cat, mouse, hamster	75
Substance P	Few cells, extrinsic nerves	Rat	85
	Chromaffin vesicles, heterogeneous	Rat	74
Vasoactive intestinal polypeptide	Few cells, chromaffin vesicles, ganglion cells, nerves	Rat	86
Calcitonin gene-related peptide	Epinephrine-NPY-containing vesicles, nerves	Rat	87
	$>1\%$ of epinephrine-containing cells, some nerves	Rat, cat, mouse, hamster	75

TABLE 1 (continued)
Cytochemical Localizations in the Adrenal Medulla

Compound	Compartment	Species	Ref.
	Neuropeptides		
Neurotensin	Nerves to norepinephrine-containing cells	Hamster	88
	65% of norepinephrine-containing cells, not with enkephalin, nerves	Cat	89
Galanin	Epinephrine-containing vesicles, with NPY	Rat, cat, mouse	75
	Proteins		
Chromogranin A	Core of chromaffin vesicles	Cow	90
	Core of chromaffin vesicles	Rat	91
	Chromaffin vesicle matrix	Cow	50
	Core of chromaffin vesicles	Man, pheochromocytoma	92
	Unfixed chromaffin vesicles	Man	93
	Chromaffin vesicles	Cow	51
Chromogranins B and C	In same chromaffin vesicles as chromogranin A and NPY	Rat	95
Pituitary polypeptide 7B2	Like chromogranins	Rat	96
GAWK	Like chromogranin B	Man	97
Tyrosine, leucine, and L-dopa	Spatial and temporal study (radioautography)	Rat	98–99
Glycoprotein D2	Surface of all chromaffin cells, nerves, Schwann cells	Rat	100
Cerebellar Purkinje cell-specific protein	Norepinephrine-containing vesicles, cytosol, some ganglion cells	Rat	102
Leu-7 (HNK-1)	Matrix of chromaffin vesicles	Man, pheochromocytoma	103
Cytochrome b_{561}	Chromaffin vesicles, with fodrin	Rat	104
Basic fibroblast growth factor	Chromaffin vesicles	Cow	105
	Contractile and Structural Proteins		
Myosin	Plasma membrane of chromaffin cells, vessels, fibroblasts	Cow, rat	50
Actin	Chromaffin vesicle membranes	Cow	108
Fodrin	Inside of plasma membrane of all chromaffin cells, nerves, Schwann cells	Rat	109
	Inside of plasma membrane	Rat	110
	Following stimulation, around small vesicles and large vacuoles	Rat	104
	Membrane Dynamics		
Nonspecific membrane retrieval	Pits at cell surface	Hamster	112
	Vesicles, tubules, multivesicular bodies, dense bodies, lysosomes	Rat	113
	As above, not in Golgi	Hamster	114

the cytosol of epinephrine- and norepinephrine-containing cells, but not in the mitochondria or nuclei; reaction product was also seen on the outside of chromaffin vesicle membranes. They later extended this finding to the house shrew *(Suncus murinus riukiuanus),* in which a similar cellular localization of tyrosine hydroxylase was found.[39,40]

Stephens et al.[41] used a monospecific antibody against tyrosine hydroxylase to examine distribution of the enzyme in the adrenal medullae of stressed and nonstressed rats. In all treatment groups, chromaffin cells were found that had tyrosine hydroxylase associated with some chromaffin vesicles. The stressed groups had higher levels of enzyme activity, as measured biochemically, and some stressed animals had fewer chromaffin vesicles that appeared to contain tyrosine hydroxylase. Semenenko et al.[42] produced a high-affinity monoclonal antibody against tyrosine hydroxylase, but did not examine the adrenal medulla in this study. Kwan et al.[43] also produced and characterized a monoclonal antibody to tyrosine hydroxylase.

Oomori et al.[44] used monoclonal and polyclonal antibodies to localize tyrosine hydroxylase immunoreactivity in rat adrenal medulla and pheochromocytoma (PC12) cells. They obtained similar results with both types of antibodies. Immunoreactivity was clearly demonstrated in epinephrine- and norepinephrine-containing vesicles, as well as in the cytoplasm, in ribosomes attached to the endoplasmic reticulum, and free ribosomes of both adrenal medullary cells and PC12 cells (Figure 2). Oomori et al.[44] explained their findings by pointing out that the more rigorous fixation used in their study prevented solubilization and redistribution of the enzyme from vesicles into the cytoplasm. Since the enzyme also appeared to be associated with attached and free ribosomes, it was suggested that enzyme molecules destined for chromaffin vesicles and the cytoplasm are synthesized, respectively, on the two types of ribosome.

In a subsequent study, Oomori et al.[45] extended their finding to neuronal processes within the adrenal gland. They found immunoreactivity to tyrosine hydroxylase in varicose nerve fibers in both the medulla and cortex, where they were near smooth muscle cells of blood vessels.

Using biochemical methods, Kuhn et al.[46] demonstrated that a fraction of tyrosine hydroxylase is an integral component of bovine chromaffin vesicle membranes. If the studies of Oomori et al.[44] and Kuhn et al.[46] stand the test of time, they will change fundamental ideas on the localization of tyrosine hydroxylase, which has generally been considered to be a cytosolic enzyme.

2. Dopa Decarboxylase

This enzyme, also referred to as aromatic L-amino acid decarboxylase, is ubiquitous and plays no role in regulating catecholamine synthesis. There are few biochemical studies measuring dopa decarboxylase in the adrenal medulla, and no cytochemical studies at the ultrastructural level. It is considered to be a cytosolic enzyme. Hökfelt et al.[47] and Verhofstad et al.[48] demonstrated this at the histochemical level. Verhofstad et al.[48] showed that tyrosine hydroxylase, dopa decarboxylase, and dopamine β-hydroxylase appeared in the medullary blastema of the developing rat adrenal gland, but phenylethanolamine *N*-methyltransferase does not appear until shortly before birth.

3. Dopamine β-Hydroxylase

The localization of dopamine β-hydroxylase within adrenal medullary chromaffin cells has not been controversial. There are many biochemical studies demonstrating that the enzyme is within chromaffin vesicles in both a soluble form and membrane-bound form, attached to the inner aspect of the chromaffin vesicle membrane.

The first study to localize dopamine β-hydroxylase at the electron microscopic level was published by Redick et al.[49] They used a peroxidase-antiperoxidase unlabeled antibody technique to localize the enzyme throughout the matrix of the chromaffin vesicle in specimens

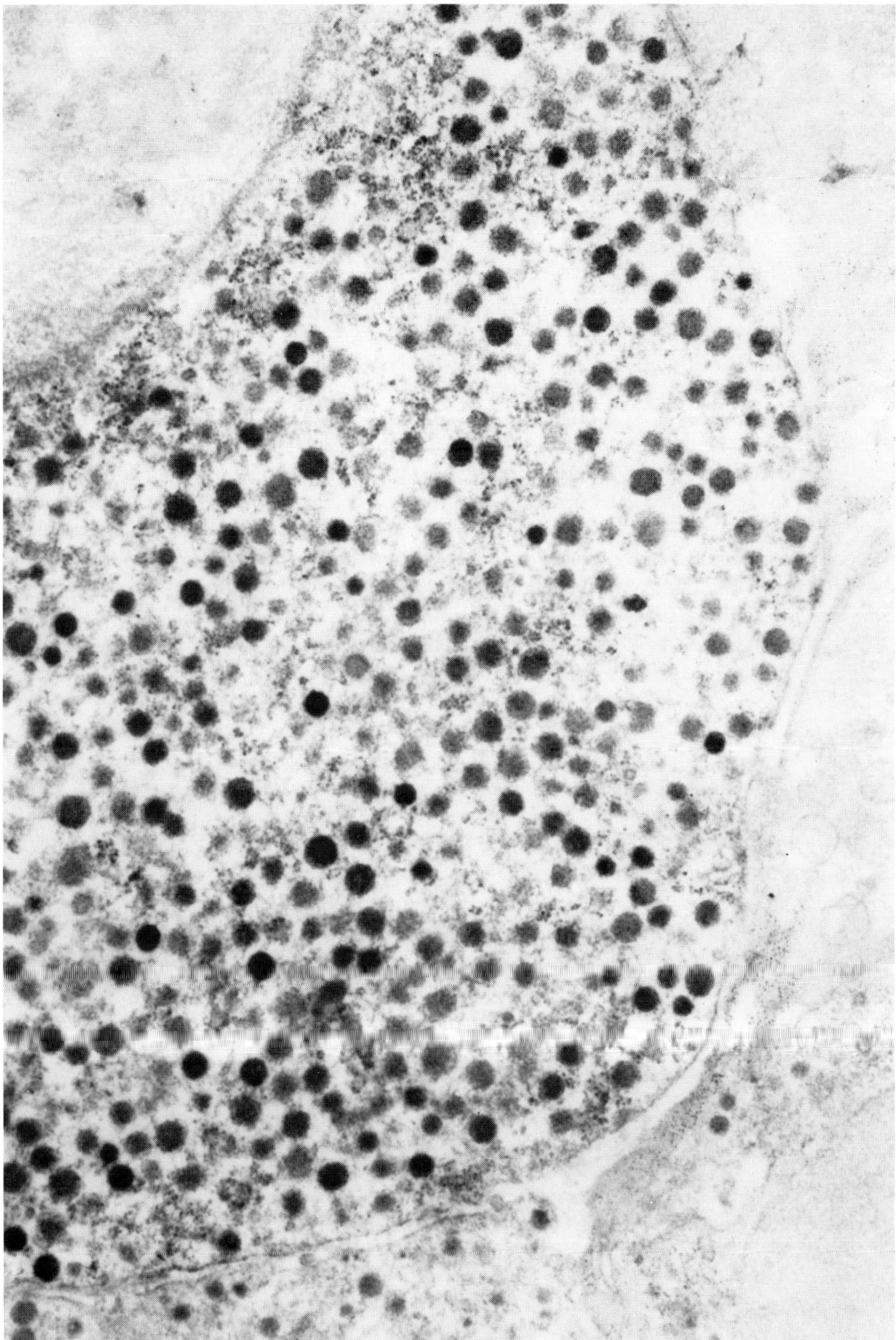

FIGURE 2. Tyrosine hydroxylase localization in chromaffin cell of rat adrenal medulla. PAP pre-embedding method with monoclonal antiserum. Unstained. (Magnification × 17,200.)

from the bovine adrenal medulla. They demonstrated more immunoreactivity when less vigorous preparation techniques were employed, suggesting that the soluble form is removed under more vigorous conditions, leaving only the membrane-bound form of the enzyme.

Nagatsu et al.[38] observed immunoreactivity for dopamine β-hydroxylase only within chromaffin vesicles in the adrenal medulla of rat and frog, and later in the house shrew.[40] Aunis et al.[50] colocalized dopamine β-hydroxylase and chromogranin A within chromaffin vesicles from bovine adrenal medulla. Interestingly, they noticed that staining for dopamine β-hydroxylase gave a more organized pattern than chromogranin A. The localization of dopamine β-hydroxylase and chromogranin A in the bovine adrenal medulla was confirmed by Matsumoto et al.[51]

4. Phenylethanolamine *N*-Methyltransferase

There are considerable biochemical data to indicate that the enzyme which converts norepinephrine to epinephrine, phenylethanolamine *N*-methyltransferase (PNMT), is located in the cytosol of epinephrine-containing adrenal chromaffin cells, and not within chromaffin vesicles. Using a peroxidase-labeled immunocytochemical technique, Nagatsu and Kondo[52] demonstrated enzyme in the ground substance of epinephrine-containing cells, but not in norepinephrine-containing cells of the bovine adrenal medulla. Some of their observations indicated that PNMT may also be located in epinephrine-containing vesicles. They later extended their findings to the rat and frog[38] and house shrew.[40]

Brownfield et al.[53] presented supporting evidence in a later immunocytochemical study. They generated a primary antibody against bovine adrenal PNMT and compared the distribution of the immunostaining in adjacent serial thin sections using a postembedding method in conjunction with peroxidase-antiperoxidase immunocytochemistry. They found immunoreactivity within chromaffin vesicles. While the presence of the enzyme within the vesicle makes intuitive sense (for norepinephrine could be converted to epinephrine without leaving and then reentering the vesicle), it is still not accepted that PNMT is localized within chromaffin vesicles, since it runs counter to so much biochemical evidence.

C. PHOSPHATASES

It has been well established that high-energy phosphates, in particular adenosine triphosphate (ATP), are colocalized with catecholamines. The physiologic significance and metabolism of these phosphates remain unknown. ATP is hydrolyzed at several locations within the adrenal chromaffin cell, and cytochemical studies have helped to elucidate these sites.

The first study to localize nucleotidase activity at the subcellular level in chromaffin cells was by Wood.[54] In a study of the rat and cat, he demonstrated triphosphatase and diphosphatase activity along membranes of chromaffin cells and nerve fibers, including spaces between chromaffin cells (canaliculi) and nerve terminals, and in endothelial cells. Diphosphatase activity was additionally seen within cells. Monophosphatase activity was present at cell membranes and related to chromaffin vesicles. Conflicting results were reported by Yokoyama.[55] No triphosphatase activity was seen between chromaffin cells and monophosphatase activity was not seen in adrenal medullary cells.

There has also been conflicting evidence regarding ATPase activity on or in the chromaffin vesicle. This is of particular significance because Mg-ATPase has been shown to translocate protons from the cytosol into the chromaffin vesicle, creating an electrochemical gradient to drive the uptake of catecholamines and other molecules into the vesicle (Figure 3). Enzyme cytochemical studies using 10 m*M* Mg in the incubation medium failed to demonstrate activity at the chromaffin vesicle, though enzyme activity was demonstrated on the plasma membrane and in the cytosol.[54-58] However, when 40 m*M* Mg was used, Mg-ATPase activity could be detected on the chromaffin vesicle membrane.[59] A striking observation was a heterogeneity of the enzyme's activity; whereas some vesicles were labeled with

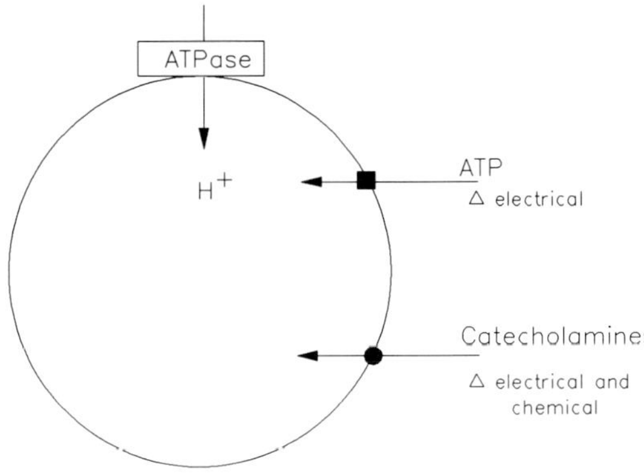

FIGURE 3. Mg-ATPase in the chromaffin vesicle membrane drives protons into the vesicle. The resultant electrochemical gradient drives the uptake of ATP and catecholamines.

reaction product, others in the same cell were not. Biochemical studies have established a functional role for Mg-ATPase activity in the chromaffin cell, but this activity does appear to be heterogeneous. It is estimated that less than ten ATPase molecules are associated with each chromaffin vesicle;[60] thus, activity of the enzyme is difficult to detect with routine methods.

Ca-activated ATPase activity was detected in the Golgi region of chromaffin cells by Benedeczky et al.[56] Biochemical studies had indicated that this enzyme was present in the microsomal fraction, but this study localized it in the Golgi apparatus of bovine adrenal chromaffin cells, as well as on the plasma membrane and on nerves. Yokoyama[55] found a weak reaction in the Golgi apparatus of rat adrenal chromaffin cells, and further suggested that it could be due to acid phosphatase, rather than ATPase. Nakai et al.[58] found Ca-activated ATPase in the Golgi apparatus and in chromaffin vesicles of mouse adrenal medulla.

The first cytochemical demonstration of acid phosphatase activity in adrenal chromaffin cells was presented by Holtzman and Dominitz.[61] In a study of the rat, they localized enzyme activity in chromaffin vesicles and the Golgi apparatus. Palkama and Rechardt[62] demonstrated enzyme activity in lysosomes of epinephrine-containing cells of the rat and in norepinephrine-containing cells of the mouse, but not in the Golgi apparatus. Palkama and Rechardt[62] suggested that lead precipitate in chromaffin vesicles was nonenzymatic, but due to high lead concentrations. The methods used by Holtzman and Dominitz[61] and Palkama and Rechardt[62] were very similar, so it is unclear why one study found evidence of enzyme activity in the Golgi apparatus and the other did not. However, Yokoyama[55] found cytochemical evidence for acid phosphatase activity in the Golgi apparatus and lysosomes in the rat adrenal medulla.

In a study of frog *(Rana esculenta),* Mastrolia and Manelli[63] demonstrated acid phosphatase activity in the membrane of chromaffin vesicles, lysosomes, and occasionally in the Golgi apparatus in both epinephrine- and norepinephrine-containing cells. The study was expanded to three other amphibian species (*Discoglossus pictus, Xenopus laevis,* and *Bufo bufo*) by Manelli et al.[64] Evidence of enzyme activity was distributed similarly in the three species: in the lysosome, on the membrane of epinephrine-containing chromaffin vesicles, occasionally on the membrane of norepinephrine-containing vesicles, and occasionally in the Golgi apparatus.

In addition to the phosphatases mentioned above, Yokoyama[55] also demonstrated activity for thiamine pyrophosphatase and inosine diphosphatase in the rat adrenal medulla. Both of these enzymes were localized in the Golgi apparatus, on axons, and on Schwann cells.

D. CHOLINESTERASES

Simultaneous demonstration of catecholamines and cholinesterases in the rat adrenal medulla was performed by Palkama.[65] Acetylcholinesterase and nonspecific cholinesterase activities were seen on the plasma membrane and nuclear membranes of chromaffin cells. Norepinephrine-containing vesicles could be clearly differentiated from epinephrine-containing vesicles. Cholinesterase activities were also seen on axonal membranes and in the rough endoplasmic reticulum of ganglion cells.

In a series of comparative studies, cholinesterase activities were detected in chromaffin cells of lizard *(Lacerta sicula)*,[66] urodelian amphibians (*Triturus cristatus, Siren lacertina*, and *Desmognathus quadrimaculatus*);[67] and an anuran amphibian *(Discoglossus pictus)*.[68] Enzyme activity was associated with the nerves in all species, confirming that the innervation is cholinergic. Some variation in enzyme activity and distribution in the chromaffin cells was seen.

Acetylcholinesterase and pseudocholinesterases in the developing chick adrenal gland were investigated by Mastrolia et al.[69] Activity of both types of cholinesterases was detected in the perinuclear space and rough endoplasmic reticulum in chromaffin cells during the first phases of organogenesis. Acetylcholinesterase activity was also on the plasma membrane. It was suggested that these enzyme activities play a role in morphogenesis.

A detailed cytochemical study of acetylcholinesterase and nonspecific cholinesterase activity in the bovine adrenal medulla was done by Somogyi et al.[70] Enzyme activity was detected on the plasma and nuclear membranes. Of particular interest was the localization of acetylcholinesterase activity within the endoplasmic reticulum of chromaffin cells and axons of the splanchnic nerve. This raised the question whether the enzyme activity released from the perfused adrenal gland might originate from chromaffin cells and/or preganglionic neurons.

In an attempt to localize acetylcholinesterase activity in chromaffin vesicles, Carmichael[71] examined intact chromaffin cells and isolated vesicles. A small amount of enzyme activity was seen on chromaffin vesicle membranes.

E. OTHER ENZYMES

In a related study, Carmichael[72] demonstrated a small amount of adenylate cyclase activity on chromaffin vesicles. In contrast to acetylcholinesterase, adenylate cyclase activity appeared to be distributed within the vesicle.

The distribution of catalase-containing peroxisomes in the rat adrenal medulla and sympathetic ganglia was examined by Arnold and Holtzman.[73] Peroxisomes were identified in chromaffin cells, ganglion cells, and the Schwann cells associated with axons in the gland.

F. NEUROPEPTIDES

Cytochemical studies are revealing the localization of peptides in vesicles and other compartments of chromaffin cells, as well as other locations within the adrenal medulla. Kondo[74] and Pelto-Huikko[75] have reviewed the literature on the localization of neuropeptides in the adrenal medulla. Their reviews underscore the *heterogeneity* of peptide distribution, not only in chromaffin cells, but also in the preganglionic sympathetic nerve endings. As will be seen below, rapid progress has been made recently in the localization of peptides in the adrenal medulla of many species, but their biological significance is not yet appreciated.

The first neuropeptide to be localized in the adrenal medulla at the ultrastructural level was enkephalin. Hervonen et al.[76] demonstrated enkephalin-like immunoreactivity in the rat using the PAP method. The label was associated with chromaffin vesicles; the number of labeled vesicles varied greatly from cell to cell. In some chromaffin cells, all the vesicles were labeled. Examining the human adrenal medulla, Hervonen et al.[77] found that enkephalin-like immunoreactivity was associated with some, but not all, chromaffin vesicles, and in a few cells, with the rough endoplasmic reticulum. No correlation with subpopulations of vesicles was observed.

Several mammalian species were examined by Varndell et al.[78] [Met]enkephalin-like immunoreactivity was localized using a modified on-grid immunogold staining method; reactivity was localized to vesicles in the adrenal medulla of cat, pig, horse, and man, including human pheochromocytoma. The identity of the vesicles as chromaffin vesicles was verified by demonstrating dopamine β-hydroxylase-like immunoreactivity.

Kobayashi et al.,[79] using a protein A-colloidal gold method, demonstrated [Met]enkephalin-Arg^6-Gly^7-Leu^8-like immunoreactivity in large-cored vesicles, but not small-cored vesicles, in the presynaptic nerve terminals on adrenal chromaffin cells in the rat. This was confirmed by Kondo et al.[129] In a direct comparison study, Kobayashi et al.[81] examined the adrenal medulla of the guinea pig and found a distribution of immunoreactivity similar to that seen in the rat. The physiologic significance of these observations is not known, but it does indicate that modulation of secretion from chromaffin cells is very complicated.

In the rat, both [Met^5]- and [Leu^5]enkephalin-like immunoreactivities were localized by Pelto-Huikko et al.[82] in a few epinephrine cells, in ganglion cells, and in a population of nerves innervating epinephrine cells. Only a few norepinephrine cells were reactive, and no nerve fibers around norepinephrine cells were reactive. These results suggested that enkephalin is colocalized with epinephrine in the rat adrenal medulla, and enkephalin-containing terminals regulate the secretion of epinephrine.

Kondo and Yui[80] demonstrated that [Met]enkephalin and epinephrine are colocalized in the chromaffin vesicles in the frog adrenal (interrenal) gland. In these animals, chromaffin cells are scattered in the gland and do not form a central medulla. Although some suggestions were offered, the physiologic significance of enkephalin in epinephrine cells remains to be established.

[Met]enkephalin was demonstrated in epinephrine-containing cells of the adrenal medulla of an insectivore, the house shrew *(Suncus murinus),* by Yamada et al.[40] They also demonstrated neuropeptide Y in the cytoplasm of both epinephrine- and norepinephrine-containing cells.

Neuropeptide Y (NPY) was demonstrated in the adrenal medulla of several mammals (horse, cat, rat, guinea pig, and mouse) by Varndell et al.[83] Immunoreactive material appeared to be localized in norepinephrine-containing cells, colocalized with enkephalin.

In a more detailed morphologic study, Kuramoto et al.[84] found immunoreactivity to NPY in approximately 50% of chromaffin cells in the rat adrenal medulla. Their study indicated that NPY is in epinephrine-containing cells. As a possible explanation for the discrepancy of the study by Varndell et al.,[83] Kuramoto et al.[84] suggested that they were able to more accurately identify epinephrine- and norepinephrine-containing cells. Kuramoto et al.[84] further found that derivatives of preproenkephalin A coexist with about 20% of the NPY-reactive cells. NPY-reactive material was localized in the core of chromaffin vesicles as well as diffusely in the cytoplasm of chromaffin cells. Ganglion cells were not intensely immunoreactive. Nerve fibers demonstrating immunoreactivity made synaptic contacts with both cortical and chromaffin cells.

Pelto-Huikko[75] also demonstrated NPY-like immunoreactivity in the adrenal medulla of rat, cat, mouse, and hamster. His findings were in agreement with those of Kuramoto et al.[84]

Substance P-like immunoreactivity was demonstrated in a relatively small number of chromaffin cells in the rat adrenal medulla by Kuramoto et al.[85] In addition, immunoreactive nerve fibers, which appeared to originate outside the adrenal gland, formed synapses on nonimmunoreactive chromaffin cells.

In a study by the same group, Kondo et al.[86] localized vasoactive intestinal polypeptide (VIP)-like immunoreactivity in the rat adrenal medulla. Immunoreactivity was revealed in a few chromaffin cells, ganglion cells, and single varicose nerve fibers. VIP in chromaffin cells appeared to be confined to the core of chromaffin vesicles; in ganglion cells and nerve fibers, it was present in vesicles and in the cytoplasm or axoplasm.

Calcitonin gene-related peptide (CGRP)-like immunoreactivity was reported in a few chromaffin cells of the rat adrenal medulla by Kuramoto et al.[87] The peptide was localized in epinephrine-containing cells that also exhibited NPY immunoreactivity. CGRP was densely localized within chromaffin vesicles. The peptide was also localized in nerve fibers, which appeared to be extrinsic in origin. Pelto-Huikko[75] demonstrated CGRP-like immunoreactivity in less than 1% of epinephrine-containing cells of the adrenal medulla of the rat, cat, mouse, and hamster.

Neurotensin-like immunoreactivity was localized in nerve fibers and terminals of the hamster adrenal medulla by Pelto-Huikko et al.[88] Only norepinephrine-containing chromaffin cells and ganglion cells were innervated by immunoreactive nerves.

In a more detailed study, enkephalin and neurotensin imunoreactivities were localized in the cat adrenal medulla by Pelto-Huikko et al.[89] Enkephalin was localized mainly in epinephrine cells, most of which were heavily labeled. Less than 10% of the norepinephrine cells were immunoreactive for enkephalin. A dense network of enkephalin-immunoreactive nerves was localized among epinephrine cells. Neurotensin was not colocalized with enkephalin, being present in about 65% of the norepinephrine cells, but not those that labeled for enkephalin. Neurotensin-immunoreactive nerves were restricted to norepinephrine cells. It was suggested that the restriction of enkephalin and neurotensin to separate populations of chromaffin cells and nerves may be a mechanism by which epinephrine and norepinephrine secretion can be differentially regulated.

Galanin was localized in the adrenal medulla by Pelto-Huikko.[75] He demonstrated immunoreactivity in epinephrine-containing vesicles, colocalized with NPY.

G. PROTEINS

The first localization of soluble protein in the adrenal medulla at the ultrastructural level was reported by Smith et al.[90] An indirect immunocytochemical technique was used to localize soluble proteins purified from lysates of bovine chromaffin vesicles. Examining sections of bovine adrenal medulla, Smith et al.[90] found nearly all of the reaction product deposited over the core of chromaffin vesicles. It is evident that they localized proteins that are now called chromogranins.

Winkler et al.[91] were the first to specifically investigate the presence of chromogranin A in the adrenal medulla at the electron microscopic level. Using a peroxidase-labeled antibody method, they clearly demonstrated staining for chromogranin A in the chromaffin vesicle in the rat adrenal gland. The vesicle membranes were unstained. This was confirmed in the bovine adrenal medulla by Aunis et al.[50]

Using a monoclonal antibody against human chromogranin, Varndell et al.[92] localized this soluble glycoprotein to the matrix of secretory vesicles from several endocrine cells, including normal human adrenal medulla and pheochromocytomas. They proposed chromogranin as a marker for the ultrastructural recognition of endocrine secretory vesicles.

The effect of specimen postfixation with osmium tetroxide on postembedding localization of chromogranin A was evaluated by Hearn.[93] Using a protein A-gold technique, he demonstrated positive chromaffin vesicles in human adrenal medulla, as well as in other tissues and tumors.

The first ultrastructural demonstration of chromogranins B and C (also known as secretogranin II; Eiden et al.[94]) was published by Steiner et al.[95] In the rat adrenal medulla, they found that all three proteins are present in the same cell. Double labeling of ultrathin sections demonstrated the presence of the three chromogranins, as well as NPY, within the same chromaffin vesicles.

A polypeptide isolated from the pituitary, designated 7B2, was identified in bovine chromaffin vesicles and localized in the rat adrenal medulla by Marcinkiewicz et al.[96] This polypeptide, apparently not derived from chromogranins A, B, or C, was shown to be similar to chromogranins with regard to distribution, subcellular localization, and physicochemical properties.

Another polypeptide isolated from the pituitary, named GAWK (from the first four amino acids in its sequence: Gly, Ala, Trp, Lys), was localized in the human adrenal medulla and other neuroendocrine cells by Bishop et al.[97] The entire 74 amino acid sequence of GAWK was found to be within chromogranin B, and, not surprisingly, GAWK was distributed similarly to this chromogranin.

Although arguably not cytochemical studies, Benchimol and Cantin[98,99] published two detailed studies following the production of proteins and catecholamines in the rat adrenal medulla. Using ultrastructural radioautography, they followed the incorporation of tritiated leucine, tyrosine, and dopa in chromaffin cells. The spatial and temporal aspects of chromogranin formation were followed with leucine, whereas catecholamine formation was followed with dopa. Tyrosine appeared to participate in the formation of proteins and catecholamines.

The ultrastructural localization of the glycoprotein D2 in the rat adrenal medulla was investigated by Langley and Aunis.[100] D2 is apparently related to, if not equivalent to, neuronal cell adhesion molecules (nCAM). D2 was found on the surface of all chromaffin cells of the rat adrenal medulla, and on nerve fibers and Schwann cells. In a related study, Langley et al.[101] found that nCAM was expressed in several endocrine cells, but not the adrenal medulla of the rat.

A cerebellar Purkinje cell-specific protein (also known as spot 35 protein) was localized in all norepinephrine-containing and some ganglion cells of the rat adrenal medulla by Kondo et al.[102] Immunoreactivity was observed in chromaffin vesicles as well as in the cytosol.

Tischler et al.[103] demonstrated that the anti-lymphocyte antibody Leu-7 (HNK-1) reacts intensely with normal and neoplastic adrenal medullary cells. The epitope was localized over the matrices of most chromaffin vesicles in the human adrenal medulla and pheochromocytoma.

Fujimoto and Ogawa[104] used an antibody to cytochrome b_{561} as a marker for the chromaffin vesicle membrane. Cytochrome b_{561} is an integral protein of the vesicle membrane, involved in electron transport for the synthesis of norepinephrine.

The presence of basic fibroblast growth factor (bFGF) immunoreactivity in the bovine adrenal medulla was demonstrated by Westermann et al.[105] The typical 18-kDa form of bFGF was localized to chromaffin vesicles, along with higher molecular mass forms, which they suspected are precursors. Presta and Rifkin[106] challenged the presence of the largest molecular form of bFGF in chromaffin vesicles, contending it is a cytosolic protein. Westermann et al.[107] replied that they localized a different protein within chromaffin vesicles. It is significant that this growth factor has been demonstrated within a secretory organelle, since this is the first suggestion that bFGF is in a location suitable for regulated release from cells.

1. Contractile and Structural Proteins

Contractile proteins represent a special category of proteins. In the adrenal chromaffin cell, contractile proteins have been considered the ''muscle'' of stimulus-secretion coupling. However, an active role for proteins in moving chromaffin vesicles toward the plasma membrane as a prelude to exocytosis has never been demonstrated. It is now suspected that contractile and structural proteins form a barrier between chromaffin vesicles and the cell surface. Upon the proper stimulus, this barrier somehow diminishes, allowing vesicles access to the plasma membrane. Whereas these dynamic events have not yet been demonstrated with cytochemical techniques, cytochemistry has played an important role in establishing the presence and location of certain contractile and structural proteins.

Myosin was localized by Aunis et al.[50] in the vasculature of the bovine adrenal medulla, in fibroblasts lining the vessels, and in chromaffin cells. In chromaffin cells, staining was associated with the plasma membrane, occasionally with Ω-figures, thought to be morphologic evidence of exocytosis. It is tempting to speculate on an active role of myosin in exocytosis, but Aunis et al.[50] agreed that such a role could not be supported by their study.

Wilkins and Lin[108] found actin, in the form of stable oligomers, on chromaffin vesicle membranes. They demonstrated high-affinity cytochalasin-binding sites on the membranes, indicating that they contain stable actin nuclei. This study indicated that actin associates with the chromaffin vesicle membrane in a specific fashion.

Fodrin, an analog of the erythrocyte structural protein spectrin, was localized in adrenal chromaffin cells by Langley et al.[109] and Fujimoto et al.[110] The protein was localized in all chromaffin cells on the cytoplasmic side of the plasma membrane, in a position where it could be a barrier between chromaffin vesicles and the cell surface. Fodrin was also located in nerves and Schwann cells. Fujimoto et al.[110] further demonstrated that when the chromaffin cell is stimulated to secrete, fodrin is also located on vesicles and vacuoles.

In an elegant study, Fujimoto and Ogawa[104] further followed the distribution of fodrin after massive stimulation of the adrenal medulla. Using insulin-induced hypoglycemia as a strong stimulus, they used immunoelectron microscopy to examine the rat adrenal medulla 1 to 4 h after insulin injection. In control cells fodrin was distributed only in the periphery. After stimulation, markers were found around small vesicles and large vacuoles. The results of Fujimoto and Ogawa[104] suggest that these organelles are derived from the plasma membrane after massive secretion, and fodrin remains associated with the vesicle and vacuolar membranes following their retrieval from the cell surface.

H. MEMBRANE DYNAMICS

Cytochemistry has played a key role in elucidating the membrane dynamics of catecholamine storage and secretion. It was suggested by De Robertis and Vaz Ferreira[111] and De Robertis and Sabatini[20] that catecholamines are synthesized and stored within chromaffin vesicles, then released in a regulated manner via exocytosis. This scenario is now well accepted. It has been apparent that the vesicle membranes could not be continuously incorporated into the plasma membrane, since the surface area of the cell would become unmanageably large within a short time. Obviously, a retrieval mechanism was operating. Until recently, the question remained whether the retrieval was specific (the same patch of vesicle membrane that was inserted into the plasma membrane was retrieved) or not (any comparable area of membrane was retrieved).

One of the first cytochemical studies to address the problem of membrane retrieval in chromaffin cells was done by Nagasawa and Douglas.[112] They introduced colloidal thorium dioxide into the extracellular space of the adrenal medulla of hamsters. The electron-dense material could be clearly seen within pits at the cell surface, demonstrating that a mechanism like endocytosis was operating. However, no quantitative information or specific identity of the membranes involved was available.

In a similar study, Abrahams and Holtzman[113] used horseradish peroxidase to follow endocytosis in the rat adrenal medulla. The tracer was taken up into vesicles, tubules, multivesicular bodies, and dense bodies. Further experiments identifying acid phosphatase suggested that many structures into which the peroxidase accumulated were lysosomes. This suggested that membranes retrieved from the surface undergo autolysis, but the specificity of the membranes could not be determined.

Benedeczky and Somogyi[114] followed the same events with similar methods in the hamster adrenal medulla. It is of interest that they did not find any tracers in the Golgi complex, leading to the conclusion that membranes are not reutilized. Demonstration of membrane reutilization would await more specific markers for membranes.

That marker turned out to be glycoprotein III, a membrane-bound molecule restricted to facing the interior of the chromaffin vesicle. Patzak and Winkler[115] conjugated antibodies to glycoprotein III with colloidal gold and exposed them to isolated bovine adrenal medullary cells. Labeled patches were present on the surface of stimulated cells. Within minutes, labeled patches became coated, giving rise to coated vesicles, and then to smooth vesicles. These vesicles were found throughout the cell, including the Golgi region, but Golgi stacks were not labeled. Some label was incorporated into multivesiclar bodies, probably representing a lysosomal pathway. After 30 min, labeled membranes were seen in what appeared to be nascent chromaffin vesicles. After 6 hours, immunolabel was seen in mature chromaffin vesicles. Thus, the selective nature of membrane reuptake and partial reutilization was established in the adrenal medulla.

Using cultured bovine chromaffin cells, Patzak et al.[116] found some differences in membrane recycling when compared to freshly isolated cells. Immunolabel was not reutilized into vesicles positive for chromogranin A, but rather the label went into lysosomes.

A cytochemical technique employing tannic acid to precipitate proteins was used to demonstrate exocytosis. This was done both in cultured bovine chromaffin cells[117] and in the rat adrenal medulla *in situ*.[118]

In a different approach to membrane dynamics during exocytosis, Rosenheck and Plattner[119] examined interactions between bovine chromaffin vesicles and adrenal medullary plasma membrane vesicles. Cytochemical demonstration of acetylcholinesterase confirmed that vesicles were derived from the plasma membrane. Other studies demonstrated that the majority of vesicles were oriented right-side out; this is unfortunate, since this is the incorrect orientation for an *in vitro* model of exocytosis. However, plasma membrane vesicles and chromaffin vesicles did aggregate in isotonic ionic media.

Unfixed chromaffin cell plasma membranes were studied by Himmelhoch and Rossi.[120] Membranes from bovine adrenal medulla were isolated on a urografin density gradient, frozen, and sectioned. Using a monoclonal antibody against its cholinergic binding site, nicotinic receptors were localized in patches.

Cholesterol is a membrane component thought to have a major role in fluidity related to exocytosis. Fujita et al.[121] used the polyene antibiotic filipin to localize cholesterol in membranes of the mouse adrenal medulla. Filipin-cholesterol complexes were visualized in freeze-fracture images. Complexes were heterogeneously distributed, but were noticeably more dense on plasma membranes and chromaffin vesicles; coated pits were lacking in complexes.

I. OTHER CYTOCHEMICAL STUDIES

Calcium-binding sites were localized in bovine adrenal medullary cells with cytochemical techniques by Burgoyne et al.[122] They used pyroantimonate to localize calcium and did control studies to test for specificity and accuracy of positional information. Since they could not exclude the possibility that some diffusion occurred during fixation, they regarded the technique as localizing binding sites rather than the ions themselves. Perhaps the most striking finding they presented is the demonstration of increased plasma membrane-associated calcium following stimulation with carbamylcholine. This observation and other evidence support a role for calcium ions as a mediator in exocytosis.

The localization of calcium in resting chromaffin cells was reported by Ravazzola.[123] Her results in the rat adrenal medulla were consistent with those of Burgoyne et al.;[122] the major sites of calcium antimonate deposits were in chromaffin vesicles and mitochondria.

Cytochemical localization of 5-hydroxydopamine in the cat adrenal medulla was reported by Prentice and Wood.[124,125] They administered 5-hydroxydopamine, a "false" adrenergic transmitter, to cats, then localized the compound using a combination of glutaraldehyde and molybdate. They were, for the first time, able to demonstrate convincingly adrenergic nerves in the adrenal medulla at the ultrastructural level. The physiologic role of these nerves remains unknown.

Serotonin (5-hydroxytryptamine) was localized in the rat adrenal medulla by Brownfield et al.[53] Serotonin immunoreactivity was found on the membrane, as well as the matrix, of epinephrine-containing vesicles. They suggested that serotonin may be synthesized by adrenal chromaffin cells.

Delarue et al.[126] localized serotonin in the frog adrenal (interrenal) gland. Using a preembedding procedure associated with gold-silver intensification of the immunoperoxidase reaction, serotonin was visualized in chromaffin vesicles at the periphery of epinephrine-containing cells. Delarue et al.[126] suggested that serotonin may be released with epinephrine under stressful conditions.

Cantin and Benchimol[127] used several cytochemical techniques to localize and characterize carbohydrates in the adrenal medulla of rat and hamster. They found that chromaffin vesicles and the Golgi complex in epinephrine-containing, but not norepinephrine-containing, cells are rich in glycoproteins.

The uranaffin reaction was introduced by Richards and Da Prada[128] as a technique for localizing adenine nucleotides in organelles such as chromaffin vesicles. Based on the observation that uranyl ions have an extremely high affinity for catecholamine-containing organelles, they incubated specimens, including rat adrenal medulla, for 18 hours in 4% aqueous uranyl acetate. Their results suggested that chromaffin vesicles containing norepinephrine, but not epinephrine, are rich in nucleotides. This conclusion was later revised: both norepinephrine- and epinephrine-containing cells possess a mixture of reactive and nonreactive vesicles, possibly reflecting variations in nucleotide contents.[128a]

ACKNOWLEDGMENTS

The author thanks Drs. Virginia Black, Susan Stoddard, Albert Verhofstad, and Joe Wood for reviewing the manuscript and making many helpful suggestions.

REFERENCES

1. **Carmichael, S. W. and Winkler, H.,** The adrenal chromaffin cell, *Sci. Am.,* 253, 40, 1985.
2. **Winkler, H., Apps, D. K., and Fischer-Colbrie, R.,** The molecular function of adrenal chromaffin granules: established facts and unresolved topics, *Neuroscience,* 18, 261, 1986.
3. **Hopwood, D.,** The histochemistry and electron histochemistry of chromaffin tissue, *Prog. Histochem. Cytochem.,* 3, 1, 1971.
4. **Colin, G.,** *Traité de Physiologie Comparée,* Paris, 1856.
5. **Vulpian, M.,** Note dur quelques reactions propres àla substance de capsules surrénales, *Compt. Rend. Acad. Sci., (Paris),* 43, 663, 1856.
6. **Virchow, R.,** *Arch. Pathol. Anat.,* 12, 481, 1857.
7. **Werner, B.,** De Capsulis Suprarenalis, *Diss. Ingorg. Dorpati,* Livonorum, 1857.
8. **Henle, J.,** Ueber das Gewebe der Nebenniere und der Hypophyse, *Z. Rat. Med.,* 24, 143, 1865.

9. **Kohn, A.,** Das chromaffine Gewebe, *Ergeb. Anat. Entwicklungsgesch.*, 12, 253, 1902.
10. **Wood, J. G.,** Positive identification of intracellular biogenic amine reaction product with electron microscopic X-ray analysis, *J. Histochem. Cytochem.*, 22, 1060, 1974.
11. **Wood, J. G.,** Analytical electron microscopic identification of cytochemical products in thin sections, *Acta Histochem.*, 53, 143, 1975.
12. **Holtz, P. and Shüman, H. J.,** Arternol — Ein neues Hormon des Nebennierenmarks, *Naturwissenschaften*, 35, 159, 1948.
13. **Bänder, A.,** Über zwei verschiedene chromaffine Zelltypen in Nebennierenmark und ihre Beziehung zum Adrenalin- und Arterenolgehalt, *Verh. Anat. Ges.*, 48, 172, 1950.
14. **Eränkö, O.,** On the histochemistry of the adrenal medulla of the rat, with special reference to acid phosphatase, *Acta Anat.*, 16 (Suppl. 17), 1, 1952.
15. **Hillarp, N.-Å. and Hökfelt, B.,** Evidence of adrenaline and noradrenaline in separate adrenal medullary cells, *Acta Physiol. Scand.*, 30, 55, 1953.
16. **Wood, J. G.,** Identification of an observation on epinephrine and norepinephrine containing cells in the adrenal medulla, *Am. J. Anat.*, 112, 285, 1963.
17. **Lever, J. D.,** Electron microscopic observation on the normal and denervated adrenal medulla of the rat, *Endocrinology*, 57, 621, 1955.
18. **Benedeczky, I. and Smith, A. D.,** Ultrastructural studies on the adrenal medulla of golden hamster: origin and fate of secretory granules, *Z. Zellforsch.*, 124, 367, 1972.
19. **Carmichael, S. W. and Blair, B. C.,** Normal ultrastructure of the day old dog adrenal medulla, *J. Anat.*, 115, 113, 1973.
20. **De Robertis, E. D. P. and Sabatini, D. D.,** Submicroscopic analysis of the secretory process in the adrenal medulla, *Fed. Proc.*, 19 (Suppl. 5), 70, 1960.
21. **Kano, M.,** [Electron microscopic study on the adrenal medulla of domestic fowl], *Arch. Histol. Jpn.*, 18, 25, 1959.
22. **Yates, R. D., Wood, J. G., and Duncan, D.,** Phase and electron microscopic observations on two cell types in the adrenal medulla of the Syrian hamster, *Tex. Rep. Biol. Med.*, 20, 494, 1962.
23. **Wood, J. G. and Barrnett, R. J.,** Histochemical demonstration of norepinephrine at a fine structural level, *J. Histochem. Cytochem.*, 12, 197, 1964.
24. **Coupland, R. E., Pyper, A. S., and Hopwood, D.,** A method for differentiating between adrenaline- and noradrenaline-storing cells in the light and electron microscope, *Nature*, 201, 1240, 1964.
25. **Tramezzani, J. H., Chiocchio, S., and Wassermann, G. F.,** A technique for light and electron microscopic identification of adrenalin- and noradrenalin-storing cells, *J. Histochem. Cytochem.*, 12, 890, 1964.
26. **Benedeczky, I., Puppi, A., Tigyi, A., and Lissak, K.,** Electron microscopic study of adrenaline and noradrenaline secretion of the adrenal medulla, *Acta Biol. Hung.*, 15, 285, 1965.
27. **Carmichael, S. W.,** Morphology and innervation of the adrenal medulla, in *Stimulus-Secretion Coupling in Chromaffin Cells*, Rosenheck K. and Lelkes, P. I., Eds., CRC Press, Boca Raton, 1987, 1.
28. **Coupland, R. E. and Hopwood, D.,** The mechanism of the differential staining reaction for adrenaline- and noradrenaline-storing granules in tissues fixed in glutaraldehyde, *J. Anat.*, 100, 227, 1966.
29. **Arnold, M. and Hager, G.,** Zum kontrast von catecholamingranula im elektronenmikroskop, *Histochemie*, 14, 297, 1968.
30. **Wood, J. G., Seelig, L. L. J., and Benjamin, C. P.,** Cytochemistry of epinephrine and norepinephrine adrenomedullary cells, *Histochemie*, 28, 183, 1971.
31. **McClung, R. E. and Wood, J. G.,** Analytical electron microscopic evaluation of the effects of paraformaldehyde pretreatment on the reaction glutaraldehyde with biogenic amines, *J. Histochem. Cytochem.*, 30, 481, 1982.
32. **Benchimol, S. and Cantin, M.,** Ultrastructural cytochemistry of the human adrenal medulla, *Histochemistry*, 54, 9, 1977.
33. **Kanerva, L., Hervonen, A., and Rechardt, L.,** Permanganate fixation demonstrates the monoamine-containing granular vesicles in the SIF cells but not in the adrenal medulla or mast cells, *Histochemistry*, 52, 61, 1977.
34. **Rechardt, L., Kanerva, L., and Hervonen, H.,** Ultrastructural demonstration of amine granules in the adrenal medullary cells of the rat using acid permanganate fixation, *Histochemistry*, 54, 339, 1977.
35. **Verhofstad, A. A. J., Steinbusch, H. W. J., Penke, B., Varga, J., and Joosten, H. W. J.,** Use of antibodies to norepinephrine and epinephrine in immunohistochemistry, *Adv. Biochem. Psychopharmacol.*, 25, 185, 1980.
36. **Verhofstad, A. A. J., Coupland, R. E., and Colenbrander, B.,** Immunohistochemical and biochemical analysis of the development of the noradrenaline- and adrenaline-storing cells in the adrenal medulla of the rat and pig, *Arch. Histol. Cytol.*, 52, 351, 1989.

37. **Kobayashi, S., Uchida, T., Ohashi, T., Fujita, T., Nakao, K., Yoshimasa, T., Imura, H., Mochizuki, T., Yanaihara, C., Yanaihara, N., and Verhofstad, A. A. J.,** Immunocytochemical demonstration of the co-storage of noradrenaline with Met-enkephalin-Arg^6-Gly^7-Leu^8 in the carotid body chief cells of the dog, *Arch. Histol. Jpn.*, 46, 713, 1983.
38. **Nagatsu, I., Karasawa, N., Kondo, Y., and Inagaki, S.,** Immunocytochemical localization of tyrosine hydroxylase, dopamine-β-hydroxylase and phenylethanolamine-*N*-methyltransferase in the adrenal glands of the frog and rat by a peroxidase-antiperoxidase method, *Histochemistry,* 64, 131, 1979.
39. **Nagatsu, I., Karasawa, N., Kondo, Y., and Inagaki, S.,** Localization of catecholamine synthesizing enzymes in the adrenal medulla of *Suncus murinus riukiuanus* — immunofluorescent and peroxidase-antiperoxidase (PAP) methods, *Acta Histochem. Cytochem.*, 12, 550, 1979.
40. **Yamada, K., Karasawa, N., Fujii, T., Isomura, G., Takagi, I., Sakai, M., and Nagatsu, I.,** Immunocytochemical localization of amines and peptides in adrenal chromaffin cells of the house shrew *(Suncus murinus), Biog. Amines,* 7, 223, 1990.
41. **Stephens, J. K., Masserano, J. M., Vulliet, P. R., Weiner, N., and Nakane, P. K.,** Immunocytochemical localization of tyrosine hydroxylase in rat adrenal medulla by the peroxidase labelled antibody method: effects of enzyme activation on ultrastructural distribution of the enzyme, *Brain Res.*, 209, 339, 1981.
42. **Semenenko, F. M., Cuello, A. C., Goldstein, M., Lee, K. Y., and Sidebottom, E.,** A monoclonal antibody against tyrosine hydroxylase: application in light and electron microscopy, *J. Histochem. Cytochem.*, 34, 817, 1986.
43. **Kwan, S.-W., Patel, N. T., Vulliet, P. R., Hall, F. L., Denney, R. M., Shen, R. S., Westlund, K. N., and Abell, C. W.,** Isolation, characterization, and application of monoclonal antibodies to rat tyrosine hydroxylase, *J. Neurosci. Res.*, 23, 316, 1989.
44. **Oomori, Y., Okuno, S., Fujisawa, H., Satoh, T., and Ono, K.,** Intracellular localization of tyrosine hydroxylase immunoreactivity in the rat adrenal chromaffin and pheochromocytoma cells, *Acta Anat.*, 136, 1, 1989.
45. **Oomori, Y., Okuno, S., Fujisawa, H., and Ono, K.,** Immunoelectron microscopic study of tyrosine hydroxylase immunoreactive nerve fibers and ganglion cells in the rat adrenal gland, *Anat. Rec.*, 229, 407, 1991.
46. **Kuhn, D. M., Arthur, R., Jr., Yoon, H., and Sankaran, K.,** Tyrosine hydroxylase in secretory granules from bovine adrenal medulla: evidence for an integral membrane form, *J. Biol. Chem.*, 265, 5780, 1990.
47. **Hökfelt, T., Fuxe, K., Goldstein, M., and Joh, T. H.,** Immunohistochemical localization of three catecholamine synthesizing enzymes: aspects on methodology, *Histochemie,* 33, 231, 1973.
48. **Verhofstad, A. A. J., Hökfelt, T., Goldstein, M., Steinbusch, H. W. J., and Joosten, H. W. J.,** Appearance of tyrosine hydroxylase, aromatic amino-acid decarboxylase, dopamine β-hydroxylase and phenylethanolamine N-methyltransferase during ontogenesis of the adrenal medulla, *Cell Tissue Res.*, 200, 1, 1979.
49. **Redick, J. A., Thomas, J. A., Van Orden, L. S., III, Van Orden, D. E., and Kopin, I. J.,** Immunocytochemical localization of dopamine β-hydroxylase in adrenal chromaffin granules, *Neuropharmacology,* 13, 1005, 1974.
50. **Aunis, D., Hesketh, J. E., and Devilliers, G. D.,** Immunohistochemical and immunocytochemical localization of myosin, chromogranin A and dopamine-β-hydroxylase in nerve cells in culture and in adrenal glands, *J. Neurocytol.*, 9, 255, 1980.
51. **Matsumoto, S., Tanaka, K., Yamamoto, A., Nakada, H., Uchida, M., and Tashiro, Y.,** Immunoelectron microscopic localization of dopamine β-hydroxylase and chromogranin A in adrenomedullary chromaffin cells, *Cell Struct. Funct.*, 12, 483, 1987.
52. **Nagatsu, I. and Kondo, Y.,** Immunoelectronmicroscopic localization of phenylethanolamine-*N*-methyltransferase in the bovine adrenal medulla, *Histochemistry,* 42, 351, 1974.
53. **Brownfield, M. S., Poff, B. C., and Holzwarth, M. A.,** Ultrastuctural immunocytochemical co-localization of serotonin and PNMT in adrenal medullary vesicles, *Histochemistry,* 83, 41, 1985.
54. **Wood, J. G.,** The relationship of nucleotidase activity to catecholamine storage sites in adrenomedullary tissue, *Am. J. Anat.*, 121, 671, 1967.
55. **Yokoyama, M.,** Ultracytochemical study on some phosphatases in rat adrenal medulla, *J. Electron Microsc.*, 23, 285, 1974.
56. **Benedeczky, I., Smith, A. D., and Dubois, F.,** A cytochemical study of the calcium-activated adenosinetriphosphatase in hamster adrenal medulla: its occurrence in the Golgi region of chromaffin cells, *Histochemistry,* 29, 16, 1972.
57. **Al-Lami, F. and Farman, N.,** Ultrastructural and histochemical study of the adrenal medulla in normal and cold-stressed Syrian hamsters, *Anat. Rec.*, 181, 113, 1975.

58. **Nakai, Y., Uchida, M., and Iwashita, T.,** Fine-structural localization of adenosinetriphosphatase in the mouse adrenal medulla under normal conditions and following treatment with reserpine, in *Chromaffin, Enterochromaffin and Related Cells,* Coupland, R. E. and Fujita, T., Eds., Elsevier, New York, 1976, 35.
59. **Benedeczky, I. and Carmichael, S. W.,** Ultrastructural demonstration of Mg-ATPase activity of the chromaffin granule, *Histochemistry,* 68, 317, 1980.
60. **Winkler, H. and Westhead, E. W.,** The molecular organization of adrenal chromaffin granules, *Neuroscience,* 5, 1803, 1980.
61. **Holtzman, E. and Dominitz, R.,** Cytochemical studies of lysosomes, Golgi apparatus and endoplasmic reticulum in secretion and protein uptake by adrenal medulla cells of the rat, *J. Histochem. Cytochem.,* 16, 320, 1968.
62. **Palkama, A. and Rechardt, L.,** Light and electron microscopic observations on acid phosphatase activity in the adrenal medulla of the rat and mouse, *Ann. Med. Exp. Fenn.,* 48, 77, 1970.
63. **Mastrolia, L. and Manelli, H.,** Localization of acid phosphatase in the suprarenal gland of Rana esculenta, *Cell Mol. Biol.,* 25, 11, 1979.
64. **Manelli, H., Mastrolia, L., and Arizzi, M.,** Comparative acid phosphatase distribution in the suprarenal gland of *Discoglossus pictus, Xenopus laevis* and *Bufo bufo* (Anurans, Amphibia), *Basic Appl. Histochem.,* 25, 189, 1981.
65. **Palkama, A.,** Demonstration of adrenomedullary catecholamines and cholinesterases at electron microscopic level in the same tissue section, *Ann. Med. Exp. Fenn.,* 45, 2295, 1967.
66. **Gallo, V. P. and Mastrolia, L.,** Acetylcholinesterase and pseudocholinesterases in the adrenal chromaffin tissue of *Lacerta sicula, Gen. Comp. Endocrinol.,* 37, 415, 1979.
67. **Mastrolia, L., Manelli, H., Gallo, V. P., and Arizzi, M.,** Acetylcholinesterase activity in the adrenal chromaffin cells of *Triturus cristatus, Siren lacertina* and *Desmognathus quadrimaculatus* (Amphibia Urodela), *Cell Mol. Biol.,* 32, 511, 1986.
68. **Gallo, V. P., Civinini, A., and Mastrolia, L.,** Comparative data on acetylcholinesterase cytochemistry in various chromaffin cell types of *Discoglossus pictus* (Amphibia, Anura), *Basic Appl. Histochem.,* 31, 135, 1987.
69. **Mastrolia, L., Bichi, R., Arizzi, M., and Manelli, H.,** Acetylcholinesterase and pseudocholinesterases in developing chick adrenal, *Basic Appl. Histochem.,* 30, 97, 1986.
70. **Somogyi, P., Chubb, I. W., and Smith, A. D.,** A possible structural basis for the extracellular release of acetylcholinesterase, *Proc. R. Soc. London, B,* 191, 271, 1975.
71. **Carmichael, S. W.,** Acetylcholinesterase activity of adrenal chromaffin vesicles, *Acta Histochem.,* 75, 183, 1984.
72. **Carmichael, S. W.,** Adenylate cyclase activity of adrenal chromaffin vesicles, *Acta Histochem.,* 75, 189, 1984.
73. **Arnold, G. and Holtzman, E.,** Peroxisomes in rat sympathetic ganglia and adrenal medulla, *Brain Res.,* 83, 509, 1975.
74. **Kondo, H.,** Immunohistochemical analysis of the localization of neuropeptides in the adrenal gland, *Arch. Histol. Jpn.,* 48, 453, 1985.
75. **Pelto-Huikko, M.,** Immunocytochemical localization of neuropeptides in the adrenal medulla, *J. Electron Microsc. Tech.,* 12, 364, 1989.
76. **Hervonen, A., Pelto-Huikko, M., and Linnoila, I.,** Ultrastructural localization of enkephalin-like immunoreactivity in the rat adrenal medulla, *Am. J. Anat.,* 157, 445, 1980.
77. **Hervonen, A., Linnoila, I., Vaalasti, A., Alho, H., and Pelto-Huikko, M.,** Electron microscopic localization of enkephalin-like immunoreactivity in the human adrenal medulla, *J. Electron Microsc. Tech.,* 12, 380, 1989.
78. **Varndell, I. M., Tapia, F. J., De Mey, J., Rush, R. A., Bloom, S. R., and Polak, J. M.,** Electron immunocytochemical localization of enkephalin-like material in catecholamine-containing cells of the carotid body, the adrenal medulla, and in pheochromocytomas of man and other mammals, *J. Histochem. Cytochem.,* 30, 682, 1982.
79. **Kobayashi, S., Ohashi, T., Uchida, T., and Yanaihara, N.,** Met-enkephalin-Arg^6-Gly^7-Leu^8-like immunoreactivity in presynaptic nerve terminals on both adrenaline-storing (A) and noradrenaline-storing (NA) cells of rat adrenal medulla, *Biomed. Res.,* 4, 151, 1983.
80. **Kondo, H. and Yui, R.,** Co-existence of enkephalin and adrenalin in the frog adrenal gland, *Histochemistry,* 80, 243, 1984.
81. **Kobayashi, S., Miyabayashi, T., Uchida, T., and Yanaihara, N.,** Met-enkephalin-Arg^6-Gly^7-Leu^8 in large-cored vesicles of splanchnic nerve terminals innervating guinea pig adrenal chromaffin cells, *Neurosci. Lett.,* 53, 247, 1985.

82. **Pelto-Huikko, M., Salminen, T., and Hervonen, A.,** Localization of enkephalins in adrenaline cells and the nerves innervating adrenaline cells in rat adrenal medulla, *Histochemistry,* 82, 377, 1985.
83. **Varndell, I. M., Polak, J. M., Allen, J. M., Terenghi, G., and Bloom, S. R.,** Neuropeptide tyrosine (NPY) immunoreactivity in norepinephrine-containing cells and nerves of the mammalian adrenal gland, *Endocrinology,* 114, 1460, 1984.
84. **Kuramoto, H., Kondo, H., and Fujita, T.,** Neuropeptide tyrosine (NPY)-like immunoreactivity in adrenal chromaffin cells and intraadrenal nerve fibers of rats, *Anat. Rec.,* 214, 321, 1986.
85. **Kuramoto, H., Kondo, H., and Fujita, T.,** Substance P-like immunoreactivity in adrenal chromaffin cells and intra-adrenal nerve fibers of rats, *Histochemistry,* 82, 507, 1985.
86. **Kondo, H., Kuramoto, H., and Fujita, T.,** An immuno-electron-microscopic study of the localization of VIP-like immunoreactivity in the adrenal gland of the rat, *Cell Tissue Res.,* 245, 531, 1986.
87. **Kuramoto, H., Kondo, H., and Fujita, T.,** Calcitonin gene-related peptide (CGRP)-like immunoreactivity in scattered chromaffin cells and nerve fibers in the adrenal gland of rats, *Cell Tissue Res.,* 247, 309, 1987.
88. **Pelto-Huikko, M., Salminen, T., Partanen, M., Toivanen, M., and Hervonen, A.,** Immunohistochemical localization of neurotensin in hamster adrenal medulla, *Anat. Rec.,* 211, 458, 1985.
89. **Pelto-Huikko, M., Salminen, T., and Hervonen, A.,** Localization of enkephalin- and neurotensin-like immunoreactivities in cat adrenal medulla, *Histochemistry,* 88, 31, 1987.
90. **Smith, D. J., Redick, J. A., Van Orden, D. E., and Van Orden, L. S., III,** Immunocytochemical localization of soluble protein in the adrenal medulla, *Histochem. J.,* 5, 255, 1973.
91. **Winkler, H., Schneider, F. H., Rufener, C., Nakane, P. K., and Hörtnagl, H.,** Membranes of adrenal medulla: their role in exocytosis, in *Cytopharmacology of Secretion, Advances in Cytopharmacology, Vol. 2,* Ceccarelli, B., Clementi, F., and Meldolesi, J., Eds., Raven Press, New York, 1974, 127.
92. **Varndell, I. M., Lloyd, R. V., Wilson, B. S., and Polak, J. M.,** Ultrastructural localization of chromogranin: a potential marker for the electron microscopical recognition of endocrine cell secretory granules, *Histochem. J.,* 17, 981, 1985.
93. **Hearn, S. A.,** Electron microscopic localization of chromogranin A in osmium-fixed neuroendocrine cells with a protein A-gold technique, *J. Histochem. Cytochem.,* 35, 795, 1987.
94. **Eiden, L. E., Huttner, W. B., Mallet, J., O'Connor, D. T., Winkler, H., and Zanini, A.,** A nomenclature proposal for chromogranin/secretogranin proteins, *Neuroscience,* 21, 1019, 1987.
95. **Steiner, H.-J., Schmid, K. W., Fischer-Colbrie, R., Sperk, G., and Winkler, H.,** Co-localization of chromogranin A and B, secretogranin II and neuropeptide Y in chromaffin granules of rat adrenal medulla studied by electron microscopic immunocytochemistry, *Histochemistry,* 91, 473, 1989.
96. **Marcinkiewicz, M., Fischer-Colbrie, R., Falgueyret, J. P., Benjannet, S., Seidah, N. G., Lazure, C., Winkler, H., and Chrétien, M.,** Two-dimensional immunoblotting analysis and immunocytochemical localization of the secretory polypeptide 7B2 in adrenal medulla, *Neurosci. Lett.,* 95, 81, 1988.
97. **Bishop, A. E., Sekija, K., Salajuddin, M. J., Carlei, F., Rindi, G., Fahey, M., Steel, J. H., Hedges, M., Domoto, T., Fischer-Colbrie, R., Winkler, H., Krausz, T., Ghatei, M. A., Bloom, S. R., and Polak, J. M.,** The distribution of GAWK-like immunoreactivity in neuroendocrine cells of the human gut, pancreas, adrenal and pituitary glands and its co-localisation with chromogranin B, *Histochemistry,* 90, 475, 1989.
98. **Benchimol, S. and Cantin, M.,** Ultrastructural radioautography of the incorporation of tritiated leucine by the rat adrenal medulla *in vivo, Cell Tissue Res.,* 193, 179, 1978.
99. **Benchimol, S. and Cantin, M.,** Ultrastructural radioautography of synthesis and migration of proteins and catecholamines in the rat adrenal medulla, *Cell Tissue Res.,* 225, 293, 1982.
100. **Langley, O. K. and Aunis, D.,** Ultrastructural immunocytochemical demonstration of D2-protein in adrenal medulla, *Cell Tissue Res.,* 238, 497, 1984.
101. **Langley, O. K., Aletsee-Ufrecht, M. C., Grant, N. J., and Gratzl, M.,** Expression of the neural crest adhesion molecule in endocrine cells, *J. Histochem. Cytochem.,* 37, 781, 1989.
102. **Kondo, H., Kuramoto, H., Iwanaga, T., and Fujita, T.,** Cerebellar Purkinje cell-specific protein-like immunoreactivity in noradrenalin-chromaffin cells and ganglion cells but not in adrenaline-chromaffin cells in the rat adrenal medulla, *Arch. Histol. Jpn.,* 48, 421, 1985.
103. **Tischler, A. S., Mobtaker, H., Mann, K., Nunnemacher, G., Jason, W. J., Dayal, Y., DeLellis, R. A., Adelman, L., and Wolfe, H. J.,** Anti-lymphocyte antibody Leu-7 (HNK-1) recognizes a constituent of neuroendocrine granule matrix, *J. Histochem. Cytochem.,* 34, 1213, 1986.
104. **Fujimoto, T. and Ogawa, K.,** Retrieving vesicles in secretion-induced rat chromaffin cells contain fodrin, *J. Histochem. Cytochem.,* 37, 1589, 1989.
105. **Westermann, R., Johannsen, M., Unsicker, K., and Grothe, C.,** Basic fibroblast growth factor (BFGF) immunoreactivity is present in chromaffin granules, *J. Neurochem.,* 55, 285, 1990.
106. **Presta, M. and Rifkin, D. B.,** Immunoreactive basic fibroblast growth factor-like proteins in chromaffin granules, *J. Neurochem.,* 56, 1087, 1991.

107. **Westermann, R., Unsicker, K., and Grothe, C.,** Immunoreactive basic fibroblast growth factor-like proteins in chromaffin granules — Reply, *J. Neurochem.*, 56, 1088, 1991.
108. **Wilkins, J. A. and Lin, S.,** Association of actin with chromaffin granule membranes and the effect of cytochalasin B on the polarity of actin filament elongation, *Biochim. Biophys. Acta,* 642, 55, 1981.
109. **Langley, O. K., Perrin, D., and Aunis, D.,** α-Fodrin in the adrenal gland: localization by immunoelectron microscopy, *J. Histochem. Cytochem.*, 34, 517, 1986.
110. **Fujimoto, T., Yoneda, K., Nihira, M., Mizuno, N., and Ogawa, K.,** Cytoplasmic occurrence of fodrin in the principal cell of the renal collecting tubule, the epidermal keratinocyte and the adrenal chromaffin cell, *Acta Histochem. Cytochem.*, 22, 119, 1989.
111. **De Robertis, E. and Vaz Ferreira, A.,** Electron microscope study of the excretion of catechol-containing droplets in the adrenal medulla, *Exp. Cell Res.*, 12, 568, 1957.
112. **Nagasawa, J. and Douglas, W. W.,** Thorium dioxide uptake into adrenal medullary cells and the problem of recapture of granule membrane following exocytosis, *Brain Res.*, 37, 141, 1972.
113. **Abrahams, S. and Holtzman, E.,** Secretion and endocytosis in insulin-stimulated rat adrenal medulla cells, *J. Cell Biol.*, 56, 540, 1973.
114. **Benedeczky, I. and Somogyi, P.,** Cytochemical localization of exogenous peroxidase in adrenal medullary cells of hamster, *Acta Biol. Acad. Sci. Hung.*, 29, 155, 1978.
115. **Patzak, A. and Winkler, H.,** Exocytotic exposure and recycling of membrane antigens of chromaffin granules: ultrastructural evaluation after immunolabeling, *J. Cell Biol.*, 102, 510, 1986.
116. **Patzak, A., Aunis, D., and Langley, O. K.,** Membrane recycling after exocytosis: an ultrastructural study of cultured chromaffin cells, *Exp. Cell Res.*, 171, 346, 1987.
117. **Brooks, J. C. and Carmichael, S. W.,** Ultrastructural demonstration of exocytosis in intact and saponin-permeabilized cultured chromaffin cells, *Am. J. Anat.*, 178, 85, 1987.
118. **Carmichael, S. W., Brooks, J. C., Malhotra, R. K., Wakade, T. D., and Wakade, A. R.,** Ultrastructural demonstration of exocytosis in the intact rat adrenal medulla, *J. Electron Microsc. Tech.*, 12, 316, 1989.
119. **Rosenheck, K. and Plattner, H.,** Ultrastructural and cytochemical characterization of adrenal medullary plasma membrane vesicles and their interaction with chromaffin granules, *Biochim. Biophys. Acta,* 856, 373, 1986.
120. **Himmelhoch, S. and Rossi, G.,** Ultracryotomy for the study of unfixed membranes, *Histochemistry,* 84, 191, 1986.
121. **Fujita, H., Ishimura, K., Matsuda, H., and Ban, T.,** Freeze-fracture images of distribution of filipin-cholesterol complexes in secretory cells of the anterior pituitary and adrenal medulla of mice, *Acta Histochem. Cytochem.*, 15, 284, 1982.
122. **Burgoyne, R. D., Barron, J., and Geisow, M. J.,** Cytochemical localisation of calcium binding sites in adrenal chromaffin cells and their relation to secretion, *Cell Tissue Res.*, 229, 207, 1983.
123. **Ravazzola, M.,** Intracellular localization of calcium in the chromaffin cells of the rat adrenal medulla, *Endocrinology,* 98, 950, 1976.
124. **Prentice, F. D. and Wood, J. G.,** Cytochemical localization of 5-hydroxydopamine in adrenergic elements of cat adrenal medulla, *Experientia,* 30, 645, 1974.
125. **Prentice, F. D. and Wood, J. G.,** Adrenergic innervation of cat adrenal medulla, *Anat. Rec.*, 181, 689, 1975.
126. **Delarue, C., Leboulenger, F., Morra, M., Héry, F., Verhofstad, A. J., Bérod, A., Denoroy, L., Pelletier, G., and Vaudry, H.,** Immunohistochemical and biochemical evidence for the presence of serotonin in amphibian adrenal chromaffin cells, *Brain Res.*, 439, 17, 1988.
127. **Cantin, M. and Benchimol, S.,** Localization and characterization of carbohydrates in adrenal medullary cells, *J. Cell Biol.*, 65, 463, 1975.
128. **Richards, J. G. and Da Prada, M.,** Uranaffin reaction: a new cytochemical technique for the localization of adenine nucleotides in organelles storing biogenic amines, *J. Histochem. Cytochem.*, 25, 1322, 1977.
128a. **Richards, J. G.,** personal communication, 1978.
129. **Kondo, H., Hirofumi, K., and Wanaga, T.,** Immunohistochemical study on Met-enkephalin-Arg-Gly-Leu-like immunoreactive nerve fibers in the rat adrenal medulla, *Brain Res.*, 310, 371, 1984.

Chapter 10

ENDOCRINE ORGANS

10.3 Cytochemistry and Immunocytochemistry of the Thyroid Gland

Hisao Fujita

TABLE OF CONTENTS

0-8493-6012-9/93/$0.00 + $.50

A. INTRODUCTION

Thyroid hormones are classified into two categories: (1) thyroxine (T_4) and triiodothyronine (T_3), and (2) calcitonin; T_4 and T_3 are not peptides but are characteristic diphenylethers with ether linkage (–O–) obtained by removing alanine from two molecules of iodinated tyrosine.[1,2] In all the peptide-secreting cells, the secretory granule, which originates from the rough endoplasmic reticulum and Golgi apparatus system, as in other protein secretory cells, is released into the pericapillary connective tissue space. However, formation of T_4 and T_3 in the thyroid requires more complicated steps. The hormones are not synthesized directly from their constituents, tyrosine and iodine, in the cell, and their synthetic process occurs in both the follicle epithelial cell and extracellular lumen, called follicle lumen.[1,2] The secretory process of T_4 and T_3 is as follows: (1) synthesis of a large glycoprotein (mol wt 660,000), called thyroglobulin, in the follicle epithelial cell; (2) release of thyroglobulin into the follicle lumen; (3) iodination of the tyrosyl residues in thyroglobulin, and coupling of iodotyrosyl residues to make T_4 or T_3 in the molecule of thyroglobulin; (4) reabsorption of thyroglobulin from the follicle lumen into the follicle epithelial cell; (5) hydrolysis of thyroglobulin to liberate T_4 and T_3; and (6) release of T_4 and T_3 from the follicle epithelial cell into the connective tissue space.[1,2] (Refer to Figure 1.)

Calcitonin is a peptide secreted from the parafollicular cell.

This chapter deals with an outline of cytochemical aspects of the secretion mechanisms of T_4 and T_3 from the follicle epithelial cell and of calcitonin from the parafollicular cell.

B. AUTORADIOGRAPHY OF ^{3}H-AMINO ACID AND ^{3}H-CARBOHYDRATE

The follicle epithelial cell synthesizes a high-molecular-weight glycoprotein called thyroglobulin. As in other protein-secretory cells, the proteinaceous part of thyroglobulin is synthesized in the rough endoplasmic reticulum and Golgi apparatus. The secretory granules of thyroglobulin, about 200 nm in diameter, arise in the Golgi apparatus and are released from the apical part of the cell into the follicle lumen. By electron microscopic autoradiography of ^{3}H-amino acid, the intracellular route for the synthesis and release of thyroglobulin has been clarified.[1,3] In mouse thyroid, the labeled amino acid first appears in the cisternae of rough endoplasmic reticulum within 15 min after the intraperitoneal injection of ^{3}H-leucine. Then the silver grains begin to appear in the Golgi apparatus and subapical vesicles (secretory granules) after 30 to 45 min and in the follicle lumen after 45 min.[1]

The proteinaceous part of thyroglobulin has to be glycosylated to become a glycoprotein. Using electron microscopic autoradiography of ^{3}H-mannose, ^{3}H-galactose, and ^{3}H-fucose, the site of glycosylation of thyroglobulin has become clear morphologically.[4,5] It is now believed that mannose is bound to protein in the rough endoplasmic reticulum, while galactose, fucose, and sialic acid are bound to the protein in the Golgi apparatus.[4,5]

C. PAS METHOD AND SILVER METHENAMINE METHOD FOR CARBOHYDRATES

It is well known that the colloid in the follicle lumen and colloid droplets in the apical cytoplasm of the follicle epithelial cell are strongly positive with the PAS (periodic acid Schiff) method which stains carbohydrates. Since the luminal colloid is thyroglobulin consisting of glycoprotein, this positive reaction is easily understood. (Refer to Figure 2.)

The silver methenamine method is useful in detecting the ultrastructural localization of carbohydrates and glycoproteins.[6,7] When this method is applied to the thyroid gland, the cisternae of the rough endoplasmic reticulum show a weak but apparently positive reaction. The cisternae of 1 to 3 inner saccules (lamellae of the *trans* side) of the Golgi apparatus are intensely stained, while those of 1 to 3 outer saccules (lamellae of the *cis* side) are not so heavily stained.[7] The Golgi apparatus has an internal polarity and its *trans*-saccules have a

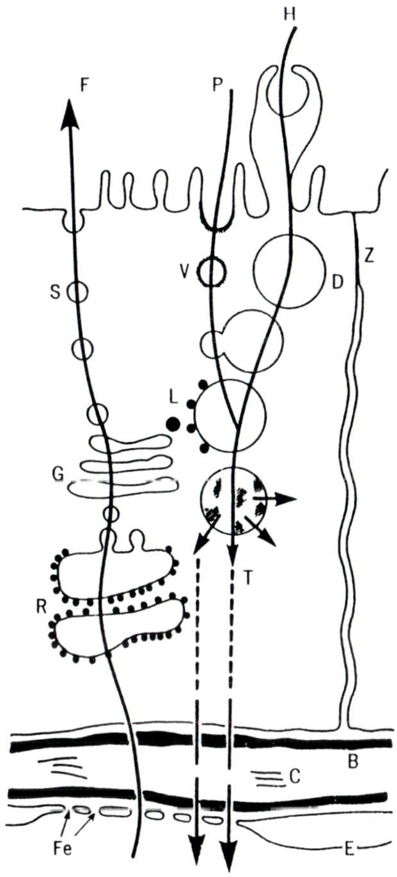

FIGURE 1. Schematic representation of the secretory process of thyroxine and triiodothyronine in the thyroid follicle epithelial cell. The protein is synthesized in the rough endoplasmic reticulum (R) and transported to the Golgi apparatus (G), where the secretory granules (S) containing thyroglobulin arise and are released into the follicle lumen (F). Thyroglobulin stored in the follicle lumen is reabsorbed into the cell by pinocytosis (P) as well as phagocytosis (H). Reabsorbed thyroglobulin is contained in colloid droplets (D) and vesicles (V). They are fused with each other and with primary lysosomes (L), and hydrolysis of thyroglobulin takes place. The thyroid hormones (T) are released from the basal part of the cell into the connective tissue space (C) passing through basal lamina (B), and then enter into the blood capillary passing through the fenestrations (Fe) of the endothelium (E). Z, zonula occludens.[7]

stronger ability for binding carbohydrates to thyroglobulin. Colloid droplets, lysosomes, and secretory granules are strongly positive to this reaction. Luminal colloid is also positive with the same intensity as the intracellular colloid droplets.[7] These findings support the autoradiographic and biochemical data that mannosylation of thyroglobulin occurs initially in the rough endoplasmic reticulum, and galactosylation, fucosylation, and sialylation take place in the *trans* side of the Golgi apparatus.

D. LOCALIZATION OF THIAMINE PYROPHOSPHATASE ACTIVITY

Thiamine pyrophosphatase (TPPase) activity is localized in the cisternae of 1 to 2 stacks of the *trans* side of the Golgi apparatus.[8] This fact is common to the Golgi apparatus of all kinds of cells. This enzyme has been implicated in galactosylation of the protein. Generally, it is well known that the galactosylation takes place in the *trans* side of the Golgi saccules, and the localization of TPPase well coincides with the site of the galactosylation.

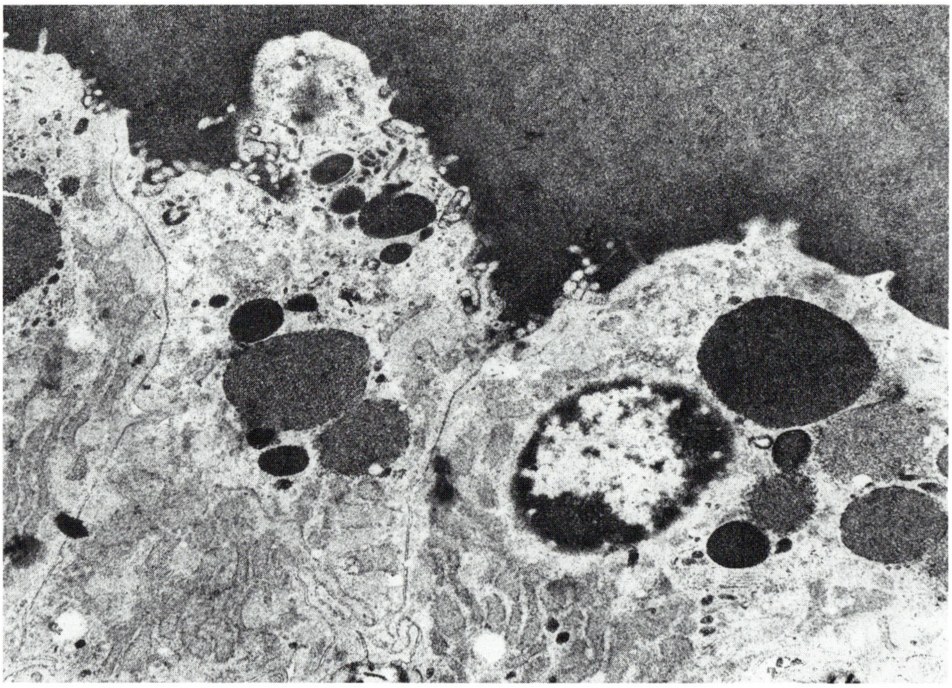

FIGURE 2. Follicle epithelial cell of a mouse stained by the periodic acid-chromic acid-silver metheriamine method. Note a strongly positive reaction in the luminal colloid, reabsorbed colloid droplets, apical small vesicles (may be secretory granules or reabsorbed vesicles containing thyroglobulin), and a weakly positive reaction in the cisternae of the rough endoplasmic reticulum. (Magnification × 12,300.)[7]

E. IMMUNOCYTOCHEMISTRY OF THYROGLOBULIN

Using an antibody of thyroglobulin, Ring and Johanson[9] demonstrated that thyroglobulin is located in the rough endoplasmic reticulum, Golgi apparatus, exocytotic vesicles, luminal colloid, reabsorbed colloid droplets, and lysosomes. They recommended the use of cold alcohol (-20°C) for the whole dehydration procedure for getting a good result.

F. AUTORADIOGRAPHY OF ^{125}I

Electron microscopic autoradiography of ^{125}I is a powerful tool for detecting the site of iodination of tyrosyl residues in the molecule of thyroglobulin.[1,2,9-13] Since inorganic iodide is washed away by autoradiographic procedures such as washing, fixation, and dehydration of the tissue, and only iodide which is bound to protein is detected, the usual autoradiographic method of ^{125}I is suitable.

In various vertebrates, it was shown that silver grains begin to appear almost entirely in the follicle lumen, especially in the periluminal region, within a few minutes after the intravenous or intraperitoneal injection of ^{125}I and are increased in number with time. In the mouse thyroid cultured in a medium containing ^{125}I, silver grains appear only in the follicle lumen within 30 seconds, and increase rapidly in number with time. (Refer to Figure 3.)

From these facts, it is considered that the iodination of thyroglobulin takes place almost entirely in the follicle lumen, especially at the periluminal region.[1,2,10] The most primitive thyroid gland of the hagfish also shows a similar pattern for the site of iodination of tyrosine in thyroglobulin-equivalent protein.[15] In the thyroid-homologous organ, called endostyle, in

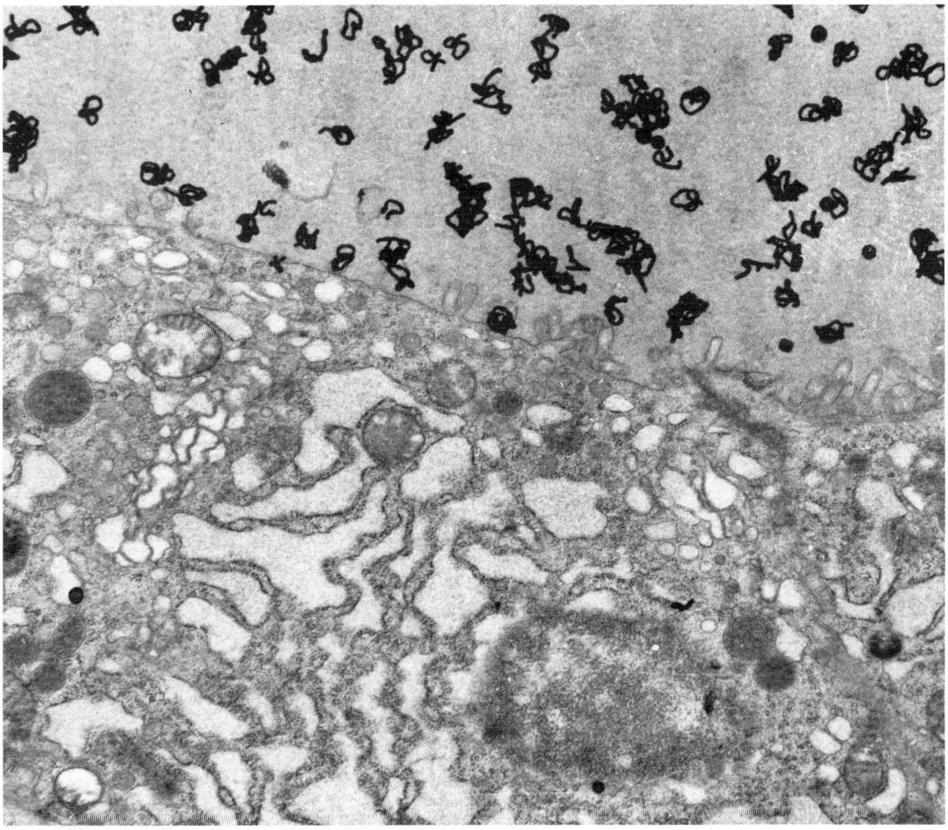

FIGURE 3. Autoradiograph of ^{125}I in the thyroid follicle of the mouse, 2 h after the intraperitoneal injection of radioactive iodine. Silver grains are localized only in the follicle lumen. (Magnification × 15,000.)[11]

the larval lamprey (ammocoetes of *Lampetra japonica*) and in the tunicate *(Ciona intestinalis),* the main site of the iodination of tyrosyl residues in thyroglobulin-equivalent protein is an apical plasma membrane region.[1,16]

For detecting the localization of inorganic iodide in the thyroid, autoradiography of freeze-dried sections is needed.[17] Light microscopic autoradiography of freeze-dried sections shows that silver grains for inorganic iodide are accumulated chiefly over the follicle lumen after the injection of ^{131}I into mercaptoimidazole-treated rats, while the silver grains disappear in the fixed, dehydrated, and paraffin-embedded sections obtained from the same animal.[17] Since mercaptoimidazole is well known to inhibit the iodination of tyrosyl residues in thyroglobulin, it is reasonable to conclude that radioactive iodide is not detected by the usual autoradiography in this experimental animal. Thus, it is concluded that inorganic iodide taken up into the thyroid is mostly transported to the follicle lumen passing through the epithelial cell and is accumulated in the lumen.

G. LOCALIZATION OF PEROXIDASE AND H_2O_2

The enzyme for iodination of tyrosyl residues of thyroglobulin is a peroxidase. Iodide ion taken up into the thyroid is oxidized to become I_2, the active form of iodine. In this reaction, H_2O_2 acts as the electron acceptor. Reactions are summarized as follows:

$$\text{NADH (or NADPH)} + H^+ + O_2 \xrightarrow{\text{peroxidase}} \text{NAD}^+ \text{ (or NADP}^+\text{)} + H_2O_2 \quad (1)$$

$$H_2O_2 + I^- \xrightarrow{\text{peroxidase}} \text{oxidized iodide} + H_2O_2 \quad (2)$$

$$\text{Oxidized iodide} + \text{tyrosine} \xrightarrow{\text{peroxidase}} \text{iodotyrosine} \quad (3)$$

The localization of endogenous peroxidase has been demonstrated histochemically using the 3,3′-diaminobenzidine tetrahydrochloride (DAB) method.[1,2,18-21] Reaction products for peroxidase are localized in the perinuclear cisternae, cisternae of rough endoplasmic reticulum, Golgi saccules, subapical vesicles, and periluminal colloid of the rat thyroid. However, Tice and Wollman,[21] who tried peroxidation of thyroid tissue using hydrogen peroxide, glucose-glucose oxidase, or potassium ferricyanide before immersion in DAB solution, emphasized that the peroxidase activity is localized on the apical plasma membrane instead of periluminal colloid. They considered that the luminal reaction is a diffusion artifact of the histochemical procedure. As mentioned above, hydrogen peroxide is also necessary for iodination of thyroglobulin. An iodide ion taken up into the follicle epithelial cell becomes oxidized to I_2, the active form of iodine. In this reaction, H_2O_2 acts as the electron acceptor. The active form of iodine binds with the tyrosyl residue in thyroglobulin. Bjorkman et al.,[22] using NADPH oxidase, showed histochemically that H_2O_2 is generated on the apical surface of the follicular epithelial cell in the rat thyroid. In comparing data from ^{125}I autoradiographic and histochemical tests for peroxidase and H_2O_2, it is considered that iodination of tyrosyl residues in thyroglobulin takes place mainly at the peripheral region of the follicular lumen, though it is obscure whether the reaction occurs strictly on the plasma membrane facing the follicle lumen.

Although about 120 tyrosyl residues are contained in a thyroglobulin molecule, only a few of them are iodinated and coupled and only a few moieties of tetraiodothyronine (T_4) and triiodothyronine are made in the thyroglobulin molecule.

H. FREEZE-FRACTURE IMAGES OF APICAL PLASMA MEMBRANE

Thyroglobulin stored in the follicle lumen is reabsorbed into the follicle epithelial cell by phagocytosis as well as by micropinocytosis. After injection of thyroid-stimulating hormone (TSH), it is recognized by electron microscopy that cytoplasmic pseudopods, suggesting an engulfment of the luminal colloid, appear on the apical surface of the cell, and the reabsorbed colloid droplets are increased in number and size in the cytoplasm. In addition, it has been shown by electron microscopy that the luminal colloid is also reabsorbed into the cell by micropinocytosis. (Refer to Figure 4.)

In the freeze-fracture image of the apical plasma membrane, intramembrane particles (IMP) are distributed randomly on the P-face. The particles on the P-face of the apical plasma membrane often form aggregates, especially in TSH-treated mice.[23] Each aggregate, about 200 nm in diameter and consisting of 15 to 25 particles, corresponds to a depression of the apical region, suggesting an initial site for micropinocytosis of luminal colloid.[23]

In sections obtained from blocks stained with tannic acid, it is clear that most of the micropinocytotic pits are coated. Some coated vesicles containing reabsorbed thyroglobulin fuse with one another or with a reabsorbed colloid droplets. Accumulation of the intramembrane particles seen in the freeze-fracture image has generally been known to be implicated with the receptor-mediated endocytosis. However, in the thyroid epithelial cell, the functional meaning of the accumulation of the intramembranous particles at the initiation site of micropinocytosis is obscure.

FIGURE 4. Freeze-replica image of PF of apical plasma membrane of follicle epithelial cell. Note micropinocytosis of the luminal colloid in a TSH-treated mouse. Accumulation of membranous particles are seen at initial sites of micropinocytosis (shallow depressions). (Magnification × 42,800.)[23]

Freeze-fracture images of filipin (antibiotic)-treated tissues have been used to detect the distribution of cholesterol in the membrane (Figure 5). Filipin is bound to membrane cholesterol to form filipin-sterol complexes that show 25- to 30-nm protuberances on the fractured plasma membrane.[24-26] The procedure is as follows. Animals are perfused with 2% glutaraldehyde solution buffered at pH 7.4 with Milloning's phosphate, and the thyroid glands removed are cut into small pieces and immersed in the same fixative for 1 hr. Then the specimens are treated with 0.02% filipin (Upjohn Co., Kalamazoo, MI) in 0.2 *M* cacodylate, pH 7.4, for 12 to 24 h at room temperature. After soaking in 30 to 40% glycerol solution for 10 to 12 h, the tissues are frozen rapidly in liquid nitrogen, cleaved, shadowed with platinum-palladium, and stabilized with carbon. The replicas made are examined in an electron microscope. By this method, it has become clear that the membrane of the pinocytotic pit where the membrane particles aggregate, as mentioned above, lacks the filipin-sterol complexes, although most other parts of the plasma membrane are very rich in cholesterol.[26]

I. IMMUNOCYTOCHEMISTRY FOR ACTIN, CALDESMON, AND CALSPECTIN, AND HEAVY MEROMYOSIN DECORATION METHOD FOR ACTIN

By immunohistochemistry, actin was demonstrated in the apical cytoplasm of the follicle epithelial cell. The microvilli as well as pseudopods contain numerous actin filaments.

It is well known that cytoplasmic actin filaments can be decorated by treatment with heavy meromyosin (HMM) to show arrowhead structures with a constant polarity.[28] Using this method, numerous actin filaments are demonstrated at the apical cytoplasm of the follicle

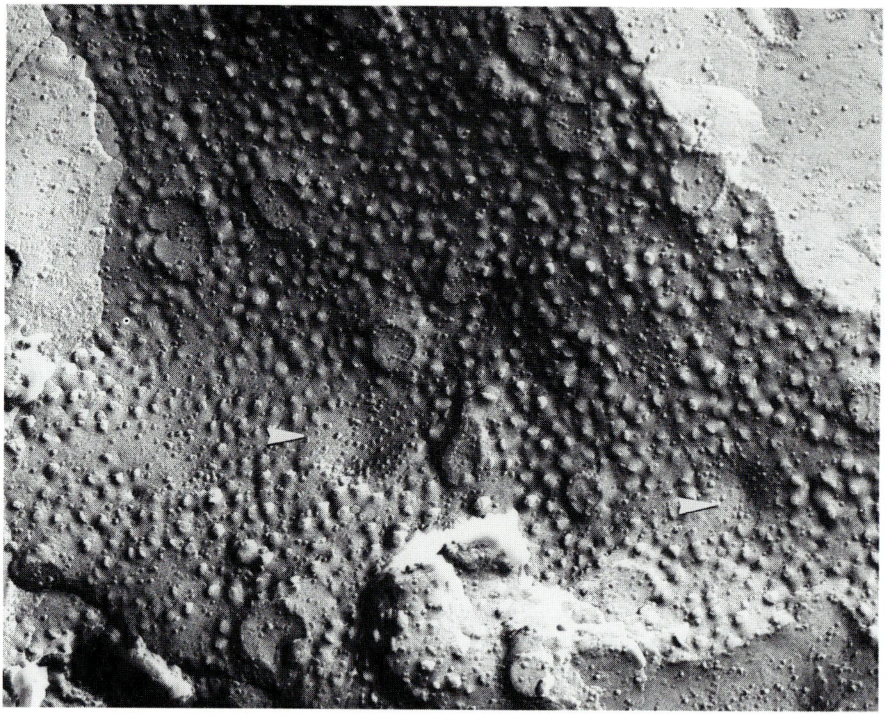

FIGURE 5. Freeze-fracture image of PF of apical plasma membrane of filipin-treated follicle epithelial cell of a mouse. Note numerous filipin-sterol complexes (large protuberances) distributed on fracture membrane, and absence of complexes at shallow depressions or pits (arrow) where aggregates of intramembranous particles (small particles) occur. (Magnification × 61,000.)[26]

epithelial cell. The actin filaments attached to the apical plasma membrane ran perpendicularly toward the network of actin filaments in the subapical region. Coated pits and coated vesicles are often connected with these filaments running toward the deeper regions of the cytoplasm.[29] These facts suggest that the actin filament system is involved in the reabsorption of the colloid by micropinocytosis as well as phagocytosis, though the detailed relationship is obscure.

There are some studies using inhibitors of polymerization of actin filaments and microtubules. The release of thyroid hormone and the phagocytosis of colloid are inhibited by cytochalasin B, an inhibitor of the polymerization of actin.[30] Microtubules may also be implicated in colloid-phagocytotic activity in this cell.[31] The phagocytosis-inducing effect of TSH in the luminal colloid is inhibited by treatment with colchicine, vinblastine, vincristine, ethanol, and deuterium oxide, which are microtubule inhibitors.[30-32] Although these data suggest a role of actin filaments and microtubules in phagocytosis, further studies are needed to clarify the detailed mechanism.

Caldesmon, discovered in the smooth muscle by Sobue et al.,[33] is located also in the apical cytoplasm of the follicle epithelial cells. Immunoreactivities for both actin and caldesmon showed almost the same pattern in localization. Caldesmon is a calmodulin-binding protein composed of 150- and 147-kDa subunit polypeptides that are able to bind to actin filaments (F-actin). The binding of caldesmon to F-actin is inhibited by Ca^{2+}-dependent binding of calmodulin to caldesmon, and the binding of caldesmon to calmodulin or F-actin can alternate, depending upon the concentration of Ca^{2+} as assumed by the flip-flop binding hypothesis.[34] Both myosin and troponin C-like phosphodiesterase activators (calmodulin) in the thyroid gland have been purified. From these results, it is speculated that caldesmon, actin, and calmodulin play a role in the regulation of apical cell motility such as micropinocytosis, phagocytosis, and exocytosis of the follicular epithelial cell.

Calspectin (fodrin, a nonerythroid, spectrin-like protein) was also demonstrated immunocytochemically in the follicle epithelial cell, especially along the basolateral plasma membrane, in normal rats.[35] As spectrin underlying the plasma membrane of the erythrocyte is known to participate in the maintenance of the cell's unique shape, calspectin in the follicle epithelial cell might function to maintain the structure of the follicle. In TSH-treated animals, the apical plasma membrane region of the follicle epithelial cell becomes positive for calspectin.[35]

J. LOCALIZATION OF ACID PHOSPHATASE ACTIVITY

Thyroid hormones T_4 and T_3 are liberated by hydrolysis of reabsorbed thyroglobulin. Acid phosphatase (AcPase) is known to be an indicator for lysosomes containing hydrolytic enzymes. In the thyroid epithelial cell, hydrolytic enzymes are necessary for hydrolysis of thyroglobulin. For clarifying the relationship between the reabsorbed colloid droplet (thyroglobulin) and lysosomes, localization of AcPase activity has been detected histochemically.[36,37] In the follicle epithelial cell, this enzyme is localized in primary lysosomes, some colloid droplets (secondary lysosomes), small vesicles, multivesicular bodies, rigid lamellae, and the cisternae of 1 to 2 saccules of the *trans* side of the Golgi apparatus (Figure 6). Small lysosomes, homogeneously dense and 100 to 300 nm in diameter, are primary lysosomes. These lysosomes are fused with the reabsorbed colloid (thyroglobulin) droplets, which become positive for AcPase activity. In the colloid droplet showing AcPase reaction, hydrolysis of thyroglobulin takes place, liberating T_4 and T_3. These hormones, which are amino acid derivatives of low molecular weight, do not show any secretory granule form. So the hormones produced by hydrolysis of thyroglobulin are considered to be liberated into the cytoplasmic matrix from the secondary lysosome (AcPase-positive colloid droplet); T_4 and T_3 are then released from the basal part of the cell into the pericapillary connective tissue space by passing through the plasma membrane, which is named diacrine secretion.

In old animals, heterogeneously dense lysosomes containing lipid-like droplets appear in the follicle epithelial cells and increase in number with age.[38] Though these lysosomes are positive for AcPase activity, the specific activities of AcPase, β-glucuronidase, and β-galactosidase show a relative decrease biochemically.[38] Some of the lysosomes observed in such older animals may be residual bodies. The localization of T_4 and T_3 in the rat thyroid gland was immunocytochemically detected using antibodies against T_4 and T_3 by Ring and Johanson.[9] The reaction products, which are localized in the luminal colloid, reabsorbed colloid droplets, and secondary lysosomes, correspond to T_4 and T_3 moieties in the thyroglobulin molecule. T_4 and T_3 liberated from the secondary lysosome may be difficult to detect by this method.

K. FREEZE-FRACTURE IMAGES OF CAPILLARY ENDOTHELIUM

The number of fenestrations and their population density in the capillary endothelium are easiy observed and calculated in freeze-fracture images.[2,39] Fenestrations of the thyroid blood capillaries are increased in number and population density in long-term TSH-treated mice, and decreased in long-term Thyradin (thyroid powder)-treated animals which are in a lower active state.[2,39] It is believed that endothelial fenestrations vary in number and population density, according to the functional activity of the gland.[2]

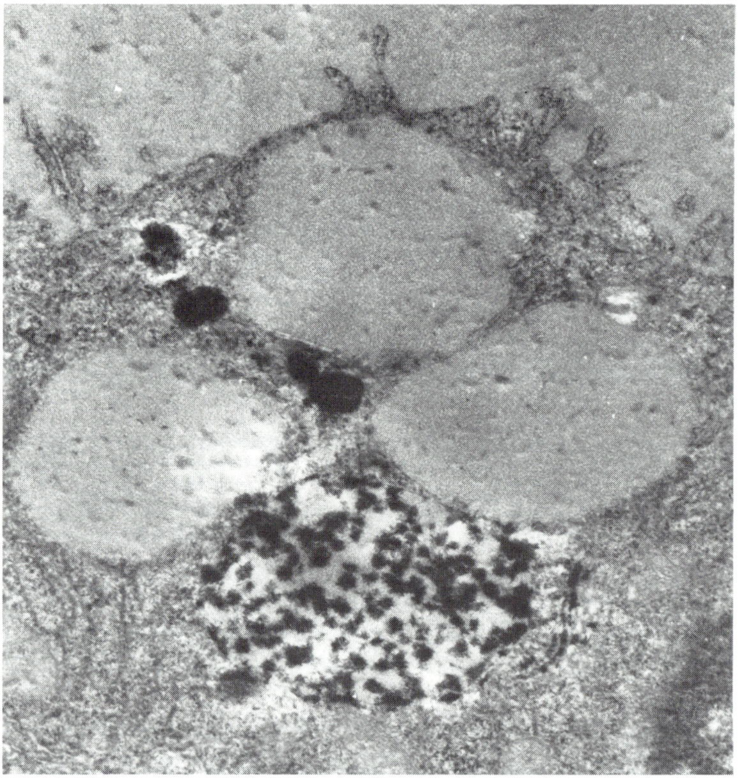

FIGURE 6. Acid phosphatase (AcPase) activity in the thyroid follicle epithelial cell of a TSH-treated mouse. Note three colloid droplets negative for acid phosphatase, primary lysosomes (small in size and localized among the colloid droplets), and a colloid droplet (secondary lysosome) positive for acid phosphatase). (Magnification × 30,000.)

L. IMMUNOCYTOCHEMISTRY FOR CALCITONIN

Calcitonin, a hormone which reduces calcium level in the blood stream, is produced in the parafollicular cell (or C-cell). The secretory granules of this cell are immunocytochemically positive for this hormone.[40,41] This cell also produces serotonin (5-hydroxytryptamine) from 5-hydroxytryptophane and dopamine from L-dopa. These amines are contained in the secretory granule. When freeze-dried materials of the thyroid gland are fixed by vapor of formaldehyde, fluorescent compounds detectable under a fluorescence microscope are produced by these amines.[42] The secretory granules are argyrophilic and stain intensely by Davenport's silver impregnation method.[43,44]

REFERENCES

1. **Fujita, H.,** Fine structure of the thyroid gland, *Int. Rev. Cytol.,* 40, 197, 1975.
2. **Fujita, H.,** Functional morphology of the thyroid, *Int. Rev. Cytol.,* 113, 145, 1988.
3. **Nadler, N. J., Young, B. A., Leblond, C. P., and Mitmaker, B.,** Elaboration of thyroglobulin in the thyroid follicle, *Endocrinology,* 74, 333, 1964.

4. **Whur, P., Herscovics, A., and Leblond, C. P.,** Radioautographic visualization of the incorporation of galactose-^{3}H and mannose-^{3}H by rat thyroids in vitro in relation to the stages of thyroglobulin synthesis, *J. Cell Biol.*, 43, 289, 1969.
5. **Haddad, A., Smith, M. D., Herscovics, A., Nadler, N. J., and Leblond, C. P.,** Radioautographic study of in vivo and in vitro incorporation of fucose-^{3}H into thyroglobulin by rat thyroid follicular cells, *J. Cell Biol.*, 49, 856, 1971.
6. **Rambourg, A., Hernadez, W., and Leblond, C. P.,** Detection of complex carbohydrates in the Golgi apparatus of rat cells, *J. Cell Biol.*, 40, 395, 1969.
7. **Tamura, S. and Fujita, H.,** Cytochemical localization of complex carbohydrates in the thyroid gland of normal and TSH-treated mice, *Histochemistry,* 58, 57, 1978.
8. **Sawano, F. and Fujita, H.,** Cytochemical studies on the internal polarity of the Golgi apparatus and the relationship between this organelle and GERL, *Histochemistry,* 71, 335, 1981.
9. **Ring, P. K. M. and Johanson, V.,** Immunoelectron microscopic demonstration of thyroglobulin and thyroid hormones in rat thyroid gland, *J. Histochem. Cytochem.*, 35, 1095, 1987.
10. **Stein, O. and Gross, J.,** Metabolism of ^{125}I in the thyroid gland studies with electron microscopic autoradiography, *Endocrinology,* 75, 787, 1964.
11. **Fujita, H.,** Studies on the iodine metabolism of the thyroid gland as revealed by electron microscopic autoradiography of ^{125}I, *Virchows Arch. B,* 2, 265, 1969.
12. **Fujita, H.,** Morphological aspects on the site of iodination of thyroglobulin in the thyroid gland, *Arch. Histol. Jpn.*, 34, 109, 1972.
13. **Ekholm, R.,** Iodination of thyroglobulin, an intracellular or extracellular process?, *Mol. Cell. Endocrinol.*, 24, 141, 1981.
14. **Ericson, L. E.,** Ultrastructural aspects on secretion and iodination in the thyroid gland, *J. Endocrinol. Invest.*, 6, 311, 1983.
15. **Fujita, H. and Shinkawa, Y.,** Electron microscopic studies on the thyroid gland of the hagfish *Eptatretus burgeri* (a part of phylogenetic studies of the thyroid gland), *Arch. Histol. Jpn.*, 37, 277, 1975.
16. **Fujita, H. and Honma, Y.,** Iodine metabolism of the endostyle of larval lampreys, ammocoetes of *Lampetra japonica.* Electron microscopic autoradiography of ^{125}I, *Z. Zellforsh.*, 98, 525, 1969.
17. **Fujita, H.,** Application of electron microscopic autoradiography of radioactive iodine for a comparative study of the thyroid gland, in *Recent Progress in Electron Microscopy of Cells and Tissues,* Yamada, E., Mizuhira, V., Kurosumi, K., and Nagano, T., Eds., Igaku-Shoin, Tokyo, 1976, 175.
18. **Nakai, Y. and Fujita, H.,** Fine structural localization of peroxidase in the rat thyroid, *Z. Zellforsch.*, 107, 104, 1970.
19. **Strum, J. M. and Karnovsky, M. J.,** Cytochemical localization of endogeneous peroxidase in thyroid follicular cells, *J. Cell Biol.*, 44, 655, 1970.
20. **Sawano, F. and Fujita, H.,** Some findings on the cytochemistry of the thyroid follicle epithelial cell in rats and mice, *Arch. Histol. Jpn.*, 44, 439, 1981.
21. **Tice, L. W. and Wollman, S. H.,** Ultrastructural localization of peroxidase activity on some membranes of the typical thyroid epithelial cell, *Lab. Invest.*, 26, 63, 1972.
22. **Bjorkman, U., Ekholm, R., and Denef, J. F.,** Cytochemical localization of hydrogen peroxide in isolated thyroid follicles, *J. Ultrastruct. Res.* 74, 105, 1981.
23. **Ishimura, K., Okamoto, H., and Fujita, H.,** Freeze-etching observations on the characteristic arrangement of intramembranous particles in the apical plasma membrane of the thyroid follicular cell in THS treated mice, *Cell Tissue Res.*, 171, 297, 1976.
24. **Tillack, T. W. and Kinsky, S.C.,** A freeze-etch study of the effects of filipin on liposomes and human erythrocyte membranes, *Biochim. Biophys. Acta,* 323, 43, 1973.
25. **Nakajima, Y. and Bridgman, P. C.,** Absence of filipin-sterol complexes from the membranes of active zones and acetylcholine receptor aggregates at frog nonmuscular junctions, *J. Cell Biol.*, 88, 453, 1981.
26. **Fujita, H., Ishimura, K., and Matsuda, H.,** Freeze-fracture images on filipin-sterol complexes in the thyroid follicle epithelial cell of mice with special regard to absence of cholesterol at the site of micropinocytosis, *Histochemistry,* 73, 57, 1981.
27. **Fujita, H., Ishimura, K., Ban, T., Kurosumi, H., Sobue, K., and Kakiuchi, S.,** Immunocytochemical demonstration of caldesmon and actin in thyroid glands of rats, *Cell Tissue Res.*, 237, 375, 1984.
28. **Ishikawa, H., Bischoff, R., and Holtzer, H.,** Formation of arrowhead complexes with heavy meromyosin in a variety of cell types, *J. Cell Biol.*, 43, 312, 1969.
29. **Miyagawa, J., Ishimura, K., and Fujita, H.,** Fine structural studies on the reabsorption of colloid and fusion of colloid droplets in thyroid glands of TSH-treated mice, *Cell Tissue Res.*, 223, 519, 1982.
30. **Neve, P., Ketelbant-Balasse, P., Willems, C., and Dumont, J. E.,** Effect of inhibitors of microtubules and microfilaments on dog thyroid slices in vitro, *Exp. Cell Res.*, 74, 227, 1972.

31. **William, J. A. and Wolff, J.,** Colchicine-binding protein and the secretion of thyroid hormone, *J. Cell Biol.,* 54, 157, 1972.
32. **Ekholm, R., Ericson, L. E., Joseffson, J.-O., and Melander, A.,** In vivo action of vinblastine on thyroid ultrastructure and hormone secretion, *Endocrinology,* 94, 641, 1974.
33. **Sobue, K., Morimoto, K., Kanda, K., Maruyama, K., and Kakiuchi, S.,** Reconstitution of Ca^{2+}-sensitive gelation of actin filaments with filamen, caldesmon and calmodulin, *FEBS Lett.,* 138, 289, 1982.
34. **Kakiuchi, S. and Sobue, K.,** Control of the cytoskeleton by calmodulin and calmodulin-binding proteins, *Trends Biochem. Sci.,* 8, 59, 1983.
35. **Ishimura, K., Senda, T., Kitajima, K., Fujita, H., Fujio, Y., and Sobue, K.,** Immunocytochemical localization of calspectin (a non-erythroid spectrin-like protein) in thyroid glands of normal and TSH-treated rats, *Histochemistry,* 6, 537, 1987.
36. **Wetzel, B. K., Spicer, S. S., and Wollman, S. H.,** Changes in fine structure and acid phosphatase localization in rat thyroid cells following thyrotropin administration, *J. Cell Biol.,* 25, 593, 1965.
37. **Seljelid, R.,** Electron microscopic localization of acid phosphatase in rat thyroid follicle cells after stimulation with thyrotropic hormone, *J. Histochem. Cytochem.,* 13, 687, 1965.
38. **Fujita, H., Tamura, S., Takano, T., Ishibashi, S., and Tanaka, T.,** Fine structural changes in the aging mouse thyroid, *J. Gerontol.,* 35, 3, 1980.
39. **Ishimura, K., Okamoto, H., and Fujita, H.,** Freeze-etching studies on ultrastructural changes of endothelial cells in the thyroid of normal, TSH-treated and Thyradin-treated mice, *Cell Tissue Res.,* 175, 297, 1976.
40. **Bussolati, G. and Pearse, A. G. E.,** Immunofluorescent localization of calcitonin in the C cells of pig and dog thyroid , *J. Endocrinol.,* 37, 205, 1967.
41. **Kameda, Y. and Ikeda, A.,** C cell (parafollicular cell) immunoreactive thyroglobulin, purification, identification and immunological characterization, *Histochemistry,* 60, 155, 1979.
42. **Falck, B., Hillarp, N.-A., Thieme, G., and Torp, A.,** Fluorescence of catecholamines and related compounds condensed with formaldehyde, *J. Histochem. Cytochem.,* 10, 348, 1962.
43. **Davenport, H. A.,** Staining nerve fibers in mounted sections with alcoholic silver nitrate solution, *Arch. Neurol. Psychiatry,* 24, 690, 1930.
44. **Kameda, Y.,** Parafollicular cells of the thyroid as studied with Davenport's silver impregnation, *Arch. Histol. Jpn.,* 30, 83, 1968.

Chapter 11

THE DIFFUSE ENDOCRINE SYSTEM

Susan Van Noorden and Julia M. Polak

TABLE OF CONTENTS

0-8493-6012-9/93/$0.00 + $.50

distinguished from another by immunocytochemistry, and even then it must be remembered that there may exist unrecognized peptides of the same family. The ultrastructure of the pyloric gastrin cell was characterized tentatively at an early stage in these studies because it was the major cell type of the pylorus and not found elsewhere. However, it was not until an antibody to gastrin had been produced that the gastrin cells could be confidently separated from the somatostatin and other endocrine cell types of the pylorus.[17,18] Once antibodies had been raised, a second type of gastrin cell was found in the small intestine. This posed a greater problem because the intestine also contained cells producing cholecystokinin (CCK), and the antibody did not distinguish between the two molecules, since it reacted with their common C-terminal pentapeptide. An antibody was therefore raised to a synthesized, unshared portion of the CCK molecule and was then used to differentiate between the two cell types and characterize them at the ultrastructural level.[19-21]

3. Peptide Precursor Localization

Another example concerns the problem of whether the peptide localized is the biologically active, free peptide or part of an uncleaved precursor molecule. The source of glucagon in the pancreatic islets had been identified as the A cells, one of at least four endocrine cell types of the islets of Langerhans. The cells (in some species) are located towards the periphery of the islets and their granules have a homogeneously electron-dense, glucagon-immunoreactive core surrounded by a less dense, nonimmunoreactive halo.[22] The N-terminally reactive antibody also recognized cells in the large intestine which were ultrastructurally different from the A cells of the islets. It was subsequently found that the intestinal glucagon-immunoreactive cell, in fact, contains glicentin, part of the glucagon precursor molecule, incorporating the glucagon sequence (hence its immunoreactivity with antisera to pancreatic glucagon) but not reactive with glucagon receptors.

Immunolabeling at the electron microscopic level demonstrated clearly that in the intestinal L or enteroglucagon (EG) cell, glicentin immunoreactivity was homogeneously distributed over the secretory granules, whereas in the pancreatic A cells, the core of the granules was glucagon-immunoreactive while the less dense halo was glicentin-immunoreactive.[16] It appears that in the intestine the glucagon precursor is not fully cleaved, but remains at the glicentin stage. This intragranular segregation of immunoreactivity was confirmed by Varndell et al.,[23] who also showed that the C-terminal extensions of the proglucagon molecule, the two glucagon-like peptides (GLP1 and GLP2), were found in the core of the A cells along with glucagon, but over the entire granule in the ER cells, further evidence for the integrity of the prohormone molecule in the EG cells. Whether glicentin has a biofunction has still not been determined.

A somewhat different segregation of prohormone and hormone has been suggested by recent evidence that a glycine-extended, nonamidated form of gastrin (G-gly) is present in the small, dense granules of the antral gastrin (G) cells, while the amidated, biologically active form is present in both large and small types of granule in the same cells. G-gly may be an intermediate form between the prohormone and hormone.[24] This finding lends support to earlier work[25] which suggested that the C-flanking peptide of gastrin, part of the prohormone, was mainly in the small, dense granules while gastrin itself was present in both the small and the larger, less dense granules, indicating a change in granule structure on maturation of the peptide by cleaving of the prohormone molecule.

4. Multiple Production of Unrelated Peptides

The studies described above show how electron microscopic immunocytochemistry can be used to map the pathways of posttranslational processing of peptides. The examples cited have dealt with structurally related peptides or peptides produced from the same precursor molecule. However, it has become the rule rather than the exception that two or more regulatory peptides, even quite unrelated in structure, may be produced within the same cell, sometimes accompanied by other regulatory substances such as amines. The packaging of cell products in secretory granules is shown by electron microscopical immunocytochemistry to be a process separate from their individual production from mRNA. To take the intestinal EG cell again — it had been suggested that pancreatic polypeptide (PP) was a product of the EG cell since PP-like immunoreactivity was colocalized with glucagon-like immunoreactivity by light microscopic immunocytochemistry. Nevertheless, extracts of mammalian lower intestine consistently failed to show much PP immunoreactivity by radioimmunoassay. When PYY (peptide with tyrosine [Y] at the N- and C-terminals) was discovered,[26] it was shown to be a member of the PP family immunoreactive with some antibodies to the C-terminal of PP. A PYY-specific antibody showed that the PP-like immunoreactivity in the EG cells was really due to the presence of PYY. Electron microscopic immunocytochemistry showed that PYY was costored with intestinal glucagon (glicentin), a completely unrelated peptide, in the same granules (Figure 1). In addition, PYY was shown to be transiently present in the granules of the morphologically primitive A cells of the pancreas in human and rat fetuses.[27] These findings have now received support from the identification of the mRNA for PYY in fetal rat pancreas and both fetal and adult rat colon.[28]

As many as four distinct hormonal peptides have been identified in some cells (mammosomatotrophs) of the cow anterior pituitary. Two of these, growth hormone and prolactin, were present either together or in separate granules within the same cell, while the other two hormones, thyroid-stimulating hormone and follicle-stimulating hormone, were similarly colocalized or separate, but in smaller granules. Three proteins of the chromogranin family, secretogranins I and II and chromogranin A, normally found in thyrotrophs and gonadotrophs, were preferentially localized to the small granule population in these cells.[29]

The chromogranins, which are present in the granules of most, though not all, endocrine cell types, are not necessarily secreted concomitantly with the peptide content of the granule. Studies on a calcitonin-producing cell line (TT cells) derived from a human thyroid medullary carcinoma, showed that calcitonin release was stimulated to a greater extent than chromogranin release by a range of secretagogues. After stimulation, the number of granules containing both chromogranin and calcitonin was decreased, with more chromogranin than calcitonin being retained by the cells. Despite their usual coexistence, differential secretory pathways or rates of synthesis of the two proteins are thus implied.[30] Although the chromogranins have as yet no proven function, they may be involved in the packaging of secretory proteins within the Golgi complex.[31]

5. Plasticity

A new concept is that of plasticity of peptide production in the cells of the diffuse endocrine system. The novel peptides, many of which do not circulate in the blood in measurable quantities, are considered to act locally in a paracrine fashion, or even within or on their own cell of origin in an autocrine fashion, to influence production and secretion of other peptides, conventional hormones, or other regulatory factors. It is thus predictable that the amount of peptide within a cell will vary with physiological status, production being dependent on demand and controlled by feedback mechanisms. Thus, one of the nervous system peptides, galanin, which is present in the normal rat anterior pituitary mainly in lactotrophs, is present in increased quantity, along with prolactin, after estrogen treatment.[32] Electron microscopic immunocytochemistry has shown that galanin is costored with prolactin in the same secretory granules in normal and estrogen-treated rats but can also be present in other cell types, e.g., corticotrophs. The proportion of galanin-positive corticotrophs increases after adrenalectomy.[33]

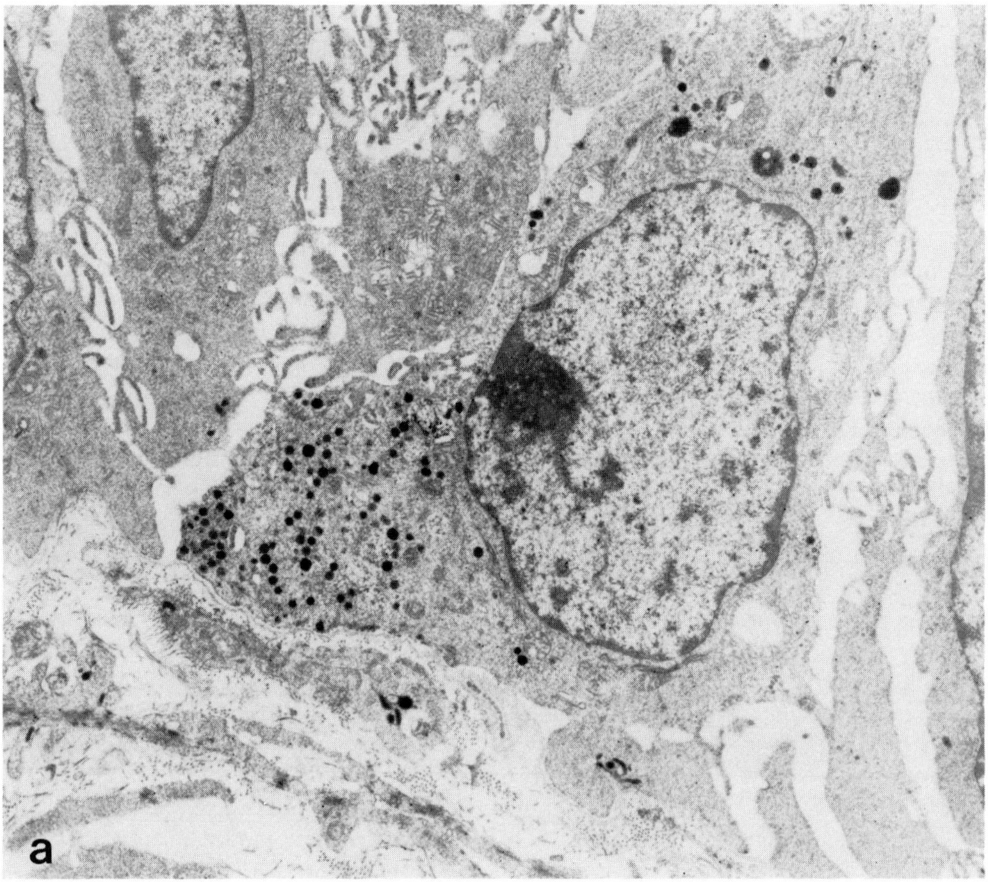

FIGURE 1. (a) Enteroglucagon (L) cell in human colon, double immunolabeled for glicentin (20-nm gold particles) and PYY (10-nm gold particles) by the reverse face method. (Magnification × 9,000.) (b) Enlargement shows that the same secretory granules are labeled for both peptides. Tissue was fixed in 2.5% glutaraldehyde; sections were contrasted with uranyl acetate. (Magnification × 33,800.)

E. ELECTRON MICROSCOPIC CYTOCHEMISTRY

Electron microscopical cytochemistry, as distinct from immunocytochemistry, has played a rather small part in studies of the diffuse endocrine system.

1. Amine Storage and Synthesis

Early work concentrated on the various silver impregnation methods that could be carried out at the ultrastructural level and which distinguished between certain types of amine-containing cell.[34,35]

Another approach combined organ culture with the uptake of tritiated amine precursors and subsequent electron microscopic autoradiography to identify the amine-synthesizing cells of the stomach.[36] The storage of biogenic amines in endocrine cell granules was identified by the uranaffin reaction for adenine nucleotides[37] and this reaction was suggested as a diagnostic ultrastructural test for neuroendocrine neoplasms.[38] However, it has not proved any more reliable than simple morphology for identifying secretory granules and does not seem to be in common use.

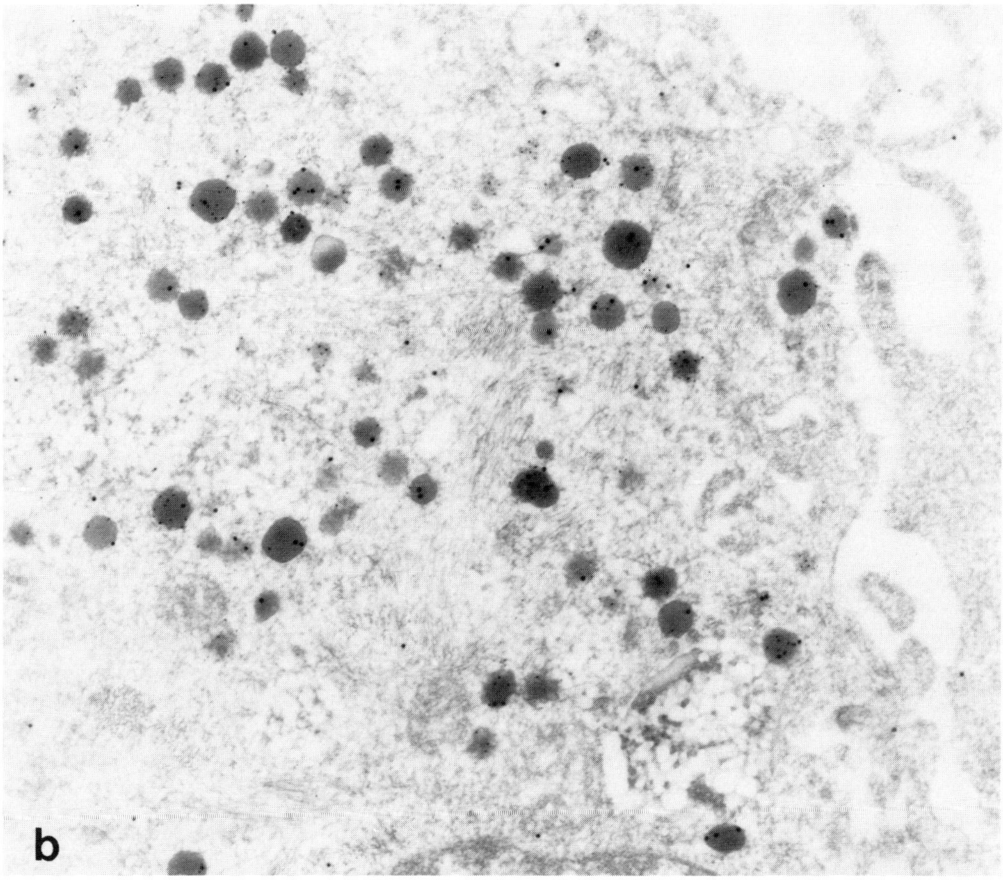

FIGURE 1 (continued)

2. Intracellular Acidity

More recently, an elegant combination of cytochemistry and immunocytochemistry has been used by Orci et al.[39] to examine the acidity of intracellular compartments, such as secretory granules and Golgi vesicles. They developed a novel type of vital "dye", a weak base that is taken up by cells and accumulates in subcellular compartments in direct proportion to their acidity. The marker was 3-[2,4,-dinitroanilino]3′-amino-*N*-methyldipropylamine (DAMP), which has a dinitrophenyl group to which monoclonal antibodies have been raised, and a primary amino group that is available for cross-linking by aldehyde fixation. Quantitative immunocytochemistry, counting the gold particles that label DAMP in a subcellular compartment after protein A-gold immunolabeling, compared with the background count from a neutral compartment such as the nucleus, gives a measure of the acidity of the compartment. The method can be combined with other morphological and immunocytochemical methods. For instance, Orci et al.[39,40] demonstrated that the maturing secretory vesicles in which the conversion of proinsulin to insulin took place (shown by an immunocytochemical reaction for proinsulin or insulin) were acidic in nature (pH 5.7). There was variation in the acidity of the secretory granules of different endocrine cell types in the pancreas, somatostatin granules being less heavily labeled for accumulated DAMP than glucagon and insulin granules.

F. QUANTIFICATION

The immunogold method, in theory, would lend itself well to the quantification of immunolabeled antigens at the electron microscopic level. In fact, the determination of the ratio of label to antibody and of antibody to antigen are not yet sufficiently standardized for any absolute measurements to be made. However, with carefully controlled experiments, attempts at comparative quantification have been made. In the example cited above of postembedding detection of acidic compartments, the accumulation of DAMP is proportional to the H^+ concentration and the relative density of gold particles can be used to measure the acidity of the compartment.

In another example,[41] pancreatic beta cells in Zucker diabetic fatty rats (a rat model of non-insulin-dependent diabetes mellitus) were immunolabeled for a high-K_m (15 to 20 m*M*) glucose transporter, GLUT-2. They showed a greatly decreased count of gold particles, correlated both with the severity of hyperglycemia and with the percentage of plasma membrane occupied by microvilli, which falls in severely diabetic rats.

At the level of the secretory granule, Shimada et al.[42] have investigated the mode of action of somatostatin in suppressing growth hormone release. They perfused rats with growth-hormone-releasing factor to stimulate the growth hormone cells, then followed this with somatostatin. They showed, by immunogold labeling of growth hormone, that somatostatin decreases the number of exocytoses from growth hormone cells but does not affect the microtubular system nor the growth hormone in the Golgi vesicles. They thus present evidence that the site of action of the release-inhibiting factor, somatostatin, is the plasma membrane and that it prevents hormone release rather than preventing hormone production.

G. RECEPTOR LOCALIZATION

With the concept of autocrine regulation of secretion from cells of the diffuse endocrine system, the identification of receptor sites on individual cells is of great interest. Lackie et al.[43] introduced a novel way of localizing such sites which avoided the hazards and difficulties of autoradiography and also alteration of the ligand by attachment of a radioactive label. The system investigated was a cell line from a small cell carcinoma of the lung, known to produce mammalian bombesin, which was thought to be a stimulator of growth of these cells. Bombesin receptors had already been shown by other methods in this material. A dimeric bombesin ligand was constructed by incubation of bombesin with glutaraldehyde, and its sites of attachment to the cell membranes were demonstrated by immunogold labeling on ultrathin frozen sections, using a mouse monoclonal antibody to bombesin and a gold-labeled anti-mouse immunoglobulin. The identification of bombesin receptor sites on the cells producing the very same peptide support the theory of autocrine or paracrine regulation. This method should be applicable to the identification of receptors for other small peptide ligands but does not seem to have been taken further, perhaps because of the difficulty of preparing ultrathin frozen sections.

H. *IN SITU* HYBRIDIZATION OF NUCLEIC ACIDS

DNA and mRNA can be identified with suitable complementary probes at the electron microscopic level, for example, to demonstrate the potential for, or level of production of, a particular peptide or hormone with greater accuracy of localization than with the light microscope. The combination of autoradiographic *in situ* hybridization with immunogold labeling is beginning to provide a powerful tool for the electron microscopist. Tong et al.[44] confirmed that prolactin was actually produced in each of two ultrastructurally distinct types of prolactin-immunopositive cells in the rat pituitary, one characterized by large granules, the other by small granules. They used a ^{35}S-cDNA probe for prolactin on Vibratome sections before embedding in resin, then immunolabeled ultrathin sections before exposing the grid to radiation-sensitive emulsion. Silver grains were seen over the nucleus of the actively secreting cells as well as over the rough endoplasmic reticulum, indicating that the method is sensitive enough to pick up nuclear pre-messenger RNA.

A more complex experiment was carried out by Morel et al.[45] on ultrathin frozen sections of rat pituitary gland after *in vivo* treatment with growth-hormone-releasing factor (GRF) or somatostatin. These authors used two probes, a cDNA for growth-hormone mRNA and a probe for intron 1 of the growth hormone gene. With the cDNA probe, after GRF or somatostatin treatment, increased labeling was seen in the nucleus and decreased labeling in the cytoplasm, corresponding with an increased rate of transcription of the growth hormone gene and accumulation of mRNA or pre-mRNA. With the growth-hormone intron 1 probe, labeling was increased in the cytoplasm and decreased in the nucleus. The suggestion was made that somatostatin inhibited the export of growth-hormone mRNA molecules from the nucleus to the cytoplasm, resulting in decreased cytoplasmic mRNA. Somatostatin could thus act at the mRNA level as well as at the cell membrane, as described above.[42] The decrease in intron 1 in the nucleus is explained by the suggestion that this intron, the first to be removed, is rapidly spliced out from the mRNA precursor before the mRNA has matured enough to leave the nucleus.

I. PATHOLOGY

In the field of tumors of the diffuse endocrine system, electron microscopy and, in particular, electron microscopic immunocytochemistry are of great value. Several regulatory peptides are often produced by these tumors, although usually one is dominant clinically, and these may be recognized by light microscopic immunocytochemistry. Even in normal endocrine cells, morphology alone is insufficient to identify the cells at the electron microscopic level,[46] and in tumors, where a variety of granule types is often found, sometimes demarcating distinct cell populations, sometimes related morphologically to the cells of origin of the tumor, but frequently quite atypical of any normal peptide granule, electron microscopic immunocytochemistry with immunogold methods is the optimal way of characterizing the tumor content (Figure 2). Thus, it can identify both the dominant and the less clinically active peptides in their particular granules, demonstrate multiple secretion from the same cells, and show the secretory state of the peptide-containing cells. Immunocytochemically identified precursors of the biologically active peptides may be correlated with granules or cells. At present, this information may be of little benefit to the patient, but the gradual build-up of knowledge about the forms and functions of these tumors must eventually lead to greater understanding of the system and, possibly, to appropriate therapy.

Electron microscopic immunocytochemistry has also played a part in tracing the morphological stages of tumor production. An ontogenic protein, Simian Virus 40 large T-antigen, was used to form part of a hybrid oncogene with the regulatory sequence of bovine arginine vasopressin and inserted into the mouse genome. Tumors developed in the pancreas and pituitary of the transgenic mice. The production of the large T-antigen was identified by immunocytochemistry in the structurally abnormal proliferating cells of the pancreatic pretumorous lesions, as well as in the tumors, which could be multihormonal. At the electron microscopic level, large T-antigen was localized to the heterochromatin of transformed cells, whether dysplastic or neoplastic, indicating that the large T-antigen interacts with DNA and/or nucleoproteins. One of the conclusions drawn from this study was that endocrine tumorigenesis begins with dysplasia.[47]

Among nontumor diseases of the diffuse endocrine system, the laying down of endocrine amyloid has been studied. The participation of the heart in the diffuse endocrine system was established by the discovery of a peptide hormone, atrial natriuretic peptide, in the

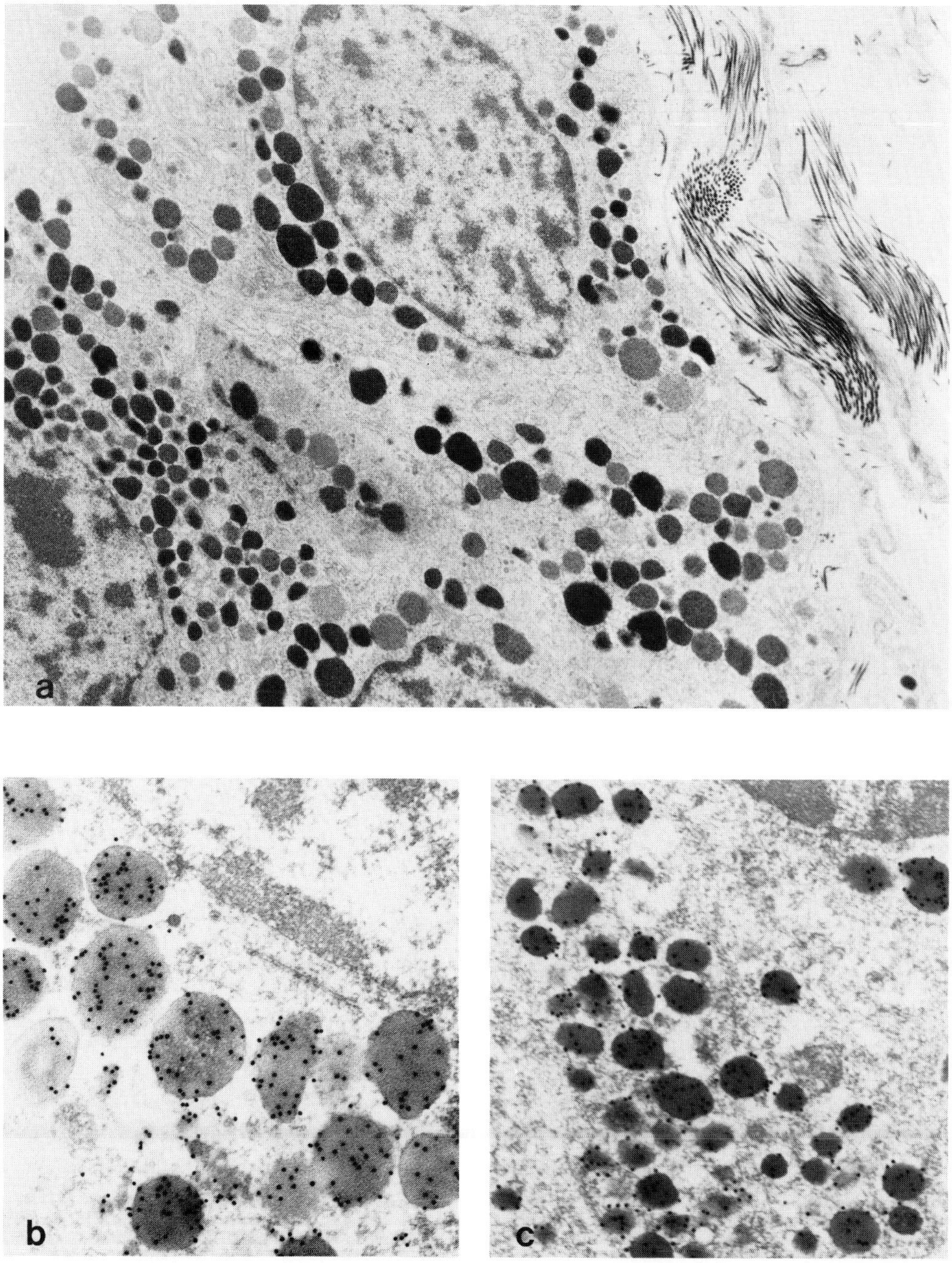

FIGURE 2. (a) Somatostatin-secreting tumor of the duodenum with large and small granules in separate cells. Tissue was fixed in glutaraldehyde followed by osmium tetroxide; sections were contrasted with lead citrate and uranyl acetate. (Magnification × 8,500.) (b and c) Both large and small granules are labeled for somatostatin by indirect immunogold labeling. Tissue was fixed in 2.5% glutaraldehyde; sections were contrasted with uranyl acetate. (Magnification × 33,600.)

atrial "specific granules". It has now been shown that the heart is subject to endocrine disease in that its peptide product can, like insulin and calcitonin in the pancreas and thyroid, be involved in the deposition of amyloid. In the heart this may be an age-related effect. The electron microscopic localization of atrial natriuretic peptide to amyloid fibrils in cardiac amyloidosis confirms previous evidence that cardiac amyloid is partly composed of this peptide.[48]

J. CONCLUSION

While this chapter is not the place to describe in detail the various exciting methods now available, the examples cited should serve to show how these new techniques can be used to provide information on the secretion of peptides, hormones, and other factors and the inter- and intracellular processes of its regulation within the diffuse endocrine system, even with the essentially "static" morphological discipline of electron microscopy.

REFERENCES

1. **Feyrter, F.,** *Uber Diffuse Endocrine Epitheliale Organe,* J. A. Barth, Leipzig, 1938.
2. **Masson, P.,** La glande endocrine de l'intestin chez l'homme, *C.R. Acad. Sci. D (Paris),* 158, 52, 1914.
3. **Hellerstrom, C. and Hellman, B.,** Some aspects of silver impregnation of the islets of Langerhans, *Acta. Endocrinol.,* 35, 518, 1960.
4. **Grimelius, L.,** A silver nitrate stain for A2 cells of human pancreatic islets, *Acta Soc. Med. Ups.,* 73, 271, 1968.
5. **Pearse, A. G. E.,** The cytochemistry and ultrastructure of polypeptide hormone-producing cells of the APUD series, and the embryologic, physiologic and pathologic implications of the concept, *J. Histochem. Cytochem.,* 17, 303, 1969.
6. **Fujita, T.,** The gastroenteric endocrine cell and its paraneuronic nature, in *Chromaffin, Enterochromaffin and Related Cells,* Coupland, R. E. and Fujita, T., Eds., Elsevier, Amsterdam, 1976, 191.
7. **Fujita, T. and Kobayashi, S.,** Current views on the paraneurone concept, *Trends Neurosci.,* 2, 27, 1979.
8. **Thorndyke, M. C.,** Immunocytochemistry and evolutionary studies with particular reference to peptides, in *Immunocytochemistry, Modern Methods and Applications,* 2nd ed., Polak, J. M. and Van Noorden, S., Eds., John Wright & Sons, Bristol, 1986, chap. 19.
9. **Solcia, E., Capella, C., Buffa, R., Usellini, L., Fiocca, R., and Sessa, F.,** Endocrine cells of the digestive system, in *Physiology of the Gastrointestinal Tract,* Johnson, L. R., Ed., Raven Press, New York, 1981, chap. 2.
10. **Osamura, R. Y.,** Pituitary adenomas investigated with pre-embedding electron microscopic immunocytochemistry, in *Endocrine Pathology Update,* Vol. 1, Lechago, J. and Kameya, T., Eds., Field and Wood, New York, 1990, chap. 1.
11. **Batten, T. and Hopkins, C. R.,** Use of protein A-coated colloidal gold particles for immuno-electron microscopic localization of ACTH on ultrathin sections, *Histochemistry,* 60, 317, 1979.
12. **Larsson, L.-I.,** Simultaneous ultrastructural demonstration of multiple peptides in endocrine cells by a novel immunocytochemical method, *Nature (London),* 282, 743, 1979.
13. **Garaud, J.-C., Eloy, R., Moody, A. J., Stock, C., and Grenier, J. F.,** Glucagon and glicentin-immunoreactive cells in the human digestive tract, *Cell Tissue Res.,* 213, 121, 1980.
14. **Bendayan, M.,** Double immunocytochemical labeling applying the protein A-gold technique, *J. Histochem. Cytochem.,* 30, 81, 1982.
15. **Tapia, F. J., Varndell, I. M., Probert, L., De Mey, J., and Polak, J. M.,** Double immunogold staining method for the simultaneous ultrastructural localization of regulatory peptides, *J. Histochem. Cytochem.,* 31, 977, 1983.
16. **Ravazzola, M. and Orci, L.,** Glucagon and glicentin immunoreactivity are topologically separated in the α-granule of the human pancreatic A cell, *Nature (London),* 284, 66, 1980.

17. **Bussolati, G. and Canese, M. G.,** Electron microscopic identification of the immunofluorescent gastrin cells in the cat pyloric mucosa, *Histochemie,* 20, 198, 1972.
18. **Greider, M. H., Steinberg, V., and McGuigan, J. F.,** Electron microscopic identification of the gastrin cell of the human antral mucosa by means of immunocytochemistry, *Gastroenterology,* 63, 572, 1972.
19. **Polak, J. M., Pearse, A. G. E., Szelke, M., Bloom, S. R., Hudson, D., Facer, P., Buchan, A. M. J., Bryant, M. G., Christophides, N., and MacIntyre, I.,** Specific immunostaining of CCK cells by use of synthetic fragment antisera, *Experientia,* 33, 762, 1977.
20. **Buchan, A. M. J., Polak, J. M., Solcia, E., Capella, C., Hudson, D., and Pearse, A. G. E.,** Electron immunocytochemical evidence for the human intestinal cell as the source of CCK, *Gut,* 19, 403, 1978.
21. **Buchan, A. M. J., Polak, J. M., Solcia, E., and Pearse, A. G. E.,** Localisation of intestinal gastrin in a distinct endocrine cell type, *Nature (London),* 277, 138, 1979.
22. **Bussolati, G., Capella, C., Vassallo, G., and Solcia, E.,** Histochemical and ultrastructural studies on pancreatic A cells. Evidence for glucagon and non-glucagon components of the α-granule, *Diabetologia,* 7, 181, 1971.
23. **Varndell, I. M., Bishop, A. E., Sikri, K. L., Uttenthal, L. O., Bloom, S. R., and Polak, J. M.,** Localization of glucagon-like peptide (GLP) immunoreactants in human gut and pancreas using light and electron microscopic immunocytochemistry, *J. Histochem. Cytochem.,* 33, 1080, 1985.
24. **Varndell, I. M., Harris, A., Tapia, F. J., Yanaihara, N., De Mey, J., Bloom, S. R., and Polak, J. M.,** Intracellular topography of immunoreactive gastrin demonstrated using electron immunocytochemistry, *Experientia,* 39, 713, 1983.
25. **Funata, N., Fukayama, M., Sugano, K., and Koike, M.,** Intracellular topography of glycine-extended pro-gastrin processing intermediates in human antral mucosa: an electron microscopic immunocytochemical study, *J. Histochem. Cytochem.,* 37, 287, 1989.
26. **Tatemoto, K.,** Isolation and characterization of peptide YY (PYY), a candidate gut hormone that inhibits pancreatic secretion, *Proc. Natl. Acad. Sci. U.S.A.,* 79, 2514, 1982.
27. **Ali-Rachedi, A., Varndell, I. M., Adrian, T. E., Gapp, D. A., Van Noorden, S., Bloom, S. R., and Polak, J. M.,** Peptide YY (PYY) immunoreactivity is co-stored with glucagon-related immunoreactants in endocrine cells of the gut and pancreas, *Histochemistry,* 80, 487, 1984.
28. **Krasinski, S. D., Wheeler, M. B., and Leiter, A. B.,** Isolation, characterization and developmental expression of the rat PYY gene, *Mol. Endocrinol.,* 5, 433, 1991.
29. **Bassetti, M., Huttner, W. B., Zanini, A., and Rosa, P.,** Co-localization of secretogranins/chromogranins with thyrotropin and luteinizing hormone in secretory granules of cow anterior pituitary, *J. Histochem. Cytochem.,* 38, 1353, 1990.
30. **Kameya, T., Kadoya, K., Iguchi, H., Abe, Y., and Shimaoka, K.,** Discordant secretion of calcitonin and chromogranin in the human medullary thyroid carcinoma cell line, *Endocr. Pathol.,* 2, 16, 1991.
31. **Rosa, P., Hille, A., Lee, R., Zanini, A., de Camilli, P., and Huttner, W. B.,** Secretogranins I and II: two tyrosine-sulfated secretory proteins common to a variety of cells secreting peptides by the regulated pathway, *J. Cell Biol.,* 101, 1999, 1985.
32. **Steel, J. H., Gon, G., O'Halloran, D. J., Jones, P. M., Yanaihara, N., Ishikawa, H., Bloom, S. R., and Polak, J. M.,** Galanin and vasoactive intestinal polypeptide are colocalised with classical pituitary hormones and show plasticity of expression, *Histochemistry,* 93, 183, 1989.
33. **Cimini, V., Van Noorden, S., and Polak, J. M.,** Modulation of galanin- and neuromedin U-like immunoreactivity in the rat anterior pituitary gland after alteration of endocrine status, in press, 1992.
34. **Grimelius, L.,** An electron microscopic study of silver stained adult human pancreatic islet cells, with reference to a new silver nitrate procedure, *Acta Soc. Med. Ups.,* 74, 28, 1969.
35. **Hakanson, R., Owman, C., Sporrong, B., and Sundler, F.,** Electron microscopic identification of the histamine-storing argyrophil (enterochromaffin-like) cells in the rat stomach, *Z. Zellforsch. Mikrosk. Anat.,* 122, 460, 1971.
36. **Rubin, W. and Schwarz, B.,** Use of EM autoradiography and organ culture to identify pathways of amine synthesis in human and murine gastric endocrine cells, in *Cellular Basis of Chemical Messengers in the Digestive System,* Grossman, M. I., Brazier, M. A. B., and Lechago, J., Eds., Academic Press, New York, 1981, 57.
37. **Richards, J. G. and Da Prada, M.,** Uranaffin reaction: a new cytochemical technique for the localization of adenine nucleotides in organelles storing biogenic amines, *J. Histochem. Cytochem.,* 25, 1322, 1977.
38. **Payne, C. M., Nagle, R. B., and Borduin, V.,** An ultrastructural cytochemical stain specific for neuroendocrine neoplasms, *Lab. Invest.,* 51, 350, 1984.
39. **Orci, L., Ravazzola, M., Amherdt, M., Madsen, O., Perrelet, A., Vassalli, J.-D., and Anderson, R. G. W.,** Conversion of proinsulin to insulin occurs coordinately with acidification of maturing secretory vesicles, *J. Cell Biol.,* 103, 2273, 1986.

40. **Orci, L., Ravazzola, M., Storch, M.-G., Anderson, R. G. W., Vassalli, J.-D., and Perrelet, A.,** Proteolytic maturation of insulin begins in clathrin-coated secretory vesicles, *Cell,* 49, 865, 1987.
41. **Orci, L., Ravazzola, M., Beatens, D., Inman, L., Amherdt, M., Peterson, R. G., Newgard, C. B., Johnson, J. H., and Unger, R. H.,** Evidence that down-regulation of β-cell glucose transporters in non-insulin-dependent diabetes may be the cause of diabetic hyperglycemia, *Proc. Natl. Acad. Sci. U.S.A.,* 87, 9953, 1990.
42. **Shimada, O., Tosaka-Shimada, H., and Ishikawa, H.,** Morphological effects of somatostatin on rat somatotrophs previously affected by growth hormone-releasing factor, *Cell Tissue Res.,* 261, 219, 1990.
43. **Lackie, P. M., Cuttitta, F., Minna, J. D., Bloom, S. R., and Polak, J. M.,** Localisation of receptors using a dimeric ligand and electron immunocytochemistry, *Histochemistry,* 83, 57, 1985.
44. **Tong, Y., Zhao, H. F., Simard, J., Labrie, F., and Pelletier, G.,** Electron microscopic autoradiographic localization of prolactin mRNA in rat pituitary, *J. Histochem. Cytochem.,* 37, 567, 1989.
45. **Morel, G., Dihl, F., and Gossard, F.,** Ultrastructural distribution of growth hormone (GH) mRNA and GH intron I sequences in rat pituitary gland: effects of GH releasing factor and somatostatin, *Mol. Cell. Endocrinol.* 65, 81, 1989.
46. **Dobbins, W. O. and Austin, L. L.,** Electron microscopic definition of intestinal endocrine cells: immunogold localization and review, *Ultrastruct. Pathol.,* 15, 15, 1991.
47. **Rindi, G., Bishop, A. E., Murphy, D., Solcia, E., Hogan, B., and Polak, J. M.,** A morphological analysis of endocrine tumour genesis in pancreas and anterior pituitary of AVP/SV40 transgenic mice, *Virchows Arch. A,* 412, 255, 1988.
48. **Pucci, A., Wharton, J., Arbustini, E., Grasso, M., Diegoli, M., Needleman, P., Vigano, M., and Polak, J. M.,** Atrial amyloid deposits in the failing human heart display both atrial and brain natriuretic peptide-like immunoreactivity, *J. Pathol.,* 165, 235, 1991.

Chapter 12

CYTOCHEMISTRY AND IMMUNOCYTOCHEMISTRY IN THE CLASSIFICATION OF BONE MARROW AND BLOOD CELLS AND IN THE DIAGNOSIS OF HEMATOLOGIC DISORDERS

Olivera Markovic-Saljinska and Nenad Markovic

TABLE OF CONTENTS

0-8493-6012-9/93/$0.00 + $.50

A. DEVELOPMENT OF TECHNOLOGY AND CONCEPTS

1. Light Microscopic Morphology (Classic)

Blood cells were first recognized in the middle of the 19th century as red and unstained corpuscles floating freely in a drop of blood investigated under the microscope. Hence, the name erythro- and leukocytes (from greek: *erythro,* red; *leukos,* unstained, white; and *kytos,* hollow vessel, cell). The unstained cells found in bone marrow were called myelocytes (*myelos,* marrow). In the later 1880s and 1890s Ehrlich, a German textile engineer, discovered tissue stains. Papencheim, Giemsa, Romanowski, Wright, and others developed methods to stain blood and bone marrow cells. These methods used stoichiometric mixtures known as neutral stains or acid/base dyes distributing inside cells according to the pH difference among intracellular organelles and other substructures. As a result, cytoplasm, nucleus, granules, vacuoles, inclusions, and nucleoli became microscopically identifiable features with distinct characteristics. Their number, hue, size, shape, and interrelations became indices for definition of criteria for categorization and classification of blood and bone marrow cells.[1,22]

Classic morphology provides about 60 subcategories for classification of hematopoietic cell progenitors in bone marrow. This list does not include stem and undifferentiated cells,which lack distinct morphologic characteristics. In certain disorders like acute leukemias, some less common hematologic malignancies, and blastic phases in myeloproliferative and myelodysplastic disorders, morphologic criteria alone cannot provide enough information for accurate classification. Morphology is also an insufficient source of information for estimation of disease prognosis in individual patients, administration of therapy, and evaluation of response. To satisfy those needs, different complementary assays are available. Some of them, helpful for diagnosis of blast cells, are presented below.

2. Cytochemistry

In the first part of the 20th century, cytochemistry provided additional morphologic criteria (chemical morphology). Chemical reactions based on intracellular catalysis of artificial substrates were used to produce deposits of insoluble color substances contrasting to structures counterstained by neutral stains.[1-10,12,14,15] Several enzymes and metabolically active intracellular constituents (myeloperoxidase, esterases, alkaline and acid phosphatase, terminal deoxynucleotidyl transferase, lipids, glycogen), otherwise invisible, have become clearly labeled by these deposits and made amenable for microscopic analysis (Figure 1).[4,6,10,20] They provided hematologists with qualitative markers to ascertain the lineage of origin for a number of abnormal cells and cells lacking distinct morphologic criteria.[1,8,10,12,14,15,17,24] These markers are particularly helpful in distinguishing blasts in acute nonlymphoblastic leukemia (ANLL) and acute phases of chronic myeloproliferative and myelodysplastic diseases.[1,8,32,41,47,50,107,150] The value of these methods has been evaluated extensively by Hayhoe et al.[1] and limited credit was given primarily because of a lack of quantitation.

In the late 1970s, development of image analysis systems allowed introduction of quantitative cytochemistry.These systems are widely available and can be used for morphometric and densitometric measurements of intracellular structures and cytochemical markers of intracellular enzyme activity.[4,7,8,54,56,90,139,158,177] They provide objective, sensitive, and exact measurement, storage of data, and automatic file management (Figure 2, A and B). Kinetic studies were used to improve standard cytochemical methods and to address some cell membrane diffusion problems (Figure 2C).[8,37,43,63,96,158] Since cytochemical markers label different types of cells, the assessment of a quantitative difference might be considered to be of limited value for diagnosis of blasts. End-point assays were better employed to study cell sensitivity to drugs targeting intracellular enzymes (Figure 2D*).[8,36,43,96,117,119,154,157] This

* See also Plate 1, following page 672.

approach may be very attractive for selection of patients for treatment with drugs such as tiazofurine, an inhibitor of inosine monophosphate dehydrogenase, and other similar ones.[79]

3. Immunocytochemistry

Discovery of membrane antigens and production of monoclonal antibodies with labels to identify their epitopes led to the introduction of immunocytochemistry in the second part of this century.[25,33,36,38,39] Cell types identifiable by morphologic and cytochemical characteristics did not benefit much by this technology. Other cell types, particularly lymphocytes and stem cells, gained criteria for classification of maturation and differentiation stages and recognition of bone marrow cell compartments, such as stem cell pool, committed stem cell pool, mitotic and postmitotic pools.[36] Bone marrow was recognized as a major source of B cells, thymus and peripheral lymph nodes as sources of T cells. B cell maturation stages, from pre-B to plasma cells, are identified only by using these techniques.

Lymphoproliferative disorders such as non-Hodgkin's lymphoma and ALL, were reclassified according to immunocytochemical criteria, and the immunophenotyping became a routine for diagnosis of acute leukemia.[40,190] The new classification of B cell leukemia is based completely on immunologic criteria (pre-B leukemia is a new name for the common ALL, early-B for the B cell ALL, intermediate-B for CLL, mature-B for hairy cell leukemia [HCL], plasma cell leukemia for the leukemic phase of multiple myeloma, and plasmacytoid lymphocyte leukemia for Waldenstrom macroglobulinemia).[33,36,38]

Identification of single antigens and measurement of the number of labeled cells inside a certain population of cells became parallel to peripheral blood counts as routine methods for patient blood examination. Flow cytometry was successfully developed to automate those assays.[189,191]

On the other hand, none of the manufacturers of antibodies have been able to demonstrate any relation between the number of antigens involved and the amount of enzyme-cytochemical label attached to target cells. Image analysis was again successful, but only to measure the change of the peroxidase label.

4. Ultrastructural (TEM, SEM, BEI) Morphology, Cytochemistry, and Immunocytochemistry

Transmission electron microscopy (TEM) provided a new insight into cell structures already visualized at the light microscope (LM) level, e.g., interphase nuclei, chromosomes, granules, mitochondria, and Golgi apparatus, and uncovered lysosomes, ribosomes, and endoplasmic reticulum, clarifying features of cell differentiation and maturation and identifying a number of abnormal structures that became additional criteria for diagnosis.[10,12,13,16,18-21,23,29,31]

Ultrastructural morphology combined with differential centrifugation, subcellular organelle separation, and biochemical microanalysis highlighted the physiology of cells and their function in disease (this includes structure and function of mitochondria: respiratory chain, oxidative phosphorylation, and energy production; ribosomes; messenger RNA function; cell membrane diffusion; lipid peroxidation; receptors; etc).[29,163] This new source of information contributed significantly to a better understanding of the origin and development of blood and bone marrow cells, particularly the structure and function of leukemic blast cells. The concept of an altered biochemistry underlying the asynchronous development of cell organelles in blast cells is based on these techniques.[13]

Introduction of scanning electron microscopy (SEM) highlighted cell membrane properties and completed the understanding of cell images obtained by light microscopy and TEM (Figure 3).[11,57,59-61,146] New emphasis was given to cell membrane projections in hairy cell leukemia.[11,60,143,165]

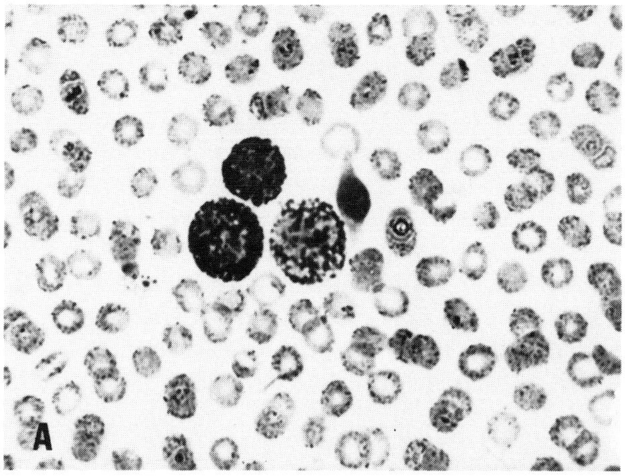

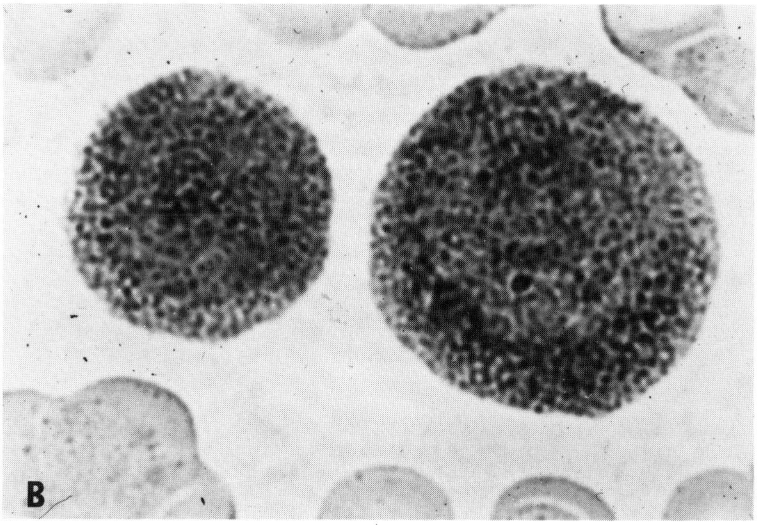

FIGURE 1. Enzyme cytochemistry of blood cells. (A) Peripheral blood, healthy donor. Myeloperoxidase, counterstained with Giemsa. Oxidized benzidine indicates peroxidase reaction. High positive PMNs, an intermediate positive monocyte, and a negative lymphocyte. (Magnification × 400.) (B) Peripheral blood, acute monoblastic leukemia (M5), α-naphthyl acetate esterase. Granular azo-dye deposit indicating high enzyme activity inside two leukemic monoblasts. (Magnification × 1,000.) (C) Peripheral blood, healthy donor, acid phosphatase (α-naphthyl AS-BI phosphatase). Azo-dye deposit indicating a single, coarse granule of phosphatase activity. (Magnification × 1,000.)

EM cytochemistry and EM immunocytochemistry provided ultrastructural localization of enzymes, metabolically active substances, and target antigens that became powerful additional tools to study the function and maturation of normal and abnormal blood cells.[10,12,14,15,17,24-27,31,170,171,188] These techniques were used to recognize early signs of myeloid maturation inside undifferentiated stem cells and blasts (identification of a myeloperoxidase inside the cytosol, a stage before its storage into lysosomal granules) and the megakaryocyte cell lineage commitment identifying a platelet peroxidase inside the perinuclear region of early megakaryoblasts (Figure 4).[26,28,30]

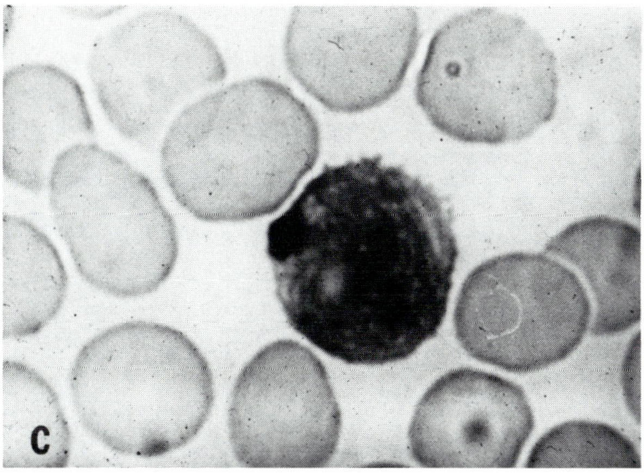

FIGURE 1 (continued)

Some criteria established at the LM level had to be changed. Ultrastructural cytochemistry of esterases revealed an abundant activity in histiocytomonocytic series.[115] Contrary to the findings at the LM level, it was found that esterases are rarely related to any cytoplasmic organelles, such as mitochondria or lysosomes. Granulocytes and plasma cells are negative. Some lymphocytes, outside of the germinal center, and most lymphocytes in the blood flow show punctate esterase reaction in the form of one to four dense granules.

Ultrastructural morphology, cytochemistry, and colloidal gold techniques were combined to study the proliferation and maturation of normal cells. This approach revealed data unavailable by LM and some of them are used in distinguishing undifferentiated blast cells.[34-36]

Ultrastructural detection revealed the presence of nucleoside phosphatase and ATPase activity in plasmalemma, cytoplasmic reticulum, Golgi complex, mitochondria, inside the nucleus, nucleolus, and perinuclear area. Lymphocytes showed a considerable degree of heterogeneity of the distribution of these enzymes that is used for subclassification of lymphoblasts.[183]

Using ultrastructural cytochemistry and immunocytochemistry, valuable data were obtained for the location of the nucleolus organizer region in interphasic nucleolar components, the structure of the nucleolar chromatin *in situ,* and the structural-functional relationship of the nucleolar components.[172]

Information was also obtained for the localization of actin-binding protein and myosin in human platelets, revealing that the disc shape of these cells is not identical to resting platelets, and helping in distinguishing platelet-productive megakaryocytes.[173] A novel technical approach was described for the visualization of lysozyme in neutrophils and precursors.[174]

The backscattered electron imaging mode of SEM (BEI) brought new application of this technology, correlating them with cytochemical and immunologic specific features of mature and precursor cells (Figure 5).[58-61,108,165,166,188,189] Erythroblasts were identified by silver methenamine staining, myeloblasts by MPO and 3C5-positive immunogold technique, lymphoblasts by positive J5. This technology is now expanding for the study of pathologic blast cells.

5. Cytogenetics and DNA

Investigation of nuclear DNA in the resting phase of the cell cycle is among the first that was directly connected with cancer research. Cancer cells were defined as those which

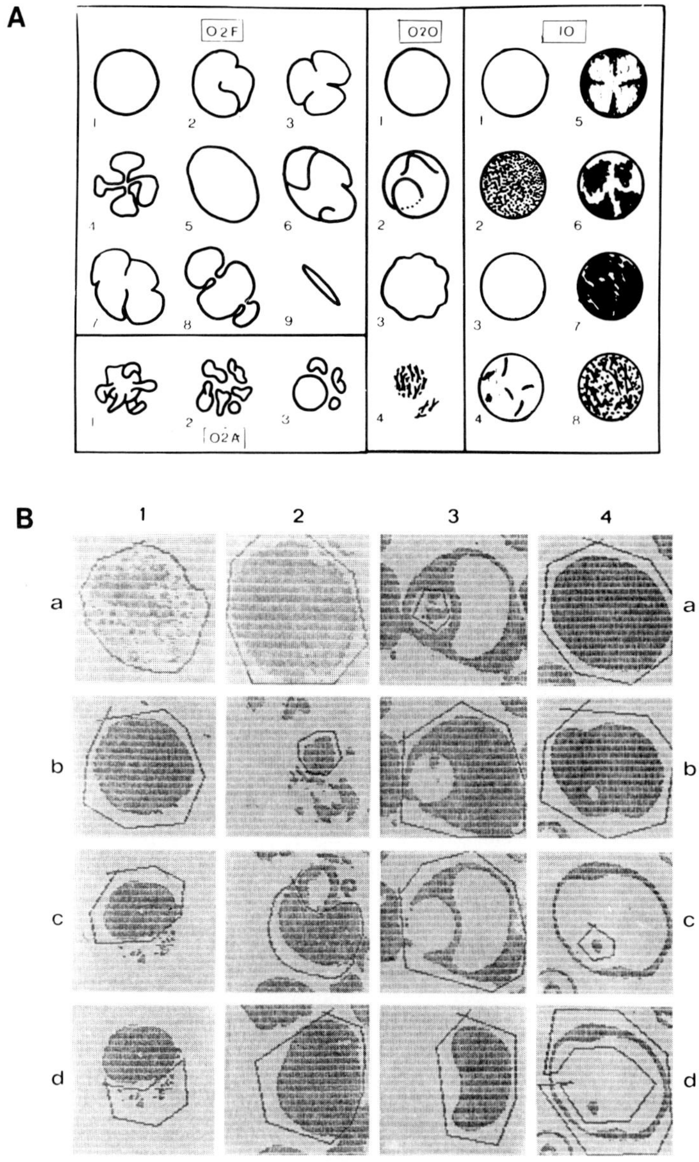

FIGURE 2. Image analysis of blood cells. (A) Morphometric criteria for blast cells: O2F, shape; O2O, outline; O2A, level of fragmentation; 10, degree of chromatin condensation. (B) Computerized cell image tomography of leukemic blasts: two-dimensional segmentation of the binary image of different blasts: 1b, c, d, APL (segmentation of granules and nucleus); 2b, c, d, AML (segmentation of an inclusion and nucleus); 3a, b, c, d, AMoL (segmentation of vacuole, nucleus, and cytoplasm); 4a, b, c, d, ALL (segmentation of nucleus, nucleolus, and cytoplasm); 2a, NAE activity in monoblasts; 1a, drug-induced inhibition of NAE activity in monoblasts. (The Imagescan, Image Analysis System, Carl Zeiss, Co., Inc. Thornwood, NY) (C) Kinetics of myeloperoxidase activity in normal PMNs as measured by the NCI "Grain Counter". Polaroid prints, taken from the monitor screen, are presenting gradual accumulation of oxidized benzidine. Numerical increase in area, mean optical density, and integrated optical density of the final reaction product can be seen. Four phases of the reaction are presented: introductory, fast development, stabilization, and saturation phase. (From Markovic, N. et al., *Clin. Chem.*, 29, 1528, 1983. With permission.) (D) Inhibition of PMNs naphthol AS-D-chloroacetate esterase by quinidine. D1, untreated cell (ctrl); D2, D3, and D4, quinidine pretreatment with increasing drug concentration. Dose dependent decrease of enzyme activity. (The Imagescan, Image Analysis System, Carl Zeiss, Co., Inc. Thornwood, NY.) (From Markovic, O. et al., *Clin. Chem.*, 34, 518, 1988. With permission.)

C

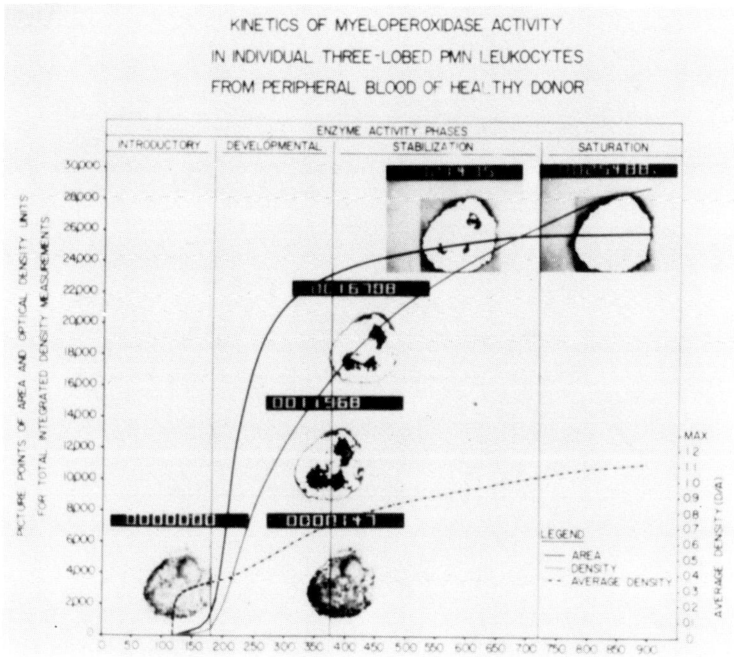

D

Computerized Cell Image Tomography of Quinidine Tolerance Test in Neutrophils with α-Naphthyl AS-D Chloroacetate Esterase

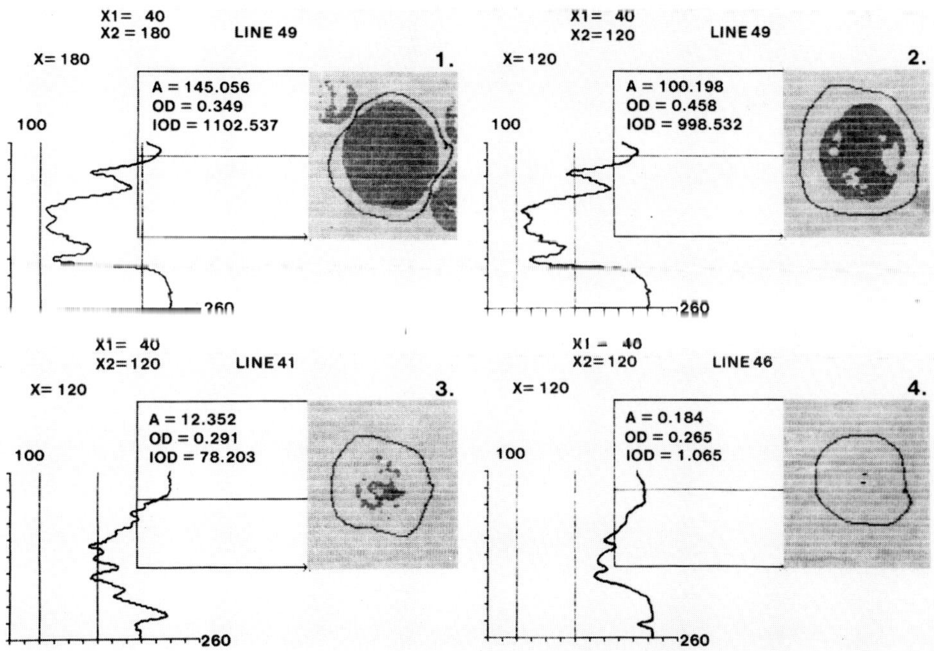

FIGURE 2 (continued)

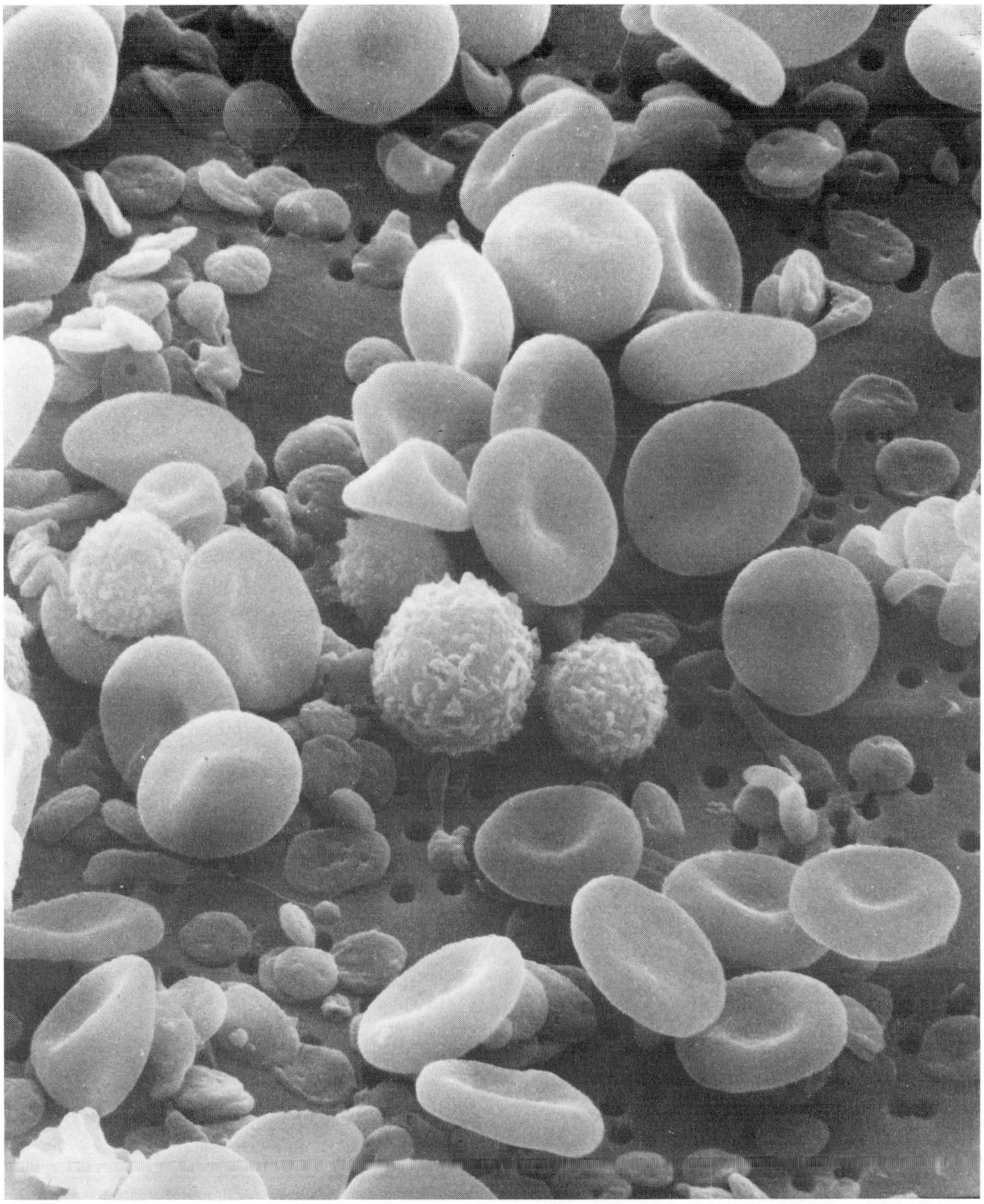

FIGURE 3. Scanning electron microscopy of blood cells. (above) Peripheral blood cells (erythrocytes, platelets, monocyte, and three lymphocytes) fixed in whole blood, SEM micrograph; (facing page) same cells, Feulgen stained and freeze-dried. (Courtesy of Dr. Bruce Wetzel.)

contain hyperchromatic, aneuploid nuclei. Feulgen stain was designed to color nuclear chromatin DNA. Many measuring systems were developed to quantify the images of the Feulgen-stained nuclei: from planimetric to image analysis techniques.[4,21]

Cytogenetic studies and chromosome analysis of cells in mitosis promised new avenues of investigation, particularly in leukemia, following the discovery of the Philadelphia

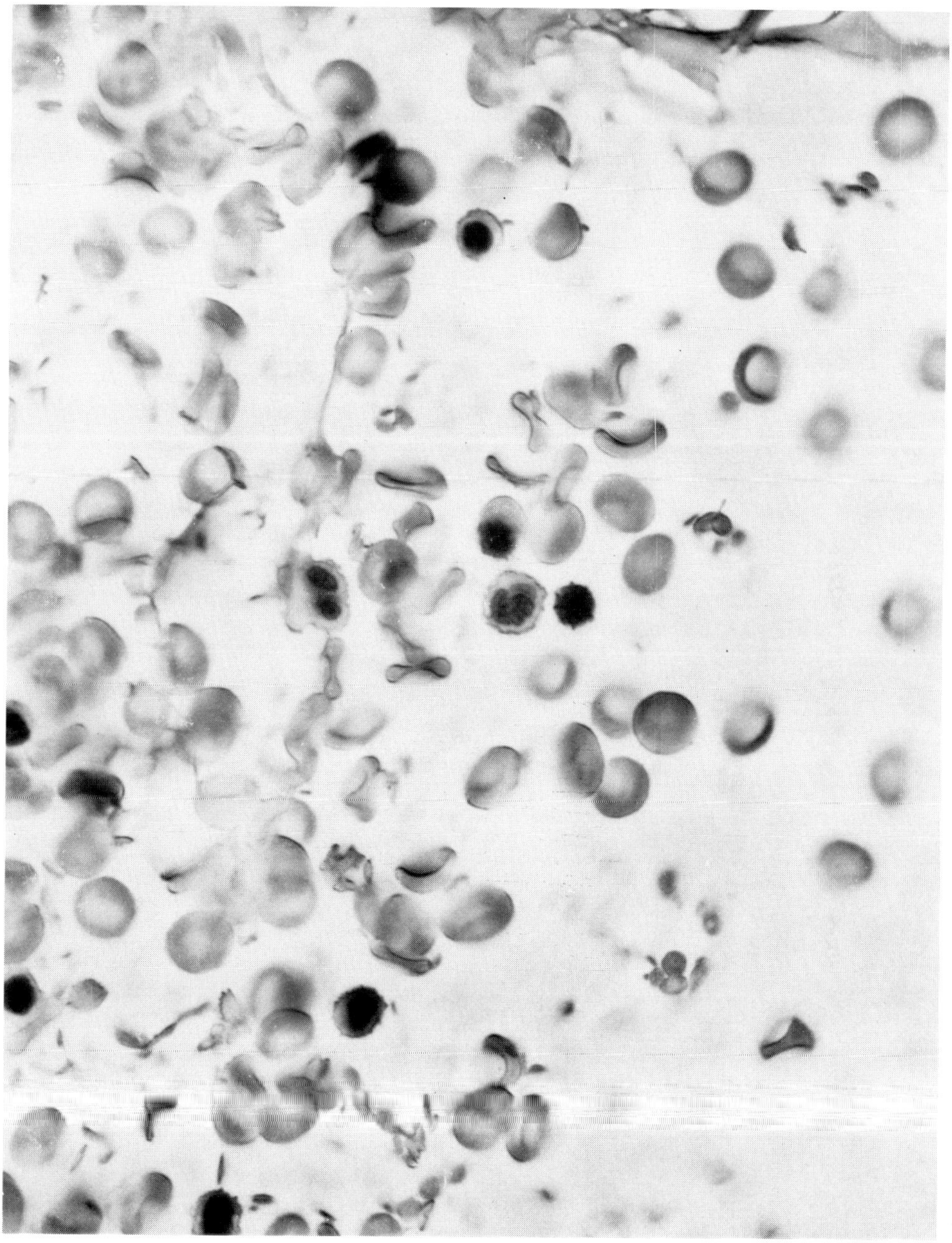

FIGURE 3 (continued)

chromosome (translocation 22/9, see Figure 4). Numerous chromosome abnormalities have been described, but only few of them significantly identified selected subtypes of leukemia (M3:t,15;17).

DNA probe assays as *in situ* hybridization technology have followed this pattern and, at present, the majority of morphologic features described at the beginning of the century

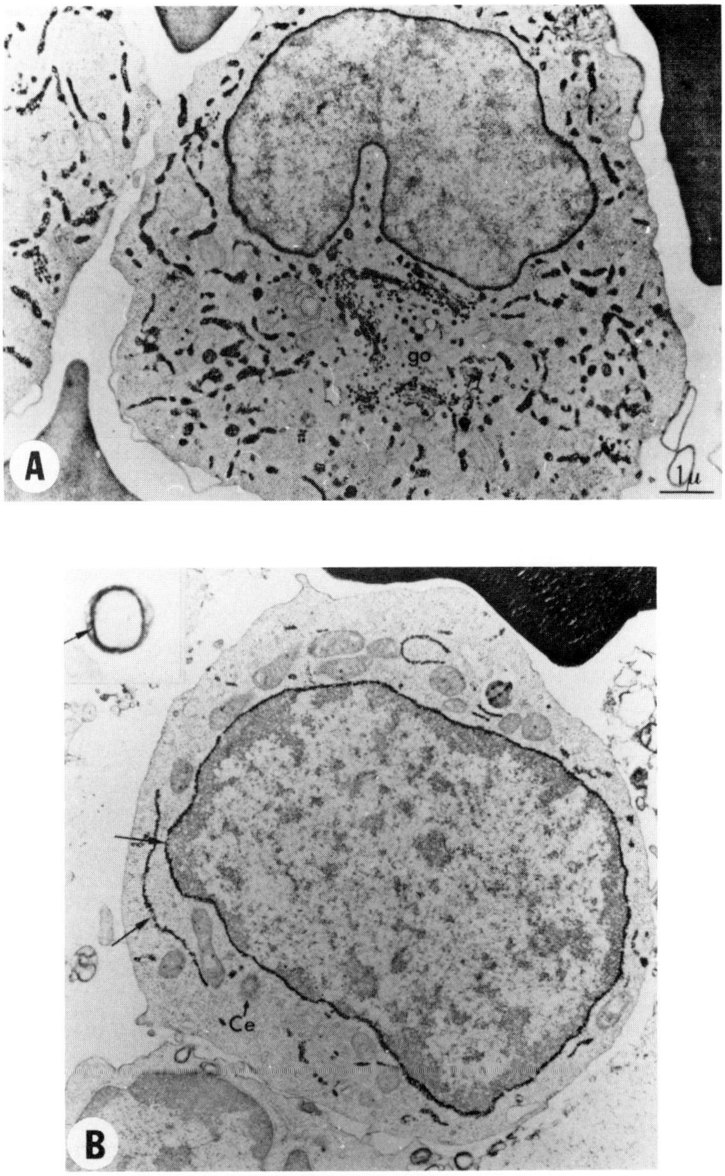

FIGURE 4. Transmission electron microscopy of peroxidases in blasts. (A) Myeloperoxidase in a myeloblast. This cell, which lacks granules, belongs to the granulocytic series because there is peroxidase activity in the endoplasmic reticulum and Golgi apparatus. (From Brenton-Gorius, J., in *Human Leukemias,* Polliack, A., Ed., Martinus Nijhoff, Boston, 1984. With permission.) (B) Platelet peroxidase in a megakaryoblast. Marrow MgBl with a "lymphoid" appearance identified by the presence of PPO. (From Brenton-Gorius, J., Reyes, F., Duhamel, A. et al., *Blood,* 51, 45, 1978. With permission.)

as leukemic blast characteristics may be interpreted as specific expressions of abnormal or altered genes. Unfortunately, such precision in diagnosis has not been followed by improvement of therapeutic skills, and this knowledge has evolved beyond the application in the area of patient care. The balance has to be established, as has happened with other techniques after their auspicious introduction.

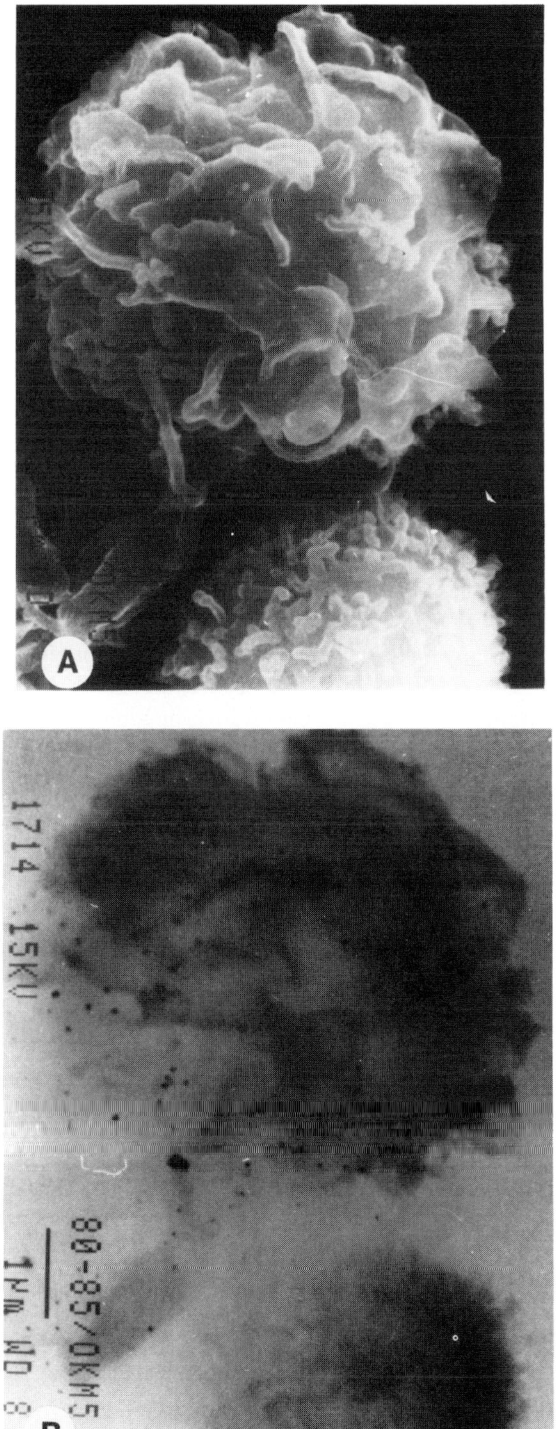

FIGURE 5. Backscattered electron imaging mode of scanning electron microscopy. (A) Secondary electron image of ruffling monocyte and villous lymphocyte; (B) same cells viewed in backscattered electron imaging mode in reverse polarity. All the colloidal gold particles (OKM5) appear in a high contrast and could be counted easily on the surface of the monocytes. (From DeHarven, E., *Ultrastruct. Pathol.*, 11, 711, 1987. With permission.)

It is obvious that ultrastructural morphology, cytochemistry, immunocytochemistry, and DNA assessment have become powerful additional tools to light microscopy in differentiation and classification of immature cells in hematologic malignancies. This is particularly true for so-called unclassifiable "difficult" cases of acute leukemia, mixed and hybrid types, blast phases of chronic myelogenous leukemia, and other myeloproliferative and myelodysplastic disorders. These tools have given a better understanding of blasts in myelodysplastic syndromes and rare types of hematological diseases. Precise classification of cells has the utmost importance for the proper administration of therapy and estimation of prognosis.

B. MULTIPARAMETER IDENTIFICATION OF LEUKEMIC BLAST CELLS

1. Classification of Leukemia

Classification of leukemias has always been a subject of dispute and conventional agreement. It was accepted that names of diseases should indicate the prognosis of an untreated disease and the cell type involved. The first classification appeared at the end of the last century and included acute and chronic forms of myeloid and lymphatic leukemias.[22,41] Almost 30% of cases were left unclassified. Morphology alone allowed recognition of acute leukemias (myeloblastic, monomyeloblastic, monoblastic, lymphoblastic, and erythroblastic variants) and chronic myeloid and lymphatic leukemia.[22,41] Most of this concept survived until today, and it has been included into the FAB (French-American-British study group) Classification.[41,47] We now search for signs more specific for prognosis of disease response to chemotherapy, hoping to plan the most appropriate therapy for individual patients. These signs are the extension of classic morphology and FAB classification and are due to the achievements accomplished by ultrastructural investigation and information obtained by cytochemical and immunocytochemical techniques.[12,15,22,48-50,62,111,125,131,138,144,180,190]

Using the current FAB classification, hematologists encounter about 10% of leukemias that are difficult or unclassifiable on the basis of light microscopical morphology and cytochemistry.[48,52,53,55] Additional techniques such as immunophenotyping, transmission and scanning electron microscopic morphology, cytochemistry, quantitative cytochemistry, immunocytochemistry, and cytogenetics are necessary to decrease this number to less than 1%.[1,4,15,38] The following set of criteria will highlight some of the important characteristics obtained by the combined approach when it is used to determine an appropriate cell type, when it might be difficult or uncertain by conventional techniques.

2. Criteria for Classification of Selected Types of Blast Cells

a. Megakaryoblasts

Megakaryoblasts (MgBl) are preponderant blasts in megakaryocytic leukemia and may be present in mixed cases of acute leukemias and the blast crisis of chronic myelogenous leukemia (CML).[30,64,67,68,101] With the aid of EM, these cells are more easily identified in the terminal phase of AML, acute myelodysplastic and preleukemic syndrome, acute myelofibrosis, and malignant myelosclerosis.[71,75,76] There are data suggesting that leukemic MgBl may be involved in the pathogenesis of myelosclerosis.[100] Proliferation of megakaryocyte precursors has also been reported in thrombocythemia, myelosclerosis, and erythroleukemia.[70,72,76]

MgBls are difficult to recognize at the LM level because they lack specific morphological features.[16,77] Cytochemically, the majority of MgBls express periodic acid-Schiff (PAS)-positive reaction sensitive to amylase digestion, the degree of which increases with maturation (Figure 6A).[78] Mature, platelet-forming megakaryocytes are highly glycogen positive.[78] Platelets are glycogen positive. In most cases, MgBls contain a sodium fluoride-sensitive α-naphthylacetate esterase (NAE).[79,80] They also possess acid phosphatase (AcPase) which increases with maturation of the cell (Figure 6B).[77] However, mature megakaryocytes contain predominantly methanol-sensitive AcPase whereas the methanol-resistant enzyme is found in the immature forms (Figure 6, C and D).[77,81-84] Acetylcholine esterase is found in rat and guinea pig MgBl, but not in humans.[86] Myeloperoxidase (MPO) is absent. Cytochrome oxidase is variably positive.[79] Naphthol AS-D chloroacetate esterase (NCE) is negative, as is α-naphthyl butyrate esterase (NBE).[79]

The ultrastructural characteristics of MgBl resemble lymphoblasts without differentiation features of the cytoplasm. Demarcation membranes and alpha granules appear later.[68,87,92,104] Cells with prominent demarcation membranes, granules, and open canalicular systems are named micromegakaryocytes.[101] MPO is absent on the EM level.

Platelet peroxidase (PPO) is present in the majority of the cases.[26,29,87,97,102] It is localized in the endoplasmic reticulum (ER) and perinuclear membrane and is negative in Golgi cisternae and granules (Figure 4B). This peroxidase differs from MPO found in monocytic and granulocytic precursors which is localized in the Golgi apparatus and azure granules.[26,29,92,93] PPO is much more sensitive to glutaraldehyde fixation than MPO and it is visible only on the EM level.[92,93] The functions of PPO also differ from MPO. The most important effects of MPO are bactericidal, viricidal, and fungicidal activity, whereas PPO is implicated in platelet prostaglandin synthesis.[94] Catalase may also be visible in EM.[98,99] Glycoprotein IIb/IIIa (the platelet-megakaryocytic antigen) is specific for the early stage of megakaryocyte lineage commitment.[95,103,105]

b. Myeloblasts (MyBl) and Monoblasts (MoBl)

It is conventionally accepted that myeloid and monocytic cell lineages have a common precursor in the uncommitted stem cell pool and that they separate with further maturation and differentiation in the committed stem cell pool.

EM is a useful tool for identifying early features of myeloid maturation. The appearance of cytoplasmic granules in normal granulocyte and monocyte precursors were described as a first morphologically distinguishable differentiation feature in these cells. Granules are larger in granulocytic series (primary granules) with MPO and AcPase activity.[27,89] The presence of MPO is a cytochemical expression of myeloid differentiation, and an important criterion for the diagnosis of AML.[1,9,10,15,42] MPO is present in granulocytes, monocytes, and their precursors. However, the enzyme differs in these cells.[8,63,65] Eosinophilic peroxidase is distinct from neutrophil peroxidase with regards to subcellular localization, substrate specificity, inhibitor sensitivity, and different sensitivity to fixatives.[63,65]

Myeloperoxidase at the EM level is considered the earliest marker of myeloid differentiation, more sensitive than MPO on the LM level. MPO appears very early in the maturation of myeloid cells in the ER and Golgi apparatus, prior to the development of granules (Figure 4A). The same pattern was seen in basophilic and eosinophilic precursors helping in the distinction of nongranular myeloid precursors from lymphoblasts.[27,89] Similarly, small MgBl can be recognized by the presence of PPO localized exclusively in the ER and nuclear membrane.[93] PPO can also be seen in MyBl, but in these cells it is primarily localized in granules.[89,92,93] MPO granules are found in monocytic precursors but they are smaller than those in MyBl and appear relatively late in monocyte maturation.[31]

Acid phosphatase activity in blast cells may not be specific for any particular type of leukemia, since a positive reaction was seen in MyBl, MoBl, lymphoblasts (LyBl), and MgBl.[1,2,8-10,15,111,118] AcPase presents coarse granular deposit in the myeloid series, a finer and more diffuse reaction throughout the cytoplasm in monocytes, and different characteristics in LyBl, MgBl, and hairy cell precursors.[1,2,10,24,130] At the EM level, AcPase was observed inconsistently in the early myeloid precursors. This enzyme appears later than MPO.[10] The appearance of small granules with AcPase activity is considered an early feature of monocytic differentiation.[31] Monoblasts contain small granules with AcPase activity and

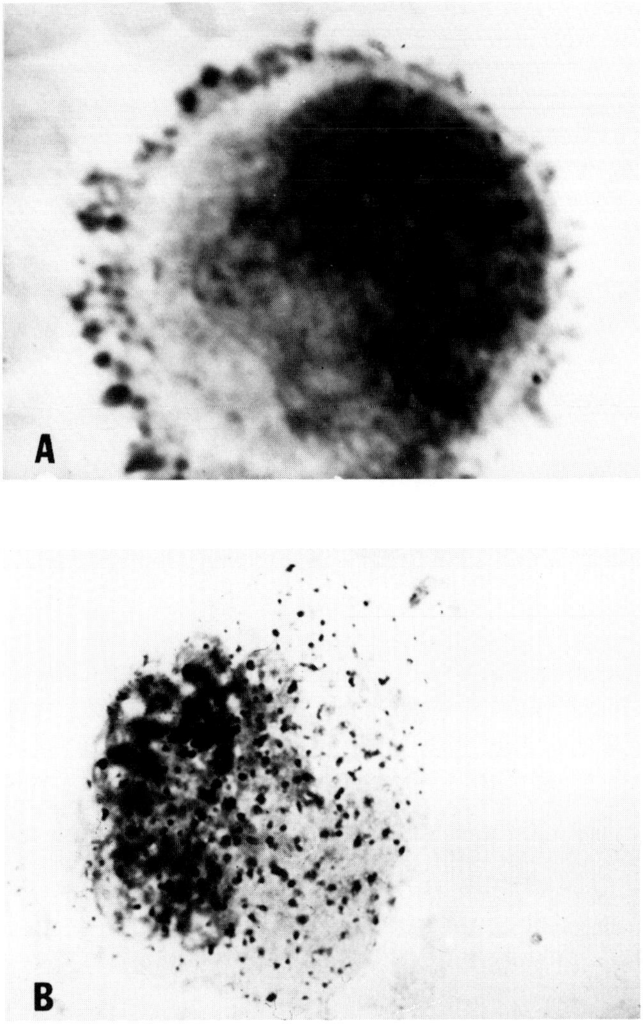

FIGURE 6. Cytochemical aspects of megakaryocyte maturation. (A) Platelet-forming megakaryocyte, PAS reaction. Areola of PAS-positive, young platelets. (Magnification × 1,250.) (B) Mature megakaryocyte, strong acid phosphatase (naphthol AS-MX phosphatase) activity. (Magnification × 1,250.) (C) Methanol-induced inhibition of the sensitive AcPase in mature megakaryocyte. (Magnification × 1,250.) (D) Methanol does not inhibit resistant AcPase in immature megakaryocyte. (Magnification × 1,250.)

lack MPO. AcPase reaction can also be demonstrated in the Golgi cisternae. With AcPase and MPO at the EM level, three types of granules can be seen in monocytic cells. They appear sequentially during maturation: primary — AcPase-positive, MPO-negative; secondary — MPO-positive, AcPase-positive; and tertiary — AcPase-negative, MPO-negative granules.

Cells of monocytic lineage show strong NAE activity sensitive to sodium fluoride at the LM and EM levels (Figures 1B and 7A).[1,15,17,24,114,155] NAE esterase is inconsistently present in MyBl. Most are negative, with the remainder showing diffuse, granular, or focal reactions of varying intensity.[8] However, this esterase is different from monocyte-associated isoenzymes.[116,178]

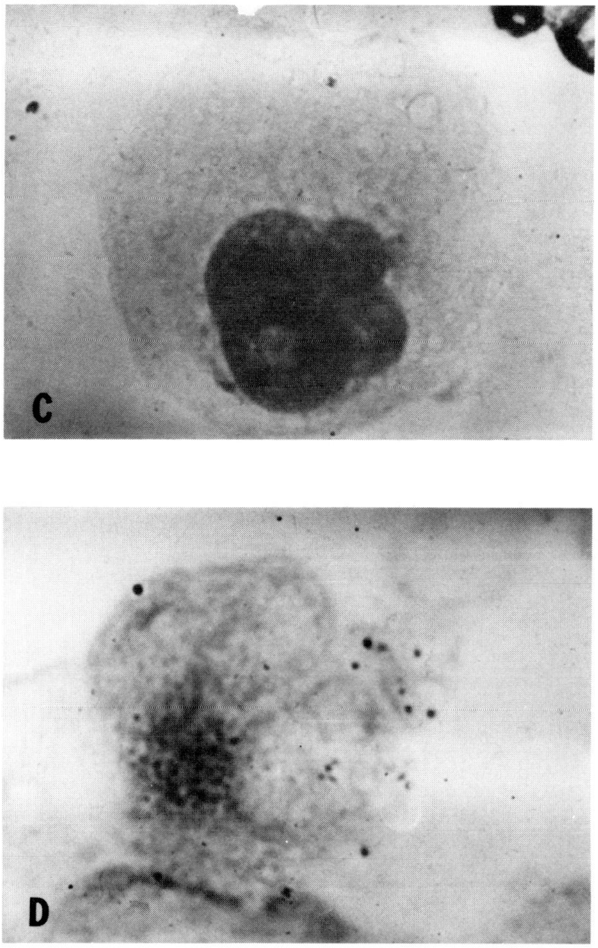

FIGURE 6 (continued)

Naphthol AS-D-chloroacetate esterase (NCE) is considered to be a specific esterase for myeloid series, although lesser activity can be visualized in many other cell types.[10,117,178] NCE activity increases with maturation, but because the reaction may be negative in more immature MyBl, identified by MPO, it is considered as less indicative. Esterase activity is low or absent in monocytic series. NCE distinguishes a granulocytic sarcoma from large-cell lymphoma.[120]

The majority of MyBls are PAS negative. Some may have a very light background or fine granules, which is quite different from the coarse PAS granular deposits typical for ALL.[1,9,10,107] This reaction is not helpful in the characterization of monocytic precursors, since some MoBls are negative, some show fine granular reaction like normal monocytes, and some show large aggregates of PAS-positive material. It is still accepted that Sudan Black B positivity should separate MyBls and MoBls from other immature cells.[1,8]

Alkaline phosphatase is negative in blast cells. This enzyme is low in mature polymorphonuclear neutrophils (PMNs) of CML and AML, which is contrary to ALL, where high or normal activity is found in those cells.[5,6,88]

Antibodies to myeloid and monocytic cell lineages have been produced, but most of them identify already committed cell types.[121-124] Some specificity to earlier stages was

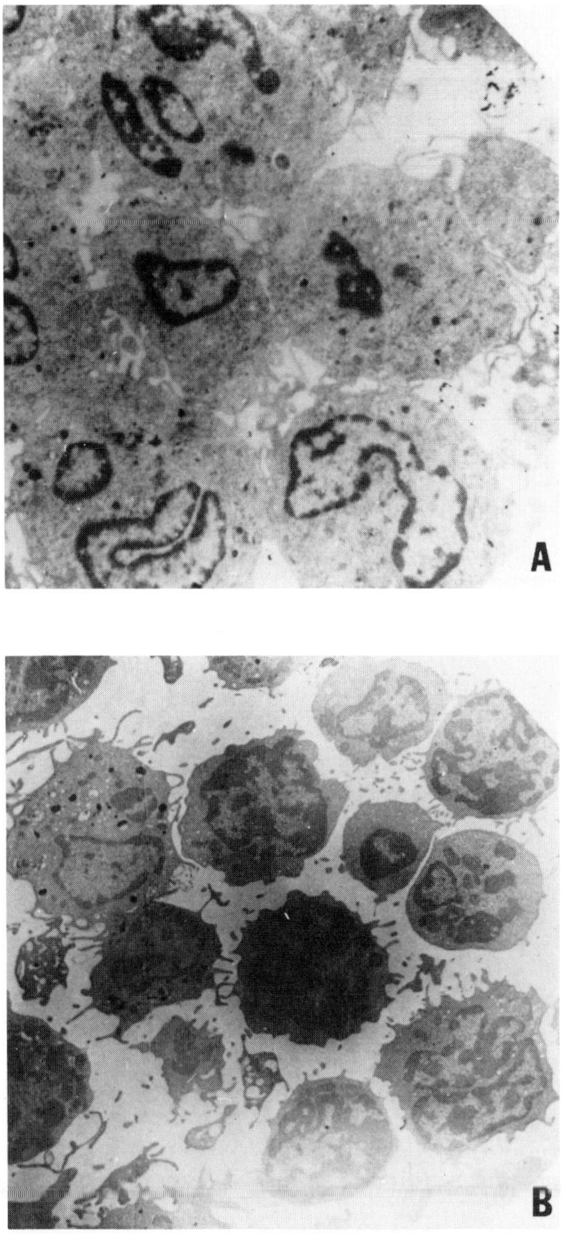

FIGURE 7. Less common types of leukemic cells. (A) Monoblast from a patient with monocytic blastic transformation, showing positive stain for nonspecific esterase. (Magnification × 5,000.) (B) Myeloperoxidase-stained, TEM micrograph of a mixed leukemia. Two cells containing MPO-positive granules are seen at the bottom, the remaining cells are mostly "lymphoid" in nature. (Magnification × 5,000.) (C) TEM, ultrastructure, hairy cell leukemia in buffy coat. Long cytoplasmic villi and pinocytic activity can be seen. (Magnification × 11,250.) (D) Peripheral blood from a patient with AML showing hybrid cells: "Burkitt"-type vacuoles inside cytoplasm of myeloblasts. (Magnification × 1,250.) (A and B from Polliack, A. et al., in *Human Leukemias,* Polliack, A., Ed., Martinus Nijhoff, Boston, 1984; C from Flandrin, G. et al., in *Human Leukemias,* Polliack, A., Ed., Martinus Nijhoff, Boston, 1984. With permission.)

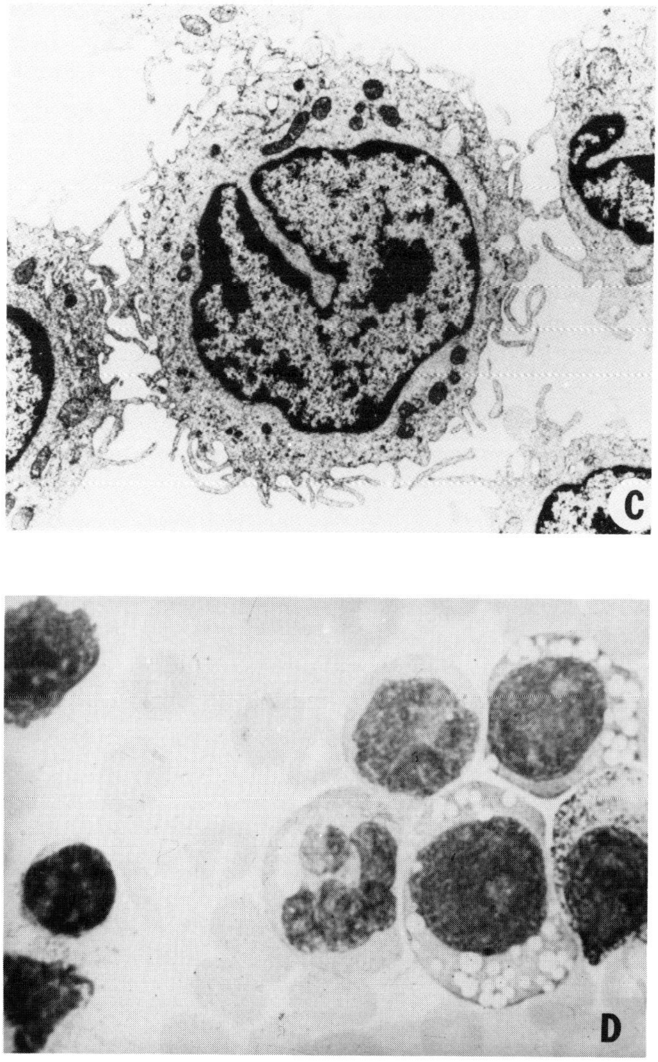

FIGURE 7 (continued)

attributed to OKM1.[186,187] There is still a need for a more specific determinant of an early granulocyte differentiation.

c. Lymphoblasts

Lymphoblasts (LyBl) lack morphologic characteristics at the LM level, and cytochemical markers are less significant than in myeloid cell precursors. Even at the EM level the assessment of lymphoblast-specific identifier is restricted. The main criteria used to distinguish lymphoid cell precursors have evolved from studies of immune markers, deficiency of cytochemical markers typical for other cell types, presence and distribution of terminal desoxynucleotidyl transferase (TdT) and acid hydrolases, and ultrastructural morphology.[109,110]

This multiparameter approach is not only helpful in distinguishing LyBls from blasts of unknown origin but also in identifying specific subgroups of lymphoid cells.

Immunological marker studies using a panel of well-characterized monoclonal antibodies are of proven value for the differentiation of lymphoid precusors from other blasts and for the subclassification of ALL.[36,38,125-129,181,192] Monoclonal antibodies (McAb) are helpful in identifying both the major lymphocyte types (B cells and T cells) and their subsets.[175] The entire group of lymphocyte-related malignant diseases has been reclassified according to these criteria.[191] The distinction between subgroups has been shown to have important therapeutic and prognostic implications.[190,192] Common ALL has a better prognosis. Consequently, pre-B ALL responds well to therapy and should be distinguished from B cell ALL, with the worst prognosis. A typical phenotype may be demonstrated in the majority of cases. However, some variations are not excluded. Immunological marker studies are beyond the scope of this chapter, and the reader is addressed to the excellent review articles cited above.

A recently introduced assay based on immunogold staining of cell surface markers provided a more distinct and reliable criterion than standard immunofluorescent assays of bone marrow aspirates and peripheral smears.[34-36,145,146] Antigen sites were detected not only on the surface membrane of lymphocytes but on cytoplasmic membrane of ER and perinuclear envelope. Combined cytochemistry with immunogold staining in the same TEM preparation permits the precise identification of subtypes. This has provided additional information to LM phenotyping.[38]

Myeloid and megakaryocyte markers such as MPO, PPO, and Sudan Black B (SBB) are lacking in LyBls.[1,12,15] Although 90% of ALL blasts (except L3) are PAS positive, this cannot be considered a distinctive pattern for ALL because other blasts may contain PAS-positive material. Only the blocks of PAS-positive material contrasting an empty cytoplasm can be used as an example for a distinct criterion in favor of ALL. Other PAS-positive reactions may be seen in erythroleukemia, monoblastic leukemia, etc. Therefore, a positive PAS reaction is indicative of ALL only in the battery of cytochemical assays where the majority of blasts are negative for specific markers like MPO, SBB, and NAE.[51,91,169]

Terminal desoxynucleotidyl transferase (TdT) and acid hydrolases have been identified in lymphocytic cell lineage more frequently than other enzyme markers. TdT is an intracellular enzyme which is generally not detectable in mature lymphocytes.[131,133,140] It is found in ALL, with the exception of B cell subtype and may be used to detect early relapse or residual leukemic cells. Approximately 5% of AML may be TdT positive and this enzyme should not be used alone for distinguishing AML from ALL.[133] TdT detection by immunofluorescent assays is the procedure of choice in clinical practice. This is also a useful predictor of the likelihood of the response to treatment with vincristine and prednisone in acute leukemia and in the blast phase of CML.[113]

It was known from long ago that several hydrolyses (AcPase, β-glucuronidase, β-glucosaminidase, and acid NAE) can be helpful in distinguishing normal from leukemic T cells. High activity is found in T cell as compared to B cell leukemias.[8,134-137,153]

AcPase appears in two distinct patterns: a single large granule inside cells with high nucleocytoplasmic ratio and/or two to three small- to medium-sized granules localized in the Golgi apparatus of lymphocytes with larger cytoplasm (Figure 1C).[8] AcPase activity is related to immunologic markers.

Ultrastructural morphology may also help to distinguish LyBls. LyBls are cells with an irregularly shaped indented nucleus, containing heavily condensed chromatin aggregates along the nuclear membrane. Contrary to LyBl, MyBl have round or oval, uniformly shaped nucleus with a fine, dispersed chromatin pattern. The amount of the cytoplasm is not a reliable point of distinction but LyBls do not contain cytoplasmic granules.[10,16,18,39,180]

B cells may present pale, small, and localized granules. They are positive with CD10, CD19, or CD26 McAbs. These lymphocytes may also have morphological features characteristic of monocyte series, such as peripherally localized heterochromatin. However, anti-T (CD2, CD5, or CD7) and anti-monocyte (CD11 or CD14) antibodies can be used as additional criteria for the classification of B cells.

The multiparameter approach provides a number of criteria for subclassification of acute leukemia. How much they can add to a more precise prognosis depends more on drugs available and specific sensitivity of leukemic cells than to any rigid subclassified cell type.

C. CHARACTERIZATION OF BLASTS IN LESS COMMON TYPES OF LEUKEMIA

Differentiation of blasts is important for diagnosis of less frequent types of leukemia.

1. Mixed- and Hybrid-Cell Leukemia

For a long time, morphologic studies have suggested that, although rare, there may be a proliferation of leukemic blasts with markers of more than one cell type or even cell lineages. Ultrastructural morphology, including cytogenetics, cytochemistry, and immunocytochemistry, combined with immunogold technique as auxillary tools, is helpful to differentiate those blasts into leukemia of mixed or hybrid type.[40,44,45,69,142]

Mixed leukemia was recognized as an entity in the case of AMMoL (M4) and erythroleukemia (M6). In these two examples, the two emerging populations of leukemic cells may belong either to a close common progenitor giving rise to myeloid and monocyte lineage (AMMoL) or, because the malignant alteration hit a more immature stage of stem cell maturation, to a more distant progenitor giving rise to different cell lineages such as myeloid and erythroid (AEL). The existence of blast cells involving malignant proliferation in more than one cell lineages is considered biphenotypic or mixed acute leukemia (Figure 7, B and D).

Recent studies suggest the existence of hybrid blasts (single cells with markers indicating different blast types). Leukemias of hybrid phenotype are those in which cells appear to express markers specific for two or more distinct cell lineages.[45] The pathogenesis of such mixed cell population may result from the loss of structure or function by some malignant cells, expansion of a small subpopulation of cells, or as a result of chemotherapy.

2. Erythroleukemia

Erythroleukemia (AEL) was described by DiGuglielmo.[141] The name was given to denote involvement of erythroblasts in the initial stage of acute leukemia. Many of these cases develop a myeloblastic type of leukemia later on. The same author distinguished another entity, a single erythroblastic involvement and, in 1956, he named it myelosis erythremica. These two entities were given the names [illegible] and [illegible], respectively.[illegible]

The FAB classification does not recognize these two types and suggests M6 category for erythroblastic involvement in blast cell proliferation. AEL blast cell proliferation needs distinction from myelodysplastic syndrome (MDS). Sometimes erythroblasts cannot be distinguished morphologically from lymphoblasts. Immunohistochemical staining for hemoglobin and EM morphology may help to provide information about hemoglobin synthesis that is characteristic only for erythroblastic maturation.[147,148]

In AEL, proliferating erythroblasts are abnormal, although some signs of disturbed maturation may be present. EBls show the presence of polyribosome clusters, well-developed Golgi, regions of smooth ER, and mitochondria. Nucleoli are present. Ferritin can be seen in most cells dispersed throughout the cytoplasm and in aggregates. Colloidal gold has been applied to visualize cell surface antigens.

The main cytochemical characteristics of EBls are the following: deposits of PAS-positive material, MPO may be positive, PPO is negative, AcPase is positive, and NaF-sensitive NAE is positive. Strong pseudoperoxidase activity provoked by hemoglobin is also present.

ACKNOWLEDGMENT

The authors are indebted to Dr. Edward Henderson, for reviewing the manuscript; Dr. Bruce Wetzel, for providing us with SEM microphotographs; Dr. Ljubisa Rakic, for continuous support; Dr. Barbara Illman, for editing this manuscript; and Mrs. Florence Rosenthal, for help with the bibliography.

The authors acknowledge support by the Fannie E. Rippel Foundation, Annandal, NJ, and the Carl Zeiss, Co., Inc., U.S./Germany.

REFERENCES

1. **Hayhoe, F. G. J., Quaglino, D., and Doll, R.,** The Cytology and Cytochemistry of Acute Leukemia, MRC Special Report Series, Her Majesty's Stationery Office, London, 1964.
2. **Barka, T. and Anderson, P.,** *Histochemistry,* Harper & Row, New York, 1963.
3. **Barka, T.,** Nomenclature in histochemistry, *J. Histochem. Cytochem.,* 14, 575, 1966.
4. **Bahr, G.,** Frontiers in quantitative cytochemistry, *Anal. Quant Cytol.,* 1, 1, 1979.
5. **Kaplow, L. S.,** The application of cytochemistry in automation, *J. Histochem. Cytochem.,* 25, 990, 1977.
6. **Kaplow, L. S.,** Leukocyte alkaline phosphatase in diseases, *Crit. Rev. Clin. Lab. Sci.,* 2, 243, 1971.
7. **Hirai, K., Ogawa, K., Wang, J. Y., and Ueda, T.,** Varied cytochrome oxidase activities of the alveolar type I, II and III cells in rat lungs. Quantitative cytochemistry, *Electron Microsc.,* 36, 449, 1989.
8. **Markovic, O., Markovic, N., and Rakic, Lj.,** *Quantitative Cytochemistry of Enzymes,* University Press, Belgrade, 1986.
9. **Stoward, P. and Polak, J.,** *Histochemistry, The Widening Horizons of its Application in the Biomedical Sciences,* John Wiley & Sons, Chichester, 1981.
10. **Polliak, A.,** *Human Leukemias, Cytochemical and Ultrastructural Techniques in Diagnosis and Research,* Martinus Nijhoff, Boston, 1984.
11. **Wetzel, B., Cannon, G. B., and Alexander, E. L.,** A critical approach to the scanning electronmicroscopy of cells in suspension, Scan Electr Micr, 1974, Part III, Proceedings of the Workshop on Advances of the SEM, IIT Research Institute, Chicago, IL.
12. **Guntz, F. and Henderson, E.,** *Leukemia,* Grune & Stratton, New York, 1983.
13. **Bessis, M.,** *Live Blood Cells and Their Ultrastructure,* Springer-Verlag, Berlin, 1983.
14. **Catovsky, D.,** *Leukemic Cell,* Churchill Livingstone, London, 1981.
15. **Flandrin, G. and Daniel, M. T.,** Practical value of cytochemical methods for the classification of acute leukemia, *Recent Results Cancer Res.,* 43, 43, 1973.
16. **Cawley, J. C. and Hayhoe, F. G. J.,** *Ultrastructure of Hemic Cells. The Cytologic Atlas of Normal and Leukemic Blood and Bone Marrow Cells,* W. B. Saunders, Philadelphia, 1973.
17. **Meistner, H.,** Applied cytochemistry in hematology, *Folia Haematol. (Liepzig),* 100, 186, 1973.
18. **Beard, M. E. J. and Fairley, G. H.,** Ultrastructure of normal and leukemic cells and the characteristic features of the various types of acute leukemia, *Semin. Hematol.,* 11, 5, 1974.
19. **Dvorak, A. and Monahan, R.,** Diagnostic electron microscopy, *Pathobiol. Annu.,* 16, 101, 1981.
20. **Hayhoe, F. G. J. and Quaglino, D.,** *Hematological Cytochemistry,* Churchill Livingstone, London, New York, 1980.
21. **Pearse, A. G. E.,** *Histochemistry. Theoretical and Applied,* Williams & Wilkins, Baltimore, 1968.
22. **Markovic, N. S.,** Morphology of Blast Cells, Possible Parameter to Estimate Prognosis of Acute Leukemia, Doctorate thesis, University Press, Belgrade, 1975.
23. **Henderson, D. and Papadimitrou, J.,** *Ultrastructural Appearance of Tumors,* Churchill Livingstone, London, 1986.
24. **Koiki, T.,** Hematological cytochemistry: current status, *Nippon Ketsueki Gakkai Zasshi (Acta Hematol. Jpn.),* 50(8), 1519, 1987.
25. **Greaves, M. F.,** Recent progress in the immunological characterization of leukemic cells, *Blut,* 34, 349, 1977.
26. **Brenton-Gorius, J. and Guichard, T.,** Amelioration techniques permettant de reveler la peroxydase plaquetaire, *Nouv. Rev. Fr. Hematol.,* 16, 381, 1976.

27. **Ackerman, G. A. and Clark, M. A.,** Ultrastructural localization of peroxydase activity in human bone marrow cells, *Z. Zellforsch. Microsc. Anat.,* 117, 463, 1971.
28. **Bainton, D. F.,** Differentiation of human neutrophilic granulocytes normal and abnormal, in *The Granulocyte: Function and Clinical Utilization,* Greenwalt, T. G. and Jamieson, G. A., Eds., Allan R. Liss, New York, 1977, 1.
29. **Brenton-Gorius, J. and Rayes, F.,** Ultrastructure of human bone marrow maturation, *Int. Rev. Cytol.,* 46, 251, 1976.
30. **Marie, J. P.,** Ultrastructural localization of peroxydases in undifferentiated blasts during the blast crisis of chronic myelogenous leukemia, *Br. J. Haematol.,* 43, 549, 1979.
31. **O'Brien, M., Catovsky, D., and Costello, C.,** Ultrastructural cytochemistry of leukemic cells: characterization of early small granules in monocytes, *Br. J. Haematol.,* 45, 201, 1980.
32. **Catovsky, D. and Costello, C.,** Cytochemistry of normal and leukemic leukocytes: a review, *Basic Appl. Histochem.,* 23, 255, 1979.
33. **Greaves, M. F.,** Monoclonal antibodies as probes for leukemia heterogeneity and haemopoietic differentiation, in *Leukemia Markers,* W. Knapp, Ed., Academic Press, New York, 1981, 19.
34. **Faulk, W. P. and Taylor, G. M.,** An immunogold colloidal method for the electronmicroscope, *Immunocytochemistry,* 8, 1081, 1971.
35. **De May, J.,** Colloidal gold probes, in *Immunocytochemistry: Practical Application in Pathology and Biology,* Polak, J. M. and van Noorden, S., Eds., Wright, Bristol, 1983, 82.
36. **Matutes, E. and Catovsky, D.,** The fine structure of normal lymphocyte subpopulation, a study with monoclonal antibodies and immunogold technique, *Clin. Exp. Immunol.,* 50, 416, 1982.
37. **Markovic, N., Markovic, O., and Young, D.,** Measurement of enzyme kinetics in single cells: concept and model, *Clin. Chem.,* 23, 1427, 1977.
38. **Hansmann, M. L., Harpprecht, J., Westphal, E., and Miller, W.,** Immunoelectronmicroscopic demonstration of antigenic sites on lymphoid cells using a human monoclonal antibody, *Eur. J. Haematol.,* 49, 35, 1989.
39. **Matutes, E. and De Oliviera, P.,** The role of ultrastructural cytochemistry and monoclonal antibodies in clarifying the nature of undifferentiated cells in acute leukemia, *Br. J. Haematol.,* 69, 205, 1988.
40. **Lavabre-Bertrand, T., Vanneran, H., Donadio, M., Escola, M., and Taib, J.,** Interest of simultaneous characterization of morphological, cytochemical and immune phenotyping of putative hybrid acute leukemia, *Eur. J. Haematol.,* 42, 202, 1989.
41. **Bennett, J. M., Catovsky, D., Daniel, M. T., Flandrin, G., Galton, D., and Gralnick, H.,** Proposal for the classification of acute leukemias, *Br. J. Haematol.,* 33, 451, 1976.
42. **Markovic, O. and Markovic, N.,** Cytochemistry of leukocytes: leukocyte peroxidase, *Zb. Med. Fak. Skopje,* 18, 365, 1972.
43. **Markovic, N., Markovic, O., and Young, D.,** Simplified method for measuring enzyme kinetics in single cells, *Clin. Chem.,* 29, 1578, 1983.
44. **Morgan, M. A.,** Biphenotypic leukemia: a case of mixed T lymphoid and myeloid leukemia, *J. Clin. Pathol.,* 38, 575, 1985.
45. **Martelsmann, R. and Kozier, B.,** Evidence for distinct lymphocytic and monocytic population in patients with terminal transferase positive acute leukemia, *Blood,* 51, 1050, 1978.
46. **Quattrin, N.,** Leucemies aigues a basophilies, *Nouv. Rev. Fr. Hematol.,* 13, 745, 1973.
47. **Boll, A. and Hippel, H.,** Use of cytochemistry and FAB classification in leukemia and other pathological status, *Am. J. Med. Technol.,* 47, 437, 1981.
48. **Bessis, M.,** Pathology of the leukemic cells or reasons why some leukemias are unclassifiable, *Blood Cells,* 1, 183, 1975.
49. **Markovic, N. and Markovic, O.,** Quantitative relation among maturational stages of granulocytes: possible parameter for subclassification of acute leukemia, *Makedon. Med. Pregl.,* 32, 141, 1978.
50. **Markovic, N. and Markovic, O.,** Comparative study between cytochemistry of blood cells and recently recommended classifications of acute leukemia, *J. Histochem. Cytochem.,* 27, 715, 1979.
51. **Jensson, S. E., Gripenberg, J., Vuopio, P., and Teerenhovi, L.,** Classification of acute leukemias by light and electronmicroscopy cytochemistry, *Sci. J. Haematol.,* 25, 412, 1980.
52. **Youness, E., Trujillo, J. M., and Ahearn, M. J.,** Acute unclassified leukemia: a clinicopathologic study with implication of electronmicroscopy, *Am. J. Haematol.,* 1980, 9, 79.
53. **Polly, N., O'Brien, M., DeCastro, T. J., and Matutes, J.,** Characterization of blast cells in chronic myeloid leukemia in transformation, acute myelofibrosis and undifferentiated leukemia, *Br. J. Haematol.,* 59, 277, 1985.
54. **Altman, F. P.,** Quantitation in histochemistry: a review of some commercially available microdensitometers, *Histochem. J.,* 7, 375, 1975.

55. **Hewson, J. W., Bradstock, K. F., Kerr, A., and Rose, R. G.,** Characterizing difficult acute leukemias. A combined electronmicroscopic and immunological marker study, *Pathology,* 18, 99, 1986.
56. **Bartels, P.,** Expert system in histopathology, *Ann. Quant. Cytol. Histol.,* 11, 1, 1989.
57. **Polliak, A. and Prokocimer, M.,** Use of multiparameter studies and SEM in the interpretation and correlation of surface morphology with cell type in 135 cases of acute leukemia, *Cancer Res.,* 41, 1171, 1971.
58. **DeHarven, E.,** Scanning electronmicroscopy in the backscattered electron-imaging (BEI) mode: application in clinical hematology, *Ultrastruct. Pathol.,* 11, 71, 1987.
59. **Lamberterghi, D. G.,** Scanning electron microscopy of blood cells, *Scan. Electron Microsc.,* 2, 817, 1985.
60. **DeHarven, E. and Solgio, D.,** Scanning electron microscopy of human leukocytes, *Scan. Electron Microsc.,* 2, 773, 1984.
61. **Polliack, A., Gurfel, D., and Leizerovitz, R.,** Surface morphology of leukemic cells as seen by scanning electron microscopy, in *Human Leukemia,* Polliak, A., Ed., Martinus Nijhoff, Boston, 1984, 405.
62. **Markovic, O. and Markovic, N.,** Enzymes in blast cells, *Proc. 3rd Congr. Yugoslavian Hematol.,* 1, 109, 1978.
63. **Markovic, O.,** Kinetics of myeloperoxidase reaction in single neutrophils, Doctorate thesis, University of Belgrade Academic Press, Belgrade, 1977.
64. **Marie, J. P., Vernand, J. P., Dreyfus, B., and Brenton-Gorius, J.,** Ultrastructural localization of peroxydase in undifferentiated blasts during the blast crisis of chronic granulocytic leukemia, *Br. J. Haematol.,* 43, 549, 1979.
65. **Williams, W. C. and Wess, G. B.,** Megakaryoblastic transformation of chronic granulocytic leukemia, *Cancer,* 49, 921, 1982.
66. **Hanada, T. and Nakarava, M.,** Megakaryoblastic transformation of chronic granulocytic leukemia in child, *Scand. J. Haematol.,* 33, 476, 1984.
67. **Marshall, D. and Duhasnel, G.,** La leucemie a megakaryocytes, *Sem. Hop.,* 15, 887, 1954.
68. **Koiki, T.,** Megakaryoblastic leukemia: the characterization and identification of megakaryoblasts, *Blood,* 64, 683, 1984.
69. **Morgan, M. A.,** Biphenotypic leukemia, a case of mixed lymphoid and myeloid leukemia, *J. Clin. Pathol.,* 38, 575, 1985.
70. **DiGulielmo, G.,** I megakaryocyti nel sangue periferico contributo. Alla patologia del sisteme megakariopoetico, *Haematologica,* 4, 182, 1923.
71. **Bergman, K. L. and VanSlyck, E. J.,** Acute myelofibrosis an accelerated variant of agnogenic myeloid metaplasia, *Ann. Intern. Med.,* 74, 232, 1971.
72. **Bergman, R. M. and Pangalis, G. A.,** Acute myelosclerosis, *Cancer,* 43, 279, 1979.
73. **Habib, A. S. and Lee, H.,** Acute myelogenous leukemia with megakaryocytic myelosis, *Am. J. Clin. Pathol.,* 74, 705, 1980.
74. **Sultan, C., Sigaux, F., Imbert, M., and Reyes, F.,** Acute myelodysplasia with myelofibrosis, a report of 8 cases, *Br. J. Haematol.,* 49, 11, 1981.
75. **Balducci, L., Weitzner, S., Beghe, C., and Morrison, F. C.,** Acute megakaryocyte leukemia. Description of a case initially seen as preleukemic syndrome, *Arch. Intern. Med.,* 138, 186, 1982.
76. **Egner, J. R., Aabo, K., and Dimitrov, N.,** Megakaryocytic leukemia as a phase of myeloproliferative disorder, *Scand. J. Haematol.,* 28, 186, 1982.
77. **Markovic, O. and Shulman, N. R.,** Megakaryocyte maturation indicated by methanol inhibition of an acid phosphatase shared by megakaryocytes and platelets, *Blood,* 50, 905, 1977.
78. **Markovic, O. and Markovic, N.,** Cytochemical investigation of megacaryocyte glycogen, in Proc. 4th Congress of Yugoslavian Clinical Chemists, Vol. 1, Sarajevo, Yugoslavia, 1975, 97.
79. **Markovic, N. and Markovic, O.,** unpublished data, 1990.
80. **Denotolander, G. J. and TeVeld, J.,** Megacaryocyte leukemia, a report of three cases, *Br. J. Haematol.,* 42, 9, 1979.
81. **Markovic, O.,** Cytochemistry of megakaryocytes. Acid phosphatase: identification of the enzyme, *Zb. Med. Fak. Skopje,* 22, 91, 1976.
82. **Markovic, O.,** Cytochemistry of megakaryocytes. Acid phosphatase. Selection of the cytochemical method, *Zb. Med. Fak. Skopje,* 22, 99, 1976.
83. **Markovic, O.,** Cytochemistry of megakaryocytes. Acid phosphatase, a new assay for cytochemical investigation, *Zb. Med. Fak. Skopje,* 22, 107, 1976.
84. **Markovic, O.,** Cytochemistry of megakaryocytes. Acid phosphatase: separation of platelet acid phosphatase isoenzymes, *Makedon. Med. Pregl.,* 30, 19, 1975.
85. **Markovic, O. and Markovic, N.,** Influence of the different fixatives on peroxidase activity in leukocytes, in Proc. 3rd Congress of Yugoslavian Hematologists, Vol. 1, Sarajevo, Yugoslavia, 1978, 114.

86. **Jackson, C. W.,** Cholinesterase a possible marker for early cells of megakaryocytic series, *Blood,* 42, 413, 1973.
87. **Brenton-Gorius, J. and Gourdin, M. F.,** Ultrastructure of the leukemic cells, in *The Leukemic Cell,* Catovsky, D., Ed., Churchill Livingstone, London, 1981, 87.
88. **Kaplow, L. S.,** Leukocyte alkaline phosphatase cytochemistry: application and methods, *Ann. N.Y. Acad. Sci.,* 155, 911, 1968.
89. **Marie, J. P.,** Ultrastructural localization of peroxidase in undifferentiated blasts during the blast crisis of chronic granulocytic leukemia, *Br. J. Haematol.,* 43, 549, 1979.
90. **Ploem, J. S.,** Quantitative fluorescence microscopy, in *Analytical and Quantitative Methods in Microscopy,* Meck, G. A. and Edler, H. J., Eds., Cambridge University Press, 1977, 55.
91. **Williams, W., Beutler, E., and Erslev, A.,** *Hematology,* McGraw-Hill, New York, 1972.
92. **Brenton-Gorius, J.,** Review of the ultrastructure and cytochemistry of megakaryoblastic leukemia, in *Human Leukemias,* Polliak, A., Ed., Martinus Nijhoff, Boston, 1984, 63.
93. **Brenton-Gorius, J. and Guishard, J.,** Ultrastructural localization of peroxidase activity in human platelets and megakaryocytes, *Am. J. Pathol.,* 66, 277, 1972.
94. **Gerrard, J. M. and White, J. G.,** Localization of platelet prostaglandin production in the platelet dense tubular system, *Am. J. Pathol.,* 83, 283, 1976.
95. **Vainchenker, W., Dischamps, J., Bastin, J. M., and Guishard, J.,** Two monoclonal antiplatelet antibodies as markers of human megakaryocytic maturation. Immunofluorescent staining and PPO detection in megakaryocyte colonies and in vivo cells from normal and leukemic patients, *Blood,* 59, 514, 1982.
96. **Markovic, O., Markovic, N., and Kaplow, L. S.,** Further studies of myeloperoxidase reaction in single neutrophils, *J. Histochem. Cytochem.,* 26, 214, 1978.
97. **Brenton-Gorius, J. and Reyes, F.,** Megakaryocyte acute leukemia: identification by the ultrastructural demonstration of PPO, *Blood,* 51, 45, 1978.
98. **Roels, F. and Wisse, E.,** Cytochemical discrimination between catalases and peroxydase using DAB, *Histochemie,* 41, 281, 1975.
99. **Anderson, W. A. and Trantalis, J.,** Ultrastructural localization of endogeneous mammary gland peroxidase during lactogenesis in rat, *J. Histochem. Cytochem.,* 23, 295, 1975.
100. **Gripman, J. E.,** The pathogenesis of myelofibrosis in myeloproliferative disorders, *Ann. Intern. Med.,* 92, 857, 1980.
101. **Egushi, M., Skaibara, H., Suda, J., Ozava, T., and Hauashi, I.,** Ultrastructural and ultracytochemical differences between transient myeloproliferative disorder and megakaryoblastic leukemia in Down syndrome, *Br. J. Haematol.,* 73, 315, 1989.
102. **Heynen, M. J., Tricot, G., and Verwilghen, R.,** A reliable method with good cell preservation for the demonstration of peroxidase activity in human platelets and megakaryocytes, *Histochemistry,* 80, 79, 1984.
103. **Enomoto, Y. and Watanate, Y.,** Analytical study of leukemic megakaryoblastic cells using simultaneous demonstration of membrane antigen with monoclonal antibodies and peroxydase cytochemistry at ultrastructural level, *Rinsho Ketsueki (Jpn. J. Clin. Hematol.,* 29, 1150, 1988.
104. **Behnke, O.,** Ultrastructural aspects of megakaryocyte maturation and platelet release, in *Platelets: Production, Function, Transformation and Storage,* Baldini, M. G. and Ebbe, B. S., Eds., Grune & Stratton, New York, 1974, 26.
105. **DeOliviera, P. E.,** The role of ultracytochemistry and monoclonal antibodies in clarifying the nature of undifferentiated cells in acute leukemia, *Br. J. Haematol.,* 69, 205, 1988.
106. **Fujimoto, T. and Ogawa, K.,** Immunoelectron microscopy of ezrin in the rat uriniferous and collecting tubular epithelium, *J. Histochem. Cytochem.,* 37, 1345, 1989.
107. **Bennett, J. M.,** Leukemia, morphology and cytochemistry, in *Leukemia,* Gunz, F. and Henderson, E. S., Eds., Grune & Stratton, New York, 1983.
108. **Wells, O. C.,** Backscattered electron image (BEI) in the scanning electron microscopy, *Scann. Electron Microsc.,* 1, 747, 1977.
109. **Kataoka, K., Sukano, I., Imamura, N., and Kanilira, S.,** The structure of T and B cell of leukemia with special reference to adult T-cell leukemia in Japan, in *Human Leukemia,* Polliack, A., Ed., Martinus Nijhoff, Boston, 1984, 281.
110. **Reys, F., Gourdin, M. F., and Farcet, J. P.,** Immunoelectron microscopic observations in human B-cell leukemia, in *Human Leukemias,* Polliack, A., Ed., Martinus Nijhoff, Boston, 1984, 383.
111. **Markovic, N. and Markovic, O.,** Cytochemistry of acute leukemia, in Proc. Biannual Meeting of Yugoslavian Hematologists, Vol. 1, Dubrovnik, Yugoslavia, 1976, 3.
112. **Markovic, O. and Markovic, N.,** Cytochemistry of leukocytes. Leukocyte alkaline phosphatase, *Zb. Med. Fak. Skopje,* 18, 365, 1972.
113. **Efrati, P., Klaiman, A., Spitz, H.,** Mast cell leukemia, *Blood,* 12, 869, 1957.

114. **Markovic, O. and Markovic, N.,** Cytochemistry of leukocytes. Leukocyte esterases, *Makedon. Med. Pregl.*, 30, 25, 1975.
115. **Payne, B. C., Kim, H., and Pangalis, G. A.,** A method for ultrastructural demonstration of nonspecific esterase in human blood and lymphatic tissue, *Histochem. J.*, 12, 71, 1980.
116. **Daniel, M. T., Flandrin, G., Lejeune, F., and Liso, P.,** Les esterases specifiques monocytaires: utilisation dans la classifiacation des leukemies aigues, *Nouv. Rev. Fr. Hematol.*, 11, 233, 1971.
117. **Markovic, O., Young, D., and Markovic, N.,** Quinidine induced inhibition of esterase activity in a hepatoma cell line, *Clin. Chem.*, 34, 512, 1988.
118. **Barka, T. and Anderson, P. J.,** Histochemical method for acid phosphatase using hexazonium pararosaniline as a coupler, *J. Histochem. Cytochem.*, 10, 741, 1962.
119. **Markovic, O., Young, D., and Markovic, N.,** Quinidine induced inhibition of leukocyte esterases, *Clin. Chem.*, 34, 518, 1988.
120. **Seo, S. and Hull, M.,** Granulocytic sarcoma of the cervix as a primary manifestation, *Cancer,* 40, 3030, 1977.
121. **Mahmond, A. A. and Kellermeyer, R. W.,** Monospecific antigranulocyte sera against human neutrophiles, eosinophiles, basophiles and myeloblasts, *Lancet,* 2, 1163, 1974.
122. **Smith, L. J. and Curtis, J. E.,** Lineage identity of acute leukaemia, *Blood,* 61, 1138, 1983.
123. **Griffin, J. D., Mayer, R. D., Weinstein, H., and Rosenthal, D.,** Surface marker analysis of acute myeloblastic leukemia: identification of differentiation associated phenotypes, *Blood,* 62, 557, 1983.
124. **Thurlow, P. J. and Kerrigan, L.,** Analysis of human marrow with monoclonal antibodies, *J. Histochem. Cytochem.*, 33, 1183, 1985.
125. **Greaves, M. F.,** Analysis of the clinical and biologic significance of lymphoid phenotypes in acute leukemia, *Cancer Res.*, 41, 4752, 1981.
126. **Huhn, D. and Thiel, E.,** Cytochemical and membrane markers in acute lymphoblastic leukemia, *Scand. J. Hematol.*, 26, 311, 1981.
127. **Armitase, R. J. and Linch, D. C.,** Morphology and cytochemistry of human lymphocyte subpopulation defined by monoclonal antibody, *Br. J. Haematol.*, 51, 805, 1982.
128. **Greaves, M. F. and Kerbi, V.,** Phenotypic characterization of pre B-cell leukemia, *Leuk. Res.*, 3, 353, 1979.
129. **Sovill, R. E. and Ronston, I.,** Adult lymphatic leukemia phenotypes defined by monoclonal antibodies, *Blood,* 65, 65, 1989.
130. **Quan, S. P. and Poolsawat, S.,** Ultrastructure and tartrate resistant acid phosphatase localization in a T-cell hairy cell leukemia cell line, *J. Histochem. Cytochem.*, 28, 434, 1983.
131. **Bollum, F. J.,** Terminal deoxynucleotidyl transferase as a haematopoietic cell marker, *Blood,* 54, 1203, 1979.
132. **Bollum, F. J.,** Antibody to terminal deoxynucleotidyl transferase, *Proc. Natl. Acad. Sci. U.S.A.*, 72, 4119, 1975.
133. **Bradstock, K. F. and Hewson, J.,** Expression of TdT in malignant myeloblasts, *Am. J. Clin. Pathol.*, 80, 800, 1983.
134. **Fehr, J. and Scherer, O.,** Nonspecific esterase as a marker for human T lymphocytes: sequential studies during states of transient blood lymphocyte redistribution, *Blut,* 44, 201, 1982.
135. **Catovsky, D., Getto, J., Okos, A., and Miliam, E.,** Cytochemical profile of B- and T- leukemic lymphocytes with special reference to acute lymphoblastic leukemia, *J. Clin. Pathol.*, 27, 767, 1974.
136. **Kuhenhampff, J., Janossy, M. G., and Greaves, M.,** Acid esterase in human lymphoid cells and leukemia blasts: a marker for T lymphocytes, *Br. J. Haematol.*, 36, 235, 1977.
137. **Pangalis, G. A., Jatagunas, Y., and Fassas, P. H.,** Beta glucuronidase activity in lymph node imprints from malignant lymphomas and chronic lymphocytic leukemia, *J. Clin. Pathol.*, 30, 812, 1977.
138. **Yam, L. T., Li, C. I., and Lam, K. W.,** Tartrate resistant acid phosphatase isoenzyme in the reticulum cells of leukemic reticuloendotheliosis, *N. Engl. J. Med.*, 284, 357, 1971.
139. **Kaplow, L. S., Schanble, M. K., and Becktel, L. M.,** Validity of hematologic data in Veterans Administration Hospital laboratories, *Am. J. Clin. Pathol.*, 71, 291, 1979.
140. **Janos, G. and Hoffbrand, A.,** Terminal desoxynucleotidyl transferase enzyme assay and immunological membrane markers in the diagnosis of leukaemia. A multiparameter analysis of 300 cases, *Br. J. Haematol.*, 44, 221, 1980.
141. **DiGuglielmo, G.,** Acute and chronic erythremic myelosis and erythroleukemia, *Rev. Hematol.*, 1, 355, 1946.
142. **Morgan, M. A.,** Biphenotypic leukemia: a case of mixed T lymphoid and myeloid leukemia, *J. Clin. Pathol.*, 38, 575, 1985.
143. **Pangalis, G. A., Rousson, P., and Kittas, C.,** Hairy cell leukemia: clinical and laboratory observation of 11 cases, *Arch. Hell. Med. Soc.*, 7, 525, 1981.

144. **Matutes, E., Robinson, D., and Catovsky, D.,** Ultrastructural analysis of B and T lymphocyte subsets with the immunogold technique, in *Human Leukemias,* Polliack, A., Ed., Martinus Nijhoff, Boston, 1984, 261.
145. **DeWaele, M.,** Colloidal gold as marker for detection of cell surface antigens, *Acta Histochem. (Jena),* 35, 53, 1988.
146. **Heiner, H. E. and Namork, E.,** A study of immunogold labeled blood group in the scanning electron microscopy, *Eur. J. Haematol.,* 41, 17, 1988.
147. **Laskin, W. B. and Cotelingam, J.,** Erythroleukemia in child: value of immunocytochemistry and transmitted electronmicroscopy in its diagnosis, *Am. J. Pediatr. Hematol. Oncol.,* 7, 99, 1985.
148. **Palmey, R. T. and Gilbert, C. J.,** Ultrastructural silver enhancement of Prussian Blue Reactive iron in hematopoietic and intestinal cells, *J. Histochem. Cytochem.,* 36, 433, 1988.
149. **Naumenko, O. I.,** Ultrastructural features of erythropoietic disorders in Rausher virus leukemia, *Exp. Onkol.,* 12, 25, 1990.
150. **Cawley, J. S., Burns, G. F., and Hayhoe, G. J.,** *Hairy Cell Leukemia,* Springer-Verlag, Berlin, 1980.
151. **Flandrin, G. and Daniel, M. T.,** Hairy cell leukemia, *Nouv. Rev. Fr. Haematol.,* 13, 609, 1973.
152. **Li, C. Y., Yam, L. T., and Lam, K. W.,** Studies of acid phosphatase isoenzyme in human leukocytes, *J. Histochem. Cytochem.,* 18, 901, 1973.
153. **Higgy, K. E. and Burns, G. F.,** Discrimination of T, B and null lymphocytes by esterase cytochemistry, *Scand. J. Hematol.,* 18, 437, 1977.
154. **Markovic, O.,** New trends in enzymology: microanalysis inside single cells, *Period. Biol.,* 86, 101, 1984.
155. **Lee, C. I., Lam, K. V., and Yam, L. T.,** Esterases in human leukocytes, *J. Histochem. Cytochem.,* 21, 1, 1973.
156. **Kurosava, H. and Egushi, M.,** Ultrastructural cytochemistry of congenital basophilic leukemia, *Am. J. Pediatr. Hematol. Oncol.,* 9, 27, 1987.
157. **Markovic, O.,** Assay of enzyme activity in single cells, *Period. Biol.,* 82, 491, 1980.
158. **Markovic, N., Lipkin, L., Markovic, O., and Wade, M.,** Device for quantitative cytochemistry: a computerized microdensitometer ''Grain Counter'', *J. Histochem. Cytochem.,* 26, 792, 1978.
159. **Schaefer, H. E., Hellriegel, K. P., and Hannekeuser, H. H.,** Eosinophil leukemia with chloroacetate positive eosinophils, *Blut,* 26, 7, 1973.
160. **Henderson, E. S.,** Treatment of acute leukemia, *Semin. Hematol.,* 6, 271, 1969.
161. **Pedersen, B.,** The blastic crises of chronic granulocytic leukemia, *Br. J. Haematol.,* 25, 141, 1973.
162. **Newman, W.,** A case of acute plasma cell leukemia, *Cancer,* 5, 514, 1952.
163. **Devlin, T. M.,** *Textbook of Biochemistry,* 2nd ed., John Wiley & Sons, New York, 1982.
164. **Rosenthal, S. and Canellos, G. P.,** Characteristics of blasts in chronic granulocytic leukemia, *Blood,* 49, 705, 1972.
165. **Wetzel, B.,** Cell kinetics: an interpretative review of the significance of cell surface form, in Proc. Workshop on Advances in Biomedical Applications of the SEM, IIT Research Institute, Chicago, IL, 1976.
166. **Dautchev, D. and Belpomme, D.,** Critical study of the monuclear leukocyte morphology based on SEM in normal subjects and in patients with lymphoid or monocitoid proliferative disorders, *Biomedicine,* 27, 202, 1977.
167. **Bain, B., Catovsky, D., and O'Brien, M.,** Megakaryocytic transformation of chronic granulocytic leukemia, *J. Clin. Pathol.,* 30, 235, 1972.
168. **Laschi, R., Pasquinelli, G., and Versura, P.,** Scanning electron microscopy application in clinical research, *Scann. Electron Microsc.,* 1, 1771, 1987.
169. **Wintrobe, M. M.,** *Clinical Hematology,* 7th ed., Lea & Febiger, Philadelphia, 1974.
170. **Kaplow, L. S.,** Cytochemical heterogeneity of human circulating monocytes, *Acta Cytol.,* 19, 358, 1975.
171. **Nakane, P. K. and Pierce, G. B.,** Enzyme labeled antibodies: preparation and application for the localization of antigens, *J. Histochem. Cytochem.,* 14, 929, 1984.
172. **Direnzini, M. and Thiry, M.,** Ultrastructural cytochemistry of the mammalian cell nucleolus, *J. Histochem. Cytochem.,* 38, 1237, 1990.
173. **Sixma, J. J. and van der Berg, A.,** Immunoelectron microscopic localization of actin, alpha actin and actin binding protein and myosin in resting and activated human blood platelets, *Eur. J. Cell Biol.,* 48, 271, 1988.
174. **Mutasa, H. S.,** Combination of diaminobenzidine staining and immunogold labeling, a novel technical approach to identify lysosyme in human neutrophils, *Eur. J. Cell Biol.,* 49, 319, 1989.
175. **Crockard, A. and Catovsky, D.,** Cytochemistry of normal lymphocyte subsets defined by monoclonal antibodies and immunocoloidal gold, *Sci. J. Haematol.,* 30, 433, 1983.
176. **Liso, V., Traccoli, G., Specchia, G., and Magno, L.,** Cytochemical normal and abnormal eosinophils in acute leukemia, *Am. J. Hematol.,* 2, 123, 1977.

177. **Baskin, D., Davidson, D. A., and Corp, E.,** An inexpensive microcomputer digital imaging system for densitometry: quantitative autoradiography of insuline receptors, *J. Neurosci. Meth.,* 16, 119, 1986.
178. **Schmalzl, F. and Braunsteiner, H.,** The application of cytochemical methods in the study of acute leukemia, *Acta Haematol.,* 45, 209, 1971.
179. **Coser, P., Quaglino, D., and DePasquale, A.,** Cytological and clinical aspects of tissue mast cells leukemia, *Br. J. Haematol.,* 45, 5, 1980.
180. **Catovsky, D.,** B, T, and null leukemias: electronmicroscopic cytochemistry and surface morphology, *Blood Cells,* 1, 115, 1975.
181. **Butenko, Z. A., Naumenko, O. I., and Sanukova, L. N.,** Ultrastructural characteristics of the blood lymphocytes in some forms of sporadic T-cell lympholeukemia and leukemia associated with HTLV virus, *Exp. Oncol.,* 8, 54, 1986.
182. **Anderson, R. E., Gay, C. V., and Schraer, H.,** Carbonic anhydrase localization by light and electron-microscopy: a comparison of methods, *J. Histochem. Cytochem.,* 30, 1135, 1982.
183. **Bonartsev, P. D. and Bukhvalov, I. B.,** Ultrastructural detection of nucleoside phosphatases activity in human peripheral blood cells, *Tsitologia,* 27, 1353, 1985.
184. **Mathe, G., Pouillart, P., Sterescu, M., and Amiel, J. L.,** Subdivision of classical varieties of acute leukemia, *Eur. J. Clin. Biol. Res.,* 16, 554, 1971.
185. **Parkins, J., McKenna, R., and Brunning, R.,** Philadelphia chromosome positive blastic leukemia, *Br. J. Haematol.,* 52, 663, 1982.
186. **Breard, J., Reinherz, E. L., Kung, P. C., Goldstein, G., and Schlossman, S. F.,** A monoclonal antibody reactive with human peripheral blood monocytes, *J. Immunol.,* 124, 1980.
187. **Brooks, P. A., Zola, H., and McNamara, P. J.,** Membrane antigens of human cells of monocyte-macrophage lineage studied with monoclonal antibodies, *Pathology,* 15, 45, 1983.
188. **Wetzel, B., Horn, R., and Spicer, S.,** Fine structural studies in the development of heterophil, eosinophil and basophil granulocytes in rabbits, *Lab. Invest.,* 16, 349, 1967.
189. **Freireich, E. J.,** New strategies in the management of acute leukemia, *Acta Haematol.,* 78 (Suppl. 1), 116, 1987.
190. **Browman, G., Neame, P., and Soamboonsrup, P.,** The contribution of cytochemistry and immunotyping to the reproducibility of the FAB classification in acute leukemia, *Blood,* 68, 900, 1986.
191. **Foon, K. and Todd, D. R.,** Immunologic classification of leukemia and lymphoma, *Blood,* 68, 1, 1986.
192. **Minowada, J.,** Marker utility in the diagnosis and management of leukemias, *Lab. Med.,* 16, 305, 1985.

Chapter 13

MUSCLE AND MUSCLE DISEASE: CYTOCHEMICAL AND IMMUNOCYTOCHEMICAL ASPECTS

Takuro Ogata

TABLE OF CONTENTS

0-8493-6012-9/93/$0.00 + $.50

A. INTRODUCTION

Electron microscopic cytochemistry and immunocytochemistry are nowadays the two most powerful tools available for the determination of the precise localization of biological materials in muscle tissue. The classification of skeletal muscle fibers has been the subject of a tremendous number of histochemical and immunohistochemical studies. This is, however, beyond the scope of the present text due to its vast dimensions. This chapter begins with the description of calcium cytochemistry and the various enzymes found in normal muscle tissues. A brief account of the pathology of muscle is also presented. In the second section, the immunocytochemical aspects of structural proteins of muscle tissue, such as myofilament proteins, myoplasm proteins, and sarcoplasmic reticulum-related proteins, and various enzymes are presented and discussed.

B. ELECTRON MICROSCOPIC CYTOCHEMISTRY

1. Calcium

Calcium plays an important role in the excitation and contraction of muscle. There are two basic methods for the ultrastructural localization of intracellular calcium: the pyroantimonate method[1-3] and the cobalt method.[5]

In the pyroantimonate method, pyroantimonate forms electron-opaque salts with calcium. Ultrastructurally, fine black reaction product, considered to be calcium pyroantimate is demonstrated with unbuffered saturated potassium pyroantimonate, pH 9.4. In human, mouse, and rat muscles, fine precipitates were localized in several organelles. In extended sarcomeres, precipitates in myofibrils were mainly localized lineally in the I-band near the A-I junction, sometimes with slight precipitate in the H-zone, excluding the M-zone of the A-band. At times, there was some precipitate in the sarcoplasmic reticulum (SR) and, rarely, in mitochondria. In human pathologic muscle fibers undergoing degeneration or degeneration-regeneration, the calcium precipitate was markedly increased in the mitochondria, fairly prominently increased in the SR and nuclei, and slightly increased in myofibrils regardless of the type of neuromuscular disease (e.g., Duchenne's muscular dystrophy, polymyositis, other myopathies, and necrotic fibers of chronic degenerating diseases).[2] These changes were identical with those seen in abnormal fibers produced experimentally by trauma of rat muscles.[2] In calcium-loaded normal rat fibers, preincubated in 136 m*M* calcium-enhanced Hanks' medium, the same pattern of increased calcium deposition was found as in pathologic human and rat muscle, with the greatest accumulation of precipitate in the mitochondria, SR, and nuclei. Preincubation of normal rat and mouse muscle in 0.1 *M* acetate buffer, pH 4.65, before calcium loading enhanced the calcim deposition at myofibrillar sites. In extended sarcomeres, the precipitate was most evident in H-zones, excluding the M-zones of the A-bands, with none or considerably less deposition in regions of A-I overlap. Within the I band, three distinct transverse lines of precipitate were evident.[2] In the dog papillary muscles, calcium pyroantimonate, identified primarily by its susceptibility to removal by chelation with ethylene glycol-bis(2-aminoethyl)-tetraacetic acid (EGTA) and ethylene diamine tetraacetate (EDTA), was consistently found densely concentrated in the lateral sacs of the SR and over the sacomeric I-bands. M-zones were virtually free of precipitate.[3]

The pyroantimonate method is useful because it produces discrete electron-dense particles that are small enough to allow an accurate localization. However, this technique has the drawback that under certain conditions, pyroantimonate will precipitate not only calcium, but also sodium, potassium, and magnesium. The application of the cobalt uptake technique, with silver intensification of the tissue, to neurons was first reported by Phillips.[4] This cobalt technique was also applied to the frog sartorius muscle.[5] The muscle was soaked overnight

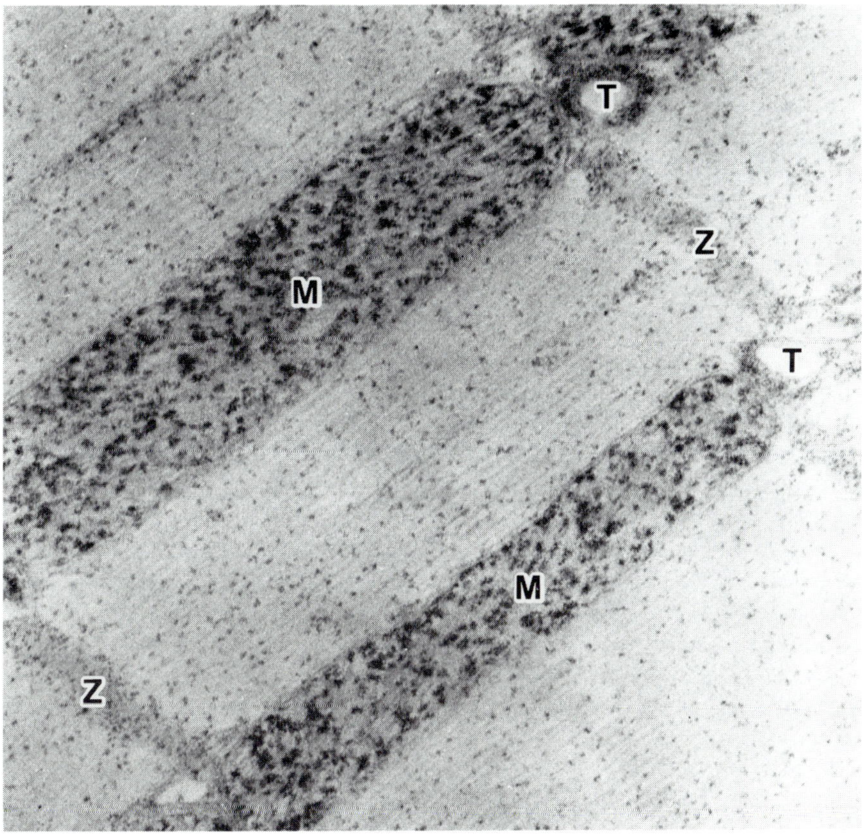

FIGURE 1. Localization of Ca^{2+}-ATPase in cardiac muscle. Rat cardiac muscle was incubated in the presence of 10 m*M* $CaCl_2$ for 10 min at 37°C. Ca^{2+}-ATPase activity is localized at the myofilaments, mitochondria (M), and SR adjacent to T-tubule (T); Z, Z-line. (Magnification × 63,000. Micrograph by courtesy of Dr. K. Fujimoto.)

in saline containing 50 m*M* cobalt, and the specimens were subsequently fixed, with or without precipitation of intracellular cobalt, by exposure to ammonium sulfide. Thick sections (3 to 4 μm) of each sample were treated with silver in order to amplify the precipitated cobalt. The samples were then cut for ultrastructural examination. In the frog sartorius muscle, silver grains were abundant over membranes of the SR, sarcolemma, mitochondria, and nucleus, and rare or absent from filaments and sarcoplasm.[5]

2. Ca^{2+}-Adenosine Triphosphatase (Ca^{2+}-ATPase)

The cytochemical procedure for Ca^{2+}-ATPase was carried out on rat cardiac muscle by the one-step lead citrate method, utilizing ATP as substrate, Ca^{2+} as the activator, and lead citrate as capture agent.[6,7] The most intense reaction was localized at the myofilaments and the matrix of mitochondria[7] (Figure 1). At higher magnification, fine particles of lead phosphate, considered to indicate the site of ATPase activity, were generally located either on or immediately lateral to thick filaments. Ca^{2+}-ATPase activity was found also in the SR (Figure 1) and the gap junctions. Within 5 min of incubation, the reaction product seemed to be associated with the cytoplasmic side of the SR. However, prolongation of the incubation time resulted in an increased coarseness and concentration of the reaction product in the luminal spaces of the SR[7] (see Figure 1).

3. Ouabain-Sensitive, K^+-Dependent *p*-Nitrophenylphosphatase

Attempts to localize the Na^+,K^+-ATPase cytochemically, using ATP as substrate, have been consistently unsuccessful. Ernst[8] developed a method for the cytochemical demonstration of Na^+,K^+-ATPase utilizing the ouabain-sensitive, K^+ *p*-nitrophenylphosphatase (K^+-NPPase) activity of the Na^+,K^+-ATPase complex.[8,9] Mayahara et al.[9] developed an alternative one-step lead method utilizing *p*-nitrophenylphosphate as substrate and lead citrate or lead nitrate as capture agent. Using this method, the localization of K^+-NPPase activity was examined in the rat cardiac muscle.[7] In the presence of K^+, the most intense reaction was localized in the sarcolemmae and the T-tubules. Preincubation in 10 m*M* ouabain, followed by incubation in the reaction medium containing 10 m*M* ouabain, considerably inhibited the reaction.

4. Adenylate (Adenyl) Cyclase

Adenylate (adenyl) cyclase forms adenosine 3′,5′-cyclic phosphate and inorganic pyrophosphate from ATP. The cytochemical localization of this enzyme was examined in chicken pectoral muscle by the lead precipitation technique.[10] The reaction product indicative of adenylate cyclase activity was mainly localized in the terminal cisternae, situated on both sides of the Z-line. The T-tubule between the terminal cisternae of the triad was free of the reaction product.[10] Adenylate cyclase activity was also examined cytochemically in the muscles of normal and muscular dystrophic chicken.[11] In both kinds of muscles, the reaction product from adenylate cyclase was present predominantly in the terminal cisternae of the SR. Adenylate cyclase activity in prefixed psoas muscle and ventricular myocardium of mice and guinea pigs was also examined by the lead precipitation technique.[12] Under suitable incubation conditions, the electron-dense reaction product (lead salt) was deposited in psoas muscle, chiefly in membranes of the lateral sacs of the SR at their points of contact with the tubules, and in heart muscle, along the plasma membrane and different parts of the sarcotubular system, including the subsarcolemmal cisternae. The depostion of the reaction product was ATP dependent and it was enhanced by fluoride. Precipitation was enhanced in skeletal and heart muscle by adrenaline, and in heart muscle by glucagon. It was abolished at all sites by zinc ions.[12]

Adenylate cyclase activity was examined in the extensor tibiae muscle of the locust to localize the sites of action of the modulatory compounds, octopamine and proctolin.[13] Reaction products, indicating the presence of adenylate cyclase activity, were found in both fast and intermediate muscle fiber types, but not in the limited number of purely slow muscle fibers, in the fan region at the proximal end of the muscle when specimens were incubated in the presence of 10 to 5 *M* DL-octopamine. In contrast, the slow muscle fibers, which are thought to produce the myogenic rhythm of contraction and relaxation, were the only ones in the muscle which exhibited proctolin-sensitive adenylate cyclase activity. In both cases, the bulk of the reaction product was characteristically localized in the dyads, with lesser amounts occurring beneath the sarcolemmal membrane, in the non-dyad SR, and in the T-tubule system.[13]

5. Acid Phosphatase

Normal soleus muscle and cardiac muscle of the rat were examined for the cytochemical detection of acid phosphatase activity, with cerium as the capture reagent.[14] The specimens were incubated in a medium containing 0.1 m*M* β-glycerophosphate, 2 m*M* $CeCl_3$, 0.00015% Tritron X-100, 5% sucrose in 0.1 *M* sodium acetate buffer, pH 5.0. In the skeletal muscle, the acid phosphatase-positive structures, which are presumed to be lysosomes, were primarily found at the periphery of the muscle fibers, although some were present within the fibers.

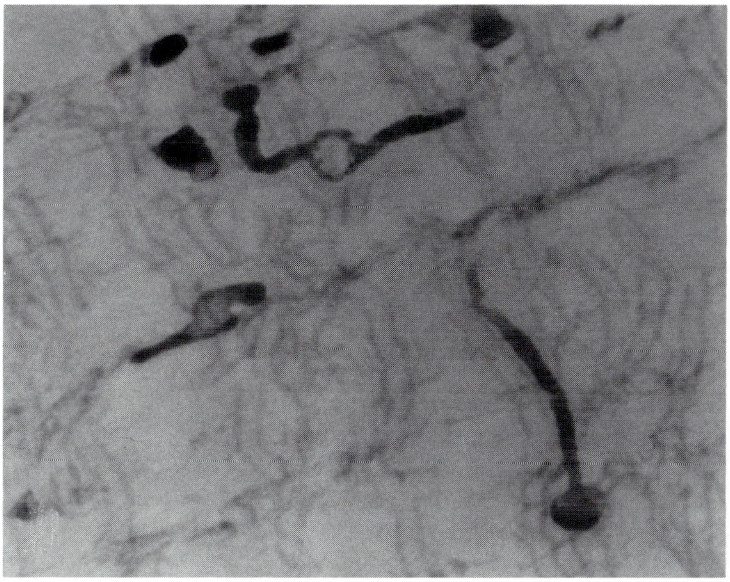

FIGURE 2. Acid phosphatase in the rat soleus muscle. Note the elongated lysosomes. (Magnification × 9,000. Courtesy of Dr. T. Okada.)

They displayed a variety of shapes, such as oval, tubular, and tortuous tubular, among the myofibrils and mitochondria (Figure 2). These profiles of lysosomes were more prominent in skeletal muscles after starvation for several days. In the cardiac muscle, most lysosomes were oval in shape and were located around the nucleus.[11]

6. Alkaline Phosphatase

The Gomori lead nitrate method has been used to examine alkaline phosphatase in rat gastrocnemius muscle.[15] The enzyme reaction was localized to the sarcolemma of striated muscle and to the membranes of endothelial cells in adjacent capillaries. In contrast to ATPase activity, alkaline phosphatase activity was inhibited by levamisole, but not by ouabain.[15]

7. Mitochondrial Cytochemistry (Cytochrome *c* Oxidase and NADH Oxidase)

Mitochondrial cytochemistry with reference to cytochrome *c* oxidase and NADH oxidase activities, as well as calcium localization at the subcellular level in experimental myopathies and in a variety of human mitochondrial disorders, were examined by Shah et al.[16,17] and van Ekeren et al.[18] For cytochrome oxidase, two methods, based on the oxidative polymerization of 3,3-diaminobenzidine to an osmiophilic product, were used.[19,20] The NADH oxidase reaction was performed using the method of Briggs et al.[21] The agents, 2,4-dinitrophenol (DNP), oleate, and crotoxin were employed to produce experimental myopathies in the rat. The DNP- and oleic acid-induced lesions showed strong cytochrome *c* oxidase and NADH oxidase activities in mitochondria. The reaction product of cytochrome *c* oxidase was observed in the outer as well as inner mitochondrial membranes and the cristae. The paracrystalline bodies also showed distinct reaction. The reaction product of NADH oxidase was observed in the outer and inner membranes and in the cristae of most mitochondria. The crotoxin lesions showed marked reduction of the enzyme activities.[21]

Human cases with mitochondrial abnormalities have also been examined by the same methods.[17] Light microscopically, lipid storage myopathies with carnitine and carnitine palmityl transferase deficiency revealed a number of muscle fibers with numerous lipid granules in addition to ragged red fibers, which displayed intense cytochrome *c* oxidase and NADH oxidase activities. Electron microscopically, strong cytochrome *c* oxidase reaction was observed in the outer and inner membrane and in the intracristal space of mitochondria. The laminar bodies and paracrystalline inclusions, when present, also showed a distinct reaction. The NADH oxidase reaction indicated the presence of strong enzyme activity in mitochondrial membranes, as well as within the concentric cristae of the altered mitochondria. However, both enzyme activities were normal in unaltered mitochondria. With regards to Ca^{2+} localization, very few dense granules were seen within the mitochondria and in the SR.[17]

Light microscopic observations of specimens from Kearns-Sayre's syndrome and chronic progressive opthalmoplegia revealed that the muscle fibers varied in size and there were ragged red fibers and necrosis.[17] The cytochrome *c* oxidase reaction showed a reduced enzyme activity as compared to the NADH oxidase reaction in the ragged red fibers. Electron microscopically, cytochrome *c* oxidase activity was conspicuously absent in the outer and inner membranes, as well as in the convoluted cristae in the mitochondrial matrix. The paracrystalline inclusions were also devoid of enzyme activity. However, normal mitochondria showed strong reaction in the outer as well as in the inner membranes. The NADH oxidase reaction was absent from mitochondrial membranes, within the concentric lamellar cristae, and from internal cristae of some mitochondria. The mitochondria in normal fibers showed enzyme reaction. As regards to Ca^{2+} localization, the altered mitochondria in the necrotic fibers frequently showed prominent large and small calcium granules. These granules were also present along the plasmalemma and basement membranes. However, the calcium granules were sparse and were only occasionally present in the mitochondria of normal muscle fibers. In polymyositis, neurogenic atrophy, and fascioscapulo humeral dystrophy, mitochondrial cytochemistry and Ca^{2+} deposition patterns were similar to those observed in Kearns-Sayre's syndrome.[17] From these results, it was suggested that the human mitochondrial disorders may be grouped into two types, one in which the morphologically altered mitochondria retain the enzyme activities and Ca^{2+} homeostasis, and the other in which altered mitochondria associated with muscle necrosis represent the loss or reduction of enzyme activities as well as Ca^{2+}-homeostasis.[17]

8. Acetylcholinesterase

The metal thiocholine technique has been widely used for the electron microscopic demonstration of acetylcholinesterase in muscle tissue.[22-24] Using the gold-thiocholine technique, acetylcholinesterase was localized at the prejunctional axoplasmic membrane and at the postjunctional sarcoplasmic membrane, including the full length of its invaginations, at the motor endoplate of the mouse intercostal muscle.[22] As an alternative method, Davis et al.[25] used substrates containing an enzyme-susceptible thiolester group and a diazonium group in the same molecule, which yields osmiophilic polymers upon enzymatic hydrolysis via diasothio ether formation linking the units. An enzyme-susceptible ester group is incorporated in the same indole molecule with a diazonium group to yield an osmiophilic polymer upon esterolysis via coupling to azo-dye formation linking the units. The thiole substrate is α-acetyl-*m*-toluenediazonium ion (ATTD) and the indoxyl substrate, 3-acetoxy-5-indolediazonium (AID) ion.[25] In the rat motor endplate, with ATTD, both acetylcholinesterase and pseudocholinesterase were demonstrated on the inner surface of the synaptic clefts, and the deposits were discrete and not linear, as has been reported with some other methods. On the other hand, deposits with AID were more finely granular and were also limited to the inner aspect of the synaptic clefts.[25]

C. IMMUNOCYTOCHEMISTRY

1. Myosin

Vertebrate skeletal muscle fibers are classified into two major groups: the twitch (fast) and slow (tonic) muscle fibers. Morphologically, the former has straight Z-lines and well-developed transverse tubular and SR systems, and is singly innervated, whereas the latter has zig-zag Z-lines, a less-developed T- and SR system, and is multiply innervated.[26] The twitch fibers are histochemically subclassified at least into three types on the basis of histochemical reactions for enzymes of aerobic oxidative metabolism, succinate dehydrogenase, cytochrome oxidase, and NADH tetrazolium reductase in mitochondria, as red, white, and intermediate fibers, respectively, or by differences in myofibrillar ATPase activity as types I, IIA, and IIB fibers.[26-28] In mammalian muscles, the red fibers generally correspond to type I fibers, the intermediate fibers to type IIA fibers, and the white fibers to type IIB fibers. Electron microscopically, the red fibers have many large mitochondria and thick Z-lines, while the white fibers have a few small mitochondria and thin Z-lines. The intermediate fibers are intermediate in content and size of mitochondria, as well as in the width of the Z-line.[26,27]

Immunohistochemically, antibodies specific for fast white myosin from the chicken pectoralis react selectively with the white and intermediate fibers of rat skeletal muscles and also with those red fibers that have high alkali-stable ATPase activity.[27,29] Antibodies against slow myosin from the chicken anterior latissimus dorsi reacted only with the remaining population of red fibers, that is, those with low alkali-stable ATPase activity. Hence, the patterns of reactivity with these two antibodies were reciprocal. Ultracytochemical studies on white (fast) myosin of the rat diaphragm revealed that in the white, intermediate fibers, there is a distinct electron-opaque band at the center of the A-band on either side of an unstained bridge-free zone.[27,29] The red fiber can be subdivided into fast and slow red fibers on the basis of positive or negative reaction with anti-white myosin antibodies visualized directly under the electron microscope. Both categories of red fibers have wide Z-lines, and thus are distinguished from white and intermediate fibers, which react with the antibody but which have narrower Z-lines.[27,29]

Immunogold labeling was used to study the distribution of newly synthesized slow muscle myosin (SM) at the ultrastructural level as it replaced fast muscle myosin (FM) in rabbit muscles undergoing stimulation-induced type transformation.[30] The primary antisera were monoclonal antibodies, anti-FM and anti-SM, raised in mice against adult rabbit fast and slow muscle myosin. Control fast muscle was labeled strongly with antibodies to FM and control slow muscle with antibodies to SM; label was confined to the A-band. After 4 weeks of continuous stimulation at 10 Hz, fibers of the tibialis anterior muscle of the rabbit reacted with antibodies to both isoforms; after 6 weeks, labeling was obtained only with antibody to SM. After a 17-week period of stimulation and a further 3 weeks of recovery, the fibers again reacted with both antibodies. In all positively labeled sections, the distribution of gold particles was characteristic of the antibody and independent of the origin or history of the fibers. This observation supports the conclusion that newly synthesized myosin can be incorporated through the length and cross section of the A-band.[30]

2. Myoglobin

Myoglobin is contained in the cytoplasm and transfers oxygen. The localization of myoglobin was ultrastructurally examined in the walrus and human muscle using the

labeling with a repeat of approximately 38 nm, a distribution characteristic of proteins associated with tropomyosin on thin filaments.[45]

6. α-Actinin

α-Actinin is an actin-binding and cross-linking protein found in both muscle and non-muscle cells at points where actin is anchored to a variety of intracellular structures.[46] This fibrous protein is a dimer of two identical polypeptides, and it is a major component of Z-disks in striated muscle, cytoplasmic plaque in smooth muscle, and intercalated disks in cardiac muscle.[46]

Using rabbit antibodies against the α-actinin prepared from porcine skeletal muscle, α-actinin was localized in the Z-lines of adult skeletal muscle, the Z-lines and intercalated discs of cardiac muscle, and in the Z-bodies and Z-plaques of embryonic heart cells from axolotls and hamsters. In addition, the dense bodies of smooth muscle cells of hamsters showed a specific staining for α-actinin.[47]

In the rabbit psoas muscle and chicken breast muscle, antibodies against α-actinin bound to the entangled Z-filaments, forming dense lines at the position of the original Z-disc.[48]

7. Vinculin

Vinculin is found in intercalated disks of cardiac muscle[49,50] and membrane-associated plaques in smooth muscle, but not in Z-disks.[49] Vinculin binds tightly to α-actinin in cell-free extracts, but the proteins with which vinculin interacts *in vivo* are unknown. Most likely it binds directly to an integral membrane protein in the intercalated disc and also to α-actinin. The localization of vinculin and α-actinin was studied in the smooth muscle cell of chicken gizzard[50] using double immunolabeling with a new type of electron-dense antibody conjugate, Imposil-antibody, along with ferritin-antibody conjugates.[50] Vinculin was sharply confined to the membrane-associated dense plaque, closer to the membrane than the α-actinin, which was also present in these dense plaques. α-Actinin was also present in the cytoplasmic dense bodies, from which vinculin was absent. These results suggest that vinculin may play an important and widespread role in the linkage of actin-containing microfilament bundles to membranes. In the chicken cardiac muscle, labeling of vinculin occurred only along the intercalated disc, but not on the Z-band nor in the interfibrillar spaces.[39]

8. Dystrophin

Dystrophin is localized to the plasma membrane of striated muscle cells and is missing in muscles from Duchene muscular dystrophy patients.[49] Dystrophin constitutes only about 0.002% of the total muscle protein. Although its function is not yet known, it has sequence homology to α-actinin and other actin-binding proteins and may be involved in anchoring actin filaments to the plasma membrane. Using ultrathin cryosections and an immunogold method, dystrophin was localized along the cytoplasmic side of the plasma membrane in the muscle of the normal mice; no labeling was seen, however, along the plasma membrane in the *mdx* mice.[51] Some gold particles were present over the region of the T-tubule system in normal mice, but not in *mdx* mice.[51]

9. Connectin (Titin) and Nebulin

Connectin (titin) and nebulin are structural components in the sarcomeres of skeletal muscle.[52-55] Connectin filaments link thick filaments to adjacent Z-lines in the sarcomere. They are considered to be mainly responsible for passive tension generation in stretched fibers and for positioning the myosin filaments at the center of the sarcomere.[52] Nebulin is a giant structural protein that is found in the sarcomere of a wide range of skeletal muscles and is considered to be an inextensible filament in the I-band region.[53] Immunoelectron microscopic observations on the semitendinous muscle of the bullfrog showed that anti-connectin antibody-binding sites return to the original position after extreme stretch and release but anti-nebulin antibody-binding sites are largely disorganized.[52] The binding sites of anti-connectin antibodies moved within an I-band with the change in sarcomere length, but those of anti-nebulin antibodies did not. Nebulin remains in the I-band at extreme stretch. Thus, connectin filaments appear to be responsible for maintaining the mechanical continuity of a sarcomere and to behave independently of thin filaments. It has been suggested that nebulin is localized in the I-band but not in the A-band and that it is associated with thin filaments, but not with the elastic structure of myofibrils.[52]

To study whether the localization of connectin and/or nebulin varies according to the sarcomere length, skinned rabbit psoas fibers of sarcomere lengths ranging from maximum shortening to overstretching were incubated with antibodies against connectin and nebulin raised in guinea pigs.[54] The results indicated that connectin belongs to the elastic filaments because its localization was changed according to stretch at sarcomere lengths greater than 2.8 μm. On the other hand, nebulin does not seem to belong to these elastic filaments, because it showed the opposite behavior.[54]

10. Projectin

Projectin is a filamentous protein isolated from crayfish claw muscle.[56] This protein has physiological properties very similar to vertebrate skeletal muscle connectin, although its apparent molecular mass (~1200 kDa) is considerably lower than that of connectin (~3000 kDa). Polyclonal as well as monoclonal antibodies against chicken skeletal muscle connectin react with the 1200-kDa protein from crayfish claw muscle. Conversely, polyclonal antibodies against crayfish 1200-kDa protein cross-react with chicken connectin. Immunofluorescence and immunoelectron microscopy reveal that crayfish projectin is localized in the A-I junction area and A-band except for its center region in crayfish claw muscle. Polyclonal antibodies against crayfish claw muscle projectin react with 1200-kDa protein of honey bee and beetle flight muscle. A monoclonal antibody against chicken skeletal muscle connectin also react with honey bee and beetle protein. Immunoelectron microscopic observations revealed that anti-crayfish projectin antibodies bound the connecting filaments linking the Z-line and the thick filaments up to the M-line of honeybee muscle sarcomere. Anti-crayfish projectin antibodies bound the I-band region near the Z-line of beetle flight muscle. From these results, it was concluded that the 1200-kDa projectin from crayfish claw muscle is an invertebrate connectin.[56]

11. C-Protein

C-protein is present in thick filaments of vertebrate skeletal and cardiac muscle.[57-59] In the rabbit skeletal muscle it comprises 1 to 2% of the total protein content of the myofibril. It binds to myosin rod, light meromyosin, myosin subfragment, and F-actin. It inhibits actin-activated ATPase, and alters myosin assembly *in vitro*. The immunocytochemical localization of C-protein, using goat antiserum to rabbit C-protein, in the glycerinated rabbit psoas muscle showed that C-protein is located on seven stripes, spaced about 43 nm apart in each half of the A-band. Between two and four C-protein molecules per filament are present at each stripe.[58] Different isoforms of C-protein were shown in striated muscles of various fiber types.[59] Using monoclonal antibodies specific for the fast (MF-1) and slow (ALD-66) isoforms of C-protein from chicken skeletal muscle, the immunocytochemical localization of C-protein in myofibers from three muscle types of the chicken, i.e., the anterior (ALD) and posterior latissimus dorsi (PLD) muscle, and pectoralis major muscle, showed that

C-protein is present in a 43-nm repeat along the filaments in all three muscles.[59] However, there were marked differences in the absolute number and position occupied by the different isoforms. The pectoralis muscle exhibited eight labeled stripes in each half of the A-band by incubation with MF-1, but did not show labeling with ALD-66. The ALD muscle demonstrated nine labeled stripes in each half of the A-band by incubation with ALD-66, but did not exhibit labeling with MF-1. The PLD muscle was labeled with both monoclonal antibodies, namely, it exhibited a pattern identical to ALD muscle by incubation with ALD-66, but showed a different pattern to pectoralis muscle with MF-1. Therefore, it was suggested that at least two C-proteins can colocalize at binding sites 43 nm apart along thick filaments of the PLD muscle.

12. T-Tubule Protein

Immunocytochemical localization of the T-tubule protein was determined in the rabbit muscle using a monoclonal antibody raised against a triad-enriched (SR-T-tubule complex) microsomal membrane fraction of rabbit skeletal muscle.[60] The avidin-biotin complex immunoperoxidase staining method was used to screen hybrid colonies. Positive antibodies exhibited a granular double-staining pattern at the A-I junction, consistent with the location of triads in rabbit muscle. One monoclonal antibody, M171, has been further characterized by ultrastructual and immunoadsorption techniques. Postembedding electron immunocytochemistry was performed on tissue sections embedded in Lowicryl K4M. Goat anti-mouse immunoglobulin adsorbed to 10-nm colloidal gold particles was used as ultrastructural label. In these studies, M171 recognized an epitope at the triads and at periodic openings along the plasmalemma. Immunoelectrophoretic blotting of isolated SR, plasmalemma, and T-tubule, and triad-enriched fractions showed that M171 reacts with a surface membrane component. Taken together, these studies suggest that M171 recognizes an epitope associated with the T-tubule at the triad and at the mouth of the T-system at the plasmalemma, but not with the plasmalemma.[60]

13. Calmodulin

Calmodulin is a Ca^{2+}-dependent modulator protein which mediates the effects of Ca^{2+} in a variety of cell reactions and processes. Calmodulin is an intracellular calcium receptor with multiple functions, one of which might involve the Ca^{2+} of the SR. Biologically active colloidal gold-labeled calmodulin has been used as a marker for identifying potential calmodulin-binding sites by electron microscopy and frozen thin sections of rat cardiac muscle have been stained with this conjugate.[61] In the presence of Ca^{2+}, gold particles indicating calmodulin-binding sites were localized on the SR, mitochondria, and gap junctions (Figure 5).

14. Calsequestrin

Calsequestrin is a calcium-binding protein present in the SR.[62,63] Ca^{2+} ions, pumped by Ca^{2+}-ATPase into the lumen of the SR, are bound to calsequestrin in the terminal cisternae. Localization of calsequestrin in chicken ventricular muscle cells was determined by indirect immunofluorescence and immunoprotein-colloidal gold labeling of cryostat and ultracryotomy sections, respectively.[62] Calsequestrin was localized in the lumen of peripheral junctional SR, as well as in the lumen of membrane-bound structures present in the central region of the I-band, while it was absent from the lumen of the SR in the A-band region. Since chicken ventricular muscle cells lack T-tubules, the presence of calsequestrin in membrane-bound structures in the central region of the I-band suggests that these cells contain non-junctional regions of SR that are involved in Ca^{2+} storage and possibly Ca^{2+} release. The

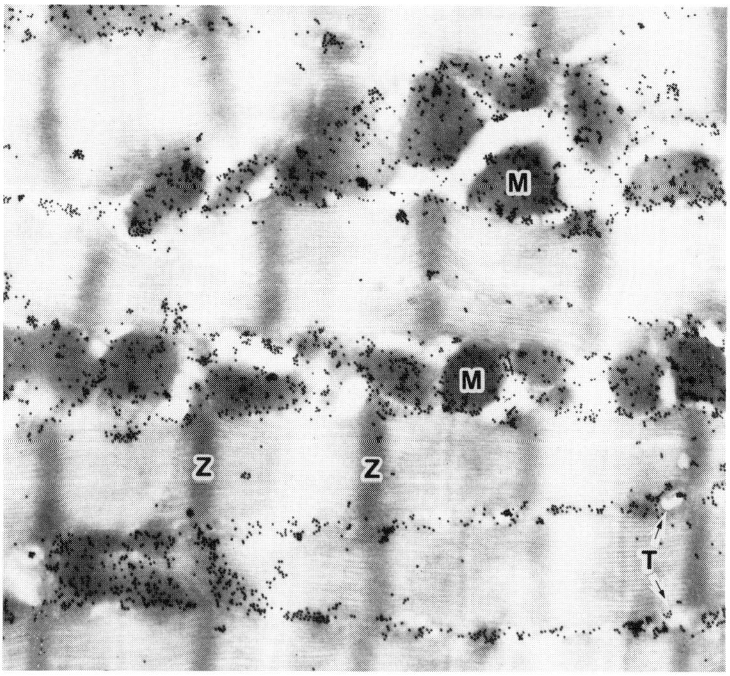

FIGURE 5. Cytochemical localization of calmodulin-bovine serum albumin-colloidal gold complex. Gold particles are observed on the myofibrillar areas, including the mitochondria (M), and SR. In addition, they are also found diffusely throughout the cells, particularly around the mitochondria or along the Z-lines (Z). (Magnification × 16,000. Micrograph by courtesy of Dr. K. Fujimoto.)

distribution of calsequestrin in rat myocardial cells was determined by indirect immunoelectron microscopic labeling of calsequestrin on ultrathin sections of rat ventricular muscle prepared by quick-freezing, freeze-drying, and direct embedding in Lowicryl K4M.[63] Calsequestrin was confined to the lumen of the junctional, cisternal, and corbular SR *(bulbous terminal swellings of the SR),* but it was absent from the lumen of the network SR. From these results, it is assumed that junctional, cisternal, and corbular SR are involved in Ca^{2+} storage and possibly release. By contrast, relocation of calsequestrin to the lumen of the network SR was observed to a variable degree in chemically fixed, ethanol-dehydrated, and Lowicryl-embedded tissue. Hence, it was concluded that for the immunoelectron microscopic localization of diffusible proteins, such as calsequestrin, tissue preparation by cryofixation, freeze-drying, and direct embedding in Lowicryl K4M is far superior to that obtained by chemical fixation, ethanol dehydration, and embedding in Lowicryl K4M.[63]

15. Calcium (Ca^{2+} + Mg^{2+})-ATPase

Striated muscle relaxes when calcium ions are removed from the sarcoplasm by the calcium ATPase in the SR membrane. The ultrastructural localization of calcium (Ca^{2+} + Mg^{2+}-dependent)-ATPase in the rat papillary muscle[64] and gracilis muscle[65] was determined by indirect immunoferritin labeling of ultrathin frozen sections. In both papillary and gracilis muscle, simultaneous visualization of ferritin particles and of adsorption-stained cellular membranes showed that the calcium ATPase is concentrated in the longitudinal SR and in nonjunctional regions of the membrane of the terminal cisternae, but is virtually absent from mitochondria, plasma membranes, T-tubules, and junctional SR. Ferritin particles were found preponderantly on the cytoplasmic surface of the membrane. Comparison of the density of ferritin particles in fast and slow fibers suggests that the density of calcium ATPase in the SR membrane in a fast fiber is approximately two times higher than in a slow fiber.[64,65]

Using monoclonal antibodies against the calcium ATPase in the SR of rabbit fast-twitch skeletal muscle, the ultrastructural distribution of the antigens was compared between the fast and slow fibers of the rabbit and rat muscle using postembedding immunolabeling.[66] The density of antibody binding to the membrane of the terminal cisternae of the fast-twitch fibers was on average seven times greater than in the slow-twitch fibers. The results provide strong evidence that the different calcium accumulation rates in mammalian fast- and slow-twitch muscles are due to different concentrations of calcium ATPase molecules in the SR membrane.

16. Carbonic Anhydrase

The subcellular distribution of carbonic anhydrase III in rat soleus and vastus lateralis muscles was immunocytochemically studied using rabbit nonspecific antiserum raised against carbonic anhydrase III prepared from human skeletal muscle.[67] The enzyme protein was found to be distributed diffusely in the cytoplasm of skeletal muscle cells. Red skeletal muscle (mainly type I fibers) revealed a very strong immunogold staining, whereas in white muscle (mainly type II fibers) gold particles were almost completely absent. No immunoreaction was observed in mitochondria or in other intracellular organelles.[67]

17. Acetylcholine Receptor

Snake venom α-bungarotoxin irreversibly binds to the acetylcholine receptor (AChRs). Thus, AChRs can be labeled for light and electron microscopy with α-bungarotoxin. For visualization, the toxin may be conjugated with horseradish peroxidase[68] or fluorescent dyes such as rhodamine,[69] or it may be labeled by ^{131}I[70] or ^{125}I.[71] Unlabeled α-bungarotoxin bound to AChRs was also visualized immunocytochemically using peroxidase-labeled antibody.[72]

By electron microscopic radioautography with ^{125}I-α-bungarotoxin, AChR was localized to the upper portions of the junctional folds where the electron-dense thickened portions of the folds are located.[73] Peroxidase-labeled bungarotoxin[72] gave a more precise localization of AChR at the light and electron microscope levels than ^{125}I-α-bungarotoxin. Electron microscopically, AChR was localized on the terminal expansions of the junctional folds and for a variable distance along the stalks of the folds.[74]

REFERENCES

1. **Clark, M. A. and Ackerman, G. A.,** A histochemical evaluation of the pyroantimonate-osmium reaction, *J. Histochem. Cytochem.*, 19, 727, 1971.
2. **Oberc, M. A. and Engel, W. K.,** Ultrastructural localization of calcium in normal and abnormal skeletal muscle, *Lab. Invest.*, 36, 566, 1977.
3. **Legato, M. J. and Langer, G. A.,** The subcellular localization of calcium ion in mammalian myocardium, *J. Cell Biol.*, 41, 401, 1969.
4. **Phillips, C. E.,** Intracellularly injected cobaltous ions accumulate at synaptic densities, *Science*, 207, 1477, 1980.
5. **Anderson, P. A. V. and Josephson, R. K.,** Cobalt uptake and localization of putative calcium binding sites in striated muscle, *J. Histochem. Cytochem.*, 31, 591, 1983.
6. **Ando, T., Fujimoto, K., Mayahara, H., Miyajima, H., and Ogawa, K.,** A new one-step method for the histochemistry and cytochemistry of Ca^{2+}-ATPase activity, *Acta Histochem. Cytochem.*, 14, 705, 1981.

7. **Fujimoto, K. and Ogawa, K.,** Enzyme cytochemical study of rat cardiac muscle. II. Ca^{++}-ATPase and ouabain-sensitive, K^{+}-dependent p-nitrophenylphosphatase, *Acta Histochem. Cytochem.*, 15, 338, 1982.
8. **Ernst, S. A.,** Transport adenosine triphosphatase cytochemistry. II. Cytochemical localization of ouabain-sensitive, potassium dependent phosphatase activity in the secretory epithelium of the avian salt gland, *J. Histochem. Cytochem.*, 20, 23, 1972.
9. **Mayahara, H., Fujimoto, K., Ando, T., and Ogawa, K.,** A new one-step method for the cytochemical localization of ouabain-sensitive, potassium-dependent p-nitrophenylphosphatase activity, *Histochemistry*, 67, 125, 1980.
10. **Raible, D. G., Cutler, L. S., and Rodan, G. A.,** Localization of adenylate cyclase in skeletal muscle sarcoplasmic reticulum and its relation to calcium accumulation, *FEBS Lett.*, 85, 149, 1978.
11. **Rodan, G. A., Rodan, S. B., Raible, D. G., Cutler, L. S., Wacholtz, M., and Sha'afi, R. I.,** Adenylate cyclase in muscular dystrophy, *Ann. N.Y. Acad. Sci.*, 317, 670, 1979.
12. **Schulze, W., Krause, E.-G., and Wollenberger, A.,** Cytochemical demonstration and localization of adenyl cyclase in skeletal and cardiac muscle, *Adv. Cyclic Nucleotide Res.*, 1, 249, 1972.
13. **Swales, L. S. and Evans, P. D.,** Histochemical localization of octopamine- and proctolin-sensitive adenylate cyclase activity in a locust skeletal muscle, *Histochemistry*, 90, 233, 1988.
14. **Okada, T., Robinson, J. M., and Karnovsky, M. J.,** Cytochemical localization of acid phosphatase in striated muscle, *Histochemistry*, 85, 177, 1986.
15. **Safadi, A., Livne, E., Silbermann, M., and Reznick, A. Z.,** Activity of alkaline phosphatase in rat skeletal muscle localized along the sarcolemma and endothelial cell membranes, *J. Histochem. Cytochem.*, 39, 199, 1991.
16. **Shah, A., Sahgal, V., Sahgal, S., Subramani, V., and Kochar, O. H.,** Mitochondrial cytochemistry in experimental myopathies, *J. Submicrosc. Cytol.*, 17, 509, 1985.
17. **Shah, A., Sahgal, V., Sahgal, S., Subramani, V., and Kochar, H.,** Variability of mitochondrial cytochemistry in human neuromuscular diseases, *J. Submicrosc. Cytol. Pathol.*, 20, 287, 1988.
18. **van Ekeren, G. J., Stadhouders, A. M., Egberink, G. J. M., Sengers, R. C. A., Daniels, O., and Kubat, K.,** Hereditary mitochondrial hypertrophic cardiomyopathy with mitochondrial myopathy of skeletal muscle, congenital cataract and lactic acidosis, *Virchows Arch. A*, 412, 47, 1987.
19. **Seligman, A. M., Karnovsky, M. J., Wasserkrug, H. L., and Hanker, J. S.,** Nondroplet ultrastructural demonstration of cytochrome oxidase activity with a polymerizing osmiophilic reagent, diaminobenzidine (DAB), *J. Cell Biol.*, 38, 1, 1968.
20. **Al-Ali, S. Y. and Robinson, N.,** Ultrastructural demonstration of cytochrome oxidase via cytochrome C in cerebral cortex, *J. Histochem. Cytochem.*, 27, 1261, 1979.
21. **Briggs, R. G., Drath, D. B., Karnovsky, M. D., and Karnovsky, M. J.,** Localization of NADH oxidase on the surface of human polymorphonuclear leukocytes by a new cytochemical method, *J. Cell Biol.*, 67, 566, 1975.
22. **Davis, R. and Koelle, G.,** Electron microscopic localization of acetylcholinesterase and nonspecific cholinesterase at the neuromuscular junction by the gold-thiocholine and gold-thiolacetic acid methods, *J. Cell Biol.*, 34, 157, 1967.
23. **Teräväinen, H.,** Localization of acetylcholinesterase in the rat myoneural junction, *Histochemie*, 17, 162, 1969.
24. **Tsuji, S.,** On the chemical basic of thiocholine methods for demonstration of acetylcholinesterase activities, *Histochemistry*, 42, 99, 1974.
25. **Davis, D. A., Wasserkrug, H. L., Heyman, I. A., Padmanabhan, K. C., Seligman, G. A., Plapinger, R. E., and Seligman, A. M.,** Comparison of ultrastructural cholinesterase demonstration in the motor end plate with α-acetylthiol-m-toluenediazonium ion and 3-acetoxy-5-indolediazonium ion, *J. Histochem. Cytochem.*, 20, 161, 1972.
26. **Ogata, T.,** Morphological and cytochemical features of fiber types in vertebrate skeletal muscle, *Crit. Rev. Anat. Cell Biol.*, 1 (3), 229, 1988.
27. **Gauthier, G. F.,** Skeletal muscle fiber types, in *Myology*, Engel, A. G. and Banker, B. Q., Eds., McGraw-Hill, New York, 1986, 255.
28. **Pette, D. and Staron, R. S.,** Cellular and molecular diversities of mammalian skeletal muscle fibers, *Rev. Physiol. Biochem. Pharmacol.*, 116, 1, 1990.
29. **Gauthier, G. F.,** Ultrastructural identification of muscle fiber types by immunocytochemistry, *J. Cell Biol.*, 82, 391, 1979.
30. **Franchi, L. L., Murdoch, A., Brown, W. E., Mayne, C. N., Elliott, L., and Salmons, S.,** Subcellular localization of newly incorporated myosin in rabbit fast skeletal muscle undergoing stimulation-induced type transformation, *J. Muscle Res. Cell Motil.*, 11, 227, 1990.
31. **Goldfischer, S.,** The cytochemical localization of myoglobin in striated muscle of man and walrus, *J. Cell Biol.*, 34, 398, 1967.

32. **Rhodin, J. A. G.,** *An Atlas of Ultrastructure,* W. B. Saunders, Philadelphia, 1963, 29.
33. **Lazarides, E., Granger, B. L., Gard, D. L., O'Connor, C. M., Breckler, J., Price, M., and Danto, S. I.,** Desmin- and vimentin-containing filaments and their role in the assembly of the Z disk in muscle cells, *Cold Spring Harbor Symp. Quant. Biol.,* 46, 351, 1981.
34. **Stromer, M. H. and Bendayan, M.,** Arrangement of desmin intermediate filaments in smooth muscle cells as shown as high-resolution immunocytochemistry, *Cell Motil. Cytoskeleton,* 11, 117, 1988.
35. **Stromer, M. H. and Bendayan, M.,** Immunocytochemical identification of cytoskeletal linkages to smooth muscle cell nuclei and mitochondria, *Cell Motil. Cytoskeleton,* 17, 11, 1990.
36. **Tokuyasu, K. T.,** Visualization of longitudinally-oriented intermediate filaments in frozen sections of chicken cardiac muscle by a new staining method, *J. Cell Biol.,* 97, 562, 1983.
37. **Tokuyasu, K. T., Maher, P. A., and Singer, S. J.,** Distributions of vimentin and desmin in developing chick myotubes in vivo. I. Immunofluorescence study, *J. Cell Biol.,* 98, 1961, 1984.
38. **Tokuyasu, K. T., Maher, P. A., and Singer, S. J.,** Distributions of vimentin and desmin in developing chick myotubes in vivo. II. Immunoelectron microscopic study, *J. Cell Biol.,* 100, 1157, 1985.
39. **Tokuyasu, K. T., Dutton, A. H., and Singer, S. J.,** Immunoelectron microscopic studies of desmin (skeletin) localization and intermediate filament organization in chicken cardiac muscle, *J. Cell Biol.,* 96, 1736, 1983.
40. **Wang, K., Ash, J. F., and Singer, S. J.,** Filamin, a new high-molecular-weight protein found in smooth muscle and non-muscle cells, *Proc. Natl. Acad. Sci. U.S.A.,* 72, 4483, 1975.
41. **Wang, K.,** Filamin, a new high-molecular-weight protein found in smooth muscle and nonmuscle cells. Purification and properties of chicken gizzard filamin, *Biochemistry,* 16, 1857, 1977.
42. **Small, J. V., Fürst, D. O., and De Mey, J.,** Localization of filamin in smooth muscle, *J. Cell Biol.,* 102, 210, 1986.
43. **Sobue, K., Muramoto, Y., Fujita, M., and Kakiuchi, S.,** Purification of a calmodulin-binding protein from chicken gizzard that interacts with F-actin, *Proc. Natl. Acad. Sci. U.S.A.,* 78, 5652, 1981.
44. **Fürst, D. O., Cross, R. A., De Mey, J., and Small, J. V.,** Caldesmon is an elongated, flexible molecule localized in the actomyosin domains of smooth muscle, *EMBO J.,* 5, 251, 1986.
45. **Lehman, W., Craig, R., Lui, J., and Moody, C.,** Caldesmon and the structure of smooth muscle thin filaments: immunolocalization of caldesmon on thin filaments, *J. Muscle Res. Cell Motil.,* 10, 101, 1989.
46. **Blanchard, A., Ohanian, V., and Critchley, D.,** The structure and function of α-actinin, *J. Muscle Res. Cell Motil.,* 10, 280, 1989.
47. **Lemanski, L. F., Paulson, D. J., Hill, C. S., Davis, L. A., Riles, L. C., and Lim, S.-S.,** Immunoelectron microscopic localization of α-actinin on Lowicryl-embedded thin-sectioned tissues, *J. Histochem. Cytochem.,* 33, 515, 1985.
48. **Takahashi, K. and Hattori, A.,** α-Actinin is a component of the Z-filament, a structural backbone of skeletal muscle Z-disks, *J. Biochem.,* 105, 529, 1989.
49. **Darnell, J., Harvey, L., and David, B.,** *Molecular Cell Biology,* Scientific American Books, New York, 1990.
50. **Geiger, B., Dutton, A. H., Tokuyasu, K. T., and Singer, S. J.,** Immunoelectron microscope studies of membrane-microfilament interactions: distributions of α-actinin, tropomyosin, and vinculin in intestinal epithelial brush border and chicken gizzard smooth muscle cells, *J. Cell Biol.,* 91, 614, 1981.
51. **Watkins, S. C., Hoffman, E. P., Slayter, H. S., and Kunkel, L. M.,** Immunoelectron microscopic localization of dystrophin in myofibres, *Nature,* 333, 863, 1988.
52. **Maruyama, K., Matsuno, A., Higuchi, H., Shimaoka, S., Kimura, S., and Shimizu, T.,** Behaviour of connectin (titin) and nebulin in skinned muscle fibres released after extreme stretch as revealed by immunoelectron microscopy, *J. Muscle Res. Cell Motil.,* 10, 350, 1989.
53. **Wang, K. and Wright, J.,** Architecture of the sarcomere matrix of skeletal muscle: immunoelectron microscopic evidence that suggests a set of parallel inextensible nebulin filaments anchored at the Z line, *J. Cell Biol.,* 107, 2199, 1988.
54. **Pierobon-Bormioli, S., Betto, R., and Salviati, G.,** The organization of titin (connectin) and nebulin in the sarcomeres: an immunocytolocalization study, *J. Muscle Res. Cell Motil.,* 10, 446, 1989.
55. **Akster, H. A., Granzier, H. L. M., and Focant, B.,** Differences in I band structure, sarcomere extensibility, and electrophoresis of titin between two muscle fiber types of the perch (Perca fluviatilis L.), *J. Ultrastruct. Mol. Struct. Res.,* 102, 109, 1989.
56. **Hu, D. H., Matsuno, A., Terakado, K., Matsuura, T., Kimura, S., and Maruyama, K.,** Projectin is an invertebrate connectin (titin): isolation from crayfish claw muscle and localization in crayfish claw muscle and insect flight muscle, *J. Muscle Res. Cell Motil.,* 11, 497, 1990.
57. **Callaway, J. E. and Bechtel, P. J.,** C-Protein from rabbit soleus (red) muscle, *Biochem. J.,* 195, 463, 1981.

58. **Craig, R. and Offer, G.,** The location of C-protein in rabbit skeletal muscle, *Proc. R. Soc. London B.*, 192, 451, 1976.
59. **Dennis, J. E., Shimizu, T., Reinach, F. C., and Fischman, D. A.,** Localization of C-protein isoforms in chicken skeletal muscle: ultrastructural detection using monoclonal antibodies, *J. Cell Biol.*, 98, 1514, 1984.
60. **Malouf, N. N., Taylor, S., Gillespie, G. Y., Bynum, J. M., Wilson, P. E., and Meissner, G.,** Monoclonal antibody specific for the T-tubule of skeletal muscle, *J. Histochem. Cytochem.*, 34, 347, 1986.
61. **Fujimoto, K., Araki, N., Ogawa, K.-S., Kondo, S., Kitaoka, T., and Ogawa, K.,** Ultracytochemistry of calmodulin binding sites in myocardial cells by staining of frozen thin sections with colloidal gold-labeled calmodulin, *J. Histochem. Cytochem.*, 37, 249, 1989.
62. **Jorgensen, A. O., Shen, A. C.-Y., and Campbell, K. P.,** Ultrastructural localization of calsequestrin in adult rat atrial and ventricular muscle cells, *J. Cell Biol.*, 101, 257, 1985.
63. **Jorgensen, A. O. and McGuffee,L. J.,** Immunoelectron microscopic localization of sarcoplasmic reticulum proteins in cryofixed, freeze-dried, and low temperature-embedded tissue, *J. Histochem. Cytochem.*, 35, 723, 1987.
64. **Jorgensen, A. O., Shen, A. C.-Y., MacLennan, D. H., and Tokuyasu, K. T.,** Ultrastructural localization of the Ca^{2+} + Mg^{2+}-dependent ATPase of sarcoplasmic reticulum in rat skeletal muscle by immunoferritin labeling of ultrathin frozen sections, *J. Cell Biol.*, 92, 409, 1982.
65. **Jorgensen, A. O., Shen, A. C.-Y., Daly, P., and MacLennan, D. H.,** Localization of Ca^{2+} + Mg^{2+}-ATPase of the sarcoplasmic reticulum in adult rat papillary muscle, *J. Cell Biol.*, 93, 883, 1982.
66. **Dulhunty, A. F., Banyard, M. R. C., and Medveczky, C. J.,** Distribution of calcium ATPase in the sarcoplasmic reticulum of fast- and slow-twitch muscles determined with monoclonal antibodies, *J. Membr. Biol.*, 99, 79, 1987.
67. **Vaananen, H. K., Takala, T., and Morris, D. C.,** Immunoelectron microscopic localization of carbonic anhydrase III in rat skeletal muscle, *Histochemistry*, 86, 175, 1986.
68. **Lentz, T. L., Mazurkiewicz, J. E., and Rosenthal, J.,** Cytochemical localization of acetylcholine receptors at the neuromuscular junction by means of horseradish peroxidase-labeled α-bungarotoxin, *Brain Res.*, 132, 423, 1977.
69. **Anderson, M. J. and Cohen, M. W.,** Fluorescent staining of acetylcholine receptors in vertebrate skeletal muscle, *J. Physiol. (London)*, 237, 385, 1974.
70. **Miledi, R. and Potter, L. T.,** Acetylcholine receptors in muscle fibres, *Nature*, 233, 599, 1971.
71. **Fambrough, D. M. and Hartzell, H. C.,** Acetylcholine receptors: number and distribution at neuromuscular junctions in rat diaphragm, *Science*, 176, 189, 1972.
72. **Daniels, M. P. and Vogel, Z.,** Immunoperoxidase staining of α-bungarotoxin binding sites in muscle end-plates shows distribution of acetylcholine receptors, *Nature*, 254, 339, 1975.
73. **Fertuk, H. C. and Salpeter, M. M.,** Localization of acetylcholine receptor by ^{125}I-labelled α-bungarotoxin binding at mouse motor endplates, *Proc. Natl. Acad. Sci. U.S.A.*, 71, 1376, 1974.
74. **Engel, A. G., Lindstrom, J., Lambert, H., and Lennon, V. A.,** Ultrastructural localization of the acetylcholine receptor in myasthenia gravis and in its experimental autoimmune model, *Neurology*, 27, 307, 1977.

Chapter 14

THE KIDNEY AND ITS PATHOLOGICAL ALTERATIONS

T. Faraggiana and A. Onetti-Muda

TABLE OF CONTENTS

0-8493-6012-9/93/$0.00 + $.50

A. INTRODUCTION

Because of its complex architectural organization, and its impressive variety of enzymatic activities, the kidney has long represented the reference tissue in the development of most histochemical methods at the light and electron microscopic level. Moreover, since the early times of immunopathology, it was clear that the kidney, and in particular the glomerulus, was the target organ in most immunologic-mediated diseases; consequently, renal pathology and immunomicroscopy became almost inseparable. These observations should explain the impressive body of literature devoted to the study of the kidney in which electron microscopic histochemistry and/or immunocytochemistry play a predominant role. This short review illustrates some of the most widely used techniques, their relevance in the understanding of basic aspects of renal physiology and pathology, and their possible future developments. One must, however, realize that the literature in this field is gigantic and that any attempt to be reasonably thorough would be futile.

For the sake of clarity, this chapter will be divided into **classical histochemical methods, immunocytochemistry,** and **enzyme histochemistry,** even though several different methods or approaches may have been used at the same time to tackle a specific problem and even if some overlapping will be unavoidable.

The impact of modern molecular biology methods, such as *in situ* hybridization, is likely to change dramatically our current views of histochemistry in the near future. However, we chose not to examine this topic since the ultrastructural applications are still too sparse and in too preliminary a stage of development to allow a meaningful review.

B. CLASSICAL HISTOCHEMICAL METHODS

1. Carbohydrates

Carbohydrate electron microscopic histochemistry has been a widely investigated field since the late 1950s, when a controversy existed over the existence of the mesangium to contemporary issues of charge distribution and of the concept of glomerular wall permselectivity. One of the first techniques to be developed, which is still widely used, is the periodic acid-silver methenamine stain (PASM),[1] a modification of the classic Gomori method. By and large, it has the same significance as the periodic acid-Schiff (PAS) reaction and it visualizes mostly *vic*-glycol groups of carbohydrates and glycoproteins with a granular, electron-dense precipitate. In the kidney it stains basement membranes (BM), cell surfaces, and thin collagen fibrils. The technique is extremely simple and can be performed on thin sections of embedded tissues, but high-purity reagents (especially water) and proper temperature control must be utilized, since the alkaline silver solution is very easily precipitated. The results largely depend on the fixative and the embedding medium used, thus making it an histochemical method of doubtful specificity in most instances. For example, glutaraldehyde will introduce a strong background staining due to free aldehyde groups, especially where highly reactive amino groups are present (e.g., nuclear chromatin); similarly, osmium fixation will oxidize a number of chemical groups and introduce reactive sites irrespective of periodate treatment. Despite these limitations, the PASM method still enjoys wide popularity among renal pathologists; certain forms of glomerulonephritis are either classified or diagnosed largely on the basis of PASM staining at the ultrastructural level. The following are a few examples.

Membranous nephropathy is a chronic form of glomerulonephritis associated with heavy proteinuria or nephrotic syndrome. The morphologic hallmark of the disease is the presence of electron-dense deposits on the epithelial side of the BM. With the progression of the disease, new BM is formed as thin projections (s.c. ''spikes'') separating immune deposits and later surrounding them completely. An overall thickening of the capillary wall takes place.

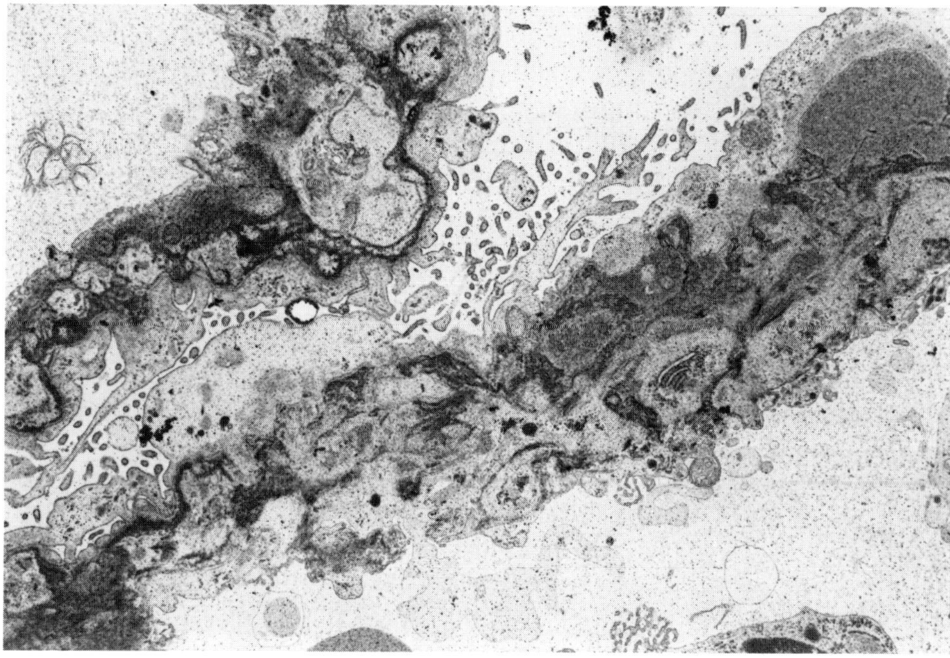

FIGURE 1. PASM reaction performed on ultrathin sections of aldehyde-osmium-fixed, Epon-embedded tissue of human glomerulus with type III membranoproliferative glomerulonephritis. PASM technique visualizes the split, disrupted basement membrane with a highly electron-dense end-reaction product; note the much lower density of the immune deposits. (Magnification × 7,700.)

Membranous nephropathy is subclassified in stages I to IV according to the extent of new BM formation.[2] Since immune deposits and BM are of similar electron density, their identification can be difficult in later stages of the disease and the PASM method can be very helpful, especially in differential diagnosis with other nonimmunologically mediated diseases where BM thickening takes place, such as diabetes mellitus.

Membranoproliferative or mesangiocapillary glomerulonephritis (MPGN) is another form of chronic glomerular disease characterized by nephrotic syndrome and progression to renal failure. Three subtypes of this entity are known with important prognostic and therapeutic differences. In all types PASM reaction contributes either to establishing the correct diagnosis, or to shedding light on the pathogenesis. As an example, various extraction methods, combined with PASM staining, were used to provide evidence for the presence of substantial lipid component in BM of the type II of this disease[3] which is characterized by very electron-dense, homogeneous deposits of unknown origin found in the lamina densa (hence the name ''dense deposits disease''). Finally in type III of the same glomerulonephritis, PASM reaction is considered essential in establishing the very diagnosis of this entity, where splitting and disruption of the BM takes place[4] (Figure 1).

Another major field of investigation where histochemistry is playing a major role, is the detection of glomerular polyanions with cationic probes. Extensive study of the properties of the glomerular filtration of proteins has provided evidence that the glomerular capillary wall has both size- and charge-selective properties:[5-7] smaller, cationic proteins are more easily filtered than larger, anionic ones.

The morphologic and biophysical basis of glomerular filtration has been very extensively investigated, and particular attention has been paid to the high density of anionic charges in

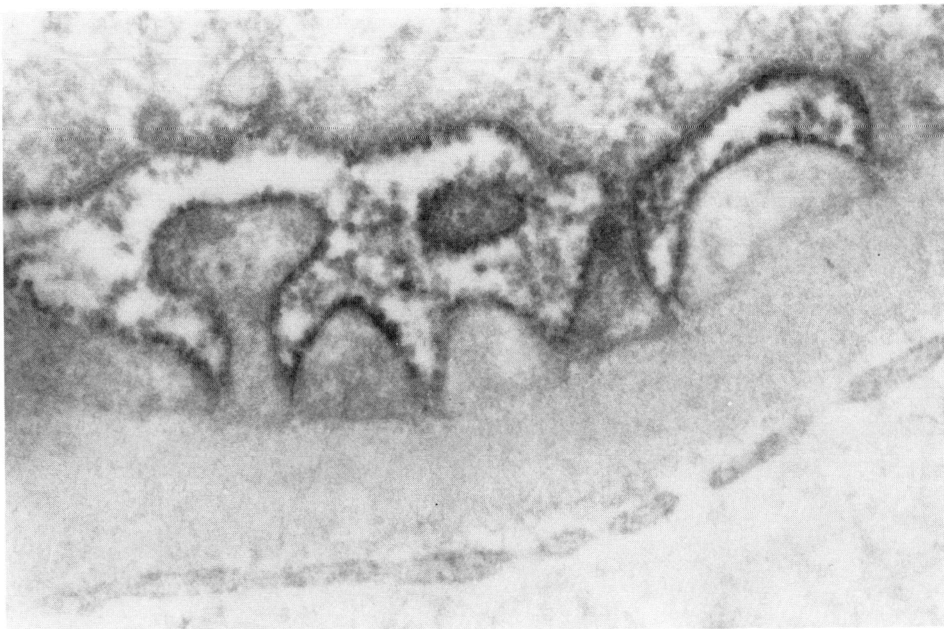

FIGURE 5. Human glomerulus treated with neuraminidase and incubated with peanut lectin-peroxidase conjugate: the exposed galactose residues are mostly present on external podocyte surface. (Magnification × 84,700. Courtesy of Dr. P. Barsotti.)

In general, all these methods show a consistent pattern of charge distribution. By combining different probes and enzymatic extraction, it became evident that the podocyte cell surface facing the urinary space is endowed with a thick, continuous coat covering plasma membrane and slit diaphragms. A thinner but equally evident coat is seen along the side facing the BM. The biochemical nature of these charges is most likely heterogeneous, but there is strong evidence that sialo groups are the most abundant. Sialic acid residues are terminally bound to the galactose moiety through a 2,4-D-Gal linkage.[21] Several light and electron microscopic studies with labeled lectins also point in this direction. For instance, lectin from *Limax flavus,*[22] selective for sialic acid, binds to the glomerular wall, but this is inhibited by neuraminidase treatment. On the other hand, peanut lectin, selective for terminal D-Gal residues, does not stain the podocyte surface since the binding site is masked by sialic acid. After neuraminidase treatment, galactose residues become terminal and a strong peanut binding occurs[23] (Figure 5).

Lectin-binding studies are performed either with peroxidase or colloidal gold probes on wet tissue (preembedding methods, especially with peroxidase tagging) or on thin sections of Lowicryl-embedded blocks (with colloidal gold probes). Both conventional transmission electron microscopy (TEM) and freeze-fracture cytochemistry[24,25] have been used. Freeze-fracture methods have the obvious advantage of making all membrane sites directly accessible to reagents.

Anionic sites are also present within the GBM, in particular along those regions traditionally indicated as *lamina rara externa* and *lamina rara interna* in view of their electron-lucent appearance in conventional TEM. One should, however, stress the fact that very recent data show that the loose texture of these regions is most likely artifactual and is due to the highly hydrated nature of their constituents, some of which are lost during preparatory steps. If chemical fixation is avoided and the tissues are quickly frozen and freeze-substituted,

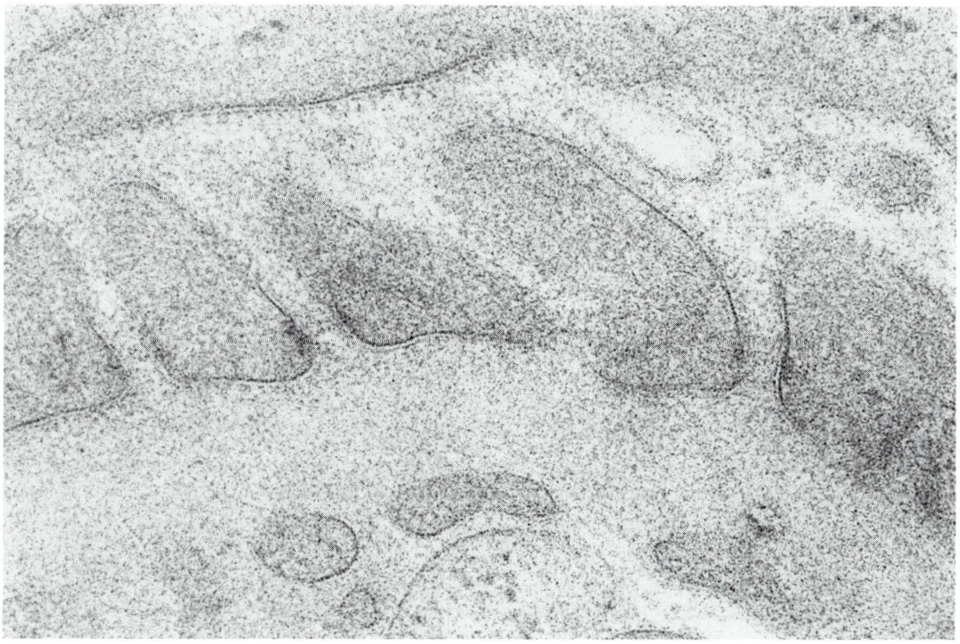

FIGURE 6. High-pressure frozen, freeze-substituted, low-temperature embedded normal rat kidney. The basement membrane lacks the traditional, three-layered appearance; the podocyte cell coat is also evident (uranyl acetate, lead nitrate). (Magnification × 107,800.)

GBM appears as a relatively homogeneous structure from podocyte to endothelial surface. In order to obtain consistent results a very fast freezing rate is necessary, as is the case with high-pressure freezing (Figure 6). The methods are, however, quite complex and require sophisticated equipment, available only in a few laboratories. Nonetheless the future of ultrastructural histochemistry will certainly depend heavily on this type of approach. More conventional methods for GBM anionic sites study are by and large those already described for podocytes. From electron microscopical and biochemical data it appears that most GBM polyanions are in fact related to heparan sulfate,[26] but carboxyl groups do probably play a role as well.[27] A few methods have therefore been proposed for the specific identification of carboxyl groups in the GBM, on the assumption that, despite the fact that heparan sulfate is the predominant charge source, it may not be the only one. Bertolatus[28] has shown that a substantial proportion of anionic sites in GBM are attributable to carboxyl groups on glycoproteins, and a method has been proposed for visualization of carboxyl groups based on previous biochemical assay. Biotin was covalently linked to sialic acid residues by a periodate oxidation-biotin hydrazide treatment, while carboxyl groups were biotinilated by the use of biotin hydrazide and carbodiimide as a cross linker. Antibiotin antibodies and colloidal gold-labeled antibodies were used for final visualization.

A somewhat different approach to the same problem is the use of classic electron microscopic immunohistochemistry, a topic to be considered later. The work pioneered by Kanwar and Farquhar[26] has conclusively shown by peroxidase or other standard immunoelectron microscopic techniques that heparan sulfate is present in the GBM. Current research still uses the same approach. For instance, Lelongt et al.[29] carried out a very detailed study of proteoglycans in aminonucleoside nephrosis by injecting into rats anti-heparan sulfate antibody directed against its core protein. The IgG fraction was either peroxidase- or

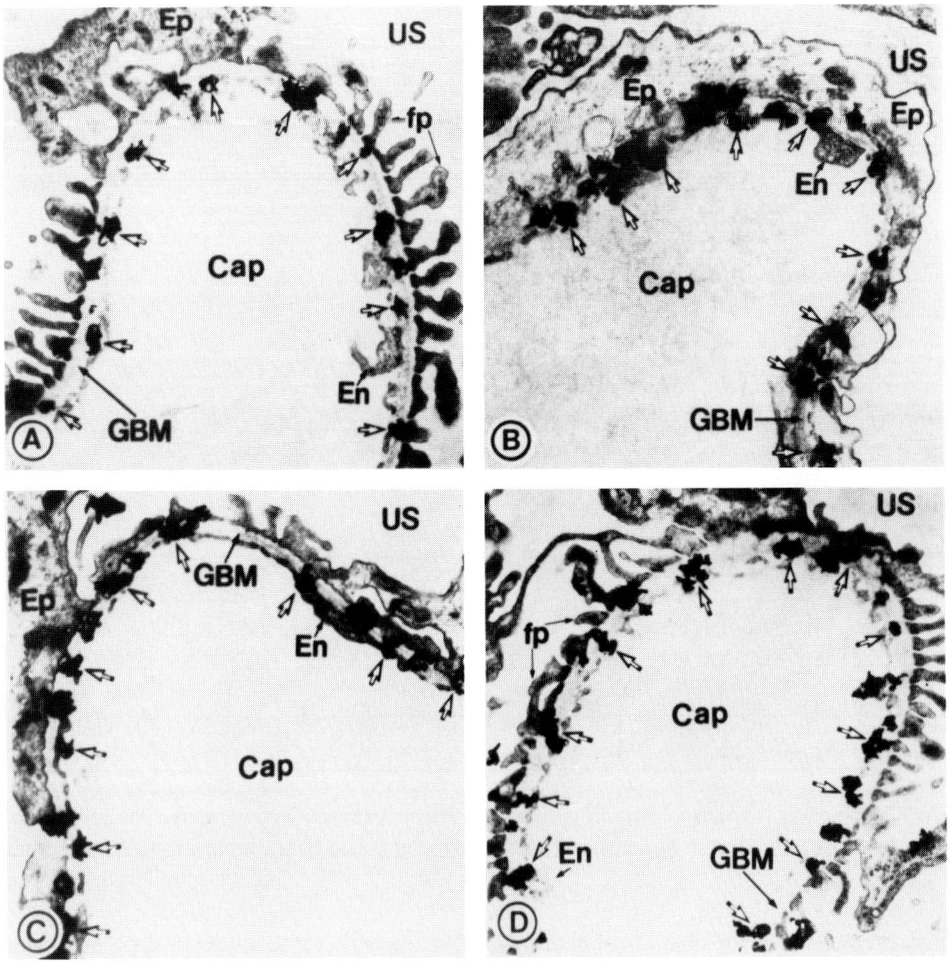

FIGURE 7. Electron microscopic autoradiographs of glomerular capillaries from kidneys of animals with PAN nephrosis, on day 0 (A), 7 (B), 14 (C), and 21 (D). After administration of radioiodinated antibody, it is seen localized to the glomerular basement membrane, as indicated by the black and white arrows. No significant differences are seen in binding of the antibody during various stages of nephrosis. US, urinary space; Ep, epithelium; fp, foot processes; En, endothelium; Cap, capillary lumen. (Magnification × 12,300. From Kanwar, Y., *Kidney Int.*, 31, 1299, 1987. With permission.)

^{125}I-labeled, and the antigen-antibody complexes were visualized by diaminobenzidine or autoradiography, respectively. The pattern of distribution on GBM of the silver grains in autoradiographs showed an impressive correlation with the known distribution of anionic charges as revealed by unrelated techniques (Figure 7). An important role of EM autoradiography seems, therefore, well established.

Much less is known of the distribution of anionic charges in peritubular capillaries. Yet, physiological studies indicate that a size and charge selectivity is also present in these regions. A few studies begin to appear on the use of cationic tracers in interstitial capillaries.[30]

2. Lipids

Lipids play a relevant role in several aspects of renal physiology. It would suffice to mention prostaglandins and related molecules. Mainly because of their solubility in practically every embedding medium, their identification is confined to ultracryomicrotomy. An indirect demonstration is, however, possible under some circumstances. By using chloroform-methanol extraction before embedding as a control, and lectin-peroxidase as label, an α-galactolipid was identified on the luminal surface of collecting ducts and endothelia of human kidney.[31] These residues are thought to be of importance in mediating bacterial adhesion (e.g., some strains of *Escherichia coli* possess specific galactolipid-binding adhesions). It is interesting that in Fabry disease, where a congenital defect of α-galactosidase takes place, there is heavy glycolipid accumulation in endothelial cells and collecting ducts.

C. IMMUNOCYTOCHEMISTRY

Immunohistochemistry and kidney diseases are so closely related that the use of immunohistochemical methods are considered an essential step in routine diagnosis of glomerular diseases. It is hardly surprising that immunology, immunopathology, and renal pathology have been developing hand in hand for the last few decades. Moreover, as stated before, the identification of most proteins and/or enzymes is currently performed by immunocytochemical methods, especially after the development of monoclonal antibodies technology. Practically every available technique of immunocytochemistry has been used, and their choice has been determined by the particular antigen to be studied, its resistance to fixatives, its intracellular, extracellular, or surface distribution, etc. Most recently, freeze-substitution and low-temperature embedding have been successfully employed.[32] It is most important to stress that no universal recipe is available at the present time and that, unless pilot experiments are carried out, no one can predict which technical approach is preferable.

Relevant use of immunocytochemistry can be made in renal research to identify (1) antigens, antibodies, or mediators of immune response in human and experimental nephritis, (2) structural antigens of the extracellular matrix (collagens, proteoglycans, noncollagenous proteins, etc.), (3) enzymes, (4) growth factors, hormones, and their receptors, and, more generally, cellular antigens in studies of almost every aspect of cell biology, as described in detail in other sections of this book, and as exemplified by Figure 8 showing the immunolocalization of renin in the juxtaglomerular apparatus of the mouse.

As will be evident, emphasis is on experimental animals, since human studies are severely hampered by limited availability of tissues and processing requirements.

According to classic concepts, it was believed that glomerulonephritis is mediated by one of two basic immunological mechanisms: deposition of circulating, preformed immune complexes or anti-glomerular basement membrane antibodies. More recently it has been shown that in the majority, if not all, cases of immune complex nephritis, the immune complexes are in fact formed *in situ*, i.e., antibodies bind to planted antigens or to normal components of the capillary wall.[33] Intense investigations in the last few years seem to indicate that the two mechanisms are not mutually exclusive but represent two aspects of a continuum of the immunological disease. Current investigation is directed to identify the antigens involved in glomerular injury (largely unknown in the vast majority of human renal diseases) and to pinpoint the fine mechanism of formation and processing of the immune deposits.

A key role has been played by ultrastructural immunocytochemistry in elucidating the nature and distribution of the antigen involved in the pathogenesis of Heymann nephritis, a very common model in recent experimental nephrology. This is a form of experimental nephritis thought to share several similarities with some forms of human glomerulonephritis, particularly membranous nephropathy. In this model, a glomerular injury is produced by injecting an autologous antigen originally obtained from proximal brush border[34,35] or by passive injection of preformed antibodies. The antigen involved has been purified and shown

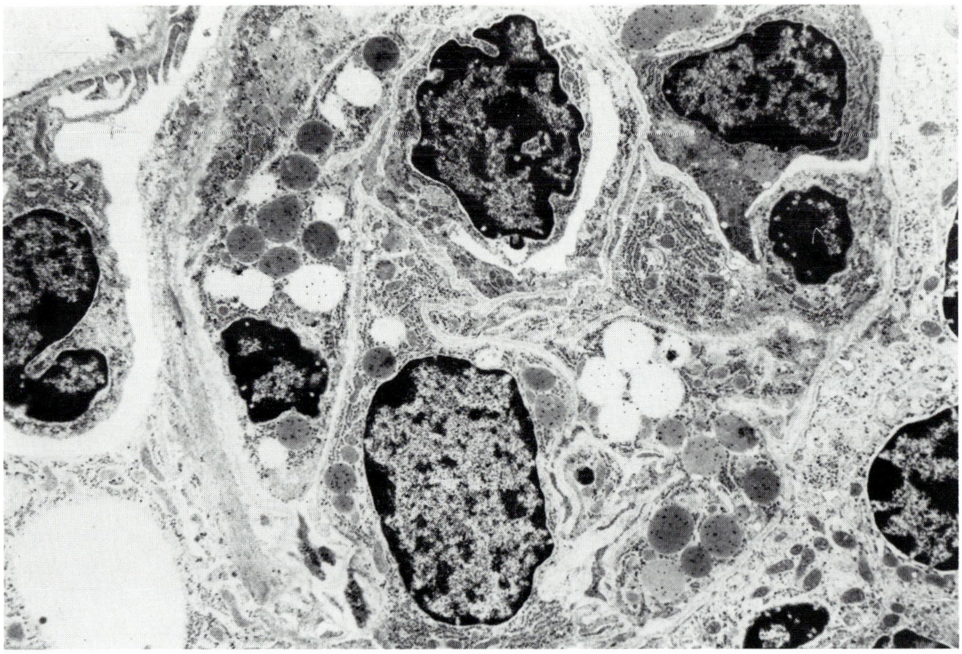

FIGURE 8. Afferent arteriole from mouse kidney treated with a rabbit anti-mouse renin antibody, followed by protein A-colloidal gold. The cytoplasmic granules of the renin-secreting cells are clearly marked by the gold particles. (Magnification × 6,200.)

to be related to a 330-kDa glycoprotein also known as gp330.[36] Several studies have shown that an antibody specific for gp330 binds to the glomerular wall when injected intravenously. By immunoelectron microscopy it was convincingly demonstrated that gp330 is localized on epithelial cells rather than on the GBM.[37] Moreover, it was mostly found in coated pits of podocytes and clathrin-coated pits of proximal tubules, thus explaining the otherwise puzzling localization in two seemingly different structures (tubular brush borders and GBM) that had been shown by immunofluorescence. Moreover, the presence of an immunogen in the endoplasmic reticulum and Golgi vesicles provided strong evidence that the antigen under investigation was indeed an autologous protein and not merely an exogenous substance.[37] The early formation and disappearance of immune deposits have also been investigated by electron microscopic immunocytochemistry and shown to "migrate" from the epithelial surface to the lamina rara externa, where they become progressively larger.

Relevant studies have also been carried out by Camussi et al.[38] on epithelial cells grown in culture in order to follow redistribution and eventual shedding of antigens under several conditions able to induce antigen modulation. Their work is relevant in that it casts some light on the process of antigen processing *in vitro,* but it also underscores possible limitations of uncritical comparisons between different experimental conditions. For instance, by immunogold technique the authors were able to show that on interaction with specific antibodies, gp330 is internalized by cells in culture rather than shed as happens at the level of coated pits *in vivo.* Among the possible explanations for such discrepancies, loss of complexes in the medium and reshaping of cells in culture are to be considered.

An equally well-studied glycoprotein is podocalyxin, the major sialoglycoprotein of the glomerulus.[39] Its immunofluorescence microscopic localization closely parallels the distribution of anionic charges as seen with cationic probes, but immunoelectron microscopic cytochemistry was instrumental in showing that its presence is limited to the podocyte surface facing the urinary space, while it is absent from the plasma membrane facing the lamina rara externa.[39] It is interesting that Ozaki et al.[40] were able to produce antibodies against neuraminidase-digested plasma membrane fraction which had been further purified by *Helix pomatia* lectin affinity. The authors, by combining immunoelectron microscopy and neuraminidase treatment on kidney tissue, showed that specific antibodies against such a purified plasma membrane fraction had a pattern of localization as well as carbohydrate composition very similar to podocalyxsin, and reached important conclusions on the crucial issue of the masking and unmasking of antigenic sites by sialic acid.

Immunoelectron microscopic studies of human renal diseases have been less numerous for obvious reasons, mostly technical, but several reports have been published showing immunoglobulin detection in human renal biopsy specimens by the use of immunogold techniques.[41] In rare cases, the antigen involved has been detected, as in hepatitis B-associated nephropathy.[42]

Besides the field of immunopathology, there are numerous studies dealing with the distribution of a variety of structural antigens in the kidney, such as collagens and other intercellular matrix constituents, mostly in the glomerulus. Mutual relationships between collagen IV, laminin, fibronectin, and proteoglycans are becoming more and more clear.[43,44] A few diseases, such as diabetes mellitus, have also been studied.[45] The reader is referred to the excellent article by Kerjaschki et al.[46] for a critical discussion of the pitfalls of available methods.

Tubular cells have also been extensively investigated, especially with monoclonal antibodies; first, in order to detect specialized cell types within nephron segments, not otherwise identifiable by morphological methods (e.g., immunodissection of the nephron), second, to investigate several cellular functions and properties, such as tubular reabsorption, cell polarity, cytoskeleton and brush border organization, etc.[47] In most cases, as in cell polarity studies, the relevance of the topic far exceeds the field of nephrology and more appropriately belongs to the very core of contemporary cell biology. A typical example is the epithelial cell line known as Madin-Darby canine kidney (MDCK) cell, probably the most thoroughly studied cell line which has been used by renal physiologists in transport studies and by cell biologists for cell polarity investigations. Since uneven distribution of highly specialized proteins is the hallmark of polarized cells, there is little wonder that these proteins have been studied over the apical and/or basolateral membranes especially by enzyme histochemistry, autoradiography, and by immunocytochemistry at the EM level.[48] In the vast majority of cases, these different approaches give similar results as far as the localization of the protein is concerned, even if some differences, due to technical reasons, are to be expected. The whole field relies heavily on fine ultrastructural localization by cytochemical methods also to detect the density of protein distribution, not merely their presence, an essential prerequisite in defining the sorting of polarized proteins.

D. ENZYME HISTOCHEMISTRY

Distribution of enzymes along the nephron, as noted earlier, is complex and varied, and it would be meaningless to review in a few pages the complex field of tubular physiology, where enzyme cytochemistry and immunocytochemistry are almost obligatory procedures in every morphofunctional study. As an example, Figure 9 shows the ultrastructural distribution of alkaline phosphatase along the luminal surface of the collecting duct of mouse kidney. Some recent reviews on the most important enzymes, such as ATPases, may give an idea of the impact of these techniques on renal physiology.[49-51] For instance, both cytochemical[52] and immunocytochemical[53] methods showed that Na^+,K^+-ATPase was

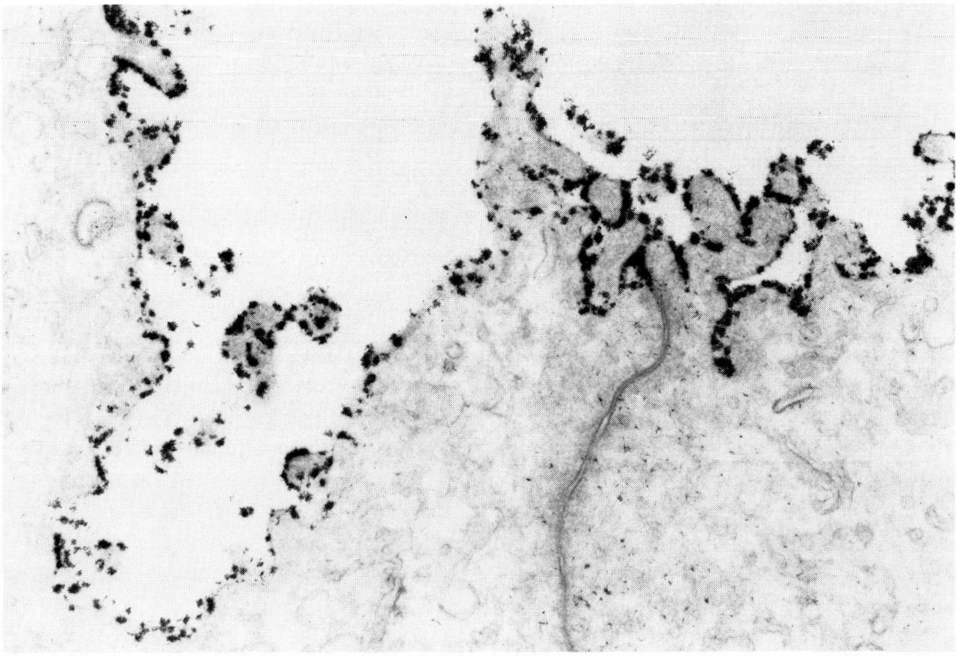

FIGURE 9. Normal mouse kidney. Alkaline phosphatase reaction (cerium chloride method). Enzyme activity is displayed along the luminal surface of a collecting duct. (Magnification × 38,500.)

exclusively found in the basolateral cell membrane. Various cytochemical assays have been used to follow changes in enzyme activity under hormonal stimulation.

A last topic we would like to mention briefly is the heterogeneity of the nephron structure. This refers to both a nonhomogeneous cell population in the same tubular segment and differences among deep vs. superficial nephrons. It has been known for a long time that subcortical nephrons are different from subcapsular ones; some diseases affect deep glomeruli more severely (as is the case with focal segmental glomerulosclerosis, a lesion associated with severe proteinuria and progressing to renal failure). Only recently has the heterogeneity of the nephron received new, closer attention. A few recent reviews have been published.[54] Although vascularization of deep vs. superficial glomeruli is the most extensively studied feature, enzyme distribution is becoming another important field of investigation. Ridderstrale et al.,[55] among others, reported on the heterogeneity of collecting ducts by localizing at the EM level carbonic anhydrase with a cobalt-precipitation method originally devised by Hansson. Na^+,K^+-ATPase, on the other hand, was localized immunocytochemically with a monoclonal antibody and peroxidase-labeled Fab of anti-mouse IgG. Na^+,K^+-ATPase, found in basolateral membranes, was more abundant in the principal than in intercalated cells in cortical collecting ducts. In the same region principal cells did not stain for carbonic anhydrase. Other segments of the collecting ducts had different patterns of enzyme activity. Species differences were also noted, which could explain functional differences between homologous segments in different animals. It has also been shown that alkaline phosphatase has higher activity in superficial than in deep proximal tubules, while the reverse is true for Na^+,K^+-ATPase.[56]

In conclusion, it is obvious that in facing the enormous task of penetrating the subtleties of an impressively complicated organ, with relatively primitive techniques (no matter how sophisticated they appear at the moment), ultrastructural histochemistry has been forced to diversify its approaches. The very existence of such a multitude of techniques should warn scientists against the danger of venturing themselves into such troubled waters with anything less than the whole array of available methods.

REFERENCES

1. **Marinozzi, V.,** Silver impregnation of ultrathin sections for electron microscopy, *J. Biophys. Biochem. Cytol.*, 9, 121, 1961.
2. **Ehrenreich, T. and Churg, J.,** Pathology of membranous nephropathy, *Pathol. Ann.*, 3, 145, 1968.
3. **Onetti-Muda, A., Barsotti, P., and Marinozzi, V.,** Ultrastructural histochemical investigations of "dense deposits disease". Pathogenetic approach to a special type of mesangiocapillary glomerulonephritis, *Virchows Arch. A*, 413, 529, 1988.
4. **Anders, D., Agricola, B., Sippel, M., and Thoenes, W.,** Basement membrane changes in membranoproliferative glomerulonephritis. II. Characterization of a third type by silver impregnation of ultrathin sections, *Virchows Arch.*, 376, 1, 1977.
5. **Kanwar, Y. S.,** Biophysiology of glomerular filtration and proteinuria, *Lab. Invest.*, 51, 7, 1984.
6. **Renke, H., Cotran, R., and Venkatachalam, M. A.,** Role of molecular charge in glomerular permeability: tracer studies with cationized ferritins, *J. Cell Biol.*, 67, 638, 1975.
7. **Brenner, B., Bohrer, M. P., Baylis, C., and Deen, W. M.,** Determinants of glomerular permselectivity: insights derived from observations in vivo, *Kidney Int.*, 12, 229, 1977.
8. **Latta, H., Johnston, W. H., and Stanley, T. M.,** Sialoglycoproteins and filtration barriers in the glomerular capillary wall, *J. Ultrastruct. Res.*, 51, 354, 1975.
9. **Caulfield, J. P. and Farquhar, M. G.,** Loss of anionic sites from the glomerular basement membrane in aminonucleoside nephrosis, *Lab Invest.*, 39, 505, 1978.
10. **Kanwar, Y. S. and Jakubowki, M. L.,** Unaltered anionic sites of glomerular basement membrane in aminonucleoside nephrosis, *Kidney Int.*, 25, 613, 1984.
11. **Whiteside, C., Prutis, K., Cameron, R., and Thompson, J.,** Glomerular epithelial detachment, not reduced charge density, correlates with proteinuria in Adriamycin and puromycin nephrosis, *Lab Invest.*, 61, 650, 1989.
12. **Mohos, S. C. and Skoza, L.,** Glomerular sialoprotein, *Science*, 164, 1519, 1969.
13. **Faraggiana, T. and Grishman, E.,** The podocyte cell coat in experimental nephrosis, *J. Ultrastruct. Res.*, 74, 296, 1981.
14. **Kanwar, Y. and Farquhar, M. G.,** Anionic sites in the glomerular basement membrane: in vivo localization to laminae rarae by cationic probes, *J. Cell Biol.*, 81, 137, 1979.
15. **Renke, H. G. and Venkatachalam, M. A.,** Glomerular permeability. In vivo tracer studies with polyanionic and polycationic ferritins, *Kidney Int.*, 11, 44, 1977.
16. **Caulfield, J. P.,** Alterations in the distribution of Alcian Blue-staining fibrillar anionic sites in the glomerular basement membrane in aminonucleoside nephrosis, *Lab. Invest.*, 40, 503, 1979.
17. **Schurer, J. W., Kalicharan, D., Hoedemaeker, P. J., and Molenaar, I.,** The use of polyethyleneimine in demonstration of anionic sites in basement membranes and collagen fibrils, *J. Histochem. Cytochem.*, 26, 688, 1978.
18. **Caulfield, J. P. and Farquhar, M. G.,** Distribution of anionic sites in glomerular basement membranes: their possible role in filtration and attachment, *Proc. Natl. Acad. Sci. U.S.A.*, 73, 1646, 1976.
19. **Hunsicker, L. G., Shearer, T. P., and Shaffer, S. J.,** Acute reversible proteinuria induced by infusion of polycation hexadimethrine, *Kidney Int.*, 20, 7, 1981.
20. **Barsotti, P., Malchiodi Albedi, F., Fabrizi, E., and Marinozzi, V.,** Ultrastructural localization of sulfate glycoconjugates in the human glomerular capillary wall using the high iron diamine method, *J. Submicrosc. Cytol. Pathol.*, 20, 549, 1988.
21. **Montreuil, J.,** Primary structure of glycoprotein glycans: basis for the molecular biology of glycoproteins, *Adv. Carbohydr. Chem. Biochem.*, 37, 157, 1980.
22. **Schulte, B. A., Spicer, S. S., and Miller, R. L.,** Histochemical localization of sialoglycoconjugates with a sialic acid-specific lectin from the slug Limax flavus, *Histochem. J.*, 16, 1125, 1984.
23. **Faraggiana, T., Malchiodi, F., Prado, A., and Churg, J.,** Lectin-peroxidase conjugate reactivity in normal human kidney, *J. Histochem. Cytochem.*, 30, 451, 1982.
24. **Brown, D., Roth, J., and Orci, L.,** Lectin-gold cytochemistry reveals intercalated cell heterogeneity along rat kidney collecting ducts, *Am. J Physiol.*, 248, C348, 1985.

25. **Chevalier, J., Appay, M. D., Wang, X. Y., Bariety, J., and Pinto da Silva, P.,** Freeze-fracture cytochemistry of rat glomerular capillary tuft. Determination of wheat germ agglutinin binding sites and localization of anionic charges, *J. Histochem. Cytochem.*, 35, 1389, 1987.
26. **Kanwar, Y. and Farquhar, M. G.,** Presence of heparan sulphate in the glomerular basement membrane, *Proc. Natl. Acad. Sci. U.S.A.*, 76, 1303, 1979.
27. **Bertolatus, J. A. and Hunsicker, L. G.,** Polycation binding to glomerular basement membrane: effect of biochemical modification, *Lab. Invest.*, 56, 170, 1987.
28. **Bertolatus, J. A.,** Affinity cytochemical labeling of glomerular basement membrane anionic sites using specific biotinylation and colloidal gold probes, *J. Histochem. Cytochem.*, 38, 377, 1990.
29. **Lelongt, B., Makino, H., and Kanwar, Y. S.,** Status of glomerular proteoglycans in aminonucleoside nephrosis, *Kidney Int.*, 31, 1299, 1987.
30. **Koshy, V. and Avasthi, P. P.,** The anionic sites at luminal surface of peritubular capillaries in rats, *Kidney Int.*, 31, 52, 1987.
31. **Faraggiana, T., Crescenzi, A., and Marinozzi, V.,** Presence of an α-galactolipid on the cell surfaces of endothelial cells of human kidney, *Histochem. J.*, 21, 235, 1989.
32. **Stirling, J. W., Coleman, M., and Brennan, J.,** The use of inert dehydration and glycolmethacrylate embedding for immunogold localization of glomerular basement membrane components, *Lab. Invest.*, 62, 655, 1990.
33. **Fleuren, G. J., Grond, J., and Hoedemaeker, Ph. J.,** In situ formation of subepithelial glomerular immune complexes in passive serum sickness, *Kidney Int.*, 17, 631, 1980.
34. **Heymann, W., Hackel, D. B., Harwood, S., Wilson, S. G. F., and Hunter, J. L. P.,** Production of nephrotic syndrome in rats by Freund's adjuvant and rat kidney suspensions, *Proc. Soc. Exp. Biol. Med.*, 100, 660, 1959.
35. **Eddington, T. S., Glassock, R. J., and Dixon, F. J.,** Autologous immune complex nephritis induced with renal tubular antigen. I. Identification and isolation of the pathogenetic antigen, *J. Exp. Med.*, 127, 555, 1968.
36. **Kerjaschki, D. and Farquhar, M. G.,** The pathogenic antigen of Heymann nephritis is a membrane glycoprotein of the renal proximal tubule brush border, *Proc. Natl. Acad. Sci. U.S.A.*, 79, 5557, 1982.
37. **Kerjaschki, D. and Farquhar, M. G.,** Immunocytochemical localization of the Heymann nephritis antigen (gp330) in glomerular epithelial cells of normal Lewis rats, *J. Exp. Med.*, 157, 667, 1983.
38. **Camussi, G., Kerjaschky, D., Gonda, M., Nevins, Th., Rielle, J-Ch., Brentjens, J., and Andres, G.,** Expression and modulation of surface antigens in cultured rat glomerular visceral epithelial cells, *J. Histochem. Cytochem.*, 37, 1675, 1989.
39. **Kerjaschki, D., Sharkey, D. J., and Farquhar, M. G.,** Identification and characterization of podocalyxin, the major sialoprotein of glomerular epithelial cell, *J. Cell Biol.*, 98, 1591, 1984.
40. **Ozaki, I., Ito, Y., Fukatsu, A., Suzuki, N., Yoshia, F., Watanabe, Y., Sakamoto, N., and Matsuo, S.,** A plasma membrane antigen of rat glomerular epithelial cells. Antigenic determinants involving N-linked sugar residues in a 140 kilodalton sialoglycoprotein of the podocytes, *Lab. Invest.*, 63, 707, 1990.
41. **Viale, G., Dell'Orto, P., Colombi, R., and Coggi, G.,** Ultrastructural localization of extracellular immunoglobulins in immune-complex-mediated glomerulonephritis. Immunoelectron microscopy of Epon-embedded human renal biopsies using the immunogold staining procedure, *Histochemistry,* 84, 1, 1986.
42. **Slusarczyk, J., Mikalak, T., Nazarewicz-de Mezer, T., Krawczynski, K., and Nowoslawski, A.,** Membranous glomerulopathy associated with hepatitis B core antigen immune complexes in children, *Am. J. Pathol.*, 98, 29, 1982.
43. **Laurie, G. W., Leblond, C. P., Inoue, S., Martin, G. R., and Chung, A. E.,** Structure of the basement membrane and immunolocalization of five basement membrane components to the lamina densa and its extension in both glomeruli and tubules, *Am. J. Anat.*, 169, 463, 1984.
44. **Courtoy, P. J., Kanwar, Y. S., Hynes, R. O., and Farquhar, M. G.,** Fibronectin localization in the rat glomerulus, *J. Cell Biol.*, 87, 691, 1980.
45. **Bendayan, M.,** Alteration in the distribution of type IV collagen in glomerular basal laminae in diabetic rats as revealed by immunocytochemistry and morphometrical approach, *Diabetologia,* 28, 373, 1985.
46. **Kerjaschki, D., Sawada, H., and Farquhar, M. G.,** Immunoelectron microscopy in kidney research: some contributions and limitations, *Kidney Int.*, 30, 229, 1986.
47. **Kenny, A. J. and Maroux, S.,** Topology of microvillar membrane hydrolases of kidney and intestine, *Physiol. Rev.*, 62, 91, 1982.
48. **Simons, K. and Fuller, S. D.,** Cell surface polarity in epithelia, *Annu. Rev. Cell Biol.*, 1, 243, 1985.
49. **Sthal, W. L. and Baskin, D. G.,** Histochemistry of ATPases, *J. Histochem. Cytochem.*, 38, 1099, 1990.
50. **Katz, A. I.,** Distribution and function of classes of ATPases along the nephron, *Kidney Int.*, 29, 21, 1986.

51. **Doucet, A.,** Function and control of Na-K ATPase in single nephron segments of the mammalian kidney, *Kidney Int.,* 34, 749, 1988.
52. **Ernst, S. A.,** Transport ATPase cytochemistry. Ultrastructural localization of K-dependent and K-independent phosphatase activities in rat kidney cortex, *J. Cell Biol.,* 66, 586, 1975.
53. **Kashgarian, M., Biemesderfer, D., Caplan, M., and Forbush, B., III,** Monoclonal antibody to Na-K ATPase: immunocytochemical localization along nephron segments, *Kidney Int.,* 28, 899, 1985.
54. **Bankir, L., Bouby, N., and Trinh-Trang-Tan, M.-M.,** Heterogeneity of nephron anatomy, *Kidney Int.,* 31, S-25, 1987.
55. **Ridderstrale, Y., Kashgarian, M., Koeppen, B., Giebish, G., Stetson, D., Ardito, Th., and Stanton, B.,** Morphological heterogeneity of the rabbit collecting duct, *Kidney Int.,* 34, 655, 1988.
56. **Schmidt, U. and Dubach, U. C.,** Differential enzymatic behaviour between single proximal segments of the superficial and juxtamedullary nephron. I. Alkaline phosphatase, (MG^{++}) ATPase and (Na^{+}-K^{+}) ATPase, *Z. Gesamte Exp. Med.,* 155, 340, 1971.

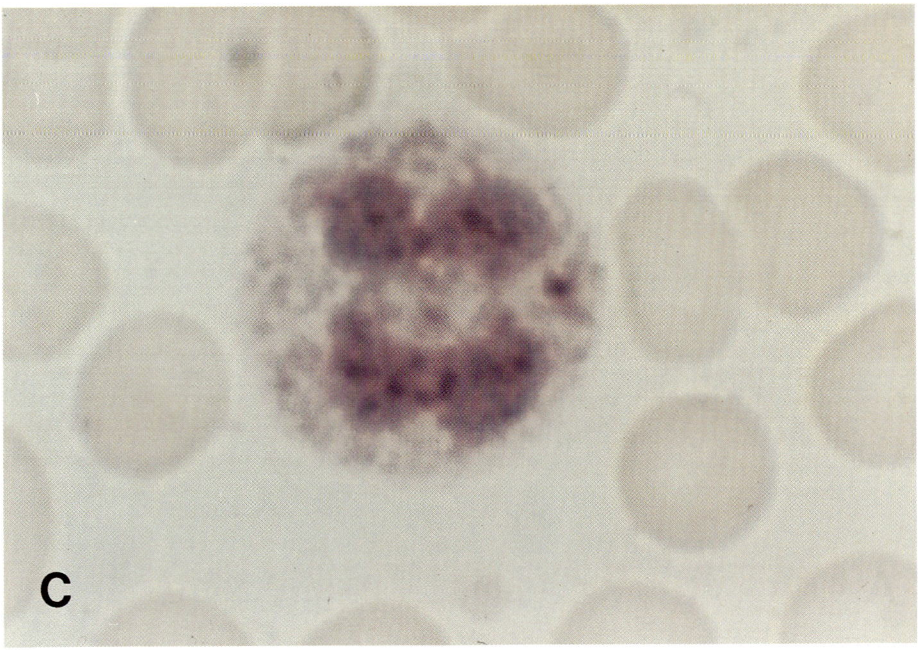

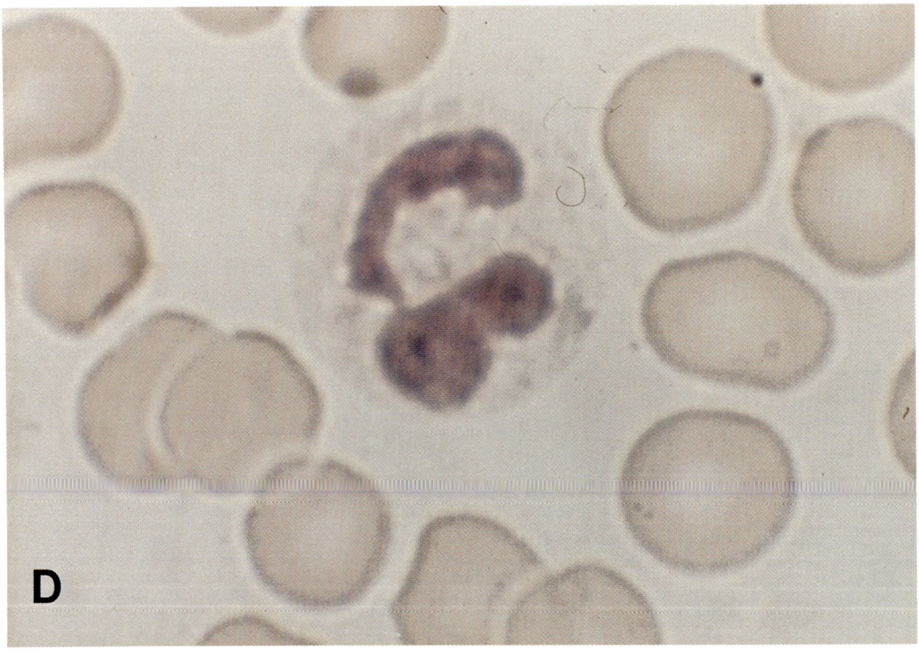

CHAPTER 12, PLATE 1 (continued).

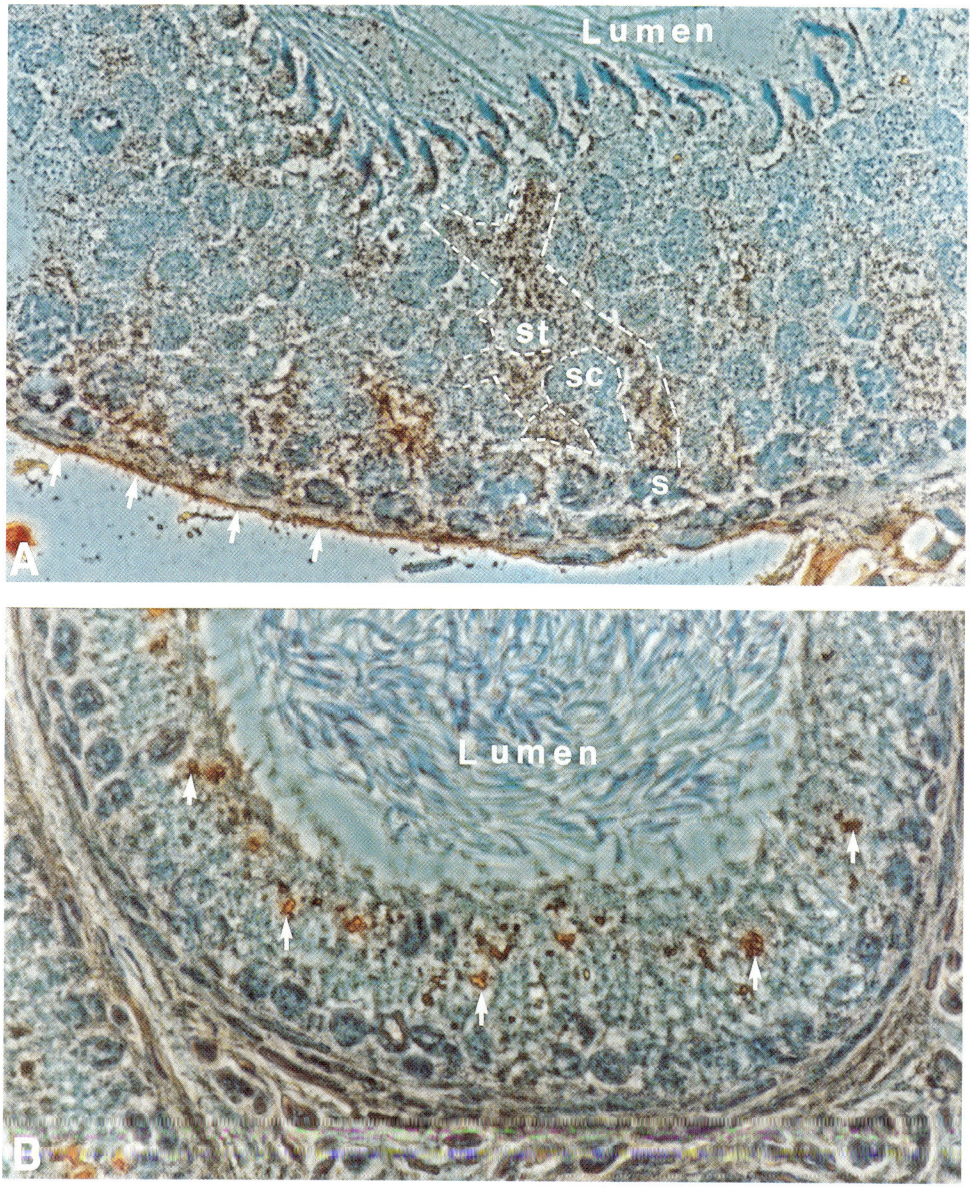

CHAPTER 15.1, PLATE 1. ABP immunolocalization in Sertoli cells and caput epididymis. ABP was immunolocalized using biotin-streptavidin peroxidase immunocytochemistry in frozen 5-μm sections that were fixed with 3.7% formalin. Within the seminiferous epithelium, ABP is homogeneously distributed throughout the Sertoli cells (outlined in white) and it can also be detected in the interstitium. In the caput epididymis, however, where ABP is internalized by receptor-mediated endocytosis, ABP is found discretely localized in vesicular structures presumed to be endosomes.

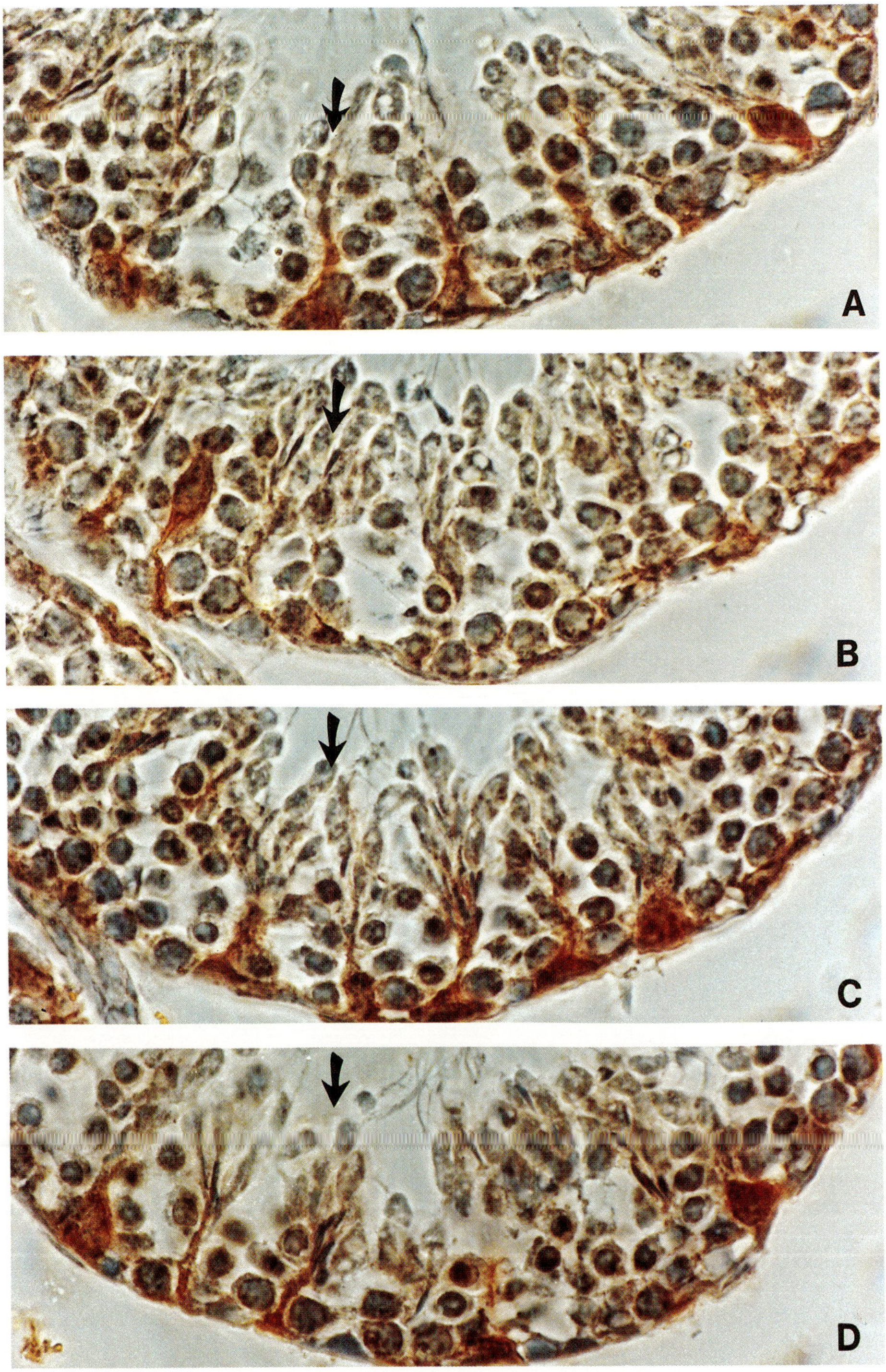

CHAPTER 15.1, PLATE 2. EGF receptor (EGFR) immunolocalization in Sertoli cells; 8-μm serial sections from polyester wax-embedded tissue were immunostained for EGFR using biotin-streptavidin immunoperoxidase. Arrows point to approximately the same spot in the different serial sections. Note the columnar immunostaining pattern in the Sertoli cells within the seminiferous epithelium. The peritubular cells and germ cells appear devoid of specific immunostaining. (From Suarez-Quian, C. A. and Niklinski, W., *Biol. Reprod.*, 43, 1087, 1990. With permission.)

Chapter 15

REPRODUCTIVE SYSTEM

15.1 Immunocytochemistry of Secretory and Membrane Proteins of the Seminiferous Epithelium

Carlos A. Suárez-Quian and Martin Dym

TABLE OF CONTENTS

0-8493-6012-9/93/$0.00 + $.50

A. INTRODUCTION

A major area of research carried out during the past 20 years in male reproductive biology was to identify blood-borne constituents involved in the regulation of spermatogenesis. Initial studies identified the gonadotropins, mainly luteinizing (LH) and follicle-stimulating (FSH) hormones working at the level of Leydig and Sertoli cells, respectively, and these have been the focus of recent reviews.[1-4] Later studies, which examined the physiological control of Leydig and Sertoli cells *in vitro,* revealed that numerous other factors, including growth factors and basement membrane components known to exert effects in other systems, also play a role in modulating certain aspects of spermatogenesis.[5-7] This wealth of information has clouded our initial, simple understanding of how LH and FSH regulate spermatogenesis and demonstrated convincingly that spermatogenesis is a complex process likely to be regulated by numerous hormones, growth factors, and other components, but which as yet is still incompletely understood.

The identification of plasma constituents, in particular hormones and growth factors, which exhibited modulatory effects on characteristic activities of Leydig and Sertoli cells *in vitro* led to the elaboration of two basic questions in testicular biology. First, what is the tissue and cell source of these hormones and/or growth factors that exert an effect on spermatogenesis? Second, given that these proteins elicit their intracellular effects via cell surface-localized receptors, what are the cell sites in the testis that respond directly to these hormones and growth factors? Whereas initial endocrine studies of hormones regulating spermatogenesis suggested extratesticular tissues as sources of these components, several laboratories began to focus their attention on the possibility that the testis itself could elaborate compounds and thereby regulate spermatogenesis via a paracrine and/or autocrine fashion.[8] Using biochemical procedures, several growth factors and growth-promoting agents were shown to be synthesized within the testis. For example, both transferrin (made by Sertoli cells) and *p*-mod-S (a differentiation factor that stimulates Sertoli cells and which is secreted from peritubular cells lining the wall of the seminiferous epithelium) have been implicated to act as paracrine regulators of spermatogenesis.[7,8] In addition, recent studies have provided evidence that germ cells themselves are capable of influencing Sertoli cell activity. This modulation is possibly brought about by germ cell secretory products serving as paracrine effectors of Sertoli cells, thereby creating a loop and exerting an effect on the spermatogenic process.[9-12] Thus, the first question posed above became an active area of investigation with respect to the cell site in the testis engaged in the synthesis of paracrine and/or autocrine modulators of spermatogenesis. To this end, as specific antibody probes became available, several laboratories employed immunocytochemical methodology to define the specific cellular distribution of additional paracrine/autocrine products. Further, attempts were initiated by different investigators to ascertain that the *in vitro* effects observed for several hormone and growth factors in Leydig and Sertoli cells, respectively, had a structural basis in their *in vivo* counterparts.[13-15] Specifically, that the presumed cell surface-localized receptors by which these hormone and growth factors are believed to exert their biological effects were in fact present in the homologous *in vivo* cell type. However, for the most part, immunocytochemical studies of secretory products have not met with optimal success, except at the light microscopic level, and the complexity of the seminiferous epithelium has often led to inconclusive interpretations in determining the precise distribution of cell surface-localized receptors.[16,17]

In this review we will present some of our immunocytochemical studies performed during the past few years and focus our attention on the seminiferous epithelium. In particular, we will address the limitations of current immunocytochemical methodology to identify with certainty the cell sites of synthesis of secretory proteins, as well as to define the specific distribution of plasma membrane components, using various examples familiar to us. In addition, we will discuss the merits of the novel confocal microscope that is capable of optically sectioning cells and tissues labeled with fluorescent probes. This new technology promises to obviate the many limitations of conventional epifluorescence immunocytochemistry.

B. IMMUNOLOCALIZATION OF SECRETORY AND MEMBRANE PROTEINS IN TESTIS

Although composed of basically two cell types, the seminiferous epithelium in the adult testis is highly complex in its organization. First, there is the nondividing Sertoli cell, a tall and columnar cell which has been ascribed a "nurse" function because of its capacity to generate a special milieu for developing germ cells.[18] Second, there are the germ cells, an actively dividing cell population which undergo a remarkable morphogenesis during their differentiation into mature spermatozoa. This complexity of the epithelium is readily appreciated in profiles of cross sections of seminiferous epithelia from plastic-embedded testis (Figure 1). The germ cells can be observed to occupy the majority of the area of the seminiferous epithelium and are characterized by distinct morphological features at different steps in their differentiation. They are also located at different levels of the epithelium as a function of their development. Sertoli cells, on the other hand, are recognized by their irregularly shaped nuclei containing mostly euchromatin and a single tripartite nucleolus, and the nuclei are often found in the basal aspects of the cell. At the ultrastructural level it is observed further that lateral projections from the Sertoli cells envelop the developing germ cells (Figure 1). These lateral projections are not easily detected with the light microscope and it is often impossible to distinguish between the cell boundaries of germ cells and Sertoli cells at this level of magnification. Thus, the complexity of the seminiferous epithelium results from the fact that germ cell morphology, as well as their location within the height of the tubules, is altered in both time and space, and that the limiting membranes of Sertoli and germ cells are not always readily apparent. In addition, the cytoplasmic "trunk" portion of the Sertoli cell may not be present in all cross sections of the seminiferous epithelium. Often, this absence causes confusion in the interpretations of histological preparations of the testis by novice investigators.

The quality of the tissue preservation presented in Figure 1 was the direct result of the development of glutaraldehyde fixatives and perfusion protocols for testis.[19] For the most part, major problems of tissue fixation faced by the pioneering electron microscopists of the testis have been overcome. In contrast, the requirements of ultrastructural immunocytochemistry, balancing the need for maintaining antigenicity vs. preserving tissue structure, have met with only limited success. One major reason for this apparent failure is that, to date, the majority of the secretion products and membrane receptors investigated, whose existence has been demonstrated by biochemical means, appear to exist at "effective" low copy numbers in the different cell types of the testis.

For example, biochemical studies performed with culture cells have demonstrated that Sertoli cells secrete numerous substances, perhaps totaling as many as 100.[18] Consistent with this abundant protein secretion is the vast amount of euchromatin found in the Sertoli cell nuclei and a well-developed Golgi apparatus. Ultrastructural studies, however, have pointed to the dearth of classical secretory vesicles present in the Sertoli cells that are generally observed in other cell types known to be actively engaged in protein secretion (Figure 1).[20] It would appear, therefore, that protein secretion by Sertoli cells occurs in the absence of packaging into secretory vesicles. This phenomenon has important implications in determining the distribution of proteins by immunocytochemistry in the seminiferous

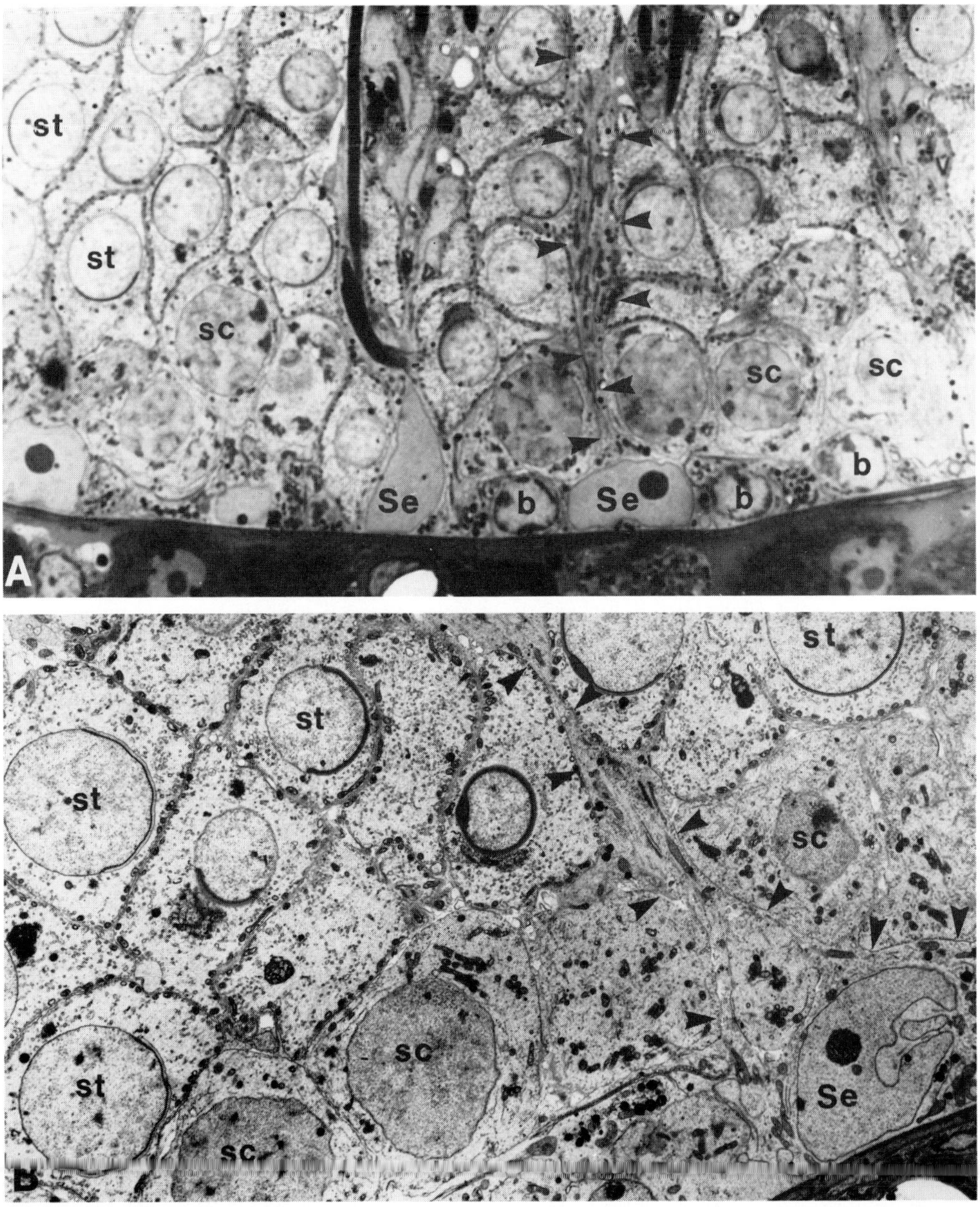

FIGURE 1. Morphology of seminiferous tubules. Light (A) and electron (B) microscopic appearance of normal adult rat seminiferous epithelium. Sertoli cells (Se), type B (b) spermatogonia, pachytene spermatocytes (sc), and round spermatids (st) are labeled. The Sertoli cell is a tall columnar cell that extends from the base to the lumen of the seminiferous tubule. The stalk portion of the cell is demarcated by arrowheads. Note that in B, lateral projections from the stalk portion, enveloping the germ cells, can be readily identified.

epithelium. Similarly, the arboreal distribution of the Sertoli cell plasma membrane, interspersed between the developing germ cells, presents only a limited volume in random sections of adult seminiferous tubules.

In this context, it is important to note that successful ultrastructural immunolocalization studies of secretory proteins have only been performed for those proteins whose secretion is regulated. That is, they are prepackaged into secretory vesicles in the *trans*-Golgi. An example of this type of packaging is insulin.[21] In contrast, although a large number of proteins are synthesized by Sertoli cells, and some of these are probably made in large quantities, absence of secretory vesicle packaging may reduce the effective concentration of proteins present at any one time in a particular area of the cell. This apparent concentration decrease in protein has often led to inadequate immunolocalization of secretory proteins in Sertoli cells. In fact, to our knowledge, only one study in the past has been successful in performing ultrastructural immunolocalization of secretory proteins in Sertoli cells. In fact, to our knowledge, only one study in the past has been successful in performing ultrastructural immunolocalization of a Sertoli cell secretory product. Anti-Mullerian hormone (AMH) was demonstrated in the endoplasmic reticulum of fetal bovine Sertoli cells.[22] Again, although known to be secreted by fetal Sertoli cells, and in relatively large quantities, AMH was not detected in secretory vesicles. That AMH was immunolocalized at all in the endoplasmic reticulum and in the fetal testis may have been fortuitous and led to its successful ultrastructural localization. First, the fetal seminiferous epithelium is composed mainly of Sertoli cell cytoplasm. Thus, the cumbersome and time-consuming process of searching for the cytoplasmic "trunk" portion of Sertoli cells rarely encountered in random sections of adult testis was mitigated significantly. Second, the endoplasmic reticulum may have served to concentrate the AMH sufficiently during its synthesis, thereby permitting its ultrastructural localization. Therefore, with respect to the first question posed in our introduction, what is the cell type in the testis responsible for production of a particular product, technical difficulties brought on by the morphology of the different cell types in the testis remain and have not yet been overcome to systematically elucidate the distribution of compounds for which antibodies are currently available.

1. Androgen-Binding Protein

A case in point with respect to the difficulties encountered in immunolocalizing secretory proteins in the Sertoli cell is that of androgen-binding protein (ABP). This protein was one of the first products to be characterized as being of Sertoli cell origin and has received considerable attention in the literature.[23,24] ABP is secreted bidirectionally by the Sertoli cells, 80% entering the lumen of the seminiferous tubules and 20% reaching the blood.[25] The luminal ABP travels to the epididymis where it is internalized by receptor-mediated endocytosis by the principal cells of the caput epididymis and degraded.[26,27] Therefore, an important question that has been raised is, what is the mechanism by which bidirectional secretion occurs? That is, is ABP packaged into different classes of vesicles in the Golgi which are destined for either apical or basal secretion? Are there morphological differences between these classes of vesicles that may be observed at the ultrastructural level? To this end, employing the peroxidase-antiperoxidase technique of Sternberger, initial immunolocalization studies demonstrated that the ABP could be localized within the seminiferous epithelium and epididymis.[26,27] In the testis, the reaction product was present in a diffused area consistent with the known distribution of the Sertoli cell cytoplasm, whereas in the epididymis, ABP immunolocalization appeared as distinct granules and vesicles now known to correspond to the endocytic compartment.[27]

Similar ABP immunolocalization to that described above is demonstrated in Plate 1.* Using biotin-streptavidin immunoperoxidase methodology in frozen sections, a diffuse reaction product can be identified in the seminiferous epithelium and in the peritubular cells lining the tubules, consistent with the bidirectional secretion of ABP by the Sertoli cell. The distribution of the reaction product within the seminiferous epithelium appears to extend from the base to the lumen of the tubules. This immunostaining pattern can only be generated by a diffuse staining of the Sertoli cell cytoplasmic ''trunk'' portion. In contrast, the uptake of ABP by endocytic organelles of principal cells of the epididymis renders an apparent concentration of this protein and facilitates its immunolocalization in vesicles and granules. Therefore, at the light microscopic level, ABP localization can be readily ascertained as being present in Sertoli cells. We have known this for over a decade.[26,27] We wish to emphasize, however, that current immunocytochemical technology has not progressed to the point where ABP localization at the ultrastructural level can be achieved with a great deal of certainty. Although different immunocytochemical techniques have been employed to determine the distribution of ABP in the Sertoli cell, to date we still do not know the identity of the secretory vesicles by which ABP exits the cell. At best, given available technology, immunocytochemical localization of secretory products of Sertoli cells can only corroborate previous biochemical studies.

2. Membrane Proteins: Epidermal Growth Factor Receptor

Recent studies have made it apparent that growth factors are likely to take part in the regulation of spermatogenesis.[6,8] Epidermal growth factor (EGF) is one such factor that has elicited much interest regarding its potential effects on spermatogenesis. For example, sialoadenectomy of adult male mice led to the elimination of circulating levels of plasma EGF, without a concomitant decrease in gonadotropin and androgen levels, and caused a marked decrease in male fertility.[29] Recent studies have been able to corroborate, in part, these findings.[30,31] Further, in experimentally induced diabetic mice that exhibit EGF deficiency which leads to the pathogenesis of oligozoospermia, daily administration of EGF significantly increased sperm count.[32] In addition, functional studies in different species of both Leydig cells and Sertoli cells *in vitro* have elucidated potential mechanisms of EGF action on spermatogenesis.[33-39] Taken together, these studies suggest that EGF could act directly in testis to bring about its effects on spermatogenesis.

Epidermal growth factor exerts its intracellular effects via a cell surface-localized 170-kDa glycoprotein molecule, the epidermal growth factor receptor (EGFR), that spans the plasma membrane.[40] Consequently, to understand fully the mechanism by which EGF exerts it effect on spermatogenesis requires first that the cell site(s) of the EGFR in the testis be elucidated with certainty.[15] As discussed above, however, the arboreal distribution of the Sertoli cell lateral appendages within the seminiferous epithelium presents significant problems with respect to distinguishing cell boundaries between germ cells and Sertoli cells.

The EGFR has been cloned and sequenced and antibody probes have been generated to peptide-specific sequences of either the extracellular or intracellular domains of the molecule. Therefore, knowing the organization of the seminiferous epithelium, we reasoned that by using an antibody to the intracellular domain and performing immunostaining for the EGFR in serial sections, we could identify whether the reaction product became limited to the Sertoli cell cytoplasmic ''trunk'' portion, or was present in germ cells also.

The procedure chosen for embedding the testis that would permit serial sectioning was low-temperature melting polyester wax.[41] Although serial sectioning of paraffin-embedded

* Plate 1 follows p. 672.

tissue is by far much easier than polyester wax, during the processing of the tissue antigenicity was apparently lost for this particular antibody. To reveal the specific sites of the EGFR, we employed the biotin-streptavidin immunoperoxidase technique.[15] Several series of 5- to 8-μm sections were prepared and immunostained for the EGFR. Each series contained at least 10 seminiferous tubule cross sections, and was composed of five to eight sections. Although specific stages of the cycle of the seminiferous epithelium are difficult to discern in polyester wax sections, representative stages of early, middle, and late stages were observed in the majority of the different serial sections immunostained. Four serial sections of a seminiferous tubule from one of these series, representing the same stage of the cycle of the seminiferous epithelium, are demonstrated in Plate 2.* At this level of magnification, the positive immunoreaction in a particular cross section of a series is limited to Sertoli cells, as evident by the immunoreactivity extending in a columnar fashion from the base to the lumen of the seminiferous tubule. However, within each cross section, the columnar immunostaining pattern was not observed in a circumferential distribution, e.g., as the spokes of a wheel. Significantly, EGFR immunostaining appears, disappears, and appears once again in the same general area of the different serial sections taken from the same tubule. This EGFR immunoreactivity is coincident with the presence or absence of the columnar cytoplasm of Sertoli cells present in the seminiferous tubule. That is, in those seminiferous tubules in which the EGFR immunocytochemistry was rigorously examined, all Sertoli cells contained within a particular stage of the cycle immunostained positively for EGFR and there was no apparent variation in the immunostaining of Sertoli cells as a function of the stages of the cycle of the seminiferous epithelium.

Had we not performed serial-section immunocytochemistry localization of the EGFR and only examined random sections, our results could have been interpreted much differently. For example, the limited immunostaining of the EGFR in Sertoli cells, as well as the apparent absence of immunostaining in every tubule cross section, may have suggested that the EGFR is expressed in a cell- and/or stage-specific fashion. Alternatively, as we demonstrate in Plate 2, the limited immunostaining is due to either poor antibody penetration into the tissue section and/or absence of sufficient Sertoli cell plasma membrane within each section to permit the efficacious accumulation of reaction product to permit detection during the immunocytochemical reaction.

3. Sertoli Cells *In Vitro*

Given that immunocytochemical localization of secretory and membrane proteins is still technologically difficult to perform, then what alternatives are available to reach similar ends? One strategy employed by us, as well as several other reproductive biologists, has been to prepare primary cultures of Sertoli cells whose morphological phenotype mimics closely their *in vivo* counterparts (see Figure 2C).[42] In addition, physiological and biochemical studies with Sertoli cells *in vitro* have demonstrated that the cultured cells function similarly to the *in vivo* Sertoli cells.[43,44] An additional advantage that this model system provides is that Sertoli cells *in vitro* can be manipulated hormonally to favor synthesis of a desired product and the presence of membrane proteins can be established in the absence of remaining germ cells.

However, one major disadvantage that the primary Sertoli cell culture system may be plagued with is the presence of germ cells and peritubular cells remaining as "contaminants". Recent studies, for example, have suggested that previously characterized biochemical responses of Sertoli cells to hormones and growth factors may have occurred indirectly; the actual cell targets for these hormones and growth factors were in reality the contaminating

* Plate 2 follows p. 672.

peritubular cells and they in turn stimulated Sertoli cells via a paracrine factor.[45] In addition, several convincing studies have provided evidence that germ cells themselves influence Sertoli cell activity.[9-12] Therefore, of principal importance in working with Sertoli cell primary cultures is the need to remove germ cells from the culture and to determine precisely the level of peritubular cell contamination.

In cases where germ cells are not desired, they can easily be removed by rinsing the cell cultures at 3 days post-plating for a brief time with a hypotonic buffer.[46,47] This wash lyses the germ cells and they slough off from the Sertoli cell monolayers. In contrast, removal of peritubular cells requires a series of enzymatic digestions in conjunction with a precise cell separation procedure and not until recently has this protocol become standard in our laboratory.[47]

The wall of the seminiferous epithelium in the rat is formed by a peritubular cell layer composed of two cell types (Figure 2). Most proximal to Sertoli cells are the myoid cells, a smooth muscle-like cell type that contains numerous cytoskeletal filaments and is capable of contracting. This layer, in turn, is surrounded by a continuous layer of lymphatic endothelial cells. Depending on the technique used to prepare the Sertoli cell primary cultures, the peritubular cell layer is often difficult to remove and several studies suggest that up to 5 to 10% of the plated cells may indeed represent peritubular cells. To determine the level of peritubular cell contamination, investigators have used a monoclonal antibody to the alpha isoform of smooth muscle actin and demonstrated that it is specific for the peritubular cells.[48] In culture, the peritubular cells can be labeled with this monoclonal antibody and their presence in the Sertoli cell monolayers quantified (see Figure 2). Quantization studies of the level of peritubular cell contamination of Sertoli cell primary cultures have clearly established that at most less than 0.5% of the cells *in vitro* express the alpha isoform of the smooth muscle f-actin filament (unpublished results). We wish to emphasize, however, that no ultrastructural immunolocalization studies have been performed with this anti-actin antibody on cells lining the wall of the seminiferous epithelium. Thus, at the limit of resolution afforded by immunofluorescence it is not possible to distinguish with certainty if only the myoid peritubular cell is immunostained with the actin antibody. Future studies should be carried out to determine if the endothelial cells are, or are not, immunostained with this antibody. If they are not immunostained, then another antibody marker must be developed which will serve to identify the potential contamination of Sertoli cell primary cultures with the lymphatic endothelial cells.

C. CONFOCAL MICROSCOPY

As discussed above, one method to determine the presence of a Sertoli cell membrane marker was to perform immunocytochemistry in serial sections of seminiferous epithelium. However, this procedure is technically demanding and may be successful only for those membrane proteins that span the lipid bilayer and for which intracellular and extracellular antibody probes are currently available. An alternative to this laborious and often risky enterprise is to use confocal microscopy to perform immunocytochemistry.

The new confocal microscope allows investigators to optically section cells and tissues which have previously been labeled with fluorescently conjugated antibodies.[49] Whereas conventional epifluorescence images are often difficult to interpret because of excessive fluorescence noise, in an optical section, out of focus fluorescence is diminished significantly. This is accomplished by illuminating only a narrow portion of the section (with a laser capable of producing a narrow and intense light source) and focusing this image on a pinhole that serves as a diaphragm. The image is then captured with a photodetector.[49]

Whereas in a homogeneous tissue such as liver, diffuse fluorescence may not present problems regarding the cell site responsible for specific immunoreactivity, in the seminiferous

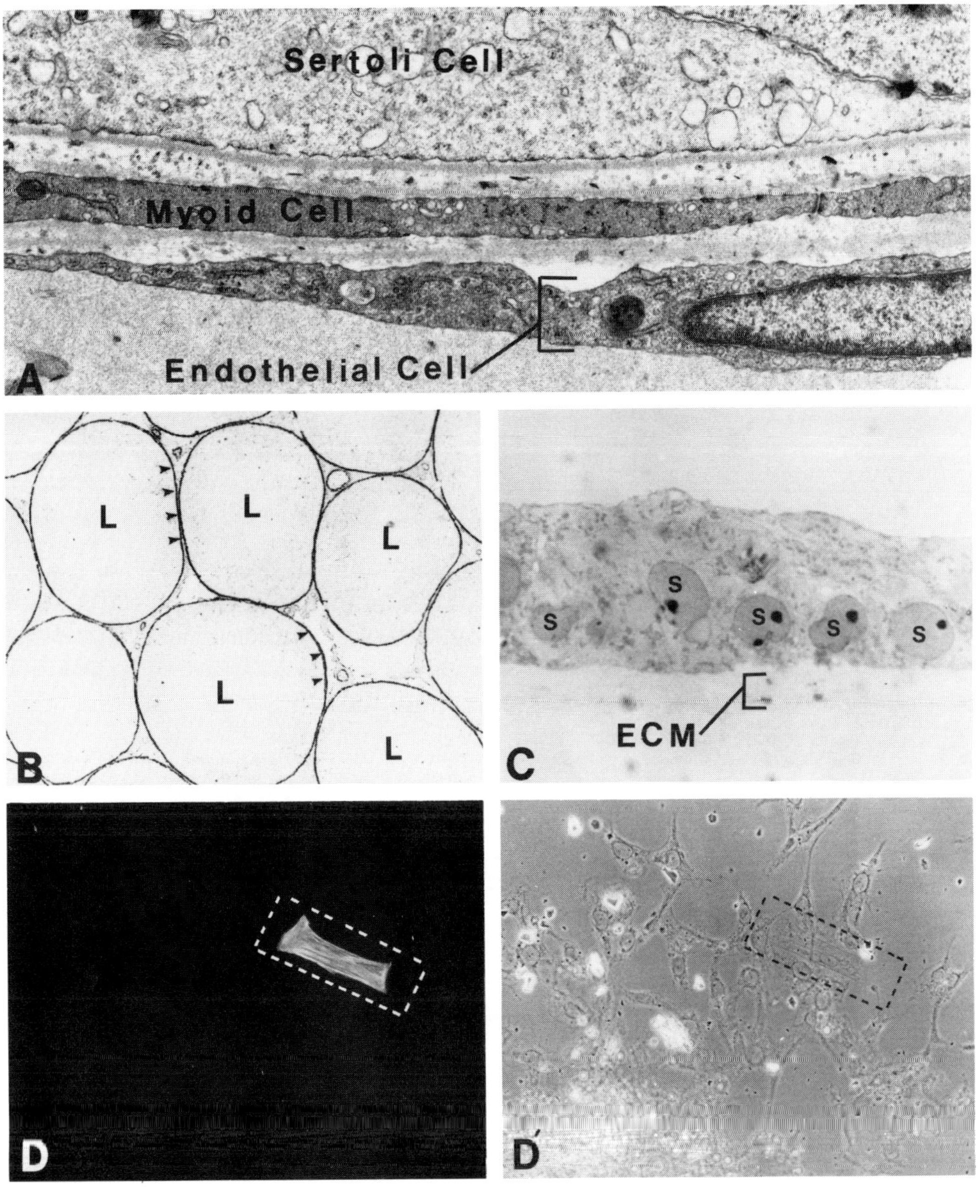

FIGURE 2. Purity of Sertoli cells *in vitro:* immunoassay of peritubular cell contamination. In panel A, the two cell types, myoid and lymphatic endothelial cells, comprising the wall of the seminiferous tubules and on which Sertoli cells rest, are demonstrated. In panel B, immunoperoxidase localization of smooth muscle myosin in the peritubular cells (arrowheads) is presented. The seminiferous epithelium is completely devoid of staining, as is the lumen (L). Note that at this level of magnification it is impossible to assign the immunoperoxidase staining as belonging to either the myoid cell or endothelial cell. Similar immunostaining is also obtained with a monoclonal antibody to the alpha-isoform of smooth muscle actin. In panel C, profiles of primary cultures of Sertoli cells (S) growing in the bicameral chamber culture system and resting on an extracellular matrix (ECM) are shown.[47] Absence of peritubular cells in this particular section is apparent. In panel D, immunostaining of a Sertoli cell primary culture with the anti-smooth muscle actin antibody by indirect immunofluorescence is shown and the corresponding phase image is presented in D′. Observe that the positive staining is limited to the cell in the enclosed rectangle whose morphology is different from the rest.

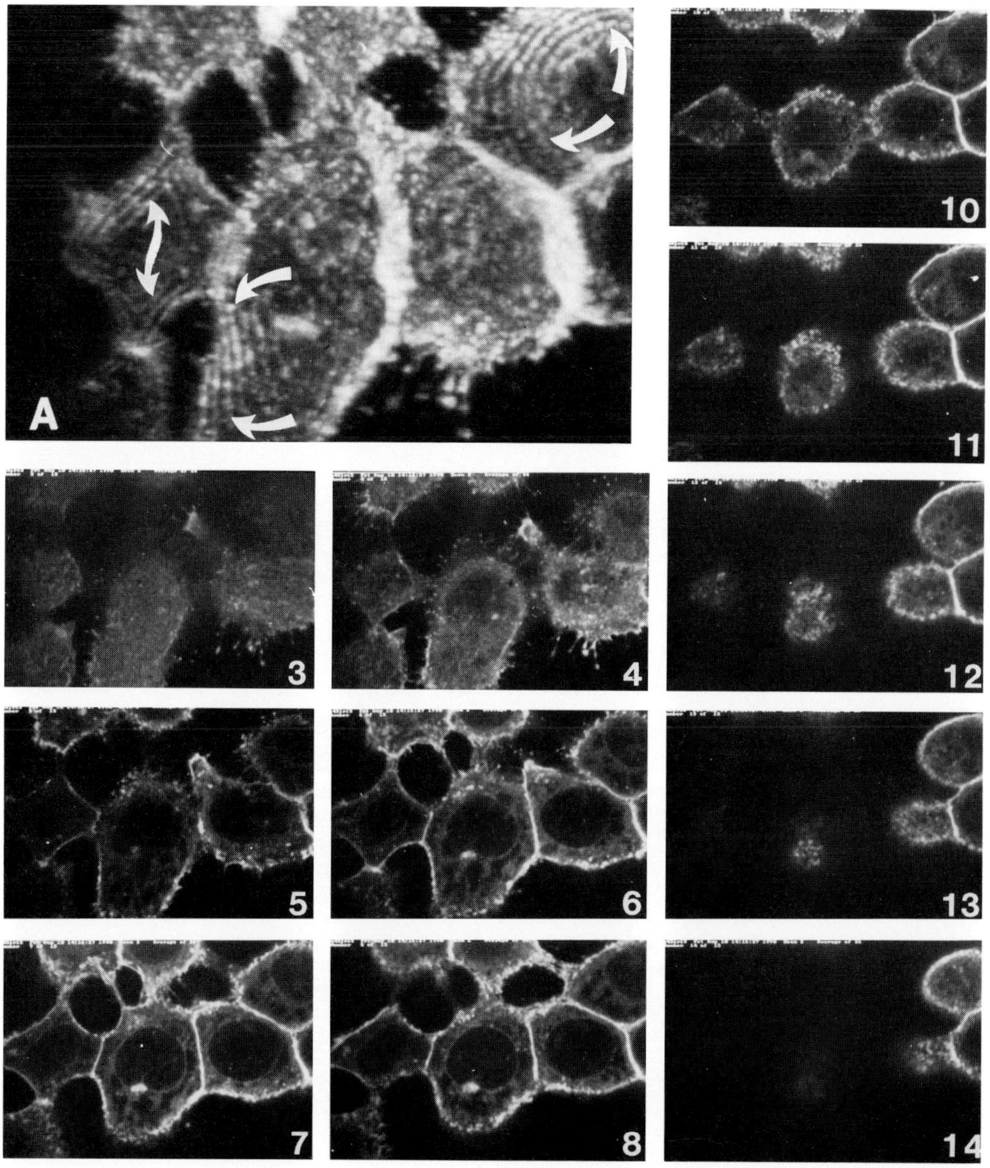

FIGURE 3. EGF receptor (EGFR) immunolocalization by confocal microscopy in A431 cells *in vitro*. Panel A demonstrates a three-dimensional reconstruction of a Z-series from 16 different optical sections of an A431 cell aggregate immunostained for the EGFR. Panels numbered 3 to 14 correspond to the respective section from the 16-section series used for the three-dimensional reconstruction illustrated in panel A. The white arrows point to samples of "ruffles" that illustrate the redistribution of the EGFR as a function of placing these cells in Ca^{2+}-free medium for as little as 15 min. Note that the ruffles cannot be discerned on the individual sections. In conventional epifluorescent images, background fluorescence noise does not permit the resolution of the ruffles and the EGFR is interpreted as being homogeneously distributed on the cell surface.

epithelium diffused fluorescence is often difficult to interpret and to assign to a specific cell type. The problem of diffused fluorescence observed with conventional epifluorescence microscopy is easily eliminated in reconstructions of optical sections obtained with the confocal microscope (Figure 3). In addition, the reconstructions often contain much more information that is not easily discerned in conventional images.

The procedure used for labeling cells presented in Figure 3 was similar to that used for conventional immunofluorescence. Cells were fixed, permeabilized with detergents, and labeled by either direct or indirect immunofluorescence. Samples were analyzed with an MRC-600 (BioRad) laser scanning confocal system equipped with a Nikon inverted microscope. The images shown were obtained using a 40× Nikon objective with a variable numerical aperture set to NA 1.1. A z-series of 0.5-μm optical sections was obtained and a three-dimensional reconstruction at average brightness projection was carried out using the Thruview software provided with the instrument.

Figure 3 illustrates confocal images of EGFR staining in the epidermoid carcinoma A431 cells that were grown to high cell density and then treated in a low-Ca^{2+} medium for 15 min. The different 0.5-μm optical sections used for the reconstruction are labeled with Arabic numbers. Since these are optical sections, it is possible to determine precisely if staining is intracellular or on the cell surface. That is, the fluorescence labeling not at the cell periphery can be confidently interpreted as being intracellular. Three-dimensional reconstructed images of this series are presented in panel A of Figure 3. While this reconstruction does not present as clear a picture of the membrane staining as that observed in the optical sections, one feature of the EGFR staining that the reconstruction demonstrated was striking. A pronounced "ruffling" of the plasma membrane EGFR immunoreactivity was observed when cells were treated with low Ca^{2+}, an effect that was not observed in cells grown in normal Ca^{2+}. Since the basis for the reconstruction is solely that of EGFR staining, these "ruffles" must reflect bona fide EGFR localization. Although we do not understand the structural basis for these "ruffles", it is possible that they represent EGFR staining associated with submembrane cytoskeletal elements in the process of a low Ca^{2+}-induced reorganization. However, of significance is the fact that these data could not have been obtained by conventional microscopy.

The power of the confocal microscope as a novel tool in immunocytochemistry has been determined only recently. Its principal use has been employed to determine the distribution of membrane antigens in cells *in vitro*. Current studies in progress will exploit this new technology to examine secretory and membrane protein distribution in seminiferous tubule sections. Although the same limitations of tissue preservation vs. maintaining antigenicity encountered in routine immunostaining protocols will remain, confocal microscopy promises to obviate serial sectioning of complex tissues to ascertain precisely the specific cell type distribution of secretory and membrane antigens.

D. FUTURE PERSPECTIVES

In this brief review, we have addressed the physical properties of the Sertoli cell with respect to the limitations of current technology in carrying out immunocytochemical localization of secretory and membrane proteins. To date, ultrastructural localization of proteins in the seminiferous epithelium has not met with tremendous success. Effective low concentration of proteins in secretory vesicles, as well as a complex epithelium, have led to these poor results. Alternative methodologies have been exploited to overcome the limitations presented by the seminiferous epithelium in carrying out satisfactory immunocytochemical localization. Future studies will require that fixation protocols and embedding materials, compatible with maintaining antigenicity, are developed to facilitate the ultrastructural immunolocalization of proteins in the seminiferous epithelium. Moreover, we wish to observe that the problems of the complexity of the seminiferous epithelium encountered in interpreting immunocytochemical results are now being faced by investigators performing *in situ* hybridization protocols. Thus, the need to develop adequate fixation and embedding materials compatible with this new technology has gained a new urgency.

ACKNOWLEDGMENTS

We wish to thank Dr. Ben Danzo of Vanderbilt University for his gift of anti-ABP antibody. Figure 3 is from work produced in collaboration with Dr. Stephen Byers, Georgetown Medical School. Some of the work presented here was funded in part by National Institutes of Health grants HD23484 to C.A.S.-Q. and HD16260 to M.D.

REFERENCES

1. **Rabinovici, J. and Jaffe, R. B.,** Development and regulation of growth and differentiation function in human and subhuman primate fetal gonads, *Endocr. Rev.,* 11, 532, 1990.
2. **Russell, L. D., Alger, L. E., and Nequin, L. G.,** Hormonal control of pubertal spermatogenesis, *Endocrinology,* 120, 1615, 1987.
3. **Sharpe, R. M.,** Follicle-stimulating hormone and spermatogenesis in the adult male, *J. Endocrinol.,* 121, 405, 1989.
4. **Ewing, L. L. and Zirkin, B. R.,** Leydig cell structure and steroidogenic function, *Rec. Prog. Horm. Res.,* 39, 599, 1983.
5. **Sanborn, B. M., Caston, L. A., Buzek, S. W., and Ussuf, K. K.,** Hormonal regulation of Sertoli cell function, in *Regulation of Ovarian and Testicular Function,* Mahesh, V. B., Dhindsa, D., Anderson, E., and Kalra, S., Eds., Plenum Press, New York, 1987, 561.
6. **Bellve, A. R. and Zheng, W.,** Growth factors as autocrine and paracrine modulators of male gonadal functions, *J. Reprod. Fertil.,* 85, 771, 1989.
7. **Griswold, M. D., Morales, C., and Sylvester, S. R.,** Molecular biology of the Sertoli cell, *Oxford Rev. Reprod. Biol.,* 10, 124, 1988.
8. **Skinner, M. K.,** Cell-cell interactions in the testis, *Endocrin. Rev.,* 12, 45, 1991.
9. **Djakiew, D. and Dym, M.,** Pachytene spermatocyte proteins influence Sertoli cell function, *Biol. Reprod.,* 39, 1193, 1988.
10. **Le Magueresse, B. and Jegou, B.,** In vitro effects of germ cells on the secretory activity of Sertoli cells recovered from rats of different ages, *Endocrinology,* 122, 1672, 1988.
11. **Janecki, A., Jakubowiak, A., and Steinberger, A.,** Effect of germ cells on vectorial secretion of androgen binding protein and transferrin by immature rat Sertoli cells in vitro, *J. Androl.,* 9, 126, 1988.
12. **Onoda, M. and Djakiew, D.,** Modulation of Sertoli cell secretory function by rat round spermatid protein(s), *Mol. Cell. Endocrinol.,* 73, 35, 1990.
13. **Sylvester, S. R. and Griswold, M. D.,** Localization of transferrin and transferrins receptors in rat testis, *Biol. Reprod.,* 31, 195, 1984.
14. **Suárez-Quian, C. A., Dai, M., Onoda, M., Kris, R. M., and Dym, M.,** Epidermal growth factor receptor localization in the rat and monkey testis, *Biol. Reprod.,* 41, 921, 1989.
15. **Suárez-Quian, C. A. and Niklinski, W.,** Immunocytochemical localization of the epidermal growth factor receptor in mouse testis, *Biol. Reprod.,* 43, 1087, 1990.
16. **Zabludoff, S. D., Erickson-Lawrence, M., and Wright, W. W.,** Sertoli cells, proximal convoluted tubules in the kidney, and neurons in the brain contain cyclic protein-2, *Biol. Reprod.,* 43, 15, 1990.
17. **Zabludoff, S. D., Karzai, A. W., and Wright, W. W.,** Germ cell-Sertoli cell interactions: the effect of testicular maturation on the synthesis of cyclic protein-2 by rat Sertoli cells, *Biol. Reprod.,* 43, 25, 1990.
18. **Bardin, C. W., Cheng, C. Y., Musto, N. A., and Gunsalus, G. L.,** The Sertoli cell, in *The Physiology of Reproduction,* Knobil, E., Neill, J., et al., Eds., Raven Press, New York, 1988.
19. **Forssmann, W. G., Ito, S., Weibe, E., Aoki, A., Dym, M., and Fawcett, D. W.,** An improved perfusion fixation method for the testis, *Anat. Rec.,* 188, 307, 1977.
20. **Fawcett, D. W.,** Ultrastructure and function of the Sertoli cell, in *Handbook of Physiology, Section 7, Endocrinology, Volume V, Male Reproductive System,* Hamilton, D. W. and Greep, R. O., Eds., American Physiology Society, Washington, D. C., 1975, 21.
21. **Orci, L., Ravazzola, M., Amherdt, M., Madsen, O., Vassalli, J.-D., and Perrelet, H.,** Direct identification of prohormone conversion site in insulin-secreting cells, *Cell,* 42, 671, 1985.
22. **Tran, D. and Josso, N.,** Localization of anti Mullerian inhibiting hormone in the rough endoplasmic reticulum of the developing bovine Sertoli cells using immunocytochemistry with a monoclonal antibody, *Endocrinology,* 111, 1562, 1983.

23. **Hansson, V. and Tveter, K. J.,** Uptake and binding in vivo of ^{3}H-labelled androgen in the rat epididymal and ductus deferens, *Acta Endocrinol.*, 66, 745, 1971.
24. **Ritzen, E. M., Nayfeh, S. N., French, F. S., and Dobbins, M. C.,** Demonstration of androgen binding components in rat epididymis cytosol and comparison with binding components in prostate and other tissues, *Endocrinology,* 89, 143, 1971.
25. **Gunsalus, G. L., Musto, N. A., and Bardin, C. W.,** Bidirectional release of a Sertoli cell product, androgen binding protein, into the blood and seminiferous tubule, in *Testicular Development, Structure and Function,* Steinberger, A. and Steinberger, E., Eds., Raven Press, New York, 1980, 291.
26. **Pelliniemi, L. J., Dym, M., Gunsalus, G. L., Musto, N. A., Bardin, C. W., and Fawcett, D. W.,** Immunocytochemical localization of androgen binding protein in the male rat reproductive tract, *Endocrinology,* 108, 925, 1981.
27. **Attramadal, A., Bardin, C. W., Gunsalus, G. L., Musto, N. A., and Hansson, V.,** Immunocytochemical localization of androgen binding protein in rat Sertoli and epididymal cells, *Biol. Reprod.*, 25, 983, 1981.
28. **Robaire, B. and Hermo, L.,** Efferent ducts, epididymis, and vas deferens: structure, functions, and their regulation, in *The Physiology of Reproduction,* Knobil, E. and Neill, J., Eds., Raven Press, New York, 1988, 999.
29. **Tsutsumi, O., Kurachi, H., and Oka, T.,** A physiological role of epidermal growth factor in male reproductive function, *Science,* 233, 975, 1986.
30. **Russell, L. D., Weiss, T., Goh, J.C., and Curl, J. L.,** The effects of sialoadenectomy (submandibular gland removal) on testicular and epididymal parameters, *Tissue Cell,* 22, 263, 1990.
31. **Seethalakshmi, L., Liu, L., Kinkead, T., Flores, C., Carboni, A. A., Menon, M., and Davis, R.,** Epidermal growth factor (EGF): an important role in testicular function, *J. Androl. Suppl.*, 12, 60, 1991.
32. **Noguchi, S., Ohba, V., and Oka, T.,** Involvement of epidermal growth factor deficiency in pathogenesis of oligozoospermia in streptozotocin-induced diabetic mice, *Endocrinology,* 127, 2136, 1990.
33. **Ascoli, M.,** Regulation of gonadotropin receptors and gonadotropin responses in a clonal strain of Leydig tumor cells by epidermal growth factor, *J. Biol. Chem.*, 256, 179, 1981.
34. **Welsh, T. H. and Hsueh, A. J. W.,** Mechanism of the inhibitory action of epidermal growth factor on testicular androgen biosynthesis in vitro, *Endocrinology,* 110, 1498, 1981.
35. **Bernier, M., Chatelain, P., Mather, J. P., and Saez, J. M.,** Regulation of gonadotropin receptors, gonadotropin responsiveness, and cell multiplication by somatomedin-C and insulin in cultured pig Leydig cells, *J. Cell. Physiol.*, 129, 257, 1986.
36. **Verhoeven, G. and Cailleau, J.,** Stimulatory effects of epidermal growth factor on steroidogenesis in Leydig cells, *Mol. Cell. Endocrinol.*, 47, 99, 1986.
37. **Mallea, L. E., Machado, A. J., Navaroli, F., and Rommerts, F. F. G.,** Epidermal growth factor stimulates lactate production and inhibits aromatization in cultured Sertoli cells from immature rats, *Int. J. Androl.*, 9, 201, 1986.
38. **Morris, P. L., Vale, W. W., Cappel, S., and Bardin, C. W.,** Inhibin production by primary Sertoli cell-enriched cultures: regulation by follicle-stimulating hormone, androgens, and epidermal growth factor, *Endocrinology,* 122, 717, 1988.
39. **Smith, F. F., Tres, L. L., and Kierszenbaum, A. L.,** Ornithine decarboxylase activity during rat spermatogenesis in vivo and in vitro: selective effect of hormones and growth factors, *J. Cell. Physiol.*, 133, 305, 1987.
40. **Carpenter, G.,** Receptors for epidermal growth factor and other polypeptide mitogens, *Annu. Rev. Biochem.*, 56, 881, 1987.
41. **Kusabe, M., Sakakura, T., Nishizuka, Y., Sano, M., and Matsukage, A.,** Polyester wax embedding and sectioning technique for immunohistochemistry, *Stain Tech.*, 59, 127, 1984.
42. **Hadley, M. A., Byers, S. W., Suárez-Quian, C. A., Djakiew, D., and Dym, M.,** Extracellular matrix regulates Sertoli cell differentiation, testicular cord formation, and germ cell development in vitro, *J. Cell Biol.*, 101, 1511, 1985.
43. **Hadley, M. A., Djakiew, D., Byers, S. W., and Dym, M.,** Polarized secretion of androgen-binding protein and transferrin by Sertoli cells grown in a bicameral culture system, *Endocrinology,* 120, 1097, 1987.
44. **Dym, M., Lamsam-Casalotti, S., Jia, M., Kleinman, H. K., and Papadopoulos, V.,** Basement membrane increases G-protein levels and follicle-stimulating hormone responsiveness of Sertoli cell adenylyl cyclase activity, *Endocrinology,* 128, 1167, 1991.
45. **Skinner, M. K., Takacs, K., and Coffey, R. J.,** Transforming growth factor-a gene expression and action in the seminiferous tubule: peritubular cell-Sertoli cell interactions, *Endocrinology,* 124, 845, 1989.
46. **Galdieri, M., Ziparo, E., Palombi, F., Russo, M. A., and Stefanini, M.,** Pure Sertoli cell cultures: a new model for the study of somatic germ cell interactions, *J. Androl.*, 2, 249, 1981.

A. OVARY

The mammalian ovary contains oogonia and oocytes, even in the embryonic period. The function of the ovary of mature animals is, first of all, to produce oocytes and to ovulate them at the process of maturation. In addition, it has an important function as an endocrine gland secreting sex hormones.

The surface of the ovary is surrounded by a part of the peritoneum, the cuboidal mesothelium, which may have an active function of local absorption and/or transepithelial transport between the abdominal cavity and ovary. In the cytochemical investigation, the surface epithelial cells of the ovary show 17β-hydroxysteroid dehydrogenase activity.[1] In the adult rat ovaries at estrous cycle, intense Ca^{2+}-ATPase activity is found at the functionally active areas of the ovarian epithelium where columnar epithelium with crypts and well-developed microvilli are developed (Figure 1). These areas are topographically related to adjacent structures of the ovarian cortex, such as preovulatory Graafian follicles and the corpus luteum.[2] It is reported that sulfated mucopolysaccharides stained with alucian blue are located in the epithelium.[1]

As an endocrine organ, the ovary is known to secrete mainly estrogen, progesterone, and some androgen. Concerning the formation site of estrogen (estrone, estradiol, and estriol), aromatase activity which participates in the aromatization during the transformation of estrogen from cholesterol via androgen, was detected as a marker ultracytochemically. Recent electron immunocytochemical investigations using antisera against aromatase cytochrome P450, revealed aromatase activity only in the granulosa cells of the preovulatory large Graafian follicles and cells of the corpus luteum just after ovulation in human, rat, and mouse[3,4] (Figure 2). However, there are some reports on aromatase activity in theca interna and interstitial cells.[5,6]

Although it is known that progesterone is secreted by the corpus luteum cells, there is no report on the exact location of synthesis of the hormone. Murakoshi et al.[7] described the immunocytochemical localization of glutathione-peroxidase (GSH-PO), which is involved in the metabolism of progesterone and thus presumably can be used as a marker indication the site(s) of progesterone synthesis in the cytosole near the endoplasmic reticulum (ER) and in the mitochondria in the granulosa cells of large follicles and in the cytosole near mitochondria and lipid droplets in lutein cells.[7]

Concerning the steroidogenesis of progesterone, 3β-hydroxysteroid dehydrogenase is found in the smooth ER in the granulosa cells.[8] Receptors for luteinizing hormone (LH) and human chorionic gonadotropin (hCG) were detected, by means of electron microscopic autoradiography, at the apical regions of the microvilli of the granulosa cells.[9-11] Follicle-stimulating hormone (FSH) receptors were reported to be present on the granulosa cells only at the light microscopic level.[9] LH receptors were also described on the plasma membrane and in the cytoplasm of lutein cells, or rather on polysomes, in small vesicles and ER.[12] Moreover, follicular cells inhibit the maturation of the oocytes; however, there is no report on the inhibitor substance from electron cytochemistry. Type IV collagen, laminin, and fibronectin are recognized as components of the basal lamina by immunofluorescence microscopy,[13] but there is no report at the electron microscopic level.

17β-Hydroxysteroid dehydrogenase participates in the synthesis of testosterone, the precursor of estrogen, which is formed in the granulosa cells. 17β-Hydroxysteroid dehydrogenase activity is found in the theca interna cells.[5,12] LH receptors have also been reported in the theca interna cells.[12]

B. OVUM

As specific components of the oocyte, cortical granules are known to have an important role in the fertilization process, and have been investigated in many kinds of animals,

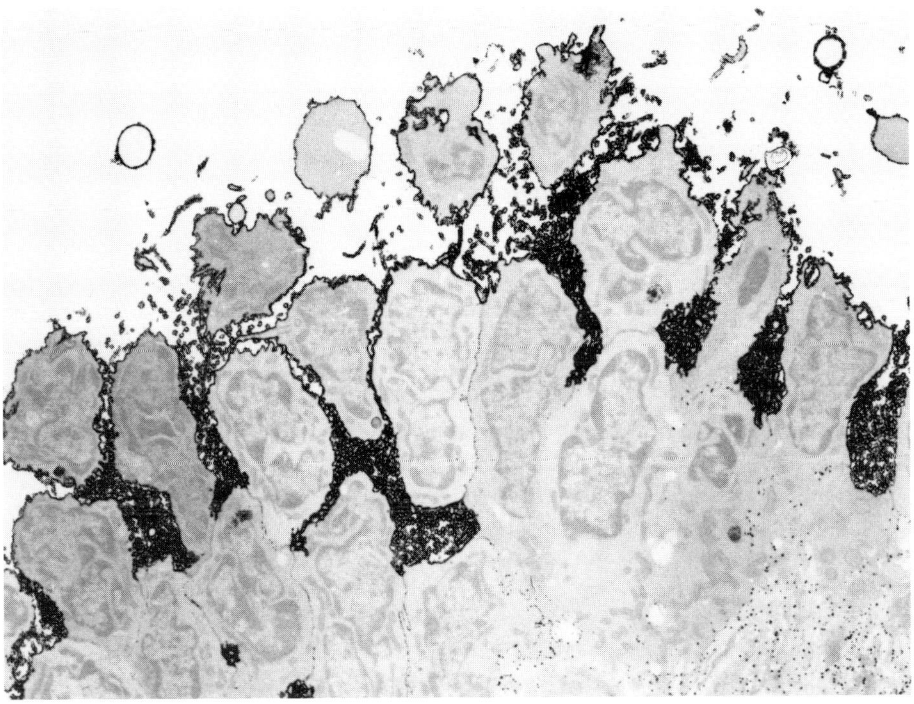

FIGURE 1. Typical columnar epithelial cells from a portion of the surface epithelium forming a crypt in the adult rat ovary. The epithelial cells have a great number of microvilli showing intense Ca^{2+}-ATPase activity on their apical membranes. The lateral membrane is also positive for the enzyme. (Magnification × 5200. From Araki, N. and Takashima, Y., *Acta Histochem. Cytochem.*, 22, 459, 1989. With permission.)

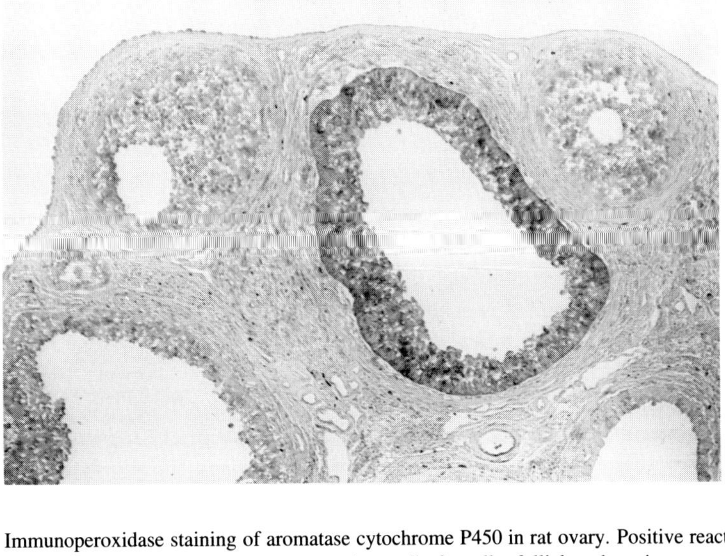

FIGURE 2. Immunoperoxidase staining of aromatase cytochrome P450 in rat ovary. Positive reaction is noticed only in the granulosa cells of a large follicle. The granulosa cell of smaller follicles, theca interna cells, interstitial gland cells, and surface epithelial cells are entirely devoid of this staining. (Magnification × 71. From Ishimura, K. et al., *Histochemistry,* 90, 413, 1989. With permission.)

especially in sea urchin.[14,15] They are found in the cortex of the mature egg. At the time of fertilization, the cortical granules undergo exocytosis and their contents are released into the perivitelline space.[14,15] Extensive morphologic and cytochemical investigations have revealed that cortical granules contain mucopolysaccharides, several proteins, proteases, ovoperoxidase, proteoliasin, β-glucanase, hyalin, and calcium (reviewed by Somers and Shapiro, 1989[15]).

The cortical granule substance was found in the nurse cell cytoplasm and presumptive cortical granules in oocyte cytoplasm in the ovary and cortical granules in mature unfertilized egg, and showed peroxidatic activity of catalase using the 3,3′-diaminobenzidine (DAB) method[16,17] at the lamellar structure of the granule (Figure 3, A and B).

Cortical granule-specific components were detected in oocytes, eggs, and accessory cells by means of antibodies, including antibodies against whole cortical granule exudate, fertilization envelope, hyaline, and β-glucanase.[18] Recent immunocytochemical investigations using polyclonal antibodies against ovoperoxidase and proteoliasin (ovoperoxidase-binding protein) and colloidal gold-conjugated secondary antibody suggested that ovoperoxidase and proteoliasin are located in the lamellar structure of the cortical granule. After fertilization, ovoperoxidase was located at the perivitelline space and proteoliasin at the fertilization envelope.[19] Materials of the hyalin layer, which surrounds and protects the fertilized eggs and early embryos, are found immunocytochemically in the homogeneous or electron-lucent portion of the cortical granule.[20-22] Experiments with ruthenium red staining and collagenase digestion indicate that the hyaline layer is a collagen-containing extracellular matrix.[23] The zona pellucida has a specific receptor for the spermatozoon.[24-27] It also has a role in protection from polyspermy.

The zona pellucida consists of a three-dimensional arrangement of fine fibrils of three kinds that are made by an assembly of glycoproteins.[26] Those three kinds of proteins, ZP1 (200 kDa), ZP2 (120 kDa), and ZP3 (80 kDa), are found cytochemically in the zona pellucida.[24,25,27-29] It is reported that, among these, the O-linked oligosaccharide α-galactose in the carbohydrate region of ZP3 protein acts as a sperm receptor.[28,30]

Hyaluronic acid was detected immunocytochemically in the oocyte-cumulus complex of the golden hamster by use of a hyaluronidase-gold complex.[31] Ultrastructural localization of several phosphatases and other enzymes, alkaline phosphatase (ALPase), 5′-nucleotidase (5′-NPase), Mg^{2+}-ATPase, Na^+, K^+-ATPase, cyclic AMP phosphodiesterase, and adenylate cyclase, were reported in preimplantation mouse embryos.[32,33]

C. OVIDUCT

The epithelium of the primate oviduct is known to change during the menstrual cycle, influenced by either endogenous or exogenous steroid hormone level.[34,35] Estradiol induces the synthesis and release of oviduct-specific glycoproteins in secretory epithelial cells; progesterone causes regression of this function.[35]

The mammalian oocyte is enclosed in an acidic glycoprotein layer, the zona pellucida, which develops during follicular growth in the ovary. After ovulation, major components of the zona pellucida are added during transit through the oviduct. Electron immunocytochemical studies indicate that the zona pellucida glycoprotein ZP-0 is secreted by the nonciliated secretory cells of the oviduct, and it is deposited on the surface of the ovulated oocyte and maturing zona pellucida.[36] The role of ZP-0 is still unclear.

The monoclonal antibodies against oviductal zona pellucida of golden hamster and protein A-gold or lectin-gold technique showed that the oviduct antigen is located in dense secretory granules, Golgi apparatus and its associated vesicles in the nonciliated secretory cells in the oviduct, and in the entire zona pellucida of the ovulated oocyte, but not in the oocyte, cortical granules, and other oocyte components.[36-38]

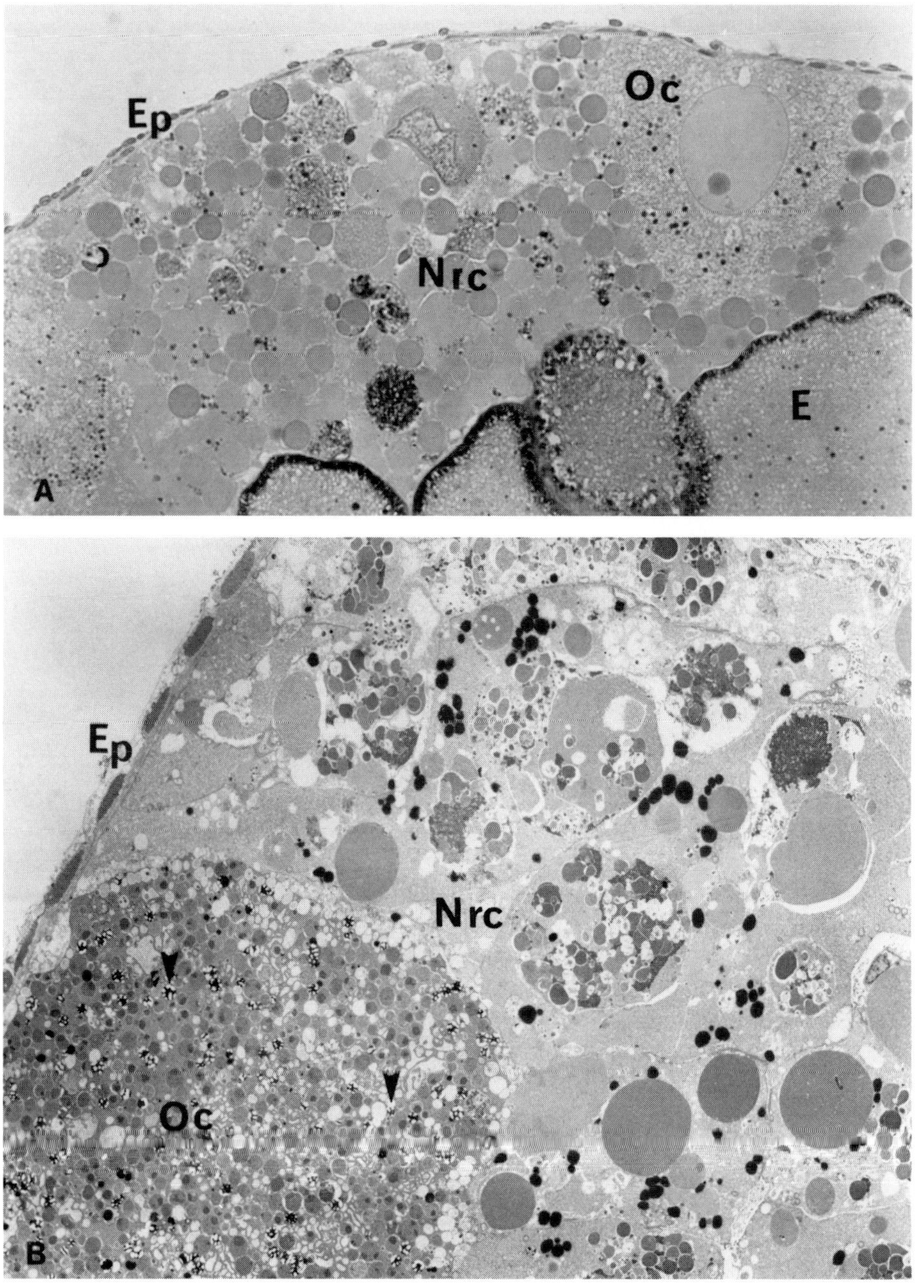

FIGURE 3. Peroxidatic activity of catalase in the sea urchin ovary in early breeding season (DAB reaction). (A) Epon section of sea urchin ovary. Under the germinal epithelium (Ep) an oocyte (Oc), nurse cells (Nrc), and mature egg (E) are shown. DAB-positive granules are shown in Nrc and on the surface of the mature eggs (cortical granules). (Magnification × 440.) (B) Sea urchin ovary (DAB reaction). Presumptive cortical granules (arrowheads), cortical-granule-like granules in Nrc are DAB positive. (Magnification × 1440.) (From Tominaga, A., *Acta Histochem. Cytochem.*, 15, 746, 1982. With permission.)

As to the major components of mature zona pellucida, glycoconjugates containing terminal *N*-acetyl-D-galactosamine residues are synthesized and secreted by nonciliated secretory cells of the oviduct, similarly to high-molecular-weight polydispersed glycoproteins.[39]

In human, ALPase activity is found at the apical and lateral plasma membrane of secretory cells, but not in that of ciliated cells. Acid phosphatase activity is observed in the lysosomes and inner surface of Golgi vesicles in both ciliated and secretory cells and the activity does not change with the menstrual cycle.[40] In the rabbit, bird, mouse, and man, ATPase activity is found at the luminal surface of secretory cells.[41]

D. UTERUS

The availability of monoclonal antibodies allowed the cytochemical demonstration of steroid receptors in the uterus. The estrogen receptor was found in the nucleus of the target cell at the light microscopic level, both in the presence and absence of estrogen hormone. Progesterone receptor showed a distribution similar to that of the estrogen receptor.[44] With an immunogold technique at the electron microscopic level, the receptor was always seen to be associated with condensed chromatin or areas surrounding condensed chromatin. A small amount of progesterone receptor antigen was also seen in the cytoplasm, where it was associated with the rough endoplasmic reticulum and free ribosome. In the chromatin, the distribution of the receptor varied according to the hormonal state. After administration of progestin R5250, the receptor was mainly distributed in the border regions between condensed chromatin and nucleoplasm.[45] These areas have been shown to be the most active sites of gene transcription.

In the rabbit uterus, synthesis of uteroglobin, a progesterone-induced protein, markedly increased at the preimplantation period in the columnar epithelium. De la Torre et al.[46] showed that the main localization of uteroglobin is the small secretory granules which were seen in the process of release into the uterine lumen. The luminal microvilli were also heavily labeled with the anti-uteroglobin immunogold complex. Uteroglobin was also seen in the secretory cells of the oviduct.[46]

In uterine smooth muscle, Ca^{2+}-ATPase activity was localized on the plasma membrane of the cell. The Ca^{2+} transport of the membrane by the Ca^{2+}-ATPase activity seems to be involved in the relaxation of the smooth muscle.[47]

Alkaline phosphatase (ALPase) activity was demonstrated cytochemically in the epithelial cells of the endometrium. The localization of heat-stable, L-phenylalanine-sensitive ALPase in the uterus seems to be noteworthy as an oncodevelopmental marker.[48-50] Heat-stable, L-phenylalanine-sensitive ALPase was rarely found in normal endometrium, but was frequently observed in endometrium cancer.[49]

E. PLACENTA

Intense ALPase activity was observed on the luminal plasma membrane of syncytiotrophoblasts by both enzyme-cytochemical and immunocytochemical techniques.[51-57] According to some investigators, ALPase activity was present not only on the luminal plasma membrane but also on the basal plasma membrane of syncytiotrophoblasts[51,55,57] and on the plasma membrane of cytotrophoblasts.[54,57] Acid phosphatase activity was localized in the lysosomes and Golgi apparatus in syncytiotrophoblasts and cytotrophoblasts.[52,55,57] Adenosine nucleotidase such as 5′-Nase, Ca^{2+}-ATPase, Mg^{2+}-ATPase, and nucleotide diphosphatase (NDPase) were demonstrated on the microvillous plasma membrane in human term placenta.[58] These enzymes may be involved in the regulation of feto-placental-maternal microcirculation. Adenosine produced by 5′-Nase hydrolysis acts as a vasodilator, and maybe also as a modulator of platelet aggregation. Adenosine may also change ACLase activity. Cytochemical localization of ACLase in the human placenta has also been demonstrated by Matsubara et al.[59]

The placenta is one of the major endocrine organs of pregnancy. The placental trophoblast is the dominant site of steroid hormone synthesis. In the human, estrogens are produced from maternal and fetal sulfated steroid precursors. Steroid sulfatase seems to be an important enzyme in the pathway of estrogen synthesis from sulfated steroid precursors. Steroid sulfatase was immunocytochemically localized in the nuclear envelope and rough ER of syncytiotrophoblasts.[60,61] The plasma membrane and the membrane of endosomes and multivesicular bodies were also immunoreactive for steroid sulfatase. This localization of steroid sulfatase may have functional implications in the placental uptake of circulating sulfated steroids.[60] Aromatase antigen was immunocytochemically localized on the surface of microvilli, the lateral membrane, and in the ER in syncytiotrophoblasts.[6] Placental progesterone is synthesized from cholesterol. Cytochrome P450 cholesterol side-chain cleavage enzyme (P450scc) was localized in bovine placentome.[62] The mononucleate cells of the cotyledon showed immunogold particles of P450scc located on the cristae of mitochondria. However, P450scc was absent in the binucleate cells. The variation in P450scc immunoreactivity is suggestive of different steroidogenetic capacities of the cells.[62] Relaxin was immunolocalized within the Golgi complex of primary and secondary giant trophoblast cells of golden hamster placenta.[63] The site of intracellular synthesis and storage of human placental lactogen (HPL) and human chorionic gonadotropin (hCG) was studied by immunogold labeling on ultracryosections.[64] The hCG and the HPL were localized throughout in the rough ER, in the Golgi cisternae, and in the infrequent small, dense granules of the syncytiotrophoblast. The result suggested that both hCG and HPL were synthesized and packaged by the classical secretory pathway, although the level of hormones stored in granules at any one time was small. In the human, passive immunity is conferred on the newborn by the transfer of maternal immunoglobulin G (IgG) across the placenta to the fetus. The pathway of IgG transport was revealed cytochemically by horseradish peroxidase-conjugated IgG (HRP-IgG) perfusion.[65] HRP-IgG was endocytosed by the syncytiotrophoblasts and released into the interstitial space and then transported through the endothelium by vesicular transport. Immunocytochemical localization of endogenous IgG on ultrathin frozen sections of human term placenta was compatible with the pathway of HRP-IgG.[66]

REFERENCES

1. **Blaustein, A. and Lee, H.,** Surface cells of the ovary and peritoneum: a histochemical and ultrastructure comparison, *Gynecol. Oncol.*, 8, 34, 1979.
2. **Araki, N. and Takashima, Y.,** Variations in the morphology and Ca^{2+}-ATPase activity of the ovarian surface epithelium in rats, *Acta Histochem. Cytochem.*, 22, 459, 1989.
3. **Ishimura, K., Yoshinaga-Hirabayashi, T., Tsuri, H., Fujita, H., and Osawa, Y.,** Further immunocytochemical study on the localization of aromatase in the ovary of rats and mice, *Histochemistry*, 90, 413, 1989.
4. **Zhong, C., Ishimura, H., Yoshinaga-Hirabayashi, T., Fujita, H., Kitawaki, J., and Osawa, Y.,** Immunocytochemical study on the localization of aromatase in the ovary of golden hamster, guinea pig and cow, *Acta Histochem. Cytochem.*, 22, 501, 1989.
5. **Yoshinaga-Hirabayashi, T., Ishimura, K., Fujita, H., Inano, H., Ishii-Ohba, H., and Tamaoki, B.,** Immunocytochemical localization of 17β-hydroxysteroid dehydrogenase (17β-HSD), and its relation to the ultrastructure of steroidogenic cell in immature and mature rat ovaries, *Arch. Histol. Jpn.*, 50, 545, 1987.
6. **Naganuma, H., Ohtani, H., Harada, N., and Nagura, H.,** Immunoelectron microscopic localization of aromatase in human placenta and ovary using microwave fixation, *J. Histochem. Cytochem.*, 38, 1427, 1990.
7. **Murakoshi, M., Ueno, K., Osamura, Y., and Watanabe, K.,** Immunocytochemical localization of glutathione-peroxidase (GSH-PO) in rat ovary, *Acta Histochem. Cytochem.*, 20, 321, 1987.

61. **Salido, E. C., Yen, P. H., Barajas, L., and Shapiro, L. J.,** Steroid sulfatase expression in human placenta: immunocytochemistry and *in situ* hybridization study, *J. Clin. Endocrinol. Metab.*, 70, 1564, 1990.
62. **Ben-David, E. and Shemesh, M.,** Ultrastructural localization of cytochrome P-450 in the bovine placentome using protein A-gold technique, *Biol. Reprod.*, 42, 131, 1990.
63. **Johns, T. C. and Renegar, R. H.,** Ultrastructural morphology and relaxin immunolocalization in giant trophoblast cells of the golden hamster placenta, *Am. J. Anat.*, 189, 167, 1990.
64. **Billingsley, S. A. and Wooding, F. B. P.,** An immunogold, cryoultrastructural study of sites of synthesis and storage of chorionic gonadotropin and placental lactogen in human syncytiotrophoblast, *Cell Tissue Res.*, 261, 375, 1990.
65. **Leach, L., Eaton, B. M., Firth, J. A., and Contractor, S. F.,** Uptake and intracellular routing of peroxidase-conjugated immunoglobulin-G by the perfused human placenta, *Cell Tissue Res.*, 261, 383, 1990.
66. **Leach, L., Eaton, B. M., Firth, J. A., and Contractor, S. F.,** Immunogold localisation of endogenous immunoglobulin-G in ultrathin frozen sections of the human placenta, *Cell Tissue Res.*, 257, 603, 1989.

Chapter 16

IMMUNOELECTRON MICROSCOPY OF NORMAL AND ABNORMAL SKIN

Ken Hashimoto and Mitsuru Setoyama

TABLE OF CONTENTS

0-8493-6012-9/93/$0.00 + $.50

A. INTRODUCTION

Immunoelectron microscopy in dermatological science has made a number of contributions in elucidating pathophysiologies of skin diseases. As in other fields of investigation, the earlier development and establishment of basic ultrastructures and histo- and cytochemistries of the normal and abnormal skin[1-3] have paved the road for the later development and fruition of immunohistochemistry and immunocytochemistry[1] and are now contributing to immunoelectron microscopy.

There are several structural elements in the skin which play vital roles in the normal function of skin and therefore are of vital importance in the diseases affecting such elements.

B. BULLOUS DERMATOSES AND BASEMENT MEMBRANE ZONE (BMZ)

The dermoepidermal junction has several structural components, all important to maintain the integrity of this junction (Figure 1). If autoantibodies are produced against these components, a separation of this junction may take place and cause blisters. Routine histopathology combined with light microscopic immunohistochemistry can determine the BMZ binding; however this BMZ pattern is common in many bullous diseases.[2-4] Therefore, in the following conditions more detailed information on the BMZ components to which autoantibodies bind is essential for the definitive diagnosis.

Epidermolysis bullosa acquisita (EBA) — This is a chronic nonhereditary blistering disease of the older individual. Blisters tend to develop in traumatized or friction areas such as hands. The direct immunohistochemical tests in the lesion are most frequently positive in BMZ for IgG, C_3, and fibrinogen, and less frequently for IgA and IgM.[5-7] However, this pattern of BMZ immunostaining is also found in bullous pemphigoid (BP) and lupus erythematosus (LE). The limited resolution of light microscopy cannot distinguish between anchoring fibrils (antigenic structure for EBA), hemidesmosomes (the major antigenic site of BP), and basal lamina/subbasal lamina fibrils (sites of immunoglobulin deposition in LE) with which autoantibodies of each disease react. Application of immunoelectron microscopy clearly demonstrated that in EBA the autoantibody immunoglobulins bind to the dermal end of anchoring fibrils, anchoring plaque, and the dermal side of basal lamina.[8,9] This pattern of immunolabeling is similar to the distribution of the carboxyl-terminal of type VII collagen[8,9] (Figure 2). By immunoblotting, it was shown that the patient's autoantibodies bind specifically with the globular carboxyl-terminus of type VII procollagen.[9]

Bullous pemphigoid (BP) — It has long been known that the patient suffering from this disease produces autoantibodies of IgG class which bind to BMZ.[1,3] However, the fine structural components of BMZ, which react with these autoantibodies, have not been elucidated. If the normal skin is incubated with 1.0 *M* saline, it is cleaved through the lamina lucida (the space between the basal lamina and basal cells). When such cleaved skin is used as substrate, most of the BP autoantibodies react only on the roof.[6] This finding at least ruled out the antigenicity of the basal lamina, anchoring fibrils, and other upper dermal components. Rather, this supported the presence of BP antigen epitopes on the basilar surface of the basal cells. The long controversy was finally settled by immunoelectron microscopic identification of the major reaction site of BP autoantibodies with hemidesmosome-tonofilament complex[10,11] (Figure 3). Recently, cDNA of 230-kDa glycoproteins were cloned and fusion proteins were produced; these reacted with BP autoantibodies.[12]

Transient bullous dermolysis of the newborn (TBDN) — Among a variety of hereditary congenital bullous diseases,[13] TBDN is unique in that the course of the disease is self-limited.[14-16] The majority of the cases are sporadic,[14-16] but dominant[17] inheritance is possible. With light microscopy, the cleavage of the skin which leads to blister formation is seen in BMZ. With periodic acid-Schiff (PAS) stain the basement membrane is demonstrated at the roof of the blister.[14] By routine transmission electron microscopy (TEM) anchoring fibrils

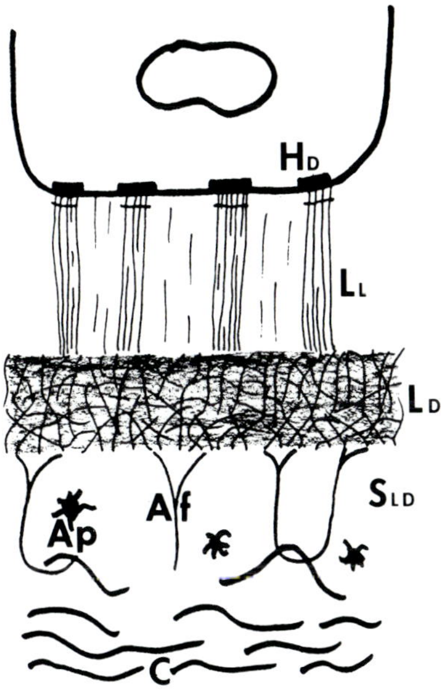

FIGURE 1. Components of BMZ and their antigenic epitopes.

are diminished or absent. These findings are helpful but not diagnostic for this disease, because similar findings are also obtained in dermal types of dystrophic epidermolysis bullosa which resemble TBDN. Multiple vesicles and vacuoles are formed in the basal cells of the epidermis which cover the blister cavity below. These are actually dilated rough-surface endoplasmic reticulum and contain peculiarly shaped inclusion bodies (Figure 4). These consist of bundled filaments with irregular striations and amorphous electron-dense laminae of various thickness and length. The former resembles anchoring fibrils and the latter the basal lamina of the epidermis (Figure 4, insert). These inclusions were subsequently found in transient, self-healing dominant dystrophic epidermolysis bullosa[17] and in a fatal case of recessive dystrophic epidermolysis bullosa.[18] The proof that these filamentous bundles with irregular striations are indeed related to anchoring fibrils came from the studies of these cases.[17,18] Light microscopic studies using type VII collagen antibody immunostained vacuoles of lesional epidermal basal cells (Figure 5), but could not resolve the fine reaction sites. The immunogold technique, however, localized the reaction sites of type VII collagen antibody to these anchoring fibril-like filament bundles (Figure 6). Amorphous laminar structures most likely represent basal lamina but so far no study has been done using type IV collagen antibody. It is obvious that some unknown disturbance of secretion of collagen types IV and VII from the epidermis causes inadequate formation of BMZ and eventually

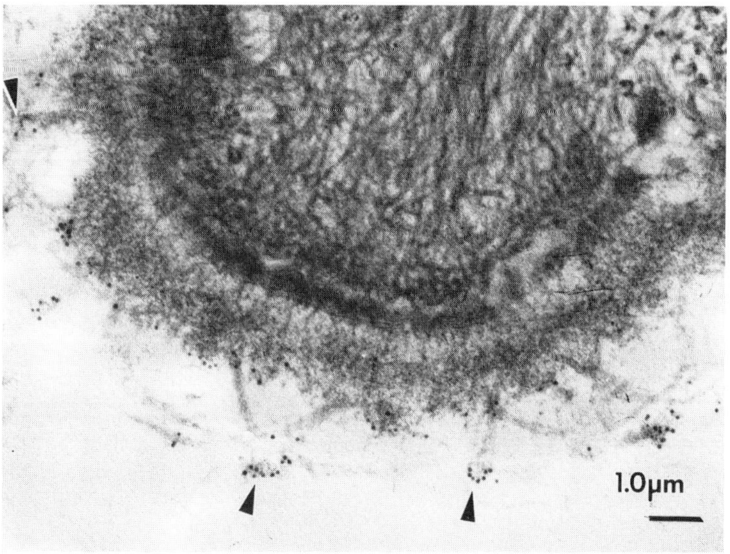

FIGURE 2. Epidermolysis bullosa acquisita. Patient's serum labeled (.5-nm gold particles) dermal end of anchoring fibrils (arrowheads) and lower surface of the lamina densa. Substrate tissue used is normal abdominal skin of adult. (Bar = 1.0 μm. Courtesy of Hiroshi Shimizu.)

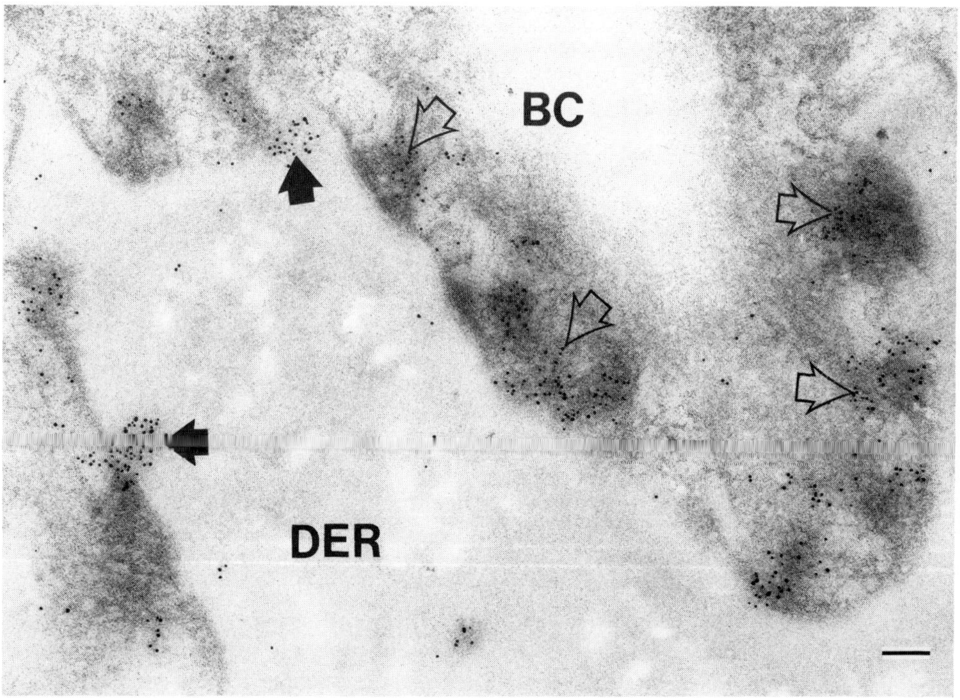

FIGURE 3. Bullous pemphigoid autoantibodies (patient's serum) have affinity to hemidesmosomes of a basal cell (BC). Gold particles are mainly localized on the hemidesmosomes (open arrows), while some labeling of extracellular antigens is noted (solid arrow). (Bar = 0.1 μm. Courtesy of Hiroshi Shimizu.)

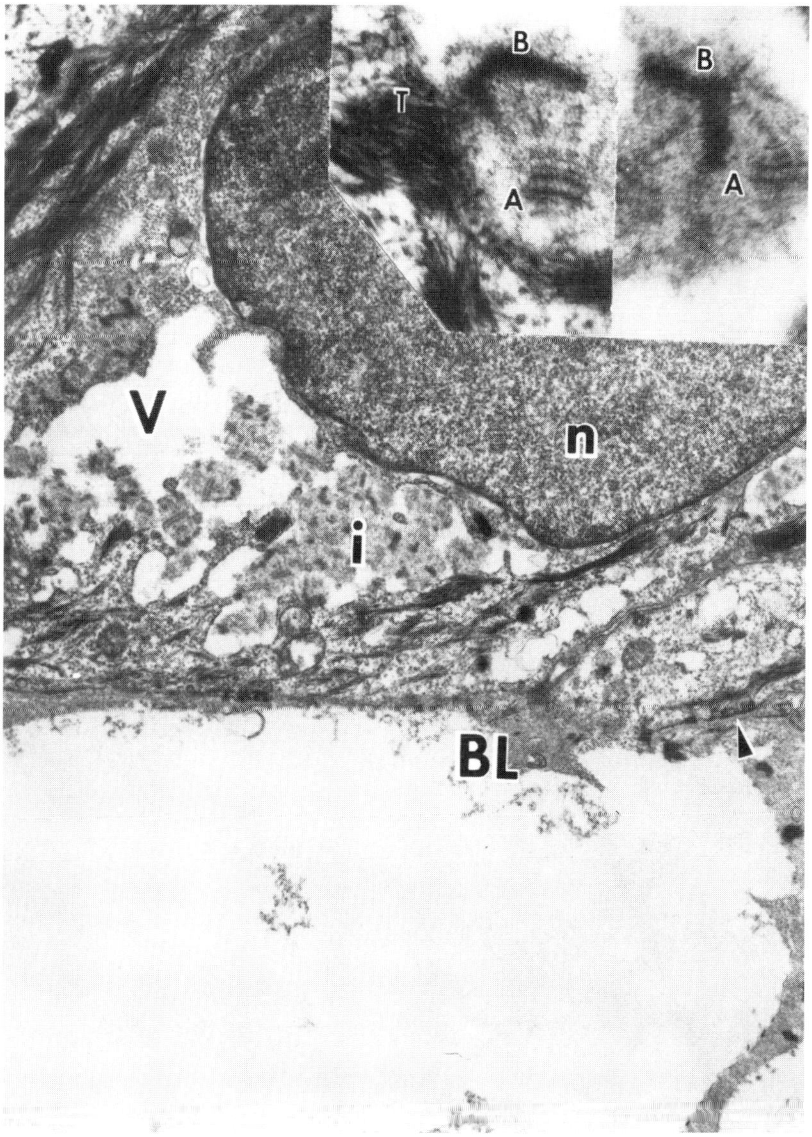

FIGURE 4. Transient bullous dermolysis of the newborn. Multiple cavities and vesicles (V) are present in a basal cell which contains multiple inclusion (i). The dermis is separated below the basal lamina (BL); n, nucleus. (Magnification × 12,000.) Insert: A high magnification of the inclusions reveals anchoring fibril-like filament aggregations (A) and a dense amorphous layer similar to basal lamina (B). Arrowhead, hemidesmosome; T, tonofilaments. (Magnification × 60,000.)

leads to separation of this junction. In this case, immunolectron microscopy was instrumental in solving the pathogenesis and will serve simultaneously as a useful diagnostic tool.

C. IDENTIFICATION OF CELL TYPES AND ORGANELLES

Probably the most frequent application of immunoelectron microscopy in dermatology is to identify cell types of cellular organelles, because immunostains at the light microscope level are not necessarily very specific. Even monoclonal antibodies label more than one cell

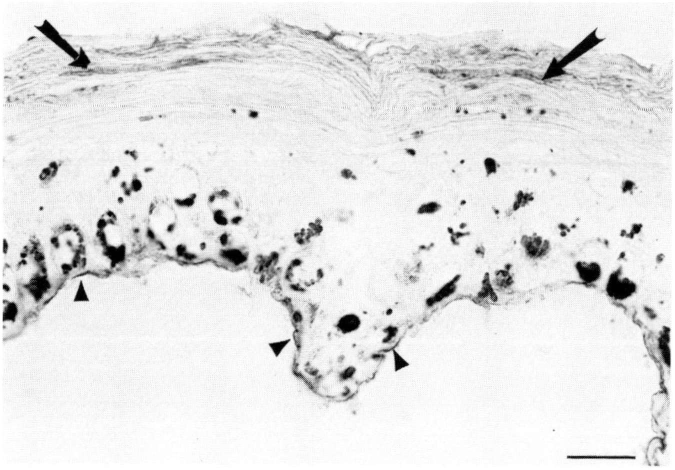

FIGURE 5. Recessive dystrophic epidermolysis bullosa (blister roof). Monoclonal antibody to type VII collagen, NP-185 (18), labeled the contents of cavities and vesicles in the lower epidermis and in some horny cells (arrows). In contrast, BMZ where anchoring fibrils should be present and stained with NP-185, is only sporadically labeled (arrowheads). (Bar = 20 μm. Courtesy of Lynne T. Smith.)

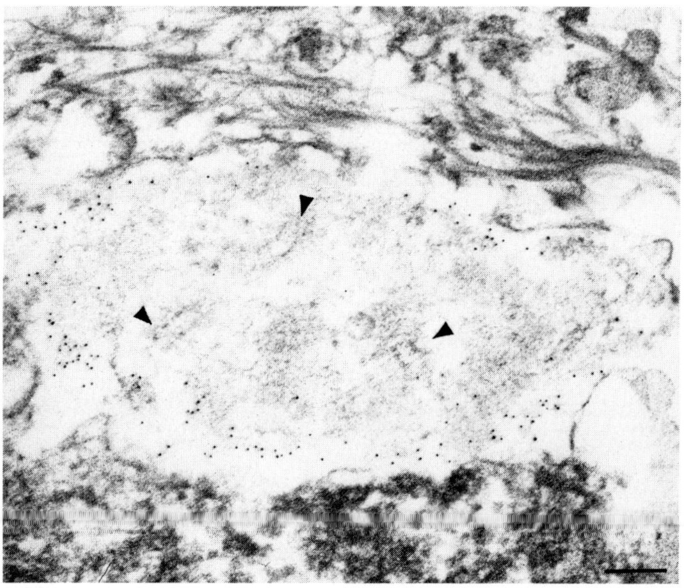

FIGURE 6. Recessive dystrophic epidermolysis bullosa. Immunogold labeling of type VII collagen of the contents of a vesicle of the lesional keratinocyte by NP-185 shows that filamentous aggregates are specifically decorated by gold particles. (Bar = 0.2 μm. Courtesy of Lynne T. Smith.)

type or organelle. For example, CD1 cell surface antigen is shared by activated thymocyte as well as Langerhans cells.

Langerhans cells and CD1 — The epidermal and dermal Langerhans cells play a critically important role in the primary processing of haptens for the development of contact (delayed or type IV) hypersensitivity. They are also the major cell type in histiocytosis-X, so much so that histiocytosis-X is now often referred to as Langerhans cell histiocytosis.[19]

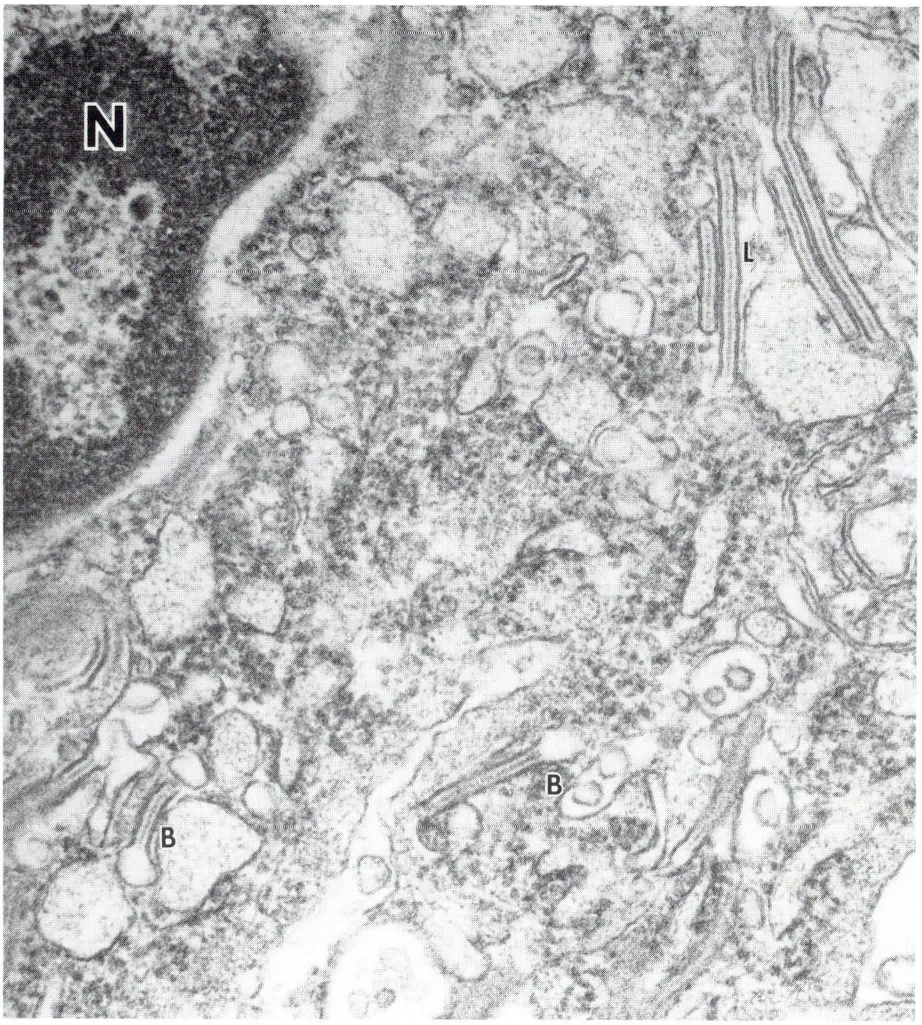

FIGURE 7. Langerhans cell of histiocytosis-X contains many tennis racket-like Birbeck's granules (B). Some granules consist of rod portion (L) only and lack swollen end; N, nucleus. (Magnification × 60,000.)

Identification of Langerhans cells relies upon the demonstration of Birbeck's granules which are tennis racket-shaped intracytoplasmic organelles (Figure 7). Although immunohistochemical methods may be faster, cheaper, and more quantitative because larger area can be scanned than using an electron microscope, immunohistochemical markers for Langerhans cells, e.g., CD1, HLA-DR, gamma-interferon (IFN-γ), and S-100 proteins, are not exclusively specific for this cell; each one of them also labels some other cell types. Before we could replace electron microscopic methods for the detection of Langerhans cell with immunostains, we had to monitor the accuracy of each monoclonal and polyclonal antibody with regard to identification of Birbeck's granule (+) Langerhans cells in normal and various pathological situations.[20] For this purpose, immunoelectron microscopy was extensively used (Figure 8). In the epidermis CD1 (+) phenotype is expressed by Langerhans cells and indeterminate cells. In histiocytosis-X, antibodies to HLA-DR and S-100 are found to stain many non-Langerhans cells, whereas antibodies to CD1 is more specific for Langerhans cell.[20]

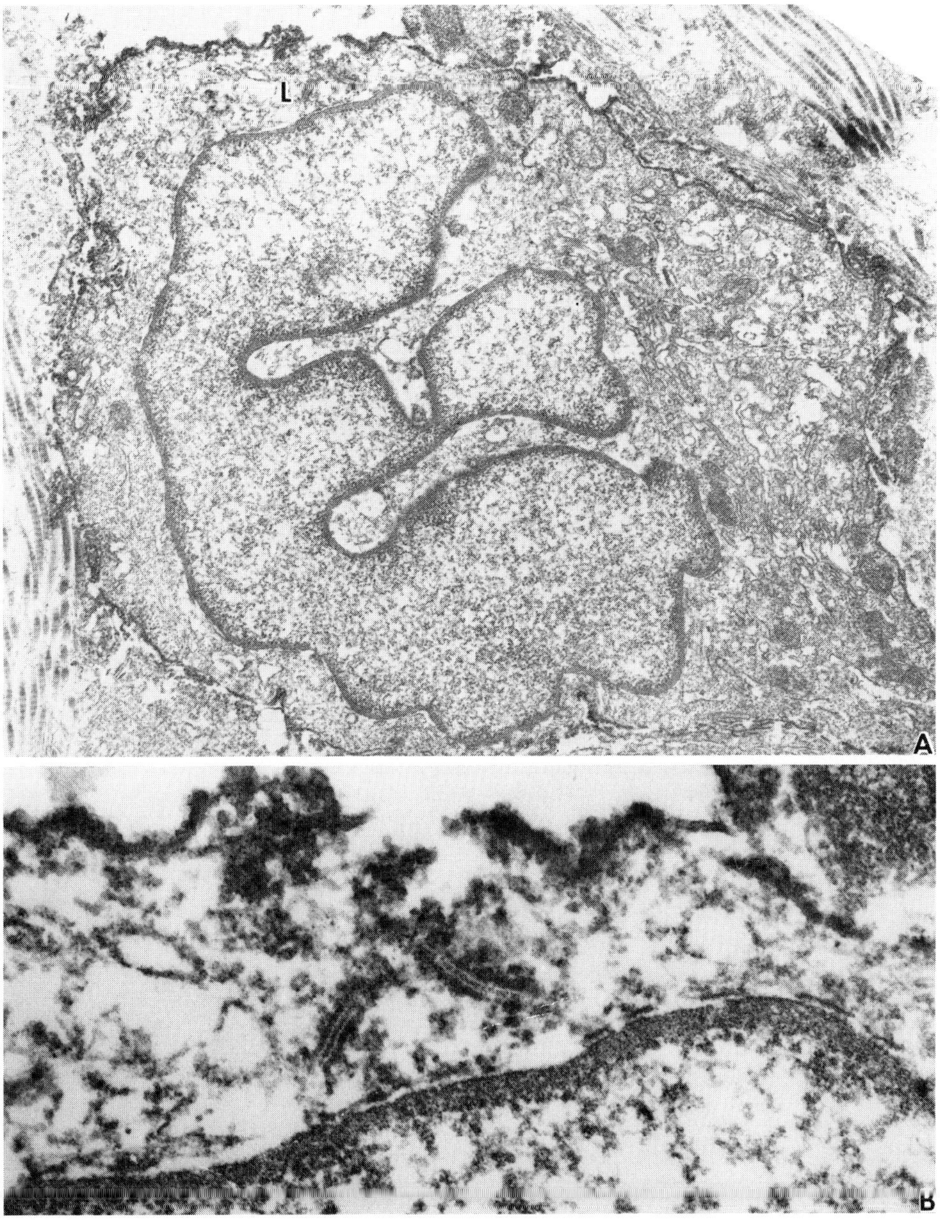

FIGURE 8. Langerhans cell. (A) This cell is labeled with CD1 and shows immunoperoxidase-labeled, electron-dense cell membrane (CD1 epitope) and simultaneously stained Birbeck's granule (L) within the cytoplasm. (B) Enlargement of CD1 labeled Birbeck's granules which are designated with L in A. (Magnifications: A, × 15,000; B, × 60,000.)

The derivation of Birbeck's granule from invagination of cell membrane had been postulated[21] on TEM evidence of continuity of these two structures. Immunoelectron microscopy using anti-CD1 antibodies as a membrane marker of Langerhans cells stained the cell membrane and Birbeck's granule and proved that these structures are related[22] (Figure 8).

Merkel cell and antikeratin antibodies — Merkel cells of the epidermis have been established as tactile receptor cells which contain specific cored dense granules. However, their ontogeny has been controversial. The theory of neural crest origin[23] has been questioned because of the presence of early Merkel cells in the fetal epidermis before either the peripheral nerve or dermal Merkel cells appear in the skin. A new theory of keratinocyte origin was based upon the observation that the epidermal Merkel cell contains tonofilaments and desmosomes, and reacts with simple epithelial keratin antibodies such as CK5 and CAM 5.2.[24,25] It was, however, necessary to identify CAM 5.2(+)/CK5(+) epidermal cells with Merkel cells by demonstrating intracytoplasmic cored granules which are specific for Merkel cell. Recently, we have demonstrated CAM 5.2-labeled tonofilaments in cored granule-positive Merkel cells by the immunogold method[26] (Figure 9). It is therefore confirmed that CAM 5.2(+) cells in fetal skin are indeed Merkel cells and this facilitated the future studies because immunostains could be used reliably, replacing the electron microscope.

D. STUDIES OF SUBCELLULAR ORGANELLES IN FUNCTIONALLY DEFINED CELL POPULATIONS

Lymphocytes have long been regarded as one type of cell, until a number of cell surface antigens were discovered. Today, the subset classification depends upon the surface markers and any morphological study should specify the cell types. If one undertakes investigation on subcellular organelles of lymphohistiocytic cells, it is required to specify the subset of those cells by CD number or other designations.

Sézary cells and nuclear contour index — In Sézary syndrome or mycosis fungoides (cutaneous T cell lymphoma) the major tumor cells are large T cells with a large nucleus. The nuclear contour is convoluted in various degrees. Some cells with a high degree of nuclear membrane invagination are called cerebriform cells because they resemble cerebral cortex. Since the degree of nuclear convolution parallels the degree of malignancy of the disease, a measure, called nuclear contour index (NCI), was devised:[27,28]

$$\text{NCI} = \frac{\text{nuclear membrane perimeter}}{\sqrt{\text{nuclear area}}}$$

If the nuclear membrane is perfectly round and smooth, NCI is 3.54; NCI will increase as the nuclear contour becomes more irregular and indented. In order to obtain NCI of tumor T-lymphocytes, immunolabeling of infiltrating cells with T cell marker (CD5), particularly helper (CD4) or suppressor (CD8) T cell surface markers, is necessary because there are many non-T cells which show various degrees of nuclear membrane invagination. The actual measurement of nuclear perimeter and nuclear area is only possible on electron micrographs and, in this case, on the immunolabeled T cells[29] (Figure 10).

Acantholytic diseases — Epidermal keratinocytes are connected with desmosomes which could be identified by an antibody against desmoplakin, a component protein of attachment plaque.[30] When desmosomes are detached (acantholysis), an intraepidermal blister is formed. The mechanisms involved in acantholysis may be due to autoantibody attack on desmoglein I,[31] an intercellular glue substance of desmosome, in pemphigus foliaceus (Figure 11A) or pemphigus vulgaris, desmoplakin is well preserved even after the acantholysis (Figures 11A and 12A). In hereditary acantholytic diseases, such as Darier's disease and Hailey-Hailey's disease, desmoplakin proteins seem to be dissolved in the lesional keratinocytes and diffused out into the cytoplasm (Figure 11B). This phenomenon has been known from light microscopic immunostains.[29] By the use of immunoelectron microscopy, it was demonstrated that desmoplakin is absent in desmosomal attachment plaques and that the immunoreactive desmoplakin is diffusely scattered in the cytoplasm[33] (Figure 12).

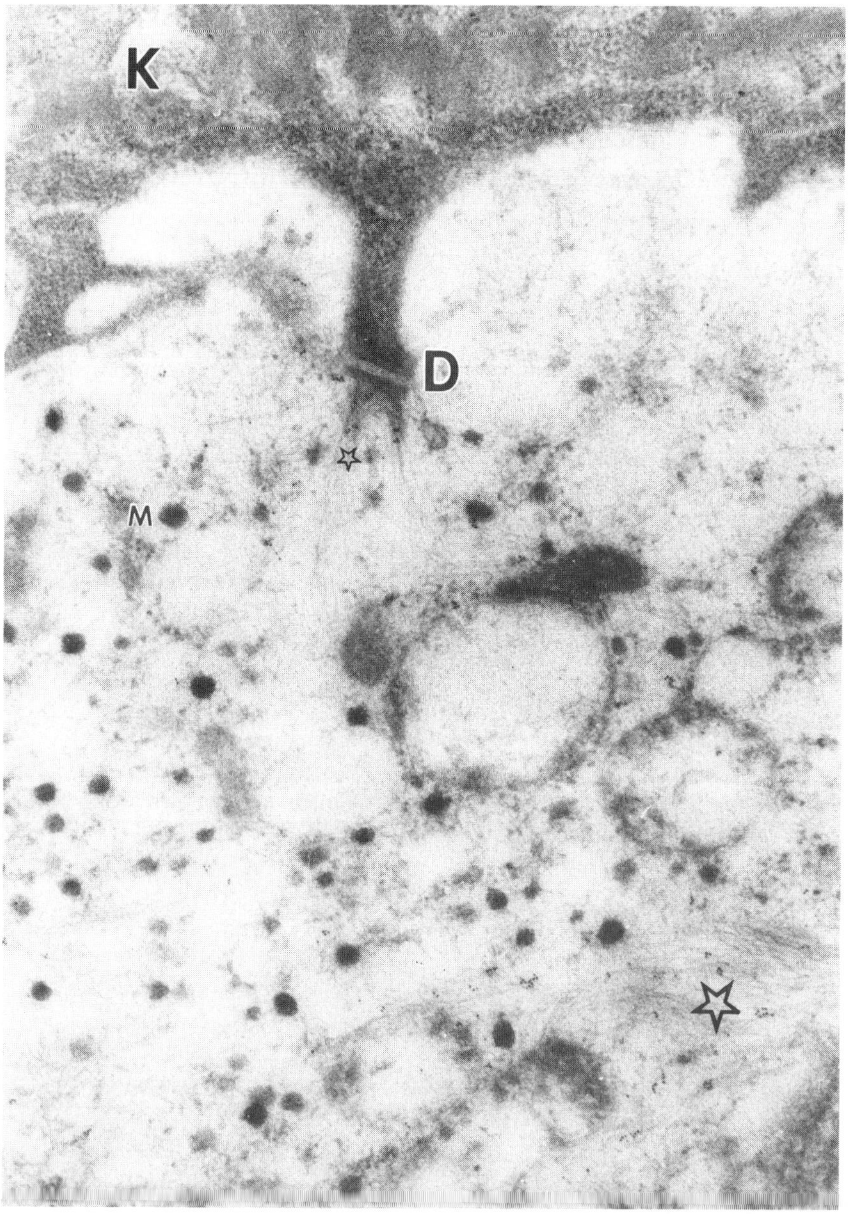

FIGURE 9. Merkel cell in the epidermis of 18-week-old fetus. Monoclonal anti-simple epithelial keratin CAM 5.2 labeled tonofilaments (☆) in a typical Merkel cell which has specific dense-cored granules (M); D, desmosome; K, keratinocyte. (Magnification × 50,000.)

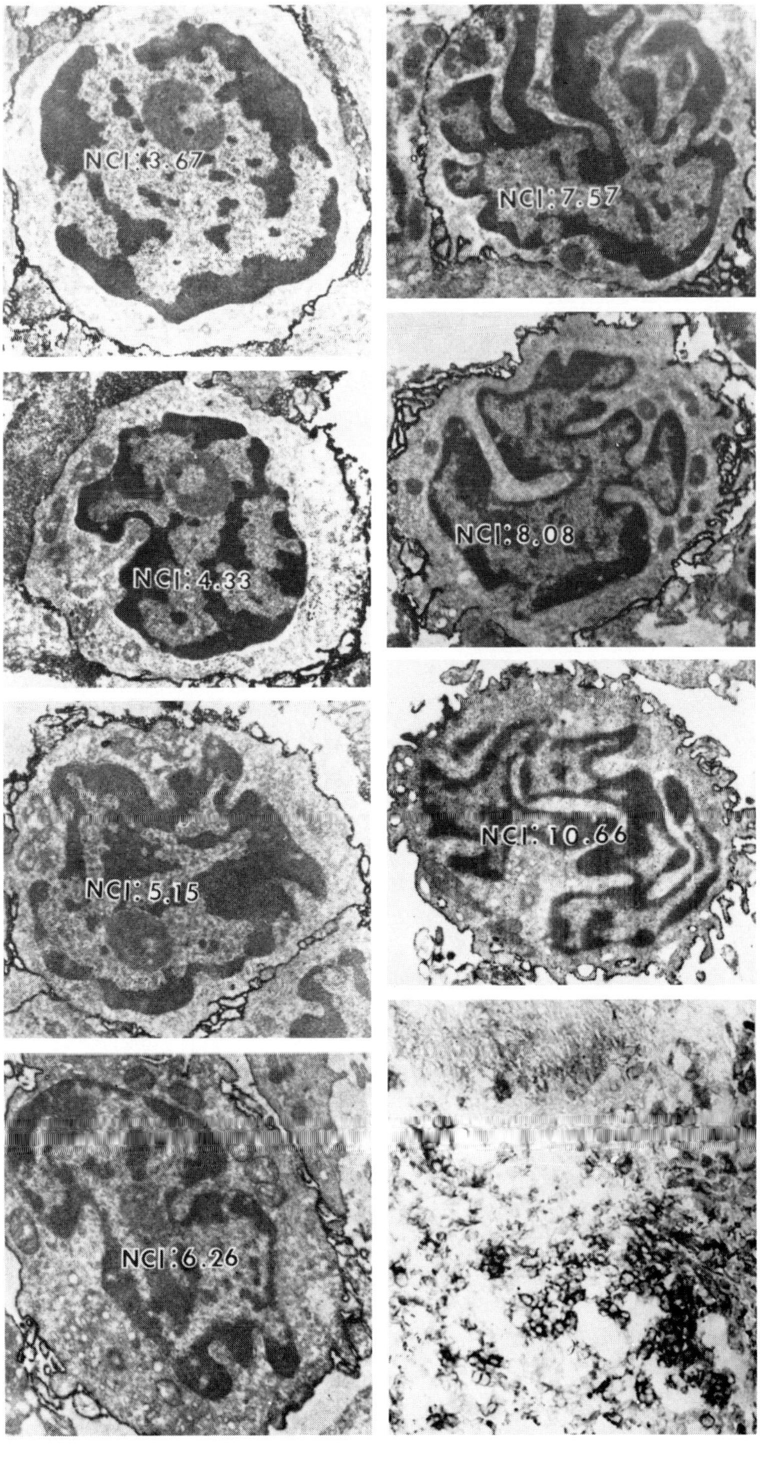

FIGURE 10. Mycosis fungoides. Tumor cells which react to CD5 antibody are selected and NCI was calculated. (Electron micrographs, magnification × 8,700; light micrograph, magnification × 170.)

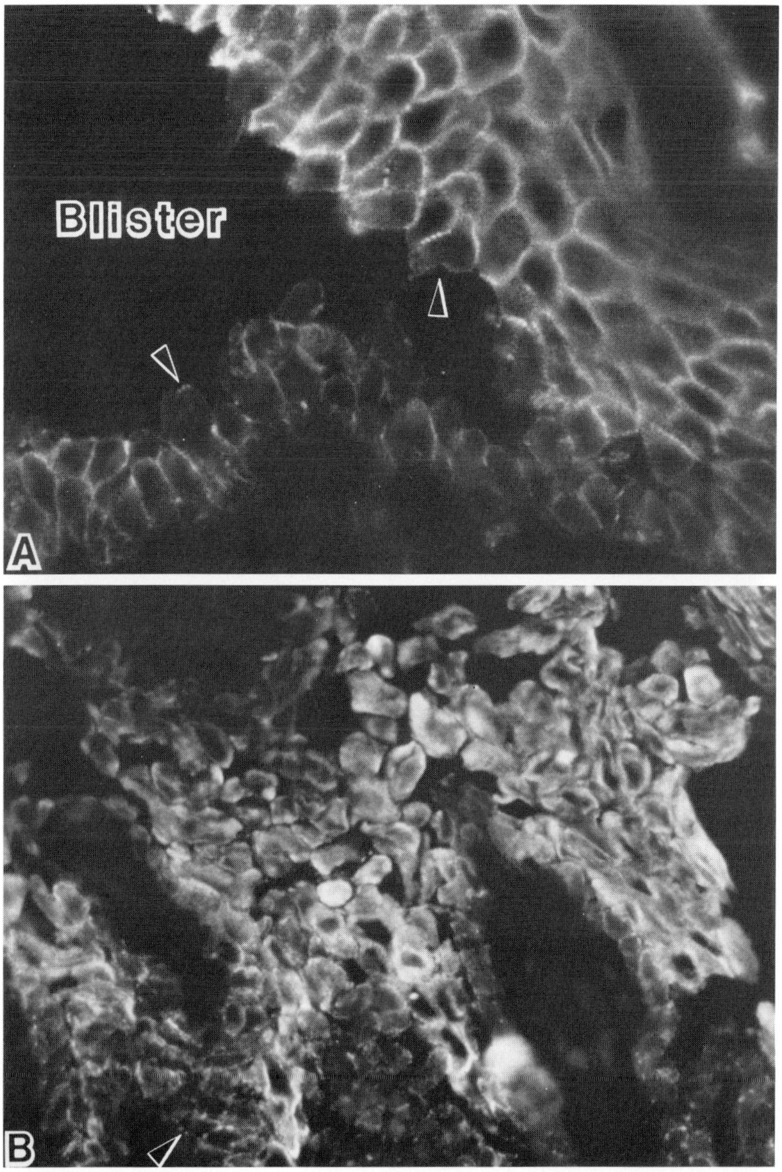

FIGURE 11. (A) Pemphigus vulgaris. IgG autoantibodies initiated the acantholysis. However, in acantholysed cells desmoplakin is mostly intact in desmosomes and demonstrated as a chain of fluorescent dots along the cell border (arrowheads). Immunofluorescence technique. (Magnification × 600.) (B) Darier's disease. Spontaneously developed acantholysis in the lesion shows that the keratinocytes contain diffusely cytoplasmic desmoplakin and lost the characteristic chain-of-dots pattern along the cell periphery which is still visible in intact areas (arrowheads). This suggests the dissolution of desmoplakin into the cytoplasm. Immunofluorescence technique. (Magnification × 600.)

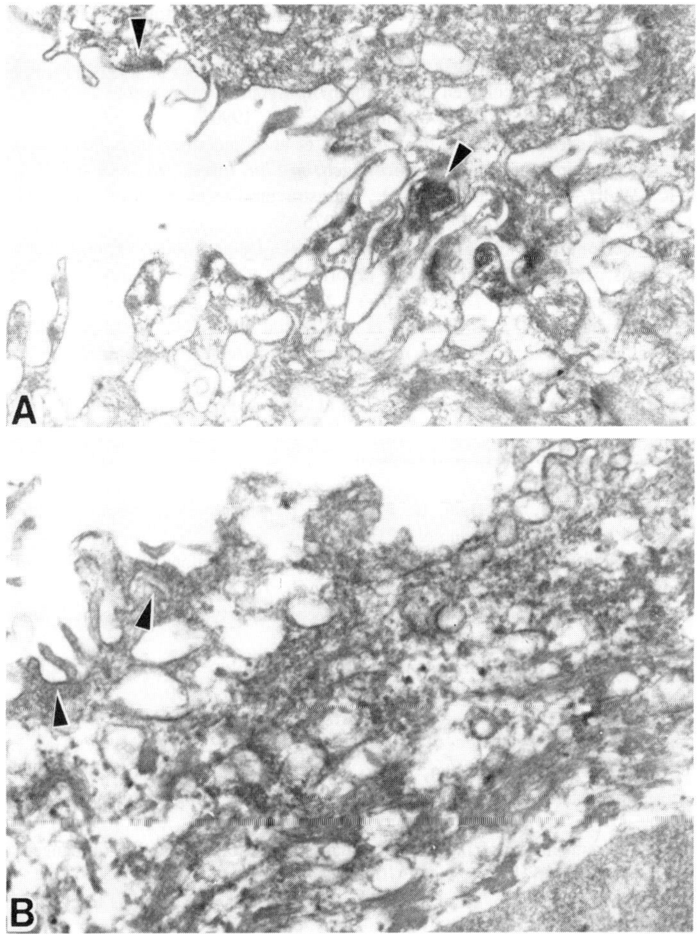

FIGURE 12. (A) Pemphigus vulgaris. Desmoplakin is demonstrated as dense plaque in both intact and acantholysed lesional keratinocytes (arrowheads). Immunoperoxidase technique, no electron stains. (Magnification × 20,000.) (B) Darier's disease. Acantholysed lesional keratinocyte shows no dense plaque where desmosomes are expected to be present (arrowheads). Instead, the density of the cytoplasm is increased. Immunoperoxidase technique, no electron stains. (Magnification × 21,000.)

REFERENCES

1. **Johannessen, J. V. and Hashimoto, K.,** *Electron Microscopy in Human Medicine,* Vol 11 (a), McGraw-Hill, New York, 1985.
2. **Lever, W. F. and Schaumburg-Lever, G.,** *Histopathology of the Skin,* 7th ed., J. B. Lippincott, Philadelphia, 1990.
3. **Mehregan, A. and Hashimoto, K.,** *Pinkus' Guide to Dermatohistopathology,* 5th Ed., Appleton & Lange, Norwalk, CT, 1991.
4. **Ueki, H. and Yaoita, H.,** *A Color Atlas of Dermatoimmunohistocytology,* Wolfe Medical Publication, London, 1989.
5. **Wilson, B. D., Birnkrant, A. F., Beutner, E. H., and Maige, J. C.,** Epidermolysis bullosa acquisita: a clinical disorder of varied etiologies, *J. Am. Acad. Dermatol.,* 3, 280, 1980.
6. **Gammon, W. R., Briggaman, R. A., Inman, A. O., III et al.,** Differentiating anti-lamina lucida and anti-sublamina densa anti-BMZ antibodies by indirect immunofluorescence on 1.0 M sodium chloride-separated skin, *J. Invest. Dermatol.,* 82, 139, 1984.

A. INTRODUCTION

The most important and characteristic function of the eye is light sensation. This function is carried out by the retina and optic nerve, but other tissues such as the cornea, iris, ciliary body, lens, etc. support the mechanism of light sensation. Numerous histo- and cytochemical studies have been carried out to elucidate the functions of the different tissues of the eye. Since it is impossible to review all the major contributions of histo- and cytochemistry to the physiology of the eye in the space available, I will concentrate on selected problems, which include the contribution of histo- and cytochemistry to the elucidation of the light-perceptive mechanism in the photoreceptor cells.

B. PHYSIOLOGICAL AND BIOCHEMICAL BACKGROUND OF THE LIGHT-PERCEPTIVE MECHANISM

The initial process in the rod photoreceptive mechanism is the bleaching of rhodopsin in the disk membranes of the photoreceptor cells by a photon. The photochemical change of the rhodopsin causes a reduction of the continuous dark sodium current via the plasma membranes of the outer segments.[1] In the darkness, a continuous diffusion of Na^+ ions into the outer segments via membrane ion channels exists and a Na^+ ion current flows steadily between the outer and inner segments. The Na^+ ions have been thought to be pumped out from the photoreceptors by the Na^+,K^+-ATPase pump. On the basis of current flow, the existence of Na^+,K^+-ATPase pumps in the inner segments was postulated.[2,3] In contrast to these physiological studies, biochemical experiments indicated that two completely different situations exist in the outer and inner segments. A high Na^+,K^+-ATPAse activity was thought to occur in the outer segments,[4] but, on the other hand, it has been claimed that this high activity of the outer segments may originate from contamination by material from the inner segments obtained during cell fractionation; the membranes of the inner segments display a high activity of this enzyme.[5] Hence, the localization of the Na,K-ATPase has so far been controversial.

Blockage of this sodium current is thought to depend on the closing of ion channels localized on the plasma membrane. As a consequence of this decrease in the sodium current, an electrical response occurs; it leads to hyperpolarization of the plasmalemma of the photoreceptor cells.[6] In the rod outer segments, the disk membranes are osmotically and electrically separated from the plasmalemma[7] by a distance of up to 200 to 300 Å.[8] Therefore, the existence of a second messenger has been postulated. Initial interest was focused on changes in cytosolic and extracellular free Ca^{2+} ions.[9] With respect to the importance of Ca^{2+} ions for generation of the electrical photoreceptor response, the localization of Ca^{2+}-ATPase, which is regarded to be one of the Ca^{2+} ion regulators, has also been studied by biochemical methods. However, in analogy to the localization of Na^+,K^+-ATPase, biochemical studies have led to diverging concepts.[10,11]

Biochemical studies revealed that the photoreceptor outer segments contain a uniquely high cGMP level different from other tissues;[12] cGMP acts mainly as an internal messenger. From biochemical studies on cGMP synthesis (guanylate cyclase) and degradation (cGMP-dependent phosphodiesterase), it appeared that the cGMP-dependent phosphodiesterase (cGMP-PDEase) activity is highly responsible for this sequential mechanism of transmission between disk membranes and plasma membranes of the photoreceptor outer segments.[13-16] Subsequently, the existence of some other important cofactors, such as G-protein (GTPase), [17]H-protein, or PDEase-inhibitory subunit was detected and their roles during light conductive mechanism have been examined. GTPase activity was also proposed to be present in the photoreceptor outer segments; special attention has been paid to this enzyme because it is partly activated by light. For elucidation of the light-perceptive mechanism, from this background, the localization, especially the ultrastructural localization, of Na^+,K^+-ATPase,

Ca^{2+}-ATPase, guanylate cyclase, cGMP-PDEase, and GTPase activities seems to be essential.

C. Na^+,K^+-ATPase

Initial histo- and cytochemical studies on the localization of Na^+,K^+-ATPase activity on the photoreceptor cells utilized techniques based on biochemical assays and on the method for Mg-ATPase as described by Wachstein and Meisel. However, the histochemical reaction, using ATP as substrate, was not inhibited by the specific inhibitor of Na^+,K^+-ATPase, ouabain.[18]

Thereafter, the localization of this enzyme was examined utilizing ^3H-ouabain autoradiography.[19] A low density of silver grains in the outer segments and relatively high density in the inner segments of photoreceptors were demonstrated. However, these studies were performed at the light microscopic level; the fine subcellular localization could not be detected because of the limited resolution power of the technique. In view of these problems, alternative ultracytochemical methods were developed by Ernst[20] and by Mayahara et al.[21] These methods are fully discussed in other chapters of this book.

In sections of the retina incubated in the standard medium of Mayahara et al., a weak K^+-*P*-nitrophenylphosphatase (K^+-NPPase) activity was found in the inner segments and outer nuclear layer of the photoreceptor cells.[22-25] In contrast, no activity was evident in the outer segments. This reaction was inhibited by ouabain and by elimination of K^+ ions from the incubating medium (Figure 1). Cytochemically, the reaction product was most abundant when the retina was fixed with paraformaldehyde alone. Fixation with 0.5% glutaraldehyde greatly diminished the activity. However, since enzyme activity survived after fixation in 0.1 to 0.25% glutaraldehyde for a short period of time, in order to obtain better morphological preservation, perfusion fixation for 5 min with a mixture of 2% paraformaldehyde and 0.25% glutaraldehyde is recommended. Ultracytochemically, the difference in the localization of the enzyme activity between the outer segments and the more proximal portions of the photoreceptors was striking. The K^+-NPPAse activity was localized along the plasmalemma of the inner segments; no reaction product was observed on the plasmalemma or the disk membranes of the outer segments (Figures 2 and 3).

From the histochemical work by Stirling and Lee,[19] and from our ultracytochemical work,[22-25] it is obvious that the major Na^+,K^+ pump site in the photoreceptor cells is localized on the plasmalemma of the inner segments, as suggested by much electrophysiological data. Moreover, our studies revealed that K^+-NPPAse activity was localized not only on the plasmalemma of the inner segments but also in the more proximal portion of photoreceptors. It is also noteworthy that this activity can also be observed on the apical plasma membrane of retinal glial Müller cells (Figure 4). Physiologically, it is known that after the peak of the light-evoked response, a decrease of K^+ ions occurs in the subretinal space.[26] The rapidity of decrease of K^+ ions suggests that it might be related to Na^+,K^+ pump activity, not only in the plasmalemma of the photoreceptors, but also in the plasmalemma of surrounding cells such as Müller cells.[27] The activity of the K^+-NPPAse, demonstrated by our ultracytochemical study, is in agreement with electrophysiological data.

D. Ca^{2+}-ATPase

Ca^{2+} ions are known to play one of the major roles in the regulation of ion permeability of the plasmalemma. The cytoplasmic Ca ion concentration is controlled by the active Ca^{2+} ion transport by ATP-dependent Ca^{2+} ion pumps, together with ion-exchange systems on the plasmalemma, the energy-dependent pump-leak system of the mitochondria, and the smooth endoplasmic reticulum under particular involvement of Ca^{2+} channels or Ca^{2+}-binding proteins.[28,29] Therefore, it seems important to analyze the subcellular localization

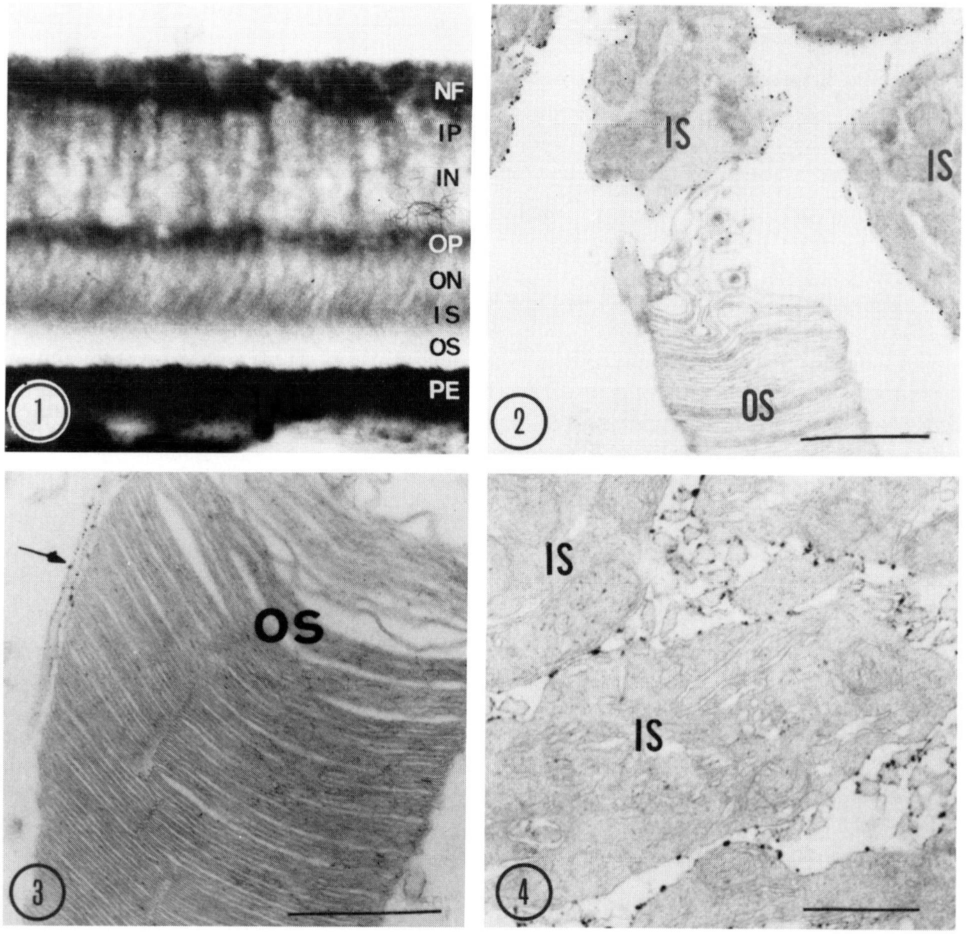

FIGURE 1. Light micrograph of guinea pig retina incubated with standard medium for K^+-NPPase. In photoreceptor cells, intense reaction was seen in outer plexiform layer (OP), moderate reaction in outer nuclear layer (ON) and inner segments (IS), whereas no obvious reaction was noted in outer segments (OS). (Magnification × 390.)

FIGURE 2. Electron micrograph of K^+-NPPase activity in the distal portion of photoreceptors. Reaction products are localized along the plasmalemma of the inner segments (IS); no reaction was seen in the outer segments (OS). (Magnification × 16,900.)

FIGURE 3. Reaction products of K^+-NPPase activity on the plasmalemma of the calyceal processes (arrow) of the inner segments, whereas no deposits mark disks and plasmalemma of the outer segments (OS). (Magnification × 20,700.)

FIGURE 4. Apical processes of Müller cells between the inner segments of photoreceptors show K^+-NPPase activity. (Magnification × 16,900.)

of the activity of Ca^{2+}-ATPase for a better understanding of the physiological function of the photoreceptors.

No cytochemical investigation of the Ca^{2+}-ATPase activity of photoreceptor cells seems to exist. The main handicap is the adequacy of available techniques for the demonstration of the activity of this enzyme, especially at the electron microscopic level. Neither two-step[30,31] nor the one-step[32] procedure is satisfactory; Ca dependency of the reaction cannot be examined since Ca^{2+} ions are utilized as the capture reagent of phosphate ions liberated from the substrate by enzyme hydrolysis; the initial reaction product, calcium phosphate, is invisible in the light microscope, and is soluble in the osmic acid solution used for postfixation in routine electron microscopy. Therefore, the calcium phosphate has to be converted into another metal phosphate in a second step.[33] The one-step method was a modification of the Mg^{2+}-ATPase procedure developed by Wachstein and Meisel. This method also exhibited some fundamental disadvantages; the reaction cannot be carried out under optimal pH[34] of the enzyme and the incubation medium yields milky precipitates.

In order to overcome these technical problems, an alternative one-step medium employing a medium which is completely clear at optimal pH, was developed by Ando et al.[35] Utilizing this method, we have demonstrated Ca^{2+}-ATPase activity in the photoreceptor cells. However, fixation with glutaraldehyde in any concentration drastically decreased activity of the enzyme. Therefore, the ultracytochemical analysis had to be performed after immersion fixation in paraformaldehyde, despite the poor morphological preservation. When sections of the retina were incubated in the standard Ca^{2+}-ATPase medium, an intense reaction was observed in the inner segment of the photoreceptors.[24] However, very little activity was found in the outer segments or proximal portion. This reaction was completely dependent on Ca^{2+} ions and could be inhibited by oligomycin or *p*-CMBS (Figure 5). Under the electron microscope, deposits of the reaction product were restricted to the cristae of the mitochondria, which are concentrated in the inner segments. In contrast, no conspicuous reaction precipitate was observed on the plasmalemma, endoplasmic reticulum, or in other cytoplasmic elements. In the outer segments, usually no reaction was detected on the plasmalemma disc membranes (Figure 6).

The almost complete lack of Ca^{2+}-ATPase activity at the outer segments is consistent with biochemical results. Mitochondria are known to take up Ca^{2+} ions into their matrix in the presence of ATP and to pool Ca^{2+} ions into their matrix in the presence of ATP and to pool Ca^{2+} ions internally. The pump-leak system of the mitochondria is thought to operate as one of the basic cell buffer systems. This buffer system, controlled by the pump-leak system of the mitochondria, may be a particularly important factor, since mitochondria are thought to be main regulators of cytoplasmic free Ca^{2+} ion concentration.[36] The increase of cytosolic Ca^{2+} ion concentration in the outer segment after light perception has so far been considered to be caused mainly by the release from the outer segment membrane or by an influx from the extracellular side through the plasmalemma. However, our results of an intensive Ca^{2+}-ATPase activity of the mitochondria suggest that the mitochondrial pump-leak system plays an important role in the Ca ion regulation of the photoreceptors.

E. CYCLASE

Cyclases are enzymes that synthesize cyclic nucleotides from nucleotide triphosphates. As previously mentioned, rod outer segments contain a uniquely high cGMP level, different from other tissues, and numerous biophotochemical works have indicated that cGMP acts mainly as an internal messenger. Therefore, the localization of guanylate cyclase (cGMP-synthesizing enzyme) activity seems to be important in the photoreceptor cells.

Initially, ATP was utilized as substrate, but using ATP, ATPase activity may contribute to the histochemical reaction, in addition to cyclase activity. In addition, the enzyme is inhibited by lead ions used as a capturing reagent. In order to overcome these pitfalls, a method was introduced by Fujimoto et al.,[37] using guanylylimidodiphosphate (GMP-PMP), instead of ATP, as a specific substrate for guanylate cyclase and dimethyl sulfoxide (DMSO) as an activator of the enzyme. (See also the chapter by K. Fujimoto in this volume.)

Utilizing this method, relatively high guanylate cyclase activity was found, restricted to the tip of the outer segments (Figures 7 and 8). No adenylate cyclase activity was observed in the outer segments.[38]

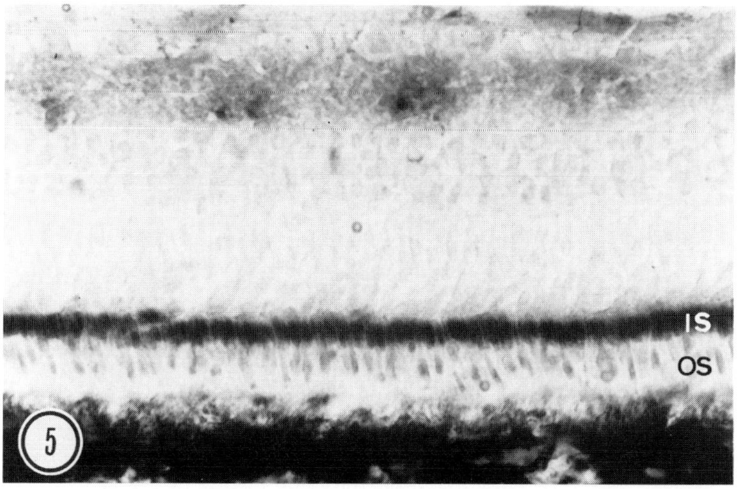

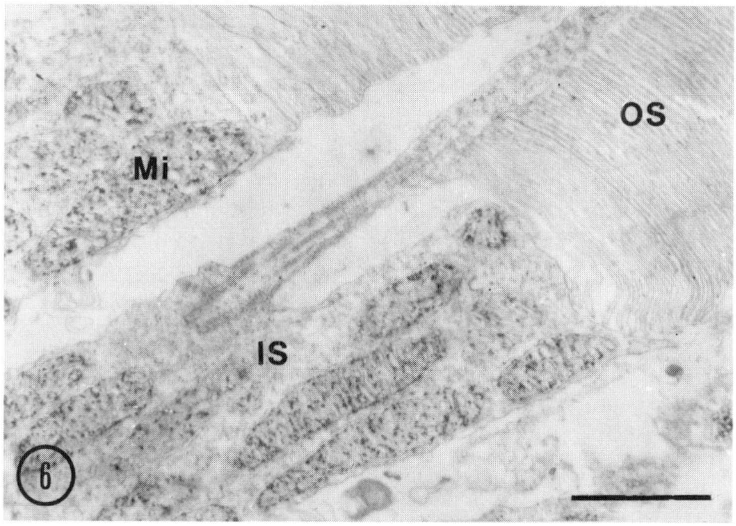

FIGURE 5. Light micrograph of Ca^{2+}-ATPase activity in the guinea pig retina. Intense reaction is restricted to the inner segments (IS) of photoreceptor cells; outer segments (OS) and basal portions of photoreceptors reveal no reaction. (Magnification × 480.)

FIGURE 6. Electron micrograph of Ca^{2+}-ATPase activity in the distal portion of photoreceptor cells. Mitochondria (Mi) in the inner segments (IS) display intense reaction; no significant reaction is seen on the plasmalemma and outer cytoplasmic elements. Also, no reaction is noted in the outer segments (OS). (Magnification × 17,000.)

F. CYCLIC NUCLEOTIDE PHOSPHODIESTERASE

A cytochemical attempt to demonstrate the activity of cyclic nucleotide-dependent phosphodiesterase (PDEase) in the photoreceptors was initially carried out by Robb,[39,40] utilizing the conventional snake venom method initially proposed by Shanta et al.[41] Although we have also tried to demonstrate cGMP-PDEase using the snake venom procedure, morphological preservation was extremely poor.[42-45] Although there was a moderate reaction in the outer segments, the sensory retina was frequently detached from retinal pigment epithelium. Under the electron microscope, the destruction of cell morphology was much more evident. Disk and plasma membranes were swollen and disrupted, and artifactual distribution of deposits of the reaction product was manifest. Moreover, Robb[39,40] could not demonstrate

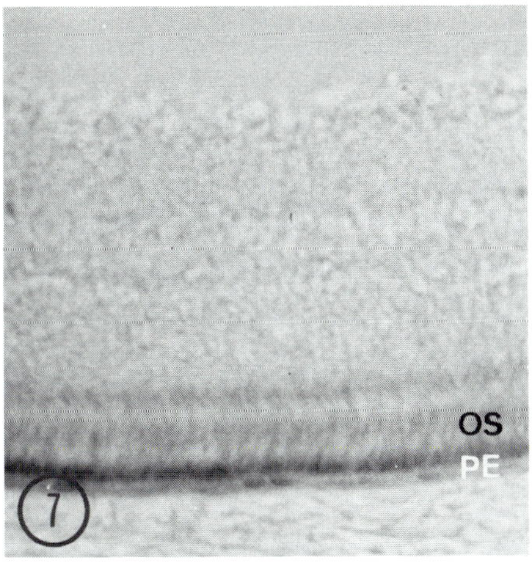

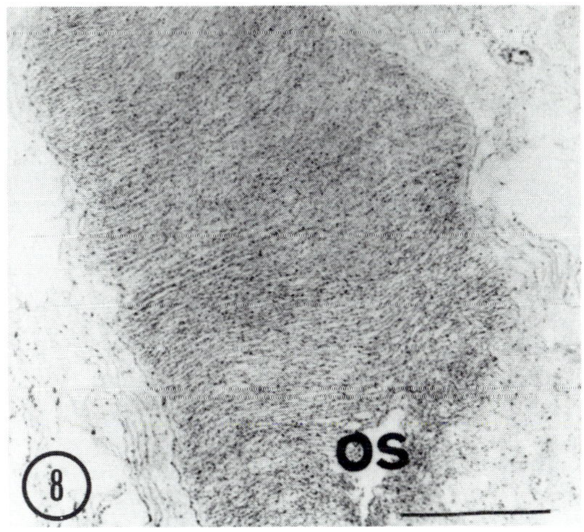

FIGURE 7. Light micrograph of guanylate cyclase activity in the rat retina; moderate reaction was seen only in the distal portion of the outer segments (OS) near the pigment epithelium (PE). (Magnification × 320.)

FIGURE 8. Electron micrograph of guanylate cyclase activity in the tip of the outer segments (OS). Large amount of reaction precipitates is noted on the disk membranes of the outer segments; moderate reaction is observed on the plasmalemma of photoreceptor outer segments and apical processes of retinal pigment epithelial cells. (Magnification × 18,000.)

substrate (cGMP and cAMP)-dependent differences utilizing the conventional snake venom procedure. Biochemical studies concerning the outer segments have revealed a more rapid hydrolysis of cGMP than cAMP by PDEase. Absorption of light by photopigment resulted in a drastic decrease of cGMP level but not that of cAMP in the outer segments. The results of Robb in the outer segments was not in complete agreement with biochemical data.

We have tried to demonstrate PDEase activity utilizing the intrinsic activity of 5Nase instead of snake venom. Details of tissue preparation, the composition of the standard incubation medium and controls for 5Nase and PDEase activities are described in Chapter 1.

Our method revealed clear-cut differences in localization, dependent on the substrate, cAMP or cGMP (see also the chapter on PDEase in Part 1). When cGMP was used as substrate, the activity was restricted to the inner and outer segments of photoreceptors and retinal pigment epithelium. The activity was much more intense uniformly in the outer segments, whereas a weaker reaction was observed in the inner segments under conditions of illumination. Under the electron microscope, heavy deposits of the reaction product were located along the disk membrane of the outer segments (Figure 9). In contrast to the localization of the 5Nase activity, no notable cGMP-dependent PDEase activity was detectable in the outer plexiform and nuclear layers. When cAMP was used as substrate, the reaction in the outer segments was much weaker than when cGMP was used. It is noteworthy that a moderate activity was noted in the outer plexiform layer. This reaction could never be observed when cGMP was used or when snake venom was utilized. In controls, these reactions were strongly inhibited by the addition of 2 m*M* 3-isobutyl-1-methylxanthine (IBMX)[46] to the standard incubation medium.

Moreover, utilizing our method, another important property of the activity was observed. When the sections were incubated in the dark (dim red light) the cGMP-dependent PDEase activity in the outer segments (Figure 4) was markedly reduced. In contrast, cAMP-dependent PDEase activity was not affected by illumination (Figure 5). Photobiochemically this cGMP dependency and light dependency of PDEase in the rod outer segments has been described.

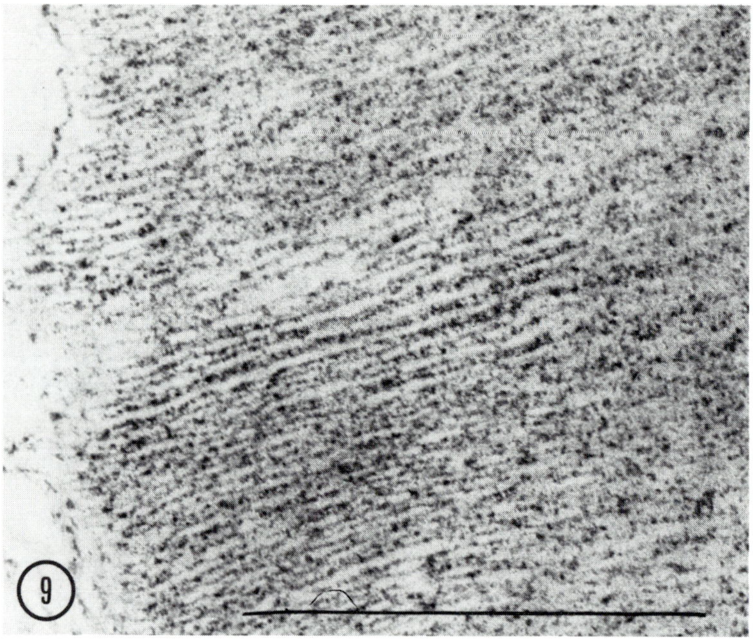

FIGURE 9. At higher magnification, reaction products of cGMP-PDEase activity are located along the disk membranes of the outer segments of photoreceptor cells. (Magnification × 64,000.)

Decrease of cGMP level by activation of PDEase was considered to be a signal to open the ion channels on the plasmalemma of the rod outer segments. A much more rapid hydrolysis of cGMP than of cAMP by PDEase was measured.[13,14] This photobiochemical difference in substrate dependency or light dependency is particularly interesting and important for checking the reliability of the histochemical procedure.

G. GTPase

GTPase activity is also described in the outer segments; this enzyme activity is given special attention because it is partly activated by light.[17] This light-dependent GTPase activity is thought to be one of the important triggers that activate cGMP-PDEase immediately after bleaching of rhodopsin by light.[47-49] Hence, studies to detect the fine structural localization of GTPase are particularly important. However, to date there have been no histo- and cytochemical reports concerning the activity of the enzyme, since no reliable cytochemical method for the demonstration of the activity exists. In 1983, we developed a method for this enzyme activity using a glycine-KOH buffer and levamisole.[50]

At the light microscopic level, an intense staining attributable to the GTPase activity was localized in the rod outer segments, outer plexiform layer, retinal pigment epithelium, and choroid. In the outer segments, the GTPase reaction is always more intense at the tip of the rods near the pigment epithelium (Figure 10). At the electron microscopic level, the precipitate was found on the disk membranes. The region adjacent to the pigment epithelial cells showed the most conspicuous deposits of the reaction product in the disk membranes. The plasmalemma of the outer segments were not stained markedly (Figure 11). In the dark phase, GTPase is thought to be associated peripherally with the disk membranes; therefore, it can be readily released from these membrane sites. In contrast, GTPase molecules become tightly bound to the membranes under the influence of light. We carried out the cytochemical

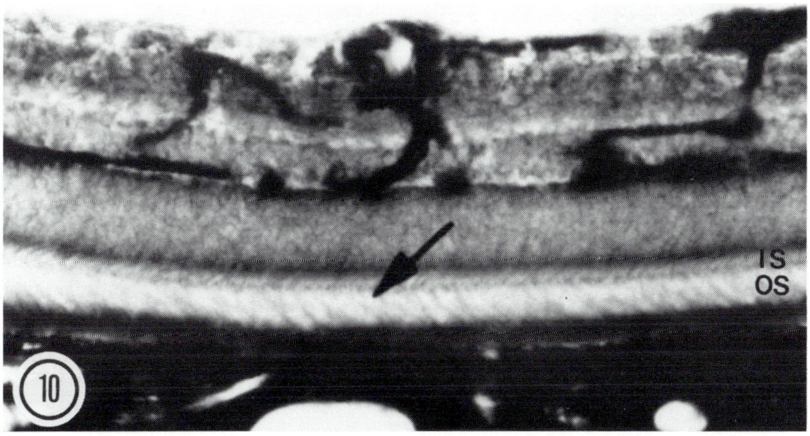

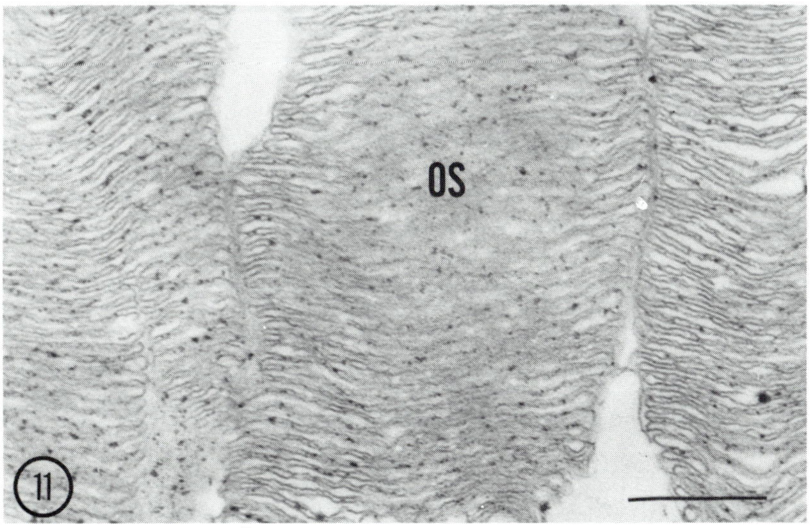

FIGURE 10. Light micrograph of GTPase activity in the rat retina. In the distal portion of photoreceptors, intense reaction is observed in the tip of the outer segments (arrow). The basal portion of the outer segments (OS) and inner segments (IS) reveal no reaction. (Magnification × 340.)

FIGURE 11. Electron micrograph of GTPase activity in the outer segments of photoreceptors (OS). Reaction deposits are located along the disk membranes of outer segments; no reaction is seen on the plasmalemma. (Magnification × 18,000.)

procedures under normal room illumination. Hence, the localization shown in our study may reflect the membrane-bound state.

Localization of the GTPase activity on the same rod outer segments was different; the reaction was always more intense at the tip of the outer segments near the pigment epithelium. Guanylate cyclase activity was also restricted to the tip of the outer segments. Although we have no simple explanation of this circumscribed enzyme location, it is interesting that both GTPase and guanylate cyclase revealed a coincidental distribution on the outer segments. However, at the fine structural level, reaction deposits of GTPase were located on the disk membrane or in the extradisk side of the membranes. In contrast, the location of guanylate cyclase was restricted to the intradisk membranes.

One more important result is the difference in distribution of GTPase and ATPase activities in the outer segments. As mentioned in the section on Ca^{2+}-ATPase, ATPase activity could not be detected in the outer segments. This result is in agreement with the results of subsequent biochemical studies suggesting that GTP is predominant in the outer segments.

H. CONCLUSION

It is hoped that this kaleidoscope of questions and histochemical findings concerning the localization and functions of enzymes in photoreceptor cells as discussed here, may be helpful for further research into what is one of the most complex and fascinating aspects of the light-perceptive mechanism. It is obvious that what is presented here is incomplete as far as the complex scheme of processes of light perception is concerned. However, we have a much better picture now than was available before. Further efforts to develop reasonable histochemical and cytochemical procedures are warranted, because they will bring much further understanding of the complicated and fascinating processes that occur in the photoreceptor cells.

REFERENCES

1. **Hagins, W. A., Penn, R. D., and Yoshikami, S.,** Dark current and photo-current in retinal rods, *Biophys. J.*, 10, 380, 1970.
2. **Frank, R. N. and Goldsmith, T. H.,** Effects of cardiac glycoside on electrical activity in the isolated retina of the frog, *J. Gen. Physiol.*, 50, 1585, 1967.
3. **Hagins, W. A., Robinson, W. F., and Yoshikami, S.,** Ionic aspects of excitation in rod outer segments, *Ciba Found. Symp.*, 31, 169, 1975.
4. **Bonting, S. L., Caravaggio, L. L., and Canady, M. R.,** Studies on sodium-potassium-activated adenosine triphosphatase. X. Occurrence in retina rods and relation to rhodopsin, *Exp. Eye Res.*, 3, 47, 1964.
5. **Bownds, D., Brodie, A., Robinson, W. E., Palmer, D., Miller, J., and Shedlvosky, A.,** Physiology and enzymology of frog photoreceptor membranes, *Exp. Eye Res.*, 18, 253, 1974.
6. **Bortoff, A. and Norton, A. L.,** An electrical model of the vertebrate photoreceptor cell, *Vision Res.*, 7, 253, 1967.
7. **Korenbrot, J. I. and Cone, R. A.,** Dark ionic flux and the effects of light in isolated rod outer segments, *J. Gen. Physiol.*, 60, 20, 1972.
8. **Hogan, M. J., Alvarado, J. A., and Weddell, J. E.,** *Histology of Human Eye*, W. B. Saunders, Philadelphia, 1971.
9. **Hagins, W. A.,** The visual process: excitatory mechanisms in the primary receptor cell, *Annu. Rev. Biophys. Bioeng.*, 1, 131, 1972.
10. **Ostwald, T. J. and Heller, J.,** Properties of a magnesium- or calcium-dependent adenosine triphosphatase from frog rod photoreceptor outer segment disks and inhibition by illumination, *Biochemistry*, 11, 4679, 1972.
11. **Berman, A. L., Azimova, A. M., and Gribakin, F. G.,** Localization of Na,K,ATPase and Ca-activated Mg-dependent ATPase in retina rods, *Vision Res.*, 17, 527, 1977.
12. **Goridis, C., Virmaux, N., Urban, P. F., and Mandel, P.,** Guanylate cyclase in mammalian photoreceptor, *FEBS Lett.*, 30, 163, 1973.
13. **Chader, G. F., Fletcher, M., Johnson, M., and Bensinger, R. E.,** Rod outer segment phosphodiesterase factors affecting the hydrolysis of cyclic AMP and GMP, *Exp. Eye Res.*, 18, 509, 1974.
14. **Chader, G. F., Fletcher, M., Johnson, M., and Bensinger, R. E.,** Cyclic nucleotide phosphodiesterase of the bovine retina: activity, subcellular distribution and kinetic parameters, *J. Neurochem.*, 22, 93, 1974.
15. **Goridis, C. and Virmaux, N.,** Light regulated guanosine 3′,5′-monophosphate phosphodiesterase of bovine retina, *Nature*, 248, 57, 1974.
16. **Pannbacker, R. G., Fleischman, D. E., and Reed, D. W.,** Cyclic nucleotide diesterase: high levels in a mammalian photoreceptor, *Science*, 175, 757, 1972.
17. **Wheeler, G. L. and Bitensky, M. W.,** A light-activated GTPase in vertebrate photoreceptors. Regulation of light-activated cyclic GMP phosphodiesterase, *Proc. Natl. Acad. Sci. U.S.A.*, 74, 4238, 1977.
18. **Glynn, I. M.,** The action of cardiac glycosides on ion movement, *Pharmacol. Rev.*, 16, 381, 1964.

19. **Stirling, C. E. and Lee, A.,** (3H) Ouabain autoradiography of frog retina, *J. Cell Biol.,* 85, 313, 1980.
20. **Ernst, S. A.,** Transport adenosine triphosphatase cytochemistry. II. Cytochemical localization of ouabain-sensitive, potassium-dependent phosphatase activity in the secretory epithelium of the avian salt gland, *J. Histochem. Cytochem.,* 20, 23, 1972.
21. **Mayahara, H., Fujimoto, K., Ando, T., and Ogawa, K.,** A new one step method for the cytochemical localization of ouabain-sensitive, potassium-dependent p-nitrophenylphosphatase activity, *Histochemistry,* 67, 125, 1980.
22. **Ueno, S., Mayahara, H., Tsukahara, T., and Ogawa, K.,** Ultracytochemical localization of ouabain-sensitive, potassium-dependent p-nitrophenylphosphatase activity in the guinea pig retina. I. Photoreceptor cells, *Acta Histochem. Cytochem.,* 13, 679, 1980.
23. **Ueno, S., Mayahara, H., Tsukahara, I., and Ogawa, K.,** Ultracytochemical localization of ouabain-sensitive, potassium-dependent p-nitrophenylphosphatase activity in the guinea pig retina. II. Neurons and Muller cells, *Acta Histochem. Cytochem.,* 14, 186, 1981.
24. **Ueno, S., Bambauer, H. J., Umar, H., Ueck, M., and Ogawa, K.,** Ultracytochemical study of Ca-ATPase and K-NPPase activities in retinal photoreceptors of the guinea pig, *Cell Tissue Res.,* 237, 479, 1984.
25. **Ueno, S., Umar, H., Bambauer, H. J., and Ueck, M.,** Localization of ATPases in retinal receptor cell, *Ophthalmic Res.,* 16, 15, 1984.
26. **Oakley, B., II and Green, D. G.,** Correlation of light-induced changes in retinal extracellular potassium concentration with c wave of the electroretinogram, *J. Neurophysiol.,* 39, 1117, 1976.
27. **Oakley, B., II and Steinberg, R. H.,** Effects of maintained illumination upon K in the subretinal space of the frog retina, *Vision Res.,* 22, 767, 1982.
28. **Schatzman, H. J. and Vincenzi, F. F.,** Calcium movements across the membrane of human red cells, *J. Physiol.,* 201, 551, 1966.
29. **Rasmussen, H. and Goodman, D. B. P.,** Relationships between calcium and cyclic nucleotides in cell activation, *Physiol. Rev.,* 57, 421, 1977.
30. **Padykula, H. A. and Herman, E.,** Factors affecting the activity of adenosine triphosphatase as measured by histochemical techniques, *J. Histochem. Cytochem.,* 3, 161, 1955.
31. **Guth, L. and Samaha, F. J.,** Procedure for the histochemical demonstration of actomyosin ATPase, *Exp. Neurol.,* 28, 365, 1970.
32. **Tice, L. W. and Barnett, R. J.,** The adenosinetriphosphatase of striated muscle, *J. Histochem. Cytochem.,* 8, 352, 1960.
33. **Mizutani, A. and Barnett, R. J.,** Fine structural demonstration of phosphatase activity at pH9, *Nature (London),* 206, 1001, 1965.
34. **Samaha, F. J. and Yunis, E. J.,** Quantitative and histochemical demonstration of a calcium activated mitochondrial ATPase in skeletal muscle, *Exp. Neurol.,* 41, 431, 1973.
35. **Ando, T., Fujimoto, K., Mayahara, H., Miyajima, H., and Ogawa, K.,** A new one-step method for the histochemistry and cytochemistry of Ca-ATPase activity, *Acta Histochem. Cytochem.,* 14, 705, 1981.
36. **Borle, A. B.,** Calcium metabolism at the cellular level, *Fed. Proc.,* 32, 1944, 1973.
37. **Fujimoto, K., Toibana, M., and Ogawa, K.,** New cytochemical method for adenylate cyclase and guanylate cyclase with dimethyl sulfoxide, *Acta Histochem. Cytochem.,* 14, 687, 1981.
38. **Toibana, M., Tsukahara, I., and Ogawa, K.,** Cytochemical demonstration of guanylate cyclase activity in retinal photoreceptors with special reference to changes under light and dark adaptation, *Acta Histochem. Cytochem.,* [illegible]
39. **Robb, R. M.,** Histochemical evidence of cyclic nucleotide phosphodiesterase in photoreceptor outer segments, *Invest. Ophthalmol. Vis. Sci.,* 13, 740, 1974.
40. **Robb, R. M.,** Histochemical demonstration of cyclic guanosine 3′,5′-monophosphate phosphodiesterase activity in retinal photoreceptor outer segments, *Invest. Ophthalmol. Vis. Sci.,* 17, 476, 1977.
41. **Shanta, T. R., Woods, W. D., Waitzman, M. B., and Bourne, G. H.,** Histochemical method for localization of cyclic 3′,5′-nucleotide phosphodiesterase, *Histochemie,* 7, 177, 1966.
42. **Ueno, S., Bambauer, H. J., Umar, H., and Ueck, M.,** A new histo- and cytochemical method for demonstration of cyclic 3′,5′-nucleotide phosphodiesterase activity in retinal rod photoreceptor cells of the rat, *Histochemistry,* 31, 445, 1984.
43. **Ueno, S., Bambauer, H. J., Umar, H., and Ueck, M.,** Localization and function of cyclic guanosine monophosphate-phosphodiesterase activity in the retinal rods of the rat by means of a newly developed cytochemical method, *Cell Tissue Res.,* 238, 453, 1984.
44. **Ueno, S. and Ueck, M.,** Localization of cyclase and cyclic nucleotide phosphodiesterase activities in retinal pigment epithelial cells of the guinea pig, *Acta Histochem. Cytochem.,* 21, 45, 1988.

A. INTRODUCTION

The gingival epithelium is considered to consist of junctional epithelium (JE), oral sulcural epithelium (OSE), and oral epithelium (Figures 1 and 2). The morphological characteristics of the JE, OSE, and oral epithelium in the rat resemble those of man, although certain differences exist in the degree of keratinization of the OSE and oral epithelium.[1] The rat is therefore suitable for the morphological study of the gingiva (junctional epithelium), and is the subject of the present report.

Of the gingival epithelium, much attention has been focused especially on the JE, because JE is permeable for bacterial and cytotic metabolic products. These substances can pass from the gingival sulcus into the connective tissue, and from the connective tissue into the sulcus through the permeable JE (see Figure 11). Therefore, JE is an important pathogenetic site in periodontal diseases.

B. MORPHOLOGICAL CHARACTERISTICS OF THE JE AND ITS RELATED EPITHELIUM

1. General Structure

As described in the excellent handbook of Schroeder,[2] prior to tooth eruption, the enamel surface is lined by reduced enamel epithelium consisting of mainly reduced ameloblasts; this epithelium is called the primary epithelial attachment. As the teeth erupt, this epithelium gradually changes into a nonkeratinizing squamous epithelium, termed the JE. The junction between the enamel surface and the JE is called the secondary epithelial attachment. The JE is thus structurally an epithelial collar that provides epithelial attachment to the enamel surface and is located between the enamel and OSE (Figure 2). JE forms the base of the gingival sulcus and terminates at the cemento-enamel junction (Figure 1), while the OSE forms the lateral wall of the gingival sulcus. Histologically, the JE consists of several layers of flattened epithelial cells which, except for basal cells, are oriented with the long axis of the cells parallel to the enamel surface. The cell layers of the JE are continuous with the OSE (see Figure 4). Ultrastructurally, the junctional epithelial cells exhibit flattened nuclei, numerous cytoplasmic processes, and a small number of rough endoplasmic reticulum (ER), lipid droplets, tonofilaments, and Golgi complexes in the cytoplasm. Intercellular spaces are much more prominent than those of the OSE and oral epithelium and correspond to almost 20% of the tissue volume (see Figure 17). Numerous leukocytes are observed in these intercellular spaces, while the narrow spaces of the OSE and oral epithelium contain no leukocytes (see Figures 4 and 5).

The OSE is thick and keratinized and lines the lateral wall of the gingival sulcus. It is continuous with the JE and also with the oral epithelium at the gingival margin (the most coronal point of the gingiva, Figure 1).

The oral epithelium is also thick and keratinized, morphologically resembling the OSE. It extends from the gingival margin to the mucogingival junction (Figures 1 and 2).

2. Epithelial Attachment of the JE

The most superficial layer of the JE is attached to the enamel surface by an internal basement lamina (IBL) and hemidesmosomes (Figure 3); the basal cell layer of the JE is attached to the adjacent gingival connective tissue by an external basement lamina (EBL) and hemidesmosomes. As termed by Schroeder,[2] the external and internal basement laminae are basement laminae which are located between the JE and the adjacent gingival connective tissue or the enamel surface, respectively. The EBL has anchoring fibrils of connective tissue origin; it can be divided into two parts, consisting of a basement lamina lucida and a basement lamina densa (see Figures 12 and 13). The IBL, however, lacks anchoring fibrils and cannot be clearly divided into these two components.

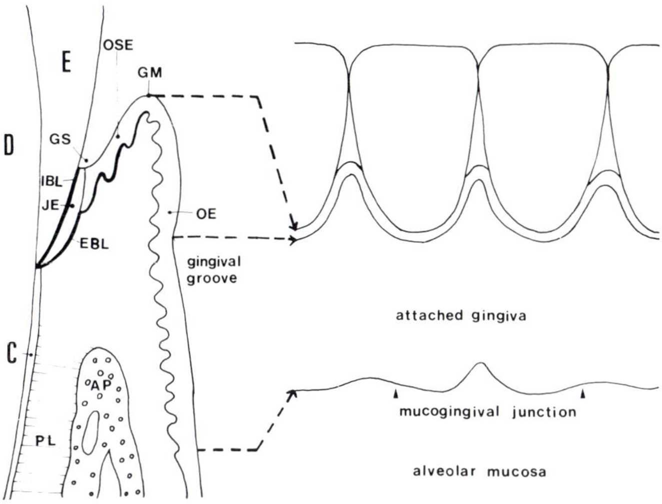

FIGURE 1. A schematic drawing of gingiva, based on the original drawing of Schroeder;[2] AP, alveolar process; D, dentin; E, enamel; GM, gingival margin; GS, gingival sulcus; JE, junctional epithelium; OE, oral epithelium; OSE, oral sulcural epithelium.

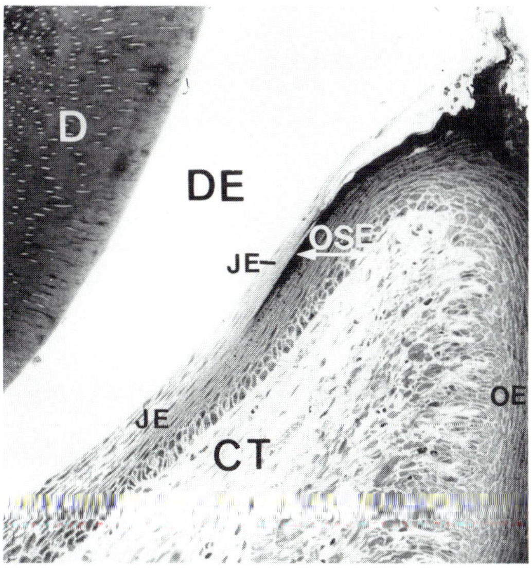

FIGURE 2. Rat gingiva; CT, connective tissue; D, dentin; DE, decalcified enamel; JE, junctional epithelium; OE, oral epithelium; OSE, oral sulcural epithelium. Toluidine blue staining. (Magnification × 150.)

3. Intracytoplasmic Vacuoles in the JE Cells

Numerous intracytoplasmic vacuoles are observed as the most characteristic organelle in the JE cells. However, they are rarely seen in the cell layer juxtaposed to the enamel surface or in the basal cells of the JE. They are generally round, surrounded by a unit membrane, and are 0.2 to 0.5 μm in diameter.[3] As the JE cells become oriented in a coronal direction, the number and size of their intracytoplasmic vacuoles increases. Finally, these intracytoplasmic vacuoles are more frequently seen in the coronal portion of the JE than in the middle and apical portions (see Figure 17).

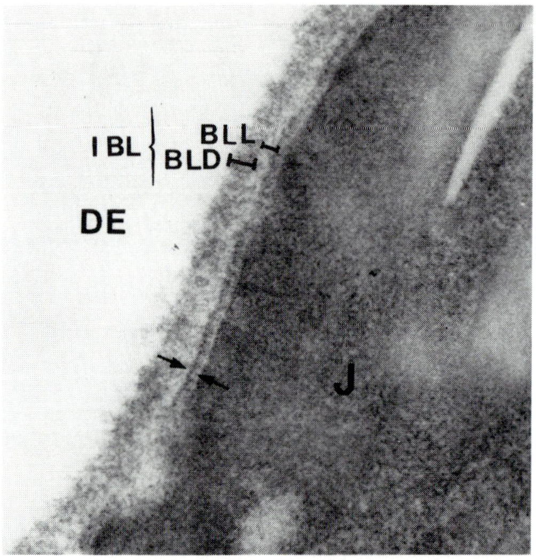

FIGURE 3. Electron micrograph of part of a JE cell (J). Internal basement lamina (IBL) consists of the basement lamina densa (BLD) and lucida (BLL). Arrows, hemidesmosome; DE, decalcified enamel. Uranyl acetate staining. (Magnification × 50,000.)

4. Intercellular Junctions Between the JE Cells

Using dog gingivae, Saito et al.[4] observed desmosomes, gap junctions, and tight junctions between the cells of the coronal and apical portions of the JE, but they occur less frequently here than in the other oral epithelium. Gap junctions are also detected between JE cells. They are usually small (0.2 to 0.7 μm in diameter) but are relatively large in the coronal portion of the JE. Tight junctions are observed only at the coronal portion of the JE and comprise maculae occludentes as leaky cell junctions rather than zonulae occludentes as tight cell junctions. It is therefore unlikely that tight junctions contribute to the barrier function of the JE.

5. Various Leukocytes Detected in the JE (Figures 4 and 5)

Various leukocytes migrate from the dento-gingival vessels into the JE and gingival sulcus. This persistent presence and continuous migration of leukocytes are ascribed to chemotactic substances from a variety of sources, e.g., bacterial products and complement such as C5a. These leukocytes migrate in an outward flow of fluid, termed ''gingival fluid'', which arises from the gingival plexus and is basically a modified and selectively diluted serum exudate. Using a smear technique,[5] gingival fluid from healthy gingivae is shown to contain few leukocytes, but neutrophils (95 to 97% of total leukocytes), lymphocytes (1 to 2%), and monocytes (2 to 3%) are regularly detected. In inflamed gingivae, leukocytes increase in number with increasing signs of inflammation, but the relative proportion of leukocytes remains the same as in healthy gingivae. In histological sections of slightly inflamed gingivae, leukocytes are numerous in the gingival sulcus but gradually decrease in number towards the apical portion of the JE. The neutrophils reside mainly in the superficial layers of the coronal portion of the JE, while monocyte or macrophages and lymphocytes are located in the basal portion.[6]

6. Nerve Fibers (Varicosities) in the JE

Byers et al.[7,8] have closely examined intraepithelial innervation in the JE. Trigeminal nerves have been demonstrated by autoradiography,[7] and numerous nerves with calcitonin

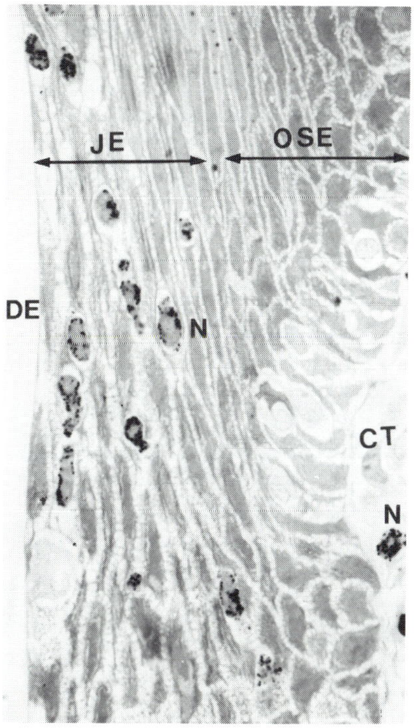

FIGURE 4. Rat gingiva after diaminobenzidine reaction. Endogenous peroxidase-positive leukocytes, probably neutrophils (N), are localized in the JE and in the connective tissue (CT); DE, decalcified enamel; OSE, oral sulcural epithelium. No counterstaining. (Magnification × 800.)

gene-related peptide (CGRP)-like immunoreactivity have been detected.[8] Also, substance P (SP)- and CGRP-like immunoreactivity coexist in trigeminal cell bodies, as well as in nerve terminals of the dental pulp.[9] Based on these findings, it is suggested that trigeminal nerve fibers in the JE contain SP.

We recently investigated the sensory innervation of the JE using an immunohistochemical method (avidin-biotin-complex[10] [ABC]) with SP antibodies, or using an anterograde transport method by the injection of wheat germ agglutinin-horseradish peroxidase (WGA-HRP)[11,12] into the trigeminal ganglion. We conclusively showed that the JE was penetrated by numerous HRP positive trigeminal nerve fibers and varicosities (Figures 6 and 7). Nerve fibers and varicosities with SP-like immunoreactivity were also observed (Figures 8 to 10). Characteristically, these HRP- and SP-positive nerve varicosities were closely apposed to JE cells or neutrophils (Figures 6, 7, 9, and 10). However, no specific junctions such as synaptic junctions between nerve varicosities and JE cells or neutrophils were seen; instead, a narrow space of at least 20 nm was evident. Thus, the morphology and distribution of SP-positive nerve fibers resembles those of HRP-positive trigeminal nerve fibers. It is therefore considered that SP-containing trigeminal nerves penetrate into the JE, and that their nerve varicosities closely approximate the JE cells or neutrophils.

C. FUNCTIONS OF JE

1. Permeability of the JE

The JE is highly permeable. JE cells are loosely connected by desmosomes and a small number of gap and tight junctions; the intercellular spaces are wide. Furthermore, in the peripheral cytoplasm of the JE, Saito et al.[4] observed dense granules, probably corresponding

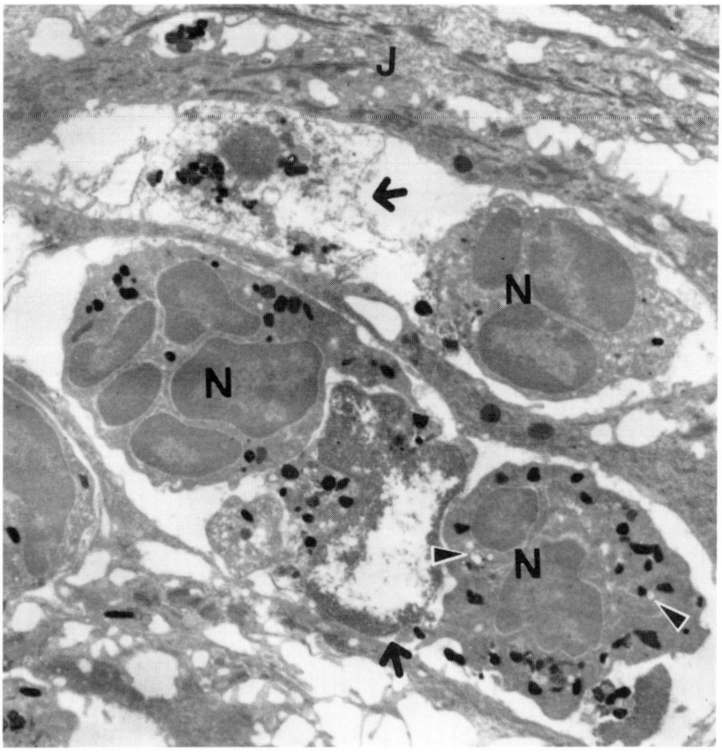

FIGURE 5. Neutrophils (N) in the intercellular spaces of the JE cells (J) after diaminobenzidine reaction. Peroxidase-positive azurophil granules and -negative specific granules (arrowheads); degenerating neutrophils (arrows). No counterstaining. (Magnification × 15,000.)

to membrane-coating granules (Odland bodies) which act as a permeability barrier blocking the penetration of substances there. However, they report that these granules do not function as a permeability barrier; JE therefore permits the inward passage of bacteria through its intercellular spaces and also the outward passage of immunoglobulins, complement, and leukocytes. Despite their high molecular weight, immunoglobulins and complement seem to easily exudate from the gingival blood vessels. To investigate the passage route of these substances, we have used tracers including microperoxidase[13] (MP, low molecular weight, mol wt, 1800); HRP[14] (mol wt, 40,000); and cationized ferritin[15] (CF, high molecular weight, mol wt, 800,000). The injected tracers were transuded from gingival capillaries, often located near the JE (Figure 11). Although MP, HRP, and CF were partially blocked by the basement lamina densa of the EBL, most penetrated into the JE through the basement lamina densa, and also the cell junctions of basal cells of the JE (Figures 12 and 13). Furthermore, they reached the gingival sulcus through intercellular spaces of the JE and along the IBL (Figures 14 and 15). Only the basement lamina densa of the EBL functioned as a rough permeability barrier, and most cell junctions of the JE were relatively permeable. Therefore, even high-molecular-weight (800,000) substances readily pass from the gingival capillaries to the gingival sulcus. In addition, EBL and IBL function as attachments which connect the JE to the adjacent connective tissue and enamel surface, respectively, rather than permeability barriers. Thus, high permeability of the JE permits easy movement of immunoglobulins and complement as a serum exudate from the gingival capillaries; these substances form a defense line against bacterial attack.

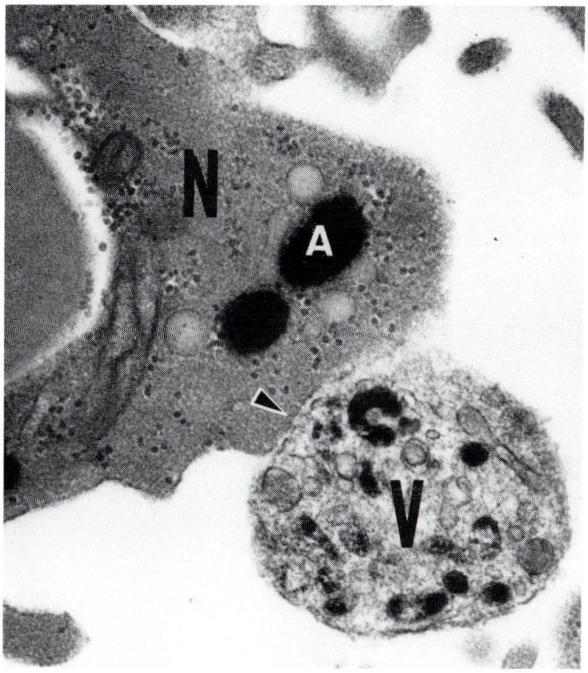

FIGURE 6. Electron micrograph of axons in the JE from rat gingivae using an anterograde axonal transport method by injection of WGA-HRP into the trigeminal ganglion. There is a narrow space (arrowhead) between an axonal varicosity (V) with HRP-labeled organelles and neutrophil (N) with endogenous peroxidase-positive azurophil granules (A). (Magnification × 50,000.)

2. Function of Neutrophils in the JE

A constant stream of migrating leukocytes and protein substances (immunoglobulins and complement) was described above. Neutrophils are the predominant leukocytes. They contain specific granules and azurophil granules and the latter contains myeloperoxidase, which reacts in the diaminobenzidine (DAB) reaction (Figure 5). Neutrophils protect the host from bacterial attack by these granular contents.[16] Neutrophils phagocytose bacteria and kill them by discharging granular contents into the bacteria-containing phagosomes. Figure 16 shows this phenomenon in the gingival sulcus, where phagosomes in the neutrophils took up numerous cationized ferritin particles which were topically applied to the gingival sulcus.[17] The cationized ferritin-containing phagosomes were often fused with the peroxidase-positive azurophil granules; it is possible that these granular contents are also secreted extracellularly. This extracellular secretion has the disadvantage of damaging adjacent gingival tissue, because many of the contents are hydrolytic substances, such as collagenase and elastase. On the other hand, these neutrophilic products may be inactive in coexistence with circulating antiproteases such as α-macroglobulin and α-antitrypsin.[16] This question awaits further investigation.

3. Endocytotic Ability of the JE Cells

JE cells are thought to have an endocytotic ability. The intracytoplasmic vacuoles may be involved in this. Mazzella et al.[18] reported that the vacuoles appear to act as morphological indications of hydropic degeneration of cells. On the other hand, Yamasaki et al.[3] suggested that the vacuoles have an endocytotic nature. In our recent report[14] using HRP as a tracer (Figure 17), it is concluded that, although the vacuoles certainly have endocytotic ability,

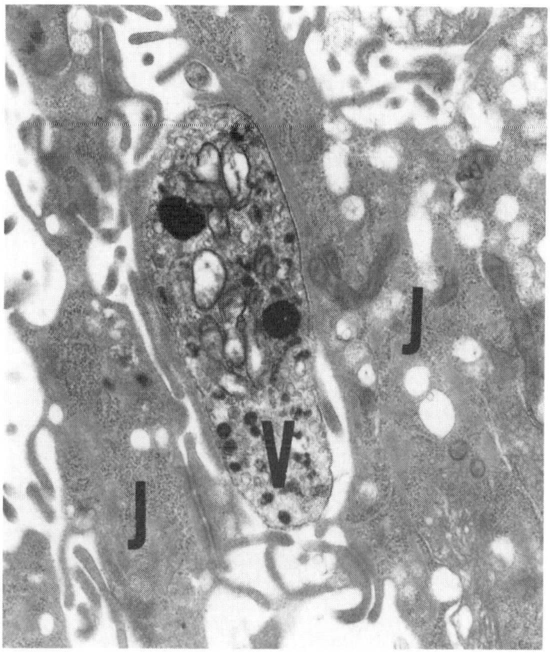

FIGURE 7. Electron micrograph of axons in the JE from rat gingivae, treated as in Figure 6. HRP-labeled organelle-containing varicosity (V) is located very close to the JE cells (J), but no specific junctions such as synaptic junctions are observed. Lead nitrate staining. (Magnification × 20,000.)

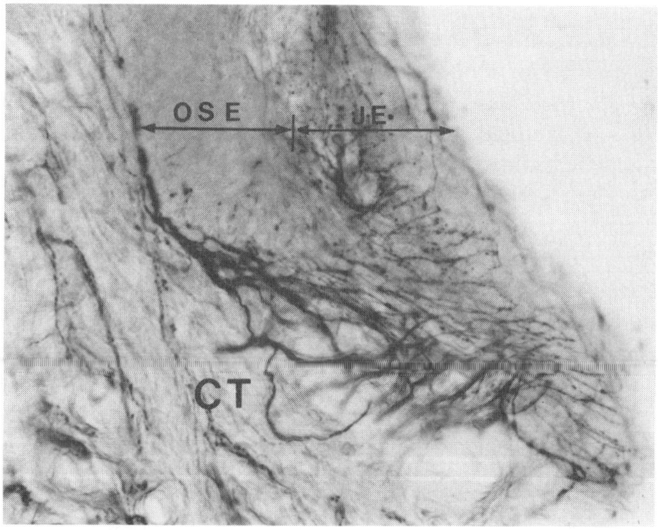

FIGURE 8. Light micrograph of JE from rat gingivae by the ABC method. A network of nerve fibers with SP-like immunoreactivity (SP-LI) is found from the connective tissue (CT) to the JE. No counterstaining. (Magnification × 300.)

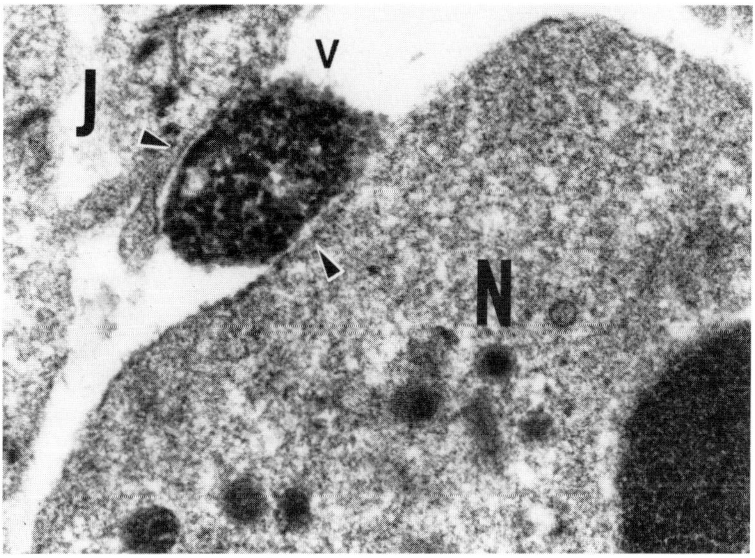

FIGURE 9. Varicosity (V) with granular and vesicular SP-LI is located very closely between neutrophil (N) and JE cell (J), but is separated from them by narrow spaces (arrowheads). Lead nitrate staining. (Magnification × 58,000.)

micropinocytotic vesicles rather than vacuoles play the main role in endocytosis in JE cells. In any case, it is certain that the JE has endocytotic ability. However, JE cells do not easily phagocytose large entities such as bacteria, therefore neutrophils, rather than JE cells, appear to function as the main protector.

4. Functions of Trigeminal Nerve Fibers (Terminals) in the JE

As described above, SP-containing trigeminal nerve fibers (terminals) seem to penetrate the JE and approximate the JE cells or neutrophils. The function of these nerve fibers must be considered. It is generally accepted that SP is present in primary afferent neurons, plays a role as a neurotransmitter, and is possibly involved in pain transmission. Furthermore, the N-terminal tetrapeptide of SP seems to enhance the endocytotic ability of human neutrophils and mouse macrophages.[19] In addition, it is possible that SP is released from nerves by antidromic stimulation.[20] SP may therefore be released also from nerve varicosities in the JE, and enhances the endocytotic ability of the JE cells. These cells are located close to SP-containing nerve varicosities and have the same endocytotic nature as neutrophils and macrophages. Of course, neutrophil function in the JE may be provoked by SP in the same way.

In summary, JE has a permeable nature, attacking bacteria from without inward by the transepithelial passage of substances such as immunoglobulins from within outward. We suggest that the endocytotic ability of JE cells and neutrophils, and the innervation of the JE, probably form a line of defense against bacteria.

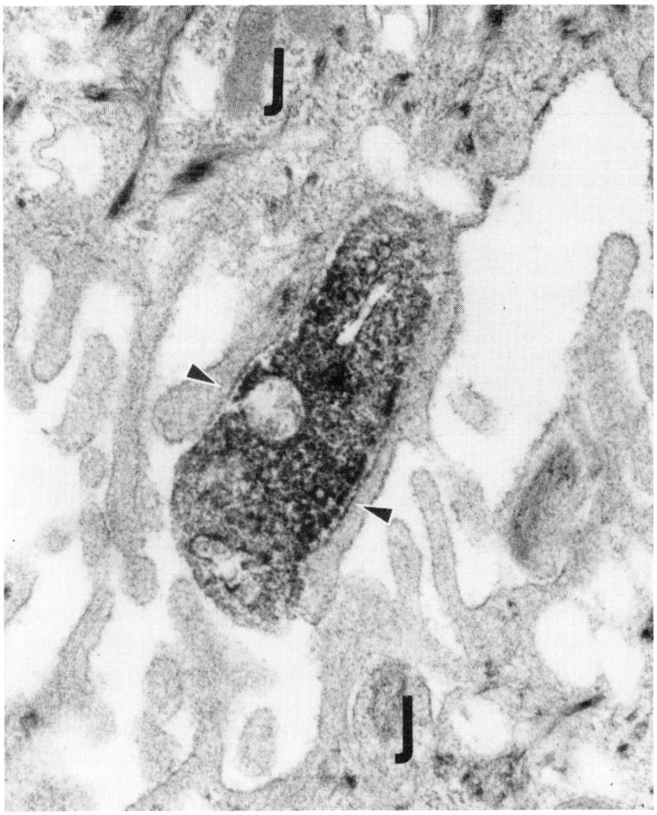

FIGURE 10. A granular and vesicular SP-LI-containing varicosity is surrounded by cytoplasmic processes of JE cells (J) but is separated from them by narrow spaces (arrowheads). Lead nitrate staining. (Magnification × 50,000.)

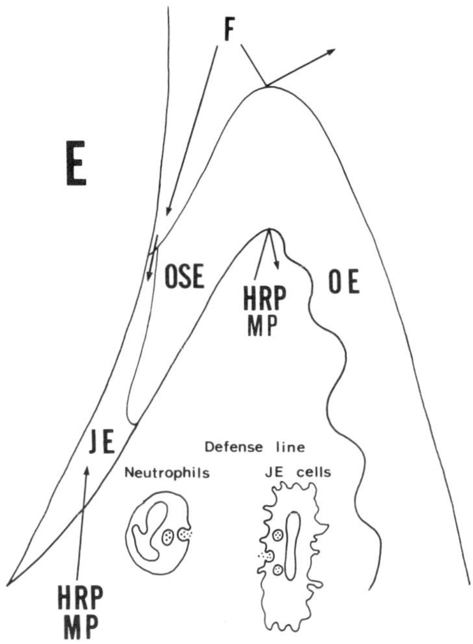

FIGURE 11. JE is permeable and permits passage of foreign substances (F) from without inward, and tracers (HRP, MP, or CF) from within outward. Oral epithelium (OE) and oral sulcural epithelium (OSE) are impermeable in both directions (angled arrows). Endocytosis of substances by JE cells and neutrophils in the JE forms a defense line.

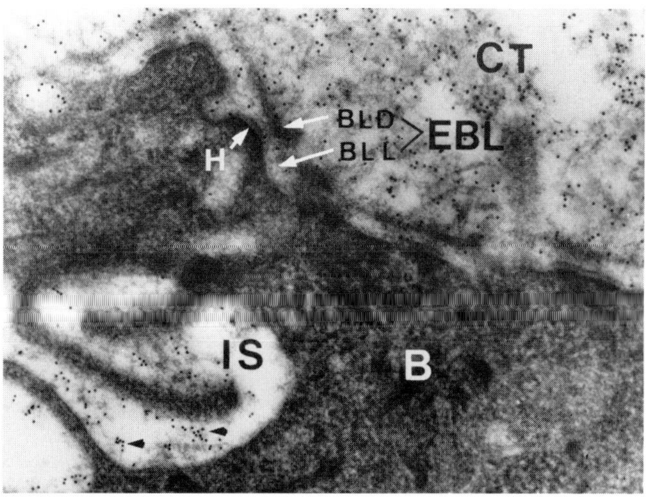

FIGURE 12. Electron micrograph of parts of basal cells (B) close to external basement lamina (EBL) from gingivae of rats injected with cationized ferritin. Tracers (arrowheads) pass from EBL and cell junctions (CJ) of basal cells into the intercellular spaces (IS). Some of the tracers are stopped by BLD; CT, gingival connective tissue; H, hemidesmosomes. Uranyl acetate staining. (Magnification × 50,000.)

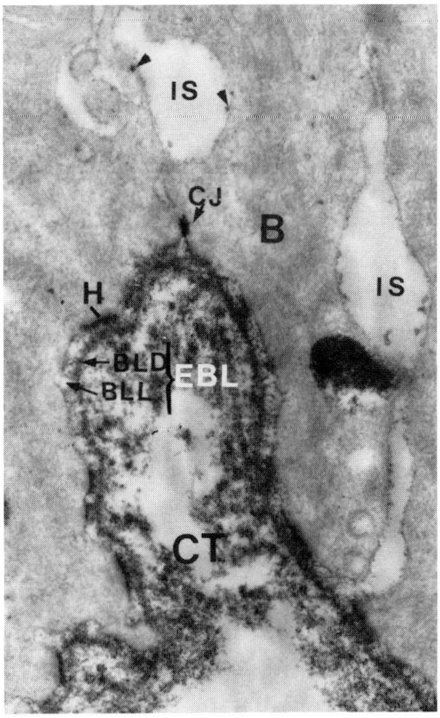

FIGURE 13. As for Figure 12, rats injected with microperoxidase. No counterstaining. (Magnification × 40,000.)

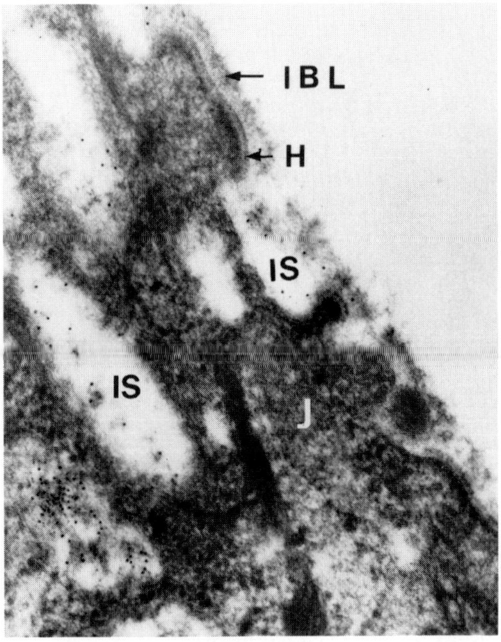

FIGURE 14. Electron micrograph of parts of JE cells close to IBL from gingivae of rats injected with cationized ferritin. Ferritin passes from intercellular spaces (IS) of JE cells (J) to IBL. DE, decalcified enamel; H, hemidesmosome. No counterstaining. (Magnification × 18,000.)

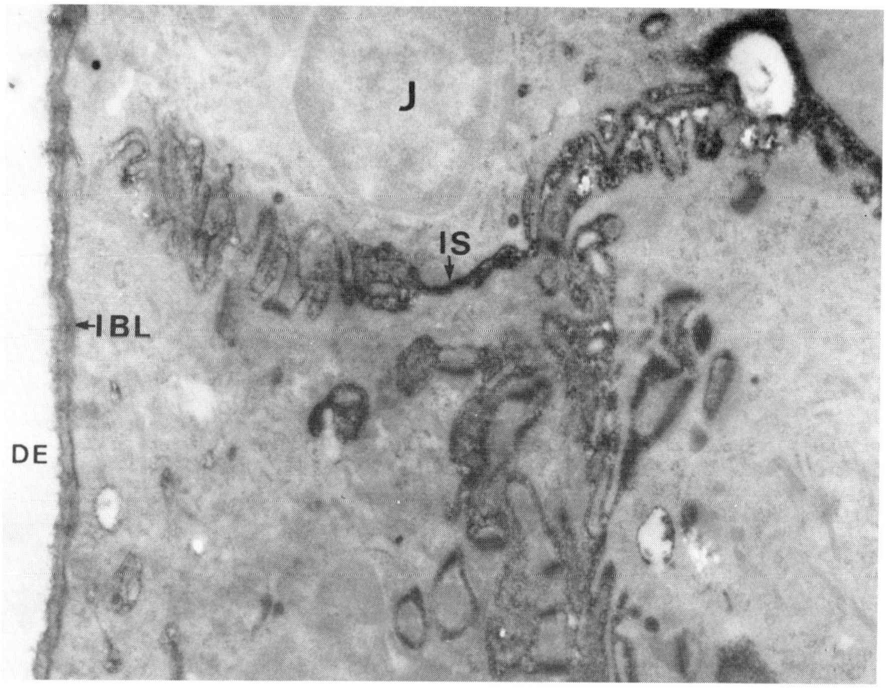

FIGURE 15. As for Figure 14, rats injected with microperoxidase. Microperoxidase passes from IS to IBL. No counterstaining. (Magnification × 18,000.)

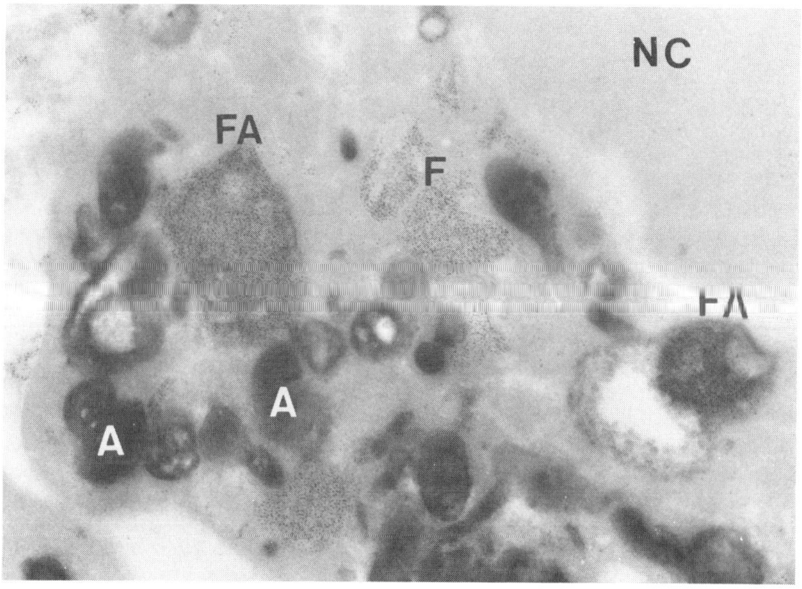

FIGURE 16. Part of a neutrophil in the gingival sulcus treated topically with hyaluronidase and cationized ferritin (F) and then reacted with diaminobenzidine. The neutrophil contains peroxidase-positive azurophil granules (A) and takes up numerous ferritin particles in the phagosomes (F). Ferritin-containing phagosomes are fused with peroxidase-positive azurophil granules (FA); NC, nucleus. No counterstaining. (Magnification × 47,000.)

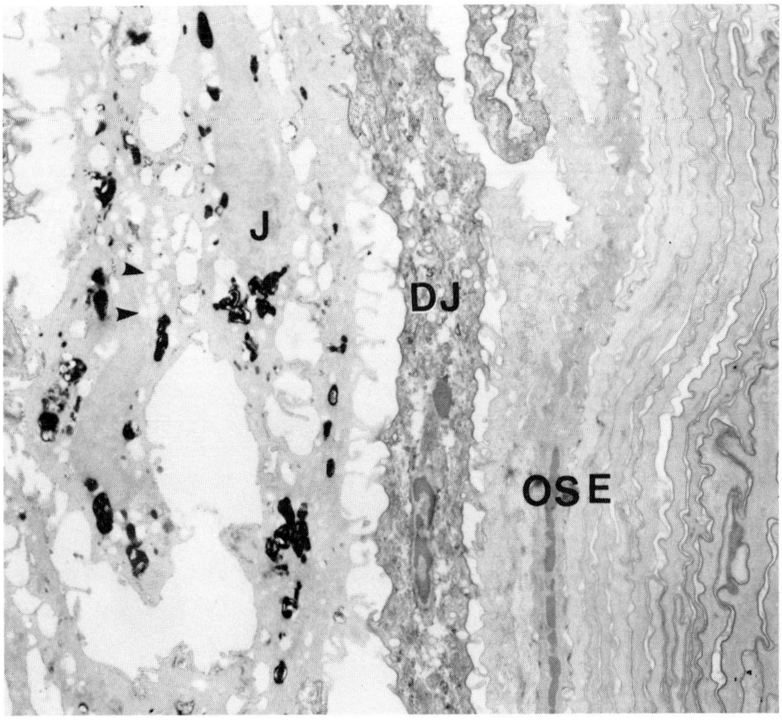

FIGURE 17. Electron micrograph of part of the coronal portion of the JE from the gingiva of a rat injected with HRP. Numerous HRP-positive organelles and HRP-negative vacuoles (arrowheads) are observed in the JE cells (J). Degenerative JE cells (DJ) labeled with HRP. Not counterstaining. (Magnification × 4500.)

REFERENCES

1. **Listgarten, M. A.,** Similarity of epithelial relationships in the gingiva of rat and man, *J. Periodontol.,* 46, 677, 1975.
2. **Schroeder, H. E.,** The periodontium, in *Handbook of Microscopic Anatomy,* Vol. V/5, Schroeder, H. E., Ed., Springer-Verlag, Berlin, 1986.
3. **Yamasaki, A., Nikai, H., Niitani, K., and Ijuhin, N.,** Ultrastructure of the junctional epithelium of germfree rat gingiva, *J. Periodontol.,* 50(12), 641, 1979.
4. **Saito, I., Watanabe, O., Kawahara, H., Igarashi, Y., Yamamura, T., and Shimono, M.,** Intercellular junctions and the permeability barrier in the junctional epithelium, *J. Periodont. Res.,* 16, 467, 1981.
5. **Attstrom, R.,** Presence of leukocytes in crevices of healthy and chronically inflamed gingivae, *J. Periodont. Res.,* 5, 42, 1970.
6. **Schroeder, H. E.,** Transmigration and infiltration of leukocytes in human junctional epithelium, *Helv. Odontol. Acta,* 17, 6, 1973.
7. **Byers, M. R. and Holland, G. R.,** Trigeminal nerve endings in gingiva, junctional epithelium and periodontal ligament of rat molars as demonstrated by autoradiography, *Anat. Rec.,* 188, 509, 1977.
8. **Byers, M. R., Mecifi, K. B., and Kimberly, C. L.,** Numerous nerves with calcitonin gene-related peptide-like immunoreactivity innervate junctional epithelium of rats, *Brain Res.,* 419, 311, 1987.
9. **Gazelius, B., Edwall, B., Olgart, L., Lundberg, J. M., Hokfelt, T., and Fischer, J. A.,** Vasodilatory effects and coexistence of calcitonin gene-related peptide (CGRP) and substance P in sensory nerves of cat dental pulp, *Acta Physiol. Scand.,* 130, 33, 1987.

10. **Hsu, S. M., Baine, L., and Fanger, H.,** Use of avidin-biotin-peroxidase complex (ABC) in immunoperoxidase techniques: a comparison between ABC and unlabeled antibody (PAP) procedures, *J. Histochem. Cytochem.,* 29, 577, 1981.
11. **Kido, M. A., Kondo, T., Ayasaka, N., Terada, Y., and Tanaka, T.,** The peripheral distribution of trigeminal nerve fibers in the rat temporomandibular joint using an anterograde axonal transport method of wheat germ agglutinin-horseradish peroxidase, *Arch. Oral Biol.,* 36, 397, 1991.
12. **Kondo, T., Ayasaka, N., Nagata, E., and Tanaka, T.,** A light and electron microscopic anterograde WGA-HRP tracing study on the sensory innervation of junctional and sulcular epithelium in the rat molar, *J. Dent. Res.,* 71, 60, 1992.
13. **Tanaka, T.,** Transport pathway and uptake of microperoxidase in the junctional epithelium of healthy rat gingiva, *J. Periodont. Res.,* 19, 26, 1984.
14. **Ayasaka, N. and Tanaka, T.,** A cytochemical study of horseradish peroxidase uptake in rat junctional epithelium, *J. Dent. Res.,* 68, 1503, 1989.
15. **Tanaka, T. and Sakano, A.,** Ultrastructural localization and passage of cationized-ferritin and microperoxidase as tracers in the outward flow of rat gingival crevicular fluid, *J. Periodont. Res.,* 22, 482, 1987.
16. **Van Dyke, T. E., Levine, M. J., and Genco, R. J.,** Neutrophil function and oral disease, *J. Oral Pathol.,* 14, 95, 1985.
17. **Tanaka, T., Ayasaka, N., and Sakano, A.,** An in vivo study of degranulation of azurophil granules in the neutrophils during phagocytosis of cationized ferritin in the gingival sulcus, *Acta Histochem. Cytochem.,* 21, 15, 1988.
18. **Mazzella, W. G. and Vernick, S. H.,** The ultrastructure of normal and pathological gingival epithelium, *J. Periodontol.,* 39, 5, 1968.
19. **Bar-Shavit, Z., Goldman, R., Stabinsky, Y., Gottlieb, M., Fridkin, M., Teichberg, V. I., and Blumberg, S.,** Enhancement of phagocytosis — a newly found activity of substance P residing in its N-terminal tetrapeptide sequence, *Biochem. Biophys. Res. Commun.,* 94, 1445, 1980.
20. **Lembeck, L. and Holzer, P.,** Substance P as neurogenic mediator of antidromic vasodilation and neurogenic plasma extravasation, *Arch. Pharmacol.,* 310, 175, 1979.

Chapter 19

CURRENT IMMUNOMORPHOLOGIC PRACTICES IN TUMOR DIAGNOSIS. A COMPREHENSIVE APPRAISAL WITH EMPHASIS ON IMMUNOLABELING AT THE LIGHT AND ULTRASTRUCTURAL LEVELS

Guillermo A. Herrera, Elba A. Turbat-Herrera, and Denise Bradley

TABLE OF CONTENTS

0-8493-6012-9/93/$0.00 + $.50

A. INTRODUCTION

Tumor evaluation probably represents the most important element in the practice of surgical and cytopathology. Clinical information, pertinent laboratory data, and gross tumor characteristics complement histologic assessment. Morphologic evaluation has been supplemented with immunologic techniques to create an immunomorphologic approach to the study and understanding of neoplasia. Hematoxylin- and eosin-stained sections remain the most important information source from which the initial diagnostic considerations are formulated. In the majority of cases in a surgical pathology practice, these routine hematoxylin- and eosin-stained section, in a few instances aided by simple histochemical stains, suffice in providing the correct diagnosis. In difficult cases other, more sophisticated, diagnostic techniques may play a key role in supplying additional information. These modalities include immunocytochemical and ultrastructural examination. While examination of the routine histological sections or cytology preparations usually permits establishing the neoplastic character of a given pathological process and definition of its benign or malignant potential, the two latter techniques aim at determination of histogenesis or differentiation in a given tumor. The amalgamation of these two specialized techniques — ultrastructural immunolabeling — has recently become an important addition to immunomorphologic evaluation. Each diagnostic technique has its innate advantages and disadvantages. It behooves the practicing pathologist to understand their specific diagnostic roles, including limitations, in order to utilize them accordingly. This chapter will deal primarily with the applications of immunolabeling techniques at the light and electron microscopic levels, emphasizing their potential in the understanding and assessment of tumor differentiation. The role of labeling techniques in the diagnostic surgical and cytopathology laboratory will be considered in detail in the context of the overall evaluation of neoplasms. Tumor speciation is becoming more critical in view of therapeutic protocols now available aiming at specific diagnostic nosologic entities. ''Lumping'' of diagnostic entities is not desirable and complete characterization of tumors is expected to be an integral part of the practice of anatomic pathology. As Dr. William L. Watson pointed out 30 years ago in reference to the practice of incomplete tumor characterization:[1] ''All that such false simplification does is to destroy information. Since the groupings are arbitrary, they are unpredictable, and rarely is it possible to compare series. More importantly, faced with ambiguous pathological diagnoses, clinicians cannot learn what course to expect in any given patient and what treatment plan to follow.''As new trainees in pathology, surgery, and oncology move into their respective practices, often outside academic centers, practicing pathologists are encouraged to update their laboratories by adding the new available techniques to their diagnostic armamentarium or to refer cases to more sophisticated centers with capabilities for more complete tumor speciation.

B. SURGICAL PATHOLOGY: HISTORICAL AND GENERAL PERSPECTIVES

Esmond R. Long said in 1928:[2] ''Nothing gives a better perspective of the subject than an appreciation for the steps by which it has reached its present state.'' This deeply revealing thought motivates the authors to briefly summarize the development of surgical pathology as it pertains to tumor diagnosis, in order to better understand the developments in the field.

The specialty of surgical pathology was born ''out of wedlock'' and out of necessity.[3] Rudolph Virchow (1821–1902), the recognized father of pathology, failed to acknowledge the subspecialty. Microscopic examination of tissue specimens was first used to diagnose cancer around the middle of the 19th century. Johannes Muller's fundamental work in 1838 on the microscopic appearance of malignant tumors provided a start for further study and awareness of the importance of microscopic evaluation.[4] Adolph Hannover (1814–1894) insisted in 1843 that there were specific characteristics of the tumor cells which permitted a diagnosis of cancer with certainty.[5] Hermann Lebert (1813–1878) also claimed that he had found the specific characteristics of cancer cells which could be identified microscopically and be used to establish a definitive diagnosis of malignancy.[6,7] This insistence on the presence of a typical malignant-appearing cell was vehemently opposed by a number of surgeons, led by Velpeau (1795–1867). In his work on breast diseases published in 1854, he relegated the histological evidence to a secondary place and insisted on clinical judgment as being much superior to that of the pathologist examining the specimens. He wrote:[8] "The intervention of the microscope is not at all necessary, to decide whether such and such a tumor, which has been removed, is or is not of cancerous nature. If we put innocent tumors on one side, and malignant on the other, I do not hesitate to affirm that, by the aid of clinical knowledge . . . an experienced practitioner will without trouble distinguish a cancerous tumor from one that is not so." At the turn of the century, surgeons were forced to become knowledgeable in the gross and microscopic examination of tumors to better understand their nature and to detect characteristics that would be helpful in guiding surgical management and, perhaps, predicting tumor behavior. Surgical pathology was initially performed by surgeons on a part time assignment. As late as 1940, the Department of Surgery at the University of Chicago maintained control of the surgical pathology laboratory functions. In 1930, Dr. Joseph A. Blake established the first surgical pathology laboratory at the College of Physicians and Surgeons. The surgical pathology and experimental surgery laboratories were fused. A formal course combining technical surgical aspects and pathology was initiated soon after at the same institution by Dr. William C. Clarke.[9] Understanding of neoplastic diseases from a surgical pathology perspective was very simplistic at the time. The majority of the discussions related to neoplastic processes were often based on philosophical speculation. Macroscopic, visual evaluation of specimens often took precedence over the microscopic findings in providing the information used to compile a diagnosis. Arthur Purdy Stout (1885–1967), initially a full time surgeon, and his pupils established surgical pathology as a separate subspecialty.[10] Stout accepted a dual appointment of instructor in surgery and assistant in surgical pathology at the Presbyterian Hospital associated with the College of Physicians and Surgeons. He helped Clarke with the reporting of surgical specimens.

Dr. Stout's approach to surgical pathology included not only identifying the nature of the neoplastic lesions and determining adequacy of margins, but also emphasized the understanding of the metastasizing potential of malignant tumors; he was extremely interested in relating his surgical pathology findings to the eventual outcome of patients. He also established criteria of operability of tumors, as well as sensitivity to radiation therapy. Some of the initial concepts of neoplastic diseases were collected by James Ewing (1866–1943) in his book *Neoplastic Diseases* published in 1919.[11] The emphasis in the book was on histopathologic findings and the treatise served as a comprehensive recollection of the known morphologic aspects of tumors in various locations. The discovery of formalin as a fixative was also an important event. Fixation of specimens in formaldehyde permitted their preservation for evaluation and reconsideration at a later time. It also allowed the use of preserved samples for teaching purposes. Formaldehyde has remained the fixative of choice and is still utilized preferentially for routine preservation of specimens. With the advent of new diagnostic techniques, other fixatives have been developed to address specific technical considerations that benefit from the characteristics of nonformaldehyde-containing fixatives.

Histochemical staining (so-called "special stains") slowly developed and even today plays an important role, albeit limited, in surgical pathology. These "special stains" are relatively quick to perform, inexpensive, suitable for most situations, and relatively easy to master. Hematoxylin- and eosin-staining of selected sections from tumors, together with one or perhaps a few "special stains" suffices for making a diagnosis in a significant number of neoplasms evaluated. However, these special histochemistry techniques simply cannot

contributing immeasurably to the success and scope of radiologic practice. Now I come neither to bury Caesar, nor to praise him. Unhappily, he does not constantly wear 'the white flower of a blameless life'.'' How accurately the pathologist can perform the task depends on the ability and knowledge to use the various techniques available that can aid in arriving at an accurate diagnosis.

This historical perspective hopefully will serve to introduce the scope of this chapter, which will be devoted to exploring the benefits and limitations of the available techniques with an emphasis on the present status of the art, which concentrates its efforts on an immunomorphologic approach, and to place them in proper perspective. Techniques such as flow cytometry, *in situ* hybridization, and other molecular biology methods will only be briefly reviewed at the end of the chapter in an effort to grasp a glimpse as to how they are perceived to interrelate with immunohistochemical methods. As will be apparent, the method of detection of the expression of a variety of markers by the utilization of molecular probes in some instances may be an immunocytochemical procedure, thus linking intimately the molecular biology revelations with immunohistochemistry.

C. CYTOPATHOLOGY: GENERAL PERSPECTIVES

Cytological examination of specimens can play a key role in tumor diagnosis. Examination of fluids, sputum, brushes from various sites, lavage specimens, and others submitted for screening for the presence of malignant cells can provide an accurate diagnosis and in some cases obviate surgical intervention. Light microscopic examination may be definitive in establishing a diagnosis; however, adjunct techniques may be needed to clarify or better define a specific diagnosis.

With the advent of fine-needle aspiration, tumors located in sites difficult to reach, inoperable neoplasms, as well as superficial masses, can be sampled to obtain a diagnosis without the need for invasive procedures (i.e., surgery). A diagnosis can be rendered at the bedside, clinic, or in the ''special procedures'' or computerized tomography (CT) scan facilities, immediately after preparation and staining of the sample. By analyzing the cellular arrangement patterns, cellular characteristics including nuclear and cytoplasmic features, as well as other findings (i.e., tumor background, necrosis), the tumors may be classified as benign or malignant and in most cases can be placed within a specific nosologic category.

In a number of instances, tumor characterization may be impossible just on light microscopic examination of the sample. In some cases even differentiation between the epithelial, lymphoid, and mesenchymal nature of a given tumor is problematic. In neoplasms deemed to belong to one of these three major diagnostic categories, further speciation may be challenging. It is in those cases that special techniques can assume a prime diagnostic role. Particularly in cytology, where the material available is often scanty, the best specimen utilization is imperative, keeping in mind the possible need for the use of special techniques.

The application of immunohistochemical techniques to cytologic specimens is, therefore, a logical step to try to resolve difficult diagnostic problems. Cytologic materials such as imprints, smears, cytospins, and cell blocks can be easily subjected to immunoperoxidase methods. Immunocytochemistry can be performed on both unstained smears, as well as in Papanicolaou-stained specimens, after destaining. The procedure can be accomplished successfully even when decolorization needs to precede immunohistochemistry evaluation. The use of cytospins affords the advantage of preparing several slides to test with a panel of antibodies. A rinse of the needle after the initial smears are prepared provides a way to stretch and make the most out of the specimen. It is also advisable that a specimen be obtained and preserved for possible ultrastructural examination, in case it is needed for tumor speciation and even for ultrastructural immunolabeling. The ''pop off'' technique is also available, whereby cells can be obtained from slides by popping them off;[26] however, the ultrastructural details are generally markedly compromised and the evaluation may be significantly hampered or the specimen may not even by worthy of examination. It is, however, of use in selected circumstances.[27]

D. CYTOPATHOLOGY/FINE NEEDLE ASPIRATION: HISTORICAL PERSPECTIVES

The history of diagnostic cytology began many years before its formally recognized inception. Since the discovery of the microscope, the cell — the unit of life — has been studied by scientists. The study of cells, from which normal and abnormal changes in the body can be inferred, has been intimately interwoven with the history of pathology. However, it was not until George N. Papanicolaou's famous manuscript on the study of vaginal cytology for the detection of cervical cancer was published in 1928[28,29] that diagnostic cytology emerged as a field of medicine.

Diagnostic cytology was for many years considered a synonym with exfoliative cytology, since it only included the evaluation of exfoliated cells. The efforts of Dr. George Papanicolaou (1883–1962), the father of modern cytology, led to a full understanding of the importance of the field of cytological diagnosis. His contributions to the cytological aspects of cancer changed the status of cytological evaluation from a theoretical field to a widely accepted laboratory procedure. Cytological evaluation of cervical smears, fluids, and other exfoliated cells from a variety of sites to detect preneoplastic and neoplastic disorders became routine in the practice of medicine since the first half of the 19th century.[30,31] In some areas, cytology progressed to become a primary source of information, obviating, in some cases, the need to obtain material by surgical procedures in order to make a definitive diagnosis.

The practice of evaluating exfoliated cells was amplified by the utilization of needle aspiration techniques to obtain material from organs and lesions that do not spontaneously yield exfoliating cells. As early as 1853, James Paget wrote a cytologic description of breast cancer.[31] He is believed to be the first to have sampled a breast tumor by the use of an aspirate. Others such as Greig and Gray in 1904, Hirschfeld in 1912 and 1919, as well as Guthrie in 1921, utilized the aspirate as a means of diagnosing hematologic conditions.[32] Perhaps most prominently, Hayes Martin and Edward Ellis from Memorial Sloan Kettering (MSK) in New York utilized the aspirate for the diagnosis of a wide range of conditions. Both Martin and Ellis, with the collaboration of Coley in the 1930s and utilizing principally an 18-gauge needle, established a long tradition of aspirates at the MSK.[33] Interestingly, these surgeons took the avenue of needle aspiration since James Ewing, then Chief of Pathology at MSK was opposed to excisional biopsies, thinking that they contributed to tumor spread and dissemination. The surgeons then resorted to needle aspiration to avoid performing surgical procedures when a preliminary diagnosis was not known. In 1933, Dr. Stewart, another American from MSK, published an important paper relating to his experience with needle aspiration.[34] Despite these early publications[34] and the great experience being obtained on the subject of tumor aspiration in some centers, the procedure was not popular in the U.S. The Europeans, however, perhaps due to the inspiration of Manheim, one of Hirschfield's students, became very adept at the practice of aspiration. The technique was further advanced in the 1940s by Lopez Cardoso and Nils Sodestrom and the utilization of a thin needle (22- to 25-gauge) was made an integral part of the procedure. The fact that these two scientists were hematologists led to the utilization of hematologic staining techniques to the aspiration smear preparations (i.e., the May-Grunewald-Giemsa stain). The field was slowly expanding and it was up to Franzen and Zajicek to make the technique more available.[35] Other Swedish physicians, such as Esposti and Lowhagen, further developed the technique of obtaining fine-needle aspirates at Karolinska Institute in Sweden, an institution that has been one of the most important centers for the study and practice of fine-needle aspiration biopsy.

The technique has now been popularized in the U.S. and is utilized routinely in some centers as a diagnostic modality. Some pathologists perform their own aspirates and immediately interpret them in a manner similar to the frozen sections in the surgical pathology laboratory.

E. DIAGNOSTIC MODALITIES — GENERAL COMMENTS, ADVANTAGES, AND LIMITATIONS

1. Light Microscopy: Hematoxylin- and Eosin-Stained Sections

It should come as no surprise to readers that light microscopy has been, and remains, the reference point for diagnosis. However, it took approximately 240 years from the time the light microscope was invented to when it was accepted as a useful tool in diagnosis. Currently, initial evaluation of tumors typically begins by a light microscopic assessment of the hematoxylin- and eosin-stained sections, which provides a reference point for decision as to whether additional diagnostic modalities are needed in an effort to arrive at a final diagnosis. The level of confidence that a specific pathologist has to formulate a final diagnosis on the basis of light microscopic assessment is quite variable and, of course, varies with the type and complexity of a given case. It is understandable that older pathologists would tend to remain more attached to light microscopy than the new generation of pathologists. The majority of tumor classifications used today, including World Health Organization (WHO) classifications, still retain light microscopic assessment as their main point of reference. It is very clear, however, that determination of the nature of a neoplastic process (whether benign or malignant) is almost always established by evaluating information provided by the clinical information, and gross and light microscopic findings in a given tumor. The advantages of "working-up" cases using a combination of techniques have been emphasized to the new generation of practitioners. These pathologists have been amply exposed to the pitfalls of light microscopic examination. This fact, as well as medico-legal considerations, have forced the new generation to approach making a diagnosis in a more versatile fashion, with the clear understanding that light microscopy is limited in so far as the diagnostic accuracy that it can achieve is concerned. The use of additional techniques needs to be evaluated in terms of patient management and benefits to be obtained vs. expense to be incurred in each individual case. Although changes may be expected in the near future, with the incorporation of data obtained from other diagnostic techniques, these will not occur unless solid evidence demonstrates their significance in patient management.

2. Histochemistry — "Special Tinctorial Stains"

"Special stains" aim at providing information regarding the nature of neoplastic cells by detecting various tissue components (i.e., lipids, glycoproteins, glycogen, mucin) which may be characteristic of various cell types. There is a great variability as to what special stains can accomplish in tumor speciation. Questions to be answered by special stains are quite variable. In some cases simple questions such as, Is there mucin in this tumor? or, Is there periodic acid Schiff-positive material? are rather simple to address. There are, however, significant degrees of complexity in the interpretation. Each case needs to be individually judged by the pathologist once the stains are examined, keeping in mind the questions addressed. In some cases, histochemistry definitely indicates whether a reaction is positive or negative. However, in other cases, interpretation of the reaction remains subjective and it may reflect the pathologist's individual biases in the case in question. Even in those cases in which the reaction is unquestionably positive, the final diagnosis has to be arrived at by deriving data from the information obtained by examining the single or panel of histochemical stains. A negative reaction is sometimes of no help whatsoever in clarifying the diagnosis. When histochemistry was the only methodology available to supplement the light microscopic findings, diagnoses were rendered based exclusively on the results obtained. As other techniques have become available, the use of ''special stains'' has been significantly reduced and in some cases if conflicting evidence arises from the examination of these and other diagnostic methods (i.e., immunocytochemistry or electron microscopy), the latter results are selectively used for arriving at the final interpretation. Another significant drawback of histochemical stains is that only a certain degree of sophistication can be attained in tumor speciation, and current practices may require more than this level of diagnostic data to be provided. Furthermore, the available histochemical methods cannot even address certain diagnostic dilemmas that other methodologies presently available can very adequately attempt to resolve.

3. Electron Microscopy

Electron microscopy has proved very useful in identifying the histogenesis (or differentiation) of various neoplasms; however, it is of little help in separating reactive, benign, and malignant cellular proliferations.

In 1968, Rosai and Rodriguez[36] stated that ''electron microscopy is a valuable tool for biologic research, yet its use in diagnostic pathology has been relatively neglected.'' This statement was made public 36 years after the first electron microscope was developed by Knoll and Ruska in 1932. Rosai's comment was taken seriously and the decade after his observation is considered the ''golden age'' of ultrastructural examination as it pertains to tumor evaluation. Diagnostic criteria were defined and limitations found. Unfortunately, some poorly informed individuals regard the electron microscope even today as an expensive and cumbersome instrument with the potential to generate confusion by providing unexpected rather than helpful data in evaluating diagnostic problems. Some have even publicly declared that the era of electron microscopy has ended and that immunocytochemistry has replaced its use. Nothing is further from reality than this statement and, as the pitfalls of immunocytochemistry have been realized, the morphologic information provided by the ultrastructural evaluation of a given case is becoming cherished. Electron microscopy is an extremely powerful diagnostic tool that, when utilized by well-trained scientists, provides information that no other available technique can at the present time. It also has limitations and any well-trained scientist with a mission to utilize this instrument for diagnostic purposes must be fully aware of these, as well as of its capabilities, in order not to over-estimate its usefulness and to avoid making unwarranted mistakes in diagnosis. Limitations are related to specimen sampling, lack of specific diagnostic features primarily in poorly differentiated tumors, heterogeneity of tumor morphology, and similarities in categories of ultrastructural cellular components. An ultrastructural diagnosis should not be based on a single finding, but rather on a comprehensive compilation of the findings present and absent in a given case.

4. Immunocytochemistry at the Light Microscopic Level

Immunocytochemical techniques have provided a new, exciting generation of ''special stains'' that aim at detecting antigenic sites, which in turn provide information regarding tissue antigens that allow characterization of tumor cells. The significance of this technique in surgical pathology is based on the assumption that different cell types can be distinguished by their immunophenotypes. Specific cell types are expected to exhibit a distinctive and more or less specific group of antigenic determinants. It has become obvious, however, that the expression of many cellular components is dependent on the stage of differentiation and functional state of the tested cellular elements. Aberrant differentiation occurs in tumors and their evaluation can be severely compromised if a final diagnosis is solely based on the results of some of these reactions. Tumors may reveal antigens not expressed by their normal cellular counterparts or vice versa, they may loose the expression of antigens commonly identified in normal cells of similar derivation or differentiation. A metastatic tumor does not necessarily reflect the same antigenic epitopes noted in its primary site. This antigenic infidelity can be a very significant source of confusion and tumor misdiagnosis. The holistic concept of tumor evaluation remains a central consideration in organizing and interpreting the data obtained by the various diagnostic techniques. In spite of the complexity and difficulties which are part of the cellular antigen detection and approach to tumor classification detailed above, immunohistochemistry has largely replaced routine histochemistry (''special stains'').

Two major immunoperoxidase techniques are widely and routinely used for localization of antigenic determinants in tissue sections. These techniques include the peroxidase-labeled antibody method (direct method)[37] and the peroxidase antiperoxidase (PAP)-unlabeled antibody method introduced by Sternberger in 1970 (indirect method).[13] Amplification methods of the latter technique to include the avidin-biotin complex (ABC),[14] protein A,[38] and streptavidin have become popularized and are utilized by many in an effort to detect smaller amounts of antigen in the tissue sections than detectable by nonamplified methodologies. In our laboratory, the ABC method is preferred over the other methodologies and the results obtained indicate that this method is highly sensitive and reliable. In general, the direct methods are less sensitive than the indirect ones and commonly require the use of higher antibody concentrations. We utilize, in our laboratory, commercially available staining kits for the ABC method and have found them reliable and easy to use for diagnostic and selected research applications.

There are a number of critical considerations that affect immunoperoxidase procedures. Tissue fixation and processing are key factors in preserving antigenic determinants to be labeled. If the antigens are not properly handled and maintained viable, their detection will fail in spite of a flawless technique. Some antigens cannot be chemically fixed at all and others require peculiar fixatives rather than the usual 10% buffered formalin. Small antigens (i.e., peptide hormones) are best immobilized by fixatives capable of cross-linkages (formaldehyde and glutaraldehyde). Conversely, extensive cross-linkage of proteins in high concentration may lead to the masking or destruction of antigens. It is indeed very fortunate that a large number of antigens survive formalin fixation, the most commonly used fixative.[39,40] Some antigenic determinants are known to be better preserved by fixing them in certain fixatives. For example, mercury-containing fixatives, such as B5 and Zenker's fixatives, have proven to be the best for demonstration of cytoplasmic immunoglobulins.[41] Intermediate filaments are better preserved in Carnoy's fixative (alcohol, formalin, and acetic acid).[42] Peptide hormones are well preserved in picric acid containing fixatives such as Bouin's or Zamboni's.[42]

It is essential to emphasize that prolonged fixation in any fixative could lead to progressive antigen denaturation and eventual loss of antigenicity.[39] This process appears to be time dependent and can sometimes be reversed by treatment of sections with proteolytic enzymes.[43] However, the result of proteolytic digestion may be variable, not only for different antigens but also for the same antigen in different circumstances.[39] Expression of some antigens may not be enhanced at all, and some may even be masked when the tissue sections are exposed to enzymatic treatment. Tissues to be immunostained must be fixed only for the minimal period necessary for adequate preservation of morphologic details and immobilization of the antigenic determinants. The latter, however, is an immediate phenomenon in the areas immediately exposed to the fixative but the overall fixation of the entire tissue fragment is, as expected, size dependent. A large tissue section exposed to the fixative for a small amount of time and later immunostained could show staining at the periphery and none in the center. In contrast, an overfixed specimen may reveal central immunostaining and absence of staining in the peripheral portions. Tissue processing, especially dehydration procedures and exposure to xylene, may also affect the ability to detect antigens by altering or masking them. Decalcification procedures commonly destroy tissue antigenicity or lead to undesirable background staining.[42] These effects are also related to the type of decalcifying agent used and the period of time that the tissue is exposed to it. High-melting-point embedding media should be avoided if at all possible, since heating may have a significant irreversible adverse effect on antigens by causing protein denaturation.

5. Monoclonal vs. Polyclonal Antibodies: Pros and Cons

An immunoperoxidase stain is only as good as the primary antibody. This primary antibody ideally must be specific, sensitive, reliable, and well characterized. Determining the proper dilution of the primary antibody is essential. This is performed by critically evaluating dilutions of antiserum to determine the titer that provides the best specific-to-background staining ratio. The issue of monoclonal vs. polyclonal antibodies continues to be a source of controversy. Monoclonal antibodies are more specific since they are targeted to a limited number of epitopes; however, if antigens are sparsely distributed, a higher level of sensitivity that some monoclonal antibodies provide may be needed. "Cocktails" of several monoclonal antibodies against different epitopes of the same antigen may be of value in these situations. If selective loss of epitopes takes place as a result of overfixation or tissue processing, a given monoclonal antibody may not be able to detect an antigenic determinant in a multivalent antigen, while a polyclonal antibody may be able to pick up other epitopes of the same antigen. Monoclonal antibodies have an intrinsic lower affinity than their polyclonal counterparts and are also more sensitive to repeated freezing and thawing. Furthermore, polyclonal antibodies can be monospecific and monoclonal antibodies are not necessarily monospecific. The choice between a polyclonal and a monoclonal primary antibody should be based on a careful consideration of the specificity, sensitivity, reliability, and characterization of the antibody in question, rather than simply on whether the antibody is monoclonal or polyclonal.

It is realistic to assume that it will not always be known when the specimen is unfixed (fresh) whether a neoplasm is to represent a diagnostic challenge and require an immunohistochemistry work-up. Due to this fact, it is recommended that tissue be handled in several ways if there is any chance that it may require special techniques to be properly evaluated. A portion should be snap-frozen and other fragments should be fixed in B-5 and/or some alcohol-based fixative, buffered formalin, and Bouin's fixative. Fixation is undoubtedly a critical factor in obtaining quality immunostaining results. Finally, a portion of the specimen should be obtained for electron microscopy.

Only the ABC-peroxidase immunostaining procedure is detailed below. Other methodologies mentioned represent relatively minor variations and will not be detailed in this chapter. The reader is referred to references that allude to the specifics of the other techniques.[13,37,38]

F. AVIDIN-BIOTIN PEROXIDASE COMPLEX IMMUNOSTAINING PROCEDURE FOR PARAFFIN SECTIONS

1. Immunocytochemistry

1. Deparaffinize and hydrate sections through xylenes or other clearing agents and graded alcohol series.
2. Rinse slides in distilled water followed by Tris-buffered saline (TBS).
3. Incubate the sections for 30 min at room temperature (RT) in 0.3% hydrogen peroxide in methanol to quench endogenous peroxidase activity. This is prepared by mixing 10 ml of 3% hydrogen peroxide and 40 ml of methanol. Incubation times may be shortened by using higher concentrations of hydrogen peroxide.

4. Wash sections in distilled water followed by TBS. *If antibody requires enzyme digestion,* place slides in 0.01% trypsin in TBS for 5 to 10 min. This is prepared by mixing 10 ml of 0.1% trypsin with 90 ml TBS. Rinse thoroughly in TBS.
5. Incubate sections for 20 min (RT) with diluted normal serum from the species in which the secondary antibody is made, i.e., normal horse serum for monoclonal antibodies or normal goat serum for polyclonal antibodies.
6. Drain excess serum from sections. No need to rinse.
7. Incubate sections for the appropriate time (refer to data sheets) in primary antibody diluted to the proper concentration in diluent. Negative control sections will be incubated in diluted normal serum.
8. Wash slides with Tris buffer. Each slide should be washed individually to prevent cross-contamination with antibodies.
9. Incubate sections for 30 min with biotinylated secondary antibody solution.
10. Wash slides in Tris buffer.
11. Incubate sections for 30 to 60 min with ABC reagent.
12. Wash slides in Tris buffer.
13. Incubate sections for 2 to 7 min in filtered diaminobenzidine (DAB) substrate solution. Monitor the reaction microscopically.
14. Wash sections briefly in tap water.
15. Counterstain for 2 min in Mayer's hematoxylin blue, dehydrate, clear, and mount. A very light counterstain is desirable.

2. Double-Labeling Techniques

Double-staining procedures consist of incubating sections with one set of reagents to stain for one antigen followed by incubation with a second set to label another antigenic determinant. One potential problem associated with double staining is that added cross-reactivity of the antibodies used for the second antigen with the first set of reagents can lead to nonspecific staining. This problem can be prevented by using primary antibodies conjugated with different enzymes, haptens or biotin, or antibodies from different species (i.e., using monoclonal and polyclonal antibodies). Elution of antibodies used in the first set of reactions is also possible, but cumbersome and of relatively limited value in a diagnostic pathology laboratory.[42] The same principles apply to double labeling at the ultrastructural level.

3. Pitfalls — False-Positive and -Negative Reactions

There are potential pitfalls in immunocytochemical analysis. The technique utilized must be meticulous to obtain valid and reproducible results. In many settings, the information obtained from immunocytochemistry may be used to favor one diagnosis over another. In some cases, the therapy of the patient will depend considerably on the results obtained. Therefore, careful scrutiny of the procedures and techniques involved is necessary to guarantee quality control and valid results. Understanding the pitfalls of the methodology becomes a major task for pathologists using it for diagnostic purposes. False-negative results may occur if the antibody used is inappropriate, as mentioned before, or if the concentration of the primary antibody utilized is incorrect. Loss of antigen through autolysis and/or diffusion may also be a source of a false-negative reaction. Even more critical is the amount of antigen necessary to be present in the tissue to generate a positive reaction. Presence of an antigen at a density below the detection threshold will give a false-negative reaction. A negative immunohistochemical reaction generally should not be used by itself to rule out a diagnosis, even in the presence of a built-in positive control. Correlation with other data available should take precedence in evaluating such a case.

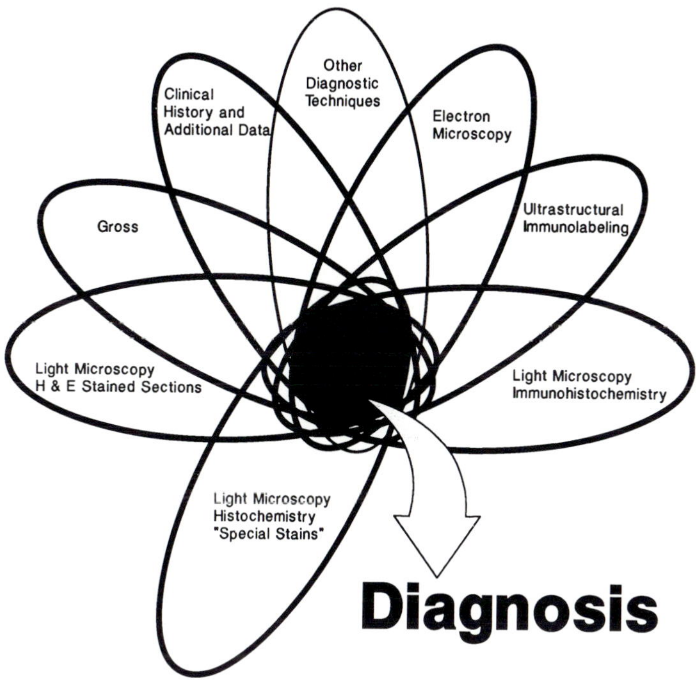

FIGURE 1. Comprehensive tumor evaluation; compiling a final diagnosis. The utilization of ancillary techniques may be required in the evaluation of difficult neoplasms. Each technique offers information that, when combined, should result in a final accurate diagnosis.

False-positive reactions also do occur and are even more dangerous, because a positive result is often used to support a diagnosis. False-positive reactions may be due to antibody cross-reactivity with antigens other than those being searched for, nonspecific binding of the antibodies to the tissues, presence of endogenous peroxidase, or avidity for the avidin-biotin complex by some cellular elements, entrapment of normal tissues, and release of soluble proteins from the cytoplasm of normal cells invaded by neoplastic cells. A false-positive result is often not immediately apparent and it may be used to support an erroneous diagnostic conclusion. The potential for misdiagnosis is far greater in false-positive reactions than in false-negative ones, since the latter are generally accepted as more capable of occurring. Pathologists in practice need to very critically judge the immunocytochemistry results obtained in a given case and interpret them in the light of all additional information available to them, rather than make a final diagnosis on the basis of stain results (Figure 1). Interpreting the stains, rather than evaluating the entire case, may lead to enormous mistakes and is not acceptable in the current practice of pathology. The pathologist interpreting the case must also be aware of the full spectrum of immunologic manifestations of the neoplasms that are considered in the differential diagnosis. The spectrum of "accepted" antigenic expressions by a given tumor, including antigenic expression resulting from antigenic infidelity and multidirectional differentiation, continues expanding as new evidence is confirmed and published in the literature.

In the discussion pertaining to ultrastructural immunolabeling, the above-mentioned reasons for false-negative and -positive reactions will be discussed in more detail. Ultrastructural labeling may play a key role in clarifying some of these false-negative and -positive results. When the immunologic parameters can be interpreted in association with morphologic markers, their specificity and sensitivity can be accurately established in many instances.

G. OTHER SIGNIFICANT TECHNICAL CONSIDERATIONS

Chromogenes — Several chromogens can be used to bring out the color reaction product. The brown-staining diaminobenzidine (DAB) is the preferred chromogen. This substance is not soluble in the organic mounting media which are routinely employed and does not easily fade with time. The major disadvantage is the fact that DAB is a potential carcinogen and needs to be handled with care. The second most commonly used chromogen is 3-amino-9-ethyl carbazole (AEC), which produces a reddish-brown reaction product that dissolves in organic solvents. Therefore, aqueous media must be used for mounting tissue sections when AEC is used. Alcohol also dissolves the reaction product and, therefore, dehydration should not be performed before mounting. One of the disadvantages of AEC is that it fades much faster than DAB, necessitating photographic documentation.[39,40]

Controls — The accuracy of the results cannot be validated unless proper controls are used in conjunction with the tested specimens. A positive control with a tissue section known to have the antigen being tested for should be fixed, handled, and immunostained similarly and concomitantly to attest for the sensitivity of the immunohistochemical method. Likewise, a negative control slide is essential to detect nonspecific reactions. When treating this slide, the primary antibody is omitted and nonimmune serum substituted. In this setting, specific staining should not take place. If staining occurs, nonspecific background is present in the immunohistochemical reaction and a positive reaction should not be interpreted in the tested material even when the chromogen can be detected. The negative control omitting the primary antibody permits evaluation of endogenous peroxidase activity and nonspecific binding of secondary antibody. Adsorbing the primary antibody with the specific antigen is another available technique to test for nonspecific positive reactions; however, this is somewhat cumbersome to do routinely in the diagnostic arena.[39] The determination of the proper dilution of the primary antibody is also essential for quality control. This is performed by critically evaluating dilutions of antisera to identify the dilution that provides the best positive stain-to-background ratio. If a high antibody concentration is used, too much background staining may occur and if a low antibody concentration is utilized, small quantities of antigen may not be detected.

H. GENERAL APPLICATIONS OF IMMUNOHISTOCHEMICAL TECHNIQUES

There are numerous applications for immunohistochemical techniques in diagnostic surgical and cytopathology and it is essentially impossible to summarize all of them in this short account. The emphasis will be placed on those diagnostically important applications that have a significant role in the current practice of immunomorphologic evaluation of neoplastic disorders in surgical and cytopathology. Immunohistochemical techniques have been utilized to

1. Demonstrate intracellular and extracellular antigenic determinants related to tissue elements
2. Better characterize specific groups of easily recognizable lesions by light microscopy, by providing additional immunologic information that would allow more complete characterization and dictate specific approaches from the therapeutic point of view otherwise not possible (i.e., lymphomas, leukemias, and plasma cell dyscrasias)
3. Identify the nature of some neoplasms that cannot be adequately classified by light microscopic examination alone
4. Determine tissue-specific antigens that may permit indication of the site of origin of a tumor (i.e., metastatic prostatic lesion tested for prostate-specific antigen)
5. Demonstrate the deletion of normal cell antigens (i.e., loss of blood-specific antigens in bladder tumors)

6. Aid in the demonstration of oncofetal antigens that are expressed by a restricted number of neoplasms (i.e., alpha-fetoprotein)
7. Identify the presence or absence of certain markers to help in the management of certain groups of patients (i.e., estrogen and progesterone receptors in breast carcinomas)
8. Identify proliferation of specific cell types within a neoplasm (i.e., insulin- or glucagon-containing cells in an islet cell tumor), which may be of significance in tumor behavior and in dictating patient's management
9. Detect indicators of early malignant change or tissue invasion by tumors (i.e., to identify by means of S-100 protein staining the presence or absence of myoepithelial cells in breast ductal carcinomas and to detect gaps in basement membranes surrounding cell clusters by means of laminin staining and thus provide evidence of early tissue invasion)
10. Identify certain groups of tumors by means of utilization of antibody testing panels aiding in the speciation of poorly differentiated neoplasms

The latter application is perhaps the most often utilized in surgical pathology. While at the beginning of immunohistochemical evaluation of tumors some years ago, the expected reactions of specific groups of tumors were portrayed as being quite restricted, the possible range of immunologic reactions in a given tumor are continuously expanding.

The interpretation of results to indicate a specific diagnosis (i.e., tumor type) needs to be analyzed critically in the clinical context of the case in question and arrived at by combining information provided from various diagnostic modalities (Figure 1). There is still much confusion as to how to separate cross-reacting antigens from antigenic expression due to multidirectional differentiation of neoplasms. Ultrastructural immunolabeling could play a key role in answering some of these existing questions. In general, those antigenic determinants that have a morphologic correlate (i.e., keratins, carcinoembryonic antigen) seem to be far more specific than those that do not (i.e., S-100 protein). The latter antigens remain of importance in the current practice of surgical and cytopathology; however, as new, more specific antigenic determinants that can answer the same questions and have morphologic correlates are discovered, they will most likely slowly be replaced and will disappear from the diagnostic arena.

I. ULTRASTRUCTURAL IMMUNOLABELING

This technique has been the latest addition to immunomorphologic diagnosis. It combines the advantages of immunocytochemistry and ultrastructure. Not only the absence or presence of antigenic sites in a given tumor can be determined, but it is also possible, if they are present, to localize them to specific cellular domains. Quantitation of the amount of antigen present is additionally possible if an immunogold labeling procedure is utilized.[22] Visual representation of the data obtained can also be readily portrayed by utilizing a microanalytical probe system (Figure 2).

In 1959, ferritin was introduced by Singer as an electron-dense marker[44] that could be conjugated to other proteins, such as antibodies, leading to the beginning of the field of ultrastructural immunolabeling. Some years later, the enzyme-labeled antibody technique became the second major landmark in immunoelectron microscopy.[45,46] Feldherr and Marshall[47] introduced colloidal particles as markers for electron microscopy. In 1971, Faulk and Taylor[48] applied colloidal gold as a specific marker for antiserum and Romano and colleagues[49] used it for labeling of isolated antibodies.

The technique relies conceptually on the same principles as light microscopy immunocytochemistry. There are only minor technical variations resulting from, or requirements

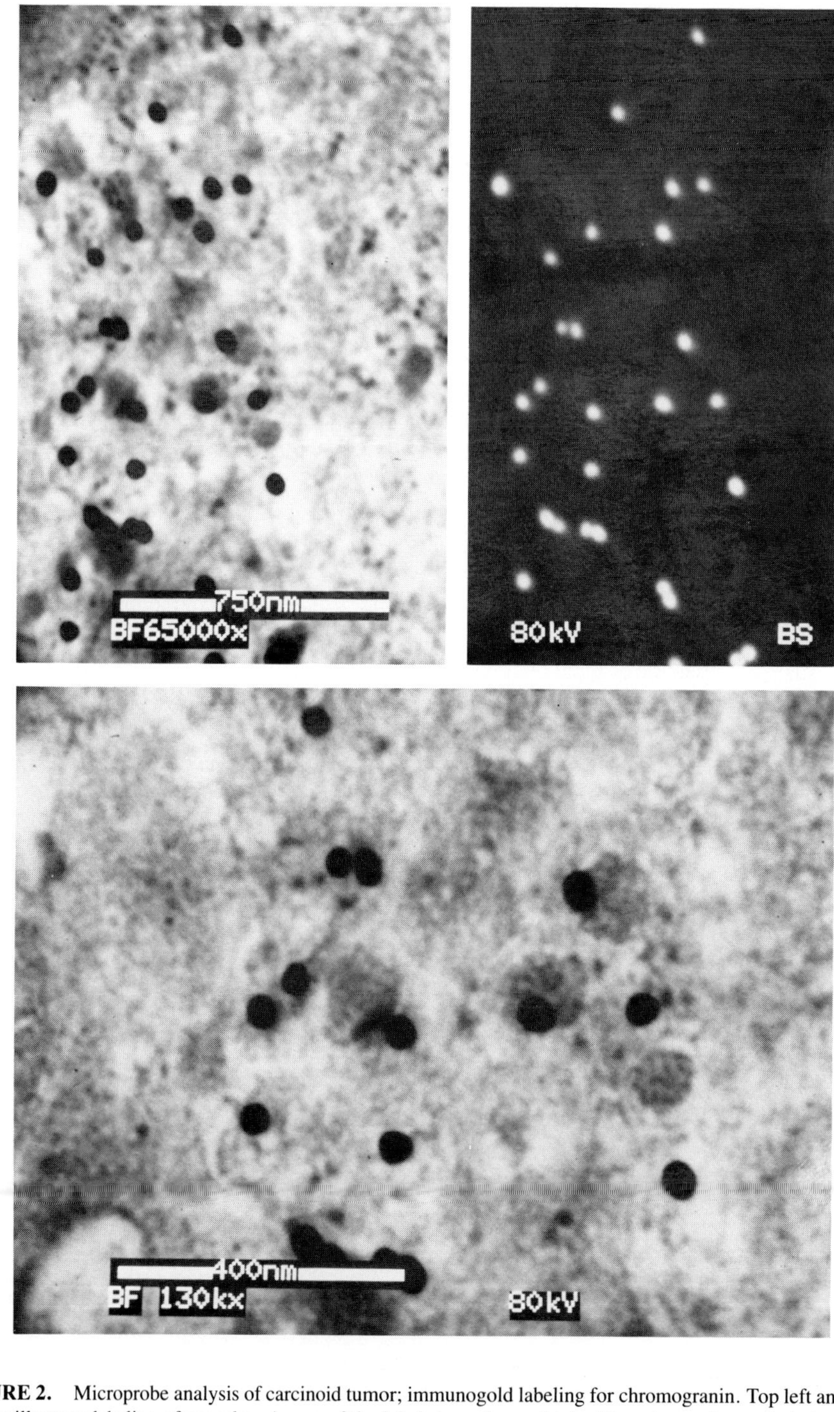

FIGURE 2. Microprobe analysis of carcinoid tumor; immunogold labeling for chromogranin. Top left and bottom figures illustrate labeling of granules. A map of the labeled specimen obtained by microprobe analysis is portrayed in the figure at top right.

created by, specific tissue-processing and labeling methods particular to electron microscopic evaluation. Immunoelectron microscopy combines the advantage of routine ultrastructural evaluation and detection of antigenic determinants as performed by light microscopy immunocytochemistry. Ultrastructural immunolabeling can play a very important diagnostic role in the evaluation of neoplasms which are not characteristic by either their immunocytochemical and/or ultrastructural manifestations. The two essential requirements to obtain excellent results in ultrastructural immunolabeling relate to preservation of antigenicity and ultrastructural morphology. Pre- and postembedding labeling techniques have been used at the ultrastructural level. Postembedding labeling is much more applicable to diagnostic work due to its simplicity as it pertains to tissue handling and the performance of the technique itself. Preembedding techniques require that the antibody incubation be performed on unfixed or mildly fixed tissue slices. Fixative-induced cross-linking may inhibit antibody penetration into the tissue. This is a particularly prominent side effect of glutaraldehyde-containing fixatives. Poor morphologic preservation of ultrastructural details usually occurs when preembedding techniques are used which may compromise the interpretation of results. Since preembedding techniques can only be realistically applied to fresh (or mildly fixed) samples, retrospective studies cannot be performed. There is also a need for large amounts of unfixed control tissues. Usually a panel testing approach is required for full characterization of a tumor. Testing for each antigenic determinant needs separate specimens for testing and for controls, making it unrealistic for this method to be utilized in a diagnostic setting where limited tissue is usually available. Furthermore, antibodies usually do not penetrate well through the tissue slices and a negative reaction may be the result of this technical problem. The only real advantage is that the antigens are not exposed, or are only minimally exposed, to chemical fixatives; therefore, sensitive antigens can be preserved.

In postembedding methods, the tissues are initially fixed, processed, and then resin embedded. Thin sections are cut and exposed to the antibodies providing direct accessibility to all cells and organelles inside. Tissue morphology is excellent and retrospective studies can be carried out at any time. Serial sections can be immunostained for a variety of antigenic determinants and double labeling is relatively easy. A new methodology, the Life-Cell process, has emerged which cryofixes tissues, maintaining antigenic integrity and eliminating the need for traditional chemical fixation which may mask, damage, or destroy extremely sensitive antigens. After cryofixation and embedding of the specimen in resin, a postembedding technique can then be applied.

There are two main labeling methods: immunogold and peroxidase. The advantages of immunogold over peroxidase methods are such that the colloidal gold methodology is selected by the great majority of scientists working in this area at the present time. Gold particles are more electron dense than the oxidation [illegible] result, easier to spot ultrastructurally. Gold particles deposit on top of the antigenic epitope detected while the peroxidase reaction products diffuse away from the main reaction site. Finally, and perhaps most crucial, is that gold particles, being discrete entities, do not mask underlying tissue morphology, permitting precise localization of the structures labeled.

There are three well-recognized issues that need to be dealt with in regards to ultrastructural labeling: methods of fixation, choice of embedding resins, and osmication. All three issues may significantly affect the ability of antigenic detection in tissues.

1. Fixation

Most electron microscopy fixatives contain glutaraldehyde in various concentrations. Glutaraldehyde has remained the fixative of choice for ultrastructural examination, due to this great ability to fix tissues while maintaining very crisp morphological details, which is directly related to cross-linking of proteins in cytoplasmic components.[50] Unfortunately, glutaraldehyde is detrimental to the preservation of antigenic viability and impairs detectability of antigenic epitopes. Only a limited number of antigenic determinants remain viable and can be labeled after glutaraldehyde fixation. These include light chains, some heavy chains, and a number of peptide hormones, among others. A significant number of antigens, however, will be affected by glutaraldehyde exposure. Fixatives such as paraformaldehyde and periodate-lysine-paraformaldehyde have been used for immunoelectron microscopy with good results, providing preservation of ultrastructural features without substantial loss of antigen reactivity. The maintenance of ultrastructural integrity has been considered to be less than optimal when paraformaldehyde alone has been used as a fixative.[50] Formalin-based fixatives, such as those used for light microscopy are appropriate for immunoelectron microscopy. Carson-Millonig's fixative (4% formaldehyde) is such an example. It is key that the fixative be properly buffered, with a pH of 7.2 to 7.4. Sodium hydroxide and phosphate are used as buffering agents. Once an acceptable fixative is found, it is recommended that it be used on a routine basis until found inadequate (i.e., for a particular antigen to be detected) and then other fixatives or even cryofixation can be tested as alternatives.

2. The Life-Cell Process

The goal of cell biologists has always been to combine the structural details offered by electron microscopic examination of tissues and information obtained from direct tissue chemical determination or analysis. Ultrastructural labeling techniques have allowed such a combination to take place to a certain degree. The conventional methods of specimen fixation and preparation using chemical fixatives, solvent dehydration, and resin embedding have in some instances provided suboptimal samples for the application of probes. The limitations imposed by these standard sample preparations have arisen from the combined effects of chemical cross-linking, aqueous extraction following the destruction of permeability barriers by the fixative, and solvent extraction during dehydration.[50] The resulting structural alterations may result in labeling difficulties or, of greater concern, artifactual localization of some antigenic sites in tissues. The Life-Cell process eliminates the effects of chemical fixation and produces a sample which is stable during the manipulations required for the application of probes and is able to withstand the effects of the vacuum and the electron beam in the electron microscope. The Life-Cell process combines cryofixation and molecular distillation drying, and even, if desired, direct resin infiltration under vacuum.[51] The end product is a stable, resin-embedded sample which is amenable to conventional sectioning and labeling employing a postembedding technique.

The most important problem that this process attempts to overcome is the formation of ice crystals which may lead to the destruction of the tissue samples. Ice crystal formation needs to be prevented during cryofixation of the specimen and the subsequent drying process. The cryofixation apparatus marketed as the CF-100 Cryofixation System includes a sample holder that consists of a metal disk attached to a magnetic plunger, followed sequentially by a 1-cm soft foam, 2-mm dense foam, 2-mm 2.5% agar, and finally a thin brass plate. A mirror-finished copper bar is cooled by flow through of liquid helium and is maintained free of contaminants during the cooling process by a surrounding ultrahigh hydrocarbon-free vacuum. At the time of the sample delivery, the vacuum system is isolated and the vacuum around the bar reversed by the introduction of dry helium gas. The sample is applied directly to the surface of the copper bar by an electromagnetic or pneumatic shutter-plunger system. Once cryofixation is completed, the sample is transferred directly to the liquid nitrogen and maintained at $-196°C$ until the molecular distillation drying process is started. The amorphous or microcrystalline state of cellular water is unstable and as the temperature is elevated, water molecule migration and ice crystal formation occurs. To avoid recrystallization during removal of the water from the cryofixed sample, a molecular distillation dryer is used. The samples are loaded under liquid nitrogen into a prechilled copper sample holder equipped with a thermocouple and a cartridge type of heater. The sample holder is then transferred to a liquid nitrogen-filled stainless steel chamber which is immersed in a Dewar containing liquid nitrogen. The liquid nitrogen is removed by a pump action and the samples are exposed to an ultrahigh vacuum generated by a helium cryopump. The equilibrium temperature of this system is $-177°C$ and the vacuum is 10^{-7} to 10^{-8} mbar. The sample is then slowly heated during the drying cycle. This process is controlled by a programmable microprocessor to permit continuous incremental temperature elevation of 1°C per hour during the most critical drying phases. Depending on the application, samples can be unloaded, infiltrated directly with resins at low temperature, or vapor fixed or osmicated in the dry state followed by direct resin infiltration, without breaking the vacuum.[51] Some prefer to retrieve the processed tissues and embed them outside the molecular distillation chamber. The specimens can then be subjected to postembedding labeling techniques. Elemental analysis can also be performed to determine the presence of various cell constituents which have been preserved in their native sites. This technique is particularly useful in the labeling of antigens which are sensitive to chemical fixation protocols.

3. Choice of Resin for Ultrastructural Labeling

If non-hydrophilic resins are used (i.e., Spurr or Epon), a preliminary step known as ''etching'' is required to permeabilize the resin so that antibodies can readily penetrate through it.[52] The process involves exposure of the section to be labeled for a short period of time (3 to 12 min at most) to oxidizing agents such as hydrogen peroxide, periodic acid, or sodium metaperiodate. These agents oxidize alkane side chains of the resins to alcohols, aldehydes, and suds, thereby increasing the hydrophilic character of the surface of the ultrathin sections.[53] As a result of this etching step, damage can occur to the thin sections and holes can be created which may in some cases compromise the viewing and photographing of the labeling results. Sections exposed to the etching step are occasionally rather unstable under the beam and fragment very easily. It is also critical to determine how much etching time is needed to create the desired effect without tissue section damage. The etching time required is quite variable for various tissues and not enough etching would make antibody penetration and, therefore, labeling impossible. The use of hydrophilic resins such as London Resin White (LR White)[54-56] and Lowicryl[56] eliminates the need for the ''etching step'', thus tissue damage and the uncertainty of the etching time required is avoided. Antibody penetration is also enhanced by these resins. Another advantage of the hydrophilic resins is that being ''porous'', the washing solutions may then control nonspecific gold trapping very efficiently. The qualities that are necessary in a resin to allow proper labeling include (1) the ability to allow easy penetration of antibody and washing solutions, (2) stability under the beam, (3) being easy to work with and demonstrating reproducibility of embedding procedures, and (4) providing adequate staining contrast for evaluation of the sections under the electron microscope beam. The schedule that we employ for embedding specimens in LR White resin is as follows:

1. Fix for a period of time needed to stabilize antigens and preserve morphology
2. Wash in buffer for 15 to 20 min
3. 60% Ethyl alcohol (ETOH) (two changes for 15 min)
4. 80% ETOH (two changes for 15 min)
5. 95% ETOH (two changes for 30 min)
6. Infiltrate with 2:1 resin-95% ETOH mixture for 1 h

and precisely documented by using immunoelectron microscopy techniques. Antibody cross-reactivity can be evaluated, as well as the presence of nonspecific binding of the antibodies to the tissues. A question that commonly arises in the evaluation of immunocytochemical reactions pertains to whether coexpression or cross-reactivity is occurring in a given neoplasm. When ultrastructural immunolabeling is utilized to address this issue, in cases with cross reactivity the same structures will be labeled for more than one antigenic determinant, usually more intensely for the true determinant than the cross-reacting antigen.[64] In coexpression, different structural components would be labeled, indicating that the tissue in question possesses more than one identifiable antigen or structural component.[65] Entrapment of normal tissues can be easily visualized by observing the ultrastructural characteristics of the labeled cells in most cases and thus confirming the presence of normal entrapped tissues. Nonspecific antibody binding to the tissue is controlled in immunoelectron microscopy by washing steps and utilization of hydrophilic resins that permit the washing procedures to be very effective. Release of soluble proteins from cells adjacent to those of interest in a given neoplasm (i.e., adjacent normal cells invaded by tumor releasing certain substances) can also be a potential problem; however, the lack of immunomorphologic correlates would help in clarifying such situations. Endogenous peroxidase and indicator dye-binding avidity to some cellular constituents are not a problem in immunogold labeling procedures and, therefore, false-positive reactions identified at the light microscopic level as a consequence of these two mechanisms would be easily clarified.

In the specific situation category fall those difficult tumor cases that require clarification of equivocal light microscopy immunocytochemistry results often originating from more than one of the situations mentioned above. Ultrastructural labeling can be of tremendous value by providing immunomorphologic correlations that allow clarification of doubtful immunocytochemistry data. Examples are many and only a few will be mentioned.

The full spectrum of immunomorphologic manifestations of melanoma continues to expand. Typical melanomas express S-100 protein and HMB-45 and lack keratin immunoreactivity.[66-68] In a spindle cell tumor, the differential diagnosis can be extensive and virtually impossible to be resolved at the light microscopic level based on morphologic alone. Figures 3 and 4 illustrate a fine-needle aspirate of a soft tissue mass. While Figure 3 reveals the cytological details, Figure 4 depicts the corresponding cell block, further illustrating difficulties in making a precise diagnosis, even when a morphologic growth pattern can be judged in a tissue fragment. The strong and unquestionable positivity for HMB-45 (Figure 4B) and S-100 protein expression (not illustrated), together with negative keratin stain, were used to render a diagnosis of melanoma. The patient had a previous history of a skin melanoma level III, excised 3 years prior. While S-100 protein expression is shared by many different cell types, HMB-45 has been stated to represent a rather specific oncofetal associated glycoprotein present in association with premelanosomal structures.[69] Recently, some melanomas have been reported to be keratin positive[70,71,72] and these tumors may or may not exhibit the other markers felt to be compatible with melanoma (i.e., expression of S-100 protein and HMB-45) (Figure 5). A metastatic neoplasm, without a primary clinically detectable tumor which is worked up for the possibility of metastatic carcinoma vs. metastatic melanoma could be a source of significant confusion to the surgeon or cytopathologist evaluating the case if the keratin stains are positive. This is further complicated by the facts that S-100 protein positivity has been reported in many carcinomas,[73] and HMB-45 has recently been reported to be present in a small number of adenocarcinomas[74-77] (Figure 6) when tested with a commercially available anti-HMB-45 antibody (Enzo Diagnostics). Ultrastructural labeling may reveal a few immature cell junctions and electron-dense bodies, either with or without diagnostic premelanosomes or melanosomes. If the melanosomal structures are diagnostic (Figure 7), and other diagnostic possibilities in which melanosomes be found

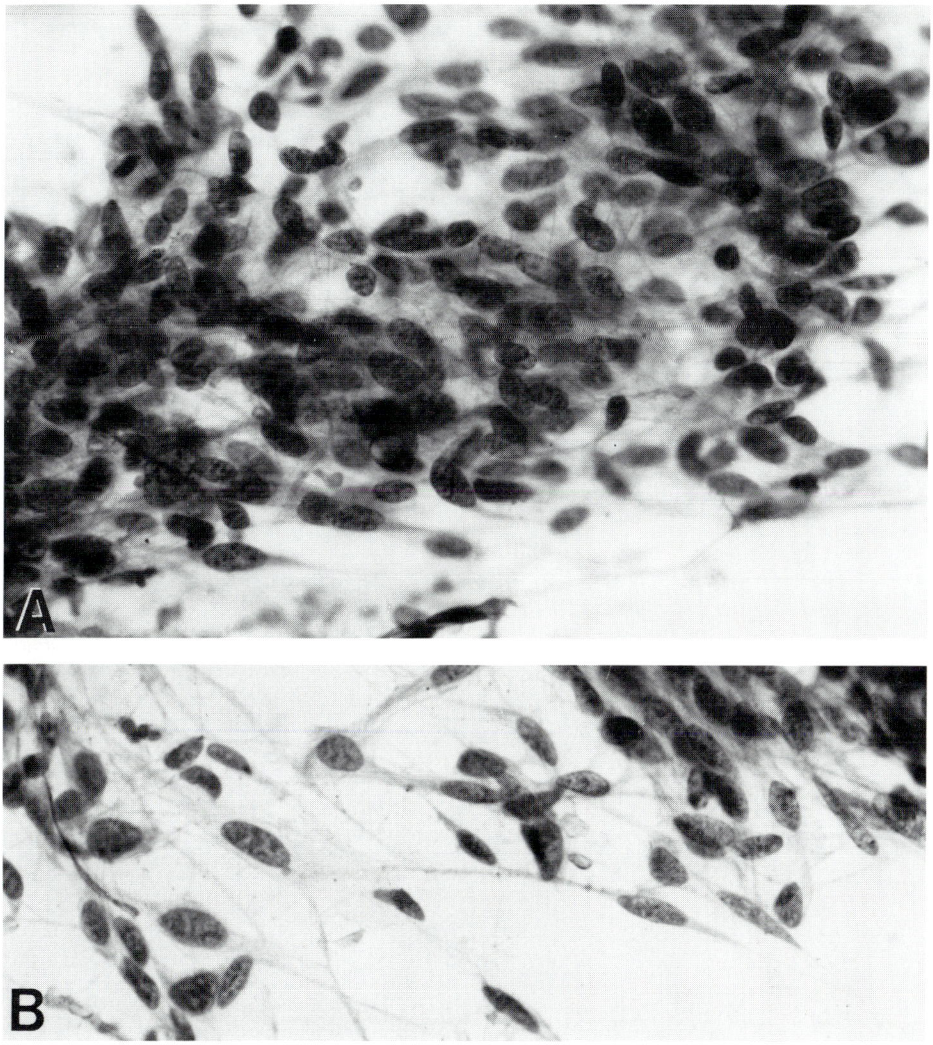

FIGURE 3. Metastatic melanoma, soft tissue mass, thigh. (A) Fine-needle aspirate. Hematoxylin and eosin stain. (B) Similar details in malignant cells in ascitic specimen. (Magnification × 330.)

can be properly ruled out, the diagnosis of melanoma can be confirmed even when the immunohistochemical profile is not typical.[72] On the other hand, the localization of the HMB-45 antibody to the dense bodies at the ultrastructural level would support the diagnosis of melanoma, in spite of a immunohistochemical profile which is not typical and dense bodies which lack an unequivocal premelanosomal internal structure.

In some situations, determination of the type of intermediate filament present in the neoplastic cells may be key in determining the proliferating cell type in a given neoplasm. An example of a tumor considered to be either a sarcomatoid mesothelioma or a sarcoma is illustrated in Figures 8, 9, and 10. Intermediate filaments may in some instances dispose themselves in such a way that their nature can be surmised from their morphology; however, in other cases they are essentially diffusely distributed in the cytoplasmic matrix without any specific arrangement (Figure 10). Ultrastructural labeling may be the indicated technique

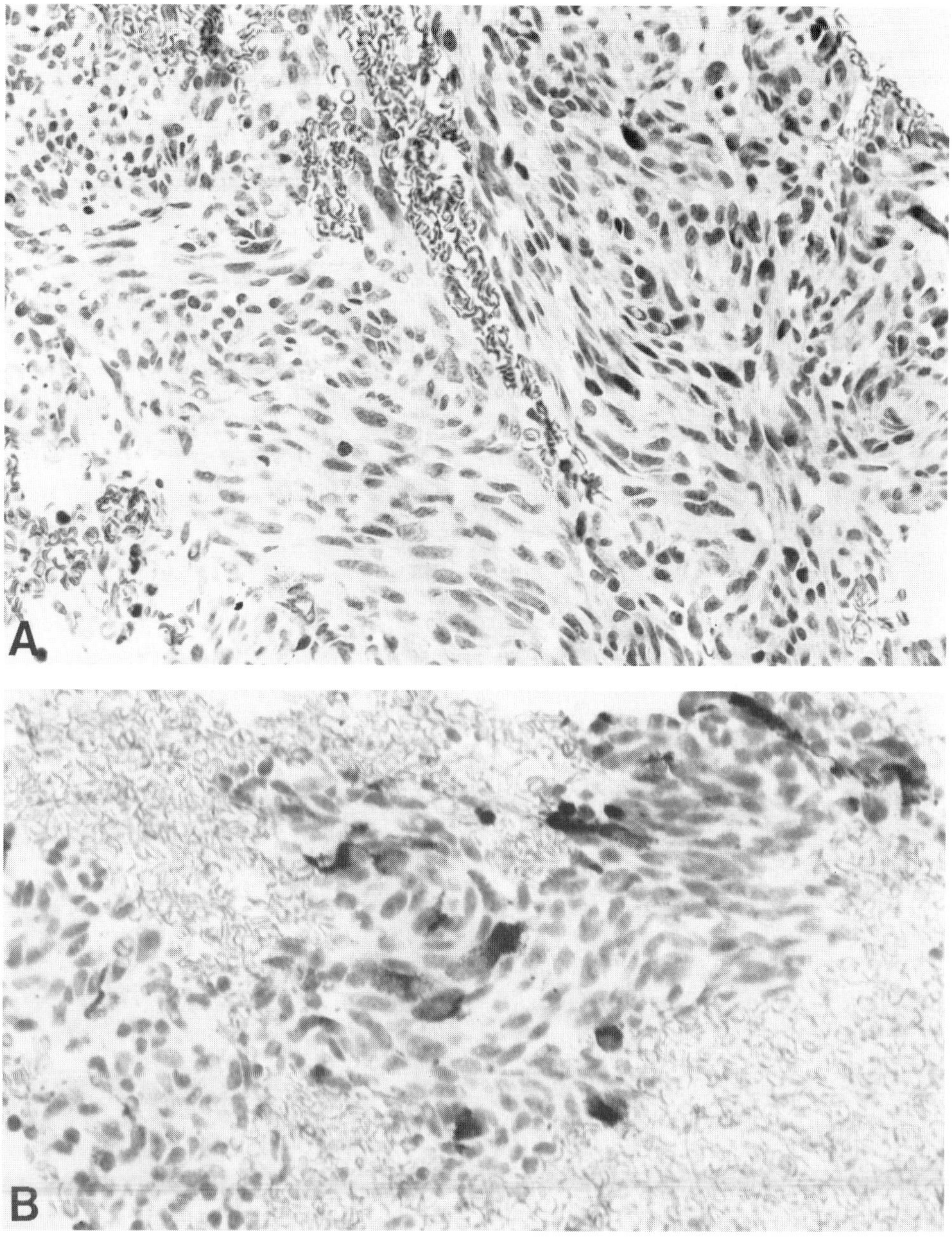

FIGURE 4. Cell block from fine-needle aspirate shown in Figure 3. Hematoxylin and eosin stain. Immunostaining for HMB-45; cell block allows a pattern of tumor growth to be recognized. (B) Strong staining of some neoplastic cells for HMB-45 is illustrated. (Magnifications: A, × 160; B, × 240.)

to address the question regarding the nature of the filaments (Figure 10C) in cases in which light microscopy immunohistochemistry does not provide a definitive assessment (Figure 9, B and C). Examples of those situations include differentiation of a sarcomatoid mesothelioma from a sarcoma, epithelial mesothelioma vs. carcinoma, angiosarcoma vs. carcinoma, spindle thymoma vs. squamous cell carcinoma vs. sarcoma, and spindle squamous cell carcinoma vs. malignant fibrous histiocytoma vs. spindle melanoma (Figures 8A and 9A). It is important to keep in mind the issue of coexpression vs. cross-reactivity in some of these situations. Ultrastructural immunolabeling can be the key technique to address this issue and thus accurately confirm the diagnosis.

The importance of detecting coexpression of markers in compiling a final definitive diagnosis is illustrated on Figures 11, 12, and 13. Figure 11A shows the light microscopic appearance of a large paranasal tumor. The possibility of a vascular tumor (i.e., hemangiopericytoma) was favored. In Figure 11, B and C show the ultrastructural findings with numerous granules in the neoplastic cells strongly suggesting a neuroendocrine neoplasm. The granules in the tumor were labeled convincingly at the ultrastructural level for chromogranin (Figure 12A), synaptophysin (Figure 12B), and serotonin (Figure 12C). Figure 13 illustrates double labeling for synaptophysin and serotonin and emphasizes the coexpression of these markers by the same neuroendocrine granules and the specific localization of the markers within the granules. Double-labeling techniques can be of extraordinary value in determining colocalization of antigenic markers.[78-80] A diagnosis of carcinoid tumor was rendered in spite of the unusual location for this neoplasm. Three months later the patient presented with a typical carcinoid syndrome and a neoplasm was discovered in the ileum. In its primary site the tumor had a typical light microscopic appearance (Figure 14).

Ultrastructural immunolabeling appears to be the next logical step in tumor analysis.[81] Immunogold labeling also permits the mapping of antigenic sites in the neoplastic cellular components by using microprobe analytical techniques, thus providing spatial definition of where they are located and confirmation of immunomorphologic correlations. In addition, quantitation of antigen expression is possible by determining relative gold-particle concentration as compared to control and other specimens similarly labeled under the same conditions.[62-64]

K. MOLECULAR BIOLOGY TECHNIQUES IN TUMOR DIAGNOSIS AND CHARACTERIZATION

The past few years have added a new aspect to the study of human neoplasia. Scientific investigations have led to new exciting insights into the molecular basis for the genesis and development of tumors. The role of oncogenes and tumor-suppressor genes, molecular genetics, detection of point mutations by restriction fragment-length polymorphism (RFLP), the development of molecular probes permitting *in situ* hybridization, and other studies, including blotting techniques and the polymerase chain reaction, have provided pathologists with powerful new tools to study human neoplasms. It is essential that the pathologists play a key role in the transfer of this new technology from the research laboratories to patient management. It is beyond the scope of this chapter to dwell on these techniques. However, they may play an important role in the future in the diagnosis and behavior prediction of some tumors. The advent of molecular genetics has greatly expanded the role of chromosome aberrations in tumors. By using RFLP and genetic linkage analysis, the precise chromosomal location of a disease gene associated with a given tumor may be determined. The presence of a certain proto-oncogene or oncogene in a given tumor may be a prognostic indicator and may alter patient management, perhaps by suggesting added therapeutic interventions that would have not normally been considered at the time. An example is the correlation between the presence of c-erb B-2 in some breast carcinomas and the propensity for these to behave more aggressively, suggesting the need for chemotherapy treatment even in cases with negative lymph nodes. Some of the proteins encoded by oncogenes stimulate growth factors or growth factor receptors. Expression of these proteins may lead to enhanced tumor cell growth. There are several questions pertinent to tumor evaluation that can be addressed,

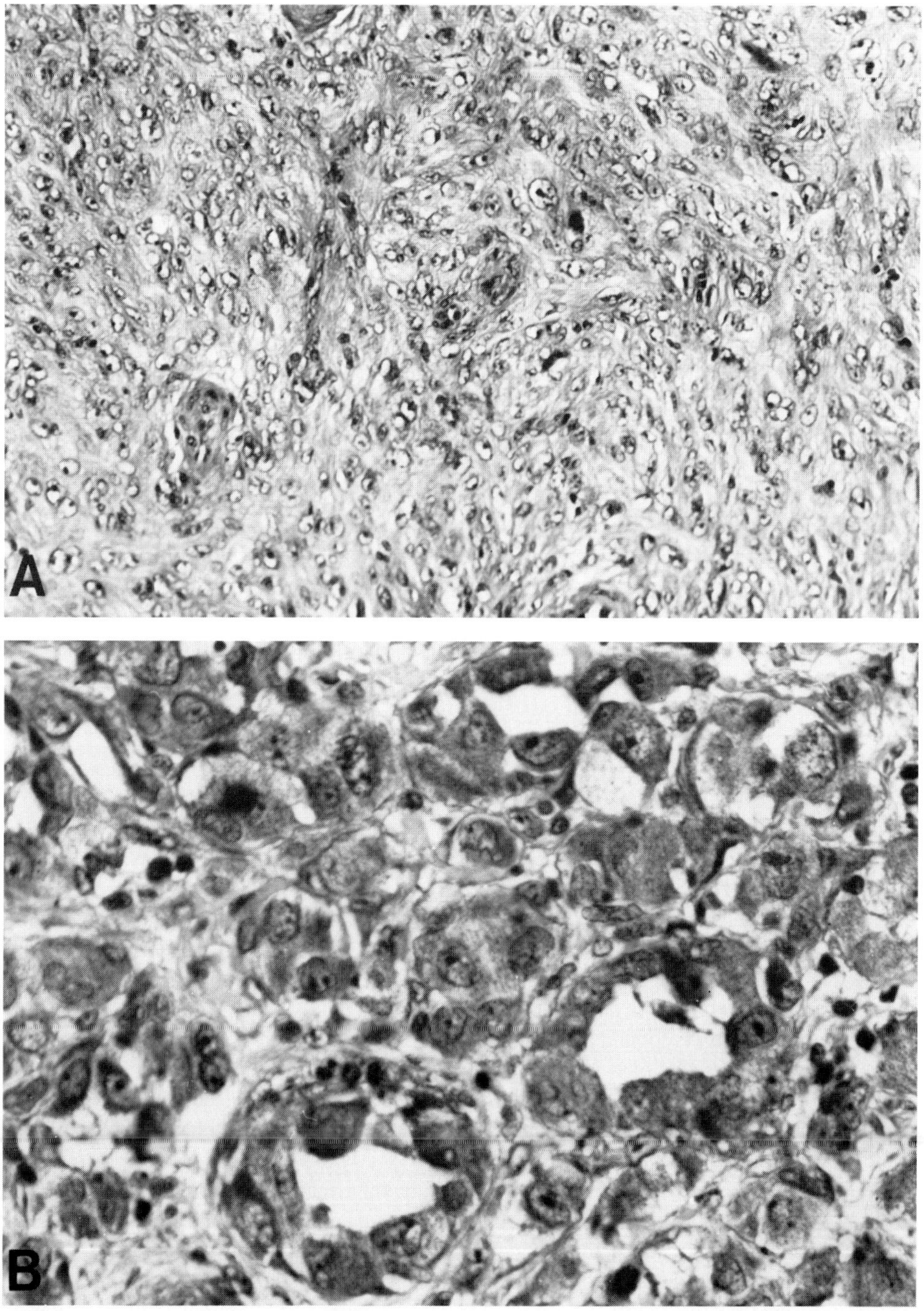

FIGURE 5. Metastatic melanoma, soft tissue mass, neck. (A) Spindle cell melanoma; hematoxylin and eosin stain. (B) Strong keratin staining in a focal area of the neoplasm with an "epithelioid" appearance; wide-spectrum keratin stain — avidin biotin complex (ABC) peroxidase technique. (C) HMB-45 stain, strong staining depicted-avidin-biotin complex (ABC) peroxidase technique. (Magnifications: A, × 160; B, × 500; C, × 350.)

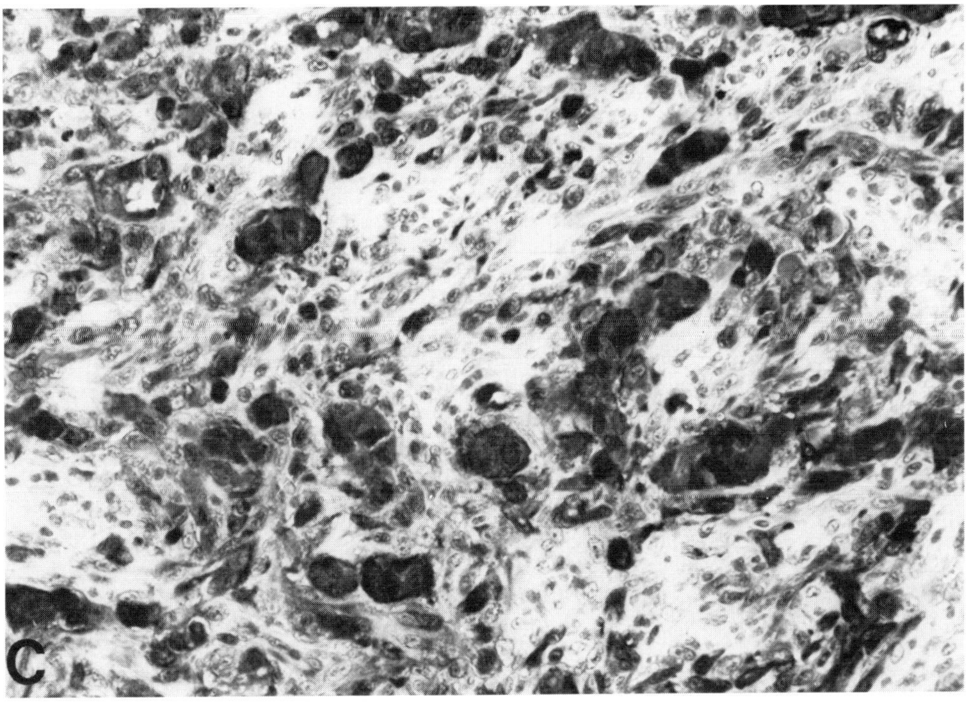

FIGURE 5 (continued).

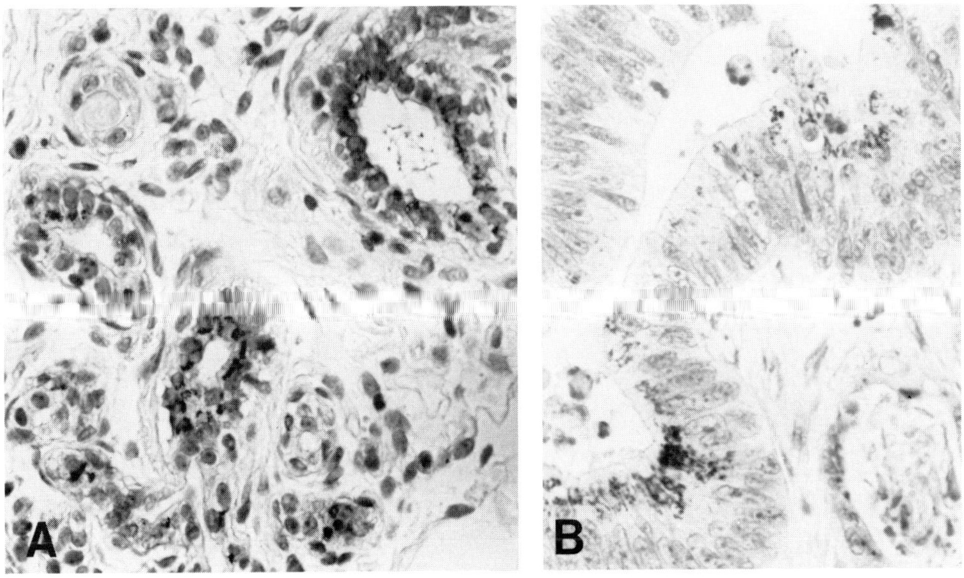

FIGURE 6. (A) Breast adenocarcinoma; HMB-45 stain, avidin-biotin complex (ABC) peroxidase stain. (B) Endometrial adenocarcinoma; HMB-45 stain, avidin-biotin complex (ABC) peroxidase stain. (A) and (B) show unexpected HMB-45 expression by two adenocarcinomas. (Magnifications: A, × 290; B, × 410.)

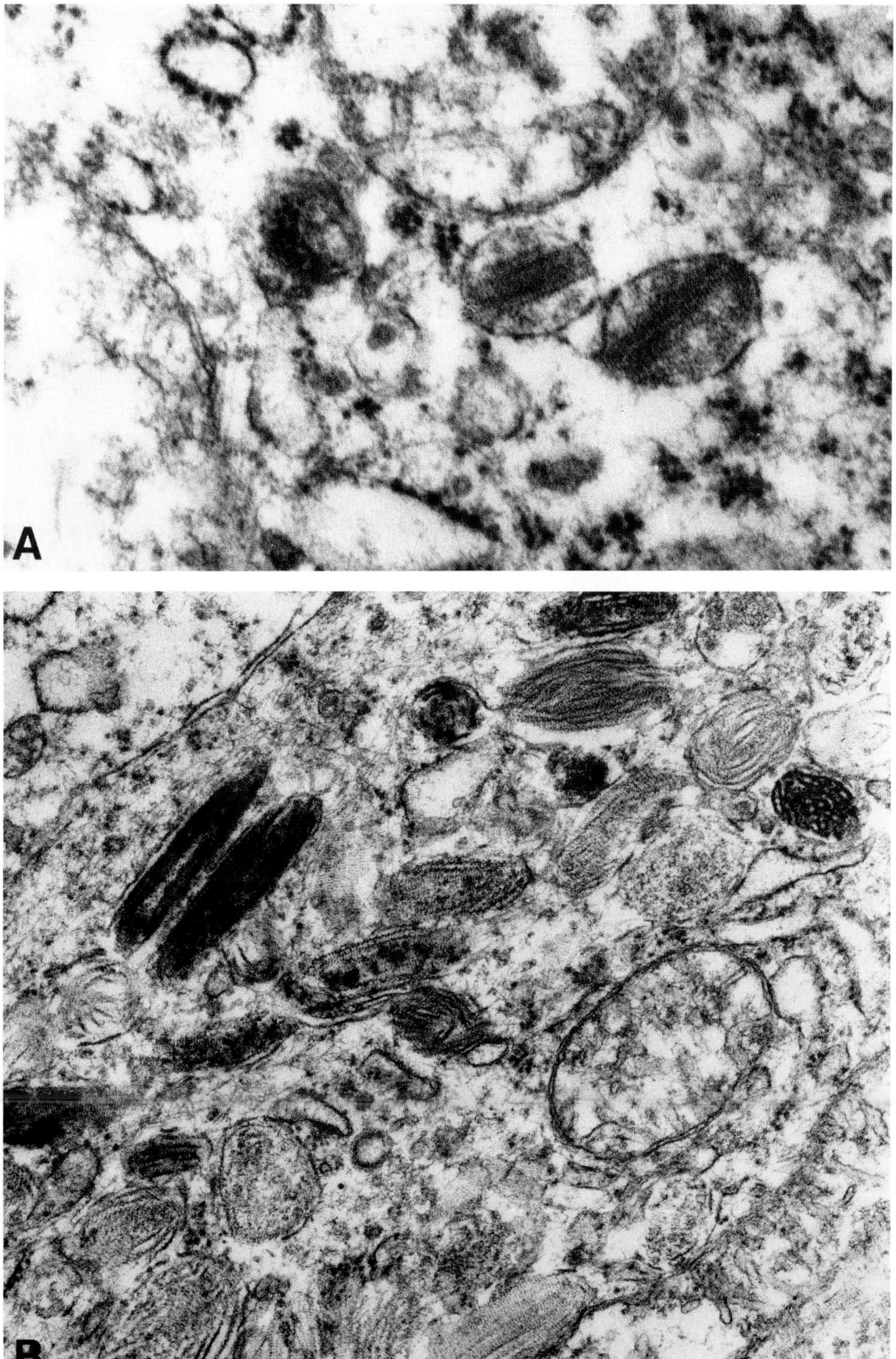

FIGURE 7. Metastatic melanoma, soft tissue mass, neck. Same case as the one illustrated in Figure 5; transmission electron microscopy. Uranyl acetate and lead citrate. Premelanosomes and melanized melanosomes are illustrated, confirming the diagnosis of melanoma. (Magnifications: A, × 1,800; B, × 21,200.)

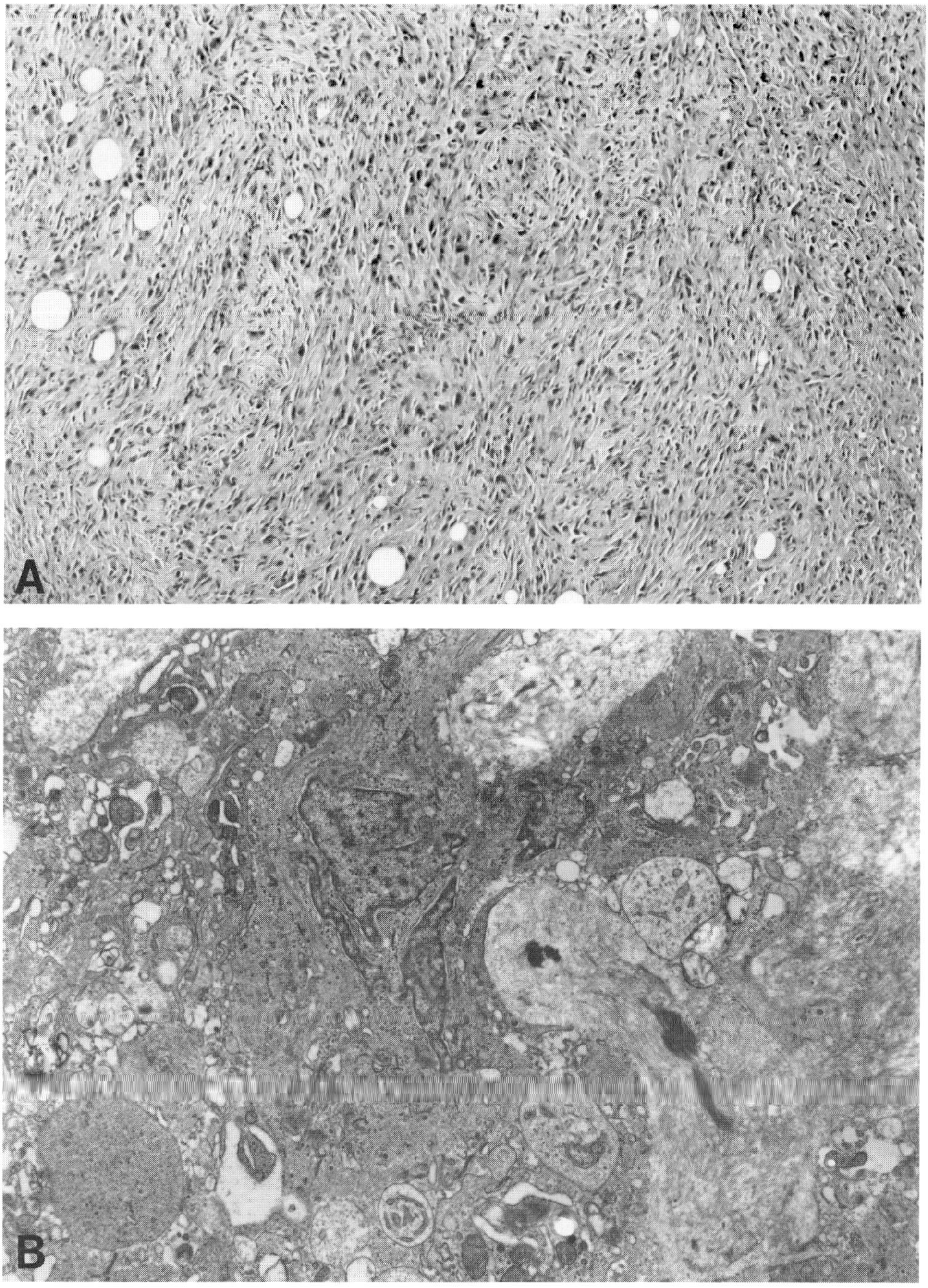

FIGURE 8. Malignant mesothelioma abdominal cavity, hematoxylin and eosin stain; transmission electron microscopy, uranyl acetate and lead citrate. (A) Shows a malignant spindle cell neoplasm difficult to accurately characterize by light microscopy; (B) reveals a typical cell in this tumor. (Magnifications: A, × 130; B, × 10,900.)

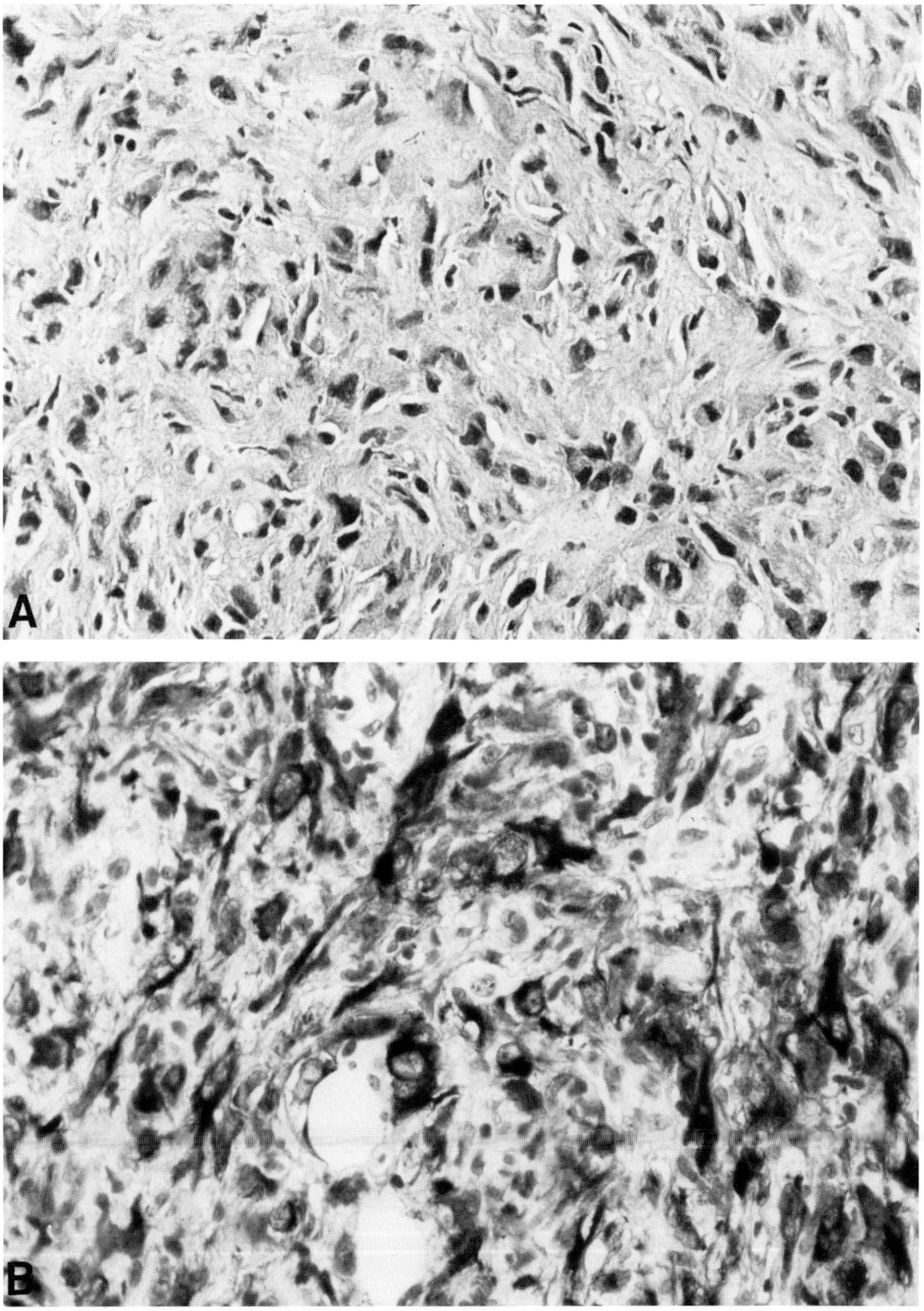

FIGURE 9. Malignant mesothelioma, abdominal cavity. (A) Hematoxylin and eosin stain; (B) keratin, wide-spectrum, stain; avidin-biotin complex (ABC) peroxidase technique; (C) vimentin avidin-biotin complex peroxidase technique. Details of neoplastic process illustrated in (A). In (B) strong staining for keratin and in (C) distinct vimentin staining is noted. (Magnifications: A, × 290; B, × 290; C, × 290.)

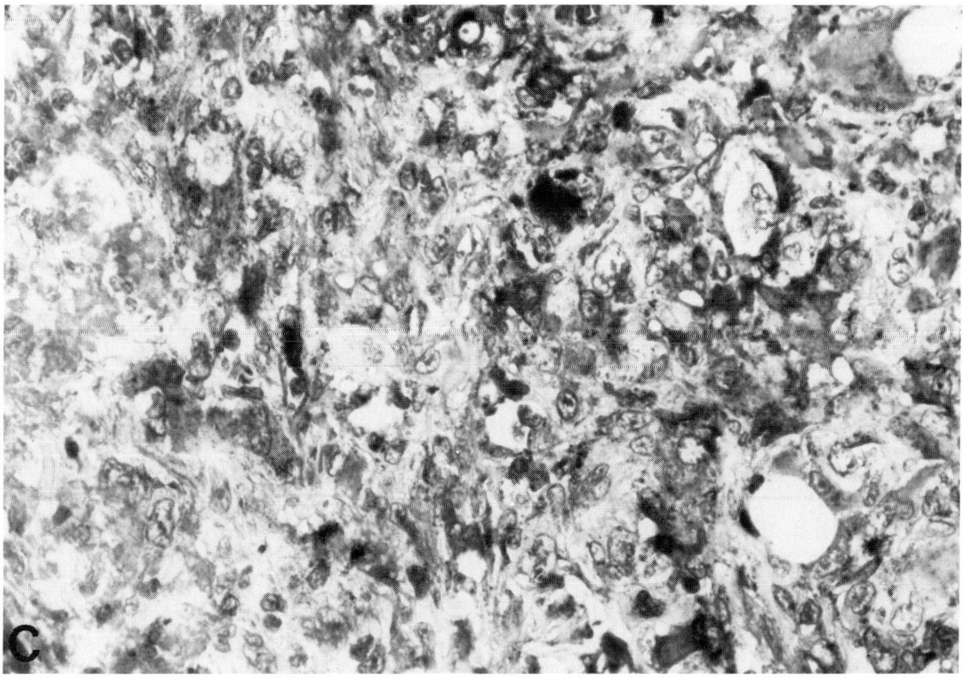

FIGURE 9 (continued).

including whether a given gene is absent, present, amplified, mutated, expressed, or over-expressed, or whether oncogene-encoded protein products are present or absent. Molecular genetic techniques are often used in the diagnosis of lymphoid processes when clinical and immunomorphologic findings are inconclusive for malignancy, aiding in the evaluation of morphologic patterns difficult to be defined by light microscopic assessment. The detection of a monoclonal process and or gene rearrangement in a given clinical setting may have crucial clinical connotations.

Immunohistochemical techniques have been used to detect the proteins encoded by oncogenes. While immunohistochemical techniques are welcome by pathologists due to their familiarity with them and their simplicity, there are several factors that should be carefully considered. Fixation techniques can alter the cellular localization of oncogene-encoded proteins. The specificity of the testing antibodies currently available is not always good and cross-reactivity with other cellular proteins may occur. Furthermore, the antigens to be detected are sensitive to degradation by endogenous proteases present in inflammatory components associated with the tumors. At the present time, with perhaps only a few exceptions (i.e., c-erb B-2), the use of immunohistochemical techniques to detect oncogene-encoded products should be restricted to frozen tissues.

The area of tumor characterization dealing with molecular biology techniques will likely expand in the coming years and may eventually become routine as the techniques are applied to diagnostic situations and perhaps automated, if they prove to add meaningful, clinically relevant information.

L. A GENERAL OVERVIEW OF TUMOR CHARACTERIZATION

As Dr. Sheldon Summers has said:[82] "After 300 years of light microscopy, some 30 years of electron microscopy, and about 10 years of immunopathologic diagnostic techniques, the specialty of pathology is now blooming in science and service. Cancer is the greatest

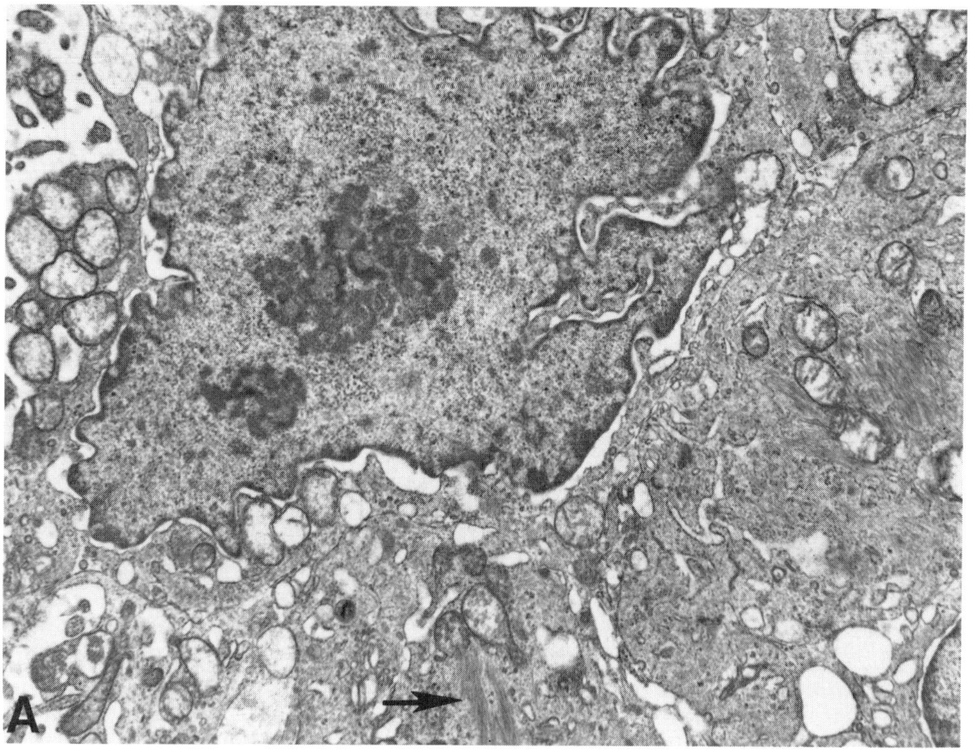

FIGURE 10. Malignant mesothelioma, abdominal cavity; transmission electron microscopy; uranyl acetate and lead citrate; immunogold labeling for keratin. (A) and (B) illustrate the ultrastructural features of the malignant cells with abundant intracytoplasmic filaments. Distinct labeling of the filaments for keratin is noted in (C). (Magnifications: A, × 21,900; B, × 25,800; C, × 25,800.)

biomedical challenge of present times.'' Innovative techniques such as flow cytometry, morphometry, cloning, autoradiography, karyotyping, DNA analysis, radioimmunoassay, tissue culture, and animal transplantation are evolving to help pathologists understand, assess, and classify neoplasms better.

The question that is often posed to the practicing pathologist is whether a given tumor needs only light microscopic evaluation, special stains, light microscopy immunocytochemistry, electron microscopic evaluation, ultrastructural immunolabeling, *in situ* hybridization, Southern or Northern blotting, or more than one of the above techniques for accurate characterization. The question, at times is not asked correctly. As Ph. U. Heitz very eloquently stated in a recent editorial:[83] ''The real problem for the pathologist is to achieve these aims (i.e., tumor characterization) with a maximum of precision in a minimum lapse of time at the lowest possible cost.'' As he concludes in his editorial: ''The problem can often be solved by using one technique or an intelligent combination of a small number of various methods, but rarely by excluding one technique in favor of another, as implied by 'or' alternatives.'' In a few accurate words the dilemma is clearly exposed. The challenge for the practicing pathologist of learning the immunomorphologic and other characteristics of various neoplasms is expanding continually, with new information becoming available essentially daily. The emphasis on cost-containment further complicates the issue. Pathologists are forced to be accurate in their tumor speciation but not to overspend. Where to stop in the work-up of a case can be quite controversial and varies from academic to private

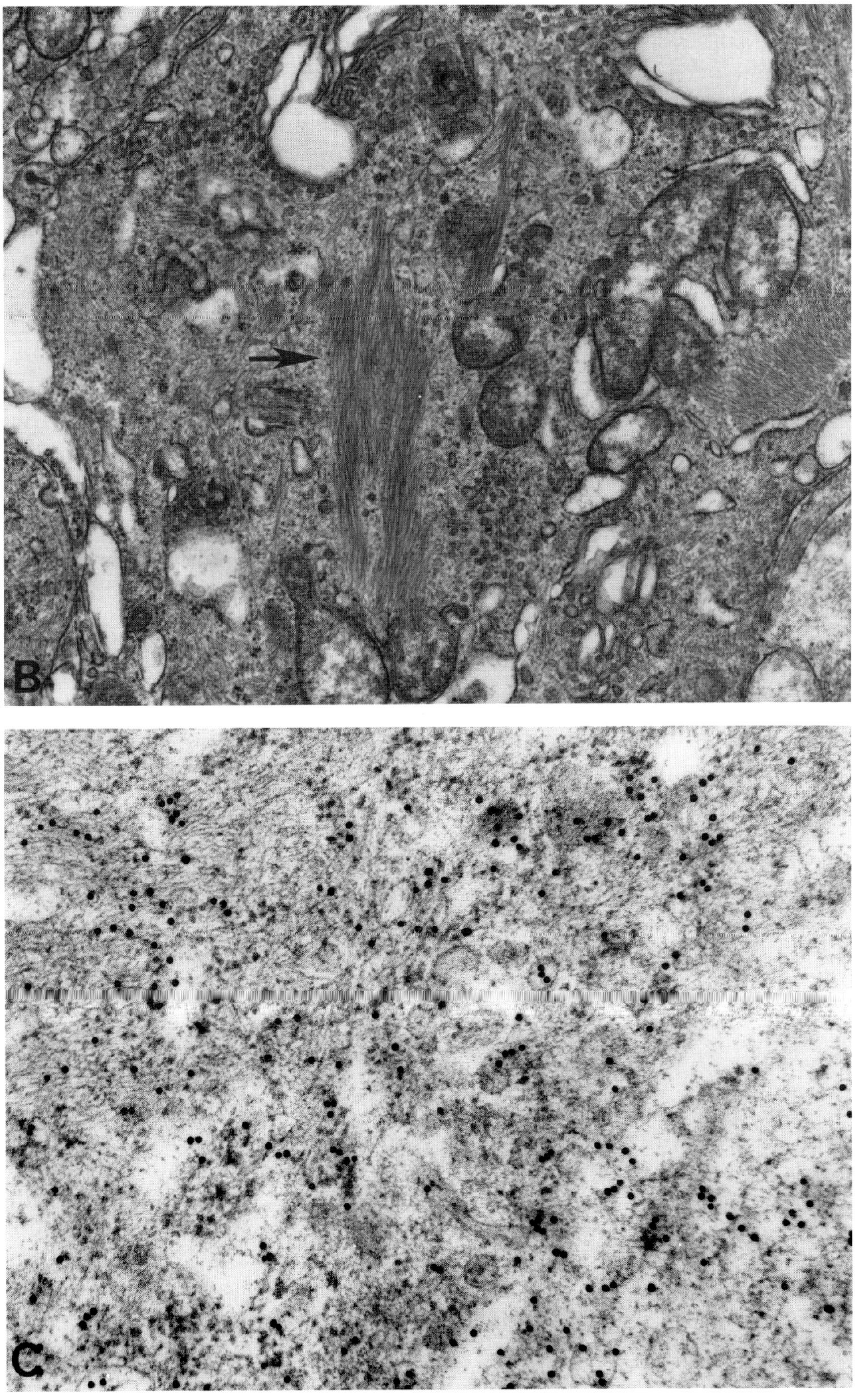

FIGURE 10 (continued).

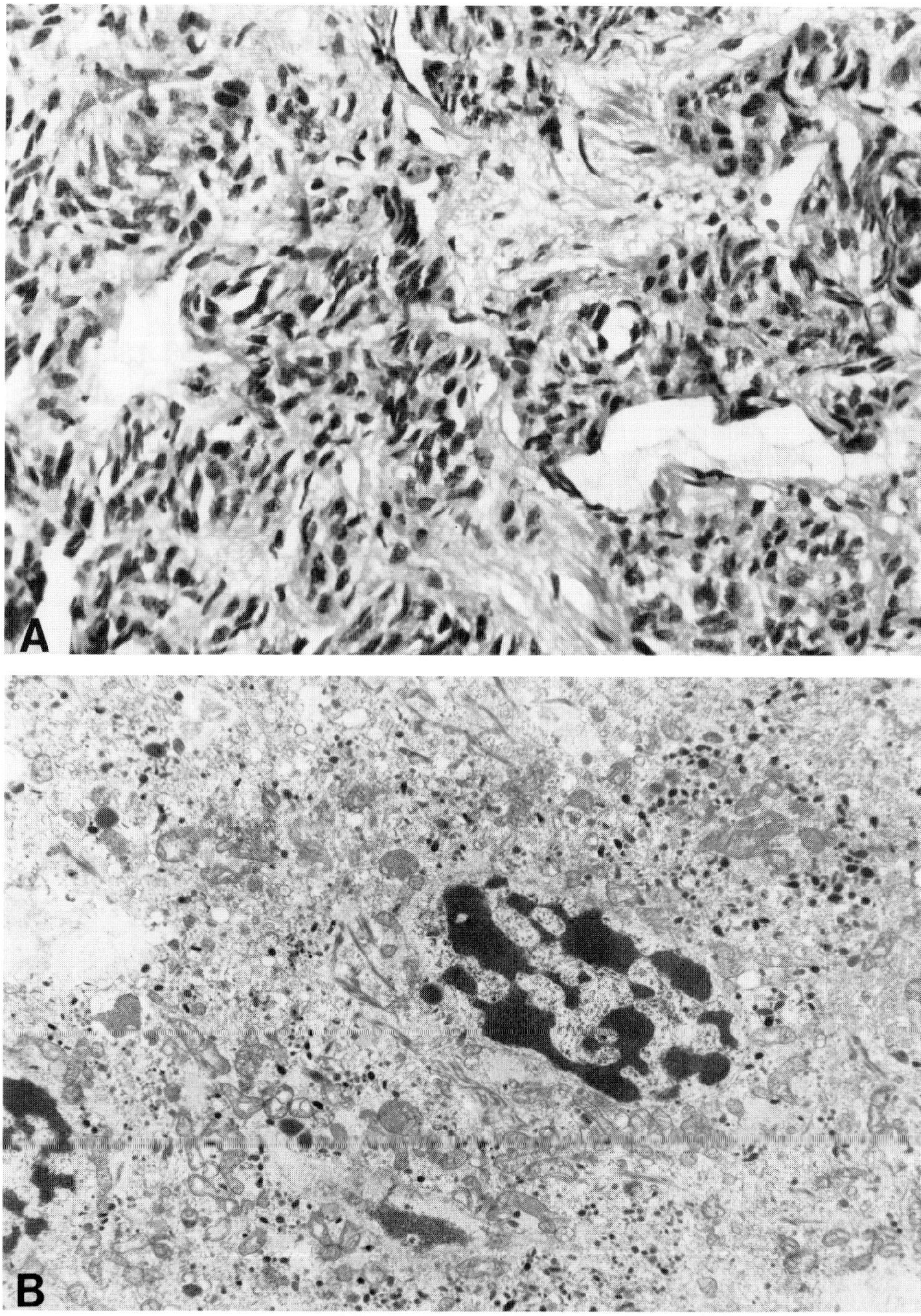

FIGURE 11. Metastatic carcinoid tumor, paranasal sinus. (A) Hematoxylin and eosin stain; (B) and (C) transmission electron microscopy; uranyl acetate and lead citrate. Overall morphologic pattern of the tumor with a prominent vascularity resembling a hemangiopericytoma is illustrated. Ultrastructural examination (B and C) reveals numerous pleomorphic cytoplasmic electron-dense granules. (Magnifications: A, × 290; B, × 10,100; C, × 18,200.)

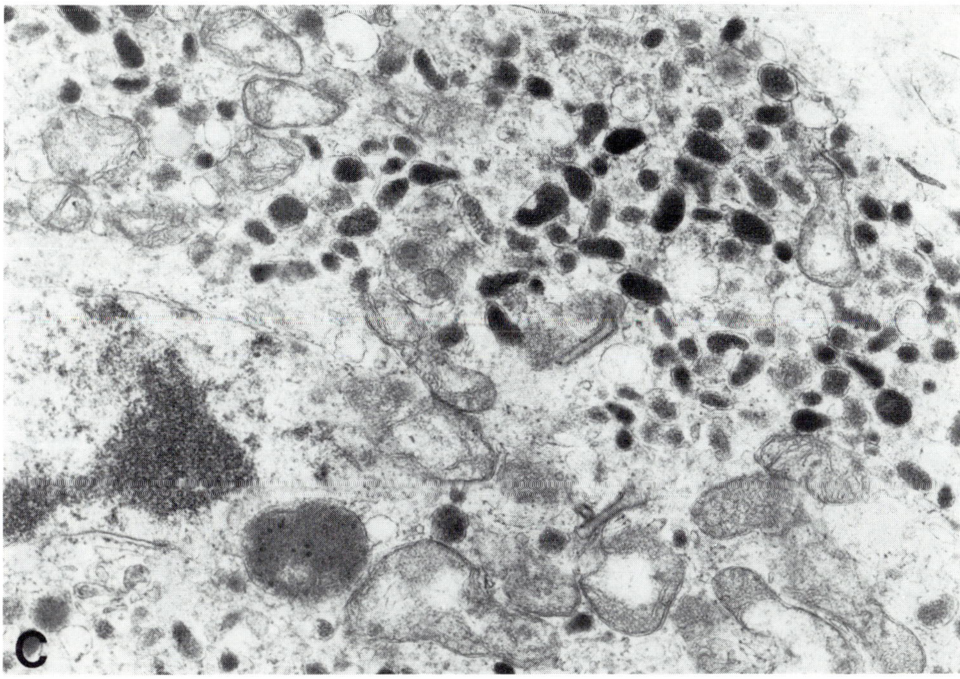

FIGURE 11 (continued).

institutions, institution to institution of the same nature, and even from pathologist to pathologist in the same institution. Guidelines can only be general, for every challenging case may be a microcosm of immunomorphologic manifestations and a new diagnostic dilemma.

Today pathology is indeed an exciting specialty full of new ideas to explore and techniques to develop. Tumor characterization remains one of the most challenging areas in diagnosis, with unquestionable importance in terms of patients management. However, as pathology develops and matures in many aspects, it is hurting in other, no less important, issues. As Dr. Sheldon Summers has stated:[82] "All is not rosy in pathology as this century nears an end. Autopsies decline, residencies are prolonged and government policy squeezes biomedical funding to the detriment of science and pathology practice. The decrease in autopsies will make evaluation of new therapeutic protocols, including specific chemotherapy and radiotherapy based protocols predicated on cancer speciation, very difficult or at least incomplete. The lack of manpower in pathology will limit the degree of sophistication that can be attained in tumor diagnosis. Incorporation of new meaningful techniques will be delayed and necessary sub-specialization will not be possible until basic service needs are met. The lack of research funding will further delay the development and sophistication of new diagnostic techniques." We must be proud of our achievements and look at the future, not with skepticism, but with controlled optimism, aware of the difficulties lying ahead but ready to conquer them.

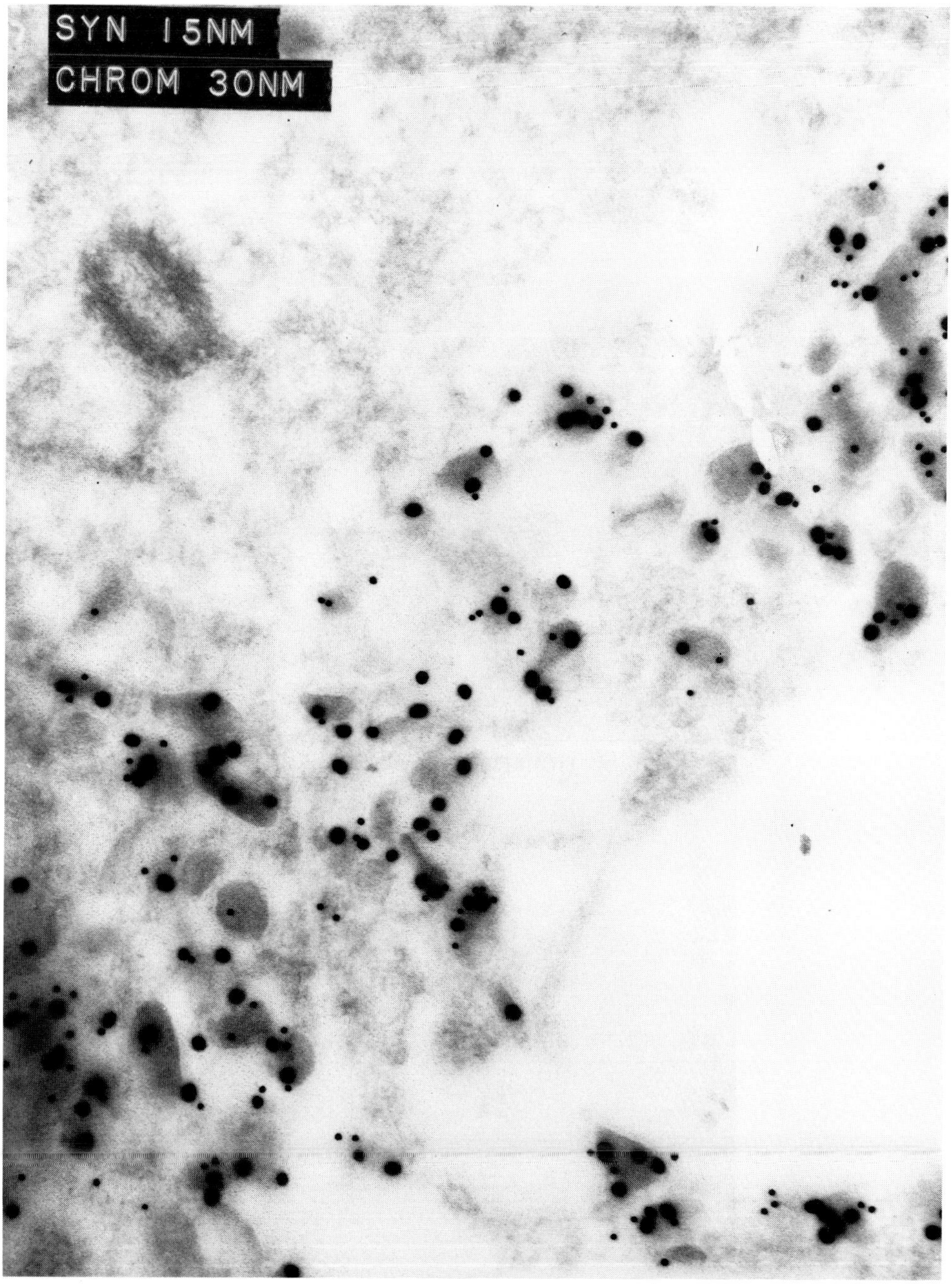

FIGURE 13. Metastatic carcinoid tumor, paranasal sinus; double immunogold labeling. Synaptophysin 15 nm and chromogranin 30 nm. Note primarily peripheral labeling of granules with 15-nm gold particles and labeling of electron-dense cores of granules by the 30-nm gold particles.

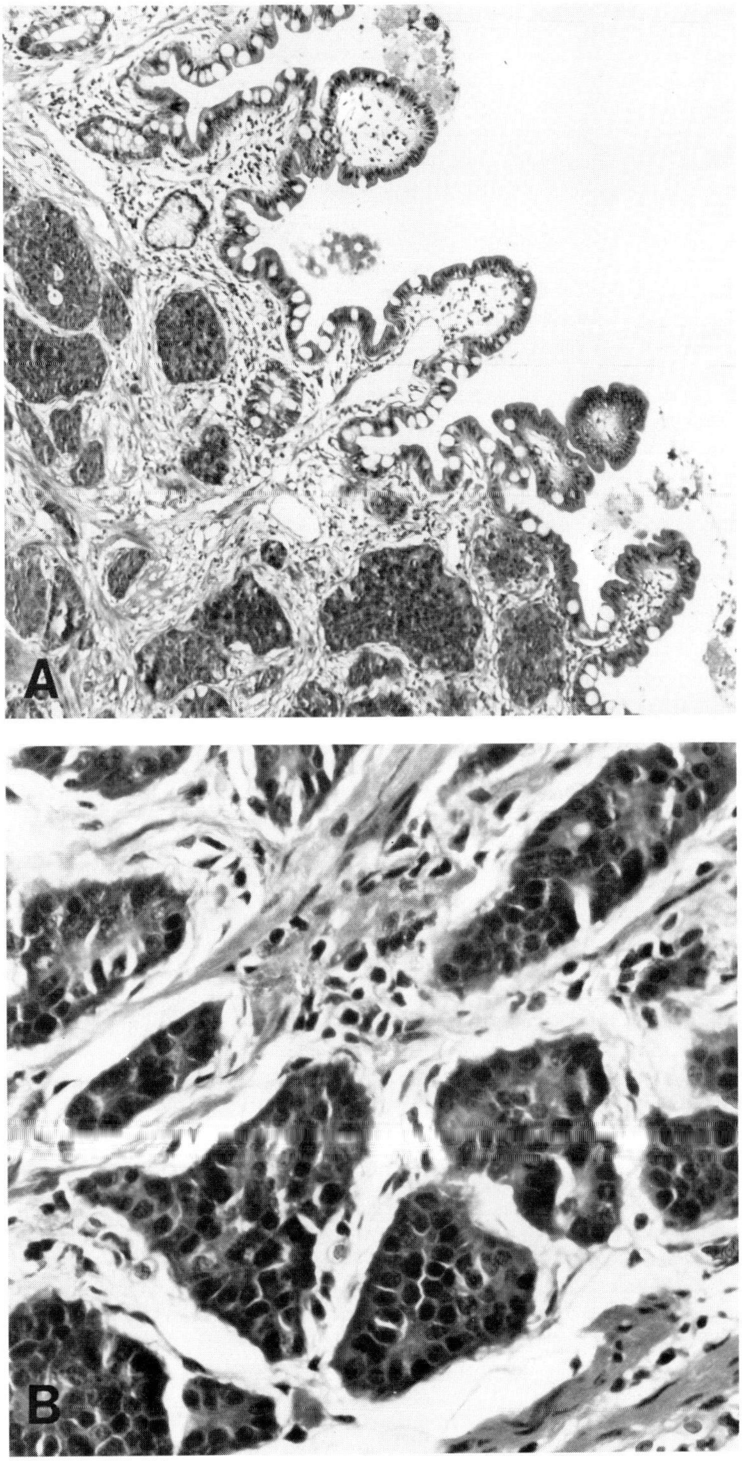

FIGURE 14. Carcinoid tumor, ileum. Note that the morphology of the primary tumor is classical for a carcinoid tumor and quite different from the appearance of the metastasis (Figure 11). (Magnifications: A, × 150; B, × 320.)

REFERENCES

1. **Watson, W. L. and Berg, J. W.,** Oat cell lung cancer, *Cancer,* 15, 759, 1962.
2. **Long, E. R.,** *A History of Pathology,* Williams & Wilkins, Baltimore, 1928, 7.
3. **Rosen, G.,** History of pathology. Beginnings of surgical biopsy, *Am. J. Surg. Pathol.,* 1, 361, 1977.
4. **Muller, J.,** *Ueber den Feinern Bau und die Formen Der Krankhaften,* Geschwulste, Berlin, 1838.
5. **Hannover, A.,** *Om Epithelioma, En Faeregen Svultst, Som Man Hidtil, Almindelighed Har Anseet for Kraeft,* C. A. Reitzels Forlag, Copenhagen, 1852, 1–9, 17–20.
6. **Lebert, H.,** *Physiologie Pathologique,* Vol. 2, Paris, 1845, 255.
7. **Lebert, H.,** *Ueber den Feinern Bau und die Formen der krankhaften Geschwulste,* Berlin, 1838.
8. **Velpeau, A. A. L. M.,** *Traite des Maladies du Sein et de la Region Mammaire,* Paris, 1853; Translated into English by M. Henry as *A Treatise on the Diseases of the Breast and Mammary Region,* London, 1856, 479.
9. **Del Regato, J. A.,** Arthur Purdy Stout, *Int. J. Radiat. Oncol. Biol. Phys.,* 16, 799, 1989.
10. **Azar, H.,** History of pathology. Arthur Purdy Stout (1855–1967). The man and the surgical pathologist, *Am. J. Surg. Pathol.,* 8, 301, 1984.
11. **Ewing, J.,** *Neoplastic Diseases,* W. B. Saunders, Philadelphia, 1919.
12. **Coons, A. H., Creech, H. J., and Jones, R. N.,** Immunological properties of an antibody containing a fluorescent group, *Proc. Soc. Exp. Biol.,* 47, 200, 1941.
13. **Sternberger, L. A., Hardy, P. H., Jr., Cusulis, J. J. et al.,** The unlabeled antibody enzyme method of immunocytochemistry: preparation and properties of soluble antigen-antibody complex (horseradish-peroxidase-antihorseradish-peroxidase) and its use in identification of spirochetes, *J. Histochem.,* 18, 315, 1970.
14. **Hsu, S. M., Raine, L., and Fanger, H.,** Use of avidin-biotin peroxidase complex (ABC) in immunoperoxidase techniques: a comparison between ABC and unlabeled antibody (PAP) procedures, *J. Histochem. Cytochem.,* 29, 577, 1981.
15. **Bendayan, M. and Zollinger, M.,** Ultrastructural localization of antigenic sites on osmium-fixed tissues applying the protein A-gold technique, *J. Histochem. Cytochem.,* 31, 101, 1983.
16. **Rantala, I. and Helin, H.,** Immunoelectron microscopic localization of immunoglobulins and complement in human renal glomeruli, *J. Histochem. Cytochem.,* 30, 932, 1982.
17. **Hearn, S. A. and Kontozoglou, T. E.,** Post-embedding immunogold electron microscopy for S-100, tubulin, and cytoskeletal proteins in an amelanotic malignant melanoma, *Mod. Pathol.,* 2, 46, 1989.
18. **Herrera, G. A., Paul, R., Turbat-Herrera, et al.,** Ultrastructural immunolabeling in the diagnosis of light-chain related renal disease, *Pathol. Immunopathol. Res.,* 5, 170, 1986.
19. **Herrera, G. A., Turbat-Herrera, E. A., Viale, G. et al.,** Ultrastructural immunolabeling in renal diseases. Past, present and future expectations, *Pathol. Immunopathol. Res.,* 6, 51, 1987.
20. **Herrera, G. A., Sanders, P. W., Lott, R. L. et al.,** The value of IF, EM and ultrastructural immunolabeling in the diagnosis of early and unusual manifestations of monoclonal light chain related renal diseases (LCRRD), *Lab Invest.,* 58, 38A, 1988.
21. **Herrera, G. A.,** Ultrastructural post-embedding, immunogold labeling. Applications to diagnostic pathology, *Ultrastruct. Pathol.,* 13, 485, 1989.
22. **Herrera, G. A., Lott, R. L., Sanders, P. W. et al.,** Ultrastructural immunogold labeling. Image processing, mapping and quantitative microanalysis of antigenic sites, *Pathol. Immunopathol. Res.,* 8, 42, 1989.
23. **Long, E. R.,** *A History of Pathology,* Williams & Wilkins, Baltimore, 1928, 185.
24. **Thiersch, C.,** *Der Epithelialkrebs Namentlich der Haut,* Wilhelm Engelmann, Leipzig, 1865, 55.
25. **Pack, G. T.,** Functions and dysfunctions of the surgical pathologist, *Surgery,* 52, 752, 1962.
26. **Bretschneider, A., Burns, W., and Morrison, A.,** "Pop Off" technique: the ultrastructure of paraffin-embedded sections, *Am. J. Clin. Pathol.,* 76, 450, 1981.
27. **Miles, P. A., Herrera, G. A., Mena, H., and Trujillo, I.,** Diagnosis of carcinoid tumor of the cervix by cytological examination, including immunohistochemistry and electron microscopy, *Acta Cytol.,* 29, 1003, 1985.
28. **Carmichael, D. E.,** *The Pap Smear: Life of George N. Papanicolaou,* Charles C Thomas, Springfield, IL, 1973.
29. **Naylor, B.,** *The History of Exfoliative Cancer Cytology,* University of Michigan Medical Bulletin, Ann Arbor, 1960.
30. **Grunze, H. and Spriggs, A. I.,** *History of Clinical Cytology, A Selection of Documents,* Verlag Ernst Gieber, Darmstadt, 1980.
31. **Webb, A. J.,** Through a glass darkly. The developments of fine needle aspiration biopsy, *Bristol Med. Chir. J.,* 89, 599, 1974.

32. **Koss, L. G., Woyke, S., and Olszewski, W.,** *Aspiration Biopsy, Cytologic Interpretation and Histologic Basis,* Igaku-Shoin, Tokyo, 1984.
33. **Martin, H. E. and Ellis, E. B.,** Biopsy by needle puncture and aspiration, *Ann. Surg.,* 62, 169, 1930.
34. **Stewart, F. W.,** The diagnosis of tumors by aspiration, *Am. J. Pathol.,* 9, 801, 1933.
35. **Franzen, S. and Zajicek, J.,** Aspiration biopsy in diagnosis of palpable lesions of the breast, *Acta Radiol.,* 7, 241, 1968.
36. **Rosai, J. and Rodriguez, H. A.,** Applications of electron microscopy to the differential diagnosis of tumors, *Am. J. Clin. Pathol.,* 50, 535, 1968.
37. **Avrameas, S.,** Coupling of enzymes to proteins with glutaraldehyde. Use of conjugates for the detection of antigens and antibodies, *Immunocytochemistry,* 6, 43, 1969.
38. **Notani, G. W., Parson, J., and Erlandsen, S. L.,** Versatility of Staphylococcus aureus protein A in immunocytochemistry. Use of unlabeled antibody enzyme system and fluorescent methods, *J. Histochem. Cytochem.,* 27, 1438, 1979.
39. **Bhan, A. K.,** Immunoperoxidase, in *Diagnostic Immunopathology,* Colvin, R. B., Bhan, A. K., and McCluskey, R. T., Eds., 1988, 437.
40. **Lewis, R. E., Johnson, W. W., and Cruse, J. M.,** Pitfalls and caveats in the methodology for immunoperoxidase staining in surgical pathologic diagnosis, *Surv. Synth. Pathol. Res.,* 1, 134, 1983.
41. **Banks, P. M.,** Diagnostic applications of an immunoperoxidase method in hematopathology, *J. Histochem. Cytochem.,* 27, 1192, 1979.
42. **Taylor, C. R.,** *Immunomicroscopy: A Diagnostic Tool for the Surgical Pathologist,* W. B. Saunders, Philadelphia, 1986.
43. **Hsutzer, N. W., Wittkuhn, J. F., and McCaughey, W. T. E.,** Trypsin digestion in immunoperoxidase staining, *J. Histochem. Cytochem.,* 28, 52, 1980.
44. **Singer, S. J.,** Preparation of an electron dense antibody conjugate, *Nature (London),* 183, 1523, 1959.
45. **Nakane, P. K. and Pierce, G. B.,** Enzyme-labeled antibodies: preparation and application for the localization of antigens, *J. Histochem. Cytochem.,* 14, 929, 1966.
46. **Ram, J. S., Nakane, P. K., Rawlins, D. G. et al.,** Enzyme labeled antibody for ultrastructural studies, *Fed. Proc.,* 25, 732, 1966.
47. **Feldherr, C. M. and Marshall, J. M.,** The use of colloidal gold for studies of intracellular exchange in amoeba Choaos Chaos, *J. Cell Biol.,* 12, 640, 1962.
48. **Faulk, W. P. and Taylor, G. M.,** An immunocolloid method for the electron microscope, *Immunochemistry,* 8, 1081, 1971.
49. **Romano, E. L., Stolinski, C., and Hughes-Jones, N. C.,** An antiglobulin reagent labeled with colloidal gold for use in electron microscopy, *Immunochemistry,* 11, 521, 1974.
50. **Bullock, G. R.,** The current status of fixation for electron microscopy: a review, *J. Microsc.,* 133, 1, 1984.
51. **Linner, J. G., Livesey, S. A., Harrison, D. S. et al.,** A new technique for removal of amorphous phase tissue water without ice crystal damage: a preparative method for ultrastructural analysis and immunoelectron microscopy, *J. Histochem. Cytochem.,* 34, 1123, 1986.
52. **Causton, B. E.,** The choice of resins for electron immunocytochemistry, in *Immunolabeling for Electron Microscopy,* Polak, J. M. and Varndell, I. M., Eds., Elsevier, Amsterdam, 1984, 37.
53. **Newman, G. R. and Jasani, B.,** Post-embedding immunoenzyme techniques, in *Immunolabeling for Electron Microscopy,* Polak, J. M. and Varndell, I. M., Eds., Elsevier, Amsterdam, 1984, 53.
54. **[illegible], D. G.,** Post embedding immunogold labeling [illegible] microscopy using LR White resin, *Am. J. Anat.,* 173, 267, 1986.
55. **Lockard, V. G.,** A simple technique for post-embedding ultrastructural immunogold labeling, *Ultrastruct. Pathol.,* 16, 47, 1992.
56. **Smith, P. S.,** Ultrastructural immuno-gold localization of immune deposits in human renal biopsies, *Pathology,* 20, 32, 1988.
57. **Knibbs, D. R.,** A simplified technique for the ultrastructural localization of cytoskeletal filament proteins by immunogold electron microscopy, *J. Histotech.,* 13, 9, 1990.
58. **Roth, J. and Heitz, P. U.,** Immunolabeling with the protein A-gold technique: an overview, *Ultrastruct. Pathol.,* 13, 467, 1989.
59. **Roth, J.,** The protein A-gold technique for antigen localisation in tissue sections by light and electron microscopy, in *Immunolabeling for Electron Microscopy,* Polak, J. M. and Varndell, I. A., Eds., Elsevier, Amsterdam, 1984, 113.
60. **Childs, G. V.,** The use of multiple methods to validate immunocytochemical stains, *J. Histochem. Cytochem.,* 31, 168, 1983.

61. **Childs, G. V., Herrera, G. A., and Lott, R. L.,** Colloidal gold labeling for diagnostic pathology, in *Colloidal Gold: Principles, Methods and Applications,* Vol. 31, Hayat, M. A., Ed., Academic Press, San Diego, 1990, 321.
62. **Herrera, G. A. and Lott, R. L.,** Ultrastructural immunolabeling in diagnostic surgical pathology, illustrative cases, *Ultrastruct. Pathol.,* 16, 71, 1992.
63. **Herrera, G. A.,** Microanalytical techniques and usage analysis in the evaluation of immunogold labeled specimens at the ultrastructural level, *Ultrastruct. Pathol.,* 16, 127, 1992.
64. **Herrera, G. A.,** Ultrastructural immunolabeling. A general overview of techniques and applications, *Ultrastruct. Pathol.,* 16, 37, 1992.
65. **Herrera, G. A., Cerezo, L., Jones, J. et al.,** Gastrointestinal autonomic nerve tumors, plexosarcomas, *Arch. Pathol. Lab. Med.,* 113, 846, 1989.
66. **Nakajima, T., Watanabe, S., Sato, Y. et al.,** Immunohistochemical demonstration of S-100 protein in malignant melanoma and pigmented nevus, and its diagnostic applications, *Cancer,* 50, 913, 1982.
67. **Gown, A. M., Vogel, A. M., Hoak, D., Gough, F., and McNutt, M. A.,** Monoclonal antibodies specific for melanocytic tumors distinguish subpopulation of melanocytes, *Am. J. Pathol.,* 123, 195, 1986.
68. **Raemaekers, F. C. S., Puts, J. J. G., Moesker, O. et al.,** Intermediate filaments in malignant melanomas, identification and use in surgical pathology, *J. Clin. Invest.,* 71, 635, 1983.
69. **Kapur, R. P., Bigler, S. J., McGrath, E. et al.,** Antimelanoma antibody HMB-45 identifies an oncofetal premelanosome associated glycoprotein, *Lab. Invest.,* 64, 105A, 1991.
70. **Miettinen, M. and Franssila, K.,** Immunohistochemical spectrum of malignant melanoma. The common presence of keratins, *Lab. Invest.,* 61, 623, 1989.
71. **Zarbo, R. J., Gown, A. M., Nagle, R. B. et al.,** Anomalous cytokeratin expression in malignant melanoma: one and two dimensional Western blot analysis and immunohistochemiocal survey of 100 melanomas, *Mod. Pathol.,* 3, 494, 1990.
72. **Allen, B. C. and Herrera, G. A.,** Phenotypic heterogeneity in melanoma: a challenge to the surgical pathologist evaluating a poorly differentiated neoplasm: an illustrative case, *Ultrastruct. Pathol.,* 15, 87, 1991.
73. **Herrera, G. A., Turbat-Herrera, E. A., and Lott, R. L.,** S-100 protein expression by primary and metastatic adenocarcinomas, *Am. J. Clin. Pathol.,* 89, 168, 1988.
74. **Hancock, C., Allen, B. C., and Herrera, G. A.,** Is HMB-45 MoAb an absolutely specific marker of melanoma?, *Am. J. Clin. Pathol.,* 94, 496, 1990.
75. **Bonetti, F., Colombari, R., Manfrin, E. et al.,** Breast carcinoma with positive results for melanoma marker (HMB-45), HMB-45 immunoreactivity in normal and neoplastic breast, *Am. J. Clin. Pathol.,* 92, 491, 1989.
76. **Kornstein, M. J. and Franco, A. P.,** Specificity of HMB-45 (Letter to the Editor), *Arch. Pathol. Lab. Med.,* 114, 450, 1990.
77. **Leong, A. S.-Y. and Milios, J.,** An assessment of a melanoma-specific antibody (HMB-45) and other immunohistochemical markers of malignant melanoma in paraffin-embedded tissues, *Surg. Pathol.,* 2, 137, 1989.
78. **Holm, R., Nesland, J. M., Attramadal, A. et al.,** Double-staining methods at the ultrastructural level applying colloidal gold conjugates, *Ultrastruct. Pathol.,* 11, 279, 1988.
79. **Bendayan, M. and Stephens, H.,** Double labeling cytochemistry applying the protein-A gold technique, in *Immunolabeling for Electron Microscopy,* Polak, J. M. and Varndell, I. M., Eds., Elsevier, Amsterdam, 1984, 143.
80. **Herrera, G. A., Lockard, V. G., and Turbat-Herrera, E. A.,** Ultrastructural labeling in the evaluation, diagnosis and characterization of neuroendocrine neoplasms, *Ultrastruct. Pathol.,* in press.
81. **Herrera, G. A.,** Ultrastructural immunolabeling: the next logical step, Editorial, *Ultrastruct. Pathol.,* 15, iii, 1991.
82. **Sommers, S. C.,** Pathology of human neoplasms, in *An Atlas of Diagnostic Electron Microscopy and Immunohistochemistry,* Foreword, Azar, H., Ed., Raven Press, New York, 1988, 7.
83. **Heitz, Ph. U.,** Special tinctorial stains of histochemistry or electron microscopy or immunocytochemistry or hybridization or . . . ?, Editorial, *Ultrastruct. Pathol.,* 12, iii, 1988.

INDEX

INDEX

A

D

E

F

G

V

XYZ